Radar Phenomenology

Maximum RCS of Simple Shapes, $\lambda \ll$ Object Size

Shape		RCS
Sphere, radius r	r	πr^2
Flat plate, edge lengths a and b	b, a	$4\pi (ab)^2/\lambda^2$
Dihedral, edge lengths a and b	b, a, a	$8\pi (ab)^2/\lambda^2$
Trihedral, square sides, edge length a	a, a, a	$12\pi a^4/\lambda^2$
Trihedral, triangular sides, edge length a	a, a	$4\pi a^4/3\lambda^2$

Swerling Models

Probability Density Function of RCS σ	Decorrelation	
	Scan-to-Scan	Pulse-to-Pulse
Exponential, $p_\sigma(\sigma) = \frac{1}{\bar{\sigma}} \exp\left[\frac{-\sigma}{\bar{\sigma}}\right]$	Case 1	Case 2
Chi-square, degree 4, $p(\sigma) = \frac{4\sigma}{\bar{\sigma}^2} \exp\left[\frac{-2\sigma}{\bar{\sigma}}\right]$	Case 3	Case 4

RCS Decorrelation

Variable	Required Change	Comment
Aspect angle (rad)	$\frac{c}{2Lf} = \frac{\lambda}{2L}$	$L =$ width as viewed along radar line of sight
Frequency (Hz)	$\frac{c}{2L}$	$L =$ depth as viewed along radar line of sight

Values of Doppler Shift

Radio frequency f		Doppler Shift f_d (Hz)		
Band	Frequency (GHz)	1 m/s	1 knot	1 mph
L	1	6.67	3.43	2.98
S	3	20.0	10.3	8.94
C	6	40.0	20.5	17.9
X	10	66.7	34.3	29.8
K_u	16	107	54.9	47.7
K_a	35	233	120	104
W	95	633	326	283

Atmospheric Attenuation

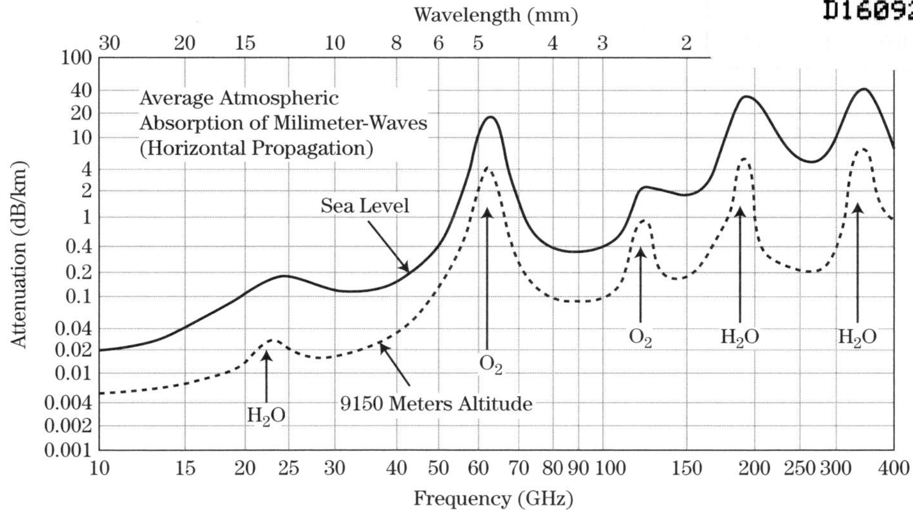

Principles of Modern Radar

Principles of Modern Radar

Vol. I: Basic Principles

Mark A. Richards
Georgia Institute of Technology

James A. Scheer
Georgia Institute of Technology

William A. Holm
Georgia Institute of Technology

Edison, NJ
scitechpub.com

Published by SciTech Publishing - An Imprint of the IET
379 Thornall Street
Edison, NJ 08837
scitechpub.com

Copyright © 2010 by SciTech Publishing, Edison, NJ. All rights reserved.

No part of this publication may be reproduced, stored in a retrieval system or transmitted in any form or by any means, electronic, mechanical, photocopying, recording, scanning or otherwise, except as permitted under Sections 107 or 108 of the 1976 United Stated Copyright Act, without either the prior written permission of the Publisher, or authorization through payment of the appropriate per-copy fee to the Copyright Clearance Center, 222 Rosewood Drive, Danvers, MA 01923, (978) 750-8400, fax (978) 646-8600, or on the web at copyright.com. Requests to the Publisher for permission should be addressed to the Publisher at the address above, or email editor@scitechpub.com.

The publisher and the author make no representations or warranties with respect to the accuracy or completeness of the contents of this work and specifically disclaim all warranties, including without limitation warranties of fitness for a particular purpose.

Editor: Dudley R. Kay
Editorial Assistant: Katie Janelle
Production Manager: Robert Lawless
Typesetting: MPS Limited, A Macmillan Company
Cover Design: Brent Beckley
Printer: Sheridan Books, Inc., Chelsea, MI

This book is available at special quantity discounts to use as premiums and sales promotions, or for use in corporate training programs. For more information and quotes, please contact the publisher.

Printed in the United States of America
10 9 8 7 6 5

ISBN: 978-1-891121-52-4

Library of Congress Cataloging-in-Publication Data

Richards, M. A. (Mark A.), 1952-
Principles of modern radar / Mark A. Richards, James A. Scheer, William A. Holm.
p. cm.
Includes bibliographical references and index.
ISBN 978-1-891121-52-4 (hardcover : alk. paper)
I. Scheer, Jim, 1944- II. Holm, William A. III. Title.
TK6575.R473 2010
621.3848–dc22

2010013808

Dedicated to the many students of Georgia Tech's professional education course *Principles of Modern Radar*, who inspired this book's development;

and

to our wives, Theresa, Ruby, and Kathleen, who inspire us.

Brief Contents

Preface xvii
Publisher Acknowledgments xxiv
Editors and Contributors xxvii
List of Acronyms xxxi
List of Common Symbols xxxiv

PART I Overview

1 Introduction and Radar Overview 3
2 The Radar Range Equation 59
3 Radar Search and Overview of Detection in Interference 87

PART II External Factors

4 Propagation Effects and Mechanisms 117
5 Characteristics of Clutter 165
6 Target Reflectivity 211
7 Target Fluctuation Models 247
8 Doppler Phenomenology and Data Acquisition 273

PART III Subsystems

9 Radar Antennas 309
10 Radar Transmitters 347
11 Radar Receivers 391
12 Radar Exciters 417
13 The Radar Signal Processor 459

PART IV Signal and Data Processing

14 Digital Signal Processing Fundamentals for Radar 495
15 Threshold Detection of Radar Targets 547
16 Constant False Alarm Rate Detectors 589
17 Doppler Processing 625
18 Radar Measurements 677
19 Radar Tracking Algorithms 713
20 Fundamentals of Pulse Compression Waveforms 773
21 An Overview of Radar Imaging 835

Appendix A: Maxwell's Equations and Decibel Notation 893
Appendix B: Answers to Selected Problems 899
Index 905

Contents

Preface xvii
Publisher Acknowledgments xxiv
Editors and Contributors xxvii
List of Acronyms xxxi
List of Common Symbols xxxiv

PART I Overview

1 Introduction and Radar Overview 3

1.1 Introduction 3
1.2 The Radar Concept 4
1.3 The Physics of EM Waves 5
1.4 Interaction of EM Waves with Matter 11
1.5 Basic Radar Configurations and Waveforms 18
1.6 Noise, Signal-to-Noise Ratio, and Detection 25
1.7 Basic Radar Measurements 27
1.8 Basic Radar Functions 33
1.9 Radar Applications 36
1.10 Organization of This Text 54
1.11 Further Reading 55
1.12 References 55
1.13 Problems 56

2 The Radar Range Equation 59

2.1 Introduction 59
2.2 Power Density at a Distance *R* 61
2.3 Received Power from a Target 62
2.4 Receiver Thermal Noise 64
2.5 Signal-to-Noise Ratio and the Radar Range Equation 66
2.6 Multiple-Pulse Effects 66
2.7 Summary of Losses 67
2.8 Solving for Other Variables 72

2.9 Decibel Form of the Radar Range Equation 72
2.10 Average Power Form of the Radar Range Equation 73
2.11 Pulse Compression: Intrapulse Modulation 74
2.12 A Graphical Example 75
2.13 Clutter as the Target 76
2.14 One-Way (Link) Equation 78
2.15 Search Form of the Radar Range Equation 79
2.16 Track Form of the Radar Range Equation 80
2.17 Some Implications of the Radar Range Equation 83
2.18 Further Reading 84
2.19 References 84
2.20 Problems 85

3 Radar Search and Overview of Detection in Interference 87

3.1 Introduction 87
3.2 Search Mode Fundamentals 89
3.3 Overview of Detection Fundamentals 95
3.4 Further Reading 111
3.5 References 111
3.6 Problems 112

PART II External Factors

4 Propagation Effects and Mechanisms 117

4.1 Introduction 117
4.2 Propagation Factor 118
4.3 Propagation Paths and Regions 119
4.4 Atmospheric Attenuation and Absorption 121
4.5 Atmospheric Refraction 130
4.6 Turbulence 137
4.7 Exploiting the Ionosphere 138
4.8 Diffraction 140
4.9 Multipath 142
4.10 Skin Depth and Penetration: Transmitting Through Walls 156
4.11 Commercial Simulations 158
4.12 Summary and Further Reading 160
4.13 References 161
4.14 Problems 163

5 Characteristics of Clutter 165
5.1 Introduction and Definitions 165
5.2 General Characteristics of Clutter 172
5.3 Clutter Modeling 202
5.4 Concluding Remarks 206
5.5 Further Reading 207
5.6 References 207
5.7 Problems 210

6 Target Reflectivity 211
6.1 Introduction 211
6.2 Basic Reflection Physics 212
6.3 Radar Cross Section Definition 219
6.4 Three Scattering Regimes 224
6.5 High-Frequency Scattering 227
6.6 Examples 236
6.7 Further Reading 244
6.8 References 244
6.9 Problems 245

7 Target Fluctuation Models 247
7.1 Introduction 247
7.2 Radar Cross Section of Simple Targets 248
7.3 Radar Cross Section of Complex Targets 251
7.4 Statistical Characteristics of the RCS of Complex Targets 253
7.5 Target Fluctuation Models 263
7.6 Doppler Spectrum of Fluctuating Targets 267
7.7 Further Reading 269
7.8 References 269
7.9 Problems 270

8 Doppler Phenomenology and Data Acquisition 273
8.1 Introduction 273
8.2 Doppler Shift 274
8.3 The Fourier Transform 276
8.4 Spectrum of a Pulsed Radar Signal 277
8.5 Why Multiple Pulses? 286

8.6 Pulsed Radar Data Acquisition 287
8.7 Doppler Signal Model 291
8.8 Range-Doppler Spectrum for a Stationary Radar 293
8.9 Range-Doppler Spectrum for a Moving Radar 296
8.10 Further Reading 303
8.11 References 303
8.12 Problems 303

PART III Subsystems

9 Radar Antennas 309

9.1 Introduction 309
9.2 Basic Antenna Concepts 310
9.3 Aperture Tapers 314
9.4 Effect of the Antenna on Radar Performance 317
9.5 Monopulse 320
9.6 Reflector Antennas 322
9.7 Phased Array Antennas 326
9.8 Array Architectures 339
9.9 Further Reading 343
9.10 References 343
9.11 Problems 345

10 Radar Transmitters 347

10.1 Introduction 347
10.2 Transmitter Configurations 351
10.3 Power Sources and Amplifiers 356
10.4 Modulators 371
10.5 Power Supplies 373
10.6 Transmitter Impacts on the Electromagnetic Environment 375
10.7 Operational Considerations 381
10.8 Summary and Future Trends 384
10.9 Further Reading 385
10.10 References 385
10.11 Problems 388

11 Radar Receivers 391

11.1 Introduction 391
11.2 Summary of Receiver Types 392

11.3 Major Receiver Functions 396
11.4 Demodulation 400
11.5 Receiver Noise Power 404
11.6 Receiver Dynamic Range 406
11.7 Analog-to-Digital Data Conversion 409
11.8 Further Reading 414
11.9 References 414
11.10 Problems 415

12 Radar Exciters 417

12.1 Introduction 417
12.2 Exciter-Related Radar System Performance Issues 418
12.3 Exciter Design Considerations 429
12.4 Exciter Components 440
12.5 Timing and Control Circuits 452
12.6 Further Reading 454
12.7 References 454
12.8 Problems 455

13 The Radar Signal Processor 459

13.1 Introduction 459
13.2 Radar Processor Structure 460
13.3 Signal Processor Metrics 462
13.4 Counting FLOPs: Estimating Algorithm Computational Requirements 464
13.5 Implementation Technology 472
13.6 Fixed Point versus Floating Point 480
13.7 Signal Processor Sizing 482
13.8 Further Reading 488
13.9 References 488
13.10 Problems 491

PART IV Signal and Data Processing

14 Digital Signal Processing Fundamentals for Radar 495

14.1 Introduction 495
14.2 Sampling 496
14.3 Quantization 504
14.4 Fourier Analysis 506

14.5 The z Transform 522
14.6 Digital Filtering 523
14.7 Random Signals 532
14.8 Integration 536
14.9 Correlation as a Signal Processing Operation 538
14.10 Matched Filters 540
14.11 Further Reading 543
14.12 References 543
14.13 Problems 544

15 Threshold Detection of Radar Targets 547

15.1 Introduction 547
15.2 Detection Strategies for Multiple Measurements 548
15.3 Introduction to Optimal Detection 552
15.4 Statistical Models for Noise and Target RCS in Radar 557
15.5 Threshold Detection of Radar Signals 560
15.6 Further Reading 584
15.7 References 584
15.8 Problems 585

16 Constant False Alarm Rate Detectors 589

16.1 Introduction 589
16.2 Overview of Detection Theory 590
16.3 False Alarm Impact and Sensitivity 592
16.4 CFAR Detectors 593
16.5 Cell Averaging CFAR 597
16.6 Robust CFARs 607
16.7 Algorithm Comparison 616
16.8 Adaptive CFARs 618
16.9 Additional Comments 619
16.10 Further Reading 620
16.11 References 620
16.12 Problems 622

17 Doppler Processing 625

17.1 Introduction 625
17.2 Review of Doppler Shift and Pulsed Radar Data 626
17.3 Pulsed Radar Doppler Data Acquisition and Characteristics 627

17.4 Moving Target Indication 629
17.5 Pulse-Doppler Processing 644
17.6 Clutter Mapping and the Moving Target Detector 665
17.7 Pulse Pair Processing 668
17.8 Further Reading 673
17.9 References 673
17.10 Problems 674

18 Radar Measurements 677

18.1 Introduction 677
18.2 Precision and Accuracy in Radar Measurements 678
18.3 Radar Signal Model 683
18.4 Parameter Estimation 685
18.5 Range Measurements 690
18.6 Phase Measurement 695
18.7 Doppler and Range Rate Measurements 696
18.8 RCS Estimation 699
18.9 Angle Measurements 700
18.10 Coordinate Systems 709
18.11 Further Reading 710
18.12 References 710
18.13 Problems 711

19 Radar Tracking Algorithms 713

19.1 Introduction 713
19.2 Basics of Track Filtering 719
19.3 Kinematic Motion Models 746
19.4 Measurement Models 751
19.5 Radar Track Filtering 757
19.6 Measurement-to-Track Data Association 760
19.7 Performance Assessment of Tracking Algorithms 766
19.8 Further Reading 767
19.9 References 768
19.10 Problems 770

20 Fundamentals of Pulse Compression Waveforms 773

20.1 Introduction 773
20.2 Matched Filters 774
20.3 Range Resolution 782

20.4 Straddle Loss 786
20.5 Pulse Compression Waveforms 787
20.6 Pulse Compression Gain 788
20.7 Linear Frequency Modulated Waveforms 789
20.8 Matched Filter Implementations 794
20.9 Sidelobe Reduction in an LFM Waveform 797
20.10 Ambiguity Functions 800
20.11 LFM Summary 808
20.12 Phase-Coded Waveforms 808
20.13 Biphase Codes 817
20.14 Polyphase Codes 824
20.15 Phase-Code Summary 829
20.16 Further Reading 830
20.17 References 830
20.18 Problems 833

21 An Overview of Radar Imaging 835

21.1 Introduction 835
21.2 General Imaging Considerations 837
21.3 Resolution Relationships and Sampling Requirements 843
21.4 Data Collection 852
21.5 Image Formation 856
21.6 Image Phenomenology 875
21.7 Summary 888
21.8 Further Reading 888
21.9 References 889
21.10 Problems 890

Appendix A: Maxwell's Equations and Decibel Notation 893

A.1 Maxwell's Equations 893
A.2 The Ubiquitous dB 895
A.3 Reference 897

Appendix B: Answers to Selected Problems 899

Index 905

Preface

Goals of the Book

As the editors of *Principles of Modern Radar: Basic Principles* (*POMR*), we had two primary goals in mind when this book was conceived. Our first goal was to design *POMR* to become the "Radar 101" textbook of choice for the next generation of radar engineers, whether students in graduate engineering courses, new hires on the job, or retraining professionals in government and industry. Our second goal was to provide a breadth of topics and modern approach that would make *POMR* the most convenient and valuable starting point for today's professionals needing to study or review a particular subject. To accomplish these twin goals, we needed to make several key trade-offs in designing the book:

1. Focus on modern techniques and systems from the start rather than historical background and legacy systems.
2. Strike a careful balance between quantitative mathematical models and tools and qualitative motivation and insight.
3. Carefully proportion the breadth of topics versus the depth of coverage of systems and external phenomenology.
4. Draw on the knowledge of a range of subject experts—and accept the intense editing effort needed to integrate their contributions into a coherent whole—versus the less comprehensive coverage but inherently consistent style and notation of just one or two authors.

What follows is a description of how these trade-offs were struck to achieve our goals.

Many in the radar community will recognize that *POMR* has evolved from the professional education short course of the same name taught to thousands of students by Georgia Tech research faculty since 1969. Some may even remember that the short course produced an earlier book, now out of print, by the same name.[1] This book is a completely new text, developed from scratch by 15 scientists and engineers working today with the most modern systems and techniques in radar technology. Each of these contributing authors brings a wealth of research and teaching experience to bear in explaining the fundamental concepts underlying all radar systems.

There are, of course, several very good books currently in use for college- and professional-level courses in radar systems and technology, so it is fair to ask why one should consider *POMR*. We believe the answer is fourfold:

- Comprehensiveness.
- Qualitative versus quantitative balance.
- Emphasis on the most modern topics and methods.
- Radar community support.

[1] Eaves, J.L., and Reedy, E.K., *Principles of Modern Radar.* Van Nostrand Reinhold, New York, 1987.

Most importantly, *POMR* provides a breadth of coverage unmatched by currently available introductory radar textbooks: chapters on fundamental concepts, propagation and echo phenomenology for targets and interference, all major subsystems of a modern radar, and all basic signal processing functions so important to modern practice. Second, these topics are presented both qualitatively and quantitatively, at a consistent level appropriate for advanced undergraduate and beginning graduate students and readers. No competing book of which we are aware strikes such a carefully constructed balance. Some competitors provide the traditional fundamental concepts but offer little on modern signal processing. Some are almost entirely descriptive, lacking the mathematical analysis students need to undertake their own analysis and modeling. A few others are highly mathematical but have limited coverage and lack the qualitative interpretation needed to develop the understanding of students new to the field. *POMR* not only provides the basic mathematical tools but also supports those tools with the explanations and insights of its experienced authors.

POMR's focus on *modern* radar is evident in its choice of topics. For example, extensive coverage is given to increasingly popular phased array antennas due to their advanced capabilities. Coherent systems, a prerequisite to most interesting signal processing, are strongly emphasized throughout the text. Last and most importantly, because so much functionality in modern systems lies in the signal processing, a significant portion of the book is devoted to methods enabled by digital radar signal processing, from pulse compression and Doppler processing to tracking and imaging. This topic choice and organization results in coverage superior to any other "Radar 101" textbook, so that *POMR* provides the most solid foundation for students progressing to "Radar 102" texts on more advanced and specialized topics.

Finally, *POMR* benefits from an extraordinary vetting by the modern radar community. It is a joint effort among the text's highly experienced authors and editors; the publisher SciTech, with its radar focus and resulting contacts; and the volunteering global community of radar experts, mostly fellow radar instructors and radar authors. As a result, the 21 chapters have been reviewed for content and style by more than 50 radar professionals representing academia, the government and military, and commercial companies. Chapters were reviewed first in their draft versions and then again after revisions. *POMR*'s editors were assisted in integrating the many reviewer suggestions by "master reviewers," each of whom read most or all of the chapters and also "reviewed the reviews" to help coordinate the improvements and perfect the emphasis, topical flow, and consistency across chapters. This extensive process of peer review iterations within the radar community ensures that *POMR* meets the needs of students, educators, and professionals everywhere.

Organization of Content

POMR is organized into four major parts: *Overview, The Radar Environment, Radar Subsystems*, and *Signal and Data Processing.* In teaching a technology area as broad as radar, it is difficult to design a topical sequence that proceeds in a straight line from start to finish without looking ahead or doubling back. The *Overview* section solves this problem by taking readers through a high-level first pass that familiarizes them with a range of fundamental radar concepts and issues, setting the stage for a more detailed examination in the remaining parts. Chapter 1 introduces basic concepts such as properties of electromagnetic waves, target and clutter echoes, monostatic and bistatic radar, and detection in noise. It also illustrates the scope of radar technology by describing a wide range of military and commercial applications. Finally, Chapter 1 introduces some radar cultural information such as the "band" terminology (e.g., L-band, X-band) and the AN Nomenclature for U.S. military systems. Chapter 2 delves into that most fundamental mathematical model in radar, the radar range equation. The basic point target range equation is derived, and its implications are explored. The chapter then develops several of the common variants tailored to specific radar modes. Chapter 3 provides a closer look at the most fundamental radar task of search and detection, describing search processes and introducing the idea of statistical detection and the resulting importance of probabilities in evaluating radar performance.

Part 2, *The Radar Environment*, is one of the truly distinguishing features of *POMR*. Few, if any, introductory radar texts provide the breadth and depth of discussion of propagation effects and target and clutter characteristics found here. Chapter 4 introduces all major electromagnetic propagation phenomenology of importance to radar, from simple attenuation in various weather conditions to more complex issues such as refraction, diffraction, multipath, ducting, and over-the-horizon propagation. Chapter 5 summarizes the extensive data on modeling the reflectivity and Doppler characteristics of atmospheric, land, and sea clutter and presents many of the common mean reflectivity and statistical models needed for clutter analysis. Chapter 6 introduces the mechanisms of scattering and reflection and the concept of radar cross section for targets, while Chapter 7 describes the common statistical models for radar cross section needed to evaluate detection performance. Chapter 8 delves more deeply into Doppler shift, concentrating on typical characteristics of Doppler spectra for stationary and moving targets and radar platforms.

Part 3, *Radar Subsystems*, describes each of the major subsystems of a typical modern radar system. Chapter 9 describes radar antenna technology, starting with basic antenna concepts and relations and then describing classic monopulse and mechanically scanned antennas. Half of this chapter is devoted to modern phased arrays, with detailed discussion of array patterns, wideband effects, and array architectures. Chapter 10 describes radar transmitter technology, including high-powered thermionic (tube-type) noncoherent and coherent transmitters, as well as solid-state transmitter technology. Again, significant coverage is devoted to transmitter modules and feed architectures for modern phased arrays. This chapter also addresses spectrum allocation and transmitter reliability issues, topics not found in other introductory textbooks. Chapter 11 presents radar receiver technology, beginning with the most basic types and extending to multistage superheterodyne receivers. Noise and dynamic range issues are discussed, and both classical analog synchronous detectors as well as the increasingly popular direct sampling digital receiver techniques for coherent systems are described. The coverage of coherent exciters in Chapter 12 is unique in an introductory textbook but important in understanding the architecture of modern systems. Exciter performance issues are presented, followed by a discussion of the technology available to implement modern coherent radar exciters. The importance of maintaining low phase noise for pulse-Doppler systems is also explained. Another topic unique to this textbook is Chapter 13, which discusses radar digital signal processor technology. Metrics and procedures for estimating processor loading are introduced, followed by discussion of alternative implementation technologies such as custom integrated circuits, reconfigurable hardware, and exciting new techniques such as the use of graphical processing units for real-time signal processing.

Part 4, *Signal and Data Processing*, concentrates on the increasingly sophisticated techniques used to extract ever more information from radar signals using advanced digital signal and data processing. The first half of Part 4 deals with signal processing basics, detection, and clutter rejection. It begins in Chapter 14 with a succinct summary of digital signal processor fundamentals such as sampling, quantization, and data acquisition, followed by a thorough review of discrete Fourier analysis, including windowing and interpolation. Other sections refresh the reader on digital filters, properties of random signals, and the all-important matched filter concept and its connection to data integration. Chapter 15 returns to the topic of threshold detection first introduced in Chapter 3. Here, much more attention is given to details of coherent and noncoherent integration and alternative ways of using the available data. Neyman-Pearson detection and the Swerling models are introduced, leading to optimum detectors for radar signals. Albersheim's and Shnidman's equations are presented as convenient computational aids. Chapter 16 continues the discussion by introducing constant false alarm rate (CFAR) threshold detection, a practical requirement in real interference environments. The properties, performance, and shortcomings of the basic cell-averaging CFAR are discussed in depth, and then many of the common "robust" and "adaptive" CFAR variants are introduced and compared. Chapter 17 covers two major forms of Doppler processing for clutter reduction: moving target indication (MTI), and pulse-Doppler processing. The discussion of MTI includes blind speeds, staggered pulse repetition frequencies, and airborne MTI. The sections on pulse-Doppler processing introduce the important topics of blind zones and ambiguity resolution. This chapter also includes a brief discussion of the pulse-pair processing method widely used in weather radar.

In the second half of Part 4, the focus turns to postdetection position measurements and tracking as well as high-resolution techniques. Chapter 18 addresses position measurements in range, angle, and Doppler. Basic concepts of precision and accuracy lead to the introduction of the Cramèr-Rao lower bound on precision. Several estimators of range, Doppler shift, and angle are then introduced, and their performance is evaluated. This chapter leads naturally into an introduction to tracking algorithms in Chapter 19. After a discussion of basic parameter estimation and some of the data association and resolution problems that complicate radar tracking, a number of tracking algorithms are introduced, from the basic $\alpha-\beta$ tracker to the Kalman filter. Chapters 20 and 21 introduce the techniques needed to achieve high-resolution radar imaging. Chapter 20 describes pulse compression for high-range resolution. The matched filter is investigated in more depth and is then applied to the most common wideband waveforms, including linear frequency modulation or "chirp" and phase-coded waveforms ranging from Barker codes to a variety of polyphase codes. Methods of range sidelobe control are described, and the ambiguity function is introduced as a means of designing and understanding waveform properties. Finally, Chapter 21 provides an overview of synthetic aperture radar (SAR) imaging. SAR data collection is described, and general, widely applicable resolution and sampling equations are derived. While the range of SAR image formation algorithms is too extensive and too advanced for an introductory textbook, descriptions are given of the two extremes: Doppler beam sharpening, one of the simplest imaging algorithms; and backprojection, the current "gold standard" for advanced imaging. The chapter closes with a discussion of the unique phenomenology of SAR imaging, including layover, shadows, and speckle. Collectively, the extensive coverage of signal processing in Part 4 of *POMR* provides an excellent springboard to study of more advanced topics such as advanced SAR, space-time adaptive processing, and multiple-input multiple-output radar.

An appendix reviews two basic electrical engineering topics that are important for understanding radar but not deemed necessary for inclusion within the chapters: Maxwell's equations and the use of decibels in describing radar values.

Features and Resources

POMR has been designed to ease the task of learning or teaching. Some of the features available to all readers include the following:

- Every chapter written by experts having "hands-on" experience in the design and development of radar systems.
- Every chapter reviewed by independent radar experts and edited by technical and publishing experts for content accuracy, level consistency, and readable style.
- Consistent notation and terminology employed throughout.
- Numerous illustrations integrated throughout, all newly drawn, clearly labeled, and carefully captioned.
- Table of common symbols and notation provided for quick reference.
- Table of acronyms, so plentiful in radar, presented alphabetically.
- Extensive, professionally prepared index facilitates reference use.
- At least 12 problems included in every chapter—over 250 total—to check and advance the student's understanding and capability. Answers to the odd-numbered problems are provided.

Several aids are available to adopting course instructors, with more being developed. The following aids can be obtained through request to SciTech at **pomr@scitechpub.com**:

- All problem answers and detailed solutions.
- All illustrations in the text in Microsoft PowerPoint sets or in high-resolution JPEG image formats for construction of custom viewgraphs.
- Copies of all equations in Microsoft Equation Editor format.

Several additional aids—tutorial simulations in MATLAB®[2] worked examples, additional problems for homework or exams—are expected to be available, and more are being developed and contributed by the radar community.

Publication of this first edition of *POMR* is just the first step in the development of a comprehensive set of resources for introducing radar systems and technology to a new generation of radar engineers. A website has been established to provide to readers these supporting materials, a complete and up-to-date list of reported errata, and an evolving set of new supplements. Visit the website periodically to check for the latest supplements and announcements:

http://www.scitechpub.com/pomr

---

[2]MATLAB is a registered trademark of The MathWorks, Inc. For MATLAB product information and cool user code contributions, go to http://www.mathworks.com, write The MathWorks, Inc., 3 Apple Hill Dr., Natick, MA 01760-2098 or call (508) 647-7101.

Companion Publications

Several remarkable publications are planned to complement, augment, and extend the material in *POMR: Basic Principles*:

Principles of Modern Radar: Advanced Techniques and Applications edited by William L. Melvin and James A. Scheer (2011)

Building on *POMR: Basic Principles*, this sequel provides extensive coverage of both advanced techniques in radar and a wide variety of specific modern applications that integrate the fundamental technologies into complete systems. Examples of advanced techniques include advanced waveforms, stripmap and spotlight synthetic aperture imaging, space-time adaptive processing, multiple-input, multiple-output radar, polarimetry, target protection, and electronic protection. Applications discussed include airborne pulse-Doppler radar, space-based radar, weather radar, air traffic control, and passive and bistatic systems. Together, the two *POMR* volumes will provide integrated and comprehensive coverage of modern radar, from basic concepts to advanced systems, all in a coherent and consistent style and notation.

Pocket Radar Guide: Key Radar Facts, Equations, and Data by G. Richard Curry (2010)

A quick reference in shirt pocket size to the very most important and commonly used facts, figures, and tables in real-world radar engineering practice.

Acknowledgments

Principles of Modern Radar could not have come into being without the dedicated efforts of many people. Each of our authors dedicated much personal time to contributing his or her individual chapters and then again to working with the entire *POMR* team to integrate the pieces into a whole that is greater than just the sum of those parts.

The authors were greatly aided by the reviewers and master reviewers. The complete list of reviewers is given in the "Publisher's Acknowledgments" section and so won't be repeated here, but every one of them had a direct hand in improving the final product in coverage, style, and correctness. Without their ability and willingness to critique the book based on their expert knowledge and their own experience with other introductory radar textbooks, we could not have achieved the level of consistency and coherency across such broad coverage and multiple authors. The authors and editors are greatly indebted to them for their efforts.

The entire *POMR* project might not have succeeded without the vision, support, and encouragement of SciTech Publishing and its president, Dudley Kay. SciTech is a wonderful asset to the radar community, and we hope this new book will add to that strength. Editorial assistant Katie Janelle managed the massive review process so important to *POMR*'s completion and quality. Production of a book is a complex endeavor requiring the efforts of many accomplished and dedicated staff. Robert Lawless is the production manager for *POMR*, responsible for managing the workflow and bringing together all the many pieces into the final product. Kristi Bennett, our copy editor, deserves great credit for bringing clarity, precision, and consistency to the writing styles of 15 authors. Freelancer Kathy Gagne conceived the eye-catching cover design. Brent Beckley has done an excellent job in marketing and promoting *POMR* so that it will reach and serve, we hope, a large audience. All of the SciTech team has been professional and, at the same time, a pleasure to work with every step of the way.

Errors and Suggestions

We have tried to bring the radar community a carefully constructed and truly valuable new introductory textbook and professional reference in *POMR*, but we recognize that there are always some residual errors and inconsistencies. In addition, experience in using *POMR* and new developments in the rapidly evolving field of radar will inevitably bring to light a need to clarify or expand some topics and introduce new ones.

The extensive review process used to develop *POMR* raised and resolved many, many such issues. Those that remain are the responsibility of the editors. We welcome the assistance of *POMR* readers in identifying errata and in making recommendations for improvements in future printings and editions. All identified errata will be posted in a timely fashion on the *POMR* SciTech web site (http://www.scitechpub.com/pomr), accessible to all users.

One of our hopes for *POMR* is that it will be adopted for use in university, professional education, and in-house training classes. There is nothing like teaching the material to newcomers to the field to quickly identify areas where the book could be improved. We invite all instructors using *POMR* to help us design the next edition by letting us know of your experience in using it and how it can be improved in the future.

Mark A. Richards
Georgia Institute of Technology
Atlanta, GA
mark.richards@ece.gatech.edu

James A. Scheer
Georgia Institute of Technology
Atlanta, GA
jim.scheer@gtri.gatech.edu

William A. Holm
Georgia Institute of Technology
Atlanta, GA
bill.holm@gatech.edu

Publisher Acknowledgments

Master Reviewers

Above and beyond the peer reviews focused on technical content, we recognized the need to bring consistency of level, notation, and writing style to an edited book composed of numerous expert contributions if we were to attain our goal of an outstanding textbook. From the initial rounds of reviews and follow-up conversations, SciTech determined which reviewers best understood our concerns, and we invited deeper involvement from them. The reviewers listed herein were called on to help resolve controversial and contradictory reviewer suggestions, to respond to editor and contributor problems, and, most of all, to work closely with SciTech as representatives of the radar community. The dedication of these "Master Reviewers" therefore merits our special recognition and appreciation.

Dr. John Milan (radar consultant): John brings more than 36 years of experience in radar systems at ITT, Gilfillan, and many years on the IEEE Aerospace and Electronic Systems Society Radar Systems Panel. He reviewed every chapter of the book for the sake of internal consistency and became the volunteer "master of figure captions," assessing every one for descriptive completeness and clarity. As impressive to us as the volume of review work undertaken, equally remarkable was John's response time to every request or question.

Mr. Byron Edde (radar/electronic warfare consultant and short course instructor, textbook author): Drawing on 40 years of experience designing and improving radar and electronic warfare systems, before reviewing materials Byron composed a hierarchal list of "What makes a textbook great?" As a successful textbook and radar study guide author who teaches courses to Navy personnel, Byron understood exactly what the *POMR* twin goals were and why they would be challenging to attain. He thus helped "set the bar" and provided broad stroke feedback on maintaining consistent depth and balance of theory, math, and practical reference to current technology. He is author of *Radar*: *Principles, Technology, Applications* and *The IEEE Radar Study Guide*.

Mr. G. Richard Curry (radar consultant and radar book author): Dick offered perceptive technical improvement comments across numerous chapters and was one of those capable of, and interested in, comparing reviewer comments objectively to help suggest the best solution. He is the author of *Radar System Performance Modeling*.

Mr. Paul G. Hannen (SAIC senior engineer, professor at Wright State University, and book author): Paul was most persistent and helpful about the first three chapters that "set the scene." He provided literally hundreds of suggested edits and worked directly with volume editor and chapter author Jim Scheer. If you perceive an especially meticulous handling of background radar facts in Part 1, give credit to Paul. He is coauthor of *Radar Principles for the Non-specialist* (3d ed.).

Dr. Randy J. Jost (senior scientist at USU Space Dynamics Lab, book author, Department of Defense consultant): Suggesting edits to various chapters for optimum organization, particularly Chapter 4, "Propagation Effects and Mechanisms," and contributing substantial content to Chapter 10, "Radar Transmitters," and always giving sound suggestions for improvements throughout, Randy proved once again why he is such an important author, reviewer, and advisor to SciTech. Randy is coauthor of *Fundamentals of Electromagnetics with MATLAB* (2d ed.).

Dr. David G. Long (professor and research director of remote sensing at Brigham Young University): David's help was invaluable in critiquing the sophisticated radar signal and data processing chapters and his contributions of content to Chapter 18, "Radar Measurements." David worked directly with volume editor and multiple chapter author Mark Richards to hone the chapters in Part 4.

Dr. Marshall Greenspan (senior systems consulting engineer, Northrop Grumman Corporation [NGC]): Marshall was not only a willing and able technical reviewer, particularly within signal processing chapters, but was also extremely helpful in suggestions and procurement of contemporary photographs via his contacts with the NGC public relations departments.

Dr. Simon Watts (deputy scientific director, Thales UK and book author): After providing excellent technical comments on several chapters, Simon was called on to assist with final edits to the organization, completeness, and notational consistency to his particular area of expertise, radar clutter, in Chapter 6. He is coauthor of the book *Sea Clutter: Scattering, the K Distribution and Radar Performance*.

Technical Reviewers

SciTech Publishing gratefully acknowledges the contributions of the following technical reviewers, who selected chapters of interest and read each carefully and completely, often in multiple iterations and often with substantive suggestions of remarkable thought and depth. Taken in aggregate, the value of their reviews was beyond measure and quite possibly unprecedented for a radar book:

Dr. Clive Alabaster, *lecturer, Cranfield University, GBR*
Chris Baker, *dean and director, ANU College of Engineering and Computer Science, Canberra, AUS*
Dr. Ronald Aloysius, *fellow engineer, Northrop Grumman Corporation*
Edward Barile, *senior principal engineer, Raytheon Corporation*
Dan Bernabei, *engineer scientist, Department of Defense*
Lee Blanton, *radar systems engineer, General Atomics Aeronautical Systems, Inc.*
Neal Brune, *vice president of countermeasures research and development, Esterline Defense Technologies*
Gerry Cain, *DSP Creations Ltd.*
Kernan Chaisson, *U.S. Air Force retired, Washington editor, Forecast International*
I.-Ting Chiang, *applicant consultant, Lorentz Solution Inc.*
Dr. Jean-Yves Chouinard, *professor, Université Laval, Quebec, CAN*
Lawrence Cohen, *electronics engineer, radar division, Naval Research Laboratory*

Carlton Davis, *senior advisory engineer, Northrop Grumman Corporation*
Dr. Muhammad Dawood, *assistant professor, New Mexico State University*
Patrick Dever, P.E., *fellow engineer, Northrop Grumman Corporation*
Robert Egri, *Cobham DES*
Dr. John J. Ermer, *engineering fellow, Raytheon Space and Airborne Systems*
Dr. Mark Frank, *principal engineer, Rohde & Schwarz Inc.*
Christophe Fumeaux, *associate professor, University of Adelaide, AUS*
Dr. Fulvio Gini, *Professor, University of Pisa, ITA*
James D. Gitre, *manager, Motorola*
Nathan A. Goodman, *associate professor, University of Arizona*
Dr. Martie Goulding, *senior radar systems engineer, MacDonald Dettwiler & Associates, CAN*
John M. Green, *senior lecturer, Naval Postgraduate School*
Dr. Hugh Griffiths, *chair of radiofrequency sensors, University College London, GBR*
Dr. Walter Gustavo Fano, *associate professor, Universidad Nacional de la Patagonia San Juan Bosco*
Dr. Stephen Harman, *radar systems technical manager, QinetiQ, UK*
Dr. Joseph Hucks, *electrical engineer, Harris Corporation*
Thomas Jeffrey, *senior engineering fellow, Raytheon Integrated Defense Systems*
Dr. Alan R. Keith, *Boeing Defense, Space and Security*
Stephane Kemkemian, *radar senior expert, Thales Airborne Systems, FRA*
Dr. Anatolii Kononov, *Senior Researcher—Dept. of Radio and Space Science, Chalmers University of Technology, SWE*
Dr. Theodoros G. Kostis, *research scientist, University of the Aegean, GR*
Dr. Richard Lane, *research scientist, QinetiQ*
François Le Chevalier, *scientific director, Thales Air Systems, FRA*
Tony Leotta, *radar consultant, ADL Associates*
Richard Lethin, *president, Reservoir Labs*
David Mackes, *senior engineer, Northrop Grumman Corporation*
Kevin McClaning, *senior radiofrequency designer, Johns Hopkins University*
Anders Nelander, *Swedish Defense Research Agency, SWE*
Natalia K. Nikolova, *professor, McMaster University, CAN*
Dr. Myriam Nouvel, *radar engineer, Radar and Warfare Technical Directorate, Thales Airborne Systems, FRA*
Dr. Chris Oliver, *CBE, technical director, InfoSAR, GBR*
Karl Erik Olsen, *senior scientist, Norwegian Defence Research Establishment, NOR*
Dr. Pinaki S. Ray, *research associate, The University of Adelaide, AUS*
Dr. Brian D. Rigling, *associate professor, Wright State University*
Firooz Sadjadi, *senior staff research scientist, Lockheed Martin Cooperation*
Dr. Earl Sager, *radar physics group chief scientist, System Planning Corporation*
Dr. Paul E. Schmid, *president, Engineering Systems, Inc.*
John Shipley, *senior scientist, Harris Corporation*
Dr. John Spurlin, P.E., *professor, Norfolk State University*
Dr. Roger Sullivan, *radar consultant*
Chin Yeng Tan, *research assistant, The University of Nottingham–Malaysia*
John Wendler, *Harris Corporation*
Dr. Andreas Wiesmann, *GAMMA Remote Sensing AG–Switzerland, CHE*
Richard Wiley, *Research Associates of Syracuse*
Ben Winstead, *principal development engineer, Honeywell International, Inc.*

Editors and Contributors

Volume Editors

Dr. Mark A. Richards

Volume editor-in-chief and multiple chapter author

Mark Richards is a faculty member in Electrical and Computer Engineering at the Georgia Institute of Technology, teaching and conducting research in the areas of digital signal processing, radar signal processing, and high performance embedded computing. He was previously Chief of the Radar Systems Division in the Sensors and Electromagnetic Applications Laboratory of the Georgia Tech Research Institute (GTRI). He is the author of *Fundamentals of Radar Signal Processing* (McGraw-Hill, 2005), as well as co-editor or contributor to four other books. He received his Ph.D. from Georgia Tech in 1982.

Mr. James A. Scheer

Associate volume editor and multiple chapter author

Jim Scheer has 40 years of hands-on experience in the design, development, and analysis of radar systems. He currently consults and works part time for GTRI and teaches radar-related short courses. He began his career with the General Electric Company (now Lockheed Martin Corporation), working on the F-111 attack radar system. In 1975 he moved to GTRI, where he worked on radar system applied research until his retirement in 2004. Mr. Scheer is an IEEE Life Fellow and holds a BSEE degree from Clarkson University and the MSEE degree from Syracuse University.

Dr. William A. Holm

Associate volume editor and multiple chapter co-author

Bill Holm is the associate vice provost for distance learning & professional education at Georgia Tech, program director for the defense technology professional education program, and is a principal research scientist at GTRI. His research in radar technology, signal processing techniques, and related subjects has resulted in over 75 technical papers, research papers, and book chapters. His 30+ years of instruction experience include the "Principles of Modern Radar" and "Basic Radar Concepts" short courses and teaching in the School of Physics. Dr. Holm holds a Ph.D. degree in physics from Georgia Tech.

Chapter Contributors

Mr. Christopher Bailey

Chapter 9 – Radar Antennas

Chris Bailey is a GTRI research engineer with experience in phased-array antenna design, analysis, and modeling, and phased-array radar-system engineering. His recent research efforts include digital beamforming, overlapped subarrays architectures, and low-power/low-cost arrays. Bailey has written numerous reports on phased array technology and regularly teaches phased array courses with the Georgia Tech Defense Technology Professional Education Program. He holds a M.S.E.E. from Johns Hopkins University.

Dr. William Dale Blair

Chapter 18 – Radar Measurements and Chapter 19 – Radar Tracking Algorithms

Dale Blair is the academic administrator of the Radar Tracking GTRI Defense course. He is a senior research engineer at the GTRI Sensors and Electromagnetic Applications Laboratory (SEAL), and has been involved in the research and development and testing of target tracking algorithms and radar signal processing for more than 14 years, and is currently involved in phased array radar and multisensor tracking.

Mr. Joseph A. Bruder, PE

Chapter 11 – Radar Receivers

Joe Bruder retired from GTRI after 25 years but is actively working part-time there and Stiefvater Consultants. He has extensive experience in radar sensor technology, including radar system design, analysis and evaluation, test planning, testing and test measurements. At the USAF Rome Laboratory his research areas were space-based radar, bistatic radar, foliage penetration, and bird hazard detection. He is an IEEE Fellow, a member of the IEEE/AESS Radar Systems Panel and is the standards representative for the Panel.

Mr. Nicholas (Nick) C. Currie

Chapter 5: Radar Clutter Characteristics

Nick Currie served on the staff of GTRI for 30 years, performing measurements of the radar backscatter of the sea, rain, snow, vegetation, and sea ice, military and civilian land vehicles, small waterborne craft, missiles, and aircraft. He has consulted with DARPA and the National Institute of Justice on concealed weapon detection and through-the-wall surveillance, and with the USAF Rome Laboratory in the development of a cylindrical, bistatic RCS range. He is a Fellow of the IEEE for work in millimeter wave measurements. He has edited and coauthored four books in the field of radar measurements and clutter.

Dr. Randy J. Jost

Chapter 10 – Radar Transmitters

Randy Jost is a Senior Scientist at the Utah State University Space Dynamics Laboratory. He also holds adjunct positions in both the Electrical and Computer Engineering Department and the Physics Department. His areas of research expertise include Computational Electromagnetics, Radar, Remote Sensing, Electromagnetic Compatibility, Wireless Communication, Electronic Materials, Electromagnetic Measurements and Metrology & Characterization of Antenna, Radar and Optical Measurement Systems. Dr. Jost is an active member and officer in the IEEE Electromagnetic Compatibility Society and the Antenna Measurement Techniques Association (AMTA).

Dr. Byron M. Keel

Chapter 16 – CFAR Processors and Chapter 20 – Pulse Compression Fundamentals

Byron Keel is a Principal Research Engineer and Head of the Signal Processing Branch within the Radar Systems Division of GTRI. He has over 20 years of experience and active research in radar waveform design, signal processing, and systems analysis. He regularly teaches in GTRI sponsored short courses including "Principles of Modern Radar" and is course director and principal lecturer in "Radar Waveforms."

Dr. David G. Long

Chapter 18 – Radar Measurements

David Long is a Professor in the Electrical and Computer Engineering Department at Brigham Young University (BYU) and is Director of the BYU Center for Remote Sensing. He has over 20 years of experience in the design of remote sensing radar systems, signal processing, and systems analysis. He is a Fellow of the IEEE.

Mr. Jay Saffold

Chapter 4 – Propagation Effects and Mechanisms

Jay Saffold is the Chief Scientist for RNI and has over 20 years engineering experience in both military and industry research in RF tags, virtual reality, digital databases, soldier tracking systems, millimeter wavelength (MMW) radar, multimode (MMW and optical) sensor fusion, fire-control radar, electronic warfare, survivability, signal processing, and strategic defense architecture. He lectures annually for GTRI on remote sensing and signal processing. He has authored or co-authored over 104 technical papers and reports. He holds a BSEE degree from Auburn University.

Dr. Paul E. Schmid

Chapter 10 – Radar Transmitters

Paul Schmid is president/owner of Engineering Systems, Inc. a Virginia consulting firm. He has fifty years industry and government experience in electromagnetic propagation, aerospace electronics, radio frequency systems, optical systems, and antenna theory that includes significant contributions to the Navy's AEGIS phased array radar, NASA's Apollo Program, and over fifty technical papers. He is a Life Senior Member of the IEEE.

Dr. John Shaeffer

Chapter 6 – Target Reflectivity

John Shaeffer has taught short courses on Radar Cross Section for over twenty years and is coauthor of *Radar Cross Section, 2nd Edition* (SciTech Publishing), the leading book on the subject. He has held senior engineering positions at McDonnell Douglas, GTRI, Lockheed Martin, and NASA, and was co-founder of Marietta Scientific and founder of Matrix Compression Technologies, LLC. He earned his PhD in Physics from Saint Louis University (1971).

Dr. Gregory A. Showman

Chapter 21 – Introduction to Radar Imaging

Greg Showman is a Senior Research Engineer at GTRI, acts as the Director of the Adaptive Sensor Technology Project Office within GTRI, and has over 20 years of experience in radar modeling, performance analysis, and signal processing algorithm development.

Mr. Tracy Wallace
Chapter 10 – Radar Transmitters

Tracy Wallace is Division Chief for the Air and Missile Defense Division of GTRI's Sensors and Electromagnetic Applications Laboratory. He supports solid-state, active-aperture radar development with focus on the frontend electronics, power systems, and system performance assessment. He has also designed and built high power tube-based transmitters for instrumentation radars. He teaches in numerous radar-related short courses: Principles of Modern Radar, Phased Array Radar Systems, Space-Based Radar, Transmit/Receive Modules for Phased Array Radar, and Coherent Radar Performance Estimation.

List of Acronyms

The following acronyms are used throughout this text. Some acronyms, *e.g.* SIR, have more than one meaning; the appropriate meaning is generally clear from the context.

Acronym	Definition
1-D	One Dimensional
2-D	Two Dimensional
3-D	Three Dimensional
A	Ampere
AAW	Anti-Air Warfare
AC	Alternating Current
ACF	Autocorrelation Function
ADC	Analog-to-Digital Converter, Analog-to-Digital Conversion
A-DPCA	Adaptive Displaced Phase Center Antenna
AESA	Active Electronically Scanned Array
AF	Array Factor
AGC	Automatic Gain Control
AGL	Above Ground Level
AL	Altitude Line
AM	Amplitude Modulation
AMTI	Airborne Moving Target Indication
AOA	Angle of Arrival
API	Application Programming Interface
AR	Autoregressive
ARMA	Autoregressive Moving Average
ASIC	Application-Specific Integrated Circuit
BIT	Built-In Test
bps	Bits per second
BPF	Bandpass Filter
BMD	Ballistic Missile Defense
BRL	Ballistics Research Laboratory (U.S. Army)
CA	Cell Averaging
CA-CFAR	Cell Averaging Constant False Alarm Rate
CBE	Cell Broadband Engine
CDF	Cumulative Distribution Function
CDL	Common Data Link
CFA	Crossed Field Amplifier
CFAR	Constant False Alarm Rate
CFLOPS	Complex Floating Point Operations Per Second
CMOS	Complementary Metal Oxide Semiconductor
CNR	Clutter-to-Noise Ratio
COHO	Coherent Oscillator
COTS	Commercial Off-the-Shelf
CPI	Coherent Processing Interval
CRLB	Cramèr-Rao Lower Bound
CRT	Chinese Remainder Theorem
CS	Censored
CUT	Cell Under Test
CW	Continuous Wave
DAC	Digital-to-Analog Converter
DARPA	Defense Advanced Research Projects Agency
dB	Decibel
dbc	Decibels relative to the Carrier
DBS	Doppler Beam Sharpening
DC	Direct Current
DCT	Discrete Cosine Transform
DDS	Direct Digital Synthesis, Direct Digital Synthesizer
DFT	Discrete Fourier Transform
DOA	Direction of Arrival
DOF	Degrees of Freedom
DPCA	Displaced Phase Center Antenna
DRO	Dielectric Resonant Oscillator
DSP	Digital Signal Processing
DSX	Direct Synthesizer
DTFT	Discrete Time Fourier Transform
EA	Electronic Attack
ECM	Electronic Countermeasures
EIO	Extended Interaction (Klystron) Oscillator
EKF	Extended Kalman Filter
EM	Electromagnetic
EMI	Electromagnetic Interference
ENOB	Effective Number of Bits
EP	Electronic Protection
ES	Electronic Support
ESA	Electronically Scanned Array
EW	Electronic Warfare
f/D	Focal length to Diameter ratio
FAR	False Alarm Rate
FCR	Fire Control Radar
FDS	Fractional Doppler Shift
FET	Field Effect Transistor
FFT	Fast Fourier Transform
FIR	Finite Impulse Response
FLOPs	Floating Point Operations
FLOPS	Floating Point Operations Per Second
FM	Frequency Modulation
FMCW	Frequency-Modulated Continuous Wave

Acronym	Definition
FOPEN	Foliage Penetration
FOV	Field Of View
FPGA	Field Programmable Gate Array
ft	Foot, feet
FWHM	Full Width at Half Maximum
GaAs	Gallium Arsenide
GaN	Gallium Nitride
Gbps	Giga bits per second
gcd	Greatest Common Divisor
GCMLD	Generalized Censored Mean Level Detector
GFLOPS	GigaFLOPS
GHz	Gigahertz
GMTI	Ground Moving Target Indication
GOCA-CFAR	Greatest-Of Cell Averaging Constant False Alarm Rate
GOPS	GigaOperations Per Second
GPEN	Ground Penetration
GPR	Ground Penetrating Radar
GPU	Graphical Processing Unit
GTRI	Georgia Tech Research Institute
HCE	Heterogeneous Clutter Estimation
HPC	High Performance Computing
HPD	High Power Density
HPEC	High Performance Embedded Computing
HRR	High Range Resolution
Hz	Hertz (cycles per second)
I	In-phase channel or signal
IC	Integrated Circuit
ICBM	Intercontinental Ballistic Missiile
ID	Identification
IEEE	Institute of Electrical and Electronic Engineers
IID	Independent and Identically Distributed
IF	Intermediate Frequency
IFF	Identification Friend or Foe
IFM	Instantaneous Frequency Measurement
IFSAR	Interferometric Synthetic Aperture Radar
IIR	Infinite Impulse Response
IMPATT	Impact Ionization Avalanche Transit Time
InP	Indium Pholsphide
IPP	InterPulse Period
I/Q	In-phase/Quadrature
ISAR	Inverse Synthetic Aperture Radar
ISR	Integrated Sidelobe Ratio
kHz	Kilohertz
kVA	KiloVolt-Ampere
kW	KiloWatt
lcm	Least common multiple
LE	Leading Edge
LEO	Low Earth Orbit
LFM	Linear Frequency Modulation
LHC	Left-Hand Circular
LNA	Low-Noise Amplifier
LO	Local Oscillator

Acronym	Definition
LOS	Line of Sight
LPD	Low Power Density
LRT	Likelihood Ratio Test
LSB	Least Significant Bit, Lower Sideband
LSI	Linear Shift-Invariant
LUT	Look-Up Table
LVDS	Low Voltage Differential Signaling
m	Meter
Mbps	Megabits per second
MB/s	MegaBytes per Second
MCM	Multichip Module
MCRLB	Modified Cramèr-Rao Lower Bound
MDD	Minimum Detectable Doppler
MDS	Minimum Detectable Signal
MDV	Minimum Detectable Velocity
MEM	Micro-Electromechanical
MESFET	Metal Semiconductor Field Effect Transistor
MFA	Multiple-Frame Assignment
MHT	Multiple-Hypothesis Tracking
MHz	Megahertz
MIPS	Millions of Instructions per Second
MIT	Massachusetts Institute of Technology
MIT/LL	Massachusetts Institute of Technology Lincoln Laboratory
MLC	Mainlobe Clutter
MLS	Maximum Length Sequence
MMIC	Monolithic Microwave Integrated Circuit
MMSE	Minimum Mean Square Error
MMW	Millimeter Wave
MoM	Method of Moments
MOPA	Master Oscillator Power Amplifier
MOTR	Multiple-Object Tracking Radar
MPI	Message Passing Interface
MPM	Microwave Power Module
MPS	Minimum Peak Sidelobe
MTD	Moving Target Detector
MTI	Moving Target Indication
MTT	Multi-Target Tracking
mW	Milliwatt
MW	Megawatt
NCA	Nearly Constant Acceleration
NCCS2	Non-Central Chi-Square of degree 2
NCV	Nearly Constant Velocity
NEES	Normalized Estimation Error Squared
NLFM	Nonlinear Frequency Modulation
NP	Neyman-Pearson
NRA	No Return Area
NRE	Non-Recurring Engineering
NRL	Naval Research Laboratory
NRE	Non-Recurring Engineering
OLA	Overlap-Add
OS	Ordered Statistic
OTH	Over the Horizon

Acronym	Definition
PA	Power Amplifier, Power-Aperture
PAG	Power-Aperture-Gain
PC	Personal Computer
PDF	Probability Density Function
PDR	Phase-Derived Range
PDRO	Phase-Locked Dielectric resonant Oscillator
PEC	Perfect Electric Conductor
PFA	Polar Formatting Algorithm
PFN	Pulse-Forming Network
PLL	Phase-Locked Loop
PLO	Phase-Locked Oscillator
PPI	Plan Position Indicator
ppm	Parts per million
PPP	Pulse Pair Processing
PRF	Pulse Repetition Frequency
PRI	Pulse Repetition Interval
PSD	Power Spectral Density
PSM	Polarization Scattering Matrix
PSR	Point Spread Response
Q	Quadrature phase channel or signal, Quality factor
QPE	Quadratic Phase Error
QRD	Q-R Decomposition
RAM	Radar Absorbing Material
RASS	Radio-Acoustic Sounding System
RBGM	Real Beam Ground Mapping
RCS	Radar Cross Section
REX	Receiver/Exciter
RF	Radiofrequency, Radar Frequency
RFLOPS	Real Floating Point Operations Per Second
RHC	Right Hand Circular
RMA	Range Migration Algorithm
rms	Root Mean Square
RPM	Revolutions per Minute
ROC	Receiver Operating Curve, Receiver Operating Characteristic
RRE	Radar Range Equation
rss	Root Sum of Squares
RTL	Register Transfer Level
rv	Random Variable
RX	Receive, Receiver
s	Second
SAR	Synthetic Aperture Radar
SAW	Surface Acoustic Wave
SBC	Single Board Computer
SBO	Shoe-Box Oscillator
SCR	Silicon-Controlled Rectifier
SFDR	Spurious-Free Dynamic Range
SiC	Silicon Carbide
SiGe	Silicon-Germanium
SINR	Signal-to-Interference-plus-Noise Ratio
SIR	Signal-to-Interference Ratio
SIR-C	Shuttle Imaging Radar-C
SLAR	Side-Looking Airborne Radar
SLC	Sidelobe Clutter
SM	Standard Missile
SMT	Surface Mount Technology
SM2	Standard Missile 2
SNR	Signal-to-Noise Ratio
SOCA-CFAR	Smallest-Of Cell-Averaging Constant False Alarm Rate
SPEC	Standard Performance Evaluation Corporation
SPST	Single-Pole, Single-Throw
SQNR	Signal-to-Quantization Noise Ratio
sr	Steradian
SSB	Single Sideband
STALO	Stable Local Oscillator
STAP	Space-Time Adaptive Processing
STC	Sensitivity Time Control
SVD	Singular Value Decomposition
TB	Time-Bandwidth product
TDRSS	Tracking and Data Relay Satellite System
TDU	Time Delay Unit
TE	Trailing Edge
TFLOPS	TeraFLOPS
THAAD	Theater High Altitude Air Defense
TI	Texas Instruments
TMR	Target Motion Resolution
TOPS	TeraOps Per Second
T/R	Transmit/Receive
TRF	Tuned Radio Frequency
TSS	Tangential Signal Sensitivity
TRF	Tuned Radio Frequency
TWS	Track While Scan
TWT	Traveling Wave Tube
TX	Transmit
UAV	Unmanned Aerial Vehicle
UDSF	Usable Doppler Space Fraction
UHF	Ultra-High Frequency
UMOP	Unintentional Modulation of Pulse
U.S.	United States
USB	Upper Sideband
V	Volt
VHDL	VHSIC Hardware Description Language
VHF	Very High Frequency
VME	VersaModule Europe
VSIPL	Vector, Signal, Image Processing Library
VSWR	Voltage Standing Wave Ratio
VXS	VersaModule Europe Switched Serial
W	Watt

List of Common Symbols

The following symbol definitions are used in multiple chapters throughout this text. Each individual chapter introduces additional notation specific to that chapter. Some symbols; *e.g.* R; have more than one usage; their meaning is generally clear from the context

Symbol	Definition
α	Attenuation coefficient
χ_1	Single-sample Signal-to-Noise Ratio
χ_N	N-sample Signal-to-Noise Ratio
δ	Grazing angle; Discrete impulse function
δ_D	Dirac (continuous-time) impulse function
Δ	Difference channel; Quantization Step Size
ΔCR	Cross-range resolution
Δf_d	Doppler spectrum width
ΔR	Range resolution
ε_r	Relative permittivity
Γ	Fresnel reflection coefficient
η	Clutter volume reflectivity; Extinction efficiency
η_a	Aperture efficiency
λ	Wavelength
Λ	Likelihood ratio
μ	Permeability
ϕ	Elevation angle (from horizontal plane); General angle or phase
ϕ_3	Elevation 3-dB one-way beamwidth
ϕ_R	Elevation Rayleigh (peak-to-null) beamwidth
ϕ_{xx}	Autocorrelation function
θ	Azimuth angle; General angle or phase
θ_3	Azimuth 3 dB one-way beamwidth
θ_B	Brewster's angle
θ_C	Critical angle
θ_{cone}	Cone angle
θ_R	Azimuth Rayleigh (peak-to-null) beamwidth
θ_{scan}	Scan angle
σ	Radar cross section
σ^0	Clutter area reflectivity
σ_n^2	Noise variance
σ_x^2	Variance of random variable or process x
$\sum$	Sum channel
τ	Pulse width (duration)
$\hat{\omega}$	Normalized frequency in radians per sample
ω	Frequency in radians per second
ω_d	Doppler frequency in radians per second
Ω	Solid angle in steradians; Impedance in ohms
A_e	Effective aperture
b	Number of Bits
B	Bandwidth in hertz
B_d	Doppler bandwidth in hertz
c	Speed of electromagnetic wave propagation
D	Antenna size; Divergence factor
D_{SAR}	Synthetic aperture size
$E\{\cdot\}$	Expected value operator
E	Energy
E_x	Energy in signal x
$\hat{f}$	Normalized frequency in cycles per sample
f	Frequency in hertz; focal length
f_d	Doppler Shift
f_s	Sampling frequency in samples per second
F	Noise factor; Noise figure; Propagation factor
FAR	False alarm rate
G	Antenna gain
G_t	Transmit antenna gain
G_r	Receive antenna gain
$(\cdot)^H$	(superscript H) Hermitian (conjugate) transpose
$h(t)$ or $h[n]$	Filter impulse response (continuous or discrete)
$H(\cdot)$	Filter frequency response
H_0	Null (target absent) hypothesis
H_1	Target present hypothesis
I	In-phase channel or signal
$\mathbf{I}$	Identity matrix
I_0	Modified Bessel function of the first kind
ISR	Integrated sidelobe ratio
$J(\cdot)$	Cramèr-Rao Lower bound
k	Boltzmann's constant
$\hat{k}_s$	Normalized total wavenumber (spatial frequency) in radians per sample
K	Discrete Fourier transform (DFT) Size
k	Total wavenumber (spatial frequency) in radians per meter
$k_x; k_y; k_z$	x; y; and z components of wavenumber (spatial frequency) in radians per meter
L	General loss; Number of range bins; Number of fast-time samples
L_a	Atmospheric loss
L_s	System loss
M	Number of pulses; Number of slow-time samples

Symbol	Definition
n	Index of refraction
N	Number of samples; Vector length
$\mathrm{N}(u; v)$	Normal (Gaussian) distribution with mean u and variance v
N_0	Noise power spectral density
$O(\cdot)$	"On the Order of"
$p_x(x)$	Probability density function of x
p_{avg}	Average power
P_D	Probability of detection
P_{FA}	Probability of false alarm
P_r	Received power
PRF	Pulse repetition frequency
PRI	Pulse repetition interval
PSR	Peak sidelobe to peak mainlobe ratio
P_t	Transmitted power
Q	Quadrature channel or signal; Power density
Q_M	Marcum's Q function
R	Range; Rain rate
$\mathbf{R_I}$	Interference covariance matrix
R_{ua}	Unambiguous range
S_{xx}	Power spectrum of random process x
SIR	Signal-to-interference ratio
$SINR$	Signal-to-interference-plus-noise ratio
SNR	Signal-to-noise ratio
$SQNR$	Signal-to-quantization noise ratio
t	Time
$(\cdot)^T$	(superscript T) Matrix or vector transpose
T	Threshold value; Pulse Repetition Interval (Interpulse Period)
T_0	Standard temperature; Period of radiofrequency (RF) sinusoid
T_{ad}	Antenna dwell time
T_d	Dwell time
T_s	Sampling interval; System noise temperature
v	Velocity
v_r	Radial velocity
$\mathrm{var}(x)$	Variance of a random variable or process x
$\boldsymbol{X}$	General vector variable
$\mathbf{X}$	General matrix variable
$\overline{x}$	Mean of a random variable or process x
$x_I(t); x_I[t]$	In-phase signal (continuous or discrete)
$x_Q(t); x_Q[t]$	Quadrature phase signal (continuous or discrete)

PART I

Overview

CHAPTER 1 Introduction and Radar Overview

CHAPTER 2 The Radar Range Equation

CHAPTER 3 Radar Search and Overview of Detection in Interference

CHAPTER

1

Introduction and Radar Overview

James A. Scheer, William A. Holm

Chapter Outline

1.1 Introduction 3
1.2 The Radar Concept 4
1.3 The Physics of EM Waves 5
1.4 Interaction of EM Waves with Matter 11
1.5 Basic Radar Configurations and Waveforms 18
1.6 Noise, Signal-to-Noise Ratio, and Detection 25
1.7 Basic Radar Measurements 27
1.8 Basic Radar Functions 33
1.9 Radar Applications 36
1.10 Organization of This Text 54
1.11 Further Reading 55
1.12 References 55
1.13 Problems 56

1.1 INTRODUCTION

Radar systems have evolved tremendously since their early days when their functions were limited to target detection and target range determination. In fact, the word *radar* was originally an acronym that stood for radio detection and ranging. Modern radars, however, are sophisticated transducer/computer systems that not only detect targets and determine target range but also track, identify, image, and classify targets while suppressing strong unwanted interference such as echoes from the environment (known as clutter) and countermeasures (jamming). Modern systems apply these major radar functions in an expanding range of applications, from the traditional military and civilian tracking of aircraft and vehicles to two- and three-dimensional mapping, collision avoidance, Earth resources monitoring, and many others.

The goal of *Principles of Modern Radar: Basic Principles* is to provide both newcomers to radar and current practitioners a comprehensive introduction to the functions of a modern radar system, the elements that comprise it, and the principles of their operation and analysis. This chapter provides an overview of the basic concepts of a radar system. The intent is to give the reader a fundamental understanding of these concepts and to identify the major issues in radar system design and analysis. Later chapters then expand on these concepts.

1.2 THE RADAR CONCEPT

A radar is an electrical system that transmits radiofrequency (RF) electromagnetic (EM) waves toward a region of interest and receives and detects these EM waves when reflected from objects in that region. Figure 1-1 shows the major elements involved in the process of transmitting a radar signal, propagation of that signal through the atmosphere, reflection of the signal from the target, and receiving the reflected signals. Although the details of a given radar system vary, the major subsystems must include a transmitter, antenna, receiver, and signal processor. The system may be significantly simpler or more complex than that shown in the figure, but Figure 1-1 is representative. The subsystem that generates the EM waves is the *transmitter*. The *antenna* is the subsystem that takes as input these EM waves from the transmitter and introduces them into the propagation medium (normally the atmosphere). The transmitter is connected to the antenna through a transmit/receive (T/R) device (usually a *circulator* or a switch). The T/R device has the function of providing a connection point so that the transmitter and the receiver can both be attached to the antenna simultaneously and at the same time provide isolation between the transmitter and receiver to protect the sensitive receiver components from the high-powered transmit signal. The transmitted signal propagates through the environment to the target. The EM wave induces currents on the target, which reradiates these currents into the environment. In addition to the desired target, other surfaces on the ground and in the atmosphere reradiate the signal. These unintentional and unwanted but legitimate signals are called *clutter*. Some of the reradiated signal radiates toward the radar receiver antenna to be captured. Propagation effects of the atmosphere and Earth on the waves may alter the strength of the EM waves both at the target and at the receive antenna.

The radar receive antenna receives the EM waves that are "reflected" from an object. The object may be a target of interest, as depicted in Figure 1-1, or it may be of no interest, such as clutter. The portion of the signal reflected from the object that propagates back to the radar antenna is "captured" by the antenna and applied to the *receiver* circuits. The components in the receiver amplify the received signal, convert the RF signal to an *intermediate frequency* (IF), and subsequently apply the signal to an analog-to-digital converter (ADC) and then to the signal/data processor. The *detector* is the device that removes the carrier from the modulated target return signal so that target data can be sorted and analyzed by the *signal processor*.[1]

The propagation of EM waves and their interaction with the atmosphere, clutter, and targets are discussed in Part 2 of this text (Chapters 4 through 8), while the major subsystems of a radar are described in Part 3 (Chapters 9 through 13).

The range, R, to a detected target can be determined based on the time, ΔT, it takes the EM waves to propagate to that target and back at the speed of light. Since distance is speed multiplied by time and the distance the EM wave has to travel to the target and back is $2R$,

$$R = \frac{c\Delta T}{2} \tag{1.1}$$

[1] Not all radar systems employ digital signal and data processing. Some systems apply the analog-detected voltage to a display for the operator to view.

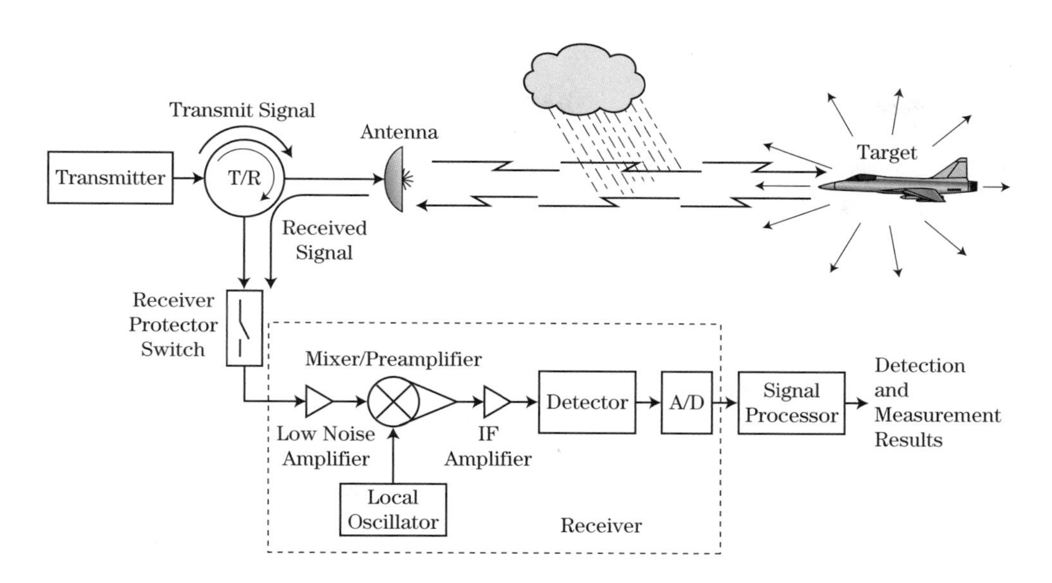

FIGURE 1-1 ■ Major elements of the radar transmission/reception process.

Here c is the speed of light in meters per second ($c \approx 3 \times 10^8$ m/s), ΔT is the time in seconds for the round-trip travel, and R is the distance in meters to the target.[2]

Received target signals exist in the presence of interference. Interference comes in four different forms: (1) internal and external electronic *noise*; (2) reflected EM waves from objects not of interest, often called *clutter*; (3) unintentional external EM waves created by other human-made sources, that is, *electromagnetic interference* (*EMI*); and (4) intentional *jamming* from an *electronic countermeasures* (*ECM*) system, in the form of noise or false targets. Determining the presence of a target in the presence of noise, clutter and jamming is a primary function of the radar's signal processor. Detection in noise and clutter will be discussed further in this and subsequent chapters; it is a major concern of a significant portion of this textbook.

EMI is unintentional, as in the case of noise from an engine ignition or electric motor brushes. Jamming signals can take the form of noise, much like internal receiver thermal noise, or false targets, much like a true radar target.

1.3 THE PHYSICS OF EM WAVES

Electromagnetic waves are electric and magnetic field waves, oscillating at the carrier frequency. The nature of electromagnetic fields is described by Maxwell's equations, presented in the Appendix. The electric, $\boldsymbol{E}$, field is in one plane, and the magnetic, $\boldsymbol{B}$, field is orthogonal to the $\boldsymbol{E}$ field.[3] The direction of propagation of this EM wave through space (at the speed of light, c) is orthogonal to the plane described by the $\boldsymbol{E}$ and $\boldsymbol{B}$ fields, using the right-hand rule. Figure 1-2 depicts the coordinate system. The $\boldsymbol{E}$ field is aligned

[2]The actual value of c in a vacuum is 299,792,458 m/s, but $c = 3 \times 10^8$ is an excellent approximation for almost all radar work. The speed of light in air is nearly the same value.

[3]Sometimes $\boldsymbol{B}$ is used to denote magnetic induction, in which case $\boldsymbol{H}$ would denote magnetic field. There are other definitions for $\boldsymbol{B}$ and $\boldsymbol{H}$; a description of these is beyond the scope of this chapter.

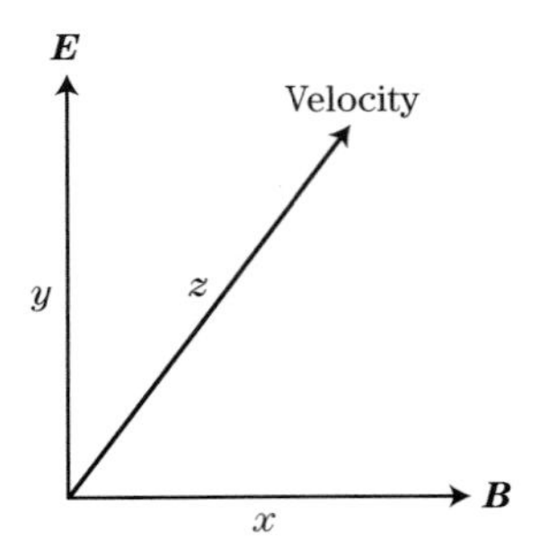

FIGURE 1-2 ■ Orientation of the electromagnetic fields and velocity vector.

along the y-axis, the ***B*** field along the x-axis, and the direction of propagation along the z-axis.

The amplitude of the x or y component of the electric field of an electromagnetic wave propagating along the z-axis can be represented mathematically as

$$E = E_0 \cos(kz - \omega t + \phi) \tag{1.2}$$

where E_0 is the peak amplitude, and ϕ is the *initial phase*.

The *wave number, k,* and the *angular frequency*, ω are related by

$$k = \frac{2\pi}{\lambda} \text{ radians/m}, \quad \omega = 2\pi f \text{ radians/sec} \tag{1.3}$$

where λ is the wavelength in meters, and f is the carrier frequency in hertz.

1.3.1 Wavelength, Frequency, and Phase

1.3.1.1 Wavelength

As the EM wave propagates in space, the amplitude of ***E*** for a linearly polarized wave, measured at a single point in time, traces out a sinusoid as shown in Figure 1-3. This corresponds to holding t constant in equation (1.2) and letting z vary. The *wavelength*, λ, of the wave is the distance from any point on the sinusoid to the next corresponding point, for example, peak to peak or null (descending) to null (descending).

1.3.1.2 Frequency

If, on the other hand, a fixed location in space was chosen and the amplitude of ***E*** was observed as a function of time at that location, the result would be a sinusoid as a function of time as shown in Figure 1-4. This corresponds to holding z constant in equation (1.2) and letting t vary. The *period*, T_0, of the wave is the time from any point on the sinusoid to the next corresponding part, for example, peak to peak or null (descending) to null

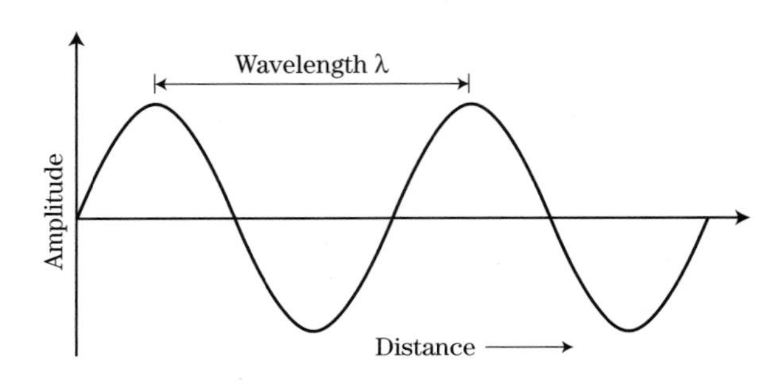

FIGURE 1-3 ■ The wavelength of a sinusoidal electromagnetic wave.

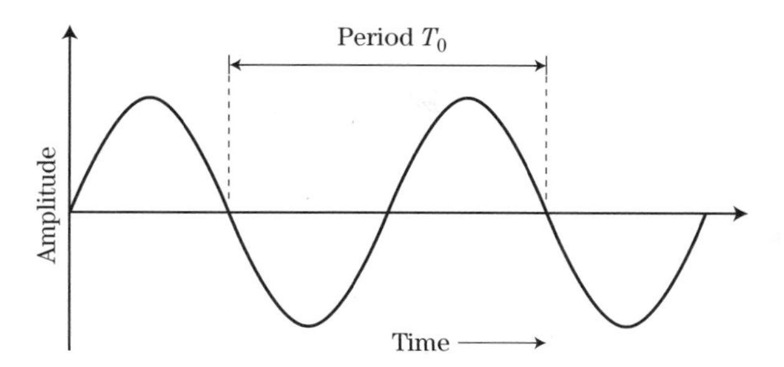

FIGURE 1-4 ■ The period of a sinusoidal electromagnetic wave.

(descending). That is, the period is the time it takes the EM wave to go through one cycle. If the period is expressed in seconds, then the inverse of the period is the number of cycles the wave goes through in 1 second. This quantity is the wave's *frequency*, *f*,

$$f = \frac{1}{T_0} \tag{1.4}$$

Frequency is expressed in hertz; 1 Hz equals one cycle per second.

The wavelength and frequency of an EM wave are not independent; their product is the speed of light (c in free space),

$$\lambda f = c \tag{1.5}$$

Therefore, if either the frequency or wavelength is known, then the other is known as well. For example, a 3 cm EM wave has a frequency of

$$f = \frac{c}{\lambda} = \frac{3 \times 10^8 \text{ m/s}}{0.03 \text{ m}} = 10^{10} \text{ Hz or } 10 \text{ GHz} \tag{1.6}$$

where "G" stands for "giga" or 10^9.

Shown in Figure 1-5 are the different types of EM waves as a function of frequency, from EM telegraphy to gamma rays. Although they are all EM waves, some of their characteristics are very different depending on their frequency. Radars operate in the range of 3 MHz to 300 GHz, though the large majority operate between about 300 MHz and 35 GHz. This range is divided into a number of RF "bands" [1] as shown in Table 1-1. Shown alongside the radar bands are the International Telecommunications Union (ITU) frequencies authorized for radar use. Note that a given radar system will not operate over the entire range of frequencies within its design band but rather over a limited range within that band. Authorization for use of frequencies as issued by the Federal Communication Commission (FCC) in the United States limits the range of frequencies for a given system. Furthermore, the FCC interacts with the ITU, a worldwide frequency coordination organization. Also, at frequencies above about 16 GHz, the specific frequencies are often chosen to coincide with relative "nulls" in the atmospheric absorption characteristics, as will be discussed shortly. The electronic warfare (EW) community uses a different set of letter band designations. Table 1-2 lists the EW bands.

1.3.1.3 Phase

Note that in equation (1.2) the wave number is in units of radians per meter and so is a kind of "spatial frequency." The quantity ϕ is often called the *fixed,* or *initial, phase*. It is arbitrary in that it depends on the electric field's initial conditions (i.e., the value of E) for

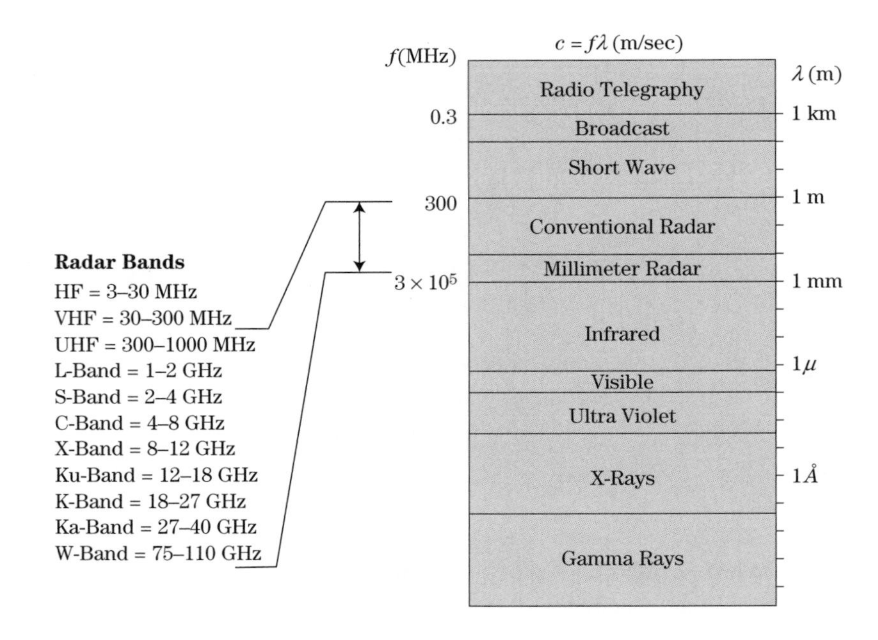

FIGURE 1-5 ■ Electromagnetic wave types.

TABLE 1-1 ■ RF and Radar Bands

Band	Frequency Range	ITU Radar Freq.
High frequency (HF)	3–30 MHz	
Very high frequency (VHF)	30–300 MHz	138–144 MHz 216–225 MHz
Ultra high frequency (UHF)	300 MHz–1 GHz	420–450 MHz 890–942 MHz
L	1–2 GHz	1.215–1.400 GHz
S	2–4 GHz	2.3–2.5 GHz 2.7–3.7 GHz
C	4–8 GHz	5.250–5.925 GHz
X	8–12 GHz	8.500–10.680 GHz
Ku (“under” K-band)	12–18 GHz	13.4–14.0 GHz 15.7–17.7 GHz
K	18–27 GHz	24.05–24.25 GHz 24.65–24.75 GHz
Ka (“above” K-band)	27–40 GHz	33.4–36.0 GHz
V	40–75 GHz	59.0–64.0 GHz
W	75–110 GHz	76.0–81.0 GHz 92.0–100.0 GHz
mm	100–300 GHz	126.0–142.0 GHz 144.0–149.0 GHz 231.0–235.0 GHz 238.0–248.0 GHz

TABLE 1-2 ■ EW Bands

Band	Frequency Range
A	30–250 MHz
B	250–500 MHz
C	500–1,000 MHz
D	1–2 GHz
E	2–3 GHz
F	3–4 GHz
G	4–6 GHz
H	6–8 GHz
I	8–10 GHz
J	10–20 GHz
K	20–40 GHz
L	40–60 GHz
M	60–100 GHz

the arbitrarily chosen spatial and temporal positions corresponding to $z = 0$ and $t = 0$. For example, if $E = 0$ when $x = t = 0$, then $\phi = \pm\pi/2$ radians. The *phase* is the total argument of the cosine function, $kz - \omega t + \phi$, and depends on position, time, and initial conditions.

The *relative phase* is the phase difference between two waves. Two waves with a zero relative phase are said to be *in phase* with one another. They can be made to have a nonzero phase difference (i.e., be *out of phase*) by changing the wave number (wavelength), frequency, or absolute phase of one (or both). Two waves originally in phase can become out of phase if they travel different path lengths. Figure 1-6 illustrates two waves having the same frequency but out of phase by $\Delta\phi = 50°$. If the waves are viewed as a function of time at a fixed point in space, as in this figure, then one is offset from the other by $\Delta\phi/\omega$ seconds.

1.3.1.4 Superposition (Interference)

The principle of superposition states that when two or more waves having the same frequency are present at the same place and the same time, the resultant wave is the complex sum, or superposition, of the waves. This complex sum depends on the amplitudes and phases of the waves. For example, two in-phase waves of the same frequency will produce a resultant wave with an amplitude that is the sum of the two waves' respective amplitudes (*constructive interference*), while two out-of-phase waves will produce a resultant wave with an amplitude that is less than the sum of the two amplitudes (*destructive interference*).

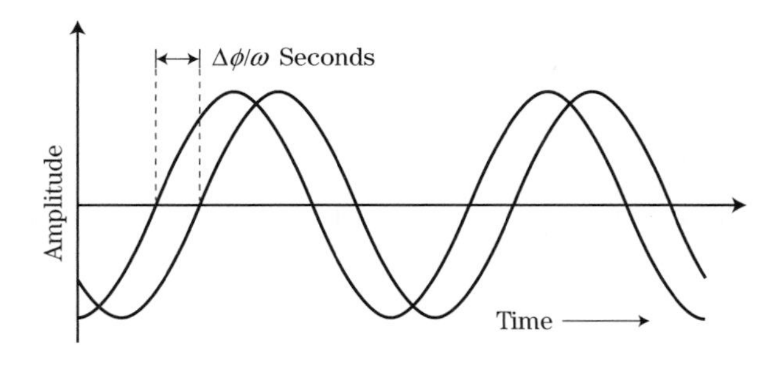

FIGURE 1-6 ■ Two sinusoidal waves with the same frequency but a phase difference $\Delta\phi$.

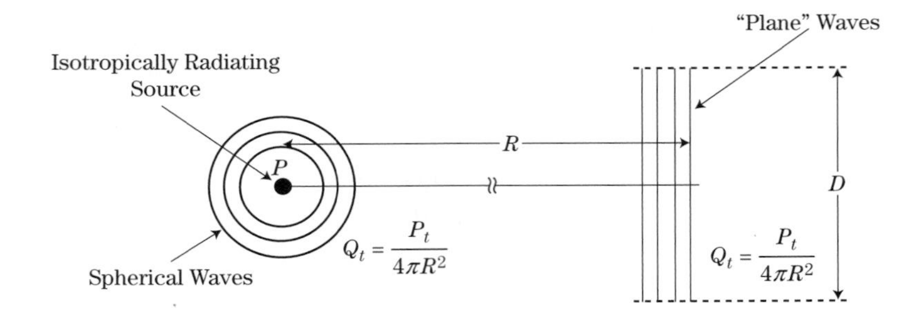

FIGURE 1-7 Intensity of spherical waves.

Two waves of equal amplitude that are π radians (180°) out of phase will produce a *null* result (i.e., no wave). The importance of the concept of superposition is seen in many topics related to radar. Among these are the formation of a defined beam produced by an antenna, the total radar cross section (RCS) of a target as a result of the many scatterers, and the effects of multipath as described in Chapter 4.

1.3.2 Intensity

The *intensity, Q,* of the EM wave is defined as the power (time-rate-of-change of energy) per unit area of the propagating wave. Thus, intensity is equivalent to *power density* (watts per square meter). Consider a single (hypothetical) antenna element emitting an EM wave of power P equally in all directions (*isotropic*) as shown in Figure 1-7. The locus of all points having the peak amplitude at a given moment in time (*wavefront*) in this wave will be a sphere; the distance between adjacent concentric spheres will be the wavelength. Since the wave is (ideally) isotropic, the power everywhere on the surface of a given spherical wavefront of radius R will be the same (because energy is conserved in a lossless medium). Thus, the transmitted power density is the total radiated transmitted power, P_t, divided by the surface area of the sphere, or

$$Q_t = \frac{P_t}{4\pi R^2} \tag{1.7}$$

The intensity of the EM wave falls off as $1/R^2$, where R is the distance from the isotropic source.

If the wave is sufficiently far from the source and a limited spatial extent of the wave is considered, then the spherical wavefronts are approximately planar, as shown in the right-hand portion of Figure 1-7. It is somewhat arbitrarily decided but universally accepted that if the wave front curvature is less than $\lambda/16$ over a given "aperture" of dimension D, then the wave is considered planar. Using relatively simple geometry, this condition is met if the distance from the source to the aperture is at least $2D^2/\lambda$. This is called the *far-field*, or plane wave, approximation.

1.3.3 Polarization

The EM wave's polarization is the description of the motion and orientation of the *electric* field vector. Suppose the wave is traveling in the $+z$ direction in a Cartesian (x-y-z) coordinate system. Then the direction of the electric field $\boldsymbol{E}$ must lie in the x-y plane. An electric field oriented along some angle in the x-y plane thus has components in both the

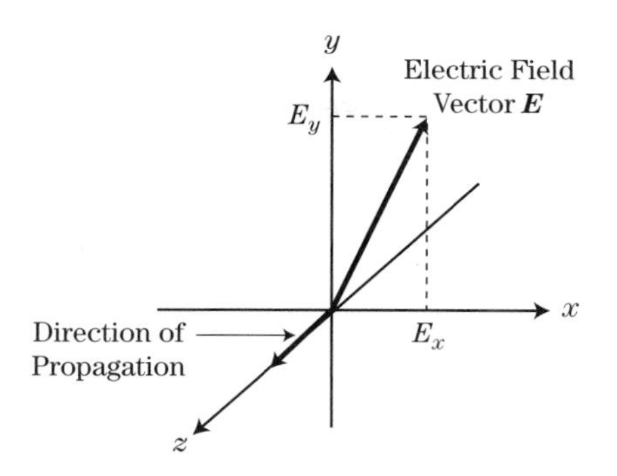

FIGURE 1-8 ■ Polarization components of a transverse EM wave propagating in the $+z$ direction.

x and y directions, say E_x and E_y, as shown in Figure 1-8, which shows the directional components of only the $\boldsymbol{E}$ field. The amplitudes of these two components will each vary sinusoidally as in equation (1.2). The relative peak amplitudes and phases of E_x and E_y determine how the orientation of the resultant vector $\boldsymbol{E}$ varies with time, and thus the *polarization* of the EM wave. For example, if the y component of the electric field is zero, then $\boldsymbol{E}$ oscillates along the x-axis and the EM wave is said to be *linearly polarized* in the x direction. If x represents a horizontally oriented axis, the wave is *horizontally polarized.* Similarly, the wave would be vertically linearly polarized if the x component is zero but the y component is not. If E_x and E_y have the same magnitude and oscillate in phase with one another ($\phi_x = \phi_y$), the field will oscillate linearly along a 45° line in the x-y-plane. In general, the polarization is linear if the x and y components differ in phase by any integer multiple of π radians; the angle of the polarization depends on the relative magnitudes of E_x and E_y.

If $E_x = E_y$ and the phases differ by an odd multiple of $\pi/2$, the tip of $\boldsymbol{E}$ traces out a circle as the wave propagates and the EM wave is said to be *circularly polarized*; one rotation sense is called "right-hand" or "right circular" polarization and the other, "left-hand" or "left circular" polarization. The polarization state is elliptical, when the tip of $\boldsymbol{E}$ traces out an ellipse as the wave propagates. This occurs when $E_x \neq E_y$.

1.4 INTERACTION OF EM WAVES WITH MATTER

The EM waves that a radar transmits and receives interact with matter, specifically, the radar's antenna, then the atmosphere, and then with the target. The relevant physical principles governing these interactions are diffraction (antenna); attenuation, refraction and depolarization (atmosphere); and reflection (target).

1.4.1 Diffraction

Diffraction is the bending of EM waves as they propagate through an aperture or around the edge of an object. Diffraction is an example of the interference phenomenon discussed in Section 1.3.1.4. The amount of diffraction present depends on the size of the aperture (antenna), *a,* relative to the wavelength, *λ,* of the EM wave. Shown in Figure 1-9 are two extreme cases.

The waves emitting from the aperture (idealized in Figure 1-9 as an opening, or "slit," in a surface) can be thought of (i.e., modeled) as being produced by many individual radiating elements separated by a wavelength (or less) and all emitting waves isotropically.

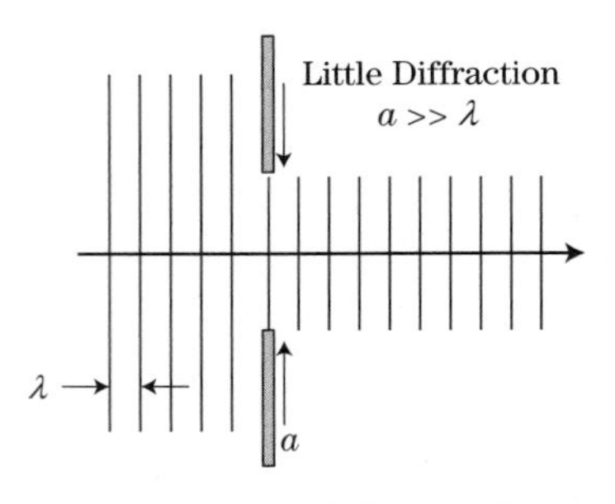

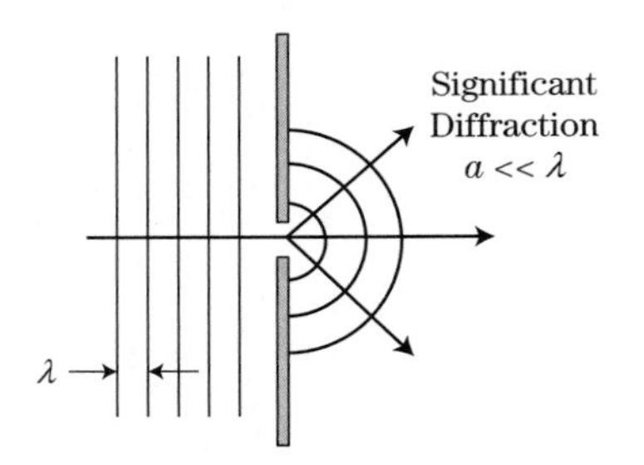

FIGURE 1-9 ■ Extreme cases of diffraction.

Many paths of width λ across the aperture, thus destructive interference occurs in all directions except forward, preventing diffraction.

Only one path of width λ across the aperture, thus diffracts in all directions on the other side, propagating isotropically.

In physics, this is known as *Huygen's principle*. The EM wave characteristics to the right of the opening in Figure 1-9 will be different from those to the left. Whereas the plane wave to the left of the aperture might be wide compared with the opening, there will be a shaped beam emerging from the opening, toward the right, including a main lobe portion, and lower amplitude, angular sidelobes, to be described later. Superposition of the waves from the individual elements using Huygen's model predicts that the radiation pattern to the right of the aperture will have a distinct main beam rather than an isotropic pattern, having a half-power beamwidth depending on the aperture size, in wavelengths. If the aperture size is much greater than a wavelength (i.e., $a \gg \lambda$), then there will be many radiating elements present and significant destructive interference in all but the forward direction. In this case, there is very little diffraction, and the antenna beamwidth will be small. Conversely, if the aperture size is much smaller than a wavelength, then there is essentially only one radiation element present, and no destructive interference takes place. In this case the EM waves propagate nearly isotropically (over only the right-side hemisphere), producing significant diffraction effects and a large beamwidth.

The angular shape of the wave as it exits the aperture is, in general, a $\sin(x)/x$ (sinc) function. The main lobe half-power (−3 dB) beamwidth, θ_3, of a sinc function is

$$\theta_3 = \frac{0.89\,\lambda}{a} \text{ radians} \tag{1.8}$$

In the case of an antenna, the same principles apply. In this case, instead of an opening in a large plate, the individual radiators are across a structure called an antenna.[4] The phenomenon of diffraction is responsible for the formation of the antenna pattern and antenna beam (or *main lobe* of the antenna pattern) as well as the sidelobes.

Consider a circular (diameter D) planar antenna made up of many (N) radiating elements, each of which is emitting EM waves of equal amplitudes over a wide range of angles. Figure 1-10 is the photograph of such an antenna. Assume that all the waves are in phase as they are emitted from the antenna elements. At a point along a line perpendicular (normal) to the plane and far away from the antenna (see Section 1.3.2 and Chapter 9 for discussions of the antenna far field), all the waves will have essentially traveled the same distance and, therefore, will all still be in phase with each other. Constructive interference will occur and, assuming that each element produced the same signal level, the resultant

[4]Often, because of this analogy, an *antenna* is called an *aperture*.

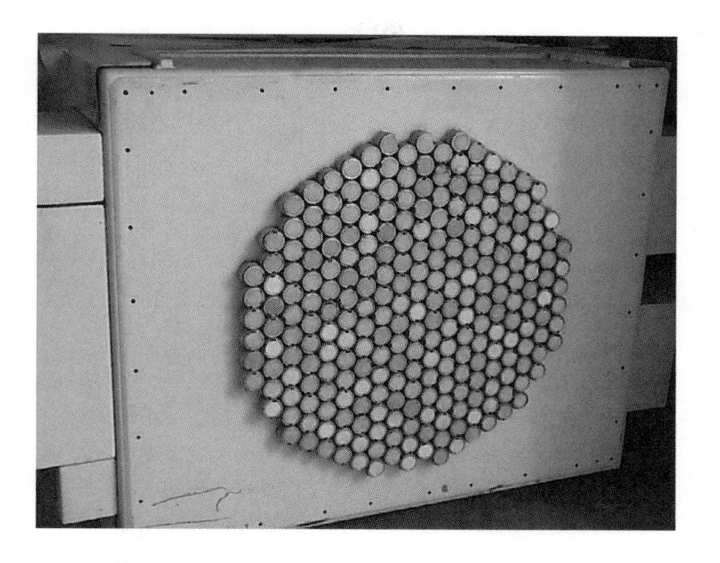

FIGURE 1-10 ■ A multi-element antenna. (Courtesy GTRI. With permission.)

wave will have an amplitude N times larger than the individual waves emitted from the elements. This represents the peak of the antenna beam. Figure 1-11 depicts the in-phase waves radiating from a linear array of elements and the resulting main beam pattern. The sidelobe pattern is not shown in the figure.

At any point off this normal, the waves will have traveled different path lengths; thus, destructive interference occurs, and the resultant wave will have an amplitude less than N times larger. As the angular distance from the normal increases, this amplitude decreases, finally reaching a perfect null (complete destructive interference). The angular region between the first null to either side of the antenna normal defines the *main beam* or *main lobe* of the antenna. Most of the radiated power is concentrated in this region. Twice the angular distance from the peak of the antenna mainbeam to the point where the EM wave power has dropped to half its peak value, or -3 dB, is the 3 dB *beamwidth*, θ_3. The exact 3 dB beamwidth depends on several things, including the shape of the antenna face, the illumination pattern across the antenna, and any structural blockage near the antenna, such as protective radomes and antenna support structures. For typical design parameters for a circular antenna,

$$\theta_3 \approx \frac{1.3\lambda}{D} \text{ radians} \tag{1.9}$$

At angles past the first null, the individual waves partially constructively interfere so that the net amplitude starts to increase, rises to a peak, and then falls again to a

FIGURE 1-11 ■ A multi-element linear array of radiating elements with in-phase signals and resulting main beam pattern.

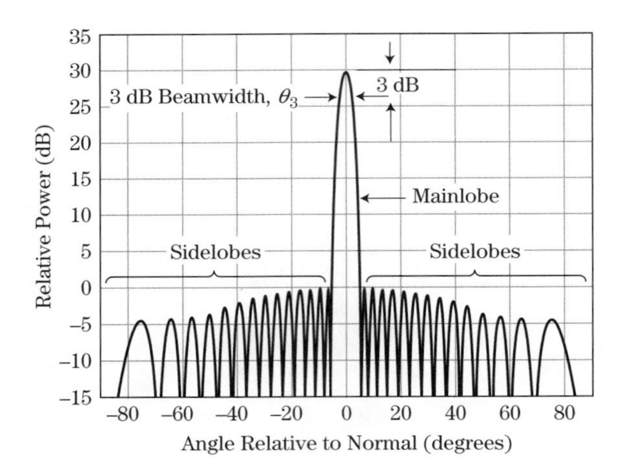

FIGURE 1-12 ■ Idealized one-dimensional antenna pattern.

second null. This pattern is repeated over and over again, forming an *antenna pattern* as shown in Figure 1-12. This figure shows a one-dimensional planar "cut" through the two-dimensional pattern of an idealized two-dimensional antenna. The lobes outside the main lobe are called *antenna sidelobes.*

If the phases of the EM waves have different values when they are emitted from the elements, then they will no longer constructively interfere in the far field in the direction of the antenna normal. If these phase values are adjusted properly, the amplitude of the far-field resultant wave can be made to peak at some angle off the normal. All the waves in this direction traveled different path lengths and, therefore, will have different path-length-induced phases. If the original phases upon emission are selected properly, they can be made to compensate for the path-length-induced phases, and all the waves will be in phase in that direction. Thus, by changing the phases of the emitted waves, the peak of the antenna beam will effectively scan from its normal position without the antenna physically moving. This is the basic concept behind a *phased array antenna* or *electronically scanned antenna* (ESA); it is discussed in more detail in Chapter 9.

The antenna can be designed to produce an ideal beamwidth for a given radar application. In fact, if the antenna is not geometrically symmetric the azimuthal and elevation angular beamwidths can be different. A circular or square antenna will produce a symmetric beam, while an elliptical or rectangular antenna will produce an asymmetric beam.

Narrow antenna beamwidths are desired in applications such as tracking, mapping, and others where good angular resolution is desired. Track precision improves as the beamwidth is narrower, as seen in Chapter 18.

Applications in which large antenna beamwidths are advantageous are (1) in the search mode and (2) in strip-map *synthetic aperture radars* (SARs). In the search mode, where high resolution is normally not required, a given volume can be searched faster with a wide beam. For an SAR, the larger the antenna beamwidth, the larger the synthetic aperture can be, and, thus, the finer the target resolution that can be achieved (see Chapter 21). However, large antenna beamwidths have negative performance effects in many radar applications. For example, the ability to resolve targets in the cross-range dimension decreases with increasing beamwidth when SAR is not used, while in air-to-ground radars, the amount of ground clutter (interfering echoes from terrain) competing with desired target signals increases with increasing antenna beamwidth. In addition, larger beamwidths result in reduced antenna gain, decreasing the signal-to-noise ratio (SNR).

1.4.2 Atmospheric Attenuation

Figure 1-13 shows the one-way attenuation (per unit of distance) of EM waves in the atmosphere as a function of frequency. There is very little clear-air attenuation below 1 GHz (L-band). Above 1 GHz, the attenuation steadily increases, and peaks are seen at 22 GHz (due to water vapor absorption), 60 GHz (due to oxygen absorption), and at higher frequencies. Curves are shown at two different altitudes to demonstrate that the different distribution of water vapor and oxygen with altitude affects the absorption characteristics. Above 10 GHz (X-band), there are troughs, or *windows*, in the absorption spectrum at 35 GHz (Ka-band), 94 GHz (W-band), and other higher frequencies. These windows are the frequencies of choice for radar systems in these higher-frequency bands that have to operate in the atmosphere. For long-range radars (e.g., surface search radars), frequencies at L-band and S-band are generally required to minimize atmospheric attenuation. Though the attenuation versus range values below 10 GHz are low, most of these systems operate at long ranges, so the loss incurred at these ranges is still significant. Chapter 4 presents more detailed information on atmospheric effects.

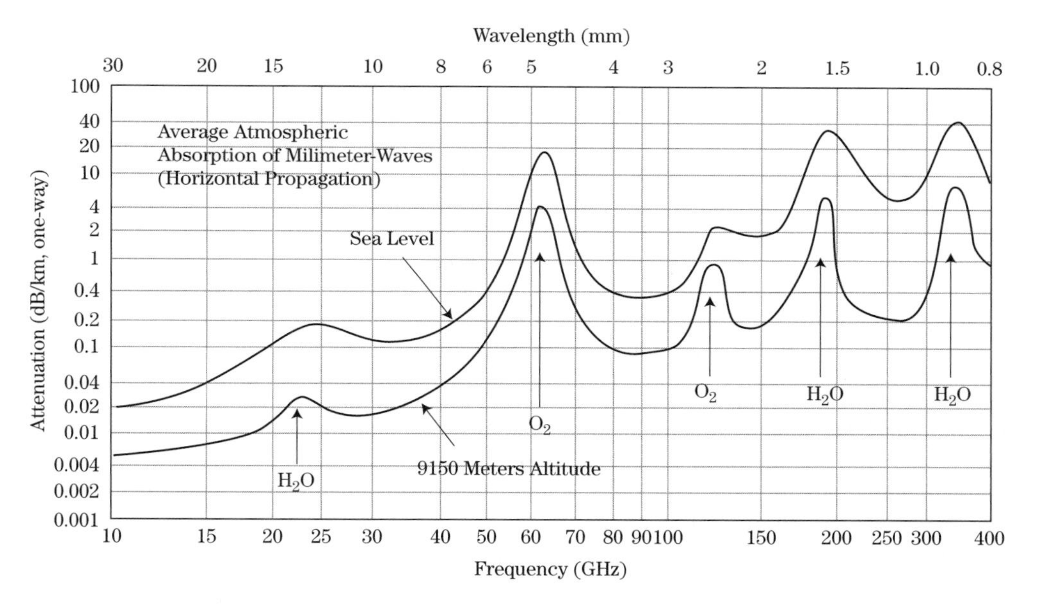

FIGURE 1-13 ■ One-way atmospheric attenuation as a function of frequency at sea level and at 9150 meters altitude. (From U. S. Government work.)

Rain, fog, and clouds further attenuate EM waves. One-way rain and cloud attenuation is shown in Figure 1-14. Rain attenuation increases with increasing rain rate and increasing frequency. At radar frequencies, rain and cloud attenuation is small, giving radar systems their famous "all weather capability" not seen in electro-optical and infrared (IR) systems. Detailed descriptions and more specific attenuation values are presented in Chapter 4.

1.4.3 Atmospheric Refraction

Refraction is the bending of EM waves at the interface of two different dielectric materials. This occurs because the speed of the EM wave is a function of the material in which it is propagating; the more "optically dense" the material, the slower the speed. Consider a wave incident on the interface to two difference materials as shown in Figure 1-15. Within the denser material (glass), the EM wave slows down due to a decrease in wavelength

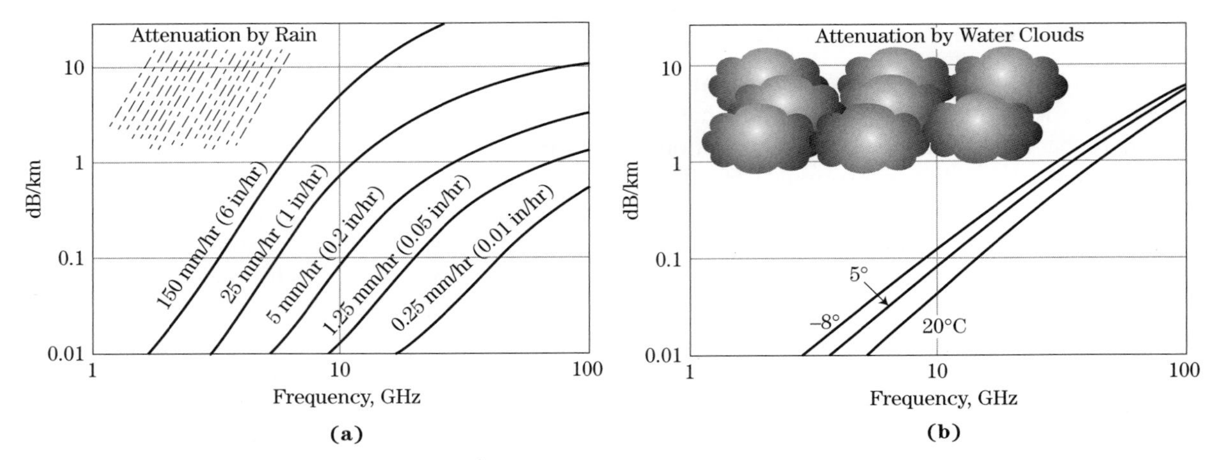

FIGURE 1-14 ■ One-way rain and cloud attenuation as a function of frequency. (a) Rain. (b) Clouds.

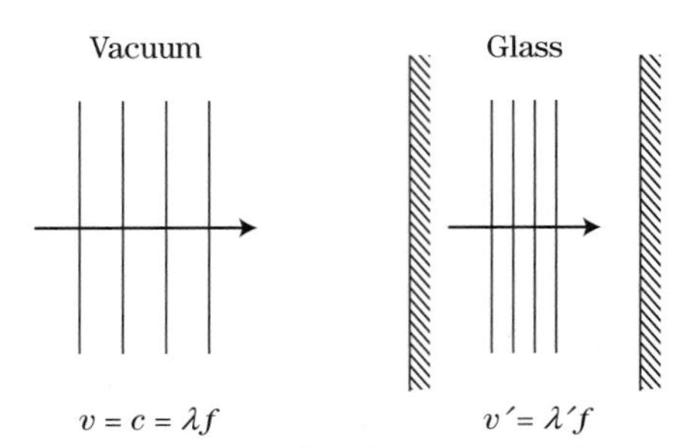

FIGURE 1-15 ■ Difference in wavelength for wavefronts in two materials.

($v = \lambda f$). The optical density of a material is quantified by the index of refraction, n, given by $n = c/v$, where v is the speed of the EM wave in the material. If this wave were incident on the interface at some angle as shown in Figure 1-16, then, given the reduction of wavelength in the material with a higher index of refraction, the only way the wavefronts can remain continuous across the interface is for them to bend at the interface. This bending is refraction.

In radar technology, refraction is encountered in radar signals directed upward (or downward) through the atmosphere at an angle relative to horizontal. Generally, the atmosphere thins with increasing altitude, causing the index of refraction to reduce. Therefore, the path of the transmitted EM wave will deviate from a straight line and bend back toward the earth. Deviations from straight-line propagation adversely affect target location and tracking accuracy unless refraction effects are accounted for.

Refraction can be beneficial for surface-to-surface radars (e.g., shipboard radars detecting other ships) since it can allow the EM wave to propagate over the horizon and detect ships not detectable if detection were limited by the geometric horizon. An extreme gradient in index of refraction with altitude causes the ray to bend more than for standard atmospheric conditions. Over the surface of the sea, this high value of refractive index with height is common. The severe ray bending is called *ducting*, and surface radar systems can "see" well past the geometric horizon.

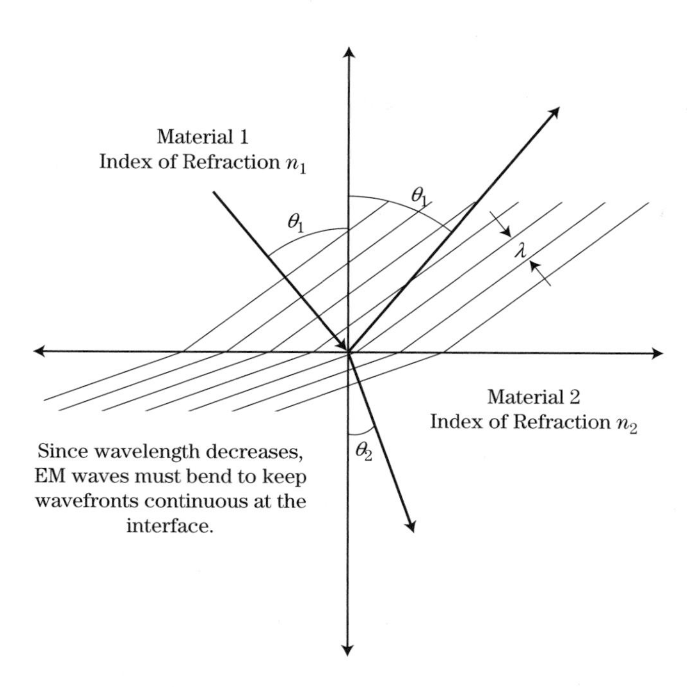

FIGURE 1-16 ■ Bending of wavefronts incident at an angle on the interface of two materials.

Over land, long-range propagation can be achieved by using the refractive effect at the earth's ionosphere. The EM wave propagates upward to the ionosphere; there the refractive bending causes the wave to travel back toward the surface of the earth, where it will intersect the earth's surface several thousand miles away from the transmitting source. The return path will experience the same effect. This condition (sometimes called *skip*) is most prominent in the high-frequency (HF) region (3–30 MHz) and is generally not encountered above 150 MHz. Radars that use this phenomenon are called *over-the-horizon* (OTH) *radars*. Chapter 4 describes the details associated with atmospheric refraction.

1.4.4 Reflection

Incident EM waves induce an electric charge on natural surfaces or the surface of a man-made object, and that object reradiates the EM wave. The reradiation of the EM wave from the surface matter of an object is called *scattering* or, more often, *reflection* of the incident wave. If the matter is a conductor so that the electric charge is free to move in the matter, then essentially all the EM wave energy is reradiated. If the matter is a dielectric material so that its electric charge is bound, some of the energy is reradiated, and some propagates into the matter where some is absorbed and some may come out the other side.

The manner in which the EM wave is reflected from the surface depends on the roughness of the surface relative to the wavelength of the incident wave. Generally speaking, roughness is the variation in surface height. It is usually quantified by the standard deviation of the surface height. If the surface is "smooth" ($\lambda \gg$ roughness), then the EM wave's angle of reflection, θ_r, equals its angle of incidence, θ_i, on the surface (see Figure 1-17). This is called *specular* scattering. Most scattering from man-made objects in radar technology is specular.

If, on the other hand, the surface is "rough" ($\lambda \ll$ roughness), then the scattering is specular only over small local regions of the surface. Macroscopically, the incident energy appears to be reflected at all angles (see Figure 1-18). This is called *diffuse* scattering. To

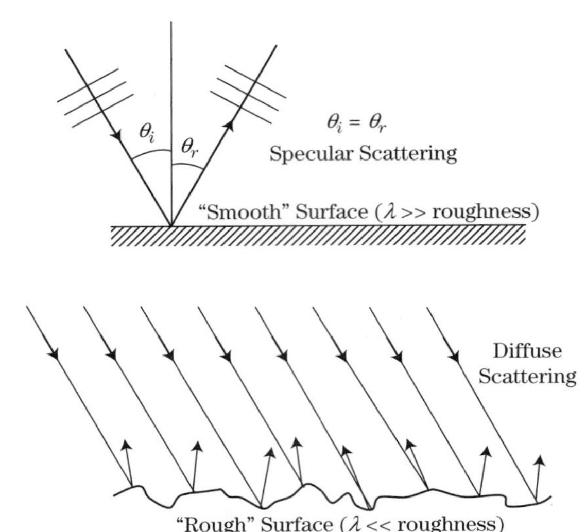

FIGURE 1-17 ■ Specular scattering.

FIGURE 1-18 ■ Diffuse scattering.

predict the scattering of EM waves from an object, both specular and diffuse scattering must be taken into consideration. Scattering from natural surfaces, especially at shorter wavelengths (higher frequencies), is often diffuse.

In radar technology, scattering phenomenology is quantified by the target parameter *radar cross section*, σ. RCS has the units of area (e.g., m^2). The RCS of a target is not a single number but is a function of target viewing angle relative to the transmitter and receiver antenna and of the frequency and polarization of the incident EM wave. RCS is a measure of not only how much of the incident EM wave is reflected from the target but also how much of the wave is intercepted by the target and how much is directed back toward the radar's receiver. Thus, these three mechanisms—interception, reflection, and directivity—all interact to determine the RCS of a target. If a target is to be made “invisible” to a radar (i.e., be a *stealth* target), then its RCS is made to be as low as possible. To do this, at least one of the three mechanisms must be addressed: (1) the amount of the EM wave energy intercepted by the target must be minimized, which is accomplished by minimizing the physical cross section of the target; (2) the amount of energy reflected by the target must be minimized, which is accomplished by absorbing as much of the EM wave as possible through the use of *radar-absorbing material* (RAM) on the surface of the target; or (3) the amount of the reflected energy directed toward the radar receiver must be minimized, which is accomplished by shaping the target. The RCS of terrain and of targets (including stealth considerations) are discussed in more detail in Chapters 5 and 6, respectively.

1.5 | BASIC RADAR CONFIGURATIONS AND WAVEFORMS

1.5.1 Monostatic versus Bistatic

There are two basic antenna configurations of radar systems: monostatic and bistatic (Figure 1-19). In the monostatic configuration, one antenna serves both the transmitter and receiver. In the bistatic configuration, there are separate antennas for the transmit and receive radar functions.

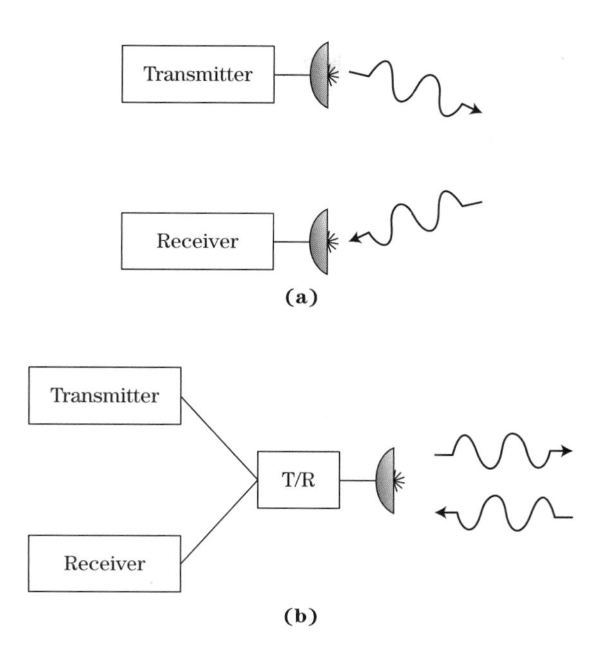

FIGURE 1-19 ■ Basic radar configurations: (a) Bistatic. (b) Monostatic.

Use of two antennas alone does not determine whether a system is monostatic or bistatic. If the two antennas are very close together, say, on the same structure, then the system is considered to be monostatic. The system is considered to be bistatic only if there is sufficient separation between the two antennas such that "... the angles or ranges to the target are sufficiently different..." [2].

The transmitter is often a high-power device that can transmit EM waves with power levels in the range of hundreds of kilowatts (10^3 watts) or even megawatts (10^6 watts). The receiver, on the other hand, is a power-sensitive device that can respond to EM waves in the range of milliwatts to nanowatts (10^{-3} to 10^{-9} watts) or less. In fact, it is not uncommon for a radar receiver to detect signals as low as −90 dBm (dB relative to a milliwatt). High-power EM waves from the transmitter, if introduced directly into the receiver, would prevent the detection of targets (self-jamming) and could severely damage the receiver's sensitive components. Therefore, the receiver must be *isolated* from the transmitter to protect it from the transmitter's high-power EM waves. The bistatic radar configuration can provide significant isolation by physically separating the transmitter and receiver antennas.

There are some applications for which the bistatic system has a significant separation between the transmitter and receiver. For example, a semiactive missile has only the receiver portion on board. The transmitter is on another platform. The transmitter "illuminates" the target while the missile "homes" in on the signal reflected from the target.

The bistatic radar can also be employed to enhance the radar's capability of detecting *stealth* targets. Recall that a target's RCS is a measure of the strength of the EM waves that are reflected from the target back toward the radar receive antenna. Stealthy targets are designed to have a low RCS, thereby reducing the distance at which they can be seen. In addition to other techniques, RCS reduction is achieved by shaping the target in a particular way. This shaping may reduce the RCS when looking at the front of a target using monostatic radar; however, it is often the case that the RF wave will scatter in a different direction, providing a large RCS in some "bistatic" direction. When the bistatic

RCS is greater than the monostatic RCS, the target is no longer "stealthy" to the bistatic radar.

Most modern radars are monostatic—a more practical design since only one antenna is required. It is more difficult to provide isolation between the transmitter and receiver since both subsystems must be attached to the antenna. The isolation is provided by a T/R device, such as a circulator or switch, as previously described. For a radar using a pulsed waveform (see the following discussion), the transmitter and receiver do not operate at exactly the same time. Therefore, additional isolation can be achieved by use of an additional switch in the receiver input path.

1.5.2 Continuous Wave versus Pulsed

1.5.2.1 CW Waveform

Radar waveforms can be divided into two general classes: *continuous wave* (CW) and pulsed. With the CW waveform the transmitter is continually transmitting a signal, usually without interruption, all the time the radar transmitter is operating. The receiver continuously operates also. The pulsed waveform transmitter, on the other hand, emits a sequence of finite duration pulses, separated by times during which the transmitter is "off." While the transmitter is off, the receiver is on so that target signals can be detected.

Continuous wave radars often employ the bistatic configuration to effect transmitter/receiver isolation. Since the isolation between the transmitter and receiver is not perfect, there is some competing signal due to the leakage, relegating CW systems to relatively low power and hence short-range applications. Since a CW radar is continuously transmitting, determination of the transmitted EM wave's round-trip time and, thus, target range, must be accomplished by changing the characteristics of the wave (e.g., changing the wave's *frequency* over time). This frequency modulated (FM) technique effectively puts a timing mark on the EM wave, thus allowing for target range determination. Though there are relatively complex CW systems employed as illuminators in fire control systems, semi-active missiles, and trackers, CW radars tend to be simple radars and are used for such applications as police speed-timing radars, altimeters, and proximity fuses.

1.5.2.2 Pulsed Waveform

Pulsed radars transmit EM waves during a very short time duration, or *pulse width* τ, typically 0.1 to 10 microseconds (μs), but sometimes as little as a few nanoseconds (10^{-9} seconds) or as long as a millsecond. During this time, the receiver is isolated from the antenna, or *blanked*, thus protecting its sensitive components from the transmitter's high-power EM waves. No received signals can be detected during this time. In addition to the isolation provided by the T/R device (shown in Figure 1-1), further protection is offered by the receiver protection switch, not shown in the figure. During the time between transmitted pulses, typically from 1 microsecond to tens of milliseconds, the receiver is connected to the antenna, allowing it to receive any EM waves (echoes) that may have been reflected from objects in the environment. This "listening" time plus the pulse width represents one pulsed radar cycle time, normally called the *interpulse period* (IPP) or *pulse repetition interval* (*PRI*). The pulsed waveform is depicted in Figure 1-20.

Pulse Repetition Frequency (PRF) The number of transmit/receive cycles the radar completes per second is called the *pulse repetition frequency* (PRF), which is properly

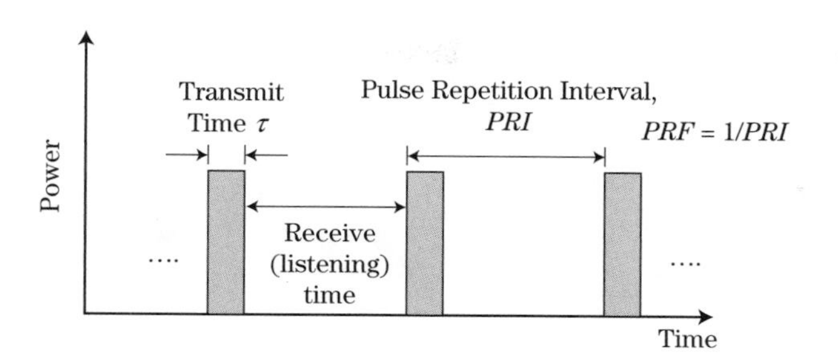

FIGURE 1-20 ■ Pulsed radar waveform.

measured in pulses per second (PPS) but is often expressed in hertz (cycles per second). The PRF and PRI are related according to

$$PRF = \frac{1}{PRI} \tag{1.10}$$

Pulse Width and Duty Cycle The fraction of time the transmitter is transmitting during one radar cycle is called the transmit *duty factor* (or *duty cycle*), d_t, and from Figure 1-20 is given by

$$d_t = \frac{\tau}{PRI} = \tau \cdot PRF \tag{1.11}$$

The average power, P_{avg}, of the transmitted EM wave is given by the product of the peak transmitted power, P_t, and the transmit duty factor:

$$P_{avg} = P_t \cdot d_t = P_t \cdot \tau \cdot PRF \tag{1.12}$$

Range Sampling Figure 1-21 depicts a sequence of two transmit pulses and adds a hypothetical target echo signal. Because the time scale is continuous, a target signal can arrive at the radar receiver at any arbitrary time, with infinitesimal time resolution. In a modern radar system, the received signal is normally sampled at discrete time intervals, using an ADC, which quantizes the signal in time and amplitude. The time quantization corresponds to the ADC sample times, and the amplitude quantization depends on the number of ADC "bits" and the full-scale voltage. To achieve detection, the time between samples must be no more than a pulse width; for example, for a 1 μs transmit pulse, the received signal must be sampled at intervals of no more than a microsecond. Usually, to achieve improved detection, oversampling is used; for example, there would be two samples for a given pulse width. A 1 μs pulse width would suggest a 0.5 μs sample period,

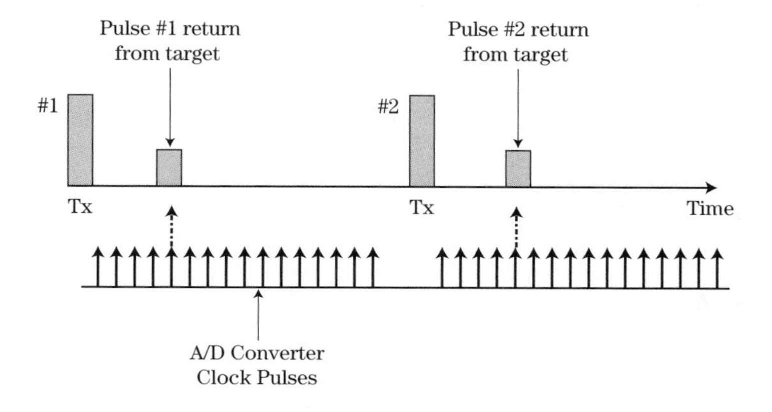

FIGURE 1-21 ■ Pulsed radar waveform showing ADC clock pulses.

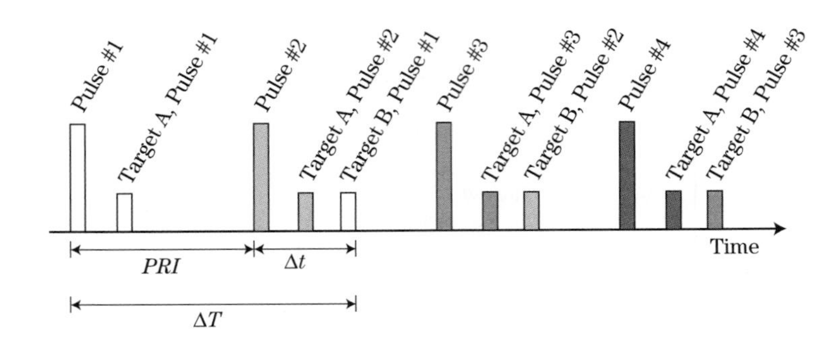

FIGURE 1-22 ■ Pulsed radar range ambiguity.

or a 2 megasample per second (Msps) sample rate. Each of these time samples represents a different range increment, often termed a *range bin*, at a range found from Equation (1.1). The target shown in the figure is at a range corresponding to sample number five.

Unambiguous Range Measurement Recall that target range is determined by measuring the delay time from transmission of a pulse to reception of the reflected signal. Problems can occur in a pulsed radar when determining the range to targets if the pulse round-trip travel time, ΔT, between the radar and the distant target is greater than the interpulse period, *IPP*. In this case, the EM wave in a given pulse will not return to the radar's receiver before the next pulse is transmitted, resulting in a time ambiguity and related *range ambiguity*. The received pulse could be a reflection of the pulse that was just transmitted and, thus, a reflection from a close-in target, or it could be a reflection resulting from a previously transmitted pulse and, thus, a reflection from a distant target.

This situation is illustrated in Figure 1-22. The tall rectangles represent transmitted pulses; the shorter rectangles represent the received echoes from two targets. The shading of the target echoes matches the shading of the pulse from which they originated. The time delay to target A and back is less than the interpulse period, so the echo from target A from a given pulse is received before the next pulse is transmitted. The time delay ΔT to target B is greater than the *PRI*; specifically, suppose $\Delta T = PRI + \Delta t$. Then the reflection from target B due to pulse #1 occurs Δt seconds after pulse #2, as shown in the figure. Consequently, it is unclear if this echo is from a short-range target Δt seconds away or a longer-range target ΔT seconds away[5].

Range ambiguities can be avoided by ensuring that the interpulse period, *PRI*, is long enough or, equivalently, the pulse repetition frequency *PRF* is low enough, such that all echoes of interest from a given pulse return to the radar receiver before the next pulse is transmitted. The round-trip time for the radar wave from equation (1.1) is given by

$$\Delta T = \frac{2R}{c} \tag{1.13}$$

Thus, to prevent range ambiguities, the following condition must be satisfied:

$$PRI \geq \Delta T_{\max} = \frac{2R_{\max}}{c} \quad \text{or} \quad R_{\max} \leq \frac{c \cdot PRI}{2} = \frac{c}{2PRF} \tag{1.14}$$

[5]It appears that the ambiguous range condition could be revealed because the target signal does not appear in the first range interval. In fact, radar systems do not usually detect a target on the basis of any single pulse; several pulses are transmitted and processed. In this case, it is not known that the target signal is "missing" for one (or more) intervals.

where R_{max} is the maximum target range of interest. Conversely, the *unambiguous range*, R_{ua}, is the maximum range at which the range to a target can be measured unambiguously by the radar. It is given by

$$R_{ua} = \frac{c}{2PRF} \tag{1.15}$$

It should be noted that not all radars satisfy this condition. Some systems cannot avoid an ambiguous range condition, due to other conflicting requirements, as is seen in the following section.

1.5.3 Noncoherent versus Coherent

Radar systems can be configured to be noncoherent or coherent. Whereas a noncoherent system detects only the amplitude of the received signal, the coherent system detects the amplitude and the phase, treating the received signal as a vector. Noncoherent systems are often used to provide a two-dimensional display of target location in a ground map background. The amplitude of the signal at any instant in time will determine the brightness of the corresponding area of the display face. Noncoherent radars can be used in cases in which it is known that the desired target signal will exceed any competing clutter signal. All early radars were noncoherent; target detection depended on operator skill in discerning targets from the surrounding environment.

For a coherent system, measurement of the phase of the received signal provides the ability to determine if the phase is changing, which can provide target motion characteristics and the ability to image a target. Though there are still applications for which noncoherent radar technology is appropriate, most modern radar systems are coherent.

A pulsed coherent system measures the phase of the received signal on a pulse-to-pulse basis. This reference sinusoid is usually implemented in the form of a local oscillator (LO) signal used to produce the transmit signal that also serves as the reference for the received signal. This process is depicted in Figure 1-23. The top line is the local oscillator signal; the solid segments represent the transmit pulse times, and the dashed segments represent "listening" times between transmit pulses. If the local oscillator signal is a fixed frequency, it can serve as a reference for measuring the phase of the received signal. The expanded "balloon" shows the phase relationship for a single transmit/receive pulse pair. The ability to measure the phase of the received signal depends on the stability of the LOs, as described in Chapter 12.

1.5.3.1 The Doppler Shift

If there is relative motion between the radar and the target, then the frequency of the EM wave reflected from the target and received by the radar will be different from the frequency of the wave transmitted from the radar. This is the *Doppler effect*, common to all wave phenomena and originally identified as an acoustic (sound wave) phenomenon. The Doppler frequency shift, f_d, or "Doppler" for short, is the difference between the frequency of the received wave and that of the transmitted wave and is approximately given by

$$f_d \approx \frac{2v_r}{\lambda} \tag{1.16}$$

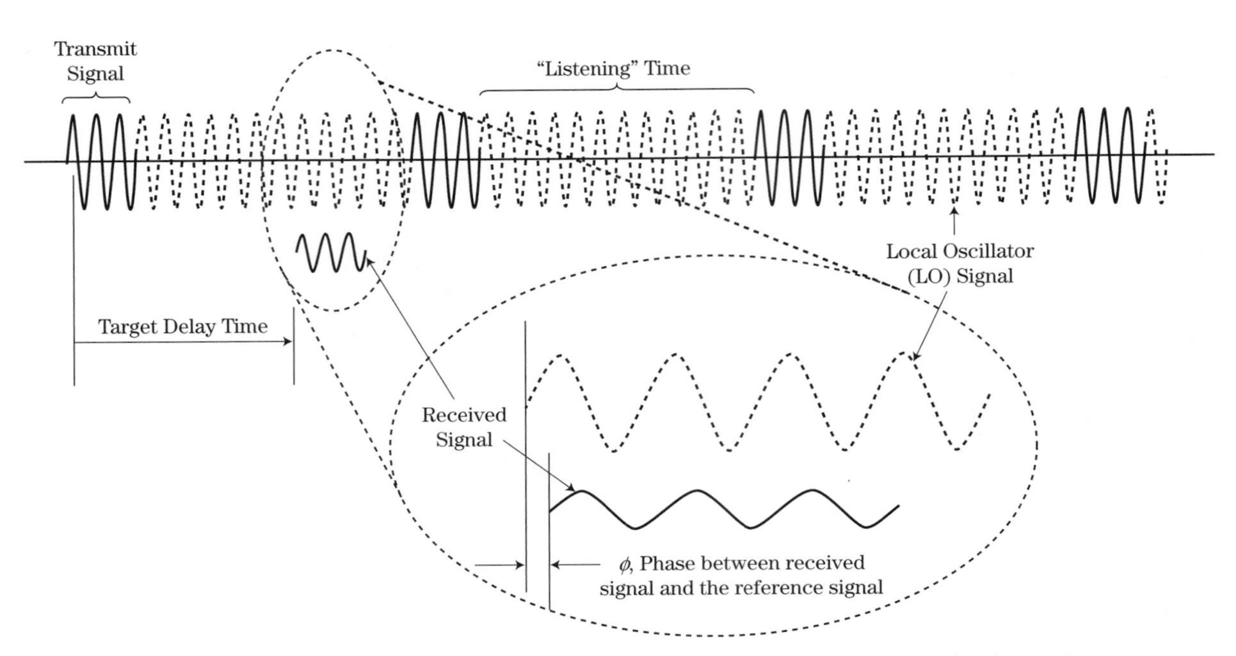

FIGURE 1-23 ■ Coherent system local oscillator, transmit, and received signals.

where v_r is the radial component[6] of the target's velocity vector toward the radar, and λ is the wavelength of the transmitted EM wave. The approximation is excellent as long as the radial component of velocity of the target is much less than the speed of light. The negative of the radial velocity is often called the *range rate*. This radial velocity component, v_r, is positive (and, thus, f_d is positive) for targets approaching the radar and negative (f_d negative) for targets receding from the radar.[7]

1.5.3.2 Unambiguous Doppler Shift Measurement

Clearly from equation (1.15), lowering the PRF of the radar will increase the radar's unambiguous range. However, lowering the radar's PRF also has a negative consequence. Most modern radars measure the Doppler frequency shift of the received EM wave. A pulsed radar samples the Doppler frequency shift at the pulse repetition frequency. This can lead to Doppler frequency ambiguities if the sampling rate (PRF) is not high enough, as discussed in Chapter 14.

One statement of the Nyquist sampling criterion or theorem is that "the maximum frequency that can be unambiguously measured is half the sampling rate." A similar statement holds for measuring negative frequencies. In a radar, Doppler shift is being

[6]The radial component is the component of velocity along the range dimension (i.e., a straight line between the radar and the target).

[7]There is often a point of confusion regarding whether the radial component of velocity is positive or negative. The Doppler frequency shift will be positive for a closing target, so a closing target represents a positive velocity. But since *range rate* is positive for a receding target (one for which the range is increasing) the sign of *range rate* will be opposite from the sign for radial component of velocity.

sampled at the radar's PRF; thus, the maximum range of Doppler shift frequencies that can be unambiguously measured is

$$f_{d_{\max}} = \pm PRF/2 \quad \text{or} \quad PRF_{\min} = 2f_{d_{\max}} = \frac{4v_{r_{\max}}}{\lambda} \tag{1.17}$$

While maximizing unambiguous range leads to lower PRFs, maximizing unambiguous Doppler shift leads to higher PRFs. In many systems, no single PRF can meet both of these opposing requirements. Fortunately, as discussed in Chapter 17, some signal processing techniques such as staggered PRFs allow radars to unambiguously measure range and Doppler shift at almost any PRF.

This conflict leads to the definition of three different *PRF regimes*: low PRF, medium PRF, and high PRF. A *low PRF* system is one that is unambiguous in range, for all target ranges of interest. Though there is no specific range of PRF values that define such a system, the PRF ranges from as low as 100 Hz to as high as 4 kHz. Of course, there may be lower or higher PRFs for low PRF systems, but a large majority of low PRF systems fall into these limits.

At the other extreme, a *high PRF* system is defined as one for which the Doppler shift measurement is always unambiguous. That is, the Nyquist sampling criterion is satisfied for the fastest target of interest. Typical values of PRF for these systems are from 10 kHz to 100 kHz (or sometimes much more).

In between these two conditions lies the *medium PRF* regime, for which both range ambiguities and Doppler ambiguities will exist. Typical values for medium PRF waveforms are from 8 kHz to 30 kHz or so.

For radar systems operating in the HF, VHF, and UHF regions, a different set of conditions apply to the definition for medium PRF. If the radar system operates in these regions it is possible that the PRF required to satisfy the Nyquist sampling criterion will be also sufficient to measure the range to the farthest target of interest unambiguously. In this case, a medium PRF system will be unambiguous in both range and Doppler.

1.6 | NOISE, SIGNAL-TO-NOISE RATIO, AND DETECTION

Because of random thermal motion of charged particles, all objects in the universe with a temperature above absolute zero will be radiating EM waves at, collectively, almost all frequencies. These EM waves, called *thermal noise*, are always present at the radar's receiving antenna and compete with the reflected EM waves from the target. In addition, the radar's receiver, being an electrical device with randomly moving electrons, generates its own internal thermal noise that also competes with the received target signal. In microwave radars, the internally generated noise usually dominates over the noise from the environment.

The noise voltage is always present in the radar receiver circuits. If the radar antenna beam is pointed in the direction of a target when the transmitter generates the transmitted signal, then the signal will illuminate the target, and the signal reflected from that target will propagate toward the receiver antenna, will be captured by the antenna, and will also produce a voltage in the receiver. At the instant in time at which the target signal is present

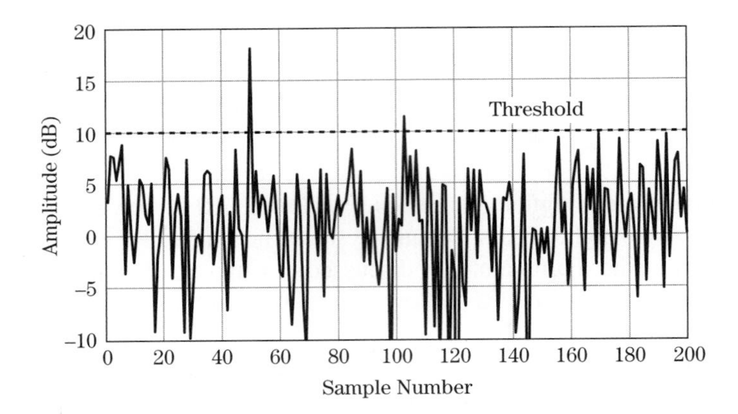

FIGURE 1-24 ■ Threshold detection of a noisy signal.

in the receiver, there will be a combination of the noise and target signal. The voltages add, but because the noise is uncorrelated with the target signal, the total power is just the sum of the target signal power S and the noise power N.

Suppose the signal power of the reflected EM wave from the target is much greater than the noise power due to environmental and receiver noise. If this is the case, the presence of a target echo signal can be revealed by setting an *amplitude threshold* above the noise level (but below the target level). Any received signals (plus noise) that are above this amplitude threshold are assumed to be returns from targets, while signals below this threshold are ignored. This is the basic concept of threshold detection. In the example of Figure 1-24 the receiver output contains 200 samples of noise with a target signal added at sample 50. The target signal is 17 dB larger than the root mean square (rms) noise power. If the threshold is set at a value of 10 dB above the noise power, the signal + noise sample easily exceeds the threshold and so will be detected, while all but one of the noise-only samples does not cross the threshold. Thus, the target signal is revealed in the presence of the noise.

Because it is a random variable, at any given time the noise alone can "spike up" and cross the amplitude threshold, giving rise to some probability that there will be a *false alarm*. In Figure 1-24, a single false alarm occurs at sample 103. In addition, the target-plus-noise signal is a random variable, so at any given time it can drop below the amplitude threshold, resulting in some probability that the target-plus-noise signal will not be detected. Because of this random nature of the signals, the detection performance of a radar must be given in terms of probabilities, usually the probability of detection, P_D, and the probability of false alarm, P_{FA}. P_D is the probability that a target-plus-noise signal will exceed the threshold and P_{FA} is the probability that the noise alone will spike above threshold. Perfect radar detection performance would correspond to $P_D = 1$ (or 100%) and $P_{FA} = 0$ (or 0%). Either P_D or P_{FA} can be arbitrarily set (but not both at the same time) by changing the amplitude threshold. When the threshold is raised, P_{FA} goes down, but unfortunately, so does P_D. When the threshold is lowered, the P_D goes up, but, unfortunately, so does P_{FA}. Thus, when the threshold is changed, P_{FA} and P_D both rise or fall together. To increase P_D while at the same time lowering P_{FA}, the target signal power must be increased relative to the noise power. The ratio of the target signal power to noise power is referred to as the *signal-to-noise ratio*. Chapter 2 develops an equation to predict the SNR called the *radar range equation* (RRE); Chapters 3 and 15 discuss the methods for relating P_D and P_{FA} to SNR; and Chapter 16 presents a description of the processing implemented to automatically establish the threshold voltage.

1.7 | BASIC RADAR MEASUREMENTS

1.7.1 Target Position

Target position must be specified in three-dimensional space. Since a radar transmits a beam in some azimuthal and elevation angular direction, and determines range along that angular line to a target, a radar naturally measures target position in a spherical coordinate system (see Figure 1-25).

Modern radars can determine several target parameters simultaneously:

- Azimuthal angle, θ
- Elevation angle, ϕ
- Range, R (by measuring delay time, ΔT)
- Range rate, $\dot{R}$ (by measuring Doppler frequency, f_d)
- Polarization (up to five parameters)

Measurements in each of these dimensions are discussed in the next section.

1.7.1.1 Azimuth Angle, Elevation Angle

The target's angular position, here denoted by the azimuth and elevation angles θ and ϕ, is determined by the pointing angle of the antenna main beam when the target detection occurs. This antenna pointing angle can either be the actual physical pointing angle of a mechanically scanned antenna or the electronic pointing angle of an electronically scanned (phased array) antenna. (See Chapter 9 for more on antenna-scanning mechanisms.) The *monopulse* technique, also described in Chapter 9, can provide a significantly more precise angle measurement than that based on the main beam beamwidth alone.

1.7.1.2 Range

The target's range, R, is determined by the round-trip time of the EM wave as discussed in Section 1.2. The range to the target is determined by measuring the time delay. Repeating equation (1.1) for convenience,

$$R = \frac{c\Delta T}{2}$$

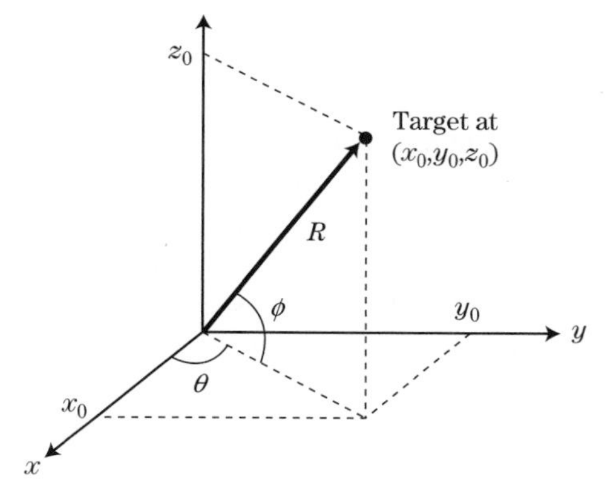

FIGURE 1-25 ■ Spherical coordinate system depicting radar-target geometry.

In most modern radar systems, the delay time, ΔT, is determined by "counting" the number of ADC clock pulses that occur between the transmit time and the target time, assuming that the first clock pulse coincides with the transmit pulse. Chapter 18 presents the details associated with range measurement results and shows that the precision of a range measurement can be much better than the resolution of the measurement.

1.7.2 Range Rate and Doppler Frequency Shift

As described in Section 1.5.4, if there is relative motion between the radar and the target, then the frequency of the EM wave reflected from the target and received by the radar will be different from the frequency of the wave transmitted from the radar. This is the Doppler effect.

The Doppler shift is measured by performing a spectral analysis of the received signal in every range increment. The spectral analysis is usually performed in modern radar systems by transmitting a sequence of several pulses, (often on the order of 30 pulses) and performing a K-point discrete Fourier transform (DFT) on this sequence of received signals for each range increment. The DFT is usually implemented in the form of the fast Fourier transform (FFT), described in Chapter 17.

Doppler shift is a very important quantity in modern radars. Measurement of the Doppler characteristics is used to suppress returns from clutter, to determine the presence of multiple targets at the same range, and to classify and identify moving targets and targets with moving components (e.g., aircraft, helicopters, trucks, tanks). In a synthetic aperture radar, the measurement of Doppler shift is used to improve the cross-range resolution of the radar.

For example, consider a stationary radar designed to detect moving targets on the ground. The EM wave return from a moving target will have a nonzero Doppler shift, whereas the return from stationary clutter (e.g., trees, rocks, buildings) will essentially have a zero Doppler shift. Thus, Doppler shift can be used to sort (discriminate) returns from targets and clutter by employing a high-pass filter in the radar's signal processor. This is the essence of *moving target indication* (MTI) radars discussed in Chapter 17.

1.7.3 Polarization

Because a typical target comprises a multitude of individual scatterers, each at a slightly different distance from the radar, the RCS of an object changes with viewing angle and wavelength, as explained in Chapter 7. It is also sensitive to the transmit and receive polarization of the EM wave. Polarization refers to the vector nature of the EM wave transmitted and received by the radar antenna. The EM wave's polarization is sensitive to the geometry of the object from which it reflects; different objects will change the polarization of the incident EM wave differently. Therefore, the change in polarization of the EM wave when it reflects from an object carries some information regarding the geometrical shape of that object. This information can be used to discriminate unwanted reflected waves (e.g., returns from rain) from those reflected from targets. Also, polarization can be used to discriminate targets from clutter and even to facilitate identifying different targets of interest.

Maximum polarization information is obtained when the *polarization scattering matrix* (PSM) **S** of a target is measured. Equation (1.18) describes the four components of the PSM. Each of the four terms is a vector quantity, having an amplitude and a phase. The subscripts refer to the transmit and receive polarization. Polarizations 1 and 2 are orthogonal; that is, if polarization 1 is horizontal, the polarization 2 is vertical. If polarization 1 is right-hand-circular, then polarization 2 is left-hand-circular.

$$\mathbf{S} = \begin{bmatrix} \sqrt{\sigma_{11}}e^{j\phi_{11}} & \sqrt{\sigma_{12}}e^{j\phi_{12}} \\ \sqrt{\sigma_{21}}e^{j\phi_{21}} & \sqrt{\sigma_{22}}e^{j\phi_{22}} \end{bmatrix} \tag{1.18}$$

Measuring the PSM requires the radar to be polarization-agile on transmit and to have a dual-polarized receiver. An EM wave of a given polarization (e.g., horizontal polarization) is transmitted and the polarization of the resulting reflecting wave is measured in the dual-polarized receiver. This measurement requires, at a minimum, the measurement of the *amplitude* of the wave in two orthogonal polarization receiver channels (e.g., horizontal and vertical polarizations) and the *relative phase* between the waves in these two channels. The transmit polarization is then changed to an orthogonal state (e.g., vertical polarization) and the polarization of the resulting reflecting wave is measured again. For a monostatic radar, this process results in five unique measured data: three amplitudes and two relative phases.[8] These data constitute the elements of the PSM. Ideally, the two transmit polarizations should be transmitted simultaneously, but in practice they are transmitted at different, but closely spaced times (typically on successive pulses). This time lag creates some uncertainty in the integrity of the PSM; however, if the two transmit times are closely spaced, the uncertainty is minor. A more detailed discussion of the PSM is in Chapter 5.

1.7.4 Resolution

The concept of *resolution* describes a radar's ability to distinguish two or more targets that are closely spaced, whether in range, angle, or Doppler frequency. The idea is illustrated in Figure 1-26, which imagines the receiver output for a single transmitted pulse echoed from two equal-strength point scatterers separated by a distance ΔR. If ΔR is large enough, two distinct echoes would be observed at the receiver output as in Figure 1-26b. In this case, the two scatterers are considered to be *resolved* in range. In Figure 1-26c, the scatterers are close enough that the two echoes overlap, forming a composite echo. In this case the two scatterers are not resolved in range. Depending on the exact spacing of the two scatterers, the two pulses may combine constructively, destructively, or in some intermediate fashion. The result is very sensitive to small spacing changes, so the two scatterers cannot be considered to be reliably resolved when their echoes overlap.

The dividing line between these two cases is shown in Figure 1-26d, where the two pulses abut one another. This occurs when

$$\Delta R = \frac{c\tau}{2} \tag{1.19}$$

The quantity ΔR is called the *range resolution* of the radar. Two targets spaced by more than ΔR will be resolved in range; targets spaced by less than ΔR will not. This equation represents the range resolution achieved using a simple, unmodulated pulse of length τ.

Note that in the discussions so far, the range resolution is proportional to the pulse width τ. A pulse width of 1 μs results in a range resolution of $(3 \times 10^8)(10^{-6})/2 = 150$ m.

---

[8]Though there are eight values in the matrix, only five are unique. σ_{12} and σ_{21} will be equal, $\angle_{11}$ is the reference angle, and $\angle_{21}$ will equal $\angle_{12}$, so the unique values are σ_{11}, σ_{12}, σ_{22}, $\angle_{12}$, and $\angle_{22}$.

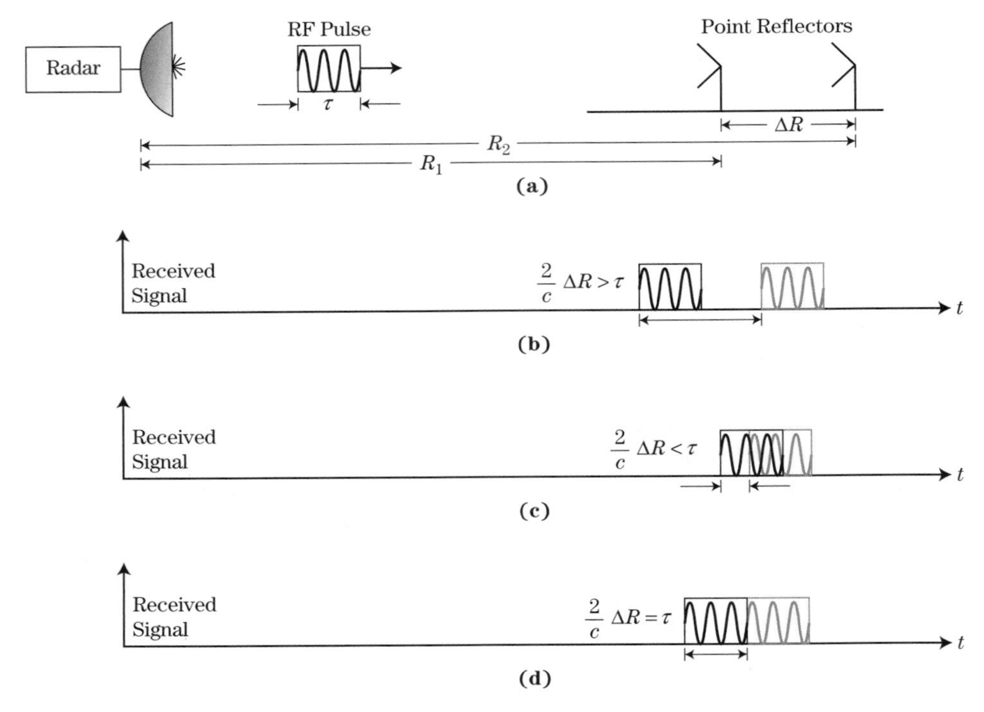

FIGURE 1-26 ■ Concept of resolution in range. (a) Transmitted pulse and two targets. (b) Receiver output for resolved targets. (c) Receiver output for unresolved targets. (d) Receiver output for defining range resolution.

This is the minimum separation at which two targets can be reliably resolved with a 1 μs simple, unmodulated pulse. If finer resolution is needed, shorter pulses can be used. It will be seen later that shorter pulses have less energy and make detection more difficult. Chapter 20 will introduce *pulse compression,* a technique that provides the ability to maintain the energy desired in the pulse while at the same time providing better range resolution by modulating the signal within the pulse.

Figure 1-20 showed the timing of a pulsed radar waveform. A number of choices are available for the actual shape of the waveform comprising each pulse. Figure 1-27 illustrates three of the most common. Part (a) of the figure is a simple unmodulated pulse, oscillating at the radar's RF. This is the most basic radar waveform. Also very common is the *linear frequency modulation* (LFM) or *chirp* pulse (Figure 1-27b). This waveform sweeps the oscillations across a range of frequencies during the pulse transmission time. For example, a chirp pulse might sweep from 8.9 to 9.1 GHz within a single pulse, a swept bandwidth of 200 MHz. Part (c) of the figure illustrates a *phase-coded* pulse. This pulse has a constant frequency but changes its relative phase between one of two values, either zero or π radians, at several points within the pulse. These phase changes cause an abrupt change between a sine function and a negative sine function. Because there are only two values of the relative phase used, this example is a *biphase-coded* pulse. More general versions exist that use many possible phase values.

The choice of pulse waveform affects a number of trade-offs among target detection, measurement, ambiguities, and other aspects of radar performance. These waveforms and their design implications are discussed in Chapter 20.

A radar also resolves a target in azimuth angle, elevation angle, and Doppler frequency. The only difference is the characteristic signal shape that determines achievable resolution. Figure 1-28a shows the Fourier spectrum of a signal consisting of the sum of two sinusoids. This could arise as the Doppler spectrum of a radar viewing two targets at the same range

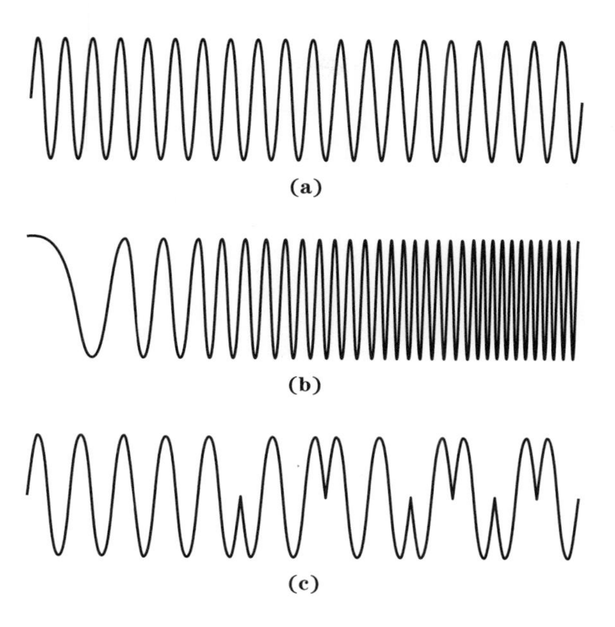

FIGURE 1-27 ■ Three common choices for a single pulse in a pulsed radar waveform. (a) Simple pulse. (b) Linear FM or chirp pulse. (c) Biphase coded pulse.

but different range rates. Each sinusoid contributes a characteristic "sinc" function shape (see Chapter 17) to the Doppler spectrum, centered at the appropriate Doppler shift. The width of the main lobe in the spectrum is proportional to the reciprocal of the time during which the input data were collected. For a "sinc" function, the main lobe 3 dB width is

$$\text{3 dB width} = \frac{0.89}{\text{dwell time}} \tag{1.20}$$

The dwell time is the time duration associated with transmitting the sequence of pulses for performing the FFT, as described in Section 1.7.2. The dwell time for this example is 0.01 seconds, so the width of the central lobe of the sinc function measured 3 dB below its peak (about 71% of the peak amplitude) is 89 Hz in this example. When the two Doppler shifts are separated by more than 89 Hz, they are clearly resolved. When the separation becomes less than 89 Hz, as in Figure 1-28b, the two peaks start to blend together. While two peaks are visible here, the dip between them is shallow. If noise were added to the signals, the ability to resolve these two frequencies reliably would degrade. At a separation of 50 Hz (Figure 1-28c), the two signals are clearly not resolved.

In this Doppler frequency example, the resolution capability is determined by the width of the sinc function main lobe, which in turn is determined by the total pulse burst waveform duration. Also, as will be seen in Chapter 17, when a weighting function is used in the Doppler processing to reduce the sidelobes, the main lobe will be further spread, beyond 89 Hz in this example, further degrading the Doppler resolution. Thus, as with range resolution, Doppler resolution is determined by the transmitted waveform and processing properties.

The signals shown in Figure 1-28 could just as easily represent the receiver output versus scan angle for two scatterers in the same range bin but separated in angle. The "sinc" response would then be the model of a (not very good) antenna pattern. Given a more realistic antenna pattern, as described in Section 1.4.1 and equation (1.9), the angular resolution would be determined by the antenna size. Two targets can be resolved in angle if they are separated by the antenna beamwidth or more.

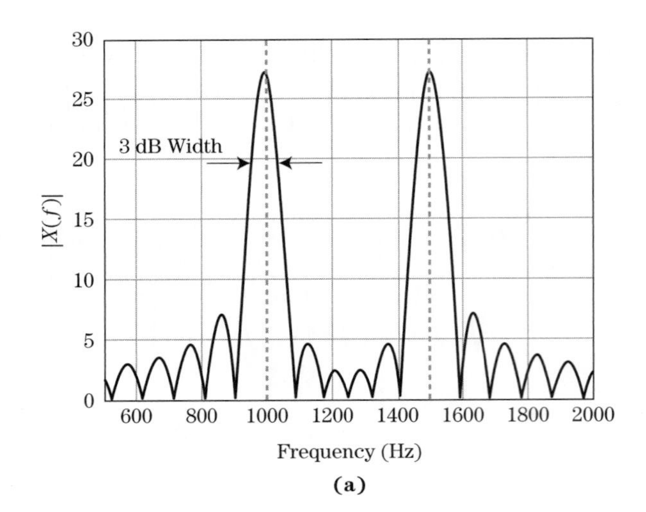

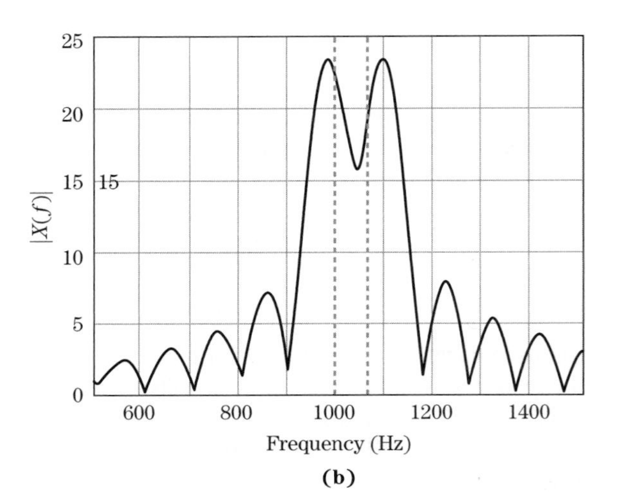

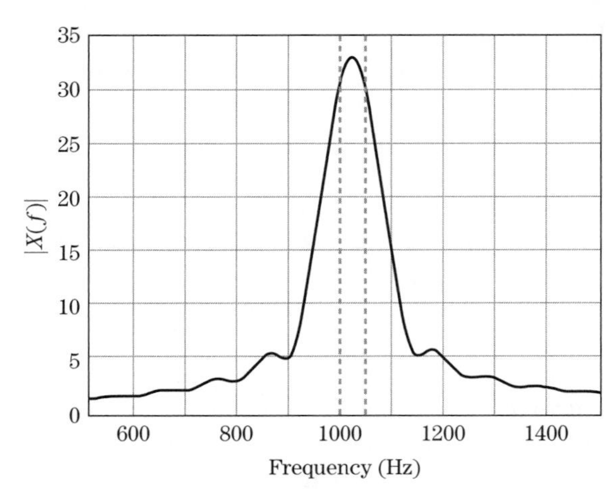

FIGURE 1-28 ■ Example of frequency resolution. (a) Spectrum of two sinusoids separated by 500 Hz. The 3 dB width of the response is 89 Hz. (b) Spectrum for separation of 75 Hz. (c) Spectrum for separation of 50 Hz. The two dashed lines show the actual sinusoid frequencies in each case.

1.8 BASIC RADAR FUNCTIONS

While there are hundreds of different types of radars in use, the large majority have three basic functions: (1) search/detect; (2) track; and (3) image. These functions are briefly discussed now, followed by a discussion of some of the many types of radar systems and how they apply these functions.

1.8.1 Search/Detect

Almost all radars have to search a given volume and detect targets without a priori information regarding the targets' presence or position. A radar searches a given volume by pointing its antenna in a succession of beam positions that collectively cover the volume of interest. A mechanically scanned antenna moves through the volume continuously. Rotating antennas are an example of this approach. An ESA is pointed to a series of discrete beam positions, as suggested in Figure 1-29.

At each position, one or more pulses are transmitted, and the received data are examined to detect any targets present using the threshold techniques described earlier. For example, 10 pulses might be transmitted in one beam position of an ESA. The detected data from each pulse might then be noncoherently integrated (summed) in each range bin to improve the SNR. This integrated data would then be compared with an appropriately set threshold to make a detection decision for each range bin. The antenna is then steered to the next beam position, and the process is repeated. This procedure is continued until the entire search volume has been tested, at which point the cycle is repeated.

A major issue in search is the amount of time required to search the desired volume once. The search time is a function of the total search volume, the antenna beamwidths, and the *dwell time* spent at each beam position. The latter in turn depends on the number of pulses to be integrated and the desired range coverage (which affects the PRF). Optimization of the search process involves detailed trade-offs among antenna size (which affects beamwidths and thus number of beam positions needed), dwell time (which affects number of pulses available for integration), and overall radar timeline. The search and detection process and these trade-offs are discussed in more detail in Chapter 3.

1.8.2 Track

Once a target is detected in a given search volume, a measurement is made of the target *state*, that is, its position in range, azimuth angle, and elevation angle, and, often, its radial component of velocity. Tracking radars measure target states as a function of time. Individual position measurements are then combined and smoothed to estimate a target *track*.

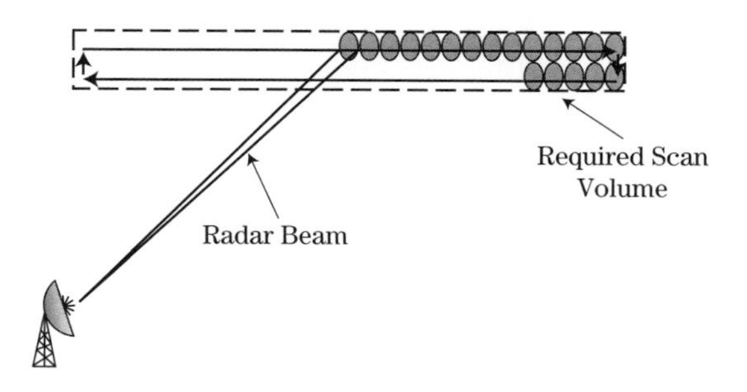

FIGURE 1-29 ■ Coverage of a search volume using a series of discrete beam positions.

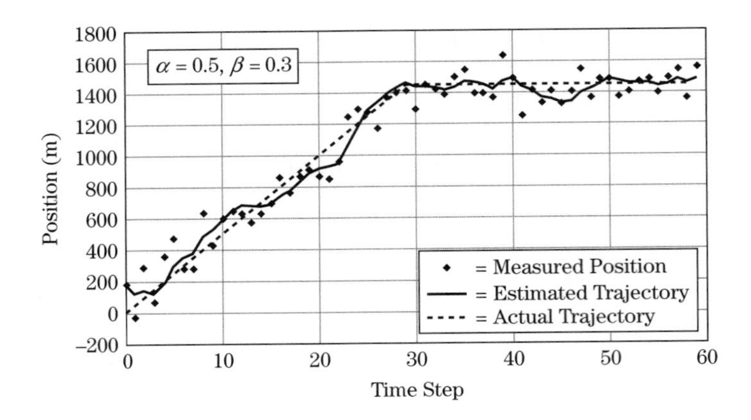

FIGURE 1-30 ■ Example of track filtering for smoothing a series of individual position measurements.

Tracking implies measuring the position and velocity of a target to an accuracy better than the radar's resolution. A variety of techniques are used to do this. For instance, the azimuth position can be estimated to a fraction of the antenna azimuth beamwidth in a mechanically scanned radar by measuring the detected target strength on several successive pulses as the antenna scans and then computing the centroid of the resulting measurements. Similar concepts can be applied in range and Doppler. These and other measurement techniques are described in Chapter 18. Individual measurements are invariably contaminated by measurement noise and other error sources. An improved estimate of the target position over time is obtained by *track filtering*, which combines multiple measurements with a model of the target dynamics to smooth the measurements. For example, the dotted line in Figure 1-30 shows the actual position of a target that is initially moving away from the radar in some coordinate at constant velocity and then at time step 30 stops moving (constant position). The small triangles represent individual noisy measurements of the position, and the solid line shows the estimated position using a particular track filtering algorithm called the *alpha-beta filter* or *alpha-beta tracker*. Advanced systems use various forms of the Kalman filter and other techniques. Track filtering is discussed in Chapter 19.

The optimum radar configurations for tracking and searching are different. Consequently, these search and track functions are sometimes performed by two different radars. This is common in situations where radar weight and volume are not severely limited (i.e., land-based and ship-based operations). When radar weight and volume are limited, as in airborne operations, the search and track functions must be performed by one radar that must then compromise between optimizing search and track functions. For example, a wide antenna beamwidth is desirable for the search mode and a narrow antenna beamwidth is desirable for the track mode, resulting in a medium antenna beamwidth compromise solution.

1.8.3 Imaging

In radar, *imaging* is a general term that refers to several methods for obtaining detailed information on discrete targets or broad-area scenes. The imaging process involves two steps: (1) developing a high-resolution range profile of the target; and (2) developing a high-resolution cross-range (angular) profile. As suggested in equation (1.19) the range resolution is proportional to the pulse width. A shorter transmitted pulse width will lead to better (smaller is better) range resolution. Improved range resolution can also be achieved

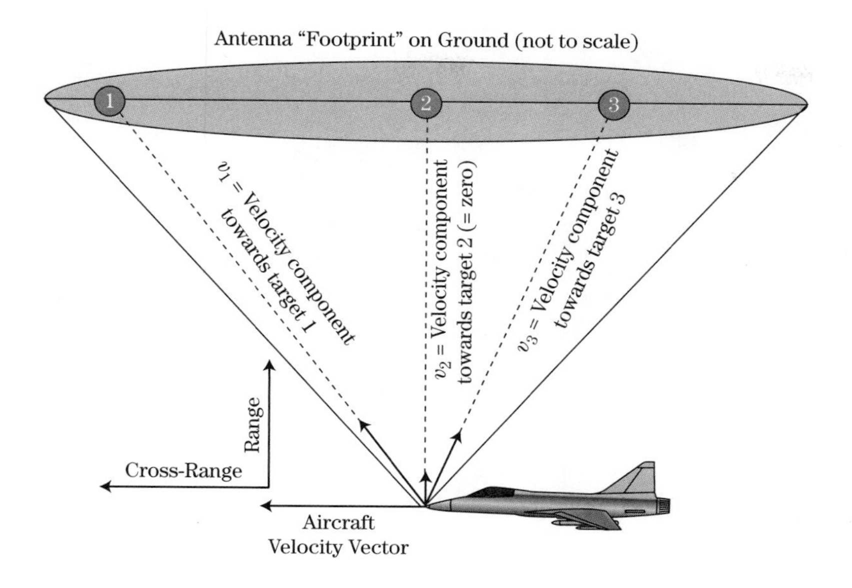

FIGURE 1-31 ■ Synthetic aperture radar geometry.

by modulating the signal within the pulse as described in Section 1.7.4 and shown in Figure 1-27. As presented in Chapter 20, there are other techniques for developing good range resolution.

With the few basics given in Section 1.7.2, one can easily understand how Doppler shift can be used to improve the resolution in an SAR. Consider an aircraft with a radar on board pointing out of the side of the aircraft down to the ground (i.e., a *sidelooking* radar). If the radar's antenna has a 1 degree beamwidth and the ground is 1,000 meters from the radar, then the antenna beam will be approximately 21 meters wide in the cross-range dimension (the direction of aircraft motion, which is also perpendicular to the range dimension) on the ground. This means that two objects on the ground at the same range and closer than 21 meters in the cross-range dimension essentially appear as one object to the radar; that is, they are not resolved. Doppler shift can be used to resolve objects within the antenna's beam in cross-range. As shown in Figure 1-31, objects in the front, middle, and back of the beam will have a positive, zero, and negative Doppler shift, respectively. In fact, each object in the beam with a different cross-range spatial location will have a different Doppler shift. Therefore, if the radar's signal processor is capable of sorting (filtering) the EM wave returns according to Doppler shift, this is tantamount to sorting the objects in the cross-range dimension, thus resolving objects at different positions. Doppler processing is discussed further in Chapters 8 and 17 and synthetic aperture radar in Chapter 21.

Synthetic aperture radars form two-dimensional images of an area at resolutions ranging from 100 m or more to well under 1 m. The first case would typically be used in wide-area imaging from a satellite, while the latter would be used in finely detailed imaging from an airborne (aircraft or *unmanned autonomous vehicle* [UAV] platform). Figure 1-32 is an example of a 1 m resolution airborne SAR image of the Washington, D.C., mall area. Two-dimensional SAR imagery is used for a variety of Earth resources and defense applications, including surveillance, terrain following, mapping, and resource monitoring (discussed in more detail in Chapter 21). In recent years, *interferometric SAR* (IFSAR or InSAR) techniques have been developed for generating three-dimensional SAR imagery.

FIGURE 1-32 ■ 1 m resolution SAR image of the Washington, D.C., mall area. (Courtesy of Sandia National Laboratories. With permission.)

To accomplish their mission, many radars must not only detect but also identify the target before further action (e.g., defensive, offensive, traffic control) is initiated. One common way to attempt identification is for the radar to measure a one-dimensional high-range-resolution "image" (often called a *high-range resolution* [HRR] *profile*) or two-dimensional range/cross-range image of the target, a high-resolution Doppler spectrum, or to determine target polarization characteristics. The radar will employ specific waveforms and processing techniques, such as pulse compression SAR processing or polarization scattering matrix estimates, to measure these properties. *Automatic target recognition* (ATR) techniques are then used to analyze the resulting "imagery" and make identification decisions.

1.9 RADAR APPLICATIONS

Given that the fundamental radar functions are search/detect, track, and image, numerous remote sensing applications can be satisfied by the use of radar technology. The uses for radar are as diverse as ground-penetrating applications, for which the maximum range is a few meters, to long-range over-the-horizon search systems, for which targets are detected at thousands of kilometers range. Transmit peak power levels from a few milliwatts to several megawatts are seen. Antenna beamwidths from as narrow as less than a degree for precision tracking systems to as wide as nearly isotropic for intrusion detection systems are also seen. Some examples are now given. The grouping into "military" and "commercial" applications is somewhat arbitrary; in many cases the same basic functions are used in both arenas. The radar applications represented here are some of the most common, but there are many more.

1.9.1 Military Applications

In about 1945, the U.S. military developed a system of identifying designations for military equipment. The designations are of the form AN/xxx-nn. The x's are replaced with a sequence of three letters, the first of which indicates the installation, or platform (e.g., A for airborne), the second of which designates the type of equipment (e.g., P for radar), and the third of which designates the specific application (e.g., G for fire control). Table 1-3 lists a subset of this "AN nomenclature" that is pertinent to radar. The n's following the

TABLE 1-3 ■ Subset of the AN Nomenclature System for U.S. Military Equipment Applicable to Radar Systems

First Letter (Type of Installation)		Second Letter (Type of Equipment)		Third Letter (Purpose)	
A	Piloted aircraft	L	Countermeasures	D	Direction finger, reconnaissance, or surveillance
F	Fixed ground	P	Radar	G	Fire control or searchlight directing
M	Ground, mobile (installed as operating unit in a vehicle which has no function other than transporting the equipment	Y	Signal/data processing	K	Computing
P	Pack or portable (animal or man)			N	Navigational aids (including altimeter, beacons, compasses, racons, depth sounding, approach, and landing)
S	Water surface craft			Q	Special, or combination of purposes
T	Ground, transportable			R	Receiving, passive detecting
U	Ground utility			S	Detecting or range and bearing, search
V	Ground, vehicular (installed in vehicle designed for functions other than carrying electronic equipment, etc., such as tanks			Y	Surveillance (search, detect, and multiple target tracking) and control (both fire control and air control)

designation are a numerical sequence. For example, the AN/TPQ-36 is a ground-based transportable special purpose radar, in this case for locating the source of incoming mortars. Another example is the AN/SPY-1, a shipboard surveillance and fire control radar (FCR) system.

1.9.1.1 Search Radars

Often, the primary functions associated with the search and track requirements are performed by two independent radar systems. One system performs the search function, and another performs the track function. This is common, though not always the case, for ground-based or surface ship systems. Some applications prohibit the use of more than one radar or more than one aperture. For example, platforms that have limited prime power or space for electronics force the search and track requirements to be performed by one system. This is common in an airborne application and for many electronically scanned antenna systems.

Two-Dimensional Search Some volume search systems employ a "fan"-shaped antenna pattern to perform the search, usually in the range and azimuth dimensions. The antenna aperture will be quite wide horizontally and somewhat narrower vertically. This leads to a narrow azimuth beamwidth and a wide elevation beamwidth. The elevation extent of the search volume is covered by the wide elevation beamwidth, while the azimuth extent is covered by mechanically scanning the antenna in azimuth. Figure 1-33 depicts a fan beam pattern searching a volume. This configuration is common in air traffic control or airport surveillance systems. A system with this beam pattern can provide accurate range and azimuth position but provides poor elevation or height information due to the wide elevation beamwidth. Consequently, it is termed a two-dimensional (2-D) system, providing position information in only two dimensions.

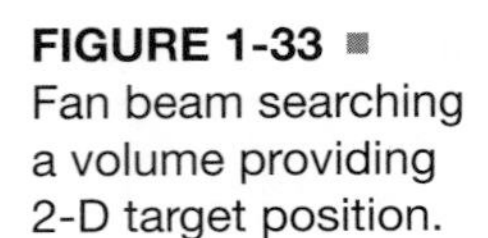

FIGURE 1-33 ■ Fan beam searching a volume providing 2-D target position.

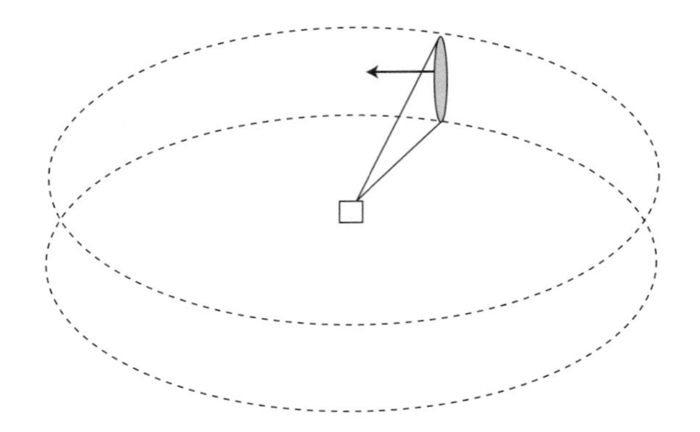

An example of a 2-D radar is the AN/SPS-49 shipboard radar, shown in Figure 1-34. The SPS-49 is a very long-range, two-dimensional air search radar that operates in the UHF band (850–942 MHz). Nominal maximum range of the radar is approximately 250 nmi. The AN/SPS-49 provides automatic detection and reporting of targets supporting the antiair warfare (AAW) mission in Navy surface ships. The AN/SPS-49 uses a large truncated parabolic mechanically stabilized antenna to provide acquisition of air targets in all sea states. Originally produced in 1975 by the Raytheon Company, the SPS-49 is a key part of the combat system on many surface combatants of several navies of the world. It has been extensively modified to provide better detection capabilities of both sea-skimming and high-diving antiship missiles.

The SPS-49 performs accurate centroiding of target range, azimuth, amplitude, ECM level background, and radial velocity with an associated confidence factor to produce accurate target data for the shipboard command and control system. Additionally, processed and raw target data are provided for display consoles.

The AN/SPS-49 has several operational features to optimize radar performance, including an automatic target detection capability with pulse-Doppler processing and clutter maps. This helps ensure reliable detection in both normal and severe clutter. A key feature of the most recent version of the radar, the SPS-49A (V)1, is single-scan radial velocity

FIGURE 1-34 ■ AN/SPS-49 2-D search radar antenna. (Courtesy of U.S. Navy.)

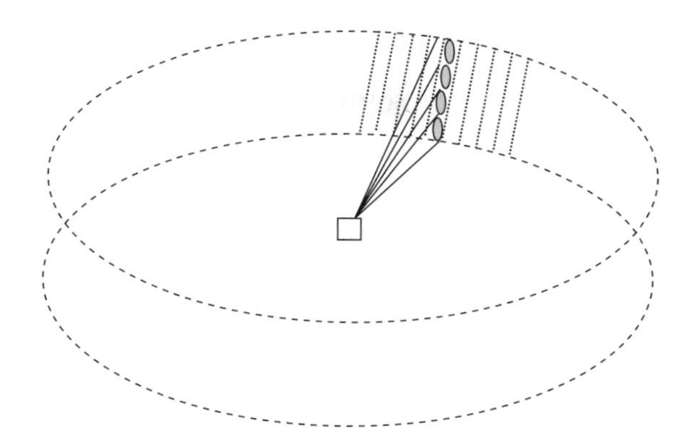

FIGURE 1-35 ■ Pencil beam searching a volume providing 3-D target position.

estimation of all targets, allowing faster promotion to firm track and improved maneuver detection.

The SPS-49 beamwidths are 3.3° in azimuth and 11° in elevation. The narrow azimuth beamwidth provides good resistance to jamming. The antenna rotates at either 6 or 12 rpm. The radar operates in a long-range or short-range mode. In the long-range mode, the radar antenna rotates at 6 rpm. The radar can detect small fighter aircraft at ranges in excess of 225 nautical miles. In the short-range mode, the antenna rotates at 12 rpm to maximize the probability of detection of hostile low-flying aircraft and missiles and "pop-up" targets. The MTI capability incorporated in the AN/SPS-49(V) radar enhances target detection of low-flying, high-speed targets through the cancellation of ground/sea return (clutter), weather, and similar stationary targets.

Three-Dimensional Search Figure 1-35 depicts a pencil beam antenna that provides accurate range, azimuth, and elevation information. A system using this approach is termed a three-dimensional (3-D) search radar.

An example of 3-D search radar used by the U.S. Navy on surface ships, including large amphibious ships and aircraft carriers, is the AN/SPS-48 produced by ITT Gilfillan and shown in Figure 1-36. The antenna is the square planar array consisting of slotted

FIGURE 1-36 ■ AN/SPS-48 3-D search radar antenna. (Courtesy of U.S. Navy.)

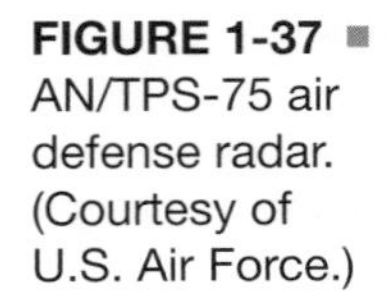

FIGURE 1-37 ■ AN/TPS-75 air defense radar. (Courtesy of U.S. Air Force.)

waveguide. The antenna is fed at the lower left into the serpentine structure attached to the planar array. This serpentine provides frequency sensitivity for scanning in elevation.

The SPS-48 scans in the azimuth plane at 15 rpm by mechanical scanning and in the elevation plane by electronic (frequency) scanning. The large rectangular antenna on top of the main antenna is for the Identification, Friend, or Foe (IFF) system.

The SPS-48 operates in the S-band (2–4 GHz) at an average rated power of 35 kW. The radar scans in elevation (by frequency shifting,) up to 65° from the horizontal. It can detect and automatically track targets from the radar horizon to 100,000 ft. Maximum instrumented range of the SPS-48 is 220 nmi.

The SPS-48 is typically controlled by the shipboard combat system. It provides track data including range, azimuth, elevation, and speed to the combat system and to the display system for action by the ships automated defense system and by the operators.

1.9.1.2 Air Defense Systems

The AN/TPS-75 air defense system used by the U.S. Air Force is shown in Figure 1-37. It has functionality similar to a multifunction 3-D search radar. It scans mechanically in the azimuth direction and forms simultaneous receive beams stacked in elevation with monopulse processing for elevation calculations. The long, narrow antenna shown at the top of the square array is an antenna that interrogates the detected targets for an IFF response. The IFF antenna angle is set back somewhat in azimuth angle so that the IFF interrogation can occur shortly after target detection as the antenna rotates in azimuth.

The AN/MPQ-64 Sentinel shown in Figure 1-38 is an air defense radar used by the U.S. Army and U.S. Marine Corps with similar functionality. This is an X-band coherent (pulse-Doppler) system, using phase scanning in one plane and frequency scanning in the other plane. The system detects, tracks, and identifies airborne threats.

1.9.1.3 Over-the-Horizon Search Radars

During the cold war, the United States wanted to detect ballistic missile activity at very long ranges. Whereas many radar applications are limited to "line-of-sight" performance, ranges

FIGURE 1-38 ■ Photo of an AN/MPQ-64 Sentinel air defense radar. (Courtesy of U.S. Army.)

of several thousand miles were desired. OTH radars were developed for this application. These radars take advantage of the refractive effect in the ionosphere to detect targets at extremely long ranges, sometimes thousands of miles, around the earth. The ionospheric refraction has the effect of reflecting the EM signal. The frequency dependence of this effect is such that it is most effective in the HF band (3–30 MHz). Given the desire for a reasonably narrow beamwidth, the antenna must be very large at such low frequencies, typically thousands of feet long. Consequently, OTH antennas are often made up from separate transmit and receive arrays of elements located on the ground. Figure 1-39 shows an example of such a transmit array. Figure 1-40 depicts the operation of an over-the-horizon system, showing the ray paths for two targets.

1.9.1.4 Ballistic Missile Defense (BMD) Radars

Radar systems can detect the presence of incoming intercontinental ballistic missiles (ICBMs) thousands of kilometers away. These systems must search a large angular volume (approaching a hemisphere) and detect and track very low-RCS, fast-moving targets. Once detected, the incoming missile must be monitored to discriminate it from any debris and

FIGURE 1-39 ■ Over-the-horizon radar system—Transmit array. (Courtesy of U.S. Air Force.)

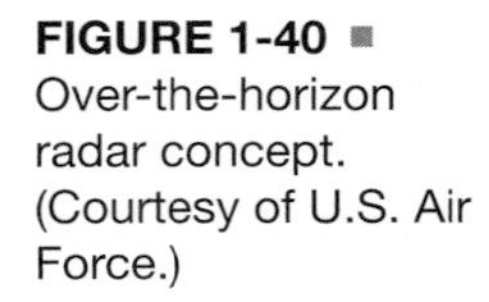

FIGURE 1-40 ■ Over-the-horizon radar concept. (Courtesy of U.S. Air Force.)

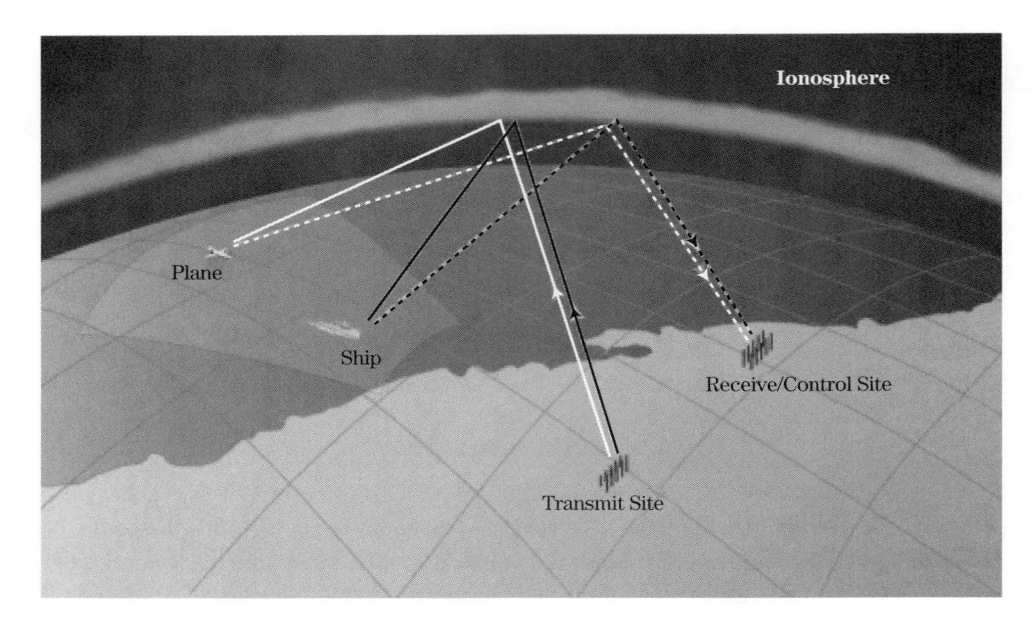

decoys from the warhead. This is accomplished with high-range resolution and Doppler processing techniques that are well suited to radar. Examples of BMD radar systems are the sea-based Cobra Judy system and X-Band (SBX) radar and the land-based Pave Paws and Terminal High Altitude Air Defense (THAAD) AN/TPY-2 systems. Figure 1-41 shows the Pave Paws (AN/FPS-115) system, featuring its two extremely large pencil beam phased array antennas.

A newer system is the THAAD radar shown in Figure 1-42. It is an X-band coherent active phased array system, with over 25,000 active array elements. As opposed to fixed-location systems such as the AN/FPS-115, it is transportable so that it can be redeployed as needed.

FIGURE 1-41 ■ Pave Paws (AN/FPS-115) ballistic missile defense radar. (Courtesy of Missile Defense Agency.)

FIGURE 1-42 ■ Photograph of the Terminal High Altitude Air Defense AN/TPY-2 radar. (Courtesy of U.S. Missile Defense Agency.)

1.9.1.5 Radar Seekers and Fire Control Radars

While many air-to-air and air-to-ground missile systems designed to attack threat targets employ infrared sensors to detect the thermal (heat) signatures of these targets, there are also missile systems that employ radars to detect and track the targets of interest. Radar systems can operate at longer ranges and in atmospheric conditions (e.g., fog and rain) that make infrared sensors ineffective.

Bistatic, *semiactive seekers* in the nose of a missile receive a reflected signal from a target that is being "illuminated" with an RF signal transmitted from a fire control radar on a stand-off platform (e.g., aircraft, ship). Such systems require that the platform maintain line of sight (LOS) to the target until it is engaged by the missile. Ship-based standard missile (SM) and NATO Seaparrow AAW missiles are examples of such a semiactive mode. The NATO Seasparrow requires a constant signal for homing. The Standard Missile does not. It requires illumination only during the last few seconds of flight. Figure 1-43 shows a Seasparrow being launched from a surface ship. Figure 1-44 shows a Navy Standard Missile 2 Block IIIA launching from a guided missile destroyer.

FIGURE 1-43 ■ NATO Seasparrow semiactive homing AAW Missile. (Courtesy of U.S. Navy.)

FIGURE 1-44 ■ Navy Standard Missile 2 Block IIIA launching from a guided missile destroyer. (Courtesy of U.S. Navy.)

The AIM-7 missile shown in Figure 1-45 is a semiactive air-to-air missile used in variety of airborne interceptors, including the U.S. Navy F-14, U.S. Air Force F-15 and F-16, and the U.S. Marine Corps F/A-18 aircraft. The radar in the aircraft illuminates the target as the missile is launched so that the seeker has a signal to which it can "home."

An active radar seeker in the nose of a missile can perform a limited search function and track the target of interest in an autonomous mode, eliminating the requirement of the platform to maintain LOS. This mode is often referred to as the *fire-and-forget* mode. The helicopter-based Longbow FCR system shown in Figure 1-46 is an example of such a system. The Longbow radar is mounted on an Apache helicopter above the main rotor. The missile has its own internal radar *seeker*. The target is acquired, located, and identified by the FCR, and target location information is sent to the missile. Once the

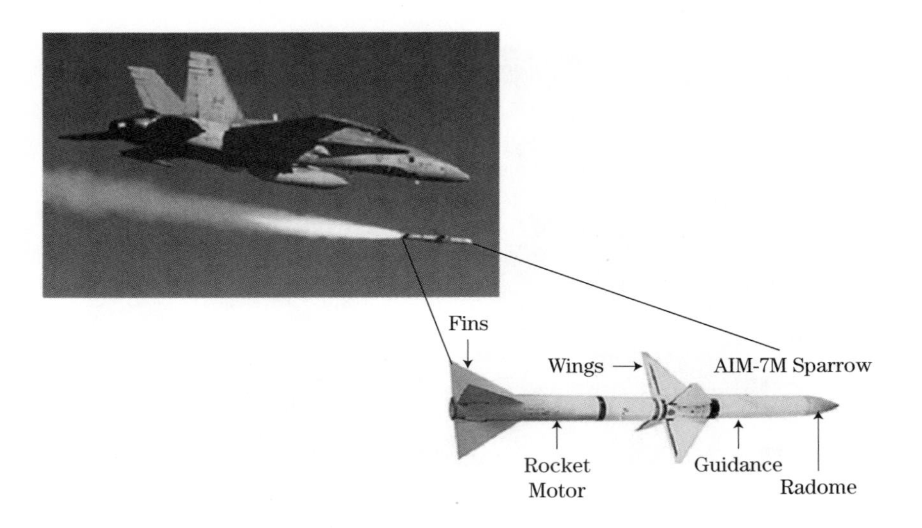

FIGURE 1-45 ■ AIM-7 Sparrow semiactive air-to-air missile. (Courtesy of U.S. Air Force.)

FIGURE 1-46 ■ Apache Longbow fire control radar and active hellfire missile. (Courtesy of U.S. Army.)

missile is launched, the helicopter can descend into a protected posture while the missile autonomously acquires and engages the target with its onboard radar seeker.

1.9.1.6 Instrumentation/Tracking Test Range Radars

Many defense department test ranges use instrumentation radars to aid in testing events. For example, missile testing at the White Sands Missile Range in New Mexico and at the U.S. Army Missile Command in Huntsville, Alabama, require that the target drones and missiles be tracked by precision tracking radars to aid in analyzing tests results and to provide for range safety. Large antennas provide a narrow beamwidth to achieve accurate track data. Long dwell times associated with these radars result in very high Doppler resolution measurements yielding target motion resolution (TMR) data for event timing analysis and phase-derived range (PDR) data for exact relative range measurements. Figure 1-47 is a photograph of an AN/MPQ-39 multiple-object tracking radar (MOTR), a C-band phased array instrumentation radar used at test ranges such as White Sands Missile Range.

Many indoor and outdoor target RCS measurement ranges are designed to measure the RCS of threat targets and provide inverse synthetic aperture radar (ISAR) images of such targets to train pattern-recognition-based target identification systems. Indoor RCS ranges measure small targets, such as missiles and artillery rounds, as well as scale models of threat vehicles and aircraft. Outdoor ranges measure the RCS characteristics of full-sized targets such as tanks and aircraft. Figure 1-48a is an example of an outdoor ISAR range located at the Georgia Tech Research Institute (GTRI). The tank is on a large turntable that

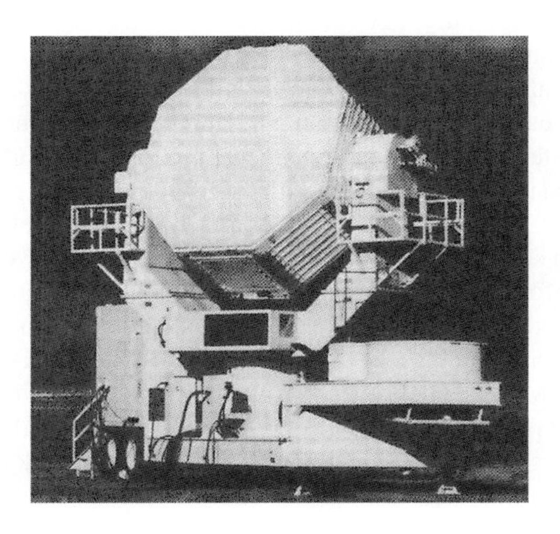

FIGURE 1-47 ■ AN/MPQ-39 C-band phased array instrumentation radar. (Courtesy of Lockheed Martin Corporation. With Permission.)

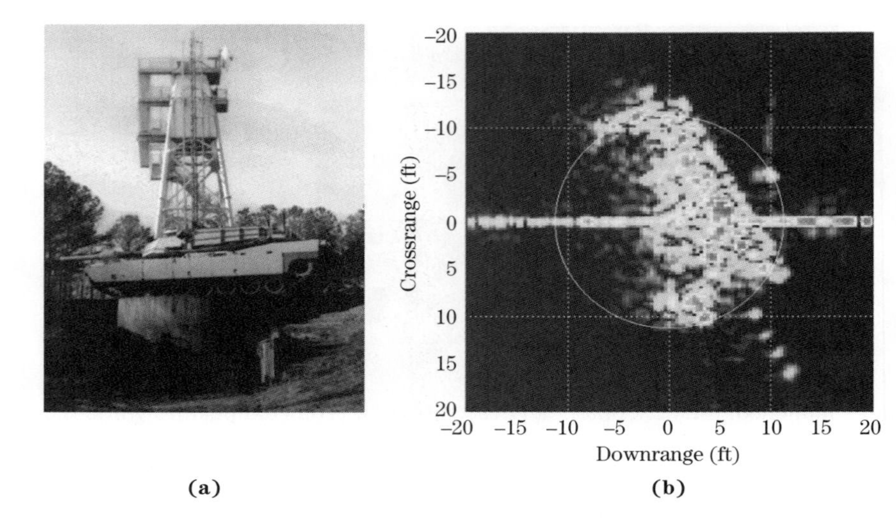

FIGURE 1-48 ■ (a) Turntable ISAR range. View is from behind the target, looking past the turntable to the radar tower in the background. (b) Quick-look image of a tank. (Photos courtesy of the Georgia Tech Research Institute. With permission.)

is flush with the ground; access to the turntable machinery is from behind the turntable. In the distance is a tower that serves as a platform for an instrumentation radar. Figure 1-48b shows a sample "quick look" ISAR image.

1.9.1.7 Tracking, Fire Control, and Missile Support Radars

Ground-based, ship-based, and airborne tracking radars support fire control missions by providing target position and velocity estimates so that an interceptor can position itself to detect and track the target, either autonomously or in a semiactive mode as it approaches the target. Early tracking radar systems could track only one target at a time. Tracking of multiple targets simultaneously required multiple radars. Modern radar tracking systems can track multiple targets using electronically scanning antennas while continuing to perform the search function. Examples of such systems are the ship-based Aegis fire control radar (AN/SPY-1), the ground-based Patriot air defense radar, and certain airborne fire control radars commonly found in fighter-interceptor aircraft such as the MiG-29, F-15, F-16, F-18, the Sukhoi SU-27 series, and the F-22 aircraft.

An aircraft fire-control radar system may include, in addition to the radar, another RF or IR source to illuminate the target with RF energy or to locate the hostile aircraft by searching for its heat signature. For example, a semiactive radar seeker in a missile can home on the target from the reflections of its own radar return signal or from the radar signal from the hostile aircraft radar system. Often, the more modern and sophisticated fire control systems include the capability to use the track data from other sources, such as land, sea, or airborne radar systems such as the Air Force E3A Sentry (AWACS) or Navy E2C AEW system. Tracking and up-linking data to an airborne interceptor in flight is another mission of a fire control radar. The airborne interceptor may be guided solely by the tracking radar or may have its own short-range radar onboard for the final phase of the engagement.

Some radars perform both the search and track functions simultaneously. One example of this type of radar is the AN/SPQ-9A Surface Surveillance and Tracking Radar, developed by Northrop Grumman Norden Systems. The "Spook-9" is an X-band track-while-scan radar used with the Mk-86 Gunfire Control system on certain surface combatants. The latest model of the "Spook-9" is the AN/SPQ-9B, designed primarily as an antiship missile defense radar. It is designed to detect hostile missiles as they break the radar horizon even in heavy clutter while at the same time provide simultaneous detection and tracking of

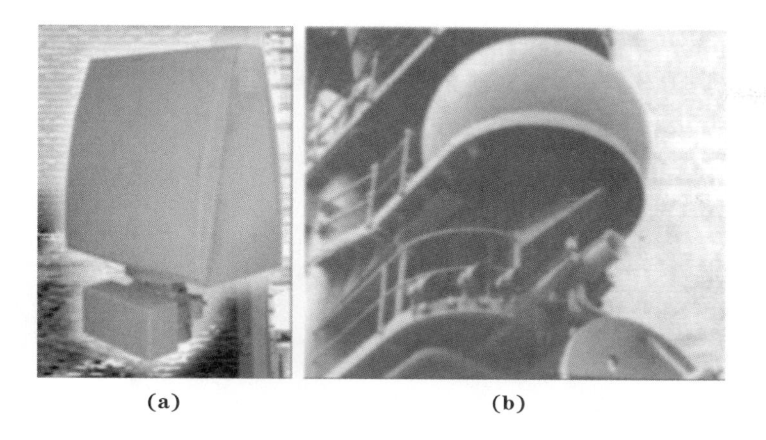

FIGURE 1-49 ■ Photograph of an AN/SPQ-9B ship-based TWS radar slotted line antenna (a) and protective radome (b). (Courtesy of U.S. Navy.)

surface targets. A mock-up of the back-to-back slotted array antenna is shown in Figure 1-49a. Figure 1-49b shows the protective radome aboard ship.

The AN/SPQ-9B scans the air and surface space near the horizon over 360 degrees in azimuth at 30 rpm. Real-time signal and data processing permit detection, acquisition, and simultaneous tracking of multiple targets.

The AN/SPQ-9B has three modes of operation: air, surface, and beacon. The AN/SPQ-9B complements high-altitude surveillance radar in detecting missiles approaching just above the sea surface. The antenna generates a 1 degree beam that, at a range of approximately 10 nautical miles, can detect missiles breaking the radar horizon at altitudes up to 500 feet.

1.9.1.8 Multifunction Radars

The advent of electronically scanned antennas using phased array antenna technology (described in Chapter 9) enables radar systems to interleave multiple functions. In particular, search and independent track modes can be implemented using one radar. The AN/SPY-1 is an example of a phased array multifunction radar used on surface ships. Figure 1-50

FIGURE 1-50 ■ USS Ticonderoga CG-47 with the AN/SPY-1 radar installed. (Courtesy of U.S. Navy.)

FIGURE 1-51 ■ AN/TPQ-36 Firefinder radar system used for weapons location. (Courtesy of U.S. Army.)

shows the AN/SPY-1 mounted on the USS Ticonderoga, the first ship to have the Aegis fire control system installed. The AN/SPY-1 is a major component of the Aegis system. Two of the four antenna faces that are required to provide full 360 degree coverage are visible in the figure.

1.9.1.9 Artillery Locating Radars

Another application of the multifunction radar is the artillery locating radar function. Artillery locating radars are designed to search a volume just above the horizon to detect artillery (e.g., mortar) rounds and track them. Based on a round's calculated ballistic trajectory, the system can then determine the location of the origin of the rounds. The U.S. Army Firefinder radar systems (AN/TPQ-36 and AN/TPQ-37) are examples of such radars. These are phased array systems employing electronically scanned antennas to perform the search and track functions simultaneously for multiple targets. Figure 1-51 shows an AN/TPQ-36.

1.9.1.10 Target Identification Radars

Early radar systems could detect a target and determine its position if the signal-to-noise ratio was sufficient. The result of a target detection was a "blip" on the display screen. Little information regarding the nature of the target was available. Modern radar systems have the ability to produce more information about a given target than just its presence and location. Several techniques are available to aid in discriminating the target from clutter, classifying it as a particular target type (e.g., a wheeled vehicle such as a truck vs. a tracked vehicle such as a tank) and even with some degree of success identifying the target (e.g., a particular class of aircraft). These techniques include high-resolution range profiles (Chapter 20), high-resolution cross-range imaging (Chapter 21), and high-resolution Doppler analysis (Chapter 17).

1.9.2 Commercial Applications

1.9.2.1 Process Control Radars

Very short-range radars can be used to measure the fluid levels in enclosed tanks very accurately or determine the "dryness" of a product in a manufacturing process to provide

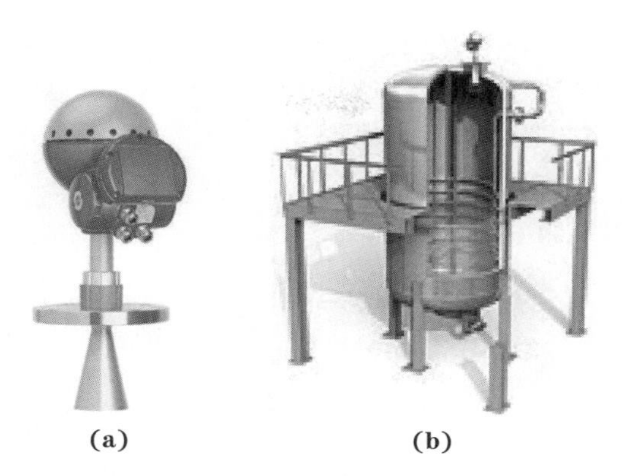

FIGURE 1-52 ■ Noncontact radar fluid level sensor. (a) Radar unit. (b) Illustration of unit mounted through the top of a fluid tank. (Courtesy of Rosemount, Inc. With permission.)

feedback to the process controller. A typical system uses a fairly high frequency such as 10 GHz and uses frequency modulated continuous wave (FMCW) techniques to measure distance to the top of the fluid in a tank. Figure 1-52a is an example of a noncontact fluid level measuring radar that mounts through the top of a tank, as shown in part (b) of the figure.

1.9.2.2 Airport Surveillance Radars

Airport surveillance radars detect and track many commercial and general aviation planes simultaneously. They are typically 2-D systems as described previously, rotating mechanically in azimuth while using a wide elevation beamwidth to provide vertical coverage. As the radar's antenna beam makes its 360 degree scan and detects an aircraft target, the target track file is updated and displayed to the operator. Often a beacon transponder on the aircraft reports the flight number and altitude back to the surveillance radar. Figure 1-53 shows the antenna of the ASR-9 air surveillance radar, a common sight at most large U.S. commercial airports.

1.9.2.3 Weather Radars

Government and news organizations keep track of weather activities using radar in conjunction with other weather station instruments. Modern Doppler weather radars measure not only the reflectivity of precipitation throughout the radar's field of view (FOV) but also

FIGURE 1-53 ■ Antenna of the ASR-9 airport surveillance radar. (Courtesy of Northrop-Grumman Corporation. With permission.)

FIGURE 1-54 ■ (a) Antenna tower for the WSR-88D (NEXRAD) radar. (b) Reflectivity image of Hurricane Katrina. (U.S. Government images.)

the wind speeds (using Doppler techniques) and a measure of turbulence called the *spectral width*. Indeed, Doppler weather radar images are ubiquitous on television, and their basic features are widely understood by the general population. Some modern weather radars can also discriminate between rain and hail using polarization characteristics of the precipitation echo, while others can detect wind shear and rotating atmospheric (tornado) events using Doppler techniques.

In the United States, the primary operational network of weather radars used by the National Weather Service is the WSR-88D ("NEXRAD"). The antenna tower for a typical installation is shown in Figure 1-54a. The contiguous 48 states are covered by a network of 159 systems. Figure 1-54b shows the reflectivity image of Hurricane Katrina from the WSR-88D in New Orleans, Louisiana, on August 29, 2005, a few minutes before the radar shut down.

A related use of radar is in radio-acoustic sounding systems (RASS), which can measure the temperature profile above the ground for several kilometers of altitude without invading the atmosphere with anything more than an acoustic wave and a radar RF signal. An acoustic wave is transmitted vertically. The acoustic wave causes compression of the air, which creates local variations in the dielectric properties of the atmosphere. A radar transmits pulses in the same vertical direction. The dielectric variations in the atmosphere result in radar backscatter from which the Doppler shift, and thus the speed of the acoustic wave can be recorded. Since the speed of sound is related to air temperature, the temperature profile can then be inferred. Figure 1-55 shows a RASS system located at the Alaska North Slope site at Barrow, Alaska. The large central square horn is the radar profiler antenna. The four surrounding circular sensors are the acoustic sources.

A very small equivalent radar cross section results from the acoustic disturbance. Normally, this small RCS would not be detectable, except for two special features of the combined system. Since the acoustic wave is spherical, and the radar wave is spherical, there is a focusing effect due to the fact that both the acoustic horns and the radar antenna are at the center of the sphere. Also, the acoustic wave is designed to produce on the order of 100 cycles at a particular wavelength. The radar frequency is chosen so that it has the

FIGURE 1-55 ■ RASS system at Barrow, Alaska. (U.S. Government image.)

same wavelength. This condition creates a constructive interference condition providing a larger received signal.

1.9.2.4 Wake Vortex Detection Radars

Large aircraft in flight produce a significant wake vortex, or turbulence, behind them in what might be otherwise laminar or still air. This vortex can persist for some time, depending on the local atmospheric conditions, and can present a dangerous flight control situation for light aircraft landing or taking off immediately behind large aircraft. Normally a separation of a minute or so is sufficient for this wake turbulence to dissipate. In some conditions, the wake turbulence persists for longer periods. Radars placed at the end of a runway can sense this wake turbulence and warn an approaching aircraft about such conditions.

1.9.2.5 Marine Navigation Radars

Radar systems can provide navigation information to a ship's captain. Shorelines, channel buoys, marine hazards (above the water surface) and other marine traffic can easily be detected at distances in excess of that required for safe passage of a ship, even in foul weather. Such systems often employ a narrow antenna azimuth beamwidth (1 or 2 degrees) and a relatively wide elevation beamwidth (10 degrees or more). The Canadian LN-66 and U.S. AN/SPS-64 radars are examples of navigation radars for military ships. Figure 1-56 shows the display and control units of a common commercial radar, the Furuno FAR2817 X-band radars.

FIGURE 1-56 ■ Control and display units of Furuno FAR2817 X-band marine radar for small ships. (Courtesy of Furuno. With Permission.)

FIGURE 1-57 ■ Artist's rendering of the RADARSAT 2 satellite mapping radar. (Courtesy of Canadian Space Agency.)

1.9.2.6 Satellite Mapping Radars

Space-based radar systems have the advantage of an unobstructed overhead view of the earth and objects on the earth's surface. These systems typically operate from satellites in low Earth orbit, which is on the order of 770 km altitude. Pulse compression waveforms and synthetic aperture radar (SAR) techniques (described in Chapters 20 and 21) are used to obtain good range and cross-range resolution.

An example of a satellite mapping radar is the Canadian RADARSAT 2 system, shown in an artist's rendering in Figure 1-57. The satellite was launched in December 2007. Table 1-4 lists the resolution modes available in RADARSAT 2. Obtainable resolutions range from 100 m for wide area imaging, down to 3 m for high-resolution imaging of limited areas. Another series of space-based mapping radars are the Shuttle Imaging Radars (SIR) A, B, and C, which operate at altitudes of about 250 km.

1.9.2.7 Police Speed Measuring Radars

Police speed measuring radars are simple CW radars that can measure the Doppler frequency shift for a target (vehicle) in the antenna beam. When the relative speed is derived from the Doppler shift using equation (1.2) and is added to or subtracted from the speed

TABLE 1-4 ■ RADARSAT 2 Resolution Modes

Beam Mode	Nominal Swath Width	Approximate Resolution (Range)	(Azimuth)	Approximate Incidence Angle	Polarization
Ultra-Fine	20 km	3 m	3 m	30°–40°	Selective Single Polarization
Multi-Look Fine	50 km	8 m	8 m	30°–50°	
Fine Quad-Pol	25 km	12 m	8 m	20°–41°	Quad-Polarization
Standard Quad-Pol	25 km	25 m	8 m	20°–41°	
Fine	50 km	8 m	8 m	30°–50°	Selective Polarization
Standard	100 km	25 m	26 m	20°–49°	
Wide	150 km	30 m	26 m	20°–45°	
ScanSAR Narrow	300 km	50 m	50 m	20°–46°	
ScanSAR Wide	500 km	100 m	100 m	20°–49°	
Extended High	75 km	18 m	26 m	49°–60°	Single Polarization
Extended Low	170 km	40 m	26 m	10°–23°	

FIGURE 1-58 ■ Photograph of a handheld, single-antenna police speed-timing radar. (Courtesy of Decatur Electronics. With permission.)

of the police cruiser, the absolute speed of the vehicle can be determined. The radars use very low transmit power and simple signal detection and processing techniques, such that they can be handheld, as shown in Figure 1-58.

1.9.2.8 Automotive Collision Avoidance Radars

Collision avoidance radars installed in automobiles are currently under development and have been deployed in some models. These short-range systems usually employ an inexpensive antenna which may be electronically scanned and a millimeter-wave radar (e.g., Ka-band or W-band) to provide a reasonably narrow azimuth beamwidth. There are challenges, however, in reducing the interpretations of nondangerous situations as dangerous, thus employing braking or steering commands unnecessarily.

1.9.2.9 Ground Penetration Radars

A ground-penetrating radar (GPR) has a low carrier RF (usually L-band and below) that can penetrate the ground (as well as other surfaces) and detect dielectric anomalies several feet deep. Almost any object that is buried will create a dielectric discontinuity with the surrounding ground, resulting in a reflection of the transmitted wave. Extremely high-range resolution (on the order of 2–3 cm or less) is important in such applications. The range resolution is achieved by using very wide bandwidth. The challenge for these systems is designing an antenna system that has a high percentage bandwidth and efficiently couples the EM wave into the ground or other material. Common uses for GPR include buried pipe detection, gas leak location, buried land mine detection, tunnel detection, and concrete evaluation and void detection in pavements.

Figure 1-59 shows a vehicular-towed system designed to locate voids in concrete highways. The resulting plot, shown in Figure 1-60, shows the void as well as the reinforcing bars (rebar) used in the fabrication of the roadbed.

1.9.2.10 Radar Altimeters

Relatively simple FMCW radars are used to determine the height of an aircraft above ground level (AGL), from nearly 0 feet to several thousand feet altitude. A strong ground reflection will be received from the surface when the radar is pointed directly downward, and the range of the ground will be the altitude of the radar/aircraft. Radar altimeters are used in commercial as well as military aircraft. Figure 1-61 is a Freeflight Systems TRA-3000 radar altimeter, showing the flush-mounted antenna and the display unit. This is an FMCW radar with about 100 MHz bandwidth, operating in the 4.2 to 4.4 GHz region. It provides altitude accuracy of about 5 to 7%.

FIGURE 1-59 ■ A ground-penetrating system designed to locate voids in a concrete highway. (Courtesy of Geophysical Survey Systems, Inc. With permission.)

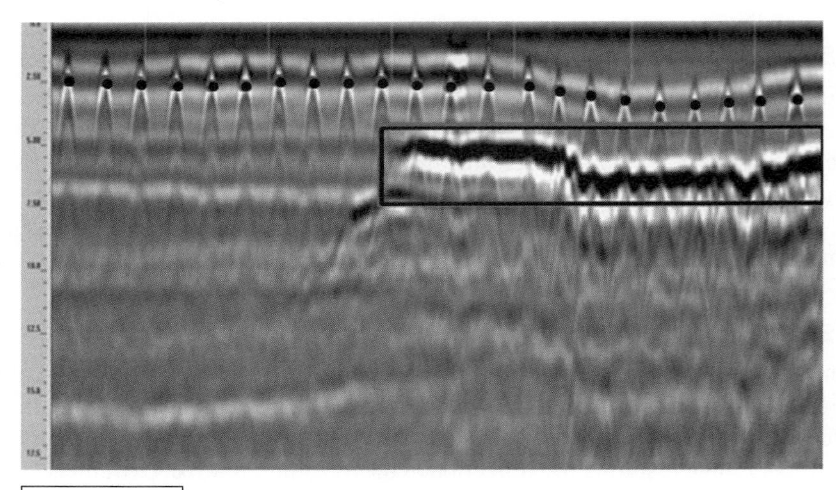

FIGURE 1-60 ■ Plot showing highway void as well as the reinforcing bars (rebar) used in the fabrication of the roadbed. (Courtesy of Geophysical Survey Systems, Inc. With permission.)

FIGURE 1-61 ■ Photograph of a radar altimeter. (Courtesy of Freeflight Systems Inc. With permission.)

1.10 ORGANIZATION OF THIS TEXT

This textbook is organized into four major parts. The first, consisting of Chapters 1–3, introduces the basic concepts and terminology of radar systems and operation, without many of the details. This part gives the reader an overview of the major issues in designing and evaluating radar systems. The remaining parts provide more detailed information about the elements of a radar system.

Part 2, consisting of Chapters 4–8, is concerned with the phenomenology of radar signals, including targets, clutter, Doppler shift, and atmospheric effects. This part provides the information needed to model realistic radar signals and thus to understand how to process them. Part 3 comprises Chapters 9–13 and represents the "hardware" section of the radar system. These chapters describe the types and characteristics of typical modern radar transmitters, receivers, antennas, and signal processors.

Chapters 14–21 comprise the fourth part, on radar signal processing. Beginning with a review of digital signal processing principles, this part of the book describes a wide variety of radar signal analysis and processing methods, ranging from basic threshold detection through Doppler processing, tracking, and an introduction to imaging.

1.11 FURTHER READING

There are a number of excellent introductory texts on radar systems and technology. The most classic is Skolnik's text [3], now in its third edition, which provides a primarily qualitative overview of a wide range of radar systems, technologies, and issues. Toomay and Hannen [4] provide an introduction to a broad range of fundamental radar topics, with supporting mathematics at a straightforward level. Kingsley and Quegan's book [5] is another good radar survey. All of the preceding textbooks, like this one, provide sample problems to aid in understanding and applying the concepts. Stimson's text [6] focuses on airborne radars but is perhaps the best-illustrated book on radar. It provides an excellent intuitive and visual discussion of many radar topics.

More advanced introductions are provided by Mahafza [7] and Peebles [8]. Mahafza provides a number of MATLAB scripts to support the textbook topics. Peebles's text is the most advanced of those discussed here, providing very thorough coverage at an advanced undergraduate or beginning graduate student level. Finally, Richards [9] provides a senior- or graduate-level text that concentrates on the signal processing aspects of radar such as Doppler processing, integration, detection, waveforms, and imaging. His text provides a good basis for study of more advanced radar signal processing sources.

1.12 REFERENCES

[1] IEEE Aerospace & Electronic Systems Society, "IEEE Standard Letter Designations for Radar-Frequency Bands", IEEE Standard 521-2002, The Institute of Electrical and Electronics Engineers, New York, 2003.

[2] IEEE Aerospace & Electronic Systems Society, "IEEE Standard Radar Definitions", IEEE Standard 686-2008, Institute of Electrical and Electronics Engineers, New York, 2003.

[3] Skolnik, M.I., *Introduction to Radar Systems,* 3d ed., McGraw-Hill, New York, 2003.

[4] Toomay, J.C., and Hannen, P.J., *Radar Principles for the Non-specialists,* 3d ed., SciTech Publishing, Raleigh, NC, 2004.

[5] Kingsley, S., and Quegan, S., *Understanding Radar Systems,* SciTech Publishing, Raleigh, NC, 1999.

[6] Stimson, G.W., *Introduction to Airborne Radar,* 2d ed., SciTech Publishing, Raleigh, NC, 1998.

[7] Mahafza, B.R., *Radar Systems Analysis and Design Using MATLAB*, Chapman and Hall/CRC, Boca Raton, FL, 2000.

[8] Peebles Jr., P.Z., *Radar Principles*, Wiley, New York, 1998.

[9] Richards, M.A., *Fundamentals of Radar Signal Processing*, McGraw-Hill, New York, 2005.

1.13 PROBLEMS

1. Find an expression for the range of a target in kilometers (km) for a reflected signal that returns to the radar ΔT μs after being transmitted.
2. Find the distance to a radar target (in meters) for the following round-trip delay times:
   a. 12 μs
   b. 120 μs
   c. 1.258 ms
   d. 650 μs
3. Find the delay times associated with the following target distances:
   a. 1 statute mile
   b. 1 km
   c. 100 km
   d. 250 statute miles
   e. 20 feet
4. Find the wavelength of the EM wave associated with the following carrier frequencies (in free space):
   a. 325 MHz
   b. 1.2 GHz
   c. 2.85 GHz
   d. 5.8 GHz
   e. 9.325 GHz
   f. 15.6 GHz
   g. 34.5 GHz
   h. 94 GHz
5. Find the carrier frequency associated with the following wavelengths for an EM wave in free space.
   a. 1 inch
   b. 0.35 cm
   c. 8.6 mm
   d. 90 cm
   e. 9.0 cm
   f. 1 foot
6. The intensity of a transmitted EM wave at a range of 500 m from the radar is 0.04 W/m^2. What is the intensity at 2 km?
7. How far from an antenna must one be positioned such that the wavefront whose source is at your position is estimated to be planar at the antenna, for the following conditions, where f is the carrier frequency, λ is the wavelength in meters, and D is the antenna dimension in meters:

	f	λ (meters)	D (meters)
a.	10 GHz	–	1.0
b.	–	0.1	1.0
c.	10 GHz	–	0.1
d.	3 GHz	–	1.0
e.	3 GHz	–	7.5

8. What is the approximate beamwidth in radians and in degrees for a circular aperture for each of the cases listed in problem 7?

9. Consider the special case of an interferometer, which can be described as a 2-element phased array antenna, consisting of two isotropic, in-phase, radiating elements separated by a distance d. Assume d is much greater than λ, the wavelength of the transmitted EM wave. Show that the first null off boresight in the far-field antenna pattern occurs at angle $\theta = \lambda/2d$ radians.

10. The peak power of 200 kW radar is reduced by 3 dB. If its duty cycle is 1.0%, what is it resulting average power in dBW?

11. Find an expression for a radar's maximum unambiguous range in kilometers if the radar's PRF is x kHz.

12. A high-PRF radar has a pulse width of 1.0 μs and a duty factor of 20%. What is this radar's maximum unambiguous range?

13. Find the Doppler shift associated with the following target motions; where v_t is the target speed, θ is the angle of velocity vector relative to LOS from the radar to the target, and f is the radar carrier frequency:

	v_t	θ	f
a.	100 mph	0°	10 GHz
b.	330 m/s	0°	10 GHz
c.	15 m/s	0°	10 GHz
d.	15 m/s	45°	10 GHz
e.	15 m/s	45°	3 GHz
f.	15 m/s	60°	10 GHz

14. What is the maximum unambiguous Doppler shift that can be measured with a radar with a PRI of 0.25 milliseconds?

15. What is the range resolution of a radar system having the following characteristics?

	Pulse length	Frequency
a.	1.0 μs	9.4 GHz
b.	1.0 μs	34.4 GHz
c.	0.1 μs	9.4 GHz
d.	0.01 μs	9.4 GHz

16. Consider a 2-D search radar having an antenna that is 6.5 meters wide. If it is rotating (in azimuth) at a constant rate of 0.8 radians per second, how long is a potential target in the 3 dB beam if the operating frequency is 2.8 GHz?

17. Consider a police speed timing radar with a circular antenna of 6 inch diameter.

a. What is the approximate beamwidth (in degrees) for an operating frequency of 9.35 GHz?

b. What is the approximate beamwidth (in degrees) for an operating frequency of 34.50 GHz?

c. What is the approximate diameter of the beam (in feet) at a distance of 0.25 miles for an operating frequency of 9.35 GHz?

d. What is the approximate diameter of the beam (in feet) at a distance of 0.25 miles for an operating frequency of 34.50 GHz?

CHAPTER 2

The Radar Range Equation

James A. Scheer

Chapter Outline

2.1	Introduction	59
2.2	Power Density at a Distance R	61
2.3	Received Power from a Target	62
2.4	Receiver Thermal Noise	64
2.5	Signal-to-Noise Ratio and the Radar Range Equation	66
2.6	Multiple-Pulse Effects	66
2.7	Summary of Losses	67
2.8	Solving for Other Variables	72
2.9	Decibel Form of the Radar Range Equation	72
2.10	Average Power Form of the Radar Range Equation	73
2.11	Pulse Compression: Intrapulse Modulation	74
2.12	A Graphical Example	75
2.13	Clutter as the Target	76
2.14	One-Way (Link) Equation	78
2.15	Search Form of the Radar Range Equation	79
2.16	Track Form of the Radar Range Equation	80
2.17	Some Implications of the Radar Range Equation	83
2.18	Further Reading	84
2.19	References	84
2.20	Problems	85

2.1 INTRODUCTION

As introduced in Chapter 1, the three fundamental functions of radar systems are to search for and find (detect) targets, to track detected targets, and in some cases to develop an image of the target. In all of these functions the radar performance is influenced by the strength of the signal coming into the radar receiver from the target of interest and by the strength of the signals that interfere with the target signal. In the special case of receiver thermal noise being the interfering signal, the ratio of target signal to noise power is called the signal-to-noise ratio (SNR); if the interference is from a clutter signal, then the ratio is called signal-to-clutter ratio (SCR). The ratio of the target signal to the total interfering signal is the signal-to-interference ratio (SIR).

In the search mode, the radar system is programmed to reposition the antenna beam in a given sequence to "look" at each possible position in space for a target. If the signal-plus-noise at any spatial position exceeds the interference by a sufficient margin, then a *detection* is made, and a target is deemed to be at that position. In this sense, detection is a process by which, for every possible target position, the signal (plus noise) is compared with some threshold level to determine if the signal is large enough to be deemed a target of interest. The threshold level is set somewhat above the interference signal level. The probability that a target will be detected depends on the probability that its signal will exceed the threshold level. The detection process is discussed in more detail in Chapters 3 and 15, and special processing techniques designed to perform the detection process automatically are discussed in Chapter 16.

In the tracking mode, the accuracy or precision with which a target is tracked also depends on the SIR. The higher the SIR, the more accurate and precise the track will be. Chapter 19 describes the tracking process and the relationship between tracking precision and the SIR.

In the imaging mode, the SIR determines the fidelity of the image. More specifically, it determines the dynamic range of the image—the ratio between the "brightest" spots and the dimmest on the target. The SIR also determines to what extent false scatterers are seen in the target image.

The equation the radar system designer or analyst uses to compute the SIR is the *radar range equation* (RRE). A relatively simple formula, or a family of formulas, the RRE predicts the received power of the radar's radio waves "reflected"[1] from a target and the interfering noise power level and, when these are combined, the SNR. In addition, it can be used to calculate the power received from surface and volumetric clutter, which, depending on the radar application, can be considered to be a target or an interfering signal. When the system application calls for detection of the clutter, the clutter signal becomes the target. When the clutter signal is deemed to be an interfering signal, then the SIR is determined by dividing the target signal by the clutter signal. Intentional or unintentional signals from a source of electromagnetic (EM) energy remote from the radar can also constitute an interfering signal. A noise jammer, for example, will introduce noise into the radar receiver through the antenna. The resulting SNR is the target signal power divided by the sum of the noise contributions, including receiver thermal noise and jammer noise. Communications signals and other sources of EM energy can also interfere with the target signal. These remotely generated sources of EM energy are analyzed using one-way analysis of the propagating signal. The one-way link equation can determine the received signal resulting from a jammer, a beacon transponder, or a communications system.

This chapter includes a discussion of several forms of the radar range equation, including those most often used in predicting radar performance. It begins with forecasting the power density at a distance R and extends to the two-way case for monostatic radar for targets, surface clutter, and volumetric clutter. Then radar receiver thermal noise power is determined, providing the SNR. Equivalent but specialized forms of the RRE are

[1]Chapter 6 shows that the signal illuminating a target induces currents on the target and that the target reradiates these electromagnetic fields, some of which are directed toward the illuminating source. For simplicity, this process is often termed *reflection*.

developed for a search radar and then for a tracking radar. Initially, an idealized approach is presented, limiting the introduction of terms to the ideal radar parameters. After the basic RRE is derived, nonideal effects are introduced. Specifically, the component, propagation, and signal processing losses are introduced, providing a more realistic value for the received target signal power.

2.2 POWER DENSITY AT A DISTANCE *R*

Although the radar range equation is not formally derived here from first principles, it is informative to develop the equation heuristically in several steps. The total peak power (watts) developed by the radar transmitter, P_t, is applied to the antenna system. If the antenna had an isotropic or omnidirectional radiation pattern, the power density Q_i (watts per square meter) at a distance R (meters) from the radiating antenna would be the total power divided by the surface area of a sphere of radius R,

$$Q_i = \frac{P_t}{4\pi R^2} \tag{2.1}$$

as depicted in Figure 2-1.

Essentially all radar systems use an antenna that has a directional beam pattern rather than an isotropic beam pattern. This means that the transmitted power is concentrated into a finite angular extent, usually having a width of several degrees in both the azimuthal and elevation planes. In this case, the power density at the center of the antenna beam pattern is higher than that from an isotropic antenna, because the transmit power is concentrated onto a smaller area on the surface of the sphere, as depicted in Figure 2-2. The power density in the gray ellipse depicting the antenna beam is increased from that of an isotropic antenna. The ratio between the power density for a lossless directional antenna and a hypothetical

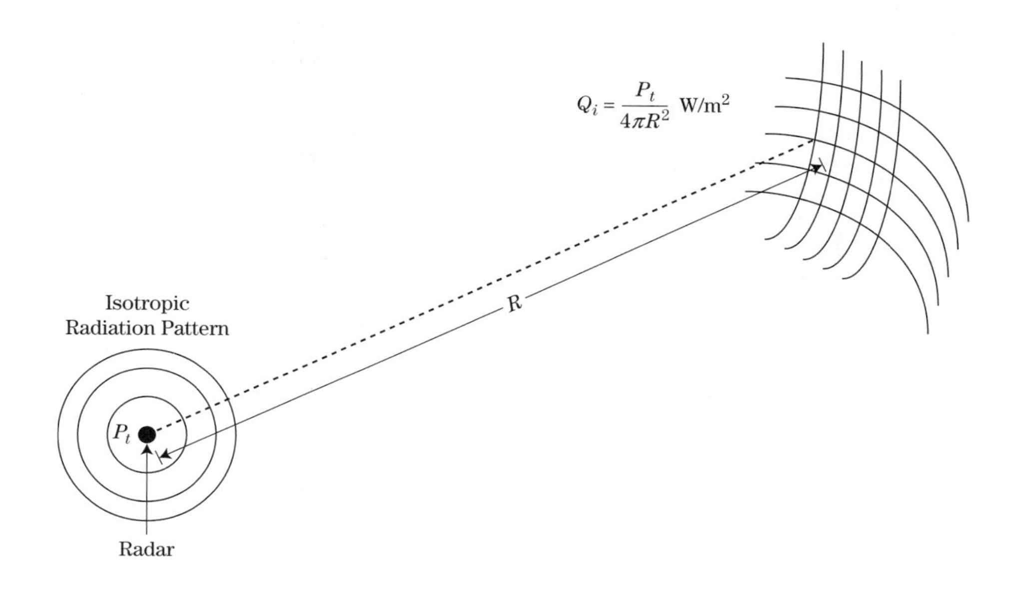

FIGURE 2-1 ■ Power density at range R from the radar transmitter, for an isotropic (omnidirectional) antenna.

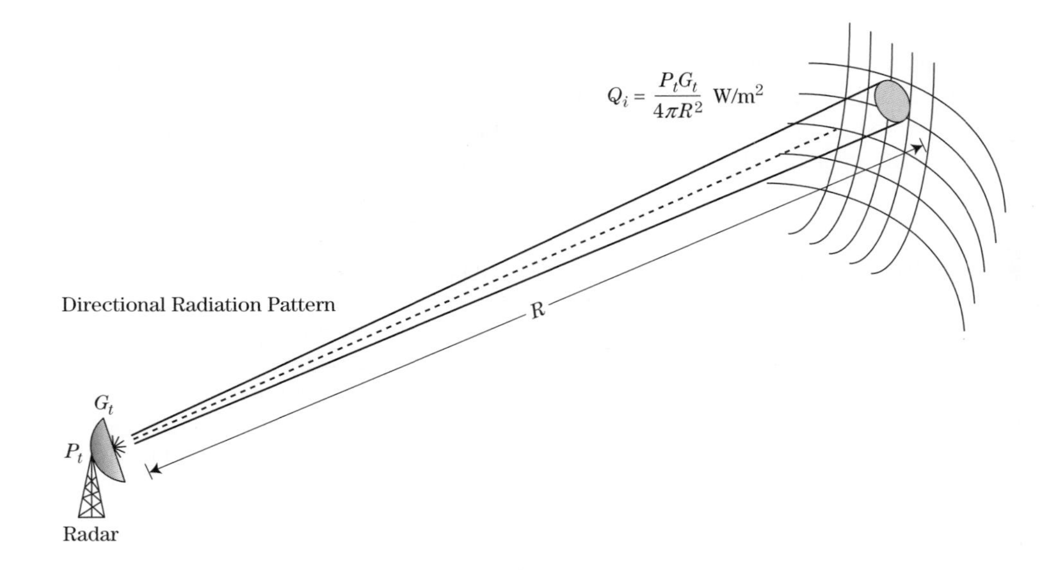

FIGURE 2-2 ■ Power density at range R given transmit antenna gain G_t.

isotropic antenna is termed the *directivity*. The gain, G, of an antenna is the *directivity* reduced by the losses the signal encounters as it travels from the input port to the point at which it is "launched" into the atmosphere [1]. The subscript t is used to denote a transmit antenna, so the transmit antenna gain is G_t. Given G_t, the increased power density due to use of a directional antenna is

$$Q_i = \frac{P_t G_t}{4\pi R^2} \tag{2.2}$$

2.3 RECEIVED POWER FROM A TARGET

Next, consider a radar "target" at range R, illuminated by the signal from a radiating antenna. The incident transmitted signal is reflected in a variety of directions, as depicted in Figure 2-3. As described in Chapter 6, the incident radar signal induces time-varying currents on the target so that the target now becomes a source of radio waves, part of which will propagate back to the radar, appearing to be a *reflection* of the illuminating signal. The power *reflected* by the target back toward the radar, P_{refl}, is expressed as the product of the incident power density and a factor called the *radar cross section* (RCS) σ of the target. The units for RCS are square meters (m^2). The radar cross section of a target is determined by the physical size of the target, the shape of the target, and the materials from which the target is made, particularly the outer surface.[2] The expression for the power reflected back toward the radar, P_{refl}, from the target is

$$P_{refl} = Q_i \sigma = \frac{P_t G_t \sigma}{4\pi R^2} \tag{2.3}$$

[2]A more formal definition and additional discussion of RCS are given in Chapter 6.

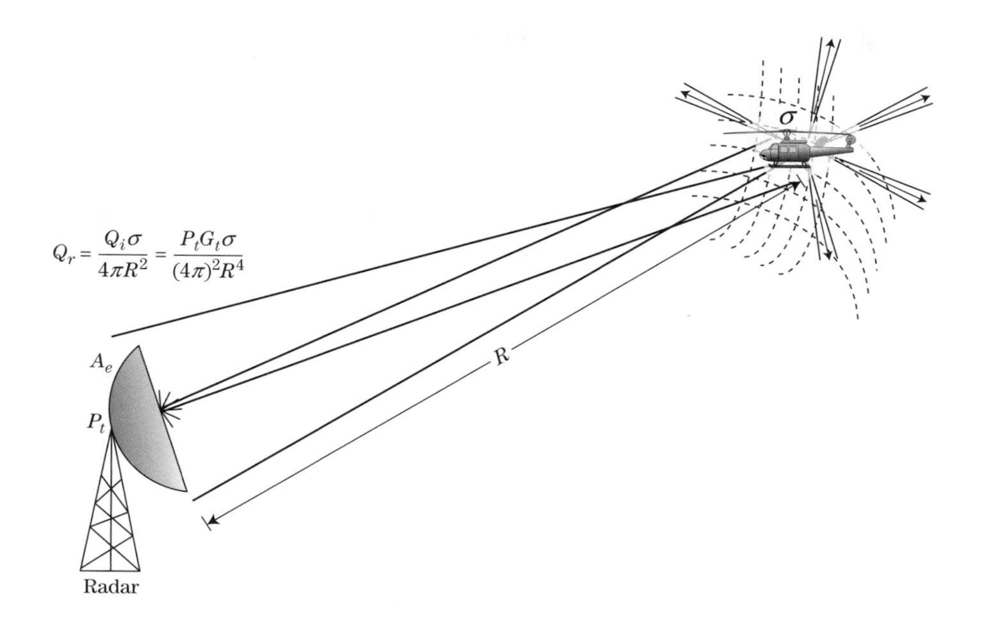

FIGURE 2-3 ■ Power density, Q_r, back at the radar receive antenna.

The signal reflected from the target propagates back toward the radar system over a distance R so that the power density back at the radar receiver antenna, Q_r, is

$$Q_r = \frac{P_{refl}}{4\pi R^2} \tag{2.4}$$

Combining equations (2.3) and (2.4), the power density of the radio wave received back at the radar receive antenna is given by

$$Q_r = \frac{Q_i \sigma}{4\pi R^2} = \frac{P_t G_t \sigma}{(4\pi)^2 R^4} \tag{2.5}$$

Notice that the radar-target range R appears in the denominator raised to the fourth power. As an example of its significance, if the range from the radar to the target doubles, the received power density of the reflected signal from a target decreases by a factor of 16 (12 dB) (ouch!).

The radar wave reflected from the target, which has propagated through a distance R and results in the power density given by equation (2.5), is received (gathered) by a radar receive antenna having an effective antenna area of A_e. The power received, P_r (watts) from a target at range R at a receiving antenna of effective area of A_e is found as the power density at the antenna times the effective area of the antenna:

$$P_r = Q_r A_e = \frac{P_t G_t A_e \sigma}{(4\pi)^2 R^4} \tag{2.6}$$

It is customary to replace the effective antenna area term A_e with the value of receive antenna gain G_r that is produced by that area. Also, because of the effects of tapering and losses, the *effective* area of an antenna is somewhat less than the physical area, A. As discussed in Chapter 9 as well as in many standard antenna texts such as [1], the relationship between an antenna gain G and its effective area A_e is given by

$$G = \frac{4\pi \eta_a A}{\lambda^2} = \frac{4\pi A_e}{\lambda^2} \tag{2.7}$$

where η_a is the antenna efficiency. Antenna efficiency is a value between 0 and 1; however, it is seldom below 0.5 and seldom above 0.8.

Solving (2.7) for A_e and substituting into equation (2.6), the following expression for the received power, P_r results:

$$P_r = \frac{P_t G_t G_r \lambda^2 \sigma}{(4\pi)^3 R^4} \tag{2.8}$$

In this expression,

P_t is the peak transmitted power in watts.
G_t is the gain of the transmit antenna.
G_r is the gain of the receive antenna.
λ is the carrier wavelength in meters.
σ is the mean[3] RCS of the target in square meters.
R is the range from the radar to the target in meters.

This form is found in many existing standard radar texts, including [2–6].

For many monostatic radar systems, particularly those using mechanically scanned antennas, the transmit and receive antennas gains are the same, so in those cases the two gain terms in (2.8) are replaced by G^2. However, for bistatic systems and in many modern radar systems, particularly those that employ electronically scanned antennas, the two gains are generally different, in which case the preferred form of the radar range equation is that shown in (2.8), allowing for different values for transmit and receive gain.

For a bistatic radar, one for which the receive antenna is not colocated with the transmit antenna, the range between the transmitter and target, R_t, may be different from the range between the target and the receiver, R_r. In this case, the two different range values must be independently specified, leading to the bistatic form of the equation

$$P_r = \frac{P_t G_t G_r \lambda^2 \sigma_{bistatic}}{(4\pi)^3 R_t^2 R_r^2} \tag{2.9}$$

Though in the following discussions the monostatic form of the radar equation is described, a similar bistatic form can be developed by separating the range terms and using the bistatic radar cross section, $\sigma_{bistatic}$, of the target.

2.4 RECEIVER THERMAL NOISE

In the ideal case, the received target signal, which usually has a very small amplitude, could be amplified by some arbitrarily large amount until it could be visible on a display or within the dynamic range of an analog-to-digital converter (ADC). Unfortunately, as discussed in Chapter 1 and in the introduction to this chapter, there is always an interfering signal described as having a randomly varying amplitude and phase, called *noise*, which is produced by several sources. Random noise can be found in the environment, mostly due

[3]The target RCS is normally a fluctuating value, so the mean value is usually used to represent the RCS. The radar equation therefore predicts a mean, or average, value of received power and, when noise is taken into consideration, SNR.

to solar effects. Noise entering the antenna comes from several sources. Cosmic noise, or galactic noise, originates in outer space. It is a significant contributor to the total noise at frequencies below about 1 GHz but is a minor contributor above 1 GHz. Solar noise is from the sun. The sun's proximity makes it a significant contributor; however, its effect is reduced by the antenna sidelobe gain, unless the antenna main beam is pointed directly toward the sun. Even the ground is a source of noise, but not at as high a level as the sun, and usually enters the receiver through antenna sidelobes.

In addition to antenna noise, thermally agitated random electron motion in the receiver circuits generates a level of random noise with which the target signal must compete. Though there are several sources of noise, the development of the radar range equation in this chapter will assume that the internal noise in the receiver dominates the noise level. This section presents the expected noise power due to the active circuits in the radar receiver. For target detection to occur, the target signal must exceed the noise signal and, depending on the statistical nature of the target, sometimes by a significant margin before the target can be detected with a high probability.

Thermal noise power is essentially uniformly distributed over all radar frequencies; that is, its *power spectral density* is constant, or uniform. It is sometimes called "white" noise. Only noise signals with frequencies within the range of frequencies capable of being detected by the radar's receiver will have any effect on radar performance. The range of frequencies for which the radar is susceptible to noise signals is determined by the receiver bandwidth, B. The thermal noise power adversely affecting radar performance will therefore be proportional to B. The power, P_n, of the thermal noise in the radar receiver is given by [4]

$$P_n = kT_S B = kT_0 FB \tag{2.10}$$

where

k is Boltzmann's constant (1.38×10^{-23} watt-sec/K).
T_0 is the standard temperature (290 K).
T_s is the system noise temperature ($T_s = T_0 F$).
B is the instantaneous receiver bandwidth in Hz.
F is the noise figure of the receiver subsystem (unitless).

The *noise figure, F*, is an alternate method to describe the receiver noise to system temperature, T_s. It is important to note that noise figure is often given in dB; however, it must be converted to linear units for use in equation (2.10).

As can be seen from (2.10), the noise power is linearly proportional to receiver bandwidth. However, the receiver bandwidth cannot be made arbitrarily small to reduce noise power without adversely affecting the target signal. As will be shown in Chapters 8 and 11, for a simple unmodulated transmit signal, the bandwidth of the target's signal in one received pulse is approximated by the reciprocal of the pulse width, τ (i.e., $B \approx 1/\tau$). If the receiver bandwidth is made smaller than the target signal bandwidth, the target power is reduced, and range resolution suffers. If the receiver bandwidth is made larger than the reciprocal of the pulse length, then the signal to noise ratio will suffer. The optimum bandwidth depends on the specific shape of the receiver filter characteristics. In practice, the optimum bandwidth is usually on the order of $1.2/\tau$, but the approximation of $1/\tau$ is very often used.

2.5 | SIGNAL-TO-NOISE RATIO AND THE RADAR RANGE EQUATION

When the received target signal power, P_r, is divided by the noise power, P_n, the result is called the signal-to-noise ratio. For a discrete target, this is the ratio of equation (2.8) to (2.10):

$$SNR = \frac{P_t G_t G_r \lambda^2 \sigma}{(4\pi)^3 R^4 k T_0 F B} \tag{2.11}$$

Ultimately, the signal-to-interference ratio is what determines radar performance. The interference can be from noise (receiver or jamming) or from clutter or other electromagnetic interference from, for example, motors, generators, ignitions, or cell services. If the power of the receiver thermal noise is N, from clutter is C, and from jamming noise is J, then the SIR is

$$SIR = \frac{S}{N + C + J} \tag{2.12}$$

Although one of these interference sources usually dominates, reducing the SIR to the signal power divided by the dominant interference power, *S/N*, *S/C*, or *S/J*, a complete calculation must be made in each case to see if this simplification applies.

2.6 | MULTIPLE-PULSE EFFECTS

Seldom is a radar system required to detect a target on the basis of a single transmitted pulse. Usually, several pulses are transmitted with the antenna beam pointed in the direction of the (supposed) target. The received signals from these pulses are processed to improve the ability to detect a target in the presence of noise by performing coherent or noncoherent integration (i.e., averaging; see Chapter 15). For example, many modern radar systems perform spectral analysis (i.e., Doppler processing) to improve target detection performance in the presence of clutter. Doppler processing is equivalent to coherent integration insofar as the improvement in SNR is concerned. This section describes the effect of coherent integration on the SNR of the received signal, and how that effect is reflected in the range equation.

Because the antenna beam has some angular width, as the radar antenna beam scans in angle it will be pointed at the target for more than the time it takes to transmit and receive one pulse. Often the antenna beam is pointed in a fixed azimuth-elevation angular position, while several (typically on the order of 16 or 20) pulses are transmitted and received. In this case, the integrated SIR is the important factor in determining detection performance. If coherent integration processing is employed, (i.e., both the amplitude and the phase of the received signals are used in the processing so that the signal contributions can be made to add in phase with one another), the SNR resulting from coherently integrating n_p pulses in white noise, $SNR_c(n_p)$, is n_p times the single-pulse SNR, $SNR(1)$:

$$SNR_c(n_p) = n_p \cdot SNR(1) \tag{2.13}$$

A more appropriate form of the RRE when n_p pulses are coherently combined is thus

$$SNR_c(N) = \frac{P_t G_t G_r \lambda^2 \sigma n_p}{(4\pi)^3 \, R^4 k T_0 F B} \tag{2.14}$$

This form of the RRE is often used to determine the SNR of a system, knowing the number of pulses coherently processed.

The process of coherent integration adds the received signal vectors (amplitude and phase) from a sequence of pulses. For a stationary target using a stationary radar, the vectors for a sequence of pulses would be in phase with one another to form the vector sum and would add head to tail, as described in [4]. If, however, the radar or the target were moving, the phase would be rotating, and the vector sum would be reduced, in some cases to zero. The SNR may be more or less than that of a single vector, or even zero. To ensure an improved SNR, the signal processor would have to "derotate" the vectors before summing. The fast Fourier transform (FFT) process essentially performs this derotation process before adding the vectors. Each FFT filter output represents the addition of several vectors after derotating the vector a different amount for each filter.

Coherent processing uses the phase information when averaging data from multiple pulses. It is also common to use *noncoherent integration* to improve the SNR. Noncoherent integration discards the phase of the individual echo samples, averaging only the amplitude information. It is easier to perform noncoherent integration. In fact, displaying the signal onto a persistent display whose brightness is proportional to signal amplitude will provide noncoherent integration. Even if the display is not persistent, the operator's "eye memory" will provide some noncoherent integration. The integration gain that results from noncoherent integration of n_p pulses $SNR_{nc}(n_p)$ is harder to characterize than in the coherent case but for many cases is at least $\sqrt{n_p}$ but less than n_p:

$$\sqrt{n_p} \cdot SNR(1) \leq SNR_{nc}(n_p) \leq n_p \cdot SNR(1) \tag{2.15}$$

It is suggested in [4] that a factor of $N^{0.7}$ would be appropriate in many cases. Chapter 15 provides additional detail on noncoherent integration.

2.7 SUMMARY OF LOSSES

To this point, the radar equation has been presented in an idealized form; that is, no losses have been assumed. Unfortunately, the received signal power is usually lower than predicted if the analyst ignores the effects of signal loss. Atmospheric absorption, component resistive losses, and nonideal signal processing conditions lead to less than ideal SNR performance. This section summarizes the losses most often encountered in radar systems and presents the effect on SNR. Included are losses due to clear air, rain, component losses, beam scanning, straddling, and several signal processing techniques. It is important to realize that the loss value, if used in the denominator of the RRE as previously suggested, must be a linear (as opposed to dB) value greater than 1. Often, the loss values are specified in dB notation. It is convenient to sum the losses in dB notation and finally to convert to the linear value.

Equation (2.16) provides the total system loss term

$$L_s = L_t L_a L_r L_{sp} \tag{2.16}$$

where

L_s is the system loss.
L_t is the transmit loss.
L_a is the atmospheric loss.
L_r is the receiver loss.
L_{sp} is the signal processing loss.

As a result of incorporating the losses into (2.14), the RRE becomes

$$SNR = \frac{P_t G_t G_r \lambda^2 \sigma n_p}{(4\pi)^3 R^4 k T_0 F B L_s} \tag{2.17}$$

The following sections describe the most common of these losses individually.

2.7.1 Transmit Loss

The radar equation (2.14) is developed assuming that all of the transmit power is radiated out an antenna having a gain G_t. In fact, there is some loss in the signal level as it travels from the transmitter to the antenna, through waveguide or coaxial cable, and through devices such as a circulator, directional coupler, or transmit/receive (T/R) switch. For most conventional radar systems, the loss is on the order of 3 or 4 dB, depending on the wavelength, length of transmission line, and what devices are included. For each specific radar system, the individual losses must be accounted for. The best source of information regarding the losses due to components is a catalog sheet or specification sheet from the vendor for each of the devices. In addition to the total losses associated with each component, there is some loss associated with connecting these components together. Though the individual contributions are usually small, the total must be accounted for. The actual loss associated with a given assembly may be more or less than that predicted. If maximum values are used in the assumptions for loss, then the total loss will usually be somewhat less than predicted. If average values are used in the prediction, then the actual loss will be quite close to the prediction. It is necessary to measure the losses to determine the actual value.

There is some loss between the input antenna port and the actual radiating antenna; however, this term is usually included in the specified antenna gain value provided by the antenna vendor. The analyst must determine if this term is included in the antenna gain term and, if not, must include it in the loss calculations.

2.7.2 Atmospheric Loss

Chapter 4 provides a thorough discussion of the effects of propagation through the environment on the SNR. The EM wave experiences attenuation in the atmosphere as it travels from the radar to the target, and again as the wave travels from the target back to the radar. Atmospheric loss is caused by interaction between the electromagnetic wave and oxygen molecules and water vapor in the atmosphere. Even clear air exhibits attenuation of the EM wave. The effect of this attenuation generally increases with increased carrier frequency; however, in the vicinity of regions in which the wave resonates with the water or oxygen molecules, there are sharp peaks in the attenuation, with relative nulls between these peaks. In addition, fog, rain, and snow in the atmosphere add to the attenuation caused by clear air. These and other propagation effects (diffraction, refraction, and multipath) are discussed in detail in Chapter 4.

Range-dependent losses are normally expressed in units of dB/km. Also, the absorption values reported in the technical literature are normally expressed as one-way loss. For a monostatic radar system, since the signal has to travel through the same path twice, two-way loss is required. In this case, the values reported need to be doubled on a dB scale (squared on a linear scale). For a bistatic radar, the signal travels through two different paths on transmit and receive, so the one-way loss value is used for each path.

Significant loss can be encountered as the signal propagates through the atmosphere. For example, if the two-way loss through rain is 0.8 dB/km and the target is 10 km away, then the rain-induced reduction in SNR is 8 dB compared with the SNR obtained in clear air. The quantitative effect of such a reduction in SNR is discussed in Chapter 3, but to provide a sense of the enormity of an 8 dB reduction in SNR, usually a reduction of 3 dB will produce noticeable system performance reduction.

2.7.3 Receive Loss

Component losses are also present in the path between the receive antenna terminal and the radar receiver. As with the transmit losses, these are caused by receive transmission line and components. In particular, waveguide and coaxial cable, the circulator, receiver protection switches, and preselection filters contribute to this loss value if employed. As with the transmit path, the specified receiver antenna gain may or may not include the loss between the receive antenna and the receive antenna port. All losses up to the point in the system at which the noise figure is specified must be considered. Again, the vendor data provide maximum and average values, but actual measurements provide the best information on these losses.

2.7.4 Signal Processing Loss

Most modern systems employ some form of multipulse processing that improves the single-pulse SNR by the factor n_p, which is the number of pulses in a *coherent processing interval* (CPI), often also called the dwell time. The effect of this processing gain is included in the average power form of the RRE, developed in Section 2.10. If the single-pulse, peak power form (Equation 2.11) is used, then typically a gain term is included in the numerator of the RRE that assumes perfect coherent processing gain as was done in equation (2.14). In either case, imperfections in signal processing that reduce this gain are then accounted for by adding a signal processing loss term. Some examples of the signal processing effects that contribute to system loss are beam scan loss, straddle loss (sometimes called scalloping loss), automatic detection constant false alarm rate (CFAR) loss, and mismatch loss. Each of these is described further in the following paragraphs. The discussion describes the losses associated with a pulsed system that implements a fast Fourier transform to determine the Doppler frequency of a detected target as described in Chapter 17.

Beam shape loss arises because the radar equation is developed using the peak antenna gains as if the target is at the center of the beam pattern for every pulse processed during a CPI. In many system applications, such as a mechanically scanning search radar, the target will at best be at the center of the beam pattern for only one of the pulses processed for a given dwell. If the CPI is defined as the time for which the antenna beam scans in angle from the –3 dB point, through the center, to the other –3 dB point, the average loss in signal compared with the case in which the target is always at the beam peak for a typical beam shape is about 1.6 dB. Of course, the precise value depends on the particular shape of the beam as well as the scan amount during a search dwell, so a more exact calculation may be required.

Figure 2-4 depicts a scanning antenna beam, such that the beam scans in angle from left to right. A target is depicted as a helicopter, and five beam positions are shown. (Often there would be more than five pulses for such a scan, but only five are shown here for

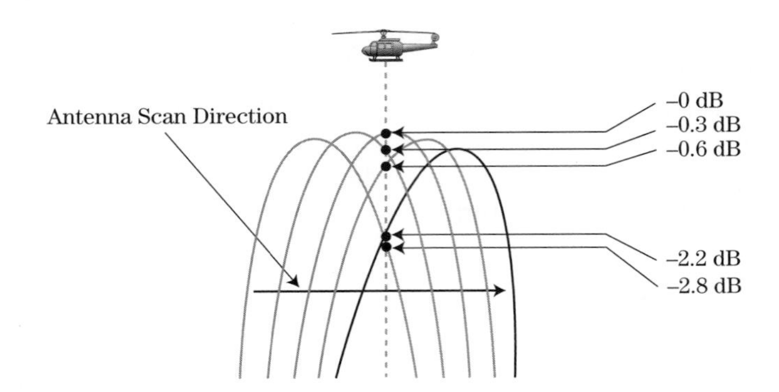

FIGURE 2-4 ■ Target signal loss due to beam scan.

clarity.) For the first pulse, the target is depicted at 2.8 dB below the beam peak, the second at 0.6 dB, the third at nearly beam center (–0.0 dB), the fourth at –0.3 dB, and the fifth at –2.2 dB. An electronically scanned antenna beam will not scan continuously across a target position during a CPI but will remain at a given fixed angle. In this case, the beam shape loss will be constant during the CPI but on average will be the same as for a mechanically scanning antenna during a search frame. Beam shape losses are discussed in more detail in Chapter 9.

In a tracking mode, since the angular position of the target is known, the antenna beam can be pointed directly at the target such that the target is in the center (or at least very close to the center) of the beam for the entire CPI. If this is the case, the SNR for track mode will not be degraded due to the beam shape loss.

The radar system is designed to search for targets in a given volume, defined by the range of elevation and azimuth angles to be considered and the range of distances from the nearest range of interest, R_{min}, to the farthest, R_{max}. Many modern systems also measure the Doppler frequency exhibited by the target. The Doppler frequency can be measured unambiguously from minus half the sample rate to plus half the sample rate. The sample rate for a pulsed system is the pulse repetition frequency (PRF) so the Doppler can be unambiguously determined from $-PRF/2$ to $+PRF/2$ hertz.

The system does not determine the range to the target as a continuous value from R_{min} to R_{max}, but rather it subdivides that range extent into contiguous range increments, often called range cells or range bins. The size of any range bin is usually equal to the range resolution of the system. For a simple unmodulated pulse, the range resolution, ΔR, is $\Delta R = c\tau/2$, where τ is the pulse width in seconds, and c is the speed of light. For a 1 microsecond pulse, the range resolution is 150 meters. If the total range is from 1 km to 50 km, there are 327 150-meter range bins to consider.

Likewise, the Doppler frequency regime from $-PRF/2$ to $+PRF/2$ is divided into contiguous Doppler bands by the action of the Doppler filters. The bandwidth of a Doppler filter is on the order of the reciprocal of the dwell time. A 2 ms dwell will result in 500 Hz filter bandwidth. The total number of Doppler filters is equal to the size of the FFT used to produce the results. If analog circuits are used to develop the Doppler measurement, then the number of filters is somewhat arbitrary.

Straddle loss arises because a target signal is not generally in the center of a range bin or a Doppler filter. It may be that the centroid of the received target pulse/spectrum is somewhere between two range bins and somewhere between two Doppler filters, reducing

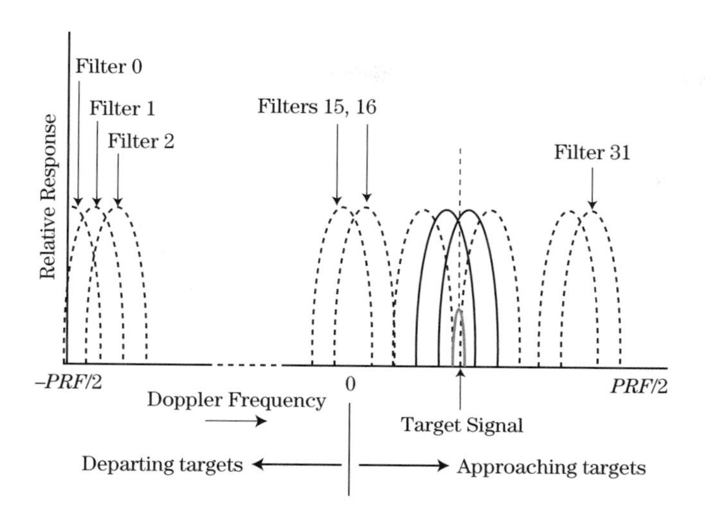

FIGURE 2-5 ■ Doppler filter bank, showing a target straddling two filters.

the target signal power in any one bin. Figure 2-5 depicts a series of several Doppler filters, ranging from $-PRF/2$ to $+PRF/2$ in frequency. Nonmoving clutter will appear between filters 15 and 16, at 0 Hz (for a stationary radar). A target is depicted at a position in frequency identified by the dashed vertical line such that it is not centered in any filter but instead is "straddling" the two filters shown in the figure as solid lines. A similar condition will occur in the range (time) dimension; that is, a target signal will, in general, appear between two range sample times. The worst-case loss due to range and Doppler straddle depends on a number of sampling and resolution parameters but is usually no more than 3 dB each in range and Doppler. However, usually the average loss rather than the worst case is considered when predicting the SNR. The loss experienced depends on the extent to which successive bins overlap—that is, the depth of the dip between two adjacent bins. Thus, straddle loss can be reduced by oversampling in range and Doppler, which decreases the depth of the "scallop" between bins. Depending on these details, an expected average loss of about 1 dB for range and 1 dB for Doppler is often reasonable. If the system parameters are known, a more rigorous analysis should be performed. Straddle loss is analyzed in more detail in Chapters 14 and 17. As with the beam shape loss, in the tracking mode the range and Doppler sampling can be adjusted so that the target is centered in these bins, eliminating the straddle loss.

Most modern radar systems are designed to automatically detect the presence of a target in the presence of interfering signals, such as atmospheric and receiver noise, intentional interference (jamming), unintentional interference (electromagnetic interference), and clutter. Given the variability of the interfering signals, a CFAR processor might be used to determine the presence of a target. The processor compares the signal power for each resolution cell with a local average of the surrounding cells, which ostensibly contain only interference signals. A threshold is established at some level (several standard deviations) above such an average to maintain a predicted average rate of false alarm. If the interference level is constant and known, then an optimum threshold level can be determined that will maintain a fixed probability of false alarm, P_{FA}. However, because the interfering signal is varying, the interference in the bin being tested for the presence of a target may be higher than the mean in some region of the sample space and may be lower than the mean in other regions. To avoid a high P_{FA} in any region, the threshold will have to be somewhat higher than the optimum setting. This means that the probability of detection, P_D, will be somewhat lower than optimum. The consequence is that the SNR for a given P_D must be higher than that required for an optimum detector. The increase in SNR required to meet the required P_D is considered to be a *loss*. Such a loss in detection performance is called a *CFAR loss* and is on the order of 1 to 2 dB for most standard conditions. Chapter 16 provides a complete discussion of the operation of a CFAR processor and its attendant losses.

For a pulse compression system, the matched filter condition is obtained only when there is no Doppler frequency offset on the target signal or when the Doppler shift is compensated in the processing. If neither of these is the case, a Doppler mismatch loss is usually experienced.

2.8 | SOLVING FOR OTHER VARIABLES

2.8.1 Range as a Dependent Variable

An important analysis is to determine the detection range, R_{det}, at which a given target RCS can be detected with a given SNR. In this case, solving equation (2.17) for R yields

$$R_{det} = \left[\frac{P_t G_t G_r \lambda^2 \sigma n_p}{(4\pi)^3 SNR \; kT_0 FBL_s} \right]^{\frac{1}{4}} \tag{2.18}$$

In using (2.18), one must bear in mind that some of the losses in L_s (primarily atmospheric attenuation) are range-dependent.

2.8.2 Solving for Minimum Detectable RCS

Another important analysis is to determine the minimum detectable radar cross section, σ_{min}. This calculation is based on assuming that there is a minimum SNR, SNR_{min}, required for reliable detection (see Chapter 15). Substituting SNR_{min} for SNR and solving (2.17) for radar cross section yields

$$\sigma_{min} = SNR_{min} \frac{(4\pi)^3 \; R^4 kT_0 FBL_s}{P_t G_t G_r \lambda^2 n_p} \tag{2.19}$$

Clearly, equation (2.17) could be solved for any of the variables of interest. However, these provided forms are most commonly used.

2.9 | DECIBEL FORM OF THE RADAR RANGE EQUATION

Many radar systems engineers use the previously presented algebraic form of the radar equation, which is given in linear space. That is, the equation consists of a set of values that describe the radar parameters in watts, seconds, or meters, and the values in the numerator are multiplied and then divided by the product of the values in the denominator. Other radar systems engineers prefer to convert each term to the dB value and to add the numerator terms and subtract the denominator terms, resulting in SNR being expressed directly in dB. The use of this form of the radar equation is based strictly on the preference of the analyst. Many of the terms in the SNR equation are naturally determined in dB notation, and many are determined in linear space, so in either case some of the terms must be converted from one space to the other. The terms that normally appear in dB notation are antenna gains, RCS, noise figure, and system losses. It remains to convert the remaining values to dB equivalents and then to proceed with the summations. Equation (2.20) demonstrates the dB form of the RRE shown in equation (2.17)

$$\begin{aligned} SNR_c\ [\text{dB}] = {} & 10\log_{10}(P_t) + G_t\ [\text{dB}] + G_r\ [\text{dB}] + 20\log_{10}(\lambda) + \sigma[\text{dBsm}] \\ & + 10\log_{10}(n_p) - 33 - 40\log_{10}(R) - (-204)\ [\text{dBW / Hz}] \\ & - F[\text{dB}] - 10\log_{10}(B)\ [\text{dBHz}] - L_s\ [\text{dB}] \end{aligned} \tag{2.20}$$

In the presentation in (2.20) the constant values (e.g., π, kT_0) have been converted to the dB equivalent. For instance, $(4\pi)^3 \approx 1{,}984$, and $10\log_{10}(1{,}984) \approx 33$ dB. Since this term is in the denominator, it results in –33 dB in equation (2.20). The (−204) [dBW/Hz] term results from the product of k and T_0. To use orders of magnitude that are more appropriate for signal power and bandwidth in the radar receiver, this is equivalent to –114 dBm/MHz. Remembering this value makes it easy to modify the result for other noise temperatures, the noise figure, and the bandwidth in MHz. In addition to the simplicity associated with adding and subtracting, the dB form lends itself more readily to tabulation and spreadsheet analysis.

2.10 | AVERAGE POWER FORM OF THE RADAR RANGE EQUATION

Given that the radar usually transmits several pulses and processes the results of those pulses to detect a target, an often used form of the radar range equation replaces the peak power, number of pulses processed, and instantaneous bandwidth terms with average power and dwell time. This form of the equation is applicable to all coherent multipulse signal processing gain effects.

The average power, P_{avg}, form of the RRE can be obtained from the peak power, P_t, form with the following series of substitutions:

$$T_d = \text{dwell time} = n_p \cdot PRI = n_p / PRF \tag{2.21}$$

where *PRI* is the interpulse period (time between transmit pulses), and *PRF* is the pulse repetition frequency. Solving (2.21) for n_p,

$$n_p = T_d \cdot PRF \tag{2.22}$$

Duty cycle is the fraction of a PRI during which the radar is transmitting:

$$\text{Duty cycle} = \tau \cdot PRF \tag{2.23}$$

Combining equations (2.22) and (2.23) gives the average power,

$$P_{avg} = P_t \cdot (\text{duty cycle}) = P_t \cdot (\tau \cdot PRF) \tag{2.24}$$

For a simple (unmodulated) pulse of width τ, the optimum receiver bandwidth, B, is

$$B = 1/\tau \tag{2.25}$$

Combining (2.22), (2.24), and (2.25) and solving for P_t gives

$$P_t = P_{avg} T_d B / n_p \tag{2.26}$$

Substituting P_t in (2.26) for P_t in (2.17) gives

$$SNR_c = \left(\frac{P_{avg} T_d B}{n_p} \right) \frac{G_t G_r \lambda^2 \sigma n_p}{(4\pi)^3 R^4 k T_0 F L_s B} = \frac{P_{avg} T_d G_t G_r \lambda^2 \sigma}{(4\pi)^3 R^4 k T_0 F L_s} \tag{2.27}$$

In this form of the equation, the average power–dwell time terms provide the energy in the processed waveform, while the kT_0F terms provide the noise energy. Assuming that all of the conditions related to the substitutions described in (2.21) through (2.25) are met—that is, the system uses coherent integration or equivalent processing during the dwell time and the receiver bandwidth is matched to the transmit bandwidth—the average power form of the radar range equation provides some valuable insight for SNR. In particular, the SNR for a system can be adjusted by changing the dwell time without requiring hardware changes, except that the signal/data processor must be able to adapt to the longest dwell. Often, for a radar in which n_p pulses are coherently processed, the dwell time, T_d, is called the coherent processing interval, CPI.

2.11 | PULSE COMPRESSION: INTRAPULSE MODULATION

The factor of N in equation (2.17) is a form of signal processing gain resulting from coherent integration of multiple pulses. Signal processing gain can also arise from processing pulses with intrapulse modulation. Radar systems are sometimes required to produce a given probability of detection, which would require a given SNR, while at the same time maintaining a specified range resolution. When using simple (unmodulated) pulses, the receiver bandwidth is inversely proportional to the pulse length τ, as discussed earlier. Thus, increasing the pulse length will increase the SNR. However, range resolution is also proportional to τ, so the pulse must be kept short to meet range resolution requirements. A way to overcome this conflict is to maintain the average power by transmitting a wide pulse while maintaining the range resolution by incorporating a wide bandwidth in that pulse—wider than the reciprocal of the pulse width. This extended bandwidth can be achieved by incorporating modulation (phase or frequency) within the pulse. Proper matched filtering of the received pulse is needed to achieve both goals. The use of intrapulse modulated waveforms to achieve fine-range resolution while maintaining high average power is called *pulse compression* and is discussed in detail in Chapter 20.

The appropriate form of the radar range equation for a system using pulse compression is

$$SNR_{pc} = SNR_u \tau \beta \tag{2.28}$$

where

SNR_{pc} is the signal-to-noise ratio for a modulated (pulse compression) pulse.
SNR_u is the signal-to-noise ratio for an unmodulated pulse.
τ is the pulse length.
β is the pulse modulation bandwidth.

Substituting (2.28) into (2.17) gives

$$SNR_{pc} = \frac{P_t G_t G_r \lambda^2 \sigma\, n_p}{(4\pi)^3\, R^4 k T_0 F B L_s} \tau\beta \tag{2.29}$$

Using (2.29) and the substitution developed in equation (2.26) for the average power form of the radar range equation, the result is

$$SNR_{pc} = \left(\frac{P_{avg} T_d}{N\tau}\right) \frac{G_t G_r \lambda^2 \sigma\, n_p}{(4\pi)^3\, R^4 k T_0 F B L_s} \tau\beta = \frac{P_{avg} T_d G_t G_r \lambda^2 \sigma}{(4\pi)^3\, R^4 k T_0 F L_s} \tag{2.30}$$

This equation, which is identical to (2.27), demonstrates that the average power form of the radar range equation is appropriate for a modulated pulse system as well as for a simple pulse system. As with the unmodulated pulse, appropriate use of the average power form requires that coherent integration or equivalent processing is used during the dwell time and that matched filtering is used in the receiver.

2.12 A GRAPHICAL EXAMPLE

Consider an example of a hypothetical radar system SNR analysis in tabular form and in graphical form. Equation (2.27) is used to make the SNR_c calculations. Consider a ground- or air-based radar system and two targets with the following characteristics:

Transmitter:	150 kilowatt peak power
Frequency:	9.4 GHz
Pulse width:	1.2 microseconds
PRF:	2 kilohertz
Antenna:	2.4 meter diameter circular antenna (An efficiency, η, of 0.6 is to be used to determine antenna gain.)
Processing dwell time	18.3 milliseconds
Receiver noise figure:	2.5 dB
Transmit losses:	3.1 dB
Receive losses:	2.4 dB
Signal processing losses:	3.2 dB
Atmospheric losses:	0.16 dB/km (one way)
Target RCS:	0 dBsm, –10 dBsm (1.0 and 0.1 m^2)
Target range:	5 to 105 km

It is customary to plot the SNR in dB as a function of range from the minimum range of interest to the maximum range of interest. Figure 2-6 is an example of a plot for two target RCS values resulting from the given parameters. If it is assumed that the target is reliably detected at an SNR of about 15 dB, then the 1 m^2 target will be detectable at a range

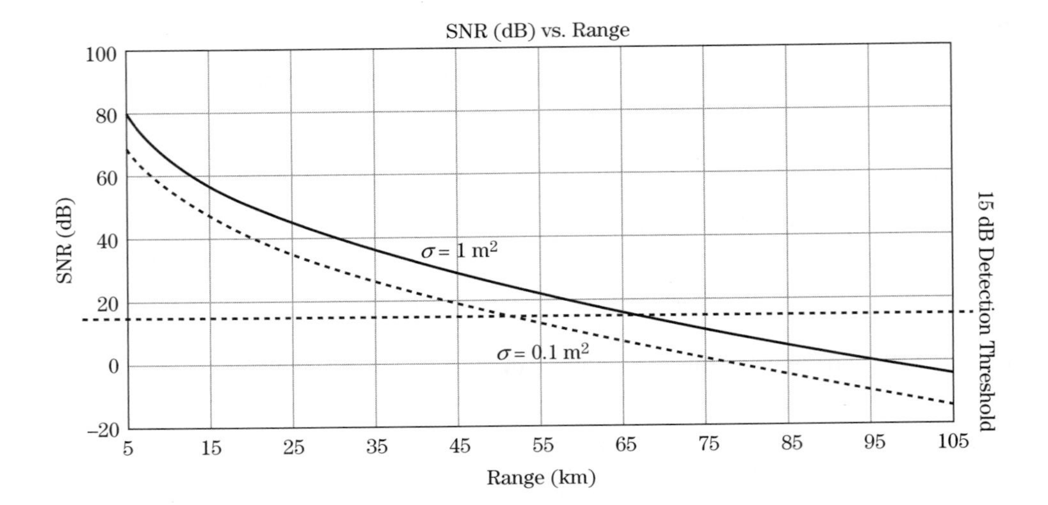

FIGURE 2-6 ■ Graphical solution to radar range equation.

of approximately 65 km, whereas the 0.1 m^2 target will be detectable at approximately 52 km.

Once the formulas for this plotting example are developed in a spreadsheet program such as Microsoft Excel, then it is relatively easy to extend the analysis to plotting probability of detection (see Chapters 3 and 15) and tracking measurement precision (Chapter 19) as functions of range, since these are dependent primarily on SNR and additional fixed parameters.

2.13 CLUTTER AS THE TARGET

Though the intent is usually to detect a discrete target in the presence of noise and other interference, there are often unintentional signals received from other objects in the antenna beam. Unintentional signals can result from illuminating clutter, which can be on the surface of the earth, either on land or sea, or in the atmosphere, such as rain and snow. For surface clutter, the area illuminated by the radar antenna beam pattern, including the sidelobes, determines the signal power. For atmospheric clutter, the illuminated volume is defined by the antenna beamwidths and the pulse length. The purpose of the radar equation is to determine the target SIR, given that the interference is surface or atmospheric clutter. In the case of either, the ratio is determined by dividing the target signal, S, by the clutter signal, S_c, to produce the target-to-clutter ratio, SCR. The use of the RRE for the signal resulting from clutter is summarized by substituting the RCS of the clutter cell into the RRE in place of the target RCS. In many cases, all of the terms in the radar equation cancel except for the RCS (σ or σ_c) terms, resulting in

$$SCR = \frac{\sigma}{\sigma_c} \tag{2.31}$$

In some cases, as with a ground mapping radar or weather radar, the intent is to detect these objects. In other cases, the intent is to detect discrete targets in the presence of these interfering signals. In either case, it is important to understand the signal received from these "clutter" regions. Chapter 5 describes the characteristics and statistical behavior of clutter in detail. A summary is provided here.

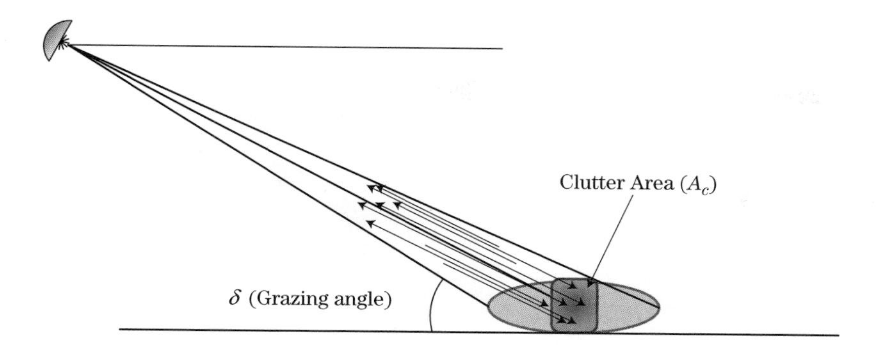

FIGURE 2-7 ■ Area (surface) clutter.

2.13.1 Surface Clutter

The radar cross section value for a surface clutter cell is determined from the average reflectivity, σ^0, of the particular clutter type, in square meters per unit area, times the area of the clutter cell, A_c, illuminated by the radar:

$$\sigma_{cs} = A_c \sigma^0 \tag{2.32}$$

where

σ_{cs} is the surface clutter radar cross section.
A_c is the area of the illuminated (ground or sea surface) clutter cell (square meters).
σ^0 is the surface backscatter coefficient (average reflectivity per unit area) (square meters per square meter).

Chapter 5 provides the formula for the calculation of the clutter area A_c. Figure 2-7 depicts the area of a clutter cell illuminated on the surface. The clutter consists of a multitude of individual reflecting objects (e.g., rocks, grass, dirt mounds, twigs, branches), often called *scatterers*. The resultant of the echo from these many *scatterers* is a single net received signal back at the radar receiver.

2.13.2 Volume Clutter

The radar cross section value for volumetric clutter cell is determined from the average reflectivity of the particular clutter type per unit volume, η, times the volume of the clutter cell, V_c, illuminated by the radar:

$$\sigma_{cv} = V_c \eta \tag{2.33}$$

where

σ_{cv} is the volume clutter radar cross section.
V_c is the volume of the illuminated clutter cell (cubic meters).
η is the volumetric backscatter coefficient (average reflectivity per unit volume) (cubic meters per cubic meter).

The volume, V_c, of the cell illuminated by the radar is depicted in Figure 2-8. Chapter 5 provides the formula for the calculation of the clutter volume V_c. The clutter in Figure 2-8 consists of a multitude of individual reflecting objects (e.g., rain, fog droplets). The resultant echo from these many scatterers is again a single net received signal back at the radar receiver.

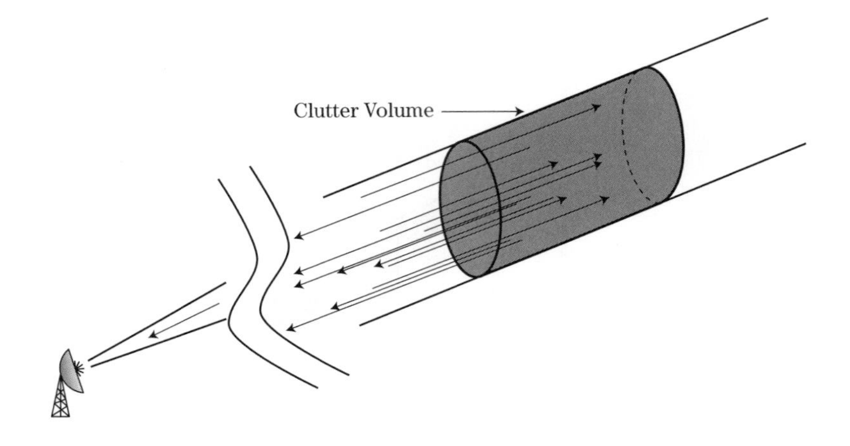

FIGURE 2-8 ■ Volumetric (atmospheric) clutter.

2.14 ONE-WAY (LINK) EQUATION

To this point, the components of the interfering signals discussed have been receiver thermal noise and clutter. In many defense-oriented radar applications, it is expected that the radar system will encounter intentional jamming. Jamming signals are one of two varieties: noise and false targets. The effect of noise jamming is to degrade the SIR of target signals in the radar receiver to delay initial detection or to degrade tracking performance. In many cases, intentional noise jamming is the most limiting interference. The intent of false target jamming is to present so many target-like signals that the radar processor is overloaded or to create false track files. In any case, the power received at the radar from a jammer is required to determine its effect on radar performance. Since the signal from the jammer to the radar has to propagate in only one direction, a simplification of the radar equation for one-way propagation is valuable.

The first step is to determine the effective radiated power out of the jammer antenna, and the power density at a distance, R_{jr}, from the jammer radiating antenna to the radar. The jammer signal power density, Q_j, in watts per square meter at a distance (range) R from a transmitting source can be determined using equation (2.1). Q_j depends on the jammer's transmitted power, P_j; the transmit path losses in the jammer equipment, L_{tj}; the range from the jammer to the radar, R_{jr}; the gain of the jammer transmitting antenna, G_j; and the losses through the propagation medium, L_{atm}. Usually, the antenna gain includes the losses between the antenna input terminal and the antenna. The power density, Q_j, from an isotropic radiating jammer source is the total power, reduced by losses and distributed uniformly over the surface area of a sphere of radius R_{jr}:

$$Q_j = \frac{P_j}{4\pi R_{jr}^2 L_{tj} L_a} \tag{2.34}$$

The power density given in (2.33) is increased by the effects of a jammer antenna, which concentrates the energy in a given direction. The power density, Q_j, at the center of the beam for a radiating source with an attached jammer antenna of gain, G_j, is

$$Q_j = \frac{P_j G_j}{4\pi R_{jr}^2 L_{tj} L_a} \tag{2.35}$$

The jammer antenna peak gain, G_j, accounts for the fact the transmitted radio waves are "focused" by the antenna in a given direction, thus increasing the power density in that direction.

Next, consider the radar receiving system at a distance, R_{jr}, from the jammer to the radar. Such a receiving system will have a directive antenna with an effective area of A_e and will have receiver component losses, L_r. The total power received at the radar from the jammer, P_{rj}, is

$$P_{rj} = Q_j A_e = \frac{P_j G_j A_e}{4\pi R_{jr}^2 L_{tj} L_a L_r} \tag{2.36}$$

Equation (2.36) is the *one-way link equation*. It is very useful in predicting the performance of a one-way system such as a communications system or, for a radar, a jammer.

Often, the antenna gain is known instead of the effective area. In this case, using equation (2.7) the area can be replaced by

$$A_e = \frac{G_{rj}\lambda^2}{4\pi} \tag{2.37}$$

which results in

$$P_{rj} = \frac{P_j G_j G_{rj}\lambda^2}{(4\pi)^2 R_{jr}^2 L_{tj} L_a L_r} \tag{2.38}$$

where G_{rj} is the gain of the radar antenna in the direction of the jammer. This is important, since the radar antenna is not necessarily pointed directly at the jammer. In this case, the main beam gain is not appropriate, but the sidelobe gain is. The sidelobes are not easily determined until the antenna is tested in a high-quality environment.

2.15 | SEARCH FORM OF THE RADAR RANGE EQUATION

Section 1.8.1 and Figure 1-29 depict a scanning antenna being used to scan a volume. Analysis of such a system designed to search a given solid angular volume, Ω, in a given search frame time, T_{fs}, is often made easier by using the so-called search form of the radar equation. The predominant figure of merit for such as system is the *power-aperture product* of the system. This section derives and describes this form of the radar equation.

The total time required to search a volume, T_{fs}, is easily determined from the number of beam positions required, M, multiplied by the dwell time required at each of these positions, T_d:

$$T_{fs} = M \cdot T_d \tag{2.39}$$

The number of beam positions required is the solid angular volume to be searched, Ω, divided by the solid angular volume of the antenna beam, which is approximately the product of the azimuth and elevation beamwidths[4]:

$$M = \frac{\Omega}{\theta_3 \cdot \phi_3} \tag{2.40}$$

[4]There are $(180/\pi)^2 \approx 3{,}282.8$ square degrees in a steradian.

For a circular antenna with diameter, D, the estimated beamwidth is about $1.22\lambda/D$, and the solid angle $\theta_3 \cdot \phi_3$ is about $1.5\lambda^2/D^2$. The area, A, of a circular antenna is $\pi D^2/4$, and, from equation (2.8) the effective antenna area, depending on the weighting function and shape, is about $0.6\pi D^2/4$, leading to

$$\theta_3 \cdot \phi_3 \approx \lambda^2/A_e \tag{2.41}$$

The antenna gain, G, is related to the effective area by

$$G = 4\pi A_e/\lambda^2 \tag{2.42}$$

Using the previous substitutions into (2.17), it can be shown that the resulting SNR can be expressed as

$$SNR = (P_{avg} A_\varepsilon) \frac{1}{4\pi k T_0 F L_s} \left(\frac{\sigma}{R^4}\right) \left(\frac{T_{fs}}{\Omega}\right) \tag{2.43}$$

By substituting the minimum SNR required for reliable detection, SNR_{min}, for SNR and arranging the terms differently, the equation can be repartitioned to place the "user" terms on the right side and the system designer terms on the left side

$$\frac{P_{avg} A_e}{L_s T_0 F} \geq SNR_{min} 4\pi k \left(\frac{R^4}{\sigma}\right) \left(\frac{\Omega}{T_{fs}}\right) \tag{2.44}$$

Since these modifications to the RRE are derived from equation (2.17), the assumptions of coherent integration and matched filtering in the receiver apply.

This *power-aperture form of the RRE* provides a convenient way to partition the salient radar parameters (P_{avg}, A_e, L_s, and F) given the requirement to search for a target of RCS σ at range R over a solid angle volume Ω in time T_{fs}. Since it is derived from the average power form of the basic RRE (2.17), it is applicable for any waveform, whether pulse compression is used, and for any arbitrary length dwell time. It does assume that the entire radar resources are used for search; that is, if the system is a multifunction radar then a loss must be included for the time the radar is in the track mode or is performing some function other than search.

2.16 | TRACK FORM OF THE RADAR RANGE EQUATION

With modern radar technology rapidly evolving toward *electronically scanned arrays* (ESAs, or phased arrays), with additional degrees of freedom, target tracking systems can track multiple targets simultaneously. As with the search form of the radar equation, analysis of a system designed to track multiple targets with a given precision is described in terms of the power-aperture cubed, or, equivalently, the power-aperture-gain form of the radar range equation. This variation is also called the track form of the RRE. These forms are used in cases in which the number of targets being tracked and the track precision or the required SNR are known.

Recalling (2.7), the relationship between an antenna's gain, G, and its effective area, A_e, is

$$G = \frac{4\pi A_e}{\lambda^2} \tag{2.45}$$

The approximate beamwidth, θ_{BW}, of an antenna in degrees is [7]

$$\theta_{BW}\ (\text{degrees}) \approx 70\frac{\lambda}{D} \tag{2.46}$$

This equation can be applied in both azimuth and elevation by using the corresponding beamwidth and antenna dimension. Since there are $180/\pi$ degrees in a radian, this is equivalent to

$$\theta_{BW}\ (\text{radians}) \approx 1.22\frac{\lambda}{D} \tag{2.47}$$

Of course, there is some variation in this estimate due to specific design parameters and their effects, but this approximation serves as a good estimate. Also, the effective area for a circular antenna of diameter D is [1]

$$A_e \approx \frac{0.6\pi D^2}{4} \tag{2.48}$$

For a more general elliptical antenna, it is

$$A_e \approx \frac{0.6\pi D_{major} D_{minor}}{4} \tag{2.49}$$

where D_{major} and D_{minor} are the major and minor axes of the ellipse, respectively.

From these equations, the solid angle beamwidth is approximately

$$\theta_{BW} \approx \frac{\pi\lambda^2}{4A_e} \tag{2.50}$$

As described in Chapter 18, a common expression for the estimated tracking noise with a precision σ_θ (standard deviation of the tracking measurement noise) in one place is

$$\sigma_\theta \approx \frac{\theta_{BW}}{k_m\sqrt{2SNR}} \tag{2.51}$$

where k_m is a tracking system parameter. Substituting (2.50) into (2.51) and solving for SNR gives

$$SNR \approx \frac{\pi\lambda^2}{8A_e k_m^2 \sigma_\theta^2} \tag{2.52}$$

Given a requirement to track N_t targets, each at an update rate of r measurements per second, the dwell time T_d per target is

$$T_d = \frac{1}{r \cdot N_t} \tag{2.53}$$

Finally, substituting equations (2.45), (2.52), and (2.53) into (2.17) and rearranging terms gives

$$\frac{\pi\lambda^2}{8k_m^2\sigma_\theta^2} = \frac{P_{avg} A_e^3 \sigma}{4\pi r N_t \lambda^2 k T_0 F L_s R^4} \tag{2.54}$$

As with the search form of the RRE, the terms are arranged so that the "user" terms are on the right and the "designer" terms are on the left, providing

$$\frac{P_{avg} A_e^3 k_m^2}{\lambda^4 L_s T_0 F} = \left(\frac{\pi^2}{2}\right)\left(\frac{k r N_t R^4}{\sigma \cdot \sigma_\theta^2}\right)\left(\frac{1}{\cos^5(\theta_{scan})}\right) \tag{2.55}$$

This form of the RRE shows that, given N_t targets of RCS σ at range R to track, each at rate r and with a precision σ_θ, the power-aperture cubed of the radar becomes the salient determinant of performance. Coherent integration and matched filtering are assumed, since these developments are based on equation (2.17).

Equation (2.55) also introduces a cosine^5 term that has not been seen thus far. This term accounts for the beam-broadening effect and the gain reduction that accompanies the scanning of an electronically scanned antenna to an angle of θ_{scan} from array normal. To a first order, the beamwidth increases as the beam is scanned away from array normal by the cosine of the scan angle due to the reduced effective antenna along the beam-pointing direction. The gain is also reduced by the cosine of the scan angle due to the reduced effective antenna and by another cosine factor due to the off-axis gain reduction in the individual element pattern, resulting in a net antenna gain reduction by a factor of $\cos^2(\theta_{scan})$. SNR is reduced by the product of the transmit and receive antenna gains, thus squaring the reduction to a factor of $\cos^4(\theta_{scan})$. Therefore, to maintain a constant angle precision, the radar sensitivity needs to increase by $\cos^5(\theta_{scan})$: $\cos^4$ due to gain effects, and another cosine factor due to beam broadening. This term is an approximation, because the individual element pattern is not strictly a cosine function. However, it is a good approximation, particularly at angles beyond about 30 degrees. Additional details on the effect of scanning on the gain and beamwidth of ESAs are given in Chapter 9.

Figure 2-9 is a plot of the loss associated with scanning an electronically scanned beam. Because of the wider antenna beam and lower gain, the radar energy on target must increase by this factor. Compared with a target located broadside to the array, a target at a 45 degree scan angle requires increased energy on the order of 7 dB to maintain the same tracking precision; at 60 degrees, the required increase is 15 dB. These huge equivalent losses can be overcome by the geometry of the problem or by changing the radar waveform at the larger scan angles. For example, once a target is in track, it will be close to the antenna normal (for an airborne interceptor); for a fixed or ship-based system, the scan loss is partially offset by using longer dwell times at the wider scan angles. Also, for a target in track, straddle losses are reduced because the target is likely to be near the center of the beam and also to be nearly centered in the range and Doppler bins. Nonetheless, the system parameters must be robust enough and the processor must be adaptable enough to support shorter than average dwell times at near-normal scan angles to allow time for longer dwells at the extreme angles to offset large scan losses.

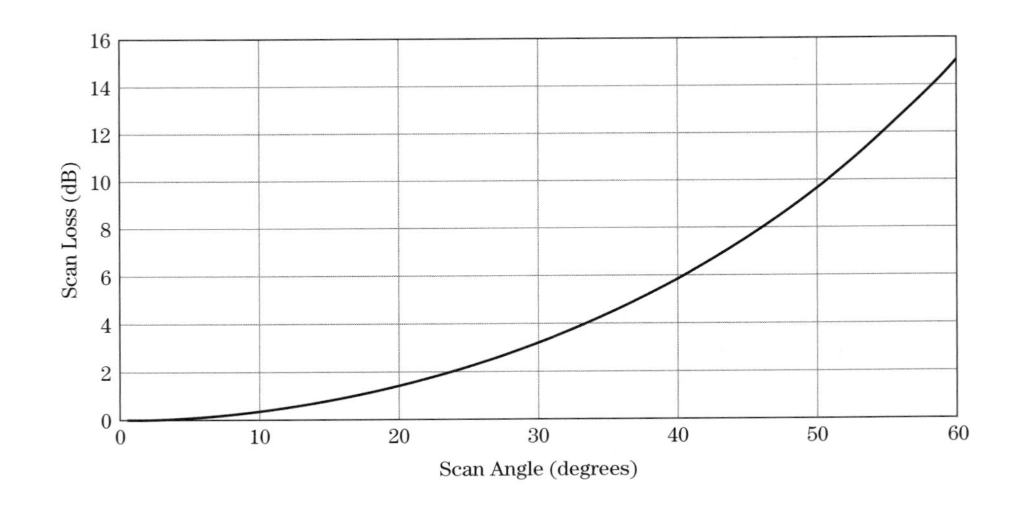

FIGURE 2-9 ■ Scan loss versus angle for an electronically scanned antenna beam.

Occasionally, another form of the radar equation as it relates to a tracking system is encountered. Beginning with equation (2.17), repeated here for convenience,

$$SNR = \frac{P_t G_t G_r \lambda^2 \sigma}{(4\pi)^3 R^4 k T_0 F B L_s} \tag{2.56}$$

making the substitution P_{avg} = average power = $P_t \cdot \tau \cdot PRF$ and solving for P_{avg} gives

$$P_{avg} = \frac{SNR\,(4\pi)^3 R^4 k T_0 F B L_s \tau PRF}{G_t G_r \sigma \lambda^2} \tag{2.57}$$

Substituting from equation (2.45) of the antenna gain gives

$$P_{avg} = \frac{SNR \cdot 4\pi R^4 k T_0 F L_s PRF \lambda^2}{\sigma A_e^2} \tag{2.58}$$

or, rearranged,

$$\frac{P_{avg} A_e^2}{L_s F \lambda^2} = \frac{SNR \cdot 4\pi R^4 k T_0 PRF}{\sigma} \tag{2.59}$$

Equation (2.59) is the power-aperture squared form of the radar range equation. This form is most useful when the required SNR is known, whereas equation (2.55) is used when the required tracking precision, σ_θ, is known. The two forms (2.55) and (2.59) are equivalent when the substitutions associated with the relationship between antenna dimensions, SNR, and tracking precision are incorporated.

A final form, also sometimes encountered is found by replacing one of the A_e terms on the left side of (2.59) with its equivalent in terms of gain,

$$A_e = \frac{G\lambda^2}{4\pi} \tag{2.60}$$

resulting in

$$\frac{P_{avg} A_e G}{L_s F} = \frac{SNR \cdot (4\pi)^2 R^4 k T_0 PRF}{\sigma} \tag{2.61}$$

which does not include wavelength, λ. Clearly some of the terms are dependent on λ, such as L and F. Equation (2.61) is often called the power-aperture-gain form of the RRE.

2.17 | SOME IMPLICATIONS OF THE RADAR RANGE EQUATION

Now that the reader is somewhat familiar with the basic RRE, its use in evaluating radar detection range performance can be explored.

2.17.1 Average Power and Dwell Time

Considering equation (2.27), the average power form of the RRE, it can be seen that for a given hardware configuration that "freezes" the P_{avg}, G, λ, F, and L_s, the dwell time, T_d, can easily be changed without affecting the hardware design. This directly impacts the SNR. For example, doubling the dwell time increases the SNR by 3 dB. As Chapter 3 shows, this increase in SNR improves the detection statistics; that is, the probability of detection, P_D, for a given probability of false alarm, P_{FA}, will improve.

For an electronically scanned antenna beam, the radar received signal power and therefore the SNR degrade as the beam is scanned away from normal to the antenna surface, as was seen in the context of equation (2.55) and Figure 2-9. This reduction in SNR can be recovered by adapting the dwell time to the antenna beam position. For example, whereas a mechanically scanned antenna beam might have a constant 2 msec dwell time, for the radar system using an ESA of similar area, the dwell can be adaptable. It might be 2 msec near normal (say, 0 to 30 degrees), 4 msec from 30 to 40 degrees scan angle, and 8 msec from 40 to 45 degrees. Depending on the specific design characteristics of the ESA, it might have lower losses than the mechanically scanned antenna system so that the dwell times for the various angular beam positions could be less.

2.17.2 Target RCS Effects

Much is being done today to reduce the radar cross section of radar targets, such as missiles, aircraft, land vehicles (tanks and trucks), and ships. The use of radar-absorbing material and target shape modifications can produce a significantly lower RCS compared with conventional designs. This technology is intended to make the target "invisible" to radar. In fact, the change in radar detection range performance is subtle for modest changes in target RCS. For example, if the RCS is reduced by 3 dB, the detection range decreases by only about 16% to 84% of the baseline value. To reduce the effective radar detection range performance by half, the RCS must be reduced by a factor of 16, or 12 dB. Thus, an aggressive RCS reduction effort is required to create significant reductions in radar detection range. Basic concepts of RCS reduction are introduced in Chapter 6.

2.18 FURTHER READING

Most standard radar textbooks have a section that develops the radar range equation using similar yet different approaches. A somewhat more detailed approach to development of the peak power form of the RRE can be found in Chapter 4 of Barton et al. [5], Difranco and Rubin [8], and Sullivan [9]. A further discussion of the energy (average power) form is found in the same references. It is sometimes appropriate to present the results of RRE analysis in a tabular form. One form that has been in use since about 1969 is the Blake chart [4]. A more complete discussion of the various sources of system noise is presented in Chapter 4 of Blake [4]. A comprehensive discussion of the various RRE loss terms is also presented in many of the aforementioned texts as well as in Nathanson [6] and Barton et al. [5].

2.19 REFERENCES

[1] Johnson, R.C., *Antenna Engineering Handbook*, 3d ed., McGraw Hill, New York, 1993, Ch. 46.

[2] Krause, J.D., *Antennas,* McGraw-Hill, New York, 1988, p. 27.

[3] Eaves, J.L., and Reedy, E.K., *Principles of Modern Radar,* Van Nostrand Reinhold, New York, 1987, equation (1–16).

[4] Blake, L.V., *Radar Range-Performance Analysis*, Artech House, Dedham, MA, 1986.

[5] Barton, D.K., Cook, C.E., and Hamilton, P., *Radar Evaluation Handbook*, Artech House, Dedham, MA, 1991, Ch. 3.

[6] Nathanson, F.E., *Radar Design Principles*, 2d ed., McGraw-Hill, Inc., New York, 1991, Ch. 2.

[7] Stutzman, W.A., and Thiele, G.A., *Antenna Theory and Design*, 2d ed., Wiley, New York, 1997.

[8] Difranco, J.V., and Rubin, W.L., *Radar Detection*, Artech House, Dedham, MA, 1980, Ch. 12.

[9] Sullivan, R.J., *Radar Foundations for Imaging and Advanced Concepts,* SciTech Publishing, Inc., Raleigh, NC, 2004, Ch. 1.

2.20 PROBLEMS

1. Target received power: Using equation (2.8), determine the single-pulse received power level from a target for a radar system having the following characteristics:

Transmitter:	100 kilowatt peak
Frequency:	9.4 GHz
Antenna Gain:	32 dB
Target RCS:	0 dBsm
Target Range:	50 km

2. Using equation (2.10), determine the receiver noise power (in dBm) for a receiver having a noise figure of 2.7 dB and an instantaneous bandwidth of 1 MHz.
3. Using equation (2.11), determine the single-pulse SNR for the target described in problem 1 if the receiver has a noise figure of 2.7 dB and an instantaneous bandwidth of 1 MHz.
4. Ignoring any losses, using equation (2.8), determine the single-pulse received power level (in dBm) for a 1 square meter target at a range of 36 km for radar systems with the following characteristics.

	P_t (watts)	G	Freq
Radar a	25,000	36 dB	9.4 GHz
Radar b	250,000	31 dB	9.4 GHz
Radar c	250,000	31 dB	2.8 GHz
Radar d	250,000	36 dB	9.4 GHz

5. Using equation (2.11), determine the SNR for the four conditions described in problem 4 for the following noise-related characteristics. Bandwidth for both frequencies is 10 MHz, the noise figure for 9.4 GHz systems is 3.2 dB, and the noise figure for the 2.8 GHz system is 2.7 dB.
6. Using equation (2.17), determine the four answers in problems 4 and 5 for the following loss conditions:

   $L_{tx} = 2.1$ dB

   $L_{rx} = 4.3$ dB.
7. If atmospheric propagation losses of 0.12 dB per km (two-way) are also considered, determine the resulting SNR values in problem 6.
8. If we desire the SNR in problem 7 to be the same as in problem 5, we can increase the SNR in problem 7 by transmitting, receiving, and processing multiple pulses. Use equation (2.14) to determine how many pulses we have to transmit to recover from the losses added in problems 6 and 7. (Hint: instead of solving the problem from the beginning, merely determine the relationship between the number of pulses transmitted and the SNR improvement.)

9. A radar system provides 18 dB SNR for a target having an RCS of 1 square meter at a range of 50 km. Ignoring the effects of atmospheric propagation loss, using equation (2.18), determine the range at which the SNR be 18 dB if the target RCS is reduced to:

   a. 0.5 square meters.
   b. 0.1 square meters.

10. A system has a single-pulse SNR of 13 dB for a given target at a given range. Determine the integrated SNR if 20 pulses are coherently processed.

11. A system SNR can be increased by extending the dwell time. If the original dwell time of a system is 1.75 msec, what new dwell time is required to make up for the loss in target RCS from 1 square meter to 0.1 square meters?

12. If the radar system in problem 1 is looking at surface clutter having a reflectivity value of $\sigma^0 = -20$ dB, dBm^2/m^2, if the area of the clutter cell is 400,000 square meters, what is the clutter RCS and the resulting signal-to-clutter ratio (SCR)?

13. If the radar system in problem 1 is looking at volume clutter having a reflectivity value of $\eta = -70$ dBm^2/m^3, if the volume of the clutter cell is 900,000,000 cubic meters, what is the clutter RCS and the resulting SCR?

14. How much power is received by a radar receiver located 100 km from a jammer with the following characteristics? Assume that the radar antenna has an effective area of 1.2 square meters and that the main beam is pointed in the direction of the jammer. Consider only atmospheric attenuation, excluding the effects of, for example, component loss. Provide the answer in terms of watts and dBm (dB relative to a milliwatt.)

Jammer peak power	100 watts
Jammer antenna gain	15 dB
Atmospheric loss	0.04 dB per km (one-way)
Radar average sidelobe level	−30 dB (relative to the main beam)

15. Using the answers from problems 2 and 14 what is the jammer-to-noise ratio (JNR)?

16. If the receiver antenna is not pointed directly at the jammer but a −30 dB sidelobe is, then what would the answer to problem 14 be?

17. What would the resulting JNR be for the sidelobe jamming signal?

18. A search system being designed by an engineering staff has to search the following solid angle volume in the stated amount of time:

Azimuth angle:	90 degrees
Elevation angle:	3 degrees
Full scan time:	1.2 seconds
Maximum range:	30 km
Target RCS:	−10 dBsm

    What is the required power aperture product of the system if the system has the following characteristics?

Noise figure:	2.5 dB
System losses:	6.7 dB
Required SNR:	16 dB

19. For the radar system in problem 18, if the antenna has an effective aperture of 1.6 square meters, and the transmit duty cycle is 12%, what is the peak power required?

CHAPTER 3

Radar Search and Overview of Detection in Interference

James A. Scheer

Chapter Outline

3.1 Introduction 87
3.2 Search Mode Fundamentals 89
3.3 Overview of Detection Fundamentals 95
3.4 Further Reading 111
3.5 References 111
3.6 Problems 112

3.1 INTRODUCTION

Though radar systems have many specific applications, radars perform three general functions, with all the specific applications falling into one or more of these general functions. The three primary functions are *search*, *track*, and *image*. The radar search mode implies the process of target detection. Target tracking implies that the radar makes measurements of the target state in range, azimuth angle, elevation angle, and Doppler frequency offset. This is not to exclude the fact that a search radar will perform target measurements to provide a cue for another sensor, for example, or that a track radar will perform the detection process.

Many tracking radar systems track a single target by continually pointing the antenna beam at the target and controlling the antenna pointing angle and range measurement to coincide with the target position. The tracking function is performed by a set of analog circuits controlling the antenna and range servo. In many modern systems, though, the tracking function is performed by processing a sequence of target state measurements made by the tracking sensor (radar). These measurements are applied to a computer algorithm that forms a target track file. The tracking algorithms, usually implemented in software, develop an accurate state vector (position, velocity, and acceleration) for the target, typically in a Cartesian coordinate system of North, East, and down. That state estimate then becomes an integral part of a fire control system, directing a weapon or cueing another sensor to the target state.

Once a target is detected and in track, depending on the application for the radar, an imaging mode may be implemented that develops high-resolution data in range, azimuth, elevation, and sometimes Doppler.[1] This would support target classification, discrimination, and or identification functions. The present chapter is designed to provide the detailed description of the radar processes associated with supporting the search and detect functions. Sensor measurements are covered in Chapter 18, the track function is described in Chapter 19, and the two-dimensional (2-D) imaging function is described in Chapters 20 and 21.

Some systems are designed to perform two or three of these tasks. One way to do this is with multiple radar systems: one optimized for search and another for track. In many cases, however, allowable space and prime power limitations do not allow for multiple radar systems. If a single radar system has to perform both the search function and the track function, then there is likely a compromise required in the design. The following sections will show that some of the features of a good search radar will not be desirable in a good track radar. These same features do not necessarily necessitate a compromise in the imaging mode, however.

The single-pulse signal-to-noise ratio (SNR) can be predicted by exercising the peak power form of the radar range equation (RRE). Seldom, if ever, is a target detected on the basis of a single pulse. Usually there is an opportunity to process several pulses while the antenna beam is pointed at a target. As presented in Chapter 2, this leads to the development of the average power form, or "energy" form, of the RRE, for which the SNR is determined by average power and *dwell time, T_d*. In this case, the dwell time is the time it takes to transmit (and receive) the n pulses used for detection. If these pulses are processed coherently using coherent integration or, more often, fast Fourier transform (FFT) processing, the time duration is often called the *coherent processing interval* (CPI). "CPI" and "dwell time" are often used synonymously. Since not all radar systems are coherent, it is not necessary that coherent processing is performed during a processing interval. The processing may be noncoherent, or the resulting signal may simply be displayed on the radar display.

As an antenna beam is scanned in angle over a designated volume in search of a target, the beam is often pointed in a given angular direction for a time that is longer than the dwell time or CPI. An electronically scanned antenna (ESA) can be pointed at a given angle for an arbitrary time, as determined (directed) by the resource management algorithms implemented in the radar processor. A mechanically scanned antenna will be pointed at a given angular point in space for the time it takes for the antenna beam to scan one half-power beamwidth (θ_3 or ϕ_3) when scanning at an angular rate of ω radians per second. It is not uncommon for the antenna to be pointed at a point for 10 CPIs, for example. To differentiate the antenna scanning dwell time from the coherent processing interval, the term *antenna dwell time*, T_{ad}, to represent this time will be adopted.

The antenna dwell time for a mechanically scanned antenna is found from

$$T_{ad} = \frac{\theta_3}{\omega} \tag{3.1}$$

and the number of CPIs, n_{CPI}, that occur during that time is

$$n_{CPI} = \frac{T_{ad}}{T_d} = \frac{\theta_3}{\omega T_d} \tag{3.2}$$

---

[1] Radars whose primary function is imaging generally do not have search-and-track modes. Some new systems combine imaging with detection and tracking of targets within the image.

The CPI dwell time, T_d, depends on the radar pulse repetition frequency (PRF) or the pulse repetition interval (PRI) and the number of pulses in a CPI, n, and is given by

$$T_d = nPRI = \frac{n}{PRF} \tag{3.3}$$

To demonstrate by example, a mechanically scanned antenna with a beamwidth of 50 mrad (about 3 degrees) scanning at a rate of 90 degrees per second (about 1.55 radians per second) will produce an antenna dwell time, T_{ad}, of about 33 milliseconds. If a radar has a 10 kHz PRF and transmits 32 pulses for a coherent processing interval, then the CPI, or T_d, is 3.2 milliseconds. There is an opportunity for up to 10 CPIs to be processed in a single antenna dwell.

Dwell time, T_d, and CPI are terms that occur regularly in the literature, whereas the term antenna dwell time, T_{ad}, does not. The use of the variable T_{ad} provides a way to unambiguously differentiate between the antenna dwell time and a signal-processing dwell time.

This chapter discusses how a radar performs a volume search, given a limited instantaneous field of view determined by the antenna beamwidth; the fundamental analysis required to relate SNR to probability of detection, P_D, and probability of false alarm, P_{FA}; and the reasons for using more than one dwell time at each beam position. It will be determined whether it is more efficient to dwell for multiple dwell periods or to use extended dwell times within a CPI for equivalent detection performance. Initially, receiver thermal noise will represent the interfering signal. This will be followed by a discussion of jamming noise and then clutter as the interfering signal.

3.2 SEARCH MODE FUNDAMENTALS

A search radar is designed to look for targets when and where there is no a priori knowledge of the target existence. It is designed to search for a given target type in a given solid angle volume out to a given slant range in a specified amount of time. These parameters are derived from the system application requirements. As an analogy, the situation is much like a prison searchlight illuminating a prison yard, looking for prisoners trying to escape over the fence. The light must scan the area in a time short enough that a prisoner can't run from the building to the fence and escape between searchlight scans. The following several sections will develop the methodology to determine radar requirements for a given search requirement.

It is important to realize that use of an ESA beam (phased array) and a mechanically scanned antenna beam lead to somewhat different search patterns. The mechanically scanned beam scans in one angular dimension over some period of time. For a system designed to search a full 360 degree azimuth sector, the antenna continually rotates in one direction. This is typical of a ground-based weather radar, an air traffic control radar, and a ship-based volume search radar, searching for threats over a full hemisphere. For a system that has to search a limited azimuth sector, such as a 90 degree sector centered in a given direction, the antenna scans in one azimuth direction and at the end of the designated sector will turn around to scan in the other direction. This is typical of an airborne interceptor system, such as an F-15, F-18, or F-22 aircraft radar. A forward-looking airborne commercial weather radar would have a similar scan pattern. In any case, for the mechanically scanned antenna the scanning motion is continuous and smoothly transitions from one beam position to the next, as the radar system performs the detection process; for an electronically scanned antenna, the beam positions will be changed incrementally.

Many modern radar systems employ phased array antenna technology, providing an ability to scan the antenna beam position electronically. An ESA beam can step incrementally from one position to the next in discrete angular steps. Since there are no limits associated with inertia, motor drives, or mechanical reliability, these time-sequential antenna pointing positions do not need to be contiguous in space. In fact, they can be somewhat arbitrary, dictated by the system requirements. For example, a radar using an ESA can search a volume, and interleaved with the search function it can track one or more targets. This is typical operation for a weapons locating radar such as the U.S. Firefinder system, which is designed to search a volume just above the horizon and to track detected artillery and mortar rounds. Legacy Naval surface ship-tracking systems were designed to track only one target, so if multiple targets had to be tracked, a separate tracking radar would be necessary for each target. These systems are currently still in service; however, the next generation of surface ships will have a phased array multifunction radar system, which will reduce the number of radars on the superstructure.

3.2.1 Search Volume

A search radar is designed to search a solid angle volume within a given time. The volume may be as small as a few degree elevation sector over a limited (e.g., 90°) azimuth sector. An example of a small search volume is the AN/TPQ-36 and AN/TPQ-37 Firefinder radars. They are designed to search just above the horizon over a 90° azimuth sector, looking for mortar and artillery rounds out to about 30 km range. If artillery rounds are detected, then the radar tracks them long enough to estimate the point of launch so that counterfire can be issued. The Firefinder is a phased array radar, so the search-and-track functions are interleaved. The search function is not interrupted for long periods of time. Another example of a limited search volume is a ship self-defense system, in which the radar is looking for incoming cruise missiles over all azimuth angles but only at elevations at or near the sea surface. The search volume for these relatively narrow elevation sectors is often referred to as a search *fence*, referring to the general shape of the search pattern.

Some search radars are designed to search a much larger volume, up to a full hemisphere (2π steradians of solid angle). This might be the case for radars searching for incoming ballistic missiles or high-flying anti-ship missiles. In this case, it is expected that the time allowed to complete a single complete search pattern would be longer than that for a Firefinder-like or ship self-defense application.

3.2.2 Total Search Time

For an ESA, the total frame search time, T_{fs}, for a given volume is determined by the number of beam-pointing positions, m, required to see the entire search volume contiguously and by the antenna dwell time, T_{ad}, required at each beam position to achieve the detection range required:

$$T_{fs} = mT_{ad} \tag{3.4}$$

The number of beam positions depends on the total solid angle to be searched, Ω, and the product of the azimuth and elevation beamwidths, θ_3 and ϕ_3. Assuming that the beam positions are contiguous and not overlapping in angle:

$$m = \frac{\Omega}{\theta_3 \phi_3} \tag{3.5}$$

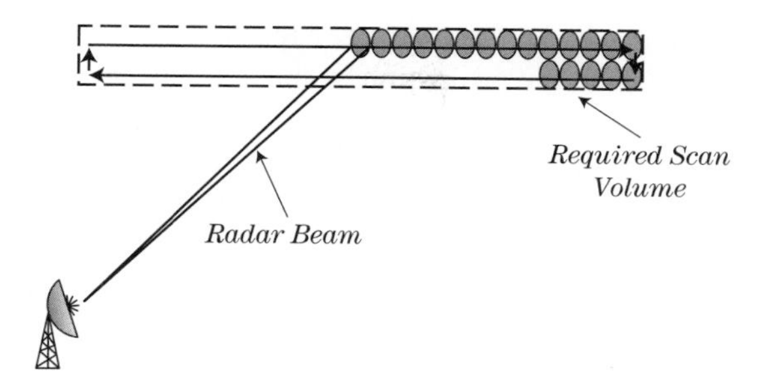

FIGURE 3-1 ■ ESA radar antenna beam scanning in the search mode.

For example, if the search volume is a sector covering 90° in azimuth and 4° in elevation and the azimuth and elevation beamwidths are both 2°, then there will be 45 beam positions in azimuth and 2 beam positions in elevation, resulting in 90 total beam positions. Figure 3-1 depicts a sequence of discrete beam positions designed to cover a specific scan volume having two rows. Note that for clarity not every beam position is depicted.

Combining equations (3.4) and (3.5), the total frame scan time, T_{fs}, is

$$T_{fs} = \frac{\Omega T_{ad}}{\theta_3 \phi_3} \tag{3.6}$$

If the antenna beam dwells at each beam position for 10 msec, then the total scan time for the 90 beam positions is 900 msec. This time ignores the time it takes for the beam to move from one position to the next, which, for a modern electronically scanned antenna, is usually short relative to the antenna dwell time. During the 10 msec antenna dwell time, there might be several CPIs. For example, if the system has a PRF of 20 kHz and 40 pulses are transmitted during a single CPI, then a single CPI, or T_d, would be 2 msec, and each beam position would have five CPIs.

In this example, antenna beams would be spaced at the –3 dB point on the beam (−6 dB round trip) which would produce a *beamshape loss*, or *scalloping loss*,[2] because the target would not necessarily be at the peak of the beam at detection. One way to reduce this loss is to reduce the beam spacing, thus increasing the number of beam positions.

For a mechanically scanned antenna, the scan rate is determined by the total angular search time and the total angle required. The antenna dwell time can be no longer than the time it takes the beam to scan past a given position and must be consistent with the detection performance in the same way as for the ESA. If, for example, the 2° beam has to be at each of the 45 beam positions for 10 msec, then the time for the beam to scan the 90 degree azimuth sector is 450 msec. The scan rate, ω, would be 90 degrees per 450 msec, or 200 degrees per second.

3.2.3 Phased Array Antenna Issues

When using an electronically scanned antenna for performing the search operation, the beamwidth and gain change with scan angle.[3] This change in beamwidth and gain must be considered in designing the search program. As the beam is scanned away from normal

[2]This effect is often called *straddle loss* in range and Doppler processing.

[3]See Chapter 9 for a more detailed discussion of these effects.

to the plane of the aperture, the antenna gain reduces because of two effects. The first is that the effective aperture (i.e., the projection of the antenna surface area in the direction of the beam scan) is reduced by the *cosine* of the scan angle. For a modest scan of, say, 10 degrees, this effect is mild. However, for a 45 degree scan angle the cosine is 0.707, resulting in a gain reduction of about 1.5 dB. For a fixed dwell time, this results in an SNR reduction of 3 dB, because the gain is squared in the radar range equation.

The second effect that reduces the antenna gain is related to each of the radiating elements in the antenna having its own beam pattern, which is at its peak broadside to the aperture but falls off with scan angle. For small scan angles this again represents a marginal reduction in gain, though for large scan angles it can be significant. This loss adds to the effective aperture loss.

The consequence of this scan loss is that the detection performance will degrade as the scan angle increases. Target detection performance is usually required to be equally good at all scan angles. If a system is designed such that it achieves the required performance at the largest scan angle, where the losses are greatest, it will perform well beyond the requirements at a small scan angle. One way to normalize performance at all scan angles is to lengthen the processing time (dwell time) at large angles and to shorten it at small scan angles to counteract the antenna gain variations. This is usually done in quantum steps. For example, a system might have a dwell time of T_d at scan angles of 0 to 15 degrees, $2T_d$ at 15 to 30 degrees, and $4T_d$ at 30 to 45 degrees. The ESA technology allows this adaptation. Mechanically scanned antennas do not experience this gain change with scan angle, so they do not have to adapt the dwell time.

In the search mode, as described in Chapter 2, the beamshape loss usually has to be accounted for since the target may be at any arbitrary angle in the beam when detection is attempted. In addition to the gain loss, the beamwidth widens as the electronically scanned beam is scanned away from normal due to the reduced effective aperture. This widening partially offsets the effect of gain loss, because the beamshape loss is reduced. The designer can take advantage of this reduced loss by decreasing the dwell time at off-normal beam positions or by increasing the antenna beam step size. Increasing the step size maintains a constant beamshape loss. In the target tracking mode, the antenna beam can be pointed directly at the target (or very close to the target), thus eliminating this beamshape loss.

3.2.4 Search Regimens

Some radar systems have to perform multiple functions nearly simultaneously. This is done by interleaving these functions at a high rate. A prime example is the case in which a radar has to continue to search a volume while also tracking targets that have previously been detected or that represent a threat. Though multifunction radars are becoming more popular, the design for such a system represents a compromise compared with a system that has to perform only a search function or only a track function.

For example, longer wavelengths are favored for search radars, and shorter wavelengths are favored for track functions, as discussed in Chapter 2. As noted earlier, available space and prime power often do not allow for more than one radar for all the required functions. A fixed ground-based air defense system might have the resources to employ multiple radars, but a mobile artillery finding radar may not. The former system may have a source of power from the national power grid, sufficient to operate both a search radar and a tracking radar. The latter system must be small and mobile and must run on a single 60 kVA generator. As another example, a surface ship-based system might have several radars onboard, but a forward-looking airborne interceptor system has space for only one radar.

With the technology improvements associated with electronically scanned antennas (phased arrays), it is now more efficient to incorporate multiple interleaved functions in a single radar system. Often, search and track functions are combined. There are two distinctly different approaches to interleaving the search and track functions in a single system: *track-while-scan* (TWS) and *search-and-track.*

3.2.4.1 Track-while-Scan

In the TWS mode, the antenna search protocol is established and never modified. When a target is detected as the beam scans in the search volume, a track file is established. The next time the antenna beam passes over this target, a new measurement is made, and its track file is updated. As an example, a 2-D air surveillance radar antenna (as described in Chapter 1) typically has a wide elevation beamwidth to provide coverage at all altitudes but a narrow azimuth beamwidth. The antenna may rotate at 10 revolutions per minute (RPM) as it searches a 360 degree azimuth sector. This means that the beam will be pointed at any given azimuth direction once every 6 seconds, and the track file for a given target will be updated with new measurements at this time interval. Since the beam scan sequence is not changed to accommodate detected targets, this technique does not require the scanning agility of an electronically scanned antenna. As the beam continues to search the volume and more targets are detected, new track files are established for these targets and are updated when the beam scans by them again. In this mode, an arbitrarily large number of targets can be tracked (1,000 or more), limited only by the computer memory and track filter throughput required. The TWS approach is used for air marshalling,[4] air traffic control, and airport surveillance applications, among others. Many of these systems use mechanically scanned antennas.

This approach provides an update rate that is adequate for benign nonthreatening targets such as commercial aircraft, but not for immediately threatening targets such as incoming missiles. An adaptive, more responsive interleaved search-and-track technique for this situation is the search-and-track mode.

3.2.4.2 Search-and-Track

In the search-and-track mode, the radar sequencer (a computer control function often called the *resource manager*) first establishes a search pattern designed to optimize the search function according to the search parameters (e.g., search volume, search frame time, prioritized sectors). If a target is detected in the search volume, then some of the radar resources are devoted to tracking this target. For instance, some percentage of the dwells each second might be assigned to target track updates. Modern search-and-track systems use an antenna beam that is electronically scanned, because the beam must be capable of being moved rapidly between arbitrary positions to optimize the tracking mode. Some legacy airborne interceptor systems used the search-and-track technique with a mechanically scanned antenna, but the time required for repositioning from the search mode to a target track position greatly extended the total search time. As a simple example, consider a radar searching a volume defined as a 90 degree azimuth sector and a 4 degree elevation

---

[4]The U.S. Navy has a requirement to monitor and control aircraft in the vicinity of an aircraft carrier. This procedure is termed *marshalling*.

sector. When a target is detected, the resource manager allocates whatever resources are required, based on, for example, target size, distance, or speed, for a high-quality track. Assume the resource manager devotes 5% of the dwells to providing measurements to the track algorithm for that target. Thus, 95% of the radar resources are left to continue the search, therefore increasing the search time somewhat. If another target is subsequently detected, another 5% of the resources may be devoted to tracking this second target, leaving 90% of the original radar dwells for searching and further increasing the search time. If 10 targets are being tracked, each using 5% of the radar resources, only 50% of the resources are left to continue the search. In this case, the search frame time will double compared with the original frame time.

This example makes it clear that in the search-and-track protocol, there is a limit to the number of targets that can be tracked simultaneously before there remain too few of the radar resources to continue an effective search. If the search frame time becomes too long, then targets of interest may get through the angular positions being searched between scans. Algorithms are developed in the system software to manage the radar resources to strike an appropriate balance between the search frame time and the number of target tracks maintained.

Some systems combine both search-and-track and track-while-scan functions. The rationale for this is as follows. The frequent updates and agility of a search-and-track system are required to accurately track targets that represent short-term threats and may have high dynamics (velocities and accelerations). Examples of such targets include incoming threats such as anti-ship missiles, fast and low radar cross section (RCS) targets, low altitude cruise missiles, and enemy artillery such as mortars and rockets. Radar systems designed to detect and track such targets are able to detect low RCS targets at long range in the presence of high RCS clutter. They also typically perform target identification functions by analyzing the target amplitude and Doppler characteristics. Such a system is likely to be a medium PRF pulse-Doppler radar and is therefore subject to the range ambiguities. That is, a long-distance low RCS target may be competing with close-in high RCS clutter.

If a target is detected and *confirmed* or *qualified* to be a true target (as opposed to a false alarm), then the target identification process is initiated. If the target is determined to be a threat rather than benign, a track file is initiated. This track initiation process consumes time for each target detected. Due to the detection range of these radar systems, target detections can occur not only for potential threat targets but also for nonthreatening targets such as friendly ground-moving vehicles and helicopters and also for birds, insects, and even turbulent air. Insects and turbulent air will be detected only at short range; however, for a range-ambiguous medium PRF system, a small target that is apparently at close range may be misinterpreted as a true threat target at a longer range. The topic of second-time-around targets, or range ambiguities, is introduced in Chapter 1 and further expanded in Chapters 12 and 17. Chapter 17 describes a technique using multiple CPIs with staggered PRFs within a given antenna dwell to resolve the range and Doppler ambiguities.

There may be many such detections in a single search scan, consuming a large fraction of the radar timeline in the target qualification process. For example, if it takes 100 msec to qualify a target and there are 100 potential target detections in a scan, then the radar will consume the next 10 seconds qualifying targets. Once the "uninteresting" targets are identified, they must still be continually tracked, or else they will be detected again on subsequent scans and have to be qualified repeatedly. However, a high-precision track is not required for these "nuisance" targets. Consequently, a track-while-scan mode can be used to maintain tracks on these targets without consuming large amounts of radar resources.

3.3 OVERVIEW OF DETECTION FUNDAMENTALS

3.3.1 Overview of the Threshold Detection Concept

The concept of *detection* of a target involves deciding, for each azimuth/elevation beam position, whether a target of interest exists in antenna beam. The technique may be as simple as an operator looking at a display and deciding if a given area of the display is "bright" enough relative to the surrounding background to be a target of interest. In most modern radar systems, detection is performed automatically in the signal/data processor. It is accomplished by establishing a threshold signal level (voltage) on the basis of the current interference (e.g., external noise, clutter, internal receiver thermal noise) voltage and then by deciding on the presence of a target by comparing the signal level in every cell with that threshold. If the signal level exceeds the threshold, then the presence of a target is declared. If the signal does not exceed the threshold, then no target is declared. This concept is shown in Figure 3-2. The detection in bin #50 may in fact be from a target, or it may be a large noise spike, creating a *false alarm*.

The detection process is performed on the received signal after whatever processing the signal experiences. It may be that a decision is made on the basis of a single transmitted pulse, though this is rare. More often, several pulses are transmitted, and the resulting received signal is integrated or processed in some way to improve the SNR compared with the single-pulse case. In any case, to detect the target signal with some reasonable probability and to reject noise, the signal must be larger than the noise. Later in this chapter the relationship between SNR and P_D will be explored.

If the interference is known to consist only of thermal noise in the receiver, then the receiver gains can be set such that the noise voltage will be at a known level. The detection threshold can then be set at a fixed voltage, far enough above that noise level to keep the probability of a false alarm (threshold crossing due to noise alone) at an acceptably low level. However, the interference is seldom this well known. In many radars the interference consists not only of receiver noise but also of clutter and noise jamming. The interference due to jamming and clutter will be quite variable as a function of range, angle, and Doppler cells in the vicinity of a target. Consequently, the interference level can vary by many dB during operation so that a fixed threshold level is not feasible. Particularly in modern radars, the threshold is often adaptive, automatically adjusting to the local interference level to

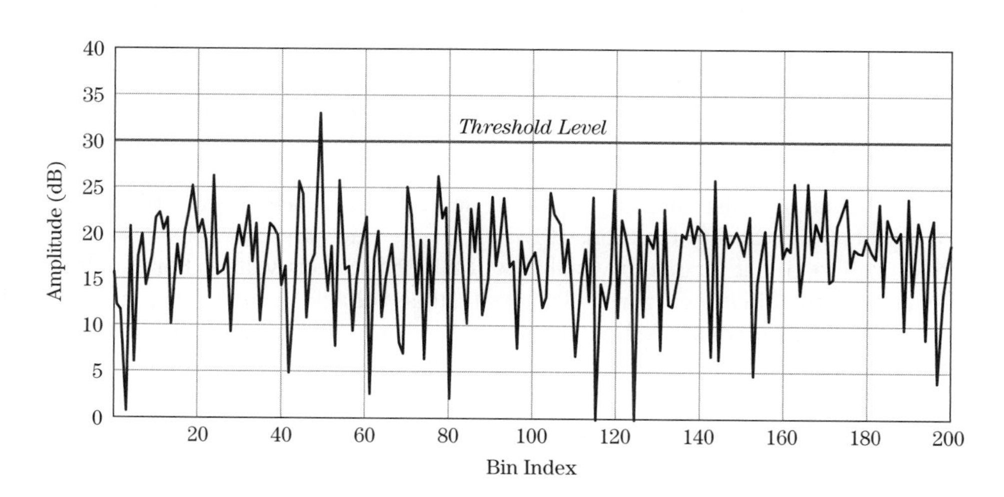

FIGURE 3-2 ■ Concept of threshold detection. In this example, a target would be declared at bin #50.

result in a constant false alarm rate (CFAR). The target signal, when present, must exceed the threshold to be detected. Details of CFAR processing are described in Chapter 16.

The description so far is of detection on the basis of amplitude alone; the amplitude of the target signal must exceed the threshold voltage determined by the interfering signal by a sufficient amount. A more severe case exists when the dominant source is clutter and especially if its amplitude exceeds that of the target signal. In this case, spectral signal processing is often employed (moving target indication [MTI] or pulse-Doppler processing) to reduce the clutter level below that of the target signal. In cases where the dominant interference is jamming and its level exceeds that of the target, often angle-of-arrival processing (e.g., sidelobe cancellation, adaptive beamforming) may be used. Systems suffering significant clutter and jamming interference may use a combination of both, called space-time adaptive processing (STAP). The detection process is performed on the output of such processors. General signal processing techniques are described in Chapter 14, and Doppler processing is described in Chapter 17.

The radar user is usually most interested in knowing (or specifying) the probability of detecting a given target, P_D, and the probability of a false alarm, P_{FA}, caused by noise. This chapter introduces the process for developing the relationship among P_D, P_{FA}, and SNR. Curves describing these relationships are often called *receiver operating curves* or *receiver operating characteristics* (ROCs).

The intent of the remainder of this chapter is to build an understanding of the issues associated with detecting a desired target in the presence of unavoidable interfering signals. The primary emphasis will be on the simplest case of a nonfluctuating target signal and noise-like (i.e., Rayleigh-distributed) interference, though some extensions will be mentioned. Chapter 15 extends the topic of detection to include the effects of the Swerling fluctuating target models, and Chapter 16 describes the implementation and performance of systems that use CFAR techniques to set the threshold level adaptively.

3.3.2 Probabilities of False Alarm and Detection

The noise at the receiver output is a randomly varying voltage. Because of the effect of multiple scatterers constructively and destructively interfering with each other, most targets of interest also present echo voltages that vary randomly from pulse to pulse, from dwell to dwell, or from scan to scan, as described in Chapters 6 and 7. However, even if the target is modeled as a constant echo voltage, the output voltage of the receiver when a target is present is the complex (amplitude and phase) combination of the target echo and noise, so it still varies randomly. Therefore, the process of detecting the presence of a target on the basis of the signal voltage is a statistical process, with a probability of detection, P_D, usually less than unity, and some probability of false alarm, P_{FA}, usually greater than zero.

The fluctuating noise and target-plus-noise voltages, v, are characterized in terms of their probability density functions (PDFs), $p_v(v)$, which are functions describing the relative likelihood that a random variable will take on various values. For example, the Gaussian or normal PDF in Figure 3-3 shows that the voltage, v, can take on positive and negative values with equal likelihood, can range from large negative to large positive values, but is more likely to take on values near zero than larger values. The likelihood of taking on values outside of the range of about -2 to $+2$ is quite small in this example.

Probability density functions are used to compute probabilities. For example, the probability that the random variable, v, exceeds some threshold voltage, V_t (a hypothetical

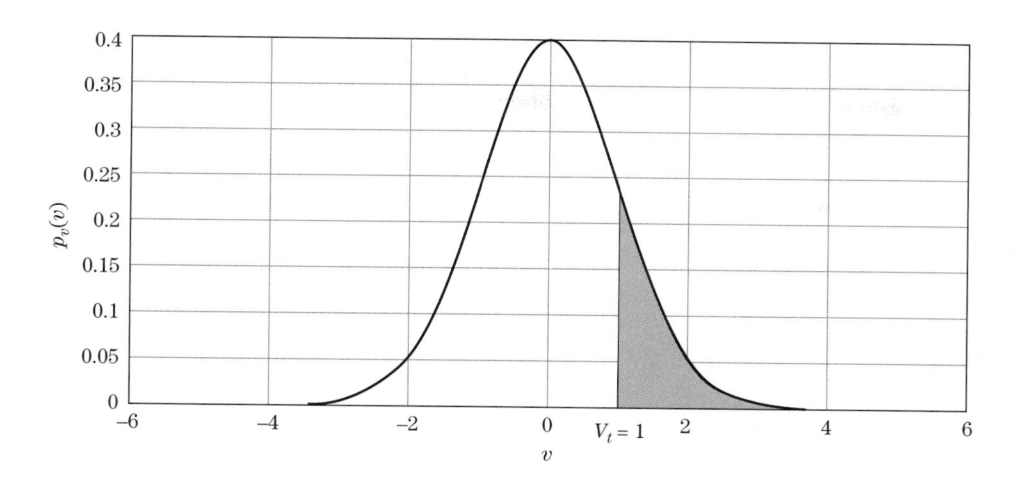

FIGURE 3-3 ■ Gaussian PDF for a voltage, v.

value of 1 for V_t is indicated in Figure 3-3), is the area under the PDF in the region where $v > V_t$, which is

$$\text{Probability}\{v > V_t\} = \int_{V_t}^{\infty} p_v(v)\, dv \tag{3.7}$$

To apply this equation to determine the P_{FA} for a given threshold voltage, suppose the PDF of the noise voltage is denoted as $p_i(v)$. A voltage threshold, V_t, is established at some level sufficiently above the mean value of the noise voltage to limit false alarms to an acceptable rate. The false alarm probability, P_{FA}, is the integral of the noise probability density function from the threshold voltage to positive infinity:

$$P_{FA} = \int_{Vt}^{\infty} p_i(v)\, dv \tag{3.8}$$

By increasing or decreasing the threshold level, V_t, P_{FA} can be decreased or increased. Thus, P_{FA} can be set at any desired level by proper choice of V_t. Determining the required threshold voltage requires solving (3.8) for V_t, given the desired P_{FA}. This process will be described in Section 3.3.3.

Once the threshold is established on the basis of the noise PDF and desired P_{FA}, the probability of detecting a target depends on the PDF of the signal-plus-noise voltage, p_{s+i}, which depends on the fluctuation statistics of both the noise and the target signals as well as on the SNR. The target detection probability, P_D, is the integral of the signal-plus-noise PDF from the threshold voltage to positive infinity:

$$P_D = \int_{Vt}^{\infty} p_{s+i}(v)\, dv \tag{3.9}$$

3.3.3 Noise PDF and False Alarms

As described already, in the absence of any target signal there is an opportunity for an interfering noise voltage to be interpreted as a target signal. A false detection of such a noise signal, caused by the noise voltage exceeding the voltage threshold, is called a false alarm. The fraction of the detection tests in which a false alarm occurs is called the probability of false alarm, P_{FA}. To analytically determine the probability of false alarm for a given noise amplitude and threshold voltage, the PDF of the noise must be known. This requires an understanding of the noise statistics and detector design.

Whereas noncoherent radar systems detect only the amplitude of the received signal, most modern radar systems are coherent and process the received signal as a vector with amplitude, v, and signal phase, ϕ. The most common detector circuit is a synchronous detector that develops the *in-phase* (I) component of the vector and the *quadrature* (Q) phase component. Even in more modern systems in which the signal is sampled at the intermediate frequency (IF) with a single analog-to-digital converter (ADC), an algorithm implemented in the system firmware constructs the I and Q components of the signal for processing. A description of the direct sampling approach is given in Chapter 11.

When the interfering signal is thermal noise, each of these signals is a normally distributed random voltage with zero mean [2]. The signal amplitude, v, is found as

$$v = \sqrt{I^2 + Q^2} \tag{3.10}$$

Equation (3.10) is referred to as a *linear detector*, and the amplitude signal, r, is often called the radar *video* signal. It can be shown that when the I and Q signals are zero mean Gaussian (i.e., normal) voltages as previously described, the resulting noise amplitude is distributed according to the Rayleigh PDF [1]. In some systems the square root function is left out of equation (3.10), resulting in a *square-law* detector. In some legacy systems, mostly noncoherent ground mapping radars, the detected signal amplitude is processed by a logarithmic amplifier. In this case the detector is called a *log detector*, and the output is sometimes called log-video. Log detectors were used to compress the wide dynamic range of the received signals to match the limited dynamic range of the display. Neither log nor square law detectors are as popular as the linear detector in modern radar systems. For this reason, the following analysis will be directed toward linear detection.

The Rayleigh PDF is

$$p_i(v) = \frac{v}{\sigma_n^2} \exp\left(\frac{-v^2}{2\sigma_n^2}\right) \tag{3.11}$$

where v is the detected envelope voltage and σ_n^2 is the mean square voltage or variance of the noise, which is simply the average noise power at the detector output. Figure 3-4 is a plot of the Rayleigh PDF when the noise power $\sigma_n^2 = 0.04$, with an arbitrary but reasonable location for a threshold voltage, V_t, indicated at $v = 0.64$. By inspection, the particular threshold voltage shown would result in a reasonably low P_{FA}; that is, it appears that a small percentage of the area under the curve is to the right of the threshold. It should be understood that, for most systems, false alarms do occur. Seldom is a threshold established that is so far above the mean noise level that no interfering signal ever exceeds the threshold.

Normally, the probability of a false alarm is low. It is typically further reduced by requiring that a target detection must occur twice: once in the initial search and subsequently in a *confirmation* or *verification* process. If there is an initial detection in the search mode, then the system is directed to immediately "look" for the target at the same range/angle/Doppler cell. If the subsequent detection is made, the detection is deemed to be a target. If not, the original detection is deemed to be a false alarm. This process greatly

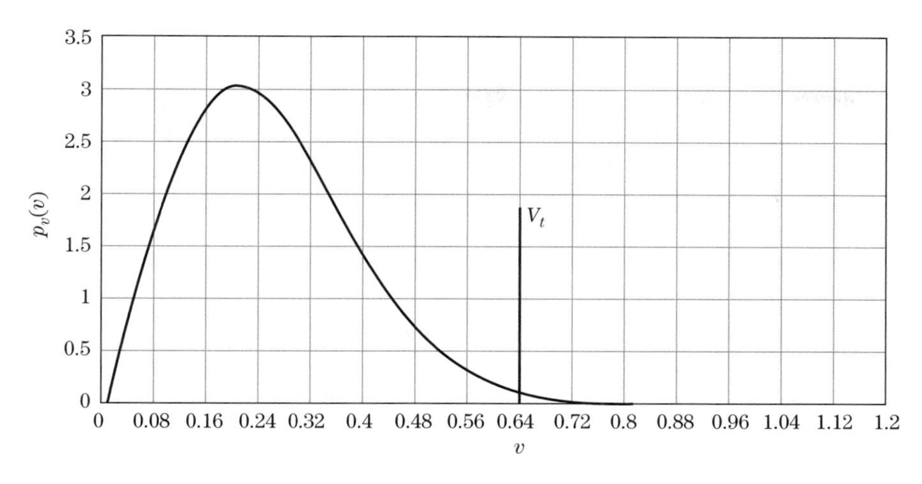

FIGURE 3-4 ■ Rayleigh distribution with an arbitrary threshold.

reduces the probability that a noise spike will initiate a track event. As an example, the system may be designed to produce, on average, one false alarm per complete antenna scan pattern.

For a single sample of Rayleigh-distributed noise, the probability of false alarm is found from equations (3.8) and (3.11):

$$P_{FA} = \int_{V_T}^{\infty} \frac{r}{\sigma_n^2} e^{-r^2/2\sigma_n^2} dr = e^{-V_T^2/2\sigma_n^2} \tag{3.12}$$

Solving for V_T provides the threshold voltage required to obtain a desired P_{FA}:

$$V_T = \sqrt{2\sigma_n^2 \ln(1/P_{FA})} \tag{3.13}$$

Thus, given knowledge of the desired P_{FA}, noise statistics, and detector design, it is possible to determine the threshold level that should be used at the detector output. Note that the target signal statistics are not involved in setting the threshold level.

As mentioned previously, search systems are designed such that if a detection is made at a given position for a given dwell, understanding that such detection may be a false alarm, a subsequent confirmation dwell at the same position is processed to see if the receiver output still exceeds the threshold. For a false alarm, the likelihood of a subsequent false detection is remote, usually settling the question of whether the detection was a false alarm or a true target. However, even this process is not perfect since there is still a small likelihood that the false alarm will persist or that a true target will not. Assume that the probability of a false alarm is independent on each of n dwells. The probability of observing a false alarm on all n dwells, $P_{FA}(n)$, is related to the single-dwell probability of false alarm, $P_{FA}(1)$, by

$$P_{FA}(n) = [P_{FA}(1)]^n \tag{3.14}$$

The likelihood that a false alarm will occur for two consecutive trials is P_{FA}^2, for three trials is P_{FA}^3, and so forth. As an example, for a single-trial P_{FA} of 10^{-4}, the two-trial P_{FA} is 10^{-8}.

Consider how the use of confirmation dwells affects the false alarm rate and search time for a hypothetical search radar. Continuing the example given earlier, suppose the radar has 90 beam positions for a given search sector, 333 range bins for each beam position, and 32 Doppler bins for each range/azimuth position. There will then be 959,040 opportunities for a false alarm in a complete search scan. For a single-dwell false alarm probability of 10^{-5}, there will be about 9.6 false alarms during the scan on average. The user normally has to determine the acceptable false alarm rate for the system, one that does not overload the signal processor or fill the display with false alarms, making it difficult to sort out the actual target detections. The use of a confirmation dwell after each regular dwell could reduce the overall P_{FA} to $(10^{-5})^2 = 10^{-10}$ so that there would be a false alarm after confirmation in only 1 in 10,000 scans, on average. Continuing with the search example given in Section 3.2.2, if each confirmation dwell (T_d or *CPI*), requires 2 ms, the same as the regular dwells, then 9.6 confirmation processes will add 19.2 ms to each nominally 900 ms search time. This is likely an acceptable expense, the alternative being to use a higher threshold voltage, which would reduce the probability of detection. However, if too many confirmation dwells are initiated, the radar search time would suffer significantly. For instance, if the P_{FA} were 10^{-4}, then there would be about 96 false alarms per scan on average. The confirmation dwells would then add about 192 msec to every 900 ms dwell, a 21.3% increase in search time. This increase would probably be considered an inefficient use of time.

3.3.4 Signal-Plus-Noise PDF: Target Detection

If, at some point during the search, the radar antenna is pointed in the direction of a target, then at the appropriate range there will be a signal resulting from the target. Of course, the noise signal is still present, so the signal in the target cell is a complex combination of target signal and the noise. This signal is the one the radar is intended to detect. Since it is composed of a combination of two signals—one a varying noise signal and the other a target signal—then there will be a variation to the combined signal. Therefore, it has a PDF that is dependent on the target signal fluctuation properties as well as the interfering signal properties. Target fluctuations are characterized in Chapter 7, and their effect on detection is described in Chapter 15. Only the simpler case of a nonfluctuating target echo is considered here to illustrate detection concepts.

For a nonfluctuating target, the PDF of the target-plus-noise signal voltage was originally shown by Rice [2] and was later discussed in [3,4] to be of the *Rician* form. This PDF is defined in equation (3.15) and plotted in Figure 3-5 along with the noise-only Rayleigh PDF for comparison.

$$p_{s+i}(v) = \frac{v}{\sigma_n^2} \exp\left[-\left(v^2 + v_{s+i}^2\right)/2\sigma_n^2\right] I_0\left(v v_{s+i}/\sigma_n^2\right) \tag{3.15}$$

where v_{s+i} is the detected signal voltage, and $I_0(\cdot)$ is the modified Bessel function of the first kind and zero order.

The procedure for determining the probability of detection given the target-plus-noise PDF is the same as for determining P_{FA} given the noise-only PDF. Using the threshold determined from the noise statistics in equation (3.13), the PDF is integrated from the threshold voltage to positive infinity:

$$P_D = \int_{V_t}^{\infty} p_{s+i}(v)dv = \int_{V_t}^{\infty} \frac{v}{\sigma_n^2} \exp\left[-\left(v^2 + v_{s+i}^2\right)/2\sigma_n^2\right] I_0\left(v v_{s+i}/\sigma_n^2\right)dv \tag{3.16}$$

This integral has no easy closed-form solution. Rather, it is defined as a new special function called *Marcum's Q function,* Q_M, discussed further in Chapter 15 and [3,5].

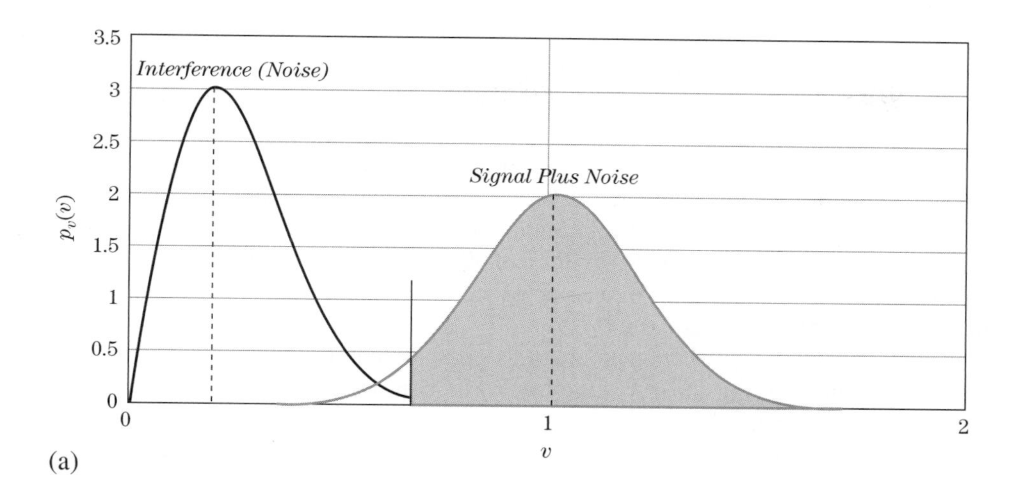

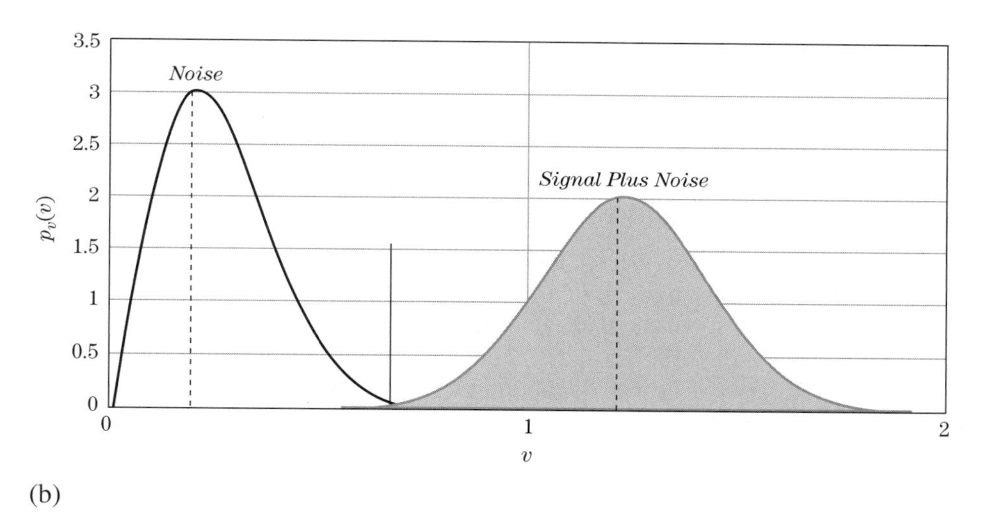

FIGURE 3-5 (a) Noise-like distribution, with target-plus-noise distribution. (b) Noise-like distribution, with target-plus-noise distribution, demonstrating the higher P_D achieved with a higher SNR.

Routines to compute Marcum's Q function are available in MATLAB and similar computational systems. In addition, various analytic approximations for the calculation of ROCs for the case of a nonfluctuating target in noise are discussed in Section 3.3.7 of this chapter.

For the case shown in Figure 3-5a, P_D will clearly be significantly greater than the P_{FA}; that is, a relatively high percentage of the area under the signal-plus-noise curve is above the threshold. This figure also again makes clear that, as the threshold is raised or lowered, both P_D and P_{FA} will be decreased (higher threshold) or increased (lower threshold).

Moving the threshold to the left (lower) or right (higher) changes the P_D and P_{FA} together for a given SNR. To get a higher P_D without increasing the P_{FA} requires that the two curves be separated more—a higher SNR is required. Figure 3-5b demonstrates that increasing the SNR while maintaining the same threshold setting (constant P_{FA}) increases the P_D. Conversely, if the SNR were to be reduced, then the curves would overlap more, and a lower P_D would result for a given threshold. This suggests that a curve of P_{FA} versus P_D for given SNR or of P_D versus SNR for a given P_{FA} would be valuable. Such curves are the topic of the next section.

By definition, the total area under a PDF curve is always unity, so P_{FA} and P_D will be less than or equal to 1. If the target signal is larger than the noise signal by a great enough margin (SNR $\gg$ 0 dB), then the threshold can be set so that P_{FA} will be nearly zero yet P_D will be nearly unity. Required values of P_D and P_{FA} are determined from higher-level system requirements and can vary greatly. However, typically P_D is specified to be in the neighborhood of 50% to 90% for a P_{FA} on the order of 10^{-4} to 10^{-6}. Clearly, if the interfering signal is of higher amplitude than the target signal (SNR $<$ 0 dB), then an unacceptably low P_D and high P_{FA} will result. This condition would suggest using some signal processing technique to reduce or cancel the noise or enhance the target signal, increasing the SNR in either case.

If the P_D and P_{FA} performance is not as good as required—that is, if the P_{FA} is too high for a required P_D or the P_D is too low for a required P_{FA}—then something must be done to better separate the noise and target-plus-noise PDFs on the plot. Either the noise has to be moved to the left (reduced in amplitude), or the target has to be moved to the right (increased in amplitude). Alternatively, the variance of the noise can be reduced, which will narrow both PDFs. These effects can be obtained only by modifying the system in some way to increase the SNR, either by changing the hardware or applying additional signal processing to adjust one of more of the terms in the RRE described in the Chapter 2. For example, the designer could increase the transmit power, the dwell time, or the antenna size (and thus gain).

3.3.5 Receiver Operating Curves

The previously shown plots depict the PDFs for noise and target-plus-noise, with an arbitrary threshold voltage plotted. The calculations already described will produce the P_D and P_{FA} for a given threshold and SNR. If the threshold is varied, a series of combinations of P_D and P_{FA} result that describe the trade-off between detection and false alarm probabilities for a given SNR. A succinct way to capture these trade-offs for a large number of radar system operating conditions is to plot one variable versus another variable for different fixed values of the third variable. The ROC is just such a curve. An example of an ROC is shown in Figure 3-6, which presents a set of curves of the SNR required to achieve a given P_D, with P_{FA} as a parameter, for a nonfluctuating target. Two alternative formats for the ROC curves are found in the literature. Some authors present plots of P_D versus SNR with

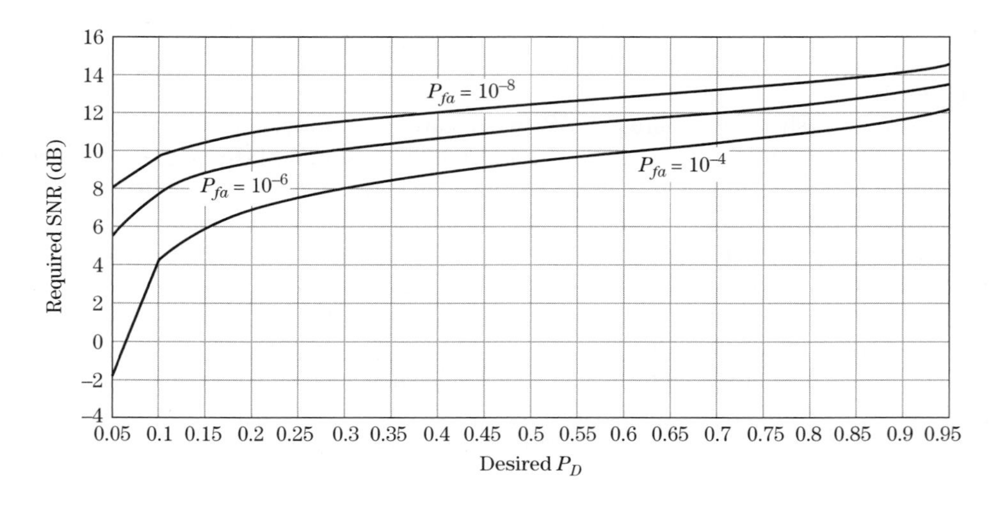

FIGURE 3-6 ■ SNR required to achieve a given P_D, for several P_{FA}'s, for a nonfluctuating (SW0) target in noise.

P_{FA} as a parameter, while others present plots of P_D versus P_{FA} with SNR as a parameter. A number of texts [e.g., 7–10] present ROC curves for a variety of conditions, most often for the case of nonfluctuating and fluctuating target signals in white noise. Results for other target fluctuation and interference-other-than-noise models are available in the literature. These results are derived using the same general strategies already described but differ in the details of the PDFs involved and thus the results obtained. Chapter 7 discusses target reflectivity statistical modeling, and Chapter 15 discusses detection performance for various target fluctuation models.

3.3.6 Fluctuating Targets

In addition to the fact that the noise signal fluctuates, it is also true that the target signal fluctuates for most real targets. Peter Swerling developed a set of four statistical models that describe four different target fluctuation conditions [8]. The four cases include two PDF models (Rayleigh and 4-th degree chi-square) and two fluctuation rates (dwell to dwell[5] and pulse to pulse.) They are labeled Swerling 1, 2, 3, and 4 (SW1, SW2, SW3, and SW4, respectively). Table 3-1 shows the PDFs and fluctuation characteristics for the four models. A nonfluctuating target is sometimes called a Swerling 0 (SW0) or a Marcum target model. These as well as other target models are discussed in detail in Chapter 7.

Table 3-2 gives some previously calculated commonly specified P_{FA} and P_D values and the required SNR in dB for the five common target models. These points were extracted from plots in [10]. For a nonfluctuating target in noise, reliable detection (90% P_D) is achieved with a reasonable (10^{-6}) P_{FA} given an SNR of about 13.2 dB. In the case of a fluctuating target signal, the PDF of the target-plus-noise is wider, leading to a lower P_D than for a nonfluctuating target at the same SNR. Thus, a higher SNR is required to achieve 90% P_D. In fact, an SNR of 17.1 to 21 dB is required for to achieve $P_D = 90\%$ at $P_{FA} = 10^{-6}$ versus 13.2 dB for the nonfluctuating case. Lower SNR values provide

TABLE 3-1 ■ Swerling Models

Probability Density Function of RCS	Fluctuation Period: Dwell-to-Dwell	Fluctuation Period: Pulse-to-Pulse
Rayleigh	Case 1	Case 2
Chi-square, degree 4	Case 3	Case 4

TABLE 3-2 ■ Required SNR for Various Target Fluctuation Models

	P_D	SW0	SW1	SW2	SW3	SW4
$P_{FA} = 10^{-4}$	50	9.2	10.8	10.5	11	9.8
	90	11.6	19.2	19	16.5	15.2
$P_{FA} = 10^{-6}$	50	11.1	12.8	12.5	11.8	11.8
	90	13.2	21	21	17.2	17.1

[5]Classically, the slowly fluctuating target was described as fluctuating from scan to scan; however, with modern system signal processing, the term *dwell to dwell* is often used.

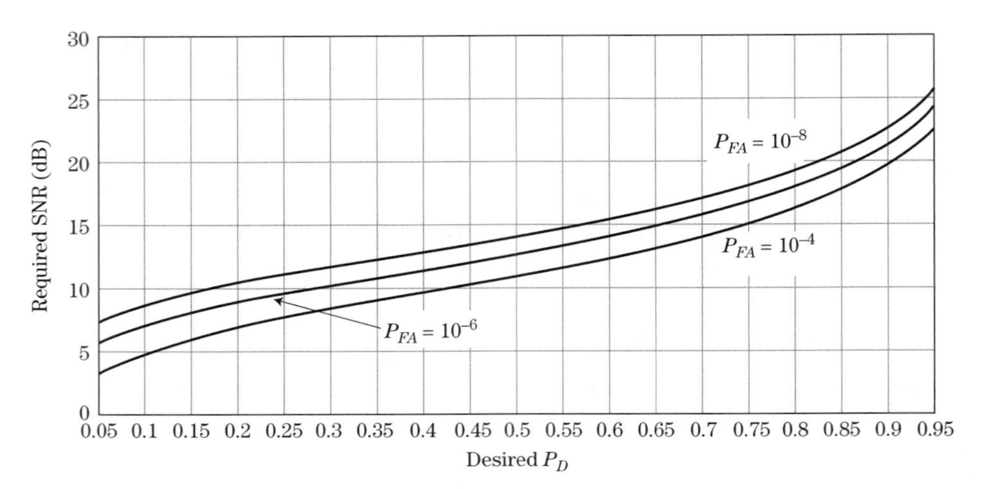

FIGURE 3-7 ■ SNR required to achieve a given P_D, for several values of P_{FA}, for fluctuating (SW1) target in noise.

lower detection probabilities, which may still be acceptable if detection can be made on the basis of several opportunities, such as over several antenna scans or several consecutive dwell periods. The system designer thus can trade radar sensitivity versus observation time in the overall system design. Figure 3-7 is an example of the ROC curves for the Swerling 1 case. This can be compared with Figure 3-6, which present the same data for the nonfluctuating target case. Chapter 7 describes the statistical nature of most target signals, showing that most targets of interest are usually fluctuating rather than of fixed amplitude. Details of the calculation of P_D and P_{FA} and corresponding ROC curves for the fluctuating target cases are given there.

Because the PDF of a fluctuating target has a higher variance (i.e., is "spread" more) than that of a nonfluctuating target, for a given SNR the target and noise PDF curves will overlap more. Therefore, for a given threshold setting, the P_D will be lower for the fluctuating target. Figure 3-8 compares the data of Figures 3-6 and 3-7 to show the SNR as a function of P_D for an SW0 (nonfluctuating) target and an SW1 target. Notice that for levels of P_D, from 50% to 95%, the SNR required for a fluctuating target RCS is significantly higher than that for a nonfluctuating target.

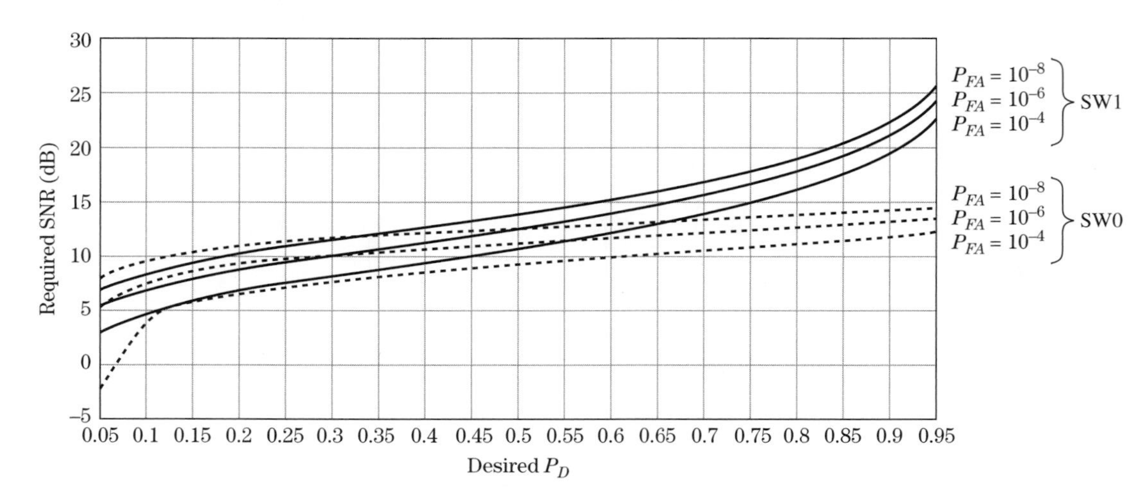

FIGURE 3-8 ■ SNR required to achieve a given P_D, several values of P_{FA}, for nonfluctuating (SW0) and fluctuating (SW1) target models.

3.3.7 Interference Other than Noise

The curves that have been plotted and presented in many standard radar texts give the P_D and P_{FA} for various values of SNR, where the interfering signal is Rayleigh-distributed after the linear detector. Thermal receiver noise in the radar and noise jamming usually satisfy this condition. Clutter interference, on the other hand, usually does not. The PDFs for clutter are proposed in many radar texts and journal articles. Some common models are Weibull, log-normal, and K-distributed [11–13]. Chapter 6 provides a summary of the clutter statistics normally encountered in modern radars.

It is not the intent of this chapter to fully describe these models; however, it is the intent to build an appreciation for the effects of these fluctuation models on target detection statistics. In general, the nature of the PDFs for distributions associated with clutter is such that they have longer "tails" in the PDFs. The effect is to increase the P_{FA} for a given threshold setting. That is, for a given mean value, there is a higher probability that the signal reaches higher amplitudes than that of noise so, compared with noise, more of the area under the curve is to the right of a given threshold. If a given P_{FA} is to be maintained if the clutter signal increases, the threshold must be increased to maintain the P_{FA} at the desired value. Doing so will lower the P_D for a given signal-to-interference ratio because less of the area under the target curve will be to the right of the higher threshold. In Figure 3-9, a hypothetical but typical clutter PDF (Weibull) is plotted in addition to the Rayleigh distribution to demonstrate the effect of the longer tail in the clutter distribution. The clutter distribution will produce a higher P_{FA} for the threshold setting shown. Therefore, the threshold will have to increase (move to the right) to recover the desired P_{FA}. As depicted in the figure, if the same false alarm probability is desired due to Weibull-distributed clutter as for Rayleigh-distributed noise, the threshold must increase, lowering the P_D for a given signal voltage. In fact, depending on the particular clutter encountered, the effect of these extended tails can be severe. In some cases, to maintain a given P_{FA} and P_D, the signal-to-clutter ratio must be 10 or even 20 dB higher than if the interference is noise-like. It is in these conditions that some method for reducing the clutter signal must be employed. Moving target indication and pulse Doppler processing techniques are the most common of these. These techniques, which are discussed in Chapter 17, reduce the clutter signal significantly.

The clutter signal is often significantly larger in amplitude than the target signal. This is because a clutter cell illuminated by the radar can result in a very large clutter area

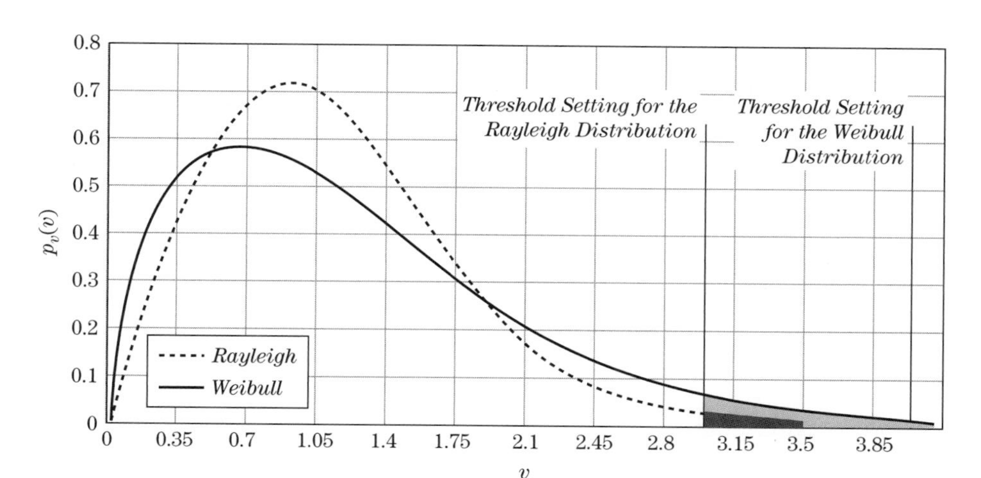

FIGURE 3-9 ■ Example clutter PDF compared with noise.

which, when multiplied by the average reflectivity of the clutter in the cell, produces a large effective RCS. The target may be quite small and, for stealth targets, smaller yet. In this case, the target signal must be separated from the clutter signal in the spectral domain, usually employing pulse-Doppler FFT processing. Since it is impractical to completely confine the clutter signal to only a few Doppler bins, there will still be some residual clutter signal persisting in the vicinity of the target signal. Thus, the question is what is the shape of the residual clutter distribution curve after the processing? The central limit theorem would suggest that since 20 or 30 samples are integrated, the new distribution might be Gaussian. This requires that the individual interference samples are statistically independent. Are the samples independent? This depends on the PRF, the decorrelation time of the clutter, and the dwell time. For most cases, the samples are not independent, suggesting that the residual clutter signal after the processor is the same shape as the original distribution (e.g., Weibull).

3.3.8 Some Closed-Form Solutions

Exact calculation of the probability of detection requires solving the integral in equation (3.16) or, equivalently, evaluation of the Marcum's Q function. While this is relatively easy using modern analysis tools such as MATLAB, it is valuable to have a simple, closed-form solution for the mathematical procedures previously described that can be solved using a spreadsheet or even a calculator. Fortunately, excellent approximations using simple formulae are available. These approximations are applicable for estimates of SNR with precision on the order of 0.5 dB and if extreme values of P_D and P_{FA} are not desired. In this chapter, Albersheim's approximation for the detection performance in the nonfluctuating target RCS case is presented. Also given are the exact results for a Swerling 1 target model when only a single sample is used for detection. The cases summarized here provide examples of detection performance and demonstrate the advantage of multiple-dwell detection techniques.

3.3.8.1 Approximate Detection Results for a Nonfluctuating Target

Albersheim's equation [14,15] is an empirically derived equation relating P_D, P_{FA}, the number of pulses noncoherently integrated, N, and the single-pulse SNR. It applies to Swerling 0 (nonfluctuating) targets and a linear detector, though it also provides good

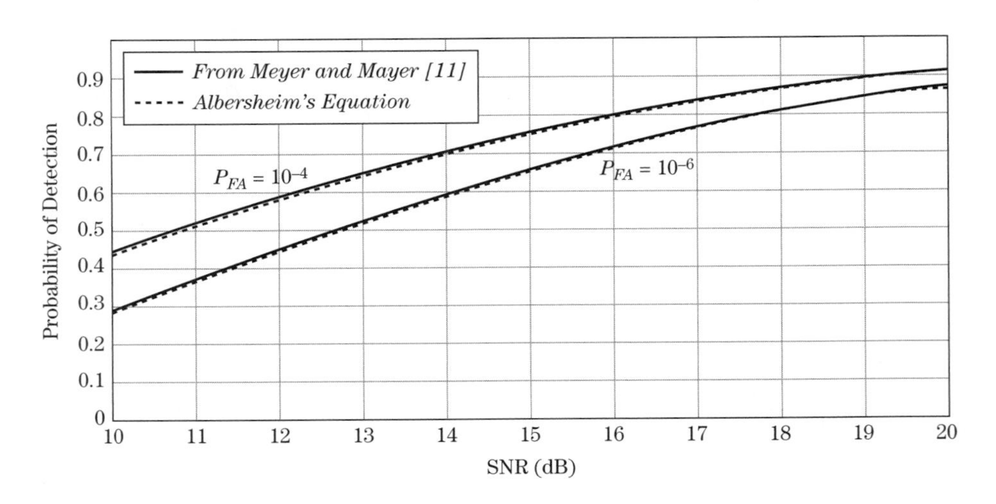

FIGURE 3-10 ■ SNR versus P_D for a Swerling 0 target using Albersheim's equation, plotted with tabulated results from Mayer and Meyer [10].

estimates for a square law detector. This is a very good approximation for values of P_D between 0.1 and 0.9 and of P_{FA} between 10^{-7} and 10^{-3}. Figure 3-10 shows ROCs computed using Albersheim's equation for P_{FA} values of 10^{-4} and 10^{-6}. The figure also shows the exact results from [10]. Note the excellent agreement between the approximated and exact results for these cases.

Albersheim's estimate of the required SNR (in dB) to achieve a given P_D and P_{FA} when N independent samples are noncoherently integrated is

$$SNR\ (\text{dB}) = -5\log_{10} N + \left(6.2 + \frac{4.54}{\sqrt{N+0.44}}\right)\log_{10}(A + 0.12AB + 1.7B) \quad (3.17)$$

where

$$A = \ln(0.62/P_{FA}) \quad (3.18)$$

and

$$B = \ln\left(\frac{P_D}{1-P_D}\right) \quad (3.19)$$

It is also possible to rearrange Albersheim's equation to solve for P_D given P_{FA}, N, and the single-pulse SNR. Details are given in Chapter 15 and [17].

Though Albersheim's equation provides simple, closed-form method for calculation, it applies only to a nonfluctuating target, which is seldom a good model in practice. In Chapter 15, an approximation similar in spirit to Albersheim's equation, but applicable to all of the Swerling models, is presented [16].

3.3.8.2 Swerling 1 Target Model

The Swerling models for describing the statistics of target fluctuations were described in Section 3.3.6. For the purpose of providing an example of an ROC for fluctuating targets, a Swerling 1 target model with a single-echo sample (no noncoherent integration of multiple samples) will be considered here. Chapter 7 presents the analysis showing that a Swerling 1 target is appropriate for targets composed of multiple scatterers of roughly the same RCS. The resulting voltage PDF for this target is a Rayleigh distribution. In fact, this is the same distribution used to describe receiver noise. When combined with the Rayleigh noise distribution, the target-plus-noise PDF is still a Rayleigh distribution [19]. Specifically, the target-plus-noise PDF is of the form

$$p_{s+i}(v) = \frac{v}{S+\sigma_n^2}\exp\left(\frac{-v^2}{2(S+\sigma_n^2)}\right) \quad (3.20)$$

where S is the mean target echo power, and σ_n^2 is the mean noise power. The probability of detection integral is

$$P_D = \int_{V_t}^{\infty} p_{s+i}(v)\,dv = \int_{V_t}^{\infty} \frac{v}{S+\sigma_n^2}\exp\left[-\frac{v^2}{2(S+\sigma_n^2)}\right]dv \quad (3.21)$$

Computing this integral gives the probability of detection as

$$P_D = \exp\left[\frac{-V_t}{1+SNR}\right] \quad (3.22)$$

where $SNR = S/\sigma_n^2$. In (3.22), V_t and SNR are linear (not dB) values. Using (3.13) it can be shown that the relationship between P_D and P_{FA} for this case is the simple relationship

$$P_D = (P_{FA})^{1/(1+SNR)} \tag{3.23}$$

Equation (3.23) applies only to the case of a Swerling 1 target in white noise, with detection based on only a single sample. This "single sample" may be developed by coherently integrating multiple pulses or the results of multiple CPIs. However, it can not include any noncoherent integration.

3.3.9 Multiple-Dwell Detection Principles: Cumulative P_D

To determine the dwell time required at each beam position, it is necessary to determine the signal-to-noise ratio required to achieve the desired P_D and P_{FA} for a target at maximum range. As an example, for a P_D of 90%, a P_{FA} of 10^{-6}, and a Swerling 2 target, the SNR required is 21 dB. Given the available average power, antenna gain, wavelength, minimum target RCS, receiver noise figure, system losses, and maximum range, equation (2.30) can be used to determine the dwell time to achieve the required SNR of 21 dB.

At first, it might seem that a P_D of 90% is not sufficient to detect a threat with adequate certainty. In fact, radar systems often combine the results from multiple opportunities to detect a target, improving the detection probability. The probability of detecting the target at least once in n dwells, $P_D(n)$, is higher than the probability of detection for a single dwell, $P_D(1)$. For example, if the probability of detection on a single dwell is $P_D(1)$, the probability of detecting the target at least once in n tries is

$$P_D(n) = 1 - [1 - P_D(1)]^n \tag{3.24}$$

provided that the detection results on each individual dwell are statistically independent. If $P_D(1) = 90\%$, then the cumulative probability for two tries will be $P_D(2) = 99\%$ and for three tries will be $P_D(3) = 99.9\%$.

The use of multiple dwells to improve detection probability presents an opportunity for a trade-off of radar detection performance for time. For example, if a 99% P_D was required on a *single* dwell for the previous Swerling 2 example, the SNR would have to be about 30 dB, about 9 dB higher than that required for 90% P_D. Using a second dwell with the 90% P_D doubles the time required to detect the target but saves about 9 dB of radar SNR requirements. This translates into some combination of reduced transmit power, reduced antenna gain, or reduced signal processing gain requirements.

Of course, the false alarm probability is also affected by the use of multiple dwells. It is normally not desirable to have P_{FA} increase due to the multiple dwells, as P_D did. The equation for cumulative P_{FA} is the same as equation (3.24), with P_{FA} substituted for P_D. Because the single-dwell false alarm probability is usually a very small number, the result is well approximated by the simpler equation

$$P_{FA}(n) \approx n \cdot P_{FA}(1) \tag{3.25}$$

Thus, for a three-look cumulative $P_{FA}(3)$ of 10^{-6}, $P_{FA}(1)$ for a single dwell would have to be 0.333×10^{-6}. This modest reduction in the allowable single-dwell P_{FA} will increase the required SNR but will be only a small amount compared with the additional SNR required to achieve a P_D of 99.9% in a single look.

3.3.10 *m*-of-*n* Detection Criterion

Instead of detecting a target on the basis of at least one detection in n tries, system designers often require that some number m or more detections be required in n tries before a target detection is accepted. If m and n are properly chosen, this rule has the effect of both significantly reducing the P_{FA} and increasing the P_D compared with the single-dwell case. The probability of a threshold crossing on at least m-of-n tries is found from the binomial theorem [1,18,19]:

$$P(m,n) = \sum_{k=m}^{n} \frac{n!}{k!(n-k)!} P^k (1-P)^{n-k} \tag{3.26}$$

Here, P is the probability of a threshold crossing on a single trial. Equation (3.26) applies both to false alarms as well as detections. If $P = P_{FA}(1)$, then it gives the cumulative false alarm probability for an m-of-n test; if $P = P_D(1)$, then it gives the cumulative detection probability for an m-of-n test. For a commonly used 2 of 3 rule ($m = 2$, $n = 3$), (3.26) reduces to

$$P(2,3) = 3P^2 - 2P^3 \tag{3.27}$$

and for 2 of 4 it becomes

$$P(2,4) = 6P^2 - 8P^3 + 3P^4 \tag{3.28}$$

Figure 3-11 is a plot of the probability of detection or false alarm versus single-dwell P_D or P_{FA}. Note that for typical single-dwell probabilities of detection (i.e., >50%) the cumulative P_D improves (increases), and for low probabilities of false alarm (i.e., <0.2) the cumulative P_{FA} also improves (is decreased). The only "cost" of this improvement is an extended antenna dwell time for each beam position to collect the required data and to conduct n detection tests instead of just one.

To get a better idea of the specific value of the effect of m-of-n detection rules, Table 3-3 presents some specific examples of the effect on P_D, and Table 3-4 presents examples of the effect on P_{FA}. A single-dwell P_D of 0.9 and P_{FA} of 10^{-3} provides 0.996 P_D and $5.92 \times 10^{-4} P_{FA}$ for 2 of 4 scans.

It is sometimes the case that the multiple dwells are not collected on a single scan but rather on successive scans. While this adds significant delay to the data collection

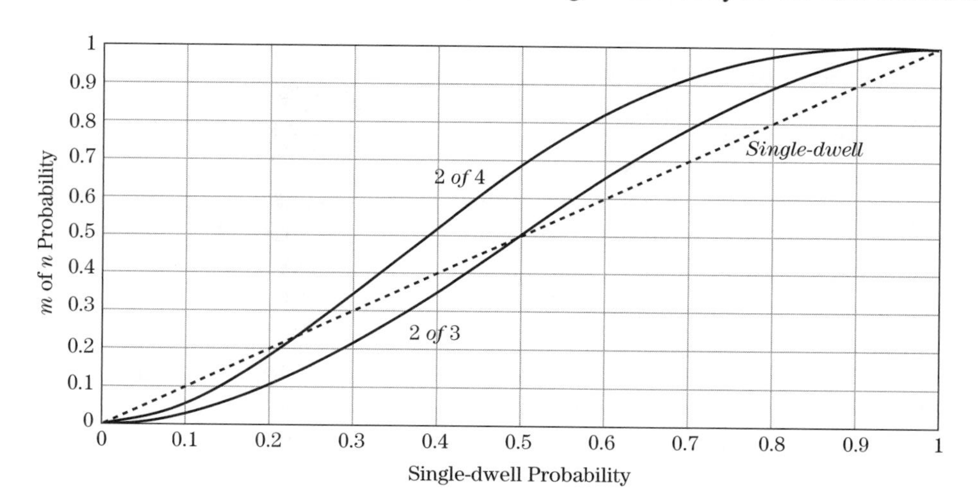

FIGURE 3-11 ■ 2-of-3 and 2-of-4 probability of threshold crossing versus single-dwell probability.

TABLE 3-3 ■ m-of-n Probability of Detection Compared with Single-Dwell Probability of Detection

$P_D(1)$	$P_D(2,3)$	$P_D(2,4)$
1	1	1
0.95	0.993	0.999
0.90	0.972	0.996
0.85	0.939	0.988
0.80	0.896	0.973
0.75	0.844	0.949
0.70	0.784	0.916

TABLE 3-4 ■ m-of-n Probability of False Alarm Compared with Single-Dwell Probability of False Alarm

$P_{FA}(1)$	$P_{FA}(2,3)$	$P_{FA}(2,4)$
1	1	1
0.1	2.8×10^{-2}	5.23×10^{-2}
0.01	2.98×10^{-4}	5.92×10^{-4}
0.001	2.998×10^{-6}	5.992×10^{-6}
0.0001	2.9998×10^{-8}	5.9992×10^{-8}
0.00001	3.000×10^{-10}	5.9999×10^{-10}

protocol and therefore the time to make a detection decision, for many applications this extra latency in the detection process is acceptable.

Some system applications require even larger numbers of dwells at a given beam position. For example, an airborne pulse Doppler radar typically performs the search function in a high PRF mode to separate moving targets from wide clutter spectral characteristics. Consequently, the radar is highly ambiguous in range, and there is a high likelihood of range eclipsing.[6] Even in the medium PRF mode, there is likely to be range and Doppler aliasing and eclipsing.[7] To improve the P_D and P_{FA} statistics in these conditions, often six or eight dwells are used. Figure 3-12 shows the results of using equation (3.26) to compute the resulting P_D and P_{FA} for 3-of-6 and 3-of-8 detection rules. A single-dwell P_D of 90% results in a processed P_D of very close to 100% for these conditions. In this application, the m-of-n rule is often combined with the use of *staggered PRFs*, discussed in Chapter 17.

The question arises as to whether it is more efficient to use a relatively large number of dwells, which costs significant time, or to increase the single-dwell time. To examine this, consider the following example. Assume a P_D of 95% and a P_{FA} of 10^{-6} is required for each complete scan. For a Swerling 1 target, the single-dwell SNR must be 24.3 dB. Using Figure 3-12, it is seen that, using 3-of-6 processing, a cumulative $P_D = 95\%$ and P_{FA} of 10^{-6} can be achieved if the single-dwell P_D is about 73% and the single dwell P_{FA}

[6] In a pulsed radar, any target that has a range delay exactly equivalent to a multiple of the interpulse period will not be detected because the target echo arrives during a subsequent transmit time. This condition is called *eclipsing*. It can be overcome by changing the pulse interval.

[7] In a coherent radar, the target can be eclipsed in the frequency (Doppler) domain, much like the time domain case already described.

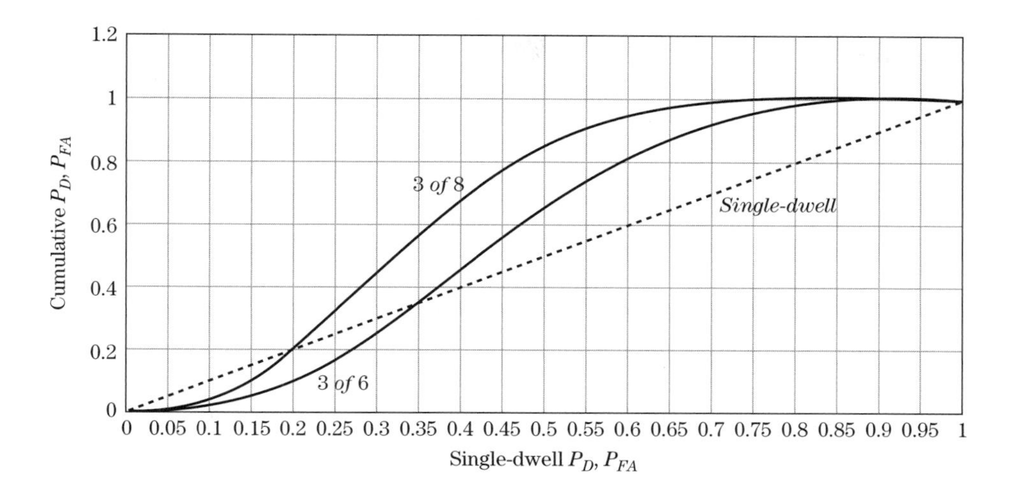

FIGURE 3-12 ■ 3-of-6 and 3-of-8 probability of threshold crossing versus single-dwell probability.

is about 0.001. Using equation (3.23), it is seen that this performance can be obtained with a single-dwell SNR of only 13.2 dB.

This reduction of the required SNR by 11.1 dB (a factor of about 10.4) means that the radar can be "smaller" in some sense by a factor of 10.4. This could take the form of a reduction in average power, antenna gain, single dwell time, or relaxed restrictions on losses. For instance, although six dwells are now required, each one could be shorter by a factor of 10.4, reducing the overall time required to meet the detection specification. Taken a step further, 99% P_D can be achieved with eight dwells and is equivalent to nearly 20 dB of additional radar sensitivity. Extending the time line by a factor of 8 can thus reduce the radar "size" by a factor of 100.

3.4 FURTHER READING

More rigorous mathematical developments of the expressions for P_D and P_{FA}, as well as the general approach of threshold detection, are found in several radar texts [e.g., 3,6,10,20]. These references also extend the results discussed here to the effect of noncoherent integration of multiple samples prior to threshold testing.

Within this volume, target fluctuation models are discussed in more detail in Chapter 7. The application of the detection concepts discussed here to fluctuating target models is discussed in Chapter 15, which also provides both the corresponding exact expressions for P_D and P_{FA} for all of the Swerling models as well as an Albersheim-like approximation.

3.5 REFERENCES

[1] Barton, D.K., *Radar System Analysis*, Artech House, Dedham MA, 1976, Section 1.2.

[2] Rice, S.O., "Mathematical Analysis of Random Noise," *Bell System Technical Journal,* vol. 23, no. 3, July 1944, pp. 282–332; vol. **24**, no. 1, January 1945, pp. 46–156.

[3] DiFranco, J.V., and Rubin, W.L., *Radar Detection*, Artech House, Dedham, MA, 1980, pp. 306–309.

[4] Long, M.W., *Airborne Early Warning System Concepts*, Artech House, Dedham MA, 1992, pp. 498–499.

[5] Farina, A., *Radar Handbook*, 3d ed., *Electronic Counter-Countermeasures*, McGraw-Hill, New York, 2008, Ch. 24.

[6] Barton, D.K., *Modern Radar System Analysis*, Artech House, Norwood, MA, 1988, Section 2.2.

[7] Blake, L.V., *Radar Handbook*, 2d ed., *Prediction of Radar Range,* McGraw-Hill, New York, 1990, Ch. 2.

[8] Skolnik, M.I., *Introduction to Radar Systems*, 3d ed., McGraw Hill, New York, 2001, p. 66.

[9] Nathanson, F.E., *Radar Design Principles*, 2d ed., McGraw-Hill, New York, 1991, pp. 54, 83–86, 89–92.

[10] Meyer, D.P., and Mayer, H.A., *Radar Target Detection, Handbook of Theory and Practice*, Academic Press, New York, 1973.

[11] Sangston, K.J., and Gerlach, K.R., "Coherent Detection of Radar Targets in a Non-Gaussian Background," *IEEE Transactions on Aerospace and Electronic Systems*, vol. 30, no. 2, pp. 330–340, April 1994.

[12] Schleher, D.C., "Radar Detection in Log-Normal Clutter," *Proceedings of the IEEE International Radar Conference,* pp. 262–267, 1975.

[13] Gini.F., Greco, M.V., and Farina, A., "Clairvoyant and Adaptive Signal Detection in Non-Gaussian Clutter: A Data-Dependent Threshold Interpretation," *IEEE Transactions on Signal Processing*, vol. 47, no. 6, pp. 1522–1531, June 1999.

[14] Albersheim, W.J., "Closed-Form Approximation to Robertson's Detection Characteristics," *Proceedings of the IEEE*, vol. 69, no. 7, p. 839, July 1981.

[15] Tufts, D.W., and Cann, A.J., "On Albersheim's Detection Equation," *IEEE Transactions on Aerospace and Electronic Systems*, vol. AES-19, no. 4, pp. 643–646, July 1983.

[16] Shnidman, D.A., "Determination of Required SNR Values," *IEEE Transactions on Aerospace & Electronic Systems*, vol. AES38, no. 3, pp. 1059–1064, July 2002.

[17] Richards, M.A., *Fundamentals of Radar Signal Processing*, McGraw-Hill, New York, 2005.

[18] Bulmer, M.G., *Principles of Statistics*, Dover Publications, New York, New York, 1967, p. 84.

[19] Barton, D.K., Cook, C.E., and Hamilton, P., *Radar Evaluation Handbook,* Artech House, Dedham, MA, 1991, pp. 4–17.

[20] Minkler, G., and Minkler, J., *CFAR, The Principles of Automatic Radar Detection in Clutter,* Magellan Book Co., Baltimore, MD, 1990.

3.6 PROBLEMS

1. For a mechanically scanned antenna having an azimuth beamwidth of 2 degrees and an elevation beamwidth of 3 degrees, how many beam positions are required to search a volume defined by a 90 degree azimuth sector and a 6 degree elevation sector?
2. If the antenna were raster scanning at 180 degrees per second (i.e., it is scanning in azimuth at 180 degrees per second), what is the maximum dwell time for each beam position? Assume it takes no time to change to a new elevation position and azimuth scanning direction.
3. For the previous conditions, what is the total frame search time?

4. For a phased array antenna, suppose the beamwidth at array normal is 2 degrees by 3 degrees (same as in problem 1). Also suppose that the dwell time used for scan angles between 0 and ± 30 degree is 4 msec, while for scan angles between ± 30 and ± 45 degrees it is 6 msec. If the beam position is stepped in equal step sizes, what is the total search frame time? Assume it takes negligible time to move from one position to the next.
5. What is the P_{FA} if the threshold voltage is set at three times the root mean square (rms) noise voltage?
6. What threshold voltage is required to effect a P_{FA} of 10^{-4} if the rms noise voltage is 150 mv?
7. Assuming $P_{FA} = 10^{-4}$, how many consecutive verifications are required to effect a P_{FA} of 10^{-9} or less?
8. For a search volume that requires: 45 beam positions, 333 range bins, 32 Doppler bins, and a P_{FA} of 5×10^{-5}, how many false alarms occur on average in a single search frame?
9. Consider a weapon locating radar having a beamwidth of 2 degrees in both azimuth and elevation that is set up to search a volume defined by a 75 degree sector in azimuth and a 4 degree sector in elevation. If the radar also has a dwell time of 2.4 msec and a plan to spend 5 dwells at each beam location, what is the total scan time?
10. If there are eight targets being tracked by the system in problem 9, each consuming 6 milliseconds per track update at an update rate of 10 Hz (10 updates per second), what is the new search scan time?
11. Your enemy, 20 km distant, fires a mortar round in your general direction. The round has a vertical component of velocity of 200 meters per second. (Assume that this does not change during the search time.) You are searching the area using a weapon-locating radar. What must your maximum scan time be to ensure at least four opportunities to detect the target before it passes through your "search fence," which is the elevation sector extending from 0 to 4 degrees above the horizon?
12. If the round in problem 11 is detected at 2 degrees above the horizon, what must the track sample rate be to get 50 track samples from between the point of detection and an elevation of 6 degrees above the horizon?
13. Given a radar system that has a single-dwell P_D of 50% and a single-dwell P_{FA} of 5×10^{-3}, what are the cumulative P_D and P_{FA} for 2-of-3 and 2-of-4 multiple-dwell processes?
14. For a radar system that has a single-dwell P_D of 75% and a single-dwell P_{FA} of 5×10^{-3}, what are the P_D and P_{FA} for 2-of-3 and 2-of-4 multiple-dwell processes?
15. Assuming that the target exhibits Swerling 1 fluctuations and that the P_D and P_{FA} that result from the multidwell (2-of-4) processing in problem 14 are adequate, what SNR improvement is required to provide the same P_D and P_{FA} in a single dwell?
16. Suppose a phased array search radar has to complete searching a volume defined by 10 degrees in azimuth, 10 degrees in elevation, and 40 km in range in 0.62 seconds. The range resolution is 150 meters, obtained with a simple 1 microsecond pulse width. At the center of the search sector, the antenna has a 2.7 degree azimuth beamwidth and a 2.7 degree elevation beamwidth. Since the target for which the system is searching is a moving target and there is surface clutter interfering with the detection, Doppler processing is used. There are 64 Doppler bins developed by the FFT processor. During a single search pattern, it is required that the probability of detecting a target is 99%, and on average one false alarm is allowed. What is the resulting P_D and P_{FA} if the single-dwell P_D is 90% and the single-dwell P_{FA} is 0.01? If 3-of-5 processing is employed, do the resulting P_D and P_{FA} meet the requirements?

PART II

External Factors

CHAPTER 4 Propagation Effects and Mechanisms

CHAPTER 5 Characteristics of Clutter

CHAPTER 6 Target Reflectivity

CHAPTER 7 Target Fluctuation Models

CHAPTER 8 Doppler Phenomenology and Data Acquisition

CHAPTER 4

Propagation Effects and Mechanisms

Jay A. Saffold

Chapter Outline

4.1 Introduction 117
4.2 Propagation Factor 118
4.3 Propagation Paths and Regions 119
4.4 Atmospheric Attenuation and Absorption 121
4.5 Atmospheric Refraction 130
4.6 Turbulence 137
4.7 Exploiting the Ionosphere 138
4.8 Diffraction 140
4.9 Multipath 142
4.10 Skin Depth and Penetration: Transmitting Through Walls 156
4.11 Commercial Simulations 158
4.12 Summary and Further Reading 160
4.13 References 161
4.14 Problems 163

4.1 INTRODUCTION

This chapter discusses the effects of the transmission medium on the propagation of electromagnetic waves traveling from a radar transmitter to a target and back to the receiver. Many factors can influence the propagating radar signal, such as the composition of the atmosphere, clouds, rain, insects, and obstacles. The wave will also be affected by the ground the wave passes over as well as other topological features, such as hills, valleys, and lakes. This chapter considers only how these topological features affect the propagating wave, while the radar return from these features, known as clutter, are discussed in the next chapter.

The effect of the propagation medium can be described as a series of mechanisms whose impact may be combined through superposition. Several of these mechanisms can exist simultaneously. Some of the major propagation mechanisms and general guidelines for assessing the primary sources of propagation impact on an application are as follows:

- *Atmospheric absorption* increases with higher frequency, longer ranges, and higher concentration of atmospheric particles (e.g., water, fog, snow, smoke).

- *Atmospheric refraction* anomalies (surface ducts), which tend to generally be less significant at higher frequencies, occur more often in conditions of high humidity, at land/sea boundaries, and at night when a thermal profile inversion exists.
- *Atmospheric volumetric scattering* increases with higher frequency, larger suspended particle sizes, and higher concentrations of atmospheric particles (e.g., water, fog, snow, smoke) and can be a strong function of wave polarization and frequency.
- *Atmospheric turbulence* is generally a high-frequency (HF) phenomenon (e.g., optical, millimeter wave [MMW], or sub-MMW) and is strongly dependent on refractive index (or temperature) variations and winds.
- *Surface diffraction* effects tend to increase with lower-frequency and higher-surface root mean square (rms) roughness specification. The effect of the earth's surface features can be roughly separated into two regions of influence: the interference region and the diffraction region (defined in Section 4.3.2)
- *Surface multipath* effects occur in the interference region and tend to increase with lower-frequency and lower-surface rms roughness specifications. Multipath forward scattering is generally confined to terrain- and target-bounced reflections in the direction of the receiver.
- *Surface intervisibility* effects (shadowing) tend to increase with higher-surface roughness specifications and lower link altitudes. Intervisibility is also affected by the presence of clutter *discretes* (specific, isolated, strong clutter scatterers) and the spherical (or effective) Earth horizon boundary limitations.

The atmospheric mechanisms of absorption, refraction, scattering, and turbulence are collectively referred to as *atmospherics*. In general, all mechanisms are present in the intermediate path for the propagating wave in the real world. In this chapter, the physics and significance of each mechanism are discussed.

4.2 PROPAGATION FACTOR

The impact of propagation mechanisms can be applied to the free-space radar range equation as a "gain" or "loss" factor. Specifically, the peak amplitude, E_0', of the one-way received electric field vector in the presence of propagation effects is related to the amplitude that would be received in free-space propagation without these effects, E_0, according to

$$\begin{aligned} E_0' &= F_v E_0 \\ &= (F e^{j\phi_F}) E_0 \end{aligned} \tag{4.1}$$

where F_v is the complex voltage *propagation factor*, F is its magnitude, and ϕ_F is its phase. F_v is composed of several component factors,

$$F_v = F_{v_1} \cdot F_{v_2} \cdot F_{v_3} \cdot \ldots \cdot F_{v_N} \tag{4.2}$$

Thus,

$$\begin{aligned} F &= F_1 \cdot F_2 \cdot F_3 \cdot \ldots \cdot F_N \\ \phi_F &= \phi_{F_1} + \phi_{F_2} + \phi_{F_3} + \ldots + \phi_{F_N} \end{aligned} \tag{4.3}$$

Since power is proportional to $|E|^2$, the propagation factor applicable to one-way received power is

$$|F_v|^2 = F^2 \tag{4.4}$$

Many texts discuss only this power form of the propagation factor.

Often the effect of propagation mechanisms on two-way propagation is needed. The fundamental assumption for two-way data is that propagation effects are path-reciprocal so that the two-way propagation factor is simply the one-way factor squared.

$$F_v^2 = F^2 e^{j2\phi_F} \tag{4.5}$$

The monostatic radar range equation of Chapter 1 then becomes

$$P_r = \frac{P_t G^2 \lambda^2 \sigma}{(4\pi)^3 L_s R^4} F^4 \tag{4.6}$$

Caution is needed in determining whether published data is based on one-way or two-way propagation. Care must also be taken not to confuse propagation factor and noise figure, since both are commonly represented by the symbol F.

4.3 PROPAGATION PATHS AND REGIONS

4.3.1 Monostatic and Bistatic Propagation

Most radio, cellular, and wireless communications links are concerned only with one-way propagation. Radar is generally concerned with two-way propagation between the transmitter and receiver. If the receiver is colocated with the transmitter (as is often the case), this is called a two-way *monostatic* case. The main difference from the communications case is that any propagation effects will be more impactive since the radar wave passes through the medium twice. Cases where the transmitter and receiver are not colocated are *bistatic*. Here, the main consideration from a propagation standpoint is that the radar wave may pass through two different propagation paths, unless the bistatic angle is relatively narrow or the target is not at a great distance. While much less common than monostatic radar, bistatic systems are not unusual. Figure 4-1 illustrates these three classes of propagation paths.

4.3.2 The Surface

Every obstacle within an expanding wave's path will impede direct propagation to an observer near the earth. Simple ray tracing illustrates the *line-of-sight* (LOS) region, also

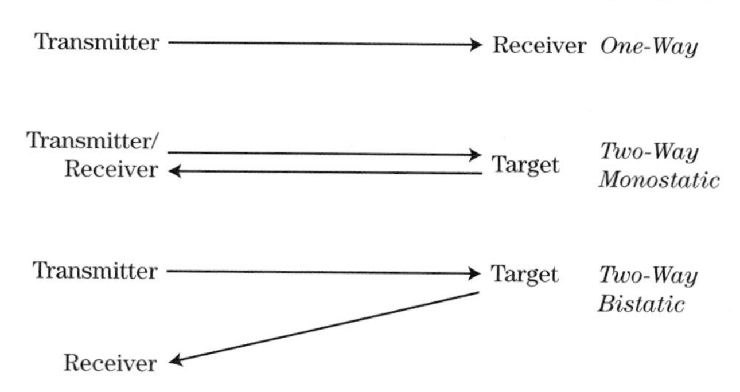

FIGURE 4-1 ■ Three classes of propagation paths.

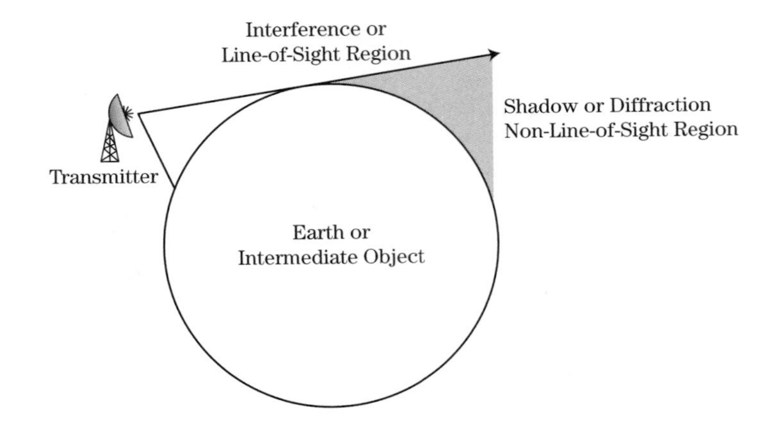

FIGURE 4-2 ■ Regions defined by the LOS propagation path near the earth's surface.

called the *interference* region, and the non-line-of-sight (*shadow* or *diffraction*) region associated with a transmitter above the surface of the earth. These regions are illustrated in Figure 4-2. A diffraction or refraction mechanism is necessary for electromagnetic (EM) waves to propagate into the shadow region.[1]

Two points in space are said to be *intervisible* with one another if there is an uninterrupted line of sight between them. In general, the rougher the surface, the higher the number of ground points that will not be visible to a given emitter (and vice versa). Clearly, local geometry and discrete location play a significant role in intervisibility estimation [1].

4.3.3 The Atmosphere

For the purposes of radar propagation analysis, the atmosphere can be characterized by a number of key radial layers. Propagation of the radiated wave is affected by the composition of these layers as well as the boundaries or interfaces between these layers according to the properties of both the wave and the boundary itself. Figure 4-3 illustrates some of the key layers associated with the atmosphere and their general altitude above sea level on the earth.

These layers can be characterized by the vertical distribution of temperature fluctuations, as illustrated in Figure 4-3, as well as the water vapor in the layer. The principal layers are the troposphere, stratosphere, mesosphere, and the thermosphere.

The *troposphere,* the region in contact with the earth, is the lowest layer of the atmosphere and extends from the earth's surface to a height of 10 to 16 km. Four fifths of the mass of the atmosphere is contained in this layer, which is characterized by a decreasing temperature as height increases in the vertical plane. The rate of decrease for this temperature, based on the 1976 U.S. Standard Atmosphere, is about 6.5 °C/km above ground level. Most weather processes (and water vapor) occur in this atmospheric layer. The *stratosphere* is the layer of the atmosphere extending above the tropopause to a height of about 55 km. Very little weather occurs in this layer due to a low water vapor content. The *mesosphere* is the atmospheric layer between the stratopause and the mesopause. Like the

[1]Of course, scattering from an LOS ray from another obstacle can cause multibounce rays to propagate into a shadow zone. This multibounce phenomenon tends to be more significant in one-way propagation (or bistatic) radar geometries.

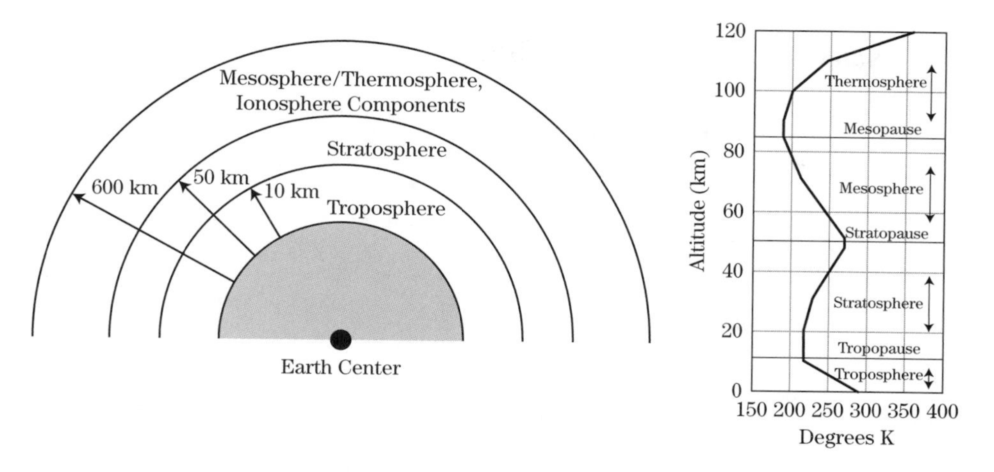

FIGURE 4-3 ■ Rays propagating through atmospheric regions. (From Eaves et al. [2] and Bogush [3]. With permission.)

troposphere, it is a layer where the temperature decreases with height. Special features of the mesosphere include a region of extremely strong winds centered near heights of 65 km.

The *thermosphere* is a high-temperature region extending upward from heights of about 80 km to the outer reaches of the atmosphere. The thermosphere contains the majority of the ionosphere, with colorful auroras often in this region. The *ionosphere* is an upper region of the earth's atmosphere in which many of the atmospheric atoms and molecules have become electrically charged by the addition or removal of electrons to produce ions. Relative to the layered regions of the atmosphere, the ionosphere begins at a base near the stratopause, rising through the mesosphere to a peak in the thermosphere. The ionosphere is subdivided into (in order of increasing altitude) the D layer, E layer, F1 layer, and F2 layer. At night the D layer vanishes, and the F1 and F2 layers merge into a single layer.

Ionization in the F2 region, primarily caused by the solar flux and cosmic radiation, serves as a low-attenuation reflector for long-wave radio signals by day. This layer has long been exploited for stable communication over long distances. For example, transoceanic navigation by ships and aircraft is accomplished with the aid of this region's reflection and attenuation properties. This is also the basis of the concept of over-the-horizon (OTH) radar used for very long-range ground-based surveillance and tracking systems, as discussed later in this chapter. See Section 4.7 for additional discussion of the ionosphere.

4.4 | ATMOSPHERIC ATTENUATION AND ABSORPTION

The *attenuation* of an EM wave through an atmosphere is caused by two major components: absorption and scattering. Absorption occurs when the atmosphere contains gases or particulates with lossy properties (e.g., oxygen molecules or raindrops). Some of the EM wave's energy is then lost to heat within the lofted particle. Scattering occurs when the particulate is of sufficient size to cause some of the wave to be reflected in directions away from the collecting receiver. The magnitude of both phenomena is a direct function of the particulate density along the EM wave propagation path.

Since atmospheric effects are volume phenomena, the potential exists for some level of scattering off the volume boundary and from the internal particles themselves, as shown

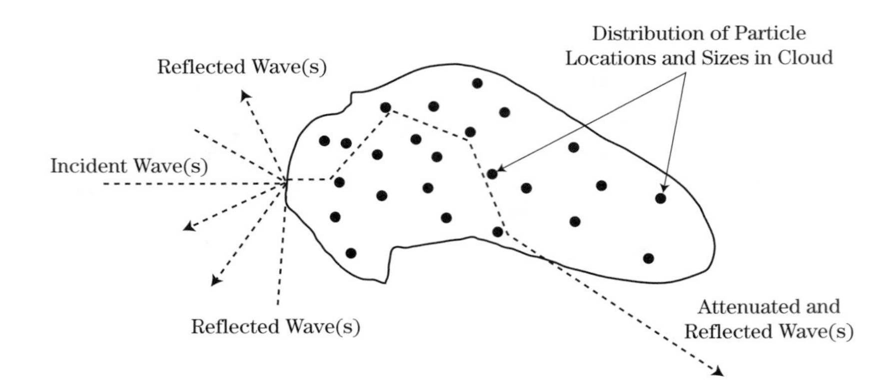

FIGURE 4-4 ■ Attenuation due to wave scattering and absorption by particulates.

in Figure 4-4. Additional scattering can also be caused by the presence of other objects in the atmosphere such as heavy dust, insects, and birds. Scattering from precipitation or other suspended particulates is an important component of total attenuation for a radar link. Further information, which can be used within this methodology, regarding volume scattering from sources other than precipitants can be found in, for example, [4,5]. Atmospheric particle scattering is generally isotropic and increases with frequency and suspended particle size and conductivity.[2] It has been shown that volumetric scattering is a strong function of incident polarization, frequency, and precipitation rates or particle concentrations.

Thus, the wave incident upon the particulate cloud will be both scattered and absorbed according to the particulate size, lossy dielectric properties, and density within the signal path. Only that signal component scattered in the direction of the collecting receiver will be measured. The ratio of the power of the one-way attenuated wave, A, to the free-space wave, A_0, defines the atmospheric loss (attenuation) in the path:

$$F^2 = \frac{A}{A_0} \tag{4.7}$$

In general, one-way attenuation of radar signals in the atmosphere can be expressed in the form

$$F^2 = 10^{\alpha L/2} \tag{4.8}$$

where α is the attenuation coefficient in units of meter^{-1} for the atmospheric type and density, and L is the path length in meters. In many references, the attenuation coefficient is given as a two-way value—hence the factor of $1/2$ in the exponent. For cases in which the α is expressed in dB/km, the one-way propagation factor can be expressed in dB as

$$F^2\ (\text{dB}) = \frac{\alpha L}{2} \tag{4.9}$$

with L (one-way propagation distance) expressed in kilometers.

For longer-range applications, the atmospheric content can be considered heterogeneous, as suggested by Figure 4-5. Using superposition, the impacts of each can be "summed" according to type, density, and path length occupancy, L_n, within the total

[2]Particles that are extremely small compared with a wavelength will scatter more isotropically.

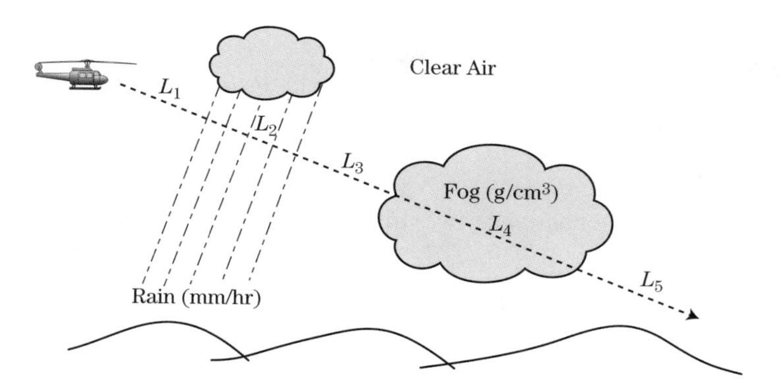

FIGURE 4-5 ■ Heterogeneous atmosphere impacts.

signal path to get the total one-way propagation factor:

$$F^2 = \frac{A}{A_0} = 10^{\alpha_1 L_1/2} \cdot 10^{\alpha_2 L_2/2} \cdot \ldots \cdot 10^{\alpha_N L_N/2} \tag{4.10}$$

Propagating waves within the troposphere will suffer some atmospheric loss due to the presence of oxygen and water vapor in the atmosphere. Other than man-made conditions, the presence of both free molecules and suspended particles such as dust grains, frozen precipitants, and water drops that condense in fog and rain are the primary absorption mechanisms. When no condensation is present (clear air conditions), the absorption is due to oxygen and water vapor molecules. Recognizing that atmospheric factors are a strong function of time, a frequency of occurrence is often used in evaluating atmospheric attenuation. For many missions atmospheric attenuation is one of the most prominent mechanisms affecting radar system performance.

Table 4-1 summarizes typical one-way loss coefficients expected for X-band (10 GHz) propagation though different atmospheric types. The factors determining the attenuation coefficient for each of these conditions are described in the following subsections. These data, derived from multiple references cited in the appropriate sections, can be used to predict the path attenuation for arbitrary path lengths.

TABLE 4-1 ■ Typical One-Way Attenuation Coefficients, α, for Some Selected Atmospherics at 10 GHz

Description	Attenuation Coefficient (dB/km)	Water Content (g/m^3)	Remarks
Clear air	0.01	7.5	Based on sea-level elevation, 42% relative humidity, and 20°C temperature
Dust	0.004	0.1	Based on sea-level elevation, 0 relative humidity, and 20°C temperature
Radiation fog	0.0688	0.1	Based on sea-level elevation, 100% relative humidity, and 20°C temperature
Fog oil (Engine smoke)	0.43	0.0001	Based on sea-level elevation, 0 relative humidity, and 20°C temperature
Rain (4 mm/hr)	0.05	n/a[1]	Based on sea-level elevation, 100% relative humidity, and 20°C temperature
(10 mm/hr)	0.17		
Snow (2 mm/hr)	0.0016	n/a[1]	Based on sea-level elevation, 100% relative humidity, and 0°C temperature
Special smokes and obscurants	8.6	0.001	Based on sea-level elevation, 0 relative humidity, and 20°C temperature

[1] Attenuation coefficients for rain and snow are based primarily on fall rate in this attenuation model.

4.4.1 Clear Air Water Vapor

In general, free space does not exist in the real world. Propagation under ideal conditions in the atmosphere will include attenuation due to water vapor, oxygen, and other normal particulates. Figure 4-6a shows the atmospheric attenuation for two temperatures at sea level, 59% relative humidity, and standard atmospheric pressure of 1013.25 kPa, and Figure 4-6b compares attenuation at two different altitudes.

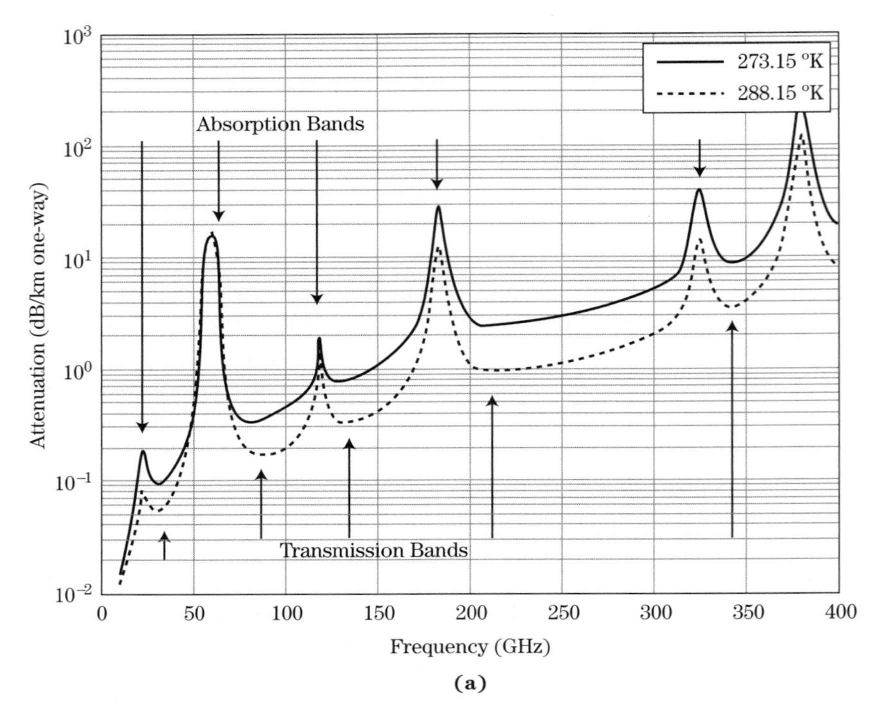

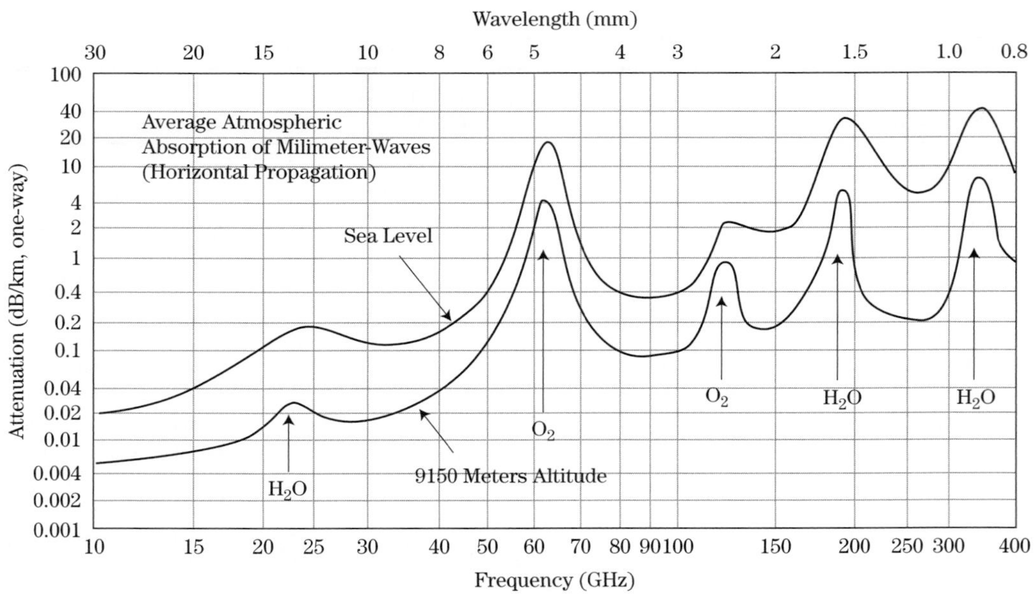

FIGURE 4-6 ■ Attenuation coefficient due to atmospheric gases and water vapor. (a) Variation with temperature. (Courtesy of Bruce Wallace, MMW Concepts. With permission.) (b) Variation with altitude. (From P-N Designs, Inc. [6].)

The dips in attenuation are typically called *transmission bands*, and the peaks are known as *absorption bands*. Radar range can be maximized by operating at a transmission-band frequency. Absorption-band frequencies are useful for applications where the designer wants the signal to reach its desired destination but also is trying to minimize the chance of it reaching farther, to undesired locations.

4.4.2 Rain

Rain is generally characterized by a fall rate, typically in millimeters per hour, known as the *rain rate*. The attenuation due to rain is a function of the rain rate and the drop-size distribution model, which differs in different areas in the world. A general model to estimate the one-way attenuation α in dB/km is given by

$$\alpha = a \cdot r^b \tag{4.11}$$

where a and b are constants based on drop-size distribution model (rain type and geographical region), temperature, frequency, and polarization, and r is the rainfall rate in mm/hr.

Polarimetric attenuation sensitivity for rain occurs where the particulate drops are aspherical and thus have preferred polarization characteristics. Numerous references indicate rain attenuation coefficients; however, only a limited amount of polarimetric data are available. Table 4-2 gives values of a and b as a function of polarization (vertical or

TABLE 4-2 ■ Summary of Empirical Coefficients for Rain Attenuation Model at Linear V, H Polarization

Frequency (GHz)	a_h	a_v	b_h	b_v
1	0.0000387	0.0000352	0.912	0.880
2	0.000154	0.000138	0.963	0.923
3	0.000650	0.000591	1.121	1.075
6	0.00175	0.00155	1.308	1.265
7	0.00301	0.00265	1.332	1.312
8	0.00454	0.00395	1.327	1.310
10	0.0101	0.00887	1.276	1.264
12	0.0188	0.0168	1.217	1.200
15	0.0367	0.0347	1.154	1.128
20	0.0751	0.0691	1.099	1.065
25	0.124	0.113	1.061	1.030
30	0.187	0.167	1.021	1.000
35	0.263	0.233	0.979	0.963
40	0.350	0.310	0.939	0.929
45	0.442	0.393	0.903	0.897
50	0.536	0.479	0.873	0.868
60	0.707	0.642	0.826	0.824
70	0.851	0.784	0.793	0.793
80	0.975	0.906	0.769	0.769
90	1.06	0.999	0.753	0.754
100	1.12	1.06	0.743	0.744
120	1.18	1.13	0.731	0.732
150	1.31	1.27	0.710	0.711
200	1.45	1.42	0.689	0.690
300	1.36	1.35	0.688	0.689
400	1.32	1.31	0.683	0.684

Source: After Nathanson [7] (With permission).

horizontal) and operating frequency [7]. Values for a and b at other frequencies can be obtained by interpolation using a logarithmic scale for a and frequency and a linear scale for b. Using these coefficients in equation (4.11) gives the expected one-way attenuation in rain as a function of rain rate for four radar frequencies, shown in Figure 4-6.

Since the magnitude of the electric field vector in circular polarization is based on the vector summation of the linear H and V field components, the circular polarization attenuation can be estimated according to

$$\alpha_{RHC} = \alpha_{LHC} = \frac{1}{\sqrt{2}}\sqrt{\alpha_{HH}^2 + \alpha_{VV}^2} \tag{4.12}$$

where α_{RHC}, α_{LHC}, α_{HH}, and α_{VV} are the attenuation coefficients for right-hand circular, left-hand circular, horizontal, and vertical polarizations. The $\sqrt{2}$ factor is used to normalize the energy in the circular wave to that of the VV or HH equivalent.

For rain rates of interest (typically less than 10 mm/hr), Figure 4-7 indicates only a slight difference between HH and VV attenuation coefficients. At higher rain rates, horizontal polarization seems to suffer slightly higher attenuation than vertical. The higher attenuation for horizontal polarization is due to a slight flattening of the falling raindrops when they become large, with the shape changing from a sphere to an oblate spheroid. With $\alpha_{VV} = k \cdot \alpha_{HH}$, $k \leq 1$, equation (4.12) becomes

$$\begin{aligned}\alpha_{RHC} = \alpha_{LHC} &= \frac{1}{\sqrt{2}}\sqrt{\alpha_{HH}^2 + k^2\alpha_{HH}^2} \\ &= \sqrt{\frac{1+k^2}{2}} \cdot \alpha_{HH}\end{aligned} \tag{4.13}$$

In rain for low rain rates there is little attenuation difference between linear and circular. As rain rate increases beyond about 10 mm/hr, the attenuation of horizontal polarization

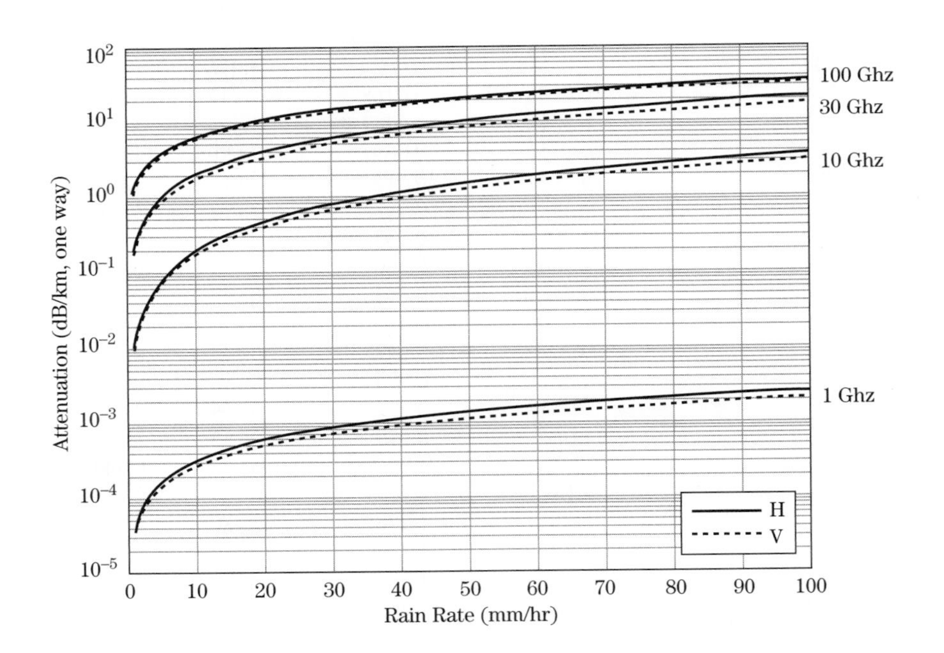

FIGURE 4-7 ■ One-way attenuation versus rain rate at four frequencies.

TABLE 4-3 ■ Summary of Fog Types and Water Concentration Ranges

Type	Water Concentrations (g/m^3)	Remarks
Steam fog	0.1–1.0	Results from cold-air movements over warm water
Warm-front fog	0.1–1.0	Evaporation of warm rain falling through cold air, usually associated with the movement of a warm front, under certain humidity, conditions yields a supersaturated air mass at ground level
Radiation fog	0–0.1	Results from radiation cooling of the earth's surface below its dew level
Coastal and inland ground fog	0.1–1.0	Radiation fog of small heights
Valley fog	0.1–1.0	Radiation fog that forms in valleys
Advection fog	0–0.1	Formed by cool air passing over a colder surface
Up-slope fog	0.1–1.0	Results from the adiabatic cooling of air up-sloping terrain

is greater than that of vertical ($k < 1$), and equation (4.13) shows that the magnitude of circular polarized energy begins to suffer measurably higher attenuation than linear magnitudes.

4.4.3 Fog

Fog is usually a result of a condensation process that occurs at or near the ground. Evaporation leads to a supersaturated condition and the formation of fog. Other occurrences of fog relate to clouds coming in contact with the ground, such as the drifting of a strata cloud into a mountain slope. Fog water content can be characterized according to Table 4-3 [3].

Fog consists of water droplets in most instances but in cold climates may be composed of ice particles. Fog formations include the presence of small drop sizes less than 100 μm in diameter. Because of the meteorological conditions associated with fog formations, the particulate size distributions may vary considerably between fog formations, even at the same or similar locations. Humidity values of 95–100% are normally associated with fog except over salt water, where values of 75–90% are common. The water concentration of fog formations is lower than that of typical water clouds, a few hundredths of a gram per cubic meter compared with values of 0.1 g/m^3 or more for clouds.

One-way attenuation (dB/km) due to fog is computed from the water concentration, M (g/m^3), frequency, f (GHz), and temperature, T (°C), according to [4,8]:

$$\alpha = M\left(-1.347 + 0.66f + \frac{11.152}{f} - 0.022T\right), \quad f > 5 \text{ GHz} \tag{4.14}$$

The fog attenuation at lower frequencies (below 5 GHz) is considered negligible for most applications. Typical values of attenuation due to fog are 0.02 to 3 dB/km at frequencies of approximately 10 GHz or higher. Higher temperatures result in higher attenuation at radar frequencies. The radar attenuation due to fog in the medium at three frequencies is illustrated in Figure 4-8.

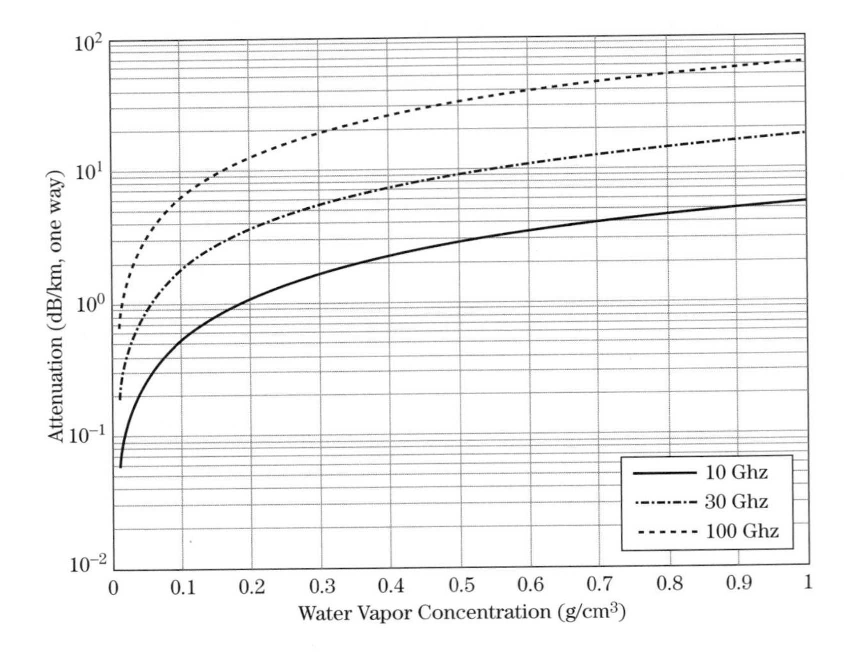

FIGURE 4-8 ■ Attenuation at three frequencies for fog versus water concentration (fog type).

4.4.4 Snow and Hail

Snow and hail differ from water particulates in two ways. The index of refraction of ice is different from that of liquid water. Snow and hail also are generally composed of aspherical, crystalline particles and are treated similarly when estimating attenuation based on equivalent fall rate. Consequently, snow and hail have very different scattering and absorption properties from rain or fog.

Table 4-4 illustrates data compiled by Nakaya and Terada [9] from their observations on Mt. Tokati, Japan, for air temperatures between −8°C and −15°C. The snow mass concentration may be converted to the equivalent rainfall rate r_e (mm/hr) according to [3]:

$$r_e = X\nu \tag{4.15}$$

where X is the mass concentration of snow (g/m^3), and ν is the velocity of the snowfall (m/s). The one-way attenuation α (dB/km) due to snow is then given by [3]

$$\alpha = 0.00349\frac{r_e^{1.6}}{\lambda^4} + 0.00224\frac{r_e}{\lambda} \tag{4.16}$$

The attenuation at four frequencies for snow based on data presented in Table 4-4 as a function of equivalent rain fall rate is illustrated in Figure 4-9. Several observations can be made from these data:

1. The attenuation increases with increasing fall rate, which is also true for rain.
2. Wet snow (snow containing liquid water) causes higher attenuation than dry snow for a given fall rate.
3. Attenuation generally increases with radar frequency.
4. Hail attenuation is traditionally computed in the same way as snow by defining the proper equivalent fall rate.

TABLE 4-4 ■ Mean Diameters, Masses, and Fall Velocities of Snow Crystals

Snow Type	Diameter (mm)	Mass (mg)	Fall Velocity (cm/s)
Needle	1.53	0.004	50
Plane dendrite	3.26	0.043	31
Spatial dendrite	4.15	0.146	57
Powder snow	2.15	0.064	50
Rimed crystals	2.45	0.176	100
Graupel	2.13	0.80	180

Source: From Nakaya and Terada [9] (With permission).

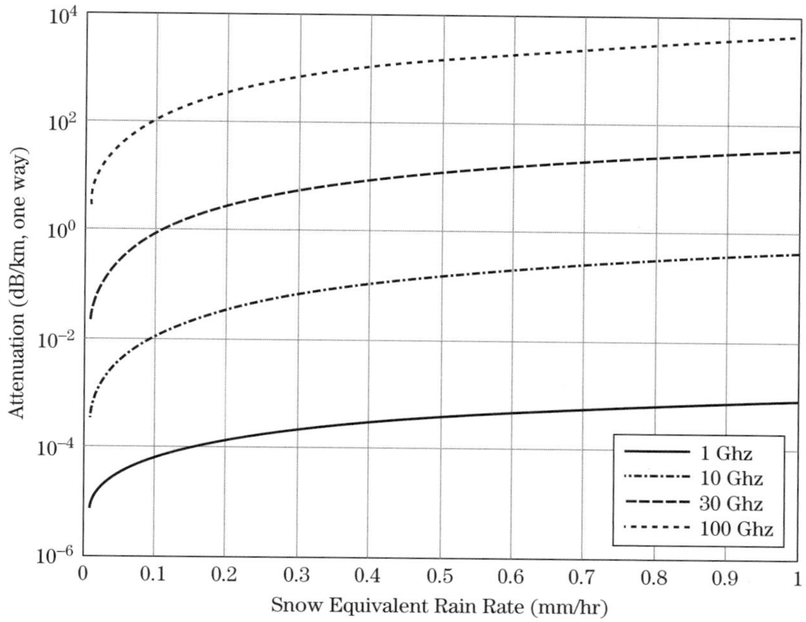

FIGURE 4-9 ■ Attenuation for snow versus equivalent rainfall rate at four frequencies.

4.4.5 Dust

For most dry soil particle concentrations in the atmosphere, the attenuation at 10 GHz is negligible (e.g., <0.0001 dB/km). This is true for dust particles up to a diameter of 300 μm. For most types of suspended particulates, the one-way attenuation α (dB/km) can be approximated as a function of *extinction efficiency,* η (m^2/g), and particulate concentration, M (g/m^3) [10,11][3]

$$\alpha = 4343 \cdot \eta M \text{ dB/km} \tag{4.17}$$

where extinction efficiency is the sum of the scattering and absorption efficiencies or the efficiency of the dust in scattering or absorbing the radar waves, respectively.

Dust particulate concentrations are generally low except in some very brief man-made instances (e.g., blowing up a dirt pile). Even the heaviest dust storm will likely have concentrations less that 0.1 g/m^3.

[3] Extinction efficiency relates to how quickly the particulates disburse when lofted and thus the reduction of attenuation through them.

4.4.6 Smoke

The smoke considered here is generally "fog oil" from a running engine. Other naturally occurring smokes can arise from burning of natural materials such as wood or rubber. For smoke, the attenuation can be expressed—similarly to dust—as a function of extinction efficiency.

Historically, smokes and obscurants have played an important role on the battlefield, and their anticipated presence on future battlefields is evidenced by the fact that some combat identification devices have requirements including operating in engine smoke. In recent years both U.S. and foreign countermeasure developers have extended their smoke developments to include radar bands. Indeed, the U.S. Army has classified the M-81 grenade for armor self-protection because it produces a cloud that is an effective obscurant in visible, infrared (IR), and MMW regions of the spectrum.

Unlike dust, fog, rain, and even fog oil, aerosols are specifically designed to affect specific wavelengths. Whereas the extinction efficiency of 4 mm/hr rain at 10 GHz is on the order of 10^{-3} m^2/gm, the extinction efficiency of average X-band aerosols is approximately 1.0 m^2/gm. Thus, it requires far less aerosol material to cause a significant attenuation effect. Recent advances in aerosol development have focused on developing multispectral aerosols that introduce attenuation in radar-, infrared-, and laser-operating bands simultaneously.

4.5 ATMOSPHERIC REFRACTION

Refraction is the change in the direction of travel of radio waves due to a spatial change in the index of refraction. The index of refraction n is equal to

$$n = \frac{c}{v_p} \tag{4.18}$$

where c is the speed of light in a vacuum (well approximated as 3×10^8 m/s), and v_p is the phase velocity of the wave (m/s) in the medium. The index of refraction is unitless. It is measured based on standard meteorological observations of air temperature, atmospheric pressure, and partial pressure of water vapor. Averaged over many locations and long periods of time, the index of refraction of the troposphere generally decreases with increasing altitude.

The refractive index variation present in the troposphere can significantly affect the propagation of radio waves. The results of these variations are divided into *standard* and *anomalous* atmosphere effects. Standard refraction implies that the relationship between refractive index and height is approximately linear: the effective Earth radar horizon model. Propagation conditions that deviate from this linear model are called *anomalous* but are not necessarily infrequent. Anomalous atmosphere conditions are in turn divided into three subcategories: *subrefraction*, *superrefraction*, and *trapping* or *ducting*.

For standard atmospheres, the net effect of refraction is a bending of a horizontally transmitted radio wave toward the earth's surface, which can cause increased errors in range and angular position measurements but also can allow propagation into masked or shadow zones. Figure 4-10 illustrates the general effect of each of these phenomena on a ray trace from a transmitter.

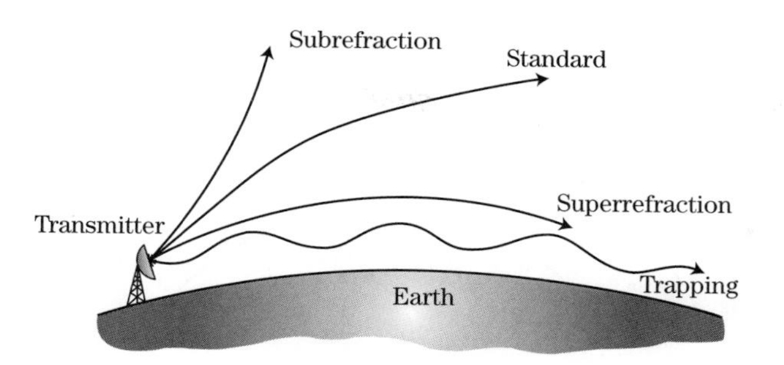

FIGURE 4-10 ■ Illustration of ray bending for standard and anomalous refraction profiles.

4.5.1 Standard Refraction

4.5.1.1 The Standard Atmosphere

For wave propagation in clear air the *standard atmosphere*, which also implies standard refraction, is most often used to describe the conditions of the medium. The standard atmosphere is actually a series of models that define the values for atmospheric temperature, pressure, and density as well as other parameters, such as sound speed, viscosity, and thermal conductivity. These parameters are presented in tables as a function of altitude, with varying resolutions ranging from 0.05 km at low altitudes to 5 km at higher altitudes. The most commonly referenced model is the U.S. Standard Atmosphere [12], but the reader should be aware that other models are available, such as the Standard Atmosphere of the International Civil Aviation Organization [13] and the International Standard Atmosphere [14]. Each model has various refinements, although most of them differ only in the region above approximately 30 km. Because of this, the U.S. Standard Atmosphere is suitable for most calculations below that altitude.

The path of a ray traveling through the atmosphere can be determined from Snell's law. For a first approximation, the earth's surface can be modeled as a plane, with the atmosphere modeled as a series of planar slabs above it, each having a constant index of refraction (see Figure 4-11). From Snell's law, one obtains

$$n_0 \cos\alpha_0 = n_1 \cos\alpha_1 = n_2 \cos\alpha_2 \ldots = n_i \cos\alpha_i \tag{4.19}$$

where n_i is the index of refraction of the i-th slab. Thus, if n_0 and the angle α_0 at which the wave is transmitted are known, the subsequent angles α_i can be found from Snell's law at each of the layer boundaries, and the path of the ray can be determined as it travels through the layers.

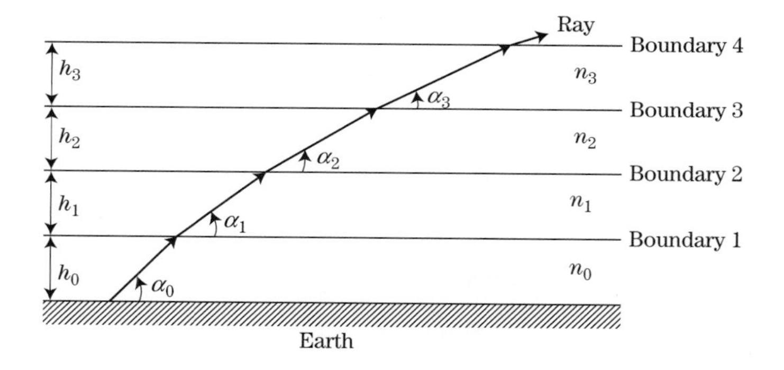

FIGURE 4-11 ■ Path of a ray through a horizontally stratified atmosphere (troposphere).

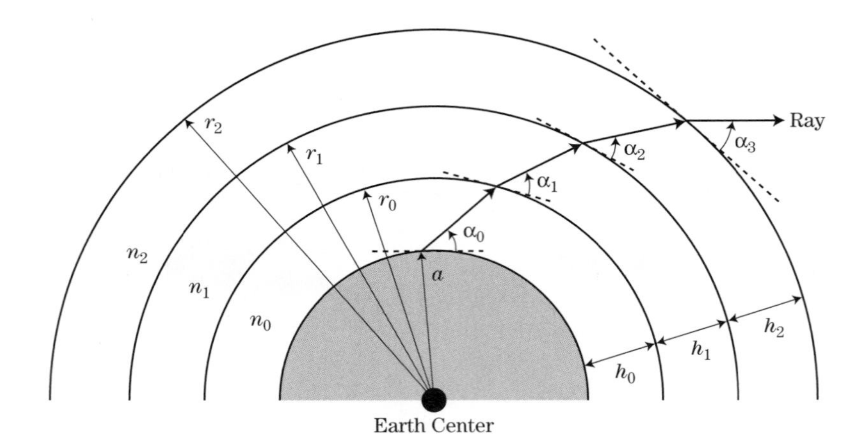

FIGURE 4-12 ■ Path of a ray through a radially stratified atmosphere (troposphere).

A more accurate solution to the refraction problem is obtained by assuming that the earth is spherical instead of planar and that the electrical properties of the atmosphere are constant between concentric spheres of outer radii, r_i (see Figure 4-12). For such a radially stratified medium, Snell's law assumes the form

$$n_0 r_0 \cos\alpha_0 = n_1 r_1 \cos\alpha_1 = n_2 r_2 \cos\alpha_2 \ldots = n_i r_i \cos\alpha_i \tag{4.20}$$

Both the planar and spherical approximations improve as the thickness of each slab is made smaller.

The curved path traveled by an EM wave due to refraction causes errors in the estimate of target location in both range and angle. This effect is illustrated in Figure 4-13. In the standard refraction case, the waves travel along the curved path bending slightly downward toward the earth at each layer. The effect is to yield measured ranges and elevation angles that are larger than the true values.

The resulting elevation angle and range errors for a spherically stratified model and standard atmosphere are shown in Figures 4-14 and 4-15. In these figures, the altitude is that of the radar system. According to Berkowitz [15], angle errors can range from approximately 0.1 to 0.8 degrees and range errors from 10 to 381 feet, depending on relative humidity and transmitter height in a standard atmosphere. These errors impact the

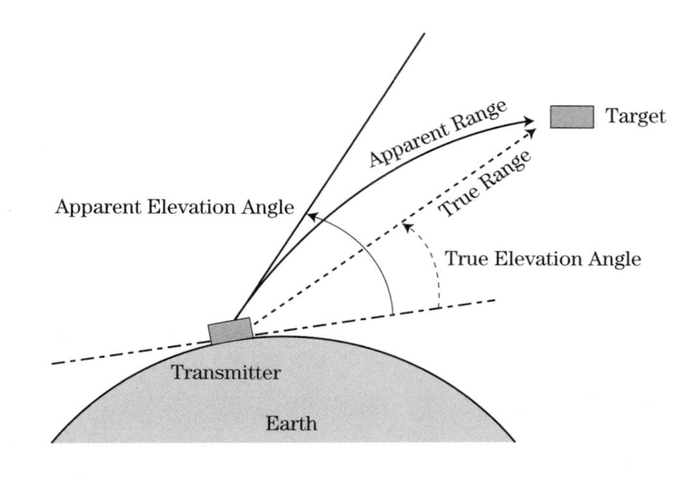

FIGURE 4-13 ■ Normal refraction effects on target location (troposphere).

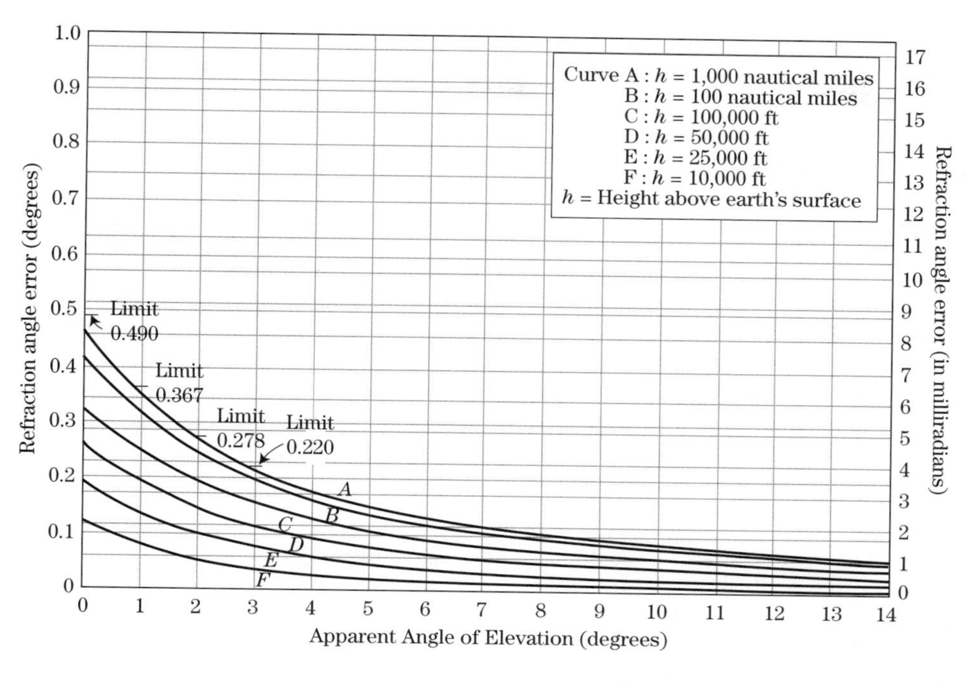

FIGURE 4-14 ■ Normal refraction effects on target location—angle estimate, 0% humidity (troposphere).

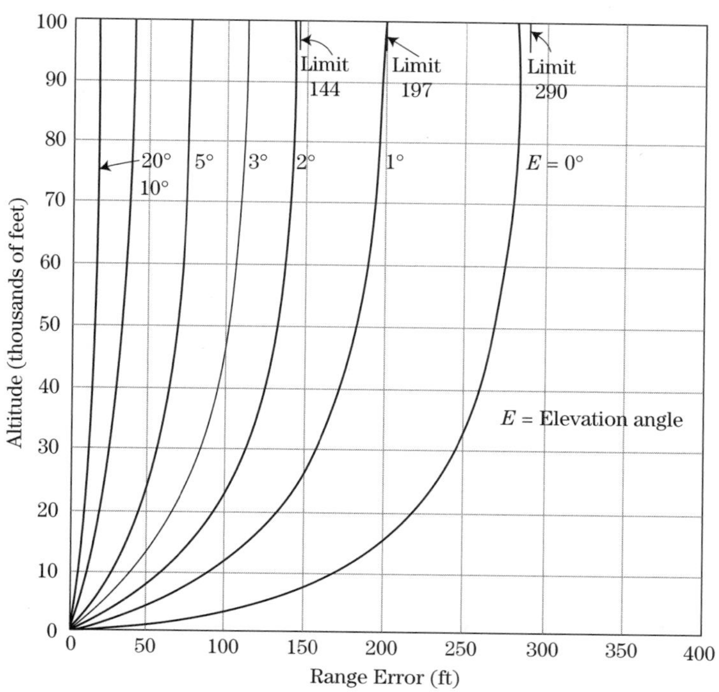

FIGURE 4-15 ■ Normal refraction effects on target location—range estimate, 0% humidity (troposphere).

ability of the radar to achieve angle error requirements and intercept receivers to achieve positional estimates. In most cases the refraction effects are significant for longer-range applications. For example, an elevation error of only 5 milliradians (approximately 0.3 degrees) at 1,000 kilometers is equivalent to a target position error normal to the radar boresight of over 5 kilometers.

While the relationships presented here are good first-order approximations, the atmosphere is not so easily stratified. The index of refraction versus height curve varies for different points on the earth. Bean and Dutton [16] compiled information about the index of retraction as a function of height for various locations on the earth's surface that can be used to refine the estimate of refraction impacts.

4.5.1.2 Defining the Horizon: The Effective Earth Model

One effect of refraction of EM waves is the extension of the apparent horizon over the "spherical" earth for ground-based radars close to the earth. The concept of an *effective Earth radius* is an alternative way to account for the effects of a standard atmosphere and in particular to predict the additional range associated with the refractive horizon for a transmitter. It is important to keep in mind that for space-based or airborne radars or for ground radars looking significantly above the horizon, these results are not as pronounced.

The radar horizon on the "spherical" earth can be shown to be

$$R_h = \sqrt{2ah_t} \tag{4.21}$$

where a is the earth's radius (m), and h_t (m) is the height of the transmitter (or receiver) above the earth's surface. The extended horizon, R'_h, due to refraction can also be predicted using equation (4.21) if an *effective Earth model* having a slightly larger radius a_e is used. This idea is illustrated in Figure 4-16.

For standard atmospheres (typical refractive index gradient $dn/dh = 3.9 \times 10^{-8}$ m) and the actual earth radius a of 6370 km, the ratio between the actual and effective radius can be shown to be approximately 4/3 [2,17]. This is the origin of the "4/3 Earth radius" or *4/3 Earth* model cited in many texts for radar horizon computations over land. From equation (4.21), the increase in radar horizon is then $\sqrt{4/3}$, or about 15%, making $a_e = (4/3)a$. Figure 4-17 illustrates the increase in the radar horizon based on ray bending in standard refraction as a function of transmitter height.

4.5.2 Anomalous Refraction

Standard refraction implies that the relationship between refractive index and height is linear. Propagation conditions that deviate from this linear model are called *anomalous*,

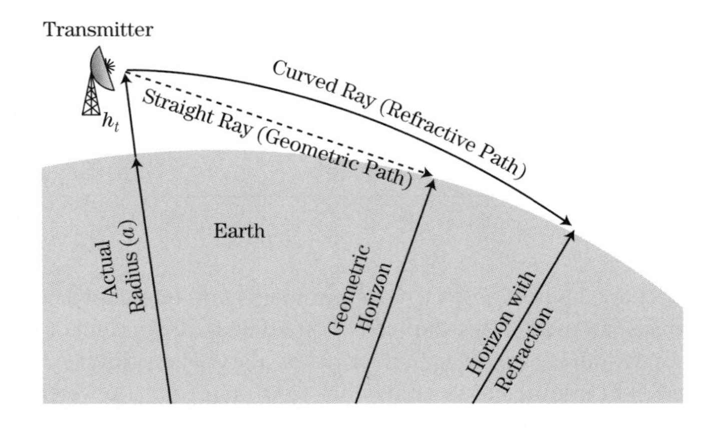

FIGURE 4-16 ■ Increased range to radar horizon due to refraction.

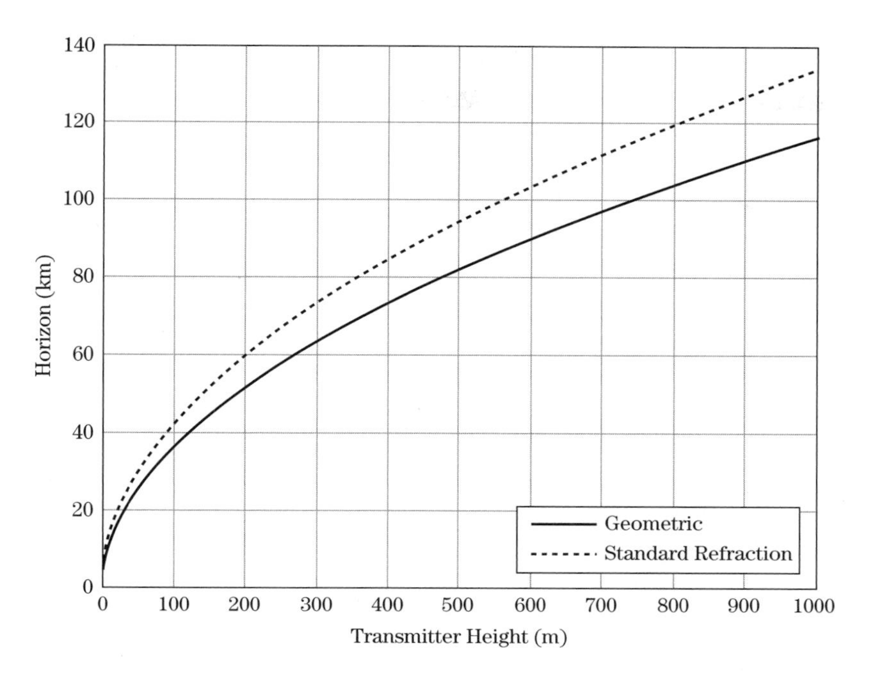

FIGURE 4-17 ■ Comparison of radar horizons in free space and a standard atmosphere.

though they are not necessarily infrequent. Three forms of anomalous deviation are possible:

1. *Subrefraction* occurs when the rate of change of the refractive index, n, as a function of altitude, h, is positive or zero ($dn/dh \geq 0$). This condition bends rays upward, thus decreasing the ground coverage of a radar.
2. *Superrefraction* occurs when dn/dh is more negative than in a standard atmosphere. This condition causes waves to bend more strongly toward the ground and can increase the range of the radar at low elevations (angles less than 1.5 degrees) but has little effect on high-angle coverage.
3. *Ducting and trapping* occurs when $dn/dh < -16 \times 10^{-8}\ \text{m}^{-1}$. Under this condition, the radius of curvature of the ray is less than or equal to the radius of curvature of the earth, trapping transmitted energy in a "parallel plate waveguide" near the earth. A duct can be formed when the temperature increases with height near the surface (temperature inversion) or the humidity decreases (moisture lapse) with height. These ducts can dramatically impair or enhance radar coverage and range depending on whether each link terminal in the radar/target or radar/receiver pair are located in or out of the ducts. Most of these effects are seen at very high frequency (VHF) and ultra high frequency (UHF). Note that in some texts a distinction is made between trapping and ducting, because trapping is a result of the same processes that cause superrefraction, as is ducting. However, in the case of trapping, the energy remains between the atmospheric boundary and the earth, creating an atmospheric waveguide.

Of these effects, ducting is the most important for radar. Ducting and trapping redistribute the radar beam over an elevation search area, causing gaps in coverage as illustrated in Figure 4-18 for ground-based radar. Under airborne radar conditions, much of the transmitted energy is redistributed into a low-elevation surface duct, significantly enhancing

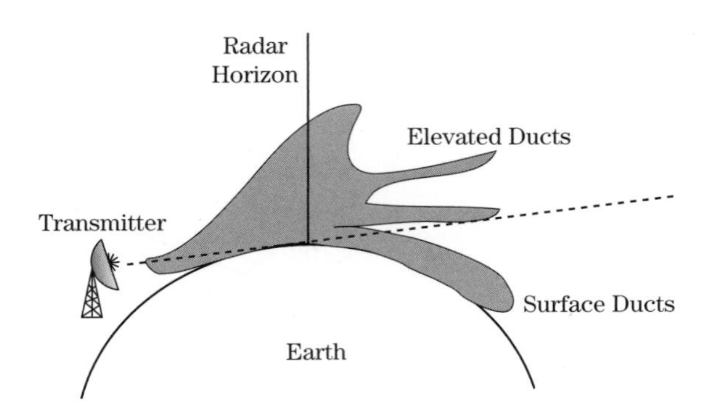

FIGURE 4-18 ■ Illustration of energy propagation over the radar horizon and gaps in coverage due to ducts.

ground coverage and range at low elevations while introducing shadow zones above the duct, as illustrated in Figure 4-19. This energy redistribution causes the performance above and inside the duct to degrade, reducing the radar's range performance for target, clutter, and weather detection.

In Figure 4-19 the incident wave is both reflected from and transmitted into the duct at point *A*. The transmitted energy in the duct reflects off the surface at point *B* back toward the top of the duct. Once at the top of the duct some of this energy transmits through the boundary and interferes with the reflected wave from *A* causing shadow zones. As the trapped energy in the duct continues to bounce around, the waves form additional interference patterns and shadow zones inside as well. In addition, for some airborne systems increased ground coverage can compromise the ability of the system to perform covert missions.

Atmospheric ducts are generally on the order of 10 to 20 meters in height, never more than about 200 meters. One type, the *evaporation duct*, is found regularly over relatively warm bodies of water. It is caused by a temperature inversion near the surface and is accentuated by the intense relative humidity near the surface due to water evaporation. These conditions can correspond to high temperatures (>70°F) and relative humidity above 75% [18]. Over land surfaces an evaporation duct is formed when an intense layer of low-lying humidity is found over a surface that is cooling more rapidly than the surrounding air (e.g., fog conditions). This condition is also representative of conditions when a large daytime temperature inversion over a locally cool surface caused by intense air temperature from heat reradiated from surrounding surfaces (e.g., over gray concrete runway surrounded

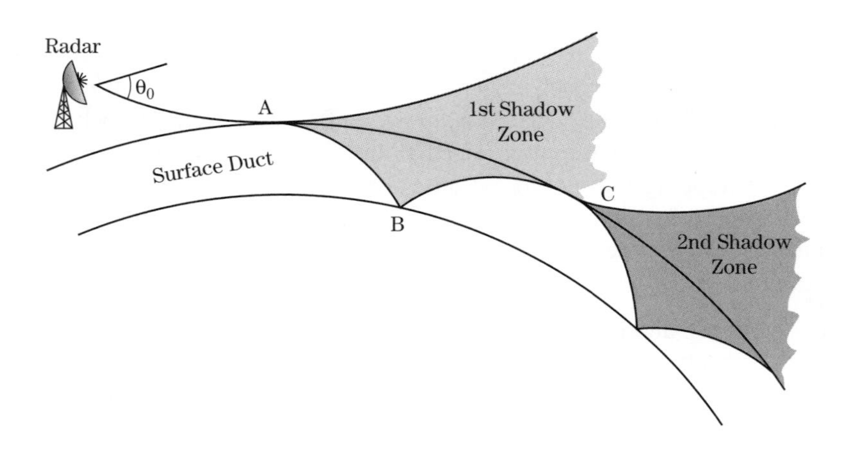

FIGURE 4-19 ■ Illustration of energy redistribution and loss in coverage when transmitter is above a surface duct.

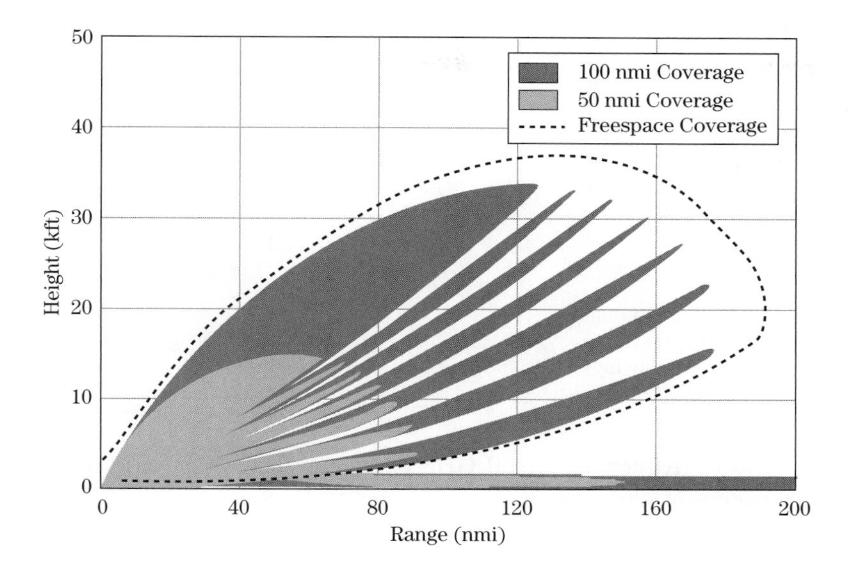

FIGURE 4-20 ■ Illustration of ducting loss in radar elevation coverage.

by black asphalt). The potential for both evaporation and elevated ducts exists in coastal regions where there are large temperature contrasts between land and water. These land–water temperature differences and their diurnal reversal produce corresponding land–water air pressure differences, resulting in a system of breezes across the shoreline that reverses its direction between day and night.

The net effect of these ducts is a redistribution of transmitted energy. The presence of extended ranges at some elevations comes at the expense of reduced ranges or radio "holes" at other elevations. Although precise elevation profiles for ducting propagation factor are not easily computed analytically, a good rule of thumb for propagation inside of a duct is to assume the one-way power falls off according to R^{-1} instead of R^{-2} [19,20]. The trapped rays may also coherently combine, causing potentially large signal fading due to the interference between the several modes.

In general, a computer model that includes an atmospheric profile is required to determine the propagation factor as a function of elevation for a particular scenario. Figure 4-20 is an example of the EREPS simulation [21] results for radar coverage based on a 3.3 GHz wave launched above an inhomogeneous evaporation duct. A list of some useful off-the-shelf simulations to determine these profiles and other propagation mechanisms are provided at the end of this chapter.

4.6 | TURBULENCE

Most people have seen the phenomenon of a blurry area appearing above a hot asphalt road. This effect is essentially a turbulence phenomenon at the eye's (optical) wavelength. The same turbulence-induced fluctuations of refractive indexes in the path through the atmospheric media can impact very short wavelength radar.[4] Atmospheric turbulence effects are generally considered to be a high-frequency phenomenon and as such are likely

[4]Typically at millimeter wavelengths or smaller.

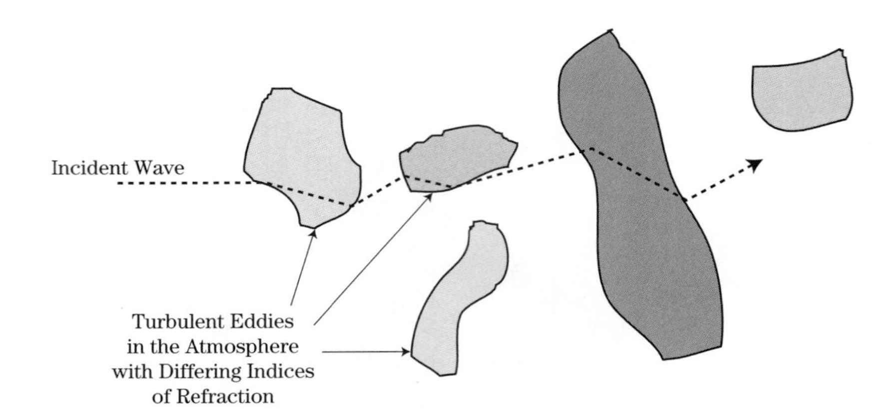

FIGURE 4-21 ■ Illustration of turbulence "pockets" in atmosphere.

significant only at frequencies of 80 GHz or higher. The worst case for atmospheric fluctuations occurs in clear, hot, humid weather and not in rain or snow as might be expected. The smallest fluctuations in both amplitude and phase tend to occur during a dense fog [22].

Radiofrequency (RF) energy is subject to phase and amplitude *scintillation* in atmospheric media, which exhibit point-to-point refractive index variations caused by small fluctuations in temperature and humidity. Intensity fluctuations probably are too small to be of consequence even at millimeter waves. Depolarization, frequency shift, and thermal blooming are other turbulence effects that can sometimes be observed at optical wavelengths but whose effects are likely negligible at RF [18].

Perhaps the major potential problem caused by atmospheric turbulence is angle of arrival (AOA) fluctuations, which are related to phase shifts by the relation $\theta = \phi/kd$, where θ is AOA, ϕ is phase shift, $k = 2\pi/\lambda$ is the wave number, and d is the separation between points in the wavefront. Index of refraction variations induce phase shifts across the wavefront of a propagating signal. These in turn can give rise to changes in the apparent angle at which the wavefront is incident on the receiver (Figure 4-21). AOA fluctuations cause a potential target to be detected at an angle different from its actual angle and thus can bring about aimpoint wander and decreased accuracy in tracking systems.

For MMW systems, amplitude fluctuations are about 1–2 dB and AOA fluctuations are about 300 microradians peak under worst-case conditions. In one experiment, 94 GHz data collected over a fairly short range path indicated that the total fluctuation in angle of arrival due to atmospheric turbulence was about 0.56 milliradians [23].

4.7 EXPLOITING THE IONOSPHERE

The ionosphere is a region of the upper atmosphere (thermosphere) that contains a sufficient number of ionized particles to affect radio propagation. The ionization of the atmospheric gases is produced primarily by the ultraviolet light from the sun. It increases with height, though not linearly, and tends to have maxima at certain heights defining layers. The shaded areas in Figure 4-22 illustrate the four main regions of high ionization.

The characteristics of the four main ionosphere layers are as follows:

1. The *D layer* exists only during daylight hours, at which time it bends and absorbs low frequency (f < 3–7 MHz) waves.

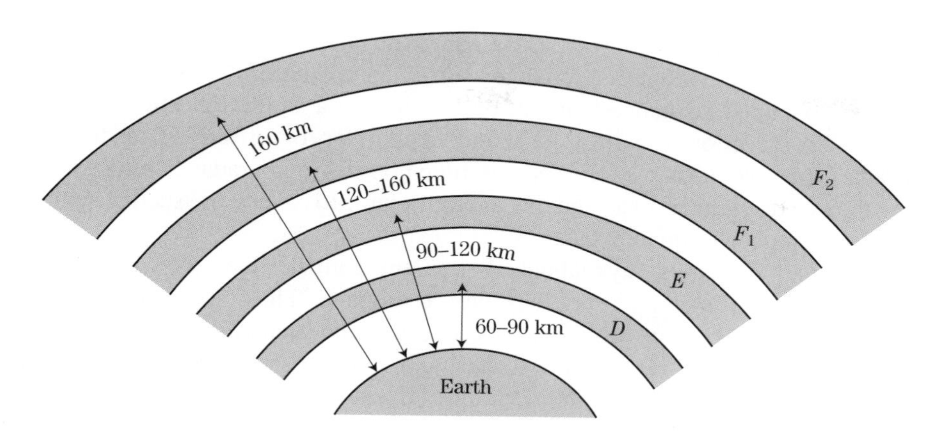

FIGURE 4-22 ■ Layers of the ionosphere (typical of noontime locations).

2. The *E layer* has characteristics similar to those of the D layer but exists at greater heights in the atmosphere.
3. The *F1 layer* is weaker and less influential on radar waves than the F_2 layer.
4. The *F2 layer* is the most important layer for long-distance transmission. It bends waves at frequencies below 30–50 MHz. Its influence is strongest during the daytime and decreases somewhat at night, when it tends to combine with the F1 layer.

The afternoon behavior of ionized layers retraces the morning values relative to local noon, providing a higher level of behavioral consistency.

The ionosphere is a dispersive medium in which the phase velocity of the waves is greater than c, resulting in a refractive index of less than 1.0. This causes rays passing into a region of increasing electron density to be bent away from the normal and back toward the ground. The bending effect of a region on EM waves can be described by Snell's law and the refractive index, n, of the layer region, as described earlier. For the ionosphere the refractive index, n, is computed from [2]

$$n = \sqrt{1 - \left(\frac{f_p}{f}\right)^2} \tag{4.22}$$

where f is the transmit frequency (Hz), and f_p is the plasma frequency for the layer based on its electron density, N_e (electrons/m^3).

$$f_p \approx 9\sqrt{N_e} \tag{4.23}$$

Typical electron density of the F2 layer during the day (noontime) ranges from 10^{11} to 10^{12} electrons/m^3.

The effect of one ionized layer on several rays is shown in Figure 4-23. A ray penetrates farther into the layer as the angle the ray makes with the horizon is increased. Rays

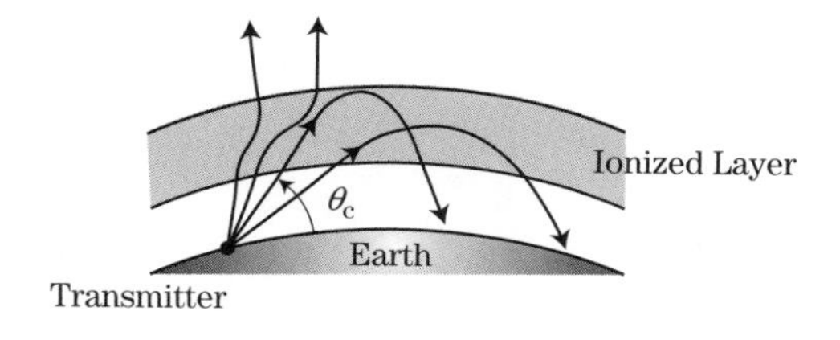

FIGURE 4-23 ■ Bending of rays passing through an ionized layer at difference incidence angles.

transmitted at angles greater than the critical angle do not return to Earth.[5] The ionosphere is therefore most influential at low angles of incidence and for low frequencies. The maximum "extended" distance that can be covered by a single reflection hop is about 1,250 miles for reflection from the E layer and 2,500 miles for reflection from the F2 layer. OTH radars use this fact and exploit one or more hops to obtain radar coverage at regions that are thousands of miles away and that are impossible to cover from such distances with microwave or higher-frequency radars.

As Figure 4-23 shows, rays launched at an angle greater than the critical angle are still affected by the ionosphere as they pass through it. Thus, space-based radars (or ground-based radars tracking objects in space) must take such effects into account when calculating, for example, ranges, range rates, and Doppler shifts. Depending on the operating frequency of the radar system, various effects must be considered. For instance, at 435 MHz one can expect approximately 1.5 complete rotations of the phase vector of the wavefront of the radar beam due to Faraday rotation. This effect diminishes as the frequency increases, so that at 1.2 GHz less than a quarter of a rotation in phase occurs.

4.8 DIFFRACTION

Diffraction is a mechanism by which waves can curve around edges and penetrate the shadow region behind an opaque obstacle. This effect can be explained by Huygens's principle, originally introduced in Chapter 1, which states that every elementary area of a wavefront can be considered a source that radiates isotropically on its front side. The new waves will interfere with each other in the shadow zone to produce an interference pattern at the observation point as illustrated in Figure 4-24.

As the incident wave diffracts around the obstacle it will recombine with scaled replicas of itself within the observation plane. The interface pattern produced is that of two new waves originating from virtual phase centers at P. These virtual phase centers are also known as *virtual sources* and are an equivalent representation of the incident wave structure *after* diffraction has occurred.

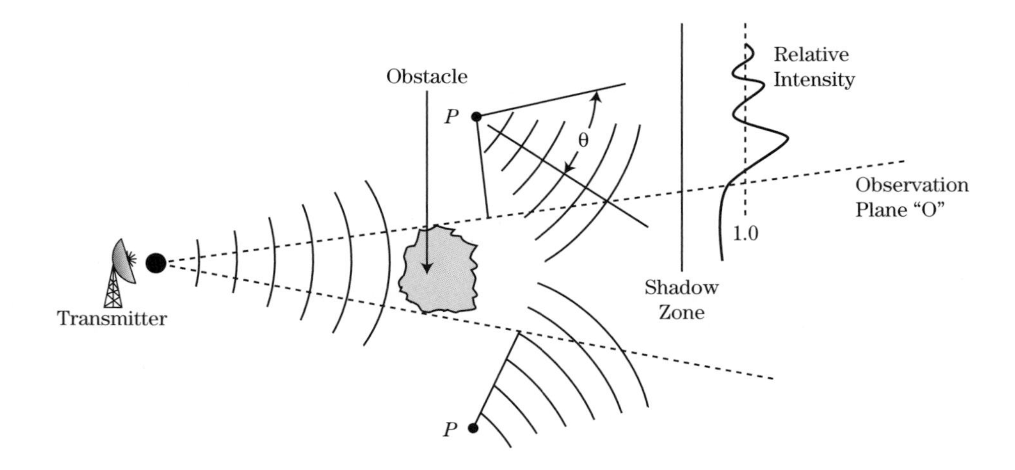

FIGURE 4-24 ■ Illustration of virtual sources for diffraction around an obstacle.

[5]Note that a signal will be totally reflected by the ionosphere if its frequency is less than the ionized layer's plasma frequency independent of incidence angle.

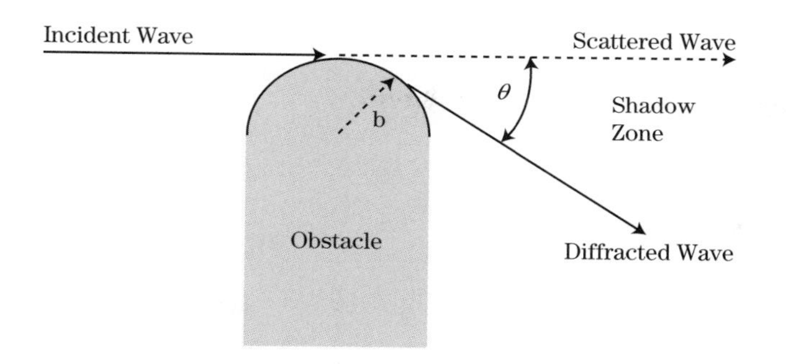

FIGURE 4-25 ■ Geometry for diffraction into shadow zones.

Depending on the transmitted wavelength, the edges of the diffracting object may be appear as a smooth, curved edge or as a sharp *knife edge* or wedge. At the boundary between the interference and diffraction regions, some signal enhancement may be realized. In general, as the observation angle falls into the shadow zones, diffracted wave attenuations increase.

The electromagnetic field in the shadow zone due to diffraction can be represented by an infinite series of terms called *modes* [24]. The effect of diffraction on the signal strength in or near the shadow region can be represented using a one-way power propagation factor, F^2, in this case called the *diffraction coefficient.*

The behavior of the diffraction coefficient is modeled by considering diffraction around a curved edge of radius b as shown in Figure 4-25. As b goes to zero, the shape becomes a knife edge. The diffraction coefficient is a function of both the edge radius times the wavenumber k and the angle, θ, into the shadow zone, $F^2(\theta, kb)$. At or near the shadow boundary, many modes must be summed to compute the diffraction coefficient and resulting signal strength. However, in the shadow zones, the first mode is often sufficient to model knife-edge diffraction.

The knife edge is considered the most frequently occurring diffraction mechanism when a wave is impeded by an obstacle. The propagation factor, F^2, into the shadow zone for knife-edge diffraction falls off with range according to $1/kR$. The falloff as a function of shadow angle is given by Sommerfield [25] as

$$F^2(\theta, kb = 0) = \frac{1}{2\sqrt{2\pi}}\left[\sec\frac{1}{2}(\theta + \pi) + \csc\frac{1}{2}(\theta + \pi)\right] \tag{4.24}$$

For other values of b greater than about $\lambda/50$, a more general form for the amplitude roll-off into the shadow region is approximated for a Fermat surface by Keller [26,27] as

$$F^2(\theta, kb) = (kb)^{1/3}\frac{C_0}{\sqrt{2}}\exp[-\tau_0(kb)^{\theta/3}]\sin(\pi/3)\sqrt{1/k} \tag{4.25}$$

where for a real dielectric surface the appropriate mode coefficients are given in [23] as $C_0 = 0.910719$ and $\tau_0 = 1.8557\exp(\pi/3)$. Note that as the frequency of the incident field increases relative to the radii of curvature (increasing kb), there is less penetration into the shadow region (decreasing F^2). Figure 4-26 illustrates the differences in the scattered signal amplitude and angle distribution for the knife and rounded-tip edges. Figure 4-27 shows, for the two edge types, the diffracted field intensity into the shadow zone as a function of shadow angle, θ. Note the near independence with respect to shadow angle for the knife-edge case. The signal falloff for the rounded-tip case in the shadow zone is significantly greater than its knife-edge counterpart.

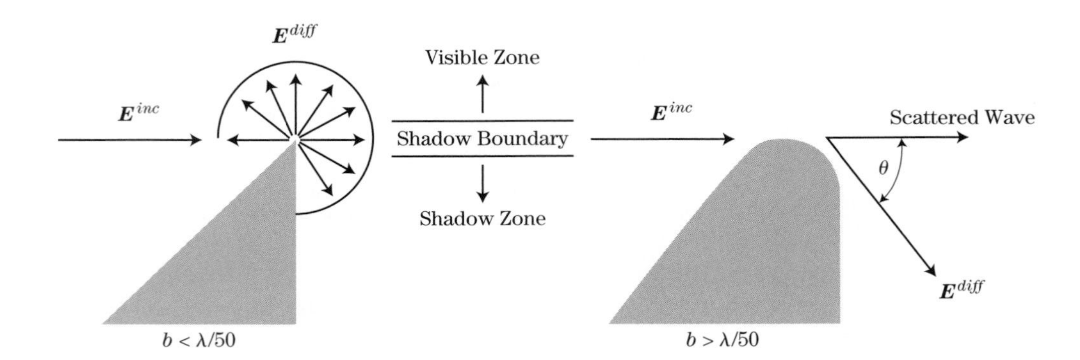

FIGURE 4-26 ■ Local diffraction coefficient, F^2, behavior for two types of edges.

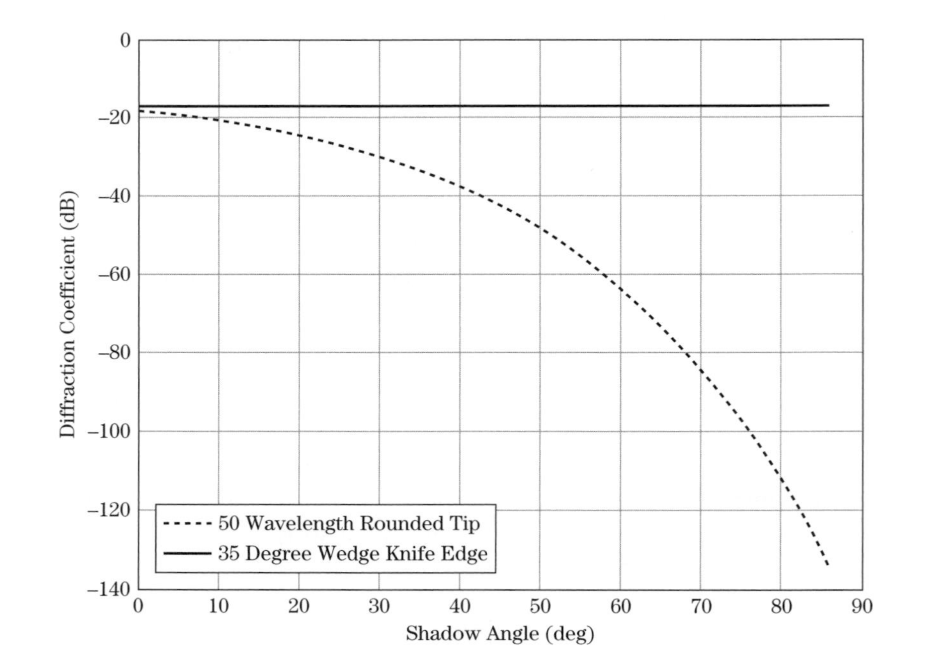

FIGURE 4-27 ■ Diffraction coefficient for rounded tip and knife edge versus shadow angle at 1 GHz.

4.9 MULTIPATH

The IEEE radar definition of multipath is "the propagation of a wave from one point to another by more than one path" [28]. When multipath occurs in radar, it usually consists of a direct path wave combined with other indirect path waves at an observation point. The signal fading (or enhancing) is a consequence of the indirect wave relative phases combining with the direct wave to produce constructive or destructive interference. Virtually everyone has experienced multipath effects on modern equipment, if not with radars. For radio and many cellular phones, sudden losses of reception in certain regions of the city may be the direct result of multipath.

Multipath wave combinations can produce complex propagation factor profiles and are highly sensitive to frequency, antenna patterns, sensor heights, and surface electrical and physical properties. Surface properties can include mountains, buildings, sand dunes,

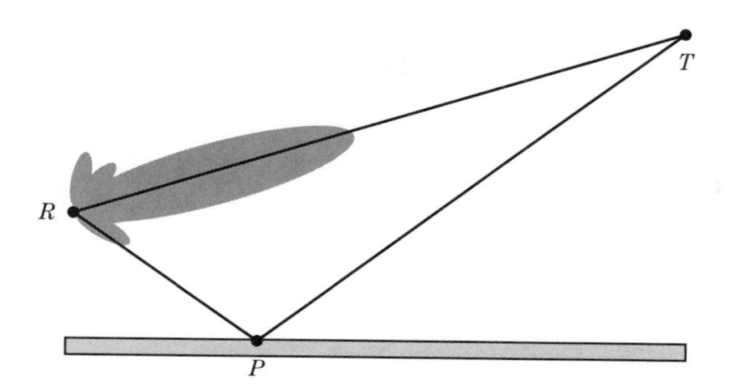

FIGURE 4-28 ■ Illustration of multipath geometry for ground reflection.

tall grass, and discrete scatters. At frequencies below about 40 MHz the multipath field at the observer may also include contributions from reflection off the ionosphere.

This section focuses on the simple case where the intermediate terrain is perfectly flat. The flat-terrain case provides the foundation for extrapolating multipath principles to more complex intermediate surfaces.

4.9.1 Propagation Paths and Superposition

When considering propagation paths it is useful to begin by defining a simple geometry. For a monostatic radar the presence of a flat ground plane adds three new two-way multipath fields. Figure 4-28 depicts a radar (R) tracking a target (T) above a ground plane (P).

The electric field amplitudes of the direct–direct (free-space) path wave and the three new field path waves are as follows:

1. E_{dd}: Path RTR (direct–direct, or DD).
2. E_{di}: Path RTPR (direct–indirect, or DI).
3. E_{id}: Path RPTR (indirect–direct, or ID).
4. E_{ii}: Path RPTPR (indirect–indirect, or II).

For the bistatic or one-way observer geometry (observer at point T), the geometry indicates that only one new path is present. E_d represents the direct field path to the receiver. The bistatic paths and vectors are as follows:

1. E_d: Path RT (D).
2. E_i: Path RPT (I).

The indirect fields will have different characteristics from the direct path due to reflection off a boundary and the additional path length they must travel to the observer. The additional path length is often denoted δR. The effect of the indirect–indirect path is commonly modeled with a virtual target, T', that exists below the ground plane. The virtual target—for analysis purposes—is at an equal distance below the plane as the real target (or observer) is above it. Figure 4-29 illustrates this concept for an air-to-ground geometry.

Neglecting the small impact on amplitude due to additional path length at the target (or observer), the indirect signal's amplitude is modified solely by the *reflection coefficient*, Γ, at reflection point P. With this in mind, the one-way power propagation factor describing the coherent sum of the direct and indirect path signals at Γ will be

$$F^2 = |1 + \Gamma \cos(k\delta R)|^2 \tag{4.26}$$

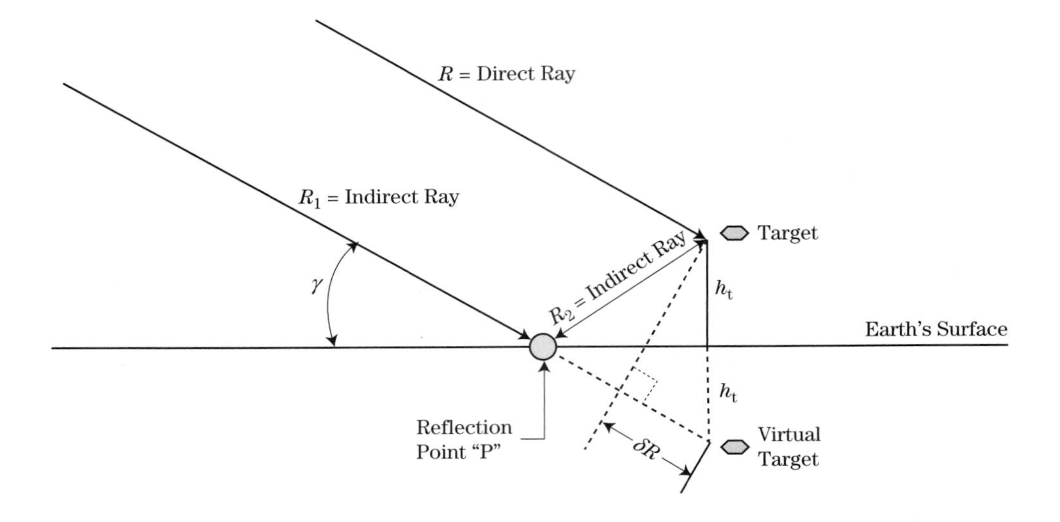

FIGURE 4-29 ■ Simplified multipath geometry for flat Earth and the virtual target.

where 1 represents the direct-path E-field magnitude, Γ is the complex reflection coefficient of the surface, k is the wavenumber, and ΔR is the difference in range between the direct path and the reflected path. For two-way geometries (e.g., radar to target and back), this factor becomes

$$F^4 = |1 + 2\Gamma\cos(k\delta R) + [\Gamma\cos(k\delta R)]^2|^2 \tag{4.27}$$

For a reflection coefficient of 1, the maximum value of F^2 (i.e., the maximum signal power enhancement due to multipath) is 6 dB (4×) for the one-way case and 12 dB (16×) for the two-way case (F^4).

For most long-range, point-to-point situations where the grazing angle, γ, is below about 5 degrees, δR may be approximated by

$$\delta R = 2h_a h_t / d \tag{4.28}$$

where $h_a$and h_t are the transmitter and target heights, respectively, and d is the ground range between them. The specular grazing angle, γ_s, defined by the simplest form of Snell's law[6] can be approximated by

$$\gamma_s = \tan^{-1}\left(\frac{h_a + h_t}{d}\right) \tag{4.29}$$

Superposition is used to determine the effect of different ray contributions to the total field. In the simplest case, the direct ray, E_d, and the forward scattered indirect ray, E_i (one-way geometry), interfere at an observation point (receiver). The composite field at any receiver in a multipath environment is the sum of the direct ray and the indirect or reflected rays (Figure 4-30), which produce significant amplitude lobing structures at the observer. The interference structure can also be classified by its lobing periodicities. The higher the frequency, the more rapidly the relative phase will change and thus the shorter the period between interference lobes. Very smooth surfaces are classified by low-frequency

[6] When the reflection is in the same medium ($n_1 = n_2$), the angle of reflection equals the angle of incidence.

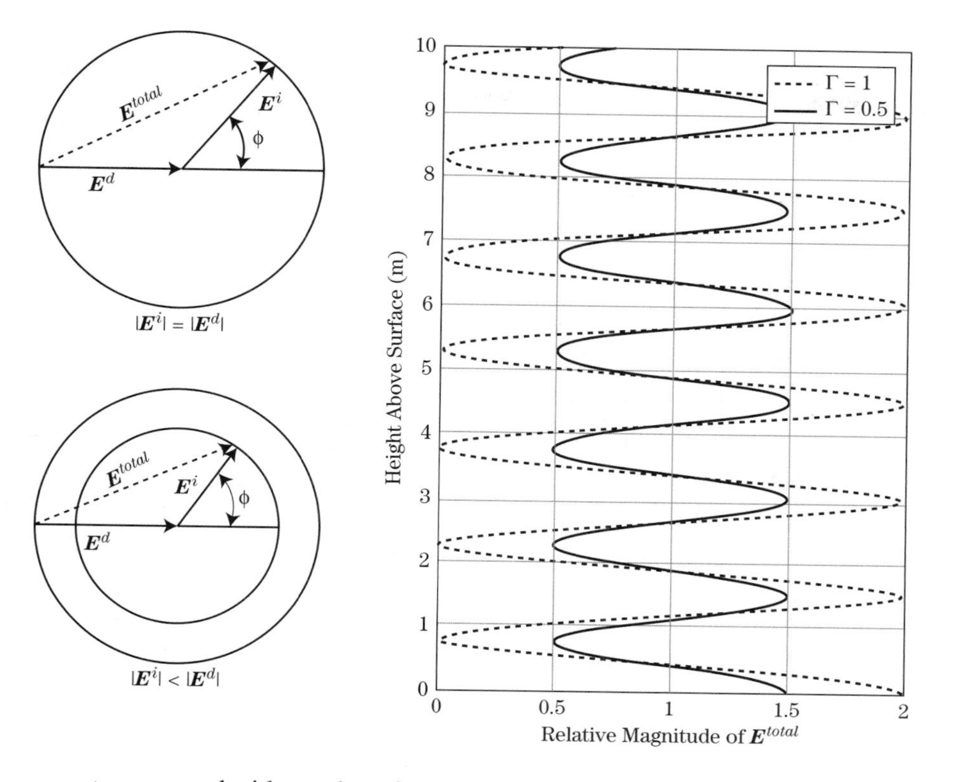

FIGURE 4-30 ■ Coherent summation of direct and indirect ray interference pattern versus observer height above ground plane (10 GHz, $h_a = 1$ m, $d = 100$ m).

components compared with rough surfaces, which have higher-frequency structures of the interference patterns lobing structure.

Figures 4-31 and 4-32 illustrate the lobing structure for the received signal as a function of range from a receiver or target in the presence of multipath. At some ranges the signal is enhanced over the free-space estimate due to constructive interference of the direct and multipath rays, allowing for extended range performance of a particular radar.

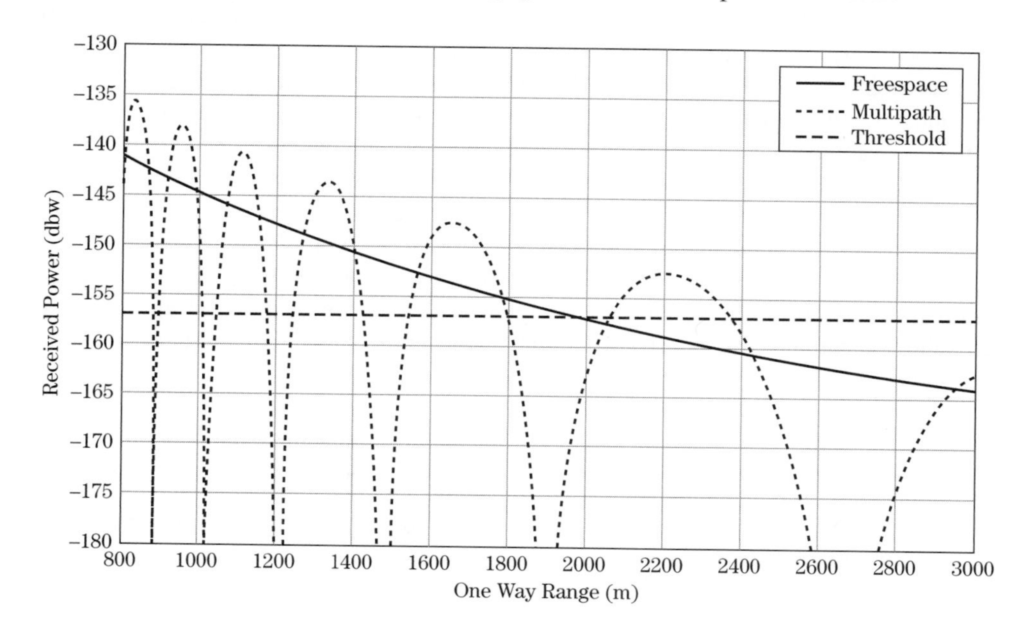

FIGURE 4-31 ■ Multipath signal lobing versus free space for one-way geometry and $\Gamma = 1$.

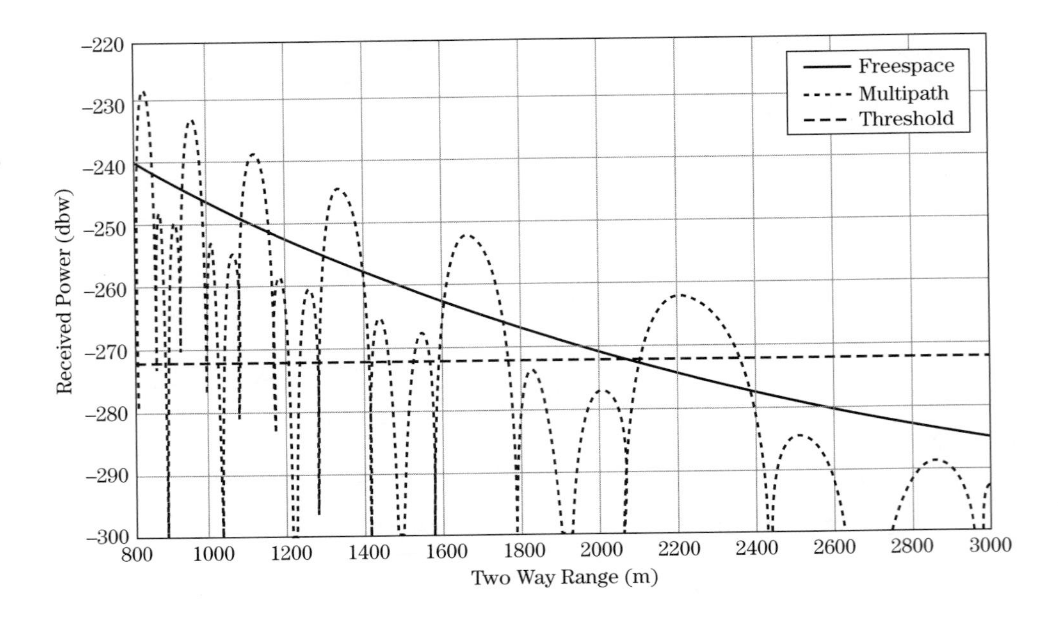

FIGURE 4-32 ■ Multipath signal lobing versus free space for two-way geometry and $\Gamma = 1$.

In the figures, note that a performance threshold is indicated for this particular radar parameter set. In both illustrations (provided for a transmitter and receiver height of 10 meters and a transmit frequency of 10 GHz), the multipath "gain" over free space actually extends the maximum range of the system's performance, by almost 300 meters in the two-way case. This gain comes at the cost of multipath losses that prevent detection at certain ranges. However, most geometries exhibit only significant losses (>10 db) over short-range separation intervals, as seen in the figures.

4.9.2 Describing the Reflecting Surface

The surface boundary can be thought of as a combination of a *macro* and a *micro* component. The macro component is described as a smooth semi-infinite dielectric or conductor with classical EM properties, whereas the micro component is described in terms of roughness and density/orientation. As an example, for a grassy field the macro component would be the general terrain profile of the field, and the micro component would be the detailed structure of grass growing on the field surface. The roughness of the micro feature modifies the ideal nature of the smooth surface boundary. Figure 4-33 shows the decomposition of a real surface structure into these two key components for analysis.

The effect of a smooth surface on a propagating wave is generally described by its Fresnel reflection coefficient. The roughness factors are derived by either discrete modeling or statistical approaches. For statistical approaches these roughness modifiers

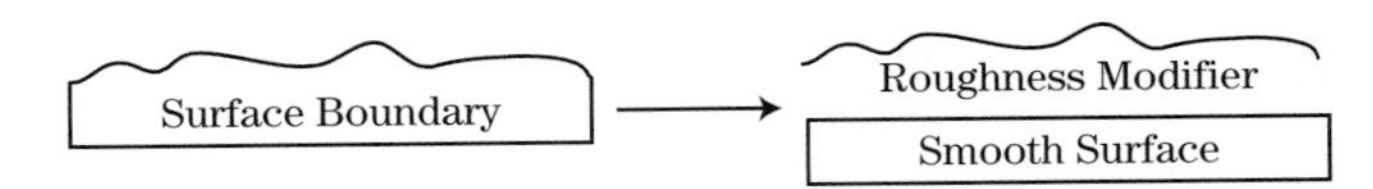

FIGURE 4-33 ■ Complex surface boundary decomposed into smooth and roughness factors.

TABLE 4-5 ■ Surface Electrical and Physical Parameters at MMW ($\mu = 1$)

Surface Type	ε_r	σ_+ (mho/m)	σ_h (m)	β_0 (rad)
Mowed grass	10	0.001	0.01	0.1
Tall grass	10	0.001	0.1	0.2
Gravel	4	0.001	0.02	0.3
Asphalt	6	0.001	0.0004	0.36
Brush	4	0.001	0.5	0.1
Snow	2.5	0.001	0.003	0.25
Desert	2.5	0.001	0.003	0.05
Trees	1.5	0.001	1.5	0.2
Sea water	80	4	*	*
Fresh water	67	0.1	*	*

relate to rms height and rms slope of elemental surface facets overlaid upon the smooth surface component [29,30]. Statistical approaches work well for bounding set prediction; however, for higher-fidelity representation of a specific scene, discrete modeling, which digitizes the ray structure on the complex surface, is best. This can be accomplished with discrete ray tracing or method-of-moments (MoM) models such as models TKMOD and MRSim (see Section 4.11).

For multipath analysis, surfaces are generally described in terms of three electrical and two physical parameters: (electrical) permeability, μ; permittivity, ε; conductivity, σ_+; (physical) rms roughness, σ_h; and rms slope, β_0. For most frequencies of interest, these parameters have been measured for specific types of surfaces. Table 4-5, compiled from many sources, lists representative values of these parameters for a number of surface classes at MMW frequencies [31,32]. Note that the permittivity listed is the relative dielectric constant, ε_r, based on the ratio of free-space dielectric to dielectric in the medium. The blank parameters marked "*" are determined from sea-state tables[7] and the relationship between surface correlation length and rms slope. For statistical models of surface features, the spatial correlation length, T, is given by [24]

$$T = \frac{2\sigma_h}{\tan \beta_0} \tag{4.30}$$

The spatial correlation length describes the minimum surface distance from a given point at which the features will have significantly changed. It can be directly measured and is often used to infer the equivalent roughness and slope of a surface. It can also be used to determine terrain sampling requirements when using digital simulations derived from measured elevation data.

Using the concepts discussed, Table 4-6 gives representative values of the physical properties for various sea states. Given a sea state (wave height), the equivalent rms roughness can be estimated. If the wave correlation length is also measured, the rms slope can then be determined from Table 4-6, the roughness, and equation (4.30).

[7] *Sea state* is a measure of rms wave height. See Chapter 5 for more information.

TABLE 4-6 ■ Water Surface Parameters by Sea State (All Wind Values)

Sea State	Type	σ_h (m)	β_0 (rad)
0	Calm	0.0001	<0.0001
1	Smooth	0.0–0.03	0.0353
2	Slight	0.03–1.0	0.3097
3	Moderate	1.0–1.5	0.1368
4	Rough	1.5–2.5	0.1064
5	Very rough	2.5–4.0	0.1021
6	High	4.0–6.0	0.1092
7	Very high	>6.0	0.1273

4.9.3 The Multipath Reflection Coefficient

The Fresnel reflection coefficient, Γ, that describes the effect of the surface on the complex voltage of the reflected wave is a combination of the smooth earth Fresnel reflection, Γ_0, the spherical earth divergence factor, D, and the sum of both specular and diffuse roughness modifier components, ρ_s and ρ_d, respectively. This relationship is

$$\Gamma = \Gamma_0 D(\rho_s + \rho_d) \tag{4.31}$$

Each of the components of Γ is discussed in the following subsections.

4.9.3.1 Fresnel Reflection Coefficients

For most surfaces of interest the Fresnel reflection coefficient, Γ_0, may be expressed for polarization parallel and perpendicular to the plane of incidence (the surface) according to [31]

$$\begin{aligned} \Gamma_0^{VV} &= \frac{Y^2 \sin\gamma - \sqrt{Y^2 - \cos^2\gamma}}{Y^2 \sin\gamma + \sqrt{Y^2 - \cos^2\gamma}} \\ \Gamma_0^{HH} &= \frac{\sin\gamma - \sqrt{Y^2 - \cos^2\gamma}}{\sin\gamma + \sqrt{Y^2 - \cos^2\gamma}} \end{aligned} \tag{4.32}$$

where Y is the boundary admittance and γ is the grazing angle. The boundary admittance is defined as $Y = \sqrt{\varepsilon_{rc}/\mu_{rc}}$, where μ_{rc} is the relative complex magnetic component of admittance and ε_{rc} is the relative complex dielectric constant. In most cases μ_{rc} is simply 1.0. ε_{rc} is given by the combination of the relative dielectric constant, ε_r, and the conductivity, σ_+, as

$$\varepsilon_{rc} = \varepsilon_r + \frac{\sigma_+}{j\omega} \tag{4.33}$$

where ω is the radian frequency of the wave.

For many surfaces of interest (e.g., desert, sea water, grass), the Fresnel reflection coefficient, Γ_0, is typically -1.0 at lower frequencies on long-range ground-to-ground links.[8] Figure 4-34 demonstrates the computed magnitude and relative phase of the

[8]This is primarily due to Brewster's angle being above the very low grazing angle in these geometries. The phase is based on Stratton's convention [33], where the reflected polarization perpendicular to the boundary (vertical) is 180 degrees out of phase with the incident wave.

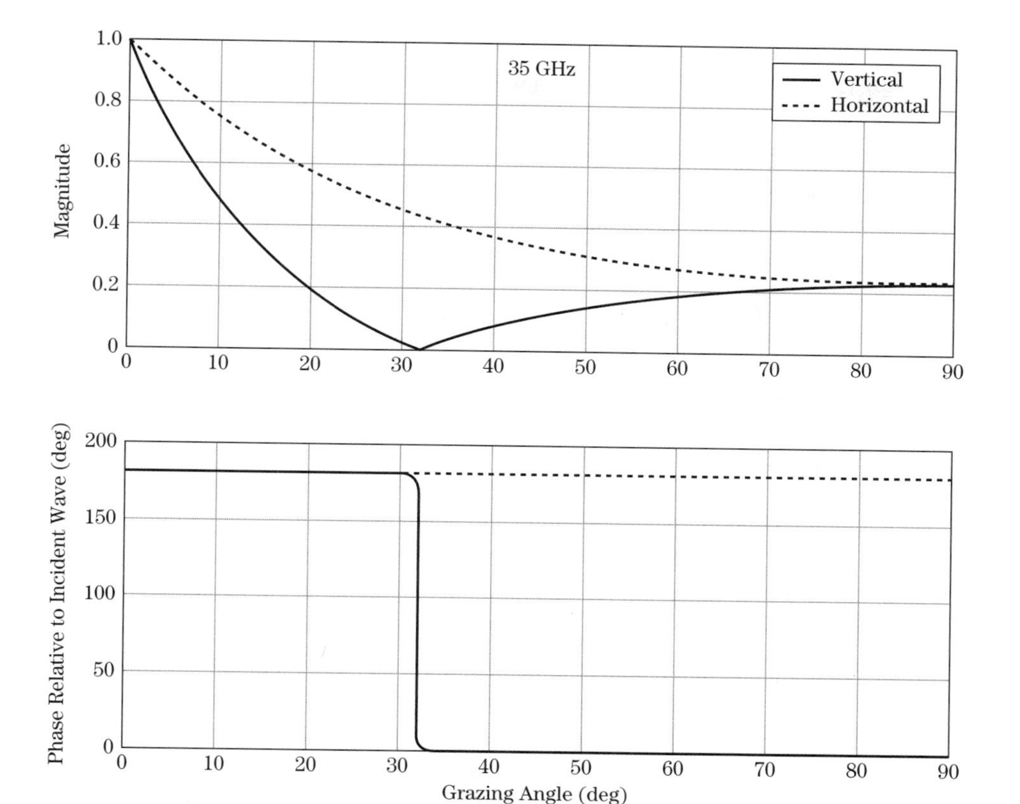

FIGURE 4-34 ■ Fresnel reflection coefficient (magnitude and relative phase) for a desert surface.

reflection coefficient to the incident wave for a desert surface ($\varepsilon_r = 2.5$) at Ka-band for the two linear polarizations.

Based on the Fresnel equations, two key angles are of interest for radar propagation: the critical angle and Brewster's angle. The *critical angle,* θ_c, allows all the energy to be reflected off the boundary layer. When considering long-range communication or OTH radar propagation, both of which operate via one or more reflections off the ionosphere, the critical angle is often used to ensure full reflection off a particular boundary when multipath is in a gain condition. It is given by

$$\theta_C = \sin^{-1}\sqrt{\frac{\varepsilon_2 v_2}{\varepsilon_1 v_1}} = \sin^{-1}\sqrt{\frac{\varepsilon_2}{\varepsilon_1}} = \sin^{-1}\sqrt{\varepsilon_r} \tag{4.34}$$

where ε_r is the relative dielectric constant between the two layer boundaries.

Brewster's angle is the grazing angle at which all of the incident energy is transmitted into the boundary with no reflection; there is no multipath field. It is given by

$$\theta_B = \sin^{-1}\sqrt{\frac{\varepsilon_2}{\varepsilon_1 + \varepsilon_2}} = \tan^{-1}\sqrt{\frac{\varepsilon_2}{\varepsilon_1}} = \tan^{-1}\sqrt{\varepsilon_r} \tag{4.35}$$

Brewster's angle is most often exploited for ground-to-ground communications and ground-to-space applications. Since the energy incident on the boundary is not reflected, it cannot interfere with the reception at a receiver. Figure 4-35 plots the critical and Brewster's angles versus relative dielectric constant of the boundary.

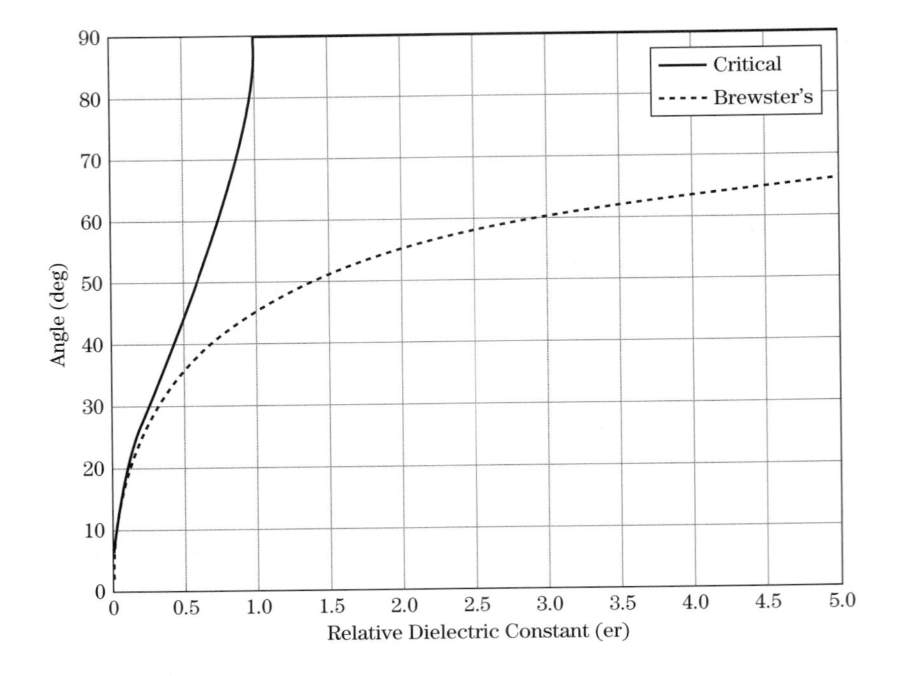

FIGURE 4-35 ■ Key reflection/transmission angles versus relative dielectric constant of the boundary.

In many ground-to-ground engagements, the grazing angle to the surface is much less than 1 degree. In air-to-ground engagements, the grazing angles can vary widely. The angle at which the vertically polarized E-field reflection coefficient reaches zero and the phase changes by 180 degrees is Brewster's angle, discussed earlier. At angles near Brewster's, the incident energy is largely transmitted into the medium, leaving less to be forward scattered to form a multipath ray. Table 4-7 provides the value of Brewster's angle for several terrain types at 35 GHz.

For grazing angles below those listed, no circular polarization sense change is observed in the scattered field. However, the closer the grazing angle is to Brewster's angle, the more the vertically polarized scattered component is attenuated, thus causing an imbalance between the V and H vector components of a circularly polarized incident wave and converting the scattered wave to elliptical polarization. For grazing angles above Brewster's, circular polarization sense flips on the reflected wave such that right-handed circular (RHC) becomes left-handed circular (LHC) and vice versa.

TABLE 4-7 ■ Computed Brewster's Angle for Selected Surfaces at 35 GHz

Surface Type	Brewster's Angle (Grazing; in Degrees)
Desert	32.3
Grass	17.5
Gravel	26.6
Snow	32.3
Sea water	6.4
Perfect conductor	0.0 (e.g., sheet of metal)

4.9.3.2 Divergence Factor

When a radar beam is reflected off a flat, perfectly conductive surface, it can be argued that the amplitude of the reflected beam equals that of the incident beam. This means that a direct and indirect ray arriving at a target will have approximately the same power. This is not the case when the reflection occurs on a smooth spherical surface, such as when a radar beam is reflected off a smooth, calm sea. Due to the curvature of the spherical surface, the reflected waves will diverge instead of following a parallel path, as seen in Figure 4-36. Therefore, the divergence factor, D, is incorporated to account for this beam divergence.

For the geometry defined in Figure 4-36, where the grazing angle, ψ, is assumed small such that $\sin(2\psi) \approx 2\psi$, it can be shown that

$$\psi \approx \frac{1}{r_1}\left[h_1 - \frac{r_1^2}{2a_e}\right] \tag{4.36}$$

This approximation for ψ can be used in an expression for D given by Kerr [24], p. 99, to obtain

$$D \approx \left[1 + \frac{2r_1^2(r - r_1)}{a_e r\left(h_1 - \left(r_1^2/2a_e\right)\right)}\right]^{-1/2} \tag{4.37}$$

which is valid for small grazing angles, ψ.

Most radar applications involve ranges that do not exceed the standard atmosphere radar horizon. For these applications, the impact of the divergence factor is negligible (e.g., $D \approx 1$). In very long-range applications, the curvature of the earth can have a

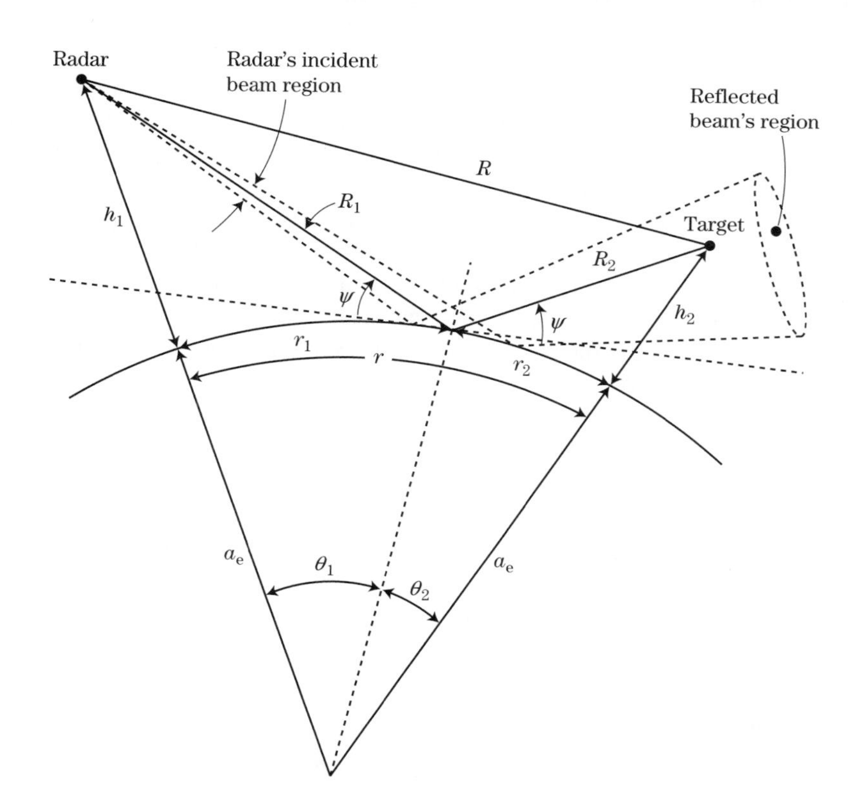

FIGURE 4-36 ■ Geometry for reflections from a spherical, smooth earth.

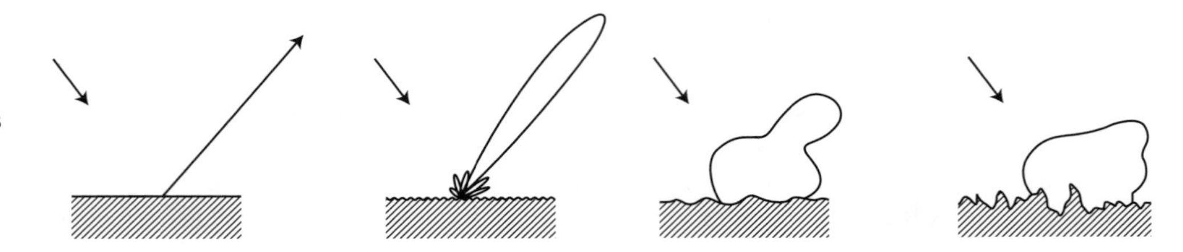

FIGURE 4-37 ■ Illustration of specular to diffuse scattering transitions with increasing surface roughness.

significant impact on the total reflection coefficient. This impact, coupled with a rough, curved surface, yields a more complex situation more fully described in Beckmann and Spizzichino [34].

4.9.3.3 Roughness Factors

The roughness factors, ρ_s and ρ_d, modify the reflection coefficient, Γ, as shown in equation (4.31). Specular scattering is generally described as a point reflection at Snell's angle from the intermediate surface. Diffusely scattered fields emanate from a spatially distributed area known as the *glistening surface* and decorrelate more rapidly than the "coherent" specular terms. Diffuse interference typically does not cause persistent signal fades but has been shown to contribute residual error in track processors.

As surface roughness increases, the dominant scattered wave transitions from specular to diffuse (Figure 4-37). In specular scattering, the reflected wave is a close replica of the incident wave and can cause coherent interference patterns at the receiver. In the limit, diffuse scattering is nearly hemispherical (also known as a *Lambertian reflectance*, or scattering in the case of light) but with lower gain in any particular direction (i.e., a fuzzy replica of the incident wave). Diffuse energy thus tends to "splatter" over a broader region of angle space.

Conservation of energy dictates that, when specular reflection is reduced, diffuse energy becomes more significant in the scattered field and also implies that surface roughness is increasing. In general, specular multipath occurs when the surface roughness is small compared with the wavelength and diffuse multipath becomes significant when roughness is higher compared with the wavelength. Beckmann and Spizzichino [34] observed that in the limit for nondirectional antenna, the maximum value of the diffuse roughness factor, ρ_d, is 0.5. Figure 4-38 illustrates the magnitude of the specular and diffuse roughness factors as a function of normalized roughness. The normalized surface rms roughness value, σ_h', takes into account the perceived roughness at a specific radar frequency and the grazing angle according to

$$\sigma_h' = \frac{\sigma_h}{\lambda} \sin\gamma \tag{4.38}$$

Specular Scattering The specular field is essentially a scaled replica of the transmit signal delayed by $\Delta R/c$ seconds. The amplitude of the specular roughness factor ρ_s is given by

$$|\rho_s| = \exp\left(-\frac{4\pi}{\lambda}\sigma_h \sin\gamma\right)^2 = \exp\left(-4\pi\sigma_h'\right)^2 \tag{4.39}$$

The phase of the specular roughness factor, ϕ_s, includes the Fresnel term and a term due to the extra path length,

$$\phi_s = \phi_{Fresnel} + k\delta R \tag{4.40}$$

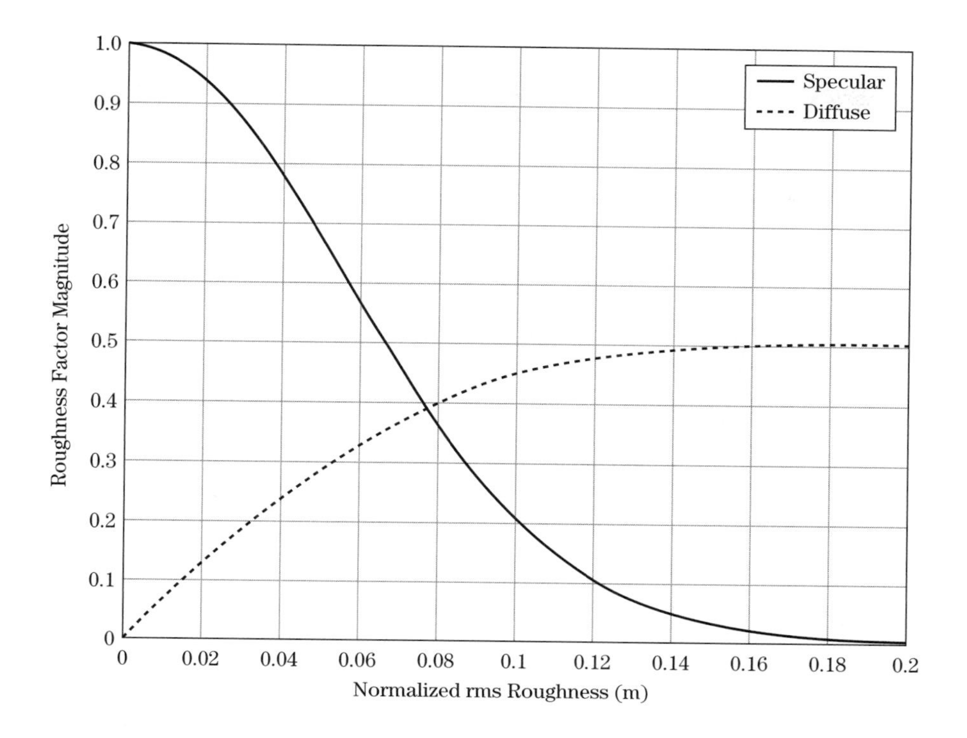

FIGURE 4-38 Specular and diffuse roughness factor transition versus normalized surface rms roughness.

For temporally varying surfaces (e.g., wind-blown trees, water), the specular roughness factor should not be modeled as a constant but as a pseudo-random process that captures the temporal, spatial, and frequency decorrelation statistics appropriate for the surface type. The power spectrum is generally modeled as a low-pass function with a cutoff frequency, ω_c, determined from the decorrelation time, T, according to

$$\omega_c = \frac{2\pi}{T} \tag{4.41}$$

An example of measured autocorrelation functions for scattered field amplitude from a moving terrain (nominal sea condition) is given in Figure 4-39. Additional examples of clutter correlation statistics are given in Chapter 5. The decorrelation point is generally considered to be the time lag at which the normalized autocorrelation function first falls to a value of $1/e$, or about 10% of its original amplitude. In Figure 4-39, the decorrelation time is evidently on the order of 1 second.

Diffuse Scattering The magnitude of the diffusely scattered field is computed from the geometry, rms surface roughness, and slope parameters. Because of conservation of energy, the magnitude of the diffuse reflection coefficient will be limited by the specular reflection magnitude [36]. Based on this conservation principle and the maximum values observed for diffuse scatter in the presence of a strong specular field [34], the magnitude of the diffuse roughness factor contribution (in the limit) can be approximated by

$$\left|\rho_{d_{\text{limit}}}\right| = 0.5\sqrt{(1 - |\rho_s|)} \tag{4.42}$$

This value was used to estimate the diffuse magnitude as a function of specular magnitude in Figure 4-38. Consider the geometry shown in Figure 4-40. The nominal

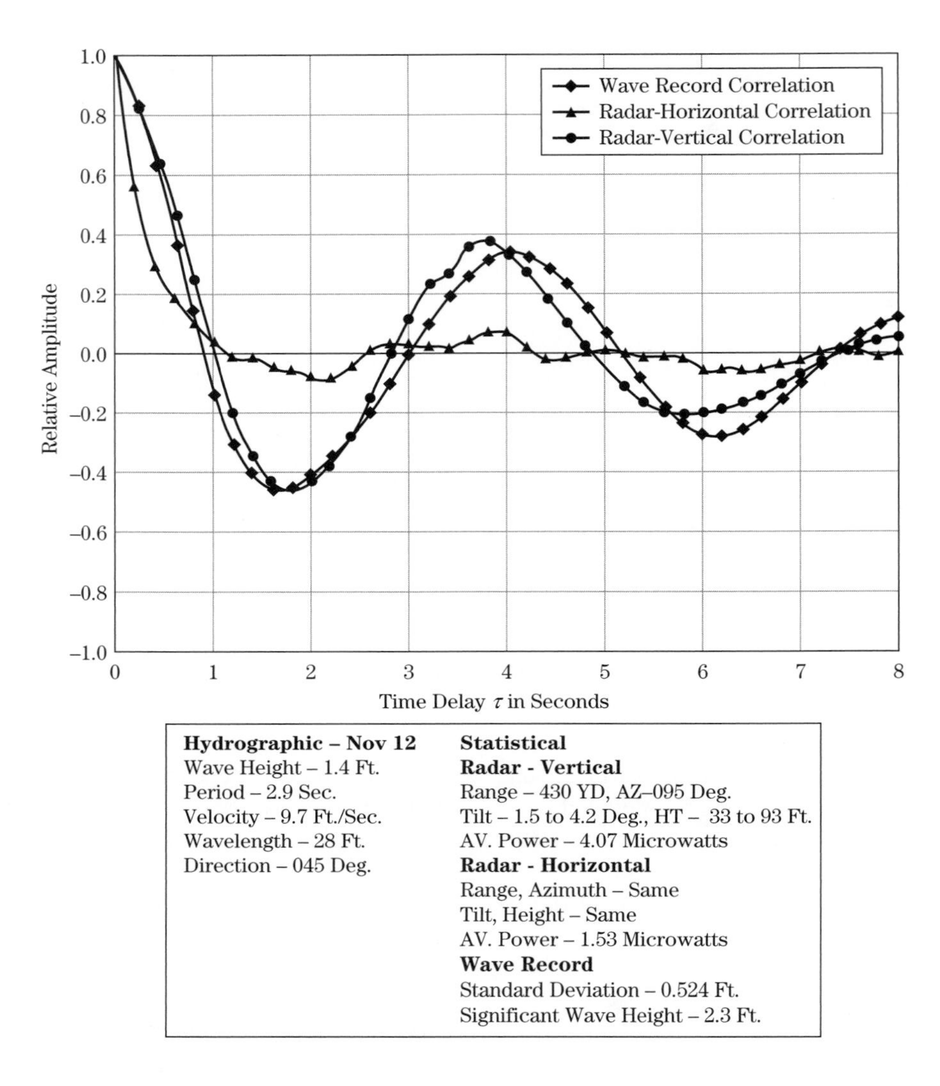

FIGURE 4-39 ■ Decorrelation data from nominal sea/wind condition. (From Meyers [35]. With permission.)

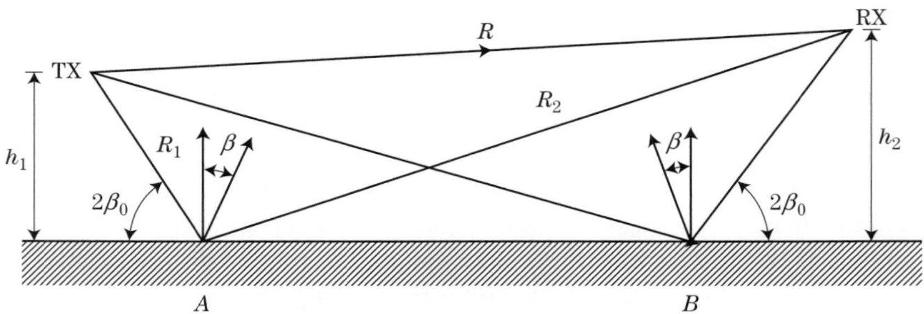

FIGURE 4-40 ■ Diffuse primary scattering geometry.

component amplitudes for each of the diffuse roughness factors, $\rho_{d_{1,2}}$, are approximated by [34]

$$\left|\rho_{d_{1,2}}\right| = \frac{R^2}{R_1^2 R_2^2 \tan^2 \beta_0} \exp\left(\frac{\tan^2 \beta}{\tan^2 \beta_0}\right) \tag{4.43}$$

where β is the difference between the surface normal and bisector angle of the incident and reflected rays (R_n,R_{n+1}), and β_0 is the rms surface slope parameter.

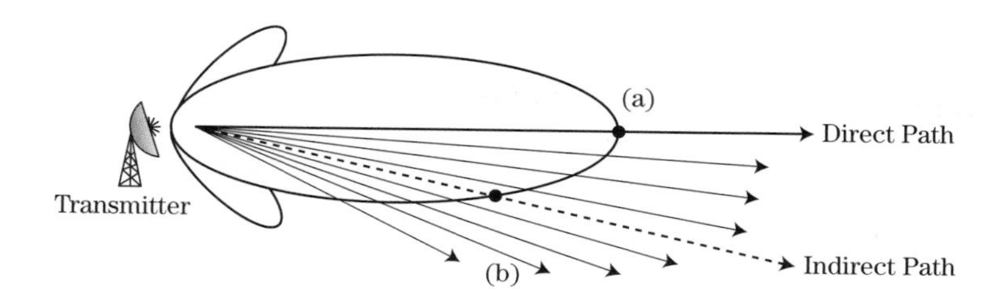

FIGURE 4-41 ■ Ratio of antenna gains at direct and indirect path angles.

The phase of the diffuse energy is based on the extended path length to two primary reflection points (points A and B in Figure 4-40) directly in front of the transmitter and receiver. The effects of the antenna gain pattern must also be included in the computation of amplitude and phase for scattered rays at the appropriate indirect angles. Similar to the specular reflection case, a random process model is needed to account for temporal decorrelation of each diffuse reflection component.

4.9.3.4 Angle Error

Specular multipath is a source of angle error biases. For a monopulse tracking system (see Chapter 18), these bias errors decorrelate very slowly and can lead to significant residual errors in the tracking filter. On the other hand, a shorter wavelength causes associated increases in diffuse multipath contributions that can be significantly reduced (but not eliminated) by simple temporal integration or averaging. Similar results can occur with narrow-band Doppler filtering. In low-angle applications where the tracker illuminates both the true target and the virtual target created by the indirect paths, the indicated angle is that of the equivalent two-scatterer target of Figure 4-29. In these cases the rms angle error caused by multipath can be computed from [36]

$$\sigma_{rms} = \frac{\rho_s \theta_3}{\sqrt{8G_{se}(\text{peak})}} \tag{4.44}$$

where θ_3 is the one-way 3 db beamwidth of the tracking antenna, and G_{se}(peak) is the ratio of the antenna gains for the direct path target and the indirect path image at angles shown in Figure 4-41a and Figure 4-41b.

Angle errors due to specular multipath reflection are summarized for two example surface roughness values in Table 4-8 for a low-angle ground-to-ground geometry where the aperture size was limited to 5.5 inches.[9] When the terrain is very smooth, all the system "sees" is a very high specular reflection. As the surface roughness increases, the higher frequency systems see the earth as an increasingly rough surface, thus reducing the magnitude of specular reflections and associated error biases. Multipath usually impacts only the elevation plane with angle errors, although diffuse multipath, which emanates from a wider surface area, can induce residual errors into the azimuth plane.

4.9.3.5 Classification Error

The presence of additional target images at range delays δR can also cause the true target to have an extended length, which can cause problems for classifiers using target length as a discriminant. For the one-way case, a replica of the true target will exist at a range

[9] This aperture size produces a 3 dB one-way beamwidth of 13 degrees at 10 GHz and 3.6 degrees at 35 GHz.

TABLE 4-8 ■ Elevation Angle Error for Two Frequencies and Surface Roughness Values (G_{se}(peak) = 1)

Roughness	10 GHz	35 GHz
0.1 m	67.2 mr	21.1 mr
1.0 m	66.5 mr	19.6 mr

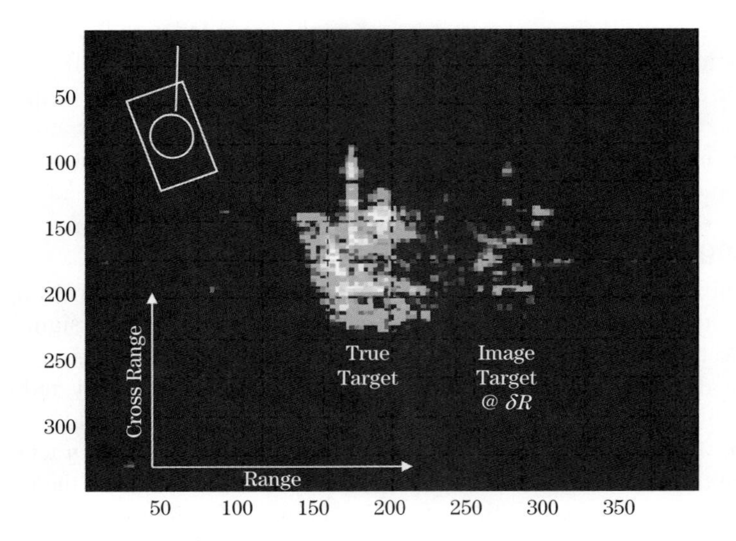

FIGURE 4-42 ■ ISAR image of a tank located over a smooth surface.

delay of δR. For the two-way case, replicas of the true target will exist at range delays of δR and $2\delta R$. The replica at δR for the two-way case can also be larger in amplitude than the true target if the reflection coefficient on the surface is high. For a distributed target (e.g., ships, tanks), each dominant scatterer may have these replicas causing the scatterer distribution to significantly change over free space. Figure 4-42 illustrates the distribution of two-dimensional scatterers on a tank target at 35 GHz and the ghost image created by the multipath bounce at range delay δR. For this two-way geometry, the third image located at $2\delta R$ is below the threshold of the gray scale, indicating that the reflection coefficient was below 0.7 on the surface.

4.10 | SKIN DEPTH AND PENETRATION: TRANSMITTING THROUGH WALLS

Most materials other than metal are poor conductors to radar waves, and, instead of the waves being reflected, the waves will penetrate into or through the material. *Skin depth* describes how far a signal penetrates past a boundary and into a material. When applied to the transmission of EM waves through walls, ground,[10] and other obstacles, it is customary

[10]Ground-penetrating radar (GPR) is another application in which these principles apply. GPR systems generally use a specialized waveform (impulse approximations or monocycles) to maximize energy into the medium.

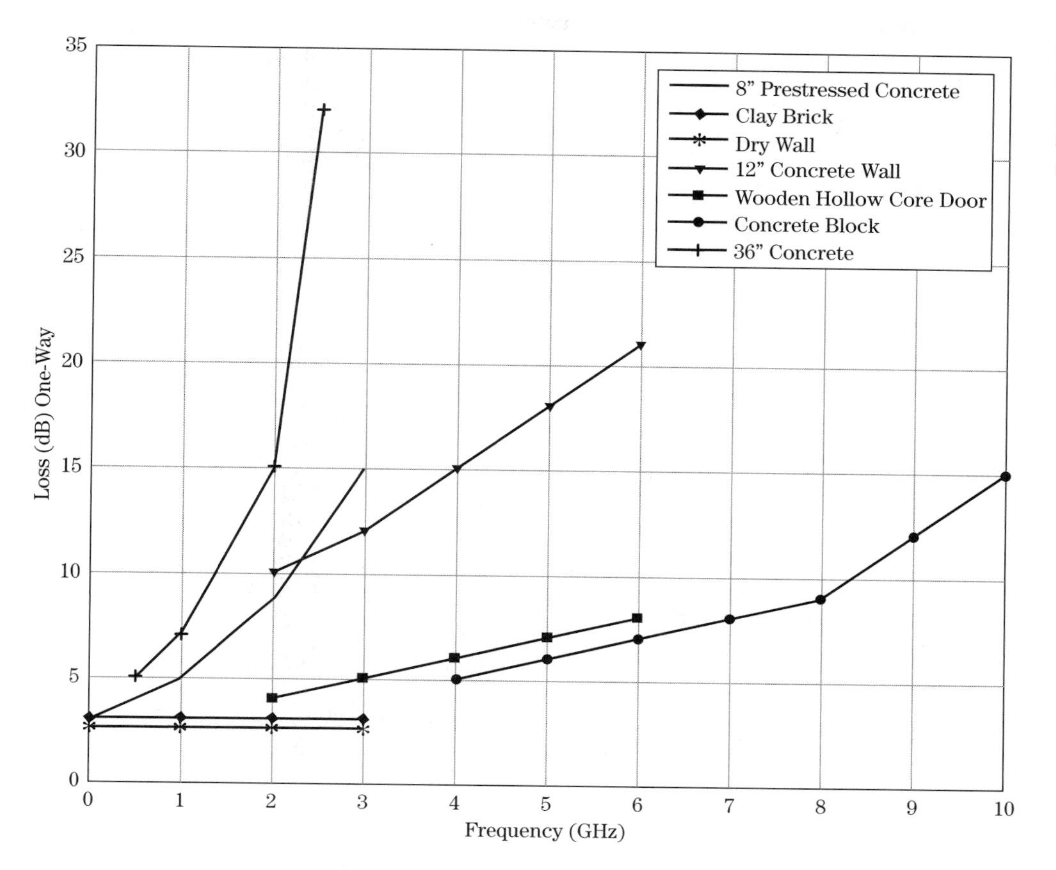

FIGURE 4-43 Radar wave penetration through some standard materials.

to define the skin depth or depth of penetration, δ, as that distance at which the transmitted wave intensity has been attenuated to $1/e$ of its value at the wall surface.

For "good conductors" the skin depth or penetration of the wave at normal incidence is

$$\delta = \sqrt{\frac{2}{\omega\mu\sigma_+}} = \sqrt{\frac{1}{\pi f \mu\sigma_+}} \tag{4.45}$$

where f is the frequency (Hz), μ is the material's permeability (H/m), and σ_+ (S/m) is the material's conductivity. A number of material properties useful for computing skin depth have been compiled by Neff [37]. If the skin depth is greater than the thickness of the material, then the attenuated wave will exit the boundary into the next medium.

Data on the penetration of radar waves into different material types and thicknesses have been published in a number of references, most notably in [38-41]. Figure 4-43 shows attenuation versus carrier frequency for several common wall materials. As expected, the losses generally increase with higher frequency.

While in the medium, the radar wave will slow down due to the refractive index change. The phase velocity in the medium is given by

$$v_p = \frac{c}{n} \tag{4.46}$$

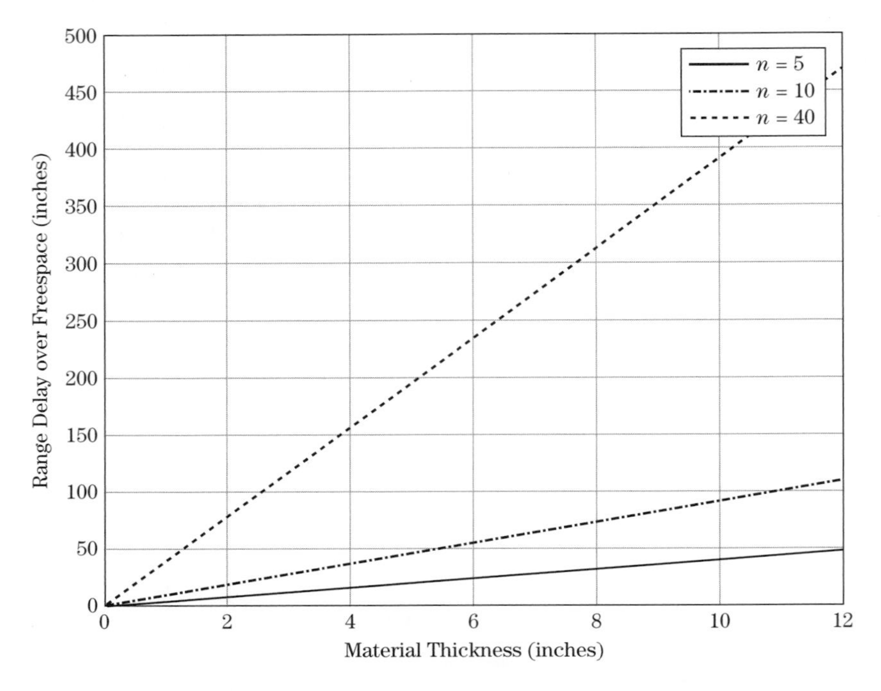

FIGURE 4-44 ■ Radar wave range delay through boundary materials with three refractive indexes.

where c is the speed of light (free-space phase velocity, or 3×10^8 m/s), and n is the refractive index of the material. Thus, the thicker the material and the higher the refractive index, the longer the propagation time through the material. Figure 4-44 illustrates the additional time (range) delay of a wave propagating through a material of various thicknesses for three refractive indexes.

This effect can be quite substantial. The more material in the path, the more delay is incurred by the propagating wave. This is a difficult obstacle for time of arrival measurement systems that must see through many walls into buildings.

4.11 COMMERCIAL SIMULATIONS

Numerous currently available computer models in government and industry may be subjectively classified as off the shelf or developmental. Following is a list of a number of such models, with links to websites providing more information on these simulations:

- TIREM (Terrain Integration Rough Earth Model)
  http://www.alionscience.com
- LEIBE MPM (Microwave Propagation Model)
  http://adsabs.harvard.edu/abs/1989IJIMW..10..631L
- TKMOD5C
  http://www2.electronicproducts.com/PrintArticle.aspx?ArticleURL=CAERP5.NOV1994
- COMBIC (Combined Obscuration Model for Battlefield-Induced Contamination)
  http://www.ontar.com
- XPatch
  http://www.saic.com/products/software/xpatch/

TABLE 4-9 ■ Propagation Model Utility and Fidelity Assessment Matrix

Model Name	Atmos Atten	Atmos Refraction	Atmos Turbulence	Surface Intervisibility	Surface Diffraction	Surface Multipath
TIREM	◑	○	○	◑	●	◑
LEIBE MPM	●	○	◑	○	○	○
TKMOD5C	◑	○	○	●	○	●
COMBIC	◑	◑	◑	○	○	○
XPatch	◑	○	○	●	◑	◑
MRSIM	◑	○	○	●	○	●
EREPS	◑	●	○	○	○	○
TEMPER	●	●	○	○	○	○
PCPEM	●	●	○	○	○	○

- MRSIM (Multispectral Response Simulation)
  http://resrchnet.com/products.htm
- EREPS (Engineering Refractive Effects Prediction)
  http://areps.spawar.navy.mil/2858/software/ereps/readerep.txt
- TEMPER (Tropospheric Electromagnetic Parabolic Equation Routine)
  http://www.jhuapl.edu/techdigest/td2502/Awadallah.pdf
- PCPEM (Personal Computer Parabolic Equation Model)
  http://www.nat-hazards-earth-syst-sci.net/7/391/2007/nhess-7-391-2007.pdf

In most cases, closed-form solutions include a number of limiting assumptions and are generally thought to be rules of thumb for quick analysis insight. Many current codes combine closed-form solutions with digital terrain and wave representations.

Table 4-9 attempts a rating of the tools based on algorithm level of fidelity for each propagation mechanism. This matrix is not intended to be complete, nor does it indicate an endorsement or criticism of each code. It is provided only to demonstrate numerous options to the reader for assessments of mechanism impacts in more complicated scenarios. Note that all the probable mechanisms are covered by at least one of the identified models. For each of the mechanisms, one of three rating levels is listed:

- ○ **No Support for Mechanism:** The simulation doesn't predict or incorporate effects of the propagation mechanism at all.
- ◑ **Low/Medium Fidelity:** The simulation or model uses rule-of-thumb solutions and presumes homogenous or uniform distributions and occurrences throughout the medium. Phenomenological variants are represented in a similar manner.
- ● **High Fidelity:** The simulation or model uses discrete models or series expansions to closely emulate the physics of a particular mechanism (e.g., MoMs, geometric optics) and can support multiple variants of a basic phenomenology.

In all of these cases, validation data of significant volume are not present to completely assess the accuracy of indicated codes. Most models are validated only against "idealized" measurement data over a few conditions since no generalized approach completely predicts anomalies in the measured data.

For additional propagation models, visit either the website of the International Telecommunications Union (http://www.itu.int), which contains many propagation models suitable for a variety of applications, or the European Communications Office (http://www.ero.dk/seamcat), which has a free download of a Monte Carlo method for propagation modeling and simulation and analysis.

4.12 SUMMARY AND FURTHER READING

This chapter addresses most, but not all, key propagation mechanisms; a number of other mechanisms and techniques not discussed include the following:

1. Atmospheric emission. Atmospheric constituents produce thermal emissions that increase the background noise temperature. The thermal emission is proportional to the absorption due to, for example, atmospheric gases, rain, clouds, fog, dust, and smoke. Emission must be taken into account not only along the path between the radar and the target but beyond the target as well. This effect depends on the frequency and elevation angle as well as on the absorber characteristics and temperature. An increase in the background noise temperature (i.e., "sky noise") can affect the performance of surface-based air surveillance, space surveillance, and ballistic missile defense radars.
2. Surface wave propagation. Surface wave (or ground wave) propagation is used by some HF radars to provide coverage to ranges well beyond the conventional radar horizon but short of the coverage provided by ionospheric propagation. Such coverage extends into the coverage gap known as the ionospheric "skip zone."
3. Ground-penetrating radar. There has been considerable research and development activity in the area of ground-penetrating radar in the last 10–15 years, including some commercial applications. Radar has also been used to probe the ice sheets of Greenland and Antarctica to depths of several kilometers. In addition, two subsurface radar sounders (VHF and UHF) are in orbit around Mars. Propagation in soil or ice presents some unique issues and challenges.
4. Atmospheric turbulence sensing. There has been considerable research and development activity in the area of sensor fusion (MMW radar and LADAR) combining to detect clear air turbulence phenomena for aircraft.
5. Trans-ionospheric propagation. Space-based radars are becoming more common in both civil remote sensing and military applications. Surface-based ballistic missile defense radars may be required to detect targets outside the atmosphere. Both of these applications must use trans-ionospheric propagation paths. Ionospheric effects can degrade the measurement accuracy of several parameters including range delay, direction of arrival, and polarization.

In addition to the books on propagation listed in the reference section [3,7,20,24,34], some additional texts are recommended for further reading and investigation for readers both new and old to the subject. Each has many strong points to recommend, and, depending on readers' specific interests, should provide a good starting point for getting into the literature of radar propagation.

Blaunstein and Plohotniuc [50] is recommended for those needing a detailed background in propagation associated with the ionosphere. It is more specific and at a much deeper level than the other books listed here. Blaunstein and Christodoulou [51] is a very complete book. Despite its orientation toward wireless communications, its core material is also useful for radar propagation and covers several topic areas the other recommended texts do not, such as statistical variations in propagation models.

Seybold's book [52] not only covers the basics of RF propagation but is also an especially good introduction to the variety of modeling techniques for RF propagation. Barclay [53] provides a very complete text on RF propagation, at a more detailed level than most of the other recommended books. There is good coverage of several of the key recommendations for further reading on propagation.

An excellent one-book compilation covering the impact of land and sea on radar propagation, covering not only the usual effects of reflection, refraction, and diffraction but also the impact of clutter on radar returns is Long [54]. Lavergnat and Sylvain [55] give a good introduction to RF propagation, with many worked examples for those wanting to get a feel for the quantities involved in propagation problems. Last but not least, Shibuya's [56] book, although older, is very complete and comprehensive.

4.13 REFERENCES

[1] Burge, C.J., and Lind, J.H., *Line-of-Sight Handbook,* Naval Weapons Center, NWC-TP-5908, January 1977.

[2] Eaves, J.L., and Reedy, E.K., *Principles of Modern Radar,* Van Nostrand Co., New York, 1987.

[3] Bogush, A.J., *Radar and the Atmosphere,* Artech House, Norwood, MA, 1989.

[4] Battan, L.J., *Radar Characteristics of the Atmosphere,* University of Chicago Press, Chicago, 1973.

[5] Austin, P.M., "Radar Measurements of the Distribution of Precipitation in New England Storms," Proceedings of the 10th Weather Conference, Boston, 1965.

[6] P-N Designs, Inc., *Electronic Warfare and Radar Systems Handbook,* http://www.microwaves101.com/encyclopedia/Navy_Handbook.cfm

[7] Nathanson, F., Radar Design Principles, 2d ed., Scitech, Raleigh, NC, 1999.

[8] Richard, V.M., Kammerer, J.E., and Reitz, R.G., "140-GHz Attenuation and Optical Visibility Measurements of Fog, Rain, and Snow," ARBRL-MR-2800, December 1977.

[9] Nakaya, U., and Terada, T., "Simultaneous Observations of the Mass, Falling Velocity, and Form of Individual Snow Crystals," *Journal of the Faculty of Science,* vol. 1, no. 7, pp. 191–200, 1935.

[10] Perry, B.P, et al., "Effects of Smoke on MMW Radar Measurements," GTRI Final Technical Report A-9007, July 1992.

[11] Pedersen, N.E., Waterman, P.C., and Pedersen, J.C., "Absorption Scattering and Thermal Radiation by Conductive Fibers," AFOSR Final Report, Panametrics, Inc., July 1987.

[12] *U.S. Standard Atmosphere, 1976,* U.S. Government Printing Office, Washington, DC, 1976. Available at: http://ntrs.nasa.gov/archive/nasa/casi.ntrs.nasa.gov/19770009539_1977009539.pdf

[13] ICAO, *Manual of the ICAO Standard Atmosphere*, Doc 7488-CD, 3d ed., 1993.

[14] International Organization for Standardization (ISO), "Standard Atmosphere," *ISO,* vol. 2533, p. 1975, 1975.

[15] Berkowitz, R.S., *Modern Radar Analysis, Evaluation and System Design,* Wiley & Sons, New York, 1965.

[16] Bean, B.R., and Dutton, E.J., "Radio Meteorology," NBS Monograph 92, U.S. Department of Commerce, 1965.

[17] Barton, D.K., *Modern Radar System Analysis*, Artech House, Norwood, MA, 1988.

[18] Ko, H.W., Sari, J.W., and Skura, J.P., "Anomalous Microwave Propagation through Atmospheric Ducts," *Johns Hopkins APL Technical Digest,* vol. 4, no. 1, pp. 12–26, 1983.

[19] Skolnik, M.I., *Introduction to Radar Systems,* 3d ed., McGraw-Hill Co., New York, 2001.

[20] Skolnik, M.I., *Radar Handbook,* 3d ed., McGraw-Hill Co., New York, 2008.

[21] "Engineering Refractive Effects Prediction System (EREPS)," http://areps.spawar.navy.mil/2858/software/ereps/readerep.txt

[22] Bohlander, R.A., McMillan, R.W., Patterson, E.M., Clifford, S.F., Hill, R.J., Priestly, J.T., et al., "Fluctuations in Millimeter-Wave Signals Propagated through Inclement weather", *IEEE Transactions on Geoscience and Remote Sensing,* vol. 26, no. 3, pp. 343–354, 1988.

[23] Churnside, J.H., and Lataitis, R.J., "Angle-of-Arrival Fluctuations of a Reflected Beam in Atmospheric Turbulence," *Journal of the Optical Society of America,* vol. 4, pp. 1264–1272, July 1987.

[24] Kerr, D.E., *Propagation of Short Radio Waves,* McGraw-Hill, New York, 1951.

[25] Sommerfield, A.J., *Optics*, Academic Press, Inc., New York, 1954.

[26] Keller, "Diffraction by a Convex Cylinder," *IRE Transactions on Antennas and Propagation,* vol. 4, no. 3, pp. 312–321, July 1956.

[27] Keller, J.B., "Geometrical Theory of Diffraction," *Journal of the Optical Society of America,* vol. 52, no. 2, February 1962.

[28] IEEE Standard Radar Definitions, IEEE Std 686-1982, Institute of Electrical and Electronic Engineers, New York, 1982.

[29] Saffold, J.A., and Tuley, M.T., "A Multidimensional Terrain Model for Low Altitude Tracking Scenarios," 1990 Summer Computer Simulation Conference (SCS), Calgary, Alberta Canada, July 18, 1990.

[30] Bruder, J.A., and Saffold, J.A., "Multipath Effects on Low Angle Tracking at Millimeter Wave Frequencies," *IEE Proceedings of the Radar and Signal Processing: Special Issue on Radar Clutter and Multipath Propagation,* vol. 138, no. 2, pp. 172–184, April 1991.

[31] Ewell, G.W., and Reedy, E.K., "Multipath Effects on Direct Fire Guidance," GTRI Final Technical Report, CR-RD-AS-87-10, June 1987.

[32] Bullington, K. "Reflection Coefficients of Irregular Terrain," *Proceedings of the IRE,* vol. 42, no. 8, pp. 1258–1262, August 1954.

[33] Stratton, J.A., "Electromagnetic Theory," McGraw-Hill, New York, 1941.

[34] Beckmann, P., and Spizzichino A., *The Scattering of Electromagnetic Waves from Rough Surfaces,* Artech House, Norwood, MA, 1987.

[35] Myers G.F. "High Resolution Radar, Part IV, Sea Clutter Analysis," Naval Research Laboratory Report 5191, October 21, 1958.

[36] Barton, D.K., "Low-Angle Radar Tracking," *Proceedings of the IEEE,* vol. 62, no. 6, pp. 687–704, June 1974.

[37] Neff, H., *Basic Electromagnetic Fields,* Harper & Row Publishers, New York, 1981.

[38] Hunt, A.R., and Hogg, R.D., "A Stepped Frequency, CW Radar for Concealed Weapons Detection and Through the Wall Surveillance," *Proceedings of the SPIE,* vol. 4708, 2002.

[39] Falconer, D.G., Ficklin, R.W., and Konolige, K.G., "Robot Mounted Through-Wall Radar for Detecting Locating and Identifying Building Occupants," *Proceedings of the 2002 IEEE International Conference on Robotics and Automation,* vol. 2, pp. 1868–1875, April 2002.

[40] Hunt, A., and Akela Inc., Briefing, "Image Formation through Walls Using a Distribution RADAR Sensor Network," CIS Industrial Associated Meeting, May 12, 2004.

[41] Stone, W.C., "Surveying through Solid Walls," National Institute of Standards and Technology (NIST) paper, Automations Robotics and Construction 14th International Symposium Proceedings, June 8–11, 1997, Pittsburg, PA.

[42] Leubbers, R., "Propagation Prediction for Hilly Terrain Using GTD Wedge Diffraction", *IEEE Transactions on Antennas and Propagation,* vol. AP-32, no. 9, pp. 951–955, September 1984.

[43] Leibe, H.J. "MPM—An Atmospheric Millimeter-Wave Propagation Model," *International Journal of Infrared and Millimeter Waves,* vol. 10, no. 6, June 1980.

[44] Queen, J.L., Stapleton, J., Kang, S., and the Army Infantry School Fort Benning GA Directorate of Combat Development, "Wideband Low-Elevation Microwave Propagation Measurements," Dahlgren Division, NSWC, NSWCDD/TR-95/18, February 1995.

[45] Reed, R., *Ultrahigh Frequency Propagation,* Chapman and Hall, Ltd., London, 1966.

[46] Currie, N.C., Dyer, F.B., and Hayes, R.D., "Analysis of Rain Return at Frequencies of 10, 35, 70, and 95 GHz", Technical Report No. 2, Contract DAA25-73-0256, Georgia Institute of Technology, Atlanta, March 1975.

[47] Pasquill, F., and Smith, F.B. (Ed.), *Atmospheric Diffusion*, 3d ed., Ellis Horwood Series in Environmental Science, 1983.

[48] Kobayashi, H.K., "Atmospheric Effects on Millimeter Radio Waves," Atmospheric Science Lab, NM, ASL-TR-0049, January 1980.

[49] Kobayashi, H.K., "Effect of Hail, Snow, and Melting Hydrometeors on Millimeter Radio Waves," Atmospheric Science Lab, NM, TR-0092, July 1981.

[50] Blaunstein, N., and Plohotniuc, E., *Ionospheric and Applied Aspects of Radio Communication and Radar,* Nathan CRC Press, Boca Raton, FL, 2008.

[51] Blaunstein, N., and Christodoulou, C.G., *Radio Propagation and Adaptive Antennas for Wireless Communication Links,* John Wiley & Sons, New York, 2007.

[52] Seybold, J., *Introduction to RF Propagation,* Wiley-Interscience, New York, 2005.

[53] Barclay, L., *Propagation of Radiowaves,* 2d ed., The Institution of Engineering and Technology, London, 2003.

[54] Long, M.W., *Radar Reflectivity of Land and Sea,* 3d ed., Artech House, Norwood, MA, 2001.

[55] Lavergnat, J., and Sylvain, M., *Radiowave Propagation,* John Wiley & Sons, New York, 2000.

[56] Shibuya, S., *A Basic Atlas of Radio-Wave Propagation,* Wiley-Interscience, New York, 1987.

4.14 | PROBLEMS

1. In a specific application the low-frequency (VHF/UHF) radar wave must propagate through continuous rain (high humidity) while remaining low to the earth's surface. The surface is very smooth (low rms roughness) and extends for great distances. Which propagation mechanisms are likely the most significant?

2. There are three significant mechanisms in a two-way path between the radar transmitter and receiver. The presence of multipath provide a propagation factor of 0.8, while atmospherics (rain and fog) provide factors of .95 and .99, respectively. What is the total propagation factor for the path?

3. A 1 watt EM wave of frequency 1 GHz travels through free space at the speed of light (3×10^8 m/s) from a source for a time of 10 μsec. What is the distance traveled and the total phase change (in radians) the wave has traversed?
4. An EM wave travels through a heterogenous atmosphere with rain, fog, and clear air. For each type the distances are 2 km for rain, 1 km for clear air, and 2 km for fog. The attenuation for each of these types is listed as 1 db/km, 0.3 db/km, and 0.7 db/km two-way, respectively. What is the total one-way propagation factor for this path in db?
5. What is the one-way attenuation from clear air at standard atmosphere for a 10 GHz radar at atmospheric temperature 273.15 Kelvin and 59% humidity?
6. What is the attenuation for a 10 GHz radar signal traveling 10 km with a 2 km dust cloud in its path? The dust cloud has a 1.0 extinction efficiency and a concentration of 0.0001 g/m^3 in the 2 km area.
7. What are the general criteria for an edge classification as a "knife" or "rounded tip"?
8. Which edge classification provides the lowest scattered energy in the shadow zone and interference regions?
9. What is the magnitude of the diffracted wave off a $\lambda/50$ radius of curvature knife edge when the incident wave has magnitude 1 and a 15 degree shadow angle from the edge?
10. Which linear polarization component offers a reduction in multipath scattering due to Brewster's angle?
11. What is the spatial correlation length of a rough surface with rms slope of 0.1 degrees and rms roughness of 0.1 meters?
12. When a transmitter and receiver (or target) are at low altitude, long range, and smooth intermediate surface, which roughness component dominates the reflection coefficient?
13. What are the specular reflection angle and range delay values for a transmitter located at 2 meters altitude and a receiver located at 4 meters altitude in a one-way link geometry when the two are separated by 100 meters?
14. Using conservation of energy principles on the roughness factors, what is the magnitude of the diffuse term if the specular reflection coefficient is 0.7?
15. What is the maximum received signal amplitude gain when the geometry is setup to coherently add all two-way reflection components from a 1 watt transmitter?
16. What is the rms angle error in elevation for a tracking radar with a one-way 3 db beamwidth of 2 degrees, a direct/indirect path gain ratio of 0.7, and a ground reflection coefficient of 0.8?
17. For a 5 GHz radar, what is the loss through a single concrete block wall?
18. What is the additional range delay over free space for a wave propagating through two 12 inch thick walls with refractive index of 5?

CHAPTER 5

Characteristics of Clutter

Nicholas C. Currie

Chapter Outline

5.1 Introduction and Definitions 165
5.2 General Characteristics of Clutter 172
5.3 Clutter Modeling 202
5.4 Concluding Remarks 206
5.5 Further Reading 207
5.6 References 207
5.7 Problems 210

5.1 INTRODUCTION AND DEFINITIONS

5.1.1 What Is Clutter?

Radar clutter is a radar return from an object or objects that is of no interest to the radar mission. For example, the mission of many radar systems is the detection and tracking of "targets" such as aircraft, ships, or ground vehicles. To these systems, clutter is considered to be an interfering return from a natural object such as precipitation, vegetation, soil and rocks, or the sea. However, to radars designed for remote sensing such as synthetic aperture radar (SAR) imagers, these objects may be the primary targets of interest. For this chapter, it will be assumed that targets of interest are man-made while natural target returns are unwanted (i.e., clutter).

5.1.2 Comparison of Clutter and Noise

Chapter 3 introduced detection in the presence of random noise, including noise properties and the effect on detection of targets. The returns from natural clutter can be strikingly similar to the effects of noise on detection yet also be quite different. So, how do clutter returns differ from noise effects? Figure 5-1 shows a high (fine) resolution (1 ft × 1 ft), two-dimensional (2-D) SAR image of a suburban terrain scene near Stockbridge, New York, collected from an airborne platform [1]. As can be seen from the figure, some areas appear uniform in nature (e.g., grassy lawns), while others appear quite nonuniform (e.g., trees and man-made structures). The high-resolution image shown in Figure 5-1 produces significantly different clutter properties from a lower-resolution radar, which averages out much of the structure shown in the figure. As an introductory text, this chapter will discuss the characteristics of lower-resolution (real-beam) systems.

FIGURE 5-1 ■ Synthetic aperture radar image of suburban terrain. (From Novak and Owirka [1]. With permission.)

Table 5-1 summarizes the primary differences between clutter returns and receiver noise. Random noise, although varying with time, exhibits a specific set of characteristics:

- The probability density function (PDF) is Rayleigh for a linearly detected (voltage) signal, or exponential[1] for a square law detected (power) signal (see Chapter 3 for a description of the Rayleigh and exponential PDFs).
- The width of the autocorrelation function (ACF) is approximately the inverse of the receiver bandwidth.
- The power spectral density (PSD) function (power spectrum) width is approximately equal to the receiver bandwidth.

Clutter statistics can be similar to those of noise when the natural targets are composed of small, nearly equal-sized scatterers but can be quite different when the nature of the scatterers change or scatterers of differing types (e.g., a tree line) are present in the radar field of view. For this case, amplitude distributions having much longer "tails" than the Rayleigh distribution have been observed. Finally, although noise is independent of transmitted frequency, spatial position, and environmental parameters, clutter varies with all of these parameters, making clutter characterization very complex.

[1]The exponential PDF is sometimes called a *Rayleigh power* PDF. Confusingly, this is sometimes shortened to just Rayleigh PDF, even though the mathematical form intended is the exponential PDF. The Rayleigh PDF describes the amplitude (magnitude) of the noise signal; the magnitude squared (power) is described by an exponential PDF.

TABLE 5-1 ■ Clutter Signals versus Noise

Noise Signal	Clutter Signal
Amplitude independent of transmitted radar signal level	Amplitude proportional to transmitted radar signal level
Wide bandwidth (limited by receiver noise bandwidth)	Narrow bandwidth (created by scatterer motion)
Statistically independent between pulses	May be highly correlated between pulses
Amplitude variation described by Rayleigh statistics	Amplitude variation may vary from none to extremely wide (log normal or Weibull statistics)
Average value is constant and independent of spatial position	Time average will differ between spatial samples as the clutter types change
Independent of transmitted frequency	Varies with changing frequency
Independent of environmental parameters	Can vary with changing environmental conditions
No spatial component	Varies with beam position and resolution

Source: Adapted from Long [2]. (© 2006 IEEE. Used with permission.)

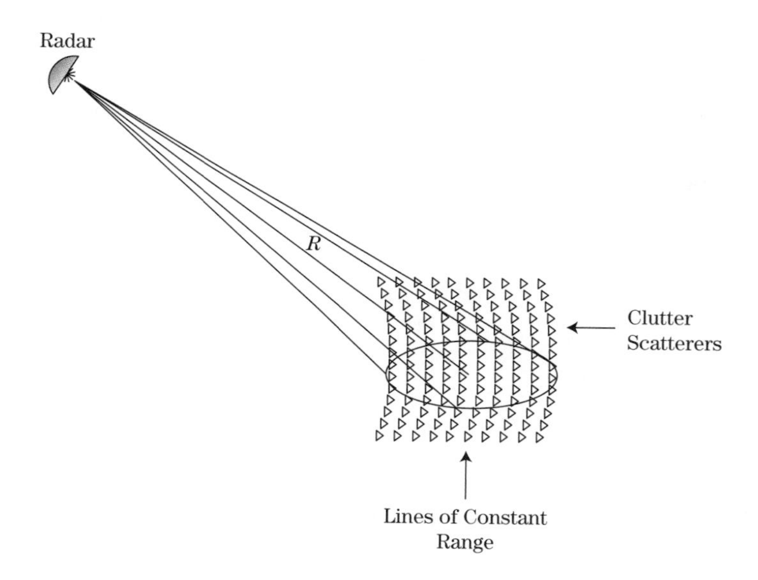

FIGURE 5-2 ■ Radar scan over clutter scatterers illustrating lines of constant range.

Figure 5-2 shows a typical situation where the radar beam illuminates a number of scatterers, each having a different reflectivity, σ_i, and distance, d_i, from the radar. Lines of constant range to the radar form ellipses on a level surface. The electric field amplitude (horizontal or vertical polarization component) measured at the radar due to the echo from the i-th scatterer will be proportional to the square root of the received power given by the radar range equation

$$|\boldsymbol{E}_i| = \left[\frac{P_t G^2 \lambda^2 \sigma_i}{(4\pi)^3 L_s d_i^4}\right]^{1/2} = k\frac{\sqrt{\sigma_i}}{d_i^2} \tag{5.1}$$

where P_t is the transmitted power, G is the antenna gain, λ is the wavelength, and L_s is the system loss (including hardware and atmospheric losses) as discussed in Chapter 2. The constant, k, absorbs all the factors that are the same for each scatterer. The phase of

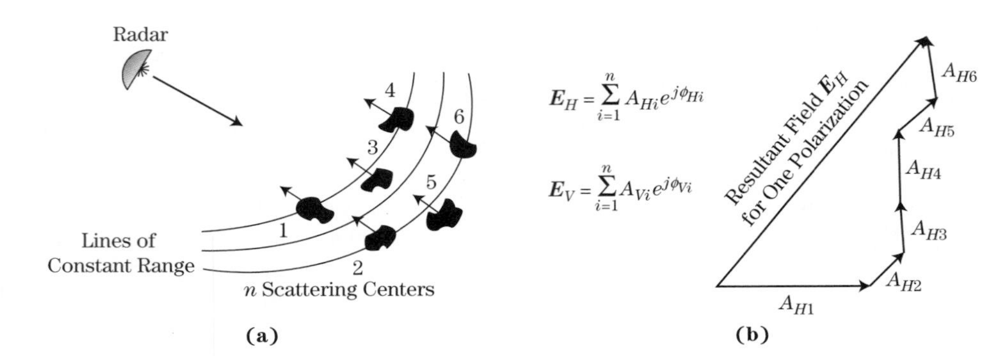

FIGURE 5-3 ■ Vector summation of scatterers at different positions and ranges. (a) Geometry of multiple scatterers. (b) Vector summation forms resultant E-field amplitude.

the received electric field, relative to the phase of the transmitted wave, is determined by the scatterer reflection phase, θ_i, and the two-way propagation distance:

$$\arg\{\boldsymbol{E}_i\} = \theta_i - \frac{4\pi}{\lambda} d_i = \theta_i + \phi_i \tag{5.2}$$

Scatterers located at the same range will generate echoes that return with the same time delay and phase at the radar, assuming that each scatterer has the same reflection phase (i.e., $\theta_i = \theta$) for some constant, θ. However, scatterers with slightly different ranges but falling within the same beam or range bin interval will have different echo phases. As shown in Figure 5-3a, the total return $\boldsymbol{E}$-field amplitude at the radar is therefore proportional to the vector summation of the electric field amplitude and phase of each group of scatterers contributing to a single radar measurement. This resultant $\boldsymbol{E}$-field is given by

$$\begin{aligned} \boldsymbol{E} &= \sum_i \boldsymbol{E}_i = \sum_i k \frac{\sqrt{\sigma_i}}{d_i^2} \exp\left[-j\left(\frac{4\pi}{\lambda} d_i + \theta_i\right)\right] \\ &\approx \frac{k}{d^2} \sum_i \sqrt{\sigma_i} \exp\left[-j\left(\frac{4\pi}{\lambda} d_i + \theta_i\right)\right] \\ &\equiv \frac{k'}{d^2} \sqrt{\sigma} \exp[j\phi] \end{aligned} \tag{5.3}$$

where d is the nominal range to the resolution cell and σ is the equivalent radar cross section (RCS) of the total clutter return, and ϕ is the equivalent phase. This process is illustrated in Figure 5-3b. The complex quantity $\sqrt{\sigma} \exp[j\phi]$ is called the *backscatter coefficient* of the clutter.

As the radar beam scans across the scatterers in the scene, the range to the particular scatterers in the beam changes so that the resulting equivalent RCS changes. This variation in return has come to be known as "speckle": the "noise-like" returns resulting from the random summation of individual scatterer echoes. The reduction of this characteristic, "despeckling," is very important in SAR processing. As the beam scans farther, new scatterers enter the beam, while other scatterers leave the beam. For a mechanically scanned system, the changes are smooth, while for an electronic scan system the changes occur abruptly as the beam is step-scanned. As long as the scatterers illuminated at any given time are similar in amplitude and phase, the return will exhibit noise-like statistics with constant parameters. However, if the beam scans across a set of scatterers that are physically different the character of the return can change. An example is scanning from a grassy field to a line of trees, which would likely result in an increase in the echo strength and thus the equivalent RCS. Also, the presence of one or two large scatterers among the smaller scatterers can cause the return to exhibit very nonnoise-like statistics.

Since the returns from clutter vary by type, polarization, environmental, and geometric conditions, it is very difficult to model clutter mathematically. Attempts to do so are briefly discussed in Section 5.3. More practically, many experiments have been conducted to measure clutter returns as functions of the various dependent parameters. Practical clutter modeling usually involves a combination of empirical and theoretical results.

5.1.3 Basic Definitions

5.1.3.1 Scattering Coefficients

To use published data on clutter, methods must be defined to normalize the returns so that data can be applied to many different radars and applications. One form of the radar equation can be written as

$$P_r = \frac{P_t G^2 \lambda^2 \sigma}{(4\pi)^3 L_s R^4} \tag{5.4}$$

where the variables are the same as in equation (5.1). This form of the radar equation is suitable for point targets (i.e., targets much smaller than the resolution of the radar) but is not convenient for distributed targets such as clutter, where many scatterers contribute at once to the total echo. Thus, equation (5.4) must be modified to account for the area or volume defined by the beamwidths and range resolution of the radar.

For surface clutter it is convenient to define the radar cross section per unit area, or *surface reflectivity*,

$$\sigma^0 \equiv \frac{\sigma}{A} \tag{5.5}$$

where σ is the total RCS of the contributing clutter, and A is the area of the contributing clutter defined by the radar beam intersection with the surface. The units of σ^0 are meters^2/meters^{2}so that σ^0 is unitless. It is often expressed in the literature in decibels, denoted dBsm/sm.

For volume clutter, the radar cross section per unit volume, or *volume reflectivity,* is defined as

$$\eta = \frac{\sigma}{V} \tag{5.6}$$

where V is the volume defined by the radar beam and range resolution cell. The units of η are meters^2/meters^3, or meter^{-1}. It is expressed in the literature in decibels per meter.

Given σ^0, the area A to be used in computing clutter RCS using equation (5.5) for surface clutter is determined by the beamwidths and the range resolution of the radar. Two situations arise: (1) the case where the range resolution is large compared with the projection of the vertical beam width onto the surface; and (2) the case where the range resolution is smaller than the projection extent. In the *beam-limited* case, the result for the beam area on the ground was

$$A = \pi R^2 \tan\left(\frac{\theta_3}{2}\right) \tan\left(\frac{\phi_3}{2}\right) \csc\delta \tag{5.7}$$

where ϕ_3 and θ_3 are the azimuth and elevation beamwidths of the antenna, respectively, and δ is the *grazing angle* of the antenna boresight with the clutter surface. For beamwidths less than about 10 degrees the small angle approximation $\tan(x) \approx x$ is valid, and equation (5.7) becomes

$$A = \frac{\pi R^2}{4} \theta_3 \phi_3 \csc \delta \tag{5.8}$$

Note that since the clutter area grows as R^2 due to beam spreading, so will the clutter RCS, σ. When used in equation (5.4), the resulting received power due to a constant clutter reflectivity will be proportional to R^{-2} instead of the R^{-4} proportionality seen for point targets. Thus, beam-limited surface clutter power does not decline with range as rapidly as does point target power.

In the *pulse-limited* case, the area defined by the beam and pulse width on the ground is given by

$$A = \left(\frac{c\tau}{2}\right) 2R \tan\left(\frac{\theta_3}{2}\right) \sec \delta = c\tau R \tan\left(\frac{\theta_3}{2}\right) \sec \delta \tag{5.9}$$

and, again, if the beam width is less than 10°, a small angle approximation gives

$$A = \left(\frac{c\tau R\theta_3}{2}\right) \sec \delta \tag{5.10}$$

Note that in this case the clutter area is proportional to R instead of R^2. Consequently, the pulse-limited echo power from constant-reflectivity surface clutter declines as R^{-3}. If the radar uses pulse compression techniques (see Chapter 20) to obtain fine range resolution, then the radar range resolution in meters should replace the factor $c\tau/2$ in equations (5.9) and (5.10).

The grazing angle at which the clutter cell area transitions from the beam-limited case to the pulse-limited case can be found by setting the pulse- and beam-limited areas equal to one another and is given by

$$\tan \delta = \frac{\pi R \tan(\phi_3/2)}{c\tau} \tag{5.11}$$

or, for small antenna beamwidths,

$$\tan \delta = \frac{\pi R\phi_3}{2c\tau} \tag{5.12}$$

When the value of $\tan \delta$ exceeds the value in equations (5.11) or (5.12), the beam-limited case applies. Conversely, when $\tan \delta$ is less than that value, the pulse-limited case applies.

Note that the beamwidths used in the previous equations are assumed to be the 3 dB two-way beamwidths. Actual radar beams are not rectangular, so some errors can occur in beam area and clutter power estimation based on the actual beam shape and sidelobes. These errors can be significant for airborne pulse-Doppler processors.

The previous discussion can be applied to area clutter (e.g., ground, sea). For computing the equivalent RCS of atmospheric clutter using equation (5.6), the reflectivity, η, and the resolution cell volume, V, defined by the radar beam and the pulse width are needed. Since air search radars are almost always narrow beam, the small angle approximation formula is usually adequate. This is given by

$$V = \pi \left(\frac{R\theta_3}{2}\right)\left(\frac{R\phi_3}{2}\right)\left(\frac{c\tau}{2}\right) = \left(\frac{\pi R^2\theta_3\phi_3}{4}\right)\left(\frac{c\tau}{2}\right) \tag{5.13}$$

Again, the actual range resolution should be used in place of the $(c\tau/2)$ term in pulse compression radars.

5.1.3.2 Clutter Polarization Scattering Matrix

The scattering properties of clutter are dependent on the transmitted and received polarization. This effect is quantified by the use of a 2-by-2 matrix known as the *polarization scattering matrix* (PSM), **S** [3]. Equation (5.14) gives a form of the matrix expressed in terms of vertical and horizontal polarization:

$$\mathbf{S} = \begin{bmatrix} \sqrt{\sigma_{HH}}\, e^{j\phi_{HH}} & \sqrt{\sigma_{HV}}\, e^{j\phi_{HV}} \\ \sqrt{\sigma_{VH}}\, e^{j\phi_{VH}} & \sqrt{\sigma_{VV}}\, e^{j\phi_{VV}} \end{bmatrix} \tag{5.14}$$

The first subscript represents the received polarization, while the second represents the transmitted polarization. For example, the lower left term in the matrix describes the vertically polarized received voltage signal component in response to a horizontally polarized transmitted signal. In fact, a matrix can be developed in terms of any two orthogonal polarizations, including circular and elliptical polarizations.

The terms in the complex matrix **S** represent the backscattering coefficients of the clutter for four polarization cases: (1) transmit and receive horizontal polarization; (2) transmit horizontal and receive vertical polarization; (3) transmit vertical and receive horizontal polarization; and (4) transmit and receive vertical polarization. For a specific frequency and geometry of the radar and a specific set of environmental parameters, the polarization scattering matrix contains all of the available information about the clutter return at the time of the measurement.

Unfortunately, a radar that can transmit and receive all four polarizations is both complex and expensive. Fortunately, for the case of a monostatic radar the *reciprocity theorem* [3] requires that the two "cross polarized" terms are equal, that is, $\sigma_{VH} = \sigma_{HV}$ and $\phi_{VH} = \phi_{HV}$. Thus, only three complex polarization values must be measured to determine the full scattering matrix.

Much work has been done to try to exploit the information inherent in the scattering matrix to identify targets in the presence of clutter. Holm [4] and others have attempted to use the scattering matrix to separate targets from clutter as well as to identify classes or types of targets on the theory that the PSMs of man-made targets and natural clutter will be significantly different. He determined that, to be effective, high-range and azimuthal resolution is required so that individual range-azimuth cells contain only target returns or only clutter returns. If cells contain a mixture of both target and clutter returns, the results are much less useful even if the clutter returns are much lower in amplitude than the targets.

In the past, attempts to use partial polarization matrix information to discriminate between targets and clutter have been tried with mixed results [3,5]. Such discriminants have included the following:

- Parallel/cross polarization ratio: $\sqrt{\sigma_{HH}}/\sqrt{\sigma_{VH}}$.
- Vertical/horizontal polarization ratio: $\sqrt{\sigma_{VV}}/\sqrt{\sigma_{HH}}$.
- Polarimetric phase: $\phi_{HH} - \phi_{VV} \equiv \phi_{H-V}$.

Polarimetric phase is usually expressed as quadrature components $\cos(\phi_{H-V})$ and $\sin(\phi_{H-V})$. These discriminants have the advantage of requiring only the polarimetric amplitude or phase instead of the entire PSM and can improve target detectability in clutter under some conditions.

5.2 GENERAL CHARACTERISTICS OF CLUTTER

5.2.1 Overview

Because clutter is one of the major limitations in target detection for practical radar, clutter measurements have been performed since the advent of radar use in World War II. The development of digital recording techniques has greatly improved the quality (and quantity) of data in the last 20 years. Because clutter exhibits noise-like fluctuations in echo strength, clutter is characterized in terms of statistical parameters describing the variability of σ^0 and η, including the following:

- Mean or median values.
- Standard deviations or variances.
- Probability density functions.
- Spectral bandwidths for temporal variability.
- Autocorrelation functions or power spectral density functions for temporal and spatial variations.

These measured values almost always include propagation factor effects (see Chapter 4). Usually, for clear air measurements of surface clutter at moderate to high angles the propagation factor effects are negligible. However, for measurements of atmospheric precipitation or for low grazing angles for surface clutter, attenuation due to multipath can be significant. These effects can be minimized by using calibration targets located near the clutter region being measured but can never be totally eliminated. For this reason, data from precipitation or surface clutter at large ranges tend to have more variability than measurements of surface clutter at steeper grazing angles. Since an operational radar will experience these same effects under the same conditions, this situation is not necessarily bad.

5.2.2 Surface Clutter

5.2.2.1 General Dependencies

Given that the radar-received clutter power in a particular application affects the detection performance and influences radar design, a simple and accurate way is needed to estimate clutter levels for a variety of scenarios. The goal is to develop models that include in a mathematical form all the known parameter dependencies of the backscattering coefficient that have been identified through experimental measurements. Fundamentally, clutter model development starts with theoretical calculations of reflectivity and comparison with the interpretation of experimental observations, which leads to an understanding of the underlying scattering mechanisms. Some of the observed dependencies are as follows:

- *Grazing angle*: Grazing angle, δ, is the angle at which the illumination energy strikes a clutter surface.
- *Vertical variation of the clutter scatterers*: Rough surfaces have a larger σ^0 than smooth ones for low grazing angles; at very high grazing angles (near 90°), smooth surfaces have a higher σ^0 than rough ones.
- *Wavelength*: σ^0 is a function of vertical texture expressed in wavelengths.

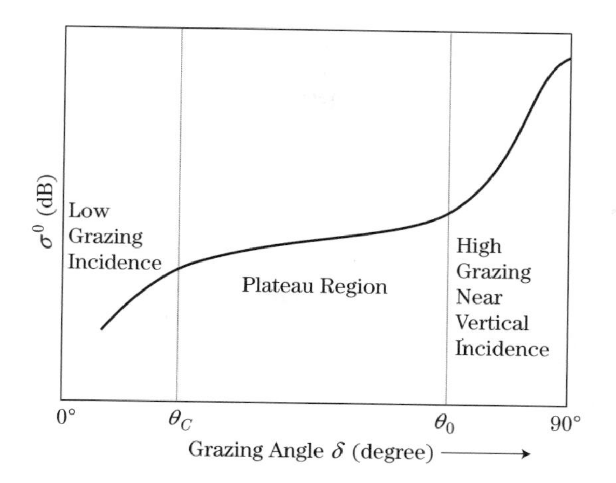

FIGURE 5-4 ■ General dependence of σ^0 on grazing angle. (Adapted from [6]. With permission.)

Depression angle is closely related to grazing angle. It is the angle below the radar's local horizontal at which the illumination energy is transmitted from the radar. For a flat horizontal clutter surface the depression and grazing angles are identical. Grazing angle is the most relevant for describing scattering, but clutter reflectivity data are often reported as a function of depression angle in the literature because the grazing angle may not be known precisely. In this chapter, the two are generally assumed to be the same unless otherwise stated, and the discussion is framed in terms of grazing angle. Depression angle is used when referring to previous published results given in terms of depression angle.

Rough surface theory and experimental measurements have determined that the backscattering for both land and sea surfaces—as a function of grazing angle—exhibit a common general dependence on grazing angle as shown in Figure 5-4. From this figure, three distinct regions of clutter behavior can be identified: (1) a low grazing angle region; (2) a plateau region; and (3) a high grazing angle region. The boundaries of these three regions, defined by θ_C and θ_0, change with frequency, surface condition, and polarization [6].

The *low grazing angle region* extends from zero to a *critical angle,* θ_C, determined by the root mean square (rms) height of surface irregularities in wavelengths. This critical angle is the grazing angle below which a surface seems "smooth" by Rayleigh's definition and above which it is "rough." Based on Rayleigh's definition, a surface is smooth if

$$\sigma_h \sin\delta < \frac{\lambda}{8} \tag{5.15}$$

where σ_h is the rms height of the surface irregularities, δ is the grazing angle, and λ is the radar wavelength. Thus, the critical angle is given by

$$\sin\delta_C = \frac{\lambda}{8\sigma_h} \tag{5.16}$$

In the *plateau region* the incident wave encounters the surface irregularities in such a way that the dependence of σ^0 on grazing angle is much less than at lower angles. Chapter 7 provides an introduction to the "rough surface" theory applicable to this regime.

For a constant σ^0 and beam-limited geometry, it is expected that the return power for surface clutter would vary as $R^{-4} \times R^2 = R^{-2}$ since the power from a point target varies as R^{-4} while the area of surface clutter increases as R^2. However, measurements have indicated that clutter power often varies as R^{-3}. For this to be the case, σ^0 must vary as

R^{-1}. The grazing angle satisfies the equation $\sin\delta = h/R$, where h is the radar altitude and R is the slant range. If the height of the radar is constant, then $\sin\delta$ is proportional to $1/R$. This fact led to the definition of the so-called *constant gamma model* for clutter reflectivity, given by

$$\sigma^0 = \gamma \sin\delta = \frac{\gamma}{R} \tag{5.17}$$

Here γ is a constant depending on terrain type, surface roughness (sea state and terrain type for sea and land, respectively), and frequency. If equation (5.17) applies, then clutter reflectivity will be proportional to R^{-1}, and the R^{-3} clutter power dependence would be expected. Note that the constant gamma model is applicable only in the plateau angular region.

In the *high grazing angle region* the scattering becomes more directional and rapidly increases to a maximum value based on the reflectivity and smoothness of the clutter, in a manner somewhat analogous to the behavior of the main lobe of a rough flat plate at near perpendicular incidence (see Chapter 6).

Figure 5-5 gives the RCS of a conducting sphere of diameter a normalized to its projected area as a function of the circumference normalized to the wavelength. As can be seen from the figure, the relative RCS increases with radian frequency $2\pi/\lambda$ until λ equals the circumference, whereupon the RCS varies rapidly until λ is about $1/10$ the circumference, at which point the RCS equals the projected cross sectional area of the sphere and is independent of frequency. Thus, for large wavelengths (low frequencies), the return from a sphere would be expected to increase with increasing frequency until the point known as the *resonance region*, where the RCS fluctuates rapidly with frequency. Cylinders exhibit similar frequency dependence relative to the ratio of circumference to wavelength. Since most clutter scatterers can be considered to be approximately either spherical or cylindrical in shape, clutter RCS should be expected to increase with decreasing wavelength up to a wavelength in the millimeter wave region for most scatterers. At high frequencies facets or ripples create resonance effects that overcome this effect. As will be seen in the data, this effect does exist, although variations among different types of clutter return can often be much greater than frequency effects.

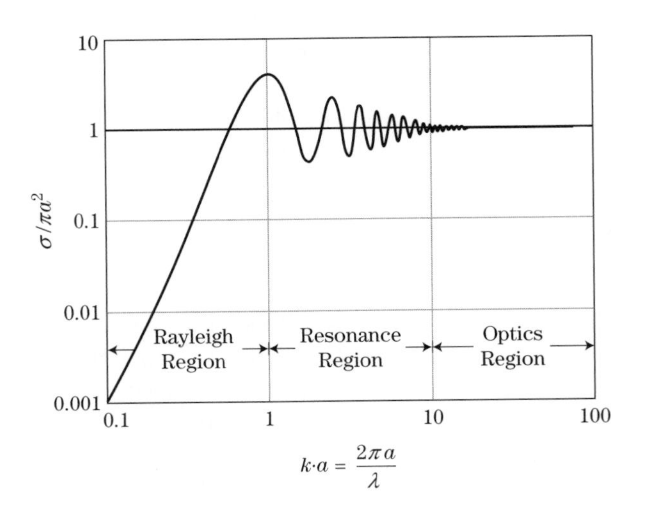

FIGURE 5-5 ■ Dependence of the RCS of a sphere on wavelength.

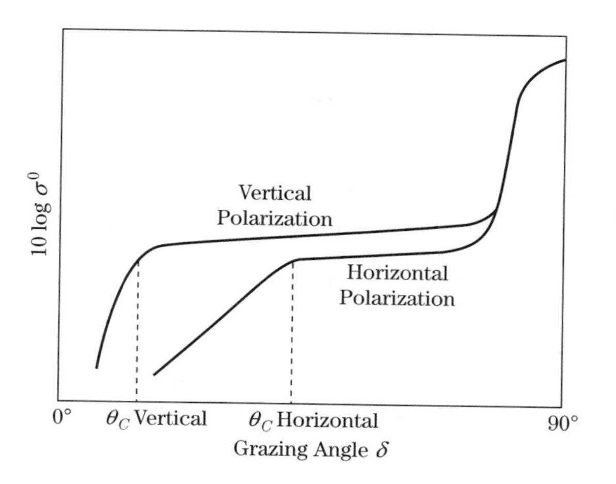

FIGURE 5-6 ■ Theoretical polarization dependence of a relatively smooth surface. (From Long [6]. With permission.)

Figure 5-6 gives the theoretical polarization dependence of σ^0 (generally verified by experiment) as a function of grazing angle for a moderately smooth surface, which tends to occur at lower frequencies. The critical angle, θ_C, occurs at different grazing angles for the two polarizations due to multipath effects so that at lower angles the difference between clutter reflectivity at vertical and horizontal polarizations can be quite large. However, for higher angles and frequencies, the horizontal polarization reflectivity is usually only a few dB lower than the vertical polarization reflectivity. Although these polarization trends on average are valid, for specific clutter patches the occasional presence of natural diplanes can result in widely varying polarization returns.

5.2.2.2 Temporal and Spatial Dependencies

Temporal and spatial variations occur due to either the motion of scatterers within the radar cell (usually due to wind) or the nature of scatterers changing as the radar cell moves in range or azimuth. Such statistical variation is described in terms of probability distributions with simple mathematical equations to facilitate modeling. Coherent receiver noise has a complex normal (Gaussian) amplitude distribution before detection and a Rayleigh distribution after detection, assuming a linear (voltage) detector, or an exponential distribution if a square law (power) detector is used. Clutter amplitudes can also appear Gaussian or Rayleigh distributed for the low-resolution case where there are a large number of scatterers within the radar cell. However, as the resolution improves, the radar cell may contain only a few clutter scatterers, resulting in a non-Gaussian or Rayleigh distribution. Also, shadowing at low grazing angles can result in hiding large scatterers some of the time. The resulting distributions are said to have long "tails" because the probability of observing large values of the clutter amplitude is greater than with Rayleigh statistics.

A commonly used family of distributions for describing clutter power is the Weibull, which is illustrated in Figure 5-7. One form of the equation for the general Weibull distribution is [6]

$$p_\sigma(\sigma) = \begin{cases} \dfrac{b\sigma^{b-1}}{\alpha} \exp\left(-\dfrac{\sigma^b}{\alpha}\right), & \sigma \geq 0 \\ 0, & \sigma < 0 \end{cases} \tag{5.18}$$

where $\alpha = \sigma_m^b / \ln 2$, and σ_m is the median of the distribution. The parameter $a = 1/b$ is called the *width* parameter; b itself is called the *shape parameter*. The parameter $\alpha^{1/b}$ is

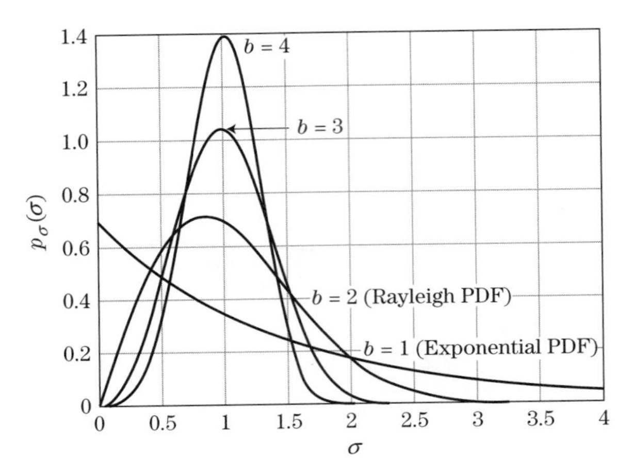

FIGURE 5-7 ■ Weibull distributions for $\sigma_m = 1$ and several values of b.

called the *scale* parameter. For a given σ_m or α, the higher the value of a and thus the lower the value of b and the longer the high value "tails" of the distribution. Figure 5-7 illustrates this for $\sigma_m = 1$. When $a = 1$ (and thus $b = 1$), the Weibull distribution reduces to the exponential PDF, while for $a = 1/2$ ($b = 2$) it reduces to the Rayleigh PDF.

Because the Weibull is, in general, a two-parameter (b and σ_m) distribution, it can be adjusted to fit both the mean and variance of experimental data. Experimenters express the variation of clutter by choosing values of b and σ_m to best fit a Weibull distribution to their data.

Other experimenters prefer to use the log-normal distribution, another two-parameter function. The log-normal distribution is simply a distribution in which the logarithm of the return is normally distributed. It is given by [6]

$$p_\sigma(\sigma) = \frac{1}{\sigma s\sqrt{2\pi}} \exp\left(-\frac{(\ln\sigma - \bar{\sigma})^2}{2s^2}\right) \tag{5.19}$$

where $\bar{\sigma}$ is the mean RCS, and s^2 is the variance of σ.

This distribution is often used when modeling radars using a logarithmic receiver. The logarithmic receiver increases dynamic range by compressing large values of returns relative to smaller returns. An advantage of the log-normal distribution is that it models even higher "tails" than the Weibull, so it may provide a better fit to severe clutter data. Figure 5-8 gives several samples of the log-normal family, all with $\sigma_m = 1$ as in Figure 5-7.

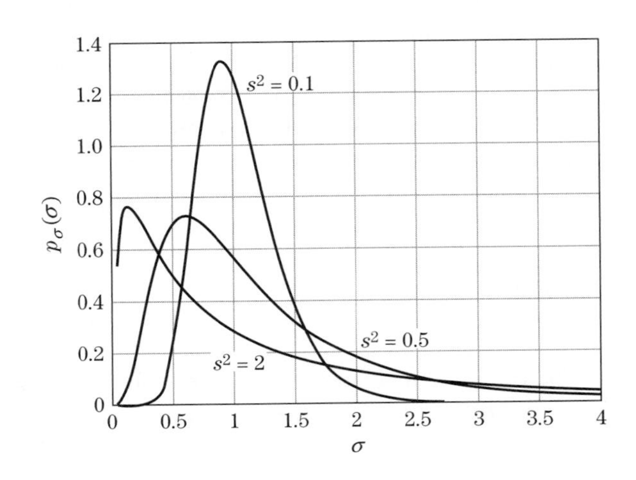

FIGURE 5-8 ■ Log-normal distributions for $\sigma_m = 1$ and several values of s^2.

When compared with Figure 5-7, it is obvious that log-normal-distributed data will exhibit a larger percentage of high values than Weibull-distributed data with the same median.

Some authors simply express the variation of clutter data in terms of the mean (or median) and the variance (or standard deviation, the square root of variance) without specifying an associated PDF. This approach is adequate for comparing the general strength and variability of different types of clutter for various conditions but does not provide enough information for analysis of radar detection performance.

5.2.2.3 Average Value Data

An extensive body of definitive experiments has been performed in the last 40 years to characterize radar clutter. Some of the key studies are as follows:

- Rain backscatter measurements from 10 to 100 GHz performed jointly by the U.S. Army Ballistic Research Laboratory (BRL) and the Georgia Tech Research Institute (GTRI) in the mid 1970s [7,8].
- Measurement of frozen precipitation by the U.S. Army Harry Diamond Laboratory millimeter waves in the late 1980s [9].
- University of Kansas measurements on terrain at high angles in the 1980s [10].
- Measurement of land clutter at low grazing angles by the Massachusetts Institute of Technology Lincoln Laboratories (MIT/LL) in the 1980s and 1990s [11].
- Naval Research Laboratory (NRL) four frequency sea clutter measurements at high grazing angles [12].
- Georgia Tech measurements of sea clutter at low grazing angles [13,14].
- The "SNOWMAN" MMW measurements performed jointly by the U.S. Army MICOM and GTRI on snow-covered ground [15].

Countless other measurement programs have been conducted in the United States and Europe. This section will attempt to summarize data from these and many other experimental programs, but the interested reader is urged to also review the references at the end of the chapter.

Land Reflectivity

Dependence on Grazing Angle Figure 5-9 shows the backscatter reflectivity of crops and short grass at X-band compiled from three data sources. The data are reported as a function

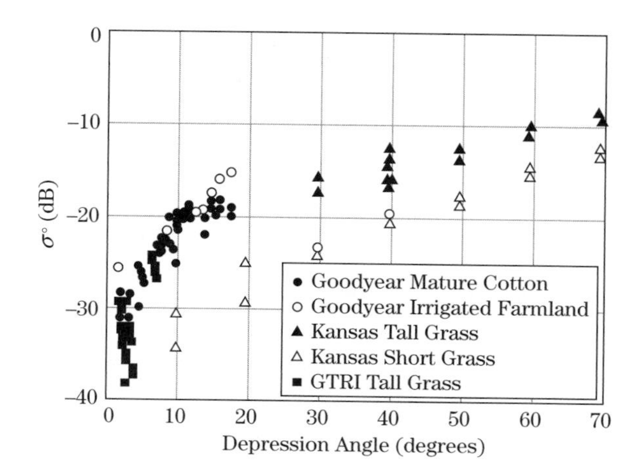

FIGURE 5-9 ■ σ^0 data for grass and crops from several sources at X-band. (Data from [16–18]. With permission.)

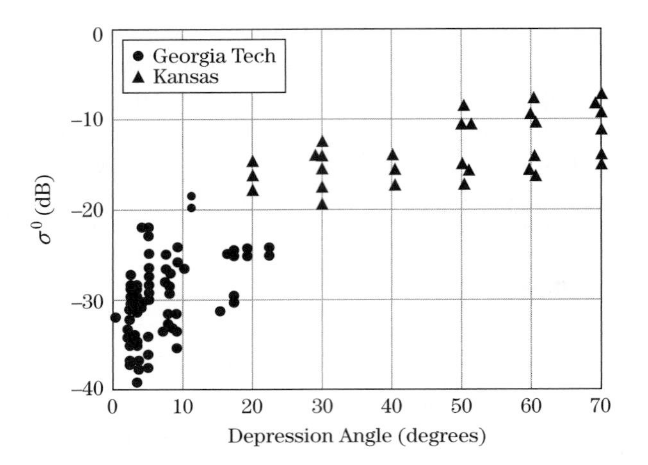

FIGURE 5-10 ■ σ^0 data for trees from two sources for X-band. (Data from [16,18]. With permission.)

of depression angle [16–18]. The data consist of time averages of individual spatial samples and exhibit the expected characteristic dependence on grazing angle shown in Figure 5-6, being relatively independent of angle above 10 degrees while rapidly decreasing in value at angles below 10 degrees. The greater spread in the data at very low depression angles is most likely due to shadowing effects.

Figure 5-10 presents backscatter reflectivity data from trees over the same angular regime as Figure 5-8. Again, the data represent time averages of spatial samples. A similar dependence is exhibited for the angular dependence as for grass and crops; however, the values for σ^0 are several dB higher, and the spatial variation is greater.

Dependence on Frequency Band Figures 5-11 through 5-14 present plots of averaged reflectivity data for five frequency bands compiled by Nathanson [15]. The data consist of σ^0 values at 0–1.5°, 3°, 10°, 30°, and 60°. Figure 5-11 gives data for relatively flat desert for L-band through X-band as a function of grazing angle. The spread in the data over the frequency range is approximately 10 dB, and the spread over 0 to 60° depression angle is more than 30 dB. Data from other sources including bare hills at L-band have yielded extremely high values for the reflectivity ($\sigma^0 > 0$ dB with corresponding RCS values of +40 dBsm) when viewing the sides of the hills at essentially 0° depression angle [16].

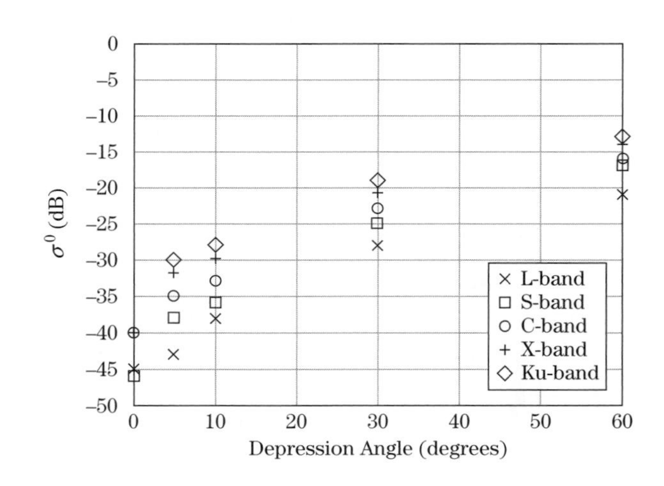

FIGURE 5-11 ■ Averaged reflectivity data for desert terrain as a function of frequency. (Adapted from Nathanson [15]. With permission.)

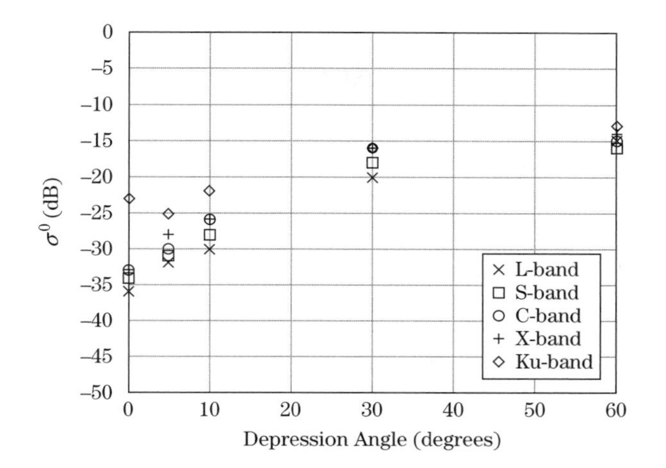

FIGURE 5-12 ■ Averaged reflectivity data for rural farmland as a function of frequency. (Adapted from Nathanson [15]. With permission.)

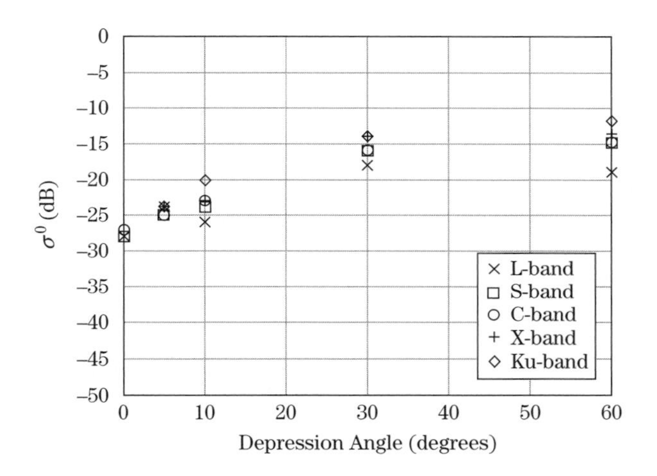

FIGURE 5-13 ■ Plot of averaged reflectivity data for heavy vegetation/jungle as a function of frequency. (Adapted from Nathanson [15]. Used with permission.)

Figure 5-12 shows averaged data for rural farmland. The spread in the data with frequency is lower than for the desert terrain, being approximately 6 dB. Also, the variation with depression angle is somewhat less, being approximately 25 dB. Figure 5-13 gives averaged data for heavy vegetation and jungle. The spread in the data with frequency is again approximately 6 dB with a similar angular variation as Figure 5-12.

Figure 5-14 gives averaged data for urban environments. The spread in the data with frequency band is lower than the previous plots, but the data are high in value as might be expected since presumably many man-made targets are included. Also, the dependence on depression angle is less than for the previous figures.

Very low angle clutter returns are of particular concern as such returns can significantly affect the detection of low flying objects such as missiles. The MIT Lincoln Laboratory performed extensive measurements in the 1980s to characterize low angle clutter over many terrain types [11]. Figure 5-15 gives σ^0 data collected by the MIT/LL for very low depression angles (0.4 to 1°) and for several frequency bands from UHF to X-band, including both vertical and horizontal polarizations. As can be seen from the figure, σ^0 appears to be at a maximum in the UHF band, decreasing at the higher frequencies, presumably due to absorption of the energy. Also, little difference is seen between

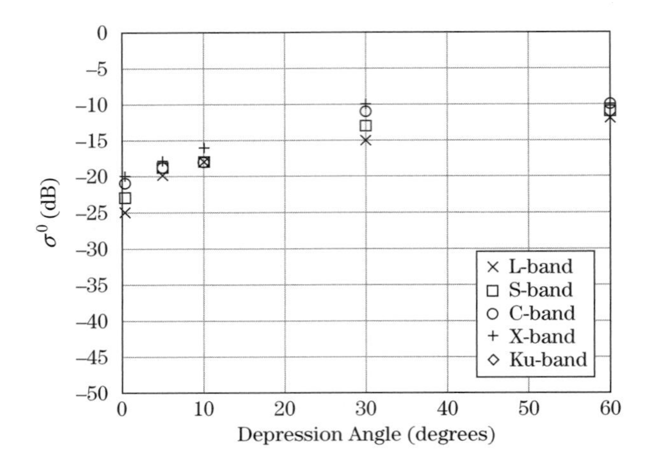

FIGURE 5-14 ■ Plot of averaged reflectivity data for urban terrains as a function of frequency. (Adapted from Nathanson [15]. With permission.)

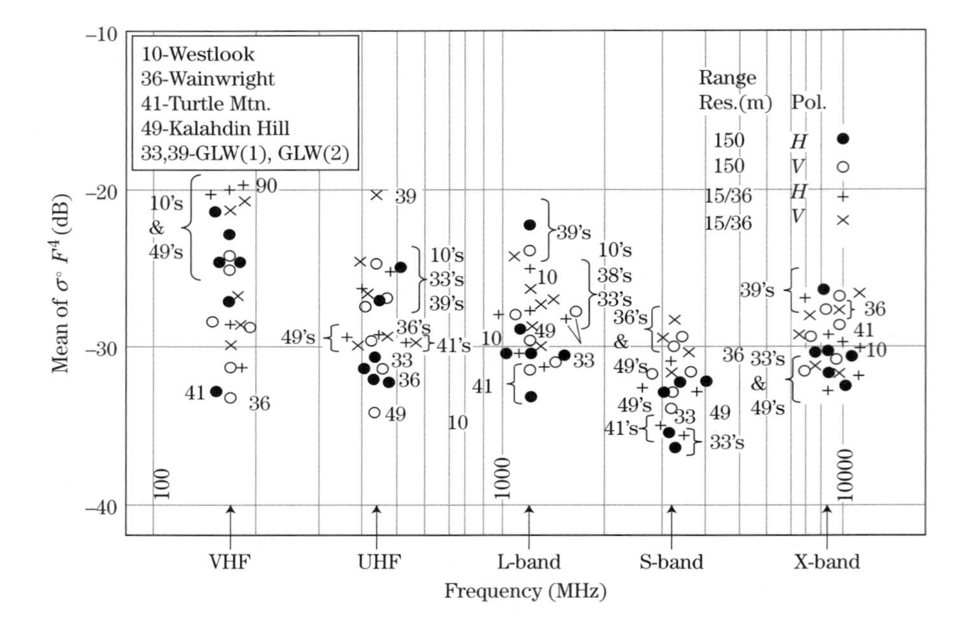

FIGURE 5-15 ■ σ^0 data for several forest/low-relief terrains at low (0.4–1°) depressions angles as a function of frequency. (From Billingsley [11]. With permission)

horizontal and vertical polarizations. Figure 5-16 gives data on mountain terrains for several frequency bands. As can be seen, the reflectivity decreases rapidly with increasing frequency, presumably caused by vegetation attenuating the returns from bare rock and ground at the higher frequencies. Note that σ^0 values of nearly 0 dB are reported at the lowest-frequency bands. Such values can often overwhelm radar MTI processors, leading to false alarms.

Sea Reflectivity The sea surface is composed of salty water with a reflection coefficient of almost -1 at microwave frequencies for small grazing angles. Thus, a smooth sea appears like an infinite flat conductive plate that scatters all of the energy impacting the surface in a forward direction so no backscatter occurs. As the wave height starts to increase, the sea begins to appear like a rough surface, and as the wave height continues to grow, organized

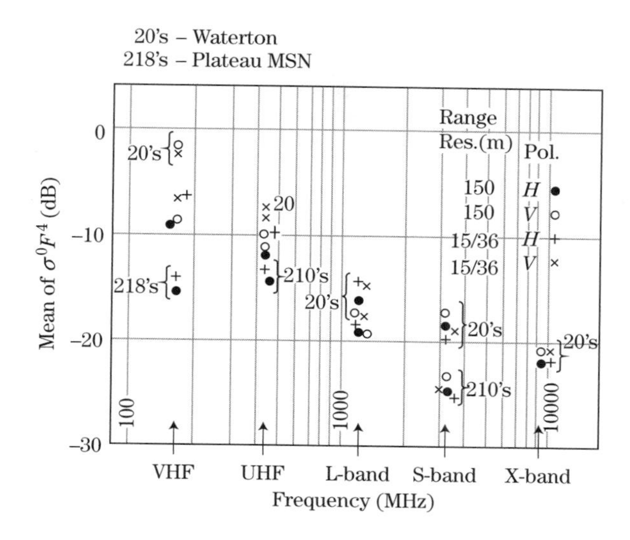

FIGURE 5-16 ■ σ^0 data for two mountain terrains at low (1.2°) depressions angles as a function of frequency. (From Billingsley [11]. With permission.)

TABLE 5-2 ■ Parameters Affecting Sea Return

PARAMETER	COMMENTS
Wave height	Strong proportional dependence
Wind speed	Dependence increases with increasing frequency
Wind/wave look direction	Significant difference between up-wave and down-wave
Polarization	Dependence decreases with increasing frequency
Grazing angle	Strong dependence at low angles, weaker dependence in the plateau region
Frequency band	Proportional to frequency in the microwave region

wavefronts occur that provide a strong directional dependence to the scattering. Table 5-2 gives the primary physical parameters that can affect sea return.

As can be seen from Table 5-2, wave height is one of the major physical parameters affecting sea return. Unfortunately, wave height is often difficult to measure during experiments. In addition, wave height is irregular. The wave height is considered to be the "significant wave height," which is an estimate of the average peak-to-trough height of the largest one-third of the observed waves.

Since it is easier to estimate a range of wave heights than a specific wave height, sea return data are often give in terms of *sea state*. The Douglas sea number [6] is a specific, widely used scale of sea states, correlated wind speeds, and subjective descriptions in which each defined sea state represents a range of wave heights as given in Table 5-3.[2] Note that sea states are defined only for a fully developed sea, that is, a sea over which a constant wind has been blowing long enough to build waves to their maximum height and the distance over which the wind has been blowing (called the *fetch*) is far enough to build waves to their maximum value.

[2]Other sea scales exist, such as the Beaufort scale and the World Meteorological Organization (WMO) scale. The WMO scale generally adopts the Douglas sea-scale definitions.

TABLE 5-3 ■ Douglas Sea State versus Wave Height and Wind Speed for a Fully Developed Sea

Sea State	Significant Wave Height (ft)	Wind Speed (Kts)
0	0 to 0.5	0 to 2
1	0.5 to 1	2 to 7
2	1 to 3	7 to 12
3	3 to 5	12 to 16
4	5 to 8	16 to 20
5	8 to 12	20 to 25
6	12 to 20	25 to 32
7	20 to 40	32 to 45
8	40+	45+

Source: Adapted from Long [6] (with permission).

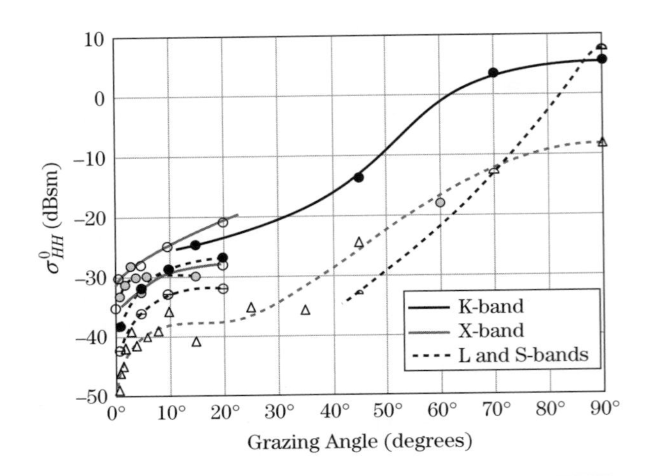

FIGURE 5-17 ■ Sea return as a function of grazing angle for four radar bands. (From Long [6]. With permission.)

Angular and Frequency Dependencies Figure 5-17 gives horizontally polarized sea return data from several original sources for four radar bands as a function of grazing angle, wind speeds corresponding to high sea states, and upwind versus downwind directions. The data illustrate strong angular, frequency, and look direction dependence. Note that very high values of σ^0 (above 0 dB) are seen at nadir. This suggests that sea clutter could be a significant limitation when searching for a small, slow-moving target on the sea surface. In addition to high backscatter, shadowing and sea Doppler accentuate the detection problem.

Figure 5-18 presents a plot of averaged data for sea clutter from Nathanson [15] for five radar bands as a function of depression angle for sea state 1. For this low sea state, σ^0 is relatively small, particularly for low depression angles. Note that there is much more

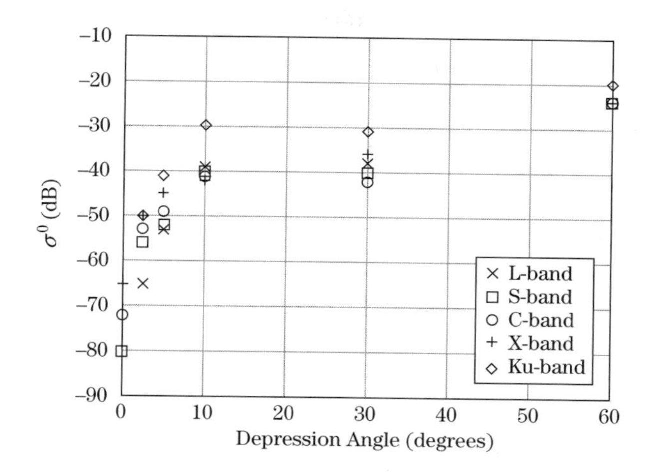

FIGURE 5-18 ■ Averaged sea return as a function of depression angle and radar band, sea state 1, VV polarization. (Adapted from Nathanson [15]. With permission.)

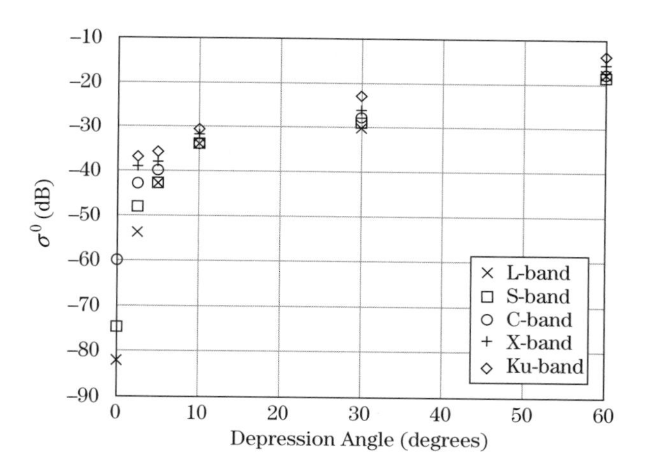

FIGURE 5-19 ■ Averaged sea return as a function of depression angle and radar band, sea state 3, VV polarization. (Adapted from Nathanson [15]. With permission.)

variation with frequency at the lower depression angles. Figure 5-19 gives averaged sea return data for sea state 3. These data are almost 10 dB higher than for sea state 1. The frequency dependencies appear to be similar to those for Figure 5-18.

Figure 5-20 compares values of σ^0 for vertical and horizontal polarized returns from sea clutter at L- and X-bands. As was discussed previously, there is a much greater difference between the vertical and horizontal values of σ^0 at L-band than there is at X-band.

One important parameter for sea return is the range fall-off of sea clutter echo power. As range increases, the grazing angle δ at which the sea surface is viewed decreases. When δ falls below the critical angle, θ_C, the surface becomes "smooth" by the Rayleigh criterion of equation (5.15). Figure 5-21 gives the measured sea data as a function of range dependence for two regions: above and below the critical angle; and low and high wave heights [19]. Note that the range for the critical angle moves in (higher depression angle), and the wave height (and thus, the rms surface roughness) increases as predicted by equation (5.16). These data show that, below the critical angle, the R^{-3} range dependence of pulse-limited sea clutter return transitions to approximately an R^{-7} dependence so that the clutter return rapidly becomes insignificant. In addition, the critical angle often appears near the first multipath null angle, further enhancing the clutter roll-off with range. (See Chapter 4 for a discussion of multipath.)

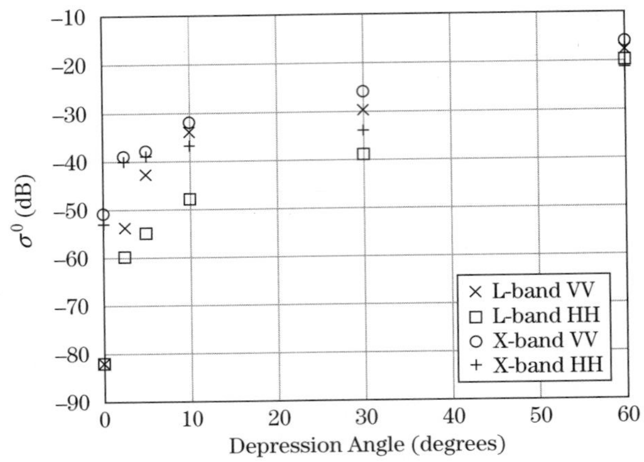

FIGURE 5-20 ■ Sea return as a function of depression angle for VV and HH polarizations, land X-bands, sea state 3. (Adapted from Nathanson [15]. With permission.)

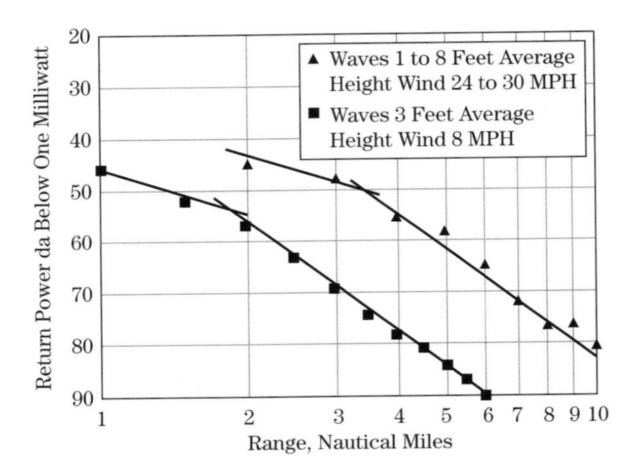

FIGURE 5-21 ■ Range dependence of sea return for two wave conditions, X-band, HH polarization. (From Dyer and Currie [19]. With permission.)

Ducting Because the sea surface has a strong reflection coefficient and water vapor is always present near the surface, conditions can occur such that a high water concentration and thus a reflective layer occur over some region above the water. When this happens, the high water concentration layer and the sea surface act like a two-dimensional waveguide, trapping the radiofrequency (RF) energy and extending the range of sea return detection. The reflection coefficient of the sea determines the duct shape and transmission efficiency. Figure 5-22 shows two sets of sea return data as a function of grazing angle taken with the same radar, the only difference being the lapse of several hours in time. The morning data show much less dependence on angle than the afternoon data. Apparently, ducting conditions were present in the morning but were absent in the afternoon. These data were collected in February in Wildwood, New Jersey, showing that ducting can occur in cold as well as tropical conditions.

Figure 5-23 gives a summary of range dependence measurements above and below the critical angle performed in Boca Raton, Florida, over a period of several years at X-band. Range dependencies quite different from the expected R^{-3} for a pulse-limited radar occurred a significant portion of the time, indicating both ducting conditions and possibly variations in the multipath field from time to time.

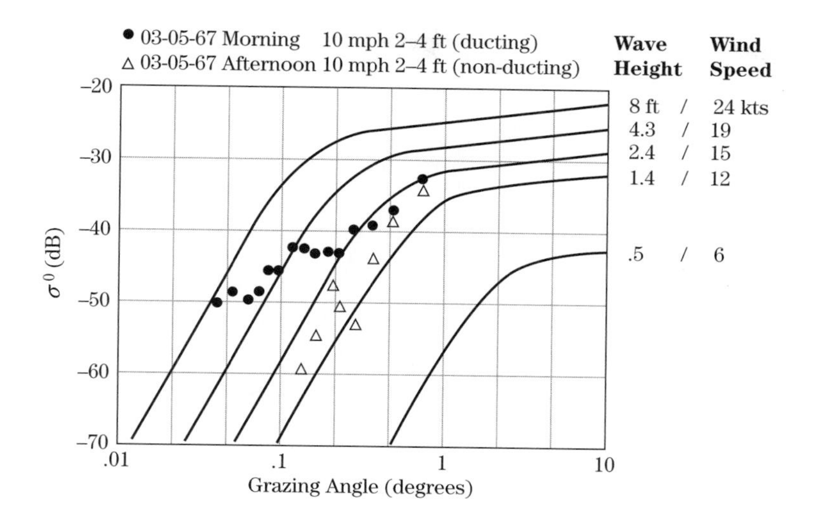

FIGURE 5-22 ■ Comparison of sea return achieved under ducting and nonducting conditions, X-band, HH polarization. (From Dyer and Currie [19]. With permission.)

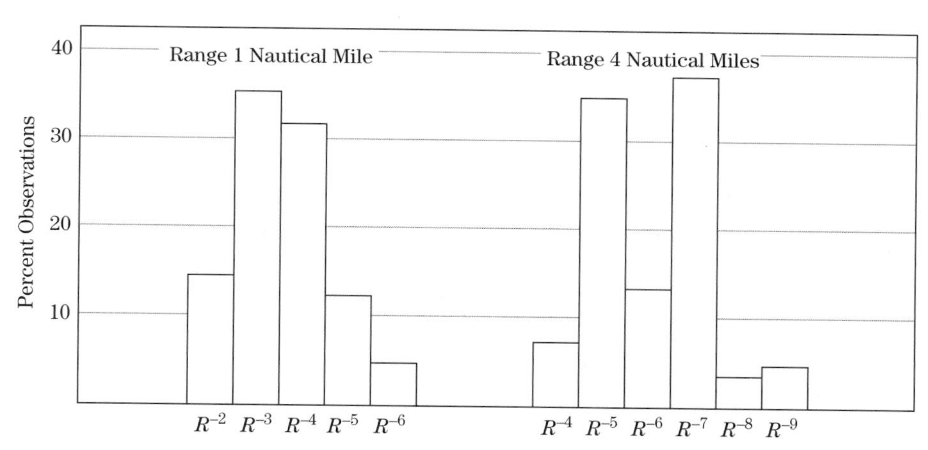

FIGURE 5-23 ■ Measured range dependencies above and below the critical grazing angle as a percentage of total measurements. (From Dyer and Currie [19]. With permission.)

5.2.2.4 Clutter Variability Properties

In the previous section average values for surface clutter were discussed, but the variability of clutter reflectivity with time and or space is as least as important, if not more so. These topics are addressed in this section.

Land Clutter Variations

Temporal Variations Land clutter generally contains some vegetation. Consequently, land clutter returns will vary with time due to wind-blown motion of leaves, needles, branches, and stalks. Since such motion makes detection more difficult, it must be described and allowances made for it when calculating the probability of detection for a target in clutter. The probability density function, $p_\sigma(\sigma)$, and the cumulative distribution function (CDF), $P_\sigma(\sigma)$, are used to describe the variation in RCS or power. Figure 5-24 illustrates an estimated unnormalized probability density function obtained as the histogram of the measured data, which is just a plot of the number of independent clutter samples that fall within a series of narrow power intervals. If the histogram is normalized by dividing the sample counts by the total number of samples times the width of an amplitude bin, an estimate of the PDF is obtained.

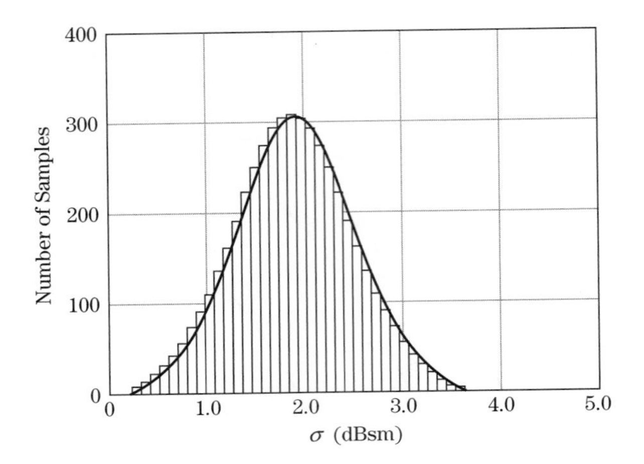

FIGURE 5-24 ■ A histogram provides an estimate of an unnormalized probability density function.

The CDF is the integral of the PDF,

$$P_\sigma(\sigma) = \int_{-\infty}^{\sigma} p_\sigma(v)\,dv \tag{5.20}$$

Values of the CDF increase monotonically from zero to one. If the area under the normalized histogram from the lowest value to some value σ is calculated, an estimate of the CDF is obtained. Cumulative distributions are quite useful because some key parameters are easily read from a plot of the CDF. For instance, the median value is the 50% point $(P_\sigma(\sigma) = 0.5)$ on the curve. Rivers [20] observed that for log-normal and Weibull distributions with parameters appropriate for modeling measured sea clutter, the mean value of the power, $\bar{\sigma}$, can be estimated from the value corresponding to the 90% point on the CDF, $\sigma_{0.9}$, to be 3.5 dB below the 90% point by the relation

$$\bar{\sigma} = (\sigma_{0.9} - 3.5\text{ dB}) \pm 0.5\text{ dB} \tag{5.21}$$

Newer land clutter data observed by Billingsley [11] is spikier than that used by Rivers, suggesting that the ±0.5 dB tolerance in equation (5.21) should be wider [6]. As another example of the usefulness of CDFs, the standard deviation in dB of a log-normal distribution is $\sigma_{0.84} - \sigma_{0.16}$.

Figure 5-25 gives a measured CDF for wind-blown trees at X-band. The dotted line approximates the CDF corresponding to an exponential PDF in this plot format. The data in Figure 5-25 appear approximately exponential in their general shape, but their distribution is seen to be wider than exponential. For example, the 90% mark on the CDF occurs at about –39 dB for the exponential distribution, but not until the larger value of –30 dB for the X-band tree data. Variable clutter complicates detection in two ways. First, since the return is changing with time, part of the time the reflectivity will be larger than the average value. Second, the rate of fluctuation can limit the effectiveness of Doppler processing.

Table 5-4 gives the standard deviations (square root of the variance) measured for wind-blown vegetation as a function of frequency and polarization. The standard deviation for an exponential distribution, converted to a decibel scale, is approximately 5.7 dB [6]. Thus, at 9 GHz the distributions appear narrower than exponential, but at 95 GHz and higher they are wider than exponential.

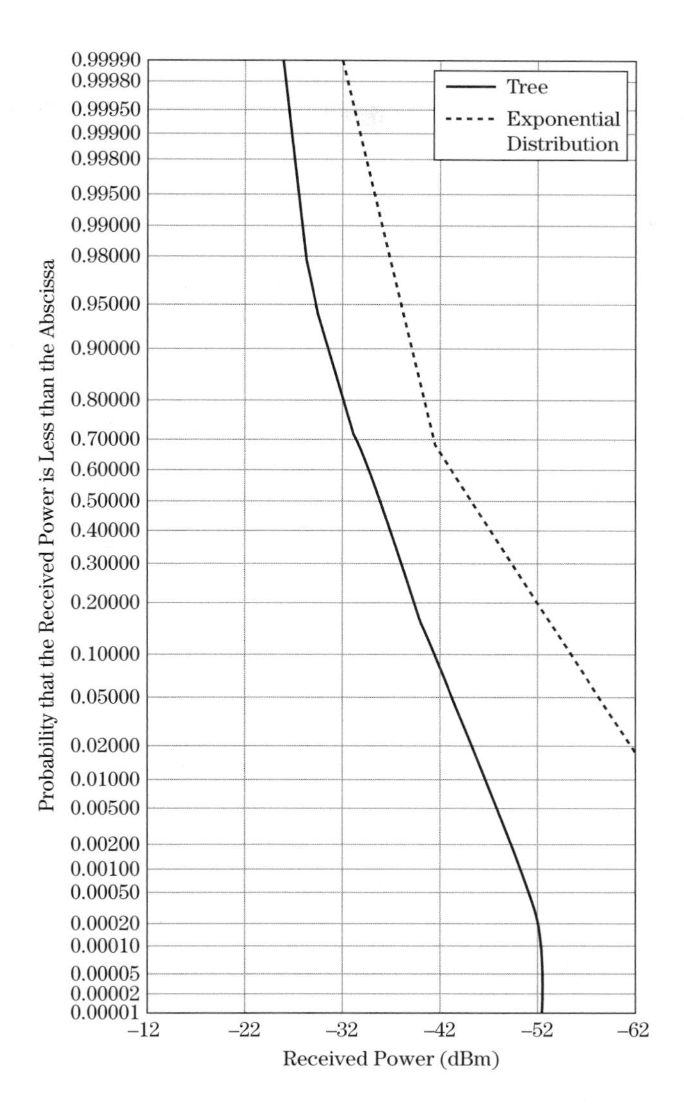

FIGURE 5-25 ■ Cumulative probability distribution for deciduous trees at X-band, 4.1°depression angle. (From Currie et al. [21]. With permission.)

Decorrelation Time An important metric of clutter temporal properties is the decorrelation time. The decorrelation time, τ_0, is defined as the time lag required for the autocorrelation function of a set of clutter samples to decay to some defined fraction, usually 0.5 or $1/e = 0.367$ of its peak value at zero lag. The autocorrelation function $\phi_{cc}(\tau)$ of the time-varying clutter return, $c(t)$, measures how similar successive data samples are to a first sample as the delay time between samples increases. The power spectral density function is the Fourier transform of $\phi_{cc}(\tau)$,

$$S_{cc}(f) = \int_{-\infty}^{+\infty} \phi_{cc}(\tau)\, e^{-j2\pi f\tau} d\tau \tag{5.22}$$

where τ is the autocorrelation lag time.

If the clutter power spectrum is "white" (constant) over the receiver bandwidth B, then the decorrelation time equals $1/B$. This is the same as the case for random noise.

TABLE 5-4 ■ Measured Standard Deviations of Temporal Variations for Trees

		Average Value of Standard Deviation (dB)			
Frequency Clutter Type	**Polarization**	**9.5 GHz**	**16.5 GHz**	**35 GHz**	**95 GHz**
Deciduous trees, summer	Vertical	3.9	—	4.7	—
	Horizontal	4.0	—	4.0	5.4
	Average	4.0	—	4.3	5.4
Deciduous trees, fall	Vertical	3.9	4.2	4.4	6.4
	Horizontal	3.9	4.3	4.3	5.3
	Average	3.9	4.2	4.3	5.0
Pine trees	Vertical	3.5	3.7	3.7	6.8
	Horizontal	3.3	3.8	4.2	6.3
	Average	3.4	3.7	3.9	6.5
Mixed trees, summer	Vertical	3.3	—	4.0	—
	Horizontal	4.6	—	4.2	—
	Average	4.4	—	4.1	—
Mixed trees, fall	Vertical	4.1	4.1	4.7	6.3
	Horizontal	4.5	4.3	4.6	5.0
	Average	4.4	4.2	4.6	5.4
Field, tall grass	Vertical	1.5	—	1.7	2.0
	Horizontal	1.0	1.2	1.3	—
	Average	1.3	1.2	1.4	2.0
Rocky area	Vertical	1.1	2.2	1.8	1.6
	Horizontal	1.2	1.7	1.7	1.7
	Average	1.1	1.9	1.8	1.7
10-in. corner reflector located in grassy field		1.0	1.0	1.2	1.2

Source: From Currie et al. [21] (with permission).

Because the receiver bandwidth also determines the Nyquist rate of its output, the received signal output will normally be sampled at a rate of about $1/B$ samples per second (see Chapter 14). Since the samples are spaced by the decorrelation time, the sampled data will appear uncorrelated with a white power spectrum. If the clutter instead has a decorrelation time greater than $1/B$ seconds, then the sampled clutter data will not appear white and will have some degree of correlation from one sample to the next.

The decorrelation time is important for detection analysis because it determines the number of uncorrelated samples, N_i, available for integration for signal-to-clutter improvement according to

$$N_i = \begin{cases} \dfrac{N_t PRI}{\tau_0}, & PRI \le \tau_0 \\ N_t, & PRI \ge \tau_0 \end{cases} \tag{5.23}$$

where N_t is the total number of clutter samples and PRI is the time between samples. As will be seen in Chapter 15, coherent or noncoherent integration of radar data can improve the detectability of targets in the presence of interference, provided that the interference samples are uncorrelated so they can be made to "average out" while the target signal is reinforced. Thus, if the total sample collection time $N_t PRI$ is shorter than the decorrelation time, no uncorrelated clutter samples are obtained, and no improvement in target detectability is gained by integration of successive samples.

If the clutter does decorrelate during the collection time, then some improvement in target detectability is possible by integrating multiple samples. In coherent integration, the complex (in-phase [I] and quadrature [Q]) data are integrated (added) to cause the target component of the samples to add in phase, whereas the uncorrelated clutter and noise components do not add in phase. In this case, the signal-to-clutter ratio (SCR) and signal-to-noise ratio (SNR) (and thus signal-to-interference ratio [SIR]) are increased, significantly improving target detectability. In noncoherent integration, the magnitude or magnitude squared of the complex receiver output data is taken and then integrated. Discarding the phase information eliminates the possibility of a gain in SCR. Nonetheless, an improvement in target detectability is still achieved, though less than in the coherent integration case. Chapter 15 discusses the effects of coherent and noncoherent integration on target detection in white interference in greater detail.

Figure 5-26 presents the autocorrelation functions for windblown trees in conditions described as a "windy day," while Table 5-5 gives the 50% ($\tau_{1/2}$) and $1/e$ ($\tau_{1/e}$) decorrelation times in seconds. As can be seen, essentially no integration improvement could be achieved at the lower frequency bands, and only limited improvement could be achieved at X-band.

Figure 5-27 gives measured decorrelation times for higher frequencies (10 GHz through 95 GHz) as a function of wind speed. As can be seen, at the higher frequencies much shorter decorrelation times are observed, an advantage for obtaining better noncoherent integration efficiency.

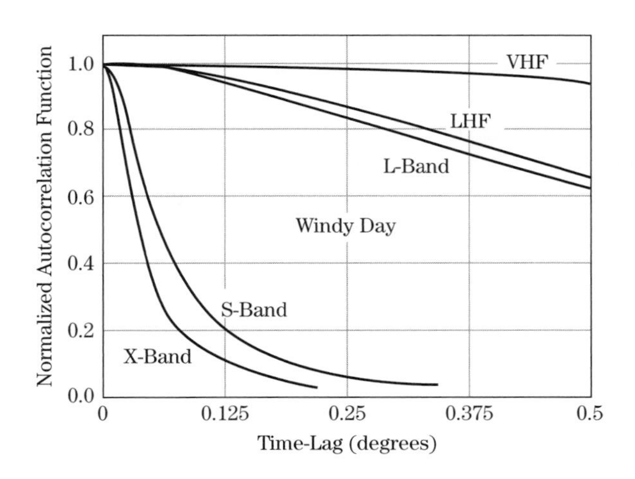

FIGURE 5-26 ■ Autocorrelation functions of the returns from windblown trees for several frequency bands on a windy day. (From Billingsley [11]. With permission.)

TABLE 5-5 ■ Measured Decorrelation Times at Five Frequency Bands

Frequency Band	Correlation Time(s)	
	$\tau_{1/2}$	$\tau_{1/e}$
VHF	4.01*	5.04*
UHF	0.69	0.94
L-Band	0.67	0.95
S-Band	0.062	0.081
X-Band	0.033	0.049

Note: * = extrapolated estimate
Source: From Billingsley [11] (with permission).

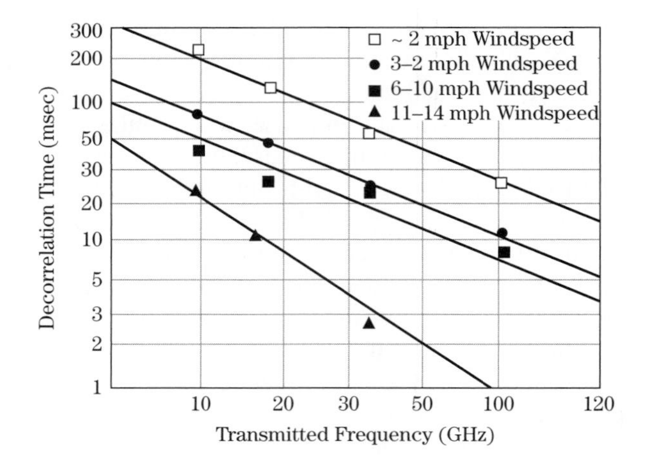

FIGURE 5-27 ■ Decorrelation time for windblown trees as a function of wind speed. (From Currie et al. [21]. With permission.)

Clutter Frequency Spectra A second way to look at temporal variations in clutter returns is in terms of clutter power spectra. Power spectra tell us how rapidly a return is varying. This is of interest since many radars use Doppler processing to improve target detection in heavy clutter. Clutter spectra are of interest for two situations: (1) very slow-moving targets such as ground vehicles or boats; and 2) high-speed targets that may be affected by aliasing ("foldover") of the clutter spectra around the pulse repetition frequency (PRF). (See Chapter 8 for a discussion of Doppler foldover.)

Theoretical formulations of clutter spectra yield Gaussian-shaped spectra, yet a number of actual clutter measurements have yielded power law-shaped spectra. These differences are of concern since power law spectral shapes roll off more gradually with frequency than Gaussian spectra. This problem was first identified by Fishbein et al. [22] when attempting to develop models for the spectral data on trees at X-band shown in Figure 5-28 [22]. They found that a Gaussian curve fit rolled off much too rapidly. A much better fit was obtained with a power law curve of the form

$$S_{cc}(f) = \sigma_{DC}\left(\frac{1}{1+\left(\dfrac{f}{f_c}\right)^n}\right) \tag{5.24}$$

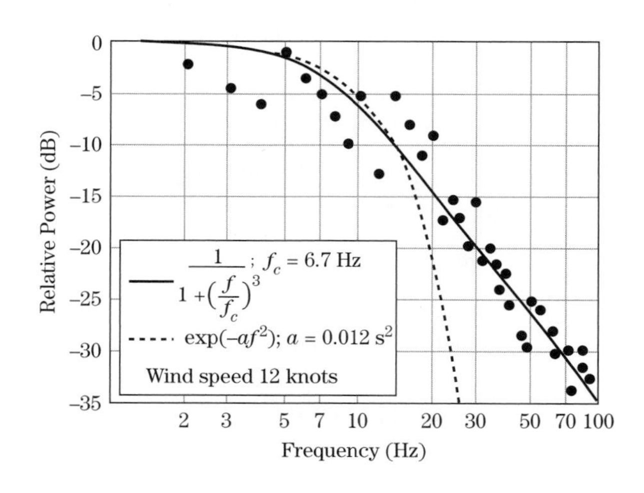

FIGURE 5-28 ■ Spectral data from trees at X-band with Gaussian and power function curve fits. (From Fishbein et al. [22]. With permission.)

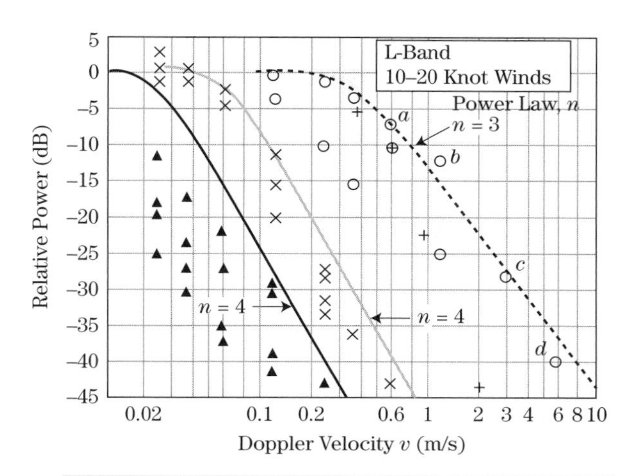

Measurements		Models
Mountains	▲	
Partially Wooded Hills	×	
HeavilyWooded Valleys (Lowlands)	○	
a, b, c, d: Extrema		
+ :Widest Lincoln Measurement		

FIGURE 5-29 ■ L-band spectral data with various power law curve fits. (From Simpkins et al. [23]. With permission.)

where σ_{DC} is the DC clutter power return, f_c is the 3 dB cutoff frequency for the power spectrum, and n is an integer selected to best fit the data. Fishbein et al. found that a value of $n = 3$ fit the data best. Other experimenters have also measured power law dependencies. Figure 5-29 presents power spectral data at L-band for woods and wooded hills measured by the U.S. Air Force's Rome Air Development Center (RADC) [23]. For these data, the best power law fits were obtained with $n = 3$ or 4.

Table 5-6 presents the cutoff and power law exponents determined for wind-blown tree data at 10 GHz though 95 GHz for both logarithmic and linear receiver transfer functions. The exponent seems to lessen with increasing frequency, implying a slower roll-off, while the logarithmic receiver provides a steeper roll-off than the linear case.

To date, theorists have not been able to determine a theoretical basis for power law frequency dependence, yet experimenters continue to observe them in measured data. One possibility is that Doppler spectra can be significantly affected by radar system imperfections such as nonlinearities, dynamic range limits, and oscillator phase noise. Recently, Billingsley [11] measured tree data with a high-quality coherent instrumentation radar having high linearity over a wide dynamic range, stabilized oscillators, and low-sidelobe narrow Doppler filters. He obtained clutter power spectra that were well modeled by a

TABLE 5-6 ■ Corner Frequencies and Power Exponents for Tree Return Spectra

Power Function Parameter	Frequency			
	9.5 GHz	16 GHz	35 GHz	95 GHz
n (linear)	3	3	2.5	2
n (log)	4	3	3	3
f_c (Hz), 6–15 mph wind speed	9	16	21	35

Source: From Currie et al. [21] (with permission).

two-sided exponential roll-off. This spectrum has a rate of decay that falls between the Gaussian and power law models.

These observations suggest that equipment imperfections may be a significant contributor to the slower decay of data that are well modeled by power law spectra. However, many fielded systems will suffer these limitations, so a practical approach to performance assessments could use both Gaussian and power law clutter power spectrum models to establish upper and lower bounds on performance in the presence of clutter.

Spatial Variations As discussed earlier, when the radar scans across the surface, the number and types of scatterers within the radar beam change, resulting in a changing return. These variations are described using the same tools as for temporal variations (i.e., amplitude distributions and spatial correlation functions).

Spatial amplitude distributions can be much wider than temporal distributions, particularly at lower grazing angles where shadowing and multipath come into play. Most experimenters resort to Weibull distributions to describe the variations. A key finding of experimenters is that at higher grazing angles, spatial distributions tend to appear exponential (Weibull width parameter $a \approx 1$) while at low grazing angles the width parameter increases greatly. The calculated spatial distributions by Booth [24] for cultivated land at X-band in Figure 5-30 illustrate this effect. They are approximately exponential-distributed (width parameter $a \approx 1$) at a grazing angle of 5° but become much wider at lower angles.

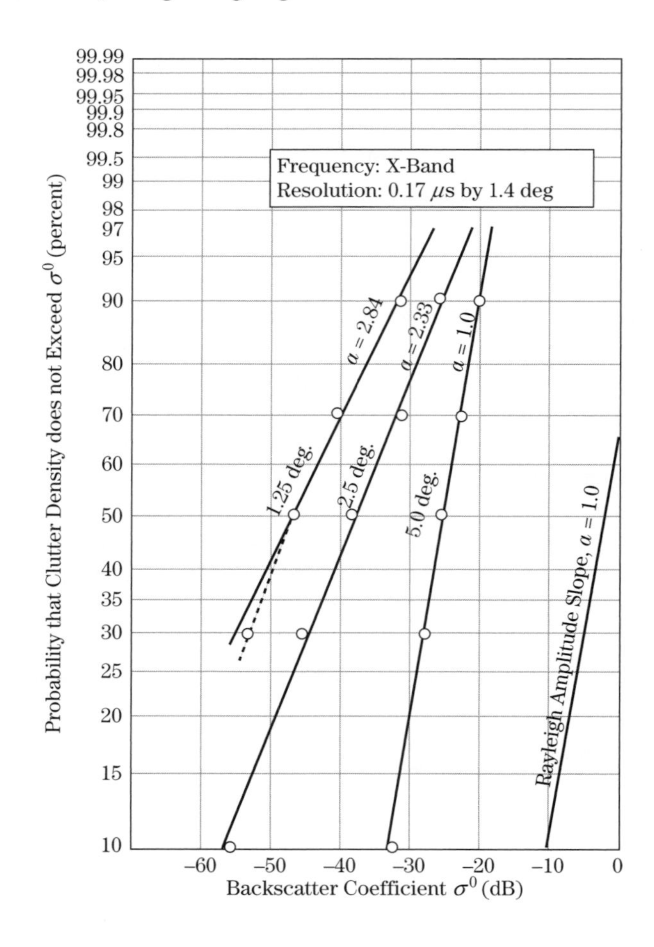

FIGURE 5-30 ■ Spatial distributions of cultivated land as function of grazing angle X-band. (From Booth [24]. With permission.)

TABLE 5-7 ■ Spatial Statistical Attributes for X-Band Ground Clutter

Terrain Type	Depression Angle (deg)	Weibull Parameters			Ensemble Mean Clutter Strength σ^0 (dB)	Percent of Samples above Radar Noise Floor	Number of Patches
		a	σ_m (dB)	σ_w^0 (dB)			
Rural/ Low-Relief	0.00–0.25	4.8	−60	−33	−32.0	36	413
	0.25–0.50	4.1	−53	−32	−30.7	46	448
	0.50–0.75	3.7	−50	−32	−29.9	55	223
	0.75–1.00	3.4	−46	−31	−28.5	62	128
	1.00–1.25	3.2	−44	−30	−28.5	66	92
	1.25–1.50	2.8	−40	−29	−27.0	69	48
	1.50–4.00	2.2	−34	−27	−25.6	75	75
Rural/ High-Relief	0–1	2.7	−39	−28	−26.7	58	176
	1–2	2.4	−35	−26	−25.9	61	107
	2–3	2.2	−32	−25	−24.1	70	44
	3–4	1.9	−29	−23	−23.3	66	31
	4–5	1.7	−26	−21	−22.2	74	16
	5–6	1.4	−25	−21	−21.5	78	9
	6–8	1.3	−22	−19	−19.1	86	8
Urban	0.00–0.25	5.6	−54	−20	−18.7	57	25
	0.25–0.70	4.3	−42	−19	−17.0	69	31
	0.70–4.00	3.3	−37	−22	−24.0	73	53

σ_m = median reflectivity
σ_w^0 = mean reflectivity
F = propagation factor (see Chapter 4)
Source: Adapted from Billingsley [11] (with permission).

The effect of grazing angle on the PDF is further illustrated by Table 5-7, which gives spatial statistics for rural and urban settings as a function of grazing angle for very low angles. For rural settings, and angles above 5 degrees, the Weibull width parameter a indicates an approximately exponential power PDF ($a \approx 1$), while the PDF becomes much wider for lower angles. As might be expected, the urban setting exhibits larger width parameters and thus distributions wider than exponential even at 4°. Thus, in general, exponential power statistics are expected in the plateau regions for all but urban clutter, but much wider distributions are seen for low grazing angles. For urban settings, wide distributions may be observed even in the plateau region because of the height of man-made structures that cause shadowing even at higher grazing angles.

Sea Clutter Variation The return from the sea varies in time due to the effects of the wind on sea waves. These effects are of several types: waves created by the wind blowing for a period of time over a given "fetch" of water, ripples that appear on the surface due directly to the wind, white caps created when the tops of waves break over the front of the waves, and airborne spray that results from ripples and white caps.

Initially, amplitude (voltage) statistics from a fixed range-azimuth cell were assumed to be Rayleigh. This model fit measured data for low-resolution radars well. However, as finer-resolution radars came into use over time, non-Rayleigh statistics were encountered. The current view is that wind-produced ripples on the sea surface produce noise-like variations in sea return or "speckle" that are approximately Rayleigh distributed in amplitude but that the large-scale moving structure of the sea swells changes the local slope of the rippled surface, imposing a time-varying change in the mean of the Rayeigh PDF in a given spatial cell. The resulting amplitude statistics of the return are non-Rayleigh, implying a non-Gaussian model of the complex I/Q data.

In general, the amplitude statistics are observed to become "spikier" with decreasing grazing angle and decreasing radar footprint area, possibly due to a decreasing average or median clutter value while the strength of the sea clutter spikes remains relatively constant. Also, horizontal polarization tends to present spikier clutter than does vertical polarization, particularly at low grazing angles. Sea clutter statistics also vary with the radar-wind/wave look direction.

Experimenters have used Weibull, log-normal, and K distributions to model sea statistics. In particular, Ward et al. [25] and Ward [26] performed extensive modeling of sea return using the K distribution, a two-parameter distribution that can be related to the Weibull and the log-normal. It is given by [6]

$$p_\xi(\xi) = \begin{cases} \dfrac{2b}{\Gamma(v)}\left(\dfrac{b\xi}{2}\right)^{v} K_{v-1}(b\xi), & \xi > 0 \\ 0, & \xi < 0 \end{cases} \tag{5.25}$$

where

$\xi = \sqrt{\sigma}$ is the clutter amplitude.
v is the "shape factor."
b is a scale factor.
$\Gamma(\cdot)$ is the gamma function.
K_n is the modified Bessel function of the third kind and order n.

Ward showed that this distribution describes the statistics of the product of two random variables, one described by a Rayleigh PDF and the other by a chi or gamma PDF [26]. Thus, the Rayleigh component can be used to model the speckle portion of the sea return, while the chi component can be used to model the time-varying mean of the local Rayleigh PDF due to large-scale sea structure, as discussed previously. The K distribution is believed to provide a better fit to sea clutter data, particularly for high resolution radars. See [25–27] for more information.

Temporal Variations

Correlation Properties Work on calculating autocorrelation functions for sea returns has demonstrated that there are three mechanisms involved: (1) return from sea spray and white caps; (2) specular returns (spikes) from wavefronts; and (3) Bragg scattering [28]. The return from sea spray and Bragg scattering decorrelates very rapidly, while the specular return from the wavefront is highly correlated in time. Figure 5-31 illustrates this effect. The figure shows the autocorrelation function for sea return at K_u-band on two time scales. The right-hand figure indicates that almost 600 ms are required for the return to decorrelate to 50% of the ACF peak. The left-hand figure shows that the return drops from a normalized correlation of 1 to 0.85 in approximately one ms. The initial 15% decorrelation is due to the rapidly moving sea spray, while the longer 50% decorrelation time is due to the much slower motion of the wavefronts. Attempts have been made to use frequency agility to decorrelate sea return, but only the decorrelation due to sea spray and speckle is affected. Thus, sea waves tend to be highly correlated over many milliseconds, so that only scan-to-scan integration will be effective in increasing target detectability.

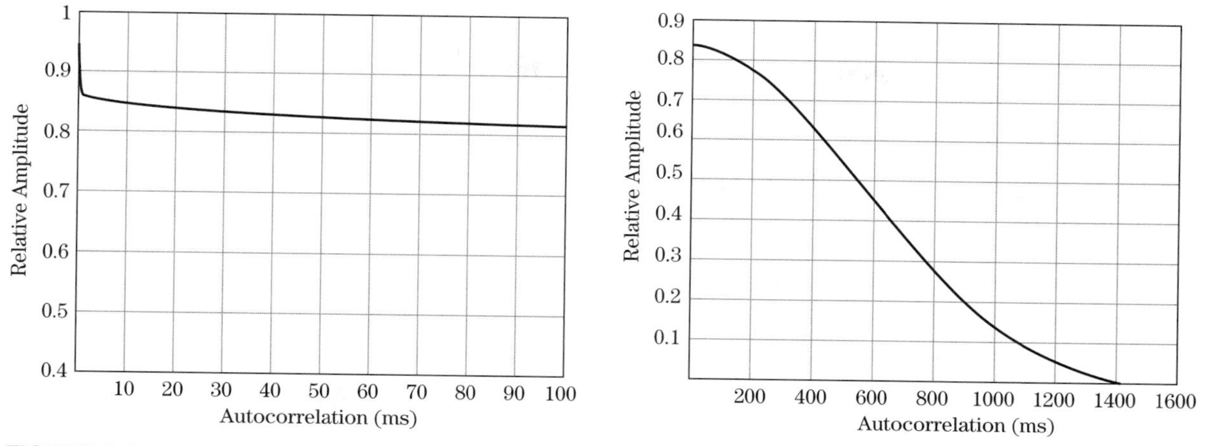

FIGURE 5-31 ■ Autocorrelation function of sea return at K_u-band. 3.9° depression angle, and downwind/down-wave look direction. (From Ward et al. [25]. With permission.)

Spectral Properties The spectrum for sea clutter differs from that of land since sea waves can move physically toward or away from the radar. Thus, sea clutter can have a nonzero average Doppler frequency, unlike land clutter where moving scatterers are anchored in place. The Doppler frequency at which the motion of scatterers at a radial velocity, v, appears is given by the equation

$$f_d = \frac{2v}{\lambda} \tag{5.26}$$

Consequently, the width, Δf_d, of the Doppler spectrum is related to the width, Δv, of the velocity spectrum according to

$$\Delta v = \frac{\lambda}{2}\Delta f_d \tag{5.27}$$

Figure 5-32 gives an example of the coherent spectrum at L-band, indicating the spectrum width for various amplitude levels [29]. At higher frequencies the center frequency and spectral width would be expected to scale proportional to the transmitted frequency in accordance with equations (5.26) and (5.27). While the spectra in this figure are approximately symmetric and Gaussian in shape, sea clutter spectra, especially at fine resolution and low grazing angle, are often distinctly asymmetric; see [28] for a number of examples.

Sea Spikes *Sea spikes* are strong returns from highly localized points or regions on the sea surface. The region generating the spike is usually, but not always, much smaller than the resolution cell area. Spikes are a problem for radar detection because they can exceed the signal level of small targets and can last for many seconds. Explanations of the causes for such returns are disputed by various researchers and have included returns from white caps, to sea spray, to returns from particular facets located on the wavefronts. Most likely all are responsible. Some general properties of sea spikes from the radar point of view reported by Werle [30] are as follows:

- Spikes occur significantly more often for horizontal than for vertical polarization.
- Spike intensity is usually greater for horizontal than for vertical polarization.
- HH/VV polarization ratios of greater than 10 dB were observed.

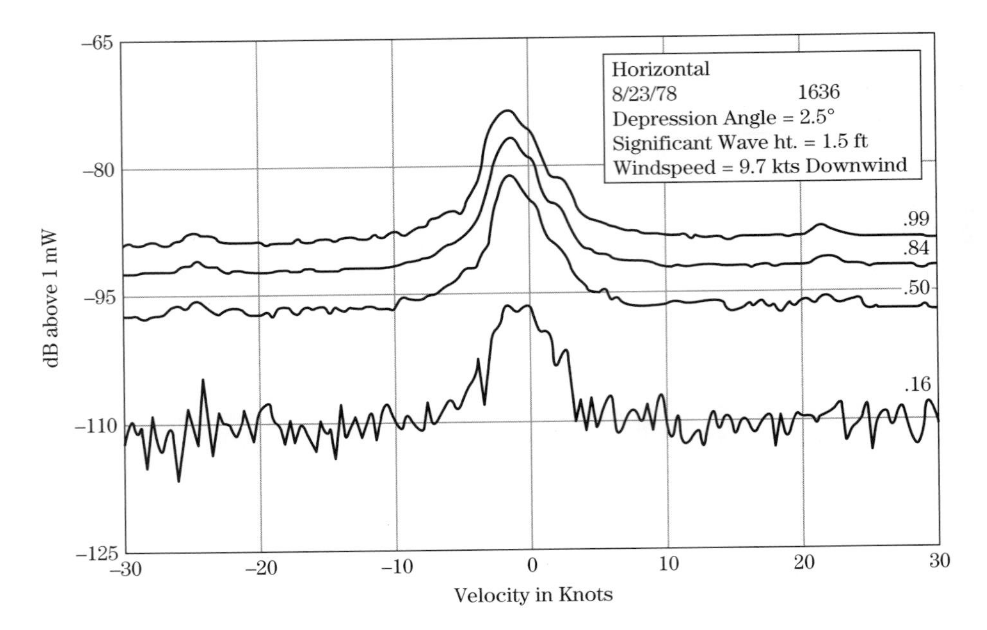

FIGURE 5-32 ■ Percentile of Doppler spectra at L-band. (From Plummer et al. [29]. With permission.)

- Spikes at both HH and VV occurred much less often than for HH alone.
- Rarely was a VV spike seen without a concurrent HH spike.

Walker [28] sates that the HH/VV intensity difference holds for the specular and Bragg scattering components but that the white cap component has roughly equal HH and VV backscatter. Given these properties, use of a dual-polarized radar and analysis of measured HH/VV polarization ratios may be one technique for recognizing and removing spikes from clutter data.

Spatial Variations Sea clutter returns are due to moving scatterers: the wind-driven spray and the gross and fine shape of the moving waves. Consequently the temporal and spatial variations of sea clutter are correlated. The only major difference between the two is due to wind-wave look direction. Generally, upwind/up-wave provides the highest radar return while downwind/down-wave provides the lowest return. This occurs because waves lean away from the wind so that the more vertical wavefronts are exposed in the upwind/up-wave direction, while the sloping backs are exposed in the downwind/down-wave direction.

5.2.3 Atmospheric Clutter

Atmospheric clutter primarily consists of hydrometeors, of which rain is the primary case of interest. Other atmospheric phenomena such as *angels* (clear air returns of unknown origin) can be of concern but will not be covered here. Frozen precipitation is also of interest, but primarily at millimeter wave frequencies because of the lower reflectivity of ice compared with water in the microwave bands. Experiments to determine average reflectivity as well as the spectral properties of rain are discussed in the following sections.

5.2.3.1 Average Value Data

Rain Reflectivity Average Values Raindrops can be modeled as dielectric spheres of differing sizes that are contained within the radar resolution cell. The radar return from the rain is, thus, the sum of the reflectivity of all the drops. As a result, the drop-size distribution is critical to the resultant reflectivity. The variability between most rain clutter

TABLE 5-8 ■ Average Rain Reflectivity versus Frequency Band

		η, dB m^{-1} Transmit frequency, GHz						
Z, dBz	Radar band: Type	S 3.0	C 5.6	X 9.3	K_u 15.0	K_a 35	W 95	mm 140
−12	Heavy stratus clouds				−100	−85	−69	−62
14	Drizzle, 0.25 mm/h	−102	−91	−81	−71	−58	−45*	−50*
23	Light rain, 1 mm/h	−92	−81.5	−72	−62	−49	−43*	−39*
32	Moderate rain, 4 mm/h	−83	−72	−62	−53	−41	−38*	−38*
41	Heavy rain, 16 mm/h	−73	−62	−53	−45	−33	−35*	−37*

* Approximate
Source: From Nathanson [15] (with permission).

models is due to differences in the drop-size distribution selected. Referring to Figure 5-5, recall that the reflectivity of a sphere is a strong function of the ratio of the circumference to the wavelength in the Rayleigh scattering region, so that rain reflectivity depends strongly on the percentage of large drops and will increase with increasing frequency until the resonance region is encountered.

Of course, every rainstorm has a different drop-size distribution, and distributions often change in differing parts of the same storm. Thus, a great deal of variation in the return with both time and space can be expected. Consequently, when discussing average values for reflectivity, it should be understood that considerable variation will be encountered in a realistic situation.

Table 5-8 gives average rain backscatter from several sources for S-band through W-band for several rain situations. As can be seen, the backscatter coefficient increases with increasing frequency (at least up to 35 GHz) and increasing rain rate. Observers of rainstorms have noted that rain drop size increases with increasing rain rate, which accounts for the direct dependence of reflectivity on rain rate. Attenuation has been eliminated from the rain data, so the apparent backscatter can be less for heavy rain rates.[3]

Table 5-8 includes a column listing an alternative scale for rain reflectivity denoted with the symbol Z. This scale, common in meteorological applications, is also called volume reflectivity. It relates radar reflectivity to the distribution of drop sizes, which is more useful for estimating rain rates. It is usually expressed in decibel units and denoted dBz. The relationship between η and Z is [31]

$$\eta = \frac{\pi^5 |K|^2}{\lambda^4} Z \tag{5.28}$$

where K is the complex index of refraction and Z is in units of m^6/m^3 = m^3. While K depends on temperature and wavelength, for most weather conditions $|K|^2 \approx 0.93$ for liquid scatterers (e.g., rain, fog) and 0.197 for frozen scatterers (e.g., snow, hail).

For use in meteorology, Z is converted to units of mm^6/m^3 (which requires multiplication of Z in m^3 by 10^{18}) and then to a decibel scale denoted dBz. Thus,

$$Z\ (\text{dBz}) = 10 \log_{10}(10^{18} Z) = 10 \log_{10}(Z) + 180 \tag{5.29}$$

The calculations of equations (5.28) and (5.29) were used to obtain the dBz values in Table 5-8.

[3] Of course, the target echo signal strength is reduced by rain attenuation as well.

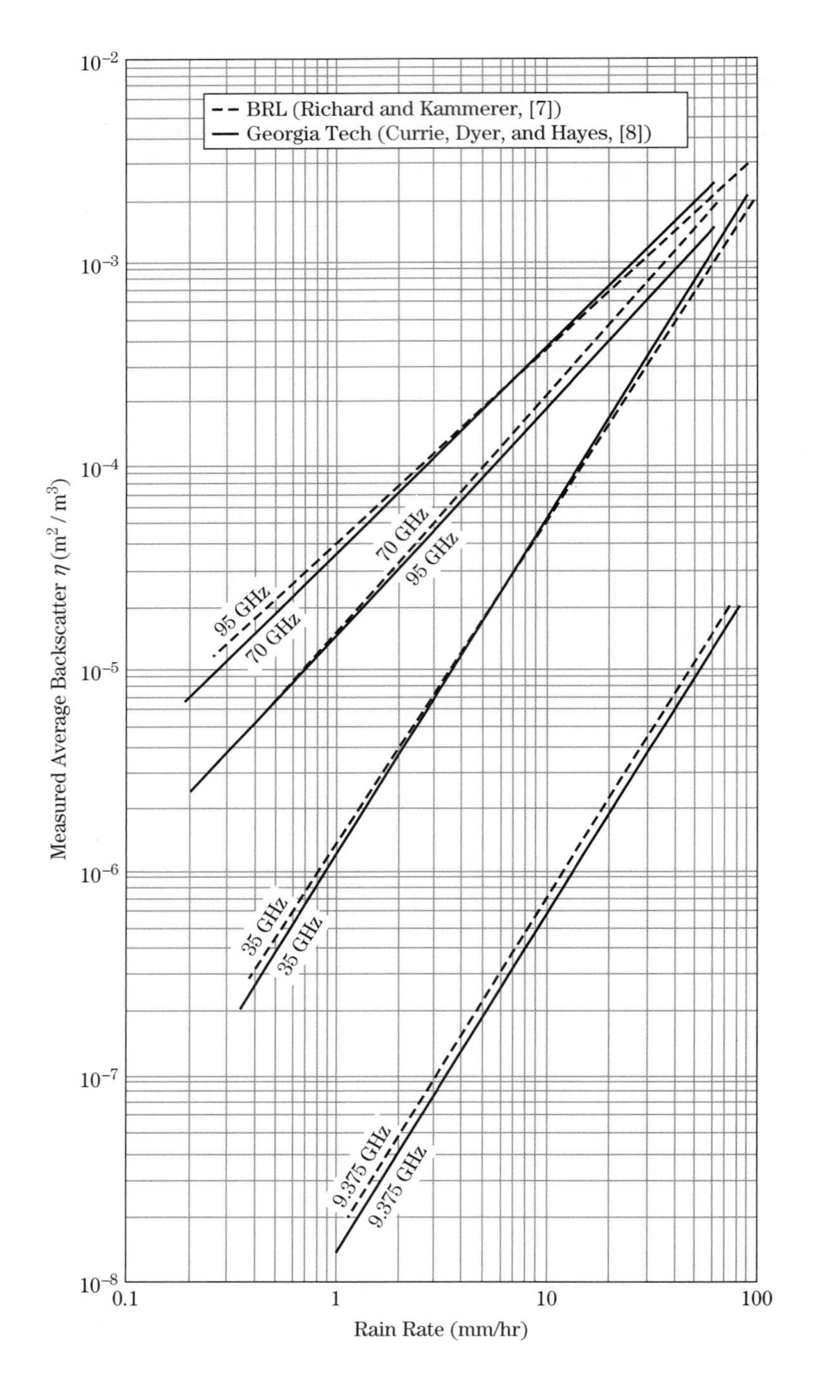

FIGURE 5-33 ■ Least squares fit to rain data at four frequency bands. (From Currie [32]. With permission.)

Figure 5-33 gives least square fit curves to rain data obtained from thunderstorms in a joint experiment between the U.S. Army BRL and GTRI in Orlando, Florida, in the 1970s [32]. Thunderstorms are known for producing large drops, so the wide variation in reflectivity with rain rate is not surprising. One interesting factor is the flattening of the curves at 70 and 95 GHz. This may indicate that the ratio of drop circumference to wavelength is approaching the resonance or optical regions, resulting in a lessened dependence on drop size. BRL and GTRI each separately analyzed the data, getting slightly different results as shown. However, the overall spread in the data dwarfs these small differences.

Frozen Precipitation As indicated earlier, the reflectivity of frozen precipitation such as snow is generally ignored at microwave frequencies. However, at higher frequencies snow reflectivity can be significant. Figure 5-34 presents data on snow reflectivity at 95,

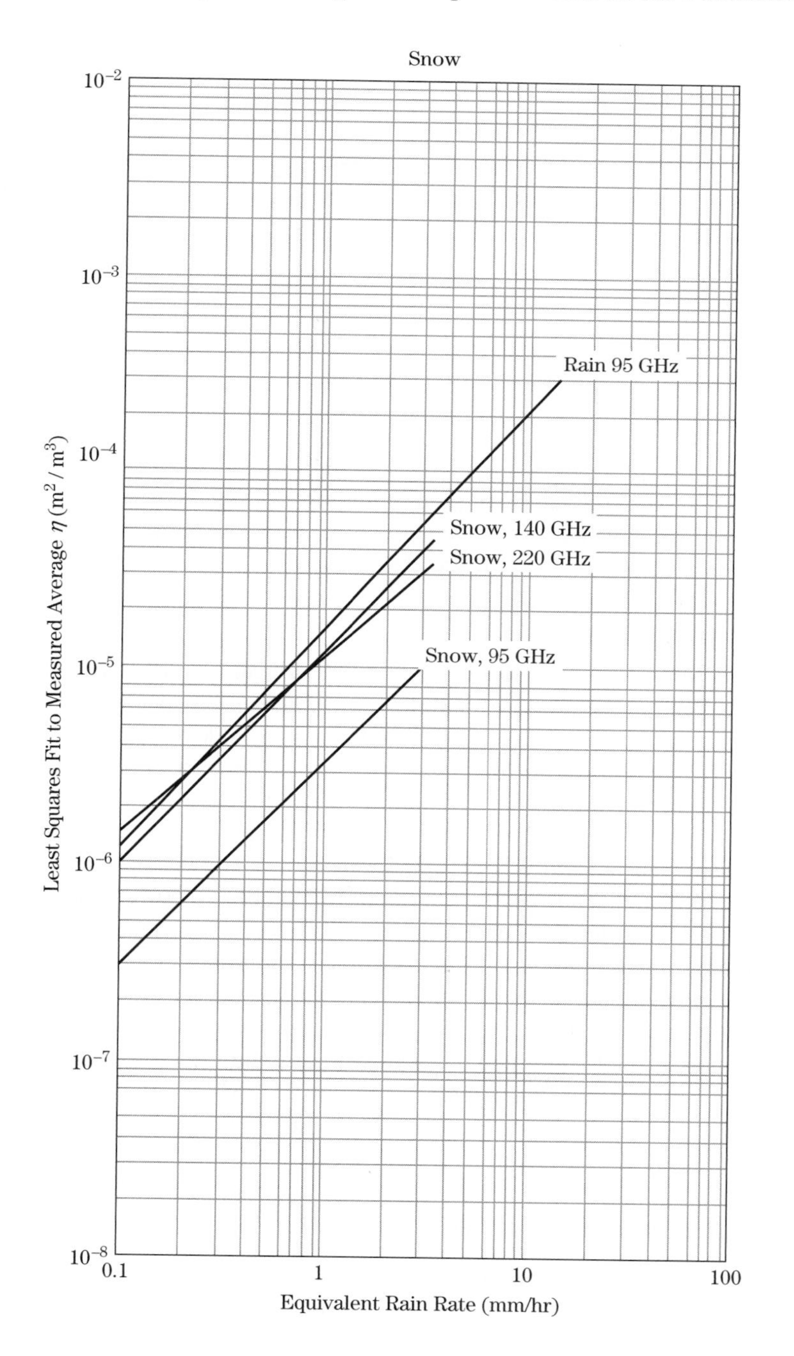

FIGURE 5-34 ■ Least squares fit to snow data at two frequency bands compared with rain data. (From Currie et al. [34]. With permission.)

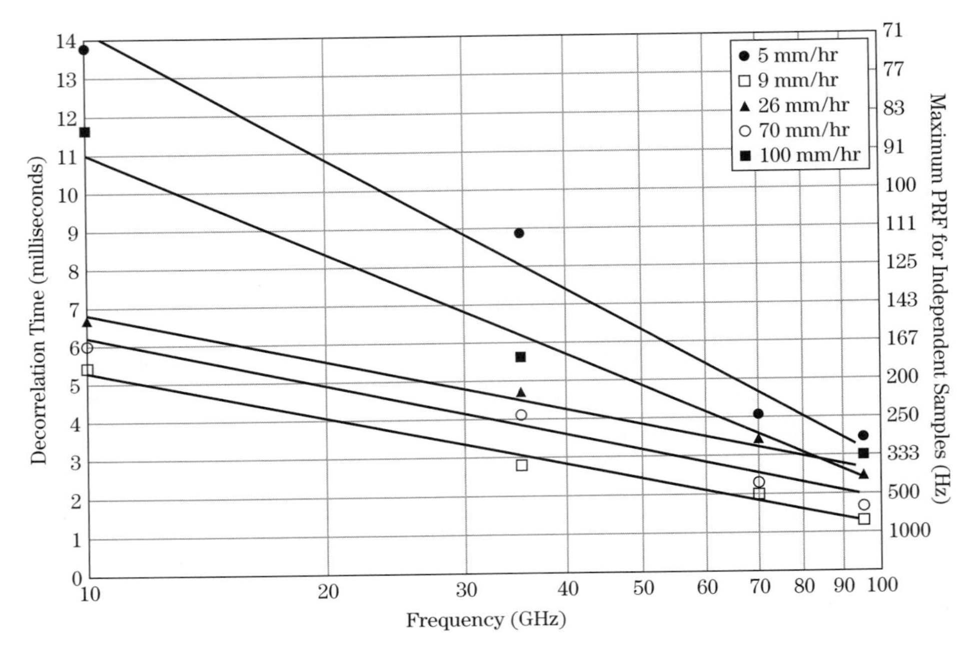

FIGURE 5-35 ■ Decorrelation time for rain backscatter as a function of frequency. (From Currie et al. [8]. With permission.)

140, and 220 GHz measured by Nemarich et al. [33,34] at the U.S. Army Harry Diamond Laboratory compared with the Georgia Tech 95 GHz rain data from Figure 5-33 The snow data are plotted versus equivalent rain rate based on snow water content. While the snow return is much lower at 95 GHz than the rain return, the snow returns at higher frequencies are comparable to the rain return for 95 GHz.

5.2.3.2 Temporal Spectra

Figure 5-35 presents data on the decorrelation time for rain return at 10 through 95 GHz. The maximum PRF for independent samples, calculated as the inverse of the decorrelation time as discussed in Section 5.2.2.4, is given on the right side. Note that for 5 mm/hr rain rate and 10 GHz, the maximum PRF for independent samples is only 71 Hz. At higher rain rates and higher frequencies, the maximum PRF increases significantly.

Figure 5-36 shows the spectral response at 10 through 95 GHz measured during the experiment displayed in Figure 5-33 [35]. Also shown is the Doppler shift for a slow-moving target at 3 mph. The spectral curves were matched to power law functions of the form of equation (5.24) in a manner similar to the data for trees discussed earlier. The figure shows that the rain return frequency spectra would be approximately 15 dB down from the peak (constant power level) at the Doppler frequency for a 3 mph target. This implies that the SCR could be improved by up to 15 dB through the use of careful Doppler processing.

Other important factors related to rain clutter are the spatial extent of a storm (both horizontal and vertical) and its Doppler characteristics, which are affected by prevailing winds, rain rates, and atmospheric turbulence. Attenuation due to rain is extremely important and was discussed in Chapter 4.

5.2.4 Millimeter Wave Clutter

The only difference between millimeter waves and microwaves is the wavelength. However, because of the change in the reflectivity properties of spheres and cylinders as the

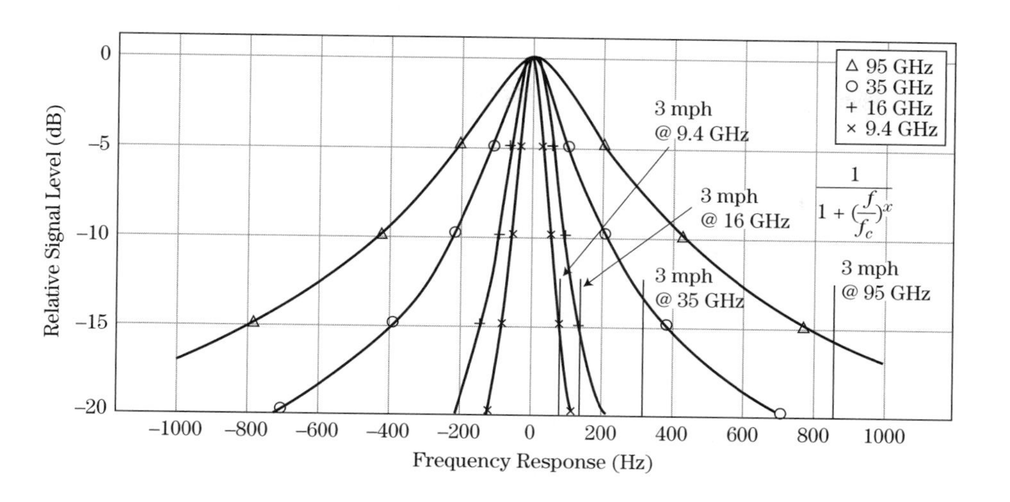

FIGURE 5-36 ■ Rain spectral response for four frequencies compared with a 3 mph moving target. (From Hayes [35]. With permission.)

circumference-to-wavelength ratio approaches unity, significant differences in reflectivity can be observed at higher frequencies. The dependence of rain reflectivity on rain rate at 70 and 95 GHz in Figure 5-33 is one example.

Table 5-9 compares the diameter-to-wavelength ratios of different natural and man-made objects at 35 through 300 GHz. As the ratio approaches 1 or larger unusual reflectivity effects will be observed. From the table it can be seen that this can occur for many types of common scatterers.

As an example of unusual effects at millimeter waves, Figure 5-37 shows the radar reflectivity of snow-covered ground at 35 GHz as a function of time of day. The air temperature was below freezing at the start of the measurement but rose above freezing between 8 a.m. and 8:15 a.m. As can be seen, the reflectivity of the snow dropped about 10 dB in 45 minutes. The temperature continued to hover around freezing, and the reflectivity varied depending on whether the temperature fell below or rose above freezing. This phenomenon has been observed at lower frequencies, but the effect was much smaller. The current theory is that when snow melts and refreezes, large resonant crystals form that reflect RF energy back to the radar, whereas, when melting occurs, surface water prevents

TABLE 5-9 ■ Comparison of the Circumference in Wavelengths of Different Natural and Man-made Items

Scatterer	Diameter (mm)	Ratio of Diameter to Wavelength (a/λ)			
		35 GHz	95 GHz	140 GHz	300 GHz
Raindrops	0.2–6	0.02–0.7	0.6–2	0.09–2.8	0.2–6
Sea Spray	0.2–10	0.02–1.15	0.6–3.3	0.09–4.7	0.2–6
Pine Needles	0.5–1.5	0.057–0.171	0.167–0.5	0.23–0.7	0.5–1.5
Screw Heads	1.5-25	0.17–2.9	0.5–8.3	0.7–11.7	1.5–25
Rivets	10	1.15	3.3	4.7	10
Grass Blades	2–8	0.23–0.92	0.7–2.7	0.93–3.7	2–8
Deciduous Leaves	6–20	0.7–2.31	2.0–6.7	2.8–9.4	6–20
Branches	5–76	0.7–8.78	2.0–25	2.8–35.5	6–76
Snow Crystals	5–50	0.58–5.77	1.7–16	2.3–23.3	5–50
Hail	1–10	0.12–1.15	0.3–3.3	0.47–4.7	1–10

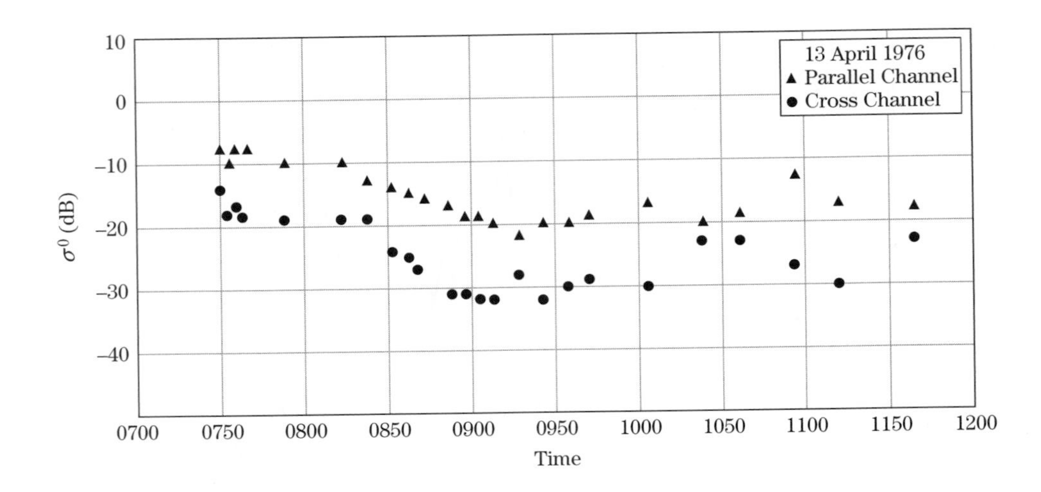

FIGURE 5-37 ■ Snow-covered ground reflectivity at 35 GHz as a function of time of day. (From Currie et al. [36]. With permission.)

penetration into the snow layers and forward scatters more of the energy. The key point of this discussion is that the radar designer must be aware of possible unusual scattering effects as the frequency increases into the millimeter wave region.

5.3 CLUTTER MODELING

5.3.1 General Approaches for Estimating Detection Performance in Clutter

Although a great deal of effort has been expended in developing theoretical models for clutter, most of these models have severe limitations because of the complexity of the real world. As a result, most radar designers either develop their own empirical models from clutter data or use models from the literature to estimate radar performance in clutter. For readers interested in pursuing theoretical modeling, several references are presented at the end of this chapter. The remainder of this section presents some empirical models that have been proven to be of use over the years.

5.3.2 Clutter Models

5.3.2.1 Surface Clutter

In the late 1970's GTRI developed an open literature empirical model for the reflectivity σ^0 of varying types of land clutter for grazing angles in the low angle and plateau regions [36]. In the 1980s this model was extended to higher frequencies, and additional data were used to refine the model [37]. The model takes into account wavelength, rms surface roughness, and grazing angle and has the form

$$\sigma^0 = A(\delta + C)^B \exp\left[\frac{-D}{1 + \dfrac{0.1\sigma_h}{\lambda}}\right] \tag{5.30}$$

TABLE 5-10 ■ Coefficients for GTRI Empirical Model

Constant	Frequency	Soil/ Sand	Grass	Tall Grass Crops	Trees	Urban	Wet Snow	Dry Snow
	3	0.0045	0.0071	0.0071	0.00054	0.362	—	—
	5	0.0096	0.015	0.015	0.0012	0.779	—	—
A	10	0.25	0.023	0.006	0.002	2.0	0.0246	0.195
	15	0.05	0.079	0.079	0.019	2.0	—	—
	35	—	0.125	0.301	0.036	—	0.195	2.45
	95	—	—	—	3.6	—	1.138	3.6
	3	0.83	1.5	1.5	0.64	1.8	—	—
	5	0.83	1.5	1.5	0.64	1.8	—	—
B	10	0.83	1.5	1.5	0.64	1.8	1.7	1.7
	15	0.83	1.5	1.5	0.64	1.8	—	—
	35	—	1.5	1.5	0.64	—	1.7	1.7
	95	—	1.5	1.5	0.64	—	0.83	0.83
	3	0.0013	0.012	0.012	0.002	0.015	—	—
	5	0.0013	0.012	0.012	0.002	0.015	—	—
C	10	0.0013	0.012	0.012	0.002	0.015	0.0016	0.0016
	15	0.0013	0.012	0.012	0.002	0.015	—	—
	35	—	0.012	0.012	0.012	—	0.008	0.0016
	95	—	0.012	0.012	0.012	—	0.008	0.0016
	3	2.3	0.0	0.0	0.0	0.0	—	—
	5	2.3	0.0	0.0	0.0	0.0	—	—
D	10	2.3	0.0	0.0	0.0	0.0	0.0	0.0
	15	2.3	0.0	0.0	0.0	0.0	—	—
	35	—	0.0	0.0	0.0	—	0.0	0.0
	95	—	0.0	0.0	0.0	—	0.0	0.0

Source: From Currie [32] (with permission).

where δ is the grazing angle in radians, σ_h is the rms surface roughness, and A, B, C, and D are empirically derived constants.

Table 5-10 gives the values for A, B, C, and D in equation (5.30) for frequencies of 3 through 95 GHz for several types of clutter. Figure 5-38 shows the model output for trees at X-band plotted against some GTRI data on tree reflectivity. Note that the model prediction appears a little high when compared to the data since the model attempts to predict an average of both temporal and spatial variations. This model has not been updated to reflect the low angle data reported by Billingsley [11].

Although several empirical models have been developed for sea return, the GTRI model has been one of the more popular ones used, particularly for low angles. The model is based on many years of radar data collection at a test site near Boca Raton, Florida. The model predicts average sea clutter values as a function of polarization, wavelength, grazing angle, radar boresight-wind direction, average wave height, and wind speed. Table 5-11 summarizes the equations for the model.

The definitions of variables are as follows:

- λ is the radar wavelength.
- δ is the grazing angle.

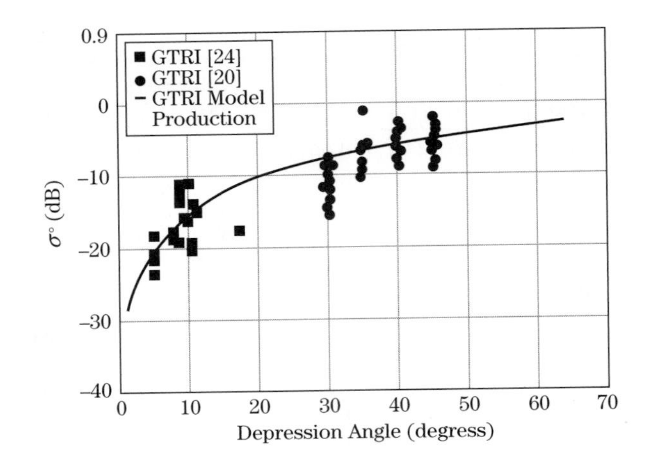

FIGURE 5-38 ■ Comparison of GTRI model output with data for deciduous trees at X-band. (From Currie [32]. With permission.)

- ϕ is the angle between radar boresight and wind direction.
- h_{av} is the average wave height.
- A_i is the "interference factor," which takes into account rms surface roughness and multipath.
- A_u is the upwind/downwind factor.
- A_w is the wind speed factor.

Figure 5-39 gives a sample output of the model. Shown is the value of σ^0 as a function of bore sight-upwind direction for three frequencies. (The model assumes that the wind and waves are proceeding in the same direction.) Note that after the original 1 to 10 GHz sea clutter model was developed, a second model was developed for 10 to 100 GHz [38].

TABLE 5-11 ■ GTRI Sea Clutter Model Equations

$\sigma^0_{HH} = 10\log[3.9\times10^{-6}\lambda\delta^{0.4}A_iA_uA_w]$
For 1 to 3 GHz $\sigma^0_{VV} = \sigma^0_{HH} - 1.73\ln(h_{av}+0.015) + 3.76\ln(\lambda) + 2.46\ln(\delta+0.0001) + 222$
For 3 to 10 GHz $\sigma^0_{VV} = \sigma^0_{HH} - 1.05\ln(h_{av}+0.015) + 1.09\ln(\lambda) + 1.27\ln(\delta+0.0001) + 9.70$
$\sigma_\phi = (14.4\lambda + 5.5)\delta h_{av}/\lambda$
$A_i = \sigma_\phi^4\left(1+\sigma_\phi^4\right)$
$A_u = \exp\left[0.2\cos\phi(1-2.8\delta)(\lambda+0.015)^{-0.4}\right]$
$qw = 1.1/(\lambda+0.015)^{-0.4}$
$V_w = 8.67h_{av}^{0.4}$
$A_w = [1.94V_w/(1+V_w/15.4)]^{qw}$

Note: Values for h_{av} and λ are given in meters, δ and ϕ are in radians.
Source: From Horst et al. [39] (with permission).

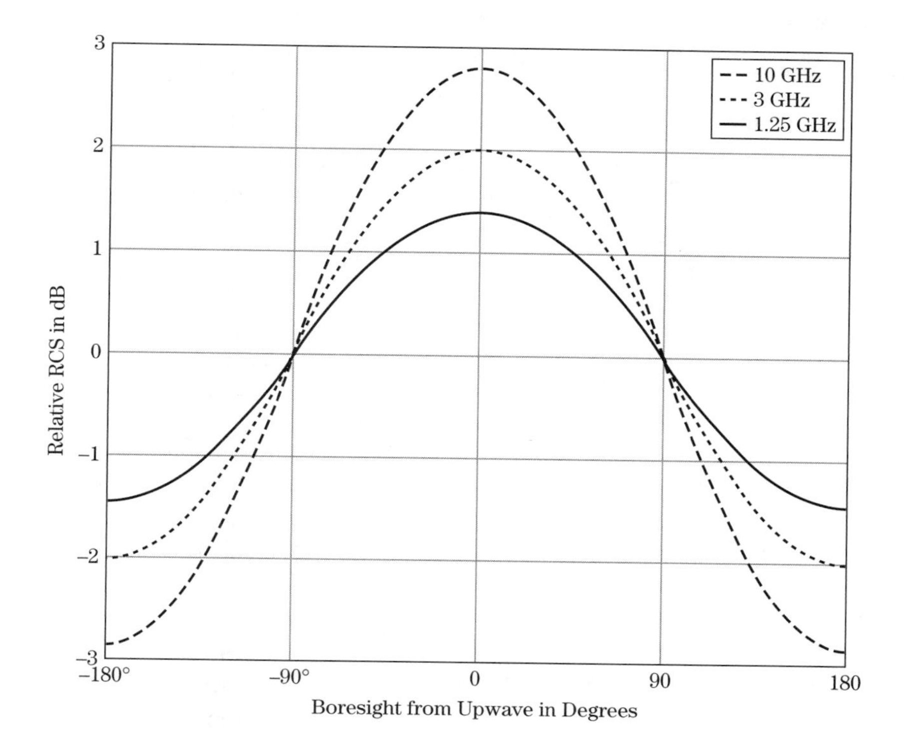

FIGURE 5-39 ■ GTRI sea clutter model output for three frequencies at 1° grazing angle. (From Long [6]. With permission.)

Unfortunately, the experimenters used a different X-band data set from that used for the original model development so that the results of the two models do not agree at 10 GHz. Some users that need to model sea return over the entire frequency range average the results from the two models at 10 GHz.

5.3.2.2 Atmospheric Clutter Models

The equations for the least squares fit to rain data presented in Figure 5-33 are presented in Table 5-12. The equations have the form

$$\eta = AR^B \text{ m}^{-1} \tag{5.31}$$

where η is the radar cross section per unit volume, and A and B are constants given in Table 5-12. Note that significant variability in the data occurred around the calculated least squares fits.

TABLE 5-12 ■ Model Coefficients for Rain

Frequency (GHz)	A	B
9.4	1.3×10^{-8}	1.6
35	1.2×10^{-6}	1.6
70	4.2×10^{-5}	1.1
95	1.5×10^{-5}	1.0

Source: From Currie et al. [34] (with permission).

5.4 CONCLUDING REMARKS

This chapter defined the common terminology for dealing with radar clutter; discussed the average, temporal, and spatial properties of clutter; and introduced some empirical models for clutter. In no way should this discussion be considered as comprehensive in nature. Dozens of books have been written on this topic from both theoretical and empirical points of view. Some of these books are listed in Section 5.5. The next section provides a summary of average clutter data taken from the sources discussed previously.

5.4.1 Reflectivity Summary

Table 5-13 gives some average values for land clutter as a function of aspect angle that are plotted in Figures 5-11 through 5-14, which were adapted from Nathanson [15]. Table 5-14 summarizes average values for sea clutter as a function of polarization and depression angle which were plotted in Figures 5-18 through 5-20, again adapted from Nathanson. These data reflect the general trends discussed previously but omit any dependence on wind direction, a weakness corrected by the GTRI model of Table 5-11.

Table 5-8 given previously summarizes average values for rain return as a function of frequency and rain rate, again, taken from Nathanson [15].

The temporal and spatial variations of land clutter were best represented by Weibull distributions. The width parameters (standard deviation or Weibull width parameter) for land clutter were given in Tables 5-4 and 5-7. Spectral bandwidth and roll-off characteristics for wind-blown trees were presented in Table 5-6. Rain spectral constants were presented in Figure 5-35.

TABLE 5-13 ■ Summary of Averaged Land Reflectivity (σ^0 in db)

		Grazing Angle (Deg.)			
Clutter Type	Frequency Band	1.5	10	30	60
Desert	L	−45	−38	−28	−21
	S	−46	−36	−25	−17
	C	−40	−33	−23	−16
	X	−40	−30	−21	−14
	K_u		−28	−19	−13
Farmland	L	−36	−30	−20	−15
	S	−34	−28	−18	−16
	C	−33	−26	−16	−15
	X	−33	−26	−16	−14
	K_u	−23	−22	−16	−13
Woods	L	−28	−26	−18	−19
	S	−28	−24	−16	−15
	C	−27	−23	−16	−15
	X	−26	−23	−14	−14
	K_u	−13	−20	−14	−12
Urban	L	−25	−18	−15	−12
	S	−23	−18	−13	−11
	C	−21	−18	−11	−10
	X	−20	−16	−10	−10
	K_u				

TABLE 5-14 ■ Summary of Averaged Sea Reflectivity (σ^0 in db)

Sea State	Frequency Band	Polarization	Grazing Angle (Deg.)			
			0.1	10	30	60
1	L	VV		−39	−38	−22
	L	HH		−56	−46	−24
	S	VV	−80	−40	−40	−24
	S	HH	−80			−25
	C	VV	−72	−41	−42	−24
	C	HH	−75	−53	−48	−26
	X	VV	−65	−42	−36	−24
	X	HH	−71	−51	−44	−24
	K_u	VV		−40	−31	−20
	K_u	HH			−38	−20
3	L	VV	−82	−34	−30	−18
	L	HH	−82	−48	−39	−20
	S	VV	−75	−34	−29	−19
	S	HH	−68	−46	−38	−20
	C	VV	−60	−34	−28	−18
	C	HH	−69	−40	−37	−20
	X	VV	−51	−32	−26	−16
	X	HH	−53	−37	−34	−21
	K_u	VV		−31	−23	−14
	K_u	HH		−32	−28	−16

5.4.2 Clutter Effect on Detection

This chapter has summarized the clutter characteristics and provided clutter data and modeling for simulating clutter. Clutter limits detection of targets due to its competing signal, which interferes with the target signal. Methods for dealing with these problems using Doppler processing are discussed in Chapter 17.

5.5 FURTHER READING

Perhaps the two best comprehensive references for understanding radar clutter are the texts by Long [6] and Ulaby and Dobson [10]. Both combine in-depth discussion of clutter phenomenology and modeling with extensive summaries of real-world clutter measurements. Ward et al. concentrate on sea clutter measurements and modeling with the K-distribution in [25].

More extensive theoretical analyses are given for sea clutter by Ward et al. in [25] and for various clutter sources in [40–43]. An in-depth analysis of land clutter data is given by Billingsley [11]. Other good reviews of land and sea clutter data can be found in [15,40–43].

Most of the preceding sources are concerned with radar clutter at microwave frequencies. Additional discussion of millimeter-wave clutter data is available in the publications by Currie and various colleagues [34,44,45].

5.6 REFERENCES

[1] Novak, L.M., and Owirka, G.J., "Radar Target Identification Using an Eigen-Image approach," *1994 IEEE National Radar Conference*, Atlanta, GA, p. 130, March 29–31, 1994.

[2] Long, M.W., "Radar Clutter," Tutorial presented at the *2006 IEEE Radar Conference,* Verona, NY, April 2006.

[3] Mott, H., *Polarization in Antennas and Radar*, Wiley, New York, 1986.

[4] Holm, W.A., "MMW Radar Signal Processing Techniques," Chapter 6 (pp. 279–310) in *Principles and Applications of Millimeter-Wave Radar*, Ed. N.C. Currie and C.E. Brown, Artech House, Norwood, MA, 1987.

[5] Echard, J.D., et al., "Discrimination between Targets and Clutter by Radar," Final Technical Report on Contract DAAAG-29-780-C_0044, Georgia Tech Research Institute, Atlanta, December 1981.

[6] Long, M.W., *Radar Reflectivity of Land and Sea*, 3d ed., Artech House, Norwood, MA, 2001.

[7] Richard, V.W., and Kammerer, J.E., "Rain Backscatter Measurements and Theory at Millimeter Wave Measurements," Report No. 1838, U.S. Army Ballistic Research Laboratory, Aberdeen Proving Ground, MD, October 1975.

[8] Currie, N.C., Dyer, F.B., and Hayes, R.D., "Analysis of Radar Rain Return at Frequencies of 9.375, 35, 70, and 95 GHz," Technical Report No. 2 on Contract DAAA 25-76-C-0256, Georgia Tech Research Institute, Atlanta, February 1975.

[9] Nemarich, J., Wellman, R.J., and Lacombe, J., "Backscatter and attenuation of Falling Snow and Rain at 96, 140, and 220 GHz," *IEEE Trans. Geoscience and Remote Sensing*, vol. 26, no. 3, pp. 330–342, May 1988.

[10] Ulaby, F.T., and Dobson, M.C., *Handbook of Radar Scattering Statistics for Terrain,* Artech House, Norwood, MA, 1989.

[11] Billingsley, J.B., *Low-Angle Radar Land Clutter*, William Andrew Publishing, Norwich, NY, 2002.

[12] Aley, J.C., Davis, W.T., and Mills, N.B., "Radar Sea Return in High Sea States," Naval Research Laboratory Report No. 7142, September 1970.

[13] Long, M.W., et al., "Wavelength Dependence of Sea Echo," Final Report on Contract N62269-3019, Georgia Tech Research Institute, Atlanta, 1965.

[14] Trebits, R.N., et al., "Millimeter Wave Radar Sea Return Study," Interim Technical Report on Contract N60921-77-C-A168, Georgia Tech Research institute, Atlanta, July 1978.

[15] Nathanson, F.E., "Sea and Land Backscatter," Chapter 7 in *Radar Design Principles,* 2d ed., McGraw-Hill, Inc New York, 1991.

[16] Currie, N.C., "Performance Tests on the AN/TPQ-31a Radar," Technical Report on Contract N00014-75-C-0228, Mod P00001, Georgia Tech Research Institute, Atlanta, 1975.

[17] "Radar Return Study," Goodyear Aircraft Corp., Final Report on Contract NOAS-59-6186-CGERA 463, Phoenix, AZ, September 1959.

[18] Stiles, W.H., Ulaby, F.T., and Wilson, E., "Backscatter Response of Roads and Roadside Surfaces," Sandia Report No. SAND78-7069, University of Kansas center for Research, Lawrence, March 1979.

[19] Dyer, F.B., and Currie, N.C., "Some Comments on the Characteristics of Radar Sea Echo," 1974 IEEE APS International Symposium, Atlanta, GA, June 1974, pp. 323–326.

[20] Rivers, W.K., "Low Angle Sea Return at 3mm Wavelength," Final Technical Report on Contract N62269-70-C-0489, Georgia Tech Research Institute, Atlanta, November 1970.

[21] Currie, N.C., Dyer, F.B., and Hayes, R.D., "Radar Land Clutter Measurements at 9.375, 16, 35, and 95 GHz," Technical Report No. 3 on Contract DAA25-73-0256, Georgia Tech Research Institute, Atlanta, February 1975.

[22] Fishbein, W., et al., "Clutter Attenuation Analysis," Technical Report No. ECOM-2808, US Army ECOM, Ft, Monmouth, NJ, March 1967.

[23] Simpkins, W.L., Vannicola, V.C., and Ryan, J.P., "Seek Igloo Radar Clutter Study," Technical Report No. RADC-TR-77-338, RADC, Rome, NY, October 1977.

[24] Booth, R.R., "The Weibull Distribution Applied to Ground Clutter Backscatter Coefficient," Report No. RE-TR-69-15, US Army Missile command, Huntsville, AL, June 1969.

[25] Ward, K.D., Tough, R.J.A., and Watts, S., *Sea Clutter: Scattering, the K Distribution and Radar Performance*, Institution of Engineering and Technology, London, 2006.

[26] Ward, K.D., "Compound Representation of High Resolution Sea Clutter," *Electron Lett.*, vol. 17, pp. 561–565, 1981.

[27] Watts, S., "Radar Detection Prediction in K-Distributed Sea Clutter and Thermal Noise," *IEEE Trans. AES*, vol. 23, pp. 40–45, 1987.

[28] Walker, D., "Doppler Modelling of Radar Sea Clutter," *IEE Proceedings—Radar, Sonar and Navigation,* vol. 148, no. 2, pp. 73–80, April 2001.

[29] Plummer, D.K., et al., "Some Measured Statistics of Coherent Radar Sea Echo and Doppler at L-Band," Final Technical Report on Contract NADc-78254-30, Georgia Tech Research Institute, Atlanta, December 1969.

[30] Werle, B.O., "Sea Backscatter Spikes, and Wave Group Observations at Low Grazing Angles," *Proceedings of the 1995 IEEE International Radar Conference*, Washington, DC, pp. 187–195, May 1995.

[31] Richards, M.A., *Fundamentals of Radar Signal Processing*, McGraw-Hill, New York, 2005.

[32] Currie, N.C., "MMW Clutter Characteristics," Chapter 5 in *Principles and Applications of Millimeter-Wave Radar*, Ed. N.C. Currie and C.E. Brown, Artech House, Norwood, MA, 1987.

[33] Nemarich, J., et al., "Comparative Near Millimeter Wave Propagation Properties of Snow and Rain," *Proceedings of Snow Symposium III*, U.S. Army CRREL, Hanover, NH, August 1983.

[34] Currie, N.C., Hayes, R.D., and Trebits, R.N., *Millimeter-Wave Radar Clutter*, Artech House, Norwood, MA, 1992.

[35] Hayes, R.D., private communication, Atlanta, GA, 1980.

[36] Currie, N.C., et al., "Radar Millimeter Wave Measurements: Part 1, Snow and Vegetation," Report No. AFATL-TR-77-92, July 1977.

[37] Currie, N.C., and Zehner, S.P., "MMW Land clutter Model," *IEE Radar 82 International Symposium*, London, September 1982.

[38] Horst, M.M., and Perry, B., "MMW Modeling Techniques," Chapter 8 in *Principles and Application of Millimeter-Wave Radar*, Ed. N.C. Currie and C.E. Brown, Artech House, Norwood, MA, 1987.

[39] Horst, M.M., Dyer, F.B., and Tuley, M.T., "Radar Sea Clutter Model," Proceedings of the IEEE Conference on Antennas and Propagation, November 1978.

[40] Skolnik, M.I., *Radar Handbook,* 2d ed., Chapters 12, "Ground Echo" and 13, "Sea Echo," McGraw-Hill Publishing Company, New York, 1990.

[41] Ulaby, F.T., Moore, R.K., and Fung, A.K., *Microwave Remote Sensing: Vol. II, Radar Remote Sensing and Surface Scattering and Emission Theory*, Addison-Wesley Publishing Company, Reading, MA, 1982.

[42] Ulaby, F.T., and Elachi, C. (Eds.), *Radar Polarimetry for Geoscience Applications*, Artech House, Norwood, MA, 1990.

[43] Bogush Jr., A.G., *Radar and the Atmosphere*, Artech House, Norwood, MA, 1989.

[44] Currie, N.C., and Brown, C.E. (Eds.), "MMW Clutter Characteristics," Chapter 5 in *Principles and Applications of Millimeter-Wave Radar*, Artech House, Norwood, MA, 1987.

[45] Button, K.J., and Wiltse, J.C. (Eds.), "Millimeter Radar," Chapter 2 in *Infrared and Millimeter Waves: vol. 4, Millimeter Systems*, Academic Press, New York, 1981.

5.7 PROBLEMS

1. A radar has a pulse length of $\tau = 10\ \mu$s, an azimuth beamwidth $\theta_3 = 3^\circ$, and an elevation beamwidth $\phi_3 = 3^\circ$. At what grazing angle δ does the transition occur between the pulse-limited and beam-limited ground clutter cases when the nominal range to the ground is $R = 10$ km? Repeat for $R = 50$ km.

2. For the same radar used in problem 1, what is the volume V in cubic meters of a volume clutter resolution cell at $R = 10$ km? Repeat for $R = 50$ km.

3. A radar is attempting to detect a point target in the presence of ground clutter. The parameters of the radar and its environment are such that the SNR at a range of $R_0 = 10$ km is 30 dB, while the SCR at the same range is 20 dB. The detection performance at this range is "clutter limited" because the clutter is the dominant interference. Assume the clutter interference is pulse-limited and that σ^0 does not vary with range. At what range will the SNR and SCR be equal? (At ranges longer than this, the detection performance for this target will be "noise limited.")

4. Consider two radar targets with polarization scattering matrices $\mathbf{S}_1$ and $\mathbf{S}_2$ as follows:

$$\mathbf{S}_1 = \begin{bmatrix} 1 & 0 \\ 0 & 1 \end{bmatrix}, \quad \mathbf{S}_2 = \begin{bmatrix} 1 & j \\ -j & -1 \end{bmatrix}$$

where $j = \sqrt{-1}$. Compute the parallel/cross-polarization ratio and the vertical/horizontal polarization ratio for each target. Which ratio could be used to discriminate between the two targets?

5. Compute the critical grazing angle for an X-band (10 GHz) radar when the surface roughness σ_h is 1.0 cm, and again when $\sigma_h = 10.0$ cm. Repeat for an L-band (1 GHz) radar and a K-band (35 GHz) radar.

6. Compute the diameter of a conducting sphere at the boundary between the Rayleigh and resonance regions at a radar frequency of 5 GHz (C-band). Repeat for the boundary between the resonance and optics regions.

7. Show that the Weibull distribution reduces to the exponential distribution when $b = 1$, and to the Rayleigh distribution when $b = 2$.

8. A radar collects $N_t = 30$ samples of clutter data having a decorrelation time τ_0 of 200 μs. What is the number of uncorrelated samples N_i if the PRF is 1 kHz? Repeat for PRF = 5 kHz and 40 kHz.

9. Consider the GTRI model for land clutter reflectivity given by equation (5.30). Show that the model predicts that σ^0 becomes independent of surface roughness when $\sigma_h \ll 10\lambda$.

10. Use equations (5.28) and (5.29) to confirm that a reflectivity $\eta = -92$ dB corresponds to a meteorological reflectivity of 23 dBz at S-band (3 GHz) as shown in Table 5-8.

CHAPTER 6

Target Reflectivity

John F. Shaeffer

Chapter Outline

6.1 Introduction 211
6.2 Basic Reflection Physics 212
6.3 Radar Cross Section Definition 219
6.4 Three Scattering Regimes 224
6.5 High-Frequency Scattering 227
6.6 Examples 236
6.7 Further Reading 244
6.8 References 244
6.9 Problems 245

6.1 INTRODUCTION

The basic motivation for this chapter is to describe a key link in the understanding of radar: how does a radar wave, more properly known as an electromagnetic (EM) wave, transmitted from some transmitter source, interact with a target to produce reflected energy at some receiver position? The goals for this chapter are as follows:

1. To understand what an electromagnetic wave is.
2. To understand its properties.
3. To describe a measure of the amount of reflected energy, a quantity known as radar cross section (RCS).
4. To understand basic scattering or reflectivity physics.
5. To focus on the typical microwave scattering mechanisms.
6. To understand how two or more scattering centers add and subtract.
7. To illustrate examples of high-frequency scattering from targets along with scattering center images.

6.2 BASIC REFLECTION PHYSICS

Basic reflection physics must start with a description of the characteristics of an electromagnetic wave and then progress to how this EM wave interacts with a target object to cause reflection.

6.2.1 Electromagnetic Wave Fundamentals

An electromagnetic wave is the self-propagating transport of energy (voltage and current) through space, without this energy being attached or directed via some external structure such as a transmission line or waveguide. James Clerk Maxwell showed in the 1860s that a time-changing electric field, $\boldsymbol{E}$ (V/m), is the source for the magnetic field, $\boldsymbol{H}$ (A/m), and, in turn, a time-changing $\boldsymbol{H}$ is the source for $\boldsymbol{E}$. Thus, once an EM wave is launched, it becomes self-propagating. EM waves propagate in free space as well as inside material media.

The three most important characteristics of an EM wave (Figure 6-1) are (1) its frequency or rate of temporal variation, f, with units of hertz (Hz or cycles/s); (2) its wavelength or spatial variation, λ, with units of meters (m); and (3) its velocity of propagation, v, with units of meters per second (m/s). These three fundamental wave quantities are not independent; they are related as

$$\lambda f = v \tag{6.1}$$

When an EM wave propagates in free space, $v = c$, the velocity of light, which is approximately 3×10^8 m/s. Mathematically it is given by the free space values of electric

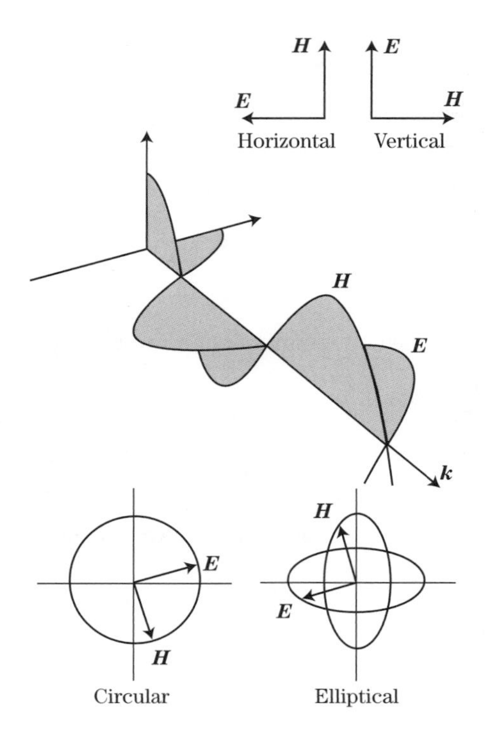

FIGURE 6-1 ■ Snapshot of an EM wave in time and allowed polarization directions (from Knott [1]).

permittivity, ε_0, and magnetic permeability, μ_0:

$$c = \frac{1}{\sqrt{\varepsilon_0 \mu_0}} \approx 3 \times 10^8 \text{ m/s} \tag{6.2}$$

which is about 1 foot per nanosecond, or about 1,000 feet per microsecond.

EM waves propagating inside a material media travel slower than in free space with a velocity

$$v = \frac{c}{n} = \frac{1}{\sqrt{\varepsilon_m \mu_m}} \tag{6.3}$$

where n is the material index of refraction, a composite measure of electric energy storage, and is a function of the material electric and magnetic energy storage properties characterized by its relative dielectric permittivity, ε_m (electric energy storage), and permeability, μ_m (magnetic energy storage). Thus, the more energy that can be stored inside a material, the slower an EM wave will propagate. Also, inside a material, the time variation of the wave is the same as in free space, but the wavelength becomes smaller to always satisfy the relation $\lambda f = v$. In free space a wave is typically characterized by either its frequency or wavelength, whereas inside a material the wave is typically characterized by its frequency.

Frequency and wavelength may also be characterized by radian frequency, $\omega = 2\pi f$, and by wave number, $k = 2\pi/\lambda$, and the relationship $\lambda f = v$ becomes $\omega/k = v$.

For the sake of analysis a fictional plane wave is often considered, given mathematically by

$$\begin{aligned} &\boldsymbol{E} = \boldsymbol{E}_0 e^{j(\omega t - \boldsymbol{k} \cdot \boldsymbol{R})} \\ &\text{or} \\ &\boldsymbol{H} = \boldsymbol{H}_0 e^{j(\omega t - \boldsymbol{k} \cdot \boldsymbol{R})} \end{aligned} \tag{6.4}$$

where the spatial phase $\boldsymbol{k} \cdot \boldsymbol{R}$ is the distance from some origin measured in units of wavelength λ in the direction of propagation $\boldsymbol{k}$. The nature of the sinusoidal temporal and spatial variation is shown within the complex phasor exponent $(\omega t - \boldsymbol{k} \cdot \boldsymbol{R})$. The constant phase fronts of this plane wave are perpendicular to the direction of propagation $\boldsymbol{k}$. Recall from geometry that a plane in space is defined by $\boldsymbol{k} \cdot \boldsymbol{R} = $ constant, where $\boldsymbol{k}$ is normal to the plane, and $\boldsymbol{R}$ is a vector to a point on the plane.

The electric and magnetic fields $\boldsymbol{E}$ and $\boldsymbol{H}$ are time-varying vector quantities. The quantities $\boldsymbol{E}_0$ and $\boldsymbol{H}_0$ are their vector "amplitudes." While these are in general three-dimensional vectors, for plane waves the component in the direction of propagation $\boldsymbol{k}$ is zero, so $\boldsymbol{E}_0$ and $\boldsymbol{H}_0$ will be considered to be two-dimensional vector quantities.

This plane wave is fictional because its intensity does not fall off as the wave propagates away from its source and because it has exactly planar wave fronts. A spherical wave at a sufficient distance from its source can be considered planar over the target dimensions. In addition, there is insignificant $1/R$ decrease in intensity over the relatively small target. Thus, the plane wave concept is a practical simplification for waves interacting with targets.

Three fundamental vectors characterize an EM wave:

- The electric field, $\boldsymbol{E}$, with units of V/m.
- The magnetic field, $\boldsymbol{H}$, with units of A/m.
- The direction of propagation, $\boldsymbol{k}$, the vector wavenumber, whose magnitude (also called the scalar wavenumber or just the wavenumber) is $2\pi/\lambda$, with units of rad/m.

Once an EM wave is launched so that $\boldsymbol{E}$ and $\boldsymbol{H}$ are no longer attached to some sort of conducting structure, the $\boldsymbol{E}$ and $\boldsymbol{H}$ fields circle and close back on themselves, and the fields are now solenoidal.

In free space, there is equal energy in the wave $\boldsymbol{E}$ and $\boldsymbol{H}$ field components so that time-changing $\boldsymbol{E}$ is the source for $\boldsymbol{H}$ and vice versa. While there is equal energy contained in each field component, the numerical values of $\boldsymbol{E}$ and $\boldsymbol{H}$ are not the same. The ratio of the norm of $\boldsymbol{E}$ to the norm of $\boldsymbol{H}$ is the wave impedance, Z, which in free space is the constant value of approximately 377 ohms. This numerical value results from using the SI system of units:

$$\frac{\|\boldsymbol{E}\|}{\|\boldsymbol{H}\|} = Z = \sqrt{\frac{\mu_0}{\varepsilon_0}} \approx 377 \text{ ohms}$$

Because this ratio is constant, this free space wave can be characterized by either its electric or magnetic field component.

Maxwell's equations for propagating waves require that the spatial directions of $\boldsymbol{E}$, $\boldsymbol{H}$, and $\boldsymbol{k}$ be perpendicular to each other (Figure 6-1) so that $\boldsymbol{E}$ and $\boldsymbol{H}$ must always be perpendicular to the direction of propagation and be perpendicular to each other. In free space $\boldsymbol{E}$ and $\boldsymbol{H}$ are in phase; that is, each peaks at the same time. However, inside a material media, where energy storage occurs, this is often not the case.

An EM wave transports electrical energy characterized as an energy flux (i.e., energy per unit cross section area, W/m^2). The magnitude and direction of the energy flux are given by the Poynting vector, $\boldsymbol{P}$, which is the vector cross product of $\boldsymbol{E}$ and $\boldsymbol{H}$. Averaged over time, $\boldsymbol{P}$ is

$$\boldsymbol{P} = \frac{1}{2}\text{Real}(\boldsymbol{E} \quad \boldsymbol{H}^*) \text{ W/m}^2 \tag{6.5}$$

where $\boldsymbol{H}^*$ is the complex conjugate of $\boldsymbol{H}$.

6.2.2 Electromagnetic Wave Polarization

Another characteristic of an EM wave is its polarization, which is the direction of its electric field vector $\boldsymbol{E}$. Maxwell's equations require only that $\boldsymbol{E}$, $\boldsymbol{H}$, and $\boldsymbol{k}$ be perpendicular. Thus, $\boldsymbol{E}$ may point in any direction in a plane perpendicular to the direction of propagation $\boldsymbol{k}$ (Figure 6-1). Polarization may be linear where the direction $\boldsymbol{E}$ is always in the same direction, or it may be circular (or more generally, elliptical) where $\boldsymbol{E}$ and $\boldsymbol{H}$ rotate as the wave propagates.

Linear polarization is specified typically as relative to one's surroundings (Figure 6-2). In the real world with the earth as a reference, horizontal or vertical are usually chosen as linear polarization directions. In the computational world with spherical coordinates, polarization is specified in the azimuth, ϕ, and polar, θ, directions.

Circular polarization is specified as left or right circular, counterclockwise (LC) or clockwise (RC) rotation of $\boldsymbol{E}$, respectively. The reference for LC or RC is determined by looking in the direction of propagation $\boldsymbol{k}$ and sensing which way $\boldsymbol{E}$ is rotating. For transmission, the sense is based on looking *away* from the antenna (outgoing wave), and for reception it is based on looking *toward* the antenna (incoming wave) [2].

Polarization information is required to describe how an EM wave is both transmitted and received. The physical characteristics of the transmitting antenna that launches an EM

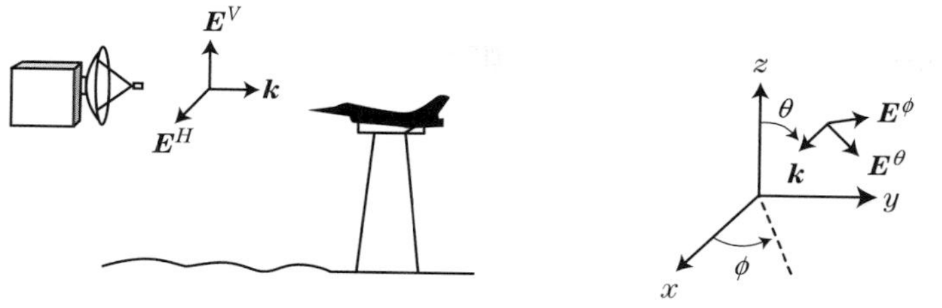

(a) Experimental

(b) Analytical Spherical Coordinate System

FIGURE 6-2 ■ Polarization convention is given by the direction of $\boldsymbol{E}$ relative to local surroundings [1].

wave determine the polarization of the outgoing wave. The receiving antenna polarization characteristic determines the amount of signal actually received from an incoming EM wave of a given polarization. Thus, RCS becomes a function of both transmitter polarization and receiver polarization. In addition, some target reflection properties such as edges and surface waves are a function of the polarization of the incident wave relative to target geometry.

For transmitting, a horizontal dipole radiates an EM wave with horizontal polarization, whereas a vertical dipole launches a wave with vertical polarization. For receiving, a horizontal dipole responds only to the horizontal component of the incoming EM wave polarization, whereas a vertical dipole is sensitive only to the wave's vertical component of polarization.

6.2.3 Electromagnetic Wave Reflection

A discussion of EM wave reflection, the familiar "echo" of radar signals, requires consideration of boundary conditions, induced currents, re-radiation, and Maxwell's equations. In optics, high-frequency scattering waves reflect according to Snell's law such that the angle of reflection is equal to the angle of incidence, much the same way a billiard ball bounces. This section discusses the physics of how this behavior arises.

For simplicity, consider a perfect electric conductor (PEC). A PEC has infinite conductivity. Its surface is always at the same electric potential and cannot support a tangential electric field; that is, the PEC surface is a short circuit to any applied electric field because charge is free to move under an applied electric field (force). When a PEC surface is completely closed, there can be no electric field inside the surface; that is, the closed surface becomes a perfect shield called a Faraday shield.

It is customary to split the total vector electric field into the *incident field* and *scattered field* $\boldsymbol{E}^{inc}$ and $\boldsymbol{E}^{scat}$. The total electric field is $\boldsymbol{E}^{total} = \boldsymbol{E}^{inc} + \boldsymbol{E}^{scat}$. The incident field comes from some distant source, whereas the scattered field is due to currents and charges on the scattering body. The PEC boundary condition of zero tangential field applies to $\boldsymbol{E}^{total}$, that is, to the tangential sum of the incident and scattered fields. It can be expressed mathematically as

$$\boldsymbol{E}^{total} \quad \boldsymbol{n} = (\boldsymbol{E}^{inc} + \boldsymbol{E}^{scat}) \quad \boldsymbol{n} = 0 \tag{6.6}$$

where $\boldsymbol{n}$ is a unit vector normal to the surface. Thus, the tangential scattered field is always equal and opposite in direction and amplitude to the incident tangential incident field *at the PEC surface*. Electric field lines of the local scattered field begin and end on surface-induced charges. At the surface the $\boldsymbol{E}^{total}$ field must be perpendicular to the surface.

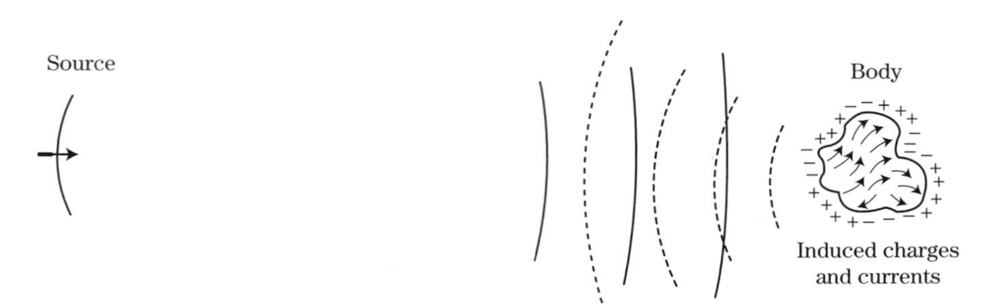

FIGURE 6-3 ■ Charges and currents: are induced on a PEC to satisfy the perfect conductor boundary conditions of zero tangential field (short circuit); and, consequently, re-radiate a scattered field $\boldsymbol{E}^{scat}$. From Knott [3].

With no applied $\boldsymbol{E}^{inc}$ field, there are no induced electric charges on the PEC, and $\boldsymbol{E}^{scat}$ is zero. Now, say at time $t = 0$, a *static* electric field $\boldsymbol{E}^{inc}$ is applied. Because the PEC is a short circuit, there is an immediate separation of electric charge on the surface, + and –, which creates a scattered field $\boldsymbol{E}^{scat}$. The nature of the induced charge separation is such that the boundary condition of zero tangential field is now satisfied everywhere on the conducting surface in accordance with equation (6.6).

For this static applied field the instant after $t = 0$, electric charges have been induced, the boundary conditions have been satisfied, and nothing more happens because the total tangential field is zero: $\boldsymbol{n} \quad \boldsymbol{E}^{scat} = -\boldsymbol{n} \quad \boldsymbol{E}^{inc}$. Now consider what happens if an EM wave is incident on the PEC rather than a static field (Figure 6-3). The PEC is exposed to a time-varying $\boldsymbol{E}^{inc}$ field as the wave propagates over the surface. Because the surface must be a short circuit, the induced scattered field, $\boldsymbol{E}^{scat}$, and the induced electric charge distribution must vary with time so as to always satisfy the boundary conditions of equation (6.6). The creation of the induced electric charge by the incident field and the resulting induced scattered field required to satisfy equation (6.6) is the fundamental physical process by which the zero tangential total electric field boundary condition is satisfied.

As a result of the applied EM wave, the induced charges are constantly in time motion. Because current is nothing more than charge in motion, the EM wave also creates induced currents. These currents and charges, which form fundamentally to create a tangential scattered field on the surface of the PEC that is exactly opposite the incident field, now act like antenna sources in that they radiate the scattered field far beyond the PEC itself.

Maxwell's equations in integral form show how the scattered field is formed from the PEC induced surface currents, $\boldsymbol{J}$ (A/m), and charge density, ρ_S (coulombs/m^2).

$$\boldsymbol{E}^{scat}(\boldsymbol{R}_f) = \int \left(-j\omega\mu \boldsymbol{J} g + \frac{\rho_S}{\varepsilon} \nabla g \right) dS \tag{6.7}$$

This form of Maxwell's equations, where time harmonic waves $\exp(j\omega t)$ have been assumed, shows that the scattered electric field vector $\boldsymbol{E}^{scat}$ at spatial field position $\boldsymbol{R}_f$ is computed as the sum over the surface currents $\boldsymbol{J}$ and charges ρ. When field point $\boldsymbol{R}_f$ is on the PEC surface, the surface tangential component of $\boldsymbol{E}^{scat}$ is exactly equal and opposite of the tangential incident field such that the total surface tangential field is zero as required for a short circuit surface.

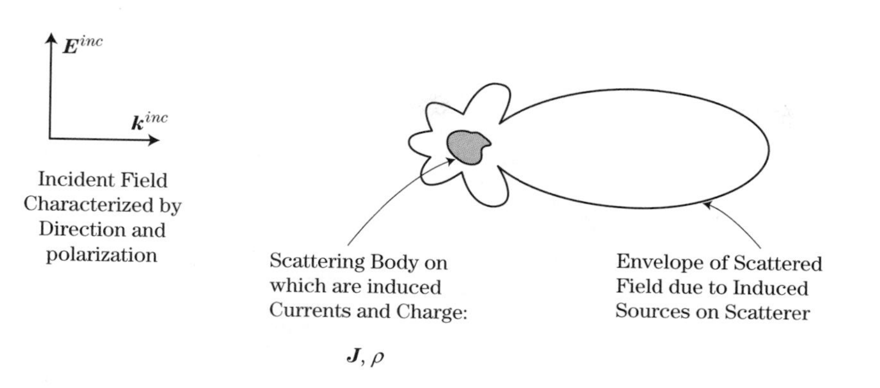

FIGURE 6-4 ■ PEC induced currents and charges reradiate a scattered field [1].

The g term in this equation is the Green's function, which relates a source quantity $\boldsymbol{J}$ or ρ at a surface source location $\boldsymbol{R}_s$ to the field $\boldsymbol{E}$ at spatial point $\boldsymbol{R}_f$. The Green's function is

$$g = \frac{e^{-jk \cdot (\boldsymbol{R}_f - \boldsymbol{R}_s)}}{4\pi |\boldsymbol{R}_f - \boldsymbol{R}_s|} \tag{6.8}$$

and can be thought of as a Huygen's wavelet in that it describes how a source at position $\boldsymbol{R}_s$ influences the field at position $\boldsymbol{R}_f$. The denominator is the $1/R$ falloff of intensity. The numerator is the phase or time delay from the source disturbance to influence the field point. The argument of the complex exponential is proportional to distance from a surface source point to the spatial point where the scattered field is being evaluated. This distance is measured in units of wavelength, $kR = 2\pi R/\lambda$.

While the primary function of the induced sources (charges and currents) is to satisfy the boundary conditions, those sources also act as antenna sources in that they radiate fields far away from the surface. This is the scattered wave (Figure 6-4).

A PEC sphere can be used as an example of how an incident wave induces sources on the surface that satisfy the boundary conditions and the resulting scattered and total fields surrounding the sphere. Figure 6-5 shows a time snapshot of currents induced on a sphere illuminated from the right with vertical polarization. The maximum current zone is where the incident wave first hits the sphere, and behind this point the surface creeping wave currents can be seen.

The induced currents and charges reradiate (i.e., become the source) for the scattered fields shown in Figure 6-6. In this figure the view is changed such that the incident wave

FIGURE 6-5 ■ Time snapshot of currents induced on a sphere due to an incident plane wave with vertical polarization.

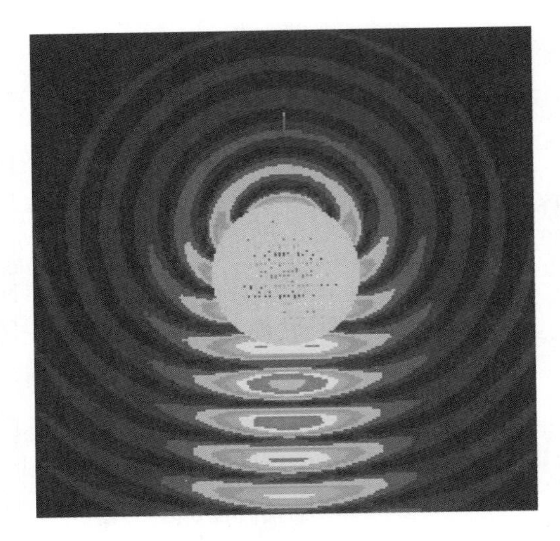

FIGURE 6-6 ■ Time snapshot of scattered fields from currents on a PEC sphere illuminated from the top.

is traveling from top to bottom over the sphere. There are three important things to note: (1) A scattered spherical wave is created traveling out from the sphere; (2) inside the sphere, partially masked by the graphic, is a strong horizontal field; and (3) behind the sphere (bottom of the figure) is a strong forward scattered field. This scattered field is propagating in *all* spatial directions away from the surface. This includes the direction back to the original source of the incoming plane wave (i.e., the backscatter direction).

Figure 6-6 shows the scattered field. Adding back the incident plane wave gives the total field (Figure 6-7). This figure shows the obvious incident plane wave and the important physics of how the scattered field and incident field add to zero inside the conducting surface and, behind the sphere, create a shadow on the side opposite the incident wave.

To illustrate this for a slightly more complicated object, Figure 6-8 shows the currents and scattered field from a PEC cylinder with an ogive nose when illuminated from 20 degrees above and from the left with the polarization in the plane of the figure. Clearly seen is the forward scattered field, which, when added to the incident field, forms a shadow behind. The figure also shows a strong scattered field in the specular direction from the

FIGURE 6-7 ■ Time snapshot of the total $\boldsymbol{E}$ field, $\boldsymbol{E}^{tot} = \boldsymbol{E}^{scat} + \boldsymbol{E}^{inc}$, when sphere is illuminated from the top with a plane wave $\boldsymbol{E}^{inc}$.

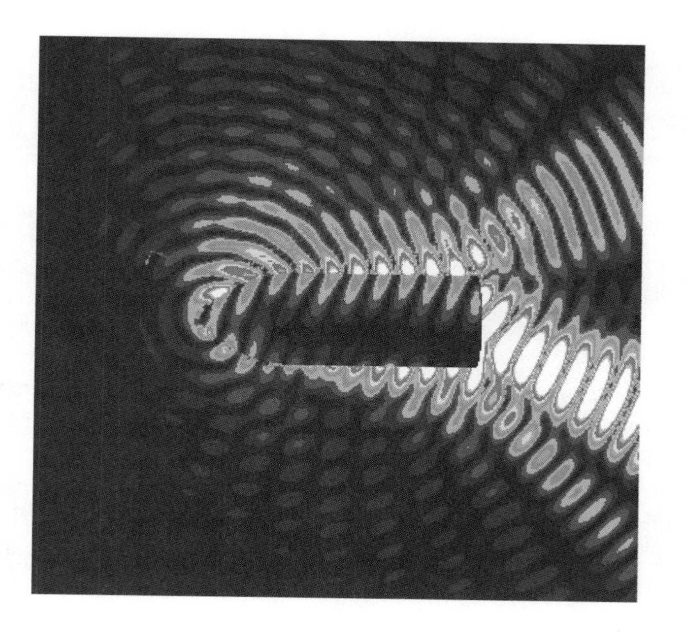

FIGURE 6-8 ■ Time snapshot of the scattered field from a PEC ogive-cylinder illuminated 20 degrees up from the axis with $\boldsymbol{E}$ in the plane of the figure. Also shown are the instantaneous surface currents producing this scattered field.

cylinder (angle of reflection equal to angle of incidence). An ogive tip diffracted field radiating spherically from the front can be observed as well.

6.3 RADAR CROSS SECTION DEFINITION

Radar cross section is a measure of power scattered in a given spatial direction when a target is illuminated by an incident wave. Another term for RCS is echo area. RCS is intended to characterize the target and not the effects of transmitter power, receiver sensitivity, and the distance or location between a target and the transmitter or receiver. Therefore, RCS is normalized to the power density of the incident wave *at the target* so that it does not depend on the distance from the illumination source to the target. This removes the effects of the transmitter power level and target distance. RCS is also normalized so that the inverse square falloff of scattered intensity back toward a receiver due to spherical spreading is not a factor; therefore, it is not necessary to know the position of the receiver.

6.3.1 IEEE RCS Definition

The Institute of Electrical and Electronics Engineers (IEEE) dictionary of electrical and electronics terms [4] defines RCS as a measure of the reflective strength of a target. Mathematically, it is defined as 4π times the ratio of the power per unit solid angle scattered in a specified direction to the power per unit area in a plane wave incident on the scatterer from a specified direction. More precisely, it is the limit of that ratio as the distance from the scatterer to the point where the scattered power is measured approaches infinity,

$$\sigma = \lim_{R\to\infty} 4\pi R^2 \frac{\left|\boldsymbol{E}^{scat}\right|^2}{\left|\boldsymbol{E}^{inc}\right|^2} \tag{6.9}$$

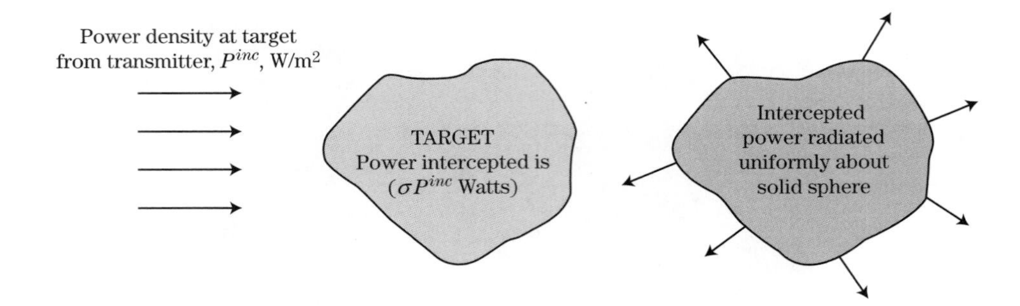

FIGURE 6-9 ■ Intuitive derivation of radar cross section [1].

where $\boldsymbol{E}^{scat}$ is the scattered electric field, and $\boldsymbol{E}^{inc}$ is the field incident *at the target*. Three case are distinguished: (1) monostatic or backscatter; (2) forward scattering; and (3) bistatic scattering.

A formal cross section may also be defined for the energy that is scattered, absorbed, or removed from the incident wave. A total cross section can also be defined that includes all of these effects. The scattered energy is of greatest practical interest because it represents the energy available for radar detection.

6.3.2 Intuitive Derivation for Scattering Cross Section

The formal IEEE definition for RCS given in equation (6.9) can be made more plausible by considering the following derivation (Figure 6-9). Let the incident power density at a scattering target from a distant radar be P^{inc} W/m^2. Considering the density at the target automatically removes from the definition transmitter power and the $1/R^2$ intensity falloff. The amount of power intercepted by the target is then related to its cross section σ, with units of area, so that the intercepted power is σP_{inc} W. This intercepted power is then either reradiated as the scattered power or absorbed as heat. Assume for now that it is reradiated as scattered power uniformly in all 4π sr of space so that the scattered power density is given by

$$P^{scat} = \frac{\sigma P^{inc}}{4\pi R^2} \quad \text{W/m}^2 \tag{6.10}$$

Equation (6.10) is then solved for σ, assuming that the distance R is far from the target to avoid near-field effects:

$$\sigma = \lim_{R\to\infty} 4\pi R^2 \frac{P^{scat}}{P^{inc}} \tag{6.11}$$

RCS is therefore fundamentally a ratio of scattered power density to incident power density. The power or intensity of an EM wave is proportional to the square of its electric or magnetic field magnitude, so RCS can be expressed as

$$\sigma = \lim_{R\to\infty} 4\pi R^2 \frac{\left|\boldsymbol{E}^{scat}\right|^2}{\left|\boldsymbol{E}^{inc}\right|^2} = \lim_{R\to\infty} 4\pi R^2 \frac{\left|\boldsymbol{H}^{scat}\right|^2}{\left|\boldsymbol{H}^{inc}\right|^2} \tag{6.12}$$

because in the far field $\boldsymbol{E}$ and $\boldsymbol{H}$ are related to each other by the impedance of free space. The units for radar cross section σ is area, usually square meters. RCS is sometimes made nondimensional by dividing by wavelength squared, σ/λ^2.

This definition is made more recognizable by examining the basic radar range equation for power received by the radar, P_r, in terms of transmitted, scattered, and received power:

$$P_r = \frac{(P_t G_t / 4\pi R^2)\sigma}{4\pi R^2} A_r \tag{6.13}$$

The term in parentheses in the numerator is the power density at the target location (measured in W/m^2). This incident power flux is multiplied by a cross section (area) and represents power captured from the incident wave and then reradiated by the target, some of which goes back toward the receiver. When this is divided by the return path spherical spreading factor, it gives the power density at the receiver for capture by the receiving antenna effective area, A_r.

The radar cross section of a target is a function of several attributes of the target, the radar observing the target, and the radar-target geometry. Specifically, RCS depends on the following:

- Target geometry and material composition.
- Position of transmitter relative to target.
- Position of receiver relative to target.
- Frequency or wavelength.
- Transmitter polarization.
- Receiver polarization.

When the transmitter and receiver are at different locations (Figure 6-10), the RCS is referred to as the bistatic radar cross section. In this case, the angular location of target relative to transmitter and receiver must be specified to fully specify the RCS. Monostatic or backscatter cross section is the usual case of interest for most radar systems. In this configuration the receiver and transmitter are collocated, often using the same antenna for transmitting and receiving (Figure 6-10). In this case only one set of angular coordinates is needed. Most experimental measurements are of backscatter cross section. Analytical RCS predictions, however, are much easier to make for bistatic cross section, with the illumination source fixed and the receiver position moved.

The radar cross section of a target may also be a function of the pulse width, τ, of the incident radiation. In the usual radar case with microsecond pulse widths, the pulse is large enough such that the pulse illuminates the entire target at once (target length assumed much less than 1,000 feet in length). This condition, which is loosely equivalent to the

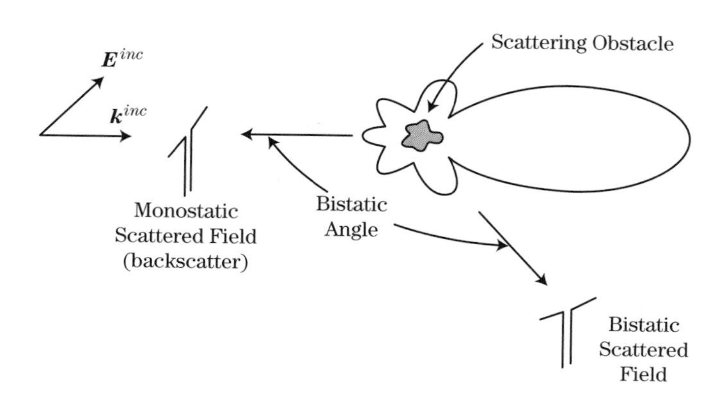

FIGURE 6-10 ■ Monostatic or bistatic target cross section cases.

target being illuminated by a continuous wave (CW) at a specific frequency, is known as long-pulse illumination. In this case, all scattering mechanisms from the target add coherently to give the net reflected signal. When short pulses are used (large bandwidth), such as nanosecond pulses with spatial extents of only several feet, then each scatterer on the target contributes independently to the return signal in time. In this case the RCS is a collection of individual scattering returns separated in time. Short-pulse radars (or their wide bandwidth pulse compression waveform equivalents; see Chapter 20) are often used to identify these scattering centers on complex targets.

6.3.3 RCS Customary Notation

The units for radar cross section are square meters. However, the RCS of a target does not necessarily relate to the physical size of that target. Although it is generally true that larger physical targets have larger radar cross sections (e.g., the optical front face reflection for a sphere is proportional to its projected area, $\sigma_{sphere} = \pi a^2$), not all RCS scattering mechanisms are related to size. Typical values of RCS can span 10^{-5} m^2 for insects to 10^{+6} m^2 for large ships. Due to the large dynamic range, a logarithmic power scale is most often used with the reference value of $\sigma_{ref} = 1$ m^2:

$$\sigma(\text{dBsm}) = \sigma(\text{dBm}^2) = 10\log_{10}\left(\frac{\sigma}{\sigma_{ref}}\right) = 10\log_{10}(\sigma) \tag{6.14}$$

Both the notations dBm^2 and dBsm are often used.

6.3.4 Polarization Scattering Matrix

6.3.4.1 Scattering Matrix for Linear Polarization

Radar cross section, as a scalar number, is a function of the polarization of the incident and received wave. A more complete description of the interaction of the incident wave and the target is given by the polarization scattering matrix (PSM), which relates the scattered electric field vector $\boldsymbol{E}^{scat}$ to the incident field vector $\boldsymbol{E}^{inc}$, component by component. In matrix notation, this is

$$\boldsymbol{E}^{scat} = \mathbf{S} \cdot \boldsymbol{E}^{inc} \tag{6.15}$$

$\boldsymbol{E}$ can be decomposed into two orthogonal directions or polarizations (because there is no component in the direction of propagation $\boldsymbol{k}$); thus, the polarization scattering matrix, $\mathbf{S}$, is a 2×2 complex matrix:

$$\boldsymbol{E}^{scat} = \begin{bmatrix} \boldsymbol{E}_V^{scat} \\ \boldsymbol{E}_H^{scat} \end{bmatrix} = \begin{bmatrix} S_{VV} & S_{VH} \\ S_{HV} & S_{HH} \end{bmatrix} \boldsymbol{E}^{inc} = \begin{bmatrix} S_{VV} & S_{VH} \\ S_{HV} & S_{HH} \end{bmatrix} \begin{bmatrix} \boldsymbol{E}_V^{inc} \\ \boldsymbol{E}_H^{inc} \end{bmatrix} \tag{6.16}$$

where $\boldsymbol{E}^{scat}$ and $\boldsymbol{E}^{inc}$ are the scattered and incident fields. $\boldsymbol{E}_H^{scat}$ and $\boldsymbol{E}_V^{scat}$ are the orthogonal vector vertical and horizontal polarization components of $\boldsymbol{E}^{scat}$, while $\boldsymbol{E}_H^{inc}$ and $\boldsymbol{E}_V^{inc}$ are the orthogonal vector vertical and horizontal polarization components of $\boldsymbol{E}^{inc}$.

The four complex elements of $\mathbf{S}$ that specify the scattering matrix represent eight scalar quantities, four amplitudes and four phases. One phase angle is arbitrary and is used as a reference for the other three. If the radar system is monostatic (backscatter), then $S_{VH} = S_{HV}$, and $\mathbf{S}$ can be specified by five quantities.

When a coherent radar that transmits and receives two orthogonal polarizations is present, the scattering matrix is determined for a given aspect angle at the radar frequency, f.

For a given target, aspect angle, and frequency, no more signal information can be extracted than what is contained in the scattering matrix for a narrow bandwidth pulse.

6.3.4.2 Scattering Matrix for Circular Polarization

In circular polarization, the electric field vector rotates in the plane perpendicular to the direction of propagation. The two orthogonal components of the electric field are called the *right-handed circular* and *left-handed circular* components (or just *right circular* [RC] and *left circular* [LC]). Consider an observer viewing a circularly polarized EM wave along the direction of propagation, and assume the wave is propagating away from the observer. This corresponds to considering the transmitted EM wave. The IEEE defines the case where the electric field vector rotates clockwise when viewed in this manner as RC rotation and the case where it rotates counterclockwise as LC rotation (Figure 6-11).

It follows that the apparent direction of rotation for a given circular polarization component is reversed for an incoming wave. To an observer viewing an incoming RC polarized wave, the electric field vector will appear to be rotating counterclockwise, while that of an incoming LC polarized wave will appear to be rotating clockwise.

Linear polarization can be transformed into circular polarization by shifting the phase of one of the two linear components by ±90 degrees. The choice between right- and left-handed circular is determined by the sign of the phase shift. Specifically, the RC and LC components of an electric field vector can be derived from the horizontal and vertical components according to [2]:

$$\begin{bmatrix} E_{RC}^t \\ E_{LC}^t \end{bmatrix} = \frac{1}{\sqrt{2}} \begin{bmatrix} 1 & +j \\ 1 & -j \end{bmatrix} \begin{bmatrix} E_{HH}^t \\ E_{VV}^t \end{bmatrix} \tag{6.17}$$

where the superscript t indicates the transmitted field components. The inverse transform to obtain linear polarization components from circular components is

$$\begin{bmatrix} E_{HH}^t \\ E_{VV}^t \end{bmatrix} = \frac{1}{\sqrt{2}} \begin{bmatrix} 1 & 1 \\ -j & +j \end{bmatrix} \begin{bmatrix} E_{RC}^t \\ E_{LC}^t \end{bmatrix} \tag{6.18}$$

as may be verified by taking the matrix inverse of the transformation matrix in equation (6.17).

Received polarization can be defined in a similar manner. The LC and RC definitions change because the observer is now looking in the direction of propagation, which is from

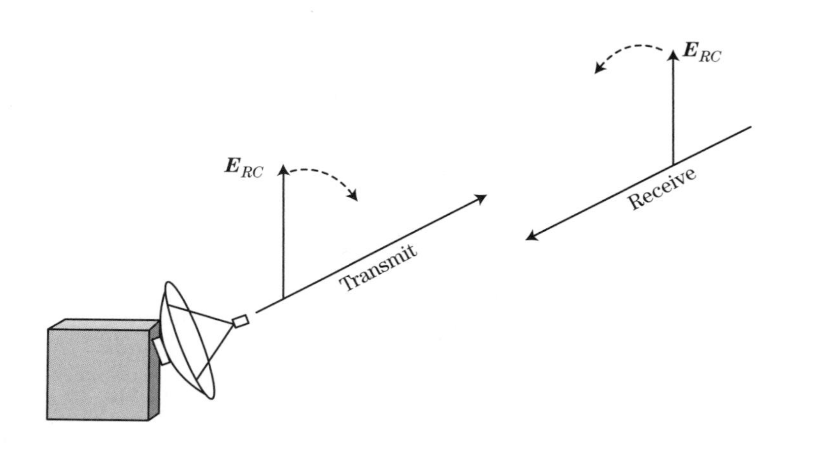

FIGURE 6-11 ■ Circular polarization sense, CCW or CW, is referenced by looking in the direction of propagation $\boldsymbol{k}$ [1].

the target toward the receiver, while the radar system has defined LC and RC as viewed from the radar to the target. Therefore, with the superscript r denoting received field components,

$$\begin{bmatrix} E^r_{RC} \\ E^r_{LC} \end{bmatrix} = \frac{1}{\sqrt{2}} \begin{bmatrix} 1 & -j \\ 1 & +j \end{bmatrix} \begin{bmatrix} E^t_{HH} \\ E^t_{VV} \end{bmatrix} \tag{6.19}$$

which is seen to be the complex conjugate of the transmitted case (6.17).

The circular polarization PSM contains no more information than the linear polarization PSM. If a linear PSM has been computed or measured, the corresponding circular PSM can be obtained by using equations (6.17) through (6.19) to obtain [2]

$$\begin{bmatrix} S_{LL} & S_{LR} \\ S_{RL} & S_{RR} \end{bmatrix} = \frac{1}{2} \begin{bmatrix} 1 & -j \\ 1 & +j \end{bmatrix} \begin{bmatrix} S_{HH} & S_{HV} \\ S_{VH} & S_{VV} \end{bmatrix} \begin{bmatrix} 1 & 1 \\ -j & +j \end{bmatrix} \tag{6.20}$$

where subscripts L and R indicate left and right circular polarization.

A characteristic feature of circular polarization is that single- or odd-bounced scattering changes the sense of the polarization, from LC to RC or from RC to LC. For linear polarization single-bounce specular scattering, the scattered energy has the same polarization as the incident polarization. This occurs because the scattered field is phase-shifted by 180 degrees relative to the incident field—that is, in the opposite direction (reflection coefficient $\Gamma = -1$).

6.4 | THREE SCATTERING REGIMES

Scattering mechanisms depend on scattering body size, L, relative to wavelength, λ. When λ is much larger than the body size, scattering is due to induced dipole moments; when λ is approximately the same as body size, surface wave effects such as edge, traveling, and creeping waves along with optical effects are important. When λ is much smaller than body size, surface wave effects become insignificant, and only optical effects are important.

The three scattering regimes are as follows:

Rayleigh	$\lambda \gg L$	Dipole-like scattering
Resonant	$\lambda \approx L$	Optics scattering + traveling, edge, and creeping surface waves
High frequency	$\lambda \ll L$	Optics scattering: angle of incidence = angle of reflection, end region, edge diffraction, multiple bounce

The classic illustration of radar cross section over these three regimes is that of a sphere as shown in Figure 6-12, where σ has been normalized to the projected area of the sphere, πa^2, plotted as a function of sphere circumference normalized to wavelength, $ka = 2\pi a/\lambda$. Figure 6-12 is a log-log plot over four orders of magnitude in σ and three orders of magnitude in sphere circumference. When the wavelength is much greater than the sphere circumference, the sphere's RCS is proportional to $a^2(ka)^4$. Although σ is small in this regime, it increases as the fourth power of frequency and sixth power of radius. When the circumference is between 1 and 10 wavelengths, the RCS exhibits an oscillatory behavior due to the interference of the front face optics specular return and the two surface creeping waves that propagate around the back of the sphere, as implied by the small sphere shown in the middle of Figure 6-12. This range of wavelengths is known as the resonant region. When the sphere's circumference is large compared with wavelength, the creeping waves

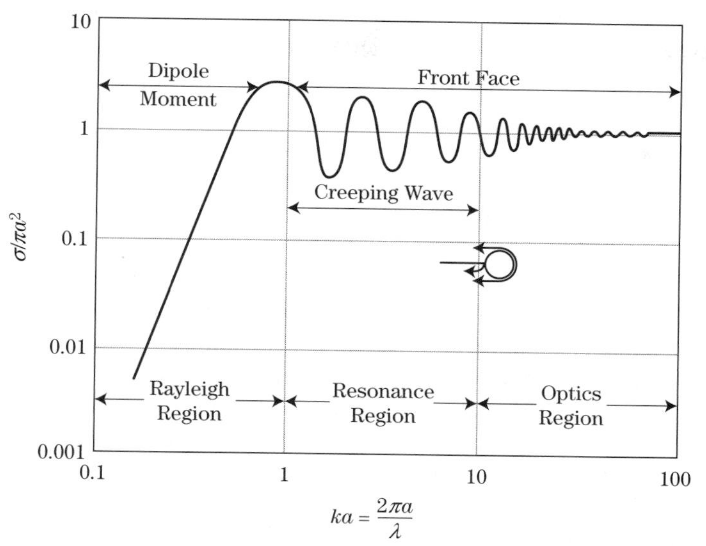

FIGURE 6-12 ■ Sphere scattering from Rayleigh, resonance, and optics regions, [1].

die out before they can radiate back toward the radar, the oscillatory behavior decays, and only the optics front face reflection is left, which for a double-curved surface is $\sigma = \pi a^2$, the projected area of the sphere. This is the optics region.

6.4.1 Rayleigh Region Dipole Scattering

When the incident wavelength, λ, is much greater than body size, L, scattering is called Rayleigh scattering, after Lord Rayleigh's analysis of why the sky is blue: the shorter blue wavelengths are more strongly scattered than the longer red wavelengths.

In the Rayleigh region, also called the low-frequency case, there is essentially little phase variation of the incident wave over the spatial extent of the scattering body: each part of the body "sees" the same incident field at each instant of time. This situation is equivalent to a static field problem, except that now the incident field is changing in time. This quasi-static field builds up opposite charges at the ends of the body; in effect, the incident field induces a dipole moment. The strength of this dipole is a function of the size and orientation of the body relative to the vector direction of the incident field. Dipole moments are defined as charge density multiplied by separation distance. The salient characteristic of Rayleigh scattering is that cross section is proportional to the fourth power of the frequency or wave number [5]:

$$\sigma \propto \omega^4 \quad \text{or} \quad k^4 \tag{6.21}$$

The low-frequency approach can be used until there is appreciable phase change of the incident wave over the length of the scatterer.

6.4.2 Resonant Region Scattering

When the incident wavelength is on the order of the body size, the phase of the incident wave changes significantly over the length of the scattering body (Figure 6-12). Although there are no absolute definitions, the resonant region is typically taken to be the range of wavelengths such that the target body is between λ and $10\,\lambda$ in size, $1 < L/\lambda < 10$. This region has two classes of scattering mechanisms: (1) optical, with the local angle of reflection equal to the angle of incidence; and (2) surface wave effects, where nonlocal regions interact.

Surface wave scattering mechanisms are distinctly different from optical scattering mechanisms. The name *resonant region* is a bit of a misnomer in that these mechanisms are not from high-quality factor sharply resonant phenomena. Rather, the physical mechanisms are due to EM energy that stays attached to the body surface. Surface wave types are *traveling*, *edge*, and *creeping*. Surface wave backscatter occurs when this surface energy is reflected from some aft body discontinuity or, as in the case of a creeping wave, when the energy flow is completely around the body.

Surface wave scattering is independent of body size. Cross section magnitudes are proportional to wavelength squared—that is, $L^0\lambda^2$. From this relation it can be seen why surface wave effects are important for resonant region body sizes. Surface wave effects are present in the optics region, but because of the smaller wavelengths the scattering magnitudes are much smaller than optical region scattering magnitudes, which most often are proportional to one of the following: $L\lambda$, $L^2\lambda^0$, $L^3\lambda^{-1}$, or $L^4\lambda^{-2}$.

Resonant region scattering mechanisms are typically due to one region of the scattering body interacting with another region. Thus, edge waves involve an entire edge in front of an edge termination aft discontinuity, surface waves over much of the entire surface in front of the surface termination aft discontinuity, and creeping waves over the shadowed regions. Overall geometry is important; however, small-scale details (relative to wavelength) are not.

6.4.3 High-Frequency Optics Region

When the wavelength becomes much smaller than the body size, $\lambda \ll L$, a localized scattering center approach can be used to represent the scattering physics; that is, local effects are more important than collective body–body interactions. In this region, collective surface edge, traveling, and creeping wave effects are very weak, so the body is now treated as a collection of independent scattering centers. Detailed geometries now become important in the scattering process, and net scattering from the body is the complex phasor sum of all the individual scattering centers.

True optics scattering is defined in the limit as $\lambda \to 0$. For most cases of microwave interest, finite body size effects must still be dealt with.

Optical regime scattering mechanisms are as follows:

- *Specular scattering*: This is true optics scattering in the sense of $\lambda \to 0$. Ray optics, where angle of reflection is equal to angle of incidence, apply. Scattering is analogous to mirror reflections in optics and is responsible for bright spike-like scattering.
- *End-region scattering*: This is scattering from the end regions of finite bodies. It produces the *sidelobes* in directions away from the direction of specular scattering. End-region sources arise from the rapid truncation of surface currents at body ends.
- *Diffraction*: This is end-region scattering in the specular direction due to edge-induced currents at leading or trailing edges, tips, or body regions of rapid curvature change.
- *Multiple bounce*: This is the separate case of mutual body interaction in the sense that one body surface specularly scatters energy to another body surface, which reflects that energy back to the observer (e.g., corner reflectors and cavities).

TABLE 6-1 ■ Scattering Mechanism and Relevant Scattering Regime

Scattering Mechanisms	Scattering Regime	Comments
Dipole	Dipole	Small scattering varies as the fourth power of frequency and the sixth power of size.
Surface waves	Resonance	Traveling, edge, and creeping waves; grazing angle phenomena; depends on polarization
Specular	Optics, resonance	Angle of reflection = angle of incidence for planar, single-, and double-curved surfaces
Multiple bounce	Optics, resonance	Few bounces (e.g., corner); many bounces (e.g., cavities)
End region	Optics, resonance	Sidelobes of a plate or cylinder from the ends of the surface
Edge diffraction	Optics, resonance	Diffraction in the specular direction; depends on polarization
Discontinuities, gaps, cracks	Optics	Surface imperfections important at higher frequencies

Table 6-1 lists the various scattering mechanisms and the scattering regimes where they apply.

6.5 HIGH-FREQUENCY SCATTERING

High-frequency scattering is defined as the case when the wavelength, λ, is much smaller than body size. Scattering centers are now typically local regions on the body and are referred to as scattering centers.

6.5.1 Phasor Addition

There is almost always more than one scattering center in view for any given aspect angle. The overall or net scattering is then the coherent phasor sum of all the scattering mechanisms from individual centers. The coherent phasor sum is the complex vector addition of scattering amplitudes depending on their electrical phase (path length measured in wavelengths) according to

$$\boldsymbol{k} \cdot \boldsymbol{R} = 2\pi \left(\frac{R}{\lambda} \right) \tag{6.22}$$

where distance, R, is the total two-way distance from the transmitter to the target scattering center and back to the receiver. Specifically, the total coherent RCS from N scattering centers active at a given viewing angle is

$$\sigma_{total} = \left| \sum_{i=1}^{N} \sqrt{\sigma_i} e^{j2\boldsymbol{k} \cdot \boldsymbol{R}_i} \right|^2 \tag{6.23}$$

where $\sqrt{\sigma_i}$ and $\boldsymbol{R}_i$ are the amplitude and spatial position of each individual scattering center. This sum depends on relative spatial position (i.e., spatial phase) of each contributor as well as its amplitude. The factor of two in the phase term results from the fact that the

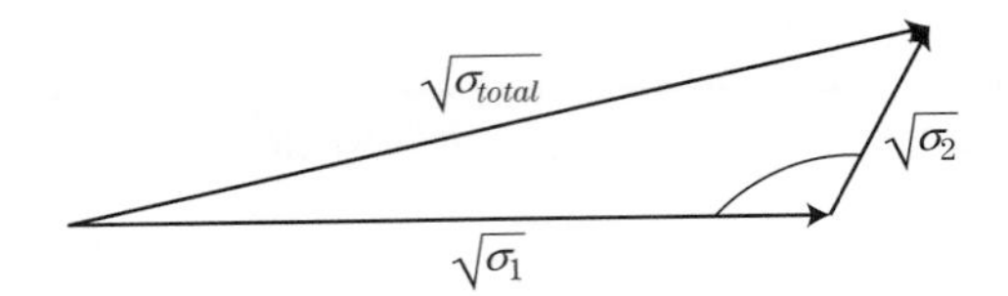

FIGURE 6-13 ■ Total scattering is the phase sum of individual scattering centers.

total distance from transmitter to target and back to receiver is twice the one-way distance. The coherent phasor sum is analogous to force vector addition in kinematics.

To illustrate, consider just two scattering centers, σ_1 and σ_2, with voltage amplitudes proportional to $\sqrt{\sigma_1}$ and $\sqrt{\sigma_2}$, and let $\sqrt{\sigma_{total}}$ be the amplitude of the phasor sum of the electric fields produced by these two scattering mechanisms. The coherent phasor sum (Figure 6-13) depends not only on the amplitude of the two scattering centers but also on their relative phase.

Figure 6-14 shows how the RCS of two individual point scatterers add and subtract to create different values of total RCS depending on their relative sizes and spatial phase. Point scatterers have no angular dependence, so the only source of phase differences is differences in their spatial location. Let the point scatterers lie on some baseline and be spaced 2λ apart. Now consider three cases.

In case 1, both scatterers have the same RCS, $\sigma_1 = \sigma_2 = 1\text{ m}^2 = 0\text{ dBsm}$. Figure 6-14a illustrates the coherent phasor sum RCS in the plane of the two scattering centers. When the transmitter/receiver is on a line perpendicular to the baseline, the distance to each scattering center is the same, so the two complex amplitudes add in phase to give $\left(\sqrt{1}+\sqrt{1}\right)^2 = 4\text{ m}^2 = 6\text{ dBsm}$ for the total RCS. When the transmitter/receiver is collinear with the baseline, the round-trip distance difference is 4λ, so again the complex amplitudes of the two scatterers add in phase to give a total RCS of 6 dBsm. At selected angular locations in between, the spatial phase difference becomes 180 degrees, and the two scatterers add out of phase. Since they are equal-RCS scatterers, the net RCS is zero, and the null depths are infinitely deep on the logarithmic dBsm scale.

In case 2, shown in Figure 6-14b, the radar cross sections of the two scatterers differ in magnitude by a factor of four, $\sigma_1 = 1\text{ m}^2 = 0\text{ dBsm}$ and $\sigma_2 = 0.25\text{ m}^2 = -6\text{ dBsm}$. As before, when the receiver is on a line perpendicular to the baseline, the distance to each scattering center is the same, and the individual voltages add in phase to give

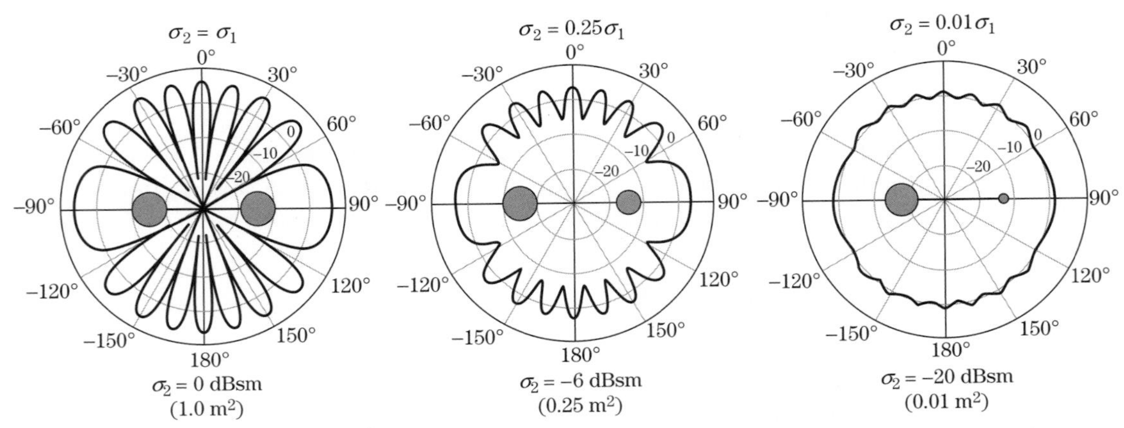

FIGURE 6-14 ■ Phase sum of two point scattering centers of different magnitudes. Scale: dBsm, 10 dB/div.

$(\sqrt{1}+\sqrt{1/4})^2 = 2.25\ \text{m}^2 = 3.52$ dBsm. When the transmitter/receiver is collinear with the baseline, the round-trip distance difference is 4λ, so again the two scatterers add in phase to 3.52 dBsm. At angles where the two spatial phases are 180 degrees apart, the scatterers add out of phase to give $(\sqrt{1}-\sqrt{1/4})^2 = 0.25\ \text{m}^2 = -6.0$ dBsm for the null depths.

In case 3 (Figure 6-14c), the two scatterers differ in magnitude by a factor of 100, $\sigma_1 = 1\ \text{m}^2 = 0$ dBsm and $\sigma_2 = 0.01\ \text{m}^2 = -20$ dBsm. As before, when the receiver is on a line perpendicular to the baseline or is collinear with the baseline, the two scatterers add in phase to give $\left(\sqrt{1}+\sqrt{1/100}\right)^2 = 1.21\ \text{m}^2 = 0.83$ dBsm. At angles where the two spatial phases are 180 degrees apart, the scatterers phase subtract to give $(\sqrt{1}-\sqrt{1/100})^2 = 0.81\ \text{m}^2 = -0.92$ dBsm for the null depths.

From these results it is seen that the interaction of two equal or nearly equal RCS scatterers can be up to four times, or 6 dB, higher than that of a single scattering center when they add in phase, and that when they subtract, the null depths can be very deep. However, when two scatterers of significantly different RCS interact, the lower-RCS scatterer does not contribute much to the net scattering on the dB scale.

6.5.2 Specular Scattering

Specular scattering from flat, single-curved, or double-curved surfaces is scattering where the angle of reflection is equal to the angle of incidence. In very simple terms, the reflected wave bounces like a billiard ball from the surface (Figure 6-15). Specular scattering is often of very high intensity. The target scattering center surface normal must point back toward the radar transmitter/receiver direction for specular scattering to occur in the backscatter direction.

A key notion in specular scattering is that of the *specular point*. This point is the surface location where the angle of incidence is equal to the angle of reflection. This location is also called the flash point, hot spot, or stationary phase point. In geometrical terms, this point determines the shortest distance between the transmitter and receiver for a location on the surface. This point is also the surface location for Fermat's minimum path length; that is, the path length is stationary with respect to variations in path. When the receiver and transmitter are collocated, as in backscatter, the specular points are those surface locations where the local normal points back toward the illuminating radar.

Figure 6-16 shows a bistatic configuration of transmitter and receiver. The surface specular point lies on a line between an image of the transmitter (below the surface) and the receiver or, equivalently, on a line between an image of the receiver and the transmitter. This line is the shortest distance between the transmitter and receiver via the surface. Currents in the vicinity of the specular point add in phase to create the high-magnitude flash in the specular direction. Currents outside this specular region add out of phase and do not contribute to the flash.

The occurrence of specular scattering is very dependent on the geometry of the scattering surface and the angles to the transmitter and receiver. A flat surface has only one backscatter specular point—that is, only one spatial direction where the surface normal

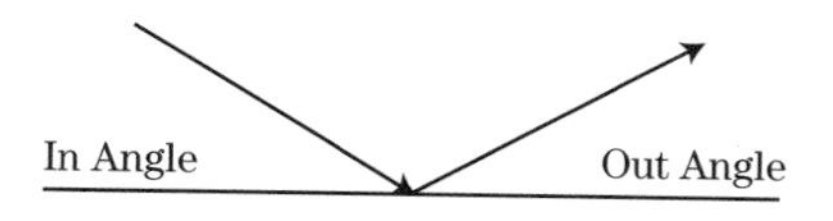

FIGURE 6-15 ■ Specular scattering, Snell's law.

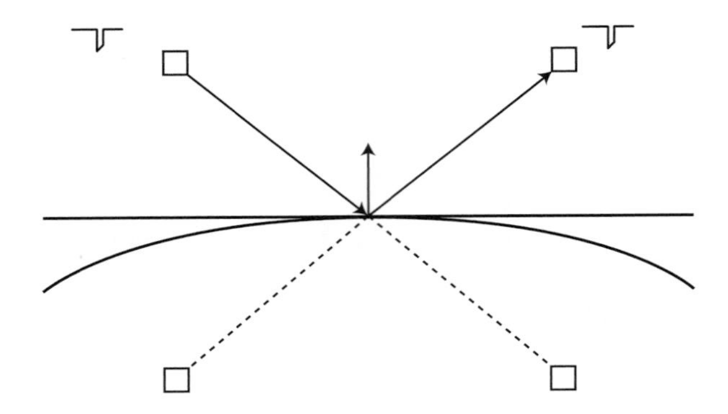

FIGURE 6-16 ■ Surface specular point for a bistatic radar.

can point back toward a radar. A sphere has an infinite number of specular points; that is, one is always perpendicular to the sphere surface regardless of view angle. A cylindrical surface has only one plane in space perpendicular to the surface where there is specular scattering.

Backscatter amplitude for specular scattering depends on the constant phase area, A_{cp}, at the specular point and on the wavelength,

$$\sigma_{specular} = 4\pi \frac{A_{\text{cp}}^2}{\lambda^2} \tag{6.24}$$

In the flat plate case, the constant phase area is the physical area of the plate. In this case all elementary regions of the plate are equidistant from the transmitter/receiver. Hence, all of these elemental areas add in phase to give the large net specular amplitude.

The double-curved surface constant phase area depends on the two radii of curvature at the specular point. As the incident wave passes over the surface, only the currents in a small region about the specular point can add in phase to the receiver/transmitter. The sharper the curvatures (smaller radii of curvature), the smaller the constant phase area dimension, L_{cp}:

$$L_{\text{cp}} \propto \sqrt{\frac{R_c \lambda}{2}} \tag{6.25}$$

For double-curved surface backscatter, the two constant phase dimensions correspond to the curvature of the surface at the specular point. The RCS is

$$\sigma = \pi \; R_1 \; R_2 \tag{6.26}$$

where R_1 and R_2 are the two specular point surface radii of curvature. Equation (6.26) is a very simple result that depends only on geometric curvature parameters and not on wavelength. As an example, the RCS of a prolate spheroid when viewed from the small end is less than when viewed from broadside due to the difference in the radii of curvature with the change in the position of the specular point.

In the cylindrical case, the surface has only one radius of curvature for the constant phase dimension, with the other dimension determined by cylindrical length. Cylindrical RCS becomes

$$\sigma_{cyl} = \frac{2\pi}{\lambda} R L^2 = k R L^2 \tag{6.27}$$

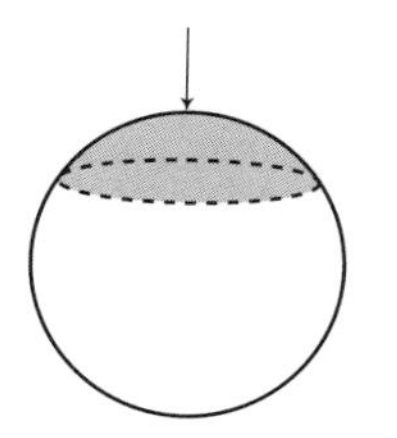

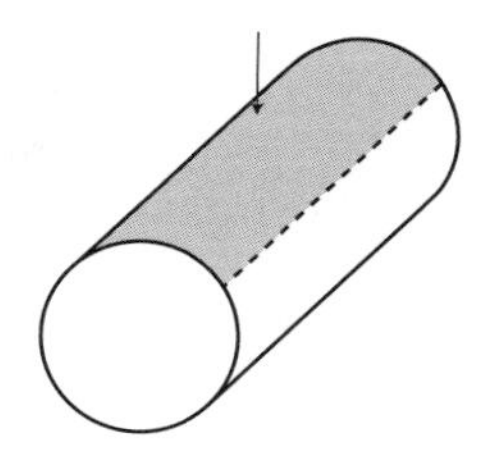

FIGURE 6-17 ■ Specular point constant phase current region for backscatter when viewed from the top.

Figure 6-17 shows spherical and cylindrical specular point areas when a backscatter radar is above the target.

6.5.3 End-Region Scattering

Currents at the ends of a surface have two types of scattering. One is called end-region scattering and is responsible for plate sidelobes. The second is edge diffraction and is due to local edge line sources. Even though they both occur at edges, the sources are fundamentally different. Each of these sources scatters energy into directions other than the specular direction of the surface.

End-region scattering occurs when surface currents do not taper smoothly to zero at an edge; rather, they abruptly change value. This abrupt change gives rise to a scattering center. This mechanism scatters energy into directions other than the specular direction. This is why the end-region sources are said to produce the sidelobes of the RCS pattern. End-region scattering does not depend on polarization.

Consider backscattering from a 10λ flat plate when viewed perpendicular to one of its edges (Figure 6-18 and the left half of Figure 6-19). Each end region becomes a scattering center source, and they are of equal magnitude. Thus, they add in or out of phase to give the oscillatory sidelobe pattern away from the specular direction. As the radar moves toward a grazing aspect angle, the projected area of the plate ends decreases, so the amplitude of the sidelobes also decreases. The end-region "source" area is determined by its projected area relative to the wavelength. When viewed perpendicular to an edge, the linear plate dimension, L, determines the effective end region size.

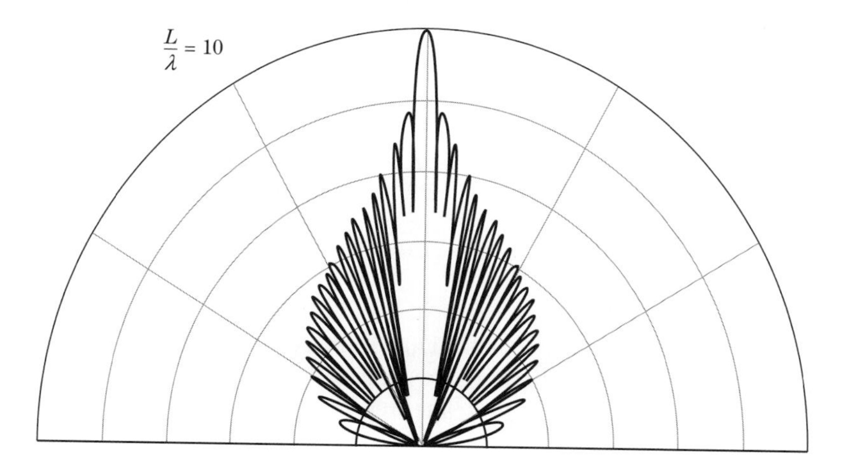

FIGURE 6-18 ■ Backscatter from a flat plate when viewed perpendicular to an edge. The side lobes are due to the truncated end region currents phase adding/subtracting. The first side lobe is 13 dB down from specular. Scale is 10 dB/div.

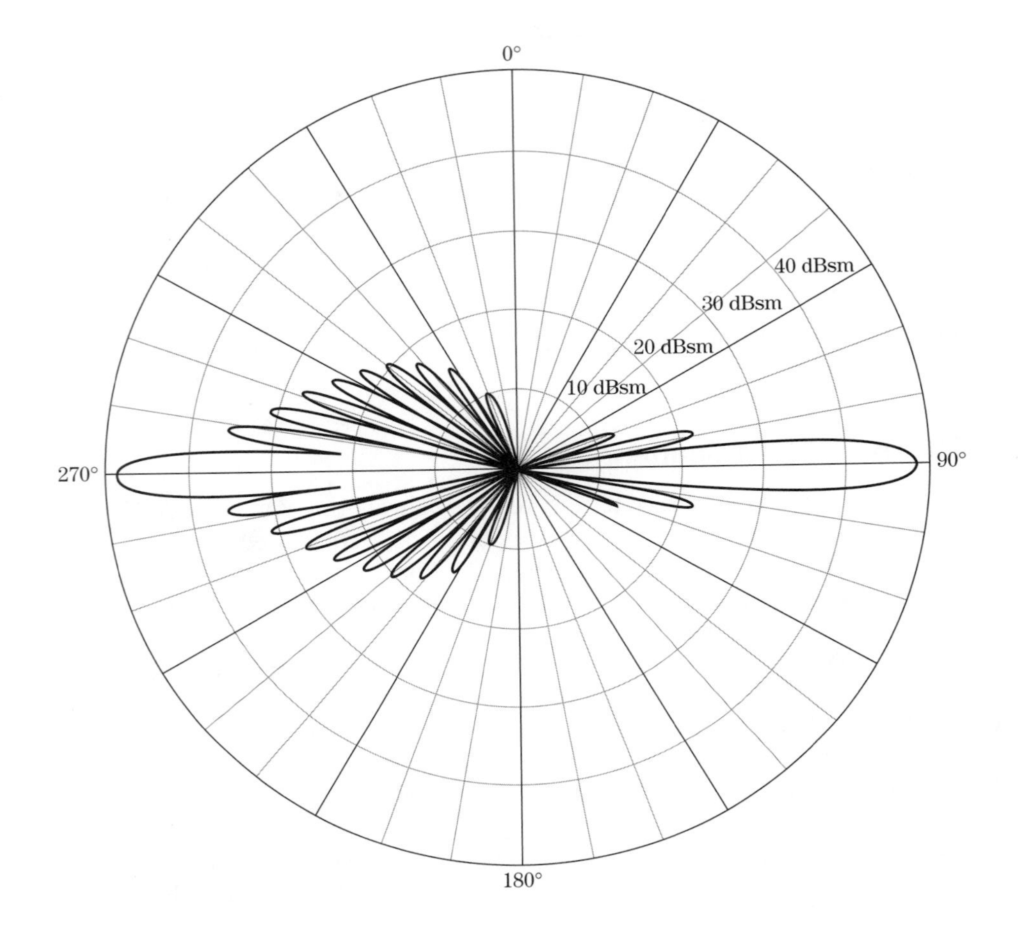

FIGURE 6-19 ■ Plate backscatter showing end region side lobe scattering: Left) Viewed perpendicular to edges; Right) Viewed along diagonal.

The end-region source strengths for this plate change if the plate is viewed in a plane along its diagonal. In this case the end regions are the four corners of the plate. These areas are much smaller than the end-region area when the plate is viewed perpendicular to its edge; thus, the sidelobe pattern is much smaller (right half of Figure 6-19). In this case, the first sidelobe is 27 dB down from the specular peak, and the entire sidelobe pattern is much lower than when the plate is viewed perpendicular to its edges.

As further illustration of end-region sources, Figure 6-20 shows the RCS scattering centers of the plate when viewed at 45 degrees perpendicular to its edges. Two end-source regions are evident. When the plate is viewed at 45 degrees along its diagonal (Figure 6-21), four much smaller scattering center end regions are seen. Thus, in the diagonal view cut, sidelobes are much smaller since the end-region area sources are smaller.

6.5.4 Edge Diffraction

Edge diffraction is due to induced localized line source (wire-like) currents at edges. Edge currents give rise to specular scattering on what is called the Keller cone of reflected rays and typically dominates grazing angle scattering (Figure 6-22).

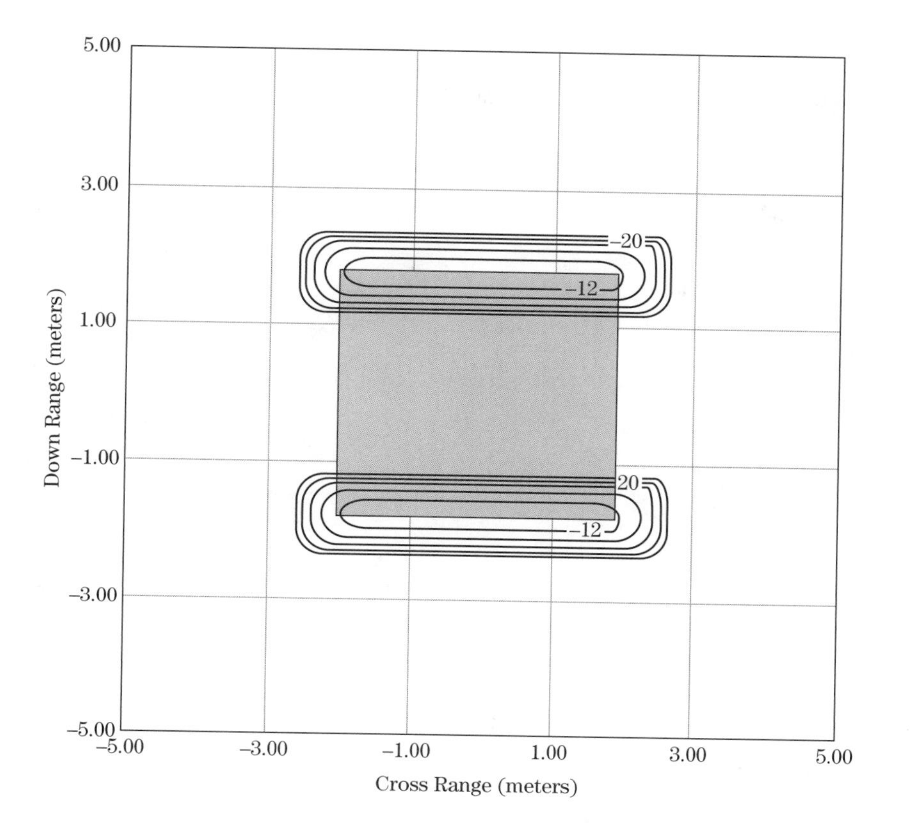

FIGURE 6-20 ■ Flat-plate physical optics end-region return for oblique 45 degree backscatter looking perpendicular to edge of plate. These end regions produce the side lobes as they phase add or subtract. Image analytically computed using physical optics currents; that is, edge diffraction is not included, which at this angle is not yet significant. Plate size is 5λ.

This "cone" of rays occurs due to cylindrical azimuthal symmetry of the local wire-like edge line source. Diffraction edge currents add in phase only in the specular direction, that is, on the Keller cone [6]. Every ray on this cone is in the specular direction. Edge normals are a disk of normal vectors perpendicular to the edge. The edge specular point is that point on the edge where the angle of reflection is equal to the angle of incidence.

When only backscatter is considered, the Keller cone degenerates into a disk of reflected radiation, and, in this case, the specular direction occurs only when the disk normal points back toward the radar. Thus, for backscatter, edge diffraction occurs only when the radar is perpendicular to an edge.

Edge diffraction depends on polarization (Figure 6-23). Two cases occur: (1) the incident electric field is parallel to the edge (leading-edge [LE] diffraction); and (2) the incident electric field is perpendicular to the edge (trailing-edge [TE] diffraction).[1]

The magnitude of edge diffraction has a weak dependence on the interior-included angle. A practical approximation for backscatter radar cross section for an edge of length L is

$$\sigma = \frac{L^2}{\pi} \tag{6.28}$$

which for a 1 ft. edge has a specular return of approximately –15 dBsm.

[1]This leading- and trailing-edge terminology is not to be confused with the LE and TE of airfoil surfaces.

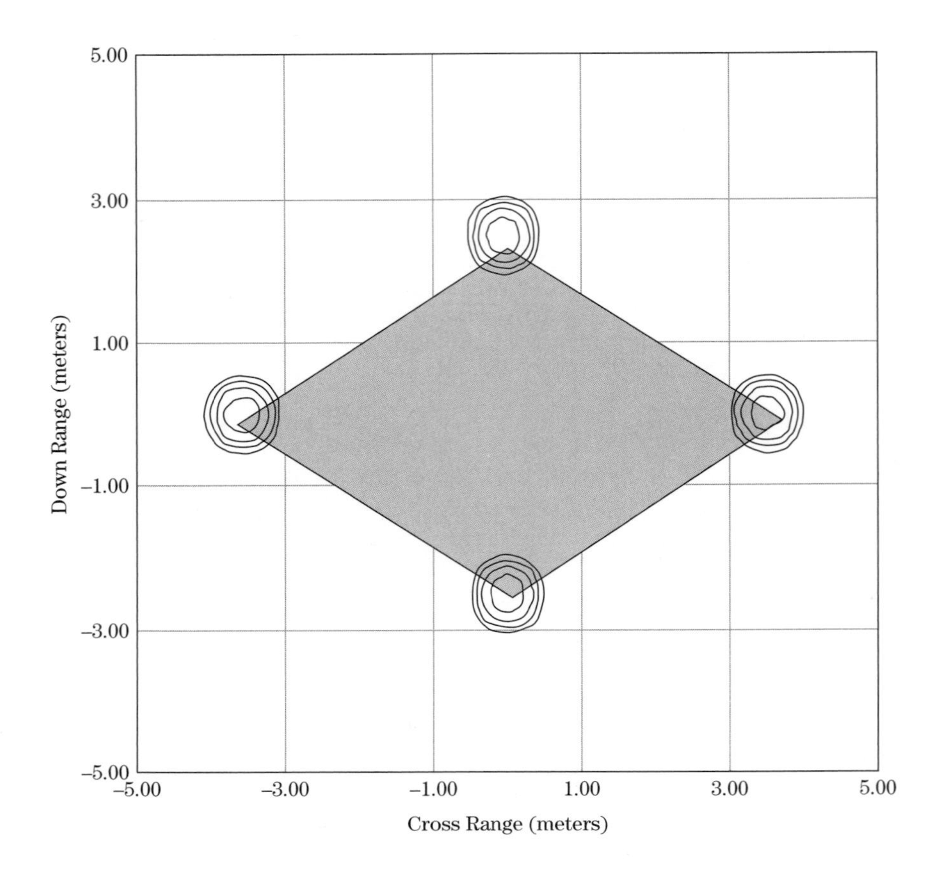

FIGURE 6-21 ■ Flat-plate physical optics end-region return for oblique 45 degree incidence looking along diagonal of plate. These much smaller end regions produce lower side lobe envelopes as they phase add/subtract compared to looking perpendicular to an edge. Image analytically computed using physical optics current; that is, edge diffraction is not included, which at this angle is not yet significant. Plate size is 5λ.

FIGURE 6-22 ■ Keller cone of edge specular reflected rays. Cone is due to symmetry of wire like local edge currents. Cone is the specular direction(s) of the incident ray, (from Knott [1]).

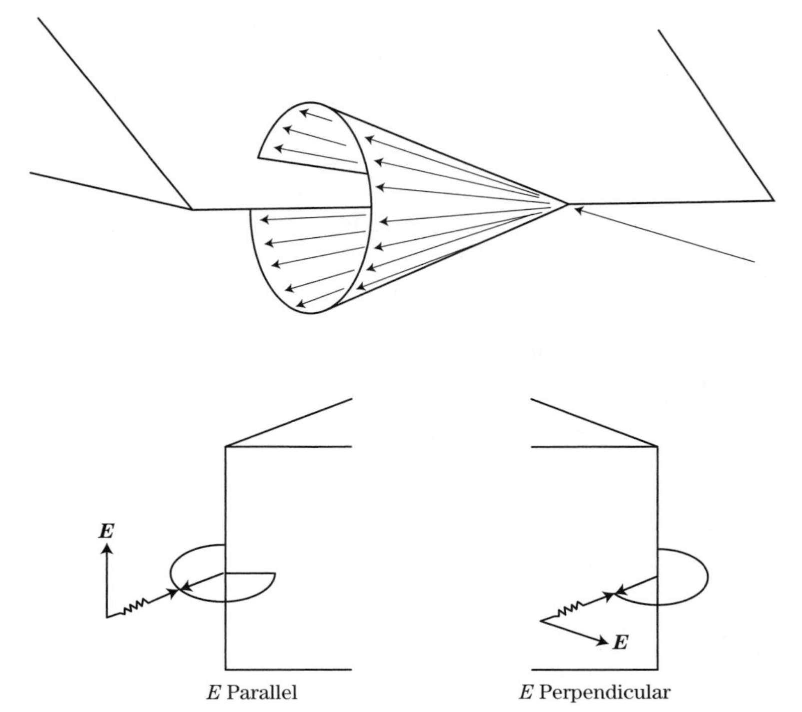

FIGURE 6-23 ■ Edge diffraction depends on polarization: E parallel or perpendicular to edge.

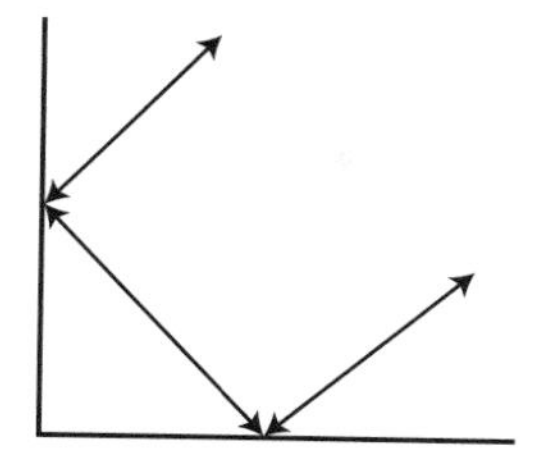

FIGURE 6-24 ■ Multiple bounce is two or more specular scatters which reflect back to a radar.

6.5.5 Multiple-Bounce Scattering

Multiple-bounce scattering occurs when two or more specular bounces act in combination to reflect incident energy back to a receiver (Figure 6-24). Multiple-bounce scattering has two broad classes: one involving only a few specular bounces, such as a corner reflector; and the other involving many bounces such as a jet engine inlet/exhaust cavity of an aircraft. Multiple-bounce scatters are often characterized as having a high-level return over a wide angular region. This is why corner reflectors are often used as RCS augmenter devices on the masts of sailboats and why traditional ship construction, with many bulkheads and decks at right angles to each other and to the sea surface, produces vessels with rather large RCS.

The simplest corner reflectors are dihedrals and trihedrals formed by the intersection of two or three surfaces, respectively, intersecting at right angles. The RCS of a dihedral at X-band with square sides of dimension 17.9 cm is shown in Figure 6-25 [1]. Note the large central return over a wide angular region. The peaks at ±45 degrees correspond to the single-bounce specular return from each of the two planar surfaces.

Another class of multiple-bounce geometries includes the jet engine inlet cavity of high-performance aircraft. This case is typically characterized by more than two bounces. Unless there is an energy absorption mechanism, a cavity that is open at only one location will reflect back all of the energy incident upon it. When numerous bounces occur, exit energy is often randomized in the exit direction.

Multiple-bounce geometries can also occur in cases where there is not an "obvious" corner reflector, such as an edge that is perpendicular to a single- or double-curved surface (Figure 6-26). Another example is a double-curved surface near any other target, such as a sphere in the vicinity of a planar surface (Figure 6-27).

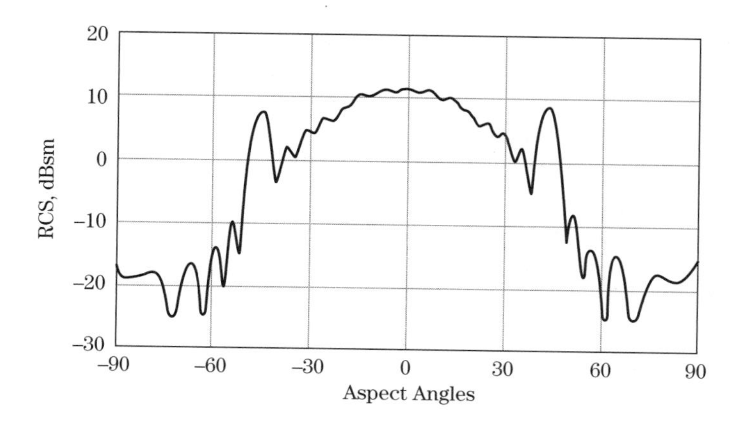

FIGURE 6-25 ■ Multiple bounce dihedral backscatter showing a large central region of scattering [Knott, 1].

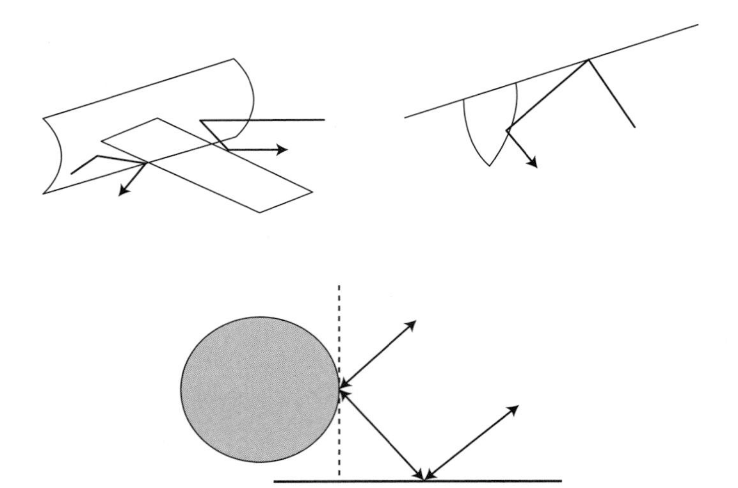

FIGURE 6-26 ■ Multiple bounces can exist for a variety of geometric arrangements.

FIGURE 6-27 ■ Doubly curved surfaces often form corner reflectors with other surfaces.

6.6 EXAMPLES

This chapter concludes with RCS examples from a flat plate, an A7 aircraft at X-band, and a fictional geometry called a "stove pipe aircraft." Far-field coherent patterns as well as scattering center images will be presented.

The backscattering from a flat plate when viewed perpendicularly to one of its edges exhibits three optics regime scattering mechanisms: specular, end region, and edge diffraction. Figure 6-28 [7] shows the measured RCS for a 6.5 inch square plate at X-band for both polarizations.

The specular spike occurs when the plate is viewed at 0 degrees, that is, perpendicular to the plate. This spike has the same amplitude, approximately 8.75 dBsm, for both polarizations. The null-to-null beamwidth, given by

$$\theta = 57\frac{\lambda}{L} \text{ degrees} \tag{6.29}$$

where L is the plate dimension, is approximately 12 degrees. End-region scattering is dominant from the first null out to about 40 degrees. The in-phase or out-of-phase addition from the two equal area end regions creates this part of the RCS pattern. The sidelobes fall off as the end-region projected area becomes smaller (cosine function).

This part of the pattern is the same for both polarizations. Edge diffraction then dominates the pattern from 40 degrees out to grazing (approaching 90 degrees), now with distinctly different results for each polarization. The case of the electric field parallel to an edge (on the left side of the figure) has a value at grazing of –20 dBsm, which is consistent with equation (6.28). The case where the electric field is perpendicular to an edge (right side of the figure) can better be characterized as a traveling wave return since the plate is only 5λ in size.

A very complex backscatter example at X-band is that of an A7C Corsair aircraft as measured at the Navy's Junction Ranch range in California. The test measurements were of a full-scale aircraft (Figure 6-29), where the vehicle is supported on a tapered low RCS Styrofoam column. The column is tapered so that its (small) specular cylinder return is not normal to the test radar. The tie-down ropes are also oriented so that they are not perpendicular to the test radar.

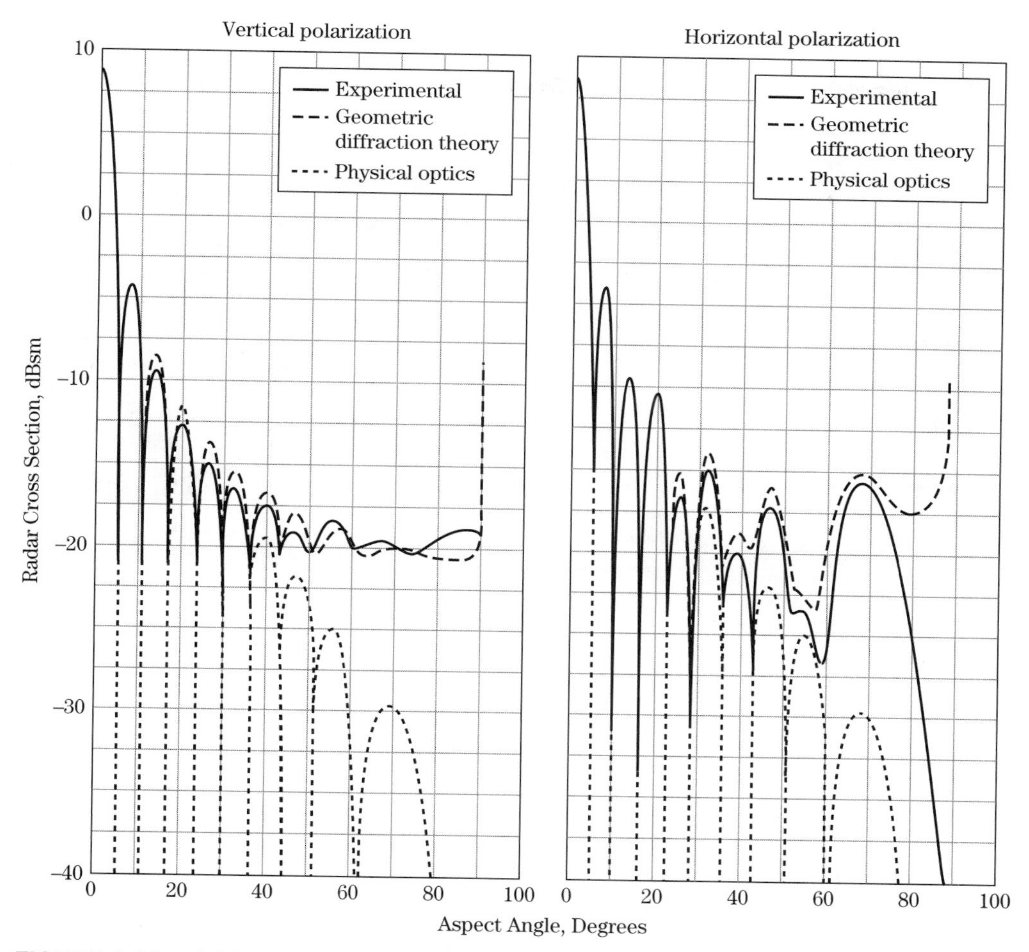

FIGURE 6-28 ■ RCS patterns of a 6.5 in. square plate at a wavelength of 1.28 inches viewed perpendicular to its edges [from 7].

This aircraft at X-band is truly in the optical scattering regime. At each viewing angle it has tens to hundreds of scattering centers that combine in or out of phase to produce the final RCS pattern. Major scattering centers include surfaces normal to the radar line of sight, engine inlet and exhaust cavities, the pilot crew station, control surface edges, avionics antennas and sensors, and the gaps and cracks of construction joints and control surface breaks. Since this vehicle is hundreds of wavelengths long, it does not take much angular movement for these scattering centers to change from adding in phase to adding out of phase with one another. The RCS pattern (Figure 6-30) is very oscillatory and has a rapidly fluctuating backscatter return. While this pattern does not display distinct features, one does see an increase near broadside due to specular scattering. Trailing edge diffraction spikes can also be identified.

Figure 6-31, an image of the scattering centers when the radar is perpendicular to the wing leading edge, shows numerous scattering centers. The major ones include the wing leading edge, the pilot crew station region, the engine inlet cowling, and the engine inlet cavity. The inlet cavity images are off the plan form view at the cross-range location of

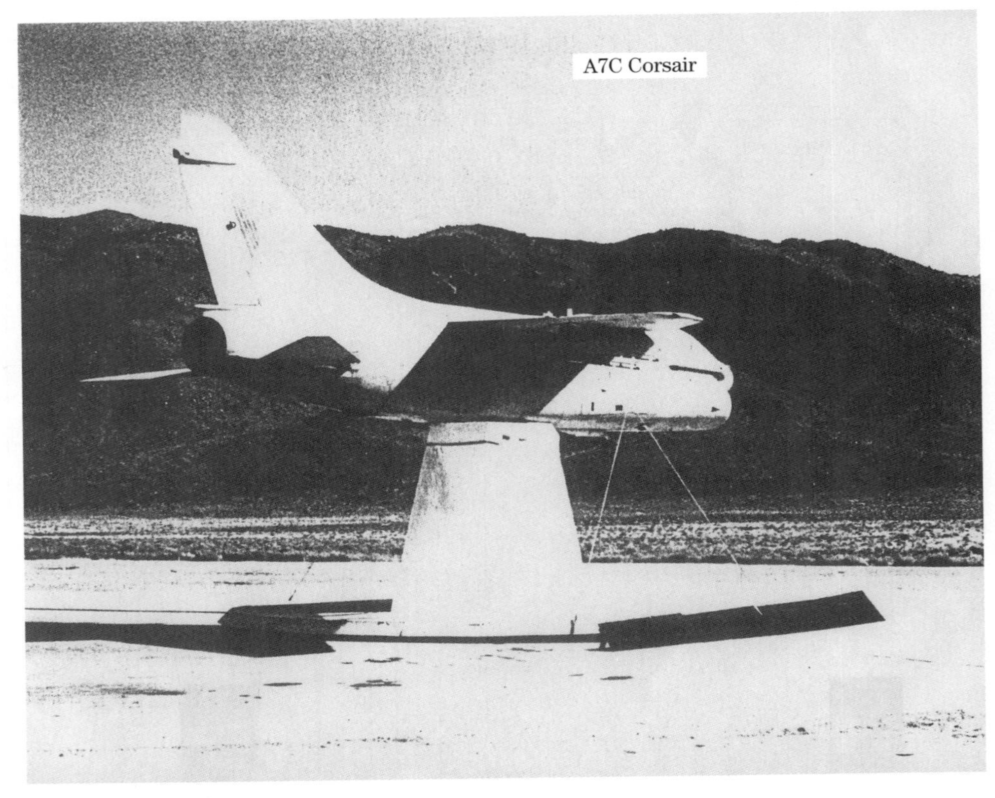

FIGURE 6-29 ■ A7C Corsair RCS measurement set up at the Navy Junction Ranch Range.

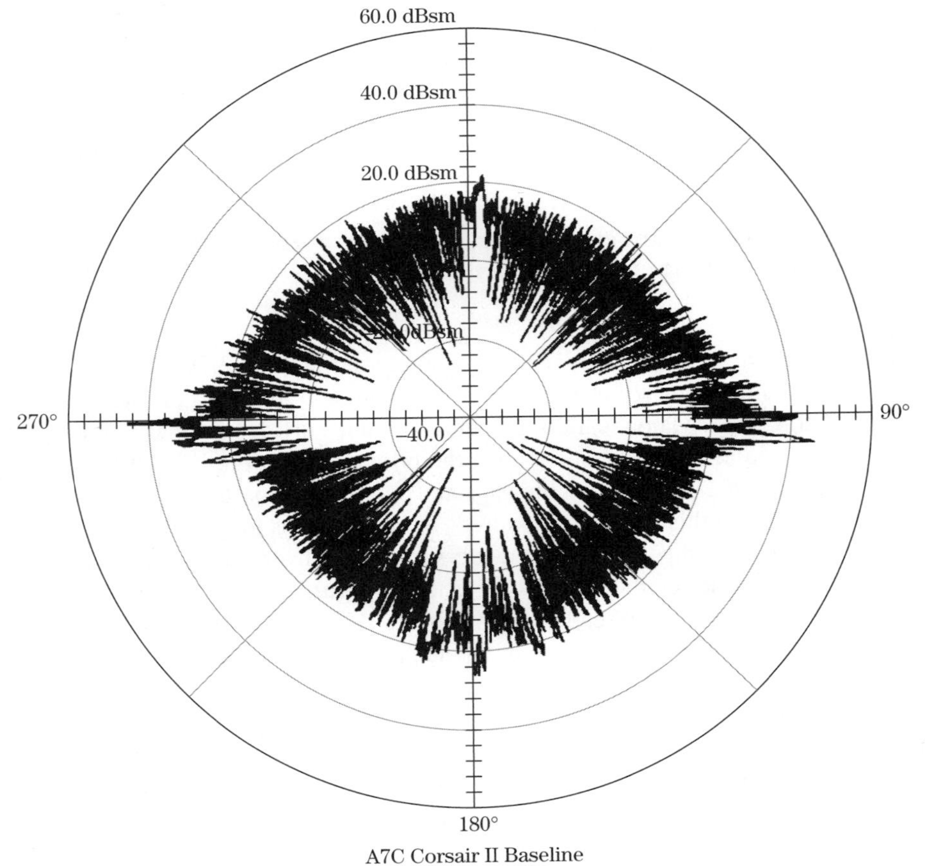

FIGURE 6-30 ■ A7 C measured backscatter for horizontal polarization at 9.5 GHz, 20 dB/div.

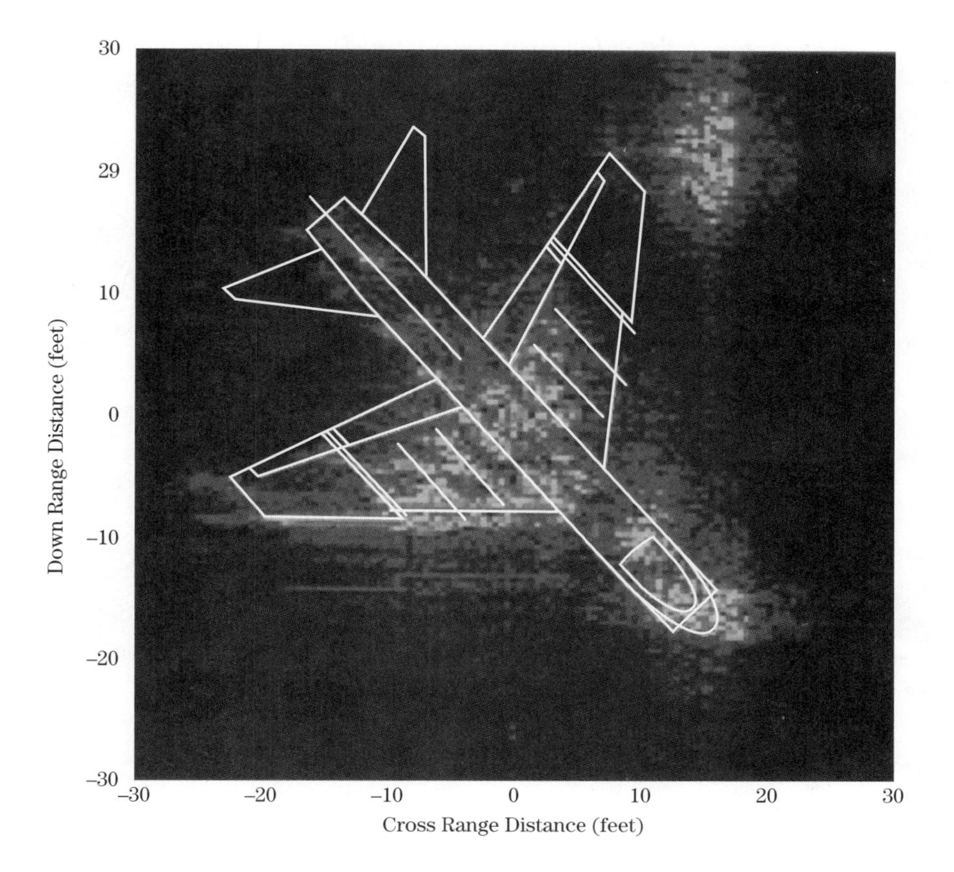

FIGURE 6-31 ■ A7 down/cross range image measurements when perpendicular to wing leading edge for horizontal polarization at X-band [Navy Junction Range Range].

the engine inlet and at a down-range location corresponding to the engine front face. Note that this cavity return is one of the major scattering centers at this viewing angle.

The last example is a "stove pipe" aircraft (Figure 6-32). The model is approximately 3 feet long and was measured at C-band, horizontal polarization, along with image measurements of the scattering centers at various aspect angles. Figure 6-33 shows the measured backscatter RCS (left half of the figure) along with a high-frequency computer physical optics code prediction (right half of the figure). The measurements and the optics prediction code show reasonable agreement, as should be the case since this RCS target measurement is in the optics regime. Specific specular scattering centers as a function of aspect angle—such as the cylinder broadside and vertical tail return, tail flat disk return, nose cone return, and wing leading-edge returns—can be seen on the polar plot.

Image measurements were also performed for a number of aspect angles (Figures 6-34 through 6-38). In these images, the radar is below the picture so that down-range (time delay) is up the page and cross-range is left or right. Target rotation allows for cross-range processing, and radar bandwidth allows for down-range image processing [8].

The nose-view scattering center image (Figure 6-34) shows the scattering centers to be the discontinuity between the nose cone and cylinder (the line of sight is perpendicular to the rim edge), the trailing edge of the vertical fin (trailing edge diffraction for horizontal polarization), and cylinder surface traveling wave energy reflecting from the forward wing

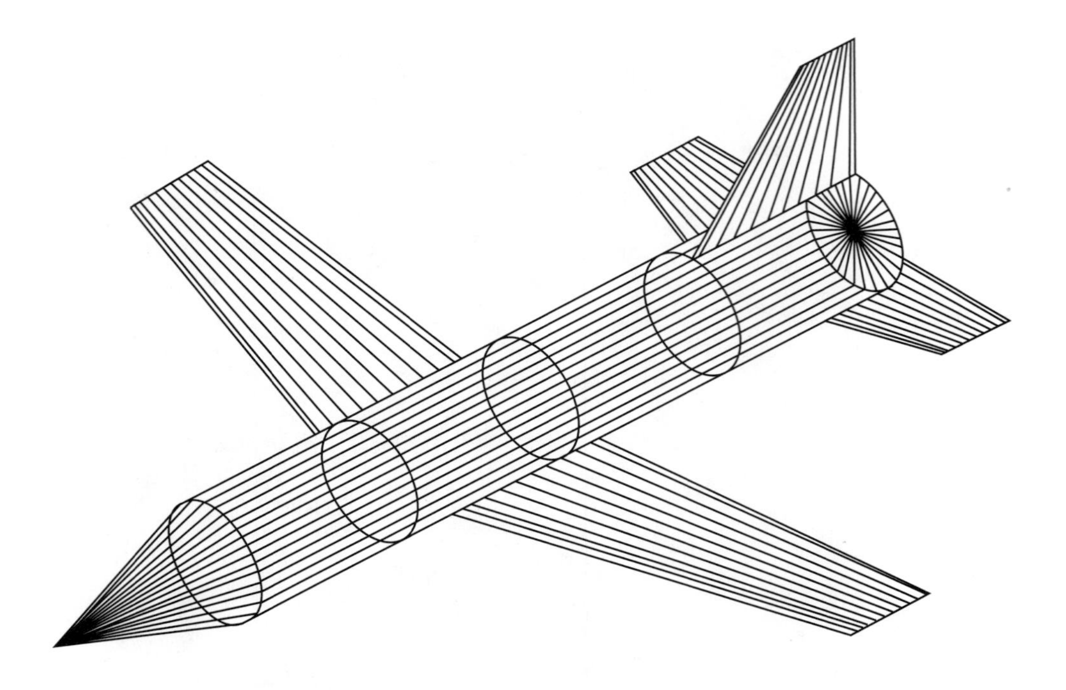

FIGURE 6-32 ■ "Stovepipe" RCS backscatter model.

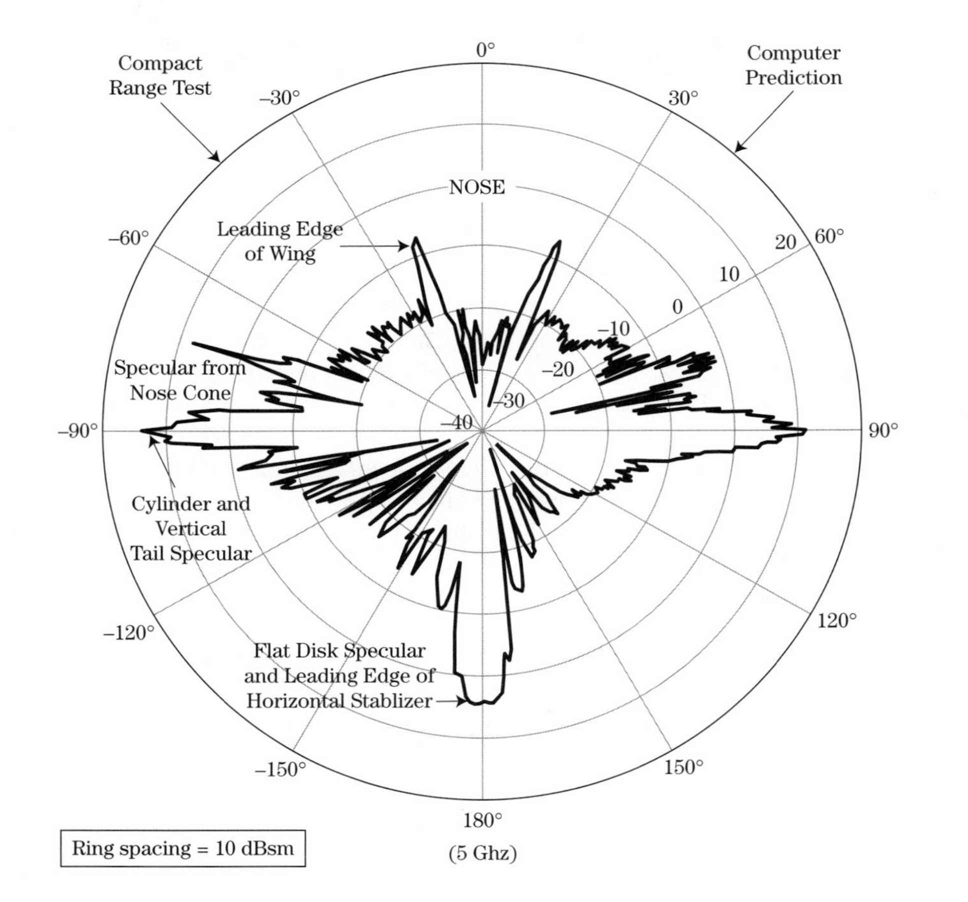

FIGURE 6-33 ■ Measured and Physical Optics predicted RCS for "Stovepipe" geometry at C band.

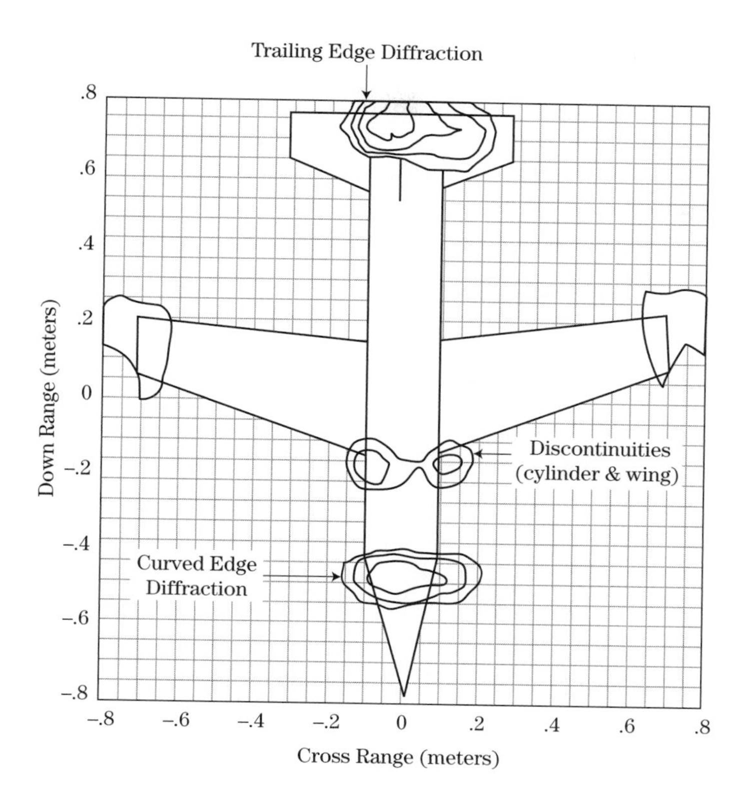

FIGURE 6-34 ■ "Stovepipe" nose view scattering centers.

root. These four major scattering centers combine in or out of phase to give the 0 degree net RCS value of approximately –25 dBsm seen in Figure 6-33.

Broadside-view scattering centers (Figure 6-35) are the cylinder sides, the vertical edge of the vertical fin, and the vertical surface of the tail fin. Note that, for horizontal polarization, the wing partially shadows the cylinder.

Tail-view scattering centers (Figure 6-36) are the flat disk specular scattering at the end of the cylinder, the edge diffraction from the horizontal stabilizer (horizontal polarization), several very small traveling-wave scattering mechanisms from energy going down the sides of the cylinder and across the vertical fin surfaces, and remnants of wing-edge diffraction.

Scattering centers when the model is viewed perpendicular to the wing leading edge (Figure 6-37) are the wing-edge diffraction (electric field parallel to the edge) and the vertical fin trailing edge (electric field perpendicular to the edge).

The last view is from an aspect perpendicular to the front nose cone (Figure 6-38). Now the scattering centers are the specular return from the single-curved nose cone surface and the edge diffraction from the trailing edge of the vertical fin (electric field perpendicular to edge).

This stove pipe geometry shows that electromagnetic backscattering is very much a geometry issue; that is, scattering centers appear when surfaces or edges are perpendicular to the radar. These images also show that scattering centers in the optical regime are local and that they are dependent on geometry, radar orientation, and polarization. If the images were formed at X-band rather than C-band, the scattering centers would be very similar since the geometry is the same. Only the magnitude of the frequency dependent scattering mechanisms would be different.

FIGURE 6-35 ■ "Stovepipe" broadside view scattering centers.

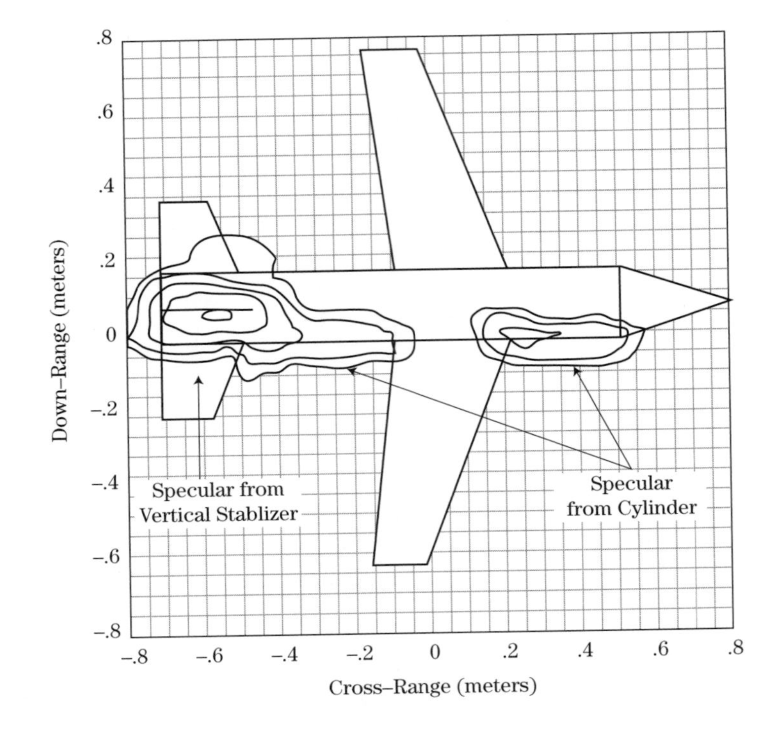

FIGURE 6-36 ■ "Stovepipe" tail view scattering centers.

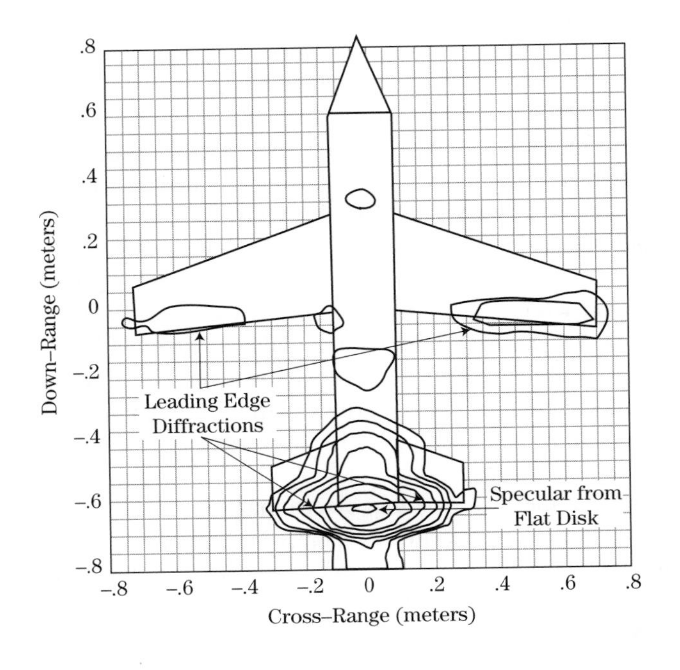

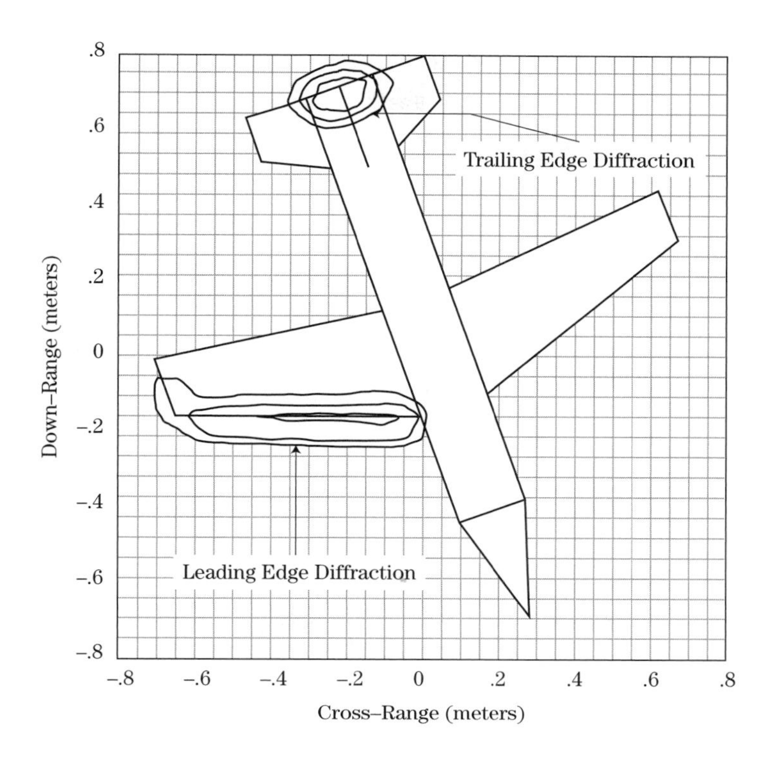

FIGURE 6-37 ■ "Stovepipe" wing leading edge view scattering centers.

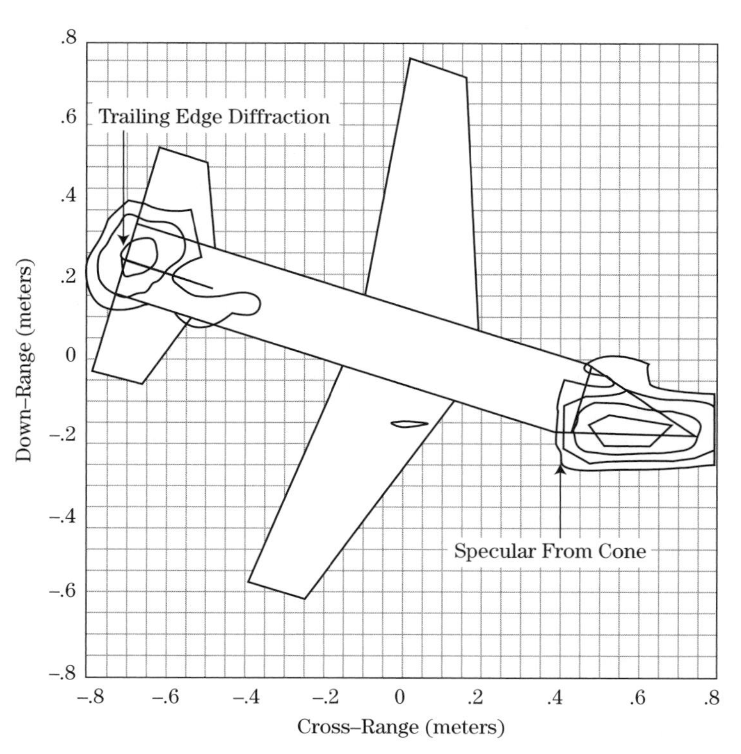

FIGURE 6-38 ■ "Stovepipe" scattering centers when viewed normal to front nose cone.

6.7 FURTHER READING

There are several classic and modern textbooks on radar cross section theory, phenomenology, measurement, and prediction. The most modern, and perhaps the best introduction to the subject, is the text by Knott et al. [1]. Other comprehensive sources include Jenn [9] and the two-volume *Radar Cross Section Handbook* [10]. More theoretical treatments are available in Crispin and Siegel [11], Ogilvy [12], Maffett [13], and Ross Stone [14].

RCS reduction is a topic of great interest in military applications. An excellent introduction is available in [1], with additional detail provided in the text by Bhattacharyya and Sengupta [15]. Vinoy and Jha [16] discuss the particular method of using radar-absorbing materials for RCS reduction.

6.8 REFERENCES

[1] Knott, E.F., Shaeffer, J.F., and Tuley, M.T., *Radar Cross Section*, 2d ed., Scitech Publishing, Raleigh, NC, 2004.

[2] Emmons, G.A., and Alexander, P.M., "Polarization Scattering Matrices of Polarimetric Radar," Technical Report RE-83-1, U.S. Army Missile Command, Redstone Arsenal, AL, March 1983.

[3] Knott, E.F., "Radar Cross Section Short Course Notes," Georgia Institute of Technology, Atlanta, 1984.

[4] Jay, F. (Ed.), *IEEE Standard Dictionary of Electrical and Electronic Terms*, ANSI/IEEE Std 100-1984, 3d ed., IEEE Press, New York, 1984.

[5] Ruck, G.T. (Ed.), *Radar Cross Section Handbook*, vols. 1 and 2, Plenum Press, New York, 1970.

[6] Keller, J.B., "Geometrical Theory of Diffraction," *Journal of the Optical Society of America*, vol. 52, p. 116, 1962.

[7] Ross, R.A., "Radar Cross Section of Rectangular Flat Plates as a Function of Aspect Angle," *IEEE Transactions on Antennas and Propagation*, vol. AP-14, pp. 329–335, May 1966.

[8] Mensa, D. L., *High Resolution Radar Imaging*, Artech House, Norwood, MA, 1981.

[9] Jenn, D.C., *Radar and Laser Cross Section Engineering*, American Institute of Aeronautics and Astronautics, Reston, VA, 1995.

[10] Ruck, G.T., Barrick, D.E., Stuart, W.D., and Krichbaum, C.K., *Radar Cross Section Handbook*, vols. 1 and 2, Plenum Press, New York, 1970.

[11] Crispin, J.W., and Siegel, K.M., *Methods of Radar Cross Section Analysis*, Academic Press, New York, 1968.

[12] Ogilvy, J.A., *Theory of Wave Scattering From Random Rough Surfaces*, Taylor & Francis, Philadelphia, 1991.

[13] Maffett, A.L., *Topics for a Statistical Description of Radar Cross Section*, John Wiley & Sons, New York, 1989.

[14] Ross Stone, W., *Radar Cross Sections of Complex Objects*, IEEE Press, New York, 1990.

[15] Bhattacharyya, A.K., and Sengupta, D.L., *Radar Cross Section Analysis and Control*, Artech House, Boston, 1991.

[16] Vinoy, K.J., and Jha, R.M., *Radar Absorbing Materials*, Kluwer Academic Publishers, Boston, 1996.

6.9 | PROBLEMS

1. For specular scattering, what is the key geometric concept involving the transmitter, receiver, and target surface?
2. What is specular backscatter?
3. Where in space do edge normals point?
4. What is the direction of target surface normals for (a) specular backscatter and (b) edge normals?
5. Why can RCS be measured using either $\boldsymbol{E}$ or $\boldsymbol{H}$ fields?
6. What scattering mechanisms do not depend on polarization?
7. What scattering mechanisms do depend on polarization?
8. In free space, how much of the EM wave energy is in the E field? In the H field?
9. Near a PEC surface, which EM wave field component contains most of the energy and why?
10. For an EM wave propagating inside a material medium, what physical properties of the material determine the velocity of propagation?
11. How does the wavelength of an EM wave inside of a material differ from that of free space?
12. If you are designing an outdoor measurement range for RCS with a specified error tolerance, what level of background RCS would you try to achieve? Why?
13. What was Maxwell's key contribution such that the EM laws of Gauss, Ampere, and Faraday became collectively known as Maxwell's Equations?
14. For an EM wave propagating toward a PEC surface, what happens to the impedance of the wave very close to the PEC surface?
15. In world of computational EM, if an aircraft model has its fuselage along the x axis and it wings parallel to the x-y plane, what spherical unit vectors correspond to horizontal and vertical polarization for an angle cut in the x-y plane?
16. When can a spherically spreading EM wave be characterized as a plane wave?
17. What are the EM fields inside a closed PEC surface?
18. What is the polarization scattering matrix?
19. Can a scattering-only measurement ever tell us more about a target at a fixed angle than the PSM?
20. For circular polarization, what is the reference for deciding whether a wave is left or right circularly polarized?
21. In Rayleigh region scattering, what is the phase variation of the incident wave over the target?
22. In high-frequency scattering, why does the backscatter RCS usually vary so rapidly with target movement?
23. What is the most important feature of specular scattering from (a) planar surfaces, (b) single-curved surfaces, and (c) double-curved surfaces?
24. If you were to design a target to have low specular RCS, how would you proceed?
25. The formula for the magnitude of specular scattering has an area term. In the backscatter case, what area is this?

CHAPTER 7

Target Fluctuation Models

Mark A. Richards

Chapter Outline

7.1 Introduction 247
7.2 Radar Cross Section of Simple Targets 248
7.3 Radar Cross Section of Complex Targets 251
7.4 Statistical Characteristics of the RCS of Complex Targets 253
7.5 Target Fluctuation Models 263
7.6 Doppler Spectrum of Fluctuating Targets 267
7.7 Further Reading 269
7.8 References 269
7.9 Problems 270

7.1 INTRODUCTION

A sample of radar data is composed of either interference alone or interference plus target echoes. The interference is, at a minimum, receiver noise and might also include clutter echoes, electromagnetic interference (EMI) from other transmitting sources (e.g., radars, television stations, cellular telephones), and hostile jamming. Most of these interfering signals are noise-like and are therefore modeled as random processes. Occasionally, a target echo will also be present in a particular sample of radar data. One of the major tasks of a radar system is to detect the presence of these targets when they occur. As was seen in Chapter 3, this is generally accomplished by *threshold detection*.

Chapter 3 introduced the *nonfluctuating* target, that is, one in which the radar cross section (RCS) remains constant over the group of samples used for detection. While this is a useful reference point, it is rarely a realistic model of real-world radar targets. Chapter 6 described how target size, materials, shape, and viewing angle contribute to determining RCS. Variations in radar-target geometry, target vibration, and radar frequency changes can lead to variations in target RCS, resulting in *fluctuating targets*. In this chapter, common models for the statistics of target echoes are discussed, with an emphasis on the traditional Swerling models. The effect of these models on radar detection performance is considered in Chapter 15.

7.2 | RADAR CROSS SECTION OF SIMPLE TARGETS

7.2.1 Basic Scatterers

The simplest radar target is a perfectly conducting sphere, such as those shown in Figure 7-1. Because of its spherical symmetry, the RCS of a sphere is independent of aspect angle. As discussed in Chapter 6 it is not, however, independent of radar frequency. Figure 7-2 shows the normalized RCS of a sphere as a function of its size relative to the radar wavelength, computed using the Mies series solution in [1]. It oscillates dramatically when the sphere is relatively small compared with the radar wavelength λ but approaches an asymptotic value of πa^2 (where a is the radius of the sphere) when the sphere is large compared with the wavelength. Another example of a relatively simple target with a known RCS is a *corner reflector* such as the dihedral or trihedral shown in Figure 7-3. Corner

FIGURE 7-1 ■ Examples of RCS calibration spheres. (Courtesy of Professor Nadav Levanon, Tel-Aviv University.)

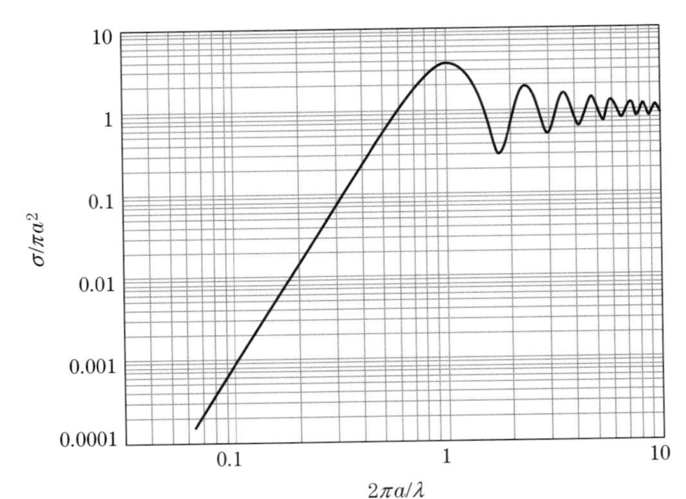

FIGURE 7-2 ■ Normalized RCS of a perfectly conducting sphere as a function of radius a and wavelength λ.

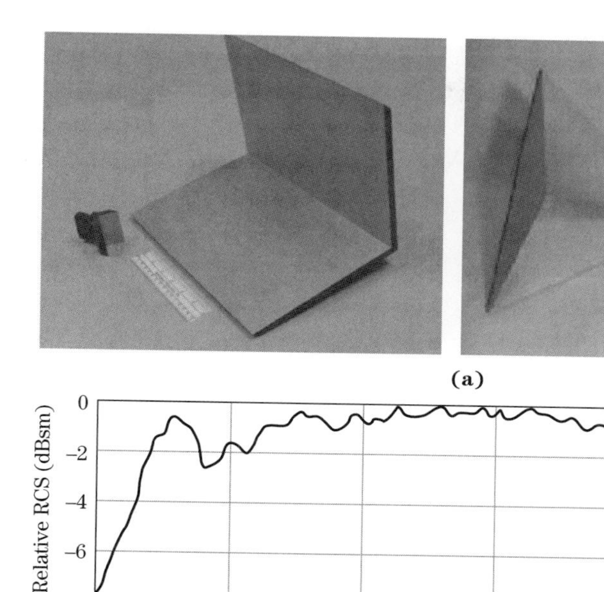

FIGURE 7-3 ■ Corner reflectors. (a) Dihedrals. (b) Trihedrals. (Courtesy of Professor Nadav Levanon, Tel-Aviv University.)

reflectors are meant to be illuminated along their axis of symmetry. For example, the RCS of a square-sided trihedral of dimension a on a side is $12\pi a^4/\lambda^2$ when viewed along the axis of symmetry [1]. The RCS of both corner reflectors (provided the radar aspect angle is near the axis of symmetry) and spherical targets can be characterized by a single value for a given wavelength and polarization. Thus, these simple targets have deterministic, not statistical, RCS models. Nonetheless, the radar received signal is the sum of these deterministic echoes and random interference and still requires a statistical characterization.

7.2.2 Aspect Angle and Frequency Dependence of RCS

The radar cross section of real targets cannot be effectively modeled as a simple constant. In general, RCS is a complex function of aspect angle, frequency, and polarization even for relatively simple targets. Furthermore, received power at the radar (target echo only, not including noise or other interference) is proportional to target RCS as seen in the radar range equation. Thus, RCS fluctuations result in received target power fluctuations.

A simple example of frequency and aspect dependence of the RCS of a "complex" target is the two-scatterer "dumbbell" target shown in Figure 7-4. If the nominal range, R, is much greater than the separation, D, the range to the two scatterers is approximately

$$R_{1,2} \approx R \pm \frac{D}{2}\sin\theta \tag{7.1}$$

If the transmitted signal is $ae^{j2\pi ft}$, the echo from each scatterer will be proportional to $ae^{j2\pi f(t-2R_{1,2}/c)}$. The voltage, $y(t)$, of the composite echo is therefore

$$\begin{aligned} y(t) &\propto ae^{j2\pi f(t-2R_1/c)} + ae^{j2\pi f(t-2R_2/c)} \\ &= ae^{j2\pi f(t-2R/c)}\left[e^{-j2\pi fD\sin\theta/c} + e^{+j2\pi fD\sin\theta/c}\right] \\ &= 2ae^{j2\pi f(t-2R/c)}\cos\left(\pi fD\sin\theta/c\right) \end{aligned} \tag{7.2}$$

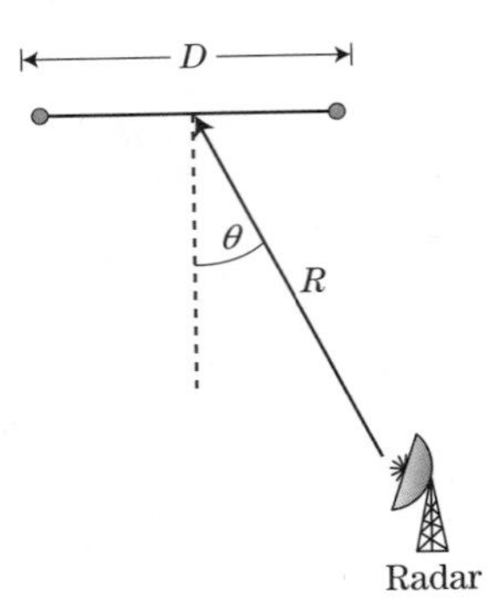

FIGURE 7-4 ■ Geometry for determining relative RCS of a dumbbell target.

RCS is proportional to the power of the composite echo, as shown in Section 6.5.1. Taking the squared magnitude of (7.2) leads to the result

$$\sigma \propto 4a^2 \left|\cos\left(2\pi f D \sin\theta / c\right)\right|^2 = 4a^2 \left|\cos\left(2\pi D \sin\theta / \lambda\right)\right|^2. \tag{7.3}$$

Equation (7.3) shows that the RCS is a function of both radar frequency and aspect angle. The larger the scatterer separation in terms of wavelengths, the more rapidly the RCS varies with angle or frequency. An exact calculation of the variation in RCS of the dumbbell target is plotted in Figure 7-5 for the case $D = 5\lambda$ and $R = 10{,}000D$. The plot has been normalized so that the maximum value corresponds to 0 dB. Notice the multi-lobed structure as the varying path lengths traversed by the echoes from the two scatterers cause them to shift between constructive and destructive interference. Also note that the maxima at aspect angles of 90° and 270° (the two "end-fire" cases) are the broadest, whereas the maxima at the two "broadside" cases of 0° and 180° are the narrowest. Figure 7-6 plots the same data in a more traditional polar format.

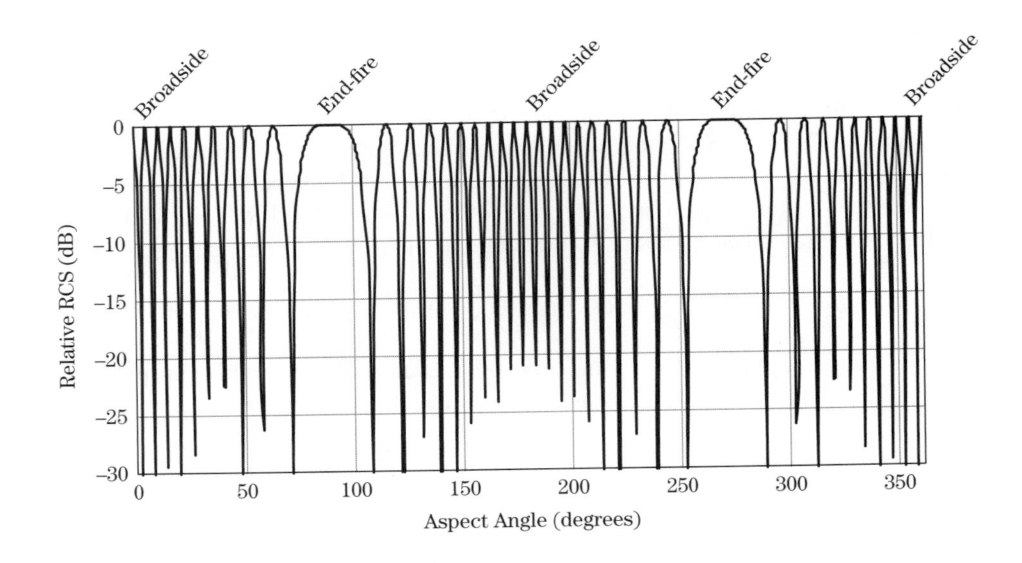

FIGURE 7-5 ■ Relative radar cross section of the dumbbell target of Figure 7-4 when $D = 5\lambda$ and $R = 10{,}000\,D$.

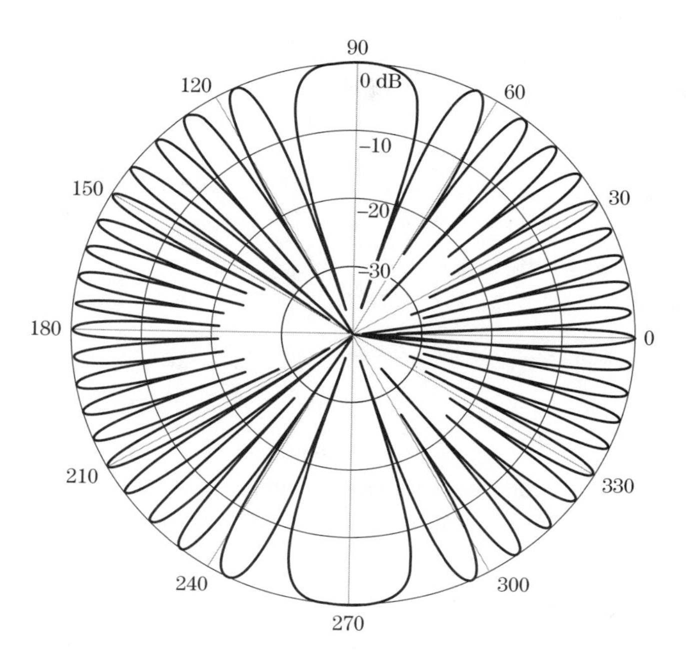

FIGURE 7-6 ■ Polar plot of the data of Figure 7-5.

7.3 | RADAR CROSS SECTION OF COMPLEX TARGETS

RCS variations become very complicated for complex targets having many scatterers of varying individual RCS. The relative RCS of a target with multiple scatterers can be computed using a generalization of (7.2). Suppose there are N scatterers, each with its own RCS σ_i, located at ranges R_i from the radar. The complex voltage of the echo will be, to within a proportionality constant,

$$\begin{aligned} y(t) &= \sum_{i=1}^{N} \sqrt{\sigma_i}\, e^{j2\pi f(t-2R_i/c)} \\ &= e^{j2\pi ft} \sum_{i=1}^{N} \sqrt{\sigma_i}\, e^{-j4\pi f R_i/c} \\ &= e^{j2\pi ft} \sum_{i=1}^{N} \sqrt{\sigma_i}\, e^{-j4\pi R_i/\lambda} \end{aligned} \tag{7.4}$$

The RCS σ is proportional to $|y|^2$. Define the echo *amplitude* as

$$\varsigma \equiv |y| = \left| \sum_{i=1}^{N} \sqrt{\sigma_i}\, e^{-j4\pi R_i/\lambda} \right| \tag{7.5}$$

and the target RCS as

$$\sigma = \varsigma^2 = \left| \sum_{i=1}^{N} \sqrt{\sigma_i}\, e^{-j4\pi R_i/\lambda} \right|^2 \tag{7.6}$$

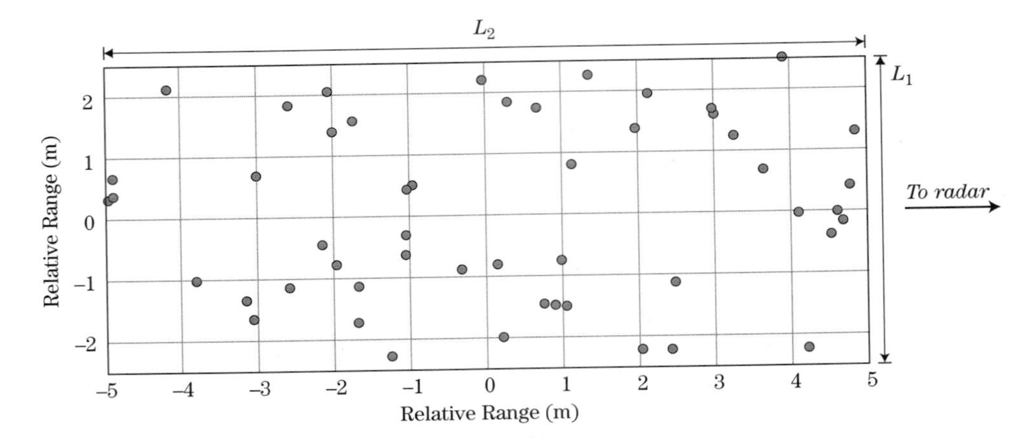

FIGURE 7-7 ■ Random distribution of 50 scatterers used to obtain Figure 7-8. See text for additional details.

While the previous equations implicitly assume a continuous wave (CW) radar, they also apply to a pulsed system so long as all the scatterers are contained in a single resolution cell so that all contribute to the receiver output simultaneously at a time corresponding to the nominal range to the target.

Figure 7-7 shows a "target" consisting of 50 point scatterers randomly distributed within a rectangle 5 meters wide and 10 meters long. The RCS of each individual point scatterer is a constant, $\sigma_i = 1.0$. Figure 7-8 shows the relative RCS, computed at 1 degree increments using (7.6), which results when this target is viewed 10 km from its center at a frequency of 10 GHz. Zero degrees corresponds to the radar being located to the right of the target, as indicated by the arrow. The dynamic range of the received power and thus the RCS is similar to that of the simple dumbbell target, but the lobing structure is much more complicated. A target whose RCS varies strongly with aspect angle or frequency is called a *fluctuating target*.

This complicated behavior observed for even moderately complex targets means that calculations of detection performance would be very sensitive to radar-target aspect angle because of the large variations of RCS and therefore of signal-to-noise ratio (SNR). Such calculations would be both complicated and of limited utility, since it would be difficult to know either the target RCS pattern or the radar-target aspect angle accurately enough to use such detailed data. It is much more practical and ultimately more useful to develop

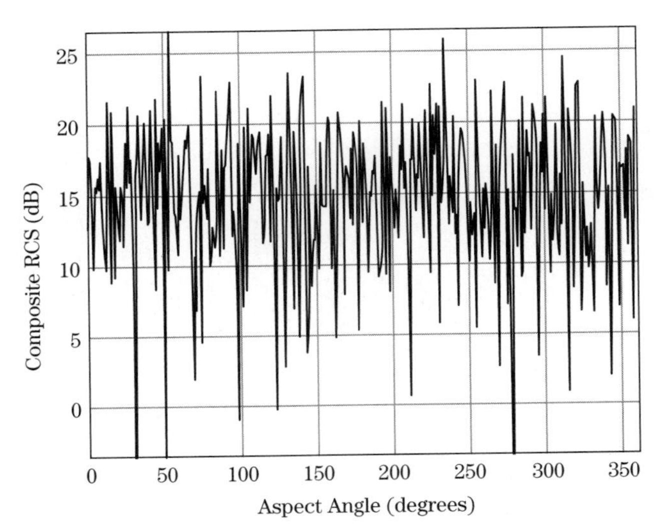

FIGURE 7-8 ■ Relative RCS of the complex target of Figure 7-7 at a range of 10 km and radar frequency of 10 GHz.

"average" performance based on a simpler model of the target RCS. This argument leads to the use of a statistical description for radar cross section [1–3] in which the composite RCS σ of the scatterers within a single resolution cell is considered to be a random variable with a specified *probability density function* (PDF). The radar range equation, or RCS modeling as described in Chapter 6, is used to estimate the mean RCS from received echo power measurements, and one of a variety of standard PDFs is used to describe the statistical variations of the RCS.

It is important to realize that using a statistical model for RCS does not imply that the actual RCS of the target is random. If it was possible to describe the target surface shape and materials in enough detail, and in addition to identify the radar-target aspect angle accurately enough, then the RCS could in principle be computed accurately using the techniques of the previous chapter. Statistical models are used because RCS behavior, even for relatively simple targets like the previous examples, is extremely complex and very sensitive to aspect angle. Combined with the uncertainty of aspect angle, particularly before a target is detected, it is much more practical to use a statistical model as a simple way to capture the complexity of the target RCS.

7.4 | STATISTICAL CHARACTERISTICS OF THE RCS OF COMPLEX TARGETS

7.4.1 RCS Distributions

Consider a target consisting of a large number of individual scatterers (similar to that of Figure 7-7) randomly distributed in space and each with approximately the same individual RCS. The phase of the echoes from the various scatterers can then be assumed to be a random variable distributed uniformly on $(0,2\pi)$. Under these circumstances, the central limit theorem guarantees that the real and imaginary parts of the composite echo can each be assumed to be independent, zero mean Gaussian random variables with the same variance, say α^2 [1,2]. In this case, the squared-magnitude σ has an exponential PDF [4]:

$$p(\sigma) = \begin{cases} \dfrac{1}{\bar{\sigma}} \exp\left[\dfrac{-\sigma}{\bar{\sigma}}\right], & \sigma \geq 0 \\ 0, & \sigma < 0 \end{cases} \tag{7.7}$$

where $\bar{\sigma} = 2\alpha^2$ is the mean value of the RCS σ.[1] The amplitude voltage, $\varsigma = \sqrt{\sigma}$ (more appropriate to a radar using a linear, rather than square law, detector), has a Rayleigh PDF:

$$p(\varsigma) = \begin{cases} \dfrac{2\varsigma}{\bar{\sigma}} \exp\left[\dfrac{-\varsigma^2}{\bar{\sigma}}\right], & \varsigma \geq 0 \\ 0, & \varsigma < 0 \end{cases} \tag{7.8}$$

While the Rayleigh/exponential model is strictly accurate only in the limit of a very large number of scatterers, in practice it can be a good model for a target having as few as 10 or 20 significant scatterers. Figure 7-9 compares a histogram of the RCS values from Figure 7-8 (after conversion from the decibel scale back to a linear scale) to an exponential curve of the form (7.7) having the same mean $\bar{\sigma}$. Even though only 50 scatterers are used,

[1] Note that this PDF applies to the RCS in linear scale units of m^2, not to decibel scale units of dBsm.

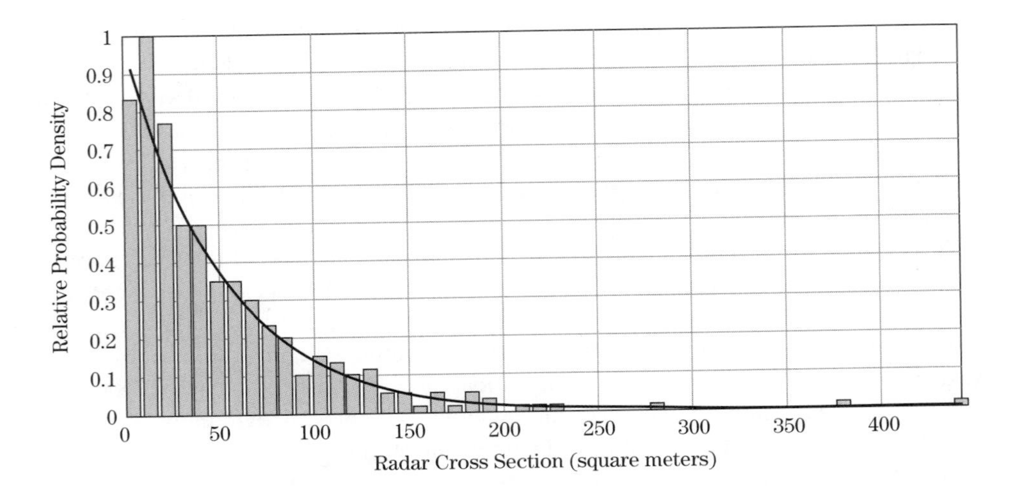

FIGURE 7-9 ■ Histogram of linear-scale RCS data of Figure 7-8.

the fit of the total RCS histogram to the exponential distribution is quite good. This same effect is observed when the randomly distributed scatterers also have random individual cross sections drawn from the same Gaussian distribution, a somewhat more general and plausible situation than the approximately fixed-RCS case.

Many radar targets are not well modeled as an ensemble of equal-strength scatterers, so a variety of other PDFs have been advocated and used for modeling target RCS. Table 7-1 summarizes several of the more common models. The mean value $\bar{\sigma}$ of RCS is given for each case in which the PDF is not written explicitly in terms of $\bar{\sigma}$. The variance $\mathrm{var}(\sigma)$ is also given for each case. The naming terminology can be confusing, because in some cases the name traditionally applied to the distribution of RCS σ is actually that of the density function of the corresponding amplitude, ζ. For example, the exponential RCS distribution of (7.7) is sometimes referred to as the Rayleigh model, because the amplitude (square root of the RCS) follows the Rayleigh PDF of equation (7.8).

Figure 7-10 is the histogram of an RCS versus aspect angle data set for a 20-scatterer target, but with an additional dominant scatterer added at a random location. The non-central chi-square distribution with two degrees of freedom is the exact PDF for this case but is considered somewhat difficult to work with because the expression for the PDF contains a Bessel function. The fourth-degree chi-square is a more analytically tractable approximation that is commonly used. This PDF is given by

$$p(\sigma) = \begin{cases} \dfrac{4\sigma}{\bar{\sigma}^2} \exp\left[\dfrac{-2\sigma}{\bar{\sigma}}\right], & \sigma \geq 0 \\ 0, & \sigma < 0 \end{cases} \tag{7.9}$$

The first two central moments (mean and variance) of the two PDFs are equal when the RCS of the dominant scatterer is $1 + \sqrt{2} \approx 2.414$ times that of the sum of the radar cross sections of the small scatterers. The data in Figure 7-10 were computed using this ratio. The fourth-degree chi-square PDF with the same mean RCS overlaid on the histogram shows that the observed RCS data are a good fit to the theoretical model provided the ratio of dominant to small scatterers is correct.

The exponential and fourth-degree chi-square PDFs are the two most traditional models for target RCS. They are the PDFs used as part of the common *Swerling models* of RCS that will be introduced in Section 7.5.1. Both are special cases of a chi-square density of

TABLE 7-1 ■ Common Statistical Models for Radar Cross Section

Model Name	PDF for RCS σ	Comment
Nonfluctuating, Marcum, Swerling 0, or Swerling 5	$p(\sigma) = \delta_D(\sigma - \bar{\sigma})$ $\text{var}(\sigma) = 0$	Constant echo power, e.g. calibration sphere or perfectly stationary reflector with no radar or target motion.
Exponential (chi-square of degree 2)	$p(\sigma) = \frac{1}{\bar{\sigma}} \exp\left[\frac{-\sigma}{\bar{\sigma}}\right]$ $\text{var}(\sigma) = \bar{\sigma}^2$	Many scatterers, randomly distributed, none dominant. Used in Swerling case 1 and 2 models.
Chi-square of degree 4	$p(\sigma) = \frac{4\sigma}{\bar{\sigma}^2} \exp\left[\frac{-2\sigma}{\bar{\sigma}}\right]$ $\text{var}(\sigma) = \bar{\sigma}^2/2$	Approximation to case of many small scatterers + one dominant, with RCS of dominant equal to $1+\sqrt{2}$ times the sum of RCS of others. Used in Swerling case 3 and 4 models.
Chi-square of degree $2m$, Weinstock	$p(\sigma) = \frac{m}{\Gamma(m)\bar{\sigma}} \left[\frac{m\sigma}{\bar{\sigma}}\right]^{m-1} \exp\left[\frac{-m\sigma}{\bar{\sigma}}\right]$ $\text{var}(\sigma) = \bar{\sigma}^2/m$	Generalization of the two preceding cases. Weinstock cases correspond to $0.6 \le 2m \le 4$. Higher degrees correspond to presence of a more dominant single scatterer.
Weibull	$p(\sigma) = CB\sigma^{C-1} \exp\left[-B\sigma^C\right]$ $\bar{\sigma} = \Gamma\left(1+1/C\right) B^{-1/C}$ $\text{var}(\sigma) = B^{-2/C}\left[\Gamma\left(1+2/C\right) - \Gamma^2\left(1+1/C\right)\right]$	Empirical fit to many measured target and clutter distributions. Can have longer "tail" than previous cases.
Log-normal	$p(\sigma) = \frac{1}{\sqrt{2\pi}\, s\sigma} \exp\left[-\ln^2\left(\sigma/\sigma_m\right)/2s^2\right]$ $\bar{\sigma} = \sigma_m \exp\left(s^2/2\right)$ $\text{var}(\sigma) = \sigma_m^2 \exp\left(s^2\right)\left[\exp\left(s^2\right) - 1\right]$	Empirical fit to many measured target and clutter distributions. "Tail" is longest of previous cases. σ_m is the median value of σ.

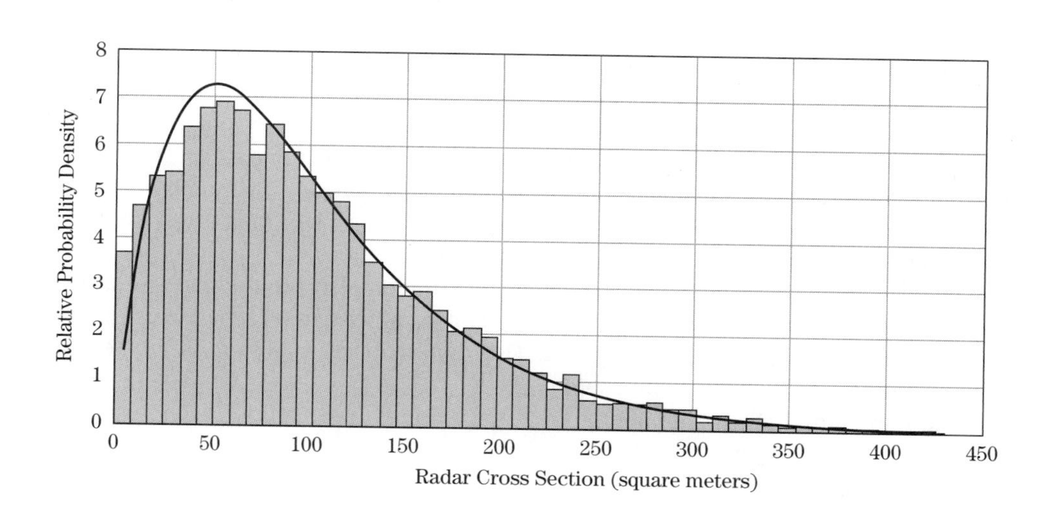

FIGURE 7-10 ■ Comparison of a fourth-degree chi-square PDF and the histogram of linear-scale RCS data for one dominant scatterer with many small scatterers. See text for details.

degree $2m$, where m is sometimes called the "duo-degree" of the density. This PDF is

$$p(\sigma) = \begin{cases} \dfrac{m}{\Gamma(m)\bar{\sigma}} \left[\dfrac{m\sigma}{\bar{\sigma}}\right]^{m-1} \exp\left[\dfrac{-m\sigma}{\bar{\sigma}}\right], & \sigma \geq 0 \\ 0, & \sigma < 0 \end{cases} \tag{7.10}$$

The exponential corresponds to $m = 1$, while the fourth-degree chi-square corresponds to $m = 2$.

The definition of the fourth-degree chi-square PDF as shown in equation (7.9) and Table 7-1 is common in radar but is otherwise somewhat nonstandard terminology. A chi-square of degree N is usually considered in the more general statistical literature to be a special case of the *gamma PDF*, which has two parameters α and β:

$$p(\sigma; \alpha, \beta) = \begin{cases} \dfrac{\sigma^{\alpha-1}}{\beta^{\alpha}\Gamma(\alpha)} e^{-\sigma/\beta}, & \sigma \geq 0 \\ 0, & \sigma < 0 \end{cases}, \tag{7.11}$$

Specifically, the conventional fourth-degree chi-square is obtained when $N = 4$, so $\alpha = 2$ and $\beta = 2$. However, the PDF in equation (7.9) is obtained from (7.11) with $\alpha = 2$ and $\beta = \bar{\sigma}/2$, which implies $N = 4$ but does not have $\beta = 2$. The more general form of the so-called shape parameter, β, is necessary to allow the mean of the distribution to be set to any desired value.

The chi-square of degree $2m$ in equation (7.10) is obtained by letting $\alpha = m$ and $\beta = \bar{\sigma}/m$ in the gamma PDF. Use of the gamma PDF to generalize the original chi-square models and to represent a wider range of target behavior was first proposed by Swerling in 1966 (reprinted in [6]). Another example of this generalization is the Weinstock models [7], which are chi-square PDFs with degrees between 0.6 and 4 that provide the ability to fit a range of dominant-to-small scatterer ratios from about 0.03 to $1 + \sqrt{2}$. For targets having stronger dominant scatterers, the non-central chi-square PDF is a better fit.

The Rayleigh/exponential function is an example of a *one-parameter PDF*; specifying only one parameter, the mean $\bar{\sigma}$, completely specifies the PDF. In particular, the variance of the PDF is related directly to its mean, as shown in Table 7-1. The chi-square models of a given degree are also one-parameter PDFs. In contrast, the Weibull and log-normal are examples of *two-parameter PDFs*. The shape of the PDF is determined by two independent parameters (B and C for the Weibull; σ_m and s for the log-normal). Consequently, their mean and variance can be adjusted independently so that two-parameter distributions can adequately fit a wider range of measured data distributions. For example, Figure 7-11 shows three variants of the log-normal distribution. All have the same mean of 1.0, but the variances are 0.2, 0.5, and 1.0. As the variance increases, the "tail" of the PDF lengthens; this is more readily evident in Figure 7-11b, which plots the PDFs on a log scale. Longer PDF tails represent a greater probability of higher RCS echoes.

Estimating the mean of a one-parameter distribution also provides the information needed for an estimate of the variance. For the two-parameter case, separate estimates of the mean and variance must be computed. This distinction is important in the design of automatic detection algorithms in Chapter 16.

The choice of PDF for modeling RCS fluctuations directly affects the estimation of detection performance, as will be seen in Chapter 15. Figure 7-12a compares the Rayleigh, fourth-degree chi-square, Weibull, and log-normal density functions when all have the same RCS variance of 0.5 and all except the exponential have a mean of 1.0.

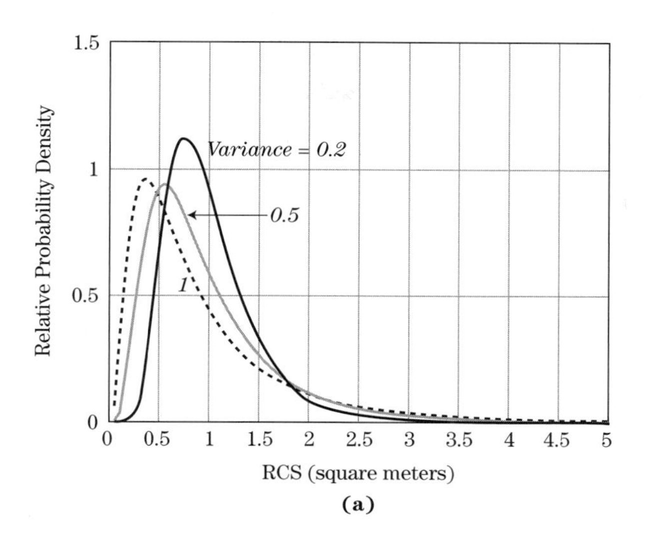

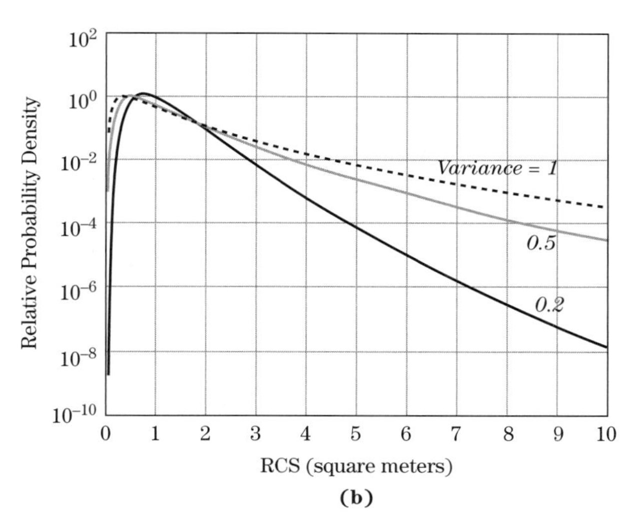

FIGURE 7-11 ■ Log-normal PDFs with mean = 1 and variances = 0.2, 0.5, and 1. (a) Linear scale. (b) Log scale, showing detail of the PDF "tails."

(Because it is a one-parameter distribution with its mean and variance always equal, the exponential distribution in the figure has a mean of 0.5). Figure 7-12b repeats the same data on a semilogarithmic scale so that the behavior of the PDF "tails" is more evident. The exponential PDF is somewhat unique, since it doesn't have a distinct peak near the RCS mean as the other three densities do. Each of the others does have a distinct peak, making each suitable for distributions with one or a few dominant scatterers. For the parameters shown, the Weibull has a broader peak and more rapidly decaying tail than does the chi-square of degree 4, while the log-normal has both the narrowest peak and the longest tail of any of the distributions shown. However, the sharpness of the peak and the length of the PDF tail can be varied for both the Weibull and log-normal by adjusting their variance.

Not all targets exhibit RCS fluctuations. Targets whose RCS is modeled as a constant, independent of aspect angle are called *nonfluctuating* targets; they are also sometimes called *Marcum* targets. Examples include the conducting sphere and, so long as the aspect angle doesn't vary too much from its axis of symmetry, the various corner reflectors.

Most radar analysis and measurement programs emphasize RCS measurements, since received power is proportional to RCS. Sometimes the corresponding amplitude voltage,

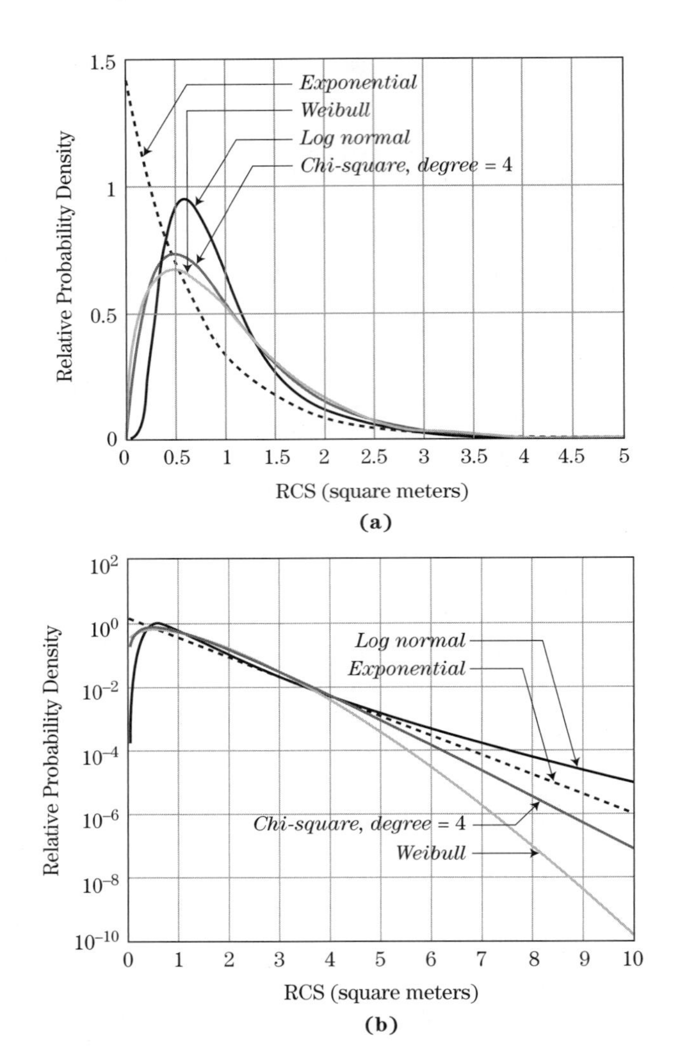

FIGURE 7-12 ■ Comparison of five models for the probability density function of radar cross section with the same mean (except for the exponential) and variance. See text for additional details. (a) Linear scale. (b) Log scale, showing detail of the PDF "tails."

ς, is of interest, particularly for use in simulations where equation (7.4) is used explicitly to model the composite echo from a multiple scatterer target. The probability density function for the amplitude is required to properly model the probabilistic variations of the complex sum. The PDF of ς is easily derived from the PDF of σ using basic results of random variables [4]. Specifically, the PDF of ς is related to that of the RCS σ by

$$p_\varsigma(\varsigma) = 2\varsigma p_\sigma(\varsigma^2) \tag{7.12}$$

Equation (7.12) can be used to write the voltage PDFs by inspection from Table 7-1. The results, given in Table 7-2, are expressed in terms of the parameters of the corresponding RCS distribution from Table 7-1. Note that the nonfluctuating, Weibull, and log-normal RCS distributions all result in distributions of the same type (but with one or more parameters changed) for the voltage. Again, note that the voltage in the Rayleigh/exponential case is Rayleigh distributed, explaining the name.

7.4.2 RCS Correlation Properties

As has been seen, the RCS of a complex target varies with both transmitted frequency and aspect angle. Another important characteristic of a target's signature is the *correlation*

TABLE 7-2 ■ Voltage Distributions Corresponding to Common Statistical Models of Radar Cross Section

RCS Model Name	PDF for Voltage ς	Description of Voltage Model
Nonfluctuating, Marcum, Swerling 0, or Swerling 5	$p(\varsigma) = \delta_D\left(\varsigma - \sqrt{\bar{\sigma}}\right)$ $\bar{\varsigma} = \sqrt{\bar{\sigma}}, \quad \text{var}(\sigma) = 0$	Also nonfluctuating model
Exponential (chi-square of degree 2)	$p(\varsigma) = \dfrac{2\varsigma}{\bar{\sigma}} \exp\left[\dfrac{-\varsigma^2}{\bar{\sigma}}\right]$ $\bar{\varsigma} = \dfrac{1}{2}\sqrt{\pi\bar{\sigma}}, \quad \text{var}(\varsigma) = \bar{\sigma}\left(1 - \pi/4\right)$	Rayleigh distribution
Chi-square of degree 4	$p(\varsigma) = \dfrac{8\varsigma^3}{\bar{\sigma}^2} \exp\left[\dfrac{-2\varsigma^2}{\bar{\sigma}}\right]$ $\bar{\varsigma} = \dfrac{3}{4}\sqrt{\dfrac{\pi\bar{\sigma}}{2}}, \quad \text{var}(\varsigma) = \left(1 - \dfrac{9}{32}\pi\right)\bar{\sigma}$	Chi-distribution of degree 4
Chi-square of degree $2m$, Weinstock	$p(\varsigma) = \dfrac{2\varsigma m}{\Gamma(m)\bar{\sigma}}\left(\dfrac{m\varsigma^2}{\bar{\sigma}}\right)^{m-1} \exp\left[-\dfrac{m\varsigma^2}{\bar{\sigma}}\right]$ $\bar{\varsigma} = \sqrt{\dfrac{\bar{\sigma}}{m}}, \quad \text{var}(\varsigma) = \bar{\sigma}\left[1 - \dfrac{1}{m}\left(\Gamma(m+0.5)/\Gamma(m)\right)^2\right]$	Chi-distribution of degree $2m$
Weibull	$p(\varsigma) = 2CB\varsigma^{2C-1}\exp\left[-B\varsigma^{2C}\right]$ $\bar{\varsigma} = \Gamma\left(1 + 1/2C\right)B^{-1/2C}$ $\text{var}(\varsigma) = \left[\Gamma\left(1+1/C\right) - \Gamma^2\left(1 + 1/2C\right)\right]B^{-1/2C}$	Also Weibull, one parameter changed ($C \rightarrow 2C$).
Log-normal	$p(\varsigma) = \dfrac{2}{\sqrt{2\pi}\,s\varsigma}\exp\left[-2\ln^2\left(\varsigma/\sqrt{\sigma_m}\right)^2/s^2\right]$ $\bar{\varsigma} = \sqrt{\sigma_m}\exp\left(s^2/8\right)$ $\text{var}(\varsigma) = \sigma_m\exp\left(s^2/4\right)\left[\exp\left(s^2/4\right) - 1\right]$	Also log-normal, both parameters changed ($s \rightarrow s/2$, $\sigma_m \rightarrow \sqrt{\sigma_m}$).

"length" in time, frequency, and angle. This is the change in time, frequency, or angle required to cause the echo amplitude to decorrelate to a specified degree. If a rigid target such as a building is illuminated with a series of identical radar pulses and there is no motion between the radar and target, each pulse will result in the same received complex voltage ς (ignoring receiver noise). If there is motion between the two, however, the relative path length between the radar and the various scatterers comprising the target will change, causing the composite echo amplitude to fluctuate, similar to the fluctuations shown in Figure 7-8. Changing the radar wavelength will also cause the relative phase of the contributing scatterers to change, producing the same effect, as will target vibration in some instances. Thus, for rigid targets, decorrelation of the RCS is induced by changes in range, aspect angle, and radar frequency.

Although the behavior of real targets can be quite complex, a sense of the change in frequency and angle required to decorrelate a target or clutter patch can be obtained by the following simple argument. Consider an idealized target consisting of a uniform line array of point scatterers tilted at an angle θ with respect to the line of sight to the radar and separated by Δx from one another, as shown in Figure 7-13. For simplicity, assume an odd number $2M + 1$ of scatterers indexed from $-M$ to $+M$ as shown. If the nominal

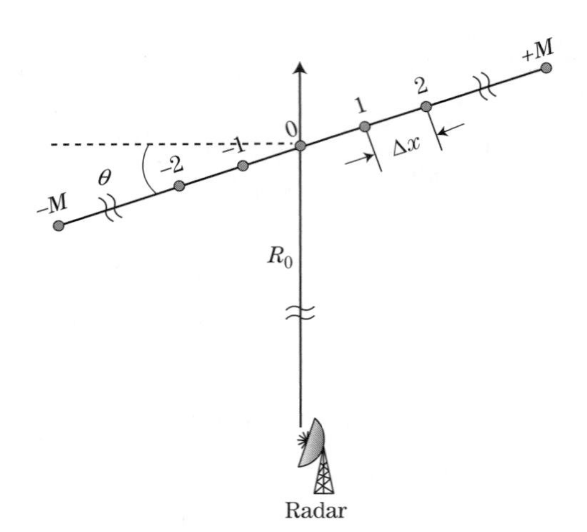

FIGURE 7-13 ■ Geometry for calculation of RCS correlation length in frequency and aspect angle.

distance to the radar, R_0, is much larger than the target extent (i.e., $R_0 \gg (2M+1)\Delta x$), then the range to the n-th scatterer is approximately

$$R_n \approx R_0 + n\,\Delta x \sin\theta \tag{7.13}$$

If the target is illuminated with the waveform $ae^{j\omega t}$, the received signal is

$$\begin{aligned}\bar{y}(t) &= \sum_{n=-M}^{M} ae^{j\omega(t-2R_n/c)} \\ &= ae^{j\omega(t-2R_0/c)} \sum_{n=-M}^{M} e^{-j4\pi n\,\Delta x\,\sin\theta\, f/c}\end{aligned} \tag{7.14}$$

To simplify the notation, define

$$\begin{aligned} z &= f\sin\theta \\ \alpha &= 4\pi\,\Delta x/c \end{aligned} \tag{7.15}$$

The new variable z includes both aspect angle and radar frequency. The signal $\bar{y}(t)$ can now be considered as a function $\bar{y}(t;z)$ of both t and z, and its autocorrelation can be computed with respect to the variable z. It can be shown that the deterministic autocorrelation function is [5]

$$s(\Delta z) = \frac{2\pi a^2}{\alpha}\frac{\sin\left[\alpha\,(2M+1)\,\Delta z/2\right]}{\sin\left[\alpha\,\Delta z/2\right]} \tag{7.16}$$

where Δz is the correlation lag in the z dimension.

One criterion for "decorrelation" is to choose the value of Δz corresponding to the first zero of the correlation function. This occurs when the argument of the numerator equals π. Using (7.15) and defining the target length $L = (2M+1)\Delta x$,

$$\Delta z = \frac{c}{2L} \tag{7.17}$$

Recall that $z = f\sin\theta$. To determine the decorrelation angle, fix the transmitted frequency f so that $\Delta z = f(\Delta\sin\theta)$. Assuming θ is small (i.e., the radar is near broadside),

$\Delta \sin\theta \approx \Delta\theta$. Equation (7.17) then becomes the desired result for the angle required to decorrelate the echo amplitude:

$$\Delta\theta = \frac{c}{2Lf} = \frac{\lambda}{2L} \tag{7.18}$$

For rigid targets, equation (7.18) estimates the amount of aspect angle rotation required to decorrelate the target echoes. Aspect angle changes will occur because of motion between the radar and the target. For instance, an airliner flying past a commercial airport presents a constantly changing aspect to the airport surveillance radar as it flies through the airspace. The amount of time over which the aircraft echoes decorrelate is simply the amount of time it takes to change the aspect angle by $\Delta\theta$ radians, which depends on the details of the relative geometry and velocity of the radar and target.

The frequency step required to decorrelate the target is obtained by fixing the aspect angle θ so that $\Delta z = \Delta f \sin\theta$. The result is

$$\Delta f = \frac{c}{2L\sin\theta} \tag{7.19}$$

This is minimum when $\theta = 90°$. Note that $L\sin\theta$ is the length of the target line array projected along the radar boresight. The value of Δf required to decorrelate the target is usually modest. For example, if the projected target size $L\sin\theta = 10$ m, $\Delta f = 15$ MHz.

This result can be generalized slightly to apply it to an $L_1 \times L_2$ meter target such as the 5 m $\times$ 10 m target used previously. The size of this rectangular target as projected along the line of sight (LOS) when the radar is at an aspect angle of θ radians measured from horizontal is the sum of the projection of each side of the target along the LOS, which is $L_1|\sin\theta| + L_2|\cos\theta|$. Using this in equation (7.19) gives the frequency decorrelation step size as

$$\Delta f = \frac{c}{2\left(L_1|\sin\theta| + L_2|\cos\theta|\right)} \tag{7.20}$$

This result assumes that the target lies entirely within a single range bin. If the range resolution, ΔR, is less than the target extent along the LOS, the required step size is $c/(2\Delta R)$.

As an example, consider a target the size of an automobile, about 5 m long. At L-band (1 GHz), the target signature can be expected to decorrelate in $(3 \times 10^8)/(2 \times 5 \times 10^9) =$ 30 mrad of aspect angle rotation, about 1.7°, while at W-band (95 GHz), this is reduced to only 0.018°. Equation (7.19) predicts the frequency step required for decorrelation with an aspect angle of 45° is 42.4 MHz. This result does not depend on the transmitted frequency.

As another example, the autocorrelation in angle of the simulated data of Figure 7-8 is shown in Figure 7-14, using only the data for aspect angles over a range $\pm 3°$. Each of the two autocorrelation functions is the average of the data from 20 different random targets, each having 20 randomly placed scatterers in a 5 m $\times$ 10 m box, similar to the simulation previously described. The black curve is the autocorrelation of the data around a nominal boresight orthogonal to the 5 m side of the target, while the gray data are the autocorrelation of the data viewed from the 10 m side. These look angles correspond to viewing the target nominally from the right and from the top in Figure 7-7. Viewed from the right, $L = 5$ m orthogonal to the LOS and $f = 10$ GHz gives an expected decorrelation interval in angle of 0.17°; viewed from the top, $L = 10$ m and the expected decorrelation angle is 0.086°. These expected decorrelation intervals are marked by the vertical dashed lines in Figure 7-14. This figure shows that in both cases, the correlation function drops to zero at the expected amount of change in the aspect angle.

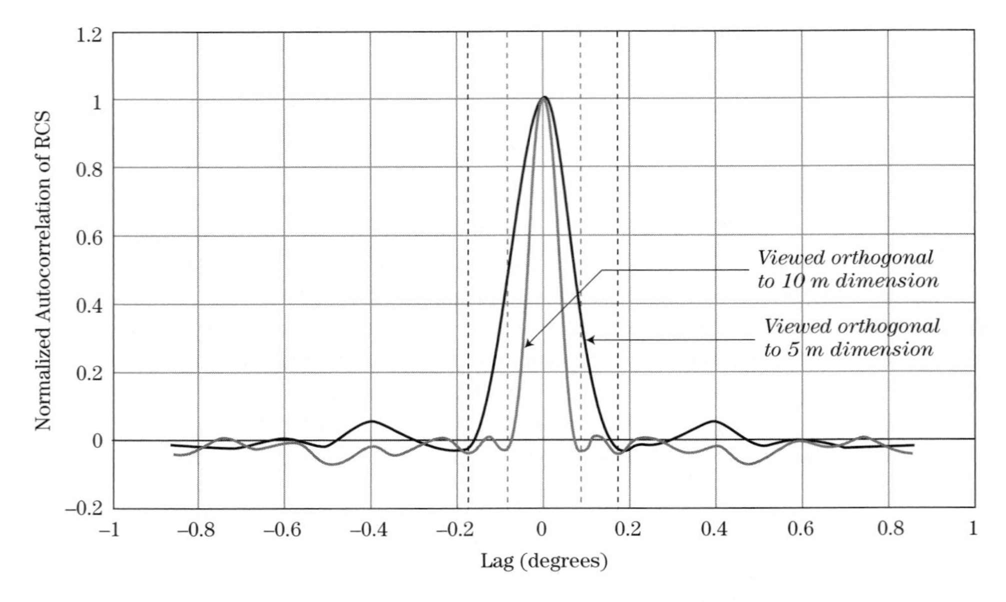

FIGURE 7-14 ■ Decorrelation in angle of RCS of target from Figure 7-8. See text for details.

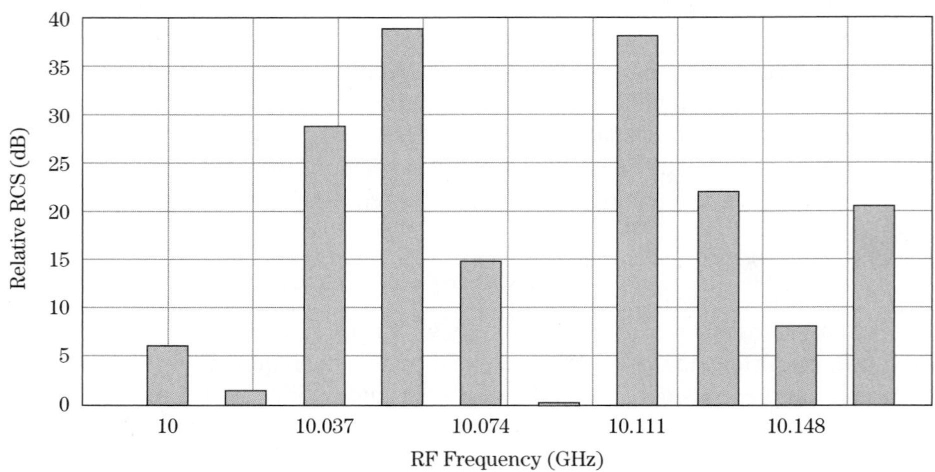

FIGURE 7-15 ■ Variation in RCS due to frequency agility for a constant viewing angle. See text for details.

Some systems deliberately change the radar frequency from pulse to pulse to decorrelate the target returns, a process called *frequency agility* [8]. As will be seen in Chapter 15, frequency agility can significantly improve the probability of detection for some systems and targets.

Figure 7-15 illustrates the ability of frequency agility to force RCS variations. A 20-scatterer, 5 m × 10 m random target similar to those previously described was observed from a fixed aspect angle of 20°. If the same radiofrequency (RF) was used for each pulse, the RCS would be exactly the same on each pulse. However, in this case the RF was increased by 18.48 MHz (calculated from equation (7.20)) from one pulse to the next, starting at 10.0 GHz. The resulting relative RCS measurements vary by 38 dB, a factor of about 6,300.

The results of equations (7.18) and (7.19) are based on a highly simplified target model and an assumption about what constitutes decorrelation. For example, defining decorrelation to be the point at which the correlation function first drops to 1/2 or $1/e$ of its peak results in a smaller estimate of the required change in angle or frequency to decorrelate the target. Also, many radars operate on the magnitude squared of the echo amplitude rather than on the magnitude, as has been assumed in this derivation. A square law detector produces a correlation function proportional to the square of equation (7.16) [9]. The first zero therefore occurs at the same value of Δz, and the previous conclusions still apply. However, if a different definition of decorrelation is used (such as the 50% decorrelation point), the required change in Δz is less for the square law than for the linear detector.

7.5 TARGET FLUCTUATION MODELS

7.5.1 Swerling Models

An extensive body of radar detection theory results have been built up using the four *Swerling models* of target RCS fluctuation [2,3,10-12]. Swerling models are intended to address the common problem of making a detection decision based on a block of N envelope-detected echo samples from a given range-angle or range-Doppler resolution cell. To see one reason why detection based on a block of N samples (instead of just one) is of interest, imagine a ground-based surveillance radar used to detect aircraft. Suppose the radar antenna rotates at a constant angular velocity Ω radians/sec with a 3 dB azimuth beamwidth of θ_3 radians and a pulse repetition frequency (PRF) of PRF Hz. The geometry is shown in Figure 7-16. Although some echo energy from the target is received in the appropriate range bin on every pulse through the antenna sidelobes, significant returns are received only when the target is in the antenna mainlobe. Every complete 360° sweep of the antenna results in a new set of $N = (\theta_3/\Omega) \times PRF$ mainbeam samples, all potentially containing significant target echoes. It is desirable to use all of this target data for detection, not just a single pulse.

This is not the only way a block of N-related pulse echoes can arise. Many modern systems are designed to transmit coherent bursts of pulses at a constant pulse repetition frequency, often with the antenna staring in a fixed direction. The time interval

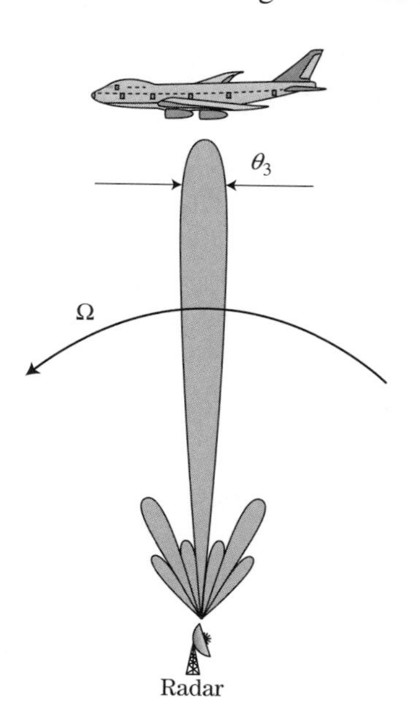

FIGURE 7-16 ■ Rotating antenna rationale for Swerling model decorrelation assumptions. Echoes from a given target are collected in blocks. Each rotation of the antenna results in a new block, and each block contains multiple pulse returns.

$N/PRF = N \times PRI$ required for this measurement is called a *coherent processing interval* (CPI).[2] The system may then repeat the entire measurement in the same or a different look direction, may change the PRF to make a related measurement, or may make any of a number of other changes in collecting the next CPI of data. This N-pulse burst is a common waveform well suited to Doppler measurements, adaptive interference suppression, and imaging applications. It will arise frequently in subsequent chapters. When multiple samples from the same range bin are combined coherently, the result is treated as a single sample for detection purposes. For example, if Doppler processing is performed as described in Chapter 17, the output would be one sample for each range-Doppler bin. However, the same data can also be combined noncoherently, in which case it again fits the model of detection based on N amplitude or power samples.

To analyze the detection performance obtainable with this block of N samples, it is necessary to model their joint statistics. The samples are assumed to be identically distributed but not necessarily independent. Thus, the amplitude of each individual pulse echo from the range bin of interest will be modeled by the same probability density function, typically one of those listed in Table 7-2. However, this still leaves the question of whether each of the N echo amplitudes is the same within a given antenna sweep, an independent random variable from the distribution, or something in between. This is a question of the *correlation* of the samples within the block of data.

The Swerling models are a combination of a specific probability density function for the echo powers and a specific assumption about the correlation of the N samples in a block. There are four Swerling models, formed from two choices for the PDF and two for the correlation behavior. The two density functions used by Swerling to describe RCS are the exponential and the chi-square of degree 4. As has been seen, the exponential model describes the behavior of a complex target consisting of many scatterers, none of which is dominant, while the fourth-degree chi-square adequately models targets having many scatterers of similar strength with one dominant scatterer, provided the ratio of the dominant to small scatterers is on the order of 2.4.

Swerling considered two bounding cases for the correlation properties of the block of N samples. The first assumes they are all perfectly correlated, so that all N echoes collected on one scan have the same value. This effectively assumes that the radar-target aspect angle varies by less than the $\Delta\theta$ of equation (7.18) over an interval of N PRIs. If the antenna scan time is greater than the target decorrelation time, the N new pulses collected on the next sweep will have the same value as one another also, but their value will be independent of the value measured on the first sweep. This case is referred to as *scan-to-scan decorrelation*. The second case assumes that each individual pulse on each sweep results in an independent random value for the amplitude, effectively assuming that the radar-target aspect angle varies by more than $\Delta\theta$ radians in one PRI, that frequency agility is used to ensure decorrelation, or both. This case is referred to as *pulse-to-pulse decorrelation*. The decorrelation properties of real data often fall between these extremes, but they are useful for bounding the detection results. It has been noted that relatively high pulse-to-pulse correlation coefficients, approximately 0.8 or more, are required before correlation has a major effect on detection performance, at least for exponential target RCS statistics [11].

[2]The term *CPI* is often used to refer to the block of data samples collected within the time interval as well as to the time interval itself.

TABLE 7-3 ■ Swerling Models

Probability Density Function of RCS	Decorrelation	
	Scan-to-Scan	Pulse-to-Pulse
Exponential	Case 1	Case 2
Chi-square, degree 4	Case 3	Case 4

The four combinations of the two choices for the PDF of the target power and the two choices for the decorrelation characteristics are denoted the "Swerling 1" through "Swerling 4" models of target fluctuation. Table 7-3 defines the four cases. In some references, the terminology is stretched to include the nonfluctuating target (Marcum) case as the "Swerling 0" or, less commonly, the "Swerling 5" model. Figure 7-17 illustrates a notional series of measurements from two different Swerling models. In each case, 10 pulses are received on each of three successive scans; the dead time between groups of

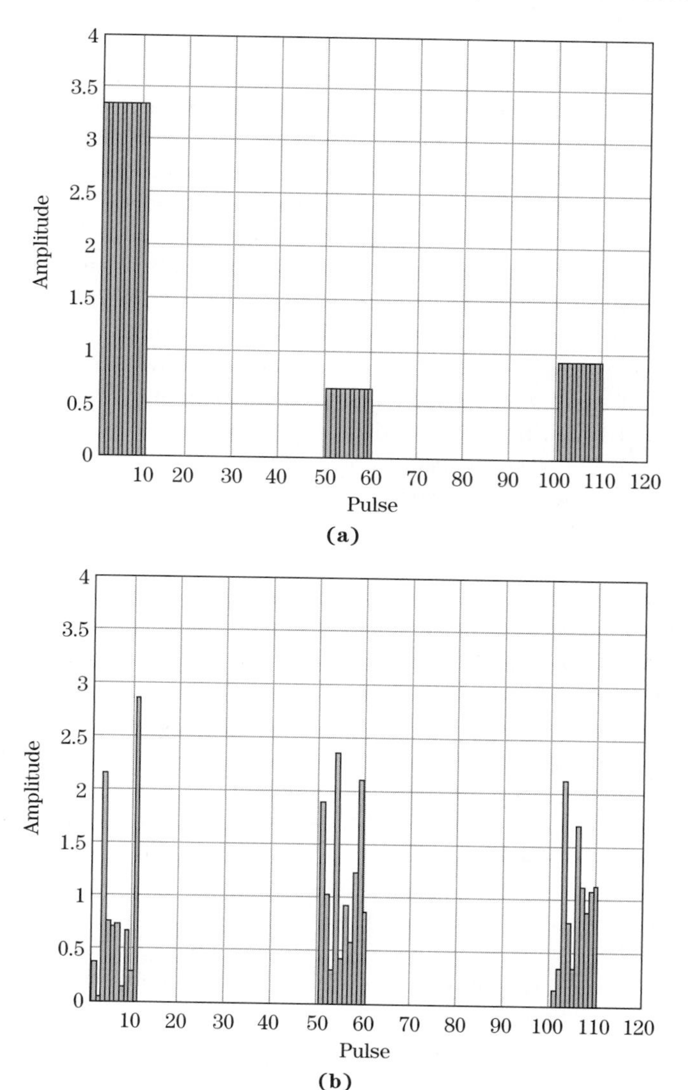

FIGURE 7-17 ■ Notional sequences of Swerling target samples. Results from three scans with 10 pulses per scan are shown. (a) Swerling case 1. (b) Swerling case 4.

pulses represents time in which the antenna is scanned in other directions so that the target is not in the antenna beam. Figure 7-17a represents Swerling case 1, which exhibits scan-to-scan decorrelation. Thus, within a scan, all 10 samples are the same, but that value changes from scan to scan according to an exponential distribution. Figure 7-17b is an example of the pulse-to-pulse decorrelation observed in Swerling case 4, where each individual sample is a new random value, in this case from a fourth-degree chi-square distribution. Chapter 15 develops exact and approximate formulas for the probability of detection and false alarm for the Swerling fluctuating target models and discusses the implications of the choice of target model.

Choosing a Swerling model for analysis requires that a choice be made between the two PDFs and between the two correlation models. To choose the PDF, the designer must have some knowledge about the RCS characteristics of the target of interest. Specifically, a judgment is needed as to whether the target RCS at the aspect angles of interest are likely to be dominated by one or two large scatterers or whether it is better described as the result of an ensemble of roughly equal scatterers. This decision is best based on measured data of the targets of interest at the appropriate frequencies, polarizations, and aspect angles, but such data are not always available. In many organizations and programs, there are well-established legacy practices for determining which Swerling or other target fluctuation model is appropriate for various classes of targets. However, care must be taken not to assume that models appropriate for use with one class of radar are necessarily valid for newer radars with different characteristics, for instance, higher RFs or much finer resolution.

The choice of the correlation model is primarily a matter of radar frequency, geometry, timeline, and the use of frequency agility. If the radar uses frequency agility to force decorrelation of the target echoes within the block of N samples, then a Swerling 2 or 4 model, depending on the selected PDF, should be chosen. If not, the designer must decide whether the aspect angle will change by more than $\Delta\theta$ radians, where $\Delta\theta$ is estimated using equation (7.18), during the collection of the block of N samples to be combined in the detection test. If not, scan-to-scan decorrelation (case 1 or 3) should be assumed. On the other hand, if the aspect angle changes by more than $\Delta\theta$ radians in one PRI, pulse-to-pulse decorrelation (case 2 or 4) should be selected. If the situation is likely to fall in between these extremes, calculations for both cases can be used to bound the range of detection results.

As an example, consider a 10 m long complex aircraft viewed with a stationary X-band (10 GHz) radar from a range of 30 km. The radar is not frequency agile. An exponential PDF is assumed due to the scattering complexity of the aircraft. The decorrelation angle estimated by equation (7.18) is 0.86° (1.5 mrad). Suppose the aircraft is flying at 200 m/s in a crossing direction (i.e., orthogonal to the radar LOS rather than directly toward or away from the radar). The angle between the radar and aircraft will change by 1.5 mrad when the aircraft has traveled $(0.0015)(30 \times 10^3) = 45$ m, which occurs in $45/200 = 225$ ms. Thus, if the radar collects a dwell of echoes from the target in less than 225 ms, the RCS would be expected to be fairly constant over the series of measurements and a Swerling 1 model should be assumed. If the PRI is longer than 225 ms, then a Swerling 2 model should be selected.

This example illustrates that very long PRIs would be required to decorrelate the data from pulse to pulse in many scenarios. For this reason, the Swerling 2 or 4 model is most often used with radars that employ frequency agility.

7.5.2 Extended Models of Target RCS Statistics

The strategy of the Swerling models can easily be extended to other target models. For example, one could define a model that uses a log-normal PDF for the target fluctuations with either pulse-to-pulse or scan-to-scan decorrelation to relate the N pulses in a block. Empirical observations have shown that such "long-tailed" distributions are often a better representation of observed radar data statistics than the traditional exponential and chi-square models, especially in high-resolution systems. Small resolution cells isolate one or a few scatterers, undermining the many-scatterer assumption of the traditional models and making large variations in observed RCS more likely as aspect or frequency changes. The approach to computing detection probabilities for such models is identical to that used with the Swerling models (see Chapter 15). However, the details are sometimes difficult to develop because the integrals of the PDFs are difficult to compute. Nonetheless, a number of results are available in the literature [e.g., 13].

Shnidman has proposed extending the Swerling models further by generalizing the gamma density underlying the traditional models to the noncentral gamma density [14]. This model allows for targets whose echo amplitudes exhibit a nonzero mean in the in-phase (I) or quadrature (Q) channels over a period of time on the order of N PRIs. Such a model is appropriate when the echoes contain one very steady component modulated by the echo from a number of smaller scatterers. This might happen, for instance, if echo from a large sea wave is modulated by rapidly varying surface ripples. A still further extension is to allow the nonzero mean component to itself fluctuate slowly, leading to what Shnidman calls the noncentral gamma-gamma density. The advantage of these extended models is that they introduce additional parameters that make it possible to more closely match the density function to experimentally observed statistics for a wide range of systems while maintaining the (relatively) tractable computational results obtainable with target models based on chi-square and gamma densities as opposed to log-normal or other density functions.

7.6 | DOPPLER SPECTRUM OF FLUCTUATING TARGETS

The correlation properties of a block of N target samples have implications for the Doppler spectrum of the target as well. Consider a possibly moving target, and ignore noise. The phase of the target echoes will change linearly from one sample to the next at a rate appropriate to the Doppler shift. If the target is either nonfluctuating or exhibits scan-to-scan decorrelation, the amplitude of the echoes will be constant or nearly so over the CPI. Thus, the slow-time target signal after demodulation to baseband will be of the form

$$y[m] = Ae^{j2\pi f_D mT}, \quad 0 \le m \le M-1 \tag{7.21}$$

where A is the nonfluctuating (constant) amplitude, f_D is the Doppler shift, T is the PRI, m is the pulse number, and M is the number of pulses in the CPI. The Doppler spectrum of this signal is a "digital sinc" function with a 3 dB width of $0.89/MT$ Hz (see Chapter 14 or [5]).

If the target echo power is fluctuating instead, the amplitude will no longer be a constant A but will vary pulse to pulse in some sequence $A[m]$ consistent with the PDF and

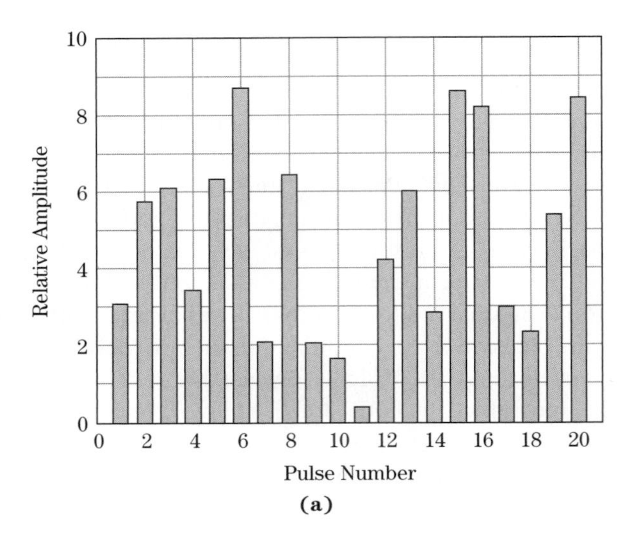

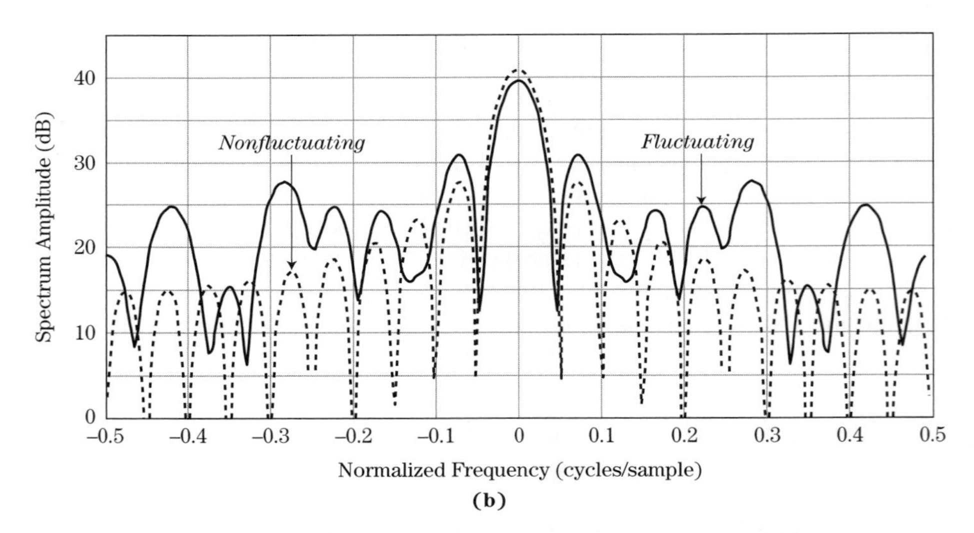

FIGURE 7-18 ■ Effect of amplitude fluctuations on target Doppler spectrum. (a) 20-pulse Rayleigh fluctuating amplitude sequence. (b) Spectrum of the fluctuating and nonfluctuating data.

correlation properties of the target amplitude. The Doppler spectrum of such a target will still be concentrated at the appropriate Doppler shift, but the nonzero bandwidth of the amplitude sequence $A[m]$ will spread the target spectrum. Figure 7-18 illustrates this effect. Figure 7-18a shows a sequence of amplitude (not power) measurements from a target with an exponential RCS distribution and pulse-to-pulse decorrelation (the Swerling 2 model). A stationary target (zero Doppler) is assumed without loss of generality. Figure 7-18b compares the spectrum of this fluctuating target with that of a nonfluctuating target with the same total energy in the data. Although the Doppler mainlobe is only slightly affected, significant energy is spread into the sidelobe regions of the response.

The Doppler spectrum of a complex target may also include discrete features due to moving parts on a target. For instance, a propeller-driven aircraft or a helicopter will have peaks in its Doppler spectrum corresponding not only to the radial velocity of the target as a whole but also to the velocity of the propeller or rotor tips as viewed from the radar. Similarly, a jet aircraft may introduce Doppler responses corresponding to the rotational speeds of the internal rotors of the jet engine. Ground vehicles may have additional Doppler components from the rotational speed of the wheels of the vehicle. Several examples of such *microDoppler* signatures are given in [15]. In some cases, they can be useful for target classification or identification as discussed in Chapter 1.

7.7 FURTHER READING

Common current introductory textbooks on RCS generally give limited attention to modeling of complex target RCS variations. Some discussion of target fluctuations and statistical models can be found in Skolnik [16], Edde [17], Levanon [18], and Mahafza [19]. Skolnik's text also has a brief but useful discussion on how to choose between models. A good discussion of the scattering characteristics of complex targets with special emphasis on correlation properties is given in Chapter 5 of the text by Nathanson [15], though the data sources used are becoming dated. Good recent sources on generalized Swerling-type models and related detection calculations in modern terminology and notation are the series of papers by Shnidman [11–14].

7.8 REFERENCES

[1] Knott, E.F., Shaeffer, J.F., and Tuley, M.T., *Radar Cross Section*, 2d ed., SciTech Publishing, Raleigh, NC, 2004.

[2] Meyer, D.P., and Mayer, H.A., *Radar Target Detection*, Academic Press, New York, 1973.

[3] Swerling, P., "Probability of Detection for Fluctuating Targets," *IRE Transactions on Information Theory*, vol. IT-6, pp. 269–308, April 1960.

[4] Papoulis, A., and Pillai, S.U., *Probability, Random Variables, and Stochastic Processes*, 4th ed., McGraw-Hill, New York, 2002.

[5] Richards, M.A., *Fundamentals of Radar Signal Processing*, McGraw-Hill, New York, 2005.

[6] Swerling, P., "Radar Probability of Detection for Some Additional Fluctuating Target Cases," *IEEE Transactions on Aerospace and Electronic Systems,* vol. AES-33(2), pp. 698–708, April 1997.

[7] Weinstock, W., "Target Cross Section Models for Radar Systems Analysis," Ph.D. dissertation, University of Pennsylvania, 1964.

[8] Ray, H., "Improving Radar Range and Angle Detection with Frequency Agility," *Microwave Journal,* vol. 9, pp. 63–68, May 1966.

[9] Birkmeier, W.P., and Wallace, N.D., *AIEE Transactions on Communication Electronics*, vol. 81, pp. 571–575, January 1963.

[10] DiFranco, J.V., and Rubin, W.L., *Radar Detection*, Artech House, Dedham, MA, 1980.

[11] Shnidman, D.A., "Radar Detection Probabilities and Their Calculation," *IEEE Transactions on Aerospace and Electronic Systems*, vol. AES-31, no. 3, pp. 928–950, July 1995.

[12] Shnidman, D.A., "Update on Radar Detection Probabilities and Their Calculation," *IEEE Transactions on Aerospace and Electronic Systems*, vol. AES-44, no. 1, pp. 380–383, January 2008.

[13] Shnidman, D.A., "Calculation of Probability of Detection for Log-Normal Target Fluctuations," *IEEE Transactions on Aerospace and Electronic Systems*, vol. AES-27, no. 1, pp. 172–174, January 1991.

[14] Shnidman, D.A., "Expanded Swerling Target Models," *IEEE Transactions on Aerospace and Electronic Systems*, vol. AES-39, no. 3, pp. 1059–1068, July 2003.

[15] Nathanson, F.E., *Radar Design Principles*, 2d ed., McGraw-Hill, New York, 1991.

[16] Skonik, M. I., *Introduction to Radar Systems*, 3d ed., McGraw-Hill, New York, 2001.

[17] Edde, B., *Radar: Principles, Technology, Applications*, Prentice-Hall, Upper Saddle River, NJ, 1995.

[18] Levanon, N., *Radar Principles, Wiley Interscience*, New York, 1988.

[19] Mahafza, B. R., *Radar Systems Analysis and Design Using MATLAB*, Chapman & Hall/CRC, Boca Raton, FL, 2000.

7.9 PROBLEMS

1. Consider a conducting sphere with a radius a_s of 1 m. What is the RCS of this sphere at an RF of 10 GHz? Give the answer in units of both square meters and dBsm. (Assume the sphere is "much larger than" the wavelength.) What must be the side length a_t of a trihedral so that it has the same RCS at 10 GHz? What is the ratio of the size of the sphere compared with that of the trihedral, a_s/a_t?

2. In terms of D/λ, what is the two-sided "mainlobe width" (angular interval between the first zero of the pattern to either side of the peak) of the dumbbell target RCS pattern of Figure 7-5 in the vicinity of a nominal aspect angle of $\theta = 0°$? Repeat for $\theta = 90°$. Assume D/λ is an integer.

3. For a given integer value of D/λ, how many RCS peaks will occur in a compass plot such as Figure 7-6? Verify your answer for the case of $D/\lambda = 5$ by counting the lobes in the figure.

4. Suppose a stationary radar illuminates a complex, but stationary, target with a series of pulses. Which probability density function is an appropriate choice to represent the series of echo power measurements? Why?

5. The fourth-degree chi-square PDF used to model the case of one dominant scatterer with many small scatterers is an approximation to the exact model for this case, which is the Rician PDF. The Rician PDF has a variance of $\bar{\sigma}^2(1+2a^2)/(1+a^2)^2$, where a^2 is the ratio of the RCS of the large scatterer to the combined RCS of the small scatterers. Show that when they both have the same mean $\bar{\sigma}$, their variances will also be the same if the Rician parameter $a^2 = 1+\sqrt{2}$.

6. Suppose that a target was modeled as consisting of one large scatterer and many small ones but that the ratio a^2 of the large scatterer RCS to the sum of the small scatterer radar cross sections is 1 (instead of $1+\sqrt{2}$ as assumed by the fourth-degree chi-square model). Assuming the means of the two distributions are the same, what degree $2m$ should be chosen for the chi-square so as to match the variance of the Rician (see Table 7-1)? Repeat for $a^2 = 10$. Note: m does not have to be an integer.

7. Part of the significance of choosing the probability density function used to model target RCS (or clutter or other interference) is that the differences in the "tails" of the PDF can have a significant impact on the probability of observing relatively large signal values, sometimes called signal "spikes." Recall that the probability that a random variable x described by a PDF $p_x(x)$ exceeds some value T is given by

$$P\{x > T\} = \int_T^{+\infty} p_x(x)\,dx$$

Consider RCS data with a mean value (linear scale) of 1.0. Compute the probability that the RCS σ is greater than 2 when an exponential PDF is a good model for the RCS statistics and again when a fourth-degree chi-square is a good model for the statistics.

8. What is the maximum estimated decorrelation frequency step size Δf for a rectangular complex target like that shown in Figure 7-7 with $L_1 = L_2 = 3$ m when viewed by a radar operating at $f_0 = 5$ GHz? If $N = 10$ pulses are collected at frequencies $f_0, f_0 + \Delta f, \ldots, f_0 + (N-1)\Delta f$, what percentage of the nominal frequency, f_0, is the total change in frequency?

9. A rectangular target has dimensions $L_1 = 3$ m and $L_2 = 10$ m. What is the largest value of Δf required to decorrelate the RCS by frequency agility, regardless of aspect angle θ? At what aspect angle does this value occur?

10. The ASR-9 is a common airport surveillance radar in the United States. It has an RF of 2.8 GHz, 3 dB azimuth beamwidth θ_3 of 1.4°, and a rotation rate Ω of 12.5 revolutions per minute. Consider an aircraft at a range of 50 nmi (nautical miles). Assume the PRF is chosen to give an unambiguous range of 60 nmi. How many pulses will be transmitted during the time the aircraft is within the 3 dB mainlobe of the antenna on a single rotation? (This will be considered a single "scan" of the aircraft by the radar.)

11. Continuing problem 10, assume the aircraft is a Boeing 757 with a length of about 47 m and a wingspan of about 38 m, flying broadside to the radar at 120 knots. In the time the aircraft is in the mainbeam of the antenna on a single scan, what will be the change in the aspect angle, $\Delta\theta$, between the radar and aircraft? Based on this result, should pulse-to-pulse or scan-to-scan decorrelation be assumed for the pulse echoes received on a single scan?

12. Show that the gamma PDF of equation (7.11) reduces to the exponential PDF of equation (7.7) when $\alpha = 1$ and $\beta = \bar{\sigma}$ and to the fourth-degree chi-square of equation (7.9) when $\alpha = 2$ and $\beta = \bar{\sigma}/2$.

CHAPTER 8

Doppler Phenomenology and Data Acquisition

William A. Holm, Mark A. Richards

Chapter Outline

8.1 Introduction ... 273
8.2 Doppler Shift ... 274
8.3 The Fourier Transform ... 276
8.4 Spectrum of a Pulsed Radar Signal ... 277
8.5 Why Multiple Pulses? ... 286
8.6 Pulsed Radar Data Acquisition ... 287
8.7 Doppler Signal Model ... 291
8.8 Range-Doppler Spectrum for a Stationary Radar ... 293
8.9 Range-Doppler Spectrum for a Moving Radar ... 296
8.10 Further Reading ... 303
8.11 References ... 303
8.12 Problems ... 303

8.1 INTRODUCTION

Many signal processing techniques used by modern radars take advantage of the differences in the Doppler frequency characteristics of targets, clutter, and noise to minimize the interference competing with the target signals, and thus to improve the probability of detection and the measurement accuracy. Consequently, it is useful to study the Doppler frequency characteristics of typical radar signals.

The chapter begins by showing how the Doppler shift predicted by special relativity reduces to the very good standard approximation commonly used in radar, including in this book. The dependence on radial velocity is described. The principal focus of this chapter is on the Doppler spectrum of pulsed radar signals. The spectrum of the received signal for idealized stationary and moving point targets viewed with a finite pulse train waveform is developed step by step with key Fourier transform relationships introduced as required. These results are used to illustrate the concept of Doppler resolution. Attention then shifts to practical measurement of Doppler shift using finite pulse trains and Fourier analysis of the pulse-to-pulse phase shift. In doing so, the idea of coherent detection, first introduced in Chapter 1, is revisited.

Finally, the contributions of noise, clutter, and moving targets are described to build an understanding of the range-Doppler or range-velocity distribution as viewed by stationary or moving (airborne or spaceborne) radars. The clutter foldover (ambiguity) effects on this distribution of range and velocity ambiguities are described and illustrated.

8.2 DOPPLER SHIFT

If a radar and scatterer are not at rest with respect to one another, the frequency, f_r, of the received echo will differ from the transmitted frequency, f, due to the Doppler effect. A proper description of the Doppler shift for electromagnetic waves requires the theory of special relativity. Consider a monostatic radar, where the transmitter and receiver are at the same location and do not move with respect to one another. Suppose a scatterer in the radar field of view is moving with a velocity component, v, toward the radar. The theory of special relativity predicts that the received frequency will be [1,2]

$$f_r = \left(\frac{1 + v/c}{1 - v/c}\right) f \tag{8.1}$$

Thus, an approaching target causes an increase in the received frequency. Substituting $-v$ for v shows that a receding target decreases the received frequency. This is in keeping with the common experience of passing train whistles or ambulance sirens. These Doppler shifts can be used to advantage in radar to detect echoes from moving targets in the presence of much stronger echoes from stationary clutter or to drastically improve cross-range resolution when there is relative rotation between a target scene and the radar. Uncompensated Doppler shifts can also have harmful effects, particularly a loss of detection sensitivity for some types of waveforms.

Equation (8.1) can be simplified without significant loss of precision because the velocity of actual radar targets is a small fraction of the speed of light, c. For example, the value of v/c for a supersonic aircraft traveling at Mach 2 (about 660 m/s) is only 2.2×10^{-6}. Expand the denominator of (8.1) in a binomial series:

$$\begin{aligned} f_r &= (1 + v/c)(1 - v/c)^{-1} f \\ &= (1 + v/c)[1 + (v/c) + (v/c)^2 + \ldots] f \\ &= [1 + 2(v/c) + 2(v/c)^2 + \ldots] f \end{aligned} \tag{8.2}$$

Discarding all second-order and higher terms in (v/c) leaves

$$f_r = [1 + 2\,(v/c)] f \tag{8.3}$$

The difference, f_d, between the transmitted and received frequencies is called the *Doppler frequency* or *Doppler shift*. For this case of an approaching target it is

$$f_d = \frac{2v}{c} f = \frac{2v}{\lambda} \tag{8.4}$$

where λ is the transmitted wavelength and positive values of v correspond to approaching targets.

Because the velocity of typical targets is so small compared with the speed of light, the numerical values of Doppler shift are small compared with the radar radiofrequencies (RFs). Table 8-1 gives the magnitude of the Doppler shift corresponding to a velocity of 1 meter per second, knot, or mile per hour at various typical RFs. The Mach 2 aircraft,

TABLE 8-1 ■ Doppler Shift as a Function of Velocity and Frequency

Radiofrequency f		Doppler Shift f_d (Hz)		
Band	Frequency (GHz)	1 m/s	1 knot	1 mph
L	1	6.67	3.43	2.98
S	3	20.0	10.3	8.94
C	5	33.3	17.1	14.9
X	10	66.7	34.3	29.8
K_u	16	107	54.9	47.7
K_a	35	233	120	104
W	95	633	326	283

observed with the L-band radar, would cause a Doppler shift of only 4.4 kHz in a 1 GHz carrier frequency.

For a monostatic radar, the Doppler shift is proportional to the relative velocity along the line of sight (LOS) between the radar and target, called the *radial velocity*. Consider the example of an aircraft flying at v m/s and illuminated by a stationary radar as shown in Figure 8-1. If the angle between the velocity vector of the aircraft and the radar LOS is ψ, the radial velocity component is $v \cdot \cos\psi$ m/s and the Doppler shift becomes

$$f_d = \frac{2v}{\lambda}\cos\psi \tag{8.5}$$

Note that a *crossing target*, which is a moving target viewed at $\psi = 90°$, will have a Doppler shift of 0 Hz. Thus, the magnitude of the Doppler shift is maximum when the target is traveling directly toward or away from the radar ($\psi = 0$ or π radians). The Doppler shift is zero, regardless of the target velocity, when the target is crossing orthogonally to the radar boresight ($\psi = \pi/2$ radians).

It is important to note that the angle, ψ, in equation (8.5) is measured with respect to the velocity vector of the radar, not the pointing direction of the antenna. The angle of

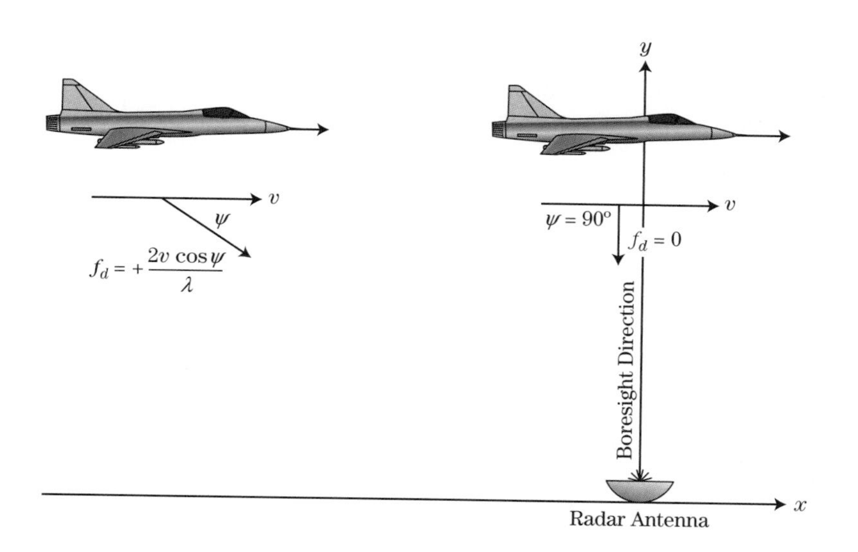

FIGURE 8-1 ■ Doppler shift is determined by the radial component of relative velocity between the target and radar.

the target with respect to the antenna determines the antenna gain in the target's direction, but not the Doppler shift of the echo. Also, note that the orientation of the aircraft body may differ from the velocity vector direction due to wind-induced crab, angle of attack, and similar effects.

8.3 THE FOURIER TRANSFORM

The frequency content of a time domain signal is analyzed by taking its Fourier transform, thus resulting in a frequency domain representation, or spectrum, of the time domain signal. Prior to discussing the Doppler frequency spectrum of radar signals, it is useful to briefly review the Fourier transform. Restricting consideration to continuous-time signals for the moment, if the time domain signal is described by the complex function $x(t)$, then its Fourier transform is given by

$$X(\omega) = \Im\{x(t)\} = \int_{-\infty}^{\infty} x(t) e^{-j\omega t} dt \tag{8.6}$$

where the operator $\Im\{\cdot\}$ represents Fourier transformation, and $j = \sqrt{-1}$. The frequency variable, ω, is in radians per second. It is related to the frequency, f, in hertz by the transformation $\omega = 2\pi f$. The Fourier transform can also be defined in terms of f as

$$X(f) = \Im\{x(t)\} = \int_{-\infty}^{\infty} x(t) e^{-j2\pi ft} dt \tag{8.7}$$

The time domain signal $x(t)$ can be regained by taking the inverse Fourier transform of $X(\omega)$ or $X(f)$:

$$x(t) = \frac{1}{2\pi} \int_{-\infty}^{\infty} X(\omega) e^{j\omega t} d\omega = \int_{-\infty}^{\infty} X(f) e^{j2\pi ft} df \tag{8.8}$$

Note that there is no factor of $1/2\pi$ on the inverse transform in terms of cyclical frequency f.

Consider an important but simple example, the infinite-length complex sinusoidal signal of amplitude A, $x(t) = A\exp(j2\pi f_0 t)$. Substituting this expression into equation (8.7) gives

$$\begin{aligned} X(f) &= \int_{-\infty}^{+\infty} (Ae^{+j2\pi f_0 t}) e^{-j2\pi ft} dt = A \int_{-\infty}^{+\infty} e^{-j2\pi(f-f_0)t} dt \\ &= A \cdot \delta_D(f - f_0) \end{aligned} \tag{8.9}$$

where $\delta_D(f)$ is the *impulse* or *Dirac delta* function [3,4]. A plot of this spectrum is shown in Figure 8-2.

In the following, it is assumed that the reader is familiar with fundamental Fourier transform theorems and properties and with common Fourier transform pairs. The texts by Papoulis [3] and Bracewell [4] are excellent references for each of the properties and transform pairs used.

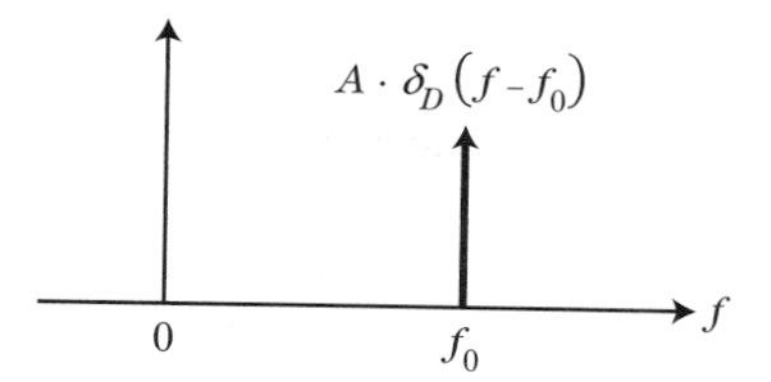

FIGURE 8-2 ■ Fourier transform of a complex sinusoid.

8.4 SPECTRUM OF A PULSED RADAR SIGNAL

It will be seen in Chapter 17 that Doppler processing relies on differences in Doppler shift between targets and interference, due to differences in velocity relative to the radar, to suppress the interference and allow target detection. For this reason, it is useful to consider the frequency spectrum of perhaps the most common radar waveform, a finite train of simple RF pulses, for both a stationary target and a moving target.

The finite RF pulse train radar signal can be described in terms of four increasingly long time scales: RF wave period, pulse width, pulse repetition interval (PRI), and coherent processing interval (CPI). The spectrum of the pulse train can be described in terms of four corresponding frequency scales: radar radiofrequency, pulse bandwidth, pulse repetition frequency (PRF), and spectral line bandwidth, respectively. A fundamental property of Fourier analysis is that the time and frequency scales have an inverse relationship to each other: the longer or more spread out a function is in time, the shorter or more compact is the corresponding frequency spectrum and vice versa. This *reciprocal spreading* relationship and its consequences are crucial to the understanding of Doppler processing techniques.

8.4.1 Spectrum of a Continuous Wave Signal

The simplest radar waveform is the real-valued infinite duration continuous wave (CW) sinusoidal signal of frequency f_0 Hz. It is defined by a single time scale, the period $T_0 = 1/f_0$. The spectrum consists of two impulse functions, one at f_0 and the other at $-f_0$. This can easily be seen by using the well-known Euler relation,

$$x(t) = A\cos(2\pi f_0 t) = \frac{A}{2}\left\{e^{j2\pi f_0 t} + e^{-j2\pi f_0 t}\right\} \tag{8.10}$$

and then taking the Fourier transform of $x(t)$ using equation (8.9) to get

$$X(f) = \frac{A}{2}[\delta_D(f - f_0) + \delta_D(f + f_0)] \tag{8.11}$$

Shown in Figure 8-3 is a plot of this waveform as a function of time and the resulting frequency-domain amplitude versus frequency plot.

Similarly, a sine function can be written by Euler's formula as

$$x(t) = \sin(2\pi f_0 t) = \frac{1}{2j}\{e^{j2\pi f_0 t} - e^{-j2\pi f_0 t}\} \tag{8.12}$$

and its Fourier transform as

$$X(f) = \frac{A}{2j}[\delta_D(f - f_0) - \delta_D(f + f_0)] \tag{8.13}$$

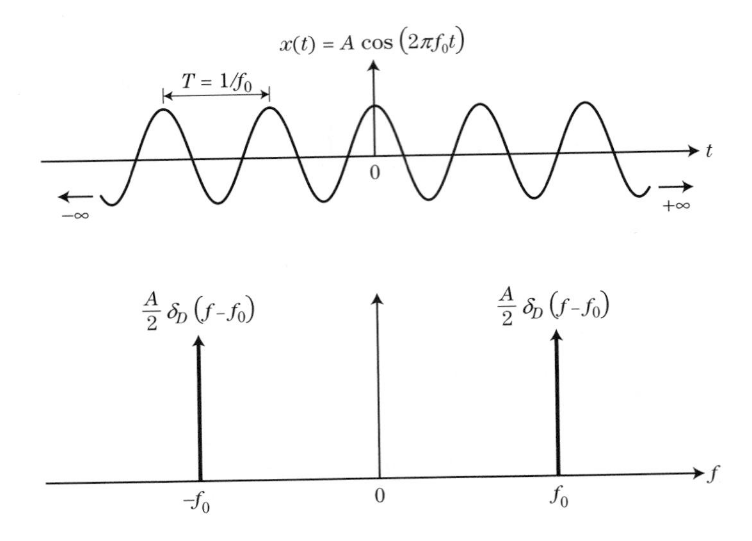

FIGURE 8-3 ■ Infinite-length continuous wave (CW) signal of frequency f_0 and its frequency spectrum.

8.4.2 Spectrum of a Single Rectangular Pulse

The spectrum of a finite-length series of modulated pulses is more complex but can be built up by a series of simple steps using the properties of Fourier transforms. Shown in Figure 8-4 is the time domain plot of a single pulse of pulse width, τ, and the magnitude of its corresponding spectrum. The waveform is defined using a single time scale parameter τ as

$$p_\tau(t) = \begin{cases} A, & -\frac{\tau}{2} \le t \le \frac{\tau}{2} \\ 0, & \text{otherwise} \end{cases} \tag{8.14}$$

Substituting this into equation (8.7), it is straightforward to show that

$$\begin{aligned} P_\tau(f) &= \int_{-\tau/2}^{+\tau/2} Ae^{-j2\pi ft} dt \\ &= \frac{A}{\pi f} \sin(\pi f \tau) = A\tau \frac{\sin(\pi f \tau)}{\pi f \tau} \\ &\equiv A\tau \operatorname{sinc}(\pi f \tau) \qquad \left(\operatorname{sinc}(z) \equiv \frac{\sin(\pi z)}{\pi z} \right) \end{aligned} \tag{8.15}$$

where the *sinc function*, which arises frequently in Fourier analysis, has been defined in the last step.

This spectrum has a $\sin f/f$ form centered at zero frequency. Notice that the nulls occur at integer multiples of $1/\tau$. The central portion of $X(f)$, between the frequencies $-1/\tau$ and $1/\tau$, is called the *main lobe* of the spectrum; the other lobes at frequencies outside of this range are called *sidelobes*. The width of the main lobe is characterized by one of several related metrics. The *Rayleigh width* is the width from the peak to the first null, which is $1/\tau$ Hz for this simple pulse. The *3 dB* or *half-power width* is the two-sided width measured at an amplitude corresponding to one-half the peak power, which occurs at $1/\sqrt{2}$ times the peak amplitude as shown in Figure 8-4. For the simple pulse, the 3 dB

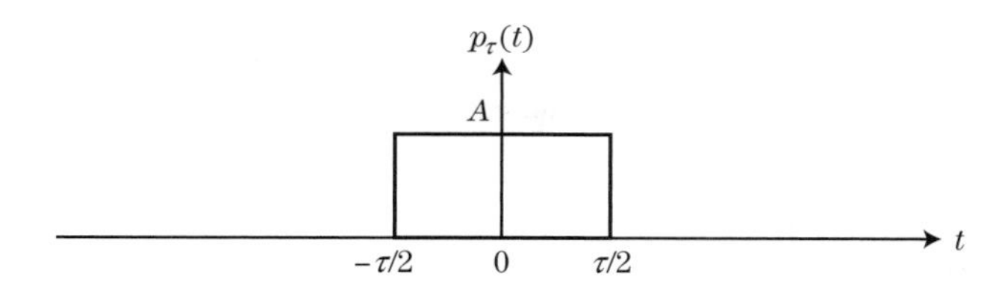

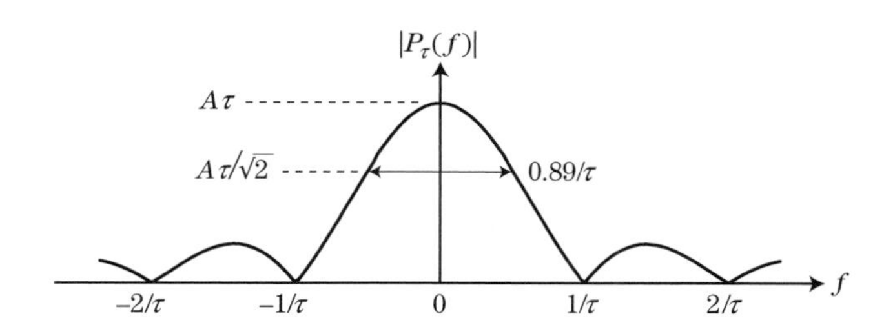

FIGURE 8-4 ■ A single simple pulse and its spectrum.

width is $0.89/\tau$ Hz.[1] Whichever metric is used, the width of the main lobe is inversely proportional to the pulse length. This is an example of the reciprocal spreading behavior of Fourier transforms previously mentioned.

8.4.3 Infinite Pulse Train

The second time scale is introduced by repeating the pulse at intervals of T seconds, creating an infinite pulse train. A convenient way to model the infinite pulse train is as the convolution of the single pulse, $p_\tau(t)$, of equation (8.14) with an infinite train of Dirac impulses spaced by the PRI of T seconds:

$$p_I(t) = \sum_{n=-\infty}^{\infty} p_\tau(t - n \cdot T) = p_\tau(t) \cdot \sum_{n=-\infty}^{\infty} \delta_D(t - n \cdot T) \tag{8.16}$$

Since $x(t)$ is the convolution of two terms, its Fourier transform is the product of the transforms of the two terms. The Fourier transform of an infinite pulse train in the time domain is another infinite pulse train in the frequency domain [4],

$$\Im\left(\sum_{n=-\infty}^{\infty} \delta_D\left(t - n \cdot T\right)\right) = \frac{1}{T}\sum_{k=-\infty}^{\infty} \delta_D\left(f - k \cdot \frac{1}{T}\right) = \frac{1}{T}\sum_{k=-\infty}^{\infty} \delta_D\left(f - k \cdot PRF\right) \tag{8.17}$$

The product of this spectrum with that of the single pulse in equation (8.15) is

$$P_I(f) = \{A\tau \operatorname{sinc}(\pi f \tau)\} \cdot \left\{\frac{1}{T}\sum_{k=-\infty}^{\infty} \delta_D\left(f - k \cdot PRF\right)\right\} \tag{8.18}$$

[1] Different bandwidth metrics are traditional in different technical specialties. While 3 dB metrics are common in radar, the Rayleigh bandwidth is common in optics. Other fields use the null-to-null bandwidth (twice the Rayleigh bandwidth). The "full width at half maximum" (FWHM), which is similar to the 3 dB width but measured at the 50% amplitude level, is common in spectroscopy.

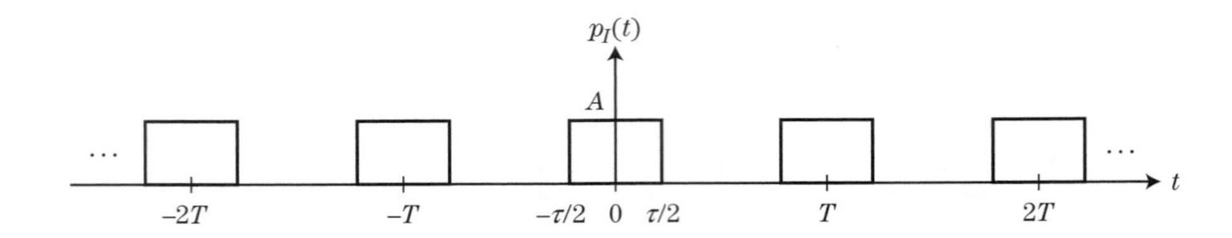

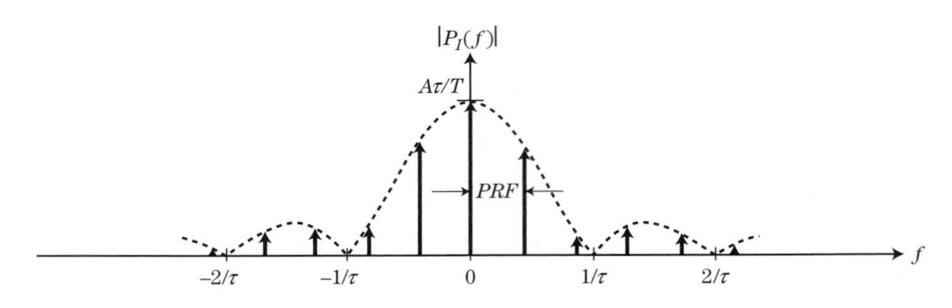

FIGURE 8-5 ■ Infinite pulse train signal and its spectrum.

The *sifting property* of impulses states that $X(f) \cdot \delta_D(f - f_0) = X(f_0) \cdot \delta_D(f - f_0)$; that is, the product of an impulse and a regular function is an impulse with a weight determined by the function. Applying this property to equation (8.18) shows that the spectrum of the infinite pulse train is an infinite series of spectral lines weighted by the spectrum of the single pulse. Specifically,

$$P_I(f) = \frac{A\tau}{T} \sum_{k=-\infty}^{\infty} \operatorname{sinc}(\pi \tau k \cdot PRF) \cdot \delta_D(f - k \cdot PRF) \tag{8.19}$$

The time- and frequency-domain sketches of an infinite sequence of pulses are shown in Figure 8-5. The broad sinc term of the single pulse has been resolved into distinct spectral lines having zero bandwidth and separated by a frequency spacing equal to the PRF. In practice, the PRI varies from just a few times to perhaps 100 times the pulse length, so the number of spectral lines in one lobe of the broader sinc weight function ranges from a few, similar to the situation in Figure 8-6, to a very dense set of lines.

8.4.4 Finite Pulse Train

The third time scale is introduced by truncating the infinite pulse train to a finite series of pulses of total duration, T_d. T_d is called the *coherent processing interval*.[2] The finite pulse train waveform is modeled as the infinite pulse train of equation (8.16) multiplied by a simple pulse like that of equation (8.14), but with duration T_d (where $T_d > \tau$) and unit amplitude ($A = 1$):

$$p_F(t) = p_I(t) \cdot p_{T_d}(t), \quad p_{T_d}(t) = \begin{cases} 1, & -\dfrac{T_d}{2} \le t \le \dfrac{T_d}{2} \\ 0, & \text{otherwise} \end{cases} \tag{8.20}$$

[2]Recall that the term CPI is sometimes used interchangeably with *dwell time*. As discussed in Chapter 3, here dwell time is considered to be the amount of time a given target is within the antenna mainbeam on a single scan. There may be several CPIs in a dwell time.

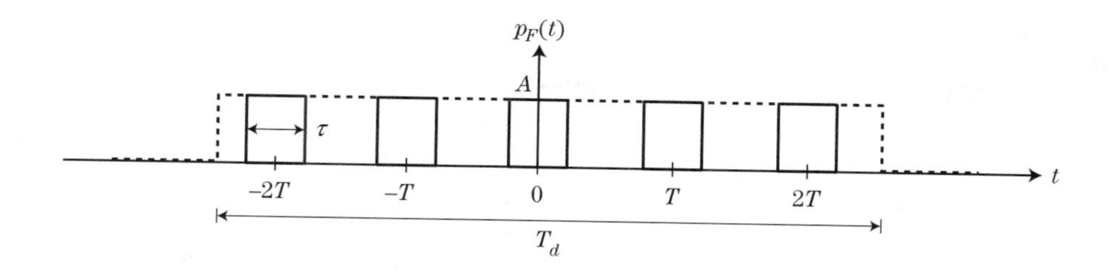

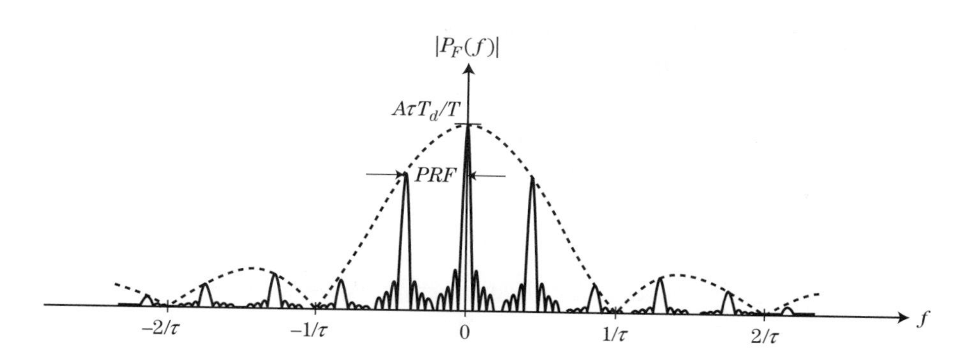

FIGURE 8-6 ■ Finite pulse train signal and its spectrum.

Since the waveform is the product of these two terms in the time domain, its Fourier transform is the convolution in frequency of the transforms of the two terms, which are given by equations (8.15) (with τ replaced by T_d and $A = 1$) and (8.19). Using the convolution property of impulses, $X(f) \cdot \delta_D(f - f_0) = X(f - f_0)$, the spectrum of the finite pulse train becomes

$$\begin{aligned} P_F(f) &= T_d \text{sinc}(\pi f T_d) \cdot P_F(f) \\ &= T_d \text{sinc}(\pi f T_d) \cdot \left\{ \frac{A\tau}{T} \sum_{k=-\infty}^{\infty} \text{sinc}(\pi \tau k \cdot PRF) \cdot \delta_D(f - k \cdot PRF) \right\} \\ &= \frac{AT_d\tau}{T} \sum_{k=-\infty}^{\infty} \text{sinc}(\pi \tau k \cdot PRF) \text{sinc}[\pi(f - k \cdot PRF)T_d] \end{aligned} \quad (8.21)$$

The resulting time-domain and frequency-domain plots are shown in Figure 8-6.

8.4.5 Modulated Finite Pulse Train

Finally, $p_F(t)$ is multiplied by a cosine function of frequency f_0 Hz to produce a finite train of modulated pulses:

$$x(t) = p_F(t) \cdot \cos(2\pi f_0 t) \quad (8.22)$$

The cosine function introduces the fourth time scale, the period $T_0 = 1/f_0$ of the modulating sinusoid.

The spectrum of the modulated pulse train can be determined using the modulation property of Fourier transforms, which states that if $x(t)$ has a Fourier transform $X(f)$, then the transform of $x_m(t) = x(t)\exp(j2\pi f_0 t)$ is

$$X_m(f) = \int_{-\infty}^{\infty} x(t)\, e^{j2\pi f_0 t}\, e^{-j2\pi f t} dt = \int_{-\infty}^{\infty} x(t)\, e^{-j2\pi(f-f_0)t} dt \\ = X(f - f_0) \tag{8.23}$$

That is, modulating a waveform by a complex exponential of frequency f_0 Hz merely shifts the waveform spectrum in frequency by f_0 Hz but does not change its shape. Applying this property along with Euler's formula to equation (8.21), the Fourier transform of the modulated pulse can be written immediately:

$$\begin{aligned} X(f) &= \frac{1}{2}\{P_F(f - f_0) + P_F(f + f_0)\} \\ &= \frac{AT_d\tau}{T}\left\{\sum_{k=-\infty}^{\infty} \operatorname{sinc}(\pi\tau k \cdot PRF)\operatorname{sinc}[\pi(f - f_0 - k \cdot PRF)T_d]\right. \\ &\quad \left. + \sum_{k=-\infty}^{\infty} \operatorname{sinc}(\pi\tau k \cdot PRF)\operatorname{sinc}[\pi(f + f_0 - k \cdot PRF)T_d]\right\} \end{aligned} \tag{8.24}$$

The magnitude of $X(f)$ is plotted in Figure 8-7. It has the form of two sets of sinc-shaped spectral lines. The sets are centered at $f = \pm f_0$ and each is weighted by a broader sinc function. Each spectral line now becomes a narrow sinc function with Rayleigh bandwidth of $1/T_d$ Hz. Thus, four frequency scales are present in the description of the spectrum of a pulse waveform: (1) the bandwidth of the spectral lines ($1/T_d$); (2) spacing of the spectral

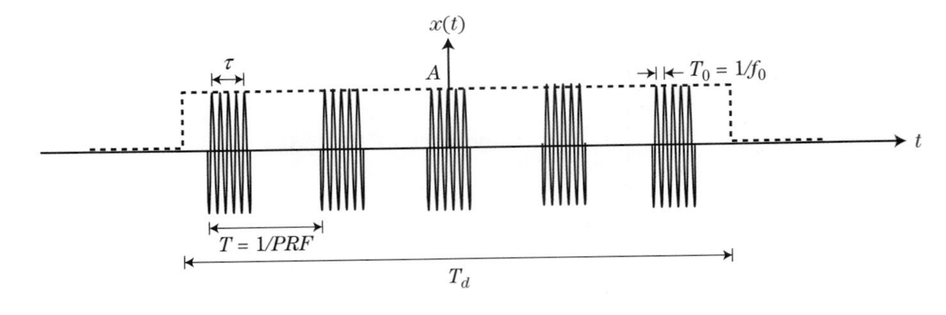

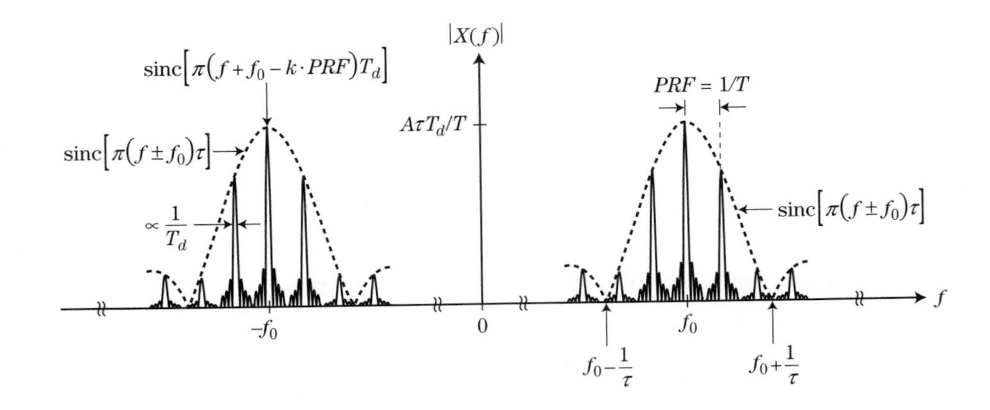

FIGURE 8-7 ■ Finite duration modulated pulse train signal and its Fourier transform.

lines ($PRF = 1/T$); (3) Rayleigh bandwidth of the single-pulse sinc envelopes ($1/\tau$); and (4) center frequencies of those envelopes ($\pm f_0 = \pm 1/T_0$). In practice, the intermediate frequency (IF) or RF f_0 would normally be from one to several orders of magnitude greater than the sinc weighting function Rayleigh width of $1/\tau$ Hz. For instance, a 1 μs pulse from a 1 GHZ radar would have $f_0 = 1$ GHz and $1/\tau = 1$ MHz, a difference of three orders of magnitude.

8.4.6 Pulsed Waveform Spectrum with Moving Targets

For a stationary radar, if the pulsed radar return is from a radar resolution cell that contains both a stationary target (clutter) and a moving target, then the return will consist of a superposition of signals. The signal from the stationary clutter will be at the transmitted RF f_0, and the signal from the moving target will be at $f_0 + f_d$, where $f_d = 2v/\lambda$ is the Doppler frequency shift, v is the radial component of the moving target's velocity, (i.e., the component along the line of sight between the radar and the target), and λ is the wavelength of the transmitted wave. The Doppler frequency shift is positive for approaching targets and negative for receding targets.

Because the effect of the moving target is simply to change the echo frequency to $f_0 + f_d$, the spectrum of the echo signal from a moving target alone (no clutter) measured with the finite pulse train waveform will be identical to that of equation (8.24) and Figure 8-7, with f_0 replaced by $f_0 + f_d$. Thus, the weighted line spectrum is simply shifted by f_d Hz. The total (moving target plus clutter) spectrum is then the superposition of the target and clutter spectra as shown in Figure 8-8. In this figure, the higher-amplitude line spectrum is due to the clutter, while the lower-amplitude, shifted spectrum is due to the moving target.

The basic *moving target indication* (MTI) approach inherent in Doppler processing techniques is now obvious. In the spectrum of the received signal, the moving targets and stationary clutter are separated. Thus, by applying appropriate filtering, the stronger stationary clutter can be removed (filtered) from the spectrum, leaving only the weaker signature of the moving target. This process is described in detail in Chapter 17.

8.4.7 Doppler Resolution

The Rayleigh main lobe width of an individual spectral line, $1/T_d$ Hz, determines the Doppler *resolution* of the finite pulse train. This is the minimum difference in Doppler shift between two targets producing equal-amplitude responses at which they can be reliably

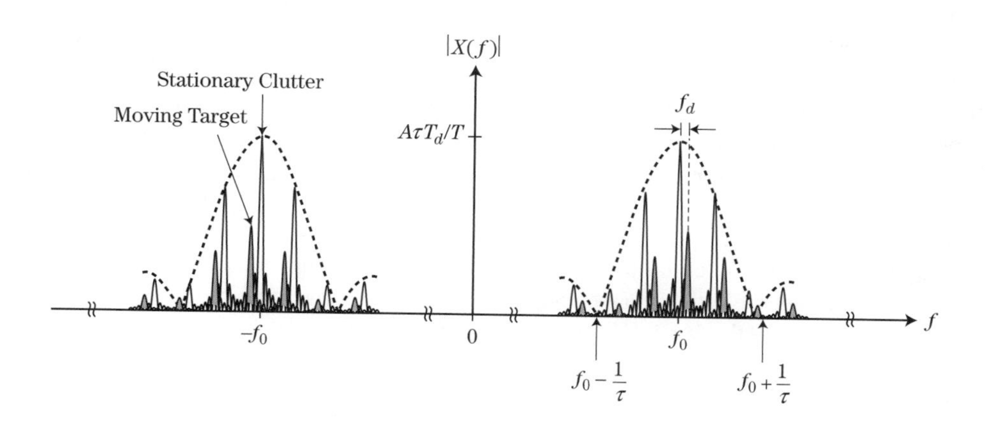

FIGURE 8-8 ■ Spectrum of the received signal from a moving target and stationary clutter.

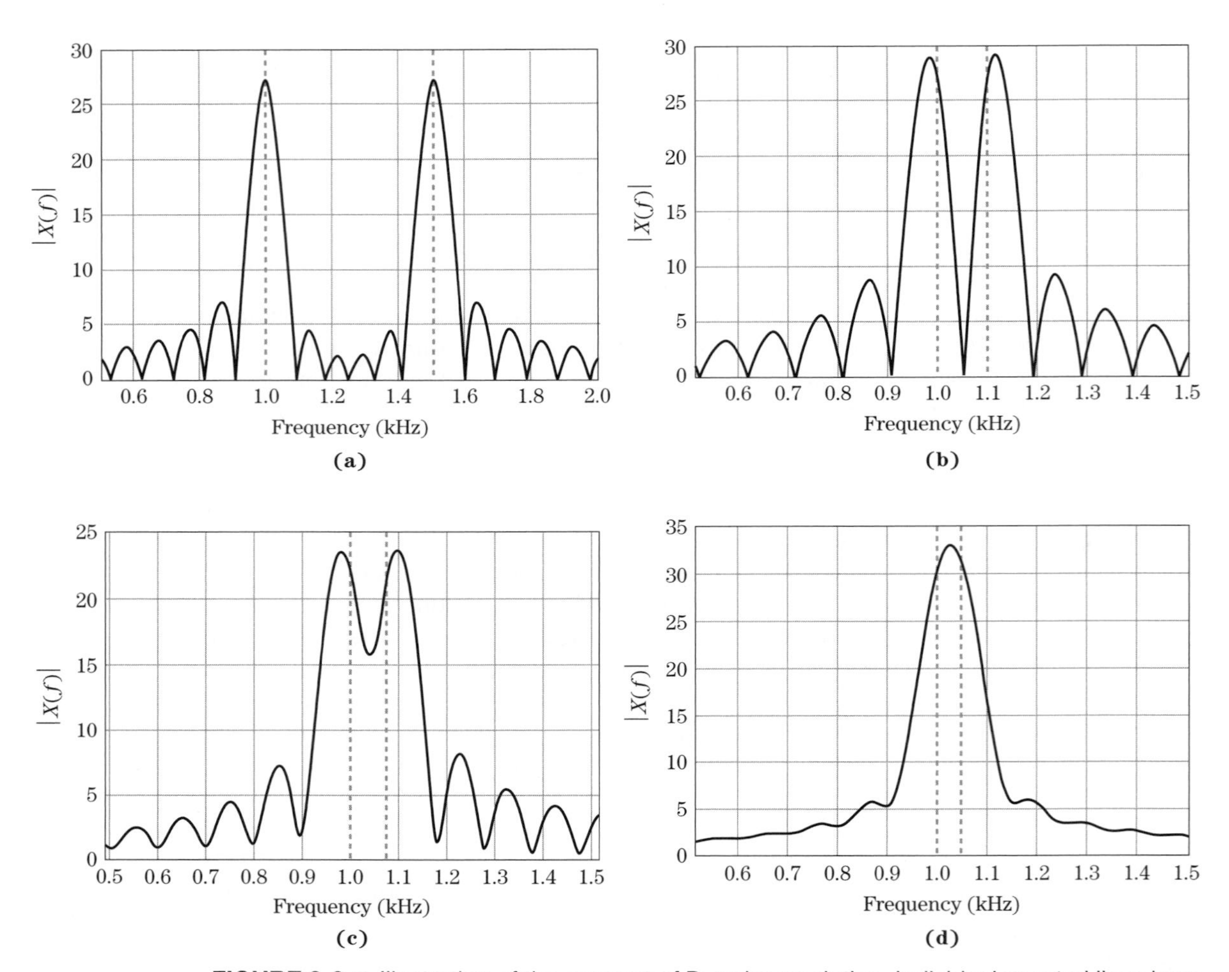

FIGURE 8-9 ■ Illustration of the concept of Doppler resolution. Individual spectral lines have 100 Hz Rayleigh bandwidth and zero relative phase. (a) 500 Hz spacing. (b) 100 Hz spacing. (c) 75 Hz spacing. (d) 50 Hz spacing.

distinguished from one another. The idea is illustrated in Figure 8-9. Part (a) of the figure shows a portion of the positive-frequency axis of the spectrum of two $T_d = 10$ ms pulse trains. A single spectral line of the spectrum from each is shown, one Doppler shifted to 1 kHz relative to the IF or RF and the other to 1.5 kHz. Because the pulse trains have a duration of 10 ms each, their individual spectral lines have Rayleigh (peak-to-first-null) bandwidths of 100 Hz. While the sidelobes of one spectrum affect the sidelobes and main lobe of the other, the two separate frequency components are obvious, and their peaks occur very close to the correct frequencies, which are marked by the vertical lines.

In the remaining parts of the figure, the pulse durations remain the same, but the spacing between their center frequencies are reduced to 100 Hz, then 75 Hz, then 50 Hz. At a spacing of 100 Hz, equal to the individual pulse Rayleigh bandwidth, there are still two easily recognizable distinct peaks. However, as the spacing is reduced to less than the pulse bandwidth, those peaks begin to blur together. When the spacing is 50 Hz, there is only a single peak, suggesting only a single-pulse frequency rather than two separate frequencies combined. While there is still some separation at 75 Hz spacing, there is significant error in the peak locations, and the addition of noise to the signal would make it difficult to recognize the presence of two distinct peaks. Thus, the two different pulse frequencies are considered to be reliably separable, or *resolved*, when they are spaced by at least the single-pulse bandwidth, which in turn is determined by the CPI, T_d. The longer the pulse train, the finer the Doppler resolution.

The specific results of Figure 8-9 depend on the relative starting phase of the two sinusoids, which was zero in this example. Other choices of initial relative phase can make the frequency spacing at which the two signals are resolved smaller or larger. For example, a relative starting phase of 90° will result in two clearly distinct peaks in Figure 8-9d, although the peaks will be at about 960 and 1,090 Hz instead of the correct 1,000 and 1,050 Hz. On the other hand, a relative starting phase of 45° will cause the partial null between the peaks in Figure 8-9c to nearly disappear. The two peaks can be reliably resolved, regardless of initial relative phase, only if separated by approximately the Rayleigh resolution or more.

8.4.8 Receiver Bandwidth Effects

The modulated pulse train spectrum is not strictly band-limited; it has some energy out to $f = \pm\infty$. When the modulated pulse train reflects from a target and is received by the radar, the radar's receiver passband bandwidth will determine how much of the target signal energy is captured by the radar. As a starting point, the receiver bandwidth might be set to the reciprocal of the pulse length, $\pm 1/\tau$, centered at f_0, a total passband width of $2/\tau$ Hz. This choice is based on the bandwidth of the main lobe of the broader sinc function term of equation (8.24). It is sufficient to capture about 91% of the total energy in the broad sinc (i.e., 9% of the energy is in the sidelobes), only 0.43 dB less than the full energy of the sinc. The fraction of the total signal energy captured will depend on the density of the spectral lines within this sinc main lobe but will be similar to the 91% figure.

As the receiver bandwidth is increased, the amount of captured target energy increases, but only slightly. At most, only an additional 0.43 dB of signal power can be obtained by increasing the bandwidth. The amount of uniformly distributed (white) noise power at the receiver output, however, is proportional to the receiver bandwidth and therefore increases linearly as the bandwidth is increased. Depending on the level of the noise spectrum, there will be some point at which the noise level increases faster than the signal energy as the receiver passband is widened. Once this point is reached, the signal-to-noise ratio (SNR) at the receiver output will decrease with further increases in receiver passband width. On the other hand, if the bandwidth is reduced from a value of $2/\tau$, the output signal and noise powers are reduced at about the same rate. Therefore, the SNR is usually maximized when the receiver passband width is approximately $2/\tau$ Hz.

Another effect of reducing the receiver bandwidth is that the fidelity of the target signal is lost. Specifically, as the passband of the receiver is reduced below $f_0 \pm 1/\tau$ Hz, the effective width of the individual time domain pulses increases and the leading and trailing edges develop a more gradual slope, as depicted notionally in Figure 8-10. The degraded pulse echo shape has the effect of reducing the range resolution of the system (see Chapter 20) and the range tracking accuracy (Chapter 18). When using a simple pulse waveform, choosing the receiver bandwidth to be about $2/\tau$ Hz provides near-optimum SNR, resolution, and tracking accuracy. If the receiver filter frequency response is not a good approximation to an ideal rectangular response, the optimum bandwidth may be somewhat wider. In many cases the optimum bandwidth may be $2.4/\tau$ or $2.6/\tau$ Hz.

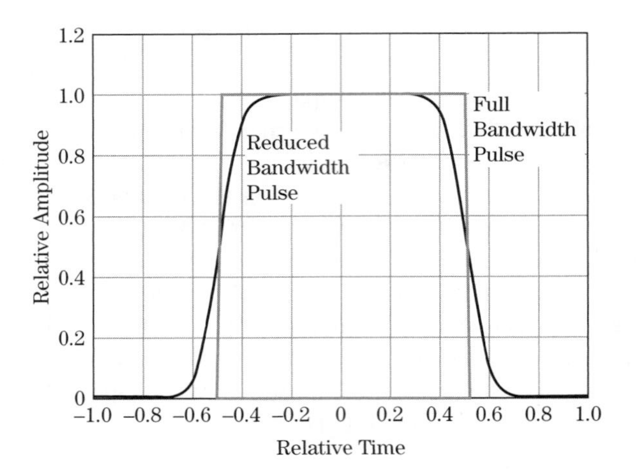

FIGURE 8-10 ■ Notional illustration of the effect of reduced bandwidth on pulse fidelity.

8.5 WHY MULTIPLE PULSES?

Why is the spectrum of a series of pulses, rather than a single pulse, important in radar? One answer to this question can be had by considering typical values of Doppler shift. As an example, compute the number of cycles in a 10 μs X-band (10 GHz) pulse, with and without a rather high Mach 1 ($\sim$340 m/s = 22.7 kHz Doppler shift at X-band) Doppler shift. With no Doppler shift, the number of cycles is $(10^{10}$ cycles/sec$)(10^{-5}$ sec$) = 10^5 = 100{,}000$ cycles. With the Doppler shift, the number is $(10^{10} + 2.27 \times 10^4)(10^{-5}) = 100{,}000.227$ cycles, a change of only about one quarter of a cycle. It seems unlikely that such a small change in the waveform could be reliably measured. Put another way, 22.7 kHz Doppler shift is not resolvable in only 10 μs because the Doppler resolution is no better than 100 kHz, similar to the example of Figure 8-9 but with all frequencies higher by a factor of 1,000. The case described here would correspond to a frequency spacing of two signals that is only about half that of Figure 8-9d. If in addition one of the signals was weaker than the other and noise was present, the two signals could not be resolved.

This example is not contrived. In most radar systems, the bandwidth of a single pulse may be a few orders of magnitude greater than the expected Doppler frequency shift (i.e., $1/\tau \gg f_d$). Thus, the spectrum of the Doppler-shifted echo from a moving target is not significantly shifted from the spectrum of the stationary (and probably stronger) clutter echoes. Remembering that the frequency resolution in the Doppler process is proportional to the reciprocal of the waveform duration, the returns from many (often 20 or 30) consecutive pulses over a CPI $T_d \gg \tau$ must be analyzed in the frequency domain so that the single-pulse spectrum will separate into individual spectral lines with bandwidths approximately given by $1/T_d$. Then, as was implied in Figure 8-10, the Doppler frequency shift of the moving object is larger than the width of the individual spectral lines. The moving target spectral lines can then be separated from any stationary object returns.

In rare cases, the Doppler shift is greater than the single-pulse bandwidth, making it possible to measure Doppler shift with a single pulse. For example, a mach 20 target (6,700 m/s) detected at Ka-band (35 GHz) will have a Doppler frequency of about 1.6 MHz, which is significantly greater than the Doppler resolution of a 10 microsecond pulse. The resulting Doppler frequency will produce several (16) cycles of Doppler within a pulse, providing the ability to detect the target motion. However, this is unusual, requiring a wide pulse, short wavelength, and very fast target. Since these conditions do not apply for most radar systems, only multiple pulse processing will be considered further.

An additional benefit of using multiple pulses for Doppler frequency estimation is that the required processing is a form of *coherent integration* that will improve target detection performance compared to a single-pulse measurement. Coherent integration and target detection are discussed in Chapter 15.

More realistic Doppler spectra are considered in Sections 8.8 and 8.9, but first it is necessary to discuss means for actually measuring Doppler shift in a modern radar system using a multiple pulse waveform. Before this can occur, however, the received signal must be detected. *Detection* in this sense means removal of the RF carrier frequency, a process also called *downconversion* or *demodulation*, to translate the echo signal energy to be centered at DC, where it is more easily processed. The downconverted signal is referred to as a *baseband* or *video* signal.

8.6 PULSED RADAR DATA ACQUISITION

8.6.1 Video Detectors and Phase Shift

Consider a single transmitted sinusoidal pulse at the radar's carrier frequency, f_0, centered at time $t = 0$:

$$\begin{aligned} x(t) &= A\cos(2\pi f_0 t + \theta), \quad -\frac{\tau}{2} \le t \le \frac{\tau}{2} \\ &= \text{Re}\{Ae^{j(2\pi f_0 t + \theta)}\} = \text{Re}\{(Ae^{j\theta})e^{j2\pi f_0 t}\}, \quad -\frac{\tau}{2} \le t \le \frac{\tau}{2} \end{aligned} \tag{8.25}$$

Assuming the RF frequency is known (as it normally would be), the pulse can be characterized by its amplitude A and phase θ, which together form the *complex amplitude* $Ae^{j\theta}$. Geometrically, the complex amplitude represents a vector of length A at an angle of θ with respect to the positive real axis in the complex plane. A coherent radar is designed to detect the rectangular components of the complex amplitude. Before discussing that process, however, it is useful to consider a single-channel detector.

Suppose the pulse of equation (8.25) reflects from a target at range, R_0. The received pulse will be

$$\begin{aligned} y_I(t) &= x\left(t - \frac{2R_0}{c}\right) \\ &= A'\cos\left[2\pi f_0\left(t - \frac{2R_0}{c}\right) + \theta\right], \quad -\frac{\tau}{2} + \frac{2R_0}{c} \le t \le \frac{\tau}{2} + \frac{2R_0}{c} \\ &= A'\cos\left[2\pi f_0 t + \theta - \frac{4\pi}{\lambda}R_0\right], \quad -\frac{\tau}{2} + \frac{2R_0}{c} \le t \le \frac{\tau}{2} + \frac{2R_0}{c} \\ &= \text{Re}\left\{A'\exp\left[j\left(\theta - \frac{4\pi}{\lambda}R_0\right)\right]\exp\left[j2\pi f_0 t\right]\right\}, \quad -\frac{\tau}{2} + \frac{2R_0}{c} \le t \le \frac{\tau}{2} + \frac{2R_0}{c} \end{aligned} \tag{8.26}$$

Compared with the envelope of the transmitted pulse, that of the received pulse has a new amplitude A' (modeled by the radar range equation) and is delayed in time by $2R_0/c$.

In addition to the delay of the pulse envelope, the phase of the received pulse is also shifted by $(-4\pi R_0/\lambda)$ radians. This phase shift, proportional to the range of the target, is

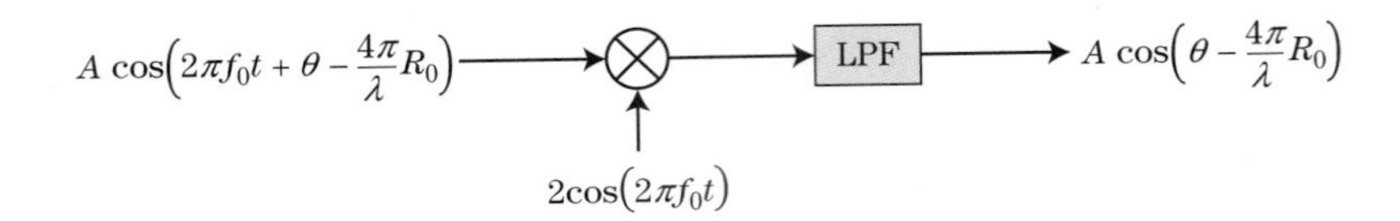

FIGURE 8-11 ■ Single-channel detector.

extremely important in coherent radar signal processing. For instance, a change in R_0 of $\lambda/4$ is sufficient to change the phase by π radians. Measuring phase changes thus provides a way to measure subwavelength range changes. This capability is the key to successful Doppler processing, adaptive interference cancellation, and radar imaging.

The goal of the radar is to measure the unknown parameters of the received pulse, namely, the amplitude A', time delay $t_0 = 2R_0/c$, and phase shift $\theta' = \theta - (4\pi/\lambda)R_0$. A block diagram of an idealized single-channel detector that achieves this is shown in Figure 8-11. The input is the sinusoidal pulse echo of equation (8.26). This signal is mixed with a reference oscillator at the RF. A simple trigonometric identity for the product of two cosine functions shows that, during the time that the input pulse is present, the signal at the output of the mixer consists of two terms, a sum frequency term $A'\cos[2\pi(2f_0)t + \theta']$ and a difference frequency term, $A'\cos\theta'$. The low-pass filter removes the sum frequency term so that the output is a sinusoidal pulse at a frequency of zero Hz, *i.e.* a constant pulse like that of Figure 8-4 with an amplitude of $A'\cos\theta'$. Note that this is just the real part of the complex amplitude $A'\exp(j\theta')$ of the received pulse. The voltage at the output of the detector is called the baseband or radar video signal. If the output of the detector is sampled at some time during the pulse, say at $t = 2R_0/c$, the measured value will be $y_I[0] = A'\cos\theta'$. The reason for the index [0] will be apparent shortly.[3]

8.6.2 Coherent Detector

The output voltage from the single channel detector of Figure 8-11 is not sufficient to identify both A' and the total phase. There are an infinite number of combinations of A' and θ' that will produce the same product $y_I[0]$. Furthermore, $\cos\theta' = \cos(-\theta')$, so the detector cannot distinguish between positive and negative phase shifts and therefore between positive and negative range changes.

This problem is solved by using the two-channel *coherent* or *I/Q detector* of Figure 8-12. This configuration splits the incoming signal into two channels. The upper channel signal is mixed with the reference oscillator $2\cos(2\pi f_0 t)$ as before, while the lower channel uses a reference oscillator of $2\sin(2\pi f_0 t)$. The lower oscillator is therefore 90° out of phase with the upper oscillator, a condition referred to as being *in quadrature* with respect to the upper channel. The upper channel is called the *in-phase* or *I channel* because it is the real part of the corresponding complex sinusoid; the lower is called the *quadrature* or *Q channel* and is the imaginary part of the complex sinusoid. A simple analysis of the Q channel similar to the previous example shows that its output voltage, when sampled at the appropriate time delay, is $y_Q[0] = A'\sin\theta'$ while the I channel output remains $y_I[0] = A'\cos\theta'$.

[3]The exact form of the result for $y[0]$ is valid under the "stop-and-hop" assumption, whereby it is assumed that the target movement while the pulse is in flight is negligible. This is an excellent approximation in most cases since the speed of propagation is much higher that target velocity. The stop-and-hop assumption is discussed in [5].

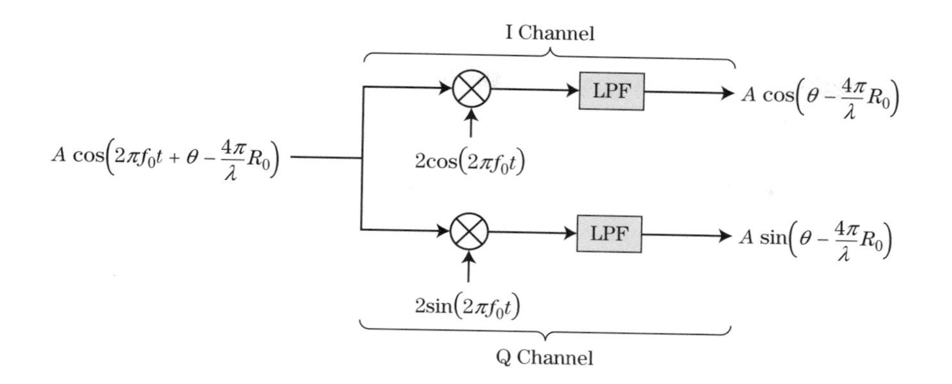

FIGURE 8-12 ■ Coherent or I/Q detector.

By itself, the Q channel does not resolve the ambiguities in measuring A' and θ'. However, a new, complex-valued signal y can be formed from the I and Q outputs:

$$\begin{aligned} y[0] &= y_I[0] + j \cdot y_Q[0] \\ &= A'(\cos\theta' + j\sin\theta') \\ &= A'\exp(j\theta') \end{aligned} \tag{8.27}$$

This complex output allows independent measurement of amplitude and phase. Specifically, $A' = |y[0]|$ and $\theta' = \tan^{-1}\theta'$.[4]

Repeating the coherent detection process for multiple samples forms a complex discrete-time signal $y[m] = y_I[m] + j \cdot y_Q[m]$ called the *analytic signal* [4]. The Q signal $y_Q[m]$ is the negative of the Hilbert transform of the I signal $y_I[m]$.

The diagrams in Figures 8-11 and 8-12 are useful models of the radar receiver and detector, but real receivers are more complex. In particular, downconversion is usually not performed with a single demodulation, but in two or more steps. The first steps translate the signal from RF to one or more IFs, while the last step translates that result to baseband. Chapter 11 discusses practical receiver architectures, and Chapter 10 describes the corresponding transmitter architectures to generate the radar signals. Chapter 11 also describes alternative receiver structures that derive the I and Q signals digitally from a single analog channel, a process often called *digital I/Q* or *digital IF* processing.

8.6.3 Range Bins

Use of the complex detector allows measurement of the received signal amplitude and phase. The time delay, t_0, is estimated by sampling the receiver output repeatedly after a pulse is transmitted and observing the time at which the echo is received. Time samples are generally taken at a spacing no greater than the time resolution of the radar pulse. For a basic pulse, this is simply the pulse length, τ. Thus, the receiver output is sampled every τ seconds from some initial time, t_1, to some final time, t_2. Since the echo received from a single scatterer is also τ seconds long, the echo will be present and contribute to only one time sample, say, at t'. Assuming the signal is strong enough to be detected, the time

[4] Care must be taken with the arctangent to place the result in the correct quadrant by taking into account the signs of y_I and y_Q.

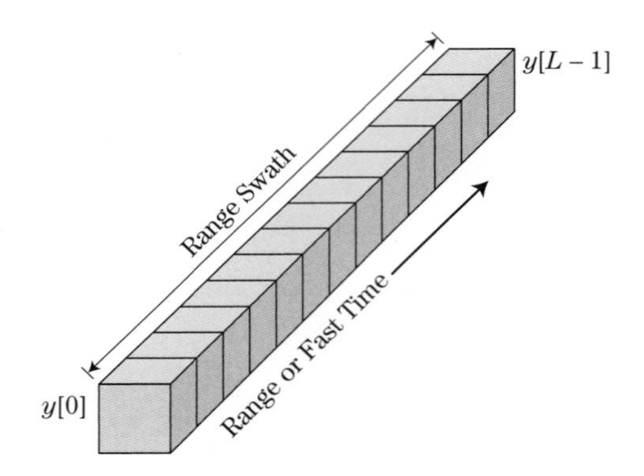

FIGURE 8-13 ■ Range bins and range swath. Each cube represents a single complex voltage measurement.

delay to the scatterer is estimated to be t' seconds, and the corresponding range estimate is $R' = ct'/2$ meters.

For the modulated pulses used for pulse compression (see Chapter 20), the sample spacing is generally is $1/B$ seconds, where B is the bandwidth of the pulse, corresponding to the compressed resolution of $c/2B$ m. Before pulse compression, the echo from a single pulse is spread out over $B\tau$ seconds. After pulse compression, the main lobe of the compressed response is limited to a single sample. (This is increased to typically 1.5 to 2 samples if windowing is used for sidelobe reduction.)

For each time sample, whether a received pulse is present, the output of the receiver will be sampled and a complex voltage, y, formed from the I and Q channel samples. This series of complex voltage measurements is typically stored in a computer memory as a one-dimensional vector as shown in Figure 8-13. The vector of samples is referred to by several names, including *range bins*, *range gates*, *range cells*, and *fast-time samples*. The interval from the range corresponding to the first sample to that corresponding to the last, that is, from $R_1 = ct_1/2$ to $R_2 = ct_2/2$ is called the range *swath*. Radars may have from as few as one to as many as several thousand range bins. The number of range bins can vary significantly in a single radar as it operates in different modes.

8.6.4 Pulsed Radar Data Matrix and Datacube

When the radar transmits M pulses in a dwell or CPI, a set of range gates like those of Figure 8-13 will be measured for each pulse. These are typically stored in memory as a two-dimensional matrix of complex voltage samples as shown in Figure 8-14a. The interval between samples in a row is the PRI, so the sampling rate in this dimension is the PRF. Because the PRF is much lower than the sampling rate in range, the pulse number axis is also called the *slow-time* dimension. Each row of the matrix represents a series of measurements from the same range bin over M successive pulses. The total amount of time MT represented by the data matrix is the dwell time or CPI.

Some radars operate multiple receivers simultaneously. Examples include phased array radars where the antenna is divided into subarrays, each with its own receiver. Another is a monopulse radar, where the antenna has three output channels (i.e., sum, azimuth difference, elevation difference), each with its own receiver. Each receiver will generate its own fast-time/slow-time data matrix for each CPI. It is common to represent the complete set of data by stacking these matrices into a *datacube* as shown in Figure 8-14b. The vertical axis is often called the receiver channel or *phase center* axis.

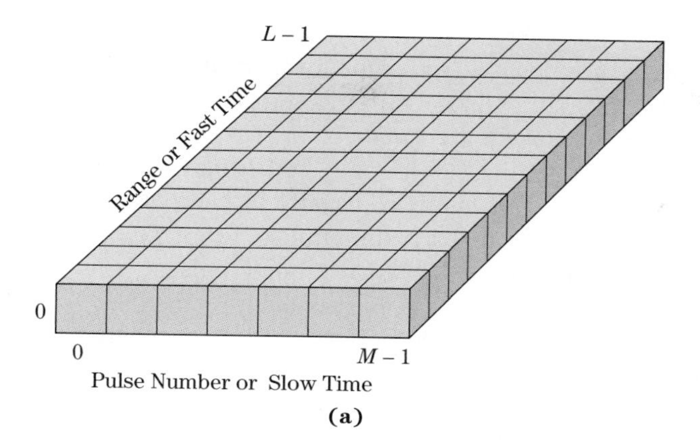

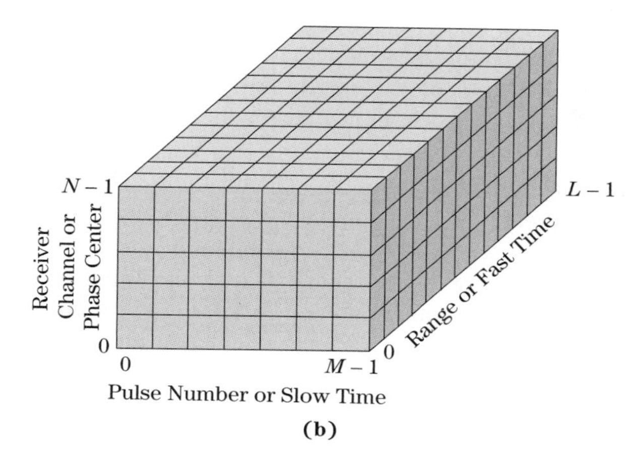

FIGURE 8-14 (a) Fast-time/slow-time CPI data matrix. (b) Datacube.

8.7 DOPPLER SIGNAL MODEL

8.7.1 Measuring Doppler with Multiple Pulses

Now suppose the same target considered in the last section is approaching the radar with a radial velocity, v, and the radar transmits a series of M pulses separated by a PRI of T seconds. The range to the target when the m-th pulse $(0 \leq m \leq M-1)$ is transmitted is $R_0 - mvT$ meters. The phase shift of the echo of the m-th pulse, following exactly the same argument as before, will be $(-4\pi/\lambda)(R_0 - vmT)$ radians. When the coherent detector output is sampled sometime during the echo (e.g., at $t = 2R_0 + mT$), the measured output for the m-th pulse will be

$$\begin{aligned}
y[m] &= A\exp\{j[\theta - (4\pi/\lambda)(R_0 - vmT)]\} \\
&= A\exp\left\{j\left[2\pi\left(\frac{2v}{\lambda}\right)(mT) + \theta - \left(\frac{4\pi R_0}{\lambda}\right)\right]\right\} \\
&= A\exp[j(2\pi f_d t_m + \theta')], \quad 0 \leq m \leq M-1
\end{aligned} \tag{8.28}$$

where $t_m = mT$ is the transmit time for the m-th pulse. Equation (8.28) shows that the sampled signal formed by measuring the phase of each successive pulse echo with a multi-pulse waveform forms a discrete-time complex sinusoid at the expected Doppler frequency! The signal is sampled at times t_m separated by the PRI. The sinusoid is the result of the changing echo phases, which in turn are caused by the changes in target range between pulses. Doppler shift measured from a series of phase measurements in this way is sometimes referred to as *spatial Doppler*.

The PRI is greater than the pulse length, sometimes by a large factor. In addition, the measurement is made with multiple (usually in the low tens, but sometimes much more) pulses. The resulting total observation time (CPI) of $T_d = MT$ seconds is typically an order of magnitude or more longer than a single pulse, giving a Doppler resolution of $1/MT$ Hz that is at least an order of magnitude finer than the $1/\tau$ Hz resolution of a single pulse. By choosing the CPI appropriately, it is possible to measure Doppler shifts on the scale needed in radar. For example, 20 pulses at a moderate PRF of 5 kHz gives a Doppler resolution of 250 Hz. Referring back to Table 8-1, this is a velocity resolution of 37 m/s (84 mph) in a 1 GHZ radar, and 3.7 m/s (8.4 mph) in a 10 GHz radar; recall that the single-pulse X-band measurement in Section 8.5 could not resolve even a Mach 1 Doppler shift. With the moving targets and clutter resolved in frequency, the detector output can now be filtered to separate moving targets from stationary targets. Moving target indication and pulse-Doppler processing techniques to do this are discussed in Chapter 17.

A common concern in measuring Doppler with finite pulse trains is that a moving target will move from one range bin to another over the course of a CPI, so its echo will not be present in all the pulses for a given range bin and equation (8.28) may not apply. That is, the target may not stay in one range bin over the entire CPI. For the velocities and CPI durations involved in Doppler processing, this *range migration* is rarely an issue. Consider a radar with relatively good 100 Hz Doppler resolution. The CPI is therefore a relatively long $1/100$ Hz $=$ 10 ms. A typical pulse length in this scenario is at least $\tau = 1\ \mu$s, so that the range bin spacing will be at least $c\tau/2 = 150$ m. An aircraft closing on the radar at a high velocity of Mach 2 (about 680 m/s) travels only 6.8 m during the CPI. Since this is much less than a range bin, it is generally safe to assume the target stays in the same range bin for the entire CPI and that equation (8.28) is a valid model. On the other hand, in Chapter 21 it will seen that range migration is very common in imaging radars due to a combination of much shorter range bins with much longer CPIs.

8.7.2 Coherent Pulses

To measure Doppler shift using multiple pulses, a deterministic phase relationship from pulse to pulse must be maintained over the CPI so that phase shifts measured by the coherent radar are due to relative radar-target motion only. Pulse trains that have this quality are called *coherent* pulse trains, and radars that can generate coherent pulse trains and measure the phase of the echoes are called coherent radars.

Figure 8-15 shows two pulse pairs—one coherent and one not. The center waveform is the output of a very stable oscillator. It provides a constantly running reference frequency for the receiver (as well as for the transmitter). The two upper pulses are both, when they are “on,” in phase with this reference oscillator. Mathematically, they both have exactly the same sinusoidal formula; only the position of the rectangular pulse envelope $A(t)$ changes to turn the successive pulses on and off at different times t_1 and t_2, respectively.

This is not true of the two pulses at the bottom of the figure. The first pulse is the same as the first pulse of the coherent pair, and is in phase with the reference oscillator. The second pulse has the same frequency but is out of phase with the reference oscillator.

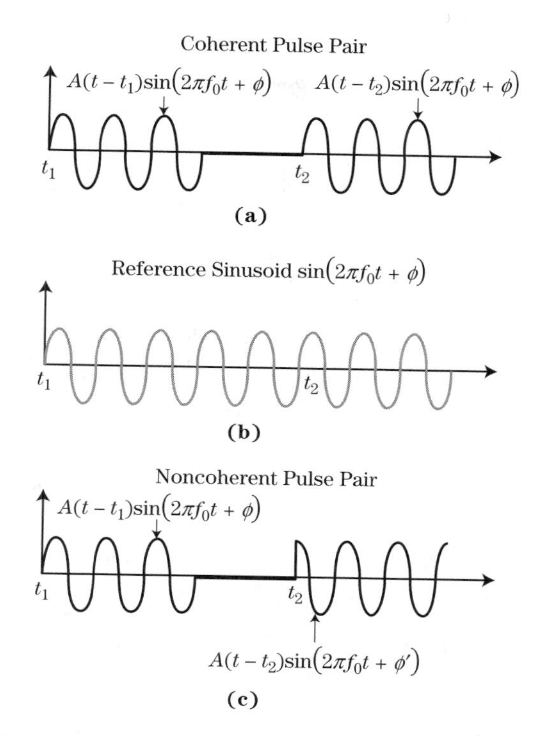

FIGURE 8-15 (a) Coherent pulse pair. (b) Reference oscillator. (c) Noncoherent pulse pair.

Thus, the sinusoidal term of the second pulse formula is different. This pair of pulses would be considered *noncoherent*.[5]

Radar coherency is obtained by deriving the transmit RF signal from very stable oscillator sources that exhibit very little phase drift. The same oscillators are used in the transmit process and in the receiver downconversion or demodulation process. Any undesired phase shift in these oscillators will impart false phase modulations on the measured echo data. These errors often appear as a false Doppler signature on all objects, including stationary objects. Since phase drifts and phase noise build up over time, they can limit Doppler processing capability at longer ranges in some systems. Practical architectures for generating and receiving coherent radar signals are described in Chapters 10 and 11, respectively.

8.8 | RANGE-DOPPLER SPECTRUM FOR A STATIONARY RADAR

8.8.1 Elements of the Doppler Spectrum

The signal at the output of a real coherent radar detector is the superposition of some or all of the following components:

- Echoes from one or more moving or stationary targets.
- Receiver noise.

[5]If the actual phase of each pulse can be measured on transmit, then the phase difference can be compensated on receive, so coherent processing is possible. Radars that do this are called *coherent on receive* systems. It is preferable to make the system coherent to begin with.

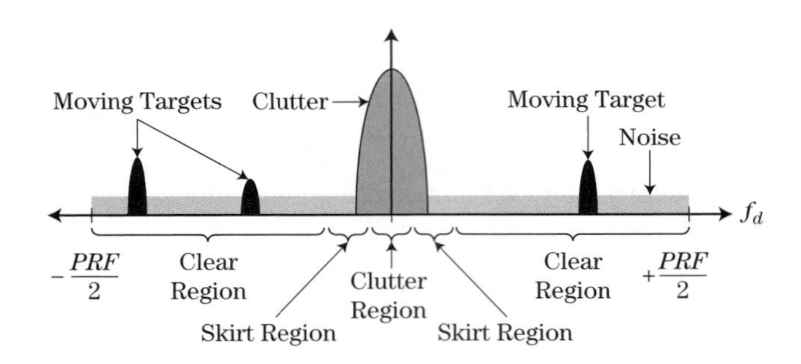

FIGURE 8-16 ■ Notional Doppler spectrum for one range bin, viewed by a stationary radar.

- Atmospheric and galactic noise.
- Spurious receiver mixer products.
- Echoes from clutter.
- Electronic countermeasures (ECM, also known as electronic attack [EA]) signals (jammers).
- Electromagnetic interference (EMI) signals, such as cell phones, other radars, or other services operating in the same or nearby frequency bands.

Not all of these are present in every case. In this chapter, attention is focused primarily on targets, clutter, and noise. Also, the radar itself may be stationary or moving (e.g., on an aircraft or satellite).

The Doppler spectrum for a range bin is the *discrete-time Fourier transform* (DTFT) of the slow-time data for that bin (see Chapter 14). Each of the aforementioned signal types will have different Doppler spectrum "signatures." The Doppler spectrum observed at the radar output will be the superposition of the contributions from each signal source.

Consider a stationary coherent radar first, and suppose its antenna is oriented such that it illuminates a scene where a particular range bin contains some ground clutter and three moving targets, one approaching the radar and two receding. A notional view of the Doppler spectrum might appear as in Figure 8-16. Only the portion of the spectrum between $\pm PRF/2$ is shown because the spectrum of the sampled data repeats periodically outside of this region. The energy around zero Doppler shift is echo from essentially stationary clutter, such as buildings, trees, and grass. As discussed in Chapter 5, real clutter has some Doppler spread that depends on the type of clutter and the weather conditions. In addition, clutter from air conditioning fans, automobile traffic, and other not-quite-stationary sources can spread the clutter spectrum.[6] The low-level energy spread uniformly across the spectrum is the white noise generated within the receiver.

Finally, the three "blips" of energy, two at negative Doppler and one at positive Doppler, represent the three moving targets. Their specific locations on the Doppler axis depend on their individual radial velocities with respect to the radar and their amplitudes relative to the noise are determined by their individual SNRs. If the target radial velocity, v, is in the interval $(-\lambda \cdot PRF/4, +\lambda \cdot PRF/4)$, then the Doppler shift will be between $(-PRF/2, +PRF/2)$ and the Doppler peaks will be indicative of the actual radial velocity.

[6] Radar platform motion can significantly spread the clutter spectrum as shown in Section 8.9, but the discussion is limited to stationary radar here.

The Doppler shift, f_d, of targets with a velocity outside this range will alias to an apparent Doppler of $f_d + k \cdot PRF$, where the integer k is chosen such that the result falls in the range $(-PRF/2, +PRF/2)$. Chapter 17 discusses resolution of these ambiguities. The amplitude of the clutter with respect to the noise floor is determined by the *clutter-to-noise ratio* (CNR). The CNR is computed using the radar range equation, but with the target RCS replaced by the total RCS of a clutter resolution cell as discussed in Section 2.13 and again in Section 5.1.

The portion of the Doppler spectrum that is free from clutter, so that noise is the dominant interference, is often called the *clear region*. This does not imply that targets at these Doppler shifts are clear of all interference, just of clutter. Noise is present at all Doppler shifts. The region of the spectrum where the clutter signal is stronger than the noise, so that clutter is the dominant interference, is called the *clutter region*. Sometimes a *skirt* or *transition region* is defined where the clutter and noise powers are approximately equal.

8.8.2 Range-Doppler Spectrum

Normally, there are many range bins. Targets may be present in several range bins, and the clutter will be distributed across many range bins. Furthermore, the clutter strength will vary with range, in part due to the fall-off in range due to the radar range equation and in part due to changes in the type of terrain producing the clutter at different ranges. Noise is present at all Doppler shifts and all ranges. Figure 8-17 is a notional illustration of the distribution of targets, clutter, and noise in both range and Doppler. The noise floor is constant through the range-Doppler space. The three targets are shown at approximately the same Doppler frequencies as in Figure 8-16 but are now in different range bins. Clutter is shown as the component centered on zero Doppler running through all of the range bins. The narrowing and lightening of the clutter ridge suggests the fall-off of the clutter power with range.

A range-velocity spectrum for a similar scenario using a 10 GHz radar, generated with a computer simulation, is shown in Figure 8-18. The left portion shows the distribution of power versus range and velocity in a three-dimensional format. The clutter ridge and

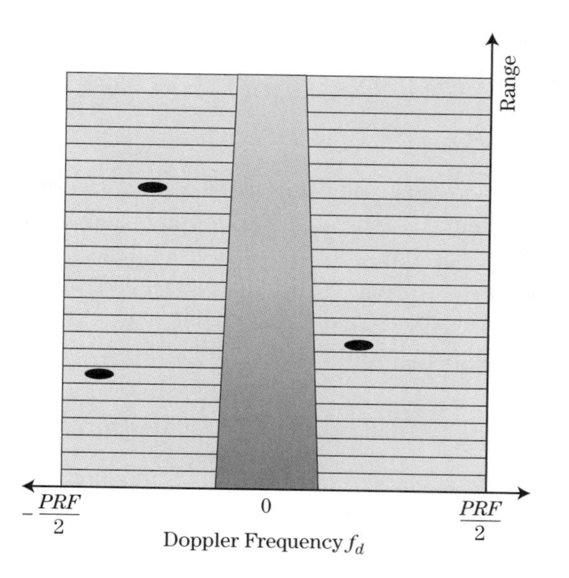

FIGURE 8-17 ■ Notional range-Doppler distribution viewed by a stationary radar.

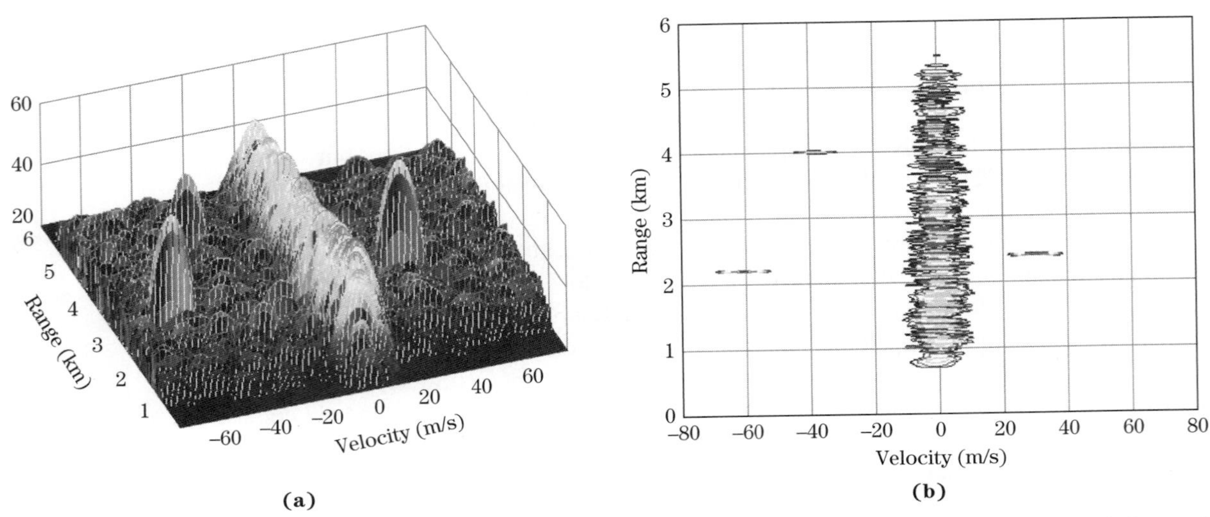

FIGURE 8-18 ■ Simulated range-Doppler distribution for a stationary radar with clutter and three moving targets. (a) Three-dimensional display. (b) Contour plot.

the three point targets are clearly visible above the noise floor. Part (b) of the figure shows the same data in a contour plot, with the lowest contour high enough to exclude the noise floor. (Figure 8-17 is essentially a stylized version of this contour plot.) This format makes it especially easy to visualize the target and clutter distribution.

The width of the point targets in Doppler is determined by the dwell time. A total of 30 pulses at a PRI of 100 μs were used, making the dwell time $T_d = 3$ ms. The Rayleigh resolution in Doppler is then $1/3$ ms $= 333$ Hz, corresponding to 5 m/s at 10 GHz, or 10 m/s null-to-null. The Doppler spectrum was computed with a Hamming window on the data to suppress Doppler sidelobes at the cost of doubling the Rayleigh bandwidth (see Chapter 14), so that the width of the peaks in Doppler is about 20 m/s, as can be seen in the figure.

8.9 | RANGE-DOPPLER SPECTRUM FOR A MOVING RADAR

8.9.1 Clutter Spreading

A moving radar not only adds a Doppler shift to the echo of stationary clutter according to equation (8.4), but it also induces a spread in the Doppler bandwidth of stationary clutter in the radar mainbeam. This is most relevant in air-to-ground radars. Figure 8-19 illustrates in two dimensions an approach to estimating the Doppler spread caused by radar platform motion. The 3 dB radar beamwidth is θ_3 radians. Recall that the Doppler shift for a radar moving at velocity v with its boresight squinted to a target ψ radians off the velocity vector is

$$f_d = \frac{2v}{\lambda}\cos\psi = f_{MLC} \text{ Hz} \tag{8.29}$$

Applied to stationary ground clutter observed from a moving platform, this formula describes the Doppler shift of the clutter on the radar boresight. The symbol f_{MLC} emphasizes that this is the Doppler shift of the main lobe clutter.

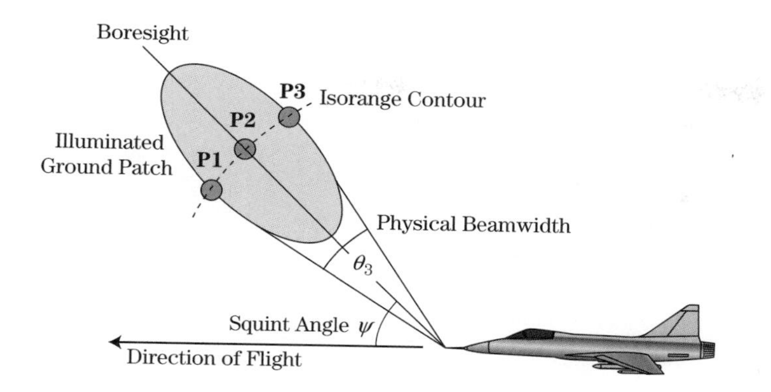

FIGURE 8-19 ■ Geometry for computing Doppler spread induced by radar platform motion.

Now consider three point scatterers **P1**, **P2**, and **P3**, each at the same range from the radar. **P1** and **P3** are at the 3 dB edges of the antenna beam, while **P2** is on boresight. Because all three are at the same range, the received echo at a delay corresponding to that range is the superposition of the echoes from all three scatterers. However, each is at a slightly different angle with respect to the aircraft velocity vector. **P2** is on the boresight at the squint angle of ψ, but **P1** and **P3** are at $\psi \pm \theta_3/2$ radians. The difference in the Doppler shift of the echoes from **P1** and **P3** is

$$\begin{aligned} B_{MLC} &= \frac{2v}{\lambda}\{\cos(\psi - \theta_3/2) - \cos(\psi + \theta_3/2)\} \\ &= \frac{4v}{\lambda}\sin\left(\frac{\theta_3}{2}\right)\sin\psi \end{aligned} \tag{8.30}$$

Radar antenna beamwidths are small, typically less than 5 degrees. Applying a small angle approximation to the sin(θ_3/2) term in equation (8.30) gives a simple expression for the variation in Doppler shift of the main lobe clutter across the beam due to platform motion:

$$B_{MLC} \approx \frac{2v\theta_3}{\lambda}\sin\psi \text{ Hz} \tag{8.31}$$

The total Doppler bandwidth is approximately the sum of the bandwidth induced by platform motion and the intrinsic bandwidth of the scene being measured. Equation (8.31) assumes the radar is squinted sufficiently that the main beam does not include the velocity vector, that is, $|\psi| > \theta_3/2$. The formula must be modified if the main beam straddles the velocity vector.

As an example, an L-band (1 GHz) side-looking ($\psi = 90°$) radar with a beamwidth of 3° traveling at 100 m/s will induce $B_d \approx 35$ Hz, while an X-band (10 GHz) side-looking radar with a 1° beam flying at 200 m/s will induce $B_{MLC} \approx 233$ Hz. This spreading of the clutter bandwidth can be either good or bad, depending on the radar's purpose. It complicates the detection of slow-moving surface targets from airborne platforms. On the other hand, this phenomenon provides the basis for obtaining cross-range resolution in imaging radars.

8.9.2 Clutter Spectrum Elements

The notional Doppler spectrum of Figure 8-16 can be greatly complicated by radar platform motion or Doppler ambiguities and by aliasing of the spectrum at low PRFs. Figure 8-20

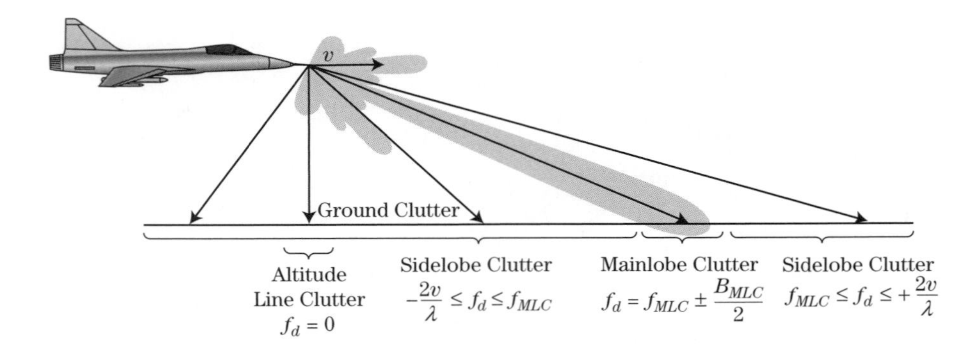

FIGURE 8-20 ■ Geometry for computing Doppler spread induced by radar platform motion.

is another view of a moving aircraft viewing stationary ground clutter, emphasizing that different patches of ground will provide echoes with different Doppler shifts due to different angles with respect to the velocity vector. In addition, these clutter echoes will have different strengths, depending on range and whether they are viewed through the antenna sidelobes or main lobe.

Figure 8-21 illustrates the resulting effects on the spectrum. For simplicity, all the signal components are shown in a single range bin's spectrum, though in reality, different components and targets will be in different range bins.

To begin, the entire Doppler spectrum is shifted by the nominal radar-to-ground Doppler shift of $f_{MLC} = 2v\cos\psi/\lambda$ Hz while the main lobe is widened by B_{MLC} Hz due to clutter spreading as described already. In addition to this broadened *main lobe clutter* (MLC), *sidelobe clutter* (SLC) and an *altitude line* (AL) are now evident as well. Sidelobe clutter results from energy radiated and received through the radar sidelobes. It is thus weaker than the main lobe clutter. Since sidelobes and backlobes exist in all directions, it is possible, depending on the aircraft direction, to receive echoes from ground or air clutter directly ahead of the aircraft as well as directly behind it. Any clutter present in these directions has a radial velocity equal to the full velocity, v, of the platform if in the direction of flight, or the negative of the velocity if behind the platform. Clutter at other angles creates echo energy at velocities between $+v$ and $-v$. Sidelobe clutter did not appear in the stationary radar case of Figure 8-16 because the sidelobe echoes occur at zero Doppler and are thus part of the main lobe clutter.

The altitude line results from the echo from energy transmitted through the sidelobes of an airborne radar straight down to the ground and back. The altitude line appears at the value of the radial velocity component toward the ground; in level flight this is zero, regardless of the platform velocity or antenna look direction. Although it is transmitted and received through the radar sidelobes, the altitude line nonetheless tends to be relatively strong due to the relatively short vertical range and the high reflectivity of most clutter at normal incidence. In a pulsed system, as will be seen shortly, the AL and MLC do not

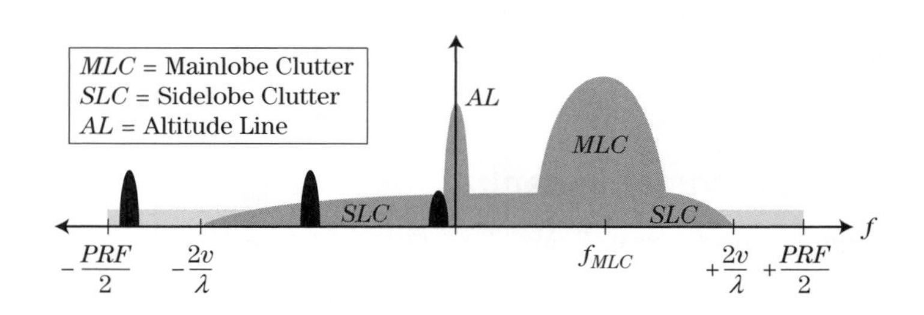

FIGURE 8-21 ■ Notional Doppler spectrum for moving radar platform (see text for details).

appear in the same range bin. The range to the altitude bounce is just the altitude of the platform, while the main lobe is usually steered to a longer range.

8.9.3 Range-Doppler Clutter Distribution

The spectrum of Figure 8-21 was used to introduce the major clutter components. However, clutter observed from a moving pulsed radar is distributed in complex patterns through the range-Doppler (equivalently, range-velocity) space. The total clutter power received in a given range-velocity cell is the sum of the power from all stationary clutter scatterers in the same range bin and having the same angle with respect to the platform velocity vector.

Since the Doppler shift of a stationary clutter scatterer is proportional to the cosine of the angle ψ between the velocity vector and the scatterer (often called the *cone angle* to the scatterer), all stationary scatterers located on a cone of half-angle ψ centered on the platform velocity vector will have the same Doppler shift. The intersection of a cone with a flat plane representing the ground is a conic section which can be either a hyperbola (if the radar travels parallel to the ground, i.e., level flight), a parabola (shallow dive), or ellipse (steep dive). Whatever its shape, this line is called an *isovelocity* or *isodoppler* contour (often *isodop* for short). Scatterers having the same range, R, from the radar are on the surface of sphere of radius, R, centered on the radar, which intercepts the ground plane on a circular locus called an *isorange* contour. Thus, scatterers at the intersection of a given isodop and isorange contour contribute power to the same range-Doppler cell.

Figure 8-22 is an example of circular isorange and hyperbolic isovelocity contours for a forward-looking radar traveling straight and level at 100 miles per hour at an altitude of 10,000 feet above the ground. Clutter scatterers at the two highlighted patches at a downrange coordinate of 68 km and cross-range of ±75 km have the same total range of 100 km and velocity of 30 m/s relative to the radar and will therefore both contribute to the same range-velocity or range-Doppler cell.

Figure 8-23 shows a complete range-velocity clutter distribution for the same radar motion and geometry, taking into account the antenna gain and the variation of clutter reflectivity with grazing angle. A sinc-squared two-way antenna pattern with a Rayleigh beamwidth of 3.5° and minimum sidelobe level of −35 dB, and a constant-gamma clutter reflectivity were assumed. Part (a) of the figure gives a three-dimensional view, while part (b) displays the distribution in a plan view.

There are several notable features of the clutter distribution. There is no clutter return at all (only noise) until a range of approximately 3 km, corresponding to the 10,000 foot

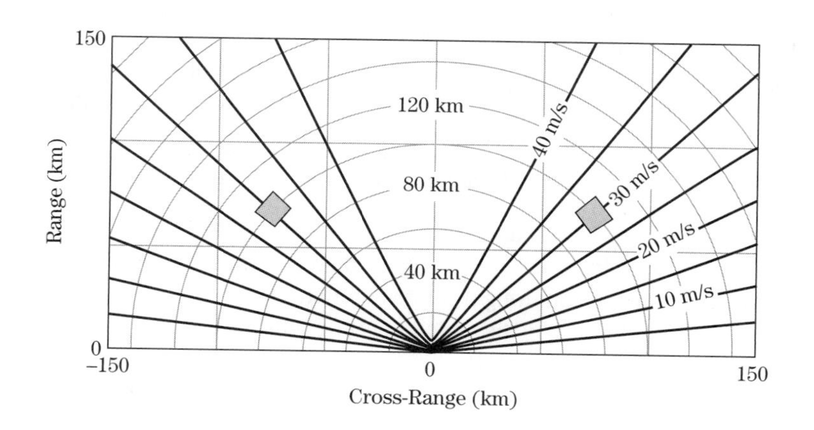

FIGURE 8-22 ■ Intersection of isorange and isovelocity contours.

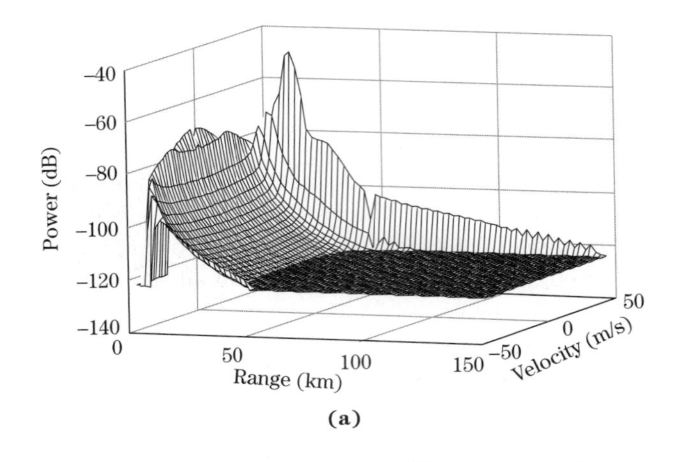

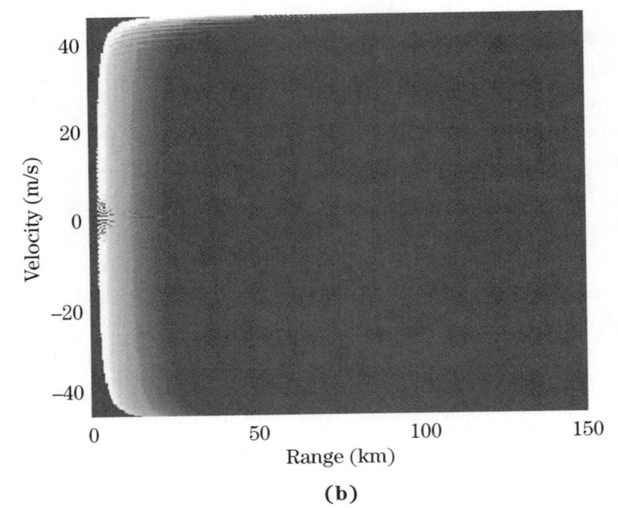

FIGURE 8-23 ■ Simulated range-velocity distribution for a stationary clutter observed from a moving radar (see text for details). (a) Three-dimensional display. (b) Plan view.

elevation of the aircraft. Note the relatively high clutter return strength at 3 km (the platform altitude) and zero velocity, which fades away rapidly as the range increases and the grazing angle decreases along the zero velocity line. This is the altitude line previously introduced; the high clutter return at short range and across all Doppler is called *near-in clutter*.

In this geometry, the boresight intercepts the ground at a range of just over 17 km and a grazing angle of only 10°. The strong peak at that range and a velocity nearly equal to the aircraft velocity is the main lobe clutter. Its greater amplitude is due to the main lobe antenna gain; the surrounding plateaus are due to the first few antenna sidelobes.

8.9.4 Range and Velocity Ambiguity Effects

The idea of a range ambiguity was introduced in Chapter 1. If the radar environment is such that the clutter echo from distances greater than the unambiguous range $R_{ua} = cT/2$ are significant, then clutter from longer ranges will "fold over" (alias) onto shorter ranges. This is more likely at higher PRFs (smaller T). Figure 8-24 illustrates this effect for a stationary radar at an altitude of 10,000 feet and a $-30°$ pointing angle, so that the boresight intercepts the ground at a range of about 5.3 km. The PRF is 20 kHz, giving

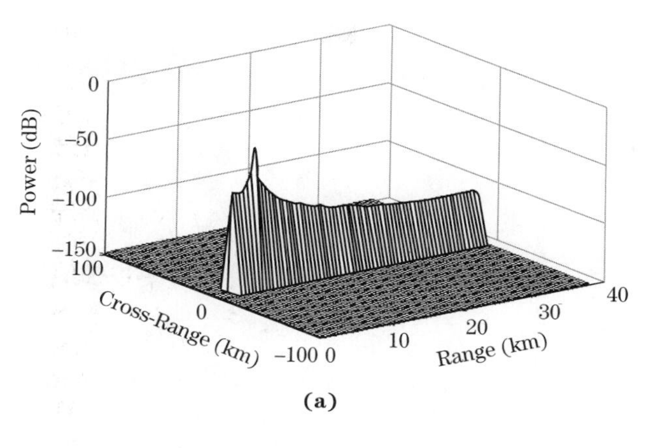

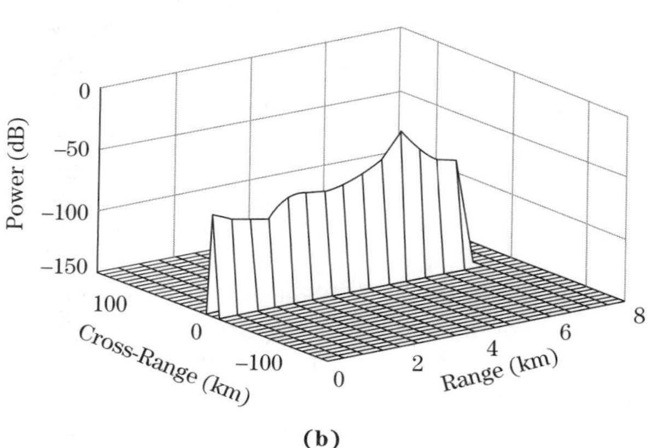

FIGURE 8-24 ■ Simulated range-velocity distribution for a stationary clutter observed from a stationary radar (see text for details). (a) Without clutter foldover. (b) With foldover.

an unambiguous range of 7.5 km. Figure 8-24a shows the actual distribution of clutter power versus range. The 3 km altitude, the main lobe peak at 5.3 km, and the fall-off in clutter power out to longer ranges are all evident. Figure 8-24b shows the clutter as would actually be measured by the radar. Note the change in range scales. Clutter in the intervals from, for example, 7.5 to 15 km and 15 to 22.5 "fold over" into the 0 to 7.5 km interval. While the main lobe peak is still apparent, the clutter power is generally flatter over range.

Ambiguities also occur in Doppler shift and thus in velocity. This is a consequence of Doppler being measured using the sampled slow-time data as was discussed in Section 8.6. The sampling rate in slow time is the PRF. The Nyquist sampling requirement (discussed in Chapter 14) limits the range of Doppler shifts that can be represented to $\pm PRF/2$, equivalent in velocity to $\pm v_{ua}/2$ where $v_{ua} = \lambda \cdot PRF/2$. If the platform velocity exceeds $v_{ua}/2$, then there will be clutter cells with Doppler shifts greater than $v_{ua}/2$. The echo from these cells will alias, wrapping around to a new value $v' = v + k \cdot v_{ua}$ for some integer k such that v' does fall within $\pm v_{ua}/2$. Velocity ambiguities are more likely to occur at low PRFs and high platform velocities.

Figure 8-25 is an example of the effect of velocity ambiguity on the range-velocity distribution of clutter. The conditions are the same as for Figure 8-23, except that the PRF has been lowered to 1 kHz. At the 10 GHz RF, this gives an unambiguous velocity of $v_{ua} = 15$ m/s, so the velocity range shown is ± 7.5 m/s. The clutter distribution is the same as in Figure 8-23b, except that now velocities greater than $+7.5$ m/s alias into

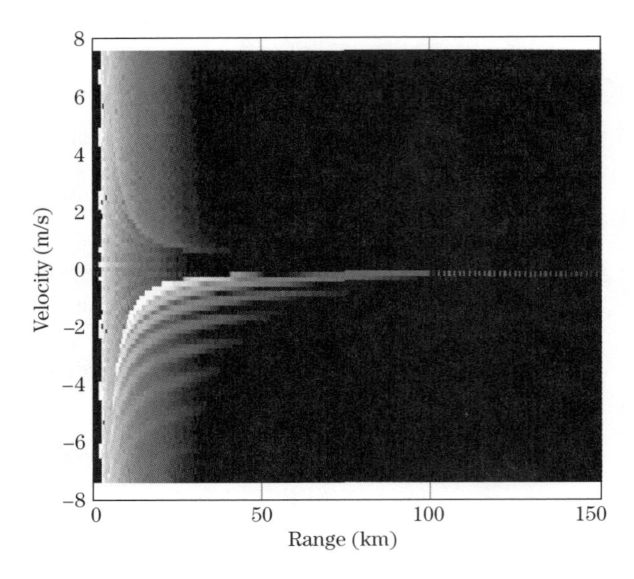

FIGURE 8-25 ■ Simulated range-velocity distribution for stationary clutter observed from a moving radar at low PRF. Compare to Figure 8-23b.

the interval ±7.5 m/s. For instance, the main lobe clutter, previously centered at about 17.3 km range and 44 m/s, still appears at 17.3 km range but now wraps around to $v' = 44 - 3 \cdot 15 = -1$ m/s.

In a so-called *medium PRF* radar, common in airborne systems, both range and velocity ambiguities may be present at the same time, resulting in complex distributions of clutter. Figure 8-26 is an example of the range-velocity clutter spectrum for an X-band radar traveling at 300 mph (134 m/s) with the antenna squinted 45° right in azimuth and 10° down. The main lobe is centered on the ground at a range of 17.3 km. The PRF is 20 kHz, giving an unambiguous range of 15 km and an unambiguous velocity interval of ±75 m/s. Thus, the clutter will fold over in both range and Doppler. To see this, notice the bright main lobe clutter signal. Although the radar is forward-looking so that the actual main lobe clutter Doppler is positive, it has folded over to a negative Doppler shift corresponding

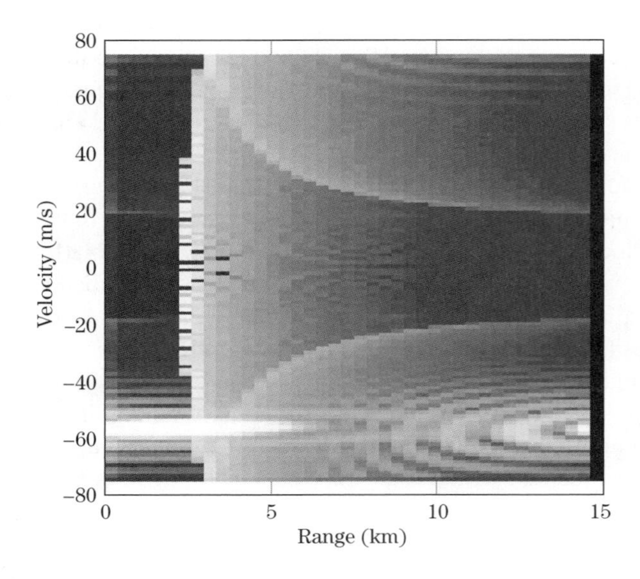

FIGURE 8-26 ■ Medium-PRF range-velocity clutter distribution (see text for details).

to about −58 m/s. The main lobe clutter is also wrapped in range: the peak occurs at the actual 17.3 km but folds over to 2.3 km. Furthermore, the main lobe peak is actually split across the 15 km unambiguous range.

As a final note, consider the area of the plot in Figure 8-26, which is a measure of the total range-velocity space in which targets and clutter can be located without encountering ambiguities. The product of R_{ua} and v_{ua} is

$$R_{ua}v_{ua} = \left(\frac{cT}{2}\right)\left(\frac{\lambda}{2}PRF\right) = \left(\frac{cT}{2}\right)\left(\frac{\lambda}{2T}\right) = \frac{\lambda c}{4} \tag{8.32}$$

which, for a given RF, is a constant. Thus, to increase the unambiguous range, the unambiguous velocity interval must be reduced and vice versa. In Chapter 17, *staggered PRF* techniques will be introduced as a way to expand the range-velocity coverage beyond the limit of equation (8.32).

8.10 | FURTHER READING

Details of the measurement of Doppler shift using multiple pulses and discrete Fourier analysis and of the stop-and-hop approximation are given in the textbook by Richards [5]. More in-depth discussions and examples of the complexities of the range-Doppler or range-velocity distribution, especially for moving platforms, are found in the books by Morris and Harkness [6] and Stimson [7]. Stimson's book deserves special mention for its comprehensive, easy-to-understand discussion of airborne radar and its extraordinary illustrations. Schleher [8] provides an in-depth analytical discussion of Doppler signal characteristics.

8.11 | REFERENCES

[1] Gill, T.P., *The Doppler Effect,* Logos Press, London, 1965.

[2] Temes, C.L., "Relativistic Consideration of Doppler Shift," *IRE Transactions on Aeronautical and Navigational Electronics*, p. 37, 1959.

[3] Papoulis, A., *The Fourier Integral and Its Applications*, 2d ed., McGraw-Hill, New York, 1987.

[4] Bracewell, R., *The Fourier Transform and Its Applications*, 3d ed., McGraw-Hill, New York, 1999.

[5] Richards, M.A., *Fundamentals of Radar Signal Processing*, McGraw-Hill, New York, 2005.

[6] Morris, G.V., and Harkness L. (Eds.), *Airborne Pulsed Doppler Radar*, 2d ed., Artech House, 1996.

[7] Stimson, G.W., *Introduction to Airborne Radar*, 2d ed., SciTech Publishing, Raleigh, NC, 2000.

[8] Schleher, D.C., *MTI and Pulsed Doppler Radar*, Artech House, Norwood, MA, 1999.

8.12 | PROBLEMS

1. Consider a target approaching an L-band (1 GHz) radar at a radial velocity of 100 m/s. Estimate the error in the approximate formula for Doppler shift given by equation (8.4). (Note: the numerical precision required for this calculation may exceed that of many calculators. The error can be estimated using tools such as MATLAB or Mathematica or by considering the terms neglected in the series expansion of (8.2).)

2. Suppose two aircraft are flying straight and level at the same altitude. One is traveling due north at 100 m/s, while the other is flying directly at the first but in the southwesterly direction, also at 100 m/s. What is the radial velocity between the two aircraft? What is the Doppler shift in hertz, including the sign, assuming an X-band (10 GHz) radar?
3. A stationary radar with a rotating antenna (typical of an airport approach radar, for instance) observes an aircraft moving through its airspace in a straight line at a speed of 200 mph. The aircraft approaches from the east, flies directly overhead of the radar at an altitude of 5 km, and continues to the west. Sketch the general behavior of the radial velocity of the target relative to the radar as it flies from east to west through the airspace. Label significant values.
4. Suppose a radar has a pulse length of 100 ns. What is the Rayleigh bandwidth of the pulse spectrum, in Hz? What is the 3 dB bandwidth in Hz?
5. Calculate the Rayleigh resolution in Doppler frequency corresponding to an RF pulse at a center frequency of 35 GHz and a pulse duration of 1 μs.
6. Consider two RF pulses at frequencies of 5.0 GHz and 5.01 GHz. What is the minimum pulse length required so that the two pulses could be reliably resolved in frequency (separated by one Rayleigh resolution)?
7. A finite pulse train waveform is composed of 20 pulses, each of 10 μs length and separated by a PRI of 1 ms. What is the coherent processing interval for this waveform? What is the peak-to-null Doppler resolution?
8. For the pulse train waveform of problem 7, sketch the spectrum in a form similar to Figure 8-6. Indicate the Rayleigh width and the spacing of the individual spectral lines. How many individual spectral lines fall within the main lobe of the sinc-shaped envelope determined by the pulse length?
9. Consider a simple pulse burst waveform with $M = 30$ pulses, each of 10 μs duration, and a PRI of $T = 100$ μs. Assuming no weighting functions are used, what are the range resolution, Doppler resolution, unambiguous range, and unambiguous Doppler shift associated with this waveform?
10. Consider a C-band (6 GHz) weather radar and suppose the desired velocity resolution is $\Delta v = 1$ m/s. What is the corresponding Doppler frequency resolution Δf required? If the PRF is 1,000 pulses per second, how many pulses, M, must be processed to obtain the desired velocity resolution, Δv?
11. Suppose a radar views a target moving away from the radar at 50 m/s. The radiofrequency is 2 GHz. If the radar PRF is 2 kHz, what is the change in the echo phase shift from one pulse to the next? How many wavelengths does the target travel between two pulses?
12. Suppose the input to the coherent detector of Figure 8-12 is changed from a cosine function to a sine function. The argument of the sine function is the same as that of the cosine shown in the figure. How are the outputs of the detector changed? Which channel (upper or lower) will be the in-phase (I) output, and which the quadrature (Q) output?
13. An aircraft has a 4° azimuth 3 dB beamwidth. The RF is 10 GHz, and the antenna is steered to a squint angle, ψ, of 30°. If the aircraft flies at 150 m/s, what is the Doppler spread, B_{MLC}, of the clutter echoes induced by the aircraft motion?
14. Suppose the aircraft in problem 13 has a PRF of 10 kHz. Sketch a Doppler spectrum similar to that of Figure 8-21, but with noise and mainlobe clutter components only (no sidelobe clutter or moving targets). What range of Doppler shifts lies in the clutter region of the spectrum?

What range of Doppler shifts lies in the clear region of the spectrum? What percentage of the total spectrum width from $-PRF/2$ to $+PRF/2$ is in the clear region?

15. If the PRF in problem 14 is changed to 1 kHz, what percentage of the total spectrum width will lie in the clear region?

16. Equation (8.31) gives the Doppler bandwidth across the radar mainbeam due to motion of the radar platform for the case where the squint angle $\psi > \theta_3/2$ so that the mainbeam does not straddle the antenna boresight. Derive a formula for Doppler bandwidth for the case where $0 < \psi < \theta_3/2$ so that the mainbeam does straddle the boresight.

17. Consider a radar with a PRF of 5 kHz. What is the maximum unambiguous range, R_{ua}, of this radar, in km? If a target is located at a range of 50 miles, how many pulses will the radar have transmitted before the first echo from the target arrives? What will be the apparent range of the target in kilometers? (The apparent range is the range corresponding to the time delay from the most recent pulse transmission time to the arrival of the target echo from a previous pulse once steady state is achieved.)

18. Consider an airborne radar traveling straight, level, and forward at 200 mph at an altitude of 30,000 feet. The antenna is pointed at an azimuth angle of 0° and an elevation angle of −20°. Sketch the approximate unaliased range-velocity distribution of the ground clutter in a format similar to that of Figure 8-23b. The range axis of the sketch should cover 0 to 100 km, and the velocity axis should cover $\pm v_{max}$, where v_{max} is the maximum possible radial velocity in m/s that could be observed from scatterers in front of the radar. Indicate where the main lobe clutter is located on the sketch.

19. Suppose the radar in problem 18 has an RF of 10 GHz and a PRF of 3 kHz. What are the unambiguous range, R_{ua}, and unambiguous velocity, v_{ua}? Sketch the approximate aliased range-velocity distribution of the ground clutter in a format similar to that of Figure 8-25. The range axis of the sketch should cover 0 to R_{ua}, and the velocity axis should cover $\pm v_{ua}$. Indicate where the main lobe clutter is located on the sketch.

PART III

Subsystems

CHAPTER 9 Radar Antennas

CHAPTER 10 Radar Transmitters

CHAPTER 11 Radar Receivers

CHAPTER 12 Radar Exciters

CHAPTER 13 The Radar Signal Processor

CHAPTER 9

Radar Antennas

Christopher D. Bailey

Chapter Outline

9.1	Introduction	309
9.2	Basic Antenna Concepts	310
9.3	Aperture Tapers	314
9.4	Effect of the Antenna on Radar Performance	317
9.5	Monopulse	320
9.6	Reflector Antennas	322
9.7	Phased Array Antennas	326
9.8	Array Architectures	339
9.9	Further Reading	343
9.10	References	343
9.11	Problems	345

9.1 INTRODUCTION

The radar antenna is responsible for transmitting electromagnetic energy into the environment and receiving energy that has reflected off a distant target. To perform this job efficiently, the radar antenna must (1) provide a matched interface between the transmitter (TX), receiver (RX), and free space, and (2) provide angular selectivity for energy being transmitted and received from the radar. These functions are illustrated in Figure 9-1.

In this chapter the primary features of the antenna will be presented, and their effect on performance will be discussed. Operational and performance issues associated with the two

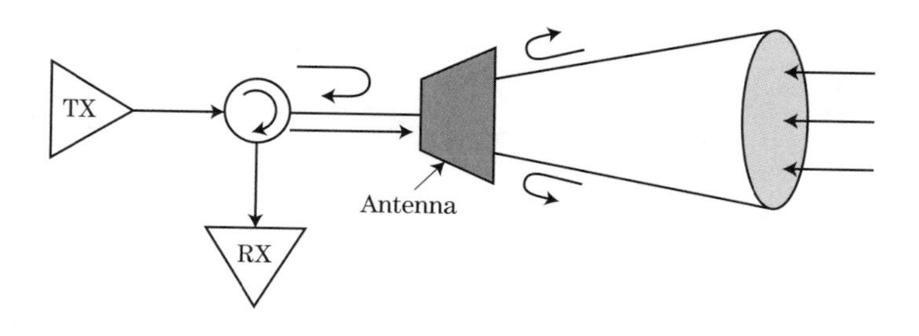

FIGURE 9-1 ■ The radar antenna must provide angular selectivity and a good electrical match.

most important classes of radar antennas, the reflector and phased array, will be described. Emphasis will not be on antenna design, which is widely covered in the literature [1–3], but on salient antenna features and characteristics that every radar engineer should understand.

9.2 BASIC ANTENNA CONCEPTS

9.2.1 The Isotropic Antenna

This introduction to the radar antenna will begin with the isotropic antenna, which is a theoretical point source that transmits and receives energy equally in all directions (Figure 9-2). This fictional antenna is an important reference for describing the radiation properties of all other antennas and will serve as a departure point.

As shown in Figure 9-2, the isotropic antenna radiates equally in all directions and therefore has no angular selectivity. The radiation intensity (watts/steradian) from an isotropic antenna can be written as

$$I = \frac{P_t}{4\pi} \tag{9.1}$$

where P_t is the total power radiated by the antenna and 4π is the steradian area of the sphere enclosing the antenna. The power density (W/m^2) measured at a distance R from the isotropic antenna is

$$Q_t = \frac{P_t}{4\pi R^2} \tag{9.2}$$

Since the area of the sphere surrounding the antenna grows as R^2, the power density must decrease as $1/R^2$. The circular lines in Figure 9-2 represent locations of equal radiated intensity and phase. The solid and dashed lines represent propagating wave fronts spaced

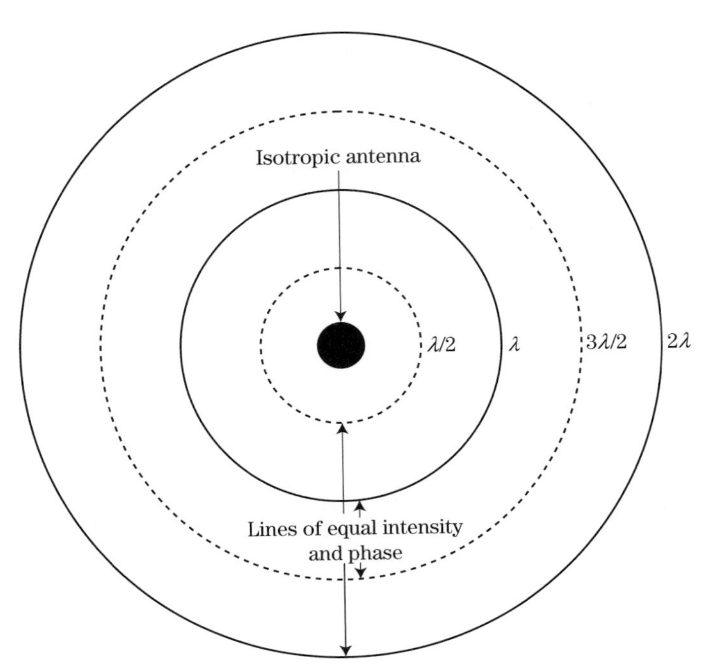

FIGURE 9-2 ■ The isotropic antenna radiates equally in all directions. The concentric rings indicate spheres of equal phase and radiation intensity.

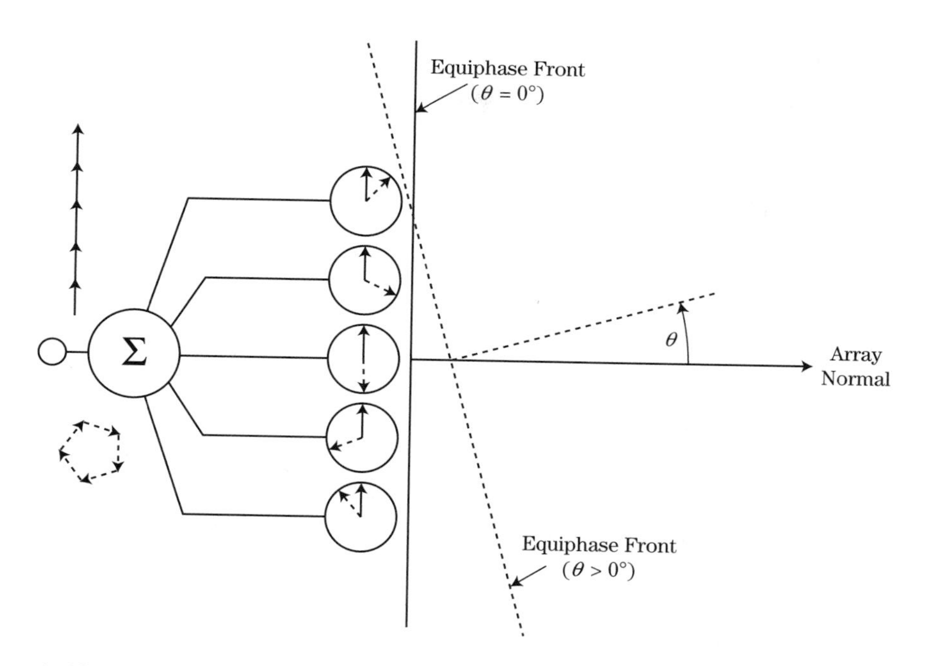

FIGURE 9-3 ■ The response of an array to an incoming plane wave is the sum of the element excitation vectors, which will combine constructively or destructively depending upon the incidence angle.

at one-half wavelength intervals. The phase of the electromagnetic wave at adjacent lines will therefore differ by 180° at any given time.

9.2.2 The Radiation Pattern

In contrast to an ideal isotropic antenna, the radiated energy from a radar antenna is highly dependent on angle. In fact, one purpose of an antenna is to concentrate the radiated energy into a narrow angular region. There are different types of radar antennas, and considerable discussion will be given to these later in the chapter. For now, a simple array of isotropic antennas will be used to explain this angular dependence. Figure 9-3 shows an array of five isotropic antennas (elements) connected at a common node referred to as the summation point. The summation point is the location where the individual element signals combine and the total antenna response is measured. This figure also depicts two plane waves incident on the radiating surface from different angles θ with respect to the normal to the array surface, or "array normal" for short. The defining property of a plane wave is that the electric field is in phase everywhere on the plane perpendicular to the direction of propagation. Thus every location on the antenna's radiating surface will detect the same phase when $\theta = 0°$. This is indicated by the solid vectors in the circles at each element location. The signal received at each antenna element will combine in phase at the summation point (assuming that the path length from each antenna element to the summation point is equal).

Now let θ be slightly greater than 0° (dashed line). This phase front reaches each element at a different moment in time, and at any instant the five elements will be experiencing different phases from the incoming plane wave. This is illustrated by the dashed excitation vectors at each antenna location. Since each element detects a different phase from the incoming plane wave, the combined energy will no longer add coherently and the amplitude of the resulting signal will be smaller than when $\theta = 0°$. As θ continues to increase there will exist an angle, or multiple angles, at which the vector sum of the phase excitations will completely cancel.

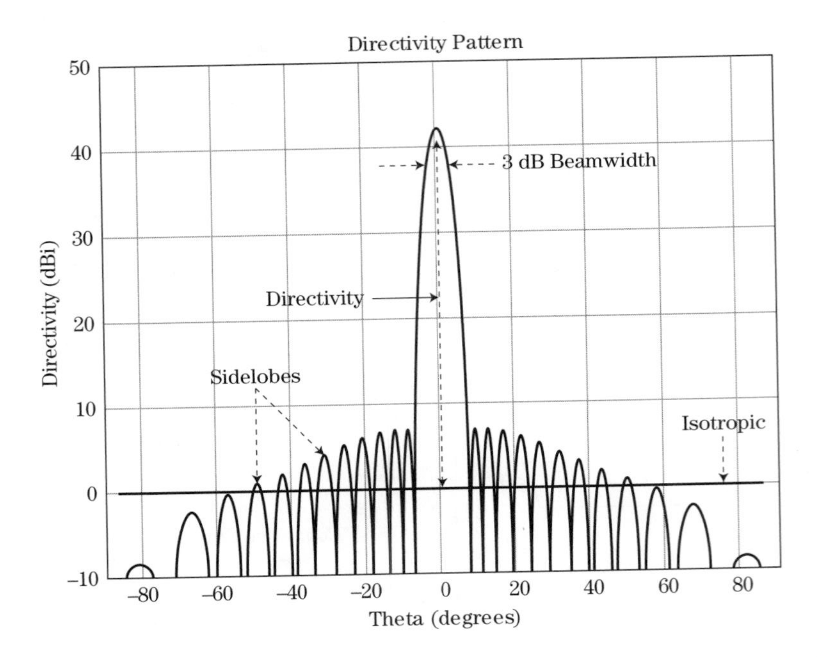

FIGURE 9-4 ■ The directivity pattern of a radar antenna and some related parameters.

Figure 9-4 shows the signal strength as a function of incidence angle. This figure shows a typical radiation pattern structure for an antenna in which energy is collected over a large surface area. The maximum of this plot, or main beam peak, occurs at the angle where all collected energy coherently adds together. For the remaining angles the collected energy will either cancel to cause nulls or partially combine to create *sidelobes*.

9.2.3 Beamwidth

The pattern in Figure 9-4 is called the *radiation* or *directivity pattern* of an antenna. The angular distance from the half power (−3 dB) point on one side of the main beam to the half power point on the other side is called the 3 dB beamwidth, or simply the beamwidth. To explain this concept further requires consideration of the phase variation across the antenna from an off-axis plane wave. The phase variation across the array surface, or *aperture*, is the total path length variation times $2\pi/\lambda$. This can be written as

$$\Delta\phi = \frac{2\pi L \sin\theta}{\lambda} \tag{9.3}$$

where the length of the antenna is D. For an array of elements, D can be replaced by $n\Delta x$, where n is the total number of elements and Δx is the distance between elements.

The total phase variation across the array increases with antenna size. Thus, as size increases the antenna response changes more rapidly with respect to θ. Consequently, the main beam becomes narrower and more directional. The beamwidth of an antenna is inversely proportional to its size. In general, the beamwidth in radians is

$$\theta_3 = \frac{\alpha\lambda}{L} \tag{9.4}$$

where α is the beamwidth factor and is determined by the aperture taper function (see section 9.3) and the aperture shape. If the aperture is circular and uniformly weighted, α is approximately equal to 1.

9.2.4 Directivity and Gain

The directivity pattern describes the intensity of the energy radiated by the antenna in all angular directions (i.e., it is a function of θ). Directivity is a unitless quantity expressed as a ratio of the radiation intensity of an antenna in a particular direction to that of a lossless isotropic antenna with the same radiated power. The vertical axis in Figure 9-4 has units of dBi, or dB relative to an isotropic antenna. The maximum directivity for a planar radar antenna is [3]

$$D_{\max} = \frac{\eta_a 4\pi A}{\lambda^2} \tag{9.5}$$

where the antenna's physical aperture area is A and the aperture efficiency is η_a. As with the coefficient α, η_a is determined by the aperture taper, which is discussed in the next section. The effective aperture A_e is the product of the physical antenna size and the aperture efficiency, or $A_e = \eta_a A$. The effective aperture is a measure of how efficiently the antenna's physical area is being used. The aperture is most efficient ($\eta_a = 1$) when uniformly illuminated.

Equations 9.4 and 9.5 can be combined to describe directivity with respect to the 3 dB beamwidth. The directivity of a uniformly illuminated rectangular antenna is

$$D_{\max} \approx \frac{4\pi(0.88)^2}{\theta_3 \phi_3} \tag{9.6}$$

where θ_3 and ϕ_3 are the azimuth and elevation 3 dB beamwidths in radians. For a uniformly illuminated rectangular array $\alpha = 0.88$ and $\eta_a = 1$. Equation (9.6) shows that the maximum directivity is proportional to the ratio of the steradian area of an isotropic antenna, which radiates equally over the full 4π steradians to the steradian area of the antenna main beam.

The terms *gain* and directivity are often incorrectly used interchangeably. These terms are closely related; however, there is an important distinction. The gain of an antenna is a measure of the ratio of the radiation intensity at the peak of the main beam to that of a lossless isotropic antenna with the same input power. Therefore, gain G is equal to the maximum directivity $D_{\max}$ minus the losses internal to the antenna. Gain can never exceed the maximum directivity.

Antenna gain should not be confused with the gain of an amplifier. An amplifier with 30 dB of gain would have 1,000 times more radiofrequency (RF) power at the output than at the input. This power increase is achieved by converting direct current (DC) power to additional RF power. An antenna with 30 dB of gain will radiate 1,000 times as much power in the direction of the main beam than will an isotropic antenna with the same input power but will deliver much less power than the isotropic antenna in most other directions. The antenna gain is achieved by concentrating the power in a preferred direction, not by creating additional power.

9.2.5 Sidelobes

The radar engineer will often want to minimize energy in the sidelobe region. Strong sidelobes will increase the clutter return, increase the number of false alarms, and make the radar more susceptible to jamming. For this reason numerous metrics are used to describe the sidelobe structure of an antenna.

The peak sidelobes are usually near the main beam and are expressed in dBi or dB relative to the peak of the main beam. The unit "dBi" means decibels relative to an isotropic antenna. In Figure 9-4, the maximum directivity is 43 dBi, and the near-in sidelobes on either side of the main beam are approximately 36 dB below the main beam peak, or 7 dBi. It is typical for a highly directive radar antenna to have peak sidelobes above the isotopic level (0 dBi). The peak sidelobe level is set by the aperture shape and the electric field distribution across the aperture.

The average sidelobe level is an important figure of merit for most radar antennas. Since the radar beam is constantly being scanned to different angular locations, the relative gain of clutter, jamming, or any other signal entering through the sidelobe region is usually assessed as a function of the average sidelobe level instead of a specific sidelobe. The average ratio of the sidelobe power to that of an isotropic antenna with the same input power is

$$SL_{ave} = \frac{\dfrac{P_{SL}}{\Omega_{SL}}}{\dfrac{P_t}{4\pi}} = \frac{(P_t - P_{MB})}{(4\pi - \Omega_{MB})} \cdot \frac{4\pi}{P_T} = \frac{\left(1 - \dfrac{P_{MB}}{P_t}\right)}{\left(1 - \dfrac{\Omega_{MB}}{4\pi}\right)} \approx 1 - \frac{P_{MB}}{P_t} \tag{9.7}$$

where P_t is the total radiated power, P_{SL} is the power radiated into the sidelobe region, P_{MB} is the power in main beam, Ω_{SL} is the steradian area of sidelobe region, and Ω_{MB} is the steradian area of main beam.

For highly directive radar antennas with a main beam of only a few degrees, Ω_{MB} is much less than 4π. Equation (9.7) then becomes a simple energy conservation equation that states the average isotropic sidelobe power is 1 minus the fraction of power in the main beam. It will be seen later that the average sidelobe level is partially determined by tolerance errors in the amplitude and phase of the electric field in the antenna's aperture. As will be discussed, aperture tapers can be used to reduce the sidelobe level; however, the minimum achievable sidelobe level is ultimately determined by errors. These errors will scatter power out of the main beam and into the sidelobe region, thereby reducing the gain and raising the average sidelobe level as indicated in equation (9.7). It can also be seen in equation (9.7) that the average sidelobe level must be below isotropic or 0 dBi.

9.3 APERTURE TAPERS

The radiating surface of an antenna is often referred to as the aperture. An antenna's aperture is defined as an opening in a plane directly in front of the antenna through which practically all the power radiated or received by the antenna passes. Figure 9-5 shows the aperture for a reflector and array antenna. The sidelobe structure of an antenna's radiation pattern can be controlled by adjusting the spatial distribution of the electric field across this aperture. This electric field distribution is called the aperture *taper*.

If the aperture taper is uniform (i.e., electric field strength is the same everywhere throughout the aperture), the antenna will have the greatest directivity and narrowest beamwidth possible; however, the peak sidelobes will be high ($\approx$13 dB below the main beam for a rectangular aperture). The peak sidelobe can be significantly reduced by using a taper that maximizes the field strength at the center of the aperture and decreases the field strength outwardly toward the aperture edges. However, with a low sidelobe aperture taper the maximum directivity is reduced and the beamwidth is increased. A fundamental trade-off exists between maximizing directivity and decreasing sidelobe levels.[1]

[1]This same trade-off between peak directivity or gain, resolution, and sidelobe level also exists in range and Doppler sidelobe control; see Chapters 14, 17, and 20.

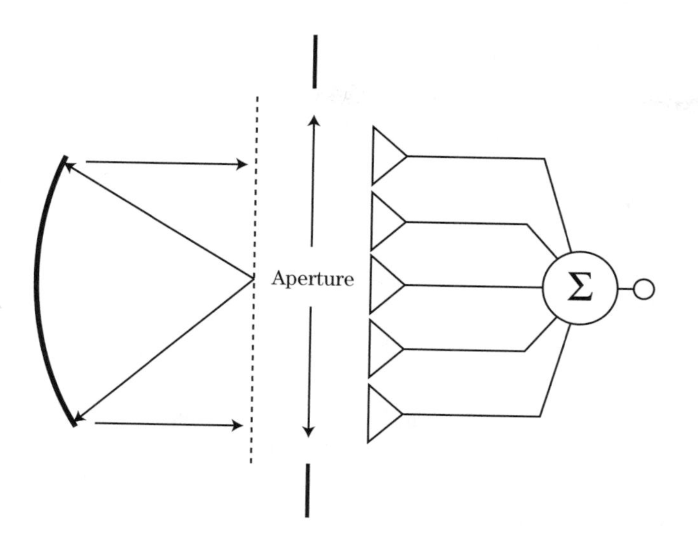

FIGURE 9-5 ■ The aperture for a reflector and array antenna.

Figure 9-6 shows uniform and low sidelobe aperture tapers and their resulting far-field radiation patterns. A mathematical function developed by Taylor provides a desirable combination of adjustable sidelobe control and minimal directivity loss [4,5]. Table 9-1 contains the aperture efficiency η_a and beamwidth factor α for several peak sidelobe levels when the Taylor function is applied to a linear array.

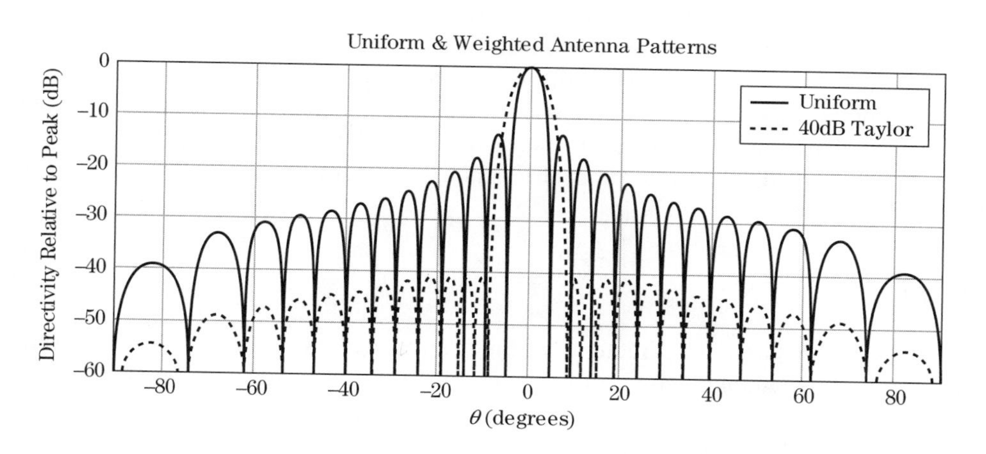

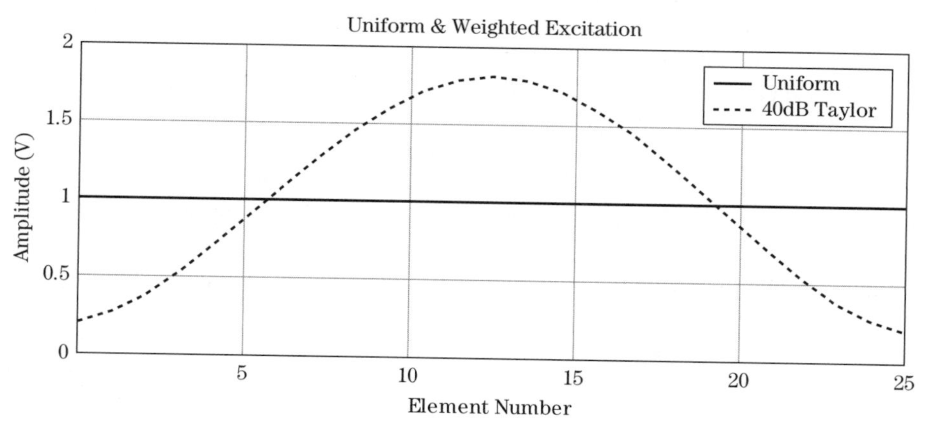

FIGURE 9-6 ■ Radiation patterns resulting from uniform and low sidelobe aperture tapers.

TABLE 9-1 ■ Aperture Efficiency and Beamwidth Factor Values for Different Taylor Distributions

Sidelobe Level (dB)	Beamwidth Factor (α)	Aperture Efficiency (η_a) dB
−13	0.88	0.0
−20	0.98	−0.22
−25	1.05	−0.46
−30	1.12	−0.70
−35	1.18	−0.95
−40	1.25	−1.18
−45	1.30	−1.39

Radar antennas are often rectangular apertures. In this case, independent Taylor tapers can be applied along the width and height of the aperture, resulting in different sidelobe levels for the azimuth and elevation radiation patterns. The beamwidth factor will then be different for the azimuth and elevation patterns, and the aperture efficiency will be the product of the horizontal and vertical efficiencies.

Taylor aperture tapers can be computed to provide arbitrarily low sidelobes, but it is very difficult to obtain first sidelobes below −40 dB in a real antenna system. This is because the ideal taper is never achieved due to errors caused by imperfections in antenna components, manufacturing tolerances, thermal effects, antenna alignment, and many other factors. Analysis of error effects is complex, is dependent on the specific type of antenna, and is treated in detail in the literature [6]. Herein, these effects are briefly discussed to illustrate how errors can affect the antenna's performance.

Aperture taper errors are typically treated as random deviations in amplitude and phase from the ideal taper function. Phase errors are typically harder to control and more destructive than amplitude errors. If there are no amplitude errors and the phase errors are uncorrelated from point to point throughout the aperture, their effect on maximum directivity can be expressed as [5]

$$\frac{D_e}{D_0} = \exp(-\delta_{rms}^2) \tag{9.8}$$

where D_0 is the maximum directivity of the antenna without errors, D_e is the maximum directivity of the antenna with errors, and δ_{rms} is the root mean square (rms) phase error in radians.

Equation (9.8) expresses the reduction of maximum directivity due to random phase errors in the aperture. This lost directivity represents power that will be scattered into the sidelobe region and will increase the average sidelobe level as described in equation (9.7). A phase error of 5° represents 5/360 or 1/72-nd of a wavelength of misalignment. At 10 GHz, this will be only 0.42 mm. Figure 9-7 shows the effect of phase errors on the directivity pattern. The 5° phase error causes significant degradation of the sidelobes.

The error analysis in Figure 9-7 treats only one type of error that can degrade an antenna's performance. There are additional sources of both amplitude and phase errors that are correlated across the aperture in different ways. These errors are best treated in computer simulations in which all errors are applied across the aperture with the proper statistics and correlation intervals. Simulations can be used to compute the far-field radiation pattern with errors added to the ideal taper to determine the combined effect. Since errors are treated as random variables, only one draw of random errors can be simulated

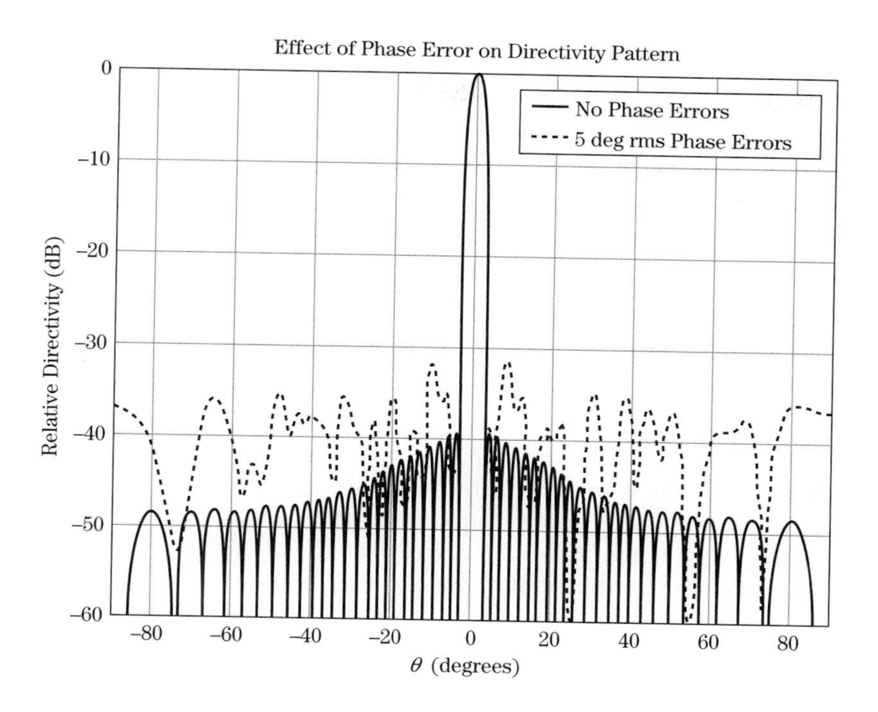

FIGURE 9-7 ■ Radiation pattern with and without random phase errors.

at a time. Therefore, the simulation should be run many times with the random number generators reseeded to get the proper statistical effects on the radiation patterns. For this reason the sidelobe metrics discussed in section 9.2 are often calculated based on the mean response of a Monte Carlo simulation that includes random errors.

9.4 | EFFECT OF THE ANTENNA ON RADAR PERFORMANCE

The antenna system has a significant impact on the overall radar performance. For instance, the radar "sees" the world through the antenna, which means a target can be detected only if it is located within the antenna's field of view (FOV). The field of view is defined as the angular region over which the main beam can be scanned. Depending on the type of antenna, the main beam can be moved throughout the field of view either by mechanically rotating the antenna or by electrically steering the beam, which will be discussed later in this chapter. In either case, the radar detection region is limited by the antenna's field of view.

The maximum range at which the radar can detect a target is highly dependent upon the antenna system. The maximum detectable range for a target of a given radar cross-section (RCS) σ is obtained from the radar range equation (see chapters 1 and 2) by determining the minimum detectable signal power $P_{\min}$ and then solving for the range at which that signal level is obtained, giving

$$R_{\max} = \left[\frac{P_t G^2 \lambda^2 \sigma}{(4\pi)^3 L_s P_{\min}} \right]^{1/4} \tag{9.9}$$

where the losses in the radar are combined into the term L_s. Note that the antenna gain is squared in this equation because the signal is transmitted and received by the same antenna. It is common for the gain to be different on transmit (G_t) and receive (G_r) due to different aperture weights, in which case G^2 is replaced by $G_t G_r$.

The ability of a radar system to resolve targets in range is inversely proportional to the system's instantaneous bandwidth. The minimum resolvable range for a compressed waveform is

$$\Delta R = \frac{c}{2B} \tag{9.10}$$

where c is the speed of light (3×10^8 m/s) and B is the radar's instantaneous bandwidth in hertz. The instantaneous bandwidth of a radar is often limited by the bandwidth of the antenna. It will be shown in section 9.7 that the instantaneous bandwidth of a phased array antenna is limited by the aperture size and the maximum electronic scan angle.

Most radar systems will either search for targets or track targets. These operating modes have slightly different antenna requirements, so key design decisions such as frequency selection and aperture size must be made with the primary radar operating mode in mind. A radar system that has to perform both of these functions is usually a compromise; that is, the search performance and track performance will be inferior to a radar designed to operate in only one of these modes.

The version of the radar range equation for tracking given in equation (2.59) can be expressed in a simplified form using peak power instead of average power as

$$\frac{P_t A_e^2}{\lambda^2} = \frac{4\pi (SNR) N_0 B R^4}{\sigma} \tag{9.11}$$

where the noise spectral power density is $N_0 = kT_0F$ and system losses L_s have been ignored. Equation (9.11) shows that the radar resources required to track a target with cross-section σ, at range R, with a given signal-to-noise (SNR) ratio SNR, can be expressed in terms of the product of the radiated power and the effective aperture area squared divided by the wavelength squared. Division by λ^2 implies that the same radar performance can be achieved with less power by increasing the frequency. The fact that area is squared indicates that it is more beneficial to increase aperture size than to raise power. In addition, it will be seen shortly that angular accuracy for tracking radars improves as the wavelength decreases. For these reasons, tracking radars tend to operate at higher frequencies with electrically large (i.e., large in wavelengths) apertures and relatively low power.

Search radars must search a solid angular sector in a given period of time to perform their mission. As presented in equation (2.44), the search form of the radar equation is

$$P_{avg} A_e = \frac{(SNR) 4\pi N_o L R^4}{\sigma} \left(\frac{\Omega}{T_{fs}}\right) \tag{9.12}$$

This equation states that a certain power-aperture product ($P_{avg} A_e$) is required to search an angular sector Ω in T_{fs} seconds to detect targets of RCS σ at range R. Unlike the tracking range equation, wavelength is not present in the search equation, which provides the freedom to operate at lower frequencies without performance degradation.

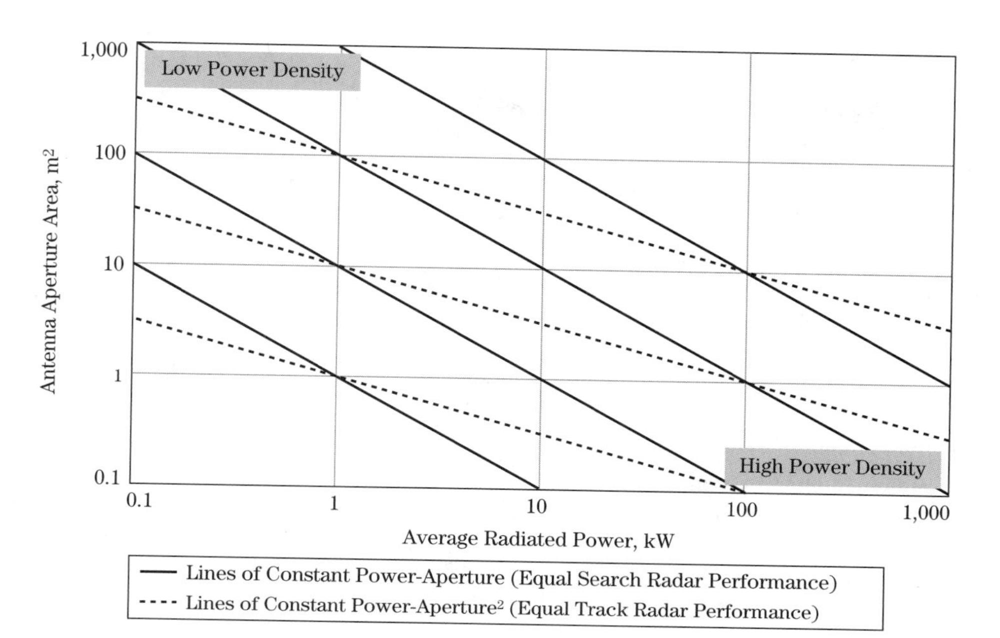

FIGURE 9-8 ■ Lines of constant track and search performance mapped onto the power-aperture space.

There are reasons lower frequencies are preferred for search radars. It is less expensive to make large apertures for lower frequencies since the tolerances, which are usually proportional to wavelength, can be relaxed. If the antenna is an array, the elements must be spaced less than a wavelength apart (see section 9.7) so fewer elements are required to fill a fixed-aperture area at lower frequencies. The beamwidth of an antenna with a fixed-aperture size is larger at lower frequencies so fewer beams are required to search the scan area in the required time. Finally, weather penetration is better at low frequencies due to reduced atmospheric attenuation.

Figure 9-8 displays the power-aperture space where the horizontal axis is average radiated power and the vertical axis is aperture area. The diagonal solid lines are lines of constant power-aperture product ($P_{avg}A_e$). As shown by equation (9.12), radar search capability will be the same for any point on a constant $P_{avg}A_e$ line. The dashed lines in Figure 9-8 are lines of constant $P_{avg}A_e^2$ for a fixed wavelength. According to equation (9.11), a radar will have the same tracking capability at any point on one of the constant $P_{avg}A_e^2$ lines. Figure 9-8 illustrates that power and aperture area can be traded to achieve the desired search or track performance. The optimum point on the curve will depend on prime power and spatial constraints imposed by the radar platform or the transportability requirements. For example, a radar in the nose of a fighter aircraft will typically have an aperture area of only 1 m^2, so it must increase power to achieve higher performance. This high-power density radar will operate in the lower right-hand corner of Figure 9-8. A space-based radar that collects solar energy to produce prime power will have a limited supply of transmit power; however, the aperture can be very large after deployment. This low-power density radar will operate in the upper left-hand corner of Figure 9-8. The power aperture requirement is a major driver on the modern radar antenna. Technologies and architectures to meet these requirements will be discussed in the following sections.

9.5 MONOPULSE

When searching for targets, it may be adequate to determine the target's angular location only to an accuracy equal to the antenna's 3 dB beamwidth. However, tracking usually requires a more accurate estimation of the target's angular position. The target's angular position can be determined more accurately than indicated by equation (9.4) by using monopulse [7]. In early systems additional accuracy was obtained by scanning the antenna beam over an angular area about the target and observing the amplitude modulation of the target return. These techniques, called conical-scan and sequential lobing [7], require several radar pulses to determine the target's position and are susceptible to errors caused by target RCS fluctuations. In addition, some jamming methods are effective against such tracking techniques.

The monopulse technique involves simultaneously generating and processing multiple closely spaced beams from the same antenna. This concept can most easily be described by considering two overlapping beams symmetrically offset from antenna normal, as shown in Figure 9-9. If the two beams are added together, an on-axis beam is formed that is commonly referred to as the sum beam (Σ). If the two beams are subtracted, the resulting pattern will have a positive lobe on one side of the axis, a negative lobe (*i.e.*, a lobe 180° out of phase) on the other, and a null that occurs on axis. This beam is referred to as the difference or delta beam (Δ).

To improve the tracking ability of a radar system, monopulse antennas create a signal proportional to the displacement of the target from the center of the Σ beam by comparing the outputs of the Σ and Δ beams. The monopulse error signal

$$v_{error}(\theta) = \frac{|\Delta(\theta)|}{|\Sigma(\theta)|} \cos \beta \tag{9.13}$$

is a function of the target's angle θ with respect to the antenna boresight and the phase angle β between the Σ and Δ beam outputs, which is designed to be nearly 0° or 180°,

FIGURE 9-9 ■ A sum and delta beam can be formed by adding and subtracting overlapping antenna beams.

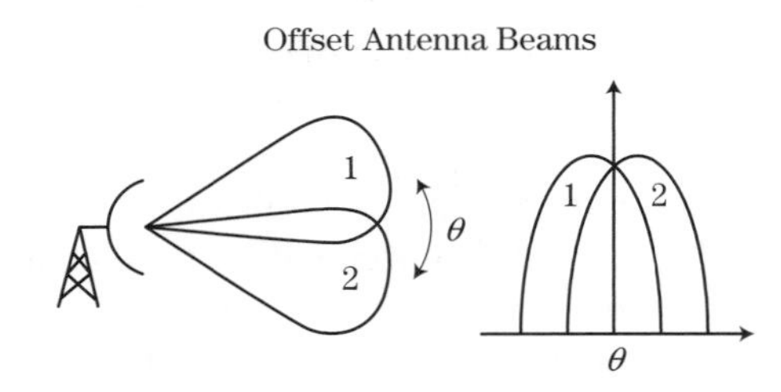

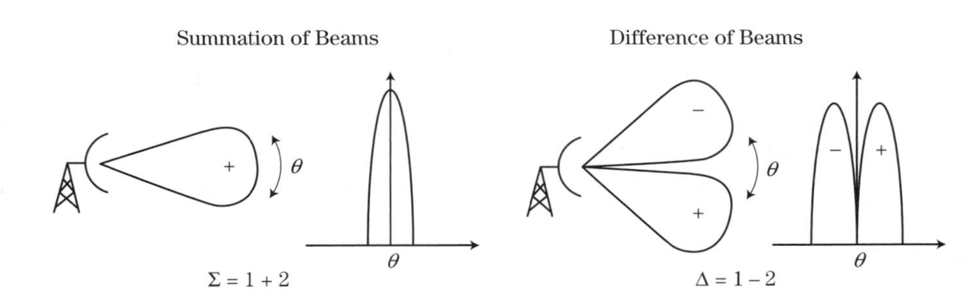

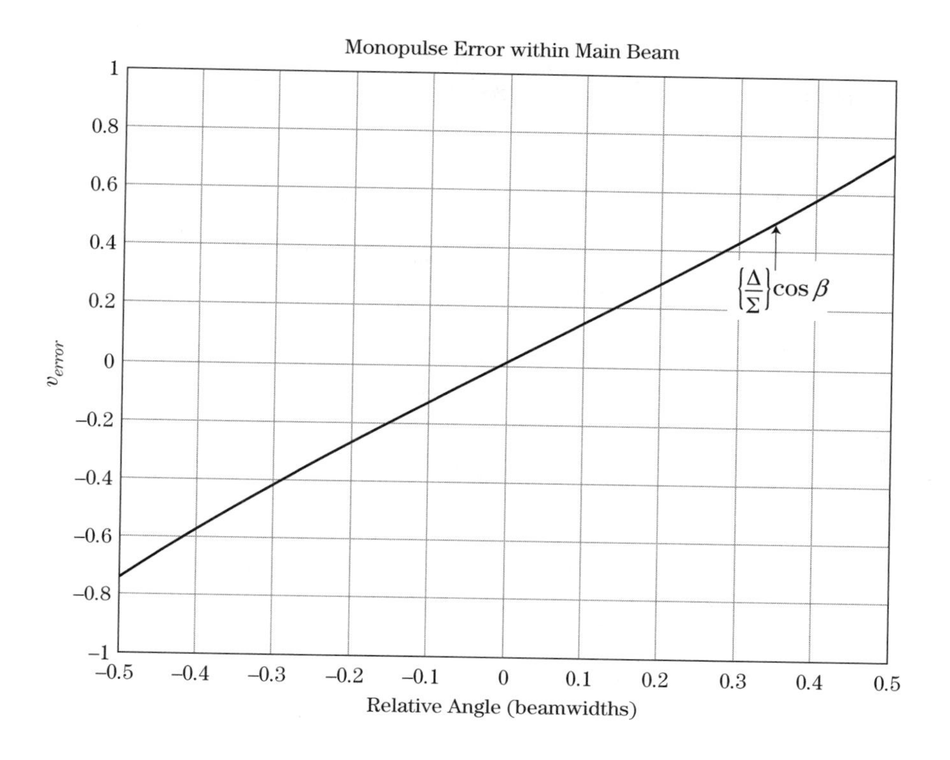

FIGURE 9-10 ■ The monopulse error signal is a ratio of the delta beam over the sum beam and is linear within the 3 dB beamwidth.

depending on from which side of center the target signal is entering. The angle θ is often normalized by the antenna beamwidth for convenience.

A typical error signal v_{error} is plotted in Figure 9-10 and is seen to be fairly linear between the 3 dB limits of the Σ beam. This error signal can be used to estimate the target's position within the Σ beam. The slope of the curve in Figure 9-10 is called the monopulse error slope k_m. With monopulse, the target's position can be measured in one pulse. Since the Δ beam is normalized by the Σ beam on each pulse, target RCS fluctuation will not create angle position errors beyond those created by changes in the SNR. The angular precision (standard deviation of angle error) that can be achieved using monopulse [7] is

$$\Delta\theta = \frac{\theta_3}{k_m\sqrt{2SNR}}\sqrt{1+\left(\frac{k_m\theta}{\theta_3}\right)^2} \tag{9.14}$$

where the monopulse error slope is k_m, and the target angle relative to the antenna normal is θ.

If a monopulse array with low sidelobes is required, a Taylor function can be used to generate the Σ pattern aperture taper, and a Bayliss [8] function can be used to generate the Δ pattern aperture taper. The Bayliss function is derived from the Taylor function and provides good sidelobe control for the Δ pattern while minimizing gain loss and beam broadening. The magnitudes of the Δ and Σ beams, using −40 dB Taylor and Bayliss functions, are shown in Figure 9-11.

The radar must transmit with the Σ pattern to put maximum power on the target and simultaneously receive with the Δ and Σ patterns. It is usually necessary to measure the target's position in both azimuth and elevation, so the previously described procedure must be performed in two dimensions. This requires a receive Σ beam and a receive Δ beam

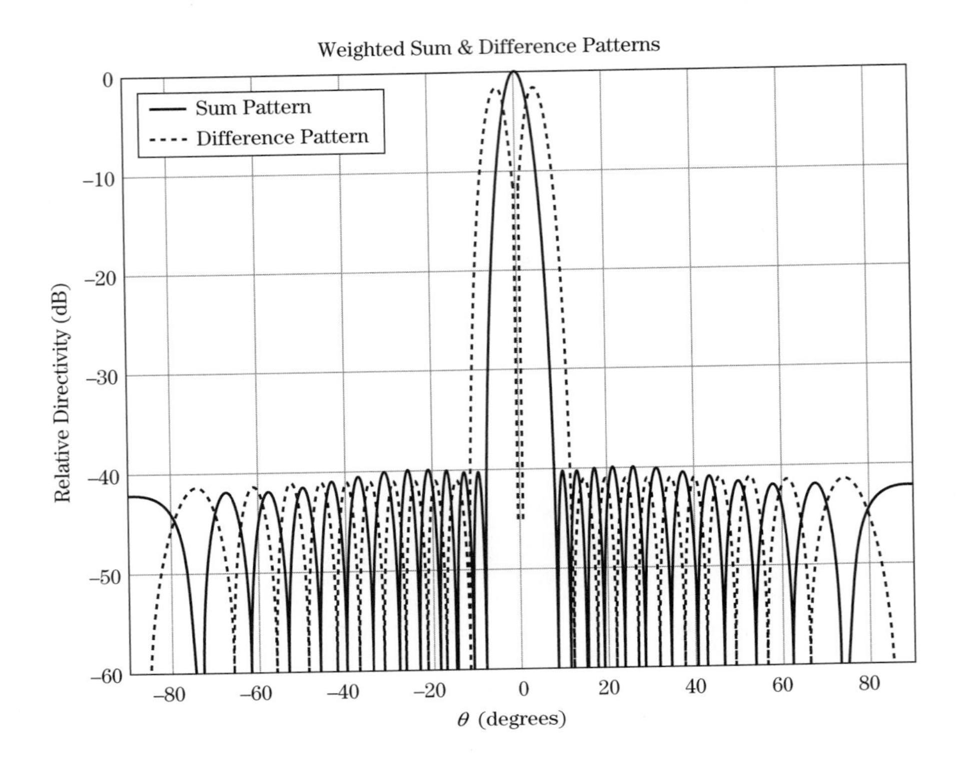

FIGURE 9-11 ■ Low sidelobe sum and delta patterns created with Taylor and Bayliss distributions.

in both the azimuth and elevation planes. Therefore, the monopulse radar requires at least three receive channels and a relatively complex antenna feed network.

9.6 REFLECTOR ANTENNAS

Mechanically positioned reflector antennas have been used in radar systems since the 1950s [9]. Today, nearly 60 years later, reflectors are still used for low-cost applications that require very high directivity with limited to no beam scanning. Most modern radar systems use phased array antennas for reasons discussed in section 9.7; nevertheless, reflectors can still be found in many radar systems. This section provides a brief introduction to the reflector antenna.

The most common reflector antenna is the parabolic reflector, which is shown in Figure 9-12. This antenna is formed by rotating a two-dimensional parabola about its

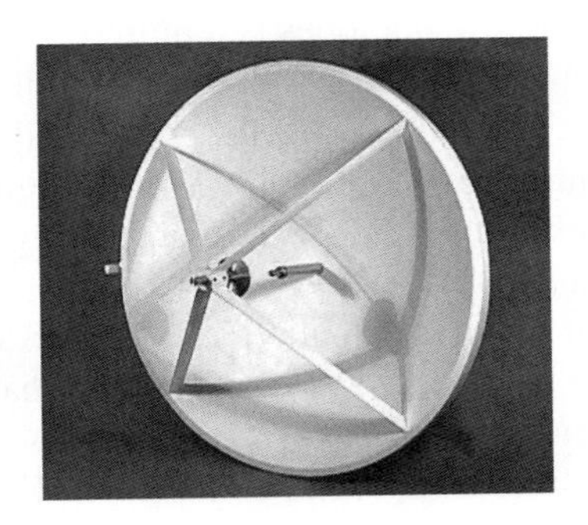

FIGURE 9-12 ■ Cassegrain reflector antenna developed by Quinstar Technology, Inc. (Used with permission.)

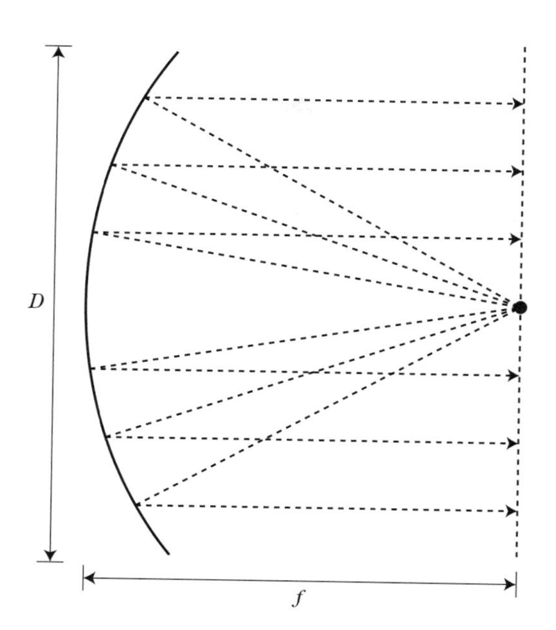

FIGURE 9-13 ■ For a parabolic reflector, all rays travel the same distance from the feed to the aperture plane.

focal axis. Geometric optics analysis (e.g., ray tracing) is a convenient way to illustrate the operation of a parabolic reflector antenna. Rays emanating from the feed point at the focus of the parabola will bounce off the parabolic reflector and travel, in parallel, to the aperture plane of the reflector. As shown in Figure 9-13, each ray will travel the same distance to the aperture plane, no matter which path is taken. Therefore, the spherically diverging wave emanating from the feed point will be collimated, or converted into a plane wave radiating perpendicular to the parabolas axis of rotation.

Conversely, if a plane wave is incident on the parabolic reflector along the axis of rotation, the energy in the plane wave will be coherently summed at the feed point. In a real, finite-sized reflector antenna, energy is concentrated to a spot located at the focal point of the parabola. A small feed antenna such as a pyramidal or conical horn is placed over the focal spot, as shown in Figure 9-14, and is connected to the radar transmitter and receiver through a circulator. The directivity pattern of the feed horn provides the aperture taper for the reflector; consequently, it is difficult to create the ideal Taylor tapers as discussed in section 9.3. If the feed pattern is too wide, significant energy will spill over (i.e., not intersect) the reflector and be wasted. On the other hand, if the feed pattern is too narrow the aperture will be under-illuminated and the aperture efficiency η_a will be reduced. The feed that provides the best compromise between spillover and aperture efficiency is one in which the -10 to -12 dB points of the feed pattern illuminate the edges of the reflector [10]. The angle to the reflector edge can be expressed in terms of the reflector diameter D and focal length f.

$$\theta_{edge} = 2\tan^{-1}\left(\frac{D}{4f}\right) \tag{9.15}$$

The focal length to diameter (f/D) ratio is an important design parameter for reflector antennas used in radar applications. This ratio is primarily determined by the curvature rate of the dish because the feed is fixed at the reflector's focal point. For example, a reflector with no curvature will have an (f/D) ratio of infinity. Reflector antennas are commonly designed to have (f/D) ratios between 0.25 and 0.5. A reflector with a high curvature

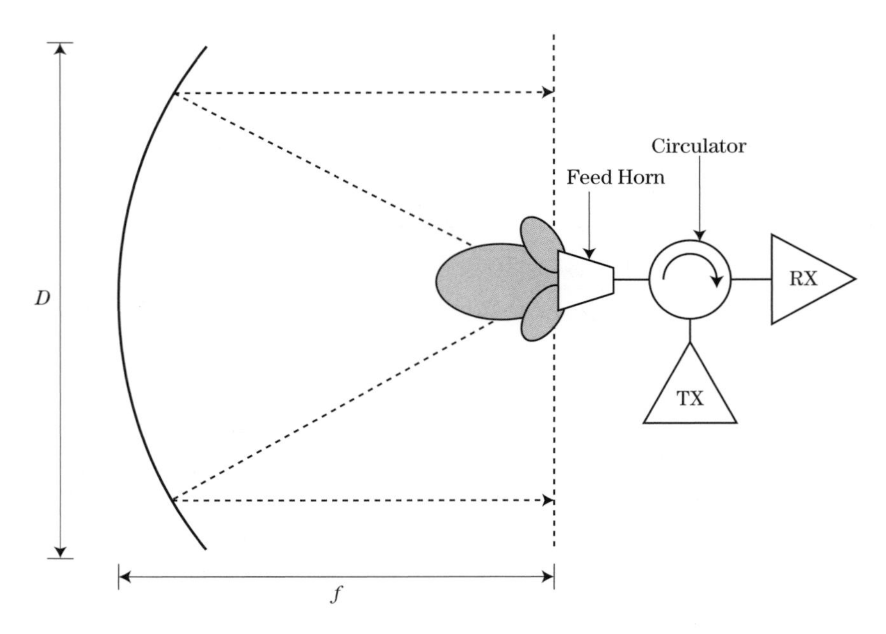

FIGURE 9-14 ■ The feed horn for a reflector antenna is sized to provide the desired aperture illumination on the reflector surface.

rate, and therefore a low value of (f/D), will suffer from polarization distortion and poor off-axis beam performance [6]. As the (f/D) ratio increases, the angle θ_{edge} will decrease, requiring a larger, more directive feed horn to provide a given aperture illumination. A large (f/D) requires that the feed be supported by long struts that are sometimes difficult to keep mechanically stable. The waveguide that connects the feed to the radar must run along one of these struts. This creates a long path between the transmitter and the feed, causing excessive losses. In addition, the feed and struts create a blockage to the antenna's aperture, which results in a gain loss and increases the sidelobe levels. This problem is exacerbated if the system is dual polarized, requiring multiple feed horns and multiple waveguide runs.

One way to alleviate the mechanical problems associated with a high (f/D) is to use a dual-reflector antenna, as shown in Figure 9-15. Here, the feed illuminates a small

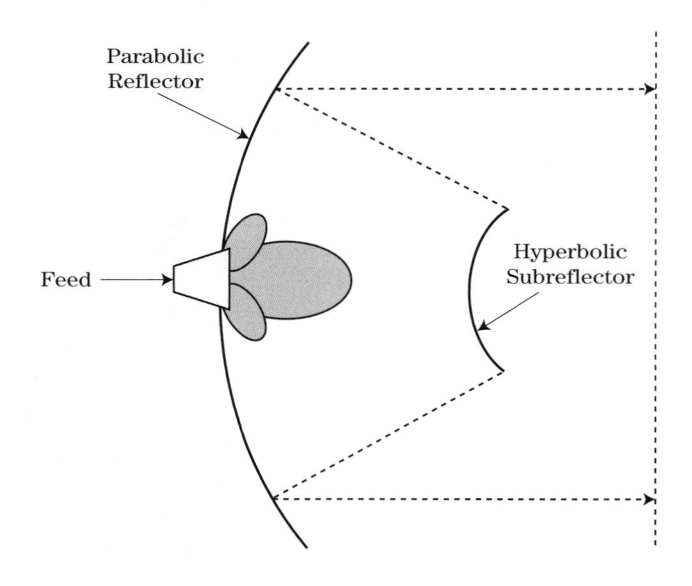

FIGURE 9-15 ■ Using a subreflector creates a long effective f/D in much less space.

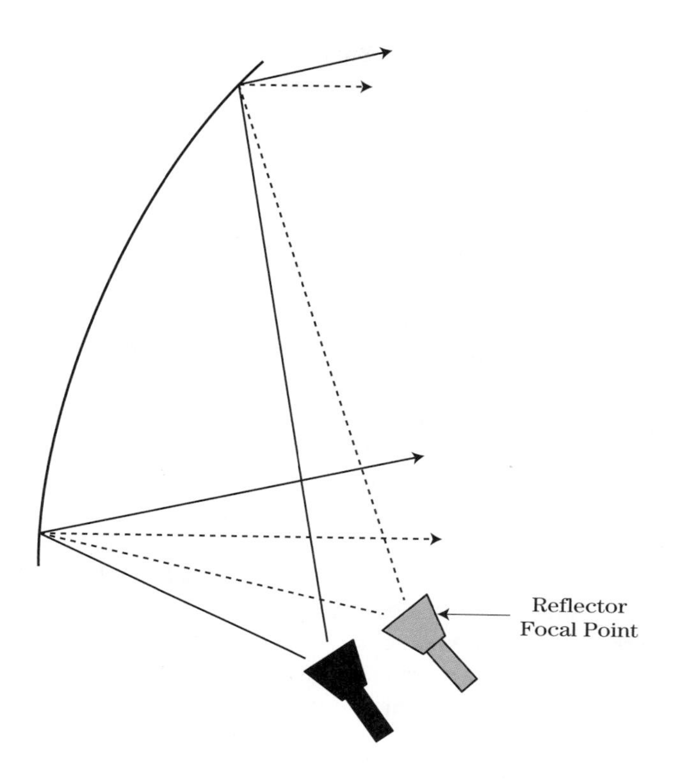

FIGURE 9-16 ■ The feed horn can be offset to eliminate blockage. Moving the horn from the focal point results in a limited amount of beam scanning.

hyperbolic subreflector with one of its foci located at the focus of the parabola [1]. This is called a *Cassegrain* configuration. This folded configuration provides an effectively long (f/D) reflector system in much less space and allows the feed waveguide to be brought in behind the reflector. The subreflector does create additional blockage of the reflector's aperture; however, the dual reflector system provides additional degrees of freedom for generating aperture tapers.

Blockage can be reduced or eliminated by using an offset or cut parabola in which the main reflector is made from a section of a parabola that is offset from the axis of rotation. From Figure 9-16 it can be seen that the focus of the parabola is now below the reflector and the radiated plane wave energy will not be blocked by the feed. The major drawback of this configuration is that the symmetry required for good monopulse performance is lost. This offset configuration is used for some search radars that do not require monopulse.

Reflector antennas can form multiple beams, or provide a limited amount of beam scanning, without mechanical motion of the reflector by displacing the feed horn from the parabola's focal point, as illustrated in Figure 9-16. However, if this method is used to scan the beam more than a few beamwidths, the resulting pattern will be severely distorted due to defocusing.

Search radars that mechanically rotate in azimuth sometimes use this limited scanning feature to form multiple elevation beams or to shape the elevation beam. If it is necessary to determine only the target's position in azimuth and range, the signals from multiple feeds can be combined to shape the elevation beam so it covers the required elevation angular area (i.e., 0° to 30°). If the target's elevation position also is needed, the radar input/output can be switched between multiple feeds to raster-scan the beam in elevation.

Reflector-based tracking radars use monopulse to enhance tracking accuracy. Sum and difference patterns are formed in the feed system by using multiple feeds displaced off the focal point to create the squinted beams and by combining them in a hybrid network or by using feed horns that can create multiple waveguide modes to form the multiple squinted beams [6].

9.7 PHASED ARRAY ANTENNAS

Phased arrays have become perhaps the most commonly used antenna in modern military radar systems. The reasons for this are plentiful. First, phased arrays provide high reliability, high bandwidth, and excellent sidelobe control. Second, there are certain applications for which phased arrays are uniquely qualified: phased arrays are ideal for stealth applications because they have no moving parts; they are ideal for airborne applications because they can electronically steer to extreme angles while maintaining a low profile, hence minimizing aircraft drag; and they are ideal for ground radar systems, which in some cases are too large for mechanical rotation, let alone rapid beam scanning. Finally, phased arrays have remarkable electronic beam agility that enables multiple functions to be performed nearly simultaneously by a single radar. Figure 9-17 is an example of a modern phased array, in this case in the nose of a fighter aircraft.

The largest disadvantage to the phased array antenna is cost. Phased arrays cost many times more than a reflector with the same gain; however, this cost gap has been decreasing. Phased array cost promises to decrease even further when technologies such as micro-electro-mechanical systems (MEMs) phase shifters and silicon-germanium (SiGe) mixed circuits replace some of the more expensive phased array components [11–13]. Nevertheless, the increased functionality of the phased array often justifies the additional cost.

9.7.1 The Array Factor

The basic concept of an array antenna was introduced in section 9.2. Essentially, an array is nothing more than collection of antennas that operates as a single unit. The individual antennas, often referred to as elements, are usually identical and uniformly

FIGURE 9-17 ■ AN/APG-81 – F-35 active electronically scanned array (AESA) radar. (Courtesy of Northrop Grumman Electronic Systems. Used with permission.)

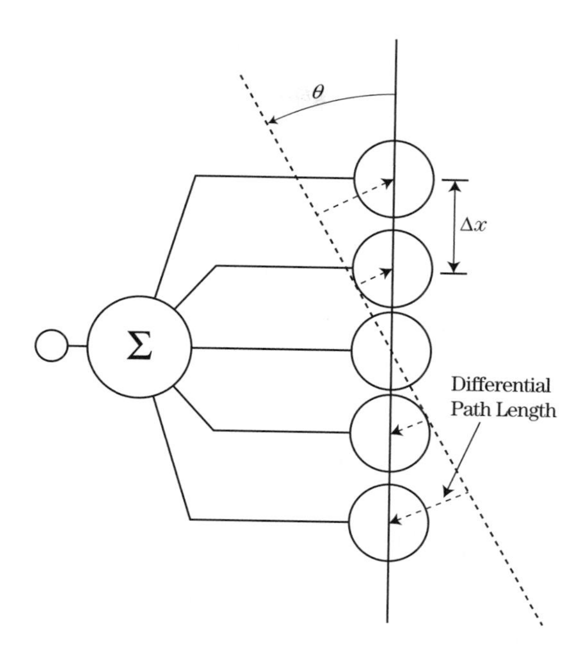

FIGURE 9-18 ■ A plane wave from angle θ intersecting with a five-element array.

spaced. Figure 9-18 shows a plane wave incident on a linear array of isotropic radiating elements from an angle θ with respect to array normal.

In the example in Figure 9-18, the interelement spacing is Δx, so the differential path length that the plane wave must propagate between one element and the next is $\Delta x \sin\theta$. This differential path length results in a relative phase shift between the signals at each element of $-(2\pi/\lambda)\Delta x \sin\theta$. The voltage response of the array to the plane wave is the sum of the individual element voltage responses. Normalized by the number of elements N, this is

$$AF(\theta) = \frac{1}{N}\sum_{n=1}^{N} \exp\left[-j\left(\frac{2\pi}{\lambda}n\Delta x \sin\theta - \phi_n\right)\right] \tag{9.16}$$

where ϕ_n is the relative phase shift between the n-th element and the array summation point. This expression is referred to as the *array factor* (AF) and is the key to understanding phased array antennas. The total radiation pattern of a phased array is the product of the array factor and the individual element pattern. If the element is assumed to be an isotropic radiator, which has no angular dependence, then the array factor and the phased array radiation pattern will be equal.

If the path length from each element to the summation point is equal, or $\phi_n = 0°$ for all n, the array factor will be maximum when the plane wave approaches from a direction normal to the array. This is confirmed in equation (9.16) where the array factor is maximum (equal to 1) when $\theta = 0°$ and $\phi_n = 0°$. As θ increases, the element signals will no longer have equal phases due to the differential path lengths and will therefore no longer combine in phase, causing the array factor to decrease. This process was described in section 9.2.2 and plotted in Figure 9-4.

It is often desirable for an array to have maximum directivity at a non-zero incidence angle of θ_s. When the n-th element is excited by a plane wave arriving from an angle θ_s the phase at that element, relative to the first element, will be $-(2\pi/\lambda)n\Delta x \sin\theta_s$. If the

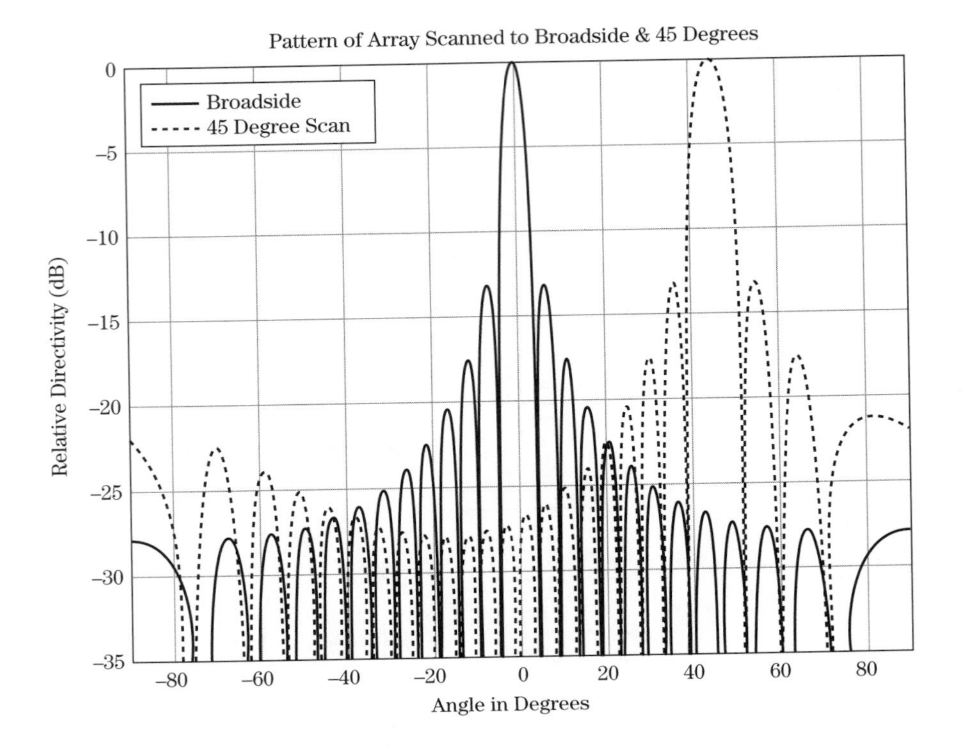

FIGURE 9-19 ■ Radiation pattern for an unsteered array and an array electronically scanned to 45°.

array elements are to coherently combine at this incidence angle, a progressive phase shift must be designed into the feed network that will essentially cancel the phase caused by the propagation delay. To accomplish this, the phase shift at element n must be

$$\phi_n = \frac{2\pi}{\lambda} n \cdot \Delta x \sin\theta_s \tag{9.17}$$

Substituting (9.17) into (9.16) yields

$$AF(\theta) = \frac{1}{N} \sum_{n=1}^{N} \exp\left[-j\frac{2\pi}{\lambda} n\Delta x(\sin\theta - \sin\theta_s)\right] \tag{9.18}$$

The array factor peak no longer occurs when $\theta = 0°$ because the feed system is now adding a progressive phase shift to the response from each element. Instead, the array factor maximum will occur at $\theta = \theta_s$. Figure 9-19 shows the array factor when the array is unscanned ($\theta_s = 0°$) and scanned to $45°$ ($\theta_s = 45°$).

By inserting the differential phase shift from equation (9.17) somewhere between the element and the summation point, the main beam can be electronically scanned to θ_s. But how is this phase shift inserted? The phase shift could be realized by adding physically longer path lengths between the elements and the summation point. The path length would need to be proportional to the desired phase delay; however, this method would be "hard-wired" and would work only at one scan angle. A smarter approach uses a device called a *phase shifter* that can dynamically change the phase delay from pulse to pulse.

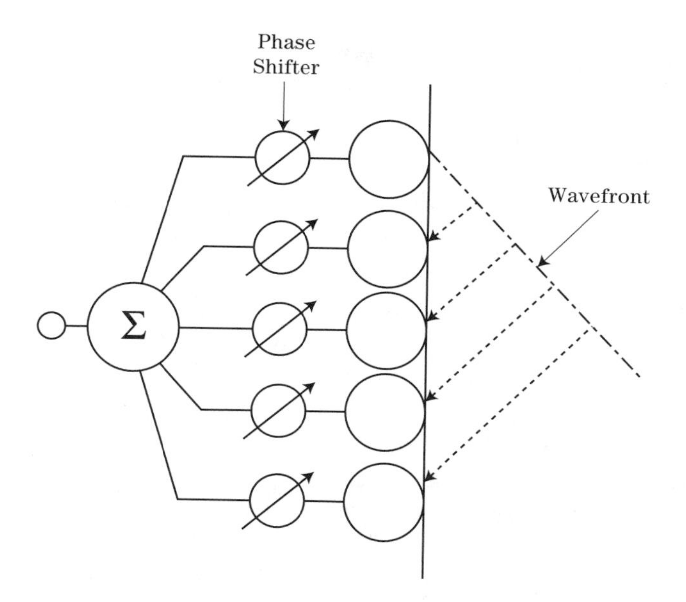

FIGURE 9-20 ■ Phase shifters can be inserted after the radiating elements to electronically scan the beam.

9.7.2 Phase Shifters

If a phase shifter is installed behind each of the radiating elements, as shown in Figure 9-20, then the array can be quickly resteered by simply adjusting the phase-shifter settings. The phase through each phase shifter will be set according to equation (9.17) for the desired scan angle. This dynamic electronic scanning capability of phased arrays makes them very attractive for modern radar applications. The antenna beam can be moved quickly in an arbitrary pattern; sequential scanning is not required of phased arrays.

Phase-shifter technology has been an energetic research area for the last 60 years, and an extensive discussion of these devices can be found in [10,14]. The discussion here will highlight key concepts that every radar engineer should understand. Figure 9-21 shows a simple switched line length phase shifter. This is commonly referred to as a 3-bit phase shifter because it has three line segments, each one-half the length of the next, that can be switched in and out to add phase shift. In this example the least significant bit causes a 45° phase delay. This leads to the first key concept, which is that phase shifters do not have arbitrarily fine phase resolution. Assume that a specific scan angle requires a phase shift of 63°. The phase shifter in Figure 9-21 would switch in its closest approximation, the 45° delay line, leaving an error of 18°. This error is referred to as a *phase quantization error* and is determined by the number of bits in the phase shifter.

Because each antenna element requires a different phase shift, the errors across the array will often appear to be random. Hence, these quantization errors are typically modeled as random errors, and, as discussed in section 9.3, they tend to increase the sidelobe

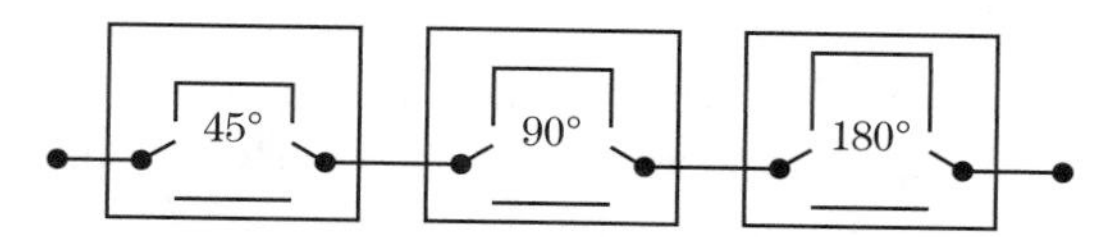

FIGURE 9-21 ■ Example of a 3-bit switched line length phase shifter.

TABLE 9-2 ■ Relationship among Phase-Shifter Bits, Phase Error, and Gain Loss

Number of Bits (N)	Least Significant Bit	RMS Phase Error	Gain Loss (dB)
2	90°	26°	0.65
3	45°	13°	0.15
4	22.5°	6.5°	0.04
5	11.25°	3.2°	0.01
6	5.625°	1.6°	0.00

levels of the radiation pattern. As the number of phase-shifter bits increases, the quantization error will decrease and the antenna sidelobe performance will improve. Also, as the sidelobe energy decreases the directivity will become larger. This relationship is shown in Table 9-2. The actual gain loss depends on the antenna size, shape, and aperture weighting, but the values in the fourth column are representative. On the other hand, insertion loss through the phase shifter will also increase with the number of bits. Therefore, it is not always beneficial to use the highest-resolution phase shifter available.

The second key concept is that phase shifters are not ideal for wideband radar applications because the required phase delay for a scanned beam is frequency dependent, as evidenced by the appearance of λ in equation (9.17). Thus, for a chirp waveform the antenna will have the right phase shift only at one frequency in the waveform. Section 9.7.6 will discuss methods to overcome this problem.

9.7.3 Grating Lobes

An examination of equation (9.18) reveals that it may be possible for more than one AF maximum to occur within the range of $-90° \le \theta \le 90°$. Specifically, maxima will occur for values of θ that satisfy

$$\frac{\Delta x}{\lambda}(\sin\theta - \sin\theta_s) = 0, \pm 1, \pm 2, \ldots \tag{9.19}$$

The maxima that corresponds to a value of zero on the right-hand side of equation (9.19) is the intended beam and has its peak at $\theta = \theta_s$. The maxima corresponding to nonzero integers are called *grating lobes* and have their peaks at values of θ other than θ_s. Grating lobes are usually undesirable for radar applications because they can induce angular ambiguities and hinder the radar's ability to locate targets. The radar may mistakenly report that a target is in the main lobe when in reality it was an object, such as clutter or a jammer, in the grating lobe.

The solid pattern in Figure 9-22 shows an array factor where the elements are spaced one wavelength apart ($\Delta x = \lambda$) and the main beam is unscanned ($\theta_s = 0°$). Since the elements are spaced one wavelength apart, a plane wave incident from either end of the array ($\theta = \pm 90°$) will travel exactly one wavelength between adjacent elements, and therefore the element signals will combine in phase. The lobes at $\pm 90°$, or $\sin\theta = \pm 1$, are the grating lobes.

As an aside, it should be noted that the array factor in Figure 9-22 is plotted as a function of $\sin\theta$ instead of θ. Since $\sin(90°) = 1$ and $\sin(-90°) = -1$, all angles in the antenna's forward hemisphere are mapped onto the $\sin\theta$ axis between -1 and 1. This coordinate system is referred to as *sine space* and is commonly used because it stresses the periodicity of the array factor.

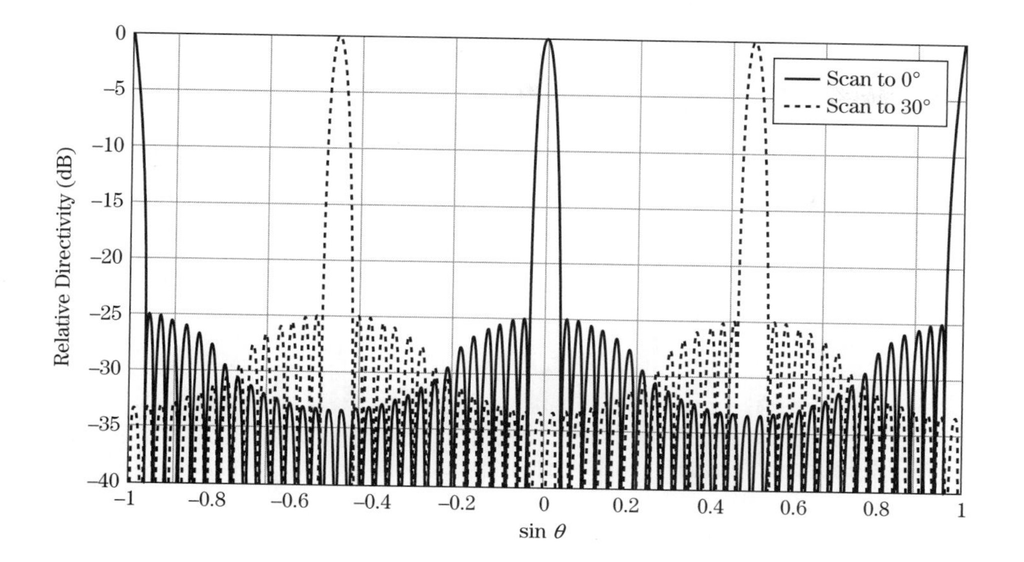

FIGURE 9-22 ■ Antenna pattern with grating lobes for a scanned and unscanned array with one wavelength element spacing.

The dashed pattern in Figure 9-22 shows the same array electronically scanned to 30° (a value of 0.5 in sine space). The grating lobe location has moved from −90° to −30°, which demonstrates that grating lobes scan with the main beam. The distance between the main lobe and the grating lobe is still one in the sine space coordinate system; hence, electronic scanning does not change the distance between the main lobe and the grating lobe.

Grating lobes can be prevented if Δx is made small enough. Equation 9.19 shows that the period between the grating lobes and the main beam is set by the element spacing in wavelengths. Specifically, in sine space, the array factor is periodic with period $\lambda/\Delta x$. Decreasing the element spacing with respect to the wavelength will push the grating lobes farther apart. Therefore, if it is required to scan the array to θ_s without grating lobes, the period of the array factor must be at least $1 + \sin\theta_s$, or

$$\Delta x \le \frac{\lambda}{(1 + |\sin\theta_s|)} \tag{9.20}$$

According to equation (9.20), Δx must be less than 0.667λ if the array is to be scanned to 30° without a grating lobe. The solid line in Figure 9-23 shows a linear array scanned to 30° with an element spacing of $\Delta x = 0.667\lambda$. The grating lobe is located at $\sin\theta = -1$. For comparison the dashed line shows a similar array with an element spacing of λ. The grating lobe is now much closer at $\sin\theta = 0.5$. By using $\theta_s = \pm 90°$ in equation (9.20), it can be seen that an element spacing of $\lambda/2$ is adequate for any scan angle and is therefore commonly assumed.

For a given aperture size there is an important trade between the number of elements and grating lobes. Grating lobes are avoided by spacing elements close together, but this requires additional elements, additional electronics, and, ultimately, incurs additional cost. Therefore, a phased array designer will often use the maximum spacing allowed by equation (9.20). This avoids grating lobes while minimizing cost. Maximum spacing is calculated using the highest operational frequency and greatest scan angle.

The grating lobe discussion is this section was limited to linear arrays. For a discussion of preventing grating lobes in a two-dimensional array, see [15].

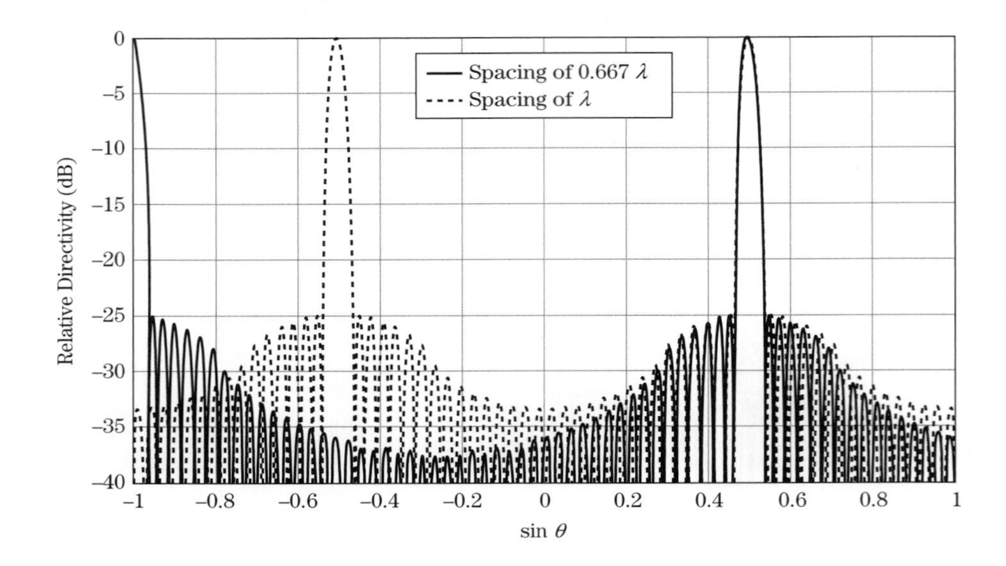

FIGURE 9-23 ■ Comparison between two arrays showing that grating lobes become less significant when the element spacing decreases.

9.7.4 Gain Loss

When a reflector antenna is mechanically scanned, the aperture and main beam are always pointing in the same direction, and the full antenna gain is realized at all scan angles. This is not the case with phased arrays. A negative byproduct of electronic scanning is gain loss. Gain loss is most significant at large scan angles and is caused by a decrease in the projected aperture of the array.

When a phased array is electronically scanned, the antenna does not move; hence, the projected aperture area in the direction of the main beam is reduced by $\cos\theta_s$. This is shown in Figure 9-24. Recall from equations (9.4) and (9.5) that the beamwidth and directivity are both related to the aperture size. If the aperture is reduced by $\cos\theta_s$ the directivity will be reduced by the same amount. The beamwidth is inversely proportional to aperture size and will be increased by a factor of $1/\cos\theta_s$. To illustrate this point further, consider a phased array scanned to 60°. At this scan angle the main beam will be twice

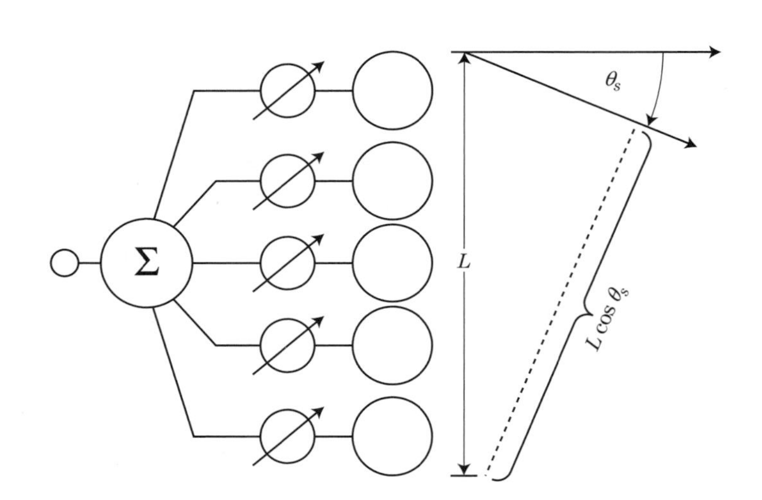

FIGURE 9-24 ■ The projected aperture of a phased array decreases with scan angle, resulting in beam broadening and a directivity loss.

as large as it is at broadside, and the directivity will be reduced by 3 dB. For monostatic radar the two-way loss at 60° will be 6 dB. This phenomenon usually limits the maximum practical scan angle of a phased array to ±60°.

As seen in equation (9.9), the maximum detection range of a radar is proportional to the antenna gain squared. Thus, at large scan angles, where gain loss is most significant, a phased array needs more sensitivity than a comparable mechanically scanned array.

Fortunately, several factors help mitigate the gain loss of a phased array. For a search radar, one of the loss terms (*scalloping loss*) is associated with the target not being at the peak of the beam at all times. Since the beam is widened at non-zero scan angles, for a constant separation of beam positions this scalloping loss is reduced, partially offsetting the gain loss. Most modern phased array systems have the ability to adapt the dwell time to compensate for gain loss. For near-normal beam positions, the dwell time can be reduced, and at far-out beam positions the dwell time can be lengthened to account for gain variation. For a track radar, the reduced gain and wider beamwidth will degrade track precision. For important threat targets, though, the dwell time can be extended to offset these effects.

9.7.5 The Array Element

So far the individual radiating elements that make up a phased array have been assumed to be fictitious isotropic radiators that do not impact the array's directivity pattern. In reality the array elements contribute significantly to the overall antenna performance. The element largely determines the array polarization as well as the impedance match and scan loss. Although the design of array elements is treated elsewhere [1–3] and is beyond the scope of this chapter, a few comments about array elements will be made.

First, combining the array factor from equation (9.18) with the element directivity pattern $E_e(\theta)$ yields

$$E(\theta) = E_e(\theta)AF(\theta) = \frac{E_e(\theta)}{N}\sum_{N}\exp\left[-j\frac{2\pi}{\lambda}n\Delta x(\sin\theta - \sin\theta_s)\right] \tag{9.21}$$

where $E(\theta)$ is the total antenna directivity pattern. Figure 9-25 shows an array with directive elements.

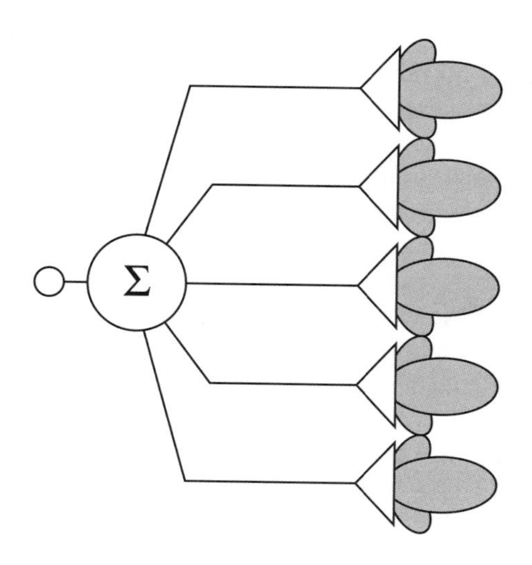

FIGURE 9-25 ■ Array elements are not isotropic and therefore have directivity.

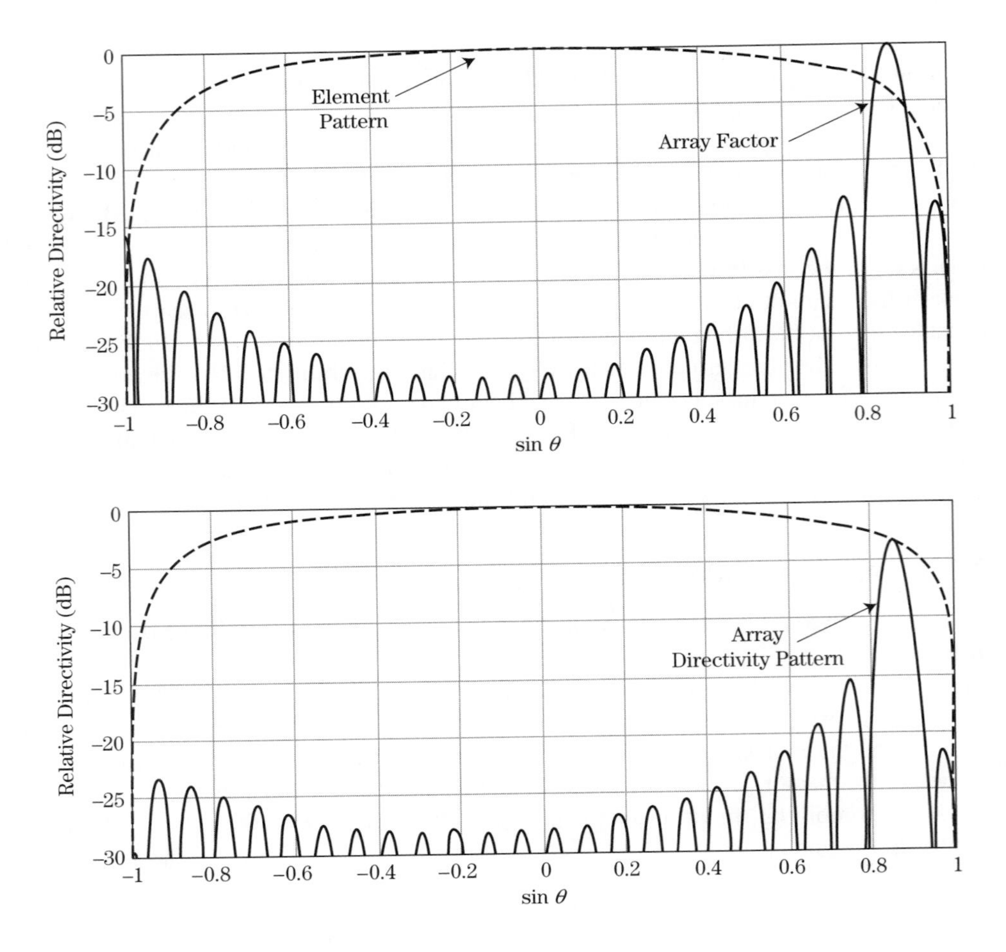

FIGURE 9-26 ■ The total antenna pattern of an array (lower plot) is the product of the array factor and the element pattern (upper plot).

According to equation (9.21) the directivity pattern of the array is the product of the array factor and the directivity pattern of a single array element. This is illustrated in Figure 9-26. When the array is electronically scanned, a phase shift is applied at each element, causing the array factor to electronically scan in the direction of θ_s, but the element pattern remains unchanged. Therefore, as the array scans it "looks" through a different part of the element pattern, and, as a result, the scanned main beam is weighted by the directivity of the element pattern at that scan angle. In an extreme scenario, if the element pattern has a null in the direction θ_s, then the array will be blind in this direction even though the array factor may be maximized. Thus, phased array gain loss is proportional to the directivity of the individual elements.

Antenna elements are usually designed such that the 3dB beamwidth is equal to the required array field of view (FOV). Therefore, $\cos\theta$ is a perfectly matched element pattern for an array with a maximum scan volume of 60°. The antenna element provides the impedance match between the feed system and free space for all scan angles and cannot be perfectly matched over all operating conditions. Elements are usually designed to have peak performance at broadside; thus, the match worsens with angle resulting in additional scan loss. For a full FOV (scan to $\pm°60$) radar, the element pattern is typically assumed to be $\cos^{1.5}(\theta)$ to account for this increased mismatch with scan angle. It is important to reiterate that the element pattern gain loss accounts for the aperture projection loss discussed previously.

There is an assumption made in equation (9.21) that is valid only for large phased arrays. This assumption is that each element has the same directivity pattern. In fact, the characteristics of an element in the array environment (surrounded by other elements typically spaced one-half wavelength apart) are quite different from what they are for an isolated element of the same type. An element near the edge of the array will "see" a different environment from what an element near the center of the array sees and will therefore have a different radiation pattern. In a 3×3 planar array, eight elements are edge elements, and only one is surrounded by others. The assumption of identical patterns for all elements is not valid for this scenario. However, if the array has 100×100 elements, 396 elements are on the edge and 9,604 are embedded in the array environment, and the assumption is valid.

Finally, polarization is specified for almost all radar antennas. For mechanically scanned systems the polarization needs to be specified only at broadside. However, with a phased array the polarization requirement will be scan angle dependent because the array "sees" through different places in the element pattern as it scans and because the element polarization can change with angle. If the polarization of the element pattern at 60° is different than at broadside, which is likely, the polarization of the main beam of the full array will also be different at a 60° scan angle. The antenna element will often exhibit the desired polarization at broadside and deteriorating polarization purity with increasing scan angle.

9.7.6 Wideband Phased Arrays

Modern radars require large instantaneous bandwidths to resolve targets in range [9,10, Chapter 20]. Regrettably, a traditional phase shifter-based phased array cannot simultaneously support wideband waveforms and electronic scanning. The phase required at element n to electronically scan a beam to angle θ_s is frequency dependent, as was seen in equation (9.17). For narrowband waveforms this is not a problem because the phase shifter can be reprogrammed between pulses to compensate for frequency hopping and new scan angles. However, for wideband operation the phase shifter can be accurate only at one frequency within the wideband spectrum and this is usually chosen to be the center frequency. As will be shown, the result is a beam-pointing error proportional to the instantaneous bandwidth.

Equation (9.21) applies to single frequency operation where λ is constant. When the radar waveform has extended bandwidth, equation (9.21) becomes

$$E(\theta, \lambda) = \frac{E_e(\theta, \lambda)}{N} \sum_{N} \exp\left[-j2\pi n \Delta x \left(\frac{\sin\theta}{\lambda} - \frac{\sin\theta_s}{\lambda_0}\right)\right] \tag{9.22}$$

where λ is the wavelength of any spectral component of the waveform and λ_0 is the wavelength at the center frequency of the waveform where the phase-shifter setting is determined.

In equation (9.21) the argument of the exponential term is zero, and all of the elements combine coherently, when $\sin\theta = \sin\theta_s$. In equation (9.22), the argument of the exponential is zero when

$$\frac{\sin\theta}{\lambda} = \frac{\sin\theta_s}{\lambda_0} \tag{9.23}$$

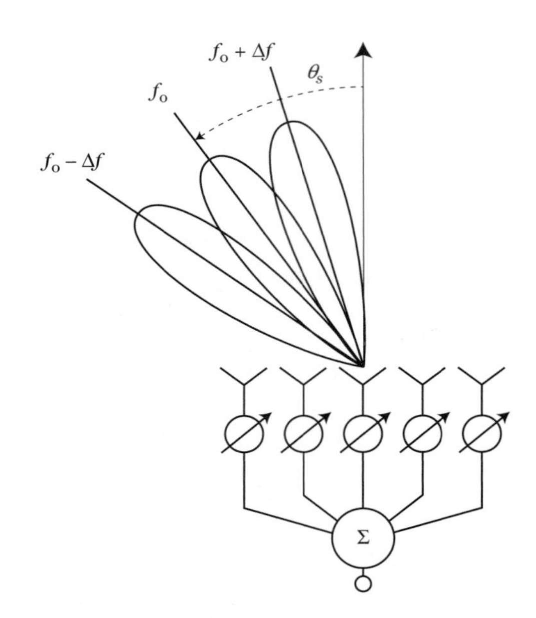

FIGURE 9-27 ■ A phase-shifter-based phased array will mispoint the beam during wideband operation for off-broadside scan angles.

From (9.23) it can be seen that a small change in frequency causes a mispointing error of

$$\Delta\theta = \frac{\Delta f}{f_0} \tan\theta_s \tag{9.24}$$

where the difference between the instantaneous frequency and the center frequency is Δf, and the resulting mispointing, or angular squint, is $\Delta\theta$. Therefore, every spectral component of the waveform will be scanned into a different direction centered about θ_s. This situation is depicted in Figure 9-27.

If it is determined that the radar can tolerate a beam squint of ±one-half of the 3 dB beamwidth, then $\Delta\theta$ can be set to $\theta_3/\cos\theta_s$ (remember the beam broadens by $1/\cos\theta_s$ with scanning), and the maximum tolerable instantaneous signal bandwidth in equation (9.24) becomes

$$B_i \equiv \left(\frac{\Delta f}{f_0}\right)_{\max} = \frac{\theta_3}{2\sin\theta_s} \tag{9.25}$$

where θ_3 is the 3 dB beamwidth in radians at broadside. The instantaneous bandwidth in equation (9.25) can be approximated with an engineering equation as

$$B_i\,(\%) \approx 2\theta_3\,(\text{degrees}) \tag{9.26}$$

where the instantaneous bandwidth B_i is expressed as a percentage, and the beamwidth has units of degrees. This simplified equation assumes a maximum scan angle of 60°. For example, an array with a 1° beamwidth can support only a 2% instantaneous bandwidth at a 60° scan angle. It is important to point out that this problem becomes exacerbated for large apertures that have smaller beamwidths. The mispointing angle does not change with aperture size; however, it becomes a larger percentage of the antenna beamwidth. Many modern radars have a beamwidth of 1° or less and require bandwidths of 10% or greater. In these cases, something must be done to overcome the limitation expressed in (9.26).

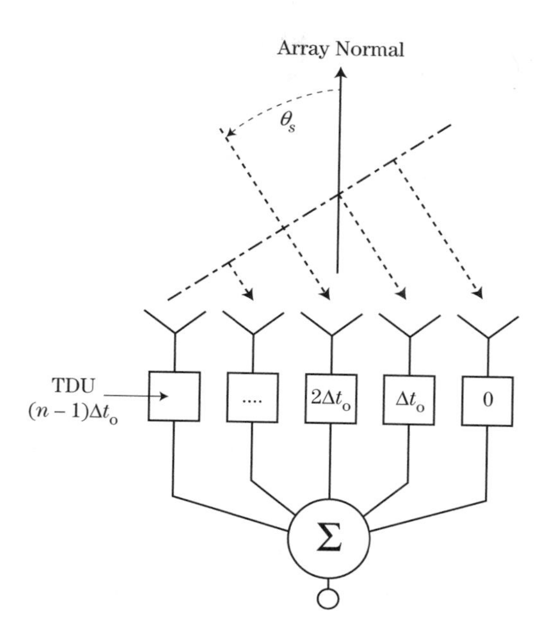

FIGURE 9-28 ■ The phase front of an off-axis plane wave will strike each element at a different time. The time of arrival difference is frequency independent. A TDU-based phased array is ideal for wideband operation.

One approach to improving bandwidth is to replace the phase shifters at each element with time delay units (TDUs). Instead of adjusting for the phase difference induced by the incident off-axis plane wave, the TDU will adjust for the difference in time of arrival at each element, as illustrated in Figure 9-28.

The differential time delay Δt at each element is

$$\Delta t_n = \frac{n \Delta x \sin \theta_s}{c} \tag{9.27}$$

which results in a phase shift of

$$\phi_n = 2\pi \Delta t_n f = \frac{2\pi}{\lambda} n \Delta x \sin \theta_s \tag{9.28}$$

Rewriting (9.22) for time delay scanning yields

$$E(\theta, \lambda) = \frac{E_e(\theta, \lambda)}{N} \sum_n \exp\left[-j\frac{2\pi}{\lambda} n \Delta x (\sin\theta - \sin\theta_s)\right] \tag{9.29}$$

By using TDUs instead of phase shifters, the phase shift at each element is automatically corrected for each spectral component and eliminates beam squinting. This is because, although each spectral component of the waveform travels a different distance in wavelengths to each element, all spectral components travel at the speed of light and therefore have equal time delay. The bandwidth problem would be solved if TDUs could be placed behind each antenna element. Unfortunately, with current technology TDUs are still too large and expensive to embed behind individual elements. Therefore, a compromise solution is gaining popularity in many large phased array systems.

The compromise solution is presented in Figure 9-29. The array is broken into small sections called subarrays, which will be discussed in section 9.8.3. Each subarray is sized to have a beamwidth determined by equation (9.25) and the desired bandwidth. Conventional phase shifters are used within the subarrays, and TDUs are used to provide time delay

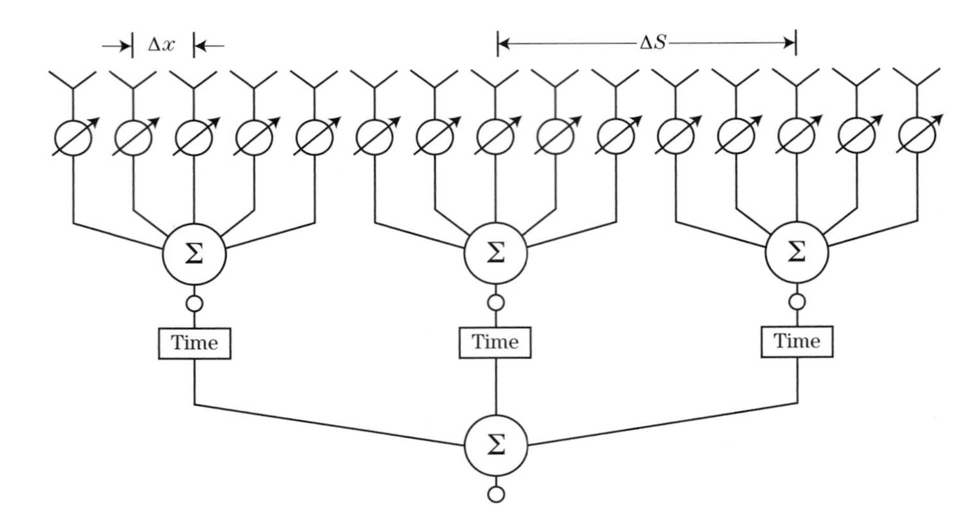

FIGURE 9-29 ■ The subarray architecture is often used for wideband operation with TDUs inserted behind each subarray.

between subarrays. The subarray pattern will squint according to equation (9.24) because it is scanned using phase. The subarray AF, which determines the location of the full-array main beam, operates according to (9.29) and will not mispoint with frequency. The resulting equation for the directivity pattern is

$$E(\theta, \lambda) = \frac{E_e(\theta, \lambda)}{MN} \tag{9.30}$$

$$= \sum_M \exp\left[-j\frac{2\pi}{\lambda} m \Delta S(\sin\theta - \sin\theta_s)\right] \sum_N \exp\left[-j2\pi n \Delta x\left(\frac{\sin\theta}{\lambda} - \frac{\sin\theta_s}{\lambda_0}\right)\right]$$

where the total number of subarrays is M, and the subarray spacing is ΔS.

The subarray pattern and subarray AF are plotted in Figure 9-30 for the spectral components at band center and band edge. The subarray centers are more than a wavelength apart so the subarray AF will have grating lobes. At band center, the subarray pattern is not squinted, and the nulls of the subarray pattern will cancel the grating lobes of the subarray AF. At band edge, the subarray pattern will squint, will create a directivity loss in the main beam, and will no longer suppresses the subarray AF grating lobes. This is the compromise created by the time delayed subarray solution to the bandwidth problem. The trade-offs associated with subarray size, number of TDUs, scan loss, and grating lobes are part of the system design of any large wideband phased array radar [16].

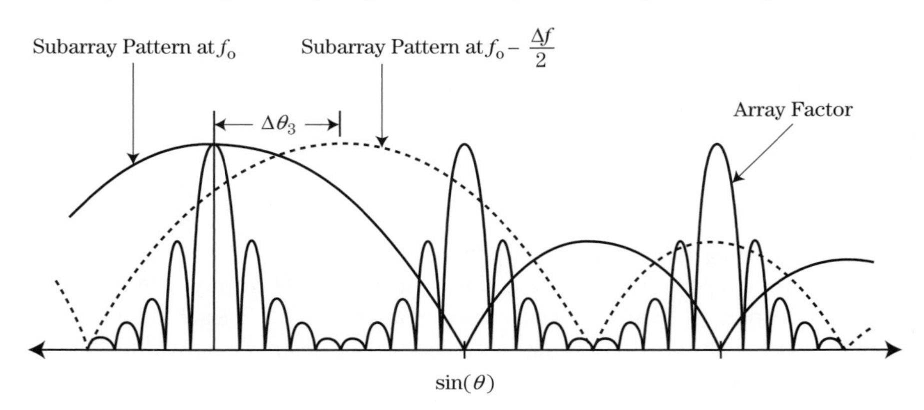

FIGURE 9-30 ■ For wideband operation the subarray pattern will squint, but the subarray AF will not.

9.8 ARRAY ARCHITECTURES

In modern radar systems the boundary between the antenna and other radar hardware is becoming less clear. With traditional systems there was a clear distinction between the receiver/exciter and the antenna. This is not so anymore. There has been a general trend to move the receiver/exciter electronics, and in some cases the signal processing, ever closer to the radiating surface of the antenna. This section will describe the most common array architectures as well as some emerging trends in antenna design.

9.8.1 Passive Array Architecture

The earliest phased array antenna systems used one large power amplifier (PA) tube to drive the entire array on transmit, and one low-noise amplifier (LNA) to set the noise figure after all array elements had been combined. This architecture is commonly referred to as a *passive phased array* because there is no amplification at the individual radiating element. The passive array architecture is shown in Figure 9-31.

The major benefit of the passive array is cost. There are, however, many pitfalls to this design that have led to a decline in its use. The passive architecture requires all array components to handle high power. Depending on the frequency and size of the array, there can be considerable loss in the combining network that will reduce the radar sensitivity. Lastly, TDUs are difficult to implement in this architecture because the array cannot conveniently be broken into subarrays; hence, passive arrays are not amenable to wideband applications.

9.8.2 Active Array Architecture

The dominant architecture at the time of this writing (2009) is the *active electrically scanned array* (AESA) where the final transmit power amplifier, receiver LNA, phase shifter, and possibly a programmable attenuator are packaged into a transmit/receive (T/R)

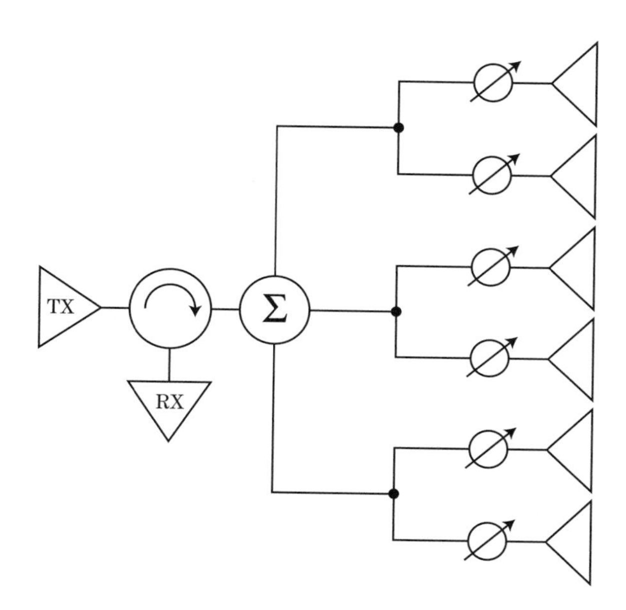

FIGURE 9-31 ■ The passive array uses one PA to drive the entire array and one LNA to set the noise figure.

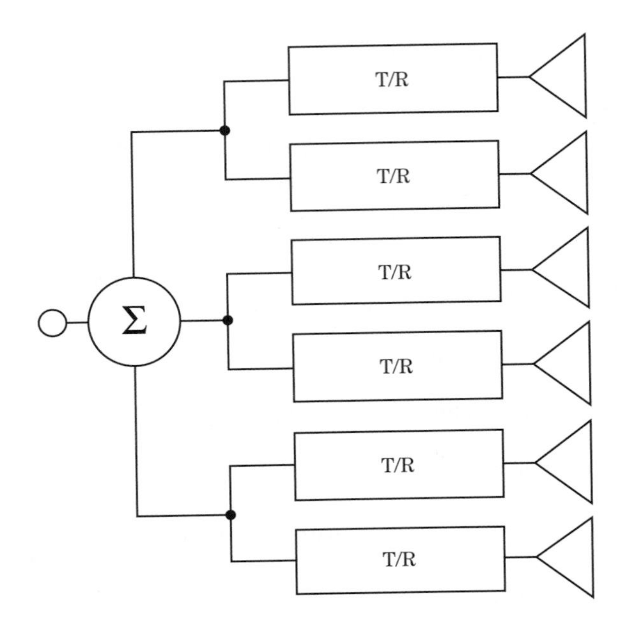

FIGURE 9-32 ■ An AESA places a PA and LNA behind each radiator. These electronics are usually packaged into a T/R module.

module behind each element [17,18]. This architecture, as shown in Figure 9-32, has several advantages over the passive architecture. Transmit losses between the PA and the radiator and receive losses between the radiator and the LNA are minimized. The AESA architecture provides additional beamforming flexibility since the amplitude and phase of each element can be dynamically controlled. This allows the aperture amplitude taper to be adjusted from pulse to pulse or from transmit to receive, whereas in a passive architecture the aperture amplitude taper is designed into the beamforming network of the array. The solid-state amplifiers can be broader band than tube amplifiers, and the received signal can be digitized or optically modulated to support digital or optical beamforming. AESAs can operate even when a small percentage of the T/R modules fail, so they degrade more gracefully than the passive architecture. Disadvantages of the AESA are cost, the requirement to cool the entire aperture to remove heat generated by the PAs, and the requirement for calibration in the field.

The cooling and cost problems may be alleviated for some applications by the emergence of the low power density (LPD) concept in which the TR module is reduced to the point where it becomes a single monolithic microwave integrated circuit (MMIC). Silicon-germanium is an ideal semiconductor technology for LPD applications because it is inexpensive, low power, and capable of excellent RF performance at typical radar frequencies [19–22].

Referring back to Figure 9-8, the required power-aperture product or power-aperture gain product of a LPD array is obtained by increasing the aperture size and reducing the transmit power. A diagonal line from the lower left corner to the upper right corner represents an aperture power density of 1 kW/m^2, which is generally considered the boundary between high and low power density. LPD AESAs have power densities low enough to eliminate the need for liquid cooling of the aperture. Although more T/R modules are required in LPD AESAs, the cost of the LPD T/R module will be significantly less than that of the high power T/R module. In LPD designs, large aperture is traded for high power, thus significantly reducing the prime power, cooling, and fuel requirements, which in turn can improve the radar's transportability.

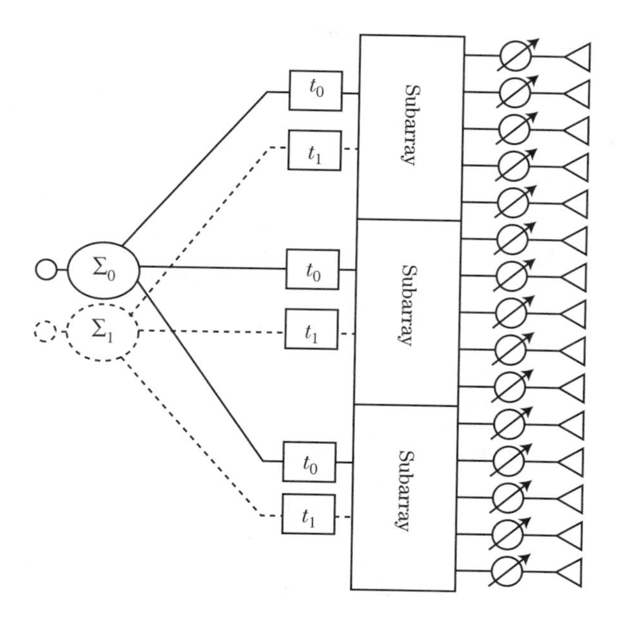

FIGURE 9-33 ■ An array divided into three subarrays. Different time delays are applied at the subarray level to form two simultaneous receive beams.

The LPD AESA is most beneficial in applications where aperture size is not fixed. For space-based or ground-based radars the antenna can be stowed for launch or transport and then erected for operation. LPD AESAs can also be used in aerostat or airship-based radars. LPD will obviously not work for radars that are severely spatially constrained, such as fighter aircraft radars.

9.8.3 Subarray Architecture

There are many variants of the subarray architecture, and one example is shown in Figure 9-33. In general, phase shift is applied at the radiating element, and amplification is applied to a grouping of elements referred to as a subarray. This makes the subarray architecture a hybrid of the passive and active architectures. Time delay is often implemented at the subarray level when the array is intended for wideband use. Time delay can be implemented with analog TDUs (as shown in Figure 9-33) or implemented digitally if the receive signal is digitized at the subarray. In addition, the subarray architecture has become a convenient method for implementing many advanced techniques such as digital beam forming and array signal processing. Many of these techniques are beyond the scope of this chapter; however, a brief introduction and relevant references will be provided [23–26].

The first advantage of the subarray is purely practical. Many modern phased arrays are manufactured using printed circuit board techniques for the radiating and beamforming layers. These techniques have size limitations that prohibit a large array from being manufactured in one piece. Subarrays are small, repeatable building blocks that can be mass produced and combined to form the full array. Figure 9-33 shows a 15-element array divided into three subarray units. The subarray units also offer a convenient location to implement time delay steering for wideband arrays as discussed in section 9.7.6.

The subarray architecture opens up additional opportunities when a digital receiver is placed behind each subarray or group of subarrays. Once the subarray outputs have been digitized, they can be weighted and combined in the digital computer to form, in principle,

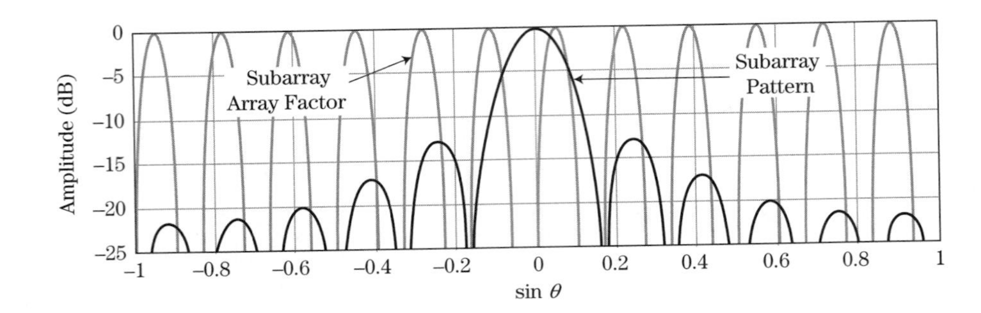

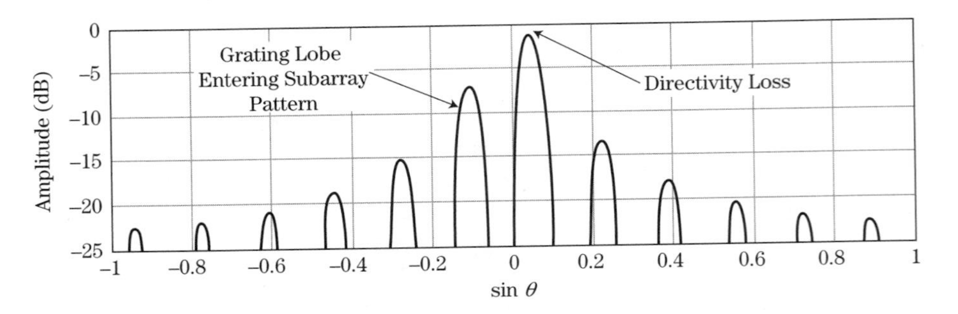

FIGURE 9-34 ■ During simultaneous beam operation the subarray pattern remains fixed, and the subarray AF will electronically scan. Grating lobes become significant when subarray AF lobes enter the subarray pattern mainlobe. Upper plot: Subarray pattern and subarray array factor. Lower: Combined antenna pattern.

an unlimited number of simultaneous offset beams without impacting radar resources or timeline.

Figure 9-33 shows the formation of two simultaneous beams using analog beamforming. For digital beamforming the analog TDUs would be replaced with digital receivers. Multiple simultaneous beams are commonly used to form clusters surrounding the main beam to improve the search timeline of a radar. The number of beams is limited only by the processing capabilities of the radar.

The extent that simultaneous receive beams can be offset from the primary, or center, receive beam is a function of the subarray beamwidth, and ultimately the subarray size. To explain this concept, remember that the radiation pattern of the array is the product of the element pattern, subarray pattern, and subarray AF. These contributors and the final radiation pattern are shown in Figure 9-34. The top figure shows the subarray AF and the subarray pattern. The bottom figure shows the combined antenna pattern.

During simultaneous beam operation, the element level phase shifters remain in the same state for all receive beams; thus, the subarray pattern does not scan. The receive beams can be digitally resteered, or offset, to anywhere within the subarray main beam by applying a phase shift or time delay at the subarray level. As discussed in section 9.7.3, grating lobes occur when array elements are spaced greater than $\lambda/2$ apart, which is certainly the case for the subarray centers. These subarray AF grating lobes are shown in the upper plot of Figure 9-34.

The majority of the grating lobes are not an issue because they are suppressed by the subarray pattern. However, during simultaneous beam operation, the grating lobes will scan and sometimes enter the subarray main beam that remains in a fixed position. Figure 9-34 shows a scenario where the subarray AF is scanned only three degrees yet the first grating lobe has entered the subarray main beam. Thus, the maximum offset angle for simultaneous beams is determined by the subarray AF grating lobes and the subarray pattern main beam.

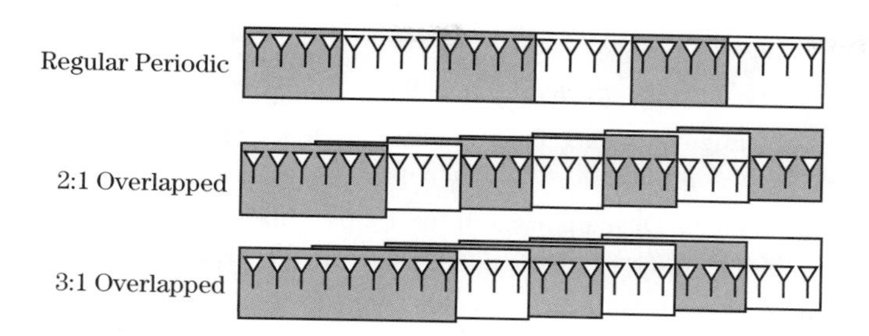

FIGURE 9-35 ■ Options for implementing the overlapped subarray architecture.

One method to increase the maximum offset angle is to create a narrower subarray beamwidth by using an overlapped subarray architecture. With overlapped subarrays each element will feed into more than one subarray receiver, hence increasing the subarray size without decreasing the number of subarrays. This concept is shown pictorially in Figure 9-35. The overlapped architecture creates larger subarray sizes, which result in narrower subarray patterns. Thus, simultaneous receive beams can be offset to larger angles without grating lobes entering the subarray pattern. This is just one of many benefits to the overlapped subarray architecture. A more through discussion of overlapped subarrays can be found in [27].

9.9 FURTHER READING

There are many superb textbooks devoted to the finer details of antenna design and analysis. Those by Johnson [1], Balanis [2], Stutzman [3] and Long [30] provide a thorough discussion of antenna fundamentals. However, these texts do not, nor were they intended to, address the design considerations of the modern radar antenna system. Most radar texts will designate a chapter specifically to the antenna system. A few excellent examples are chapter 9 of Skolnik [9] and chapters 37 and 38 of Stimson [28]. Finally, entire texts such as Mailloux [29] and Hansen [5] are devoted solely to the phased array antenna system.

The modern radar antenna is increasingly becoming a multidisciplinary system, and there are numerous emerging technologies of interest to the antenna engineer. References for these new technologies are placed throughout this chapter and can be found by monitoring the publications from the IEEE Antennas and Propagation Society and the IEEE Microwave Theory and Techniques Society.

9.10 REFERENCES

[1] Volakis, J., *Antenna Engineering Handbook*, 4d ed., McGraw-Hill Company, New York, 2007.

[2] Balanis, C.A., *Antenna Theory Analysis and Design*, 3d ed., John Wiley & Sons, Hoboken, NJ, 2005.

[3] Stutzman, W.L., and Thiele, G.A., *Antenna Theory and Design*, 2d ed., John Wiley & Sons, New York, 1998.

[4] Taylor, T.T, "Design of Line-Source Antennas for Narrow Beamwidth and Low Sidelobes," *IRE Transactions on Antennas and Propagation,* vol. AP-3, pp. 16–27, January 1955.

[5] Hansen, R.C., *Phased Array Antennas*, 2d ed., John Wiley & Sons, New York, 2010.

[6] Skolnik M.I., *Radar Handbook*, 3d ed., McGraw-Hill, New York, 2008.

[7] Sherman, S.M., *Monopulse Principles and Techniques*, Artech House, Dedham, MA, 1984.

[8] Bayliss, E.T., "Design of Monopulse Antenna Difference Patterns with Low Sidelobes," *Bell Systems Technical Journal,* vol. 47, pp. 623–650, 1968.

[9] Schell, A.C., "Antenna Developments of the 1950s to the 1980s," *IEEE Antennas and Propagation Society International Symposium,* vol. 1, pp. 30–33, July 8–13, 2001.

[10] Skolnik, M.I., *Introduction to Radar Systems,* 3d ed., McGraw Hill, New York, 2001.

[11] Maciel, J.J., Slocum, J.F., Smith, J.K., and Turtle, J., "MEMS Electronically Steerable Antennas for Fire Control Radars," *IEEE Aerospace and Electronic Systems Magazine,* vol. 22, pp. 17–20, November 2007.

[12] Brookner, E., "Phased Arrays around the World—Progress and Future Trends," *2003 IEEE International Symposium on Phased Array Systems and Technology,* pp. 1–8, Boston, MA, October 14–17, 2003.

[13] Brookner, E., "Phased-Array and Radar Breakthroughs," 2007 IEEE Radar Conference, pp. 37–42, April 17–20, 2007.

[14] Koul, S.K., and Bhat, B., *Microwave and Millimeter Wave Phase Shifters: Semiconductor and Delay Line Phase Shifters,* vol. 2, Artech House, Norwood, MA, 1992.

[15] Corey, L.E., "A Graphical Technique for Determining Optimal Array Geometry," *IEEE Transactions on Antennas and Propagation,* vol. AP-33, no. 7, pp. 719–726, July 1985.

[16] Howard, R.L., Corey, L.E., and Williams, S.P., "The Relationships between Dispersion Loss, Sidelobe Levels, and Bandwidth in Wideband Radars with Subarrayed Antennas," *1988 IEEE Antennas and Propagation Society International Symposium*, pp. 184–187, June 6–10, 1998.

[17] Cohen, E.D., "Trends in the Development of MMIC's and Packages for Active Electronically Scanned Arrays (AESAs)," *1996 IEEE International Symposium on Phased Array Systems and Technology,* pp. 1–4, October 15–18, 1996.

[18] Kopp, B.A., "S- and X-Band Radar Transmit/Receive Module Overview," *2007 IEEE Radar Conference,* pp. 948–953, April 17–20, 2007.

[19] Mitchell, M.A., Cressler, J.D., Kuo, W.-M.L., Comeau, J., and Andrews, J., "An X-Band SiGe Single-MMIC Transmit/Receive Module for Radar Applications," *IEEE 2007 Radar Conference,* pp. 664–669, April 17–20, 2007.

[20] Comeau, J.P., Morton, M.A., Wei-Min, L.K., Thrivikraman, T., Andrews, J.M. Grens, C., et al., "A Monolithic 5-Bit SiGe BiCMOS Receiver for X-Band Phased-Array Radar Systems," *IEEE Bipolar/BiCMOS Circuits and Technology Meeting, 2007,* pp. 172–175, September 30–October 2, 2007.

[21] Tayrani, R., Teshiba, M.A., Sakamoto, G.M., Chaudhry, Q., Alidio, R., Yoosin Kang, A., et al., "Broad-band SiGe MMICs for Phased-Array Radar Applications," *IEEE Journal of Solid-State Circuits,* vol. 38, no. 9, pp. 1462–1470, September 2003.

[22] Kane, B.C., Geis, L.A., Wyatt, M.A., Copeland, D.G., and Mogensen, J.A., "Smart Phased Array SoCs: A Novel Application for Advanced SiGe HBT BiCMOS Technology," *Proceedings of the IEEE,* vol. 93, no. 9, pp. 1656–1668, September 2005.

[23] Compton, R.T., *Adaptive Antennas: Concepts and Performance*, Prentice-Hall, Upper Saddle, River, NJ, 1988.

[24] Zatman, M., "Digitization Requirements for Digital Radar Arrays," *Proceedings of the 2001 IEEE Radar Conference,* pp. 163–168, May 1–3, 2001.

[25] Combaud, M., "Adaptive Processing at the Subarray Level," *Aerospace Science and Technology,* vol. 3, no. 2, pp. 93–105, 1999.

[26] Nickel, U., "Subarray Configurations for Digital Beamforming with Low Sidelobes and Adaptive Interference Suppression," *IEEE International Radar Conference,* pp. 714–719, 1995.

[27] Lin, C.-T., and Ly, H., "Sidelobe Reduction through Subarray Overlapping for Wideband Arrays," *Proceedings of the 2001 IEEE Radar Conference,* pp. 228–233, May 1–3, 2001.

[28] Stimson, G.W., *Introduction to Airborne Radar*, 2d ed., SciTech, Raleigh, NC, 1998.

[29] Mailloux, R.J., *Phased Array Antenna Handbook,* 2d ed., Artech House, Norwood, MA, 2005.

[30] Blake, L. and Long, M.W., *Antennas: Fundamentals, Design, Measurement*, 3d ed., SciTech, Raleigh, NC, 2009.

9.11 | PROBLEMS

1. Assume an isotropic antenna has been invented that can transmit 10 watts. (a) Assume there is a 20 dBm^2 target 1 km away. What is the power density at this target? (b) Assume a large clutter object (RCS of 30 dBm^2) is located at the same range as the target but 20° away. Calculate the power density at the clutter location. (c) Why would this be a problem for the radar system?

2. Consider a radar antenna that has an aperture 3.0 m high by 3.5 m wide. Imagine a 10 GHz plane wave arriving from an angle of elevation $= 0°$ and azimuth $= 15°$. (a) What is the total phase gradient across the aperture surface? (b) What if the azimuth angle is changed to 60°?

3. A fighter jet has a circular radar antenna in the nose of the aircraft that operates at 10 GHz. (a) Assuming the diameter is 1 meter and the aperture is uniformly illuminated, calculate the maximum directivity and beamwidth of the antenna. (b) How would these values change if the antenna operated at 32 GHz?

4. Consider a uniformly illuminated rectangular antenna with peak directivity of 35 dBi. What will the first sidelobe be in dBi? Now assume a -40 dB Taylor weighting is applied in the x-axis and a -25 dB Taylor weighting is applied in the y-axis. In dBi, what will the new peak directivity and peak sidelobe be? Assume the antenna aperture is error free.

5. A circular antenna with a diameter of 2 ft operates at 16 GHz. There are 2 dB of RF losses within the antenna and an RMS phase error of 15° on the aperture surface. Calculate the maximum directivity and gain of the antenna.

6. A radar system has been designed to track targets at range R. At the last minute the customer decides that she needs to track targets at $2R$, twice the initial range, and the remaining requirements must remain the same. The project manager decides to meet this new requirement by modifying the antenna. If aperture size is unchanged, by how much will transmit power be increased? What if the total transmit power remains constant and the aperture is increased? How much larger must the aperture be?

7. Assume a parabolic reflector is illuminated such that the edges of the reflector see 3 dB less power than the center of the reflector. Given a reflector diameter of 20 ft. and a focal length of 18 ft., how large must the feed aperture be if it is circular in shape and uniformly illuminated? The system operates at 10 GHz.

8. Consider a five-element linear array with element spacing of 0.5 wavelengths. Assume the path length from each element to summation point (φ_n) is equal. Using computational software (e.g., Matlab, Excel), plot the normalized array factor for incidence angles from $-90°$ to $+90°$.

9. A linear array is to be designed that operates at 10 GHz and has a 1° beamwidth (assume uniform illumination). This array must electronically scan to 65° without grating lobes. What is will the element spacing be? How many elements will be used?

10. Consider a phased array that has a 3° beamwidth and 35 dBi of directivity at broadside. What are the beamwidth and directivity when the array is electronically scanned to 55°?

11. A 10 m space-based antenna is in low Earth orbit (LEO) at 1,000 km. The radar has an instantaneous bandwidth of 500 MHz and uses a phased array without time delay. Assuming that the antenna is calibrated in the center of the band, what is the pointing error at the edge of the instantaneous bandwidth when the antenna is electrically scanned to 50°? How many meters of pointing error does this translate too when mapped to the earth's surface? (Assume a flat earth to simplify this problem.)

12. For the scenario in question 11, assume time delay units are implemented instead of the phase shifters. How much time delay is required? What does this equal in phase?

13. Determine the mispointing error caused when a phase-shifter-based phased array electronically scans to 65°. Assume 10% instantaneous bandwidth and a center frequency of 3.5 GHz. What fraction of the beamwidth will this be if the antenna is 1 m long? What fraction of the beamwidth will this be if the antenna is 10 m long? Assume uniform aperture taper.

14. A phased array is to be designed that satisfies the following requirements at 10 GHz: (1) The antenna must have a broadside directivity of 35 dBi; (2) the array must have a square shape, uniform illumination, and a rectangular element grid; (3) the array must be able to scan to 30° in azimuth and 30° in elevation without grating lobes occurring; and (4) the array must meet the directivity requirement with 10° of random phase errors. How many elements are required to meet these requirements?

15. Assume that the requirements in question 14 have been changed such that low sidelobes are required for this aperture and it must electronically scan to more extreme angles. If a 40 dB Taylor weighting is applied to the aperture in both dimensions and the maximum scan angle is increased in both dimensions to 60°, how many elements must be used? Assuming $500 per element, how much additional cost is incurred?

CHAPTER 10

Radar Transmitters

Tracy V. Wallace, Randy J. Jost, and Paul E. Schmid

Chapter Outline

10.1 Introduction 347
10.2 Transmitter Configurations 351
10.3 Power Sources and Amplifiers 356
10.4 Modulators 371
10.5 Power Supplies 373
10.6 Transmitter Impacts on the Electromagnetic Environment 375
10.7 Operational Considerations 381
10.8 Summary and Future Trends 384
10.9 Further Reading 385
10.10 References 385
10.11 Problems 388

10.1 INTRODUCTION

The radar transmitter subsystem generates the radiofrequency (RF) energy required for the illumination of a remotely situated target of interest. Targets may include aircraft, ships, missiles and even weather phenomena such as rain, snow, and clouds. The radar transmitters described in this chapter includes three basic elements: (1) a radiofrequency oscillator or power amplifier, (PA); (2) a modulator (either analog or digital); and (3) a power supply that provides the high voltages (HVs) and currents typically required in modern radar transmitters. Depending on the specific application, the peak powers generated by the radar transmitter can range from milliwatts to gigawatts. The carrier frequency can range from 3 to 30 MHz (high-frequency [HF] over-the-horizon [OTH] radars) to frequencies as high as 94 GHz (millimeter wave [MMW] radars). However, the majority of today's civilian and military search-and-track radar systems operate in the frequency range from 300 MHz to 12 GHz and typically generate an average power ranging from tens to hundreds of kilowatts. Both thermionic tube-type transmitters and solid-state transmitters are used, depending on the application.

10.1.1 The Radar Transmitter as Part of the Overall Radar System

10.1.1.1 Basic Pulse Radar

Figure 10-1 is a block diagram of a typical pulsed radar [1]. The subunit identified as the RF amplifier in practice can be either a high-power oscillator, such as a *magnetron*, or a high-power amplifier, such as the *traveling-wave tube* (TWT) or multicavity *klystron*. The characteristics of these and related high-power microwave vacuum tubes are presented in Section 10.3.

The first radar transmitters date back to World War II and used pulsed magnetrons [2]. These systems radiated a simple pulse of microwave energy and measured the time delay to the target and back (range) and the angular position of the antenna at which the detection was made (angle). The magnetron oscillator, often attributed to World War II developments, was demonstrated as early as 1928 by Professor Hidetsugu Yagi of Japan when he used a magnetron to generate RF at frequencies as high as 2.5 GHz [3]. In fact, all presently used high-power thermionic radar tubes are based on fundamental physics known well over 70 years ago. With the development of new materials and advanced manufacturing techniques, the newer high-power radar tubes (discussed in Section 10.3) exhibit much more bandwidth, higher peak power, and greater reliability than their predecessors. The modulator block shown in Figure 10-1 essentially turns the high-power oscillator or amplifier on and off at some predetermined rate termed the pulse repetition frequency (PRF; see Section 10.4).

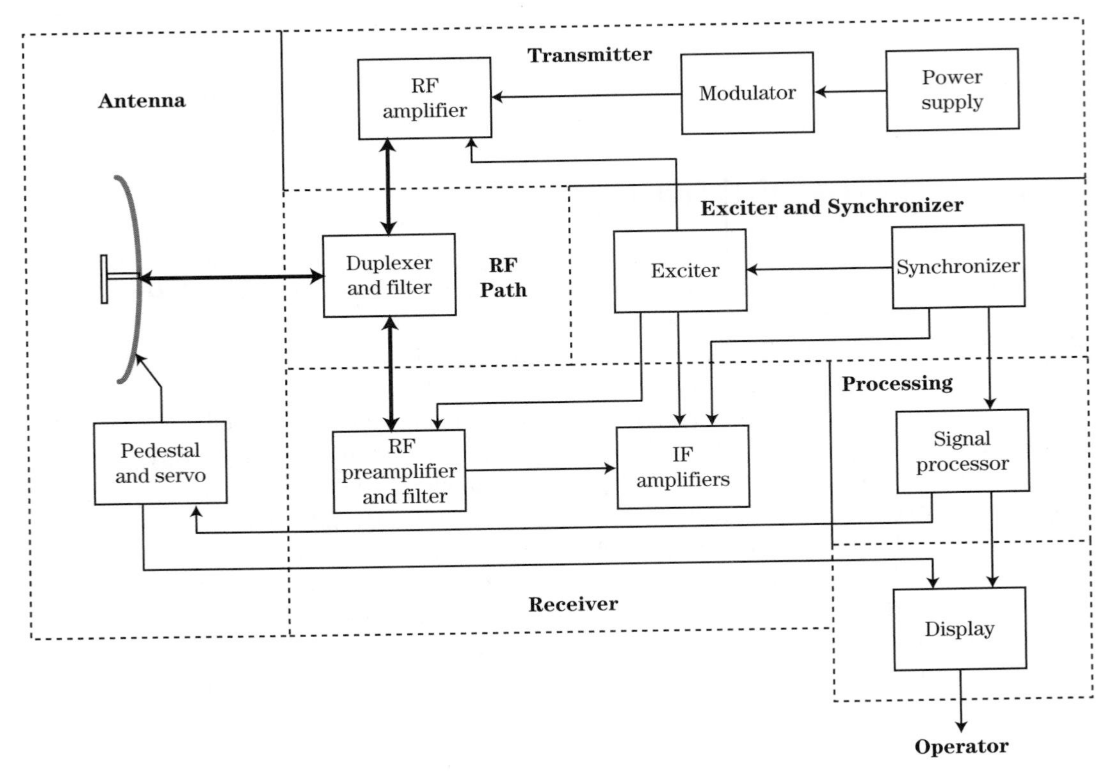

FIGURE 10-1 ■ Block diagram of typical pulsed radar. (From [1]. With permission.)

The last module of the pulsed radar transmitter in Figure 10-1 is the high-voltage power supply. The power supply typically provides thousands of volts at very high peak currents with negligible ripple while maintaining a high degree of stability and reliability. More details are described in Section 10.5.

10.1.1.2 Phased Array Radar

There are many instances where a pulsed radar with a single dish antenna is not the best way to collect the required data. Such is the case with the OTH skywave long-range missile tracking radar systems that operate in the 3 to 30 MHz HF band [4]. This application, as well as others, requires a different approach to generating the required radar signals and beam pattern. That approach is the modern phased array pulse radar.

Figure 10-2 depicts the block diagram of a typical phased array radar [1]. A brief examination makes it immediately apparent that many elements of the basic pulsed radar also exist within the phased array pulsed radar concept. The principle difference is that the phased array system, instead of using a dish antenna, incorporates an antenna aperture consisting of a large number of computer-controlled elemental antennas. Typically, these elemental antennas are spaced a wavelength or less apart to minimize unintended main lobes called *grating lobes* (see Section 9.7.3).

The pointing of individual antenna elements, or sometimes subarrays of a cluster of elements, is achieved by using computer control of digital RF phase shifters. Note by comparing Figures 10-1 and 10-2 that the transmitter subsystems could be the same in either case. The only difference is that in the case of the phased array, the high-power oscillator or amplifier output is uniformly divided and distributed to each of the computer phase-controlled antenna elements or subarrays by means of a matrix of RF power

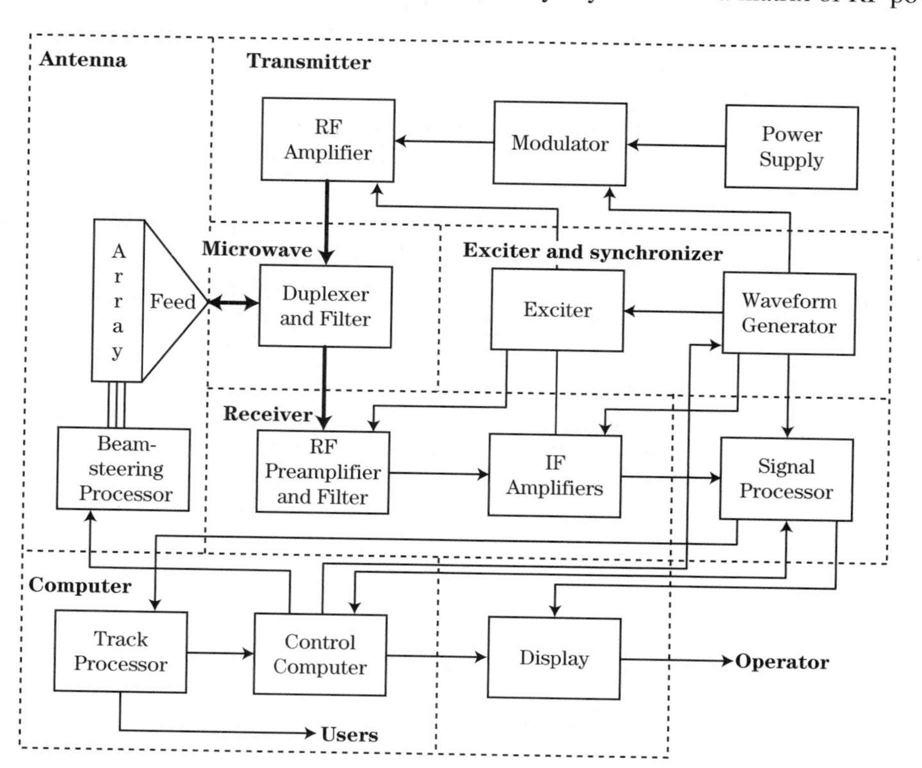

FIGURE 10-2 ■ Block diagram of a typical phased array radar. (From [1]. With permission.)

dividers. Pointing of the antenna in azimuth and elevation angle is achieved by means of the computer-directed phase shifts at each element or subarray.

The phased array concept just described is termed a *passive phased array* since the individual antenna elements do not incorporate any RF generation function. A major disadvantage of a passive phased array radar is that the RF losses in the power-dividing hardware can be large. One way of overcoming this disadvantage is to obtain the high levels of RF power needed by placing active lower-power solid-state or thermionic tube transmitting sources at each of the elements or subarrays. The total radiated power is the sum of the radiation from the aggregate of elements. This implementation is termed an *active phased array*.

Antenna pointing is again achieved using computer-controlled RF phase shifters. Often, solid-state power amplifiers are used at each element, in which case the individual radar modulators and associated direct current (DC) power supplies for each element or subarray can operate at fairly low voltages, an advantage when considering voltage breakdown issues. Also associated with each antenna element is the solid-state transmit/receive (T/R) module, which assures separation of the transmitted signal from the received radar echo. In addition to potential efficiency improvements over the passive phased array, active arrays offer the benefit of graceful degradation. When just a single high-power radar transmitter tube is used the radar becomes totally inoperative if the tube fails. In contrast, in the active phased array radar many individual low power RF transmitters can fail without totally shutting down radar operations. Solid-state power amplifiers are discussed in Section 10.3.3.1.

10.1.2 Radar Transmitter Parameters

The average radiated power, P_{av}, of a radar system in combination with the effective antenna aperture, A_e, establishes the maximum operating range, R_{max}, of a given radar [5]. That is, the maximum range is proportional to $(P_{av}A_e)^{1/4}$ (see Chapter 2). The product $P_{av}A_e$ is known as the *power-aperture product* and is a measure of performance for search radars. (A related measure is the *power-aperture-gain* product (PAG) which is often used as a measure of tracking or measurement performance.) Thus, maximum radar range can be increased by increasing either antenna size or radar transmitter net average power output. As discussed previously, this net average RF power can be derived from a single high-power tube or, as in the case of the active phased array, from the spatial summation of the output of many low to moderate power transmitters. Much of the discussion of transmitter parameters that follows is modeled after the discussions in Skolnik [5] and Curry [6].

A key specification of any radar transmitter is the RF power generated. For pulse-type radars this is usually specified by peak RF power, P_p, and average RF power, P_{av}, generated. The average power of a pulsed radar equals the peak power times the *duty cycle*, which is the fraction of the total time that the transmitter is on. That is,

$$P_{av} = P_p \tau PRF = P_p \cdot dc \text{ (watts)} \tag{10.1}$$

where

P_p = peak power (watts).

τ = maximum pulse duration (seconds).

PRF = pulse repetition frequency (hertz).

dc = duty cycle.

The duty cycles for klystron amplifiers, TWTs, and magnetrons are limited by tube element heat dissipation factors and typically range from 1% to 30%, although it is possible to obtain these devices in CW (100%) variants, depending upon the application.

The average transmitted power is limited by such factors as available prime power, heat removal from the transmitter, and computer scheduling limitations. The definition of transmitter efficiency, η_t, is given by

$$\eta_t = \frac{P_{av}}{P_{DC}} \tag{10.2}$$

where

P_{DC} = DC prime power (watts)

Typical transmitter efficiencies range from 15% to 35%.

Overall radar efficiency, η_r, is the ratio of RF power actually radiated by the antenna to DC prime power input. That is,

$$\eta_r = \frac{P_{av}}{P_{DC} L_m L_\Omega} \tag{10.3}$$

where

L_m = transmitter to antenna loss factor > 1.0
L_Ω = antenna ohmic loss factor > 1.0

Overall radar efficiencies may run from 5% to 25% or more, depending on the type of transmitter. Next, detailed transmitter configurations and the impact of the transmitter on the electromagnetic (EM) environment are discussed.

10.2 TRANSMITTER CONFIGURATIONS

There are many ways to characterize radar systems, and hence the transmitter used in them. Depending on the concerns of the designer or the end user, no one approach captures the wide variety of compromises and decisions that must be made in the final transmitter design. One way to characterize them is by end use. For example, a designer of a search radar in an air traffic control system will seek to optimize detection range, accuracy in Doppler determination, speed in covering the designated search volume, and rejection of clutter within that search volume. In this case a transmitter with a high average power and a medium to high PRF will be required. On the other hand, the designer of an instrumentation grade radar for a target measurement system on a measurement range or in an anechoic chamber would consider such issues as signal coherency for imaging purposes or high-range resolution for determining the location of scattering sources to be of great importance. In this case a solid-state transmitter may provide adequate power, and a low PRF may be all that is required.

Another way to characterize radar transmitters is by the power level required for operation. For a typical handheld police radar, the average power output is measured in tens of milliwatts. At the other end of the power spectrum, the U.S. Air Force's Space Surveillance System AN/FPS-85 radar uses a transmitter with a peak power greater than 30 MW. The two ends of this power spectrum call for very different approaches in generating the necessary power required for proper operation. A radar that emits an average power in the milliwatt to watt range can be constructed using only solid-state construction techniques, and the upper limit of output power is approaching tens of kilowatts of average power. However, radars radiating an average power in the range of hundreds of kilowatts or more will require vacuum tube sources to reach those power levels for the foreseeable future.

Finally, when considering the way the output power of the radar is generated, transmitters can be subdivided into two broad categories. The power source either directly generates the radar signal, or else it amplifies a lower-level signal generated in the exciter subsystem (see Chapter 12). In the former case, an oscillator is used, while in the latter case a power amplifier is used.

There are several basic configurations for a radar transmitter. There are many variations of these basic types, but the two most common types of transmitters are the free-running oscillator and the amplifier. Magnetrons and several types of solid-state oscillators are typically used in free-running oscillator transmitters, while TWTs, klystrons, and solid-state amplifiers are used in amplifier configurations. In a phased array transmitter, the amplifiers in the transmitter may be distributed across the face of the array and their outputs combined in space rather than combined at a single feed point as, for instance, with a dish antenna.

Before embarking on a discussion of the hardware of these different transmitter types, it is useful to review the concept of coherent versus noncoherent radar operation, which was introduced in Chapter 1 and is also discussed in Chapters 8 and 12. In a noncoherent system, the phase of the transmitted pulse relative to a reference oscillator is not known (random), so there is no significance to the phase of the received pulse relative to that same reference. Transmitters that use free-running oscillators tend to be noncoherent, with some exceptions discussed as follows.

In a coherent system, the transmit signal is derived from an ensemble of stable oscillators, so the phase of the transmitted signal is known. A key attribute of a coherent radar is the ability to maintain over time that known phase relationship between transmitted and received waveforms, which implies that the transmitter has the ability to amplify and replicate the waveform to be transmitted with minimal phase distortion, either across the pulse or from pulse to pulse. This allows for measurement of the effect of the target on the received signal phase and thus of Doppler frequency shift effects, indicative of target radial velocity relative to the radar. Coherent systems are more complicated from both hardware and signal processing perspectives, but they also generate more information about the target. These issues will be discussed in more detail later in this chapter.

Figure 10-3 presents a simplified diagram showing a simple transmitter using a free-running oscillator as the transmitter source and signal source for the radar, with the signal turned on and off at a given PRF by the modulator. A pulsed transmitter using a magnetron is shown, but the source could also be continuous wave (CW) or pulsed, solid state, or tube. A common example of a CW radar using an oscillator as the source is the infamous police speed-timing radar, which uses a solid-state source (usually a Gunn oscillator). Many maritime navigation radars use pulsed solid-state sources or, for longer-range systems,

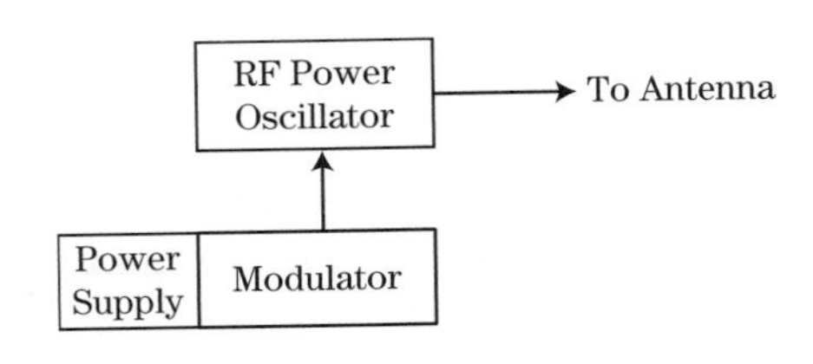

FIGURE 10-3 ■ Free-running oscillator-based transmitter.

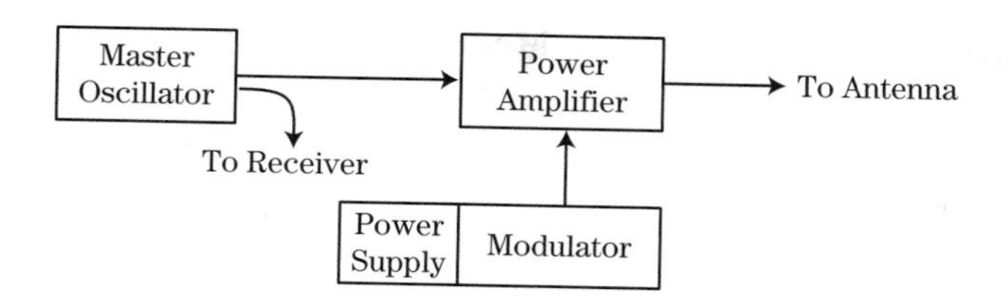

FIGURE 10-4 ■ Master oscillator/power amplifier transmitter.

pulsed magnetrons. This type of radar is somewhat simplistic when compared with more advanced coherent radars but is also much cheaper to implement. For cost reasons, this type of architecture is prevalent in high-volume, lower-performance consumer applications as opposed to low-volume, high-performance military applications.

Transmitters based on a free-running oscillator are usually operated in a noncoherent fashion but can be operated in a quasi-coherent mode in some cases. Consider a pulsed magnetron that starts each pulse at a random phase. A CW oscillator can be phase-locked to a sample of the magnetron output, resulting in what is called a *coherent-on-receive* system. Alternatively, a signal can be injected into the output port of a free-running oscillator to *injection-lock* or *injection-prime* the device [7]. This helps the device to start at a phase that is coherent with the injected signal, which can then be related to the received signal phase, resulting again in a quasi-coherent system.

Figure 10-4 shows a simplified diagram of a coherent radar transmitter using a pulsed power amplifier. Again, the source can be a tube or a solid-state amplifier. In some cases, more than one stage of power amplification is needed, depending on the gain available from a single stage. For instance, *crossed-field amplifiers* (CFAs) exhibit rather low gain per device, on the order of 10 dB to 15 dB, whereas klystrons and TWTs exhibit gains in the 35 dB to 50 dB range. In addition, the output amplifier stage could consist of a parallel combination of amplifiers to boost the output level. This configuration is commonly called the *master oscillator-power amplifier* or MOPA configuration for obvious reasons. The signal generator can range from a single-frequency oscillator to a complex, tunable, wide-bandwidth digital waveform generator under control of the radar control software. Modern systems typically use a digital waveform generator to take advantage of advances in computer processing power and the ability to change waveforms (e.g., frequencies, bandwidths, pulse widths, modulation codes) to maximize either detection or tracking performance, especially in the presence of countermeasures.

The preceding diagrams focused on simple transmitters connected to a single-feed antenna such as a conventional dish antenna. Especially in the military arena, many newer radars under development are phased arrays. Transmitters for active phased array radars are, almost by definition, of the coherent master oscillator/power amplifier type. This is because the conventional phased array antenna steers its beam via the phase relationship between radiating elements across the array face; therefore, precision is needed in knowing the phase of the signal on both transmit and receive at each element relative to some reference point in the array.

There are several ways to characterize a phased array antenna system. One approach to characterization is based on whether the array architecture uses a *passive array* or an *active-aperture array*. An active-aperture array contains an active power amplifier as well as a phase shifter at each element; a passive array contains only a phase shifter at the element. The other approach is to characterize the array by the way the energy is distributed to the radiating elements. An understanding of the array feed is fundamental in transmitter design for phased array systems because there is an intimate relationship between the RF power source and the RF distribution system. Achieving required power levels, efficiencies, bandwidths, operational reliabilities, and so forth can be accomplished only by properly matching the transmitter output to the feed system. There are many possible transmitter configurations for a phased array radar. The following paragraphs review possible feed approaches and then tie them to various approaches to generating the required RF energy for the proper operation of the phased array radar system.

In any type of phased array, the RF energy has to be distributed to the radiating elements that comprise the array. This energy is guided to the elements via a feed system or RF manifold. This feed manifold uses one of two major approaches: distributing the energy either via a guided, or constrained, approach or via a radiated or space-fed approach [8,9]. Constrained feeds use a transmission system such as a waveguide to transport the energy to the radiating element. Constrained feeds are usually classified as either series feeds (Figure 10-5a), where the radiating elements are in a series configuration with the feed system, or shunt feeds, where the radiating elements are fed in parallel via the feed system.

Series feed systems can be further subdivided into resonant feeds, which exhibit a higher efficiency at the cost of narrower bandwidths, or traveling wave feeds, which trade off a lower efficiency to achieve a broader bandwidth. A good example of a constrained series feed is the waveguide-based linear array. Linear arrays based on microstrip patches are also possible, but there are potential limitations in the array's operation due to the characteristics of microstrip structures.

Parallel feed systems can be subdivided into *corporate feeds* (Figure 10-5b) or *distributed feeds* (Figure 10-5c). Corporate feeds depend on successive power divisions as the RF energy moves through the feed network, until each radiating element is excited. Although the power division ratio at each junction is usually by a factor of two, it can range from two to five, depending on the number of elements to be fed and the type of power divider to be used. In distributed feed systems, each radiating element of a group of elements is connected to its own T/R module. Corporate feeds, as illustrated in Figure 10-5b, are known as passive arrays, as all the energy is provided by the radar transmitter. On the other hand, distributed feeds, as illustrated in Figure 10-5c, are also known as active arrays since active power sources, be they amplifiers or T/R modules, are located within the array structure and feed either individual elements or subarrays.

A *space-fed feed* system uses a feed antenna to excite the main phased array much like exciting a lens antenna system with one or more antennas. One form of the space-fed array system is the direct feed or in-line space-fed array (Figure 10-5d), where the radiation drives an array of phase shifters that feed the eventual radiating elements. The Patriot air defense radar uses this approach. The other form of the space fed array is the reflect array (Figure 10-5e), where the energy from the primary feed antenna illuminates a reflecting surface made up of array elements with a phase shifter behind them. While there are some advantages of this configuration over the in-line configuration (primarily the fact that the phase shifters, bias, and control circuitry are conveniently located behind the reflecting array elements), there is also the disadvantage of aperture blockage due to the primary feed antenna being located in front of the array.

Having reviewed the most common approaches to feeding phased array systems, the following observations can be made concerning the interaction between RF sources and the way energy is fed to the array elements. For instance, the corporate fed array of Figure 10-5b has several advantages, including a simple design approach and a reduced acquisition cost, since the RF power source is often a major cost driver for the radar transmitter and a single amplifier can be used for the source. On the other hand, using

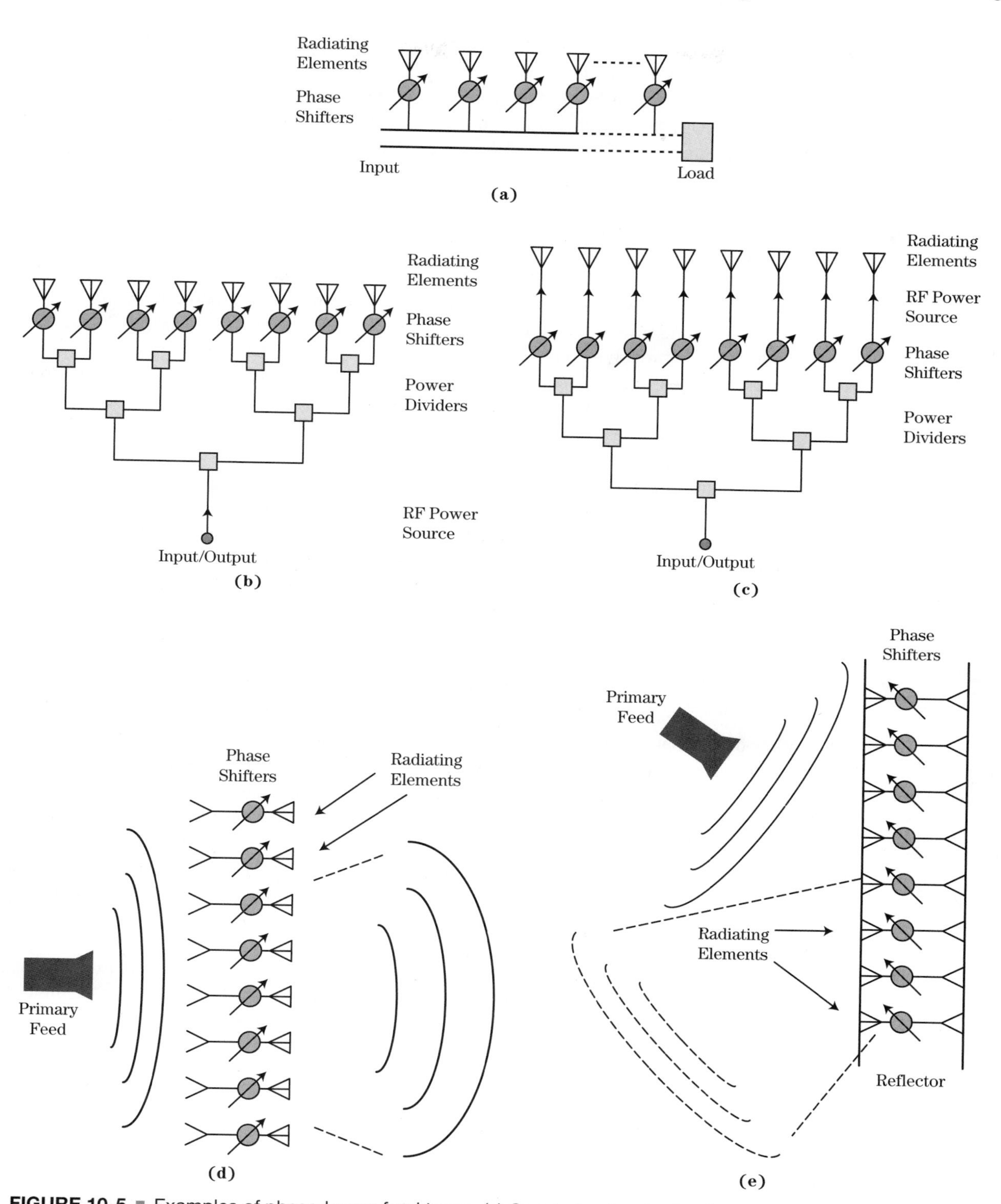

FIGURE 10-5 ■ Examples of phased array feed types. (a) Constrained series feed. (b) Constrained corporate feed. (c) Constrained distributed feed. (d) In-line space-fed array. (e) Reflect space-fed array.

a corporate feed with a single RF source means that all radar transmitter functionality will be lost if the source fails, reducing system reliability and eliminating the possibility of a graceful degradation of the radar system. One way to mitigate this situation is to replace some of the power dividers in Figure 10-5b with RF sources, resulting in a series of subarrays, each with its own power source. This is the approach used in the Cobra Dane early warning radar in Alaska, which uses 96 high-power TWTs, each feeding a single subarray [10]. This architecture allows for much greater overall reliability and graceful degradation, since the radar can still operate after losing one transmit subarray and the failed subarray can still be used in receive mode. There will be a small impact to sensitivity and perhaps an appreciable change in sidelobe levels, but the system can still operate in a degraded manner. Typical sources used for this configuration are medium- to high-power TWTs and some solid-state amplifiers.

Another approach is to feed the elements directly as indicated in Figure 10-5c, using an active array approach. Not only does this approach support a graceful degradation of the transmitter radar beam when individual elements fail, but it also has the added benefit that low-power phase shifters can be used prior to the power amplifier. This can greatly minimize ohmic losses on both transmit and receive, since on receive the phase shifter can be placed after the receiver's low-noise amplifier (LNA). Generally, solid-state amplifiers are used in the T/R modules at the element level. Solid-state active-aperture arrays will be covered in greater detail in the next section on power sources and amplifiers. Finally, it should also be pointed out that it is important to minimize the transmit losses. This can be accomplished by using a space-fed feed instead of a corporate feed, including its accompanying microwave plumbing and power dividers.

10.3 | POWER SOURCES AND AMPLIFIERS

A key decision in the transmitter design process is the selection of the type of power source to be used. Even if the architecture is predetermined, some decisions still need to be made. For instance, if the radar is to be an active-aperture solid-state type, it is still necessary to perform detailed design of the *transmit/receive module* as well as select the type of solid-state technology to be used (e.g., silicon [Si], gallium arsenide [GaAs], gallium nitride [GaN]). If the transmitter is specified to generate high average power, than an amplifier based on vacuum tube technology will be required. To make appropriate design trade-offs an understanding of the relative advantages and disadvantages of the various power sources available is needed.

For the discussion that follows, power sources will be classified into two groups: oscillators and amplifiers. The choice of which approach to take in transmitter design will be determined by the properties of the sources that make up the specific group. Within each group, there are both solid-state and tube devices as well as high- and low-power possibilities. One of the key parameters that will influence which approach to use is whether the system will be coherent. As stated earlier, noncoherent radar systems tend to be built around oscillator-based sources, whereas coherent systems tend to be based on amplifier sources.

In the subsections that follow, a basic description will be provided of the tubes and devices available to the transmitter designer for generating RF power. For more detailed information on each of these devices, the reader is directed to Gilmour's texts [2,11] as well as to the excellent survey text by Barker et al. [12].

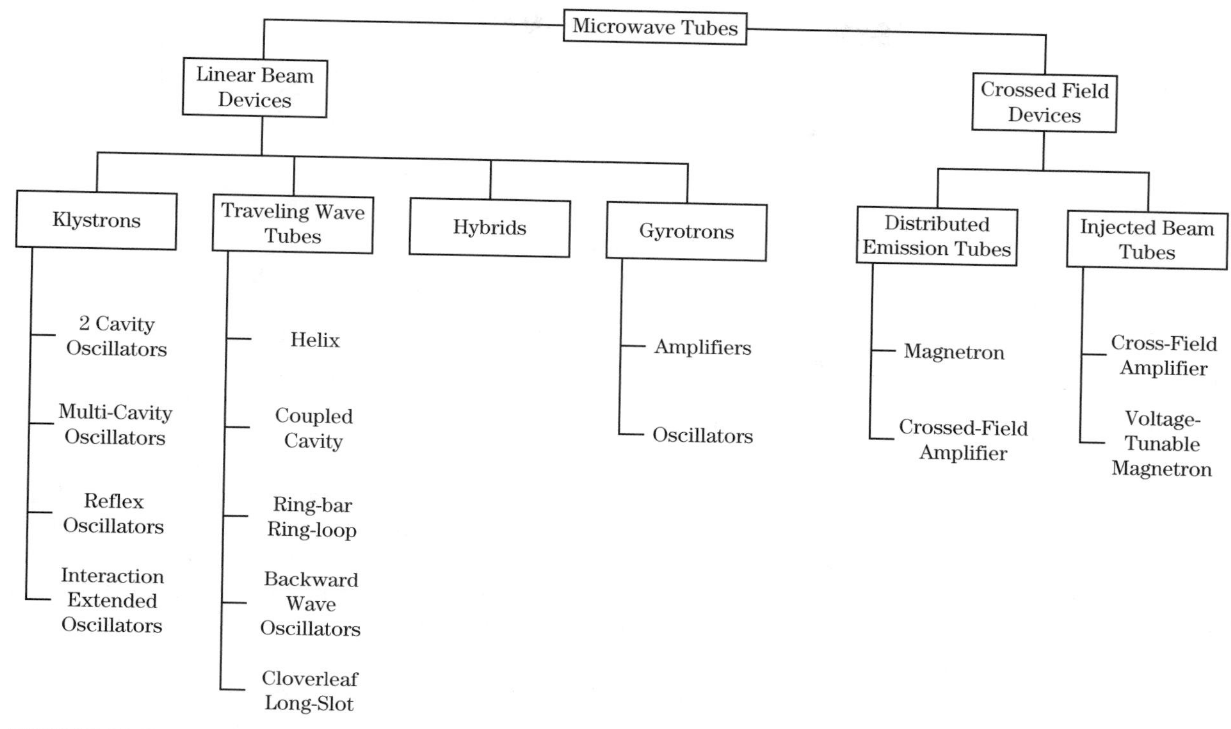

FIGURE 10-6 ■ Microwave tube family.

Figure 10-6 shows the relationship between the common tubes that are used or have been used in radar transmitters as either oscillators or power amplifiers. As can be seen by the figure, there are two main types of microwave tubes: *linear beam tubes* and *crossed-field tubes*, each with its advantages and disadvantages.

As the name implies, in a linear beam tube the electron beam and the circuit elements with which it interacts are arranged linearly. A simple schematic of a notional linear beam tube is shown in Figure 10-7. In a linear beam tube a voltage is applied to the anode, which accelerates the electrons given off by the cathode. The resultant electron beam has a kinetic energy determined by the anode voltage. A portion of the kinetic energy contained in the electron beam is converted to microwave energy as RF waves input into

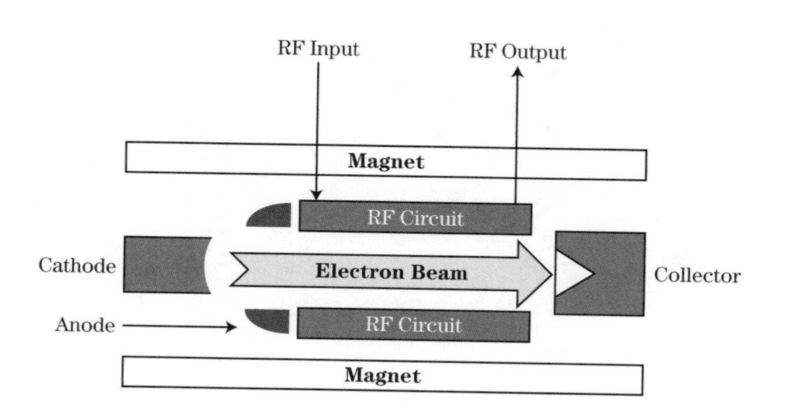

FIGURE 10-7 ■ Schematic diagram of a generic linear beam tube.

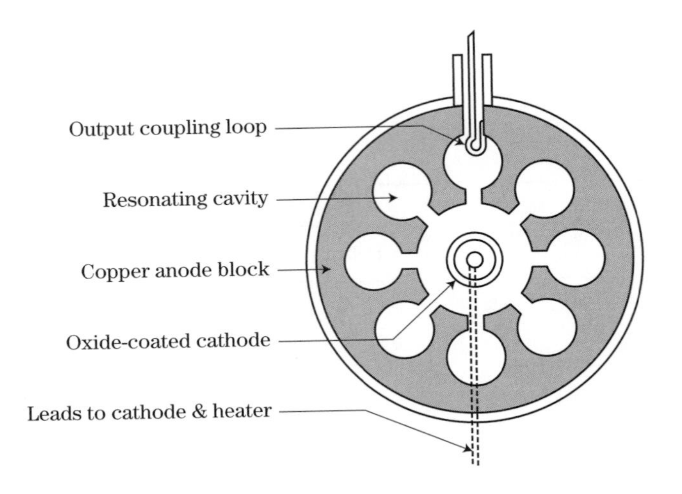

FIGURE 10-8 ■ Cross section of a magnetron tube.

the tube interact with the electron beam. The microwave energy is extracted at the RF output port, and the remainder of the electron beam is dissipated as heat from the tube or is returned to the power supply circuit at the collector. Because the electrons in the electron beam have a tendency to repel each other, a magnetic field, provided either by permanent magnets or an electromagnet, is used to focus the electrons into a beam going from cathode to collector.

Crossed-field tubes differ both in appearance and operation from linear beam tubes. The major difference is that the interaction between the electrons generated at the cathode and the anode requires a magnetic field at right angles to the applied electric field. The original device in the crossed-field tube family is the magnetron. Figure 10-8 shows a cross section of a magnetron tube. As can be seen from the drawing, a magnetron is basically a diode, with a cathode and anode. However, in this case the anode consists of a series of resonant cavities placed symmetrically around the cathode. Figure 10-9 shows a cutaway drawing of a complete magnetron assembly, showing the permanent magnet,

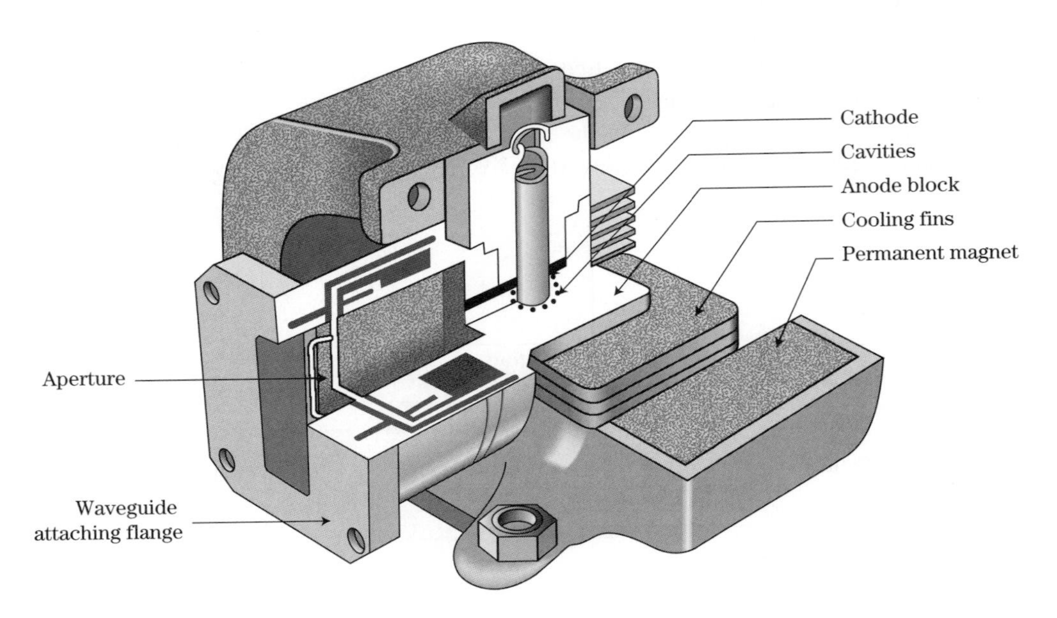

FIGURE 10-9 ■ Cutaway drawing of a typical magnetron.

which generates the focusing field, as well as the cooling fins attached to the anode block to dissipate the heat generated by the tube's operation, and the output aperture which couples the RF energy to the microwave plumbing going to the antenna.

With this general introduction to the tubes used in oscillators and high-power amplifiers, the specific tubes used in the typical radar configurations will be examined next. In addition, the basic solid-state devices used in lower-power configurations will be briefly considered as well as ways to achieve at least moderate power levels using power combining techniques.

10.3.1 Oscillators

Oscillator devices are typically used in lower-cost and hence generally lower-performance applications. These devices are easier to fabricate, are built around less expensive components, and are therefore cheaper than their stable amplifier counterparts. The primary tube oscillator for radar is the magnetron, while solid-state oscillator examples include the *Gunn* and *IMPATT diodes*.

10.3.1.1 Magnetron Oscillators

Magnetron tube oscillators, which originated prior to World War II, were the first high-power microwave radar sources developed [2]. They are crossed-field devices in that the electric and magnetic fields present in the device are orthogonal. Magnetrons are particularly useful as pulsed oscillators in simple, low-cost, lightweight radar systems.

Figure 10-10 is a photograph of a typical X-band pulsed magnetron. The large horseshoe-shaped frame is a permanent magnet that induces a magnetic field across the internal microwave cavity. When a large (20 kV) voltage pulse is applied to the electrodes at the top of the stem, an electric field is created orthogonal to the magnetic field, creating

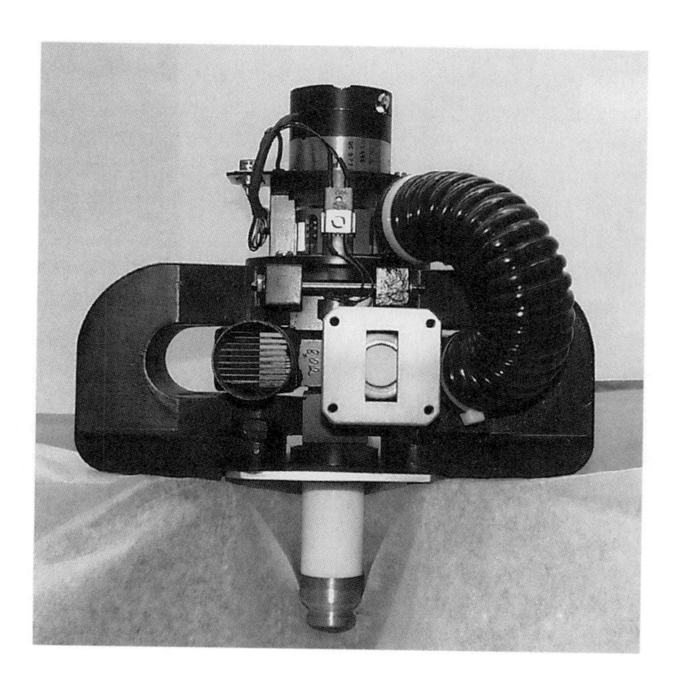

FIGURE 10-10 ■ X-band magnetron. (Photo courtesy of CPI. With permission.)

the required electromagnetic field. The frequency of oscillation depends on the mechanical characteristics of the internal cavity.

As discussed earlier, when a magnetron is used in a pulsed radar, there is no fixed relationship between the starting phase on one pulse relative to the next pulse, and hence the radar is termed noncoherent. However, injection locking can be used to allow a magnetron-based transmitter to emulate coherent operation. In injection locking, a signal is injected into the output of the magnetron prior to pulsing the tube. This microwave signal causes the energy buildup within the tube to concentrate at the frequency of the injected signal and also to be in phase with it. If the injected signal is coherent with the local oscillator (LO) of the receiver—perhaps offset from the LO by the intermediate frequency (IF)—the resulting transmitted signal is also coherent with the receiver LO, and hence the resulting radar system can measure the relative phase on receive, providing coherent operation. The degree of coherency is generally not as good as with an actual coherent amplifier chain, but such transmitters can be substantially cheaper than fully coherent systems such as TWT transmitters.

Magnetrons suffer from several undesirable operating characteristics: moding, arcing, missing pulses, and frequency pushing and pulling [13]. Many of these problems are related to how a magnetron generates microwave energy: the oscillations begin as noise, and the resonant structure of the device then forces the oscillations into a very narrow frequency band. *Moding* occurs when the tube oscillates at more than one frequency, or mode. This problem tends to be more prevalent if the rate of rise of the modulator voltage pulse is very fast. Modulator pulse shape control can help prevent moding. Since the buildup of oscillations from noise is a random process, statistically there will be instances where the pulse does not form, resulting in missed pulses. Finally, akin to phase pushing in linear beam amplifiers, the oscillation frequency is dependent on the cathode-to-anode voltage so that undesired voltage variations result in undesired frequency variations. Output load variations can affect the resonant cavity and so can also affect the operating frequency.

10.3.1.2 Gyrotron Oscillators

Gyrotrons are high-powered vacuum tubes that emit millimeter-wave beams by bunching electrons with cyclotron motion in a strong magnetic field. Output frequencies range from 20 to 250 GHz, covering wavelengths from 15 mm to less than 1 mm, with some tubes approaching the terahertz gap. Typical output powers range from tens of kilowatts to 1–2 megawatts. Gyrotrons can be designed for pulsed or continuous operation and can be used in either oscillator or amplifier applications. Although gyrotrons are more commonly used in fusion research and industrial heating applications, they have been used in radar systems that operate at millimeter wave frequencies. The Naval Research Laboratory has recently developed a high-power, coherent radar system at W-band using a gyroklystron amplifier tube with an average output power of 10 kW and a peak power of 100 kW [14]. This represents a 20-fold increase over previous systems. Although gyrotrons represent the best approach to achieving high output powers at millimeter wavelengths, much work remains to be done before gyrotrons will be commonly used in radar applications.

10.3.1.3 Solid-State Oscillators

The solid-state sources used in oscillator-based radars are primarily based on two devices: Gunn oscillators and impact ionization avalanche transit time (IMPATT) diodes. Gunn oscillators operate based on the principle of differential negative resistance within a bulk

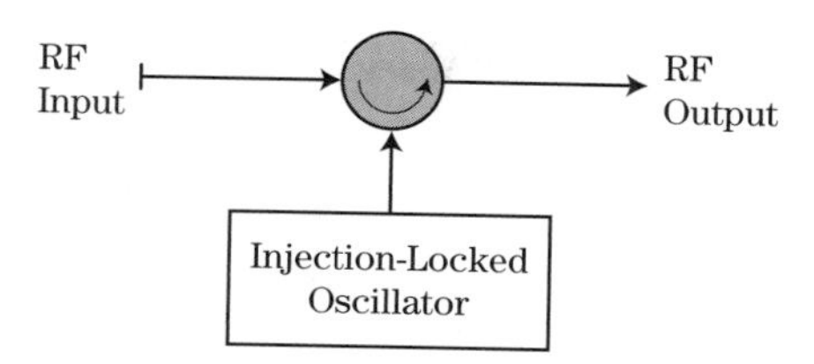

FIGURE 10-11 ■ Injection-locked amplifier.

semiconductor material, such as GaAs or indium phosphide (InP) [7]. Gunn diode oscillators are low-noise sources but are capable of only low-output power levels—tens to hundreds of milliwatts. They are useful as radar local oscillators or as the output source of low-power transmitters such as short-range frequency modulated continuous wave (FMCW) radars or altimeters. Gunn diode oscillators are available well into the MMW frequency regime.

IMPATT diode oscillators, in contrast with the Gunn diode oscillator, are fairly noisy but are capable of higher output powers, reaching into the tens of watts. They can be power-combined for even higher powers and are therefore more common as power sources. They are available up to MMW frequencies. They can also be injection-locked for amplifier operation [7] as illustrated in Figure 10-11. An RF signal is input into one port of a circulator as shown and then enters the output of the diode oscillator circuit. The signal interacts with the diode oscillator, locking the oscillation frequency to that of the input signal. The resulting output is then transferred to the load through the third port of the circulator. A reasonable MMW power amplifier can be constructed from injection-locking a power-combined set of IMPATT diodes.

10.3.2 Tube Amplifiers

The case where the radar transmitter is designed using a power amplifier to achieve the required output power for proper operation is considered next. There are many different kinds of radar transmitter amplifiers, and it is important to understand their basic differences. As with radar oscillators, tube amplifiers can be designed using either linear beam tubes or crossed-field tubes.

10.3.2.1 Linear Beam Tubes: Klystrons

Klystrons are linear beam tubes, which means the interaction between the RF field and the electron beam occurs longitudinally along the length of the tube. The klystron was the first microwave tube invented to overcome the transit-time effects that early triode and tetrode tubes experienced when used at higher frequencies. Klystron tubes are the most efficient of the linear beam tubes, are capable of the highest peak and average powers, and can be used over an extremely broad frequency range, from low ultra high frequency (UHF) (200 MHz) to W-band (100 GHz) [12]. Klystrons essentially consist of a series combination of high-Q cavities through which an electron beam passes, exchanging energy with an RF wave inserted into the input cavity. The RF is coupled from cavity to cavity via the electron beam itself, until it is amplified and extracted in the output cavity as shown schematically in Figure 10-12.

Each cavity is a resonant circuit at a particular frequency. Tuning the cavities in different ways changes the overall characteristics of the amplifier. A given design can be tuned to give broader bandwidth at reduced gain or higher gain at reduced bandwidth. The

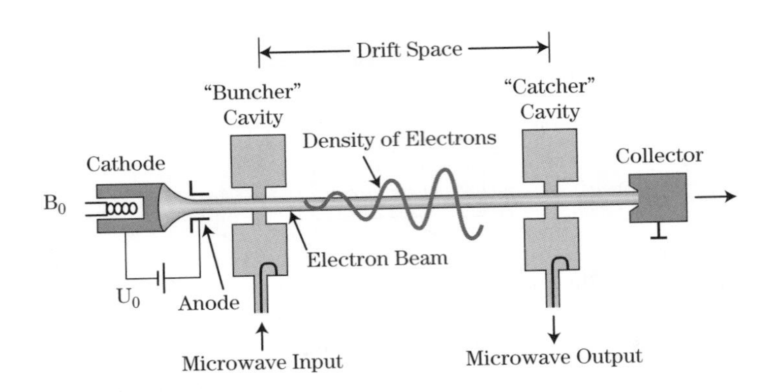

FIGURE 10-12 ■ Schematic view of a two-cavity klystron tube.

center frequency can be changed by mechanically adjusting the cavity characteristics via the tuning adjustment (typically on the side of the tube). Figure 10-13 is a photograph of a number of different klystron tubes.

The fact that the klystron uses high-Q cavities results in very low additive phase noise in the amplifier. Klystrons tend to have significant gain (40–60 dB) and good efficiency (40–60%) but suffer from inherently narrow bandwidth capability (about 1–10%) when compared with other tube types. Hence, for applications requiring a high-power source with low phase noise but not much bandwidth, the klystron is usually the proper choice.

Low-power klystrons such as reflex klystron oscillators have long been replaced by solid-state devices. However, they are unlikely to be replaced by solid-state devices at MMW frequencies any time soon. Solid-state devices cannot yet economically produce the several tens of watts of average power achieved by klystrons in those bands.

10.3.2.2 Linear Beam Tubes: Traveling Wave Tubes

Like the klystron, the traveling-wave tube is a linear beam device. Other than the magnetron used in microwave ovens, the TWT is the most commonly used microwave tube, serving in such diverse applications as the final stage amplifier in satellite communication systems, as wide bandwidth, high power, high gain, high efficiency power sources for electronic countermeasure (ECM) systems, and the driver for crossed-field amplifiers in high-power radar systems. Traveling-wave tubes are also a major component in many

FIGURE 10-13 ■ A variety of klystron tubes. (Photo courtesy of CPI. With permission.)

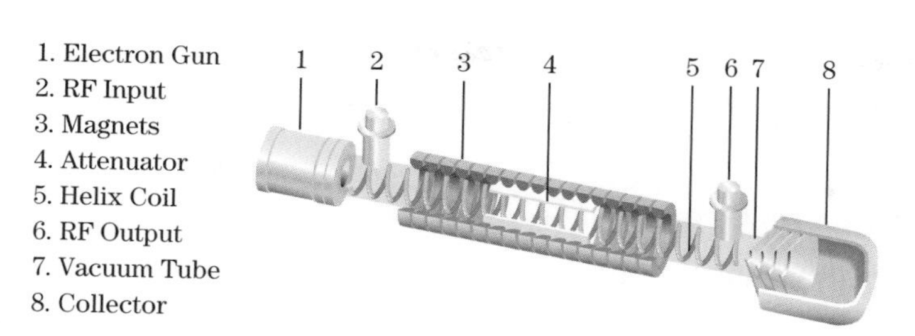

FIGURE 10-14 ■ Functional diagram of a traveling wave tube.

microwave power modules (MPMs), which combine some of the desirable attributes of solid-state devices with the power output of tubes. While many different RF circuits have been developed to use in TWTs, the two most common are the helix approach, which is well suited for broadband operations, and the coupled cavity approach, which is best for high power applications.

Figure 10-14 is a functional diagram of a TWT showing its major components. An electron gun (1) emits a beam of electrons that passes through a slow-wave structure such as a helix (5) or coupled cavity. RF energy is injected into the tube via an input port (2) and removed via an output port (6). The velocity of the electron beam is set by the cathode-to-body voltage. The slow-wave circuit slows up the longitudinal component of the velocity of the RF signal so that it travels roughly in synchronism with the electron beam. Magnets (3) are used to keep the electron beam focused as it travels down the tube. As the beam and the signal traverse the length of the tube (7), the interaction causes velocity modulation and bunching of the electrons. The bunched beam causes induced currents to flow on the RF circuit, which then causes further bunching of the electron beam. Through this regenerative process, energy is transferred from the beam to the RF signal, and amplification of the RF signal occurs. The electron beam then strikes the collector (8), which dissipates the thermal energy contained in the beam. To reduce the impedance mismatches that will occur at the RF ports, and to decrease the backward wave that can flow in the tube, it is common to use an attenuator (4) to reduce these effects. Figure 10-15 shows the various current flows that are involved in TWT operation.

Beam control is accomplished by pulsing a modulating anode or grid. Some older high-power tubes are cathode-pulsed. Grid-controlled tubes allow for beam control with lower voltages than any other means, providing for very fast rise and fall times for short pulse operation and for high-PRF operation for pulse-Doppler radars. Grids tend to intercept a small amount of beam current, which can limit the power capability of the tube. Modulating anode- or cathode-pulsed tubes do not suffer this limitation. However, their peak or average

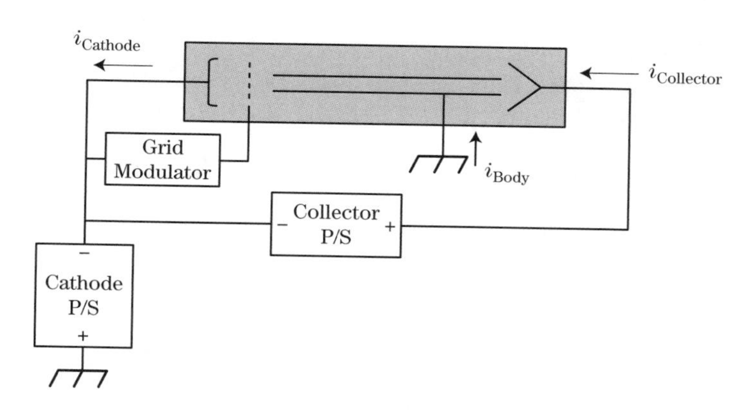

FIGURE 10-15 ■ Current flows associated with traveling-wave tubes.

powers are limited by other factors, such as RF circuit or collector heat dissipation. An excellent beginning reference for detailed information on the design and operation of TWTs is Gilmour's text [11].

10.3.2.3 Cross-Field Tubes: Crossed-Field Amplifier

The crossed-field amplifier (CFA) is similar to the magnetron in that the electric and magnetic fields are perpendicular to each other. CFAs are characterized by relatively low gain, typically 10 dB to 17 dB, and relatively high additive phase noise, especially when compared with klystrons or TWTs. The low gain is a disadvantage when generating high power levels, as it requires increased drive levels and hence more expensive driver amplifier stages. For this reason, TWTs are often used to drive CFAs. For a given amount of average output power, however, the CFA is very cost-effective, especially when considering the relatively simple power supply/modulator system required to operate it. As an example, CFAs are used in the U.S. Navy AN/SPY-1 phased array radar. Like the gyrotron, the CFA can also be used as an oscillator, although this is not a common application in current radar systems.

In summary, because of their output power, frequency range, and, in the case of TWTs, broad bandwidth characteristics, vacuum tubes will be used for the foreseeable future in high-power radar applications. For comparison purposes, the characteristics of the various tubes previously discussed are collected in Table 10-1, extracted from a U.S. Department of Defense (DoD) report on the status of the vacuum tube industry as of the late 1990s [15].

10.3.3 Solid-State Sources

There has been a strong push to replace vacuum tube-based RF sources with solid-state devices because of the many perceived advantages, including reliability and maintainability, modularity, and potentially performance. While there are many applications where solid-state devices meet all the system requirements for a radar transmitter, there are many applications where solid-state devices cannot yet compete with vacuum tube devices in terms of output power, efficiency, and cost. In fact, it can safely be predicted for the foreseeable future that solid-state devices will not be able to replace tubes in many radar applications as well as in such related fields as electronic warfare equipment, that require hundreds of kilowatts of average power. However, this does not mean that solid-state amplifiers and power modules built around solid-state devices do not have a role to play in radar technology. In fact, many applications, ranging from the radar guns used by law enforcement agencies to the radar systems being integrated into automotive systems as safety features, are best addressed using solid-state technology. The transmitter designer must be aware of the actual advantages and disadvantages of solid-state devices versus vacuum electronic devices and be able to select the best technology as appropriate.

10.3.3.1 Solid-State Amplifiers

Phased array antennas are increasingly being used in radar systems due to their many advantages. At the same time, there is a trend toward using *monolithic microwave integrated circuits* (MMICs) based on GaAs technology. According to Brookner [16], the majority of phased array antenna element PAs are fabricated using GaAs MMICs using a metal semiconductor field-effect transistor (MESFET) process, although these devices are being superseded by pseudomorphic high-electron mobility transistor (PHEMT) technology.

TABLE 10-1 ■ Compilation of Characteristics of Common Vacuum Devices

Tube Type	Frequency / Bandwidth	Power Out (Typical)	Attributes / Drawbacks	Applications
Klystron	0.1–300 GHz 5–10%	10 kW CW ** 10 MW Pulse	High Power 40–60% Efficient Low Noise Narrow Bandwidth	Radar Television Industrial Heating Satellite Uplinks Medical Therapy Science
Traveling Wave Tube (Helix)	1–90 GHz Wide Bandwidth 2–3 Octaves*	20 W CW 20 kW Pulse	Broad Bandwidth Power Handling Limitations Efficiency	Electronic Warfare Communications Commercial Broadcasting Industrial Applications
Coupled-Cavity TWT	1–200 GHz 10–20%	300 W CW 250 kW Pulse	Average Power Capability Complex & Expensive Slow Wave Structure	Airborne Radar Satellite Communications AEGIS FC Illuminator
Magnetron	1–90 GHz N/A	100 W CW 10 MW Pulse	Simple–Inexpensive Rugged Noisy	Radar/Medical Industrial Heating
Crossed-Field Amplifier	1-30 GHz 10–20%	1000 W CW 5 MW Pulse	Compact Size 30–40% Efficient Complex and Expensive Slow Wave Structure	Transportable Radars Shipboard Radar Seeker Radar Industrial Heating
Gyrotron	30–200 GHz 10% Max	0.2–3 MW Pulse	High Power at High Frequencies High Voltage Required	High-Frequency Radar Fusion Accelerators Industrial Heating

*One octave is the range defined where the highest frequency is twice the lowest (e.g., 2–4, 4–8).
**DOE's APT klystrons will run at 1 MW CW.
Source: From [15] (with permission).

In addition to GaAs, other compound semiconductors are being used in PAs for phased array antenna systems, including GaN and silicon carbide (SiC) in both discrete and MMIC form [17].

Specific device technology is undergoing constant change and improvement. Here, only the general trends in solid-state amplifiers and transmitters are discussed, providing comparisons between tubes and solid-state devices. Compared with tubes, solid-state devices posses the following advantages:

- No hot cathode is required for electron generation. Thus, there is no delay for device warm-up and no power required for a cathode heater.
- Solid-state amplifiers operate at much lower voltages. Because of this, power supply voltages are on the order of tens of volts instead of kilovolts. This has several advantages, including smaller and less expensive components and a smaller power

supply size, since large spacing between components is not required to prevent voltage breakdown and arcing between power supply components and the components do not require encapsulation for high voltage potting. Also, the lower voltage eliminates or minimizes the generation of x-rays, which are a potential health hazard in high-voltage vacuum tubes.

- Transmitters designed with solid-state devices may exhibit improved *mean time between failures* (MTBF) compared with tube transmitters. However, this assumes that the solid-state transmitter is properly matched to the surrounding subsystems and can handle the high peak-to-average power ratio that is typically present in high-power transmitters.
- Solid-state transmitters can be designed with wide bandwidths, exceeding the 10–20% bandwidths typically achievable with high-power tubes and instead reaching bandwidths up to 50%. However, to date no solid-state amplifier can achieve the 2–3 octave bandwidths of the TWT at equivalent power levels.
- Modules based on solid-state devices can exhibit a large degree of flexibility in the implementation of amplifier designs. In the next section the impact of this flexibility on the design of transmit/receive modules will be examined.

However, it should be pointed out that transmitters based on solid-state devices have their own drawbacks. For instance, a solid-state transmitter may operate with a high duty cycle, which means it will generate long pulses that require the use of pulse compression. Long pulses also result in a long minimum range, which means targets at shorter ranges might be masked by the long pulses.

Solid-state amplifiers for use in transmitters are often characterized by their class of operation. Amplifiers can operate in any of the following classes: Class A, B, AB, C, D, E, F, G, or H. Classes A, B, AB, and C are used in analog amplifier circuits where linear operation is required, while Classes D, E, F, G, and H are used in switching-mode amplifiers.

For analog amplifier circuits the class of operation is defined by the way the transistor is biased. For instance, Class A amplifying devices operate over the whole of the input cycle such that the output signal is an exact scaled-up replica of the input with no clipping. Class A amplifiers are the usual means of implementing small-signal amplifiers. They are not very efficient; a theoretical maximum of 50% is obtainable with inductive output coupling and only 25% with capacitive coupling. In a Class A circuit, the amplifying element is biased so the device is always conducting to some extent and is operated over the most linear portion of its characteristic curve. Because the device is always conducting, even if there is no input at all, power is drawn from the power supply. This is the chief reason for its inefficiency.

Contrast this with the Class C amplifier, which conducts less than 50% of the input signal. The distortion at the output is high, but efficiencies up to 90% are possible. The most common application for Class C amplifiers is in RF transmitters, where the distortion can be greatly reduced by using tuned loads on the amplifier stage. The input signal is used to roughly switch the amplifying device on and off, which causes pulses of current to flow through a tuned circuit. Collector current is drawn only when the input voltage exceeds the reverse bias across the input and the output voltage is developed across the tuned load. Thus, there is no power dissipation in the amplifier when the transmitter is switched off during receive mode.

Class D, E, F, G, and H amplifiers are switching amplifier configurations with high efficiencies that also require specialized filtering of the signal harmonics to maximize the amplifier efficiency. These can be very complicated hardware implementations and are usually warranted only if the incremental improvement in efficiency brings a significant benefit to the transmitter system.

Many different circuit configurations can be employed to implement these various classes. A detailed discussion of those configurations is beyond the scope of this chapter. The interested reader is referred to one of the many texts available on the design of RF amplifiers, including books by Krauss et al. [18], Cripps [19], and Grebennekov [20] or the articles by Raab, et al. [21] or Gao [22]. There is also an excellent chapter on solid-state transmitters by Borkowski in the third edition of the *Radar Handbook* [23].

10.3.3.2 Solid-State Transmitter/Receiver Modules

Solid-state T/R modules have received much investment and hence research and development attention over the last 20 years. Many new military radar development programs are using solid-state technology as opposed to tubes. Solid-state T/R modules are a broad category by themselves, since there are many different technologies, applications, and configurations. A typical T/R module architecture is shown in Figure 10-16. Each module generally employs an attenuator for control of receive gain and receive antenna sidelobes (via tapering across the aperture) and a low-power phase shifter for beam steering control. Several amplifiers can be power-combined to increase the power per element at the expense of increased cooling and prime power requirements. A circulator is typically used to provide a good match between the amplifiers and the antenna element, since the voltage standing wave ratio (VSWR) at the element can vary greatly as a function of scan angle. A receiver protector of some sort (e.g., a diode switch or diode limiter) is usually employed to prevent burnout and damage of the sensitive low-noise amplifier input.

Modules using high-peak power (over 100 W) silicon amplifiers for low duty cycle waveforms at UHF are used in radars such as the PAVE PAWS missile warning radar. In contrast, space-based radar applications, due to severe prime power limitations, require only milliwatts of power per module, typically operate at much higher frequencies, and tend to use GaAs devices. Most solid-state radars are somewhere in between these extremes, although to date GaAs has seen the most use in solid-state systems operating above

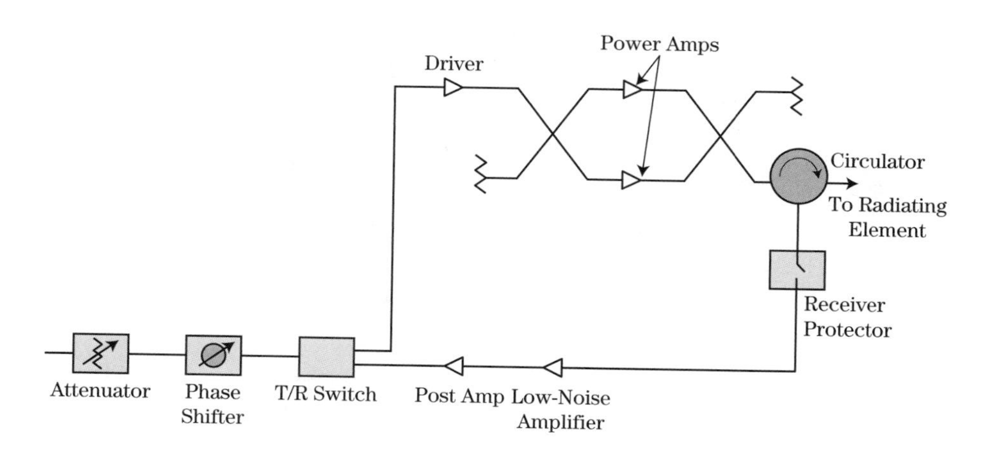

FIGURE 10-16 ■ Example T/R module architecture.

S-band. Current T/R modules using GaAs technology can have output powers in the tens of watts range. Newer materials currently under development such as gallium nitride are of interest for even higher-power amplifiers. GaN has 5–10 times the power density of GaAs and can operate at higher voltages due to higher breakdown capability.

10.3.3.3 Solid-State Active-Aperture Arrays

Solid-state active-aperture arrays can be separated into high-power density and low-power density arrays. Figure 10-17 shows a plot of aperture area and power illustrating lines of constant power aperture, a key parameter for search radars, and lines of constant power-aperture squared, a key parameter for tracking radars (see Chapter 2). Low-power density arrays are those arrays with very low power per element. These types of arrays maintain sensitivity on a target by increasing aperture size and hence transmit and receive gain while minimizing output power, which is usually done to reduce prime power, cooling requirements, and cost. Such arrays have been investigated recently for space applications as well as for ground-based applications.

High-power density arrays try to increase the amount of output power that can be generated and cooled at each element in an effort to reduce array size, for instance to reduce the footprint for storage or transport. Such arrays require much more prime power and cooling than do low-power density arrays of equivalent power-aperture-gain (PAG) product. Of course, if the volume available for the antenna is limited, then increased radar sensitivity must be obtained via increased transmitter power and reduced noise temperature.

One of the attractive features of a low-power density array is that for the same level of sensitivity, a properly designed array can be built that requires much less prime power than a high-power density array of equivalent performance. The savings in terms of the cost of the power itself is not usually the main advantage. For tactical military applications, reducing prime power requirements can mean reduced logistics requirements because of reduced fuel consumption by diesel-powered generators. Another advantage for low-power density active-aperture arrays is that the reduction in prime power results in a reduction of waste heat at the array face (where most of the waste heat is generated). This can help improve reliability of the electronics in the array as well as decrease cooling requirements at the array face, easing the thermal and mechanical engineering challenges.

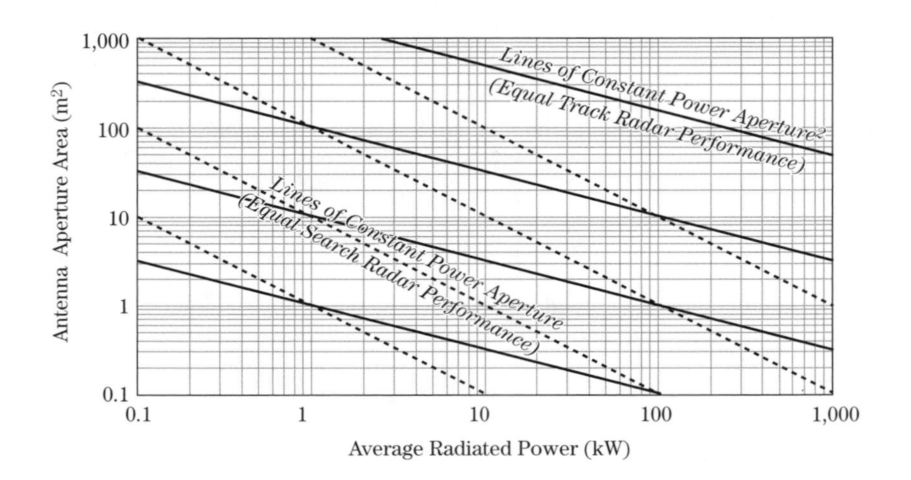

FIGURE 10-17 ■ Aperture area and radiated power for constant PA and PA^2.

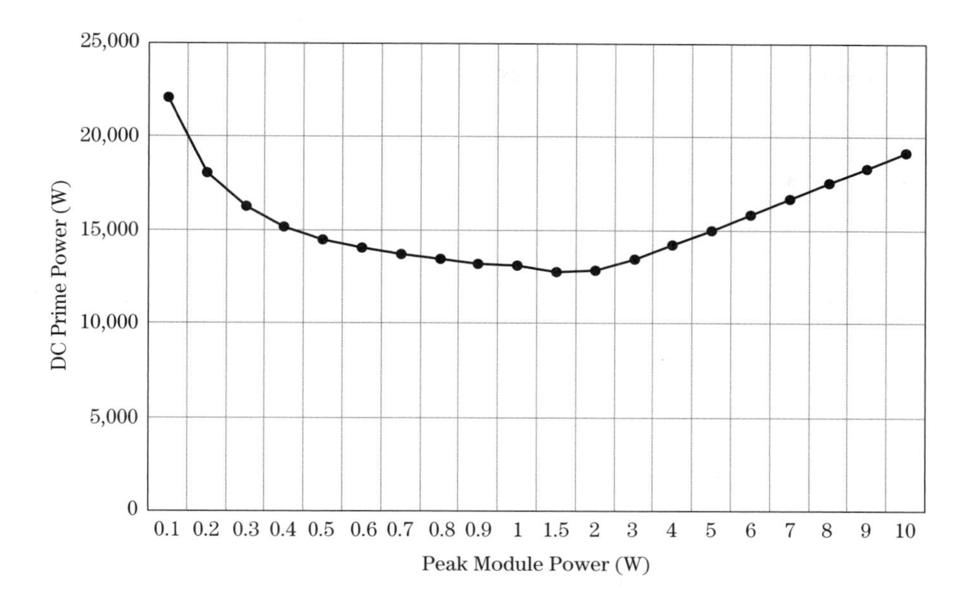

FIGURE 10-18 ■ Example phased array radar prime power requirement as a function of peak module power for a given level of power-aperture-gain product.

The variation in array prime power required as a function of peak module power is shown in the example in Figure 10-18. This example assumes a single-pulse radar PAG requirement of 90 dBWm^2; a power amplifier power-added efficiency of 33% in the transmit mode over the peak output power range of interest; prime power consumption of 1 W in the receive mode; background DC power of 0.5 W in the transmit mode (in addition to the final power amplifier DC power requirements); and a 10% transmit duty cycle. The figure then plots the average DC prime power required to operate the T/R modules for an array that meets the PAG requirement.

At extremely low module transmit power levels, the array prime power requirements are dominated by the receive-side power required by the large number of elements needed. As the peak module power increases, the array size and hence the number of elements decreases. It can be seen from the figure that there is a fairly broad minimum in the relationship, such that module output powers in the range of 0.7 W to 3 W result in reasonably low levels of prime power for this example. This range will vary as the array and module parameters change. However, as the peak module power continues to increase, the prime power required rises again as the radar sensitivity is being attained increasingly through power (P) rather than by area (A) and gain (G). This exercise neglects issues associated with power-aperture product (a key performance metric for the search mode) and any physical size constraints. Modern techniques such as beam spoiling on transmit (broadening the beam by defocusing the array) coupled with multiple simultaneous receive beams can be used to improve the search performance of a narrow-beam, large-aperture radar.

10.3.4 Microwave Power Modules

Microwave power modules combine the best attributes of both solid-state sources and tubes, particularly helix TWTs, in an attempt to create a more compact transmitter than is possible using either technology alone. In its most basic form, an MPM consists of three major components: (1) a solid-state amplifier driver; (2) an integrated power conditioner;

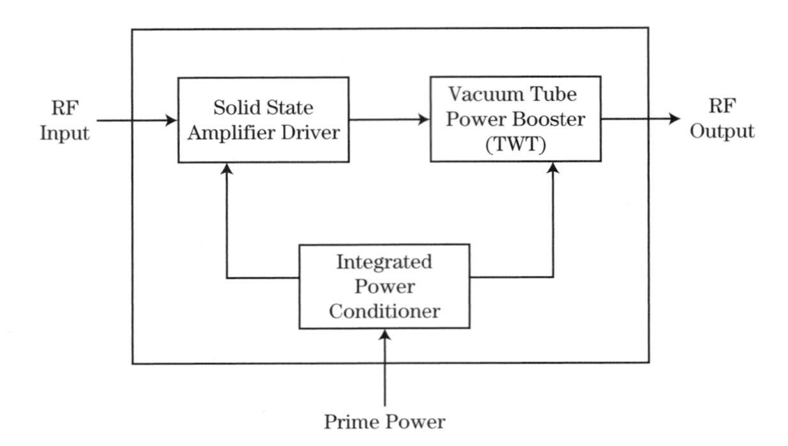

FIGURE 10-19 ■ Simple MPM block diagram.

and (3) a vacuum tube-based power booster. A simple block diagram of an MPM is shown in Figure 10-19.

MPMs take advantage of the fact that the physical length of a TWT is a function of the gain required from the RF circuit; if that gain can be minimized, the tube can be made physically smaller, assuming that the device does not generate or dissipate so much power that its mass is dominated by heat sinks. Thus, instead of requiring a 50 dB gain TWT, a solid-state driver is used to provide the first 25 dB or so of gain (where power levels are reasonably low), and the TWT is then used for the final 25 dB or so of gain where the power levels are high. Since the solid-state device can generate reasonable powers at wide bandwidths and the TWT is inherently wideband, the MPM is also capable of wide bandwidths. Typical output powers are on the order of 100 W CW for octave-bandwidth devices from S-band through K-band. Figure 10-20 shows an example MPM. The module also contains the low-voltage power supply circuitry for the solid-state amplifier as well as the high-voltage power supply for the TWT. Although MPMs have traditionally found more applications in electronic warfare (EW) systems, they are becoming more common as components of radar systems, especially on unmanned aerial vehicles (UAVs), where minimum weight and volume are key requirements. Additional information concerning MPMs is available in the paper by Smith et al. [24].

FIGURE 10-20 ■ Microwave power module (MPM). (Photo courtesy of CPI. With permission.)

10.4 | MODULATORS

A CW radio wave carries no information. Modulation of the radio wave in radar is necessary to convey intelligence, just as is modulation in radio communications. In high-power (kilowatts to megawatts), long-range detection and tracking radars, pulse modulation in some form is the norm. Although not discussed in this section it should be mentioned that certain radar transmitters make use of frequency modulation (FM). This category includes low power (i.e., milliwatts to watts) FMCW radars, which are often associated with altimeters, proximity fuzes, and traffic surveillance. Also, many high-resolution, high-power radars employ a concept called linear FM *pulse compression* in which the transmitter frequency is linearly swept throughout the pulse duration (see Chapter 20).

The following subsections discuss some of the basic concepts associated with line-type pulse modulators and active-switch pulse modulators [5]. Modern microwave power vacuum tube technology is now commonly referred to as *vacuum electron device* (VED) technology [25]. VEDs include magnetrons, klystrons, TWTs, and crossed-field amplifiers. The type of VED determines, to some extent, the type of modulator required. If the VED has a control grid or modulating anode, a low-power modulator can be used. A widely used switching element of low-power modulators is the MOSFET transistor [5]. In addition to VEDs, *solid-state power amplifiers* (SSPAs) are often used in *active electronically scanned array* (AESA) radar systems. In the latter case, low-voltage, high-current-distributed solid-state pulse modulators are indicated.

10.4.1 Line-Type Modulators

Some of the earliest pulse radar modulators used were the so-called *line-type modulators* where energy is stored in a transmission line or its equivalent, a *pulse-forming network* (PFN). The PFN simulates a transmission line by means of lumped constants consisting of many low-loss inductors and capacitors. Figure 10-21 shows the PFN being charged by a high-voltage power supply through a charging diode and discharged through a triggered switch. The switch can be a vacuum tube (typically a gas-filled triode called a thyratron in older circuits) or a stacked series of solid-state switches in more modern systems. The resultant pulse is transferred to the transmitting device (e.g., a magnetron or klystron) through a pulse transformer. Pulse transformers are used because of their impedance-matching properties and because they allow isolation between primary and secondary circuits. To operate effectively with narrow pulses, specially designed pulse transformers are required. Pulse transformers have ferromagnetic cores, are closely coupled, and have relatively few turns in each winding.

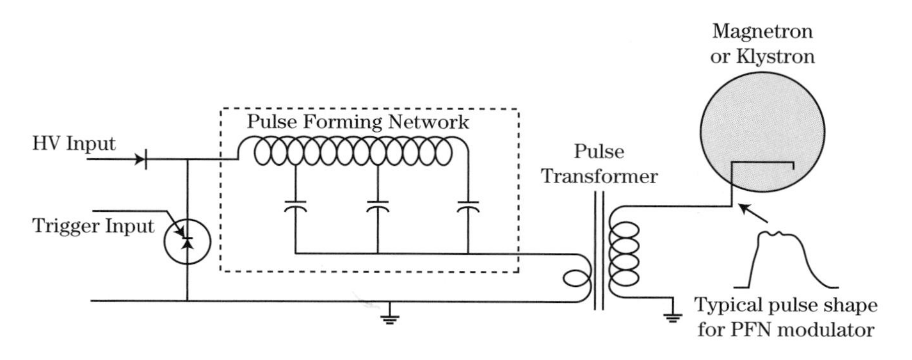

FIGURE 10-21 ■ Simplified diagram of a PFN modulator.

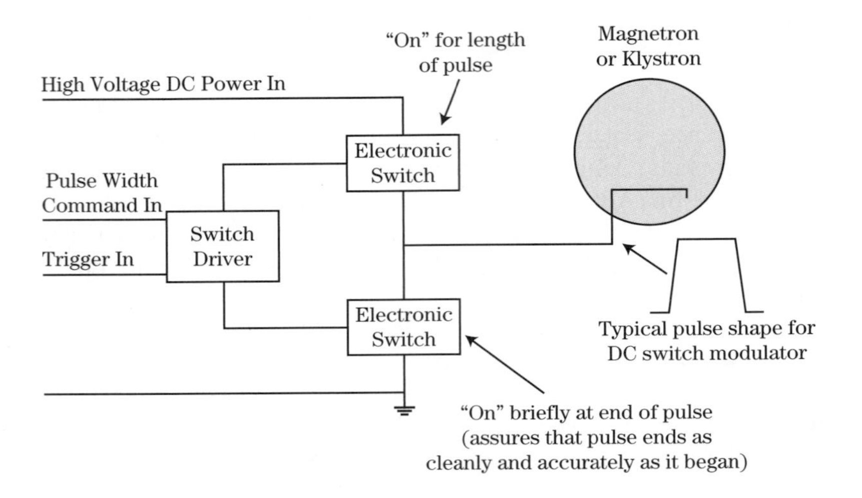

FIGURE 10-22 ■ Simplified diagram of an active switch modulator.

The switching rate sets the radar PRF while the pulse shape and duration are determined by the PFN characteristics. The switch has no effect on the pulse shape. The pulse ends when the PFN has discharged to a level that the transmitting tube is cut off. Line-type modulators tend to be less expensive than active-switch modulators (refer to Section 10.4.2) and are commonly used to operate pulsed magnetrons. One major disadvantage of this type of modulator is that the trailing edge of the pulse is usually not well defined since it depends on the discharge characteristics of the PFN. Other drawbacks to this configuration are less pulse generating flexibility and less pulse precision when compared with other types of modulators.

10.4.2 Active-Switch Modulators

Figure 10-22 is a simplified diagram of an *active-switch modulator* where the pulse has to be turned off as well as turned on. Originally, the switch was a vacuum tube and the modulator was called a *hard-tube modulator* to distinguish it from the gas tube thyratron switch often used in line-type modulators.

The "hard-tube" designation has been replaced by the more generalized term "active-switch" to allow for the use of both solid-state as well as vacuum tube switches [5]. With the advent of high-voltage, high-current solid-state switch devices such as field-effect transistors and *silicon-controlled rectifiers* (SCRs), active-switch modulators are realizable without the use of vacuum tubes. In many cases such solid-state devices must be stacked in series to handle the high voltages involved. Care must be taken in the design of series-stacked devices when switching high voltages to ensure equal voltage sharing among the devices during turn-on and turn-off transients. In a high-power radar transmitter the voltage swing across the on/off switches often exceed tens of kilovolts (kV). As shown in Figure 10-22, the switches of an active-switch modulator control both the beginning and end of the pulse. This is in contrast to the PFN modulator described previously where the pulse shape and duration is primarily determined by PFN characteristics. The active-switch modulator permits greater flexibility and precision than the line-type modulator. It provides excellent pulse shape with varying pulse durations and pulse repetition frequencies. It also provides the opportunity of using closely spaced coded bursts of pulses. A more recent outgrowth of the active-switch solid-state modulator is the solid-state cathode switch modulator [26], which can provide pulses as narrow as 50 nanoseconds at PRFs up to 400 kHz [5].

10.5 POWER SUPPLIES

The power supplies associated with any particular radar system provide the prime DC power to operate all radar system electronics. For the basic radar transmitter employing such devices as magnetrons, klystrons, and TWTs, a high-voltage power supply is required. High-voltage (HV) power supplies for all applications (not just radar) require special attention to factors such as HV insulation of wires, use of components properly rated for HV, prevention of arc-over due to ionization, and overload protective circuitry. In basic radar the HV supply must also meet the extra requirement to supply very high currents under pulsed conditions. In contrast, the power supplies for solid-state amplifiers such as associated with AESAs are relatively low-voltage, high-current DC supplies.

10.5.1 High-Voltage Power Supplies

High-power transmitter tubes (i.e., VEDs) require very high voltages for operation, generally in the range of 10 kV to 50 kV. Even so called lower-power VED transmitters can require from 4 kV to 6 kV. Since the tube may only be 20% to 40% efficient, the high-voltage power supply must produce 2.5 to 5 times the average and peak power output of the transmitter. For most applications, the high voltage supplied to the VED must be highly regulated and have extremely low ripple content to minimize phase and amplitude distortions in the RF output pulse.

Figure 10-23 illustrates a typical high-voltage power supply and shows many of the required elements of the circuit. The input alternating current (AC) voltage is preregulated, either via a variable transformer or by a high-frequency switching regulator. The resulting

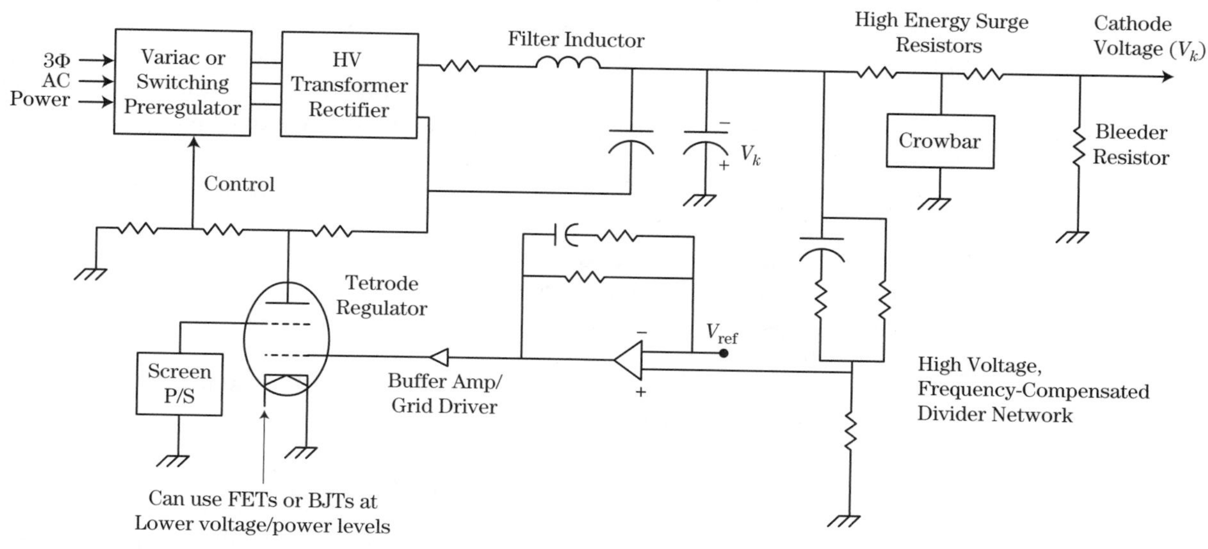

FIGURE 10-23 ■ Typical high voltage power supply for a radar transmitter.

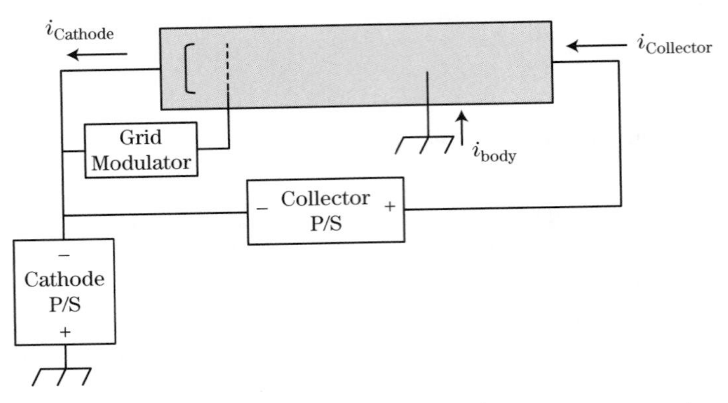

FIGURE 10-24 ■ Traveling-wave tube power supply circuitry.

voltage is then either supplied directly to the tube or, if tighter regulation is needed, is applied between the tube and a series regulator. The series regulator can be either a pass tube such as a triode or tetrode or a solid-state device such as an FET, if lower voltage swings can be maintained at the drain terminal. The standard input line frequencies are 60 Hz and 400 Hz.

Prototype multi-phase high voltage power conditioning systems using input line frequencies of 20 kHz while generating up to 140 kV or 11 megawatt pulses to drive high power accelerator klystrons have been demonstrated at the Los Alamos National Laboratory [27]. The high voltage transformers operating at 20 kHz are only 1% of the size and weight of the 60 Hz versions!

Many TWTs require so-called *depressed collector* operation, which entails the collector voltage being depressed below ground. Some TWTs have multiple collectors that require several different voltages. Such tubes are costly but exhibit much higher power efficiency than conventional TWTs. Figure 10-24 shows a block diagram of a TWT with a single depressed collector and its associated power supplies.

As an example, assume the following values in Figure 10-24: $V_{cathode} = 30$ kV, $V_{collector} = 10$ kV, $I_{collector} = 0.9$ A, $I_{helix} = 0.1$ A, and RF output = 5 kW. Without collector depression, the input DC power would be 30 kV × 1 A or 30 kW. The efficiency would then be 5/30, or 17%. With collector depression, the input DC power would be 30 kV × 0.1 A + 10 kV × 0.9 A, or 3 kW + 9 kW = 12 kW. The efficiency would then be 5/12, or 42%, a substantial improvement. Collector depression reduces thermal dissipation at the tube collector element and reduces the amount of power that must be processed, filtered, and delivered by the high-voltage power supply subsystem.

Almost without exception, high-power microwave tubes and their accompanying high-voltage switches occasionally arc over, essentially placing a short circuit across the modulator (and hence the high-voltage power supply). Since the resulting discharge of 50 joules (or more) will usually damage an RF tube (VED) or switching device, some means must be provided to divert the stored energy when an arc discharge occurs. Such a protective circuit is a called a *crowbar* since it is equivalent to placing a heavy conductor (like a crowbar) directly across the power supply to divert the energy and prevent its discharge through the tube or switching device.

10.5.2 Power Supplies for Solid-State Amplifiers

Solid-state amplifiers are associated with AESA radars such as the X-band U.S. Navy AN/SPY-3, which was designed to meet all horizon search and fire control requirements

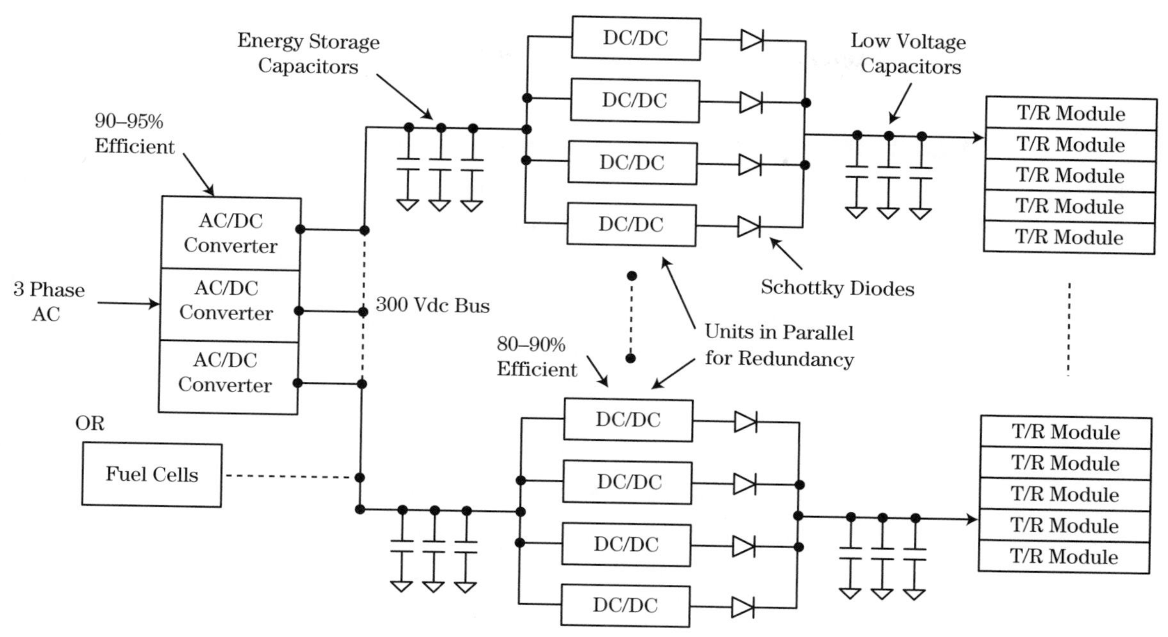

FIGURE 10-25 ■ An active aperture power supply configuration.

for the Navy. It uses three fixed arrays, each containing approximately 5,000 active elements. The aggregate of the DC power supplies for solid-state amplifiers associated with thousands of active elements must produce high currents (100s of amperes) at low voltages (5 to 10 volts). Since the transmitters of an AESA type system are distributed over the array, then the power supply system should be distributed as well. For large phased arrays, power distribution at higher voltages than required by the individual solid-state amplifiers can minimize heat generation due to ohmic losses. For example, GaAs type microwave transmitter modules typically require DC input voltages in the range of 5–10 VDC. It would be advantageous to produce these low voltages at a subarray level near the individual solid-state amplifiers by means of local DC/DC converters operating with inputs at hundreds of volts. Figure 10-25 illustrates a potential active aperture power supply configuration.

10.6 | TRANSMITTER IMPACTS ON THE ELECTROMAGNETIC ENVIRONMENT

Because of the large power levels typically associated with high-power radar systems, it is incumbent upon the designer of radar transmitters to ensure not only that the RF energy generated has the proper waveform and power levels but also that the radiated energy occupies that portion of the EM spectrum allocated to it. Additionally, because the spectrum must be shared with other radar systems within the band, the transmitted radiation should occupy only that amount of the band required to operate properly and not cause interference with other radar systems or any other electronic systems that operate within or adjacent to the designated bands for radar. High-power transmission also presents other potential negative impacts on the equipment and personnel that are in the environment located around radar transmitters.

10.6.1 Transmitter Design and Spectrum Issues

When designing the hardware that makes up a radar system, the designer attempts to balance competing requirements to maximize detection range, resolution, and system reliability while trying to minimize, for example, noise, size, weight, volume, and power requirements. Much study and effort has gone into developing design principles that meet these conflicting hardware and system requirements. An equally important area of radar engineering that has not received as much attention by radar designers is the area of spectrum management and engineering. The purpose of *spectrum management* is to coordinate and control the usage of the electromagnetic spectrum between and within countries. This encompasses a multitude of activities, including the following:

- Coordination, organization, and optimization of the use of the RF spectrum.
- Allocation of spectrum among various users.
- Controlling and licensing the operation of radio and radar systems within the spectrum.
- Controlling, avoiding, and solving interference problems between radiators and receivers,
- Advancing and incorporating new technology that impacts spectrum users.

Spectrum engineering is the complement to spectrum management. It seeks to design and develop equipment and create practices and procedures that maximize the efficient use of the EM spectrum while allowing the RF system to carry out its required functions within its frequency allocation. This means radiating the minimum amount of RF energy and using the minimum amount of spectrum that will allow the system to carry out its task, while not interfering with other equipment that lawfully operates within the same frequency band allocated to the radiating RF system.

It is relatively easy to make a determination as to how efficient the radar hardware is. There are several well-established measures of radar efficiency, including the efficiency of the radar system and the efficiency of the transmitter. This is not the case when determining the efficiency of spectrum usage. In the United States the National Telecommunications and Information Administration (NTIA) has undertaken a study of ways of measuring spectrum efficiency (SE) with the tasks of (1) developing metrics to define SE for the various radio services, (2) developing methods to implement such a metric, and (3) producing recommendations on ways to enhance SE. Until the NTIA finishes this task, the best guidance available to the radar engineer is the Radar Spectrum Engineering Criteria (RSEC) [28]. The purpose of the RSEC is to regulate the amount of bandwidth a radar is permitted to use, based on its modulation type. Additionally, it specifies a spectral roll-off requirement and a limit on spurious emissions, depending on the category a given radar falls within. Currently all primary radars are classified in one of five categories, Group A through E. Broadly speaking, Group A contains low-power (less than 1 kW peak power) radars; Group B contains medium-power (1 kw to 100 kW) radars; Groups D and E are certain special frequencies or applications; and Group C is everything else. Radars that fall within Group A are currently exempt from any RSEC specifications. Radars that fall within Groups B through E must meet the appropriate criteria for their group.

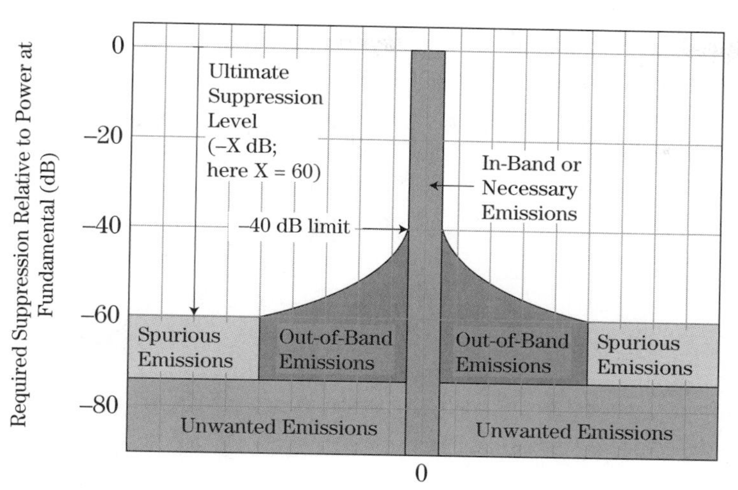

FIGURE 10-26 ■ Various signal domains considered by the RSEC.

The RSEC approach is generally to ensure that a radar's emissions fit within a spectral "mask" that is defined with respect to the radar's fundamental frequency of operation and peak power. The various types of emissions that are considered by the RSEC are illustrated in Figure 10-26, and a generic emissions mask that sets the upper bound on the radar spectral emissions is shown in Figure 10-27. Figure 10-28 shows a measured emission and the appropriate mask for checking compliance with the RSEC. Examining the figure shows that at about 3050 MHz the system exceeds the allowable emission limits.

Radar technology has advanced considerably since the RSEC was last modified in 1977. While the RSEC has bandwidth equations for pulsed, FM pulsed, phase coded, and FMCW radars, many other types of radars have no standard criteria. Thus, radars employing more than a single emitter—including phased array radars, variable PRF radars, radars whose modulation changes on a pulse-to-pulse basis, and other special types of radars—require a case-by-case examination to certify them as being compliant with spectrum engineering criteria. For the engineer tasked with designing a new radar system, the

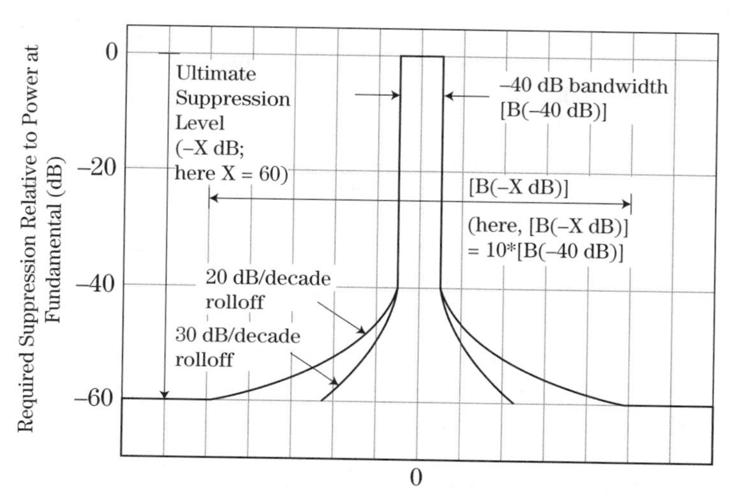

FIGURE 10-27 ■ Generic RSEC emissions box.

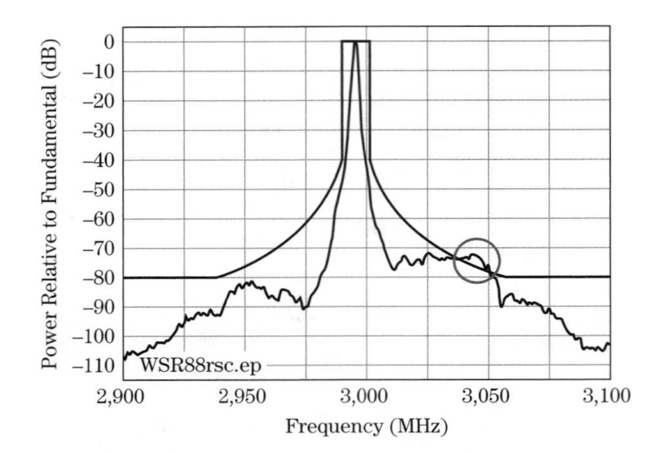

FIGURE 10-28 ■ Figure shows a measured emission within the RSEC box. At about 3050 MHz the system exceeds the allowable limits for the subject group.

best way to avoid operational problems related to spectrum issues is to work closely with the NTIA in the earliest stages of design to guarantee that the planned design approach is consistent with current regulations. This is especially important with the current effort to incorporate more waveform design and diversity into newer radar systems.

Another approach to using the spectrum more efficiently and minimizing interference to other radiating systems within a band is to consider making the radar "smarter." One way to do this is to make the radar adaptive, or to use a knowledge-based approach to transmitting the radar signals. The basic idea of this approach, also known as *cognitive radar*, is to monitor the electromagnetic environment that the radar will be operating in and to choose those portions of the band not currently occupied by other emitters or sources of interference, potentially modifying either the frequency or modulation scheme employed by the transmitter to avoid interfering with other users of the band. While this requires a much more complex transmitter design, it may be the only way to use existing radar bands efficiently. Knowledge-based radar is an active area of research. Good sources for initial study in this area are [29,30].

10.6.2 Transmitter Impacts on Spectral Purity

Transmitter impacts on radar spectral purity can be categorized as either time-varying or nontime-varying. Errors that are repeatable from pulse to pulse are considered nontime-varying. Errors that are not repeatable from pulse to pulse are considered time-varying. Time-varying errors are more serious because it is difficult to calibrate them out. In addition, they affect multipulse processes such as moving target indication (MTI) and pulse Doppler, whereas nontime-varying errors do not. It is interesting to note that the nontime-varying *unintentional modulation of pulse* (UMOP) [31] is often used as a "signature" by U.S. defense systems to identify and track foreign radars.

Transmitter output phase and amplitude are generally sensitive to the voltages applied to the power amplifier or oscillator. This is the case for both solid-state and tube-type devices. Transmitter phase and amplitude specifications in turn establish requirements on power supply voltage droop, regulation, and ripple to maintain the required spectral purity. Power supply-induced errors are usually time-varying unless the power supply frequency of operation (i.e., switching frequency or AC line frequency) is synchronized to the radar PRF or some multiple thereof.

Another source of spectral error is the nonlinear phase characteristic of transmitters, which results in frequency dispersion that in turn degrades the range sidelobes [32] associated with pulse compression systems. These effects tend to be nontime-varying as a function of waveform bandwidth for a given center frequency providing that temperature can be maintained constant. Since these effects are repeatable, they can be measured and characterized and for the most part can be calibrated or processed out. Another example of nontime-varying error is solid-state amplifier phase and amplitude sensitivity to junction temperature. These errors can vary across a pulse or from pulse to pulse depending on the waveform, but the effects can be predicted, or at least measured, and should be repeatable.

10.6.2.1 Time-Varying Errors

Within the transmitter, power supply ripple induces phase and amplitude ripple at the amplifier output due to amplifier phase/amplitude *pushing factors* [33]. Pushing factors are measures of how sensitive a device is to voltage variations at one of its electrodes. In earlier radars *frequency pulling* of magnetrons was often caused by RF load variations. This problem has been solved by means of improved RF isolation through the use of microwave circulators. Power supply ripple can cause pulse-to-pulse phase/amplitude errors, which result in Doppler sidebands that degrade coherent radar performance. High-frequency ripple can cause intrapulse modulations that can adversely affect linear frequency modulated (LFM) or chirp systems and degrade range sidelobe performance.

For example, a sinusoidal power supply ripple at 100 kHz will result in a pair of spectral sidebands that are offset from the carrier by 100 kHz. The power in these sidebands depends on the amount of voltage ripple and the phase sensitivity of the device in degrees/volt as well as amplitude modulation (AM) sensitivity in dB/V. Typical values of phase sensitivity for FETs are 1–2 deg/V (drain voltage) and 100–200 deg/V (gate voltage). The spurious sideband amplitude in decibels relative to the desired carrier (dBc) can be estimated as $20\log_{10}(\Delta\phi_{peak}/2)$, where $\Delta\phi_{peak}$ is the peak phase deviation in radians [34]. The phase deviation is calculated by multiplying the peak voltage ripple by the phase sensitivity. Note that switching power supplies do not produce ideal sinusoidal ripple waveforms, so the spectral content is complex and spread across multiple offset observed Doppler frequencies.

Figure 10-29 illustrates the effects of intrapulse ripple on the pulse-compressed receiver output for the echo from a single point target. Figure 10-29a shows the compressed ideal return with no modulation error. Figure 10-29b shows the same return with 10 degrees root mean square (rms) sinusoidal phase error, with three cycles of error across the transmitted pulse. Figure 10-29c shows the same return but with a lesser error of 2 degrees rms sinusoidal phase error, again with three cycles of error across the transmitted pulse. In this figure it is seen that the error-induced time sidelobes are three compressed range cells away from the main response. Figure 10-29d is similar to Figure 10-29b except that the 10 degree rms error waveform is now random instead of sinusoidal. The random error tends to spread out the peak time sidelobes, whereas the total integrated sidelobe floor remains the same as that in Figure 10-29b.

For several reasons, solid-state active arrays exhibit less main-beam additive spurious modulation than tube transmitters. One reason is that solid-state power amplifiers have lower noise figures. Also, if each element of an AESA has an individual transmitter amplifier and the additive noise from each amplifier is independent element to element, the resulting modulation component is suppressed by the inverse of the number of elements. (An exception would be for error components common to each element, due to common

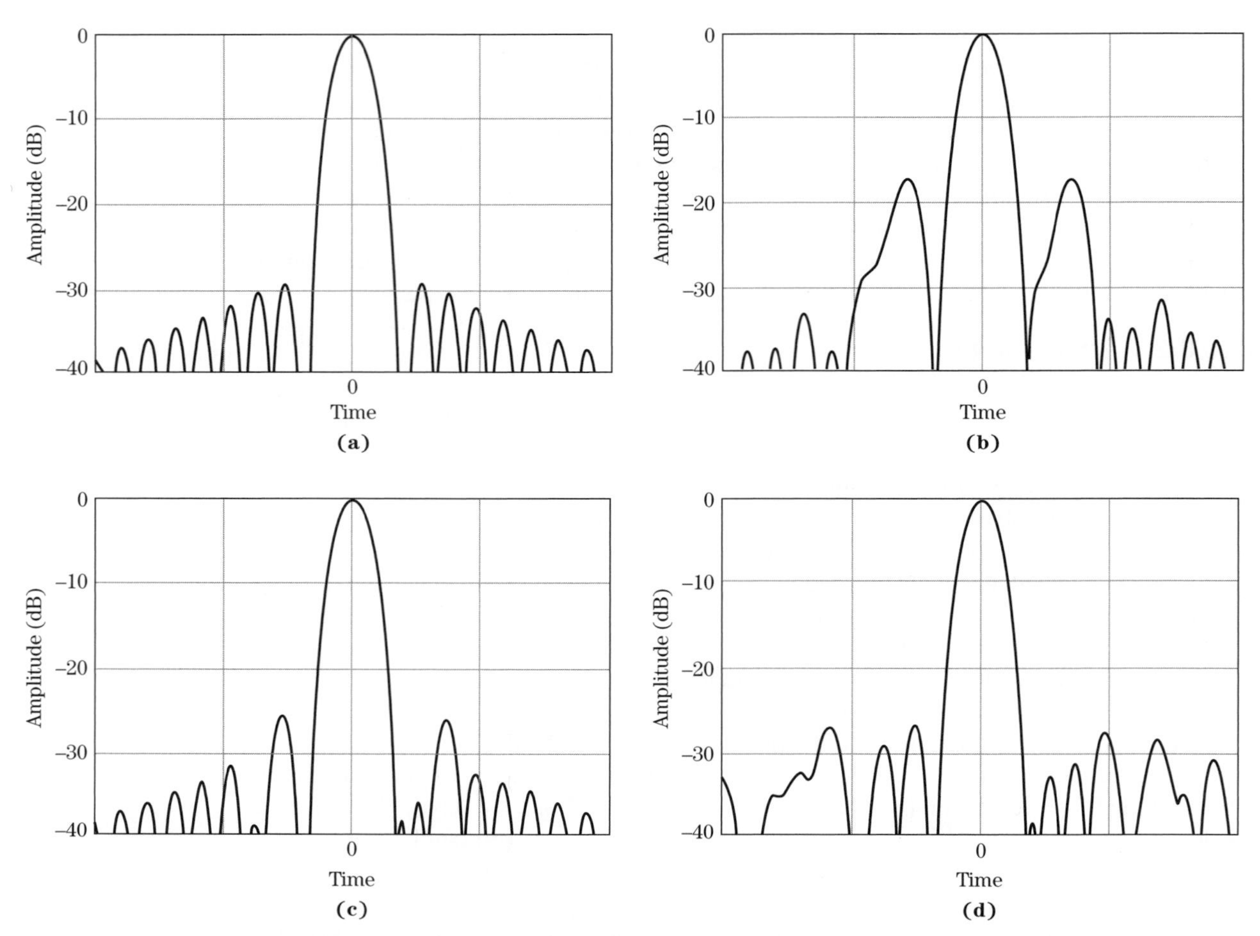

FIGURE 10-29 ■ Time sidelobe response with intrapulse modulation errors. (a) Idealized response with no intrapulse modulation error. (b) 3 cycles of 10 degrees rms sinusoidal modulation error. (c) 3 cycles of 2 degrees rms sinusoidal modulation error. (d) 10 degrees random modulation error.

power supply pushing and main signal excitation. Such independent error components will not experience the array factor gain since they will not add coherently.) They will appear as higher spurious modulation in the sidelobes of the array antenna pattern, but not in the main beam [35]. Spurious modulation that is correlated across the array achieves the same antenna directivity as the desired signal. Some examples include transmitter exciter/local oscillator phase noise, effects of power supply ripple, and noise that is common to all the elements. Generally effects with a given correlation interval (e.g., number of affected elements) will experience the spatial gain of that set of elements.

10.6.2.2 Nontime-Varying Errors

In pulse compression systems using LFM waveforms, variations in insertion phase and output power through an amplifier as a function of frequency produce range sidelobe degradation. The amount of degradation and range sidelobe spacing depends on the magnitude and spectral content of the phase and amplitude variations and follows *paired echo*

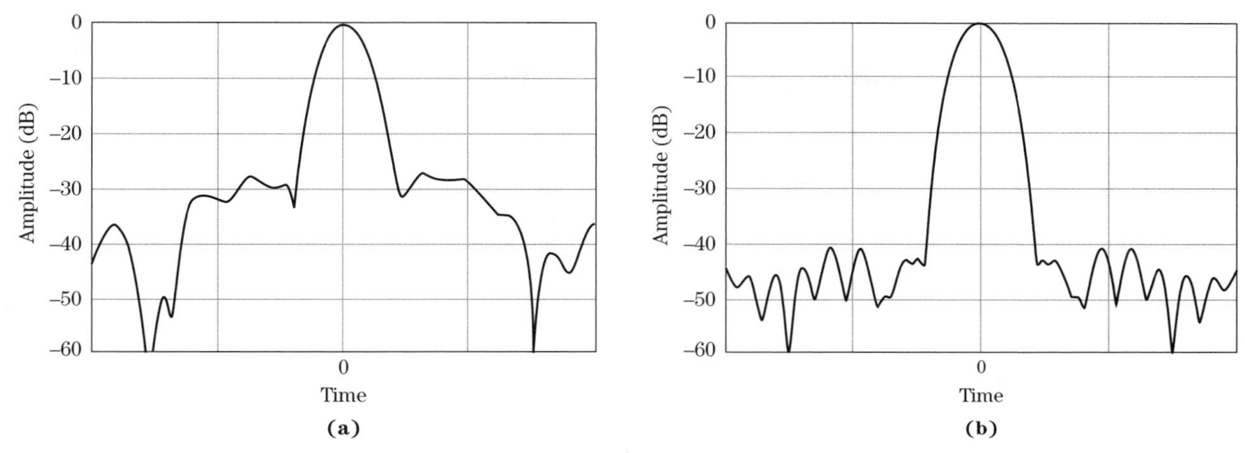

FIGURE 10-30 ■ (a) Range sidelobe response prior to TWT phase equalization. (b) Range sidelobe response after TWT phase equalization.

theory [36], which states that the stronger the modulation, the higher the sidelobes. Device nonlinearities can be viewed as mapping a complex error waveform on top of the ideal LFM waveform. Fourier analysis can be used to decompose the error waveform into individual sinusoidal terms. To some extent such errors can be corrected by means of phase equalization, which is commonly done externally for wideband TWTs or measured and corrected via predistortion in the transmitter exciter or through digital processing in the signal processor.

Figure 10-30a is processed data that show the wideband range sidelobe limitations of a high-power TWT amplifier. After correcting only for phase errors (not amplitude), these sidelobes can be brought down as in Figure 10-30b to the level inherent in the pulse compression technique that would exist even if the power supply subsystem were perfect and introduced no additive spurious signals or noise.

10.7 OPERATIONAL CONSIDERATIONS

Once a transmitter design has been optimized for a given application, the actual lifetime of the transmitter and the radar system itself will be dependent on the way the system is operated. The lifetime of the transmitter will be maximized by operating the transmitter within the tube or amplifier manufacturer's specified parameters. One of the major issues that affect transmitter reliability is overheating of the high-power components. Consequently, the radar designer must understand cooling requirements for transmitters. Safety issues must also be addressed during transmitter design.

10.7.1 Transmitter Reliability

A reliable radar is one that is available when it is needed other than scheduled down time due to, for instance, maintenance or system checks. Reliability implies that the transmitter and its critical components should have a long MTBF. The transmitter is typically the primary concern in determining radar reliability because the receiver processes radar signals at very low power levels, whereas the transmitter generates and processes signals at very high power levels or high voltage or current levels, causing high temperatures that stress components and subassemblies. These high voltages and currents also often exist in the power supply and the modulator. Thus, making an accurate estimate of the reliability and life time of a radar transmitter will depend on the voltage, current, and power levels in the transmitter subsystems; the materials used in component construction; and the way the transmitter is operated. In addition, there are some distinctive traits of both tube-based and solid-state transmitters that affect reliability estimates.

Tubes are thermionic sources; the cathode operates at a high temperature. The expected lifetime of the electron gun design is determined by the cathode operating temperature, current density, and materials [2]. As a general rule, high-power ground-based tubes are designed with expected lifetimes on the order of 10,000 hours to 40,000 hours. However, it is possible to operate properly designed and constructed tubes for much longer periods of time. In one example, a klystron in operation in a ballistic missile early warning system (BMEWS) was still operating after 30 years, or 240,000 hours [37]. Tubes designed to operate on spaceborne platforms are highly derated to allow anticipated lifetimes of hundreds of thousands of hours. Such tubes (typically TWTs) are also designed with multiple collector stages to greatly improve efficiency and reduce prime power requirements. For more detailed information on tube reliability, see chapter 18 of Gilmour [11].

The limiting component in a tube-based, high-power transmitter may not be the tube itself, however. There will be many components with high voltage/high thermal stresses, and, if not carefully designed with adequate margin, faults can occur at multiple locations. Protective circuits must be included in the design to ensure that a fault in one location does not avalanche and cause faults at other locations in the circuit. For instance, in the event of an arc in the output waveguide system, RF must be inhibited quickly, and the tube should be turned off. Temperature sensors should be employed to remove power from the tube if temperatures become excessive. Key voltages and currents should be monitored and power shut down if they fall outside of normal ranges. Properly designed interlock circuits can help extend transmitter life expectancy by reducing tube failures that are caused by power supply or waveguide system faults.

T/R modules can be highly reliable (hundreds of thousands of hours) and solid-state systems can benefit from graceful degradation, whereby some percentage of modules can fail yet the radar can still operate. Sensitivity degrades with such failures, although sidelobe levels degrade at a faster rate (see Chapter 9). As with tube-based systems, however, in many cases the reliability-limiting components or subsystems may not be power amplifiers or oscillators themselves but instead may be the supporting subsystems, such as the power supplies. Power supplies for solid-state systems must usually provide very high current levels, which can result in severe thermal stresses on critical components. Even if the T/R modules themselves are highly reliable, if the power supply feeding the modules fails that subset of modules is effectively removed from the system. Power supplies can be interconnected or bussed together to provide redundancy, at the cost of complications to the design. The larger the number of modules fed by an individual power supply, the larger the effect on radar sensitivity and sidelobe levels from the failure of that power supply. The same argument holds for any subsystem on which a group of T/R modules depends, such as a driver amplifier at the subarray level in a phased array antenna.

10.7.2 Transmitter Cooling

One of the criteria for the radar designer is to make the system as efficient as possible. Losses occur at all stages of the process where prime power is turned into RF energy that is to be radiated toward the target. These losses are usually manifested as heat, which further reduces efficiency, since additional effort must be expended to eliminate the heat and prevent it from damaging or destroying the radar. Thus, cooling the components in a radar system, especially those comprising the transmitter, is paramount to ensuring the reliability of the system.

Removing heat from the radar system is typically accomplished in one of three ways: (1) through normal air-convection currents; (2) through forced-air cooling; or (3) through liquid cooling. Cooling by normal air-convection currents is suitable for low-power radar systems, such as the "speed guns" used by law enforcement agencies. At the other extreme of a high-power microwave transmitter, the cooling system may need to dissipate as much as 70% of the input AC power in the form of waste heat. In these cases, a liquid cooled system is the only way to remove the large quantities of heat generated. In all cases, the manufacturer's specification will determine how much heat must be removed from the RF source, be it solid-state or vacuum tube.

Temperature control is important for microwave tube operation because the properties of the materials used to construct the tube will change as the temperature increases. For instance, electrical resistivity usually decreases as temperature increases, while the dielectric constant of several common tube materials increases as temperature increases. From a reliability standpoint, one of the most common failure points in tubes due to high temperatures is the metal-to-glass or metal-to-ceramic interface or seals in the tube. Above 250°C the seals can begin to deteriorate [38]. To prevent damage to the radar system, temperature monitors should be incorporated into key areas of the transmitter, including air or water inlets and outlets, near vacuum tubes, and also close to other power dissipating components such as transformers and loads.

10.7.3 Safety Issues

An important operational consideration is the safety of the personnel operating the radar system. In any system employing high voltages and currents there are the usual dangers associated with accidental electrocution. Because of this, high-power radar systems are designed with electrical interlocks to minimize danger to operations and maintenance personnel. These are often in addition to those used to provide overvoltage and overcurrent protection for the major subsystems of the radar, such as the power supply and the power amplifier.

Another potential hazard caused by high voltages and currents is the production of x-rays, which occurs when the electrons produced by a hot cathode are accelerated by a high voltage to impact on a metal anode within a vacuum tube. Because of cooling considerations, the usual anode material for high-power microwave tubes is copper. This leads to x-ray spectral lines at wavelengths of 0.139222 nm, 0.138109 nm, 0.154056 nm, and 0.154439 nm [39].

The maximum energy of the x-ray photons produced is determined by the energy of the incident electron, which is equal to the voltage on the tube. Thus, a 100 kV tube can create x-rays with energies up to 100 keV. For comparison, the voltages used in typical diagnostic x-ray tubes, and thus the highest energies of the x-rays, range from 20 to 150 kV. Higher-energy x-ray achieve greater penetration and therefore require more shielding to protect operators from the x-rays generated. For instance, a tube with a maximum voltage of 100 kV will require a minimum of 1.5 mm of lead to stop the x-rays that may be generated at that voltage [40].

Another safety issue arises from the fact that many high power vacuum tubes incorporate hazardous materials. Depending on the type of tube and the construction approach, the tubes may incorporate such materials as beryllium oxide (BeO) and antimony (Sb) as well as alkali materials such as potassium (K), cesium (Cs), and sodium (Na). As all of these are toxic, care must be exercised when working with the tubes so that personnel are not accidentally exposed to these materials, and they must be disposed of in a safe manner consistent with environmental regulations for hazardous materials.

Finally, there are the safety issues associated with the high RF power levels generated by the radar system. The specific concerns depend on the radar configuration and location. As an example, radar systems located on naval vessels must take into consideration the presence of personnel, ordinance, and other materials that absorb RF radiation. During on-loading or off-loading of ammunition, there is a danger that RF electromagnetic fields could accidentally activate electro-explosive devices (EEDs) or electrically initiated ordnance. This is a very real hazard to the ordnance, the ship, and the crew. Several occasions have been recorded when vessels of the U.S. Navy suffered severe damage due to RF radiation triggering explosions of ordnance aboard the vessels.

Several effects due to RF radiation have been observed in personnel. Some, such as the formation of cataracts at 10 GHz, are frequency dependent. The most common effect observed is tissue heating due to absorption of RF energy. Safe limits for RF exposures of personnel are based on the power density of the radiation beam and exposure time of the person being radiated. A common measure of this heating effect is known as the *specific absorption rate* (SAR) [41]. The SAR is related to the electric field at a point by the expression

$$SAR = \frac{\sigma |E|^2}{\rho} \tag{10.4}$$

where

σ = conductivity of the tissue (S/m).

ρ = mass density of the tissue (kg/m^3).

$|E|$ = rms electric field strength (V/m).

Acceptable limits on the SAR, which apply for many types of RF devices including cell phones and mobile radios, are given in ANSI/IEEE C95.1 [42] for the United States and in the International Commission on Non-Ionizing Radiation Protection [43] for Europe and most of the rest of the world.

Finally, voltages of enough potential to cause an RF burn can be induced on metallic items, such as railings or fences, that are near radar-transmitting antennas. However, there must be actual physical contact for the burn to occur. This contact can be prevented by ensuring that warning signs are properly placed to keep people away from locations where such voltages can be induced.

10.8 SUMMARY AND FUTURE TRENDS

This chapter has described radar transmitters as an integral subsystem of pulsed, pulsed compression, and active phased array radars. The transmitter consists of an RF power amplifier or high-power oscillator, a modulator, and a power supply. In addition to the technology descriptions, several transmitter system issues were discussed, including transmitter configurations, impacts on the electromagnetic environment, and operational considerations.

The technology linked to radar has evolved over many years. For example, the magnetron microwave cross-field oscillator tube used today in many low-cost ship radars was invented in 1921 by A. W. Hull [44]. The klystron amplifier was described as early as 1939 [5]. The traveling-wave tube was invented in the 1942–1943 time frame. All of these tube types (now commonly referred to as VEDs) have been improved over time for higher efficiency, improved reliability, wider bandwidth, and higher output powers. By comparison, solid-state microwave power RF amplifiers are relatively recent developments. The first commercially available GaAs MESFET X-band (10 GHz) amplifiers with good performance emerged in the 1970s [45].

Future conventional radars will certainly take advantage of the continuing evolution of high-power VEDs such as klystron amplifiers, traveling-wave tubes, and magnetrons. However, future radar systems employing active electronically scanned array antennas [46] will make increasing use of sophisticated digital hardware and software coupled with state-of-the-art solid-state RF amplifier and switching hardware. High-power solid-state switches will permit the increasing use of active-switch modulators. The crowbar circuits that are necessary to protect transmitter components during arc overs but are a major source of radar failure will be a thing of the past. Next-generation high-efficiency GaN-based power amplifiers [47,48] will be available for AESA applications.

10.9 FURTHER READING

There are relatively few texts dedicated to radar transmitter technology. The text by Ewell [13] is very focused and contains a great deal of detailed design information. Ostroff et al. [49] focused on solid-state transmitters in their work. An excellent treatment of microwave power tubes, along with some good historical context, can be found in Gilmour's work [2,11].

There are considerably more individual papers on various subjects than complete texts. Examples of conferences that are good sources for radar transmitter literature are the Microwave Power Tube Conference and the IEEE Power Modulator Symposia. The primary journals for published research on radar transmitters are the *IEEE Transactions on Aerospace and Electronic Systems* (*AESS*) and the *IEEE Transactions on Microwave Theory and Techniques* (*MTT*).

10.10 REFERENCES

[1] Barton, D.K., *Radar System Analysis and Modeling*, Artech House, Norwood, MA, 2005.

[2] Gilmour, A.S., *Microwave Tubes*, Artech House, Norwood, MA, 1986.

[3] Yagi, H., "Beam Transmission of Ultra Short Waves-Part II Magnetron Oscillators," *IRE Proceedings*, pp. 729–741, June 1928.

[4] Kolosov, A.A. et al., *Over the Horizon Radar* [tr. W.F. Barton; original in Russian], Artech House, Norwood, MA, 1987.

[5] Skolnik, M., *Radar Handbook,* 3d ed., McGraw Hill, New York, 2008.

[6] Curry, R.G., *Radar System Performance Modeling*, 2d ed., Artech House, Norwood, MA, 2005.

[7] McMillan, R.W., "MMW Solid-State Sources," Chapter 8 in *Principles and Applications of Millimeter Wave Radar*, ed. N.C. Currie and C.E. Brown, Artech House, Norwood, MA, 1987.

[8] West, J.B., "Phased Array Antenna Technology," pp. 6-156–6-179 in *The RF and Microwave Handbook,* ed. M. Golio, CRC Press, Boca Raton, FL, 2001.

[9] Mailloux, R.J., *Phased Array Antenna Handbook,* 2d ed., Artech House, Norwood, MA, 2005.

[10] Brookner, E., *Aspects of Modern Radar*, Artech House, Norwood, MA, 1988.

[11] Gilmour, A.S., *Principles of Traveling Wave Tubes*, Artech House, Norwood, MA, 1994.

[12] Barker, R.J., Booske, J.H., Luhmann Jr., N.C., and Nusinovich, G.S. (Eds.), *Modern Microwave and Millimeter-Wave Power Electronics*, IEEE Press, Piscataway, NJ, 2005.

[13] Ewell, G., *Radar Transmitters,* McGraw Hill, New York, 1981.

[14] Linde, G.J., Mai, T.N., Danly, B.G., Cheung, W.J., and Gregers-Hansen, V., "WARLOC: A High-Power Coherent 94 GHz Radar," *IEEE Transactions on Aerospace and Electronic Systems,* vol. 44, no. 3, pp. 1102–1117, July 2008.

[15] Report of Department of Defense Integrated Product Team, "Industrial Assessment of the Microwave Power Tube Industry," Washington, DC, April 1997, Accession No. ADA323772. Available at: http://www.dtic.mil/srch/doc?collection=t3&id=ADA323772

[16] Brookner, E., "Phased Arrays and Radars—Past, Present and Future," *Microwave Journal*, vol. 49, no. 1, pp. 24–46, January 2006.

[17] Edwards, T., "Semiconductor Technology Trends for Phased Array Antenna Power Amplifiers," *Proceedings of the 3rd European Radar Conference*, pp. 269–272, Manchester, UK, September 2006.

[18] Krauss, H.L., Bostian, C.W., and Raab, F.H., *Solid State Radio Engineering*, New York, Wiley, 1980.

[19] Cripps, S.C., *RF Power Amplifiers for Wireless Communication*, Artech House, Norwood, MA, 1999.

[20] Grebennekov, A., *RF and Microwave Power Amplifier Design*, McGraw-Hill, New York, 2004.

[21] Raab, F.H., Asbeck, P., Cripps, S., Kennington, P., Popovic, Z., Pothecary, S., et al., "Power Amplifiers and Transmitters for RF and Microwave," *IEEE Transactions on Microwave Theory Technology*, vol. 50, no. 3, pp. 814–826, March 2002.

[22] Gao, S., "High Efficiency Class F RF/Microwave Power Amplifiers," *IEEE Microwave Magazine*, vol. 7, no. 1, pp. 40–48, February 2006.

[23] Borkowski, M., "Solid-State Transmitters," Chapter 11 in *Radar Handbook*, 3d ed., ed. M. Skolnik, McGraw Hill, New York, 2008.

[24] Smith, C.R., Armstrong, C.M., and Duthie, J., "The Microwave Power Module: A Versatile RF Building Block for High-Power Transmitters," *Proceedings of the IEEE,* vol. 87, no. 5, pp. 717–737, May 1999.

[25] Aichele, D., "Next-Generation, GaN-based Power Amplifiers for Radar Applications," *Microwave Product Digest,* January 2009.

[26] Gaudreau, M.P.J., Casey, J.A., Hawkey, T.J., Kempkes, M.A., and Mulvaney, J.M., "Solid State High PRF Radar Modulators," *Record of the IEEE 2000 International Radar Conference*, pp. 183–186, May 7–12, 2000.

[27] Reass, W.A., Baca, D.M., Gribble, R.F., Anderson, D.E., Przybyla, J.S., Richardson, R., et al., "High-Frequency Multimegawatt Polyphase Resonant Power Conditioning," *IEEE Transactions on Plasma Science*, vol., 33, no. 4, pt. 1, pp. 1210–1219, August 2005.

[28] Radar Spectrum Engineering Criteria (RSEC), section 5.5 of Manual of Regulations and Procedures for Federal Radio Frequency Management, U.S. Dept of Commerce, National Telecommunications and Information Administration, Office of Spectrum Management, January 2008, September 2009 Revision. Available at: http://www.ntia.doc.gov/osmhome/redbook/redbook.html

[29] *IEEE Signal Processing Magazine*, vol. 23, no. 1, January 2006.

[30] Gini, F., and Rangaswamy, M., *Knowledge Based Radar Detection, Tracking and Classification*, John Wiley & Sons, 2008.

[31] Neri, F., *Introduction to Electronic Defense Systems*, Artech House, Norwood, MA, 2001.

[32] Rabideau, D.J., and Parker, P., "Achieving Low Range Sidelobes & Deep Nulls in Wideband Adaptive Beam Forming Systems," 10th ASAP Workshop, MIT Lincoln Lab, March 2002.

[33] Abe, D.K., Pershing, D.E., Nguyen, K.T., Myers, R.E., Wood, F.N., and Levush, B., "Experimental Study of Phase Pushing in a Fundamental-Mode Multiple-Beam Klystron," *IEEE Transactions on Electron Devices*, vol. 54, no. 5, pp. 1253–1258, May 2007.

[34] Graves Jr., W., "RF Integrated Subsystems: Managing Noise and Spurious within Complex Microwave Assemblies," www.rfdesign.com, July 2003. Available at: http://mobiledevdesign.com/hardware_news/radio_managing_noise_spurious/

[35] Iglehart, S.C., "Noise and Spectral Properties of Active Phased Arrays," *IEEE Transactions on Aerospace and Electronic Systems*, vol. AES-11, no. 6, pp 1307–1315, November 1975.

[36] Cook, C.E., and Bernfield, M., *Radar Signals: An Introduction to Theory and Application,* Academic Press, New York, 1967.

[37] Symons, R.S., Tubes: Still Vital After All These Years," *IEEE Spectrum,* pp. 52–63, April 1998.

[38] Whitaker, J.C., "Tubes," pp. 7-169–7-183 in *The RF and Microwave Handbook,* ed. M. Golio, CRC Press, Boca Raton, FL, 2001.

[39] Lide, D.R. (Ed.), *CRC Handbook of Chemistry and Physics,* 75th ed., CRC Press, Boca Raton, FL, 1994.

[40] Recommendations by the Second International Congress of Radiology, Stockholm, July 23–27, 1928

[41] Seabury, D., "An Update on SAR Standards and the Basic Requirements for SAR Assessment," *Conformity Magazine,* April 2005, Available at: http://www.ets-lindgren.com/pdf/sar_lo.pdf

[42] IEEE, "Standard for Safety Level with Respect to Human Exposure to Radio Frequency Electromagnetic Fields, 3KHz to 300GHz," *IEEE Std* (IEEE) **C95.1**, October 2005.

[43] International Commission on Non-Ionizing Radiation Protection, "Guidelines for Limiting Exposure to Time-Varying Electric, Magnetic, and Electromagnetic Fields (Up to 300 GHz)," 1998, Available at: http://www.icnirp.org/documents/emfgdl.pdf

[44] Hull, A.W., "The Effect of a Uniform Magnetic Field on the Motion of Electrons between Coaxial Cylinders," *Physical Review,* vol. 18, no. 1, pp. 31–57, 1921.

[45] Colantonio, P., Giannini, F., and Limiti, E., *High Efficiency RF and Microwave Solid State Power Amplifiers,* Wiley & Sons, New York, 2009.

[46] Kinghorn, A.M., "Where Next for Airborne AESA Technology?" *IEEE Aerospace and Electronic Systems Magazine*, vol. 24, no. 11, pp. 16–21, November 2009.

[47] Aichele, D., "Next-Generation, GaN-based Power Amplifiers for Radar Applications," *Microwave Product Digest,* January 2009.

[48] Information Handling Services, Inc., "Military Apps, GaN Propel RF Power Semiconductor Market," September 14, 2009. Available at: http://parts.ihs.com/news/2009/rf-power-semiconductors-091409.htm.

[49] Ostroff, E., Borkowski, M., and Thomas, H., *Solid-State Radar Transmitters*, Artech House, Norwood, MA, 1985.

10.11 PROBLEMS

1. A 95 GHz oscillator tube (extended interaction klystron oscillator) is operated at a beam voltage of 21 kV. The modulation sensitivity of the tube is 0.2 MHz/volt, and the high-voltage power supply has 0.3% peak ripple during the pulse. How much FM is caused at the RF output by the power supply ripple? What must the ripple be to keep this FM down to 1 MHz?

2. A power amplifier tube has a small signal gain of 50 dB, noise figure of 40 dB, output power of 10 kW, and 500 MHz of bandwidth. With no input RF but with beam current flowing, how much noise will be produced by the tube in dBm/MHz? What will the total noise power output be in dBm? (Hint: Thermal noise power $= kT = -114$ dBm/MHz (at 290 K). Input tube noise per MHz $= (F - 1)kT$, where $k =$ Boltzmann's constant $= 1.3807 \times 10^{-23}$ Joules/K, $T =$ temperature in K, and $F =$ noise figure.)

3. A radar needs a power amplifier for a transmitter with the following characteristics: 50 kW peak power output, 5 kW average power output, 10% bandwidth, 5 W drive power available. What type of transmitter tube would be a suitable choice? Explain why.

4. A noncoherent search radar needs a transmitter tube with the following characteristics: 250 kW peak power, 1 μs pulse width, 2,000 Hz PRF. Cost is a major consideration. What type of transmitter tube would be a suitable choice, and explain why. What type of modulator configuration might be used?

5. A solid-state module has an output spectral purity requirement of -100 dBc of additive spurious noise. If the drain voltage and gate voltage sensitivities of the power amplifiers are 1 degree/volt and 100 degrees/volt, respectively, what are the voltage ripple requirements at the power amplifiers? Assume power amplifier additive noise is the dominant component, and budget for equal contribution from gate voltage and drain voltage ripple/noise.

6. A solid-state amplifier is operated at a drain voltage of 10 V. During transmit, the voltage ripple on that voltage is 100 mV peak (one-way, not peak to peak) concentrated at a sinusoidal ripple frequency of 1 MHz (since it is created by a 1 MHz resonant power converter). The amplifier drain voltage phase sensitivity is 1 degree/V. What is the approximate sideband amplitude relative to the carrier due to this voltage ripple?

7. This same solid-state amplifier also has a gate voltage sensitivity of 200 degrees per volt. The 10 V to the T/R module is regulated down to provide the gate voltage, reducing the ripple to 0.1 mV. What is the approximate sideband amplitude relative to the carrier due to this voltage ripple?

8. A TWT amplifier has the following electrode sensitivities: cathode voltage -0.1 deg/V, grid-cathode voltage of 0.2 deg/V, and collector voltage -0.01 deg/V. The nominal voltages applied are -30 kV (cathode), 10 kV (collector with respect to the cathode), and 1 kV (grid

with respect to cathode). What does the power supply ripple and noise need to be kept to maintain −75 dBc spurious additive noise? Allocate one-half of the total noise contribution to the cathode voltage and the remaining noise equally split between the remaining two voltages.

9. A radar uses a transmitter with a 5,000 W TWT amplifier operating at a transmit duty cycle of 10%. The cathode current is 1 A peak, the body current is 100 mA peak, the cathode voltage is −30 kV, and the collector voltage is 10 kV with respect to cathode. The collector can also be operated at ground if a separate power supply is not available. Calculate the prime power required for both the depressed collector and nondepressed collector configurations assuming an AC/DC conversion efficiency of 80% for the high voltage power supplies. What is the overall transmitter/power supply combined efficiency for both?

10. For the TWT described in problem 9, recalculate the prime power required if the AC/DC conversion efficiency is increased from 80% to 85% for both modes. Compare the results.

11. A MESFET microwave power transistor has a phase sensitivity of 10 deg/volt of bias change. The DC power supply peak ripple voltage is 100 mV. Calculate the spurious sideband amplitude expressed in dBc.

12. Which of the following statements are true regarding solid-state arrays versus tube-based arrays?

    a. Solid-state arrays result in elimination of high-power waveguides.

    b. Solid-state arrays generally result in lower antenna ohmic losses that have to be accounted for in radar sensitivity budgets.

    c. Tube-based arrays don't degrade as gracefully as solid-state arrays (as a function of component failures).

    d. Solid-state arrays eliminate some of the high-voltage and x-ray concerns associated with tube transmitters.

    e. All of the above.

13. An X-band ($\lambda = 3$ cm) T/R module uses a power amplifier that exhibits a power added efficiency on transmit of nominally 33% over a peak output power range from 0.1 W to 10 W at 20% transmit duty cycle. In the transmit mode the module also requires 250 mW of DC power to power the background control electronics. In the receive mode the module requires 500 mW of DC power, including background control electronics. A radar is required to have a power-aperture-gain product of 80 dBWm^2. Assume nominally 4,000 elements per square meter as an element density to satisfy scan volume requirements. Tabulate and plot the DC prime power required by the array for peak module powers from 0.1 W to 10 W and examine the results.

14. For the module and radar of problem 13, what module power should be used to minimize the array aperture area? What would be the array area and number of elements?

15. Which of the following are typically measured to characterize the performance of a high-power transmitter? Choose all that apply.

    a. Peak and average power.

    b. Dynamic range.

    c. Output harmonics.

    d. AM/PM conversion.

    e. All of the above.

16. Which of the following are true statements with respect to the use of solid-state modules for a phased array antenna versus a power tube-type transmitter?

    a. Each module has a low-power phase shifter.
    b. The solid-state active aperture array exhibits graceful degradation.
    c. The solid-state active aperture array will have lower acquisition cost.
    d. The solid-state active aperture array will have higher ohmic losses, reducing effective radiated power and increasing noise temperature.
    e. All of the above.

CHAPTER 11

Radar Receivers

Joseph A. Bruder

Chapter Outline

11.1 Introduction 391
11.2 Summary of Receiver Types 392
11.3 Major Receiver Functions 396
11.4 Demodulation 400
11.5 Receiver Noise Power 404
11.6 Receiver Dynamic Range 406
11.7 Analog-to-Digital Data Conversion 409
11.8 Further Reading 414
11.9 References 414
11.10 Problems 415

11.1 INTRODUCTION

The receiver is an integral part of the radar system. As shown in Figure 11-1, it provides the necessary downconversion of the receive signal from the antenna and the inputs required to the signal and data processors.

In a typical pulsed radar system (see Figure 11-2), the transmit signal is coupled to the antenna via some form of a duplexer (transmit/receive [T/R] switch). The function of the duplexer is to couple the transmit signal into the antenna with low loss while providing high isolation during transmit time to prevent saturation of the receiver by the high-power transmit signal. Any radar signals present at the receiver during the transmit time are rejected by the duplexer. Following the completion of the transmit pulse, the received signals from the antenna are coupled into the receiver with low loss.

Modern receivers often have a low-noise amplifier (LNA) at the receiver input followed by a band-pass filter to reduce the noise figure of the receiver. The band-pass filter eliminates out-of-band frequency components from the mixer input. The mixer then downconverts the received signal to an intermediate frequency (IF) signal using local oscillator (LO) signals. The mixer output is subsequently amplified and narrowband filtered to reject unwanted intermodulation components. After this, detection of the intermediate frequency signal provides the output or outputs of the receiver subsystem.

The final receiver outputs are sometimes called the radar video signal. In radar, a video signal is the received signal after all carrier or intermediate frequencies are removed

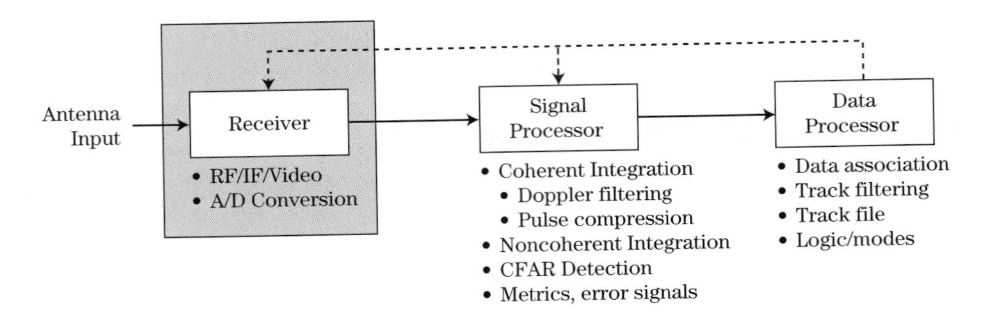

FIGURE 11-1 ■ General receiver functions.

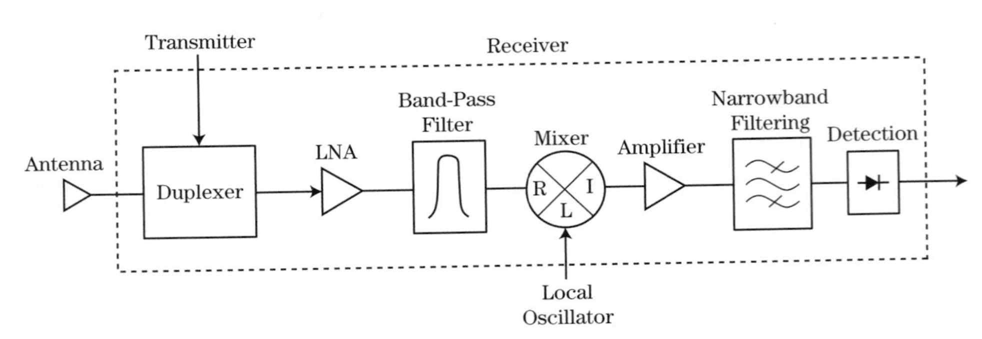

FIGURE 11-2 ■ Major receiver elements.

by demodulation to baseband. Historically, this voltage would be used to drive the radar operator's video display. In modern systems, the voltage is more likely to be digitized and subjected to digital signal processing.

The advantages and disadvantages of component placement and configurations will be addressed in this chapter. Continuing advances in analog-to-digital converters (ADCs) and digital signal processing (DSP) technology are driving receiver development. As converters improve in speed and resolution, the digitization moves closer to the antenna. Improvements in DSP resolution, speed, and cost are pushing traditional analog receiver functions into the digital domain. While these will be detailed in Chapter 14, some of the main digitization components affecting receivers will be discussed in this chapter.

Modern radar receivers are often required to perform a variety of tasks including change of frequency, bandwidth, and gain functions to support the radar modes. These more complex receivers often include digital control networks to select the appropriate receiver depending on the particular radar mode. In addition, these complex receivers often include built-in-test (BIT) functions to enable automated detection of receiver faults.

11.2 SUMMARY OF RECEIVER TYPES

There are several basic types of receiver configurations, including crystal video, superregenerative, homodyne, and superheterodyne. Most modern radars primarily use the latter, but there are other applications for which simpler receiver architectures are better suited. There has been increasing interest in digital receiver configurations for radar applications. Other types of receivers—those used primarily for electronic warfare (EW) receiver applications rather than for radar—include instantaneous frequency measurement (IFM) receivers and channelized receivers.

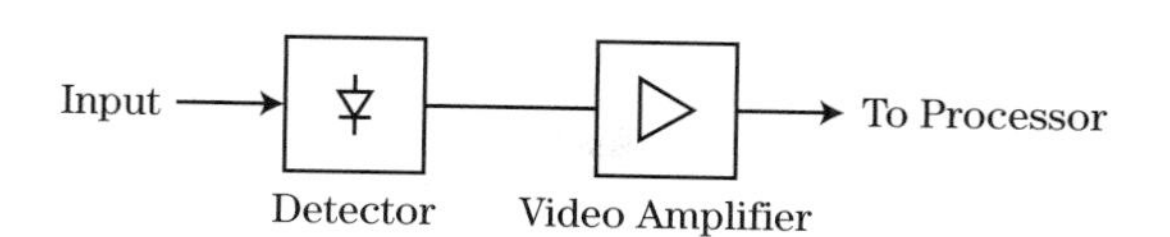

FIGURE 11-3 ■ Crystal video receiver.

11.2.1 Crystal Video Receiver

The crystal video receiver, shown in the block diagram of Figure 11-3, is inherently the simplest type of receiver configuration. The input signal to the receiver is coupled to a detector to convert the radiofrequency (RF) signal directly to video. The video output from the detector is then amplified before providing the output signal to the processor. One disadvantage of this type of receiver is that its sensitivity is 30 dB to 40 dB less than that of a typical superheterodyne receiver because the detector also processes the broadband noise at the detector input. The other disadvantage with this type of receiver is that all the amplification, typically 110 dB or more, is performed by the video amplifier. As a result, the received pulse shape is normally distorted. Because of these limitations, the use of this type of receiver is generally limited to short-range systems. Automotive collision avoidance radar might be a possible application for this type of receiver.

An alternate form of the crystal video receiver, sometimes called a tuned radio frequency (TRF) receiver, is shown in Figure 11-4. The RF input is amplified before detection. The sensitivity and associated noise figure of the receiver is improved by the selectivity and the gain of the RF amplifier, thus resulting in a higher signal to noise at the detector output. Another advantage of this configuration is that the video amplifier gain required is reduced because of the gain of the RF amplifier. Disadvantages of this type of receiver include the added cost of the RF amplifier and the considerably reduced sensitivity compared to that achievable with a superheterodyne receiver.

11.2.2 Superregenerative Receivers

Superregenerative receivers use a principle of positive feedback to cause them to oscillate periodically at the desired RF [1]. The self-quenching oscillator consists of a single tube or transistor circuit and can be used both as a transmit source as well as a receiver. The main feature of this type of receiver is the extremely high gain achieved by the single stage, thus making it attractive for applications requiring simple implementation. The drawbacks of this type of receiver include poor sensitivity compared with a superheterodyne receiver as well as inferior selectivity and gain stability, lack of frequency stability since it is not phase locked, and inherent reradiation. One of the main problems with this type of receiver is that the output has high noise content due to the regeneration noise. The noise is predominantly at the superregeneration oscillation frequency. Since the regeneration frequency is higher than the received bandwidth, the resulting bandwidth is wider due to sum and difference mixing between the input frequency and the superregeneration frequency. Some beacon implementations use this architecture, and it can also be employed for noncritical low-cost applications.

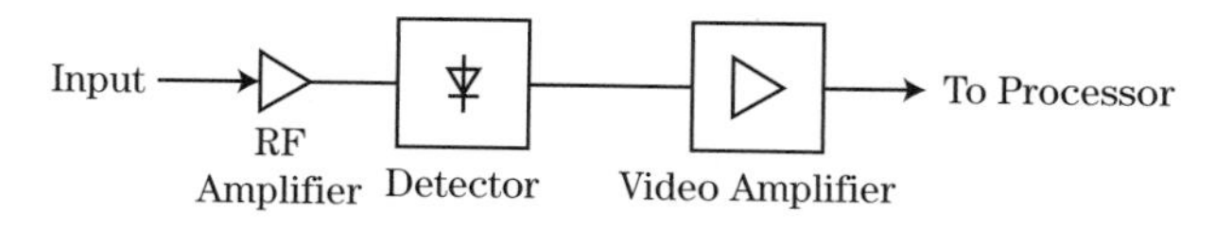

FIGURE 11-4 ■ Modified crystal video receiver.

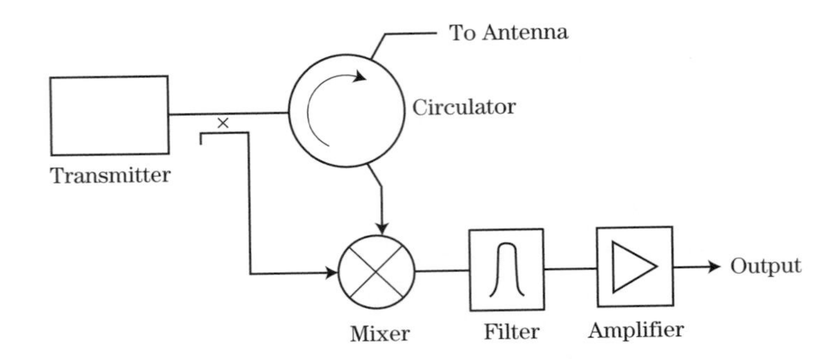

FIGURE 11-5 ■ Homodyne receiver.

11.2.3 Homodyne Receivers

An example of a homodyne receiver is shown in the block diagram of Figure 11-5. For this type of receiver, a portion of the transmit signal is coupled from the transmitter and is used as the local oscillator input to the receiver mixer. A circulator, which is a ferromagnetic device, couples the transmit signal to the antenna while isolating the transmit signal from the receiver. The received signal at the antenna is coupled to the receiver while providing isolation to the transmit port of the circulator. It thus functions as a duplexer for this radar. For this type of receiver to work, the transmitter must still be transmitting when the signals from the target are received. This type of receiver is rather simple to implement in that no local oscillator is required yet is still capable of providing coherent receive signals to the radar processor. Examples of the types of radars that can use a homodyne receiver are low-cost continuous wave (CW) radars (e.g., police speed radars) and radars using frequency modulated CW (FMCW) waveforms. Because of the long waveforms, this type of radar often uses relatively low-power solid-state transmitters, and the sensitivity is considerably improved compared with that of a crystal video receiver.

11.2.4 Superheterodyne Receivers

The superheterodyne receiver, shown in the block diagram of Figure 11-6, was originally developed as a radio receiver and used a tunable local oscillator signal to mix the signal down to a common intermediate frequency. A heterodyne receiver is essentially the same as a superheterodyne receiver, except that the LO frequency is fixed. Most current radar systems have the capability of changing the RF frequency to enable frequency diversity or specialized processing waveforms, and thus the LO frequency is then tuned to follow the RF frequency. These same features are applicable to radar receivers, except that in most cases the local oscillator is at a fixed frequency, offset from the transmit frequency

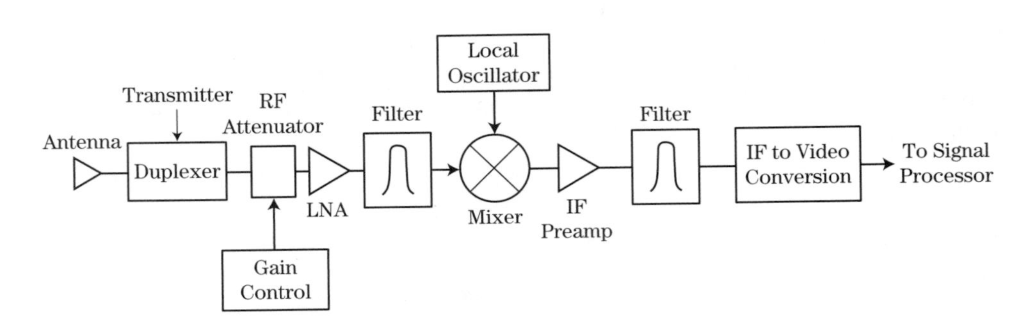

FIGURE 11-6 ■ Generic superheterodyne receiver configuration.

by a fixed IF. The receive signal from the antenna is coupled through the duplexer to the mixer. The receive signal may be fed through an attenuator to provide increased dynamic range and to protect the receiver from saturation from close-in targets. This is called Sensitivity Time Control (STC). In many current applications the receive RF signals are amplified by an LNA prior to the mixer. Recent developments of LNAs with low-noise figures enable improved sensitivity and, with the help of a band-pass filter following the LNA, can overcome the effects of the double sideband noise figure of mixers. For radar receivers requiring extreme sensitivity, cryogenic receiver front ends are used in place of the LNA. The mixer downconverts the input RF signal, using the local oscillator signal, to provide the received IF signal. The mixer output signal is commonly amplified, and in many cases the amplifier is included as part of the mixer assembly. Band-pass filtering is then used to limit the IF signal to that of the desired receive signal while providing rejection of unwanted mixer products and out-of-band signals. Following preamplification and filtering, the received signals are generally downconverted from IF to video and are detected. Some applications, such as dual polarized or monopulse radars, use multiple receiver channels, all connected to a common LO frequency.

11.2.5 Digital Receivers

The advances in high-speed A/D converters have led to their increased use in radar receiver applications. Fully digital receivers that directly digitize the RF signal are rarely used for current radar applications, although an example of such a radar receiver is included in the discussion of spurious-free dynamic range [2]. However, most modern radar receivers include A/D converters somewhere in the receiver chain to provide digital signals to the radar processor. Digitization at IF is increasingly being used for modern radar receivers, especially for coherent systems. Digital processing is also being employed to supplant analog functions in the receiver, for example, replacing analog filter functions with finite impulse response (FIR) digital filters. Software control of receiver functions via the use of embedded field programmable gate arrays (FPGAs) allow the receiver the flexibility to adapt to different radar modes of operation.

11.2.6 Instantaneous Frequency Measurement Receivers

IFM receivers can be used to determine the frequency of an incoming radar signal. Pace [3] describes the implementation of an IFM receiver, in which the incoming signal is split into N channels, with each of the N channels providing a direct path and a delayed (in time) path. The frequency of the signal is determined by measuring the phase between the direct and delayed paths. The implementation is wideband and can provide near instantaneous measurement of the receive signal frequency. However, this implementation works only if a single RF frequency is present, and it yields flawed measurements if two or more RF signals are simultaneously present.

11.2.7 Channelized Receivers

A channelized receiver separates the incoming signals into a multitude of superheterodyne receiver channels, channelized in frequency or time [4]. A frequency channelized receiver uses multiple receiver channels, each provided with separate LO frequencies. Note that many radars have multiple receiver channels; however, the term *channelized receivers* implies bands of frequency selective filtering with parallel receiver chains. The band-pass filters on the RF input signals would be selected to limit the incoming bandwidth of the individual receive channel, but, depending on the number of channels, the overall receiver bandwidth could be extremely large. The advantage of this type of receiver is that it can cover an extremely wide bandwidth while providing good receiver sensitivity. It also has the capability of separating out the returns from different radars. The disadvantage of this implementation is the increased hardware required.

11.3 | MAJOR RECEIVER FUNCTIONS

11.3.1 Receiver Protection

The sensitive components of a receiver must be protected from high-power RF signals that can leak in from the radar transmitter or come from high-power interfering signals in the environment. During transmit time, the transmit energy must be isolated from the receiver, since the relatively high transmit power could either burn out the sensitive receive components or cause the receiver to saturate. The recovery time from receiver saturation could prevent the radar from detecting close-in targets to the radar. For a high-power system, the duplexer may be a waveguide device with a radioactive gas-discharge T/R switch activated by the transmit pulse to short out the signal to the receiver and reflect the transmit energy to the antenna. For lower-power radar systems, the duplexer may be a circulator that directs the transmit energy to the antenna port while the signal received by the antenna is coupled into the receiver components. In certain cases, the isolation provided by the circulator itself is insufficient to prevent receiver saturation, so a receiver protector switch may be connected in series with the circulator to short out the input to the receiver during transmit time.

Typical radar duplexers employ one or a combination of the following: pre-T/R tubes; T/R tubes; ferrite limiters; and diode limiters. Pre-T/R tubes and T/R tubes have radioactive material to enable fast operation at the onset of the transmitter pulse, but they also have limited life. Ferrite limiters have moderate power-handling capability, fast turn-off, and inherently long life. Diode limiters provide moderate power-handling capability and fast turn-off. Multipactors have high power-handling capability and fast turn-off. While an all solid-state approach to receiver protection is being sought, it currently has limitations, and catastrophic failure of a solid-state protector could result in damage to the receiver.

At the present time, the most economical approach for high- and moderate-power radar systems is a combination of pre-T/R, T/R switch, and diode limiter [5]. For moderate- or low-power radars, a ferrite circulator can be used as receiver protector. Generally speaking, a four-port circulator is normally used to prevent transmit power from being reflected into the transmitter. This can be done with either two three-port ferrite circulators or a differential phase shift circulator, which is inherently a four-port device. Diode limiters or PIN switches must often be used in addition to circulators in the receiver to provide additional isolation during transmit time.

11.3.2 RF Preselection

RF preselection is required in most receiver designs to minimize or eliminate interfering signals. In a benign background, the most important reason for RF preselection is to reduce the exposure of the receiver to spurious signals. However, most radar returns are competing with the radar signals reflected not only from targets of interest but also from

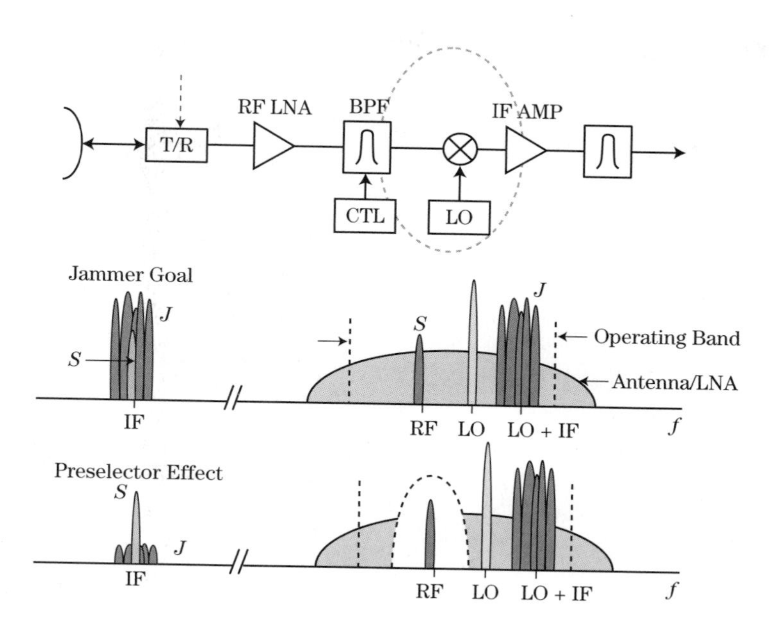

FIGURE 11-7 ■ Effects of preselection on rejection of jammer signals.

other frequency sources. These other frequency sources may be unintentional returns from other radars or communication sources or from jammers deliberately trying to jam the radar to avoid target detections. For barrage noise-type jammers, the jammer signal is spread in frequency so that only a portion of the jammer signal is in the bandwidth of the desired radar signal.

As shown in Figure 11-7, the desired radar signal (S) and the jammer signals (J) both lie within the receiver passband. So when mixed with the LO, both signals end up at the IF of the receiver due to mixer intermodulation products (see Section 11.3.3 for a discussion of mixer intermodulation components); thus, the J signal falls on top of the S signal at IF. By installing a band-pass filter on the RF signal prior to the mixer, the mixer input sees only the input signals that are present in the radar's receiver band. The RF low-noise amplifier prior to the IF improves the sensitivity of the receiver and overcomes the losses in the filter prior to the mixer. By mixing the RF signal with the LO frequency and filtering the mixer IF output to the frequency bandwidth of the transmit pulse, a substantial portion of the jammer returns not within the desired signal bandwidth are rejected by the filter. This has the effect of pulling the target returns not coincident with the desired signal bandwidth out of the jammer background. Not only does the LNA and filter combination help remove broadband jammer signals and other unintentional interference, but it also eliminates the noise and jammer signals at the image frequency from the mixer input. Normally, the LNA is located as closely as possible to the antenna (following the duplexer) so that it helps to considerably reduce additions to the noise figure due to transmission line, filter, switch, and attenuator losses prior to the mixer.

11.3.3 Frequency Downconversion and Mixers

Most radars must downconvert the RF signals to a lower intermediate frequency for detection and processing. Figure 11-8 shows an example of a single downconversion from a 10 GHz RF to a 200 MHz IF. The final IF frequency must be low enough for the input frequency to the logarithmic amplifier for downconversion to video or to enable conversion

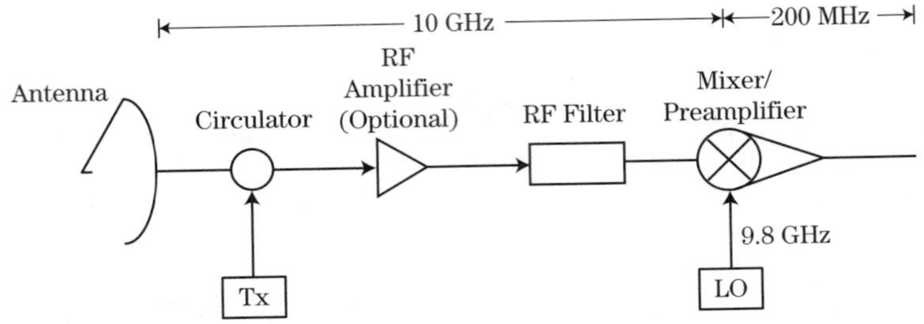

FIGURE 11-8 ■ Single downconversion process.

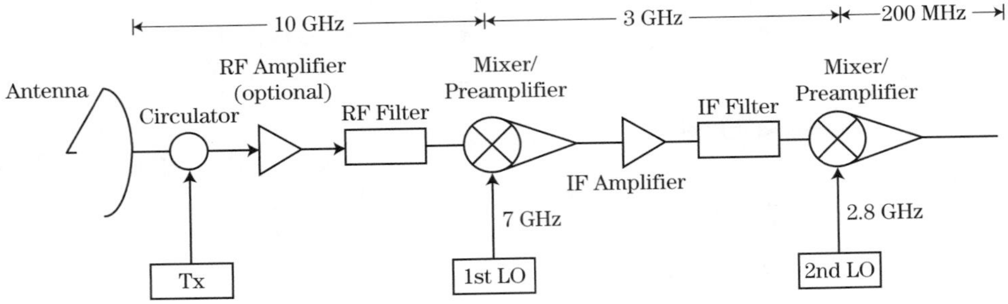

FIGURE 11-9 ■ Double downconversion process.

of the IF signal to digital format. For a radar with a single downconversion to IF, the LO frequency has to be fairly close to that of the RF frequency. Mixer intermodulation components are fairly close to the desired signal, which makes possible introducing spurious frequency components into the IF passband. It is generally difficult to incorporate an RF filter with a narrow enough bandwidth at RF to filter out the unwanted spurious frequencies due to the small offset of the LO from the transmit frequency. In addition, the image frequency component can also be in the receiver band, and, while it can be suppressed by a single-sided mixer, its level may still be in excess of that required.

Most microwave and millimeter wave radar receivers use a double downconversion process. Figure 11-9 shows this process, from a 10 GHz RF to a 3 GHz first IF and then from the 3 GHz first IF down to a 200 MHz second IF. With the first downconversion, the mixer intermodulation components are further separated from the RF, making it easier to filter them out at RF. For both, the mixer is at the heart of the downconversion process.

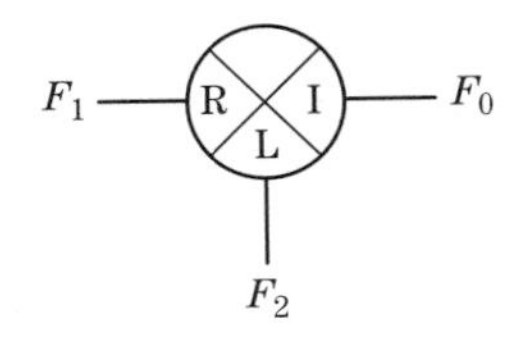

FIGURE 11-10 ■ Mixer model.

The mixer, shown in the model of Figure 11-10, is a nonlinear device used in receivers to convert signals at one frequency to a second frequency. In a receiver the mixer is normally used to mix the RF signal, f_1, with the LO signal, f_2, to a lower IF signal, f_0. The nonlinear mixer products result not only in the desired difference frequency but also in products of the harmonics of the frequencies mixing with each other resulting in undesired mixer outputs. These undesired mixer outputs are spurious signals, commonly referred to as *spurs*. These spurs have the disadvantage that they effectively reduce the spurious-free dynamic range of the receiver.

The outputs from the nonlinear mixing process can be described by the following equation [6]:

$$I_0 = F(V) = a_0 + a_1 V + a_2 V^2 + a_3 V^3 + \cdots + a_n V^n + \cdots \tag{11.1}$$

where I is the device current, and V is the voltage. For a mixer the voltage, V, is a combination of the RF voltage, $V_1 \sin \omega_1 t$, and the LO voltage, $V_2 \sin \omega_2 t$. Thus, V is of the form

$$V(t) = V_1 \sin \omega_1 t + V_2 \sin \omega_2 t \tag{11.2}$$

where ω_1 is the RF angular frequency, and ω_2 is the LO angular frequency. This results in an infinite Taylor power series:

$$\begin{aligned} I_0 = a_0 &+ a_1(V_1 \sin \omega_1 t + V_2 \sin \omega_2 t) + a_2(V_1 \sin \omega_1 t + V_2 \sin \omega_2 t)^2 \\ &+ a_3(V_1 \sin \omega_1 t + V_2 \sin \omega_2 t)^3 + a_n(V_1 \sin \omega_1 t + V_2 \sin \omega_2 t)^n + \cdots \end{aligned} \tag{11.3}$$

The desired mixing product (normally $f_1 - f_2$) results from the second-order term, and the remainder of the mixing products results in undesired spurious signals. The derivation is intended to reflect the current response of the sum of two voltages across the nonlinearity of the diode $V - F$ characteristic. Most modern radars use bridge rectifiers, where one input is of sufficient level to cause the bridge to act as double-pole, double-throw to the other signal changing the mixing function. This derivation illustrates the nonlinear mixing recognizing that in practice a different function would usually be used. The desired signal can be either $(f_1 + f_2)$ or $(f_1 - f_2)$, but for downconversion in a receiver the difference component $(f_1 - f_2)$ is generally preferred. Intermodulation products are usually considered to be the resulting products from two or more signals at the input that are each converted by the LO rather that higher-ordered products of one signal and the LO [7]. Thus, given two input frequencies f_1 and f_2 into the input of the mixer, with f_L being the local oscillator frequency, the spurs of f_1 would be of the form $mf_1 + nf_2$ where sign(m) = −sign(n) for in- or near-band products.

11.3.4 Selection of LO and IF Frequencies

Selection of LO and IF frequencies are driven by the need to provide the required IF and video bandwidths and to minimize the effect of spurious products due to the down-conversion process. Software analysis tools are available on the Internet to enable rapid determination of spurious (intermodulation) products and thus to facilitate selection of the LO and IF frequencies with the lowest spurious residuals. The levels of spurious components depend on the specific mixers and should be available from respective mixer manufacturers. In addition to avoiding or minimizing spurs, careful attention should be given to avoiding the presence of harmonics or LO frequencies in the receiver passband.

Mixer specifications normally list the double sideband noise figure for mixers. For an RF input signal f_{RF1}, the mixer output of interest is the intermediate frequency f_{IF} $(= f_{RF1} - f_{LO})$, assuming that the RF frequency is above the LO. The noise output from the mixer will contain contributions from both the RF frequencies $f_{RF1} = f_{LO} + f_{IF}$ and $f_{RF2} = f_{LO} - f_{IF}$. In addition, unwanted signals at f_{RF2} would also be downconverted to IF unless they were filtered out prior to the mixer. Low-noise amplifiers are currently available with low noise figures and are increasingly being used to amplify the signal prior to the mixer. If the LNA output is filtered to reject the unwanted f_{RF2} noise signal component, the noise out of the mixer is due only to that at f_{RF1}. If the noise figure of the LNA is equivalent to that of the mixer double sideband noise figure and if the gain of the LNA is sufficient, the effective signal-to-noise ratio out of the mixer is increased by up to 3 dB. As a bonus, the unwanted signals at f_{RF2} are also filtered out prior to the mixer, and the filter losses do not degrade the noise figure, assuming sufficient LNA gain. Generally speaking, the LNA is situated close to the antenna to minimize transmission line (waveguide or coaxial) losses and further improves the receiver noise figure since it cancels the effect of post-LNA losses into the receiver, especially if the receiver is located some distance from the antenna. Modern radar receivers generally incorporate double or triple balance stripline mixer modules to provide superior LO rejection and to decrease the effect of spurs.

11.4 DEMODULATION

Conversion of the signal to video (baseband) and detection are normally required prior to radar signals in the signal processor being used for target acquisition, tracking, and radar displays. This is true for both coherent and noncoherent systems, although some coherent receivers directly sample the IF signal and perform the required detection and processing in the digital signal processor.

11.4.1 Noncoherent Demodulation

Noncoherent detectors provide the conversion of IF signals to baseband to provide the video signal format required for radar displays and subsequent processing. This video signal is the amplitude-only envelope of the IF signal. It is called *noncoherent* because it does not preserve the phase information of the in-phase (I) signal. Noncoherent detection is used for the simplest radar configurations because it does not require complex coherent transmitter and receiver implementations.

The diode detector, such as that shown in Figure 11-11, provides the simplest form of detection. The diode is accompanied by a low-pass filter, normally consisting of a resistor–capacitor combination, to remove the IF signal component. This type of detector is known as an envelope detector or peak detector. Because of the nonlinear nature of diodes, care must be exercised in the selection of the resistor–capacitor combination to enable the video output to effectively follow the envelope of the IF signal. Another form of video detection is the square law detector, shown in Figure 11-12. The signal is split and mixed with itself to form the square law video output, and the low-pass filter is used to remove the IF signal components from the mixer outputs. This type of video detector provides a true square law relationship between the IF signal input and the video output. Figure 11-12 is just one example of a low-pass filter implementation. Selection of the

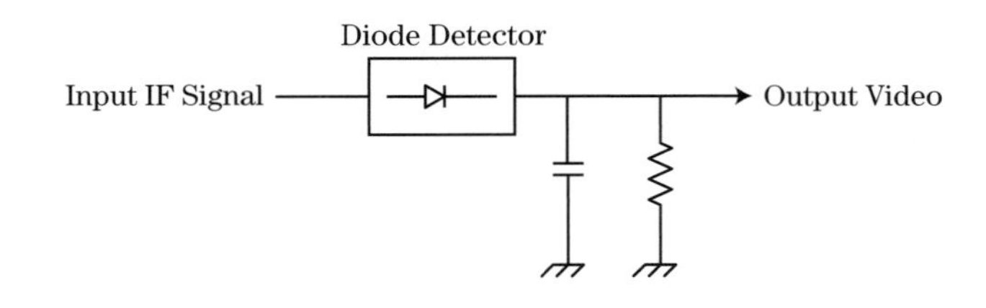

FIGURE 11-11 ■ Basic diode detector.

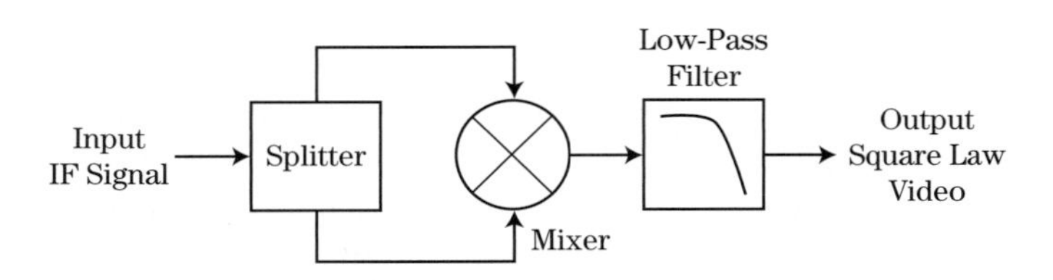

FIGURE 11-12 ■ Square law detector.

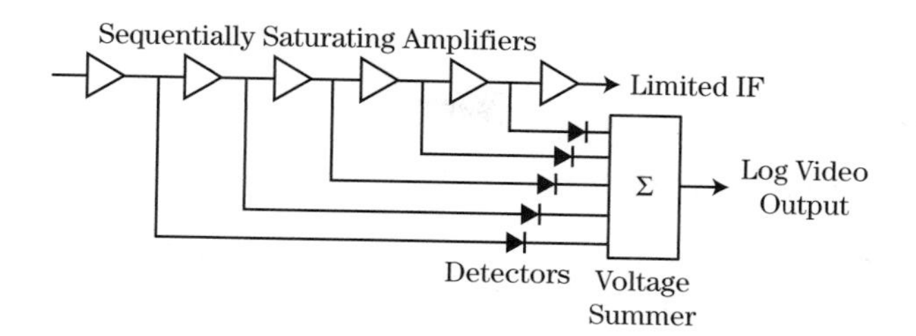

FIGURE 11-13 ■ Log amplifier block diagram.

low-pass filter resistor–capacitor combination is based on the overall circuit impedance, including the diode impedance and the characteristic impedance of the preceding circuit, the highest video frequency required, and the IF. Generally, the IF is much higher than that of the highest video frequency, so the selection is usually not critical. Often, the diode detector is followed by a video amplifier or logarithmic amplifier whose bandwidth effectively acts as a low-pass filter.

Logarithmic amplifiers are commonly used with noncoherent radars to provide amplification with wide dynamic range and subsequent video detection. Log amplifiers, shown in the block diagram of Figure 11-13, normally consist of series of limiting amplifiers, which form the logarithmic signal in the amplifier chain. The detected outputs from the limiters are summed to form the output signal, whose voltage output is proportional to the logarithm of the input IF signal level. Log amplifiers typically maintain the log relationship within approximately ±1 dB over a 70–80 dB dynamic range, as shown in the example of Figure 11-14. This makes them ideally suited for applications such as plan position indicators (PPIs) radar displays, which require good sensitivity to resolve weak targets while not overloading on stronger targets.

Tangential signal sensitivity (TSS) is one measure of sensitivity of a receiver. The received power observed as TSS is often considered the receiver's minimum detectable signal prior to processing. To measure TSS, a pulse signal is inserted into the receiver while observing the detected signal output. The input signal level is increased until the base of the detected signal-plus-noise waveform is at the peak of the detected noise-only signal. This measure is somewhat subjective, since it depends on the observer's estimate of the peak of a noise-like signal, and may vary several dB depending on the particular

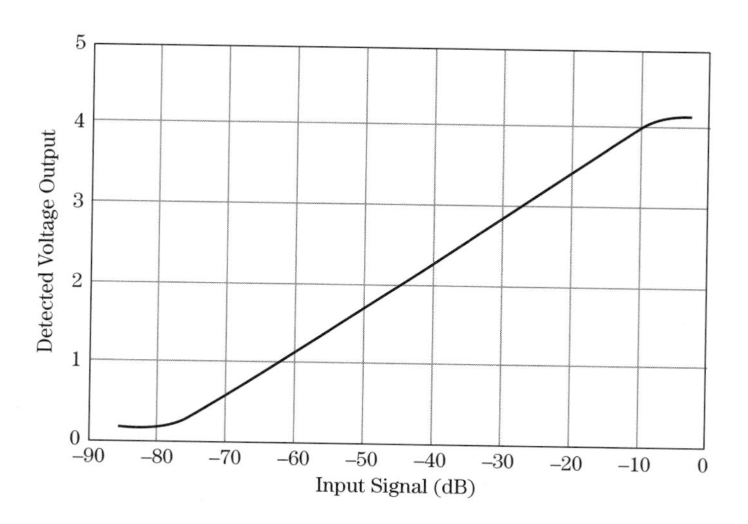

FIGURE 11-14 ■ Typical log amplifier output characteristic.

observer. Generally speaking, this measure of sensitivity is primarily used for noncoherent receiver systems.

11.4.2 Coherent Demodulation

In addition to providing amplitude information on the detected target signal, coherent detectors also provide information on the phase of the received target signal relative to the transmitter phase. The preservation of the phase of the received signal relative to that of the transmitter enables the radar processor to perform tasks such as moving target indication (MTI), Doppler processing (determination of target velocity relative to the radar), synthetic aperture radar (SAR) imaging and space-time adaptive processing (STAP). To enable coherent processing, the phase of the LO signals must be locked to that of the transmit signal. This is generally accomplished by using a highly stable frequency source to determine the frequencies of both the transmitter and the LO frequencies. Coherent-on-receive radars employ a noncoherent transmitter but sample the transmit frequency during the transmit time and use it to lock the phase of the receive signal to that of the transmit signal. In general, the performance of coherent-on-receive radars is somewhat inferior to that of fully coherent radars and is generally not adequate for the needs of modern coherent radar processing

The phase of the receive signal can be obtained in coherent radars by converting the IF signal to in-phase and quadrature (Q) phase video (Figure 11-15). The in-phase and quadrature phase signals can be represented as

$$I = A\cos\theta \tag{11.4}$$

and

$$Q = A\sin\theta \tag{11.5}$$

where A is the amplitude of the signal, and θ is the phase angle between the transmit and receive signal phases.

The amplitude and phase of the signal can be obtained from the I and Q signals, since

$$A = \sqrt{I^2 + Q^2} \tag{11.6}$$

and

$$\theta = \tan^{-1}\left(\frac{Q}{I}\right) \tag{11.7}$$

where the four-quadrant arctangent function is used.

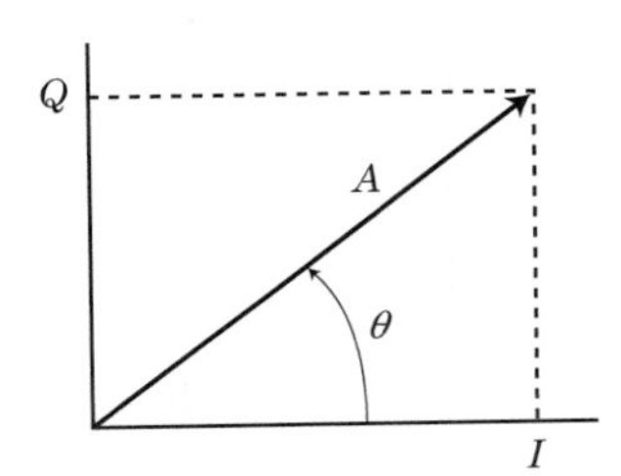

FIGURE 11-15 ■ Analog I and Q detection.

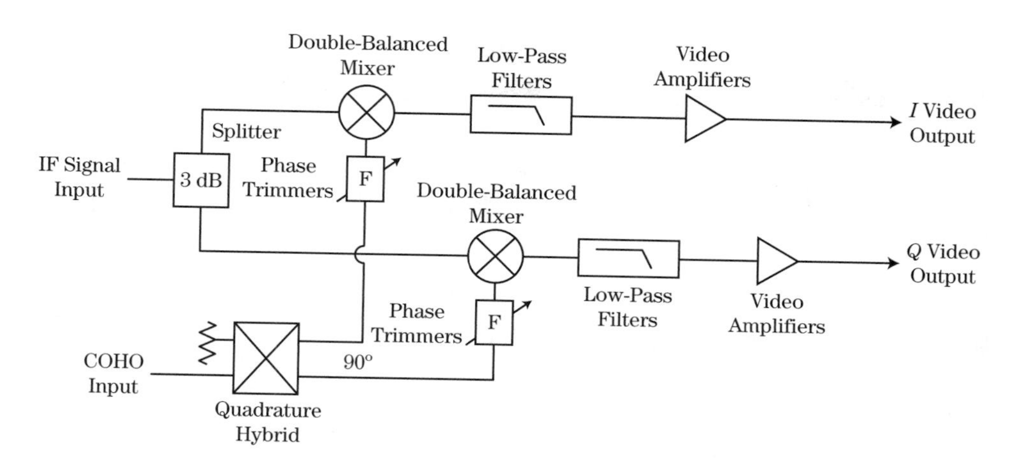

FIGURE 11-16 ■ Circuit for analog coherent I and Q processing.

11.4.3 Analog Coherent Detection Implementation and Mismatch Errors

A typical circuit for analog conversion of the IF signal to coherent I and Q signals is shown in Figure 11-16. The IF signal is split equally and is input to two identical double balanced mixers, while the local oscillator signal is normally provided to a quadrature hybrid that provides equal amplitude signals but with a phase difference of 90° to each other. The quadrature hybrid signals are then coupled to the mixers and the mixer outputs are then in phase quadrature to each other. The mixer outputs are low-pass filtered to pass the video components, which are normally amplified prior to further processing.

In performing the analog I and Q conversion, it is vital that the amplitude balance and the 90° phase difference is maintained by the circuit. Gain and phase mismatches in analog coherent processors have the potential to introduce false Doppler images (ghosts). These ghost images are apparent as false targets, showing up at a Doppler frequency that is the negative of the true target's Doppler frequency. The effect of amplitude and phase mismatches in generating undesired ghost images can be determined from Figure 11-17. In a radar with analog coherent detection, such as that shown in Figure 11-16, careful

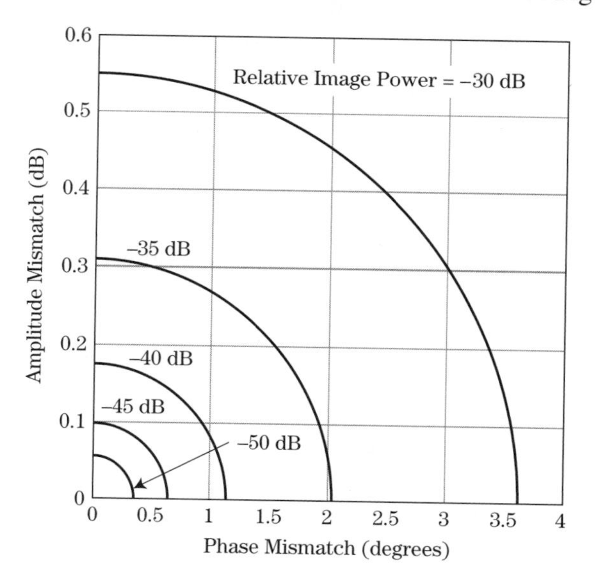

FIGURE 11-17 ■ Effect of gain and phase error mismatches.

calibration of the phase and amplitude trimmers is required to reduce the effect of phase and amplitude errors.

Direct digital coherent detection is increasingly being used to eliminate not only the physical components associated with analog I and Q detection but, more importantly, also the phase and amplitude errors and associated calibration requirements associated with analog coherent detection. A discussion of direct digital coherent detection is given in Section 11.7.2.

11.5 RECEIVER NOISE POWER

Thermal noise is inherent in receivers and limits the ability to amplify the RF spectrum. The inherent noise in receiver components is Gaussian white noise and limits the ability of a radar to detect low-level targets. The ability of a radar to detect targets of a particular radar cross section (RCS, σ) is a function of the transmitter power, the antenna, receiver parameters, and signal processing. The radar equation was developed for a point target in Chapter 2:

$$SNR = \frac{P_t G A_e \sigma}{(4\pi)^2 k T_0 B_n F R^4} \tag{11.8}$$

A modified form of the radar equation substituting for the antenna aperture area, $A_e = G/4\pi$, and including losses, L_s, results in a modified form of the radar equation:

$$SNR = \frac{P_t G^2 \sigma}{(4\pi)^3 k T_0 B_n F L_s R^4} \tag{11.9}$$

where T_0 is the temperature of the receiver (290 K nominally used), B_n is the receiver bandwidth, F is the receiver noise figure, L_s is the system losses, k is Boltzmann's constant (1.38×11^{-23} Joules/K), and R is range (m).

The receiver bandwidth, B_n, is generally taken as the final IF bandwidth of the receiver since this is normally the narrowest bandwidth in the receiver and therefore sets its noise bandwidth. In some cases video filtering is used either instead of or in addition to IF filtering. Video filtering is not as effective as IF filtering in reducing the noise bandwidth due to the double-sided nature of the filtering, so care must be exercised in evaluating its effect on the receiver noise bandwidth.

The system losses, L_s, associated with the radar equation are transmitter, receiver, and signal-processing losses. The losses associated with the receiver, L_r, are losses up to the receiver input and are normally the transmission line losses from the antenna to the receiver.

The *noise figure*, F_n, for the n-th receiver stage is defined as

$$F_n = \frac{S_{in}/N_{in}}{S_{out}/N_{out}} = \frac{1}{G_n} \frac{N_{out}}{N_{in}} \tag{11.10}$$

where S is the signal power, N is the noise power, and G_n is the gain of the n-th stage. The overall noise figure for a receiver is then

$$F = F_1 + \frac{F_2}{G_1} + \frac{F_3}{G_1 G_2} + \cdots \tag{11.11}$$

The noise figure of the first stage is determined by the input signal level to the receiver and is primarily dominated by the noise of the LNA or mixer (if an LNA is not present). However, attenuation of the signal by components prior to the LNA or mixer reduces the signal into that stage, thus increasing the noise figure F_1. Therefore, the noise figure F_1 is

$$F_1 = L_c + F_{s_1} \tag{11.12}$$

where F_{s_1} is the noise figure of the first stage, excluding components prior to the LNA or mixer, and L_c is the sum of the component attenuations prior to the LNA or mixer.

For a mixer, the noise figure stated in the mixer specifications is normally the double-sided noise figure, assuming both the signal and noise are present in both RF for the upper sideband ($f_{RF1} = f_{LO} + f_{IF}$) and RF for the lower sideband ($f_{RF2} = f_{LO} - f_{IF}$). However, for a superheterodyne receiver, this noise figure must be increased by 3 dB, since the signal is present only in one of these RF sidebands whereas the noise is present in the RF for both sidebands. Also, in most cases, the mixer is immediately followed by a mixer preamplifier (and in many cases in the same package) so that the overall noise figure is normally the mixer–preamplifier combination.

Following the first stage, the contribution of noise from the second stage is reduced due to the gain of the first stage, and the contribution of noise from the third stage is reduced even further by the combination of the gains from the first and second stages. As a result, the first stage of amplification (or mixing) is generally the predominant contribution to the noise figure. With the development of LNAs at microwave (and even millimeter wave) frequencies, most new radars use LNAs to improve the noise figure of the receiver. However, care must be exercised with component placement before or after the LNA. A band-pass filter is often placed before the LNA to reject out-of-band signals, but as depicted in Figure 11-18 the noise out of the amplifier will still contain components at both f_{RF1} and f_{RF2}. However, if the band-pass filter is placed after the LNA, as shown in Figure 11-19, the noise component out of the LNA at f_{RF2} is rejected prior to the mixer input, thus resulting in about a 3 dB lower noise figure for the receiver.

Instead of using noise figure to define receiver noise, an alternate method of characterizing receiver thermal noise is to use the receiver noise temperature. The noise power at the receiver output, P_n, is then expressed as

$$P_n = kT_s B_n \tag{11.13}$$

where T_s is the system noise temperature. The system noise temperature is given as ([8], pp. 26–31)

$$T_S = T_a + T_{tr}(L_r - 1) + L_r T_0(F - 1) \tag{11.14}$$

where T_a is the effective antenna temperature, T_{tr} is the transmission line thermal temperature (K), T_0 is the receiver thermal temperature (K), and L_r is the receiver transmission line loss factor (signal power at the antenna terminal divided by signal power at the receiver input).

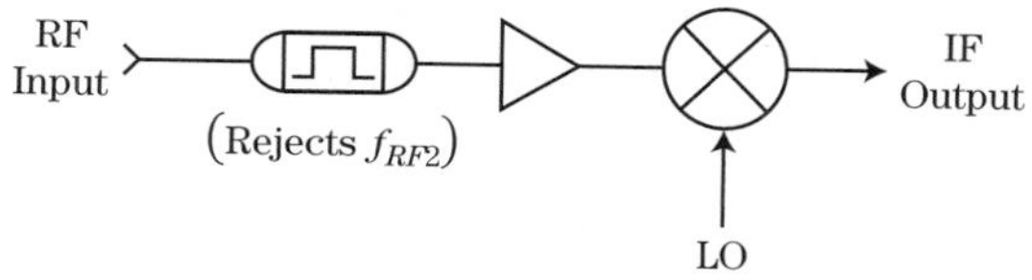

FIGURE 11-18 ■ Filter situated before LNA.

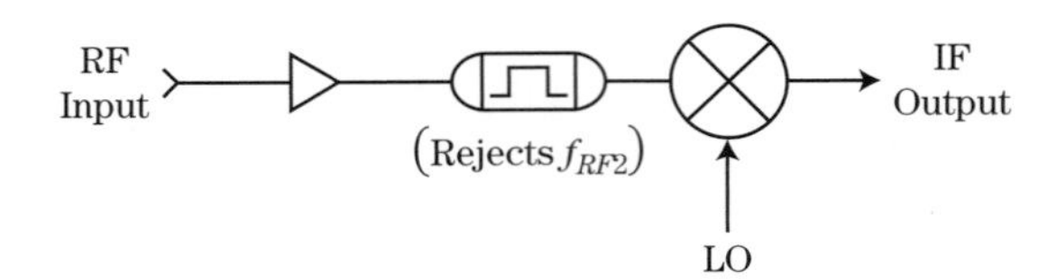

FIGURE 11-19 ■ Filter situated after LNA.

The thermal noise at the receiver input is characterized by a Gaussian probability density function with a white power spectral density. The thermal noise at the receiver output is then characterized by a Gaussian probability density function with a power spectral density of kT_s W/Hz, band limited to B_n Hz.

11.6 RECEIVER DYNAMIC RANGE

Radar returns from targets can vary over a wide dynamic range due to differences in RCS between large and small targets and due to $1/R^4$ decrease in returns as a function of range. The receiver dynamic range is limited by the components in the system, the analog components, as well as the A/D converters. For the analog portions of the receiver, the most notable contributors to limiting the dynamic range include the mixers and the amplifiers. The lower limit on the dynamic range is the noise floor, while the upper end is limited by the saturation of the amplifiers, mixers, or limiters. For linear amplifiers, the upper end of the usable range is generally the 1 dB compression point. That is, in the linear range the amplifier output will increase by a constant dB increment for a given dB increase in the input, thus forming a line of constant slope such as that shown in Figure 11-20. The 1 dB compression point for an amplifier is the point at which the output signal level departs from that of the linear slope by 1 dB.

Unwanted signal components can be generated by nonlinearity in the receiver. Second- and third-order intercept points, as shown in Figure 11-20, are measures of receiver linearity, and these distortions become dominant at the upper end of the receiver curve. The second-order intercept point is due to second-order distortions, while the third-order intercept point is due to third-order distortions. Generally, the third-order distortions are the most dominant, so most amplifiers provide information on the third-order distortions. The distortions that occur due to nonlinearity of the receiver create additional signal

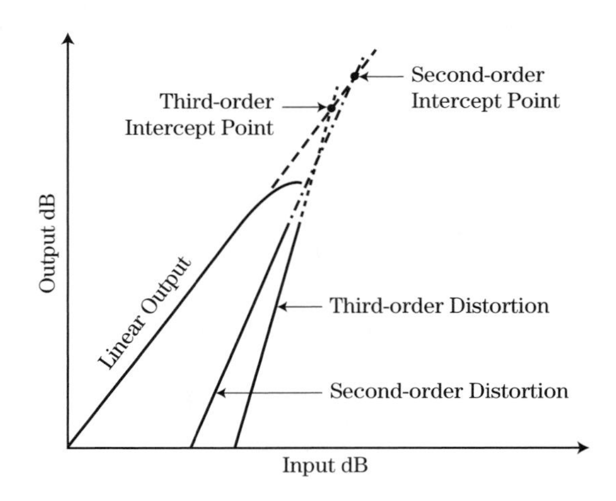

FIGURE 11-20 ■ Receiver distortion versus input power intercept point.

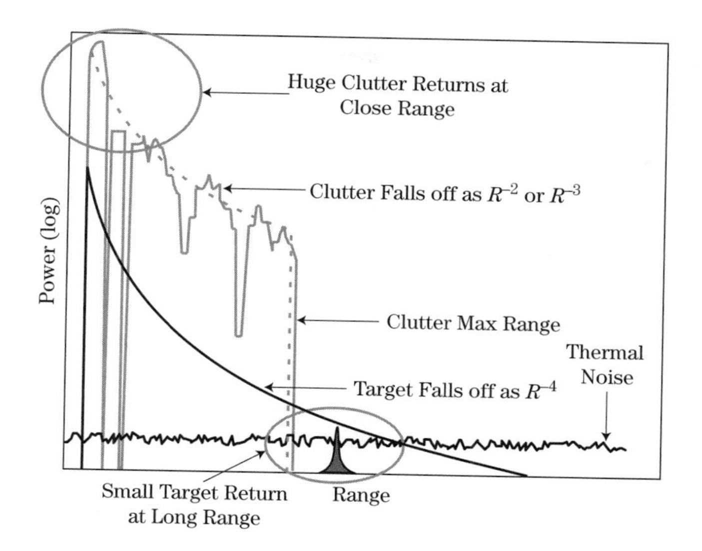

FIGURE 11-21 ■ Return signal amplitudes from targets and clutter.

components for different frequencies, such as $(2f_2 - f_1)$ and $(2f_1 - f_2)$. The intercept points are determined by inserting two different frequency signals of equal input levels that are linearly combined. The distortion products on the receiver output are measured and compared with the original signal inputs. The distortion products are generally measured at relatively low signal levels and are extrapolated to intersect at the extension of the linear gain line.

In designing a radar receiver, each stage of mixing and amplification must be given careful attention regarding degradation of the noise temperature and dynamic range of the receiver. Taylor [9] suggest using a tabular format to enable the designer to keep track of the noise temperature and dynamic range throughout the receiver. Also, the RF design guide by Vizmuller provides valuable insight into the aspects of receiver design [10]. For modern receiver design, software using spreadsheets is currently available to assist in keeping track of the gains and dynamic range throughout the receiver.

Often a radar receiver cannot achieve as large an instantaneous dynamic range needed by the system for all operational situations, and gain control components must be incorporated to achieve system performance goals. Manual gain control, automatic gain control, and sensitivity time control are tools that enable radars to achieve desired overall system dynamic range. The received signals into a radar can have a dynamic range of 120 dB or more. For a constant radar cross section target, the receive signal falls off by R^{-4} as a function of range, so the target returns decrease by 12 dB each time the range to a target increases by a factor of 2. In addition, for surface radars, returns from near-in clutter such as that shown in Figure 11-21 can exceed the target returns at similar range by tens of dB. The returns from surface clutter tend to fall off by only R^{-2} or R^{-3} (see Chapter 5) until the clutter returns become masked due to the radar horizon or other nearer obstructions. As such, the wide range of input signal levels generally exceeds the linear range of most receivers, and even receivers with log amplifiers.

11.6.1 Sensitivity Time Control

One method to increase the dynamic range of receivers is to decrease the sensitivity of the radar for near-range returns. Sensitivity time control (STC) is normally accomplished by

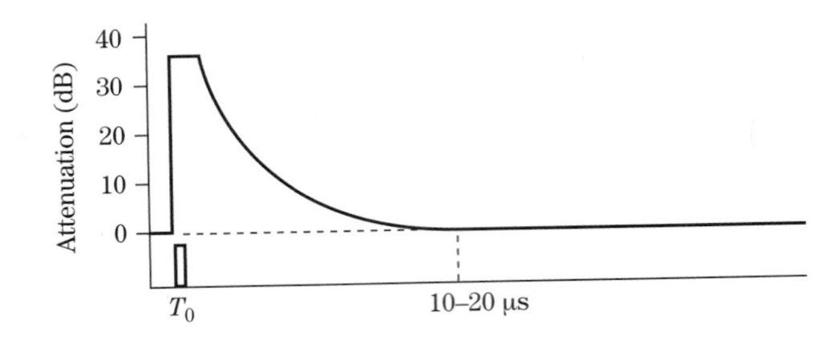

FIGURE 11-22 ■ Sensitivity time control attenuation.

inserting an attenuator between the duplexer and the receiver front end or LNA. Typically around 30 dB, the maximum attenuation (Figure 11-22) is switched in either at or just prior to transmit time. A short time following pulse transmission, the attenuation is decreased as the range increases. Decreasing the attenuation as a factor of R^{-4} would maintain constant amplitude for a given target RCS, whereas decreasing it as a function of R^{-3} would hold it constant for pulse-limited clutter RCS. Reducing the attenuation following an $R^{-3.5}$ rule provides a compromise between both. The STC attenuation present at short ranges helps prevent receiver saturation from strong targets or clutter, and then as the range is increased the attenuation is reduced to enable detection of small RCS targets.

11.6.2 Gain Control

Gain control can be accomplished either with a manual or automated gain control (AGC). Linear amplifiers have limited dynamic range compared with log amplifiers. Gain control is normally required to enable the detection of smaller targets. Typically, gain control in radar is accomplished using gain-controlled IF amplifiers. Some applications control the gain using feedback from the detected video signal.

Manual gain control allows an operator to set the overall gain of the receiver. It is not normally used for adjusting the gain based on target returns since this would require frequent adjustments.

Automated, or automatic, gain control enables the radar to control the gain based on the strength of the target returns. One form of AGC employed when tracking targets samples the detected returns from the radar and continually adjusts the gain to provide an almost constant detected output. This is commonly referred to as *slow AGC* in that the gain adjustment time constant is several radar pulses in duration. This type of gain control is normally used in monopulse radar receivers for tracking single targets. For example, the output from the monopulse sum channel is detected and used to set the gain in both the sum and the angle difference channels. In this manner the amplitude of the angle error signals in the difference channels maintains a constant relationship with the off-boresight angle. Currently, most automated gain control circuits use analog feedback to adjust the gain control voltage into the IF amplifier. Another form of AGC is to have a switchable attenuator on the RF input to prevent receiver saturation and to increase overall receiver dynamic range. This is usually a rather coarse adjustment, for example, to switch in 20 or 30 dB of attenuation on the receiver front end.

Instantaneous AGC can be used, for example, in a monopulse receiver to maintain constant angle indication independent of target amplitude returns. In an instantaneous AGC receiver, the gain is adjusted separately for each target separated in range. With this type of gain control, the detected video signal is used in an analog feedback to the gain control input of the IF amplifier. A delay line is normally incorporated to provide the receiver time to adjust the IF gain for each individual detected target. Care must be taken in this type of receiver to prevent the receiver from adjusting the gain on the basis of noise.

11.6.3 Coupling Issues

A number of radar receivers use multiple receiver channels. One example is dual polarized radars, in which two receiver channels process the orthogonal polarized signals. Inadvertent cross-coupling of the signal from one polarization channel into the orthogonal polarization channel would contaminate the polarization determination process, especially if the detected polarization ratio was being used for target discrimination purposes. The cross-coupling of the receiver channels must be kept below that of the polarization isolation of the antenna to minimize polarization contamination.

Another example is monopulse radar systems, which commonly use two or three channels for sum (Σ) and difference (Δ) angle sensing (see Chapters 9 and 19). Also, certain instrumentation radars include beacon channels, which could add an additional two or three channels to the receiver. With a monopulse receiver, the Σ signal is maximum and the Δ signal is nulled when the tracked target is on boresight. Any cross-coupling of the Σ signal into the Δ channel could result in a boresight bias and thus an angle tracking error. Isolation of at least 60 dB is generally required for monopulse receivers. Finally, channelized receivers, where the receiver channels may be signals at different RF frequencies, also suffer from cross-coupling effects.

Another form of coupling to be avoided is feedback from alternating current (AC) and direct current (DC) power lines. This is normally prevented by decoupling the receiver circuits using a combination of capacitors and resistors connected to the DC power lines feeding the circuits. The introduction of digital circuits into modern receivers makes them susceptible to transient overvoltage spikes. These transient spikes can occur when a length of cable between the power supply and the circuit acts as an inductor resulting in a voltage spike at the circuit due to instantaneous load changes.

11.7 | ANALOG-TO-DIGITAL DATA CONVERSION

Even though the radar signals are analog, most radar systems today use digital signal processing to perform radar detection, tracking, and target display functions. To convert the analog radar signal to a digital representation of that signal, ADCs are used. In the past, the signals were normally downconverted to video prior to analog-to-digital conversion, but many of the newer radar systems directly convert the IF signals to digital, reducing the amount of analog components.

Selection of the ADC sampling speed is determined based on the IF and IF bandwidth as well as on requirements to avoid aliasing of unwanted signal and noise components. Lyons ([11], Chapter 1) shows that, for a sinusoidal input to an ADC, the sampled output data obeys

$$x[n] = \sin(2\pi(f_0 + kf_S)nT_s) \tag{11.15}$$

where f_0 is the baseband signal frequency, f_S is the sampling frequency, n is the sample number, T_s is the sample interval ($1/f_S$), and k is any positive or negative integer.

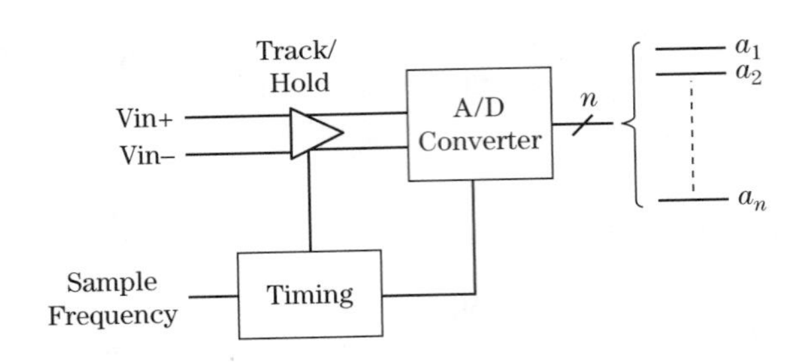

FIGURE 11-23 ■ Typical ADC configuration.

Equation (11.15) indicates that not only the baseband signal frequency f_0 but also frequencies at $f_0 + kf_S$ produce exactly the same sampled output $x[n]$. Thus, the output sampled waveform will be ambiguous unless the input frequencies are low-pass filtered with some cut-off frequency f_{co} and the sampling frequency is greater than $2f_{co}$. More generally, a non-baseband, band-limited IF signal can be sampled uniquely as long as the sampling rate of the IF signal is at least equal to twice the IF bandwidth B_n. Additional discussion of sampling and aliasing is given in Chapter 14.

The direct digital implementation is becoming increasingly popular due to the continuing advances in ADCs with high sampling speeds and increased bit resolutions [3]. Figure 11-23 shows a generic block diagram for a typical high-speed ADC. The IF or video signal, V_{IN}, is normally connected to differential input track/hold circuit (possibly through a buffer amplifier). The track/hold circuit samples the signal and holds the sampled signal constant until the analog-to-digital conversion is performed. The timing circuit is triggered by the sample frequency and samples the V_{IN} signal during the aperture time. The length of the aperture window is limited by the sampling speed and the number of bits in the ADC, and, for high-speed ADCs, the aperture window is generally less than a picosecond. The timing also signals the internal ADC after the aperture window is closed so that it can convert the sampled analog signal to digital format. The output digital signal is then provided to either complementary metal oxide semiconductor (CMOS) or low-voltage differential signaling (LVDS) drivers to provide the b-bit digital signal format.

The relationship between the analog and digital representations of a voltage V_a for an b-bit unipolar ADC is

$$V_a = V_{FS}\left(\sum_{i=1}^{b} a_i 2^{-i}\right) + q_e \tag{11.16}$$

where V_{FS} is the full-scale (saturation) voltage of the ADC, a_i is the value of the i-th bit of the digital representation (0 or 1), and q_e is the quantization error (also known as quantization noise). The digital bits are chosen so that the quantization error is no greater than the ADC quantization error (± ½ LSB [least significant bit]):

$$\left| V_a - V_{FS}\sum_{i=1}^{b} a_i 2^{-i} \right| = |q_e| < \frac{1}{2}LSB \tag{11.17}$$

where LSB is the minimum voltage step size of the ADC and equals $V_{fs}/2^b$.

Digitizing IF signals require ADCs with high sampling speeds, particularly if direct coherent sampling is implemented. In addition, ADCs with a large number of effective bits are desired. Table 11-1 lists the results of a limited survey of currently available ADCs.

TABLE 11-1 ■ Sample of Analog-to-Digital Converters

Part No.	Manufacturer	Bits	Sampling Speed Msamples/sec	SFDR[a] dBc	SNR dB
ADC083000	National Semiconductor	8	3,000	57	45.3
MAX19692	Maxim	12	2,300	68@1.2 GHz	NS
AT84AS004	Atmel	11	2,000	55	51
ADC081500	National Semiconductor	8	1,500	56	47
Model 366	Red Rapids[b]	2/8	1,500	57	47
TS860111G2B	Atmel	11	1,200	63	49[c]
ADC10D1000	National Semiconductor	10	1,000	66	57
MAX5890	Maxim	14	600	84@16 MHz	NS
MAX5888	Maxim	16	500	76@40 MHz	NS
Model 365	Red Rapids[b]	2/14	400	84	70
ADS62P49	Texas Instrument	14	250	85	73
LTC2208	Linear Technology	16	130	83	78
AD9446	Analog Devices	16	110	90	81.6

Note: NS, not specified.
[a]Spurious-free dynamic range.
[b]Dual sampler.
[c]Noise power ratio.

Table 11-1 lists, in addition to sampling speeds, the spurious-free dynamic range (SFDR) and the signal-to-noise ratio (SNR) for the devices when available. Spurs associated with the downconversion process are undesirable in that they could appear as false targets in the receiver, so maintaining a high SFDR is essential to maintaining high sensitivity for low-level targets. While the number of bits is indicative of the resolution of the ADC, the SNR of the device is an important parameter since it indicates the effective number of bits and is generally one to three bits less than that obtained from $20 \log_{10}(2^n)$ (see Chapter 14). As with all of the ADCs listed in the table, the manufacturer's product specification sheets need to be examined in detail to determine the input frequencies for which the SFDR and SNR are listed.

Random jitter in clocks sampling ADC inputs will induce additional noise on the output of the ADC, resulting in additional noise on the ADC output. For high-resolution ADCs very low clock jitter is required to prevent the effects of clock jitter from reducing the output SNR on the ADC output [12,13]. In many cases, the current trend is to sample the analog signal at the IF, and the presence of jitter can induce phase errors in the sampling process as well as affect the overall SNR [14]. This effect is increasingly evident with very high-speed sampling frequencies.

Dynamic range in a digital receiver (prior to processing gain in the signal processor) is limited by the dynamic range of the ADC. Figure 11-24 show the effective dynamic range for an ADC of 12 bits, although currently ADCs of 14 to 16 bits are readily available. Normally, the noise level into the receiver is set to be about 1 to 2 LSB. This is because the signals from small targets at far ranges are often buried in the noise, and unless gain is set so that the noise gets digitized, the follow-on signal processor will not be able to integrate the signal to improve the SNR of target returns. It is also important to keep the maximum received signal from exceeding the full-scale value of the ADC, and normally about 1 dB of headroom is maintained on clutter and targets to prevent ADC saturation effects.

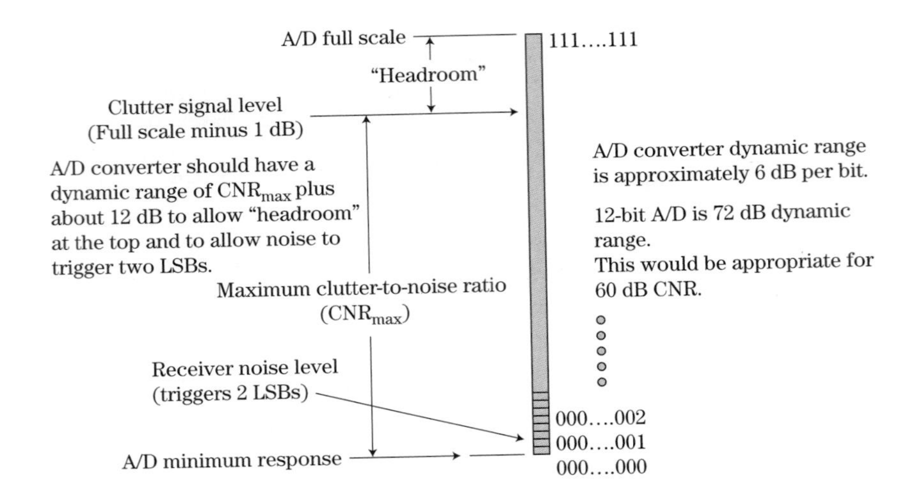

FIGURE 11-24 ■ Dynamic range of a signal-following ADC.

11.7.1 Spurious-Free Dynamic Range

The dynamic range of a receiver is limited not only by the dynamic range of the ADC and associated components but also by the spurious-free dynamic range. Spurs can cause signals in the receiver, which appear as false targets or can mask small targets. Spurs can be generated in the mixing (downconversion) process, the ADC, and other possible nonlinear sources in the receiver. It is necessary to suppress these spurs as much as possible to maintain a useable dynamic range for the receiver. Software is currently available to assist the designer in determining the overall spurious-free dynamic range of the receiver.

Figure 11-25 shows an example implementation of a direct digital receiver where the RF following band-pass filtering and amplification is directly digitized to enable the digital signal processor to form the I/Q signals [15]. Figure 11-26 shows the spurious signals associated with the 2 Gigasample/second ADC multichip module (MCM). This spurious response at twice the test signal frequency limits the dynamic range of the receiver to about 40 dB.

11.7.2 Direct Digital Coherent Detection Implementation

Direct digital in-phase/quadrature (I/Q) sampling is being used in most new radar implementations to obtain coherent video representation of the IF signal. This is primarily enabled because of advances in the sampling speeds of current analog-to-digital converters. The main advantage of direct digital sampling is that it eliminates the errors associated with phase and amplitude offsets associated with analog I/Q downconversion. Figure 11-27 shows a typical implementation of a direct digital sampler [11]. It is worth

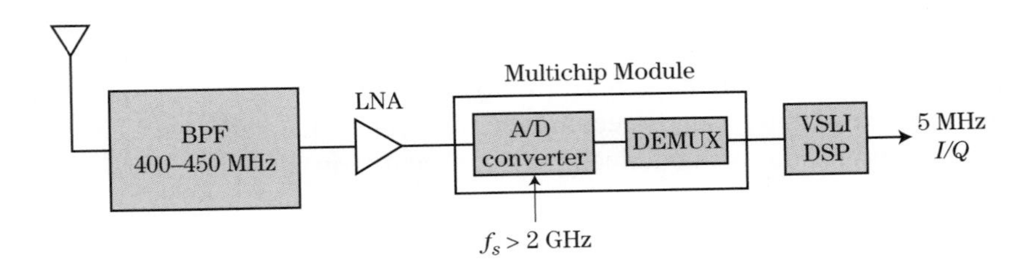

FIGURE 11-25 ■ Block diagram of a direct digital radar receiver. (From Thompson [15]. With permission.)

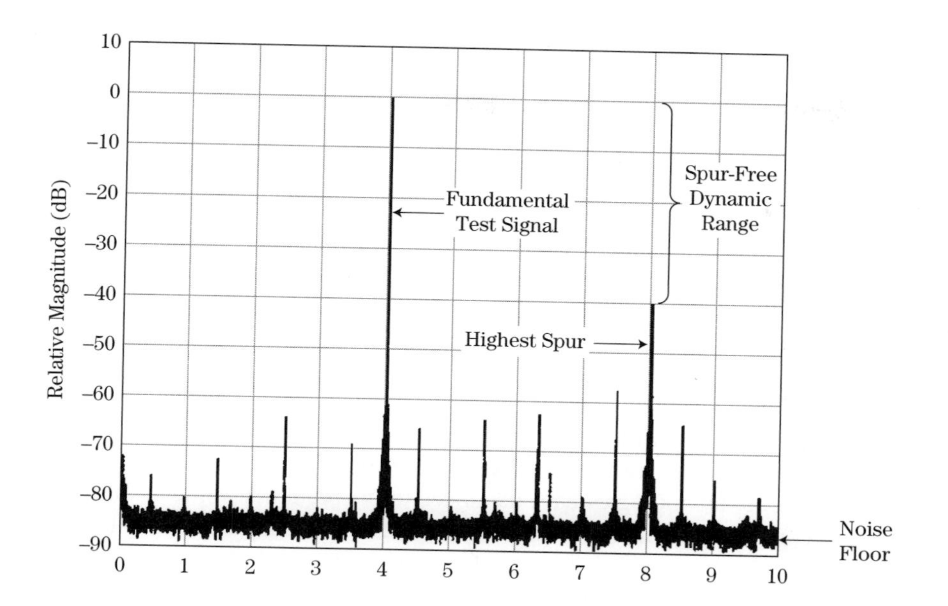

FIGURE 11-26 ■ Spurious-free dynamic range associated with the ADC MCM. (From Thompson [15]. With permission.)

pointing out that the $\cos(2\pi n/4)$ and $\sin(2\pi n/4)$ multipliers in Figure 11-27 are simply sequences of 1's, 0's, and -1's, and the downstream complex processing rate is the IF bandwidth. Also, there are other implementations with 1's, –1's, and 0's multipliers that can reduce the ADC sampling requirement slightly over two times the IF bandwidth.

The sampling frequency f_S needs to be chosen to be four times the IF center frequency, f_0, but in most cases this would require an extremely high sample frequency. Lyons ([11], Chapter 8) points out that since the IF signal is normally band-limited, the sampling can be performed at a lower frequency. For example, suppose the IF center frequency is 60 MHz and the IF bandwidth is $\leq$ 12 MHz; instead of sampling the IF at 240 MHz, the sample frequency could be chosen as 48 MHz. The sampled IF signal replicates in the frequency domain with one of the sets of spectral components centered at 12 MHz. The associated sine and cosine signals shown in Figure 11-27 could then be at a frequency of 12 MHz, which satisfies the requirement that the sampling frequency be four times the frequency of the sine and cosine multipliers. The digital I and Q signals can be derived by low-pass filtering the in-phase ($x_I[n]$) and quadrature phase ($x_Q[n]$) signals.

Urkowitz [16] describes an alternate method for generating the real and imaginary components of a signal using a Hilbert transformer, such as that shown in Figure 11-28. A Hilbert transformer has the property of providing a phase shift of $\pi/2$ (90°) to the digital representation of a signal. The IF signal is sampled, as shown in Figure 11-28, by a high-speed ADC at a sample rate equal to or greater than two times the IF. The direct signal out of the ADC is then usually called the real component, while the Hilbert transformed

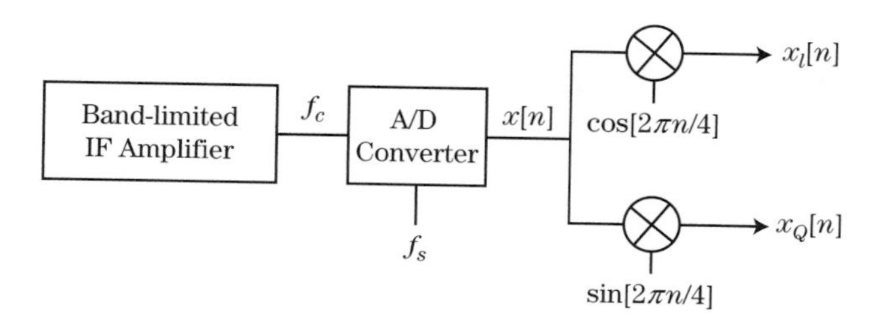

FIGURE 11-27 ■ Direct digital I/Q converter.

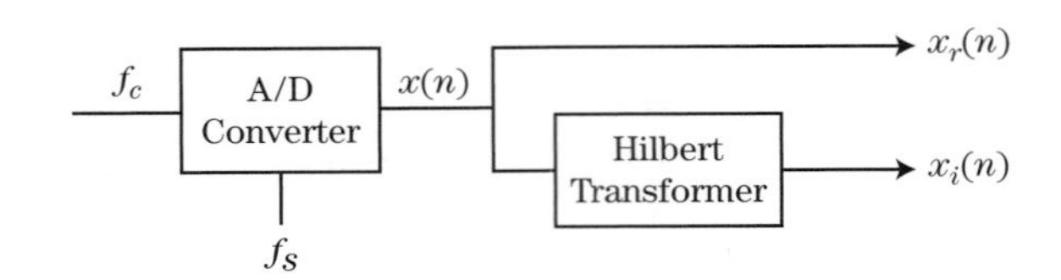

FIGURE 11-28 ■ Derivation of real and imaginary components using Hilbert transform.

output is the imaginary component. These signals can then be used to form the in-phase (I) and quadrature phase (Q) components. A detailed description of Hilbert transformer implementation is provided in ([11], chapter 9).

11.7.3 Digital Up/Down Frequency Conversion

In legacy systems, frequency up/down conversions at RF and IF were normally performed using analog signals with mixers and associated analog filters and amplifiers. Digital receiver and exciter technologies have advanced to the point that some of these frequency conversions can currently be performed digitally, thus reducing the cost and complexities associated with their analog equivalents. Oppenheim and Schafer [17, p. 798] describe an implementation for performing single-sideband frequency up/down conversions using Hilbert transformers.

11.8 FURTHER READING

The design of analog radar receivers has been fairly consistent over the past several decades, evolving primarily with improvement in analog components. This chapter is a general overview of radar receivers, and the reader is advised to investigate the references cited in the chapter to gain a more in-depth understanding of the particular aspects of receiver design and use. The effects of systematic error effects on receiver performance can be found in *Coherent Radar Performance Estimation* by Scheer and Kurtz [18].

With the rapid advances in signal processing components, more of the basic receiver functions are being implemented digitally. Although there is a tendency to consider anything past the ADC as signal processing, many of the functions have similar concerns as their analog counterparts. It is strongly recommended that engineers interested in receiver design become familiar with digital implementation of these functions. One strongly recommended book to help a designer become familiar with digital receiver design is R. G. Lyons's [11] *Understanding Digital Signal Processing*, which thoroughly explains the digital concepts in an easy-to-understand manner and without requiring advanced mathematics. This text also deals with many of the component items involved in radar receivers. After becoming familiar with his book, the advanced concepts in Oppenheim and Schafer's [17] *Discrete-Time Signal Processing* will become clearer to the reader.

11.9 REFERENCES

[1] Hall, G.O., "Super Regenerative Receivers," Chapter 20 in *Microwave Receivers*, Ed. S.N. Voorhis, MIT Radiation Laboratory Series, McGraw Hill Book Company, New York, 1948.

[2] Tsui, J., *Digital Microwave Receivers*, Artech House, Inc., Boston, MA, 1989.

[3] Pace, P.E., *Advanced Techniques for Digital Receivers*, Artech House, Inc., Boston, MA, 2000.

[4] Namgoong, W., "A Channelized Digital Ultrawideband Receiver," *IEEE Transactions on Wireless Communications,* vol. 2, no. 3, pp. 502–510, May 2003.

[5] Bilotta, R.F., "Receiver Protectors: A Technology Update," *Microwave Journal,* August 1997.

[6] Pound, R.V. (Ed.), *Microwave Mixers* (MIT Radiation Laboratory Series), vol. 16, McGraw-Hill Book Co., New York, 1948.

[7] Manassewitsch, V., *Frequency Synthesizers, Theory and Design*, John Wiley and Sons, New York, 1987.

[8] Skolnik, M.L., *Radar Handbook*, 2d ed., McGraw Hill Book Company, New York, 1990.

[9] Taylor, J.W., Jr., "Receivers," Chapter 3 in *Radar Handbook,* 2d ed., Ed. M.I. Skolnik, McGraw-Hill Book Company, New York, 1990.

[10] Vizmuller, P., *RF Design Guide: Systems, Circuits and Equations*, Artech House, Inc., Boston, MA, 1995.

[11] Lyons, R.G., *Understanding Digital Signal Processing*, 2d ed., Prentice Hall, Upper Saddle River, NJ, 2004.

[12] Brand, D.M., and Tarczynski, A., "The Effect of Sampling Jitter in a Digitized Signal," 1997 IEEE International Symposium on Circuits and Systems, June 9–12, Hong Kong.

[13] Goyal, S., Chatterjee, A., and Atia, M., "Reducing Sampling Clock Jitter to Improve SNR Measurement of A/D Converters in Production Test," pp. 165–172 in *Proceedings of the Eleventh IEEE European Test Symposium (ETS '06)*, IEEE Computer Society, Washington, DC, 2006.

[14] Zeijl, P.T.M., van Veldhoven, R.H.M., and Nuijten, P.A.C.M., "Sigma-Delta ADC Clock Jitter in Digitally Implemented Receiver Architectures," Proceedings of the 9th European Conference on Wireless Technology, September 2006, Manchester, UK.

[15] Thompson, D.L., Degerstrom, M.J., Walters, W.L., Vickberg, M.E., Riemer, P.J., Amundsen, L.H., et al., "An 8-Bit 2 Gigasample/Second A/D Converter Multichip Module for Digital Receiver Demonstration on Navy AN/APS-145 E2-C Airborne Early Warning Aircraft Radar," *IEEE Transactions on Components, Packing, and Manufacturing Technology*, part B, vol. 21, pp. 447–462, November 1998.

[16] Urkowitz, H., *Signal Theory and Random Processes*, Artech House, Inc., Boston, MA, 1983.

[17] Oppenheim, A.V., and Schafer, R.W., *Discrete-Time Signal Processing*, 2d ed., Prentice Hall, Upper Saddle River, NJ, 1999.

[18] Scheer, J.A., and Kurtz, J.L., *Coherent Radar Performance Estimation*, Artech House Inc., Boston, MA, 1993.

11.10 | PROBLEMS

1. Why is receiver protection necessary for radar receivers?
2. Consider the case of a pulsed radar at RF of 9.4 GHz that is being jammed by a broadband jammer covering the bandwidth from 9.0 to 10.0 GHz with uniform random noise. If the radar has a 20 MHz bandwidth preselect filter, what is the potential reduction in jammer power due to the preselect filter?
3. For a radar with an RF center frequency of 9.6 GHz and a desired first IF of 2.4 GHz. What is the LO frequency if it is below the RF? Determine if any of the intermodulation components fall into the IF for $n = 1$ to 3 and $m = 1$ to 4.

4. For Problem 4, consider changing the IF center frequency to 3.0 GHz. What is the new LO frequency (LO below the RF)? For $n = 1$ to 3, and $m = 1$ to 5, what is the closest intermodulation component to the IF center frequency? Will an IF with a 1 GHz bandwidth with a sharp cut-off at $\pm$ 500 MHz from the IF center frequency reject the intermodulation component?
5. For a radar with a logarithmic amplifier that provides a 0.05 volt/dB increase in the video output voltage of the logarithmic amplifier per dB increase in input signal power, what is the voltage increase for a 35 dB increase in input signal level?
6. A coherent radar for a particular target has video output voltage of $Q = +1.2$ volts and $I = -0.5$ volts. What is the magnitude and phase of the target signal?
7. Use the radar equation from Chapter 2 to solve the following problem. For a radar with a 100 kW peak power transmitter with an antenna gain of 30 dB, noise figure of 4 dB, losses of 6 dB, and receiver bandwidth of 1 MHz, find the maximum range a radar can provide a 13 dB SNR on a target with 10 m^2 radar cross section (assume that $kT_0 = -204$ dBW).
8. Consider a receiver with three stages: The first stage has a noise figure of 3.0 and a gain of 23 dB; the second stage has a noise figure of 10 and a gain of 30 dB; and the third stage has a noise figure of 15. Determine the overall noise figure for the receiver.
9. Why is it necessary to increase the noise figure stated in mixer specifications by 3 dB when used in superheterodyne receivers?
10. What is the purpose of sensitivity time control?
11. Assume that the effective antenna temperature is 50 °C (Note: K $\approx$ °C + 270), $T_{tr} = T_0 =$ 290 K. When the received signal at the antenna is 1 mW the signal at the receiver input is 0.9 mW and the receiver noise figure is 2.5 (4 dB). What is the effective system noise temperature, T_S?
12. Why is it necessary to set the input noise level to provide one to two least significant bits? Why not just set the noise level below the least significant bit to provide maximum dynamic range from the ADC?
13. Assume that the maximum signal into a 12-bit ADC is set to 1 dB below its saturation and the input noise level is set to about 2 LSBs. Assuming 6 dB/bit, what is the effective dynamic range from the ADC?
14. An ADC has a noise floor about 80 dB down relative to a peak signal near its maximum and spurious responses down below the peak signal by 70 dB, 65 dB, and 45 dB. What is the spurious-free dynamic range of the ADC?
15. Why does the sampling frequency for the in-phase and quadrature phase circuit in Figure 11-27 have to be four times the frequency of the sine and cosine multiplier functions?

CHAPTER

12

Radar Exciters

James A. Scheer

Chapter Outline

12.1	Introduction	417
12.2	Exciter-Related Radar System Performance Issues	418
12.3	Exciter Design Considerations	429
12.4	Exciter Components	440
12.5	Timing and Control Circuits	452
12.6	Further Reading	454
12.7	References	454
12.8	Problems	455

12.1 INTRODUCTION

The earliest radar systems had the ability to detect a target and determine the amplitude of the received signal and the distance to the target by measuring the time delay between the transmit pulse and the received signal. Although the intrapulse phase of such a signal was usually quite consistent, the starting phase of the transmitted signal was random, so there was no significance to the phase of the received signal. As a consequence, there was also no fixed or predictable relationship of the phase of the target echoes from one pulse to the next. Because of this lack of pulse-to-pulse coherence in the signal, these systems were termed *noncoherent* radars.[1] Most (though not all) modern radar systems are *coherent*; that is, they detect the phase of the received signal, relative to a well-controlled reference, as well as the time delay and the amplitude. The received signal is treated as a vector having an amplitude and a direction (phase angle). Whether there is a need for coherence depends on the specific application for the radar system; however, any system that needs to cancel clutter, to measure the Doppler characteristics of targets, or to image a target needs to be coherent.[2]

[1] Sometimes the term *incoherent* is used, which has the connotation of a coherent system somehow losing its coherence. The more accurate term is *noncoherent*.

[2] Some techniques used with noncoherent systems, such as coherent-on-receive processing or clutter-referenced moving target indication (MTI) to detect moving targets, provide a crude measurement of the phase of the received signal, but modern radars seldom implement these techniques, opting instead for full coherence.

To implement the ability to detect the phase of the received signal, the transmitted signal has to have a known phase. By far the most common technique is to develop a transmit signal from a set of very stable continuously operating oscillators. These oscillators are then used as the phase reference for the received signals. Such a system transmits a signal that is "in phase with" (having a fixed phase difference with) a reference oscillator or combination of oscillators. There is then significance to the phase of the received signal in the sense that any change in the received signal phase relative to the reference oscillator phase can be attributed to the target characteristics, principally its range. The radar subsystem that develops the transmit signal and supplies coherent local oscillator (LO) signals to the receiver is (usually) called the *exciter*. Often, it is physically combined with the receiver subsystem, in which case the subsystem is called the receiver/exciter (REX). However, in this chapter the exciter is treated as a separate subsystem, independent of the physical architecture of the system.

In addition to supplying the transmit and LO signals, the exciter usually supplies the timing and control signals for the radar. These include, but are not limited to, transmit timing, receiver protection timing, analog-to-digital converter (ADC) sample timing, pulse repetition frequency (PRF), and pulse repetition interval (PRI) timing.

12.2 | EXCITER-RELATED RADAR SYSTEM PERFORMANCE ISSUES

Coherence is required in three major categories of applications: clutter reduction, Doppler processing, and imaging. The reduction of strong, close-in clutter signals is the most demanding of these applications in terms of the system instabilities that can be tolerated. The major design issues involve phase noise, timing jitter, and spurious signal (spur) generation.

Ideal oscillators operate at the specific exact frequency desired from the oscillator. In practice, though, the average frequency over time might be close to the specified frequency, from one instant to the next the frequency will drift due to thermal, mechanical, and aging effects. Oscillator exactness in terms of center frequency is usually expressed in parts per million (ppm). One Hz drift for a 1 MHz oscillator would be specified as a drift of 1 ppm. A 100 MHz oscillator with 6 ppm drift would vary 600 Hz.

Due to the thermal effect on the dimensions of the frequency-determining elements in an oscillator, the frequency of an oscillator will drift with changes in the ambient temperature. The thermal coefficient of expansion for the critical components is somewhat predictable; therefore, the frequency's sensitivity to temperature is likewise predictable. Though design details can minimize the thermal effects, it is difficult to eliminate the effects entirely. Therefore, most oscillators will specify the maximum thermal drift in frequency. Usually this is specified in terms of ppm per degree (Celsius) of temperature change (ppm/degree).

Slow frequency drift in an oscillator does not adversely affect radar performance. One effect would be to produce an incorrect Doppler frequency measurement. The amount of the error, expressed in ppm, would be the same as the oscillator frequency error in ppm. For instance, with a 6 ppm oscillator error, a 100 mph target would look like a 100.0006 mph target. Such small errors in Doppler measurements are usually not considered a problem.

Another effect is to spread the spectrum of the received signal if the frequency were to change during a coherent processing interval (CPI). In the context of this discussion, a CPI is the time it takes to transmit and receive the sequence of pulses used in a pulse-Doppler process. It is usually on the order of several milliseconds. Again, the scale of the error is nearly insignificant for most radar applications. For example, with the same 6 ppm frequency error, a Doppler bin with an ideal bandwidth of 1,000 Hz would appear 1000.006 Hz wide.

12.2.1 Phase Noise Issues

Although slow frequency drift does not have an adverse effect on coherent radar performance, fast frequency modulation does. The objective of a coherent radar system is to measure the phase of the received signal. Analysis of the signal phase leads to the ability to separate targets from clutter in the frequency (Doppler) domain and to image targets by providing high range and cross-range resolution. In all these cases, the integrity with which the phase can be measured depends on the stability of the oscillators that generate the transmit signal and provide local oscillator references for the downconversion process.

The purpose of the exciter is to generate an ensemble of coherent signals used to produce the transmitted signal (not including the final amplifier) and to provide the coherent reference signals for the coherent receiver (downconverter.) These are all phase-stable signals such that they form an accurate reference for determining the phase of the received signal. The received phase is determined by comparing the phase of the received signal with that of the reference signal from which the transmitted signal was generated, as depicted in Figure 12-1. The dark sine wave segments in the figure represent the times for which the oscillator signal is applied to the transmit amplifier and radiated out of the antenna system. The dashed sine wave signal represents the local oscillator signals, which are continuously available for use as the phase reference for the receiver circuits. The segment labeled “Received Signal” is a delayed version of the transmitted signal, resulting from the reflected wave from the target at some range R, leading to the delay time $2R/c$. This received signal phase is compared with the phase of the reference oscillator to determine the target-induced phase shift. If the target changes range from one pulse to

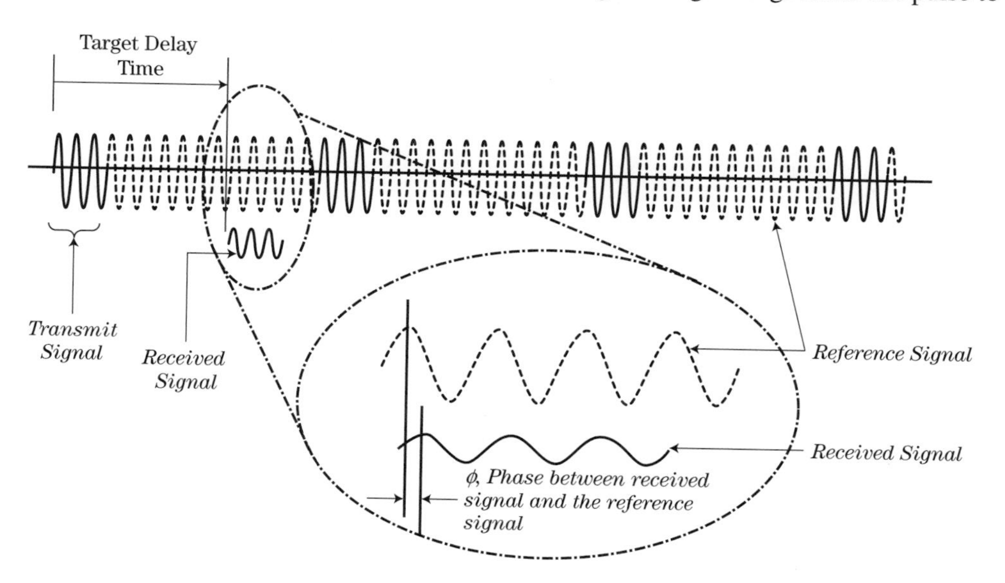

FIGURE 12-1 ■ Measurement of the received signal phase.

the next, this phase will change at a rate proportional to the radial component of target velocity, identifying the target as a moving target.

Classically, the received signal is synchronously detected, producing the two rectangular coordinate system components of the received signal: the in-phase (I) and the quadrature (Q) components. The expressions for the I and Q signals as functions of range to the target, R, and wavelength, λ, are

$$I = A\cos\left(\frac{4\pi R}{\lambda}\right) \tag{12.1}$$

$$Q = A\sin\left(\frac{4\pi R}{\lambda}\right) \tag{12.2}$$

The integrity with which the I and Q signals can be measured depends on the phase stability of the transmit signal reference between the times when the transmit signal is generated and the received signal is detected. If there is any unintentional phase modulation, $\delta\phi$, of the reference signal, then the measurement will be flawed by that amount. In this case, the expressions for I and Q become

$$I = A\cos\left(\frac{4\pi R}{\lambda} + \delta\phi\right) \tag{12.3}$$

$$Q = A\sin\left(\frac{4\pi R}{\lambda} + \delta\phi\right) \tag{12.4}$$

Though the goal is to have a stable reference signal ($\delta\phi = 0$), it is never perfectly phase stable. At a minimum thermal noise will cause a phase modulation, termed phase noise.

Phase noise is not "white" in its spectral characteristics. That is, it does not have a uniform power spectral density. Rather, the phase noise power spectrum is highest at frequencies close to the carrier frequency of the oscillator and reduces with increased frequency offset. The general characteristics of the phase noise (spectrum) are depicted in Figure 12-2, which shows a typical, hypothetical oscillator's single-sideband phase noise power spectrum, $S(f_m)$, as a function of the modulation frequency, f_m, expressed in dB relative to the carrier power (dBc) in a 1 Hz bandwidth. The vertical axis is labeled in units

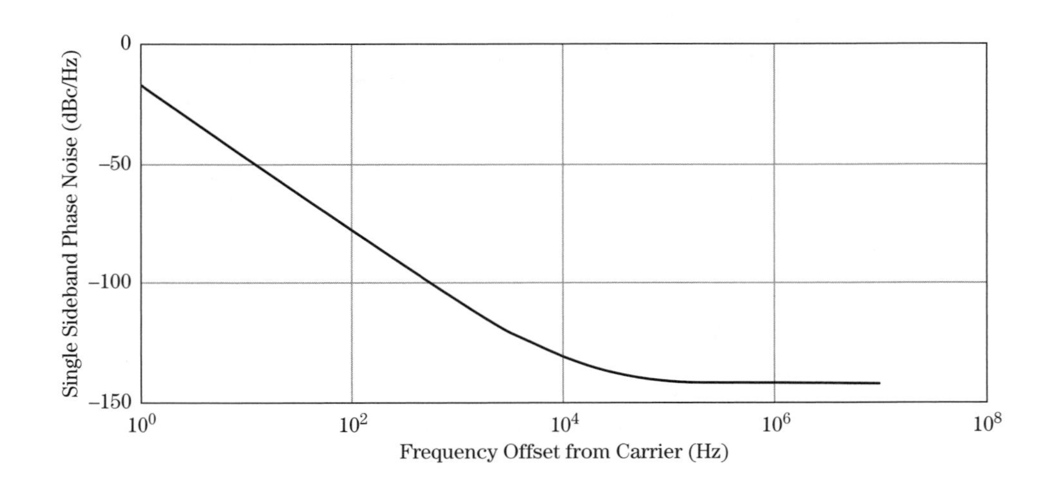

FIGURE 12-2 ■ Typical phase noise spectral characteristics.

of dBc/Hz. The figure depicts the offset frequencies above the carrier, The same curve exists at offset frequencies below the carrier.

One of the most important consequences of phase noise is that in a clutter cancellation system, such as a moving target indication or pulse-Doppler processor, some of the clutter signal energy will be spread throughout the passband of the processor, limiting the target's detectability.

The typical oscillator phase noise spectral characteristic of Figure 12-2 does not represent any particular oscillator but does demonstrate the general shape expected of many stable oscillator phase noise curves. Analysis of a typical oscillator circuit [1,2] reveals that there will be a low-frequency region in which the phase noise power falls off as $1/f^3$ and a region farther from the carrier that falls off as $1/f^2$ before the phase noise reaches a noise floor. A simple mathematical model that can be used to approximate the power spectral density of such an oscillator is

$$S(f) = 10\log 10\left\{\left[10^9\left(\frac{S_{1k}}{f^3} + \frac{2.5S_{50k}}{f^2}\right)\right] + S_{min}\right\} \tag{12.5}$$

where

$S(f)$ is the single-sideband power spectral density of the phase noise in dBc/Hz.

f is the offset frequency.

S_{1k} is the phase noise at 1 kHz offset frequency (in power relative to the carrier power/Hz).

S_{50k} is the phase noise at 50 kHz offset frequency in (in power relative to the carrier power/Hz).

S_{min} is the minimum phase noise at high offset frequencies in (in power relative to the carrier power/Hz).

Figure 12-2 is a plot of (12.5), for the following conditions:

$S_{1k} = 1 \times 10^{-11}$ per Hz (equivalent to –110 dBc/Hz)

$S_{50k} = 1 \times 10^{-13}$ per Hz (equivalent to –130 dBc/Hz)

$S_{min} = 1 \times 10^{-14}$ per Hz (equivalent to –140 dBc/Hz)

Notice that this model would produce an infinite value at 0 Hz, so use of the model is subject to some maximum single-sideband phase noise power, which usually occurs in the vicinity of 5 to 10 Hz offset frequency. Close-in phase noise can be characterized by the spectral "line width." The 3 dB line width is typically on the order of a few Hz, depending on the oscillator frequency. Therefore, plotting or analyzing the effects of phase noise this close to the carrier is usually not considered. Any target with a Doppler frequency offset this low will not be detectable in a clutter environment.

12.2.1.1 Spectral Folding Effects

Two additional calculations are required for a complete analysis of the phase noise requirements. First, the phase noise spectrum extends from a receiver bandwidth below the carrier frequency to a receiver bandwidth above the carrier frequency. Since the radar samples this signal at the much lower radar PRF, there is often significant spectral folding (aliasing). All spectral components must be included in calculating the total phase noise at any given Doppler frequency. The folded components must be added to the phase noise at a Doppler frequency of interest, $S(f_d)$. In general, an estimate of the sum of all folded spectral components, $S(f)_{total}$, can be found by assuming that the phase noise is at the noise floor, $S(f)_{min}$, for all spectral components above the PRF. The number of spectral folds between the single-sideband receiver bandwidth BW and the PRF is $n = BW/PRF$, so the sum of all these components is simply the phase noise in the first PRF interval plus n times the noise floor, as given in equation (12.6):

$$S(f)_{total} = S(f_d) + S(PRF - f_d) + 2\left(\frac{BW}{PRF}\right) S(f)_{min} \tag{12.6}$$

This must be evaluated at every Doppler frequency of interest.

Figure 12-3a is a plot of equation (12.5) for the same parameters as used previously. This plot is different from that of Figure 12-2 in that the horizontal axes (frequency) for Figures 12-3a and 12-3b are linear, whereas the horizontal axis for Figure 12-2 is logarithmic. Also, Figures 12-3a and 12-3b are plotted for offset frequencies from near 0 to only 10 kHz.

Figure 12-3b is the same plot with the phase noise components below the carrier folded into the spectral region above the carrier. This folding is due to the sampling effects of a pulsed radar.

The total phase noise is the sum of the curves in Figures 12-3a and 12-3b, plus the frequency components from the PRF out to the receiver bandwidth. This is found by exercising equation (12.6). The result is plotted in Figure 12-3c. This shows the total phase noise spectrum including the components below and above the carrier, as well as all folded components from near zero to the receiver bandwidth.

In developing these plots, it was assumed that the phase noise spectrum above the PRF is flat, so the effect of folding all components from the PRF to the receiver bandwidth is simply the number of PRF intervals times the noise floor. In most cases, however, the phase noise spectrum is not "flat" beyond the first PRF line, so the actual calculated values must be used. In this case, the total phase noise at each Doppler frequency is the sum of the noise from all PRF intervals from $-BW$ to $+BW$. Equation (12.7) shows how this summation is performed:

$$S(f)_{total} = S(f_d) + \sum_{i=1}^{n} S(f_d + i \cdot PRF) + \sum_{i=1}^{n} S(f_d - i \cdot PRF) \tag{12.7}$$

where f_d is the offset frequency of interest.

12.2.1.2 Self-Coherence Effect

The second factor affecting phase noise provides some help in meeting detection requirements. If the same local oscillator is used in the receiver downconversion process as is used to develop the transmit signal, there is some "self-coherence" in the system, reducing the clutter residue. Specifically, the noise in the received signal is partially correlated with the noise in the local oscillator, reducing the phase noise out of the receiver downconverter. The effective signal-to-clutter ratio (SCR) improvement depends on the time delay, t_c, to

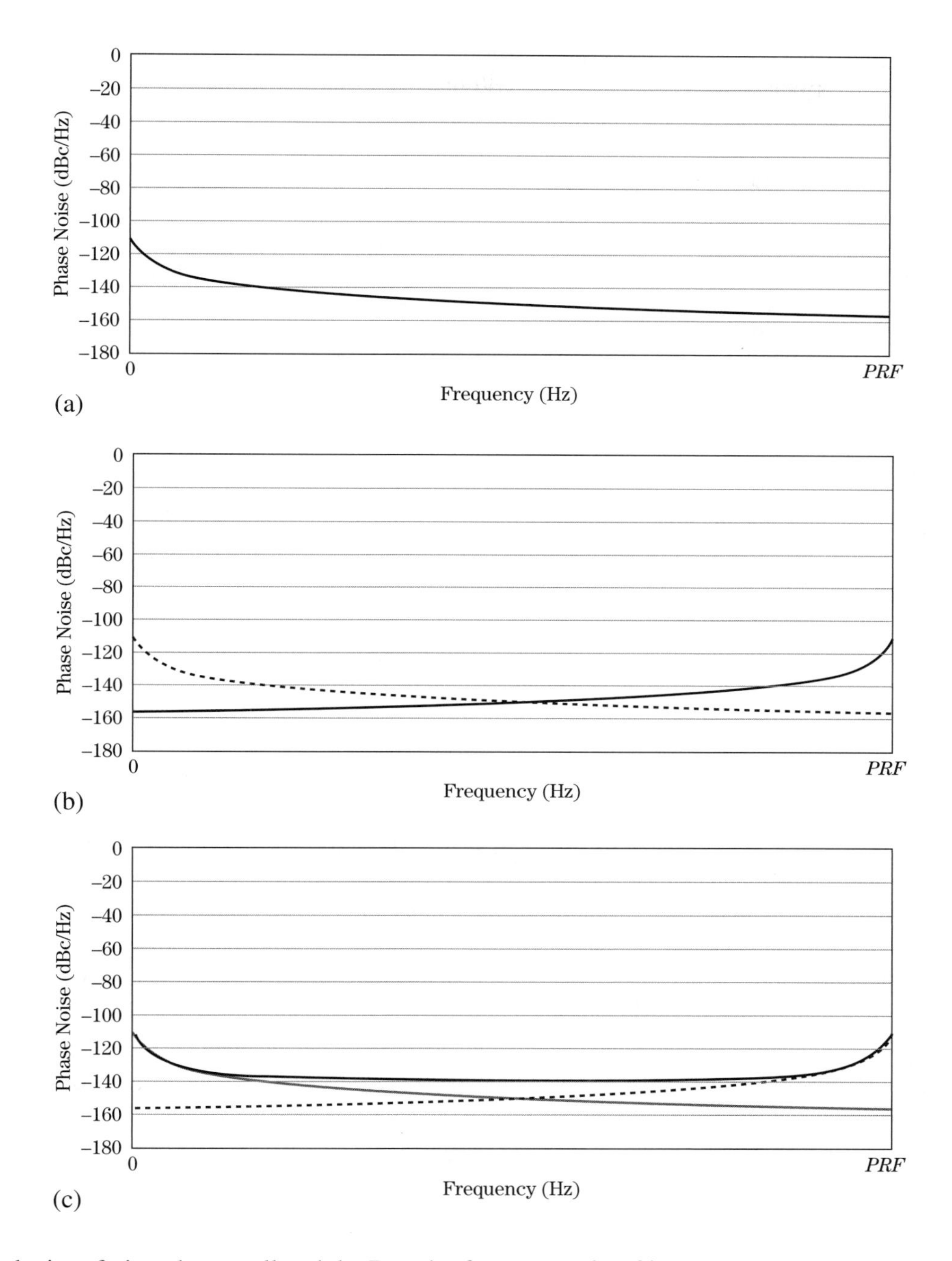

FIGURE 12-3 ■ (a) Phase noise plot. (b) Components of phase noise below the carrier frequency. (c) The sum of all folded components. Parameters are the same as Figure 12-2.

the interfering clutter cell and the Doppler frequency, f_d, of interest. Raven [3] develops the following expression for the improvement, y:[3]

$$y\,(t_c,\, f_d) = 4\sin^2\,(\pi t_c f_d) \tag{12.8}$$

[3]Note that an equivalent expression found in the literature can be derived from equation (12.8) using the double angle formula, resulting in $y(t_c,\, f_d) = 2[1 - \cos(2\pi t_c f_d)]$.

For small Doppler offset frequencies and close-in clutter the factor is less than unity, reducing the effect of phase noise in conditions for which it is needed the most. This equation predicts an improvement of about 7.6 dB at a Doppler frequency of 10 kHz and 1 km range. The factor must be applied for each range ambiguity and Doppler ambiguity before using the summation given in equation (12.7).

The previously described self-coherence factor applies only in cases for which the same oscillator is used in the upconversion and downconversion processes. If a different oscillator is used in the receiver, then its phase noise adds to the stable local oscillator (STALO) phase noise, with no self-coherence factor applied. Since the effect of self-coherence is significant—and often is the difference between a system meeting detection requirements and not—this is a very important design consideration.

12.2.2 Effect of Phase Noise on Clutter Reduction—MTI Processing

The simplest clutter reduction process is the MTI system, which is implemented as a time-domain canceller that is functionally equivalent to a periodic notch filter having notches at 0 Hz, at the PRF, and at multiples of the PRF (see Chapter 17). Many legacy hardware (analog or digital) systems use a single-delay filter having a frequency response, $H_1(f)$, of

$$H_1(f) = 2\sin\left(\frac{\pi f_d}{PRF}\right) \tag{12.9}$$

or a double-delay filter, having a frequency response of

$$H_2(f) = 4\sin^2\left(\frac{\pi f_d}{PRF}\right) \tag{12.10}$$

These two filter characteristics are depicted in Figure 12-4. For a stationary radar, the clutter will be centered at DC (0 Hz), with some spectral spread due to wind-induced motion of the individual scattering centers. The spectral width of the clutter is typically on the order of 100 Hz or less, as discussed in Chapter 5. The clutter residue, C_0, which is the amount of clutter power at the output of the MTI filter, is found by integrating the product of the frequency response of the filter and the spectrum of the clutter over the PRF interval

$$C_0 = \int_{-PRF/2}^{PRF/2} |H(f)C(f)|^2 \tag{12.11}$$

where $C(f)$ is the power spectrum of the clutter. Were it not for hardware instabilities, the clutter spectrum might be modeled by the exponential function

$$C(f) = k_c e^{-af^2} \tag{12.12}$$

However, the phase noise characterized by equation (12.5) spreads the clutter spectrum out to a bandwidth of 1 MHz or more, depending on the receiver instantaneous bandwidth. The clutter residue is then significantly higher than that due to intrinsic spectral spread of the clutter alone. The transmitted spectrum for a pulsed radar system is an ensemble of spectral lines spaced by the radar PRF and having a $\sin(x)/x$ envelope as was shown in Chapter 8. Given these sampled data system characteristics of a pulsed radar, both upper

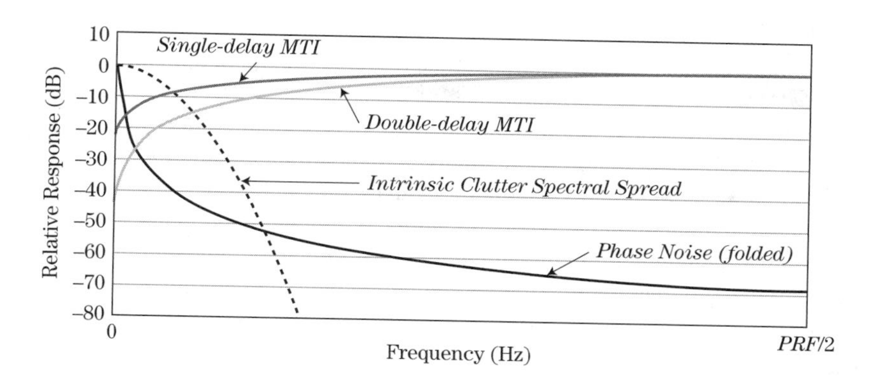

FIGURE 12-4 ■ MTI filter response, showing clutter spectrum and phase noise spectrum.

and lower phase noise sidebands for each of the PRF lines—above as well as below the carrier—fold into the PRF region.

Figure 12-4 depicts the frequency responses of a single- and double-delay MTI filter for a 1 kHz PRF MTI system, along with the natural clutter spectrum and double-sideband phase noise, demonstrating the effect of the phase noise in increasing the amount of clutter power in the MTI filter passband. Notice that the logarithmic plot distorts the MTI filter response shapes from the sine or sine-squared shape of equations (12.9) or (12.10).

12.2.3 Effect of Phase Noise on Pulse-Doppler Processing

Probably the most demanding application regarding phase noise specifications is found in a radar system designed to detect a (small) target in the presence of clutter. The first step in the performance analysis is to determine the signal-to-noise ratio (SNR) for the target signal, as described in Chapter 2. Assuming this is sufficient for detection, given the improvements obtained by whatever signal processing is employed, the next step is to determine the SCR using the radar range equation. The SCR is simply the ratio of the power received from a target to the power received from a clutter cell. The target radar cross section (RCS), σ_t, depends on the size, shape, and materials of the target. Target reflectivity characteristics were presented in Chapter 6. Assuming the interference is due to surface clutter, the clutter RCS, σ_c, depends on the area of the surface clutter being illuminated, A_c, and the average reflectivity per unit area of the clutter, σ^0. Chapter 5 provides a thorough description of the characteristics of various types of clutter.

For a pulse-Doppler system, the waveform is usually range-ambiguous; that is, the distance to the clutter cell may be significantly different from the distance to a target appearing in the same range bin due to foldover of the target, clutter, or both. This can be challenging to the radar in two ways. Not only is clutter RCS larger than the target RCS, but also, for a medium PRF radar, close-in high RCS clutter may appear in the same range bin as a faraway low RCS target. In this case, the target's echo power relative to that of the competing clutter will be decreased by ratio of their ranges raised to the fourth power. In addition, the clutter contributions from several range ambiguity intervals must be summed to determine the true effective ratio between the clutter and target signal. Figure 12-5 depicts a target at some long range, R_t, and several range-ambiguous clutter cells that compete with the target signal. The ranges to the clutter cells are designated Rc_1 through Rc_5. The areas of the clutter cells are determined by the range resolution, which

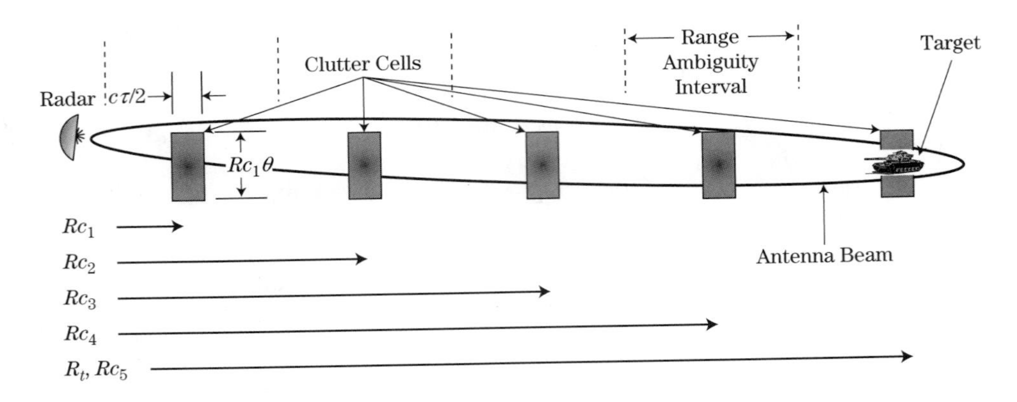

FIGURE 12-5 ■ Far-away target signal competes with close-in clutter signal in a range-ambiguous system.

is proportional to the pulse length for a simple unmodulated pulse, and the azimuthal beamwidth and range to the cell,

$$A_c = \frac{c\tau}{2} R_c \theta_3 \tag{12.13}$$

where θ_3 is the 3 dB azimuth beamwidth of the antenna.

The target signal-to-clutter ratio due to clutter at range Rc_1 is found from

$$SCR = \frac{\sigma_t}{A_c \sigma^0} \left(\frac{R_{c1}}{R_t}\right)^4 \left(\frac{G_t}{G_{c1}}\right)^2 \tag{12.14}$$

where

SCR is the target-to-clutter ratio for a clutter cell at range R_{c1} and a target at range R_t.

σ_t is the target radar cross section.

σ^0 is the average clutter reflectivity per unit area.

A_c is the area of the clutter cell.

G_t is the gain of the antenna in the direction of the target. Assuming that the main beam is centered or nearly centered on the target, this is the peak antenna gain.

G_{c1} is the antenna gain in the direction of the first clutter cell. If the clutter cell is in a direction that is, for example, below the target, it may not be in the main beam of the antenna or may be somewhat down the slope of the main beam. In this case, the antenna gain in the direction of the clutter is less than the antenna gain in the direction of the target. To include the effects of multiple clutter cells "folded" into the same range bin, as depicted in Figure 12-5, the summation of several range-ambiguous clutter cells must be considered.

The net SCR resulting from these effects is

$$SCR = \frac{G_t^2 \sigma_t}{\sigma^0 R_t^4} \sum_{i=1}^{n} \frac{R_{ci}^4}{A_{ci} G_{ci}^2} \tag{12.15}$$

where n is the number of range ambiguities between the nearest clutter cell and the target. Notice that for a radar whose clutter cell area, A_c, is proportional to the range to the clutter cell, R_c, one factor of R_c cancels, leaving R_{ci}^3 in the numerator.

As an example, typical values for these parameters might be as follows:

$\sigma_t = 1 \text{ m}^2$

$\sigma^0 = .01$

$\tau = 1\ \mu\text{ sec}$

$R_c = 5$ km

$R_t = 40$ km

$\theta_3 = 50$ mrad

The resulting SCR is –62 dB. Clearly, pulse-Doppler processing (discussed in Chapter 17) is required to detect the target in the presence of the clutter energy.

Ideally, the remaining (residual) clutter power is zero in the Doppler bin of interest after the pulse-Doppler fast Fourier transform (FFT) processing is performed. Phase noise, however, spreads the spectrum of the clutter signal into essentially all Doppler bins. As already described, the amount of residual clutter in a given Doppler bin depends on the intrinsic clutter power and spectrum, the power spectral density of the phase noise, and the bandwidth of the Doppler bin. The allowable residual clutter power in a Doppler bin depends on the required probabilities of detection and false alarm—P_D and P_{FA}, respectively—and the probability density function (PDF) of the clutter.

For conventional FFT processing, the Doppler filter bandwidth is proportional to the reciprocal of the dwell time, T_d. If no weighting function were used, the 3 dB bandwidth of the filter would be $0.89/T_d$; however, since a weighting function is usually required to reduce Doppler sidelobes, an aggressive weighting function (e.g., Blackman[4]) leading to a bandwidth of about $1.6/T_d$ is more common. Thus, a typical Doppler bin bandwidth will be

$$B_d = 1.6/T_d \tag{12.16}$$

For a 2 ms dwell time, the resulting Doppler filter bandwidth is 800 Hz (29 dBHz).

Suppose that a P_D of 90% and P_{FA} of 10^{-4} are considered to provide reliable detection. If the interference is noise only and the target exhibits Swerling 1 or 2 fluctuations, the required signal-to-interference ratio (SIR) is about 19 dB. If the interference is clutter having a probability density function PDF with longer "tails," then the SIR needs to be significantly higher (e.g., 28 dB, depending on the specific clutter PDF) for equivalent performance.

The required power spectral density (PSD) of the phase noise, $S(f)_{reqd}$, at any offset frequency of interest is found from

$$S(f)_{reqd} = SCR\,(\text{dB}) - B_d\,(\text{dBHz}) - SIR_{reqd}\,(\text{dB}) \tag{12.17}$$

[4] *Blackman* is the name of one of a number of commonly used signal processing weighting functions. These functions have various degrees of sidelobe suppression and main lobe spreading. The Blackman function provides a high degree of sidelobe suppression, with the attendant large amount of main lobe spreading. Detailed descriptions of many of these functions can be found in digital signal processing texts.

For the previous example,

$$SCR\ (\text{dB}) = -62\ \text{dB}$$

$$B_d\ (\text{dBHz}) = 29\ \text{dB}$$

$$SIR_{reqd} = 28\ \text{dB}$$

which results in a phase noise requirement of –119 dBc/Hz at any Doppler frequency of interest. This numerical example assumed that only one clutter cell, at 5 km, coincided with the target signal at 40 km. If the effect of range folding and spectral folding had been included in the analysis, the required phase noise would have been even lower. Typical coherent radar systems with modest target detection requirements will require that the total phase noise power density from all contributors be no higher than perhaps –110 dBc/Hz at any Doppler frequency of interest. Detection of a stealth target in the presence of range-folded clutter will require a more aggressive system, which may necessitate that the phase noise be no more than, say, –140 dBc/Hz at any Doppler frequency of interest.

The previous example was given for a particular target range, clutter range, and target RCS. The result scales directly with target RCS. For example, for a 10 dB smaller target, the phase noise has to be 10 dB lower. The result also scales with space loss, R^4. If the target range is doubled, the phase noise must be 12 dB lower.

The analysis of target detectability can be reduced to a step-by-step procedure, or algorithm. If the target RCS and range, clutter characteristics and range, and phase noise power spectrum are known, then the following procedure can be used to determine target detectability while accounting for the effects of phase noise:

- Determine the target range.
- Determine clutter RCS from the clutter cell area and average reflectivity.
- Fold in clutter from multiple range ambiguities, adjusted for R^4, to obtain total clutter power in the range bin of interest.
- Determine the space loss, R^4, differences between clutter and target echo power.
- Develop a model for the system phase noise, or obtain a point value from the vendor data sheet.
- Determine the Doppler filter bandwidth for the operating parameters of interest.
- Fold in phase noise from multiple Doppler ambiguities, including negative frequencies, to obtain the total phase noise power in the Doppler bin of interest.
- Multiply the phase noise PSD and the filter bandwidth to determine clutter reduction at a given Doppler frequency.
- Adjust clutter reduction by the self-coherence term (range-dependent) of equation (12.8).
- Determine the resulting signal-to-interference ratio.
- Determine the minimum SIR required to obtain the desired P_D and P_{FA} for the expected interference PDF.
- Determine the minimum detectable target RCS from the required SIR.

12.2.4 Effect of Phase Noise on Imaging

The unavoidable phase noise of the exciter oscillators manifests itself as a spectral spreading of the transmit signal and of the local oscillators used in the receiver downconversion process. This spectral spreading has several effects, including spectral spreading of the desired target signal and its spectral (Doppler) characteristics, spectral spreading of the interfering clutter signal, and increased sidelobe levels in imaging processes, such as used with a stepped frequency system or synthetic aperture system. If the goal of a target imaging system is a recognizable set of scattering centers such that a pattern recognition system can be used to identify or classify the target, then the image sidelobe structure will alter that image. Most target imaging systems use only the top 40 dB or so of image dynamic range, so the sidelobes in the image need be suppressed only about 40 dB. While this is not as demanding as the clutter cancellation applications, phase noise still needs to be considered.

12.3 | EXCITER DESIGN CONSIDERATIONS

12.3.1 Transmit Signal

Modern coherent radar systems require that the phase of the receiver LO represents the phase of the transmitted signal, delayed by the round-trip propagation time for the radiofrequency (RF) signal. In the master oscillator-power amplifier (MOPA) approach, which is the most common architecture for coherent radars, the transmitted signal is developed by modulating the signals developed in a master oscillator and amplifying the modulated signal with a power amplifier. The master oscillator usually comprises a set of very stable oscillators, each of which is often phase-locked to a common stable crystal oscillator. This network of oscillators and the associated timing and control circuits are called the exciter. Figure 12-6 shows a top-level block diagram of a coherent radar using a MOPA technique. The exciter is the part indicated by the dashed box.

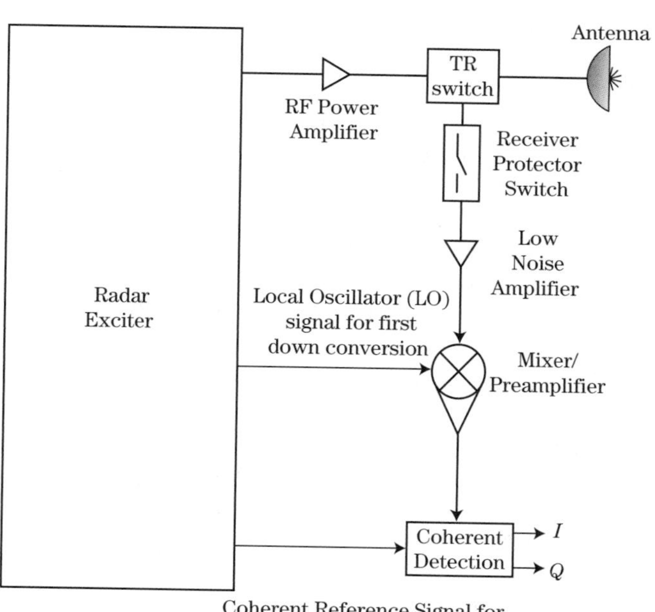

FIGURE 12-6 ■ General diagram of exciter as part of the radar system.

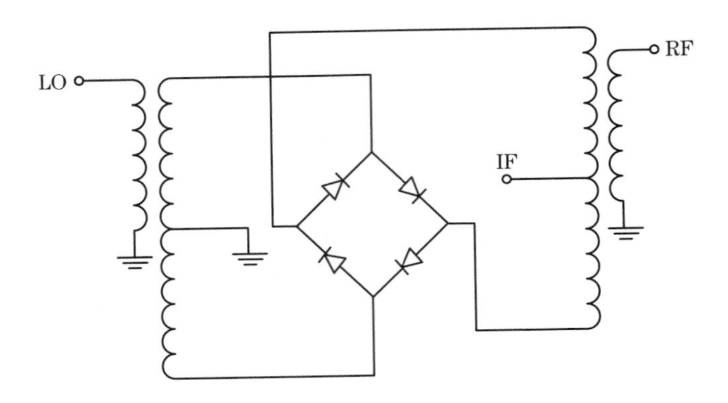

FIGURE 12-7 ■ Diagram of a double-balanced mixer.

Typically, the transmit signal is developed by mixing two or more oscillator signals to produce the sum of the frequencies of these two (or more) oscillators. Though the RF transmit signal could be generated using a single oscillator, the received signal must be downconverted to a lower frequency than the RF for high-fidelity processing. For example, an X-band radar having a carrier frequency of 10 GHz would have a received signal frequency that would be difficult to filter with a narrow bandwidth filter. This is because filter technology is such that it is difficult to produce filters with less than, for example, 1% bandwidth. This means an X-band filter would have a bandwidth of 100 MHz or more. A matched filter for a 1 microsecond pulse, for example, would be a bandwidth of about 1 MHz, a narrower bandwidth than available at X-band. Also, current coherent radar system designs include A/D conversion for digital signal processing. Current analog-to-digital technology does not provide for high-fidelity conversion of a 10 GHz signal. Therefore, the received frequency must be converted down to an intermediate frequency (IF) that is easier to process. Typical IFs for current systems are in the range from 100 MHz to 300 MHz. To downconvert the received RF to an IF, a local oscillator is required in the receiver. This oscillator must be coherent with the transmitted signal. The typical technique for producing an IF in the received signal is to upconvert an IF signal in the transmitter (exciter) and to use the same oscillators in the receive process.

Figure 12-7 is the schematic diagram of a frequency mixer, commonly called simply a double-balanced mixer. A mixer is a three-port device that produces the sum and difference of the input frequencies. The RF and LO ports normally operate at a high frequency (HF) relative to the IF port. For example, a C-band mixer might have a signal capability of from 2 to 4 GHz at the RF and LO ports and an IF port operational band from, say, 100 MHz to 500 MHz. As discussed in Chapter 11, in a receiver a mixer is usually used to downconvert the RF input to an IF output. In this case, the RF might be 3.2 GHz and the LO might be 2.9 GHz, resulting in an IF of 300 MHz. In a radar exciter, the intent is to produce the desired RF frequency from the IF and LO inputs. Usually, this is the sum of the LO frequency and the IF (LO + IF); however, in some cases the difference frequency (LO – IF) is desired. For example, it would be typical to have an LO frequency of 9.0 GHz and an IF input frequency of 300 MHz, resulting in an RF of 9.3 GHz.

Figure 12-8 depicts the use of a mixer to produce a signal at a frequency determined by the frequencies of two oscillators. One signal is input to the local oscillator, or "L" port of the mixer. This signal is normally at a frequency on the order of the desired output frequency, or at least in the same band. For an X-band radar, this might be on the order of 9 GHz. The other input is a signal into the intermediate frequency, or "I" port of the mixer. This signal in the exciter is usually the same frequency as the final IF of the radar receiver.

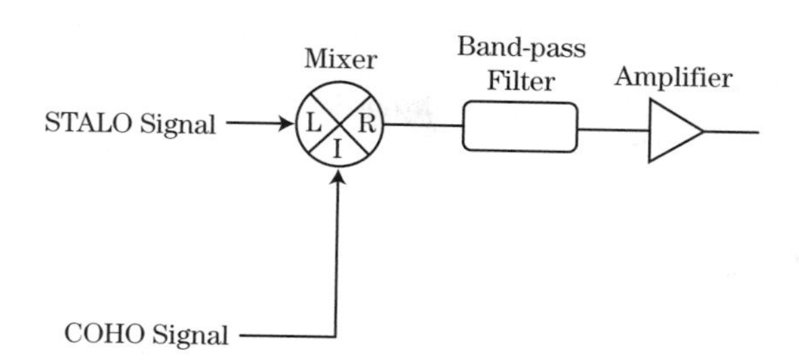

FIGURE 12-8 ■ Method for generating a signal whose frequency is the sum of two other signals' frequencies.

Though potential IFs vary widely, one might choose 300 MHz. The radiofrequency (RF), or "R" port of the mixer, will produce a signal whose frequency is the *sum* of the frequencies of the two inputs. For example, if the L port has a signal at 9.0 GHz and the I port has a signal at 300 MHz, then the output of the mixer at R will be at 9.3 GHz. For the typical coherent radar exciter architecture, the local oscillator is usually called a STALO, and the IF oscillator is usually called the coherent oscillator (COHO). Though these terms are most common, some major radar contractors may have other (proprietary) names for these functions.

The mixer also produces other unwanted mixing products, such as the two individual input frequencies (9.0 GHz and 300 MHz in this example) and the difference frequency (8.7 GHz). Additional unwanted mixing products are generated at harmonics (multiples) of the input signals and sums of these harmonic components and differences between these harmonic components. Figure 12-9 depicts the frequencies that would be generated in such a case, showing only the first and second harmonic terms. Table 12-1 lists the frequencies that are generated when mixing a 9 GHz STALO with a 300 MHz COHO. These frequencies are depicted graphically in Figure 12-9. A band-pass filter, chosen to pass only the desired sum frequency (9.3 GHz), is placed after the mixer to ensure that only that frequency is processed by the transmitter amplifier.

Figure 12-10 depicts the frequencies that are produced by the mixer, up to the second harmonic signals. Note that the filter must pass the desired $(f_2 + f_1)$ frequency and reject all others, some of which may be close to the desired frequency, particularly if higher-order harmonics are considered.

The bandwidth of the band-pass filter must be sufficient to include the bandwidth of the transmitted signal. The RF and IF selected for the process must be carefully considered to avoid the possibility of an unwanted signal (mixing product) existing in the passband of the filter. For a fixed-frequency radar, this is usually not a significant problem because the passband filter can have a relatively narrow bandwidth (though a bandwidth less than about 1% of the center frequency is difficult to realize).

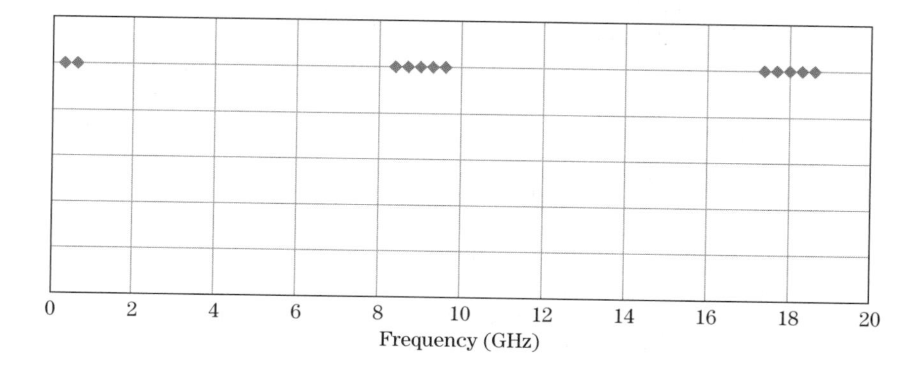

FIGURE 12-9 ■ Mixer output frequencies given two inputs, at 9.0 and 0.3 GHz.

TABLE 12-1 ■ List of Mixer Output Frequencies

Input Frequencies	Output Frequencies
f_1	9.0 GHz
f_2	300 MHz
$f_1 + f_2$	9.3 GHz (Desired, the closest undesired frequency is 300 MHz away.)
$f_1 - f_2$	8.7 GHz
$2f_1$	18.0 GHz
$2f_2$	600 MHz
$f_1 - 2f_2$	8.4 GHz
$f_1 + 2f_2$	9.6 GHz
$2f_1 + 2f_2$	18.6 GHz
$2f_1 + f_2$	18.3 GHz
$2f_1 - f_2$	17.7 GHz
$2f_1 - 2f_2$	17.4 GHz

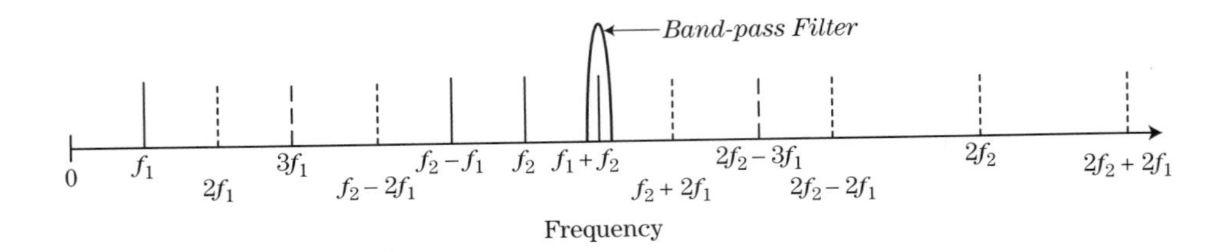

FIGURE 12-10 ■ Frequencies out of the mixer, and filter characteristics.

The amplitude of the signal is reduced as it is processed by the mixer and filter, so an amplifier is placed after the filter to recover the signal loss. The conversion loss of most mixers is on the order of 6 to 7 dB, and the filter will attenuate the signal by an amount on the order of 1 dB, so only modest gain (typically 10 dB) is required of this amplifier.

For the exciter, the desired output is usually the sum of the two input frequencies.[5] Consequently, a single-sideband upconverter is sometimes used in place of the mixer. This device is an assembly of two mixers—signal in-phase and quadrature splitters—and combiners that produces only the desired sum frequency (upper sideband [USB] or lower sideband [LSB]) and greatly suppresses the undesired sideband and carrier frequencies. The undesired signals are not fully suppressed, but the filtering process is made easier with the partial suppression. Figure 12-11 depicts a circuit diagram of a single-sideband frequency converter, which has both the USB and LSB outputs available. The user would terminate the unused output port using a matched load. The desired output would have the undesired frequency suppressed by about 20 dB, relative to a typical mixer shown in Figure 12-7.

For a pulsed radar, the transmit signal exists for only a short period of time, consistent with the pulse length. Except during the transmit pulse, it is desirable not to have any signal out of the mixer. To achieve this, two switches controlled by the timing and control circuits

[5] Some radars, such as those operating in the HF band (3–30 MHz) may employ a high-side local oscillator with the desired output frequency being the *difference* between the two input frequencies.

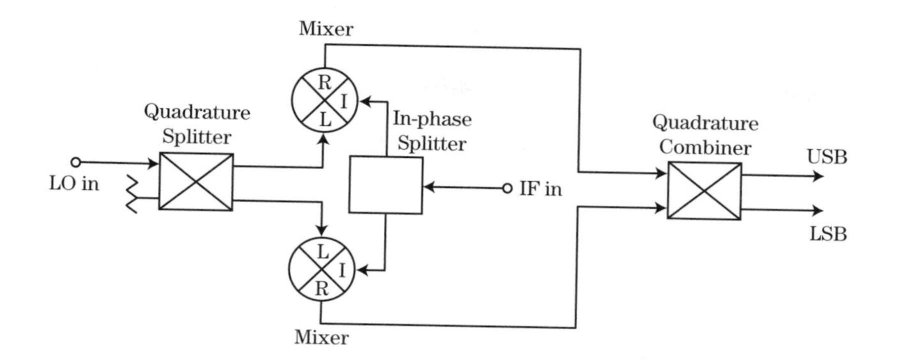

FIGURE 12-11 ■ Diagram of a single-sideband frequency converter.

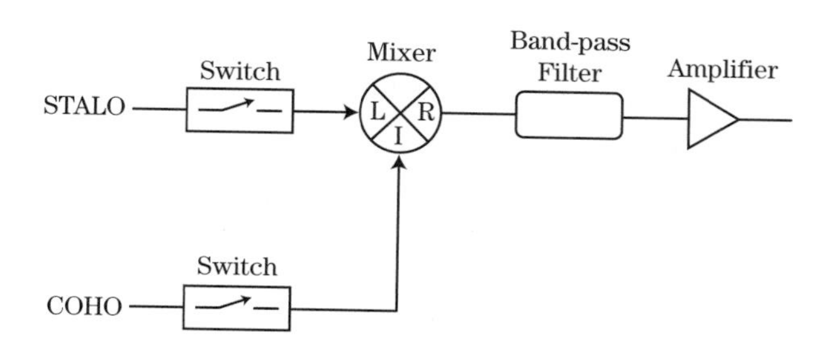

FIGURE 12-12 ■ Method for pulsing the transmit frequency using switches.

are used to pulse the signals into the mixer. Certainly, only one of the two signals needs to be pulsed; however, better isolation is achieved through the mixer if both signals are gated "off" during the receive time. Figure 12-12 depicts the use of the two switches with the mixer circuit. Usually the transmit amplifier is also pulsed on only during the transmit time. Gridded traveling-wave tube (TWT) amplifiers, such as described in Chapter 10, are a common type of amplifier for this application.

Figure 12-13 depicts the diagram of a more complete exciter circuit, which produces a transmit signal to be applied to the transmit amplifier, and the LO and COHO signals for the receiver downconversion and coherent detection process. In typical designs, the STALO is at a frequency somewhat close to the transmit RF, and the COHO is at a significantly lower frequency. For example, for an X-band radar, the STALO might be at 9 GHz and the COHO at 300 MHz. The combination would produce a transmit frequency of 9.3 GHz. Both of the stable oscillators (STALO and COHO) are phase-locked to an extremely stable low frequency crystal oscillator. The most stable of such oscillators is a stress-compensated-cut (SC-cut) quartz crystal oscillator.[6] These are available at frequencies up to about 100 MHz. It is desirable to use as high a frequency as available for this reference oscillator, so that the final output frequency is no larger a multiple of the reference oscillator than necessary.

The STALO and COHO are each followed by a signal splitter, ahead of the switch, so that continuous wave (CW) versions of these signals are available for the receiver LO and COHO requirements. If the desired transmit signal bandwidth B is wide due to the use of intrapulse coding such as a phase code or linear frequency modulation (LFM), then

[6]There are many other techniques for stabilizing the frequency or phase of an oscillator. The use of a quartz crystal is common for oscillators in the HF, VHF, and UHF ranges.

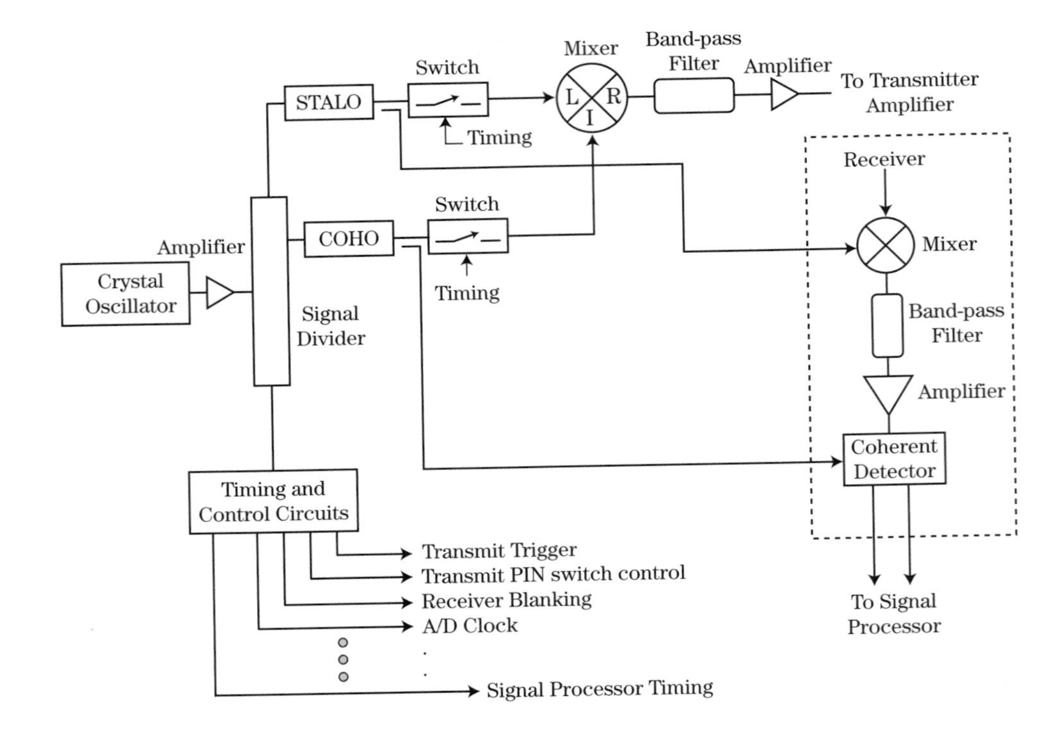

FIGURE 12-13 ■ Diagram of internal components of a simple exciter.

the selection of RF and IF is more critical, as depicted in Figure 12-14. The percentage bandwidth of the band-pass filter must be wide enough to pass the desired signal, with only marginal amplitude reduction over that bandwidth. In this case, for the example shown in Figure 12-14, one of the unintentional mixing products ($f_2 + 2f_1$) is beginning to invade the passband of the filter. As shown in the figure, the condition is probably acceptable as it is shown; however, if the bandwidth were any wider, or if f_1 were any lower, then a different IF would be chosen. In general, for a wideband signal, the use of a very low IF compared with the RF creates a problem. In this case, more than one upconversion stage (and likewise multiple stages of downconversion in the receiver) will alleviate the problem. Figure 12-15 depicts the signals, up through the second harmonics, for a 9.0 GHz and 1.0 GHz pair. Compared with Figure 12-8, the separation between the desired 10.0 GHz signal and the closest neighbor is wider. The instantaneous bandwidth of the filter at 10 GHz and therefore the bandwidth, B, of the signal could be as much as 1 GHz. This frequency plan for the upconversion process would result in a 1.0 GHz IF in the downconversion process. This may be a higher IF than desired, so a second downconversion to a lower IF would be required. To maintain coherence with the transmitter, a similar second upconversion in the transmit process would be required.

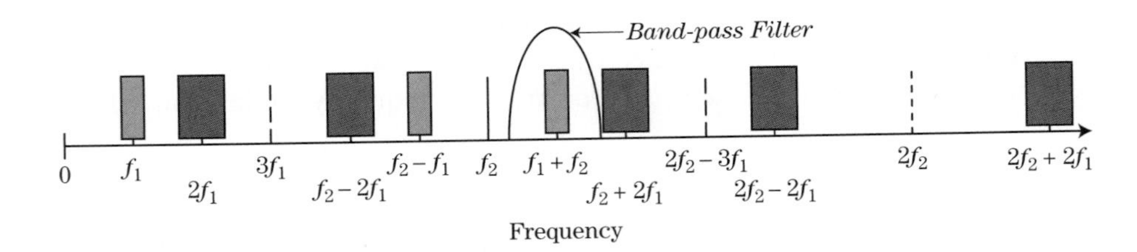

FIGURE 12-14 ■ Frequency plan with a wideband IF signal.

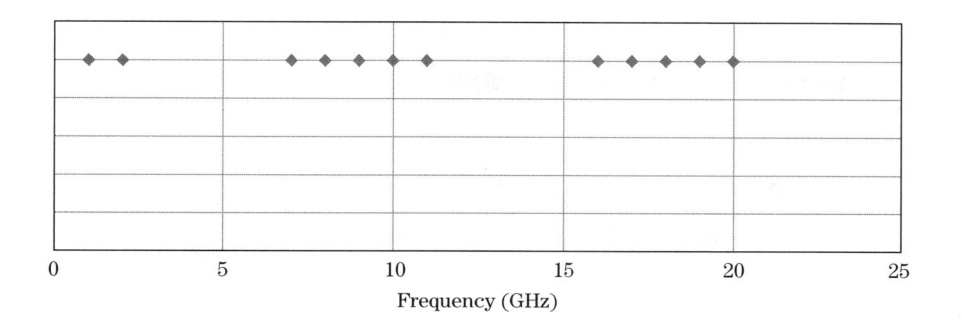

FIGURE 12-15 ■ Frequencies resulting from 9 GHz and 1 GHz pair.

Notice that the bandwidth of the filter selecting the desired signal shown in Figure 12-13 is wider than the equivalent filter shown in Figure 12-10. For a wider bandpass, the so-called skirts of the filter are wider and will pass signals farther from the desired signal than will the narrowband filter. Usually, this condition requires that there be a lower ratio of frequencies into the mixer than for the narrow band system. Whereas the narrowband system may allow a ratio of up to about 50:1 between the RF and IF, the wideband system may allow only a ratio of about 3:1 to 4:1 for each mixing stage. Therefore, two mixing stages might allow for a final ratio of 10:1 to 20:1, and three mixing stages would be required for a ratio on the order of 50:1. Table 12-2 lists the frequencies that are generated when mixing a 9 GHz STALO with a 1 GHz COHO. These frequencies are depicted graphically in Figure 12-15.

Figure 12-16 is a photograph of several typical microwave filters [4]. The devices in Figures 12-16a and 12-16b of the figure are connectorized versions, and Figure 12-16c is a surface-mounted version. Since these filters are passive devices, they typically have only an input and output coaxial connection.

Figure 12-17 shows the band rejection characteristics for a typical tubular filter having a bandwidth of 5% of the center frequency, for example, 500 MHz (3 dB) bandwidth centered at 10 GHz [4]. This filter will pass a signal in the band from 9.750 to 10.250 GHz. If the undesirable mixing products from a mixer ahead of the filter are specified to be at least 40 dB down from the desired signal, then for a five-section filter their frequencies

TABLE 12-2 ■ Output Frequencies

Input Frequencies	Output Frequencies
f_1	9 GHz
f_2	1 GHz
$f_1 + f_2$	10 GHz (Desired, the undesired frequency is 1,000 MHz away.)
$f_1 - f_2$	8 GHz
$2f_1$	18 GHz
$2f_2$	2 GHz
$f_1 - 2f_2$	7 GHz
$f_1 + 2f_2$	11 GHz
$2f_1 + 2f_2$	20 GHz
$2f_1 + f_2$	19 GHz
$2f_1 - f_2$	17 GHz
$2f_1 - 2f_2$	16 GHz

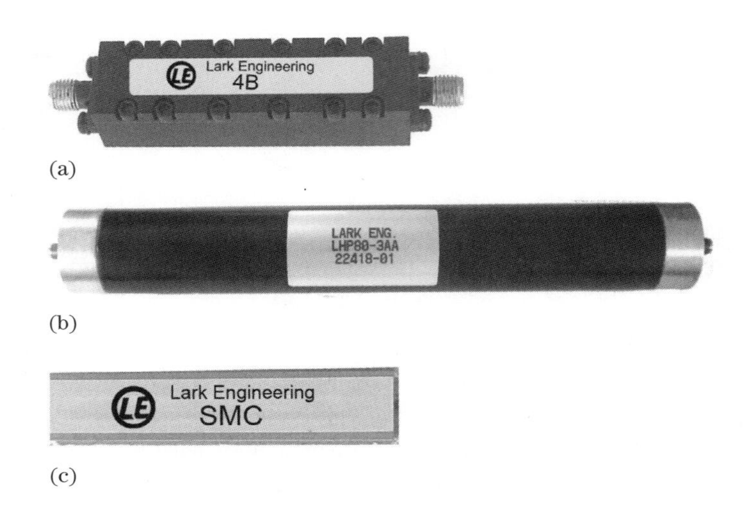

FIGURE 12-16 ■ Typical IF filters. (a) Connectorized. (b) Tubular style. (c) Surface mount style. (Courtesy Lark Engineering [4]. Used with permission.)

must be at least 1.6 bandwidths (800 MHz) above 10.0 GHz, or 1.7 bandwidths (850 MHz) below 10.0 GHz—that is, above 10,800 MHz or below 9,150 MHz. The data points for this calculation are shown as two dark dots on the figure.

There is a trade-off in the selection of filter order between frequency and time-domain characteristics. A higher-order filter, such as a six- or eight-section filter, will allow undesired mixing product signals to be closer to the center frequency (but no less than 1 bandwidth, or 500 MHz away, for the previous example and the filter characteristics shown). However, when a high-order filter (above, e.g., five sections) is used, the time-domain response usually exhibits "ringing" or a decaying oscillation in the signal. Fewer filter sections, typically three or four, provide a better response from a time-domain point of view, but the frequency restrictions are then worse; in the previous example, the undesired signals would have to be 1 or 1.5 GHz away from 10 GHz. No matter how many filter sections are used, as a practical matter the filter passband bandwidth must be somewhat less than the separation between the closest unwanted signals, by a factor of at least 1.5.

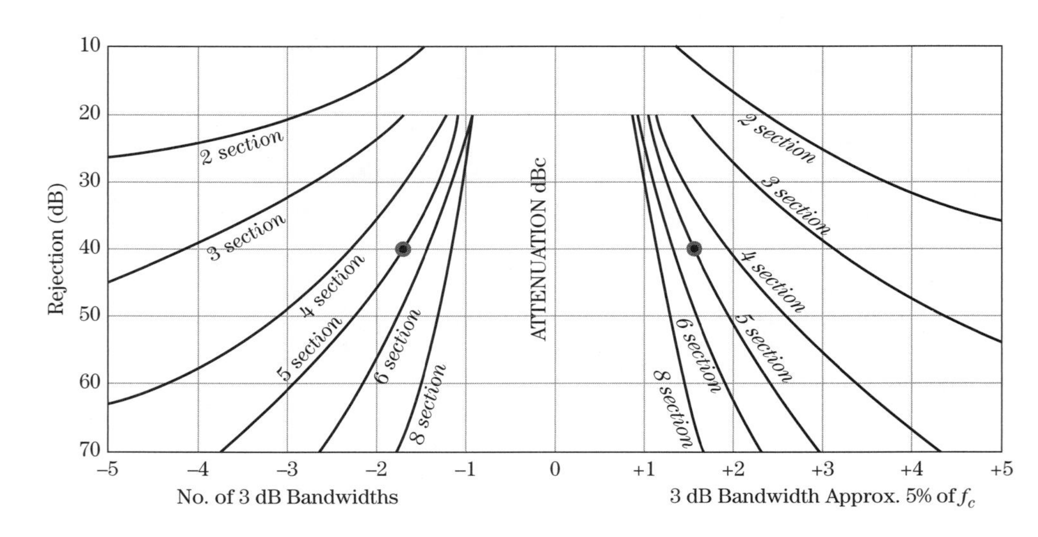

FIGURE 12-17 ■ Filter characteristics for a typical 5% bandwidth tubular filter. (Courtesy Lark Engineering [4]. Used with permission.)

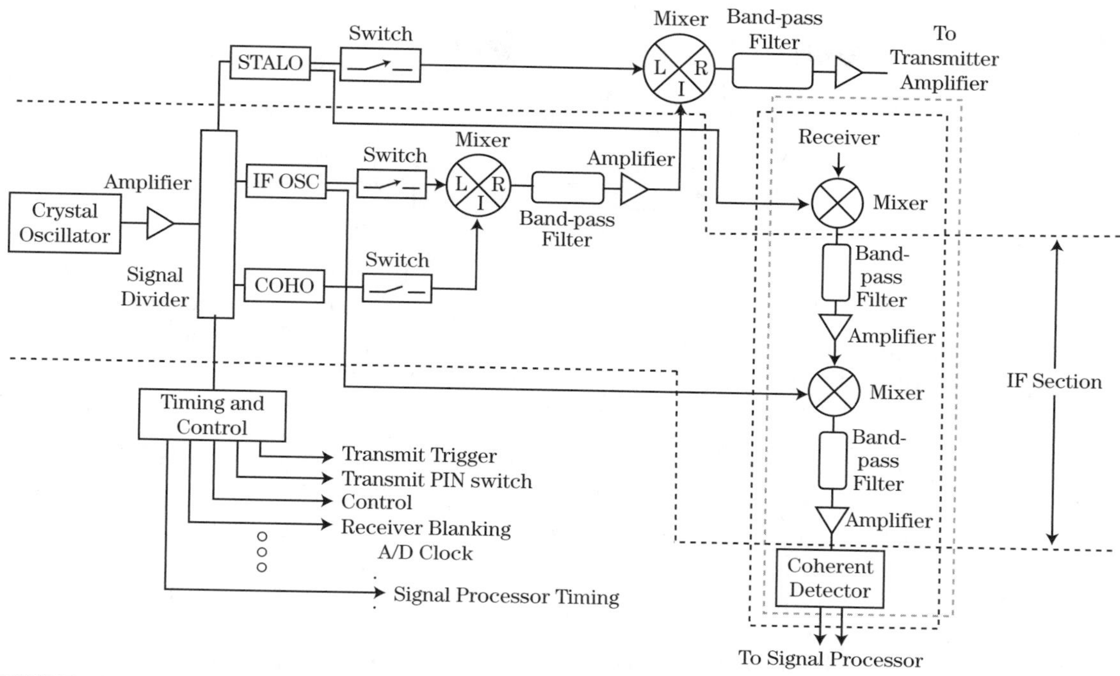

FIGURE 12-18 ■ Diagram of an exciter with two upconversion stages.

Figure 12-18 is a block diagram of an exciter that has two upconversion stages. The additional coherent oscillator required for this architecture is often called the IF oscillator. This is the most typical exciter configuration because it allows for a somewhat wideband signal, such as that used in a pulse compression waveform. The selection of frequencies is based on the need to prevent unwanted mixing products from lying within the passband of the band-pass filters at each mixing stage. Which mixer ports are used for a given oscillator signal depends on the bandwidth specified for each of the mixer ports. Typically, the IF port will support a lower range of frequencies than either the LO or RF ports. The specific ports are chosen depending on the relative frequencies of the COHO and the IF oscillator. Note that there is a filter following every mixer in the diagram.

12.3.2 Waveform Generation

There is a wide variety of techniques for generating a wideband waveform for a coded pulse system. The most common modern radar techniques use either a wideband LFM within the pulse or a pseudorandom biphase code. The properties and application of these and other wideband waveforms are discussed in Chapter 20. The wideband waveform is usually developed in the IF section of the exciter in Figure 12-18. From the point of view of the exciter, the IF section is the set of components that operates at frequencies near those of the COHO rather than at frequencies near the RF. The block diagram of Figure 12-18 shows the section designated as the IF section.

12.3.2.1 Intrapulse, Pseudorandom Biphase Modulation

If pseudorandom biphase code pulse compression is used, the phase code is applied to the IF or COHO signal using a biphase modulator controlled by the waveform generation

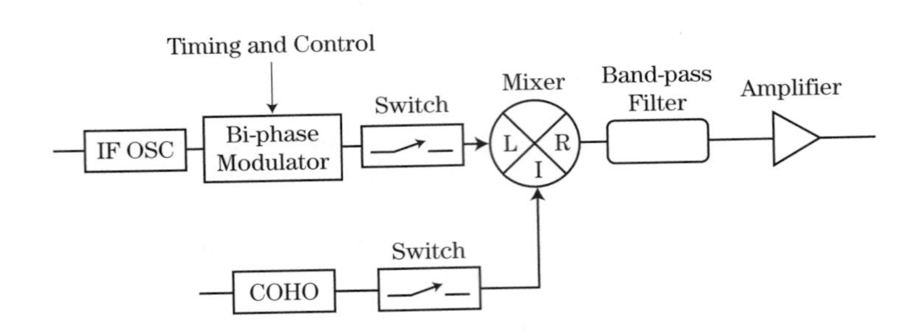

FIGURE 12-19 ■ Use of a biphase modulator to create pseudorandom phase modulation.

function of the radar computer. Figure 12-19 depicts the addition of a biphase modulator in the IF circuit. The phase either changes from one state (e.g., 0 degrees) to a complementary state (e.g., 180 degrees) or is left unchanged for each short time period consistent with the required bandwidth. Each such time period, which represents a fraction of the total pulse length, is called a *chip*. For example, a 200 MHz bandwidth would be consistent with a 5 nsec chip period. The choice to change the phase or leave it unchanged from one chip to the next would be made randomly every 5 nsec for, say, a 1 microsecond pulse. Since the computer processor develops the random sequence, the phase sequence is actually pseudorandom in nature. The computer stores the specific code sequence that was transmitted so it can perform the correlation process during the receive function. Chapter 20 provides a thorough mathematical description of the characteristics of the biphase modulated technique. Concerning implementation of the biphase modulation technique in an exciter, it is valuable to mention here that the spectrum of the biphase modulated waveform is a $\sin(x)/x$ shape, requiring a linear phase (vs. frequency) filter to properly maintain the integrity of the waveform. Bessel filters, having a "rounded" passband characteristic and linear phase, are often used in conjunction with biphase modulation techniques.

12.3.2.2 Intrapulse Linear Frequency Modulation (LFM)

If a frequency agility waveform or intrapulse LFM is required, then a synthesized signal generator is used in the IF circuit to produce the frequency modulated signal. The LFM waveform is characterized by a linear increase (or decrease) in frequency during the pulse. Often, several hundred megahertz of frequency modulation is imposed onto the pulse. Chapter 20 provides a thorough mathematical description of the operation of the LFM mode. The spectrum of the LFM waveform is flat, requiring a relatively flat filter response. Historically, an LFM waveform would be implemented using a surface acoustic wave (SAW) pulse expansion device. The received signal would use a complementary device called a SAW pulse compressor to produce a high-resolution pulse-compressed signal in the time domain. In modern systems, an LFM waveform is usually generated using a direct digital synthesizer (DDS). Figure 12-20 depicts where the DDS would be inserted into the exciter architecture. The baseband output of the DDS is then mixed with the IF oscillator. The DDS device has a random access memory in which samples of a sine wave are stored. The values are read out of the memory at a varying rate determined by the clock pulse rate (usually on the order of several hundred MHz), producing a sinusoidal signal at a controlled frequency. The sine wave is quantized in amplitude to an accuracy determined by the number of bits in the output digital-to-analog (D/A) converter. As with the biphase modulator, the DDS is controlled by the signal processor, which defines the waveform in terms of the start frequency, stop frequency, and rate of change of frequency.

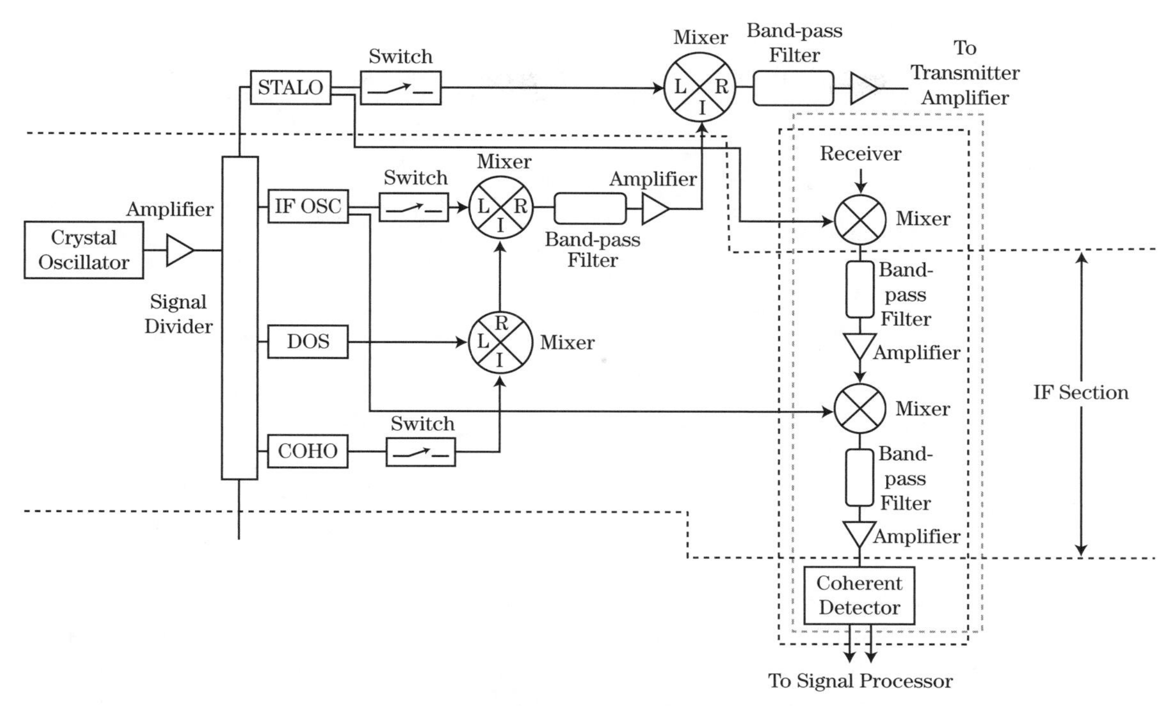

FIGURE 12-20 ■ Exciter drawing showing incorporation of a DDS.

Current DDS technology exhibits phase noise significantly worse than that required for detection of a long-range target in severe clutter. Therefore, the wideband waveform is used only after detection is made and target *imaging* or *range-profiling* is required. For example, in the search mode, when detection performance needs to be optimized the radar will generally not use a wideband waveform but rather will use either an uncoded pulse or modest frequency modulation bandwidth. In this case, high range resolution is not required, but low phase noise is. When the mode is switched to one requiring high range resolution—for instance, in a target identification application—then the wideband waveform generator is employed. Phase noise will be increased; however, low phase noise is not required at this stage. If the radar has an operational mode where low phase noise is required at the same time as high range resolution imaging, then a low-noise method for waveform generation will be required.

12.3.2.3 Dwell-to-Dwell Frequency Change

Better phase noise performance is attainable using a more sophisticated frequency synthesizer. There exist a large number of commercial synthesizers operating at frequencies from hundreds of MHz to 18 GHz and above. A typical high-performance synthesizer will change frequency in as little as a microsecond to as much as several hundred milliseconds. The phase noise performance of a frequency synthesizer is better than a DDS-based synthesizer but not as good as that of a fixed-frequency oscillator. Synthesizers do not switch frequency fast enough to implement intrapulse frequency modulation but can provide the IF signal for dwell-to-dwell frequency hopping. This class of synthesizer is large, usually rack-mounted, or on the order of the size of a shoe box. Figure 12-21 depicts an exciter architecture with a stepped frequency synthesizer, capable of changing transmit frequency

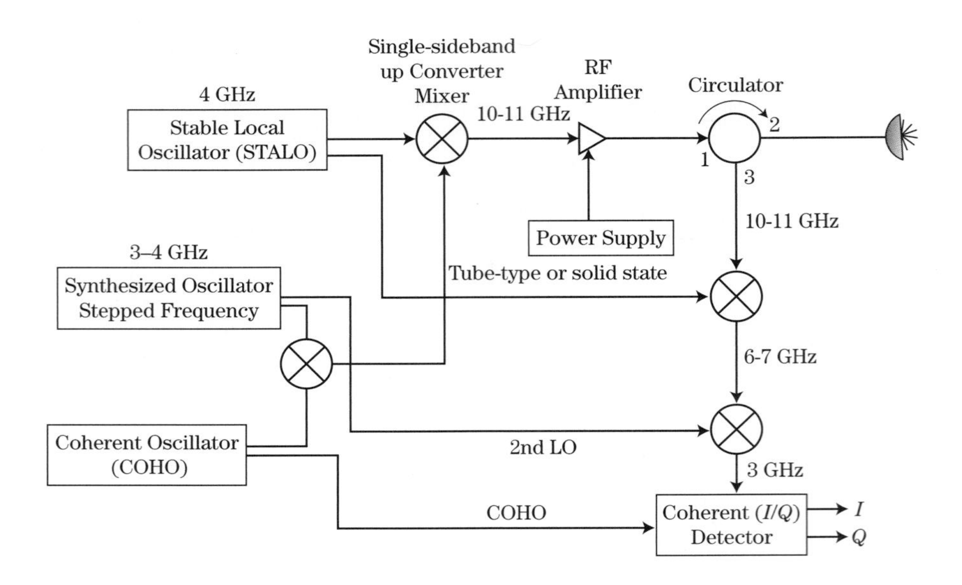

FIGURE 12-21 ■ Exciter diagram, showing stepped frequency synthesizer.

on a pulse-to-pulse or dwell-to-dwell basis. In this example, the stepped frequency synthesizer operates from 3 to 4 GHz, resulting in a 1 GHz range of frequencies for the transmit frequency. (Note that for simplicity, the switches, filters, and amplifiers are not drawn in this figure; however, they would be included in the design.)

12.4 EXCITER COMPONENTS

12.4.1 Stable Oscillators—Oscillator Technology

12.4.1.1 Crystal Reference Oscillator

The reference oscillator to which all other exciter oscillators are phase-locked must exhibit very low phase noise. Note that when the phase-locked oscillator (PLO) is locked to a harmonic of the reference oscillator frequency, or when the PLO includes a frequency divider between the PLO output and the phase detector, within the phase-locked loop bandwidth the phase noise out of the PLO will be increased relative to that of the reference oscillator.

The phase noise spectrum of the signal out of a PLO, $S_{PLO}(f)$ is related to the reference oscillator phase noise, $S_{ref}(f)$, and the frequency multiplication ratio, n, by [2]:

$$S_{PLO}(f) = n^2 \cdot S_{ref}(f) \tag{12.18}$$

The phase noise spectrum of a typical low-noise oscillator made by Wenzel Associates is shown in Table 12-3. The particular device is an SC-cut crystal oscillator with a center frequency of 100 MHz. Figure 12-22 is a picture of such an oscillator. In addition to an output coaxial connection, it has pins to which the DC voltage is provided.

When the signal out of the crystal oscillator is multiplied up to another frequency, or when another oscillator is phase-locked to the crystal oscillator, then the phase noise of the resulting signal is worse than that of the reference oscillator.

In the case of frequency multiplication, the phase noise at any given offset frequency increases by a factor of n^2, where n is the frequency multiplication ratio [1]. For example,

TABLE 12-3 ■ Phase Noise of a 100 MHz Wenzel Crystal Oscillator

Offset Frequency (Hz)	SSB Phase Noise (dBc/Hz)
100	–130
1,000	–160
10,000	–175
20,000	–176

Source: From Montress and Parker [5]. Used with permission.

FIGURE 12-22 ■ Typical crystal oscillator. (Courtesy Wenzel Associates [5]. Used with permission.)

if a 100 MHz oscillator is multiplied up to 8 GHz, then the phase noise curve shifts upward by a factor of 80^2, or 6,400 (equivalent to 38 dB).

In the case of a phase-locked oscillator, a high-frequency oscillator is phase-locked to a harmonic of a stable crystal oscillator. At offset frequencies within the locking bandwidth, the output phase noise will be determined by the reference oscillator. At offset frequencies above the locking bandwidth, the output phase noise will be determined by the output oscillator.

12.4.1.2 Stable Local Oscillator

Figure 12-23 is the measured phase noise of an extremely stable 8 GHz LO [6], the dielectric resonant oscillator (DRO) developed by Poseidon Scientific and depicted in

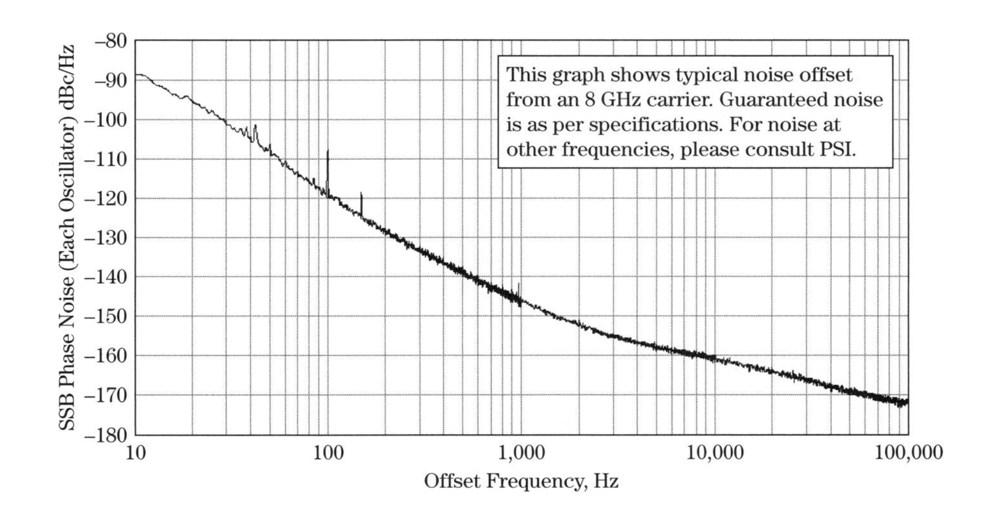

FIGURE 12-23 ■ Shoe-box oscillator phase noise. (Courtesy Poseidon Scientific Instruments [6]. Used with permission.)

FIGURE 12-24 ■ State-of-the-art X-band dielectric resonant oscillator. (Courtesy Poseidon Scientific Instruments [6]. Used with permission.)

Figure 12-24. The phase noise at 1 kHz offset frequency is about –146 dBc/Hz and at 10 kHz is about –160 dBc/Hz. This technology provides the best phase noise available in 2009 at LO frequencies. Local oscillators using a sapphire resonator at nominally 3, 6, and 8 to 10 GHz are available from PSI.

Low phase noise comes at a cost of a high-Q circuit that cannot change frequency. Any frequency changes in a radar using this device must be implemented in the IF section. Normally, any circuit that provides the ability to change frequency or that generates a wideband waveform will not have as low a phase noise as shown here. The high-Q dielectric (sapphire) loaded cavity used in the DRO design is obvious in the photograph of the so-called shoe-box oscillator (SBO) shown in Figure 12-23.

12.4.1.3 Phase-Locked Oscillators for Stable Local Oscillators

If the ultimate in performance is not required, then a more modest design can be considered for the STALO. An example of a typical oscillator is shown in Figure 12-25. This is a phase-locked dielectric resonator oscillator (PDRO) produced by Herley-CTI, Inc., and available at frequencies from 3 to 45 GHz. The phase noise characteristics when phase-locked to a very stable crystal oscillator and operated at 5 GHz are shown in Figure 12-26.

FIGURE 12-25 ■ Herley-CTI PDRO dielectric resonator oscillator. (Courtesy Herley-CTI [7]. Used with permission.)

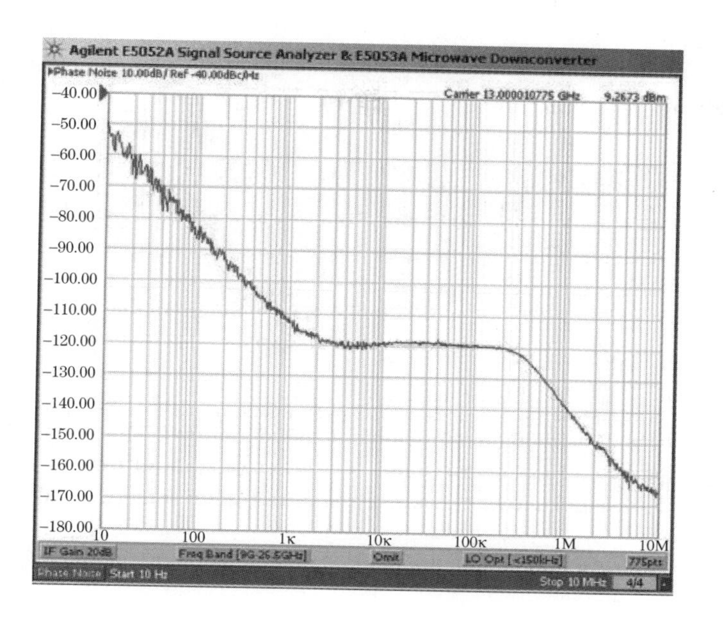

FIGURE 12-26 ■ CTI PDRO oscillator phase noise. (Courtesy Herley-CTI [7]. Used with permission.)

When an oscillator is phase-locked to a reference oscillator, the phase noise within the loop bandwidth (of the phase-locked loop) is driven by the reference oscillator. The phase noise outside (above) the loop bandwidth is driven by the free-running oscillator phase noise. In the example shown in Figure 12-26, the portion of the curve at offset frequencies less than ~200 kHz reflects the multiplied phase noise of the crystal reference oscillator within the phase-locked loop (PLL) bandwidth. The part of the curve at offset frequencies above ~300 kHz represents the unsuppressed phase noise of the DRO.

12.4.1.4 Surface Acoustic Wave Oscillators

Many system developers have the manufacturing ability to develop oscillators using SAW technology for the resonator. These are physically small devices yet with sufficient Q to produce a relatively low phase noise, particularly given their small size. The characteristics of SAW oscillators are reported in [8,9]. Table 12-4 summarizes the phase noise of two SAW oscillators and the phase noise that would be experienced when multiplying the frequency up to 10 GHz, a representative frequency for an X-band system. The last

TABLE 12-4 ■ Summary of SAW Oscillator Phase Noise

f_{offset} (Hz)	400 MHz SAW	L-Band SAW	400 MHz SAW at 10 GHz	L-Band SAW at 10 GHz	CTI PDRO at 10 GHz
	dBc/Hz	dBc/Hz	dBc/Hz	dBc/Hz	dBc/Hz
1	−50		−22		
10	−82	−70	−54	−50	
100	−112	−100	−84	−80	−85
1k	−140	−127	−112	−107	−110
10k	−167	−155	−139	−135	−120
100k	−180	−171	−152	−151	−120
1M	−184	−170	−156	−150	−135

FIGURE 12-27 ■ Herley-CTI series BBS synthesizer. (Courtesy Herley-CTI [7]. Used with permission.)

column is the phase noise of the previously discussed Herley-CTI PDRO, showing relative performance. The PDRO has better phase noise performance at lower offset frequencies, and the SAW oscillators have better performance at higher offset frequencies.

12.4.2 Synthesizers

Direct digital synthesizers develop a signal by clocking the values of a sine wave from a digital memory at a rate dependent on the desired output frequency. Though these devices are very flexible, because of quantization effects they are not very stable, as judged by phase noise performance, and they produce larger spurious signals than most other devices. Spurious signals are typically below the carrier level by approximately 9 dB for each bit of quantization in the synthesizer, with a floor of about –70 dBc [1]. Figures 12-27 through 12-29 illustrate three DDS devices from Herley-CTI. Though the basic shape and sizes of the synthesizers shown in Figure 12-27 and Figure 12-28 are similar, note that the mounting and connectors arrangements are somewhat different. Both have two coaxial connectors—one for the reference frequency input and one for the synthesized output. Power is supplied via the multipin connector on the broad-band synthesizer (BBS) synthesizer (Figure 12-27), while DC power is supplied through dedicated feedthroughs on the fast switching synthesizer (FSS) design (Figure 12-28). The additional circuits required for the very fast switching of the direct synthesizer (DSX) design require a larger package.

FIGURE 12-28 ■ Herley-CTI FSS series synthesizer. (Courtesy Herley-CTI [7]. Used with permission.)

FIGURE 12-29 ■ Herley-CTI series DSX fast-switching synthesizer. (Courtesy Herley-CTI [7]. Used with permission.)

Table 12-5 summarizes the performance of the synthesizers shown in Figures 12-27 and 12-28, listing frequency range, frequency step size, switching speed, and phase noise for a nominally 5 GHz output. Notice that at higher offset frequencies (above 10 kHz) the model FSS, which switches frequency in 50 microseconds, suffers from somewhat more phase noise than the model BBS, which switches in about 60 milliseconds. There is often a compromise between switching speed and phase noise performance. The wider bandwidth required for fast switching comes at a price of more noise. That said, the architecture of the design will also dictate the phase noise, too. The Herley-CTI FSX series, which switches in nanoseconds, and the Miteq LNS series synthesizers, which switch at speeds on the order of the Herley-CTI BBS series, produce less noise than the smaller-sized examples shown in Figures 12-27 and 12-28.

Another low phase noise design is available from MITEQ Inc. [12]. Figure 12-30 is the photograph of a MITEQ model LNS60806580, which operates from 6.08 to 6.58 GHz. Other 500 MHz-wide RF bands are available. Table 12-6 lists the phase noise achieved at various offset frequencies. The synthesizer switches in as little as 10 msec, with a maximum of 100 msec.

TABLE 12-5 ■ Phase Noise and Pertinent Parameters for Several Herley-CTI Frequency Synthesizers

Herley-CTI Model	BBS	FSS	DSX
Frequency	Up to 5.12 GHz	2–18 GHz	0.5 to 18 GHz
Frequency step size	2 Hz	1 Hz	1 Hz
Switching speed	60 msec	50 μsec	200 nsec
Phase Noise at Indicated Offset Frequency: dBc/Hz			
100 Hz	−77	−80	−93
1 kHz	−78	−90	−113
10 kHz	−88	−92	−123
100 kHz	−114	−100	−123
1MHz	−134	−114	−133

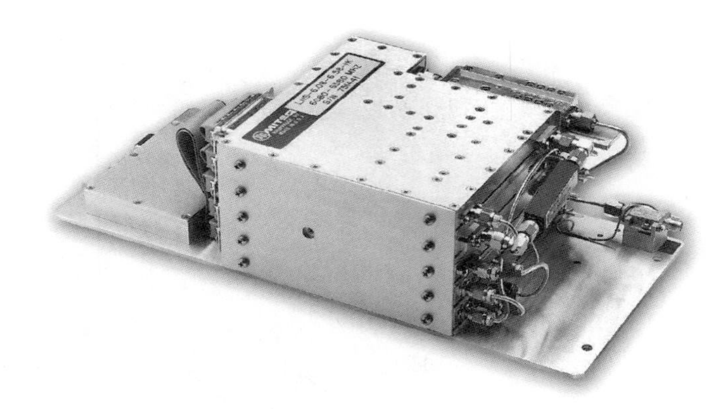

FIGURE 12-30 ■ Photograph of a MITEQ LNS60806580 frequency synthesizer. (Courtesy of MITEQ Inc. [12]. Used with permission.)

12.4.3 Other Devices

In addition to the oscillators, other devices are required to implement exciter functions. Mixers, amplifiers, direct digital synthesizers, frequency multipliers, and frequency dividers are used to generate the various signals desired. These devices all add some phase noise to that of the oscillators, so care must be taken to select devices that are specified to minimize this consequence. For the lower microwave bands—for example, from UHF to S-band—both silicon carbide (SiC) and gallium arsenide (GaAs) technology is available for the required devices. At C-band, X-band, and above, GaAs is used. It produces very low levels of phase noise, so it is not difficult to maintain good system performance as long as attention is paid to the phase noise characteristics of each device.

12.4.3.1 Assembly Approaches

Whereas oscillators and synthesizers are assemblies in themselves, many of the other components such as mixers, filters, and amplifiers required to make an exciter are individual components that are physically small compared with the previously shown oscillators and synthesizers. These devices can be assembled to meet the design requirements. There are two different approaches for assembling individual components for an exciter assembly: (1) connecting individual components via interconnecting coaxial cables; and (2) attaching surface mount technology (SMT) devices onto a microstrip circuit or printed board.

Figure 12-31 demonstrates the assembly technique for connectorized components. The advantage of this technique is that subsystems can be developed quickly and can be

TABLE 12-6 ■ Phase Noise of the MITEQ Model LNS60806580 Frequency Synthesizer

Offset Frequency (Hz)	SSB Phase Noise (dBc/Hz)
100	−80
1,000	−100
10,000	−105
100,000	−105
300,000	−105
1,000,000	−115
10,000,000	−140

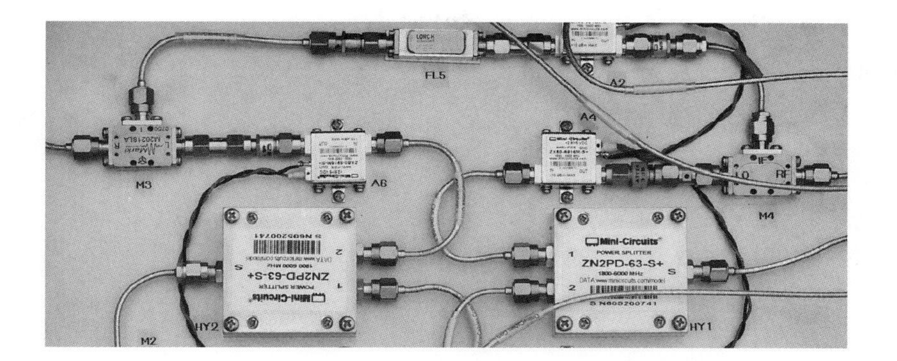

FIGURE 12-31 ■ RF subassembly showing connectorized assembly technique. (Courtesy T. L. Spangler [10]. Used with permission.)

reconfigured easily by just disconnecting the components and inserting desired components. This approach is suitable for one-of-a-kind or proof-of-principle systems, which may need to be reconfigurable. The design requires only standard tools and materials, such as aluminum plates, drills, and machine hardware. The interdevice cabling uses standard 50 ohm semi-rigid coaxial cables. Copper semirigid cabling is shown in the figure. The disadvantages of the connectorized approach are that the subsystems developed using this method are larger and heavier and often have somewhat poorer RF performance than microstrip and SMT techniques.

There is usually a loss associated with the cabling and connectors between components. Also, an impedance mismatch is inevitable, creating a voltage standing wave ratio (VSWR), which represents another loss source. The VSWR will generally change with frequency, so for a wideband system the VSWR introduces amplitude and phase ripple into the system. Without compensation these ripples create range sidelobes in a high-resolution system.

These effects can be reduced by placing attenuating "pads" between the devices. These attenuators reduce the amplitude of the reflected signals more than the forward propagating signals because the reflected signals pass through the attenuators more times. Of course, somewhat higher initial signal levels are required to offset the effects of the attenuators; however, this is usually not a problem if this need is anticipated. The use of these pads is seen in some of the critical locations in the circuit shown in Figure 12-31.

Once a design is stabilized, surface mount technology might be used to implement smaller assemblies with better performance. Figure 12-32 is an example of an assembly using SMT components. The design time is longer than for the connectorized parts because a printed microstrip or microwave integrated circuit must be designed and fabricated. Design changes take more time for the same reasons. Changes that do not require redesigning the board are limited to the values of passive components (e.g., resistors) and substitution of active components with components having the same pin configuration and form factor. However, once the design is finalized, the RF performance is usually significantly better than the performance of the connectorized technique because there are no intermediate connectors and a good impedance match is achievable using the microstrip design technique.

Figure 12-32 shows several SMT components typically used for exciter designs. These devices are intended for use in the final design, once the design has been validated using connectorized parts. Usually a given device is available in both connectorized and SMT packages. Often the specifications for the connectorized part are somewhat relaxed compared with the equivalent SMT part for the previously cited reasons.

FIGURE 12-32 ■ RF subassembly showing SMT assembly technique. (Courtesy T. L. Spangler [10]. Used with permission.)

12.4.3.2 Mixers

As passive devices, mixers add negligible phase noise; however, the output phase noise will be the sum of the phase noise power of the input signals. Usually the resulting phase noise is close to that of the highest frequency oscillator in the exciter—often the STALO.

Figure 12-33 is a photograph of a Hittite Microwave Corp. HMC_C014 mixer [11]. This is a passive device with three ports: a LO port, IF port, and an RF port. Though most mixer ports are labeled with RF, LO, and I, the particular mixer shown is labeled RF1, RF2, and IF. This is because, for this mixer, the two RF ports can be interchanged with no effect on performance. The local oscillator signal would be inserted into one of the RF ports,

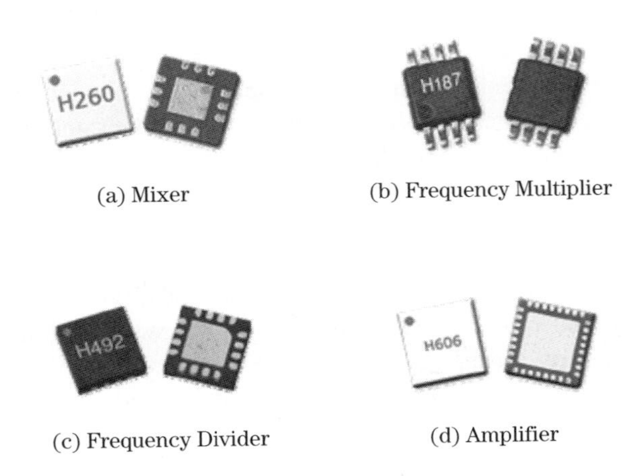

(a) Mixer (b) Frequency Multiplier

(c) Frequency Divider (d) Amplifier

FIGURE 12-33 ■ Ensemble of SMT components. (Courtesy Hittite Microwave Corporation [11]. Used with permission.)

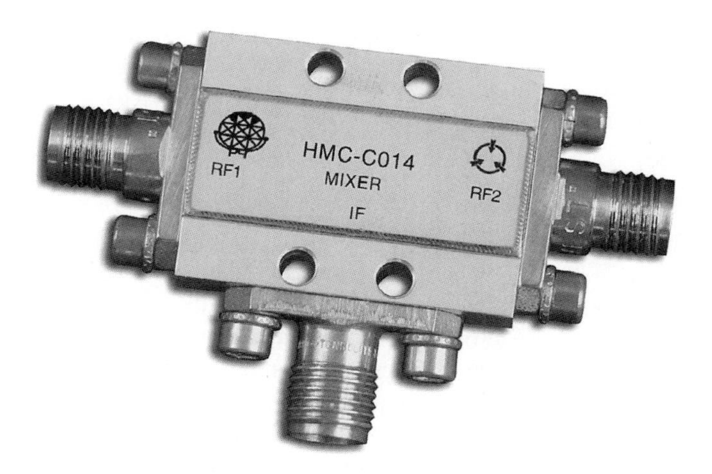

FIGURE 12-34 ■ Typical mixer, showing RF, LO, and IF ports. (Courtesy Hittite Microwave Corporation [11]. Used with permission.)

and the RF signal would come out of the other RF port. When used as an upconverter, as is the case with the exciter, the RF port is the output. (When used as a downconverter in the receiver, the IF port is the output; see Chapter 10.) Salient parameters for the Hittite HMC_C014 mixer are as follows:

Frequency range:	16–32 GHz
IF:	DC-8 GHz
LO power:	13 dBm
Conversion loss:	8 dB
Isolation:	35–40 dB, 25 dB (RF to IF)

Figure 12-33a is a photograph of the same mixer shown in Figure 12-34 except in the SMT format. This device is mounted to a microstrip or stripline circuit card by laying the gold-plated leads down against the board and flowing solder onto the connections, as is visible in Figure 12-31.

12.4.3.3 Switches

Figure 12-35 is a photograph of a connectorized single-pole single-throw (SPST) RF switch. It has two RF ports (i.e., input and output) and solder terminals for DC power and

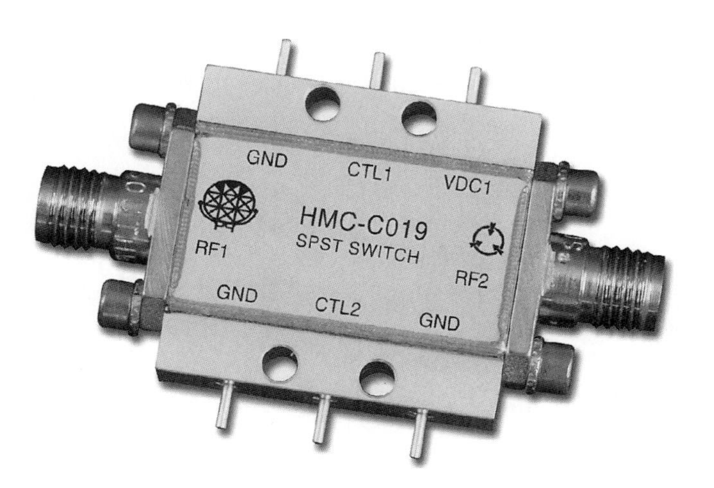

FIGURE 12-35 ■ Typical microwave SPST switch, showing input, output, power, and control connection points. (Courtesy Hittite Microwave Corporation [11]. Used with permission.)

the switching command signal. Salient parameters for this switch are as follows:

Insertion loss:	3–6 dB
Isolation:	60 to 100 dB
Switch speed:	2–10 nsec

12.4.3.4 Frequency Multipliers

When generating a set of frequencies for a coherent radar, a small number of oscillators is used, and the various frequencies are developed by multiplying, mixing, and dividing the frequencies from the few basic oscillators. Frequency multipliers use nonlinear devices (typically diodes) to generate harmonics of the input signal. Selection of the desired harmonic provides the correct output frequency. Typical GaAs frequency multipliers have a phase noise contribution of –136 dBc/Hz at 100 kHz offset frequency. The phase noise of the signal out of a multiplier, $S(f_m)_{out}$, is related to the input signal phase noise $S(f_m)_{in}$ and the frequency multiplication ratio n by [2]

$$S\left(f_m\right)_{out} = n^2 \cdot S\left(f_m\right)_{in} \tag{12.19}$$

Additional phase noise is added by the multiplier device itself. As an example, the spectral shape for the additive noise from a Hittite X-band output, 8× multiplier is typically nearly flat (white). Figure 12-33b is a Hittite HMC187MS8 SMT 1.7 to 4 GHz (output) frequency doubler. This device is surface-mounted to the stripline or microstrip board by orienting the device such that the eight leads bend down onto the board, where the solder connection is made. Figure 12-36 is a photograph of a connectorized MITEQ MX2M010060 passive frequency doubler [12]. The input frequency range is from 500 to 3,000 MHz, and the output frequency range is from 1,000 to 6,000 MHz. Since it is passive (i.e., has no active components), there is only an input and output connector—no power supply terminals. The signal loss through the doubling process is between 15 and 19 dB.

12.4.3.5 Frequency Dividers

Frequency dividers are also used to generate desired output frequencies. Digital flip-flop-like circuits divide the signal frequency by 2. Cascaded stages provide other divide ratios. Typical GaAs frequency dividers, such as the Hittite X-band output, 2× divider have a phase noise contribution of –150 dBc/Hz at a 100 kHz offset frequency. Again, the spectral shape for the additive noise is typically nearly flat (white). Figure 12-33c is a photograph of a Hittite HMC492LP3 SMT 2× frequency divider with 0.2 to 18 GHz input frequency.

FIGURE 12-36 ■ Photograph of a MITEQ MX2M010060 passive frequency multiplier. (Courtesy of MITEQ Inc. [12]. Used with permission.)

Salient parameters for this frequency divider are as follows:

P_{in}:	−15 dBm
P_{out}:	−4 dBm
Phase noise (f_{in} = 4.8 GHz, f_{out} = 2.4 GHz):	−150 dBc/Hz at 100 kHz

12.4.3.6 Amplifiers

Several kinds of amplifiers are required in the exciter, operating at the different frequencies in the subsystem. As a minimum, the COHO frequency and the STALO frequency—as well as the output RF signal to be applied to the transmit amplifier—usually require amplification. These signals are derived from very stable oscillators, so it is important to not contaminate these signals upon amplification. The additive phase noise (sometimes called "residual phase noise") of the amplifiers should be significantly less than the phase noise of the signal itself to avoid significant increase in phase noise. Fortunately, amplifier additive phase noise is usually very low, since the devices can be operated in a narrow range of conditions (e.g., limited frequency range, amplitude variations).

Signals at the intermediate frequency range (typically 0.1 to 4 GHz) are required for signals from the COHO and other IF oscillators as well as at mixer and multiplier outputs. Figure 12-37 is a photograph of a MITEQAMF-4B-02000400-20-33P-LPN amplifier [12] that operates from 2 to 4 GHz. It has a gain of 40 dB and a noise figure of 2 dB. The maximum signal output power is usually limited to approximately 10 dB below the amplifier's 1 dB compression point to reduce nonlinearity effects that lead to spurious signal generation. This amplifier is rated at an output 1 dB compression point of 2 watts (+33 dBm). Notice the cooling fins that allow the heat generated in the higher-power device to dissipate better.

Figure 12-33d is a photograph of a Hittite HMC606LC5-SMT low phase noise GaAs HBT RF amplifier that operates from 2 to 18 GHz. Its characteristics, including additive phase noise, are as follows:

Frequency:	2–18 GHz
Gain:	13.5 dB nominal
Noise figure:	5–7 dB
P_{out}:	10–13 dBm

FIGURE 12-37 ■ Photograph of a MITEQ AMF-4B-02000400-20-33P-LPN amplifier, showing the input and output connectors. Note the heat sink that provides convective cooling. (Courtesy of MITEQ Inc. [12]. Used with permission.)

FIGURE 12-38 ■ Photograph of a CTT IF amplifier. (Courtesy of CTT Inc. [13]. Used with permission.)

FIGURE 12-39 ■ Photograph of typical low-noise CTT Inc. X-band amplifiers showing drop-in and connectorized versions. (Courtesy of CTT Inc. [13]. Used with permission.)

Phase noise at 12 GHz carrier frequency:	–140 dBc/Hz at 100 Hz
	–150 dBc/Hz at 1 kHz
	–160 dBc/Hz at 10 kHz
	–170 dBc/Hz at 1 MHz

Notice that the phase noise of the amplifier is significantly lower than most of the sources available, even at X-band.

Figure 12-38 is a photograph of a CTT Inc. [13] connectorized ASM/050-1636-123A amplifier. Characteristics of this amplifier include the following:

Frequency:	2–18 GHz
Gain:	20 dB nominal
Noise figure:	3 dB
Pout:	10 dBm

Figure 12-39 is a photograph of several CTT RF amplifiers, showing the "drop-in" version at the bottom and the method of assembling that device into a connectorized frame.

12.5 TIMING AND CONTROL CIRCUITS

In addition to supplying the transmit signal and the LO signals for the receiver, the exciter also generates the timing and control signals for the system. For each pulse repetition interval, high-speed timing signals control the transmit time, receiver blanking, and ADC sample timing. The high-speed timing controls the start and end of each CPI and the FFT processing timing. The signal processor determines the transmit frequency, PRF, and waveform length for each CPI.

With recent improvements in ADC technology, it is practical to sample the IF signal directly, as described in Chapter 10. This technique mitigates analog circuit errors such as I/Q gain and phase mismatch. However, the timing of the analog-to-digital sampling is very critical because timing jitter will lead to sampling jitter noise in the digitized signal. The amount of noise developed depends on the root mean square (rms) timing jitter relative to the input signal bandwidth. Current applications require that the timing jitter be on the order of 0.1 ps or less. This jitter is caused by the timing in the ADC sample pulses as well as internal effects in the ADC itself. If the timing jitter is random, then the jitter noise will be white; that is, it will spread throughout the Doppler spectrum. If the timing jitter is sinusoidal or periodic, such as that caused by power supply ripple, then the jitter will produce spurious signals that appear at specific Doppler frequencies. The SNR resulting from an rms timing jitter of δ_t seconds when sampling a signal of effective bandwidth B_e Hz is [14]

$$SNR\,(\text{dB}) = -10\log_{10}(2\pi^2 B_e^2 \delta_t^2) \tag{12.20}$$

where $B_e = \sqrt{B^2 + f_{if}^2}$, B is the signal bandwidth, and f_{if} is the intermediate frequency.

Figure 12-40 shows the pulses that are required to develop a transmit pulse for each PRI. The first line depicts the desired transmit time for a given pulse. The pretrigger, which precedes the transmit time, is usually required by a high-power transmitter, such as a TWT, to prepare the transmitter for developing a pulse. The IF and STALO switch commands apply the transmit signal to the transmitter at precisely the right time.

Receiver blanking blanks the signal into the receiver to protect it from unintentional but unavoidable transmit signal leakage. The ADC clock pulses define the range bin timing. Intrapulse processing commands instruct the signal processor to begin pulse compression processing. The timing signals shown represent the minimum requirements; many radar systems may require more signals than those depicted in Figure 12-40.

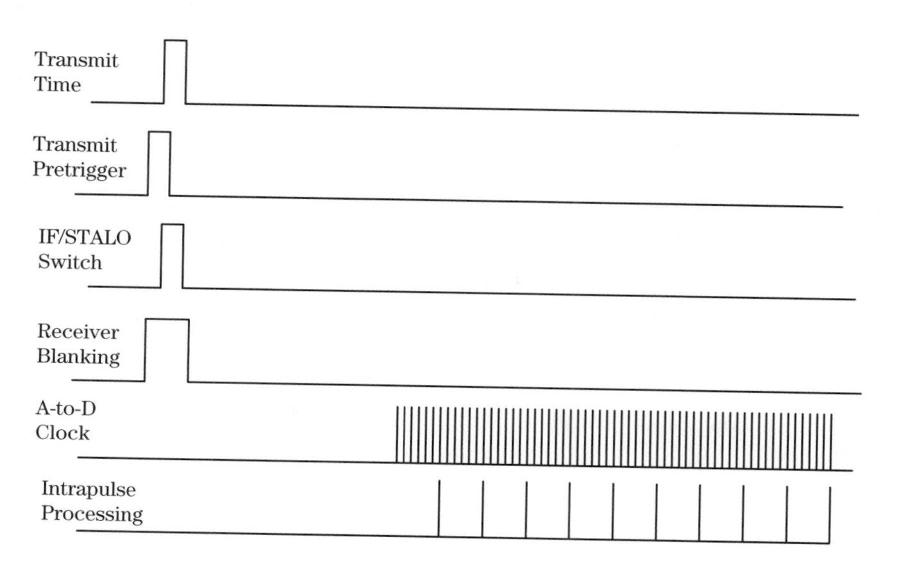

FIGURE 12-40 ■ Timing for a single-transmit pulse.

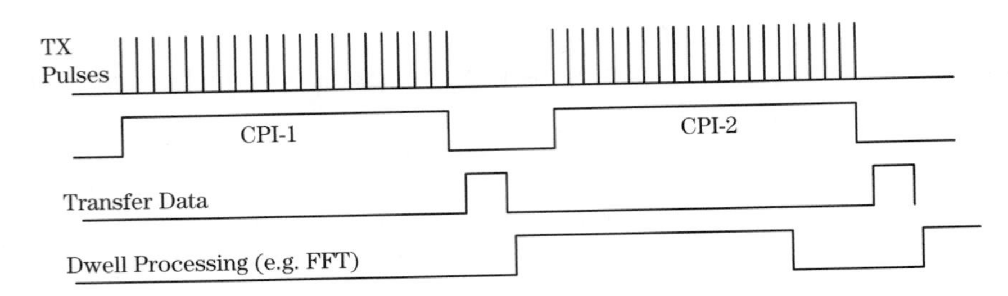

FIGURE 12-41 ■ Timing for multiple dwells.

Figure 12-41 depicts the timed events for two consecutive dwell times, or CPIs. Details will vary from system to system, but this represents the fundamental requirements. After the ADC digitizes the measured voltages in all the received range bins, these data are transferred to the processor as a group. Then, dwell processing such as weighting and FFT processing is begun and is carried out while data for the next dwell are being collected. In a real-time system, the dwell processing must be completed in an interval of no more than one dwell time or the processor will be unable to keep up with the input data.

Timing and control signals derived in the exciter have historically been developed using discrete digital integrated circuit (IC) technology such as counters, dividers, and logic gates. More recently, field programmable gate arrays (FPGAs) and similar technologies which allow reprogramming of the timing in the field have been used.

12.6 FURTHER READING

This chapter is unique in that most radar texts do not have material related to the design characteristics of the coherent radar exciter. Many of the design practices for stable and exciters are proprietary design practices held by the major radar contractors. The technical literature on the exciter technologies of, for example, low phase noise oscillators, low spurious signals, and low timing jitter is found primarily in such periodicals as *IEEE Transactions on Microwave Techniques*, *IEEE Transactions on Aerospace and Electronic Systems*, and *Proceedings of the IEEE International Frequency Control Symposia.* Two textbooks that treat the subject of phase noise in radar systems are [14,15].

12.7 REFERENCES

[1] Ewell, G.W., "Stability and Stable Sources," Chapter 2 in *Coherent Radar Performance Estimation*, Ed. J.A. Scheer and J.L. Kurtz, Artech House, Dedham MA, 1993.

[2] Leeson, D.B., "A Simple Model of Feedback Oscillator Noise Spectrum," *Proceedings of the IEEE*, vol. 54, pp. 329–330, February 1966.

[3] Raven, R.S., "Requirements for Master Oscillators for Coherent Radar," *Proceedings of the IEEE*, vol. 54, no. 2, pp. 237–243, February 1966.

[4] Lark Engineering catalog. Available at http://www.larkengineering.com.

[5] Wenzel Associates catalog. Available at http://www.wenzel.com.

[6] Poseidon Scientific Instruments catalog. Available at http://www.psi.com.au.

[7] Herley-CTI catalog. Available at http://www.herley-cti.com.

[8] Montress, G.K., and Parker, T.E., "Design and Performance of an Extremely Low Noise Surface Acoustic Wave Oscillator," *Proceedings of the 48th IEEE International Frequency Control Symposium,* pp. 365–373, June 1–3, 1994.

[9] Parker, T.E., and Montress, G.K., "Low Noise SAW Resonator Oscillators," *Proceedings of the 43rd Symposium on Frequency Control,* pp. 588–595, 1989.

[10] Spangler, T.L., Georgia Tech Research Institute, Atlanta, GA, private e-mail communication, August 12, 2008.

[11] Hittite Microwave Corporation catalog. Available at http://www.hittite.com.

[12] MITEQ Inc. catalog. Available at http://www.mitequationcom.

[13] CTT Inc. catalog. Available at http://www.cttinc.com.

[14] Scheer, J.A., "Timing Jitter Effects," Chapter 4 in *Coherent Radar Performance Estimation,* Ed. J.A. Scheer and J.L. Kurtz, Artech House, Dedham MA, 1993.

[15] Goldman, S.J., *Phase Noise Analysis in Radar Systems Using Personal Computers,* Wiley, New York, 1989.

12.8 PROBLEMS

1. a. Determine the total phase noise spectral density, in dBc/Hz, at an offset frequency of 12 kHz, without the effects of self-coherence, using the following parameters: Assume the phase noise is at the minimum level at frequencies above the PRF.

   $s(12\text{ kHz}) = -100$ dBc/Hz

   $s_{min} = -117$ dBc/Hz

   Receiver bandwidth = 1.2 MHz

   Radar PRF = 24 kHz

   Clutter cell distance = 1.0 km

   b. Determine the phase noise reduction factor associated with self-coherence.

2. If a coherent radar system is designed to detect and measure the velocity of ground moving targets at speeds up to 50 meters per second, what is the maximum error in measuring this velocity due to frequency drift of the system? The frequency drift is expected to be a maximum of +/ − 100 ppm over all thermal conditions.

3. The original design for a fixed-frequency coherent radar system is for operation at 9.375 GHz, using a STALO at 9.125 GHz and a COHO at 250 MHz. The object of the radar is to determine the radial component of velocity of ground-moving targets with an accuracy of +/ − 1 meter/second. The targets can have a maximum velocity of 50 meters per second. If the STALO is subject to frequency drift at a rate of +/ − 5 ppm per degree C, the COHO experiences a maximum of +/ − 20 ppm/degree C, and the system experiences a thermal range of +/ − 10 degrees C, does the carrier frequency stay close enough to the design to make the velocity measurement within specified limits? How much total frequency drift would be allowed?

4. A coherent ground-based radar system has a pulse length of 1 microsecond, a PRF of 8 kHz, and an antenna azimuth beamwidth of 3 degrees. There is a target at a distance of 60.0 km. This system is range-ambiguous. What is the apparent range to the target?

5. For the case described in problem 4:

a. What is the surface area of a clutter cell at the true range of the target?

b. What is the area of a clutter cell at the apparent range to the target?

c. At least the two clutter cells described in (a) and (b) provide clutter signal that competes with the target signal. Assuming a flat Earth, at least for the purpose of this problem, are there any other range ambiguities that compete?

6. For the case described in problems 4 and 5, if the average reflectivity for the clutter is -24 dB (m^2/m^2), what are the RCS values for each of the clutter cells that compete with the target?

7. For the conditions described in problems 4, 5, and 6:

a. Using the effect of only the closest clutter cell, what is the signal-to-clutter ratio for a 1 square meter target?

b. If the SCR at the detector must be 13 dB for the desired detection probability, how much must the clutter signal be reduced relative to that of the target (i.e., what is the desired CIF)?

8. For the case described in problem 7, Doppler processing is used to detect the target. If the dwell time is 3 ms, what is the nominal bandwidth of a Doppler bin, in Hz? What must the phase noise level (in dBc/Hz) be in the Doppler bin of interest to achieve the desired CIF? (Recall that the phase noise power spectral density in dBc/Hz must be multiplied by the filter bandwidth to determine the complete CIF.)

9. For a 10 GHz radar, what is the Doppler frequency shift for a ground target with a radial component of velocity of 50 meters per second? For a clutter cell at a range of 2 km, what is the self-coherence improvement (in dB) for this Doppler frequency?

10. If a 120 MHz crystal oscillator has a maximum thermal frequency drift of +/ − 2 ppm/degree C, then what would the maximum frequency change be from a 25 degree C environment to a 45 degree C environment?

11. If a COHO with a carrier frequency of 300 MHz has a maximum thermal frequency drift of +/ − 20 ppm/degree C and the STALO with a carrier frequency of 9.2 GHz has a maximum thermal frequency drift of +/−2 ppm/degree C, which oscillator contributes the most frequency drift when the ambient temperature changes from 72 degrees F to −20 degrees F?

12. a. What frequencies are produced at the RF port of a mixer if the two input frequencies are 1.5 GHz into the IF port and 7.8 GHz into the LO port? Consider all harmonics of both input signals up to and including the second harmonic and sums and differences of all harmonics.

b. If the desired output signal is the one at 9.3 GHz, how close is the closest undesirable signal above the desired frequency and below the desired frequency (in MHz)?

c. If the instantaneous bandwidth of the signal at 7.8 GHs is 1 GHz, do any of the undesirable signals enter an ideal 1 GHz wide band-pass, filter centered at 9.3 GHz?

d. If a filter having the characteristics shown in Figure 12-17 is used to suppress undesired signals by at least 24 dB, how many sections will be required for the filter for 3 dB bandwidth of 1 GHz? Assume all signals are less than 10 MHz bandwidth.

13. Using Figure 12-17, if you need to pass signals in the 3 dB passband from 5.85 to 6.15 GHz and suppress unwanted signals below 5.55 GHz and above 6.45 GHz by 50 dB, what is the fewest number of sections the filter must have? If the rejection requirement is reduced to –40 dB, can a filter with fewer sections be used? (How many sections would be required for this case?)

14. a. Using Figure 12-12, if the STALO or COHO signal into the network is at +7 dBm, what is the amplitude of the signal at the output of the filter? The following lists specify the salient features of the devices.

Switches:

Insertion loss	0.7 dB	
Isolation	23 dB	
Switching speed	25 ns maximum	
Maximum power	100 dBm	

Mixer:

Conversion loss	IF-to-RF	6 dB
	LO-to-RF	6 dB
IF	2.0–4.0 GHz	
LO frequency	2.0–10.0 GHz	

Filter:

Passband insertion loss	1.5 dB

b. If an amplifier is used at the output of the filter in (a), how much gain must the amplifier have to provide at least +10 dBm signal into the following transmit amplifier?

CHAPTER 13

The Radar Signal Processor

Mark A. Richards

Chapter Outline

13.1 Introduction 459
13.2 Radar Processor Structure 460
13.3 Signal Processor Metrics 462
13.4 Counting FLOPS: Estimating Algorithm Computational Requirements 464
13.5 Implementation Technology 472
13.6 Fixed Point versus Floating Point 480
13.7 Signal Processor Sizing 482
13.8 Further Reading 488
13.9 References 488
13.10 Problems 491

13.1 INTRODUCTION

The signal processor is the portion of the radar system responsible for extracting actionable information about the signal (target, clutter, and jamming) environment from raw radar signals. The signal processor is composed of two major elements: (1) the algorithms that analyze the radar data; and (2) the hardware on which those algorithms are hosted.

Historically, signal processing hardware has been constructed using a variety of analog and digital technologies. For instance, two-pulse moving target indication (MTI) has been implemented in older systems using analog mercury delay lines, and mixed analog and digital surface acoustic wave (SAW) devices have been used for pulse compression [1]. Early synthetic aperture radar (SAR) systems used remarkably elegant optical processors for image formation [2]. Many excellent radar systems have been fielded using analog signal processing techniques. The principal advantage of analog signal processing, whether performed in electronic circuits, microwave circuits, or even optical devices, is speed. Functions are computed at the speed of signal propagation, literally the speed of light or nearly so in most cases. However, analog techniques suffer from limited dynamic range, temperature sensitivity and aging, lack of flexibility, limited memory, and other disadvantages.

To counter these problems, digital signal processing (DSP) began to be applied to some radar functions in the 1960s. By the 1970s and 1980s the digital revolution enabled by Moore's Law (see Chapter 14) had progressed to the point where many radar signal processing operations could be implemented digitally in real time for a range of important systems. Today, radar designers can capitalize on another 20 years (10+ generations of Moore's Law improvements in digital technology) to implement signal processing techniques far beyond what can currently be done with analog circuitry.

DSP offers designers several important benefits. Digital processing results are more repeatable than those obtained with analog circuits. Both the accuracy and dynamic range are controllable through the choice of digital word lengths. It is often easier in digital processors to implement controllable or adaptive parameters, making these implementations more flexible than their analog counterparts. Digital processors are inherently more compatible with other digital portions of the radar system such as displays, data links, and the data processor.

By far the most important benefit of digital processing is a direct result of it being essentially programmable. Digital processors can implement a vastly greater range of functions than can analog systems, limited only by the imagination of the algorithm developers and the real-time constraints of the application. While very high-quality analog MTI filters and linear FM pulse compression filters have been fielded, digital techniques enable the development of much more advanced processing methods. Examples include adaptive MTI filters, waveform generators and matched filters for nonlinear frequency modulations, modern spectral analysis techniques, and elaborate space-time adaptive processors. These techniques are difficult and, in many cases, effectively impossible to implement in practical analog hardware. Thus, digital processing has made it possible to undertake more advanced algorithms than is feasible with only analog processing. This increased breadth of signal processing functions, and the increased radar performance it enables, is the deciding advantage for digital signal processing, and virtually all modern radar systems use digital processing.

In this chapter, common approaches to the architecture of radar signal processors are described, along with metrics for evaluating those architectures. The role of both the hardware and the software in determining the effectiveness of a processor is described.

13.2 RADAR PROCESSOR STRUCTURE

Figure 13-1 illustrates, at a very high level, the general structure of radar signal processing algorithms from the point of view of the computational resources required. Raw radar data from the receiver arrive at the input to the signal processor. These data are sampled at rates determined by the signal bandwidth; these rates can range from several hundred thousand samples per second to several gigasamples per second. Most modern radar systems use coherent receivers, so the input data will be complex-valued. Generally, the data will be at baseband, but in systems using digital intermediate frequency (*digital IF*) or digital in-phase/quadrature (*digital I/Q*) techniques it may be on a relatively low intermediate frequency carrier.

The early stages of a radar signal processing algorithm generally involve fixed functions that are applied to all of the data and that are not varied based on the actual content of the data. This computing style is often referred to as *streaming* processing. Examples include matched filtering/pulse compression, MTI and pulse Doppler filtering, coherent and noncoherent integration, SAR image formation, space-time adaptive processing (STAP), and constant false alarm rate (CFAR) threshold detection. The purpose of such processing is generally to improve the signal-to-interference ratio. The operations required

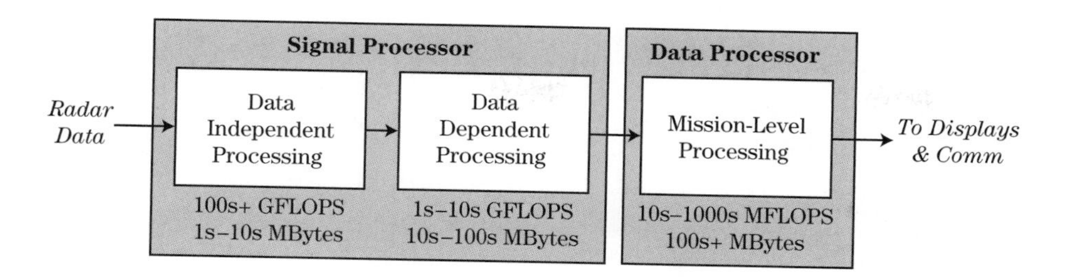

FIGURE 13-1 ■ Computational types, rates, and storage in a generic radar signal and data processor.

are primarily finite impulse response (FIR) digital filters, correlation, fast Fourier transforms (FFTs), and matrix-vector algebraic operations. Because of the high data rates and the relatively high computational complexity of the basic functions, this data-independent processing stage can require (in 2009) computational rates on the order of 1 billion to 1,000 billion operations per second (1 gigaops [GOPS] to 1 teraops [TOPS]). Because these arithmetic rates are so high, some systems use fixed-point arithmetic in these early operations due to its higher speed for a given size, weight, and space relative to a floating-point processor. Systems at the low end of the computational requirement, or those having larger-capacity signal processors, may use floating-point operations for greater arithmetic precision and ease of development. The computational rate is then measured in billions or trillions of floating-point operations per second (GFLOPS or TFLOPS).

While the signal processor stages nearest the receiver may have high computational rate requirements, their storage requirements tend to be modest, on the order of ones to tens of megabytes. This is because of the streaming nature of the calculations, which take a block or datacube of data, apply a filter or transform to it, and pass the modified data to the next operation. Intermediate results are not retained, so memory requirements at any one stage of the process are moderate.

After the initial signal conditioning operations, the data are next passed to a series of operations that do depend on the actual content of the data. Examples include single-dwell track measurements and SAR autofocus. For instance, the number of track measurements required depends on the number of targets detected by the CFAR processing. This may be only one or a few in air-to-air radars but may be thousands in some air-to-ground radars. Traditionally, the computational rates are reduced at this stage, but memory requirements may be increased.

The outputs from this data-dependent processing, such as detections ("hits") with associated position and SNR estimates, are then passed to higher-level processing that typically operates on time scales corresponding to multiple dwells. A typical example is track filtering with α-β or Kalman filters, which operates over many dwells. Processing at this level is often called *data processing*, although some systems include it in the signal processing. The computational rates at this stage may be still lower, in the tens of MFLOPS to ones of GFLOPS range, but the storage requirements may increase to hundreds of megabytes or more.

Emerging system designs in 2009 incorporate increasingly elaborate postdetection processing, for instance, for tracking, classification, and recognition of thousands of potential targets. In some newer proposed systems the computational load at the back end of the processor may actually exceed that of the front end.

Nonetheless, the typical radar signal processor reduces the computational rate but increases the storage requirements in each successive stage. The implementation technology for digital radar signal processors mimics this flow, as shown in Figure 13-2. The early,

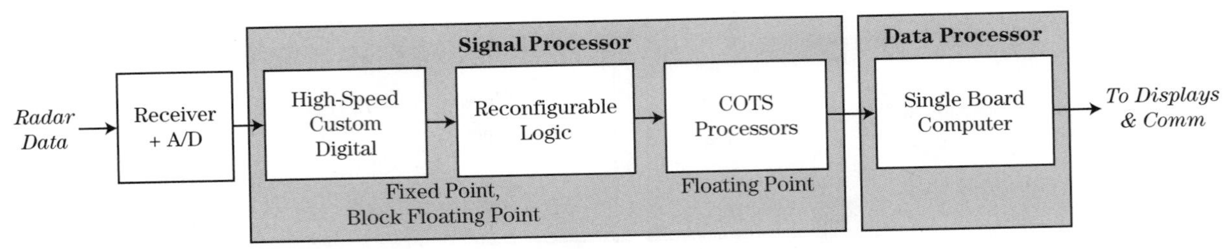

FIGURE 13-2 ■ Signal processor implementation technologies.

highest-speed data-independent operations may require the use of custom *application-specific integrated circuits* (ASICs) to obtain the required processing rates under severe size, weight, and power constraints. In many applications, this hardware may use fixed-point or block floating-point arithmetic; the latter is a compromise between the logic complexity and arithmetic precision of fixed-point and full floating-point arithmetic. The data-dependent operations can often be hosted on specialized *commercial off-the-shelf* (COTS) programmable processors designed for high-speed signal and data processing. Such processors typically use floating-point arithmetic. Finally, the relatively low computational rates, high memory requirements, and control complexity of mission-level processing can be accommodated with *single board computers* (SBCs), which are essentially equivalent to standard workstations. The various types of digital computing technology are discussed further in section 13.5.

13.3 SIGNAL PROCESSOR METRICS

The fundamental metric in evaluating a signal processor implementation is *time to solution*, which is the time required to complete all specified processing once the required data become available. In real-time radar signal processors, processing of one batch of data must be finished in time to accept the next batch of data, a so-called hard real-time requirement. If a processor does not have a hard real-time requirement, the designer may have additional flexibility to trade performance versus processor size or cost.

The time to solution depends on both the processor software and hardware. The hardware speed is determined by the type of digital processing technology being used, the amount of it available, and the speed at which it is operated. Software influences include the particular algorithms used to implement signal processing operations and the software tools such as languages, compilers, and libraries used to map those algorithms to the hardware.

13.3.1 Hardware Metrics

Signal processor metrics fall into two classes: (1) those describing the algorithms; and (2) those describing the implementation hardware. Consider hardware metrics first. The two most important are *throughput* and *latency*.

Throughput describes the rate at which a processor performs arithmetic operations. It is measured in floating-point operations per second; at radar rates, units of MFLOPS and GFLOPS are most common. If a processor uses fixed-point arithmetic, units of operations per second (MOPS, GOPS) are used.

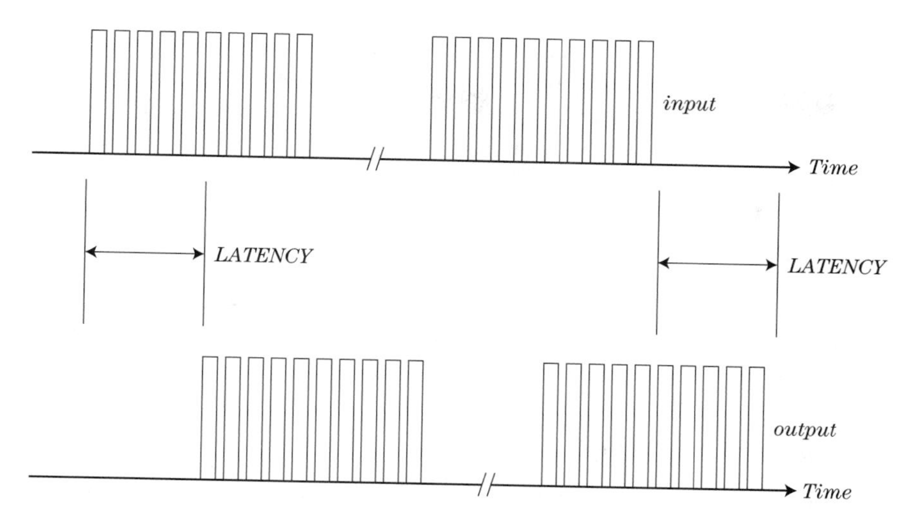

FIGURE 13-3 ■ Illustration of latency.

Another measure of processor speed is millions of instructions per second (MIPS). While MIPS is appropriate for assessing processor speed on general operations, it is not appropriate for quantifying floating-point arithmetic operations. Although some processors are designed to complete multiple floating-point additions or multiplications in a single cycle, others require many cycles, so MIPS are not a reliable indicator of floating-point arithmetic speed. However, MIPS may be appropriate for assessing arithmetic speed in fixed-point processors or custom logic.

Many signal processing operations, especially in the data-independent stages, take a finite block of data at their input and produce another block of data at their output. Examples include FFTs and FIR filters, among many others. The elapsed time from when the first element of the input block arrives at the processor to when the first element of the output block is available is called the latency of the operation.[1] This idea is illustrated in Figure 13-3. If throughput measures speed, then latency measures quickness.

High throughput is a requirement for nearly all radar signal processors. Minimizing latency is of greater importance in some systems than in others. For instance, ground MTI (GMTI) radars using adaptive processing to cancel jammers need short latencies to adapt quickly to the interference environment. Latency is less important in SAR systems, which generate imagery that may not be consumed, or even fully formed, until significantly after the data are collected. Generally, short latency requires high throughput, but the converse is not true; pipeline architectures can achieve very high throughputs but still have long latencies.

Some of the most commonly cited hardware metrics are related to peak throughput, but this value alone does not adequately characterize the technology. A processor that achieves 100 GFLOPS might be a single specialized high-performance circuit board or a large cluster of more generic workstations. Consequently, achievable throughput is most useful when normalized by the size, weight, power consumption, or cost of the processor. The corresponding metrics are GFLOPS per cubic centimeter (or cubic inch or foot or liter), kilogram (or ounce or pound), watt, or dollar.

[1]Latency can also be measured from last input to last output, or between any other start and end point definitions, as long as a consistent definition is used in comparing implementations.

Another class of hardware metrics relates to interconnection bandwidth. Examples include the available data rates in bits per second (bps, Mbps, Gbps) to and from a chip or board. Another metric is defined by dividing a processing system into two halves of equal capacity, in such a way that the interconnect bandwidth across the boundary is minimized. The *bisection bandwidth* is the rate of data transfer across that boundary. Bisection bandwidth provides a measure of the interconnect richness and speed in a multiprocessor board or system. As with throughput, interconnect metrics can be normalized with respect to size, power, weight, and cost.

Other metrics describing processor hardware relate to development costs. For example, systems using floating-point hardware typically require less software development effort than those using fixed-point arithmetic (see section 13.6)

13.3.2 Algorithm Metrics

Algorithm metrics traditionally focus on the actual number of arithmetic operations required to compute the specified functions. The quantity of floating-point operations is denoted by the acronym FLOPs (MFLOPs, GFLOPs); note the difference between this and the rate of FLOPs per second, or FLOPS. If an algorithm requires one MFLOPs to compute, and it is implemented on a processor with a throughput of one GFLOPS, the time to completion might be expected to be 1 ms. Reality is more complex, particularly in modern processors.

Another important algorithm metric is the working memory size required to hold the data being processed, any intermediate results, and any algorithm coefficients or parameters required (e.g., the impulse response coefficients in an FIR filter). Algorithms with larger working sets will require larger hardware memories and, if large enough, will suffer performance penalties whenever required data must be fetched from nonlocal memory.

Closely associated with memory size are the concepts of *spatial locality* and *temporal locality*. An algorithm with high spatial locality is one where successive data values used by the algorithm are likely to be stored in adjacent or nearby memory locations, making memory access more efficient. High temporal locality means that all accesses to the same data value tend to be clustered in time (i.e., the data are used for some period of time and then not reused later); such data values do not have to be repeatedly reaccessed from memory, again improving efficiency.

13.4 | COUNTING FLOPs: ESTIMATING ALGORITHM COMPUTATIONAL REQUIREMENTS

13.4.1 General Approach

A first estimate of the required throughput of the radar signal processor can be developed by considering each of the major modes of the radar in turn. For each mode, the major operations to be performed on the data are identified and broken down into basic signal processing functions such as linear filters and discrete Fourier transforms (DFTs). The number of arithmetic operations for each function is totaled. This total will be a function of the various parameters such as data vector lengths, impulse response lengths, and transform size. To convert this number to a computational *rate*, which is more relevant to

TABLE 13-1 ■ Number of Complex Arithmetic Operations per Output Point in Basic Signal Processing Functions[a]

Operation	Multiplications Required per Output Point	Additions Required per Output Point
FIR filter, length L	L	$L-1$
IIR filter, numerator order L, denominator order P	$L+P+1$	$L+P-1$
Autocorrelation, kernel length N	N	$N-1$
Addition of two vectors	0	1
Multiplication of two vectors	1	0
Fast Fourier transform, Cooley-Tukey radix 2	$(1/2)\log_2 N$	$\log_2 N$

[a]Complex data and coefficients assumed.

signal processor sizing, the elapsed time allotted for the operations must also be considered. The final result is a computational rate in FLOPS for that mode. The process is repeated for each mode of the particular radar, such as surveillance, stripmap imaging, spotlight imaging, GMTI, airborne moving target indication (AMTI), target tracking, or target identification. The resulting analysis of computational rates provides an initial estimate of the required signal processor throughput and also identifies the major modes driving the processor size.

As part of the same analysis, the working memory requirements can be estimated by considering the size of the data block at the input and output of each operation as well as any coefficients or auxiliary data required for that block. Provided data word lengths can be specified at each stage, the data rates into and out of each block can similarly be estimated. A spreadsheet is a common tool for building up the computational, memory, and data rate requirements because of the ease with which mode parameters can be varied to see their effect on processing requirements.

The computational requirements of the low-level signal processing operations into which the major operations are decomposed are usually easily determined from the operation formulas or from standard DSP texts such as [3]. Table 13-1 lists these operation counts for the most basic functions. In each case, the formula given is the number of operations per output point computed. For an FIR filter with an impulse response of length L (order $L-1$), computation of the complete output for an N-point input vector requires computing $L+N-1$ output points, while for an N-point FFT (N assumed to be a power of 2 in the table), the output vector is also N points. Element-wise operations on N-point vectors also return N-point results. The length of the autocorrelation of a sequence of length N is $2N-1$ lags.

Many radar signal processing operations, particularly in the earlier stages of the processing, are performed on coherent data and are therefore complex-valued. Thus, the formulas in Table 13-1 should be interpreted as complex multiplication and addition counts. To estimate hardware requirements, it is necessary to convert this to equivalent real-valued operations. For example, addition of two complex numbers requires two real additions, one for the real parts and one for the imaginary parts, a total of two real operations. Multiplication of two complex numbers using the most common algorithm requires four real multiplications and two real additions, a total of six real operations. Table 13-2 lists conversion factors for these and several other complex floating-point operations (CFLOPs) to real floating-point operations (RFLOPs). The last three rows of the tables are estimates

TABLE 13-2 ■ Conversion from Complex to Real FLOPs

Operation	Number of Real Floating-point Operations Assumed
complex-complex multiply	4 real multiplies + 2 real adds = 6 RFLOPs
complex-complex add, subtract	2 real adds + 2 real subtracts = 2 RFLOPs
magnitude squared	2 real multiplies + 1 real add = 3 RFLOPs
real-complex multiply	2 real multiplies = 2 RFLOPs
complex divided by real	1 real inverse + 1 real-complex multiply = 8 RFLOPs
complex inverse	1 conjugate (not counted) +1 magnitude-squared + 1 complex divided by real = 11 RFLOPs
complex-complex divide	1 complex inverse + 1 complex multiply = 17 RFLOPs

because the number of operations for a real inverse, which is required in each of those lines, can vary significantly on different machines, with numbers from four to eight times the number of operations for a real multiply being common [4]. In the table, a value of six RFLOPs per real inverse has been assumed.

The most compute-intensive signal processing operations, such as convolution, correlation, fast Fourier transforms, and vector-matrix operations, tend to require approximately equal numbers of additions and multiplications. As a simple example, the dot product of two complex vectors of length N requires N complex multiplications and $N-1$ complex additions. Because complex additions require two RFLOPs and complex multiplications require six RFLOPs, a good rule of thumb for front-end signal processing algorithms is that the number of real operations is approximately four times the number of complex operations.

Figures 13-4 and 13-5 illustrate this approach for estimating computational loading. This example is for a fairly complex wideband GMTI mode incorporating reduced-dimension STAP. Each box in the flowgraph of Figure 13-4 represents a major processing operation, such as filtering the input into narrow subbands or pulse compression. Each in turn is decomposed into more basic signal processing operations. For example, the subband filtering would probably be implemented using polyphase FIR filtering, while the pulse compression is implemented using fast convolution, that is, with FFTs. Figure 13-5 is a detailed spreadsheet for the same example in which the user specifies major system parameters in the left portion of the spreadsheet. The middle portion computes a number

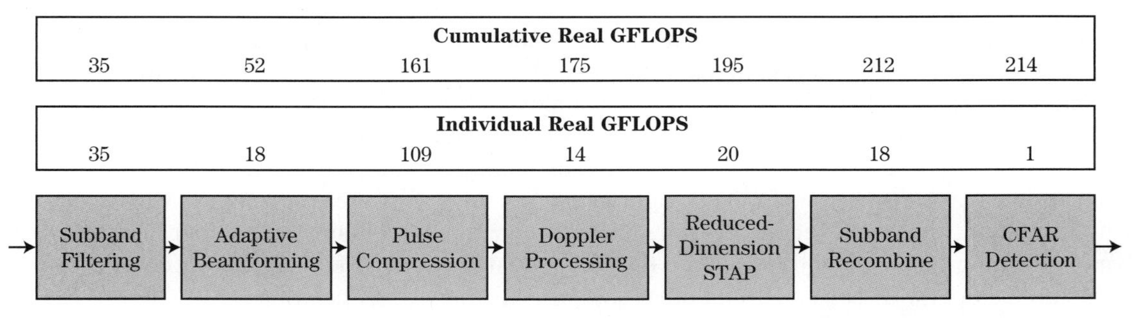

FIGURE 13-4 ■ Estimated computational load (real operations) for a wideband STAP-GMTI radar.

Input parameters:

Parameter	Value
Signal bandwidth (MHz):	50
Pulse length (us):	100
IF sampling rate (Msamp/s):	125
No. of antenna channels:	6
RMB rule averaging factor:	5
RMB rule averaging factor for STAP:	5
No. of subbands (subbanding FFT size):	16
Receiver duty factor (single pulse):	0.80
Decimation factor:	12
Subbanding prototype filter length:	240
No. of canceller beams formed:	8
No. of beams combined for STAP:	3
No. of Dopplers combined for STAP:	4
No. of STAP beams formed:	4
PRF (pps):	2000
PRI (sec):	5.00E-04
Number of pulses per CPI:	30
CPI duty cycle:	0.8
CPI (sec):	0.019
Adaptive weight update interval (s):	0.01
STAP weight update interval (s):	0.1
CFAR window size (cells):	20
CFAR guard cells:	4

Derived parameters:

Parameter	Value
Polyphpase subbanding:	
Subband filter length:	15
IF oversampling factor:	2.50
Subband sampling rate (Msamp/s):	10.42
Subband bandwidth (MHz):	10.42
Subband spacing (MHz):	7.81
Subband overlap percentage:	25.00%
Number of subbands processed:	7
Suggested subband filter length:	12
Number of range bins:	50000
Number of range bins per subband:	4167
# of subband filter output samples:	4181
Adaptive Beamforming:	
Number of DOF:	6
Number of snapshots averaged:	30
Pulse Compression:	
Pulse samples:	1042
No. of Overlap-add segments:	5
Length of OLA segments:	837
Length of filtered segments:	1878
Pulse compression FFT size:	2048
Doppler Processing:	
Doppler FFT size:	32
STAP:	
Number of STAP DOF:	12
Number of snapshots averaged:	60
Number of STAP steering vectors:	128
Polyphase Recombining:	
Recombine filter length:	15
Recombine FFT size:	8
Output sampling rate (Msamp/s):	83.3
Number of output range bins:	33334

Computational Load:

Processing Step	GMULTS	GADDS	Combined GFLOPS	GFLOPS	Cumulative GFLOPS
Polyphase subbanding:					
Polyphase FIR filtering:	12.0	11.2		23.3	23
Two-channel FFT:	3.2	4.8		8.0	31
Unravel two-channel FFT:		0.8		0.8	32
Subband modulation:	1.4	1.4		2.8	35
Polyphase subbanding GFLOPS:				34.9	
Adaptive Beamforming:					
SMI weight estimation:			0.26	0.3	35
Diagonal loading:		1.26E-04		0.0	35
Beam formation:			17.2	17.2	52
Adaptive Beamforming GFLOPS:				17.5	
Pulse Compression:					
Forward FFT:	20.2	30.3		50.5	103
Vector multiply:	3.7	3.7		7.3	110
Inverse FFT:	20.2	30.3		50.5	161
OLA additions:		0.7		0.7	161
Pulse Compression GFLOPS:				109.0	
Doppler Processing:					
Weighting:	0.7			0.7	162
Range curvature compensation:	1.5	1.5		3.0	165
Doppler FFT:	4.0	6.0		10.0	175
Doppler Processing GFLOPS:				13.7	
STAP:					
SMI Weight estimation:			0.7	0.7	176
Diagonal loading:		8.40E-07		0.0	176
Beam formation:			18.8	18.8	195
STAP GFLOPS:				19.5	
Polyphase Recombining:					
Subband modulation:	0.8	0.4		1.2	196
IFFT:	1.4	2.0		3.4	199
Polyphase FIR filtering:	6.9	6.4		13.2	212
Polyphase recombining GFLOPS:				17.9	
CFAR:					
Squared-magnitude detection:	0.5	0.2		0.7	213
Threshold estimation:	0.2	0.5		0.7	214
CFAR GFLOPS:				1.4	
TOTAL GFLOPS:				214.0	

FIGURE 13-5 ■ Example of spreadsheet for estimating computational load for a wideband STAP-GMTI system. Counts are for real operations.

of derived parameters, and the right-hand portion is the actual FLOPS computation. Note that the underlined subheadings in this section correspond to the stages of the signal flowgraph in Figure 13-4.

Returning to Figure 13-4, the row of numbers immediately above the operations is the spreadsheet estimate of the computational rate in real GFLOPS required for each individual operation for the particular parameters used. The upper row of numbers is the cumulative operation rate through that stage.[2] For instance, the first three operations require an estimated 35, 18, and 109 GFLOPS, respectively. The upper set of numbers reflects the total of 161 GFLOPS for those three operations.

Additional examples of this type of operation-counting analysis of complex radar signal processing modes are given in chapters 6 and 15 of [5].

13.4.2 Mode-Specific Formulas

The radar literature provides parametric formulas for rough estimates of computational loading for certain modes. Two common examples are SAR imaging and STAP.

An example of a simple formula applicable to stripmap SAR is developed in [6]. The formula begins by factoring the total computational rate F_t in FLOPS into two terms, the number of image pixels generated per second (R_p) and the number of FLOPs required per pixel (F_p)

$$F_t = F_p R_p \tag{13.1}$$

Consider the pixel rate first. This can in turn be expressed as the number of pixels generated per pulse (N_p) times the pulse repetition frequency (*PRF*),[3]

$$R_p = N_p PRF \tag{13.2}$$

The number of pixels per pulse is very design-specific but is at most the unambiguous range interval, which is the maximum SAR swath depth divided by the range resolution

$$N_p \leq \frac{(c \cdot PRI/2)}{(c/2B)} = B \cdot PRI = \frac{B}{PRF} \tag{13.3}$$

where B is the waveform bandwidth. This result assumes that the range sample spacing equals the range resolution, that is, one range sample per range pixel. Combining equations (13.2) and (13.3) shows that the pixel rate is bounded by a number on the order of the waveform bandwidth,

$$R_p \leq \left(\frac{B}{PRF}\right) PRF \quad \Rightarrow \quad R_p \sim O(B) = O\left(\frac{c}{2\Delta R}\right) \tag{13.4}$$

where ΔR is the range resolution of the SAR. The notation $O(x)$ means "on the order of x."

[2] Slight differences between the listed cumulative GFLOPS and the running sum of the individual GFLOPS are due to rounding.

[3] This assumes single-look SAR image products but is sufficient for rough processor sizing. The approach is extendable to multilook images.

F_p, the number of FLOPs per pixel, is a strong function of the image formation algorithm used. A middle-of-the-road estimate can be developed from the viewpoint that SAR imaging requires a matched filtering (correlation) operation in cross-range. The number of samples N_{CR} of the correlation kernel equals the number of cross-range samples (pulses) that contribute to a single pixel; this number is upper bounded in stripmap SAR by the cross-range spread of the antenna beam expressed in SAR pixels

$$N_{CR} = \frac{R\theta_{az}}{\Delta CR} = \frac{R\lambda}{D_{az}\Delta CR} \tag{13.5}$$

where θ_{az} and D_{az} are the antenna beamwidth and size in the cross-range (width) dimension, ΔCR is the cross-range resolution of the image, and λ is the radar wavelength. Since a correlation operation requires a multiplication and an addition for each sample of the kernel function, the number of operations per output pixel, F_p, is twice N_{CR}. The lower bound on ΔCR for a fully focused stripmap SAR is $D_{az}/2$, so that $D_{az} = 2\Delta CR$ [7]; using this result in equation (13.5) gives

$$F_p = 2N_{CR} \leq \frac{R\lambda}{(\Delta CR)^2} \quad \Rightarrow \quad F_p \sim O\left(\frac{R\lambda}{(\Delta CR)^2}\right) \tag{13.6}$$

The estimate of equation (13.6) is most appropriate for a simple range-Doppler imaging algorithm or similar. More complex algorithms, such as the ω-k (also called range migration) algorithm with Stolt interpolation may require four times as many operations [8]. On the other hand, FFTs can be used to implement the brute-force correlation operation implied by this discussion if range curvature is insignificant; in this case, F_p may be reduced significantly.

Combining equations (13.4) and (13.6) gives the final result

$$F_t \sim O\left(\frac{R\lambda c}{\Delta R(\Delta CR)^2}\right) = O\left(\frac{R\lambda c}{(\Delta CR)^3}\right) \tag{13.7}$$

where the last step assumes "square pixels," $\Delta R = \Delta CR$. Other variations on this result are given in [6–9]. Note that the computational load rises rapidly with increasingly fine resolution: Improving (reducing) the resolution by a factor of 2 produces an 8× increase, nearly an order of magnitude, in the computational rate.

As an example, consider applying this formula to the Shuttle Imaging Radar-C (SIR-C). This system operates at 1.28 GHz from an orbit altitude of 225 km. At 25 m resolution, the resulting estimated computational rate is 506 (complex) MFLOPS. For a more aggressive system, consider a low Earth orbit (LEO) satellite operating at X band. Using an orbit altitude of 770 km and an RF of 10 GHz, the computational rate is estimated at 55 GFLOPS at a resolution of 5 m. This number is probably too low, since a higher-quality algorithm would be required at that resolution. Thus, the true rate is more likely to be three to four times higher, on the order of 150–200 GFLOPS.

Similar estimates have been developed for STAP, another compute-intensive signal processing operation. Derivation of this result requires analysis of the computational load of matrix algebra operations such as the Q-R decomposition, which is beyond the scope of this chapter. A typical result for the "voltage domain" approach is [7]

$$\text{RFLOPS} = \begin{cases} F_u\left[(8L_sR^2 - 2.67R^3) + 12KPR^2 + 26KPR - 4KP\right] & \text{(weight computation)} \\ (8R-2)KPLF_{CPI} & \text{(weight application)} \end{cases} \tag{13.8}$$

where

- L_s is the number of data snapshots averaged to estimate the interference statistics;
- R is the number of sensor degrees of freedom (DOF, product of the number of slow-time pulses and number of antenna phase centers);
- K is the number of Doppler beams tested for targets;
- P is the number of angles of arrival (AOA) tested;
- F_u is the weight update rate; and
- F_{CPI} is the number of coherent processing intervals (CPIs) per second

Equation (13.8) is expressed in units of RFLOPS and includes both multiplications and additions. Note that, as with SAR, the computational load increases as the cube of a key parameter, in this case the number of degrees of freedom R of the STAP system.

As an example, consider a system with $N = 8$ phase centers and $M = 20$ pulses in the CPI, giving $R = MN = 160$ DOF. Limiting the covariance estimation matrix loss to 3 dB requires $L_s = 2R = 320$ reference cells [7]. Assume that the weights are updated at rate of $F_u = 20$ updates per second. Assume also that $K = 20$ Doppler frequencies and that $P = 12$ AOAs are tested in each range bin. Equation (13.8) then gives a computational rate of 2.6 GFLOPS to compute the adaptive weights. Assuming $F_{CPI} = 375$ CPIs per second then gives 115 GFLOPS to apply the weights and a total computational load of 117.6 GFLOPS for this system. Similar results for rough estimates of STAP processing loads are given in [10].

For both SAR and STAP-GMTI, it is important to remember that many algorithm variations exist. Some of the limitations to the SAR estimates were previously discussed. In STAP, much attention has been devoted to reducing the number of degrees of freedom to reduce both the data needed for interference statistics estimation and the computational load [11]. Thus, the formulas in this section, while useful as examples of an approach to developing rough estimates of computational loads, must be used with caution. Serious analysis of processor requirements calls for more detailed analysis specific to the particular algorithms of interest, typically starting with a spreadsheet analysis as discussed earlier.

13.4.3 Choosing Efficient Algorithms

It is frequently the case that a given signal processing operation can be implemented with more than one algorithm. A simple example is the DFT. A brute-force implementation of an N-point DFT requires N^2 complex multiplications and a similar number of complex additions. As is well known, however, the DFT can be computed much more efficiently with one of the many FFT algorithms. For the oft-used case where N is a power of 2, the Cooley-Tukey radix 2 FFT [12] computes the DFT using $(N/2)\log_2 N$ complex multiplications, a reduction by a factor of $2N/\log_2 N$. This speedup factor is illustrated in Figure 13-6, which also makes clear that, as is the case with most fast algorithms, the impact of the FFT algorithm increases with the problem order. A common benchmark is the $N = 1{,}024$ (1K) DFT, for which the speedup is a factor of just over 200.

Modern FFT algorithms of order $N\log N$ exist for almost all lengths, not just powers of 2, as well as for special cases such as real-valued inputs, or those having certain symmetries. In addition, modern FFT libraries have the capability to adaptively construct algorithms optimized for particular processors, matching the algorithm structure to the memory hierarchy and other characteristics of the host processor [13]. In fact, the

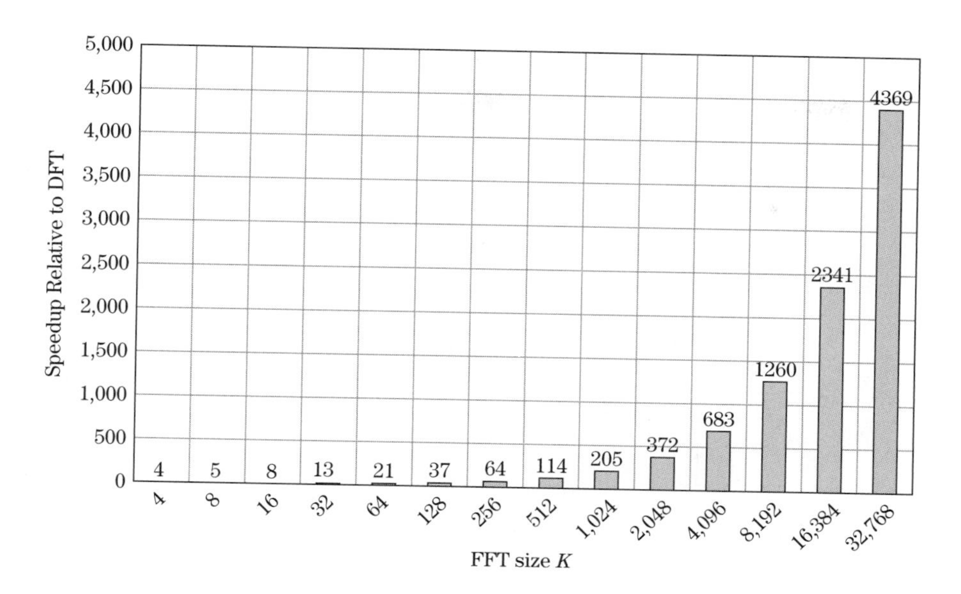

FIGURE 13-6 ■ Speedup of Cooley-Tukey radix-2 FFT relative to brute-force DFT.

complexity of modern processors makes it essential to optimize the FFT algorithm (and other algorithms as well) for the particular processors being used.

Another example of a core signal processing algorithm that can be computed using different algorithms is convolution. The brute-force algorithm for convolving an N-point data sequence $x[n]$ and an L-point filter impulse response $h[n]$ is

$$y[n] = \sum_{m=0}^{L-1} h[m]x[n-m] \tag{13.9}$$

As noted earlier, this requires L multiplications for each of the $N+L-1$ nonzero output samples, a total of $L(N+L-1)$ multiplications; if the data and impulse coefficients are complex, these are complex multiplications.

The same computation can be done using the multiplication property of DFTs; specifically

$$\begin{aligned} y[n] &= \text{DFT}^{-1}\{Y[k]\} = \text{DFT}^{-1}\{H[k]X[k]\} \\ &= \text{DFT}^{-1}\{H[k]\cdot \text{DFT}\{x[n]\}\} \end{aligned} \tag{13.10}$$

$H[k] = \text{DFT}\{h[n]\}$ is generally precomputed, so the DFT to obtain $H[k]$ is not shown in equation (13.10). Consequently, one forward and one inverse DFT are required to compute the convolution output. The minimum size of the DFT is $N+L-1$ (see chapter 14). The advantage to this *fast convolution* approach is that efficient FFT algorithms can be used to compute the forward and inverse DFTs.

Figure 13-7 illustrates the number of multiplications required to compute $y[n]$ by brute-force convolution (13.9), fast convolution (13.10), or the overlap-add (OLA) method [3] using segment sizes of $L = 100$ and 500. In the latter cases, fast convolution is used for the small segment convolutions with power-of-2 FFTs. The "stair steps" in all of the results except the direct convolution occur when it is necessary to increase the FFT size to the next power of 2, for example, from 512 to 1,024.

Of the algorithms considered, the brute-force convolution is the least efficient for most, but not all, input data lengths. The most efficient choice of those shown is the

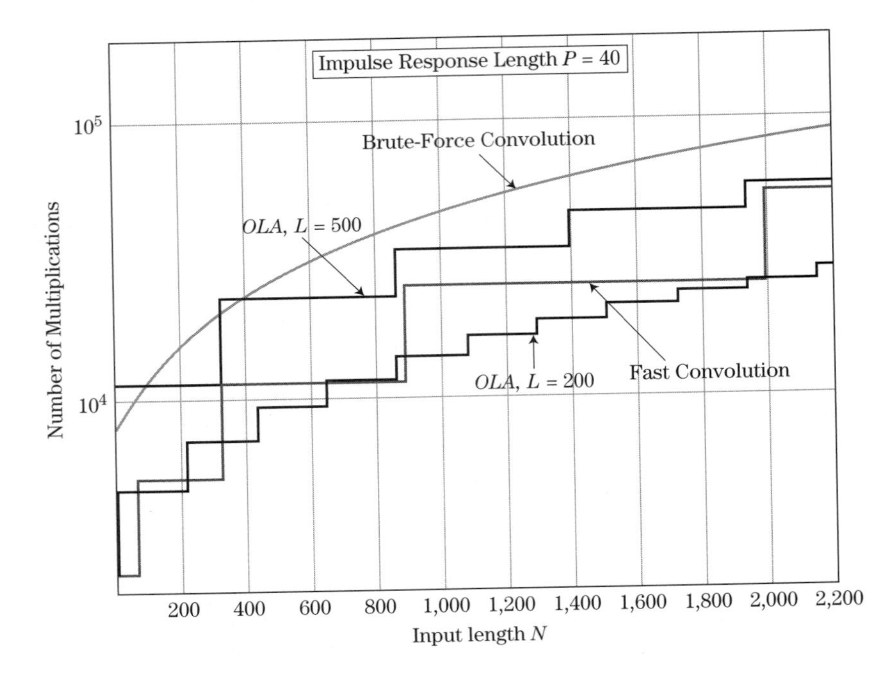

FIGURE 13-7 ■ Number of multiplications required to compute a linear convolution using various algorithms.

OLA algorithm with short $L = 100$ sample segments for most, but again not all, lengths. The difference in efficiency of the various algorithms is significant. For instance, with $N = 1{,}000$, the brute-force convolution requires 41,560 multiplications, while the OLA with $L = 100$ requires only 13,824, a reduction by a factor of three. For most input lengths, the ratio between the number of multiplications required for the least and most efficient algorithms is between 3 and 3.5. Thus, the overall processing workload can be significantly affected by the choice of the specific algorithm used for each of the major operations.

These examples show that implementing the series of operations that comprise a radar mode with other than the most efficient algorithms can result in large increases in the total number of required operations. This translates into either an increase in the time to solution or an increase in the size, weight, power, and cost of the required signal processor to compensate for the increased workload. If real-time processing is required, the time to perform the operations required on one block of data cannot exceed the block time or the processor will not be ready for the next block of data.

The previous discussion has focused on minimizing the number of operations in each algorithmic step. As noted earlier for FFTs, in modern processors the optimization of data movement and hierarchical memory is often more important than the number of arithmetic operations in determining run time. Thus, in some cases the best algorithm may not be the one with the fewest operations.

13.5 IMPLEMENTATION TECHNOLOGY

13.5.1 Analog-to-Digital Conversion

At some point in every modern radar system, the receiver output will be converted from an analog signal to a digital signal by an *analog-to-digital (A/D) converter*. The required sampling rate is determined through the Nyquist criterion by the instantaneous bandwidth of the receiver output, while the number of bits in the digital sample is determined by the instantaneous dynamic range of the data and signal-to-quantization noise ratio (SQNR) requirements of the system. Nyquist sampling and the effect of the number of bits on dynamic range and SQNR are both discussed in chapter 14.

The required sampling rate is generally equal to the radar instantaneous waveform bandwidth B, perhaps increased by a safety margin of 10% to 20% to allow for the imperfect bandlimiting of real waveforms and anti-aliasing filters. In high-resolution radars the resulting sampling rates can be very high, often tens to hundreds of megasamples per second and in some cases even reaching rates of gigasamples per second. The use of digital I/Q or digital IF techniques for non-baseband sampling can increase the required rate relative to the waveform bandwidth by a factor of 2.5 to 4, further exacerbating the problem [7]. On the other hand, many very wideband systems use a specialized linear FM waveform processing method called *stretch processing* (see [7]) to reduce the analog signal bandwidth prior to A/D conversion, often by an order of magnitude or more.

Except in a few specialized situations, the minimum number of bits required is at least 6 and preferably 8, while 12 bits or more are desired in many applications [14]. Interest is increasing in the use of sigma-delta converters, which trade very small quantizer word lengths (as little as one bit) for much higher sampling rates [15].

Figure 13-8 summarizes the state of the art in A/D converters in 2006 [16]. The quantity plotted is the effective number of bits (ENOB) versus sampling rate. ENOB is inferred from measured SQNRs as described in [17], which also discusses many other A/D metrics and limiting factors. That data show that A/D converter word lengths tend to drop about two bits per decade of sampling rate. However, an ENOB of 8 bits is achievable at rates up to about 1 Gsamples/sec, and an ENOB of 12 bits is achievable at rates up to approximately 100 Msamples/sec. These are quite adequate for most radar applications. Improvement in achievable ENOB over time is somewhat slow, constrained by progress in overcoming design limitations such as sample timing jitter and fundamental circuit limitations. The data in [17] estimate that ENOB increased about 3.5 bits at a given sampling rate between 1999 and 2006 and project a further increase of about 2 bits by 2015.

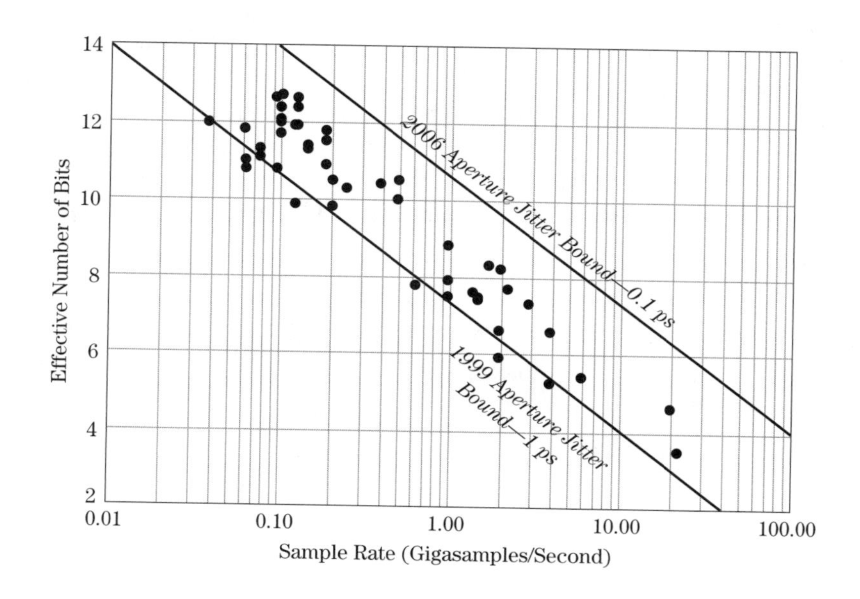

FIGURE 13-8 ■ Analog-to-digital converter performance. (Data courtesy of Dr. Robert Walden and The Aerospace Corporation. Used with permission.)

13.5.2 Processor Technologies

As was suggested in Figure 13-2, the major types of computing hardware used in implementing radar signal processors can be grouped into four broad classes: ASICs; reconfigurable modules; programmable signal processing modules; and general-purpose computing hardware. The first three classes of hardware are used for signal processing and the last for data processing.

Here, *programmable DSP module* refers to both microprocessors specialized for digital signal processing, such as those manufactured by Texas Instruments, Inc. (TI) and Analog Devices, Inc., and modular board-level products such as those manufactured by Mercury Computers, Sky Computers, CSPI, and Radstone. The board-level products typically are multiprocessors based on microprocessors such as the TI DSPs, the IBM PowerPC series, or the Analog Devices SHARC series. *General-purpose processors* refer to workstations and SBCs that are effectively workstations on a single modular circuit board, intended to be used as part of a larger system.

Reconfigurable hardware modules refers to computing elements, typically at the modular board level, usually based on *field programmable gate arrays* (FPGAs) as the core computing elements [18]. The FPGAs are manufactured by companies such as Xilinx and Altera and are integrated into board-level devices by companies such as Annapolis Microsystems and Nallatech. Classically, FPGAs provided a "sea of gates" (actually, logic cells organized into registers and multi-input look-up tables) arranged in a grid layout. An externally supplied control data stream could reconfigure the interconnection of the various registers and blocks, allowing the designer to construct semicustom complex logic functions. Newer FPGAs often add fixed-function blocks (e.g., adders or multipliers), memory, and even small programmable microprocessor cores to the basic reconfigurable logic.

Table 13-3 compares the signal processing technology classes in broad terms in several areas. In general, the performance (considered as throughput per unit volume, power, or weight) declines as the implementation technology progresses from ASICs to programmable modules. However, so does the cost, both in acquisition cost and in nonrecurring engineering effort (NRE).

Ease of functional change refers to the ability of the processing technology to adapt to an upgrade or change in processing algorithms. An ASIC can be designed to meet exactly the requirements of a particular system, but, once fabricated, it cannot be modified to meet the needs of an upgraded or different system. The programmable DSP modules are programmable in both assembly language and higher-level languages such as C or C++. As such, they can be readily adapted to new algorithms. The programmable DSP might be considered somewhat less flexible than a traditional general-purpose CPU in

TABLE 13-3 ■ Comparison of Signal Processor Implementation Techniques (after [18])

Technology	Performance	Cost	Power	Ease of Functional Change	Design Effort
ASIC	High	High	Low	Low	High
Reconfigurable Hardware Module	Medium	Medium	High	Medium	Medium
Programmable DSP Module	Low	Medium	High	High	Low

that some effort is needed to structure the source code to make sure the DSP's specialized computing resources are used effectively; simple single-thread, serial code is not generally adequate. To get the maximum efficiency from the device, the programmer must take care to identify and manage data and task parallelism so that all computing resources are kept busy, that computation is overlapped with communication, and so forth. Reconfigurable hardware is also programmable, though usually at a much lower level such as register transfer level (RTL) or through hardware description languages such as VHDL, requiring more specialized skills. It is more difficult to express signal processing algorithms at this level than in high-level languages, accounting for the greater design effort.

The arithmetic format also tends to vary between each technology type. Both fixed-point and floating-point designs exist in each of these classes. However, at least as applied to radar, it tends to be the case that reconfigurable hardware is more likely—and ASICs more likely still—to use fixed-point than programmable hardware. This is because the more specialized hardware is used for portions of the signal processing flow that require high throughput; are tightly constrained in size, weight, and power; or both. These same constraints tend also to favor the faster, smaller, and lower-power fixed-point arithmetic over floating point. Systems that combine multiple classes of hardware may thus also combine fixed-point processing in earlier stages with floating point in later stages.

The first decade of the 21st century has seen the emergence of new *multicore processor* architectures. Two examples of great interest within the radar signal processor community are the Sony/Toshiba/IBM Cell Broadband Engine (CBE) and graphical processing units (GPUs). The CBE combines a PowerPC microprocessor and on-chip random-access memory (RAM) with an array of eight signal processing elements into a single chip capable of about 200 GFLOPS in single precision [19]. The CBE was designed as the processor for the Sony Playstation 3 game machine. The primary manufacturers of GPUs are Nvidia and ATI (now part of AMD, Inc.). Designed to generate the graphics for high-end personal computers, current (2009) high-end GPUs offer performance of up to one teraflops in a single device having 240 core processors [20]. These devices hold promise of providing performance rivaling or even exceeding that of reconfigurable modules, but with development costs closer to that of more standard programmable devices because they can be programmed in standard high-level languages. For instance, an improvement of about 35×, compared to a standard microprocessor, in the execution time of a two-dimensional (2-D) interferometric imaging radar 2-D phase unwrapping algorithm was demonstrated in [21]. Furthermore, the CBE and GPUs potentially offer extremely good performance per dollar because both are commodity chips developed for very large markets, thus reducing the device cost.

The primary difficulty in using these devices, as well as many other emerging multicore systems and multiprocessor systems in general, is programming them to operate at peak rates. Partitioning algorithms among many processors and choreographing the data to keep all of the processing units busy is a difficult task. In addition, the developer must choose algorithms that are well matched to the architecture of the device being used [21]. Tools such as Nvidia's CUDA development environment aid in applying GPUs to more general high-performance computing applications. OpenCL [22] is an emerging open standard for parallel programming on heterogeneous, many-core architectures that may improve productivity and portability across GPUs from multiple vendors, the CBE, "traditional" dual- and quad-core processors, and other emerging devices.

A thorough discussion of the many classes of current and emerging processors that are candidates for radar signal processing is given in chapter 26 of [5].

13.5.3 COTS Technology and Modular Open Architectures

Since at least the 1980s, it has been increasingly common to implement radar signal processors whenever possible using commercial off-the-shelf components. This has been the natural result of the "leading edge" in electronics development moving from military systems to the much larger consumer products market. The benefits of COTS components include the following [23]:

- Incorporation of the rapid advances in commercial technology into radar systems more quickly than with conventional military acquisition practices.
- Reduction or avoidance of the research and development (R&D) costs for new processors.
- Cost savings associated with products developed for large consumer markets and multivendor competition.

For military systems, there are several drawbacks to COTS as well. These include the difficulty in obtaining components that:

- Meet military specifications, or in some cases reduced "ruggedization" specifications, on environmental characteristics such as temperature and vibration.
- Meet military requirements for safety, security, certification, hard real-time performance, and so forth.
- Do not become obsolete or unsupportable too quickly or fail to provide a certifiable upgrade path.

The rapid evolution of COTS processing technology is thus both a blessing and a curse. It allows rapid improvements in processor capability but makes it difficult to support deployed military systems for their typically multidecade life spans.

A small example from the author's experience illustrates some of the benefits of transitioning to COTS processor implementations. In 1991, an implementation of a particular real-time airborne SAR image formation processor required nine custom 12" × 14" 12-layer custom circuit boards. Of these nine boards, six were vector signal processors, and three were specialized "corner-turn" memories. The arithmetic was implemented in a combination of 16-bit fixed-point and 24-bit block floating-point formats; floating point could not be supported. All of the interfaces were custom designed. Six years later, the processor was reimplemented. This time, the entire system was hosted on four COTS 6U VME (6.3" × 9.2") boards, each containing four IBM PowerPC microprocessors. All of the arithmetic was 32-bit IEEE 754 floating point, and the major interfaces were the open standard versa module europe (VME) and RACEway protocols. Thus, the size of the processor was greatly reduced; the numerical quality of the algorithm implementation was improved; and the design cost was also greatly reduced. If the same processor were reimplemented today, it would require at most two COTS boards, and quite probably only one.

COTS-based signal processors are usually *open architecture* system designs as opposed to proprietary architectures [24]. Open architecture signal processors generally use a modular design approach. In a modular processor, a standard form factor chassis provides a physical enclosure, and the chassis backplane provides a control bus for the modules. The VME form factor, which uses circuit boards in a number of different sizes, is a frequent choice. Commonly used sizes are denoted (in order of increasing size) 1U, 3U, 6U, and 9U. The computing resources required for a particular application are obtained by inserting modules of the required type and number into the backplane. The "menu" of example module types might include the following:

- Single-board general-purpose computer modules.
- Fixed- and floating-point arithmetic modules.
- General or specialized memory modules.
- High-speed interconnect modules.
- Sensor interface modules.
- Timing and control generators.
- Power supply and conditioning modules.
- System-specific interface and control-signal modules.

The reliance on open, public standards for the mechanical, electrical, and logical interfaces allows successive models of equipment from a vendor, or competing equipment from another vendor, to be inserted into a modular system to replace or upgrade existing modules with minimal redesign and cost. Open architecture approaches do not obviate the need for testing and certification, but they provide a manageable approach to maintaining equipment over a long life cycle in an environment of rapid commercial technology product cycles.

Open modular architectures also enable the development of systems that combine multiple styles of computing devices in a single chassis. For example, the FPGA-based reconfigurable computing boards discussed previously are implemented in VME and other standard form factors. They can thus be integrated in a VME chassis with conventional processors, memories, interfaces, control computers, and power supplies.

There are numerous examples of open architecture interconnection standards. The VME (IEEE P1014–1987) and VME64 (IEEE P1014 Rev D) control bus standards for modular system backplanes remain very common in commercial products. RACEway (ANSI/VITA 5–1994) and SCI (IEEE 1596–1992) provide high-speed intermodule and intramodule data interconnects for multiprocessor systems. Newer high-speed interconnect standards include Serial RapidIO, PCI Express, InfiniBand, and Gigabit Ethernet. These buses provide effective data rates in the range of 940 to 980 MB/s. A thorough discussion of interconnection fabrics and standards, as well as of the emerging VME switched serial (VXS) standard (VITA 41.2) for VME-compatible packaging of these interconnect schemes, is given in chapter 14 of [5]. Other standard bus architectures and protocols exist for test and maintenance (e.g., high-performance serial bus (FireWire), JTAG), and for input/output (Fiber Channel, MIL-STD-1553B).

Figure 13-9 illustrates a typical COTS 6U VME signal processor. This particular system from Mercury Computer Systems, current in late 2009, was said to provides 761 GFLOPS of computational capability using quad MPC7448 PowerPC modules as well as 42 GB/s sustained throughput using the Serial RapidIO interconnect fabric. The system is compliant with the VXS form factor standard.

The most important open software standards for radar signal processors are probably Vector Signal Image Processing Library (VSIPL), VSIPL++, and message passing interface (MPI) [25]. VSIPL is an application programming interface (API) for functions such as FFTs, FIR filters, and linear algebra commonly used in signal processing [26]. Manufacturers of commercial signal processing modules typically also provide signal processing libraries optimized for their products' architectures. The proprietary libraries

FIGURE 13-9 ■ Example of commercial off-the-shelf modular signal processor. (Copyright Mercury Computer Systems. Used with permission.)

of all such vendors would include, for example, a one-dimensional FFT function, but the specific name of the function and the calling sequence of its arguments will vary from vendor to vendor. Thus, application code written for one vendor's boards is not portable to another's, even if the hardware is interchangeable. VSIPL provides a standardized interface to these functions. Products from any vendor that supports the VSIPL API will then be more software compatible at the source code level. VSIPL++ provides a C++ binding of VSIPL with additional functionality [27].

MPI plays a similar role for interprocessor communication [28]. It provides a portable, open standard for the functions needed to support the message passing communication protocol on high-performance multiprocessor machines, especially those using distributed memory.

13.5.4 The Influence of Moore's Law

Gordon Moore of Intel first articulated what later became known as *Moore's Law* in a prescient 1965 paper [29]. Moore's Law, as it is now commonly understood, states that the number of transistors in a single integrated circuit doubles approximately every 18 to 24 months. This is evidenced by the history of Intel microprocessors, shown in Figure 13-10 [30]. Other microprocessor metrics such as clock speeds or functionality per unit cost also tend to increase exponentially. Indeed, the rapid increase in computing power per unit cost is evident to every consumer of personal computers, cell phones, and other consumer devices every time he or she visits the local electronics store. Moore's Law has a dark side as well: Power consumption, design costs, and infrastructure costs also increase exponentially. The history and implications of Moore's Law are discussed in [31], and its impact on signal processing is examined in [32].

Moore's Law describes improvements in individual integrated circuits. Obviously these carry over to complete computing systems, but the performance of complete systems

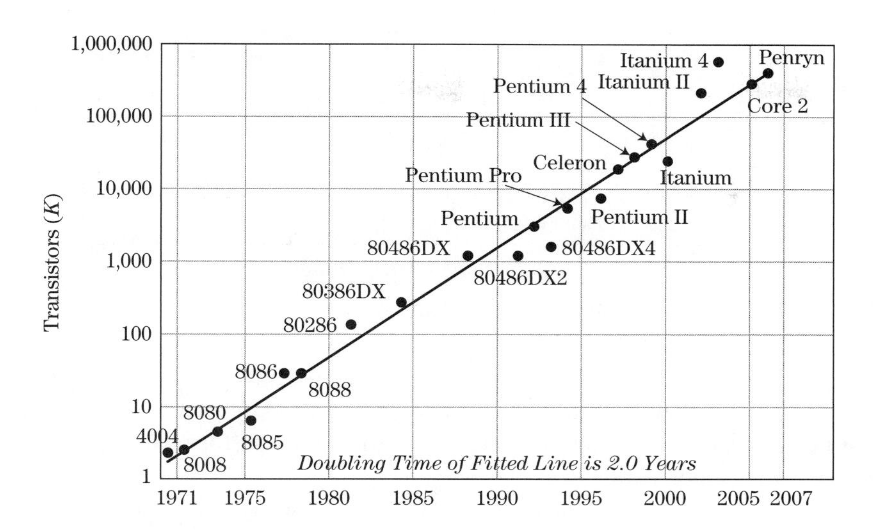

FIGURE 13-10 ■ Moore's Law, as illustrated by Intel microprocessors. (Data courtesy of Intel, Inc.)

is also impacted by the performance of other components, for example, system bus speeds, storage access speeds, and general system architecture. In fact, in recent decades the speed of computing systems as a whole has progressed more quickly than has the speed and density of individual integrated circuits. One illustration of this is the "Top 500" supercomputer rankings, which twice annually updates a list of the 500 fastest computers, based on a specific floating-point LINPACK benchmark [33]. Figure 13-11 shows the increase in speed of the fastest, 100th fastest, and 500th fastest computer over a 15-year period. The average rate of growth in performance corresponds to a doubling every 13 to 15 months, quicker than the 18- to 24-month doubling period of Moore's Law.

The fastest supercomputers are not candidates for embedded processing because they are typically building-sized systems. In mid-2008, the top-ranked system in the Top 500 list was the IBM "Roadrunner" system designed for the U.S. Department of Energy and

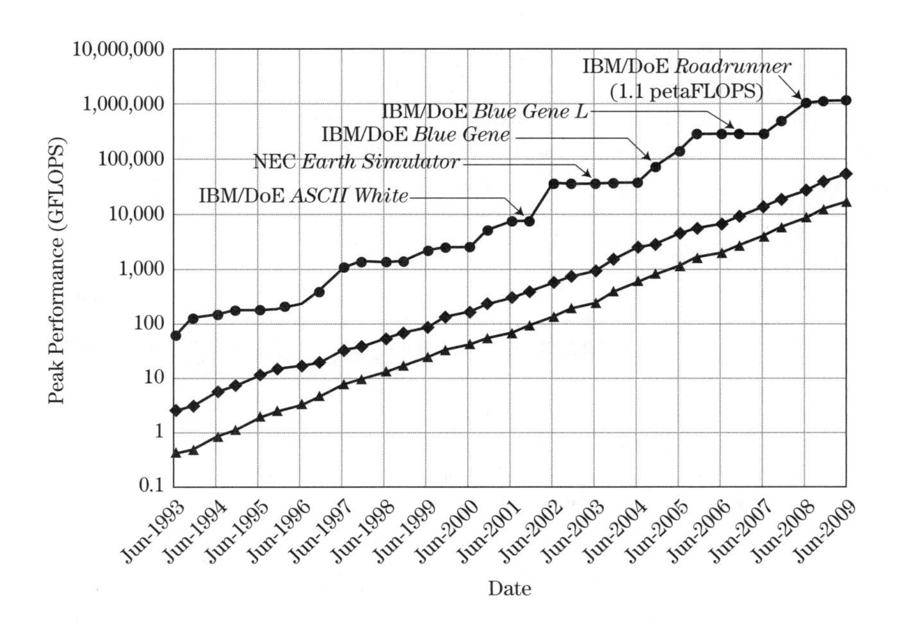

FIGURE 13-11 ■ Peak throughput of the fastest, 100th fastest, and 500th fastest computers in the Top 500 list over time.

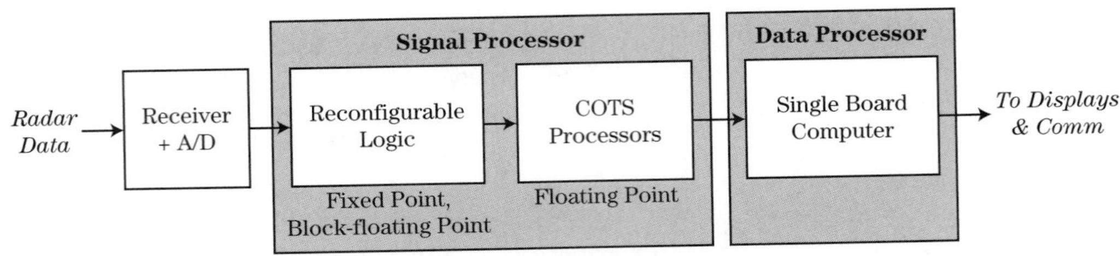

FIGURE 13-12 ■ Simplified processor architecture.

used in weapons research.[4] The machines at the #100 and #500 spots on the list are typically workstation and personal computer (PC) clusters in commercial usage. These more general-purpose machines are better indicators of the performance improvements of computer systems in general.

Moore's Law ensures that the capability of all three classes of signal processing hardware increases rapidly with time. It follows that, for a fixed processing requirement, the need for the more specialized processing technologies such as ASICs will decline over time. Specifically, as FPGAs become faster they become increasingly able to host the more demanding front-end stages of high-throughput algorithms as well as to implement floating-point instead of fixed-point arithmetic, to provide more on-chip memory, and so forth. Substituting reconfigurable hardware for ASICs allows major reductions in design cost and time as well as an increase in the ability of the processor to adapt to new algorithms. Thanks to Moore's Law, the signal processor in a wide variety of systems can be simplified from the design of Figure 13-2 to that of Figure 13-12, where only COTS reconfigurable hardware and programmable processors are needed to implement the radar signal processing algorithms [34]. Frequently, such designs can also use floating-point arithmetic throughout, even in the FPGAs. Signal processors based on reconfigurable and programmable COTS processors with floating-point arithmetic realize major savings in algorithm development cost and time as well as superior numerical performance.

Nonetheless, for the fastest, most tightly constrained processors, ASICs will remain a necessary part of the design. Space-based radars and unmanned aerial vehicle (UAV) radars with high-resolution imaging and ground moving target indication modes are most likely to require ASICs as part of the processor implementation due to high computational loads and very tight payload constraints. An example of the capability that can be obtained in an ASIC designed for radar front-end processing is given in [35]. This bit-level systolic fixed-point design achieves over 1 trillion operations per second in about one-eighth of a cubic foot, 12 kilograms, and 50 watts in a 0.25 μm CMOS technology in the year 2000.

13.6 FIXED POINT VERSUS FLOATING POINT

The choice of arithmetic format in a radar signal processor is a significant design issue, affecting numerical attributes such as accuracy, dynamic range, stability; software development costs; and hardware attributes such as speed, size, weight, and power

[4]This machine is the first to break the petaflops barrier (10^{15} FLOPS) using the Top 500 metric.

consumption. The principal choice is between fixed-point and floating-point arithmetic, though compromise formats such as block floating point [36] are used in some cases.

Digital logic for implementing fixed-point arithmetic in common word lengths may allow reductions of 30% to 50% in silicon area compared with floating-point implementations in custom digital circuits or FPGAs. Because it does not have to process exponents, fewer logic stages are needed in a fixed-point arithmetic unit, allowing it to execute computations at a faster rate. Thus, fixed-point implementations are more efficient than floating point in size, weight, and power consumption. However, when combined with memory and other functions on a complete processor chip, the difference between fixed-point and floating-point devices is reduced.

The chief disadvantages of fixed-point arithmetic are its limited dynamic range and signal-to-quantization noise ratio, both increasing at 6 dB per bit [7], and the problems of underflow and overflow of arithmetic computations. Preventing or controlling underflow and overflow requires additional scaling operations between arithmetic functions and can increase both the development cost and the implementation requirements of a fixed-point implementation of an algorithm. For instance, a 16-bit fixed-point data representation limits the dynamic range of the data to approximately 96 dB. While this may seem adequate, arithmetic operations on the data increase the dynamic range. For example, the sum of two positive numbers is larger than either alone. If two numbers that are near the maximum representable in a given format are added, the result will be twice that maximum.

To guarantee that the sum does not overflow, either the operands must be limited to a range one bit less than the maximum dynamic range or the word length must be extended by one extra bit. Similarly, the sum of N numbers has a range potentially $\log_2 N$ bits larger than that of the individual summands. The product of two operands can double the required word length. Combining these results, an operation such as a dot product of N-element vectors of M-bit numbers can require as much as many as $2M + \log_2 N$ bits. Overflow can be prevented by scaling the data down prior to an operation by enough bits to ensure that overflow cannot occur, but this approach reduces the number of bits of precision of the result. More sophisticated approaches scale the data in stages, minimizing precision loss while reducing or avoiding overflow [4,36]. Even with scaling, algorithms that explicitly or implicitly require matrix inversion, such as those common in STAP and other adaptive processing, can be very challenging to implement with adequate numerical accuracy in fixed-point arithmetic. While tools exist to aid in fixed-point algorithm design, traditionally they have been relatively limited and immature compared with the design environments available for floating-point design.

Floating-point arithmetic eliminates essentially all dynamic range and scaling issues. While a variety of formats are used, increasingly the most common is the IEEE P754 32-bit format [37]. Using a 24-bit mantissa and an 8-bit exponent in the single precision version, this 32-bit format can represent numbers over a range of about 72 orders of magnitude. The double precision version uses a 52-bit mantissa and an 11-bit exponent, increasing the representable range to about 610 orders of magnitude!

Another advantage of floating-point arithmetic is that most algorithms are developed in floating-point environments such as C or C++ source code or using simulation and analysis tools such as MATLAB, Mathematica, Maple, or MathCAD. Porting of the algorithms from the development environment to a fixed-point target requires additional development to convert the algorithm to a fixed-point equivalent while avoiding loss of accuracy or, in some cases, even numerical instability. A related advantage is that floating-point algorithms, especially when using standardized formats such as IEEE P754, are more easily ported to new hardware platforms.

The gap between the characteristics of floating-point and fixed-point design is narrowing [38]. Improving design tools and available libraries of optimized fixed-point functions for many processors aid in efficiently developing fixed-point implementations of numerical algorithms. The increased capacity of chips in accordance with Moore's Law reduces the significance of the area and power impacts of larger floating-point logic circuits. For example, it is increasingly practical to implement FPGA arithmetic in floating point rather than fixed point. Fixed-point devices retain an advantage in very high-volume, cost-sensitive applications such as cell phones, while the numerical accuracy and ease of use of floating point remains important in numerically difficult but low-volume applications such as radar.

13.7 SIGNAL PROCESSOR SIZING

The preceding sections have discussed ways to estimate the arithmetic operation counts of radar signal processing algorithms and some of the implementation technologies available to host these algorithms. It remains to translate these operation counts into estimates of time to solution for a given processor.

13.7.1 Considerations in Estimating Timing

In the early days of DSP, it was possible to estimate algorithm run time by simply multiplying multiply and add operation counts by the time the processor of interest required to perform a single addition or multiplication. This approach, while useful for rough initial processor sizing, effectively assumes 100% use of the arithmetic unit. In many cases, users would "de-rate" the processor by a factor of $2\times$ to $4\times$ to obtain more realistic results.

Caution is also needed in relating FLOPS estimates to advertised capabilities of commercial processing hardware. Advertised computational rates usually represent peak theoretical values that may be difficult or impossible to sustain in practice, because they assume that all data and instructions are in level 1 (L1) caches, that pipelines are filled, and so forth. Fetching data and instructions from level 2 (L2) cache or from off-chip can significantly slow the processor while it waits for the data transfer.

This approach is not adequate in modern processors with complex memory hierarchies and multiple cores. For example, interprocessor communication, both intrachip (for multicore processors) and interchip, often causes actual throughputs to be much lower than advertised peak rates. Most processors large enough for modern radar algorithms require the use of multiprocessor architectures. Maximum throughput requires keeping the compute units busy as much as possible, but processor compute units frequently have to idle while waiting on data transfers to or from other processors. Designers must seek to partition the data flowgraph to divide the workload evenly among the compute units, a process knows as *load balancing*.

Another major constraint is memory access. In modern processors the time required for a floating-point operation is often much less than the time required to fetch the data for the operation and to store the results unless the data are in the L1 cache; a memory access to main memory can easily require 1,000 times as long as a floating-point math instruction. Thus, the designer must overlap computation and communication as much as possible. This choreography of the data movement through levels of the memory hierarchy is frequently more important in determining throughout than operation counts. Indeed, the developers of the Fastest Fourier Transform in the West (FFTW) library stated in 2005 that "... there is no longer any clear connection between operation counts and FFT speed, thanks to the complexity of modern computers" [13].

A related issue in multiprocessor architectures is *scaling*. It is often implied that if a single processor achieves a certain throughput, then a multiprocessor composed of N such processors will achieve N times the throughput. While this is certainly the goal, such *linear speedups* are difficult to achieve, particularly as N increases, due both to *Amdahl's Law* [4], which describes the limit to the attainable speedup when only a portion of a program is parallelizable, and to the same issues of load balancing and interprocessor communication already cited.

The analysis needed to obtain time-to-solution estimates requires detailed processor architecture modeling and simulation and is too intricate for hand calculation. Many companies have internally developed tools to conduct these analyses. In addition, a number of university and commercial tools exist. An example of a university tool is the Ptolemy system developed at the University of California–Berkeley [39]. Examples of commercial tools include Simulink by The MathWorks, Inc., Signal Processing Designer by CoWare, Inc., and Gedae by Gedae, Inc., among many others. Typical capabilities of these tools include representation of algorithms using flowgraphs and user-extensible libraries of predefined functions; mapping of flowgraphs to multiprocessor architectures; simulation of flowgraphs and architectures; and visualization of processor loading, data flows, and signals. Some systems also include generation of C or C++ code for certain hardware processors.

13.7.2 Benchmarks

Another way to at least partially address the problem of estimating processor speed is to test actual hardware components using nontrivial benchmark codes that are representative of the applications of interest. Good benchmarks should include a mix of computation, communication, storage, and input/output operations. The computational operations should have a realistic mix of operation types for the application class of interest. For instance, a benchmark for evaluating general processing capability should include a wide variety of integer and floating-point operations with varied indexing and memory access patterns, while a benchmark for FFT hardware should concentrate on complex multiplications and additions and strided memory accesses.

General-purpose benchmark codes can be obtained publicly for a variety of purposes. For example, the Standard Performance Evaluation Corporation (SPEC) commercial benchmarks are widely used for evaluating general-purpose processors [40]. The SPEC suite also includes benchmarks for such areas as graphics and Web servers. Traditional arithmetic benchmarks such as the High Performance Computing (HPC) Challenge and many more are accessible through the BenchWeb site at the NETLIB repository [41]. Benchmarks for embedded processors at various scales are also available commercially. In addition, the defense community in the United States has developed several benchmark suites, including the High Performance Embedded Computing (HPEC) Challenge [42]. Table 13-4 provides information on a number of publicly available benchmarks with at least some applicability to radar signal processing. While these general-purpose benchmarks can be very useful, the best benchmarks for evaluating a processor architecture are

TABLE 13-4 ■ Selected Benchmark Suites

Name	Source	URL	Nature	Software/Platform	Comments
SPEC	Standard Performance Evaluation Corp.	www.spec.org	Broad spectrum embedded microprocessor evaluation	C	Commercial. Currently CPU2006.
EEMBC	Embedded Microprocessor Benchmark Consortium	www.eembc.org	Broad spectrum embedded microprocessor evaluation	ANSI C	Commercial.
BDTI Benchmark Suites	BDTI	www.bdti.com	DSP kernels, video, communication	C, assembly, test vectors	Commercial. Leans somewhat to communications.
HPCS Benchmarks	DARPA HPCS program	www.highproductivity.org/	HPC kernels and applications	HPC challenge requires MPI and BLAS	Not specific to signal processing.
HPC Challenge	University of Tennessee	www.hpcchallenge.org/	HPC parallel kernels	Requires MPI and BLAS	Public release.
HPEC challenge	MIT LL	www.ll.mit.edu/HPECchallenge/	Sensor signal processing	ANSI C (kernels only)	Public release
LINPACK (HPL)	University of Tennessee	www.netlib.org/benchmark/hpl/	Dense linear algebra	FORTRAN, Java, C Requires MPI and either BLAS or VSIPL	Public release. Basis of Top500 rankings. Incorporated into more comprehensive HPC challenge.

BDTI: Berkeley Design Technology, Inc.
HPCS: High Productivity Computing Systems
DARPA: Defense Advanced Research Projects Agency
HPEC: High Performance Embedded Computing
HPC: High Performance Computing
MIT LL: Massachusetts Institute of Technology Lincoln Laboratory

those that mimic the important characteristics of the actual application mix for the specific architecture of interest.

13.7.3 Software Tool Impacts

Even when an implementation of a flowgraph that minimizes operations has been developed, the actual efficiency of the executable code depends on the software tools used. The same source code, compiled to the same target machine with different compilers, is almost certain to have different execution times due to differences in compiler optimizations and in libraries used. The three-part series of articles [43–45] dramatically illustrates the effect of source code optimization, spatial and temporal locality, and compiler behavior on the execution time of a typical "inner loop" numerical computation. Execution time was reduced by a factor of over 30× compared with the original naïve code.

Radar signal processing algorithms rely heavily on certain core operations: spectral transforms such as FFTs, discrete cosine transforms (DCTs), and wavelet transforms; FIR filters; and linear algebraic calculations ranging from simple matrix-vector operations to more complex operations such as Q-R decomposition (QRD), singular value decomposition (SVD), and eigenvalue/eigenvector decomposition. Software developers rely on libraries to provide these critical functions. For instance, the BLAS – LINPACK – LAPACK – ScaLAPACK series of libraries provides versions of common linear algebra manipulations optimized for different classes of machines; information and code for all of these are available at the Netlib repository [46]. Numerous FFT libraries exist. VSIPL [26] and VSIPL++ [27] provide a portable library specification that includes a selection of the most common unary, binary, FFT, filtering, linear algebraic, and related signal processing operations.

Many vendors provide implementations of libraries optimized for their specific processors. One example is Intel's Math Kernel Library (MKL) [47], which offers the various classes of routines previously described along with statistics functions and sparse solvers in a package optimized for Intel microprocessors. A recent trend in signal processing and scientific libraries is the development of *autotuning* libraries. These are open source libraries that use deep mathematical and practical knowledge of the structure of a class of algorithms to evaluate many different implementations of a specific function on a specific target machine to experimentally find the most efficient code. The autotuning approach can provide very good performance over a wide range of target machines, often besting vendor-specific libraries. The FFTW library uses this approach for FFTs [48]. ATLAS takes a similar approach to the BLAS and some LAPACK linear algebra routines [49].

13.7.4 Data Rates

Not all radar systems need to perform all of the signal processing in an *onboard processor* colocated with the radar sensor, as has been implied so far. An alternative is to digitize the receiver outputs and transmit them to another location for subsequent processing, either using real-time communication links or by storing the data for later processing. Such *offboard* processing might be appropriate if significant latency in obtaining the final data products can be tolerated and the size of the onboard processor is severely limited by constraints of size, weight, or power. As an example, satellite systems and small unattended autonomous vehicles (UAVs) usually have very tight payload budgets. Satellite SARs that generate imagery for non-real-time uses such as cartography and earth resources monitoring generally downlink the data to a ground station, where it can be processed in a facility with much greater resources and on whatever timeline is needed. Alternatively, the data may be cross-linked to a communications satellite for further distribution. In contrast, an airborne tactical SAR used for targeting must generate imagery and display it in the cockpit in real time. This system must perform all of the required processing onboard and with minimum latency.

The data rates at the output of a radar receiver can be very high. Consider a SAR system. Using an approach similar to that of section 13.4.2, an estimate can be developed for the data rate at the receiver output. The data rate DR_{in} in bits per second can be expressed as the product of three factors: the number of receiver samples per pulse N_{sp}; the pulse repetition frequency, PRF; and the number of bits per sample N_b:

$$DR_{in} = 2N_{sp}N_b PRF \text{ bps} \tag{13.11}$$

The factor of two results from assuming a coherent receiver so that both the in-phase and quadrature (I and Q) channels must be sampled. N_{sp} in turn is the product of the fast time sampling rate and the time duration T_f of the sampling window. The fast time sampling rate is determined by the waveform bandwidth B and will normally be about B samples/sec; thus

$$N_{sp} = T_f B \text{ samples} \tag{13.12}$$

T_f must be less than the pulse repetition interval (PRI) to avoid range ambiguities. Consequently, $T_f PRF \leq 1$, and, using $\Delta R = c/2B$,

$$DR_{in} \leq 2N_b B \quad \Rightarrow \quad DR_{in} \sim O\left(\frac{cN_b}{\Delta R}\right) \text{bps} \tag{13.13}$$

Taking the SIR-C example again with $\Delta R = 25$ m and assuming eight bits per sample produces a rate of 96 Mbps. A radar with a range resolution of 10 m would increase this to 240 Mbps; 1 m range resolution would raise it to 2.4 Gbps.

Note that equation (13.13) applies to each coherent channel. Thus, a standard three-channel monopulse antenna would require three times this rate to capture the sum and both difference channels. A phased array system used for imaging and including adaptive processing for interference rejection can be particularly stressing. For example, a system having six-bit A/D converters, 10 m range resolution, and six phase centers will have a raw data rate of 1.08 Gbps.

The data rate at the output of a SAR processor is typically lower than at the input to the processor by a modest factor. The pixel rate was found in equation (13.4); multiplying by the number of bits per pixel N_p gives the bit rate at the output of the image formation:

$$DR_{out} \sim O\left(\frac{cN_p}{2\Delta R}\right) \text{bps} \tag{13.14}$$

Equation (13.14) is of the same form as equation (13.13), just reduced by a factor of 2. This result assumes the range swath is at or near its maximum, the unambiguous range. Note that since image pixels are real, there is no need to double this rate to account for I and Q channels. On the other hand, a system doing interferometric SAR (IFSAR) does retain the pixel phase. A "one-pass" IFSAR system would have four times this output rate: one doubling for I and Q channels and another for the fact that two SAR images are required for IFSAR.

Images are often represented at eight bits per pixel in an uncompressed format. Assuming this precision, equation (13.14) becomes

$$DR_{out} \sim O\left(\frac{4c}{\Delta R}\right) \text{bps} \tag{13.15}$$

The resulting output bit rate is 120 Mbps at 10 m range resolution, 400 Mbps at 3 m, and 1.2 Gbps at 1 m.

Data compression can be applied to SAR data or imagery. A variety of techniques are used, depending in part on the stage of processing at which the compression is applied [50,51]. Block adaptive quantization is common on raw SAR data; this technique should be applicable to raw data for other wideband radar systems as well, such as STAP and GMTI systems. Vector quantization has been investigated for partially processed data. Wavelet and trellis coding have been applied to complex SAR imagery, with compression factors on the order of 4 for the phase but up to 64 on the magnitude. Standard image compression algorithms such as JPEG can be used on final detected imagery, as can wavelet and trellis coding; savings by factors of 4 to 8 have been achieved. It seems safe to say that reductions of at least 4× in data rate are achievable using data compression, sometimes much more.

In contrast to the SAR example, some radar signal processing systems have output date rates significantly lower than the input data rates. Consider again the STAP-GMTI system of Figures 13-4 and 13-5. The input consists of six channels (one for each phase center) sampled at 125 Msamples/second. These data are real, not complex, because digital I/Q formation will be used. Assuming eight bits per sample the input data rate will be 6 Gbps. After the subband filtering, there are 16 complex subband channels at a 10.42 Msample/second rate. Still assuming eight bits, this lowers the data rate to 2.7 Gbps. Subsequent stages will tend to have similar data rates unless the processing is structured to reduce the amount of data, for example, by processing only some subbands or applying a reduced-dimension STAP algorithm to a subset of the Doppler filters. In practice, the data rate at the input to the detector will be between one and two orders of magnitude below that at the radar input, in this case between 60 and 600 Mbps.

At the output of the detector the data rate will likely be lower because it will consist only of a list of threshold crossings with associated range/angle/Doppler coordinates and other metadata. The actual data rate will be scenario dependent. In a ground-to-air or air-to-air system, there may be only a few tens of targets per CPI, sometimes few or none. However, an air- or space-to-ground GMTI system may detect thousands of targets per CPI.

13.7.5 Onboard versus Offboard Processing

An example of a data link used by the space shuttle radars is the National Aeronautic and Space Administration's (NASA) Tracking and Data Relay Satellite System (TDRSS). This system provides channels with data rates of 80 to 800 Mbps. The U.S. Military's Common Data Link (CDL) provides on the order of 250 Mbps per channel but also allows multiple channels to be combined to create links approaching 1 Gbps. While optical communication links under development may raise achievable rates by one to two orders of magnitude, current technology appears limited to link rates of a few hundred Mbps to, at most, about 1 Gbps.

Figure 13-4 showed the growth of the cumulative computational load for a wideband STAP-GMTI signal processor as each successive stage was executed. If the radar platform is tightly constrained in size, weight, and power, as, for instance, on a satellite or small UAV, it may not be possible to host a processor of adequate capacity to perform all of the signal processing onboard. In systems that do not require that all signal processing be completed in real time, it may simply be unnecessary to complete the processing onboard. There may, therefore, be a need for an engineering trade-off between onboard processor capacity and data link capacity. For instance, if the platform can host a 200 GFLOPS processor, it can perform all of the processing of Figure 13-4 onboard, and only a low-rate data link will be needed to downlink the detection data. If, on the other hand, the platform can host only a 160 GFLOPS processor, then the data must be downlinked at the output of the pulse compression stage and the processing completed in a ground station or other facility. For this to be possible, a data link must be available with adequate capacity for the data rate at the pulse compression output. Assuming this data rate is on the order of 1 Gbps and that appropriate data compression can reduce it by another factor of 4, the data could be downlinked with a TDRSS or CDL data link such as already discussed.

If the data rates are such that this trade-off is not possible, the system design must be revisited until a realizable combination of onboard signal processing capacity and downlink bandwidth is found. If a larger processor can be hosted, the downlink can be deferred to a later processing stage. If a large enough processor to perform the entire processing chain can be hosted, a more narrowband data link will suffice for delivery of the final image products.

13.8 FURTHER READING

The recent edited volume by Martinez, Bond, and Vai [5] gives an excellent overview of the full range of hardware, software, and system issues in high-performance embedded computing, with an emphasis on sensing applications, including specifically advanced radar techniques.

13.9 REFERENCES

[1] Skolnik, M.I., *Introduction to Radar Systems*, 3d ed., McGraw-Hill, New York, 2001.

[2] Cutrona, L.J., Leith, E.N., Porcello, L.J., and Vivian, W.E., "On the Application of Coherent Optical Processing Techniques to Synthetic-Aperture Radar," *Proceedings of the IEEE*, vol. 54, no. 8, pp. 1026–1032, August 1966.

[3] Oppenheim, A.V., and Schafer, R.W., *Discrete-Time Signal Processing*, 3d ed., Prentice-Hall, Englewood Cliffs, NJ, 2009.

[4] Goldberg, D., "Computer Arithmetic," Appendix A in *Computer Architecture: A Quantitative Approach*, 3d ed., Ed. J.L. Hennessy and D.A. Patterson, Morgan Kaufmann, San Francisco, 2002.

[5] Martinez, D.R., Bond, R.A., and Vai, M.M., Eds., *High Performance Embedded Computing Handbook: A Systems Perspective*, CRC Press, Boca Raton, FL, 2008.

[6] Elachi, C., *Spaceborne Radar Remote Sensing: Applications and Techniques*, IEEE Press, New York, 1988.

[7] Richards, M.A., *Fundamentals of Radar Signal Processing*, McGraw-Hill, New York, 2005.

[8] Cumming, I.G., and Wong, F.H., *Digital Processing of Synthetic Aperture Radar Data: Algorithms and Implementation*, Artech House, Boston, MA, 2005.

[9] Curlander, J.C., and McDonough, R.N., *Synthetic Aperture Radar: Systems and Signal Processing*, Wiley, New York, 1991.

[10] Cain, K.C., Torres, J.A., and Williams, R.T., "RT_STAP: Real-Time Space-Time Adaptive Processing Benchmark," MITRE Technical Report MTR 96B0000021, February 1997.

[11] Ward, J., "Space-Time Adaptive Processing for Airborne Radar," MIT Lincoln Laboratory Technical Report 1015, pp. 75–77, December 13, 1994.

[12] Cooley, J.W., and Tukey, J.W., "An Algorithm for the Machine Computation of Complex Fourier Series," *Mathematics of Computation*, vol. 19, pp. 297–301, April 1965.

[13] Frigo, M., and Johnson, S.G., "The Design and Implementation of FFTW3," *Proceedings of the IEEE*, vol. 93, no. 2, pp. 216–231, 2005.

[14] Merkel, K.G., and Wilson, A.L., "A Survey of High Performance Analog-to-Digital Converters for Defense Space Applications," *Proceedings of the 2003 IEEE Aerospace Conference,* vol. 5, pp. 2415–2427, March 8–15, 2003.

[15] Aziz, P.M., Sorensen, H.V., and van der Spiegel, J., "An Overview of Sigma-Delta Converters" IEEE *Signal Processing Magazine*, vol. 13, no.1, pp. 61–84, January 1996.

[16] Walden, R.H., "Analog-to-Digital Conversion in the Early 21st Century," in *Encyclopedia of Computer Science and Engineering*, Ed. B. Wah, John Wiley & Sons, Inc., New York, 2008.

[17] Walden, R.H., "Analog-to-Digital Converter Survey and Analysis," *IEEE Journal on Selected Areas in Communications*, vol. 17, no. 4, pp. 539–550, April 1999.

[18] Tessier, R., and Burleson, W., "Reconfigurable Computing for Digital Signal Processing: A Survey," *Journal of VLSI Signal Processing,* vol. 28, no. 7, pp. 7–27, 2001.

[19] IBM, "Cell Broadband Engine Technology." n.d. Available at http://www.ibm.com/developerworks/power/cell/docs_articles.html.

[20] Owens, J.D., Houston, M., Luebke, D., Green, S., Stone, J.E., and Phillips, J.C., "GPU Computing," *Proceedings of the IEEE*, vol. 96, no. 5, pp. 879–899, May 2008.

[21] Karasev, P.A., Campbell, D.P., and Richards, M.A., "Obtaining a 35x Speedup in 2D Phase Unwrapping Using Commodity Graphics Processors," *Proceedings of the IEEE Radar Conference 2007,* Boston, MA, pp. 574–578, 2007.

[22] Khronos Group, "OpenCL—The Open Standard for Parallel Programming of Heterogenous Systems." August 2009. Available at http://www.khronos.org/opencl/.

[23] Alford, L.D., Jr., "The Problem with Aviation COTS," *IEEE AES Systems Magazine*, vol. 16, no. 2, pp. 33–37, February 2001.

[24] Wilcock, G., Totten, T., Gleave, A., and Wilson, R., "The Application of COTS Technology in Future Modular Avionic Systems," *Electrical & Communication Engineering Journal*, vol. 13, no. 4, pp. 183–192, August 2001.

[25] Shank, S.F., Paterson, W.J., Johansson, J., and Trevito, L.M., "An Open Architecture for an Embedded Signal Processing Subsystem," *Proceedings of the IEEE 2004 Radar Conference,* pp. 26–29, April 2004.

[26] Defense Advanced Research Projects Agency (DARPA) and the U.S. Navy, "Vector Signal Image Processing Library (VSIPL)." April 23, 2009. Available at http://www.vsipl.org.

[27] Georgia Tech Research Institute, "High Performance Embedded Computing Software Initiative (HPEC-SI)." n.d. Available at http://www.hpec-si.org.

[28] University of Illinois, "Message Passing Interface (MPI) Forum." n.d. Available at http://www.mpi-forum.org.

[29] Moore, G.E., "Cramming More Components onto Integrated Circuits," *Electronics*, vol. 38, no. 8, pp. 114–117, April 19, 1965.

[30] ICKnowledge, LLC, "Microprocessor Trends," 2000–2008. Available at http://www.icknowledge.com/trends/uproc.html.

[31] Schaller, R.R., "Moore's Law: Past, Present, and Future," *IEEE Spectrum*, vol. 32, no. 6, pp. 52–59, June 1997.

[32] Shaw, G.A., and Richards, M.A., "Sustaining the Exponential Growth of Embedded Digital Signal Processing Capability," *Proceedings of the 2004 High Performance Embedded Computing Workshop,* MIT Lincoln Laboratory, September 28–30, 2004. Available at http://www.ll.mit.edu/HPEC/agenda04.htm.

[33] Top 500 Supercomputer Sites, "Homepage," 2000–2009. Available at http://www.top500.org.

[34] Le, C., Chan, S., Cheng, F., Fang, W., Fischman, M., Hensley, S., et al., "Onboard FPGA-Based SAR Processing for Future Spaceborne Systems," *Proceedings of the 2004 IEEE Radar Conference,* pp. 15–20, April 26–29, 2004. Available at http://hdl.handle.net/2014/37880.

[35] Song, W.S., Baranoski, E.J., and Martinez, D.R., "One Trillion Operations per Second On-Board VLSI Signal Processor for Discoverer II Space Based Radar," *IEEE 2000 Aerospace Conference Proceedings,* vol. 5, pp. 213–218, March 18–25, 2000.

[36] Ercegovec, M., and Lang, T., *Digital Arithmetic*, Morgan-Kauffman, San Francisco, 2003.

[37] IEEE. "IEEE Standard for Binary Floating-Point Arithmetic," ANSI/IEEE Standard 754-1985, August 12, 1985.

[38] Frantz, G., and Simar, R., "Comparing Fixed- and Floating-Points DSPs," Texas Instruments White Paper SPRY061, 2004. Available at http://focus.ti.com/lit/wp/spry061/spry061.pdf.

[39] The University of California at Berkeley, Electrical Engineering and Computer Science Department, "The Ptolemy Project." n.d. Available at http://ptolemy.eecs.berkeley.edu/.

[40] Standard Performance Evaluation Corporation, "Homepage," n.d. Available at http://www.spec.org.

[41] NETLIB BenchWeb, "Homepage," n.d. Available at http://www.netlib.org/benchweb/.

[42] MIT Lincoln Laboratory, "HPEC Challenge," 2001–2006. Available at http://www.ll.mit.edu/HPECchallenge/.

[43] Pancratov, C., Kurzer, J.M., Shaw, K.A., and Trawick, M.L., "Why Computer Architecture Matters," *IEEE and American Institute of Physics Computing in Science and Engineering*, vol. 10, no. 3, pp. 59–63, May–June 2008.

[44] Pancratov, C., Kurzer, J.M., Shaw, K.A., and Trawick, M.L., "Why Computer Architecture Matters: Memory Access," *IEEE and American Institute of Physics Computing in Science and Engineering*, vol. 10, no. 4, pp. 71–75, July–August 2008.

[45] Pancratov, C., Kurzer, J.M., Shaw, K.A., and Trawick, M.L., "Why Computer Architecture Matters: Thinking Through Trade-offs in Your Code," *IEEE and American Institute of Physics Computing in Science and Engineering*, vol. 10, no. 5, pp. 74–79, September–October 2008.

[46] University of Tennessee, Computer Science Dept, "Netlib Repository." n.d. Available at http://www.netlib.org.

[47] Intel, "Intel Math Kernel Library (Intel MKL) 10.2," n.d. Available at http://www.intel.com/cd/software/products/asmo-na/eng/307757.htm.

[48] M. Frigo, S. G. Johnson, and the Massachusetts Institute of Technology, "FFTW [Fastest Fourier Transform in the West]," n.d. Available at http://www.fftw.org.

[49] SourceForge, "Automatically Tuned Linear Algebra Software (ATLAS)," n.d. Available at http://math-atlas.sourceforge.net/.

[50] El Boustani, A., Brunham, K., and Kinsner, W., "A Review of Current Raw SAR Data Compression Techniques," *2001 Canadian Conference on Electrical and Computer Engineering*, vol. 2, pp. 925–930, May 13–16, 2001.

[51] Francaschetti, G., and Lanari, R, *Synthetic Aperture Radar Processing*, CRC Press, Boca Raton, FL, 1999.

13.10 PROBLEMS

1. Show that the number of real FLOPs required to compute all N points of a radix-2 FFT (N is assumed to be a power of 2) is $5N \log_2 N$.
2. Suppose a system convolves a 1,000-point complex-valued input sequence with a 40-point complex-valued filter impulse response. What is the length of the filter output sequence? How many complex multiplications are required to compute the complete output? Check your answer against Figure 13-7.
3. Repeat problem 2, but assume that the filtering is done using fast convolution, that is, using FFTs. Assume that the FFT length is required to be a power of 2 and that $H[k]$, the FFT of the impulse response $h[n]$, has been precomputed so that its multiplications need not be counted. Again, check your answer against Figure 13-7.
4. Now suppose the 1,000-point input vector is broken into 5 200-point segments. Each segment is filtered separately to produce a 239-point output, using fast convolutions with 256-point FFTs. Again, assume that $H[k]$ has been precomputed. What is the total number of multiplications to compute all of the required FFTs and inverse FFTs?
5. Suppose the input to a lowpass digital filter is a real-valued continuous stream of 12-bit data samples at a rate of 10 Msamples/second. The filter impulse response is 40 samples long and is also real-valued. The sample rate at the output of the filter is reduced to 5 Msamples/second due to the lowpass filtering. What is the filter input data rate in Mbps? What is the filter output data rate in Mbps if the outputs are represented as 16-bit fixed-point numbers? Repeat for the cases where the output is represented as single-precision floating point (32 bits per sample) and as double-precision floating point (64 bits per sample).
6. Consider the parametric formula for SAR resolution in section 13.4.2. If the SAR range swath is $^1/_2$ of the unambiguous range interval, how will equation (13.3) be altered?
7. In section 13.4.2, it was argued that F_p, the number of complex FLOPs per pixel, is two times the size of the cross-range data length N_{CR} (one multiply plus one add per cross-range sample). Assuming the system parameters are such that the correlation can be done using FFTs instead of a brute-force correlation calculation, calculation of the cross-range correlation would require an FFT, a vector multiply with the reference function, and an inverse FFT, all of size N_{CR}. What would be the new value of F_p in this case, and how would equation (13.6) and equation (13.7) be altered?
8. The first petaflops computer (10^{15} FLOPS), as measured by the Top 500 criterion, appeared in June 2008. The first terascale (10^{12}) computer appeared on the list in June 1997. When should we expect the first exaFLOPS (10^{18} FLOPS) computer to appear on the Top 500 list?

PART IV

Signal and Data Processing

CHAPTER 14 Digital Signal Processing Fundamentals for Radar

CHAPTER 15 Threshold Detection of Radar Targets

CHAPTER 16 Constant False Alarm Rate Detectors

CHAPTER 17 Doppler Processing

CHAPTER 18 Radar Measurements

CHAPTER 19 Radar Tracking Algorithms

CHAPTER 20 Fundamentals of Pulse Compression Waveforms

CHAPTER 21 An Overview of Radar Imaging

CHAPTER 14

Digital Signal Processing Fundamentals for Radar

Mark A. Richards

Chapter Outline

14.1 Introduction 495
14.2 Sampling 496
14.3 Quantization 504
14.4 Fourier Analysis 506
14.5 The z Transform 522
14.6 Digital Filtering 523
14.7 Random Signals 532
14.8 Integration 536
14.9 Correlation as a Signal Processing Operation 538
14.10 Matched Filters 540
14.11 Further Reading 543
14.12 References 543
14.13 Problems 544

14.1 INTRODUCTION

Radar technology was first seriously developed during the 1930s and through World War II. Radar systems evolved to use a variety of signal processing techniques to extract useful information from raw radar echoes. Examples include moving target indication (MTI), signal integration to improve the signal-to-noise ratio (SNR), pulse compression for fine range resolution, and angle and delay estimation for range and angle tracking.

As discussed in Chapter 13, early radars used analog processors and techniques to implement these functions. However, the ability to implement far more sophisticated algorithms, supported by the exponential improvements in digital processor speed and capacity, have caused digital signal processing (DSP) methods to displace analog processing in virtually all modern radar systems.

There are many excellent references on basic digital signal processing theory; see, for instance, [1,2]. These texts typically base most of their examples on speech and image signals rather than on radar or communications. A good reference for radar-specific signal processing is [3].

While the fundamental principles of DSP apply equally to all of these signals, some of the details important in radar do not arise in speech and images. One of the most fundamental differences is that radar signals, when detected with a coherent (in-phase/quadrature, or I/Q) receiver such as discussed in Chapter 11, are complex-valued; speech and images are real-valued signals. Care must be taken in correctly extending theoretical results such as formulas for correlation or transform symmetry properties to the complex case. Processors must provide mechanisms for complex arithmetic, and the storage per sample is doubled.

Another major difference is that speech and image signals are baseband; that is, they have no carrier. Radar signals are on a carrier frequency and must be demodulated. This fact requires some care in the application of the Nyquist theorem but also opens the door to the digital intermediate frequency (IF) (also called digital I/Q) methods discussed in Chapter 11.

Radar signals also exhibit high dynamic ranges. The dynamic range of the received signal is often on the order of 40 to 80 dB due to the combined effects of large variations in target radar cross section (RCS) and the R^4 variation in received power as a function of range R. Finally, radar signals often suffer low signal-to-interference ratios, falling well below 0 dB for the raw echoes before any processing is done.

Bandwidth, however, may be the most significant difference. Radar signal bandwidths are very large compared with the signals that characterize many other application areas. Telephone-quality speech signals are conventionally limited to about 4 kHz of bandwidth, while music signals exhibit about a 20 kHz bandwidth. In contrast, instantaneous bandwidths of pulsed radars are routinely on the order of hundreds of kilohertz to a few megahertz, similar to video signals. High-range resolution radar systems can have bandwidths from the tens to hundreds of megahertz, with very high-resolution systems occasionally reaching bandwidths in the low gigahertz!

The wide bandwidth of radar signals requires a high sampling rate, as will be seen in Section 14.2.1. This in turn means that a large number of sampled data values, often millions per second, must be processed in real time. Because of this flood of data, radar DSP was originally limited to relatively simple algorithms, requiring only a few operations per data sample. Thus, older radar processors made heavy use of the fast Fourier transform (FFT) and finite impulse response (FIR) filters. Only relatively recently have they adopted correlation methods common in sonar processing or model-based spectral estimation algorithms such as autoregressive or autoregressive moving average methods, which require the solution of sets of linear or nonlinear equations.

14.2 SAMPLING

The "digital signals" of DSP are obtained by discretizing analog signals in two independent ways, as illustrated in Figure 14-1. Discretization in the independent variable, usually time, is called *sampling*. Discretization in amplitude is called *quantization*. Sampling is discussed first; quantization will be discussed in Section 14.3 Whether a signal is sampled is denoted by enclosing its independent variable in parentheses $(\cdot)$ if it is continuous (nonsampled) and in square brackets $[\cdot]$ if it is discrete (sampled). Thus, $x(t)$ represents a continuous signal, whereas $x[n]$ represents a sampled signal. Also, a subscript a will often be used to denote an analog (continuous independent variable) signal.

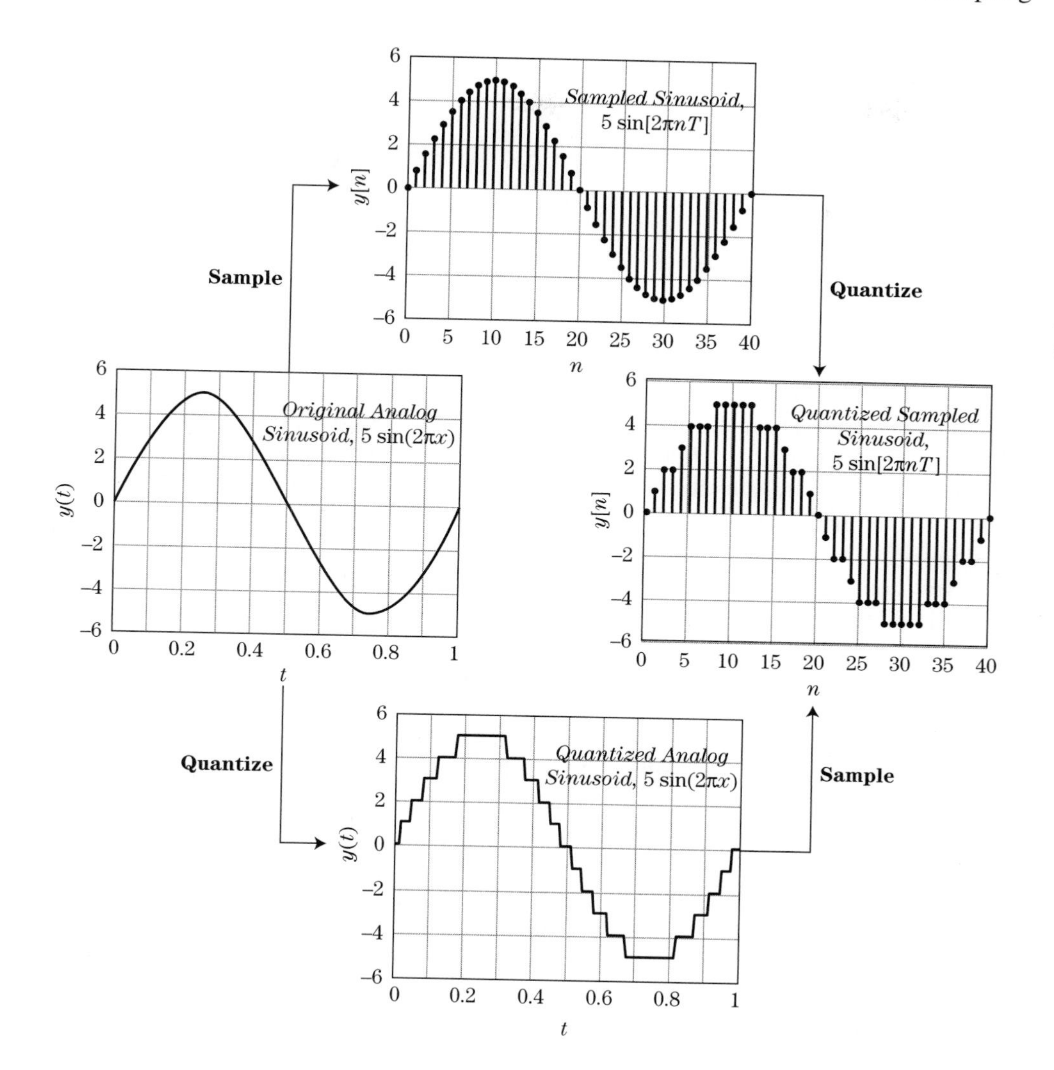

FIGURE 14-1 ■ Illustration of the difference between sampling (discretization in time) and quantization (discretization in amplitude) of an analog time-domain signal.

A sampled signal, $x[n]$, is obtained by taking the value of $x_a(t)$ at multiples of some *sampling interval,* T_s:

$$x[n] = x_a(nT_s), \quad n = -\infty, \ldots, +\infty \tag{14.1}$$

14.2.1 The Nyquist Sampling Theorem

The Nyquist sampling theorem addresses the most important question in sampling: How frequently must samples be taken to adequately represent the analog signal? Consider a signal $x_a(t)$, and assume its Fourier transform $X_a(f)$[1] is band-limited to the

[1] The symbols f and ω are used to represent frequency in cycles per sample (hertz) and radians per second, respectively, for continuous-variable signals. The symbols $\hat{f}$ and $\hat{\omega}$ are used for normalized frequency in cycles per sample and radians per sample for discrete variable signals.

interval $-B/2 \le f \le +B/2$ Hz. Now consider the continuous-time signal $x_s(t)$ defined by

$$\begin{aligned} x_s(t) &= x_a(t)\left\{\sum_{n=-\infty}^{+\infty} \delta_D(t-nT_s)\right\} \\ &= \sum_{n=-\infty}^{+\infty} x_a(nT_s)\delta_D(t-nT_s) \\ &= \sum_{n=-\infty}^{+\infty} x[n]\delta_D(t-nT_s) \end{aligned} \tag{14.2}$$

where $\delta_D(\cdot)$ is the Dirac delta (impulse) function. The "sifting property" of the impulse function has been used to obtain the second line in equation (14.2) [4], while the third line was obtained by substituting from equation (14.1). $x_s(t)$ is a sequence of impulse functions occurring at the sample intervals, nT_s, each weighted by a sample of the original analog signal.

Now consider the Fourier transform of $x_s(t)$. An important property of Fourier transforms is that sampling in one domain causes a periodic replication of the signal in the complementary domain. Thus, sampling $x(t)$ in time to obtain $x_s(t)$ (and ultimately $x[n]$) causes a periodic replication of $X_a(f)$. Specifically [1,2],

$$X_s(f) = \frac{1}{T_s}\sum_{k=-\infty}^{+\infty} X_a\left(f-\frac{k}{T_s}\right) = \frac{1}{T_s}\sum_{k=-\infty}^{+\infty} X_a(f-kf_s) \tag{14.3}$$

The spectrum $X_s(f)$ of the sampled signal $x_s(t)$ is thus formed by replicating the original spectrum every $1/T_s = f_s$ Hz. This effect is illustrated in Figure 14-2, which shows an original and sampled signal and the corresponding original and replicated spectra.

The Nyquist criterion follows immediately from the diagram in Figure 14-2 of the replicated spectrum in the lower right. So long as the spectral replicas do not overlap, the original spectrum, and therefore the original signal, can be recovered from the replicated spectrum by an ideal low-pass filter (sinc interpolator) and gain adjustment. This process is shown in Figure 14-3, which also makes clear the necessary conditions on the original spectrum. The replicas will not overlap so long as $x_a(t)$ was band-limited to some finite bandwidth B Hz as shown, and the sampling frequency satisfies

$$f_s = \frac{1}{T_s} > B \tag{14.4}$$

Equation (14.4) is known as the *Nyquist criterion*; it is the source of the widely quoted statement that a band-limited signal can be recovered from its samples so long as they are taken at a rate equal to "twice the highest frequency" (here, $B/2$ Hz). The minimum sampling rate of B samples/second is called the *Nyquist rate*. As (14.4) shows, it is equivalent and somewhat more general to restate the Nyquist theorem as requiring that a signal be sampled at a rate equal to or greater than the total width of the original spectrum in hertz.

Most DSP texts present the Nyquist sampling theorem in terms of real-valued, baseband signals, as just described. Coherent radar signals are neither: Before demodulation, they are on a carrier, so the nonzero portion of the spectrum is not centered at 0 Hz,

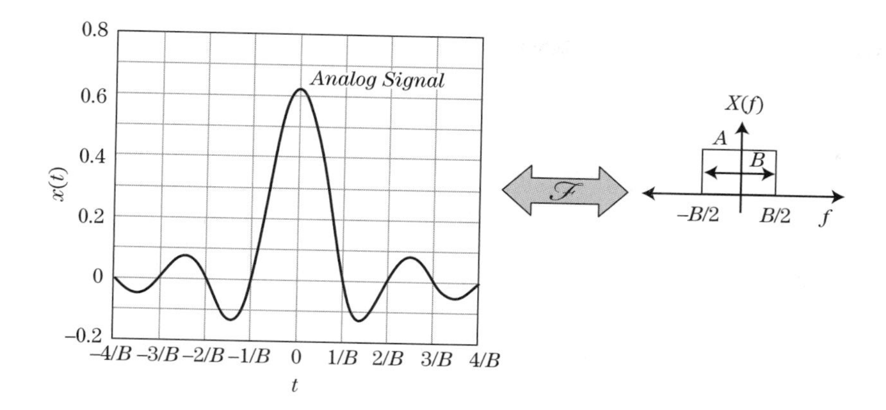

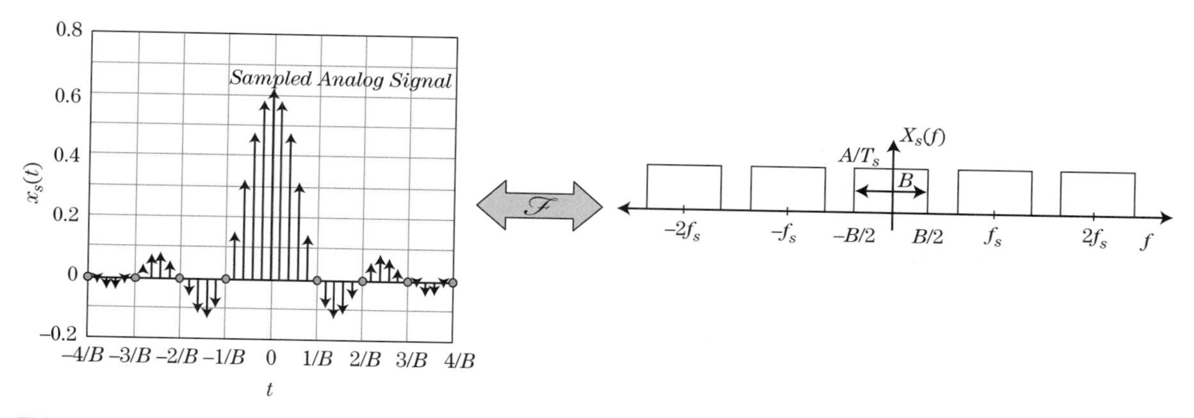

FIGURE 14-2 ■ Effect of sampling on a signal and its spectrum. Top: A band-limited analog signal $x(t)$ and its Fourier transform $X(f)$. Bottom: Analog sampled signal $x_s(t)$ and its Fourier transform $X_s(f)$, illustrating replication due to sampling.

and if a coherent receiver is used, the I and Q output signals are combined to form a complex-valued signal. Nonetheless, the Nyquist theorem still applies, because nothing in its derivation relies on the signal being real-valued or having a baseband spectrum. If the signal is complex-valued, this simply means that the samples taken at B samples/sec or faster will be complex-valued also. The spectrum replication caused by sampling, which is the basis of the Nyquist theorem (14.3), is a property of Fourier transforms and holds for any signal. Thus, if the original spectrum is still band-limited to a total width of B Hz but is centered on a carrier f_0, it is still possible to reconstruct the original signal from samples taken at $f_s = B$ samples/sec or faster. This is illustrated in Figure 14-4, which shows spectrum replication due to sampling a signal having a bandwidth of 40 MHz but not centered at zero Hz. So long as $f_s > 40$ MHz, the replicas shown in the lower part of the figure will not overlap. The difference between this case and the case where the spectrum is centered at zero Hz is that the reconstruction formula will be somewhat different.

This raises another important point. The Nyquist theorem established conditions sufficient to reconstruct an analog signal from its samples. In radar DSP, the analog signal

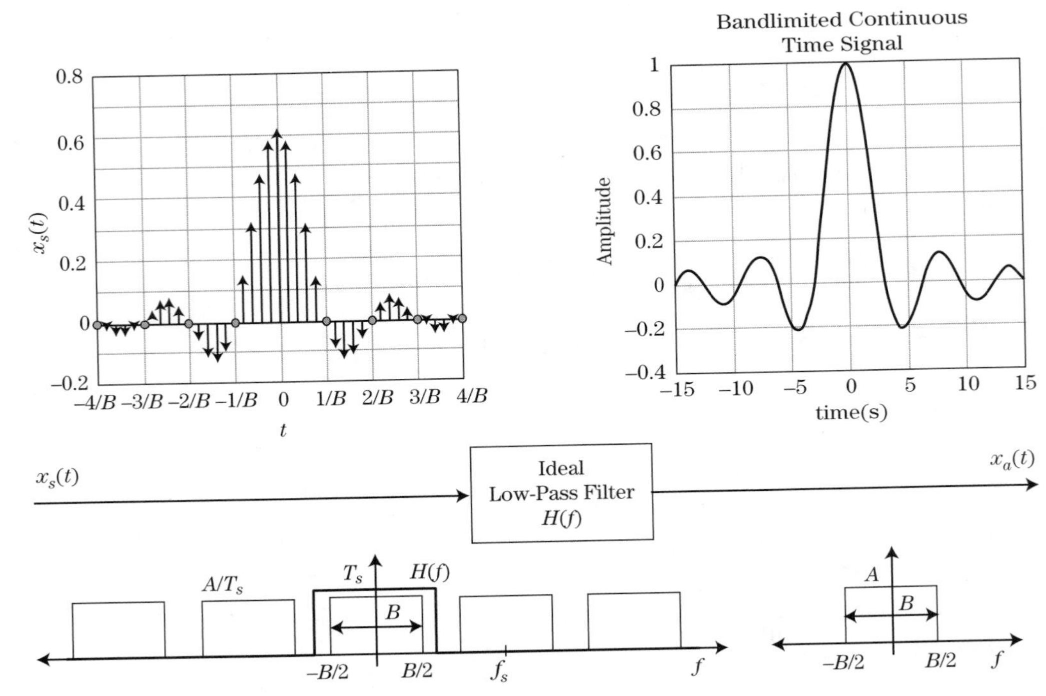

FIGURE 14-3 ■ Reconstruction of analog signal from its sampled equivalent.

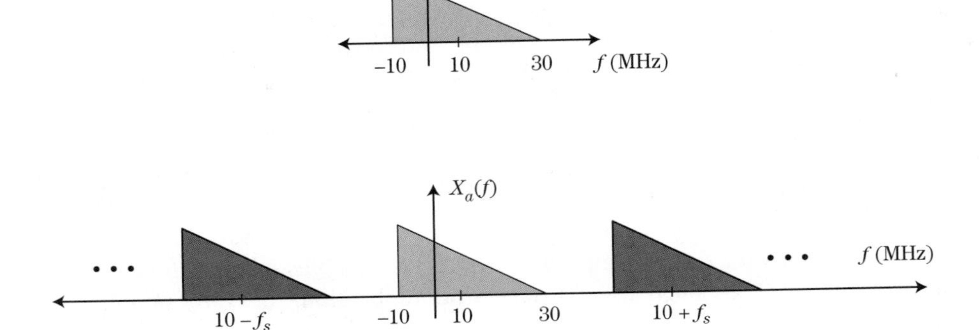

FIGURE 14-4 ■ Effect of sampling on spectrum of a complex, non-baseband signal.

is very rarely, if ever, reconstructed from its samples. Rather, the samples are processed to obtain other information, such as detection reports, track data, or radar images. One could thus ask if the Nyquist criterion is appropriate for radar. In fact, other sampling rules based on other criteria can be developed, such as limiting maximum straddle loss (see Section 14.4.5). However, the ability to reconstruct the original signal implies that all of the information in the signal has been retained, and, for this reason, the Nyquist criterion is generally used to establish required sampling rates.

Finally, note that the Nyquist sampling theorem can be applied in domains other than the time domain. Although the details are not given here, the theorem can be used

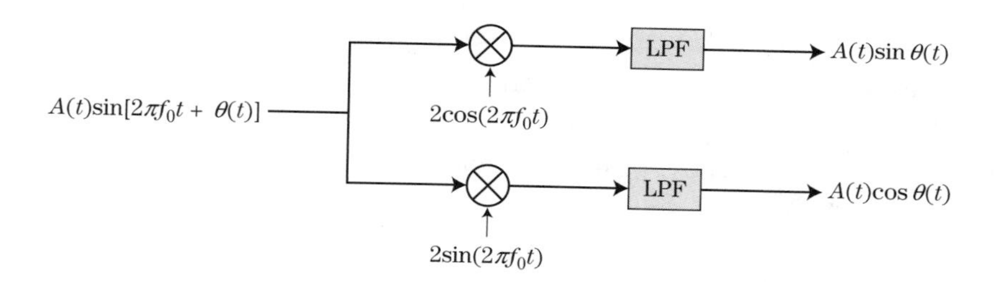

FIGURE 14-5 ■ Coherent (I and Q) receiver.

to establish the minimum-size discrete Fourier transform (DFT) required to represent a finite-length sequence or to determine the density of spatial samples needed in designing an array antenna or an imaging radar [3].

14.2.2 Sampling Nonbaseband Signals

Though it is possible to sample a signal while it is still on a carrier, most conventional systems use a coherent receiver such as the one shown in Figure 14-5 to move the signal energy to baseband prior to sampling. The coherent receiver was discussed in Chapter 11. If the spectrum of the baseband pulse $A(t)$ has a two-sided bandwidth of B Hz then the two output signals from the coherent receiver will each be a real-valued baseband signal with a frequency spectrum occupying the range $(-B/2, +B/2)$ Hz. Each can then be conventionally sampled at its Nyquist rate of $f_s > B$ samples/second.

This coherent receiver approach has been very successful in many fielded systems. Nonetheless, it has limitations. The block diagram of Figure 14-5 implies some stringent requirements on the receiver hardware. For example, the two reference oscillators must be exactly 90° out of phase, while the gain and delay through each of the two output channels must be identical across the signal frequency band.[2] This can be difficult to achieve to adequate precision in high-quality systems and has resulted in increasing interest in so-called *digital IF* or *digital I/Q* receivers. In these receivers, the signal is demodulated to an IF, but not all the way to baseband, with analog hardware but is then sampled while still on the IF carrier, with the remaining baseband conversion completed digitally. Additional detail on both conventional and digital I/Q coherent receivers are given in Chapter 11.

14.2.3 Vector Representation of Sampled Signals

It is sometimes convenient to represent the sampled signal $x[n]$ as a vector. Suppose $x[n]$ is a finite-length N-sample signal, that is, $x[n] = 0$ for $n < 0$ and $n > N - 1$. Form the N-point column vector

$$\boldsymbol{X} = [x[0] \;\; x[1] \;\; \cdots \;\; x[N-1]]^T \tag{14.5}$$

where the superscript T represents matrix transpose. Uppercase italic boldface variables such as $\boldsymbol{X}$ represent vectors, while uppercase non-italic boldface variables such as $\mathbf{R}$ represent matrices. This vector representation will be used in Section 14.10.

[2]The effects of imperfect gain and phase matching are discussed in Chapter 11.

14.2.4 Data Collection and the Radar Datacube

Advanced radar signal processing often operates on data that are sampled in more than one dimension. Consider a pulsed radar. For each pulse transmitted, a series of complex (I and Q) samples of the echo corresponding to successive range intervals will be collected at the output of the receiver. By the Nyquist criterion, these range samples are collected at a rate equal to the pulse bandwidth or greater; this dimension is often referred to as *fast time*. Range samples are also referred to as *range gates* or *range bins*.

The times of the first and last range samples are determined by the desired starting and stopping ranges R_1 and R_2 and are $2R_1/c$ and $2R_2/c$, respectively.[3] The range extent of the samples, known as the range *swath*, is $R_2 - R_1$, and the number of range samples, L, is the swath time divided by the fast time sampling interval, $2(R_1 - R_2)/cT_s$. Assuming a coherent receiver is used, each sample is a single complex number.

R_1 and R_2 are chosen based on a variety of system and mission considerations, but there are limitations on both. The largest R_2 can be is $R_{max} = cT/2$, where T is the radar's pulse repetition interval (PRI). This case corresponds to continuing to collect samples from one pulse until the next one is transmitted. In monostatic radars, the receiver is turned off during the first τ seconds of the PRI (the time during which the pulse is being transmitted) so that the high transmit power does not leak into and damage the receiver. For these radars, the minimum range is $R_{min} = c\tau/2$ m. The echo from targets at range of R_{min} or less is partially or completely ignored and is said to be *eclipsed*. The maximum swath length is then $c(T - \tau)/2$ m.

This process is repeated for each of M pulses in a *coherent processing interval* (CPI; also sometimes called a *dwell* or simply a pulse burst or group), forming a two-dimensional range–pulse number matrix of sampled data, as shown in Figure 14-6. The pulse number dimension is often called *slow time* because the sampling interval in that dimension, which is the radar's pulse repetition interval T, is much greater than the sampling interval in fast time (range). Note that the slow-time signal formed by examining values from pulse to pulse for a fixed range cell, as illustrated by the shaded row, is also a time series.

Now suppose the radar has multiple receiver channels. Multiple channels most often result from having an antenna with multiple phase centers, such as a monopulse antenna

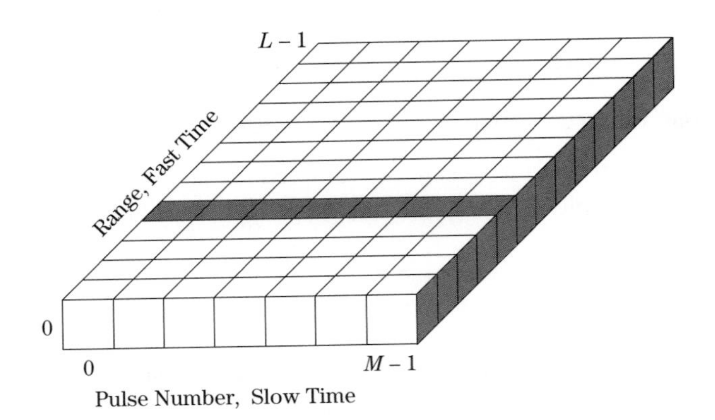

FIGURE 14-6 ■ Fast time–slow time matrix of pulse radar data.

[3]The echo signal is passed through a matched filter in many radars, as discussed in Chapter 20. In this case, the minimum and maximum range sampling times are increased by the delay through the matched filter.

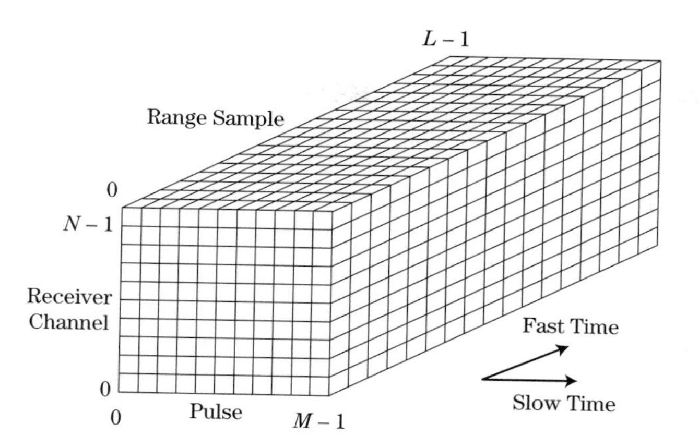

FIGURE 14-7 ■ The radar datacube.

with sum and difference channels or a phased array antenna with a number of subarrays (see Chapter 9). After demodulation and sampling, the data collected over the CPI can be represented by the radar *datacube* shown in Figure 14-7.

Various digital processing algorithms correspond to operating on the datacube in various dimensions. Figure 14-8 illustrates many of the major processing types. For example, pulse compression (see Chapter 20) is linear filtering (or correlation) in the fast-time dimension and can be performed independently for each pulse and receiver channel. Similarly, Doppler processing (MTI or pulse-Doppler; see Chapter 17) operates across multiple samples for a fixed range bin (i.e., in the slow-time dimension) and can be performed independently on each range bin and receiver channel. Synthetic aperture imaging (Chapter 21) is an example of a process that operates on two dimensions of the datacube, in this case both slow and fast time. This view of the datacube is sometimes helpful in thinking about the relationship between differing processing techniques.

Not all radar systems require real-time processing of the data, but, in those that do, any pulse-by-pulse fast-time processing such as pulse compression must be completed in T seconds or less so the processing of the samples of one pulse is completed before the samples of the next pulse arrive. Operations such as beamforming must be completed for each range bin in one PRI or less. Similarly, CPI-level processing such as space-time adaptive processing (STAP) must be completed before the next CPI of data is received, and slow-time processes such as pulse-Doppler processing must be completed for each range bin in one CPI.

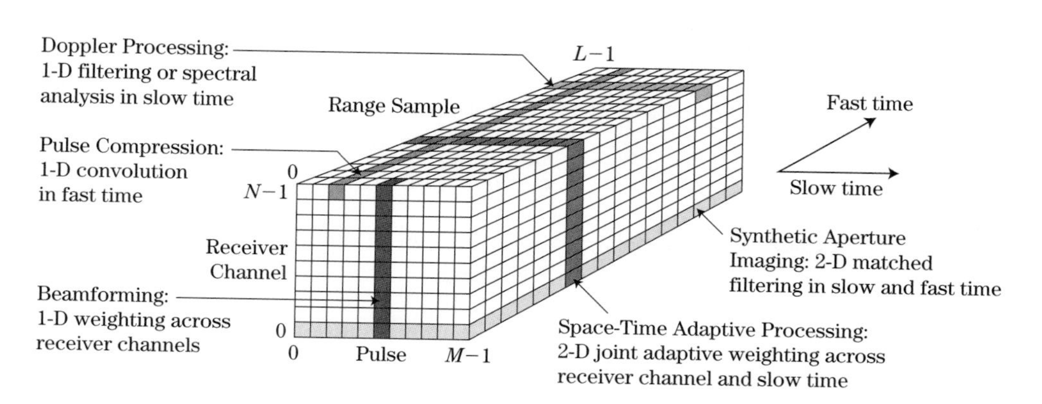

FIGURE 14-8 ■ Datacube showing the dimensions on which various radar signal processing algorithms act.

14.3 QUANTIZATION

The discretization of analog signals in amplitude is called *quantization*. Quantization is necessary so that the amplitude of each sample can be represented with a finite number of bits in a computer or digital hardware.

Binary representations of numeric data can generally be categorized as either fixed point or floating point.[4] In fixed-point representations, the b bits of a binary word are used to represent 2^b distinct and evenly spaced numerical values. In a floating-point representation, some number e of the b total bits is used to represent an exponent, and the remaining $m = b-e$ bits are used to represent the mantissa. Here, fixed-point representations are of primary interest, as this is the representation output by most analog-to-digital converters (ADCs).

The signal voltage value corresponding to each of the 2^b possible binary words in a fixed-point encoding is determined by the arithmetic coding scheme that is used and the quantization step size, Δ. The step size is the change in input value in volts for each increment or decrement of the binary word value. Thus, the value of the data sample is simply Δ times the binary number.

The two most common encodings are called *sign-magnitude* encoding and *two's complement* encoding. In sign-magnitude encoding, the most significant bit of the binary word represents the sign of the data sample; usually a value of zero represents a positive number, while a value of binary one represents a negative number. The remaining $b - 1$ bits encode the magnitude of the sample. Thus, sign-magnitude encoding can represent numbers from $(-2^{b-1} - 1)\Delta$ to $(+2^{b-1} - 1)\Delta$. Note that there are two codes for the value zero, corresponding to $+0$ and -0.

Two's complement is a somewhat more complex encoding that has advantages for the design of digital arithmetic logic. Details are given in [5]. For the present purpose, it is sufficient to note that there is only one code for zero. The extra code value allows the representation of one more negative number, so the range of values becomes $-2^{b-1}\Delta$ to $(+2^{b-1}-1)\Delta$. For any reasonable number of bits b, the difference in range is not significant.

The choice of the number of bits and quantization step size govern the trade-off between the dynamic range and quantization error of the digital signal. Dynamic range is the ratio of the largest representable magnitude to the smallest nonzero magnitude; for the two's complement case, this is

$$DR = \frac{2^{b-1}\Delta}{\Delta} = 2^{b-1} \tag{14.6}$$

For the sign-magnitude case, $DR = 2^{b-1} - 1 \approx 2^{b-1}$ for b more than just a few bits. Expressed in dB, equation (14.6) becomes

$$\begin{aligned} DR\,(\text{dB}) &= 20\log_{10}(2^{b-1}) \\ &= (b-1)20\log_{10}(2) \\ &= 6.02b - 6.02 \text{ dB} \end{aligned} \tag{14.7}$$

Equation (14.7) shows that the dynamic range that can be represented at the ADC output without saturation (overflow) increases by 6 dB per bit.

[4] Additional variations, such as block floating point, are also used but are beyond the scope of this chapter.

Conversion from analog, with its infinite allowed amplitude values, to a b-bit digital word with its finite number 2^b of possible values, necessarily entails rounding or truncating the analog sample value to one of the allowed quantized values. The difference between the unquantized and quantized samples is the *quantization error*. Although it is in fact a deterministic function of the input data and ADC parameters, the behavior of the quantization error signal is usually complex enough that it is treated as a random variable that is uncorrelated from one sample to the next. Thus, the sequence of quantization errors is modeled as a white random process called the *quantization noise*. Quantization noise is an independent noise signal that adds to the receiver noise already present in the raw analog data. The quantization error for each sample can vary between $\pm\Delta/2$, and it is commonly assumed that errors anywhere in this range are equally likely. Thus, the quantization noise process is modeled as a uniform random process over this range so that the quantization noise power is $\Delta^2/12$ [2].

Assume that the quantizer is calibrated to cover a range of $\pm A_{sat}$ without saturation. Then $\Delta = A_{sat}/2^{b-1}$. Assume also that the input signal to the ADC is modeled as a random process with some power σ^2. This could model the receiver thermal noise, for instance, which is the minimum signal expected at the ADC input. The signal-to-quantization noise ratio (SQNR) is then

$$SQNR\,(\text{dB}) = 10\log_{10}\left(\frac{\sigma^2}{\Delta^2/12}\right) = 10\log_{10}\left(\frac{2^{2b-2}12\sigma^2}{A_{sat}^2}\right) = 10\log_{10}\left(\frac{2^{2b}3\sigma^2}{A_{sat}^2}\right)$$
$$= 6.02b - 10\log_{10}\left(\frac{A_{sat}^2}{3\sigma^2}\right) \tag{14.8}$$

For a given signal power level, the SQNR will improve by 6 dB per bit, similar to the dynamic range.

There remains the issue of how to set the quantizer step size Δ relative to the signal level. If Δ is too small, large input signal values will exceed the range of the ADC and will be clipped to a value of $\pm A_{sat}$, a situation called *saturation*. If Δ is made larger to avoid saturation, then small signal variations may not cause the ADC to count up or down; very small signals may produce only a constant output of zero, thus being suppressed entirely in a condition called *underflow*. In either case, the SQNR will not follow the 6 dB per bit rule. McClellan and Purdy [6] modeled this behavior by deriving the probability density function of the ADC output using the previously given model. Ideally, the output noise level should be the same as the input so that the ADC neither adds nor detracts from the input signal power.

The results of the analysis in [6] are shown in Figure 14-9 as a function of the step size Δ, normalized to the input noise signal standard deviation. This figure suggests that, when the input is the receiver noise, a normalized step size of between ± 6 dB, meaning that Δ is between one-half and two times the quiescent noise standard deviation at the ADC input, will result in no more than 1 dB of additional quantization noise at the ADC output, while avoiding saturation or underflow effects. Many fielded systems set Δ in the lower end of this range, effectively devoting one or two bits to representing the receiver noise. Note that a normalized Δ of greater than about 12 dB (four times the noise standard deviation) results in a rapidly falling output power due to underflow because the input signal fails to exceed the relatively large ADC step size. As a result, the output is mostly zero. For a small number of bits, step sizes less than one-half the noise standard deviation result in an output

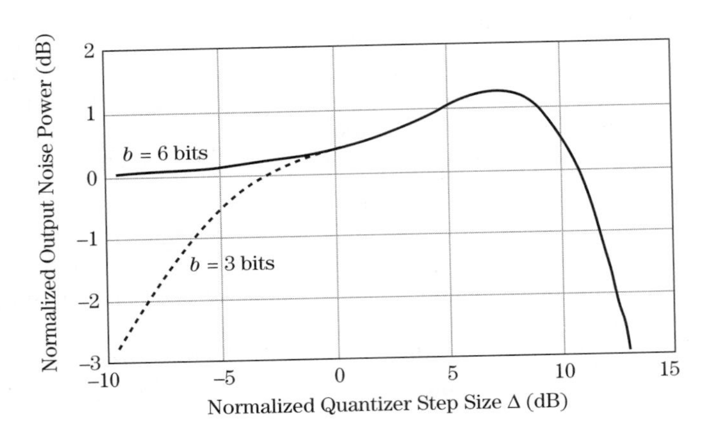

FIGURE 14-9 ■ Ratio of output power to input power for an ADC with a noise signal at its input.

power less than the input power due to saturation, since the small number of ADC levels and the small step size severely limit the range of input signals that can be represented.

Once the data are quantized, a choice must be made between implementing the processing in fixed- or floating-point arithmetic. Generally, floating-point arithmetic requires more digital logic to implement and is therefore slower. However, mathematical algorithms are easier to develop for floating-point arithmetic because numerical overflow is much less likely. Historically, most radar digital processors have relied on fixed-point arithmetic, at least in the early processing stages that tend to be more computationally intensive, because it is faster. Increasing numbers of modern systems are using floating-point processing because the dramatic increases in processor power have made it possible to implement the desired algorithms in real time.

14.4 FOURIER ANALYSIS

Just as in other signal processing application areas, Fourier analysis is instrumental to radar signal processing, both for analysis and for actual algorithm implementation. In this section, the two major forms of Fourier transforms applicable to sampled signals are reviewed. Their relation to each other and to the Fourier transform of the analog signal is illustrated. Finally, important properties of the Fourier transform of some key signals are illustrated.

14.4.1 The Discrete-Time Fourier Transform

The *discrete-time Fourier transform* (DTFT) of a discrete signal $x[n]$ is defined as

$$X(\hat{\omega}) = \sum_{n=-\infty}^{+\infty} x[n]e^{-j\hat{\omega}n}, \quad \hat{\omega} \in (-\infty, +\infty)$$

or

$$X(\hat{f}) = \sum_{n=-\infty}^{+\infty} x[n]e^{-j2\pi \hat{f}n}, \quad \hat{f} \in (-\infty, +\infty) \tag{14.9}$$

where the frequency variable is a normalized frequency, $\hat{\omega}$, in rad/sample (not rad/s) or $\hat{f}$, in cycles/sample (not cycles/s, or Hz). It will be seen momentarily why the frequency variable is considered to be normalized. The transform of equation (14.9) will be called the

DTFT even when the independent variable is not time. Note that $X(\hat{\omega})$ is periodic with a period of 2π in $\hat{\omega}$ or a period of 1 in $\hat{f}$. Thus, $X(\hat{\omega})$ is normally examined only in an interval of width 2π, usually $(-\pi, +\pi)$ or $(0, 2\pi)$. When working in units of cycles/sample, $X(\hat{f})$ is typically examined in the interval of $(-0.5, +0.5)$ or $(0, 1)$. The inverse transform is

$$x[n] = \frac{1}{2\pi}\int_{-\pi}^{\pi} X(\hat{\omega})e^{+j\hat{\omega}n}d\hat{\omega}, \quad n \in (-\infty, +\infty)$$

or

$$x[n] = \int_{-0.5}^{0.5} X(\hat{f})e^{+j2\pi\hat{f}n}d\hat{f}, \quad n \in (-\infty, +\infty) \tag{14.10}$$

Even though the signal $x[n]$ has a discrete independent variable, it is important to realize that the frequency variable of the DTFT ($\hat{\omega}$ or $\hat{f}$) is continuous: the DTFT is defined for all values of frequency. This fact will prove important in understanding the behavior of the discrete Fourier transform, or DFT, in Section 14.4.4.

Consider a discrete signal $x[n]$ obtained by sampling an analog signal $x_a(t)$ at intervals of T_s seconds. To relate the DTFT of $x[n]$ to the Fourier transform of $x_a(t)$, consider the Fourier transform of the sampled signal $x_s(t)$ from (14.2):

$$\begin{aligned} X_s(f) &= \int_{-\infty}^{+\infty} x_s(t)e^{-j2\pi ft}df \\ &= \sum_{n=-\infty}^{+\infty} x[n]\left\{\int_{-\infty}^{+\infty} \delta_D(t-nT_s)e^{-j2\pi ft}df\right\} \\ &= \sum_{n=-\infty}^{+\infty} x[n]e^{-j2\pi fnT_s} \end{aligned} \tag{14.11}$$

Comparing the last line of (14.11) to the definition of the DTFT $X(\hat{f})$ in (14.9) shows that $X(\hat{f}) = X_s(f)$ when $\hat{f} = fT_s = f/f_s$. Using (14.3) then gives the desired relation between the DTFT and the original analog signal spectrum:

$$X(\hat{f}) = \frac{1}{T_s}\sum_{k=-\infty}^{+\infty} X_a(\hat{f}\cdot f_s - kf_s) \tag{14.12}$$

Equation (14.12) shows that the periodic DTFT $X(\hat{f})$ is the replicated (due to sampling) and rescaled (in amplitude and frequency) analog spectrum $X_a(f)$. Features that appear in the original spectrum at some frequency f_0 will appear in the DTFT at the normalized frequency $\hat{f}_0 = f_0/f_s$.

The relation $\hat{f} = f/f_s$ (and equivalently, $\hat{\omega} = \omega/f_s = \omega T_s$) is the reason the DTFT frequency variable $\hat{f}$ is called *normalized* frequency: it is the original analog frequency normalized to the sampling frequency $f_s = 1/T_s$. Since the DTFT is periodic in $\hat{f}$ with a period of 1, it follows that one period of the DTFT is equivalent to f_s Hz.

A particularly important yet simple example is the DTFT of a complex sinusoid, a signal that arises in many areas of radar signal processing. Choose

$$x[n] = Ae^{j2\pi\hat{f}_0 n}, \quad n = 0, \ldots, N-1 \tag{14.13}$$

The DTFT of $x[n]$ is then

$$X\left(\hat{f}\right) = A\sum_{n=0}^{N-1} e^{-j2\pi(\hat{f}-\hat{f}_0)n} = A\frac{1-e^{-j2\pi(\hat{f}-\hat{f}_0)N}}{1-e^{-j2\pi(\hat{f}-\hat{f}_0)}} \tag{14.14}$$

where the last step was obtained by applying the very useful geometric sum formula

$$\sum_{n=N_1}^{N_2} \alpha^n = \frac{\alpha^{N_1} - \alpha^{N_2+1}}{1-\alpha} \tag{14.15}$$

with $\alpha = \exp(-j2\pi(\hat{f} - \hat{f}_0))$, $N_1 = 0$, and $N_2 = N - 1$. Equation (14.14) can be put in a more useful form as follows:

$$\begin{aligned} X(f) = A\frac{1-e^{-j2\pi(\hat{f}-\hat{f}_0)N}}{1-e^{-j2\pi(\hat{f}-\hat{f}_0)}} &= A\frac{e^{-j\pi(\hat{f}-\hat{f}_0)N}\left(e^{+j\pi(\hat{f}-\hat{f}_0)N} - e^{-j\pi(\hat{f}-\hat{f}_0)N}\right)}{e^{-j\pi(\hat{f}-\hat{f}_0)}\left(e^{+j\pi(\hat{f}-\hat{f}_0)} - e^{-j\pi(\hat{f}-\hat{f}_0)}\right)} \\ &= Ae^{-j\pi(\hat{f}-\hat{f}_0)(N-1)}\frac{\sin(\pi(\hat{f}-\hat{f}_0)N)}{\sin(\pi(\hat{f}-\hat{f}_0))} \\ &\equiv NAe^{-j\pi(\hat{f}-\hat{f}_0)(N-1)}\text{asinc}(\hat{f}-\hat{f}_0, N) \end{aligned} \tag{14.16}$$

The last line defines the "aliased sinc" or "asinc" function, the discrete-time signal equivalent to the sinc function prevalent in continuous-time Fourier analysis. Still another name for this function is the Dirichlet function. Figure 14-10 illustrates the magnitude of the DTFT of an $N = 20$ sample, unit amplitude complex sinusoid of normalized frequency $\hat{f}_0 = 0.25$. Note that the peak magnitude is equal to N. The asinc function can be evaluated numerically to show that the peak sidelobe is approximately 13.2 dB below the central peak amplitude, that the 3 dB mainlobe width is approximately $0.89/N$ cycles/sample, and that the peak-to-first null mainlobe width (also called the *Rayleigh* width) is $1/N$ cycles/sample.

The inverse dependence between the length of the signal observation, N, and the width of the mainlobe of the DTFT is called the *reciprocal spreading* property of the Fourier transform: the wider a signal is in one domain (time or frequency), the narrower it is in the complementary (frequency or time) domain. This property is illustrated explicitly in Figure 14-11, which compares the DTFTs of 20 and 40 sample complex exponentials of the same frequency $\hat{f}_0 = 0.1$. Only the real part of the complex exponentials is shown.

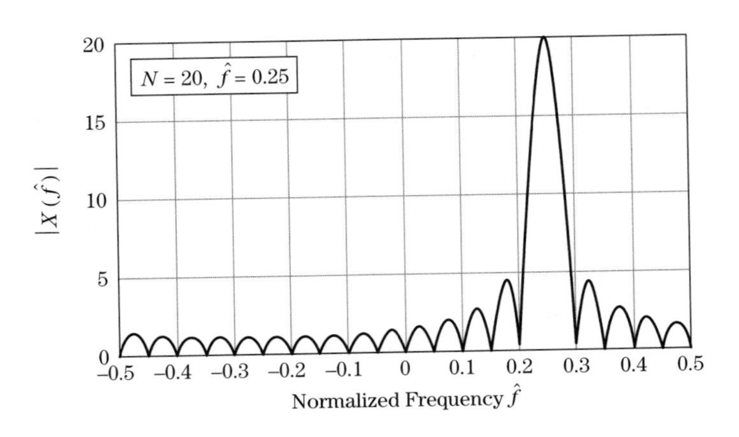

FIGURE 14-10 ■ Magnitude of the DTFT of a pure complex exponential.

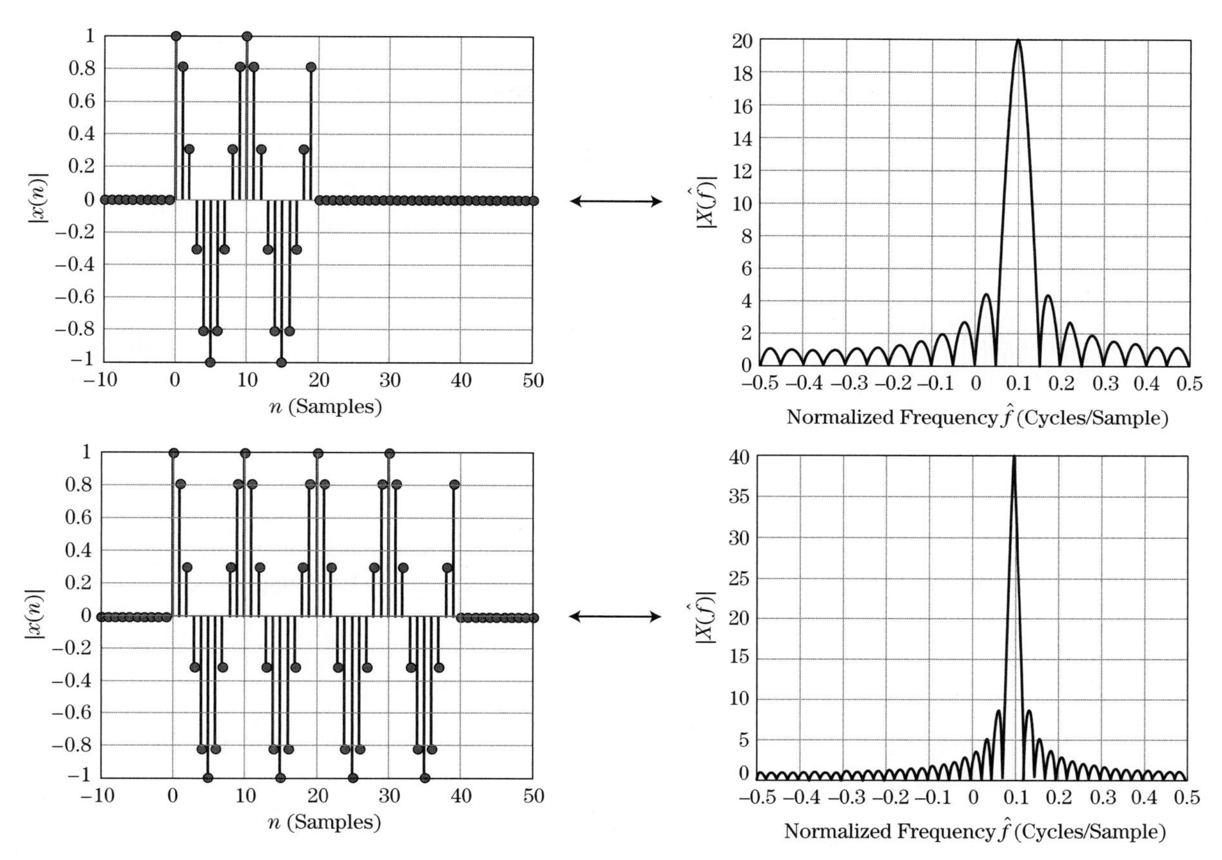

FIGURE 14-11 ■ Illustration of "reciprocal spreading" in the DTFT. (a) $N = 20$ signal samples, and DTFT. (b) $N = 40$ signal samples, and DTFT.

Clearly, doubling the length of the exponential pulse halves the width of the DTFT asinc function.[5]

Specifically, suppose N samples of data are available. The frequency resolution is often taken to be the Rayleigh width of $1/N$ cycles/sample. The $1/N$ criterion is somewhat arbitrary; another common criterion is the two-sided 3 dB width $0.89/N$. The same relationships hold in analog units: if the data are collected at a sampling interval of T_s seconds, the signal duration is NT_s seconds and the DTFT Rayleigh width is $1/NT_s$ Hz whereas the 3 dB width is $0.89/NT_s$ Hz.

14.4.2 Windowing

The –13.2 dB sidelobes of the asinc DTFT in Figure 14-10 are unacceptable in many applications. In radar, multiple targets are frequently observed with received power levels that vary by several tens of decibels due to differences in RCS and range. Consider the Doppler spectrum of a signal containing echoes from two targets. If the peak of a weak

[5]A more formalized expression of this phenomenon is the Fourier uncertainty principle described in [4].

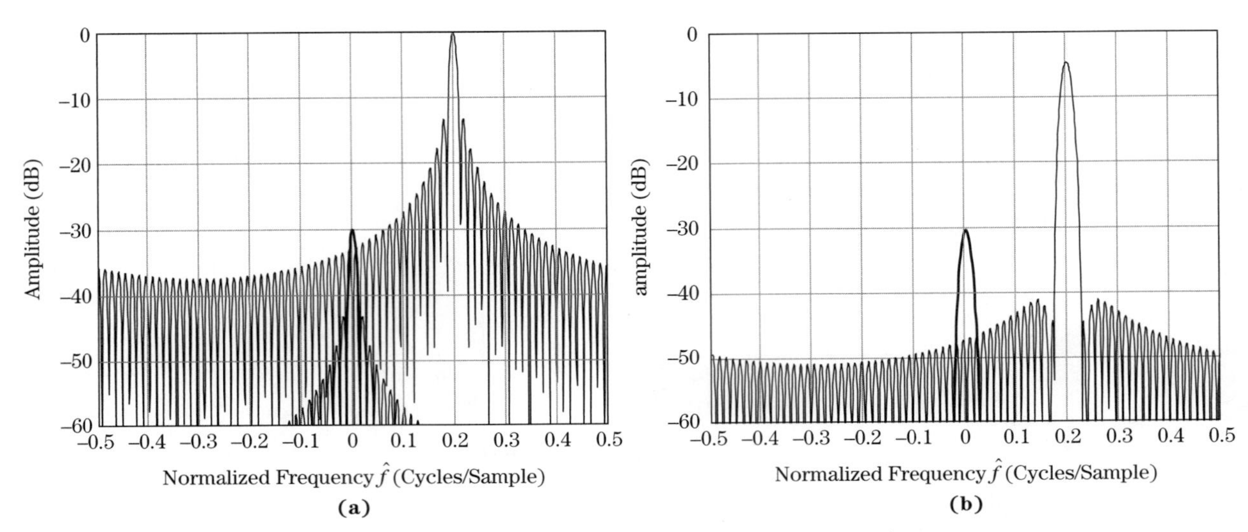

FIGURE 14-12 ■ (a) Masking of a weak target response at $\hat{f} = 0$ by the sidelobes of a 30 dB stronger response at $\hat{f} = 0.2$. (b) Same two responses with Hamming window applied before the DTFT.

target's DTFT is on a par with or is lower than the sidelobes of a strong target's DTFT, the sidelobes of the stronger target will *mask* the DTFT of the weaker target, and only the stronger target will be observed. Figure 14-12a illustrates this effect for a case where the weaker target is 30 dB lower in amplitude than the stronger target.

Windowing the data prior to computing the DTFT is the most common method of ameliorating this problem. Repeating the example of Figure 14-12a with a Hamming window applied to the data before the DTFT results in the spectrum of Figure 14-12b. In this example, the weaker target is easily detectable above the sidelobes of the stronger target.

The windowing procedure is illustrated in Figure 14-13. The upper left plot is a data sequence representing 4 microseconds of a real sinusoid. The lower left sequence is a typical window function, in this case a Hamming window [2]. The two are multiplied point by point to produce the windowed sinusoid shown on the right.

There are many window functions in use; a classic paper by F. J. Harris [8] provides an extensive list of windows and properties.[6] Typical windows are real-valued and are usually smoothly tapered from a maximum value in the center to a lower level (not necessarily zero) at the endpoints. Thus, windows enforce a smooth tapering of the data at the edges of the data sequence.

The principal motivation for windowing is to reduce sidelobes in the DTFT. Figure 14-14 illustrates the effect of a Hamming window on the DTFT of a 20-sample square pulse. The window reduces the energy in the data and its effective time-domain duration by tapering the data near the edges. The reduction in energy is reflected in the DTFT as a reduction in the peak value of the DTFT, in this case by about 5.4 dB. By the reciprocal spreading property, the reduced effective width of the data in the time domain is reflected

[6]Definitions of most common windows are also available at http://www.wikipedia.org and are embedded in many software systems such as MATLAB®.

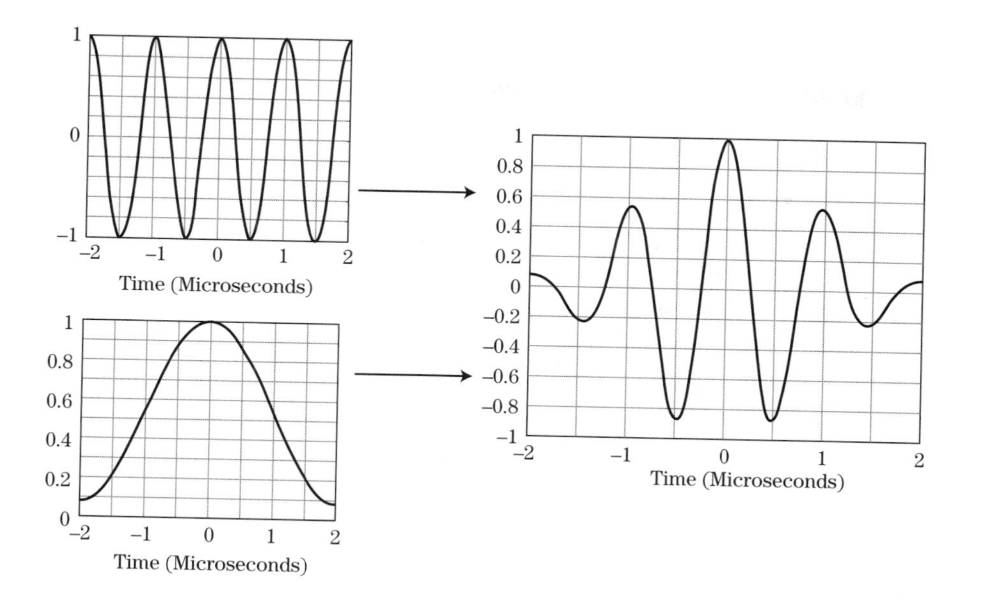

FIGURE 14-13 ■ Illustration of windowing of data.

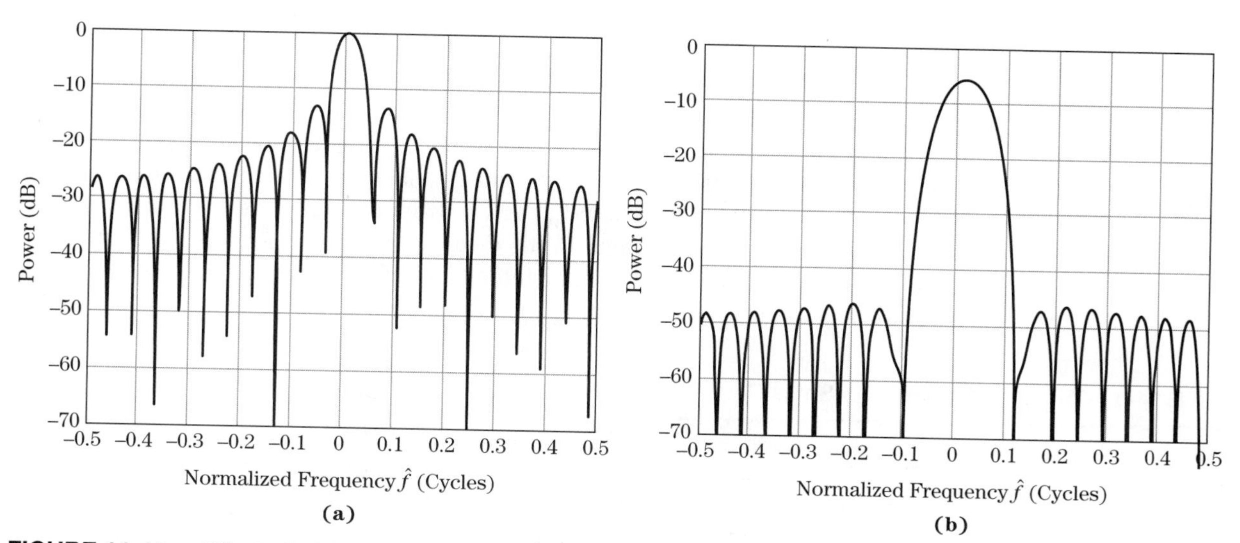

FIGURE 14-14 ■ Effect of a Hamming window on the DTFT of a rectangular pulse. (a) No window. (b) Hamming window.

in a wider mainlobe of its DTFT in the frequency domain. For this example, the 3 dB mainlobe width is increased by 46% compared with the unwindowed case; if measured by the Rayleigh (peak-to-null) width, the increase is 100%. Thus, the frequency resolution of the DTFT has been degraded by 46% to 100%, depending on the metric used. However, these costs of windowing are offset by a dramatic 30 dB reduction in sidelobes, to 43 dB below the DTFT peak. This greatly improves the dynamic range of the DTFT measurement and is the principal reason that windows are routinely used in Fourier analysis.

It is straightforward to quantify some of these window metrics as a function of the window shape. For example, the factor by which the peak power of the DTFT of the windowed data is reduced compared with the DTFT of the unwindowed data is [8]

$$\text{DTFT peak power reduction factor} = \frac{1}{N^2}\left|\sum_{n=0}^{N-1} w[n]\right|^2 \tag{14.17}$$

In (14.17), $w[n]$ is a window function of duration N samples and is assumed to be normalized so that $\max\{w[n]\} = 1$. While this reduction can be significant, the window also tapers any noise present in the data, so the reduction in signal-to-noise ratio is not as great. The factor by which the SNR is reduced due to windowing is given by [8]

$$\text{SNR reduction factor} = \frac{\left|\sum_{n=0}^{N-1} w[n]\right|^2}{N \sum_{n=0}^{N-1} |w[n]|^2} \tag{14.18}$$

In both (14.17) and (14.18) the "reduction" or "loss" is defined to be a number less than 1 on a linear scale and thus a negative number on a decibel scale. The value in dB is obtained by taking $10\log_{10}$ of the values given by the equations.

Most of these metrics are weak functions of the window length, N. As an example, for the Hamming window the loss in signal-to-noise ratio is -1.75 dB for a short ($N = 8$) window, decreasing asymptotically to about -1.35 dB for long windows. Table 14-1 summarizes the five key properties of loss in peak power, peak sidelobe level, 3 dB mainlobe width, SNR loss, and straddle loss (discussed in Section 14.4.5) for several common windows. The values given are for $N = 32$, a typical window length for slow-time pulse-Doppler processing. The "rectangular window" is an untapered window having a constant value of 1.0 for all samples; thus, it actually represents no windowing at all (other than enforcing a finite length to the data) but serves as a reference for the other entries. A much more extensive table, including more metrics as well as the definitions of these and many more types of windows, is given in [8].

TABLE 14-1 ■ Selected Properties of Some Common Window Functions

Window	3 dB Mainlobe Width (Relative to Rectangular Window)	Peak Gain (dB Relative to Peak of Rectangular Windowed Signal)	Peak Sidelobe (dB Relative to Peak of Windowed Signal)	SNR Loss (dB, Relative to Rectangular Windowed Signal)	Maximum Straddle Loss (dB)
Rectangular	1.0	0.0	−13.2	0	3.92
Hann	1.68	−6.3	−31.5	−1.90	1.33
Hamming	1.50	−5.6	−41.7	−1.44	1.68
Kaiser, $\beta = 6.0$	1.63	−6.3	−44.1	−1.80	1.42
Kaiser, $\beta = 8.0$	1.85	−7.5	−57.9	−2.35	1.11
Taylor, 35 dB, $\bar{n} = 5$	1.34	−4.4	−35.2	−0.93	2.11
Taylor, 50 dB, $\bar{n} = 5$	1.52	−5.7	−46.9	−1.49	1.64
Taylor, 50 dB, $\bar{n} = 11$	1.54	−5.8	−49.8	−1.55	1.60
Dolph-Chebyshev (50 dB equiripple)	1.54	−5.6	−50.0	−1.54	1.61
Dolph-Chebyshev (70 dB equiripple)	1.78	−7.2	−70.0	−2.21	1.19

Note: Length $N = 32$ samples in all cases.

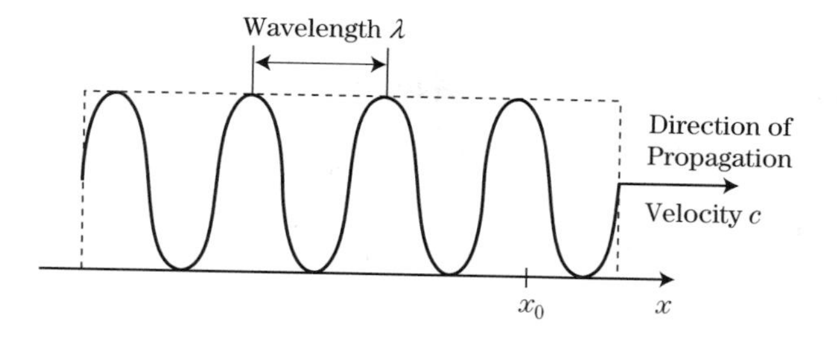

FIGURE 14-15 ■ Temporal and spatial frequency for a sinusoidal pulse.

14.4.3 Spatial Frequency

In discussing the concepts of frequency and Fourier analysis, the signal is usually thought of as being in the time domain. Its Fourier transform is then in the temporal frequency domain, usually just referred to as "frequency." However, in any application such as radar or communications that is concerned with propagating waves, *spatial frequency* is an important concept that will be needed for analyzing spatial sampling, imaging, and space-time adaptive processing. Here a simplified, intuitive introduction to the concept is given. For a more complete discussion, see [3,9].

Consider a radar pulse formed by electromagnetic plane waves propagating in the $+x$ direction with wavelength λ and velocity c as shown in Figure 14-15. An observer at a fixed spatial position x_0 will see successive positive crests of the electric field at a time interval (period) of $T = \lambda/c$ seconds; thus, the temporal frequency of the wave is $f = 1/T = c/\lambda$ Hz or $\omega = 2\pi c/\lambda$ rad/s.

A spatial period can also be defined; it is simply the interval between successive crests of the plane wave in space for a fixed observation time. From the figure, the spatial period of the pulse is obviously λ meters. The spatial frequency is therefore $1/\lambda$ cycles/m or $2\pi/\lambda$ rad/m. It is common to call the latter quantity the *wavenumber* of the pulse and to denote it with the symbol k.[7]

Because position in space and velocity are three-dimensional vector quantities in general, so is the wavenumber. For simplicity of illustration, consider the two-dimensional version of Figure 14-15 shown in Figure 14-16. The plane waves forming the radar pulse are now propagating at an angle θ relative to the $+y$ axis in an x-y plane.[8] The plane wave still has a wavenumber $k = 2\pi/\lambda$ in the direction of propagation. However, projected onto the x axis, the crests are now $\lambda/\sin\theta$ meters apart, and therefore the x component of the wavenumber is $k_x = (2\pi/\lambda)\sin\theta$. Similarly, the y component of the wavenumber is $k_y = (2\pi/\lambda)\cos\theta$. Note that as $\theta \to 0$, $k_x \to 0$ and the wavelength in the x dimension tends to ∞. Similar results hold for the y dimension when $\theta \to \pi/2$.

The extension to three dimensions of space is straightforward. The total wavenumber is related to the components as

$$k = \sqrt{k_x^2 + k_y^2 + k_z^2} \tag{14.19}$$

[7] In keeping with the notation for analog and normalized temporal frequency, the symbol k will be used for analog wavenumber and the symbol $\hat{k}$ for normalized wavenumber.

[8] Incidence angle will often be measured with respect to the normal to the x-axis (i.e., the y-axis) because this is convenient and conventional in analyzing antenna patterns. If the antenna aperture lies in the x dimension, then an incidence angle of $\theta = 0$ indicates a wave propagating normal to the aperture, that is, in the boresight direction.

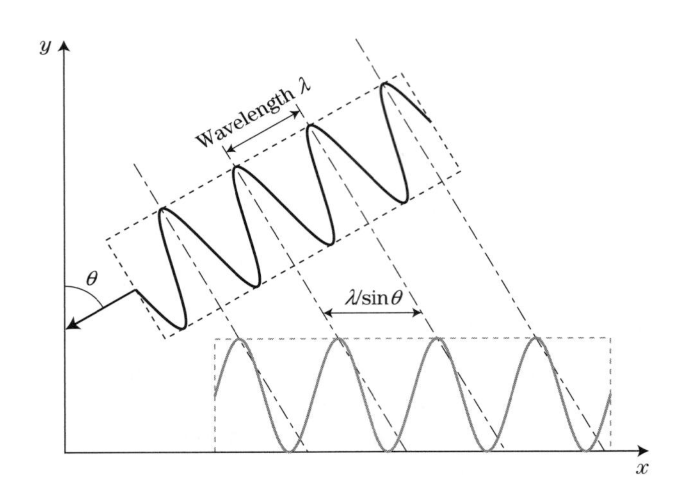

FIGURE 14-16 ■ Spatial frequency and wavelength in the x-direction, expressed in terms of the spatial frequency in the direction of propagation and the incidence angle with respect to the y-axis.

and always equals $2\pi/\lambda$. Note that the temporal frequency remains c/λ Hz regardless of the direction of propagation.

The units of wavenumber are radians per meter. An equivalent but less commonly used cyclical variable is *range frequency, p,* in units of cycles per meter. Thus, $p = k/2\pi$. In monostatic radar, of course, range usually appears as $2R$. The *two-way* range frequency is defined as $f_r = 2p$. When spatial frequencies are expressed in two-way range frequency units, signal properties such as relationships between bandwidth and resolution, sampling and aliasing, and so forth can be analyzed using the same rules that apply to temporal signals and their Fourier transforms [10].

Equivalent results hold for spatially sampled signals and the corresponding normalized spatial frequencies. If a signal is sampled in space at some interval P_s meters, the DTFT of the spatially sampled data sequence will be in normalized units of cycles/sample or rad/sample. If in rad/sample, the units can be considered to be a normalized wavenumber $\hat{k}$. Normalized wavenumber can be related back to the standard analog wavenumber in rad/m using the relation $\hat{k} = P_s k$. If the DTFT is labeled in cycles/sample $\hat{p}$, then the corresponding spatial frequency p satisfies $\hat{p} = P_s p$.

14.4.4 The Discrete Fourier Transform

Recall that the discrete-time Fourier transform is a function of a continuous normalized frequency variable $\hat{\omega}$ or $\hat{f}$. While the DTFT is an essential tool for analysis, it is not directly useful for computation because of the continuous frequency variable: the DTFT cannot be computed for all $\hat{\omega}$ with a finite number of computations. The solution to this problem is the same as used in the time domain: develop a representation that is sampled in the frequency (as well as in the time or spatial) domain. This representation is called the *discrete Fourier transform* and is not to be confused with the DTFT.

The DFT can be derived by mimicking the steps used in deriving the Nyquist criterion for sampling in time, applying them instead to the DTFT frequency domain. Given a signal $x[n]$ with DTFT $X(\hat{f})$, consider the sampled spectrum $X_s(\hat{f})$, still a function of $\hat{f}$, obtained by multiplying $X(\hat{f})$ by an impulse train of K evenly spaced Dirac impulse functions in the interval [0,1) (repeated periodically as with any DTFT):

$$
\begin{aligned}
X_s(\hat{f}) &= X(\hat{f})\left\{\frac{1}{K}\sum_{k=-\infty}^{\infty}\delta_D\left(\hat{f}-\frac{k}{K}\right)\right\} \\
&= \frac{1}{K}\sum_{k=-\infty}^{\infty} X\left(\frac{k}{K}\right)\delta_D\left(\hat{f}-\frac{k}{K}\right) = \frac{1}{K}\sum_{k=-\infty}^{\infty} X[k]\delta_D\left(\hat{f}-\frac{k}{K}\right) \qquad (14.20)
\end{aligned}
$$

where $X[k]$ is defined to be the DTFT samples $X(k/K)$. The inverse transform of $X_s(\hat{f})$, obtained by inserting (14.20) into (14.10) and using the periodicity of $X(\hat{f})$ to shift the limits of integration to the interval (0, 1) for convenience, is

$$
\begin{aligned}
x_s[n] &= \int_0^1 \left\{\frac{1}{K}\sum_{k=-\infty}^{\infty} X[k]\delta_D\left(\hat{f}-\frac{k}{K}\right)\right\} e^{j2\pi n\hat{f}}\,d\hat{f} \\
&= \frac{1}{K}\sum_{k=0}^{K-1} X\left(\frac{k}{K}\right) e^{j2\pi nk/K} \qquad (14.21)
\end{aligned}
$$

Note that the finite limits on k resulted from the finite integration limits on $\hat{f}$. To relate $x_s[n]$ to $x[n]$, substitute the definition of $X(\hat{f})$ into (14.21):

$$
\begin{aligned}
x_s[n] &= \frac{1}{K}\sum_{k=0}^{K-1}\left\{\sum_{m=-\infty}^{\infty} x[m]e^{-j2\pi mk/K}\right\} e^{j2\pi nk/K} \\
&= \frac{1}{K}\sum_{m=-\infty}^{\infty} x[m]\left\{\sum_{k=0}^{K-1} e^{j2\pi(n-m)k/K}\right\} \qquad (14.22)
\end{aligned}
$$

The inner summation can be evaluated as

$$
\sum_{k=0}^{K-1} e^{j2\pi(n-m)k/K} = K\sum_{l=-\infty}^{\infty}\delta[n-m-lK] \qquad (14.23)
$$

so that finally

$$
x_s[n] = \sum_{l=-\infty}^{\infty} x[n-lK] \qquad (14.24)
$$

Equation (14.24) states that if the DTFT of $x[n]$ is sampled in the frequency domain, the signal that corresponds to the inverse transform of the sampled spectrum will be a replicated version of the original signal. This should not be surprising. It was seen early in this chapter that sampling a signal in the time domain caused replication of its spectrum in the frequency domain. In dual fashion, sampling in frequency causes replication in the time domain. The replication period in time is K samples, inversely proportional to the sampling interval of $1/K$ cycles/sample in normalized frequency.

Continuing with the analogy to time-domain sampling, (14.24) shows that $x[n]$ cannot be recovered from the inverse transform of the sampled spectrum $X(k/K)$ unless $x[n]$ is time-limited (analogous to band-limited for time-domain sampling) and the replicas do not overlap. The replicas will not overlap provided that the support of $x[n]$ is limited to $[0, K-1]$. Thus, a signal can be represented by K uniformly spaced samples of its DTFT provided it is of finite length K samples or less.

Defining the discrete Fourier transform as the sampled DTFT, the K-point DFT and its inverse are

$$X[k] = \sum_{n=0}^{K-1} x[n] e^{-j2\pi nk/K}, \quad k = 0, \ldots, K-1$$
$$x[n] = \frac{1}{K} \sum_{k=0}^{K-1} X[k] e^{j2\pi nk/K}, \quad n = 0, \ldots, K-1 \tag{14.25}$$

The DFT $X[k]$ is a finite, computable frequency domain representation of the signal $x[n]$. Because the DTFT is periodic, so is the DFT, with period K. Examples of the DFT of specific signals and properties of the DFT can be found in most DSP textbooks, such as [1,2].

The DFT values $X[k]$ are simply samples of the DTFT of $x[n]$:

$$\begin{aligned} X[k] &= X(\hat{f})\big|_{\hat{f}=\frac{k}{K}} \\ &= X(\hat{\omega})\big|_{\hat{\omega}=k\frac{2\pi}{K}} \end{aligned} \tag{14.26}$$

Using the relation $\hat{\omega} = \omega/f_s = \omega T_s$ or $\hat{f} = f/f_s$, the DFT frequency samples can be related to frequency in analog units. Specifically,

$$\begin{aligned} f_k &= \frac{k}{KT_s} = k\frac{f_s}{K} \text{ Hz}, \quad k = 0, \ldots, K-1 \\ \omega_k &= \frac{2\pi k}{KT_s} = k\frac{2\pi f_s}{K} \text{ rad/s} \quad k = 0, \ldots, K-1 \end{aligned} \tag{14.27}$$

Because of the periodicity of the DTFT, the principal period in $\hat{f}$ can be considered to extend from -0.5 to $+0.5$ instead of 0 to 1, allowing for the representation of negative frequencies. If the DFT is interpreted as representing both positive and negative frequencies, then the samples from $k = 0$ through $\lfloor (K-1)/2 \rfloor$ represent positive frequencies kf_s/K, while the samples from $k = \lceil (K+1)/2 \rceil$ through $K-1$ represent negative frequencies $f_s(k-K)/K$. If $K/2$ is an integer, then the sample at $k = K/2$ corresponds to the Nyquist frequency $f_s/2$.

While an N-point DFT is adequate to represent an N-point discrete sequence, it is often desirable to compute more than N samples of the DTFT of a signal, for example, to obtain a better approximation to the continuous-frequency DTFT or to reduce straddle loss (see Section 14.4.5). Suppose the desired number of frequency samples $K > N$. The K-point DFT of equation (14.25) assumes that $x[n]$ is defined for $0 \leq n \leq K-1$, but in fact $x[n]$ is only N points long. The solution is to simply append zeroes to the end of the N available data values of $x[n]$, creating a new sequence $x'[n]$ that is K samples long. This process is called *zero padding*. Since no new data values have been added, only zeroes, the DTFT of $x'[n]$ is identical to that of $x[n]$. Computing the K-point DFT simply samples this DTFT at more locations along the frequency axis.

14.4.5 Straddle Loss

The sampling of the DTFT implied by the DFT raises the possibility that the DFT samples may not capture all of the important features of the underlying DTFT. So long as the DFT size, K, is equal to or greater than the signal length, N, the DFT samples capture all of

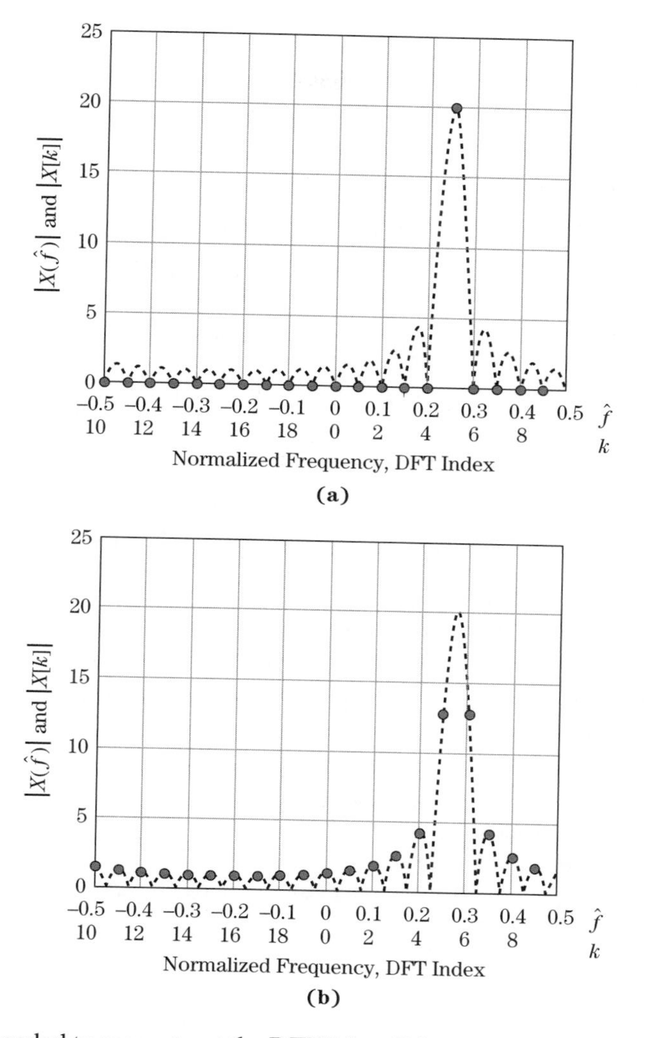

FIGURE 14-17 ■ Illustration of DFT sampling effects and straddle loss. (a) DFT samples falling on the peak and zeroes of the underlying DTFT. (b) DFT samples falling on sidelobes and straddling the peak of the DTFT.

the information needed to reconstruct the DTFT for all frequencies. Nonetheless, the DFT samples may not directly sample important signal features, such as signal peaks. This is illustrated in Figure 14-17. Consider the 20-point complex exponential sequence

$$x[n] = \begin{cases} \exp\left[j2\pi(0.25)n\right], & 0 \le n \le 19 \\ 0, & \text{otherwise} \end{cases} \tag{14.28}$$

The dashed line in Figure 14-17a illustrates the magnitude of the DTFT $X(\hat{f})$ (this is the same function shown in Figure 14-10). The solid dots indicate the magnitude of the 20-point DFT of $x[n]$. Note that in this particular case, the DFT samples all fall on zeroes of $X(\hat{f})$, except for the sample at $k = 5$, which falls exactly on the peak of $X(\hat{f})$.

This outcome for the DFT seems ideal: one sample on the peak of the DTFT, with all of the others identically zero.[9] It correctly suggests that the input signal was a single sinusoid

[9]Chapter 17 shows that the DFT is ideal in the sense of maximizing the signal-to-noise ratio for a sinusoid at one of the DFT frequencies in the presence of white noise.

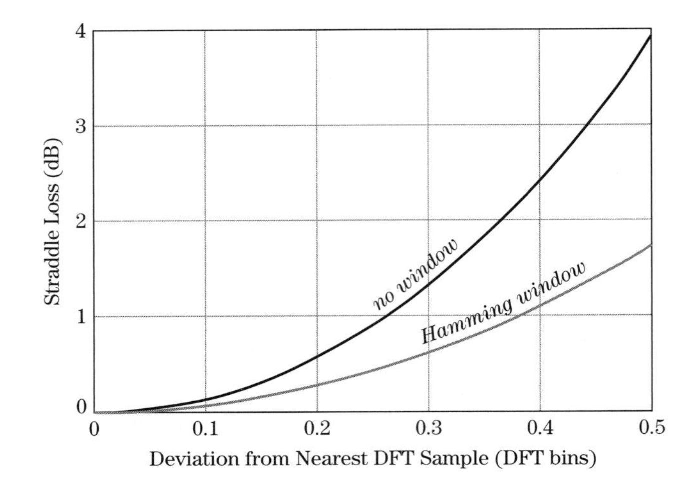

FIGURE 14-18 ■ Worst-case DFT straddle loss for a single sinusoid as a function of the deviation in bins of the sinusoid frequency from the nearest DFT sample frequency. The DFT size equals the data sequence length.

at the frequency corresponding to the nonzero DFT sample. Figure 14-17b shows the same two functions for the case where the exponential frequency is changed from $\hat{f}_0 = 0.25$ to $\hat{f}_0 = 0.275$, a frequency midway between two of the DFT frequency samples. The DTFT has exactly the same functional form but has been shifted on the frequency axis so that it now peaks at $\hat{f} = 0.275$. However, the DFT samples still occur at the same 20 frequencies. These samples now fall approximately on the peak of the DTFT sidelobes instead of the zeroes. More importantly, there is no longer a DFT sample at the peak of the DTFT. Instead, two DFT samples straddle the DTFT peak. Their numerical values are 12.75, in contrast to the peak sample value of 20 in Figure 14-17a.

This example illustrates the sensitivity of the DFT to the alignment of features of the underlying DTFT with the DFT sample frequencies. The two signals in Figures 14-17a and 14-17b are very similar, differing only slightly in frequency. Their DTFTs are identical except for a slight offset in the frequency domain. However, the DFTs look dramatically different.

The apparent reduction in peak signal amplitude in the frequency domain in the second case compared with the first is called *straddle loss*, because it results from the DFT samples straddling, and thus missing, the true peak of the data's DTFT.[10] Straddle loss represents an error in measurement of the amplitude of the underlying spectrum. If the DFT data are being used for signal detection, for example in Doppler processing, this reduction in measured signal amplitude can cause a significant reduction in signal-to-noise ratio and thus in probability of detection. Figure 14-18 shows the straddle loss in decibels for measurement of a single sinusoid as a function of the difference between the signal frequency and the nearest DFT sample frequency, assuming the DFT size equals the signal length (the worst case). For unwindowed data, the loss can be as high as 3.9 dB (the case illustrated in Figure 14-17). The average loss, assuming all values of frequency misalignment are equally likely, is 1.5 dB.

Figure 14-18 also illustrates an underappreciated benefit of windowing. With a Hamming window, the peak straddle loss is reduced from 3.9 dB to 1.7 dB and the average to 0.65 dB. This is a beneficial effect of the widening of the DTFT mainlobe caused by

[10] In many digital signal processing texts, this phenomenon is referred to as *scallop loss* [1].

windowing; the wider DTFT peak does not fall off as fast, so the DFT sample amplitude is not reduced as much when it is not aligned with the DTFT peak. While this does not compensate for the 5.6 dB reduction in peak gain caused by the Hamming window, it does reduce the variability of the apparent peak amplitude as the signal frequency varies.

Even when a peak in the DFT is strong enough to be detected, the misalignment of the DFT frequency samples with the underlying DTFT peak will still result in an underestimate of the signal amplitude and an error in the estimate of the signal frequency. The most obvious strategy for reducing these straddle-induced errors is to increase the DFT size by zero padding the data. This increases the number of frequency samples, reducing their spacing and ensuring that one will fall closer to the DTFT peak. The larger the DFT, the smaller the DFT sample spacing in frequency and the less the maximum straddle loss can be for a given data sequence length. The cost of this procedure is, of course, the increased computation required to obtain the larger DFT. It also interpolates the entire DFT, even if improved detail is desired only in limited regions of the frequency axis.

Increasing the DFT size is equivalent to a band-limited interpolation of the original K-point DFT with a set of K asinc functions [3]. A less accurate but much simpler approach to reducing straddle losses is to perform a local interpolation in the vicinity of each apparent peak [11]. The concept is shown in Figure 14-19 for the example of a quadratic interpolation. The magnitude of the DFT is searched for local peaks. For each local peak found, say, at $k = k_0$, a parabola is fitted through it and the two neighboring samples of $|X[k_0]|$. The equation for the fitted parabola is differentiated and set equal to zero to find the peak of the quadratic, which is then taken as an estimate of the location k' and amplitude A' of the true peak. It is possible to develop closed-form equations for the estimated amplitude and peak location without having to explicitly solve for the equation of the interpolating parabola. The result is [3]

$$k' = k_0 + \Delta k = k_0 + \frac{-\frac{1}{2}\{|X[k_0+1]| - |X[k_0-1]|\}}{|X[k_0-1]| - 2|X[k_0]| + |X[k_0+1]|} \tag{14.29}$$

$$\begin{aligned} |X[k']| = \frac{1}{2}\{&(\Delta k - 1)\,\Delta k\,|X[k_0-1]| - 2\,(\Delta k - 1)\,(\Delta k + 1)\,|X[k_0]| \\ &+ (\Delta k + 1)\,\Delta k\,|X[k_0+1]|\} \end{aligned} \tag{14.30}$$

The frequency estimator of equation (14.29) is used more commonly than the amplitude estimator of (14.30). There are a number of similar frequency estimators [11]. These differ

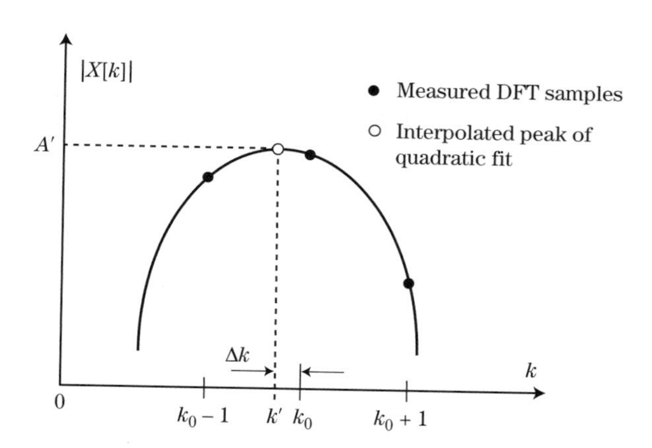

FIGURE 14-19 ■ The concept of quadratic interpolation of a DFT peak.

in the order of the estimate, the coefficients used, and whether they depend on the particular data window used (if any) in the DFT calculation. An example very similar to the previous estimator uses the complex DFT coefficients instead of the magnitude:

$$k' = k_0 + \Delta k = k_0 - \mathrm{Re}\left\{\frac{Y[k_0+1] - Y[k_0-1]}{2Y[k_0] - Y[k_0-1] - Y[k_0+1]}\right\} \tag{14.31}$$

The amplitude estimator of (14.30) can also be used with this frequency estimator. An improved amplitude estimation technique, at the cost of significantly greater computation, uses (14.29) or (14.31) to estimate the frequency of the peak but then fits an asinc centered at that location to the data [3].

The frequency estimation error using this technique is shown in Figure 14-20. Part a shows the error in the estimate of Δk as a function of the actual value of Δk for both estimators [(14.29) and (14.31)] for the unwindowed, minimum-size DFT case. On this scale, the error using the complex-data version of the estimator is nearly zero. Part b of the figure displays the residual error only for equation (14.31) on a scale magnified by a factor of 1,000.

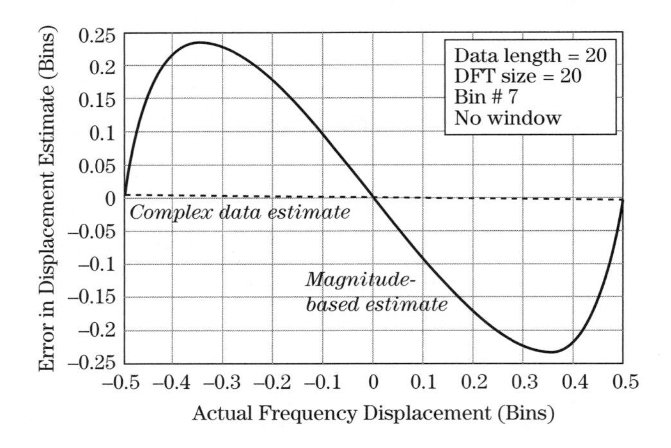

(a)

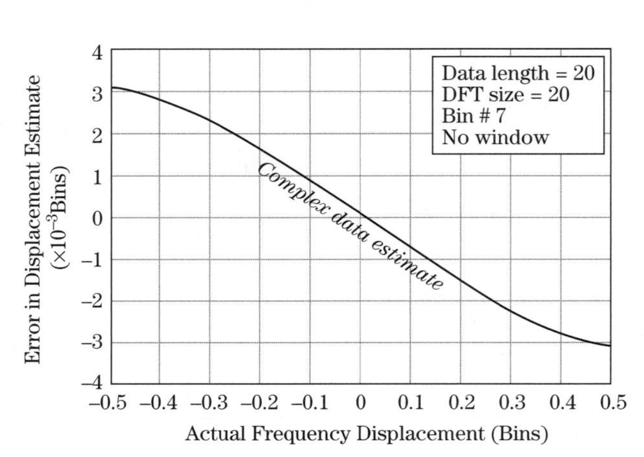

(b)

FIGURE 14-20 ■ Error in quadratic interpolation estimation of frequency in the DFT. (a) Residual error in Δk for magnitude- and complex-data estimators. (b) Complex-data estimation error on expanded scale.

14.4.6 The Fast Fourier Transform

The DFT is important because it provides a computable, sampled version of the DTFT. However, it did not become a practical tool for real-time processing until the advent of the fast Fourier transform. The FFT refers to any of a class of algorithms for efficient computation of the DFT; it is not a different mathematical transform. The first widely recognized FFT was the radix-2 algorithm developed by Cooley and Tukey [12]. Used when the DFT length K is a power of two, it reduces the number of complex multiplications required to compute the DFT from K^2 to $(K/2)\log_2 K$, a reduction by a factor of $2K/\log_2 K$. If the data sequence is not a power of 2, it can be zero padded to the required length. The savings in complex additions is a factor of $K/\log_2 K$.

As is typical of fast algorithms, the computational savings due to the FFT algorithm is larger for larger problems (higher values of K) and can become very substantial. These points are illustrated in Figure 14-21. For $K = 16$, the radix-2 Cooley-Tukey FFT achieves a modest but useful savings of $8\times$ compared with the DFT, but for $K = 1{,}024$, the savings is over two orders of magnitude ($204.8\times$, to be exact). The large speedups made possible by publication of the Cooley-Tukey FFT algorithm and its many variants radically increased the variety and order of algorithms that could be applied in real-time radar signal processing.

There are fast algorithms for computing the DFT for other values of K. Cooley-Tukey algorithms are known for radices 4 and 8. The "prime factor algorithm" can be applied when K is a product of prime numbers. Hybrid algorithms exist for other cases. In addition, there are modified algorithms offering additional savings for cases where the data is real-valued or symmetric. A comprehensive summary of DFT and FFT algorithms is available in [13].

14.4.7 Summary of Fourier Transform Relationships

Figure 14-22 summarizes the relationship between the various versions of the Fourier transform discussed in this section. The discrete-time Fourier transform is the appropriate transform for sampled data. It can be derived by sampling the data in the classical Fourier transform for continuous-variable signals. The DTFT frequency variable is continuous, not discrete, but the transform is periodic in frequency. Consequently, the DTFT is primarily an analytic tool.

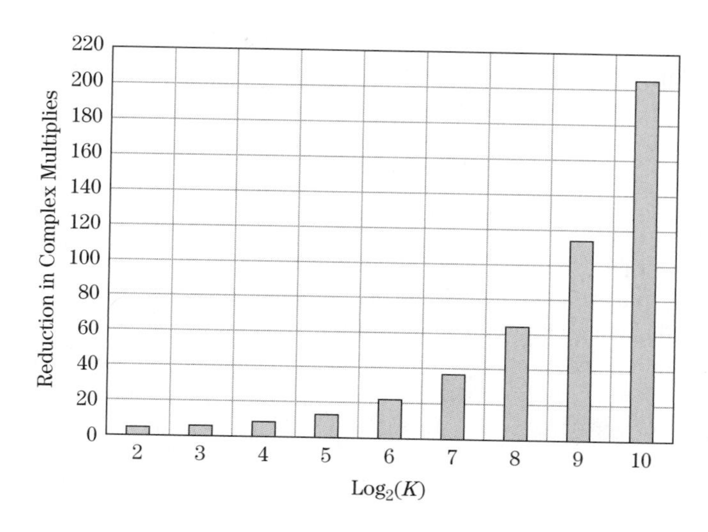

FIGURE 14-21 ■ The $2K/\log_2 K$ reduction in computational load for the radix-2 Cooley-Tukey fast Fourier transform as a function of FFT size.

FIGURE 14-22 ■ Relationship between the various Fourier transforms discussed in this chapter.

Signal Variable	Transform Variable	Transform
continuous	continuous	Fourier Transform (FT)
discrete	continuous	Discrete-Time Fourier Transform (DTFT)
discrete	discrete	Discrete Fourier Transform (DFT)
discrete	discrete	Fast Fourier Transform (FFT)

Sampled data ⇒ replicated spectrum

Sampled spectrum ⇒ finite data sequence

Fast algorithm for DFT

To obtain a computable transform, a transform with a discrete frequency variable is needed. Such a transform was obtained by sampling the DTFT in frequency, leading to the discrete Fourier transform. Finally, practical applications require a fast algorithm for actually computing the DFT. This is provided by the various fast Fourier transform algorithms that efficiently compute the DFT for various lengths K. As discussed in Chapter 13, the speed at which an algorithm executes on modern processors often has as much or more to do with the structure of cache memory hierarchies and the efficiency of multiprocessor and multicore data movement as it does with the number of arithmetic operations. The Fastest Fourier Transform in the West (FFTW) library of FFT codes [14] provides efficient implementations of $O(K \log K)$ algorithms for arbitrary values of K on many complex processors.

14.5 THE z TRANSFORM

The z transform of a signal $x[n]$ is defined as

$$X(z) = \sum_{n=-\infty}^{\infty} x[n]\, z^{-n} \tag{14.32}$$

where z is a complex variable. In general, this summation may converge only for certain values of z. A sufficient condition for convergence is [2]

$$|X(z)| \leq \sum_{n=-\infty}^{\infty} |x[n]| \left|z^{-n}\right| \leq B_x \tag{14.33}$$

for some finite bound B_x; that is, the summation must be finite. Those areas of the complex z plane (i.e., values of z) where $X(z)$ converges constitute the *region of convergence*, or ROC. The z transform plays the same role in discrete signal analysis that Laplace transforms play in continuous signal analysis, for example, the analysis of system stability.

Note that the DTFT of equation (14.9) is a special case of equation (14.32) for $z = \exp(j\hat{\omega}) = \exp(j2\pi \hat{f})$. In this case, $|z| = 1$, while the argument (phase angle) of z is $\hat{\omega}$. Thus, as $\hat{\omega}$ ranges from 0 to 2π (or $\hat{f}$ from 0 to 1), z traverses the "unit circle" in the complex z plane.

Of particular interest are rational z transforms, where $X(z)$ is the ratio of two finite polynomials in z^{-1}:

$$X(z) = \frac{b_0 + b_1 z^{-1} + \cdots + b_M z^{-M}}{1 + a_1 z^{-1} + \cdots + a_N z^{-N}} = \frac{\sum_{k=0}^{M} b_k z^{-k}}{1 + \sum_{k=1}^{N} a_k z^{-k}} \tag{14.34}$$

The roots of the denominator polynomial are called the *poles* of $X(z)$, while the roots of the numerator polynomial are called the *zeroes*.

An important property of the z transform is the effect of a delay on the z transform of a signal. If the z transform of $x[n]$ is $X(z)$, then the z transform of $x[n - n_d]$ is $z^{-n_d} X(z)$. In combination with (14.34), this fact leads to one of the three major methods for implementing digital filters, to be described in Section 14.6.3.

14.6 | DIGITAL FILTERING

Linear filtering is another major category of digital signal processing pertinent to modern radar. Filters are systems that accept an input signal and perform a useful operation or transformation to generate a modified output signal, as illustrated in Figure 14-23. A digital processor might implement a filter using custom integrated circuits, field programmable gate arrays (FPGAs), or software in a programmable processor.

Linear, shift-invariant (LSI) systems are of greatest interest. A system is linear if, when presented with a linear combination of input signals, the output is the same linear combination of the outputs observed for each input considered separately. Specifically, if the inputs $x_1[n]$ and $x_2[n]$ produce the outputs $y_1[n]$ and $y_2[n]$, then a filter is linear if the input $ax_1[n] + bx_2[n]$ produces the output $ay_1[n] + by_2[n]$. Linearity is sometimes broken down into two component properties, homogeneity and superposition. The system is homogeneous if scaling the input produces a like scaling of the output. It obeys superposition if the response to the sum of two inputs is the sum of the individual outputs. A filter is shift-invariant if delaying or advancing any input signal causes an equal delay or advance in the corresponding output but no other change. Thus, if the input $x[n]$ produces the output $y[n]$, the input $x[n - n_0]$ produces the output $y[n - n_0]$ for any n_0.

LSI filters are completely described by their impulse response $h[n]$, which is the output obtained when the input is the impulse function $\delta[n]$:

$$\delta[n] = \begin{cases} 1, & n = 0 \\ 0, & \text{otherwise} \end{cases} \tag{14.35}$$

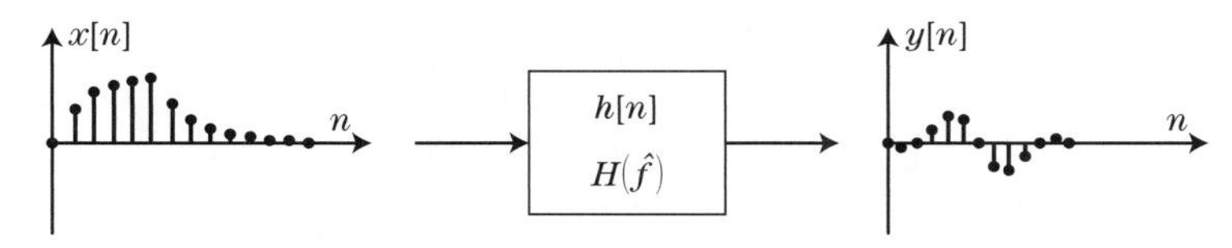

FIGURE 14-23 ■ A linear, shift-invariant system, or filter, is described by its impulse response $h[n]$ and frequency response $H(\hat{f})$.

Given the impulse response of an LSI system, its output can be computed for any input via the *convolution* sum [1,2]:

$$y[n] = \sum_{m=-\infty}^{+\infty} x[m]\,h[n-m] = \sum_{m=-\infty}^{+\infty} h[m]\,x[n-m] \tag{14.36}$$

where the second version of the sum is obtained from the first by a change of variable. Equation (14.36) is useful both for analysis and for actual computation of the filter output. It is often the case that the input, impulse response, or both are of finite duration. If both are finite duration, N and L samples respectively, then (14.36) can be used to show that the duration of the filter output will be limited to $N+L-1$ samples. This fact has importance for implementing filters using DFTs.

A system is said to be *causal* if, when starting from a state of initial rest, the output is nonzero only for indices n equal to or greater than the first index n_0 at which a nonzero input is applied. In other words, no output occurs before the input is applied. It can be seen from equation (14.36) that an LSI system will be causal if and only if $h[n] = 0$ for all $n < 0$.

It is sometimes convenient to represent linear filtering and convolution by a vector operation. Consider computing the output at time n_0 for a finite impulse response of duration L samples. Define the impulse response and signal column vectors

$$\begin{aligned} \boldsymbol{H} &= [h[0] \quad h[1] \quad \ldots \quad h[L-1]]^T \\ \boldsymbol{X}_{n_0} &= [x[n_0] \quad x[n_0-1] \quad \ldots \quad x[n_0-L+1]]^T \end{aligned} \tag{14.37}$$

Note that the samples of $\boldsymbol{X}_{n_0}$ are in time-reversed order, while those of $\mathbf{h}$ are in normal order. With these definitions, the output sample $y[n_0]$ can be computed as the vector dot product

$$y[n_0] = \boldsymbol{H}^T \boldsymbol{X}_{n_0} \tag{14.38}$$

Equation (14.38) produces only a single scalar value of the output sequence $y[n]$. To produce additional output samples, the definition of the signal vector $\boldsymbol{X}_{n_0}$ must be updated for the desired output sample.

14.6.1 Spectral Representations of LSI Systems

The impulse response $h[n]$ is a complete description of an LSI system, and the convolution sum can be used to compute the output for any input if $h[n]$ is known. However, it is sometimes more convenient to work in the frequency domain or the z domain. It is straightforward to show that the z transform of the convolution equation (14.36) gives the relation

$$Y(z) = H(z)X(z) \tag{14.39}$$

$H(z)$ is the z transform of $h[n]$ and is referred to as the *system function* or *transfer function* of the filter. Since the DTFT is simply the z transform evaluated on the unit circle in the z plane, it also follows that

$$Y(\hat{f}) = H(\hat{f})X(\hat{f}) \tag{14.40}$$

$H(\hat{f})$, the DTFT of the impulse response, is called the system's *frequency response*.

14.6.2 Digital Filter Characteristics and Design

Digital filters are divided into two broad classes, *finite impulse response* filters (also called *nonrecursive* filters) and *infinite impulse response* (IIR) filters (also called *recursive* filters), depending on whether the impulse response $h[n]$ is of finite or infinite duration. Any digital filter is fully described by any one of its impulse response, frequency response, or system function. Either class can implement common filters such as low-pass, band-pass, and high-pass designs, but each class has important advantages and disadvantages.

14.6.2.1 FIR Filters

FIR filters[11] have a finite duration impulse response. If the impulse response is nonzero for $0 \le n \le M-1$, the filter *order* is $M-1$, and the length of the impulse response is M samples. The convolution sum (14.36) is then a weighted combination of a finite number M of input samples. So long as the input data remains bounded (finite) in value, so will the output. Consequently, FIR filters are always stable.

In radar signal processing, much of the information of interest is in the phase of the received signals. Phase information is essential to Doppler processing, pulse compression, synthetic aperture imaging, and space-time adaptive processing, for example. It is important that the signal processor not distort the phase information. The frequency domain relation (14.40) suggests that this would require that the phase of $H(\hat{f})$ be identically zero. However, it is sufficient to restrict $H(\hat{f})$ to have a linear phase function. Linear phase corresponds to pure delay in the time domain [2] but no other distortion. FIR filters can be designed to have exactly linear phase. If the impulse response is either exactly symmetric or exactly antisymmetric,

$$
\begin{gathered}
h[n] = h[M-1-n] \\
\text{or} \\
h[n] = -h[M-1-n]
\end{gathered}
\tag{14.41}
$$

then the frequency response $H(\hat{f})$ will have linear phase [2]. As an added benefit, symmetric or antisymmetric filters also offer computational savings compared with an arbitrary FIR filter. Conventional IIR filters do not have linear phase.[12] Thus, FIR filters are popular in many radar applications.

There are many algorithms for designing FIR filters to meet a set of specifications, usually given in the form of a desired frequency response. Only two are briefly mentioned here. Much more detail is given in most DSP texts such as [1,2]. Whatever the algorithm, the resulting filter is represented by its impulse response $h[n]$.

The window method is a particularly simple technique that is well suited to general band-pass/band-stop frequency responses. The user specifies the desired ideal frequency response $H_i(\hat{f})$ and then computes the corresponding impulse response $h_i[n]$ using the inverse DTFT formula (14.10). In general, this impulse response will be infinite in duration.

[11] The acronym in the term *FIR filter* is often pronounced as the word *fir* (like the type of tree), resulting in the term *fir filter*.

[12] It is possible to achieve zero-phase filtering using IIR filters by filtering the signal twice, once in the forward and once in the reverse direction. Also, some advanced design techniques can achieve almost-linear phase IIR filters. Only FIR filters can readily achieve exactly linear phase without multiple filter passes.

A finite impulse response is then obtained by multiplying $h_i[n]$ by a finite length window function $w[n]$:

$$h[n] = w[n]h_i[n] = w[n] \cdot \text{DTFT}^{-1}\{H_i(\hat{f})\} \tag{14.42}$$

Because $h[n]$ is only an approximation to $h_i[n]$, the final realized frequency response $H(\hat{f})$ will be an approximation to $H_i(\hat{f})$. The choice of window function offers the designer a trade-off between the transition bandwidth between passbands and stopbands (a measure of the sharpness of the frequency response cutoffs) versus the sidelobe level in the stopbands.

FIR filters designed by the window method have generally equal peak approximation errors in the various passbands and stopbands. In addition, ripples in the frequency response are largest near the band edges. The Parks-McClellan design method provides more control over the frequency response at the expense of greater computational effort. The algorithm minimizes the maximum error between the actual and ideal frequency response for a given filter order. The resulting filters have an equiripple characteristic: the approximation error is spread evenly throughout the passband and stopbands rather than being concentrated at the edges. In addition, the Parks-McClellan algorithm allows the user to specify the relative amount of error in the various passbands and stopbands instead of being limited to equal error in each band as with the window method. For a given peak sidelobe level, the Parks-McClellan design will have a sharper transition between passbands and stopbands than a filter designed with the window method.

14.6.2.2 IIR Filters

IIR digital filters do not have the guaranteed stability of FIR filters; rather, they must be designed to be stable. Furthermore, IIR filters cannot readily be designed to have exactly linear phase. However, IIR filters generally achieve a more selective frequency response (sharper cutoff, lower sidelobes) for a given filter order than a comparable FIR filter. Linear phase is not required in some radar applications, such as MTI filters [15] and some digital IF sampling approaches [16]. For such applications, IIR filters can provide a given level of frequency selective filtering at less computational cost than FIR filters, provided attention is paid to the filter stability.

IIR filters are designed by either applying a transformation to one of the many well-documented analog filter designs or by using one of many direct computer-aided optimization techniques. For example, the *bilinear transformation* converts an analog filter specified by the analog system function $H_a(s)$ into a digital filter specified by the system function $H(z)$ by applying the substitution

$$s = \frac{2}{T_s}\left(\frac{1 - z^{-1}}{1 + z^{-1}}\right) \tag{14.43}$$

to $H_a(s)$, where T_s is the sampling period of the digital filter. The bilinear transformation has many useful properties, such as preserving the stability of analog filters in the corresponding digital design and being applicable to all types (low-pass, band-pass, high-pass, and arbitrary frequency response) of filter designs. Furthermore, rational analog transfer functions map into rational digital transfer functions of the form (14.34).

Because the impulse response is infinite in duration by definition, IIR filters are normally not represented by $h[n]$, since that would require an infinite number of coefficients. Instead, IIR filters are represented by their discrete system function $H(z)$, which will have only a finite number of coefficients.

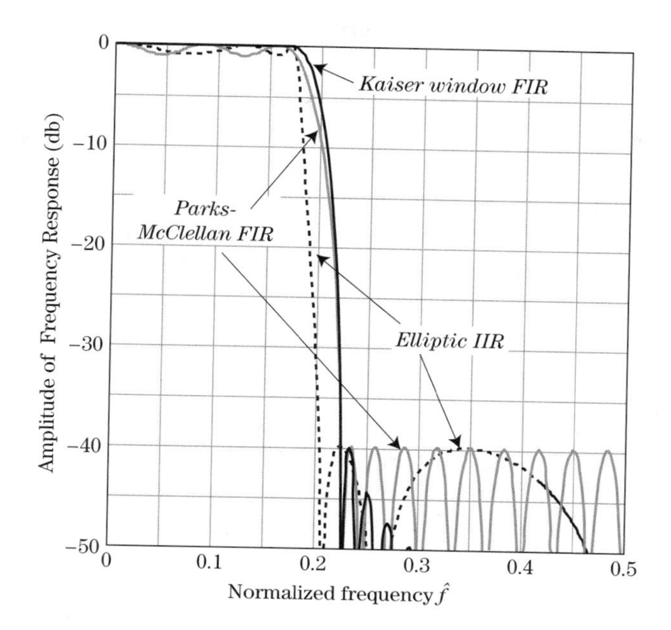

FIGURE 14-24 ■ Comparison of three classes of digital filters with similar design requirements (see text for details).

14.6.2.3 Filter Design Comparison Example

Some of the advantages and disadvantages of IIR and FIR filters can be summarized by comparing designs that result from comparable specifications. Consider a low-pass filter with a normalized frequency cutoff of $\hat{f} = 0.2$ cycles/sample, a transition bandwidth $\Delta\hat{f} = 0.1$ cycles/sample, a passband gain that does not exceed unity (0 dB) with a ripple of 1 dB, and a stopband attenuation of 40 dB. Figure 14-24 compares the frequency responses of an elliptic IIR filter and two FIR filters: one designed with the window technique using a Kaiser window and the other with the Parks-McClellan algorithm. Routines for the automated design of all of these filters classes, and several more, are readily available in MATLAB® and other mathematical software packages.

Elliptic filters are a class of recursive designs that have equiripple errors in both the passband and stopband; this behavior is clearly evident in Figure 14-24. A sixth-order elliptic filter achieves the desired specifications and can be implemented at a cost of 11 multiplications and 10 additions per output sample. As designed, the filter is stable, but if the coefficients were quantized for a fixed-point implementation, care would be required to ensure that stability is maintained.

A Parks-McClellan filter is an FIR design that also exhibits equiripple behavior in both the passband and stopband. A 29th order filter meets the same specifications as the elliptic filter. The FIR coefficients are symmetric, so that the frequency response of the filter inherently exhibits exactly linear phase over the entire frequency spectrum. By taking advantage of this symmetry, the filter can be implemented at a computational cost of 15 multiplications and 28 additions per output sample, still significantly more than for the elliptic design. The transition bandwidth, which is the change in frequency over which the frequency response transitions from the passband to the desired attenuation in the stopband, is significantly wider for the Parks-McClellan design than for the elliptic filter.

Finally, a 43rd-order Kaiser window design also meets the stopband attenuation requirement. This filter is also exactly linear phase. It requires 22 multiplications and 42 additions per output sample. Like all filters designed via the window method, this design is not equiripple in either the stopband or passband. Because the passband and stopband ripples are approximately equal, the passband appears smooth when plotted on a dB scale. In fact, the Kaiser window filter in Figure 14-24 exhibits a passband ripple of about 0.15 dB, while the Parks-McClellan and elliptic filters were designed to have a passband ripple of approximately 1 dB. The stopband ripples of the window design filter decay as approximately $1/f$. The Kaiser window gives the narrowest transition bandwidth of any window design for a given filter order and stopband attenuation. In this example, the transition bandwidth is slightly narrower than the Parks-McClellan design, though still significantly wider than that of the elliptic filter.

14.6.3 Implementing Digital Filters

There are three basic methods for implementing digital filters: time-domain convolution, time-domain difference equations, and frequency-domain convolution, also called *fast convolution*. In addition, there are a number of variants for long signals, such as the overlap-add and overlap-save techniques [2]; however, these build on the three basic methods.

The first method is the time-domain convolution of equation (14.36). This method requires that the impulse response $h[n]$ of the filter be known but is very straightforward to implement. Convolution relies on a series of multiply-accumulate (add) operations. If the impulse response is M samples long, then implementation of convolution requires M multiplications and $M - 1$ additions per output sample. In radar, the data are often complex-valued, and the filter coefficients are sometimes complex-valued as well, so these are complex multiplications and additions. Note that direct convolution is practical only for FIR filters, for which M is finite.

The second method uses the convolution theorem of Fourier transforms given in (14.40). When using the DTFT, this technique is applicable in principle to both FIR and IIR filters. However, the DTFT, while good for analysis, is not a computable transform due to the continuous frequency variable. Thus, the discrete frequency DFT version is used instead. This states that the product of two DFTs, say, $Y[k] = H[k]X[k]$, is the DFT of the *circular* convolution of the corresponding sequences $x[n]$ and $h[n]$ [2]. "Circular convolution" is the name given the result of convolving[13] periodic sequences; recall that sampling the DTFT to create the DFT implies a periodic replication of the input signal as derived in equation (14.24). Circular convolution produces results at the beginning and end of the output sequence that differ from the standard convolution result, sometimes significantly.

The standard convolution, not a circular convolution, is usually required. Despite the replication effects, this can be accomplished using DFTs if the DFT size is chosen correctly. Specifically, if the lengths of the sequences $x[n]$ and $h[n]$ are N_x and N_h samples, respectively, then their linear convolution has a length of $N_x + N_h - 1$ samples. The circular convolution produced by the inverse of the product of their DFTs is identical to the linear convolution provided that the DFT size $K \geq N_x + N_h - 1$. Thus, for finite-length data sequences, an FIR filter can be implemented using

$$y[n] = \text{DFT}_K^{-1}\{H[k]X[k]\} \tag{14.44}$$

[13]Note that the base verb for describing the operation of convolution is "to convolve," not "to convolute."

where DFT_K^{-1} represents an inverse DFT of length K and K satisfies the previous constraint.

The process of FIR filtering using multiplication of DFTs is referred to as *fast convolution*. Whether it is in fact faster than direct convolution depends on the parameters involved. Calculation of $y[n]$ using equation (14.44) requires computation of two K-point DFTs, a K-point vector multiplication to form the product $H[k]X[k]$, and a K-point inverse DFT. Thus, the number of (complex, in general) multiplications and additions is approximately $(3(K/2)\log K + K)$, where K is approximately the sum of the lengths of $x[n]$ and $h[n]$. While this is often fewer than in direct convolution, there are exceptions, particularly when one of the sequences (typically $h[n]$) is much shorter than the other. If the same filter will be applied to many different data sequences, the filter frequency response $H[k]$ need be computed only once; the number of additional operations per data sequence filtered is then reduced to $(K \log K + K)$.

Very long or continuously running input sequences can be filtered using FIR filters and fast convolution applied to the overlap-add or overlap-save methods previously mentioned. The basic idea of both methods is to divide a long input signal into shorter segments, to filter each segment individually, and then to use linearity to reassemble the filtered segments and form the filtered output. Filtering of the individual segments can be done using time-domain or fast convolution. Details are given in [2] and in most DSP texts.

The final implementation method is the one most used for IIR filters. It begins with equation (14.39), which relates the z transforms of the filter input and output and the system function of the filter. For a rational $H(z)$ of the form of (14.34), this becomes

$$Y(z) = \frac{\sum_{k=0}^{M} b_k z^{-k}}{1 + \sum_{k=1}^{N} a_k z^{-k}} X(z)$$

$$\Rightarrow \left(1 + \sum_{k=1}^{N} a_k z^{-k}\right) Y(z) = \left(\sum_{k=0}^{M} b_k z^{-k}\right) X(z) \tag{14.45}$$

Applying the z transform property that the z transform of $x[n - n_d]$ is $z^{-n_d} X(z)$ to each term on both sides of (14.45) and rearranging gives a *difference equation* for computing the filter output $y[n]$:

$$y[n] = \sum_{k=0}^{M} b_k x[n-k] - \sum_{k=1}^{N} a_k y[n-k] \tag{14.46}$$

Difference equations are the discrete equivalent of differential equations in continuous analysis.

Equation (14.46) expresses each output sample as the weighted sum of $M + 1$ input samples and N previous output samples. Thus, IIR filters, unlike FIR filters, incorporate feedback of previous outputs to the current output. For this reason, as mentioned earlier, they are also called recursive digital filters; FIR filters are nonrecursive. The presence of feedback is also the reason that IIR filters can become unstable if improperly designed. Despite the infinite impulse response, the difference equation approach allows computation of each output sample using a finite number of operations, namely, approximately $(M+N)$ multiplications and additions per output sample.

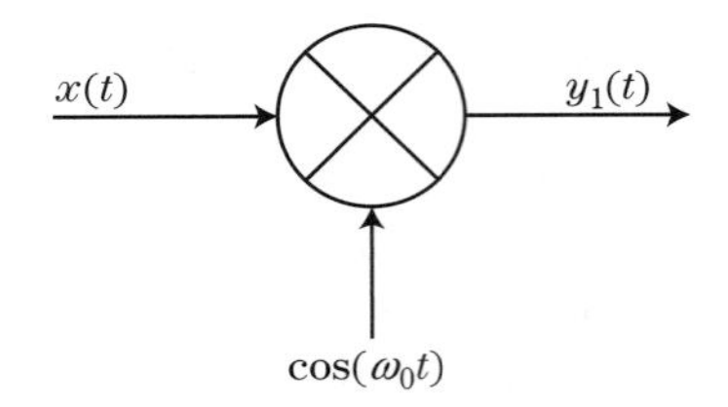

FIGURE 14-25 ■ A modulator.

14.6.4 Shift-Varying and Nonlinear Systems

FIR and IIR filters are linear, shift-invariant systems. Though not nearly so ubiquitous as LSI systems, nonlinear and shift-varying operations are commonly used in certain radar signal processing operations, and their use can be expected to grow over time.

The most important example of a shift-varying operation in radar is simple modulation, that is, multiplication of a signal by a sinusoid to shift its Fourier transform on the frequency axis. Modulation is illustrated in Figure 14-25. Suppose the input to the system is $x(t)$. The output is $y_1(t) = x(t)\cos(\omega_0 t)$. If the input to the system is delayed by t_d seconds, becoming $x(t - t_d)$, the output will be $y_2(t) = x(t - t_d)\cos(\omega_0 t)$. Because $y_2(t) \neq y_1(t - t_d)$, the system is not shift-invariant. This occurs in this case because delaying the input does not also delay the time origin of the oscillator used in the modulator. Note that the modulator is still linear, however. It is also worth noting that the output of an LSI system can contain energy only at frequencies at which there was energy in the input signal. Modulators, which move signal energy to new locations on the frequency axis, must therefore be non-LSI.

Modulators are so common in coherent receivers that they rarely draw any notice. A more interesting example of a shift-varying operation is synthetic aperture radar (SAR) image formation. SAR processing is discussed in Chapter 21. The core operation is a *matched filter* (see Section 14.9) implemented as a two-dimensional correlation using one of a variety of algorithms. However, in general (and in practice in many newer systems), it is a shift-varying correlation because the response of the SAR to a point target is a function of the range to that target. Focusing at different ranges consequently requires different matched filter correlation kernels. In lower (coarser) resolution SAR systems, the variation in range may be insignificant so that the system is effectively LSI. However, in modern high resolution SAR systems, the shift-varying nature is not negligible, leading to complicated correlation algorithms.

The previous examples, while shift-varying, are still linear. *Median filters*, and more generally *order statistics filters*, are a class of nonlinear operations that are used in radar for such functions as detection and image enhancement. A median filter is a sliding-window operation, very similar to an FIR filter. The difference is that instead of computing the output as a sum of the input samples weighted by the FIR filter impulse response, the output is the median of the input samples. Thus, equation (14.38) is replaced by

$$y[n_0] = \text{median}(\boldsymbol{X}_{n_0}) \tag{14.47}$$

A simple way to find the median of a list of N data values is to sort the list into numerical order and then to simply choose the middle element of the list (or the average of the two middle elements if N is even). An order statistics filter is a generalization of this idea in which the m-th element of the ordered list is output. For instance, the second-order statistic filter would output the second-largest element of the list. A median filter is simply the $N/2$ order statistic filter.

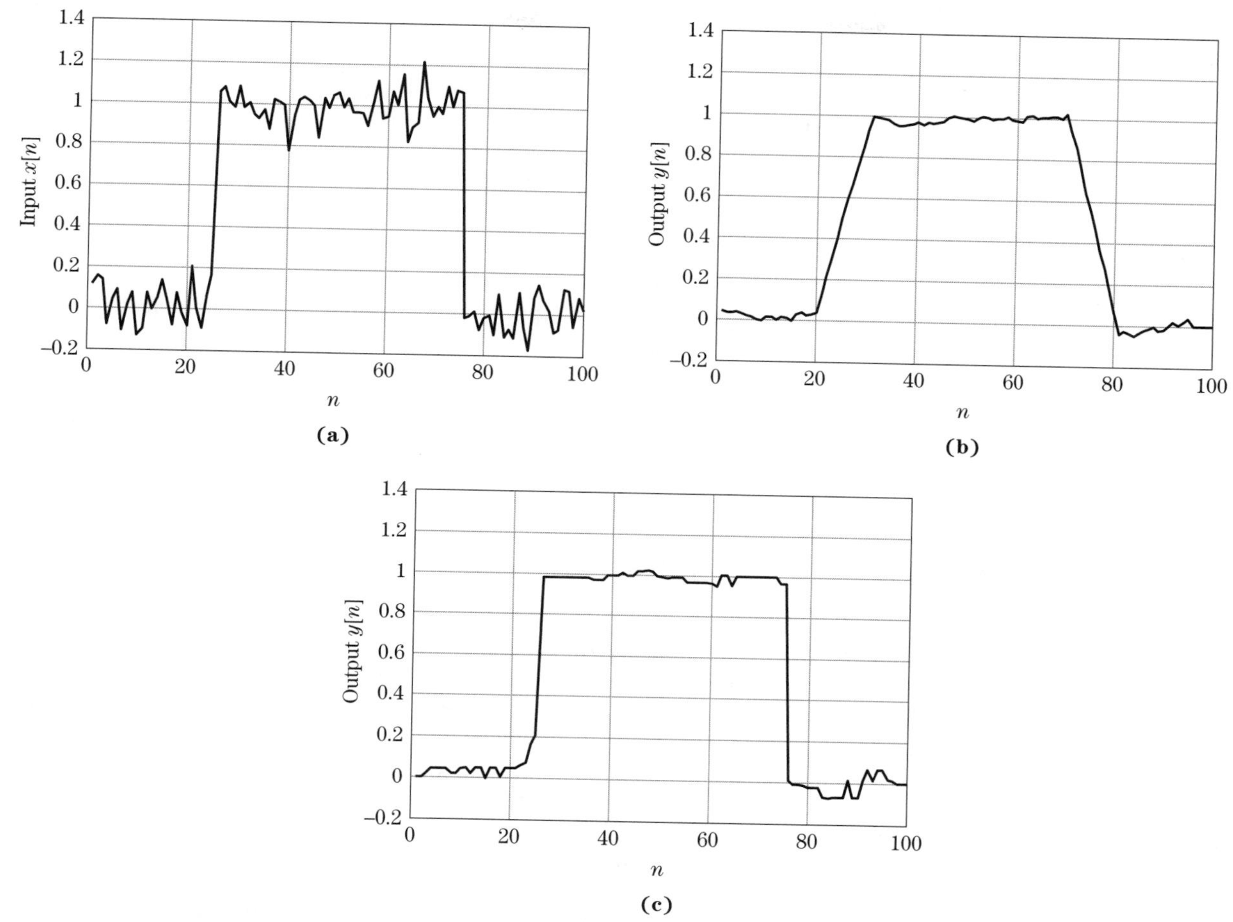

FIGURE 14-26 ■ Comparison of FIR and median filter. (a) 50-point square pulse with noise added. (b) Output of 11-point moving average FIR filter. (c) Output of 11-point median filter.

Order statistic filters are nonlinear. They exhibit the property of homogeneity, but not of superposition. To see an example of this, consider a median filter acting on the two three-point input sequences $\boldsymbol{X}_1 = \{1,4,0\}$ and $\boldsymbol{X}_2 = \{5,3,3\}$. The output of the median filter in response to the first sequence is $y_1 = 1$; this is found by sorting the list $\{1,4,0\}$ into numerical order, giving the new sequence $\{0,1,4\}$, and then taking the middle value of the list. Similarly, $y_2 = 3$. If a new sequence $\boldsymbol{X} = \boldsymbol{X}_1 + \boldsymbol{X}_2 = \{6,7,3\}$ is formed, the output of the median filter will be $y = 6 \neq y_1 + y_2$. Thus, the median filter does not exhibit the superposition property and is therefore not linear.

Figure 14-26 compares a median filter with a moving average filter in smoothing a noisy square pulse. Figure 14-26b is the output of an 11-point moving average filter, which is simply an FIR filter in which all of the filter coefficients have the same value of $1/L = 1/11$. While the filter greatly reduces the noise, it also smears the vertical edges of the pulse over an 11-sample transition region. Figure 14-26c is the output of a median filter for the same input. This nonlinear filter reduces the noise by a similar amount without introducing any smearing of the edges. A two-dimensional version of this filter is sometimes useful for reducing noise in imagery without blurring the edges of objects in the image.

The other major area of radar signal processing where order statistic filters are commonly used is in constant false alarm rate (CFAR) detection. This topic will be discussed in Chapter 16 in detail. The key operation is estimating the interference mean from a finite set of interference samples; this mean is then used to set a detection threshold. Both averaging (FIR filters) and order statistic filters can be applied to this task; that is, either the sample mean of the available data or its median (or other order statistic) can be used to estimate the true interference mean. Order statistic CFARs may be more useful when the data are likely to contain "outliers."

Nonlinear systems are more difficult to analyze than LSI systems. In particular, the concepts of impulse response and convolution do not apply, and Fourier analysis is much less informative. Another interesting aspect of order statistics filters in particular, and many nonlinear systems, is that they put much greater emphasis on sorting and counting operations than on the multiply-accumulate operations of LSI systems. This has a significant impact on processor design.

14.7 RANDOM SIGNALS

Many signals of interest are modeled as *random processes*. The most obvious example is noise. However, many deterministic signals are so complicated that they can be reasonably modeled only as random processes. Examples in radar include reflections from complex targets or from clutter, as described in Chapters 8 and 5, respectively. Here it is sufficient to consider only stationary random processes; a more general description is given in [2] and in many other DSP and random process textbooks.

14.7.1 Probability Density Functions

A stationary discrete random process is a discrete function $x[n]$ where the value x_n of x at each index n is a random variable described by some *probability density function* (PDF) $p_{x_n}(x)$, and the relationship between values x_n and x_m of $x[n]$ at different times n and m is governed by a set of joint PDFs. The stationarity restriction implies that the PDF of x_n is the same for every n; that is, the statistics do not change with time. It also implies that the joint PDF of x_n and x_m depends only on the spacing $n - m$, not on the absolute indices n and m.

Formally, a PDF describes the probability that $x_n = x[n]$ will take on a value in a particular range:

$$\text{Probability that } x_1 \le x_n \le x_2 = \int_{x_1}^{x_2} p_{x_n}(x)dx \tag{14.48}$$

For example, Figure 14-27 shows an example of a Gaussian PDF. The probability that the random variable x_n will be between 0.5 and 1.5 is the integral of the PDF from $x = 0.5$ to $x = 1.5$, which is simply the gray shaded area. It follows that the PDF can be thought of as indicating the relative likelihood that x_n will take on various values. The higher the value of the PDF at some point $x = x_0$, the more likely that x_n will take on a value near x_0. In the Gaussian example of Figure 14-27, x_n is more likely to take on values near zero than to take on values near 2, for instance.

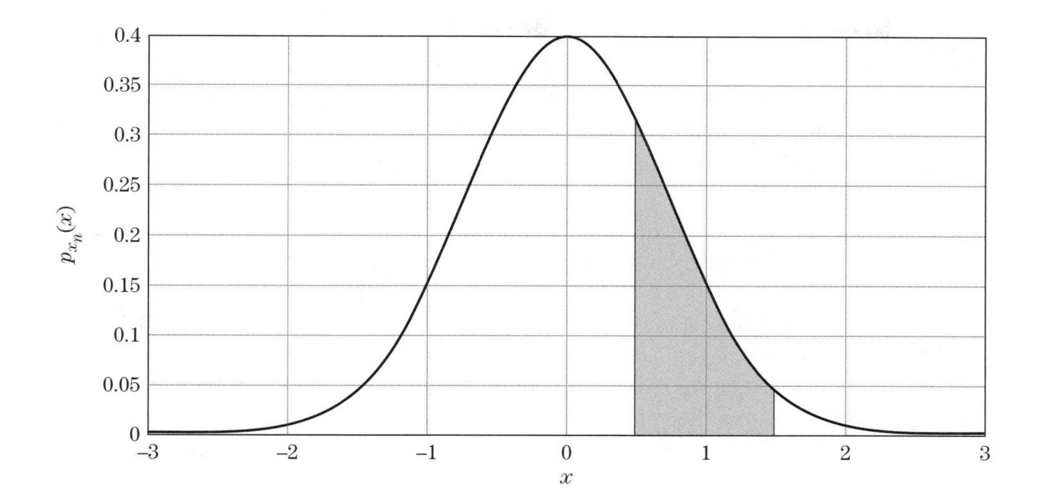

FIGURE 14-27 ■ Use of PDF to compute probabilities.

14.7.2 Moments and Power Spectrum

Random processes are also characterized by their *moments*. Most common are the mean, mean square, and variance:

$$\text{mean: } \bar{x} = m_x = \text{E}\{x\} \equiv \int_{-\infty}^{\infty} x p_x(x)\, dx$$

$$\text{mean square: } \overline{x^2} = p_x = \text{E}\{x^2\} \equiv \int_{-\infty}^{\infty} x^2 p_x(x)\, dx$$

$$\text{variance: } \sigma_x^2 = \text{E}\{(x - \bar{x})^2\} \equiv \int_{-\infty}^{\infty} (x - \bar{x})^2\, p_x(x)\, dx \tag{14.49}$$

The notation $\text{E}\{\cdot\}$ means "expected value" and is defined by the integrals shown. For any random variable, regardless of PDF, it is true that $\sigma_x^2 = \overline{x^2} - (\bar{x})^2$, which also means that the mean square and variance are equal if the mean of the random process is zero. For stationary signals, these moments are independent of the index n. That is, the mean, mean square, and variance are the same at all points within the sequence.

The mean is the average value of a process. The mean square is the average value of the square of the process and is thus the average power of the process. The variance is a measure of how wide a range of values the process takes on around its mean. A small variance means that most of the values are concentrated near the mean; a large variance means that the process takes on values over a wide range about the mean.

The moments in equation (14.49) describe behavior of $x[n]$ at a single index, n. Correlation functions are moments that relate values of random processes at different indices. The *cross-correlation* function compares values at two different times in two different random processes x and y, while the *autocorrelation* function compares values of the same random process at two different times. They are defined as

$$\begin{aligned} &\text{cross-correlation: } \phi_{xy}[m] = \text{E}\{x[n]y^*[n+m]\} \\ &\text{autocorrelation: } \phi_{xx}[m] = \text{E}\{x[n]x^*[n+m]\} \end{aligned} \tag{14.50}$$

Stationarity ensures that these correlation functions depend only on the interval m between the two samples, called the *correlation lag* or just *lag*, and not on the absolute indices n and $n+m$. Note that $\phi_{xx}[0] = \overline{x^2}$. Thus, the zero lag of the autocorrelation function gives the power in a random process. If the process is zero mean, $\phi_{xx}[0] = \sigma_x^2$.

An alternative definition of the cross- and autocorrelation is

$$\begin{aligned} \text{cross-correlation: } \bar{\phi}_{xy}[m] &= \mathrm{E}\{x[n]y^*[n-m]\} \\ \text{autocorrelation: } \bar{\phi}_{xx}[m] &= \mathrm{E}\{x[n]x^*[n-m]\} \end{aligned} \tag{14.51}$$

Comparing with equation (14.50) shows that $\bar{\phi}_{xy}[m] = \phi_{xy}[-m]$. $\bar{\phi}_{xy}[m]$ exhibits the same basic properties as $\phi_{xy}[m]$, but results derived using one definition often differ from the same results derived using the other by reflections, conjugations, or both; readers must be alert to differences in definitions when comparing results from different sources.

The autocorrelation indicates how well knowledge of a random process at one index predicts its value at another index. Specifically, it can be shown that if the autocorrelation function is normalized to have a maximum value of 1, then the minimum mean square error (MMSE) linear estimate of $x[n+m]$ given the value of $x[n]$ is [17]

$$\widehat{x[n+m]} = \bar{x} + \phi_{xx}[m](x[n] - \bar{x}) \tag{14.52}$$

Thus, if $\phi_{xx}[1] = 1$ (maximum correlation), the value of $x[n]$ at index n_0 is also the best MMSE linear estimate of what its value will be at time $n_0 + 1$. If $\phi_{xx}[1] = 0$, then $x[n_0]$ provides no information regarding the value of $x[n_0 + 1]$, so the best estimate is simply the mean $\bar{x}$. Intermediate values of $\phi_{xx}[m]$ indicate that the estimated value at $n_0 + m$ will be partially related to, and partially independent of, the value at n_0.

The auto- and cross-correlation functions have numerous applications in radar signal processing. As discussed in Chapter 20, the autocorrelation and cross-correlation functions are integral to understanding matched filtering and signal detection. They are also important in designing a variety of adaptive algorithms such as adaptive beamforming and STAP.

The DTFT of a random process does not exist mathematically. Instead, frequency domain representations of a random process are obtained by considering the *power spectrum* $S_{xx}(\hat{f})$, which is the DTFT of the autocorrelation function. The symmetry properties of the autocorrelation function are such that the power spectrum is guaranteed to be real and non-negative for all frequency. Thus, it is typically interpreted as a measure of the relative weight of various frequency components in the random process. For instance, if $\phi_{xx}[m] = \overline{x^2}$ for all m, then $S_{xx}(\hat{f}) = 2\pi\overline{x^2}\delta_D(\hat{f})$, an impulse at the origin. This indicates that the only frequency component is the direct current (DC), component, so that the random process is actually a constant for all m. A more important and realistic example is given next.

14.7.3 White Noise

Many random processes in radar are modeled as zero mean *white noise*. In many cases the noise is also assumed to have a Gaussian PDF. This model is common because it is a realistic model for such ubiquitous signals as thermal noise.

White noise is any random process whose autocorrelation function is of the form

$$\phi_{xx}[m] = \sigma_n^2\delta[m] \quad \text{(white noise)} \tag{14.53}$$

Note that this implies a zero mean process. It follows that the power spectrum is a constant for all frequencies:

$$S_{xx}(\hat{f}) = \sigma_n^2 \quad \text{(white noise)} \tag{14.54}$$

Equation (14.54) states that all frequency components are present in the random process with equal weight. Such a process is called "white noise" in analogy to white light, which contains all colors in equal measure. Because the autocorrelation function is zero for all m except $m = 0$, a white noise process is "totally random" in that knowing the value of a white noise process at one index provides no information predicting its value at any other index.

14.7.4 Time Averages

In processing real data, the precise PDFs and expected values are not known but instead are estimated from the data by computing various averages of a measured data sequence $x[n]$ that is assumed to be a realization of a random process. Given an L-point finite data sequence, common estimators are as follows:[14]

$$\begin{aligned}
\text{mean: } \hat{m}_x &= \frac{1}{L}\sum_{n=0}^{L-1} x[n] \\
\text{mean square: } \hat{p}_x &= \frac{1}{L}\sum_{n=0}^{L-1} |x[n]|^2 \\
\text{autocorrelation: } \hat{\phi}_{xx}[m] &= \frac{1}{L}\sum_{n=0}^{L-1} x[n]x^*[n+m]
\end{aligned} \tag{14.55}$$

The cross-correlation is estimated analogously to the autocorrelation. Also, alternative normalizations of the cross- and autocorrelation are sometimes used that differ in handling the end effects for finite data sequences; see [2] for details.

Note that each of these quantities is the sum of random variables or products of random variables, so each estimated moment is itself a random variable with some PDF and some mean, mean square, and variance. An estimator is *asymptotically unbiased* if the expected value of the estimated moment equals the true expected value of the corresponding moment of the random process as the amount of data available for the estimate becomes infinite, that is, as $L \to \infty$. It is *consistent* if it is asymptotically unbiased and the variance of the estimate approaches zero as $L \to \infty$. Each of the previous estimators is unbiased, and the mean and mean square estimators are consistent (provided that $\sigma_x^2 < \infty$). For example, $E\{\hat{m}_x\} = m_x$ and $\lim_{L\to\infty}\{\text{var}(\hat{m}_x)\} = 0$.

[14]These estimators are valid only if the random process is *wide sense stationary* and *ergodic*. The former property says that the mean and mean square are independent of time and that the autocorrelation depends only on the lag m, not the particular absolute times n and $n+m$. The latter property guarantees that time averages of a single realization of the random process equal the corresponding ensemble averages. See [17] for a very accessible introduction to these topics.

14.8 INTEGRATION

Integration is the process of combining multiple samples of a signal, each contaminated by noise or other interference, to "average down the noise" and obtain a single combined signal-plus-noise sample that has a higher SNR than the individual samples. This high-SNR sample is then used in a detection or tracking algorithm to obtain better performance than was possible with the individual low-SNR samples. Integration can be either coherent, meaning that the signal phase information is used, or noncoherent, meaning that only the magnitude of the signal is processed.

14.8.1 Coherent Integration

Assume a measured signal x consists of a signal component, $Ae^{j\phi}$, and a noise component, w. For example, x could be a single range sample from a single pulse, or a particular Doppler spectrum sample $X[k]$, or a single pixel from a complex SAR image. If the measurement that gave x is repeated N times, a sequence of measurements $x[n]$ can be formed. The noise $w[n]$ in each sample is assumed independent and identically distributed (i.i.d.) with variance σ_w^2, but the signal component is the same in each sample. The SNR of each individual sample is therefore $SNR_1 = A^2/\sigma_w^2$. Now consider the SNR of the integrated signal

$$\begin{aligned} x_N &= \sum_{n=0}^{N-1} (x[n] + w[n]) = \sum_{n=0}^{N-1} (Ae^{j\phi} + w[n]) \\ &= NAe^{j\phi} + \sum_{n=0}^{N-1} w[n] \end{aligned} \tag{14.56}$$

Because the signal samples all add in phase with one another, the amplitude of the coherently integrated signal component is now NA and the signal power will be $(NA)^2$. The power in the noise component is

$$\begin{aligned} \mathrm{E}\left\{\left|\sum_{n=0}^{N-1} w[n]\right|^2\right\} &= \mathrm{E}\left\{\left(\sum_{n=0}^{N-1} w[n]\right)\left(\sum_{l=0}^{N-1} w^*[l]\right)\right\} \\ &= \sum_{n=0}^{N-1}\sum_{l=0}^{N-1} \mathrm{E}\{w[n]w^*[l]\} \\ &= N\sigma_w^2 \end{aligned} \tag{14.57}$$

where in the last step the common assumptions that the noise process w is zero mean, white, and stationary have been used for simplicity. The SNR of the coherently integrated data is therefore

$$SNR_N = \frac{N^2A^2}{N\sigma_w^2} = N\frac{A^2}{\sigma_w^2} = N \cdot SNR_1 \tag{14.58}$$

Thus, coherent integration of N data samples increases the SNR by a factor of N. This increase is called the *integration gain*.

Many radar signal processing operations are coherent integration in disguise. Suppose the signal component of the series of samples consists of echoes from a target moving with constant radial velocity, resulting in a Doppler frequency of $\hat{f}_d$ on a normalized frequency scale. The signal samples are then

$$x[n] = Ae^{j(2\pi \hat{f}_d n + \phi)}, \quad n = 0, \ldots, N-1 \tag{14.59}$$

If the data are coherently integrated as is, the signal samples will not combine in phase with one another (except when $\hat{f}_d = 0$), and the factor of N gain in signal amplitude will not be realized. However, if the sum

$$\begin{aligned} x &= \sum_{n=0}^{N-1} x[n] e^{-j2\pi \hat{f}_d n} = \sum_{n=0}^{N-1} (Ae^{j\phi} e^{+j2\pi \hat{f}_d n}) e^{-j2\pi \hat{f}_d n} \\ &= Ae^{j\phi} \sum_{n=0}^{N-1} (1) = NAe^{j\phi} \end{aligned} \tag{14.60}$$

is formed, then the signal component of the sum is the same as in equation (14.56). The integration of equation (14.60) includes a phase rotation or phase compensation of the data so that the signal components will add in phase. This phase compensation will have no effect on the noise power, so there will again be an N-fold increase in SNR. Obviously, it is necessary to know $\hat{f}_d$ or, equivalently, to repeat the computation for several values of $\hat{f}_d$ to search for the one that works best, to achieve this gain.

Finally, suppose that $\hat{f}_d$ happens to be of the form k/K for some integer k. Then the summation in the first step of (14.60) is identical to the definition of the DFT given in (14.25). Thus, the DFT implements ideal coherent integration of complex sinusoids with frequencies equal to the DFT frequencies. Furthermore, a K-point DFT effectively computes K coherent integrations at once, each matched to a different signal frequency. The SNR of the DFT sample that actually corresponds to an input signal frequency will be N times that of the input data. In this manner, the DFT provides a mechanism to test multiple candidate frequencies to maximize the integration gain of an input signal; the FFT algorithm enables this search to be done very quickly.[15]

14.8.2 Noncoherent Integration

Noncoherent integration does not use phase information in the data. Thus, (14.56) is replaced by

$$|x| = \sum_{n=0}^{N-1} |x[n] + w[n]| = \sum_{n=0}^{N-1} |Ae^{j\phi} + w[n]| \tag{14.61}$$

or

$$|x|^2 = \sum_{n=0}^{N-1} |x[n] + w[n]|^2 = \sum_{n=0}^{N-1} |Ae^{j\phi} + w[n]|^2 \tag{14.62}$$

In these equations, $x[n]$ and $w[n]$ are the outputs of a coherent receiver, so they are complex (magnitude and phase) values. However, in noncoherent integration the phase information is discarded by taking the magnitude of the data.

Equation (14.61) is called noncoherent integration with a *linear* detector, while (14.62) uses a *square law* detector. Other detector laws (e.g., logarithmic) are also sometimes used. Noncoherent integration also provides an integration gain in SNR, but it is much more difficult to compute because of the cross-products between signal and noise implied by

[15]In older systems, multiple Doppler frequencies were tested for the presence of a target by subdividing the Doppler spectrum using an explicit bank of band-pass filters, or *filterbank*, within a detector at the output of each filter. The relationship between DFTs and filterbanks is discussed in Chapter 17.

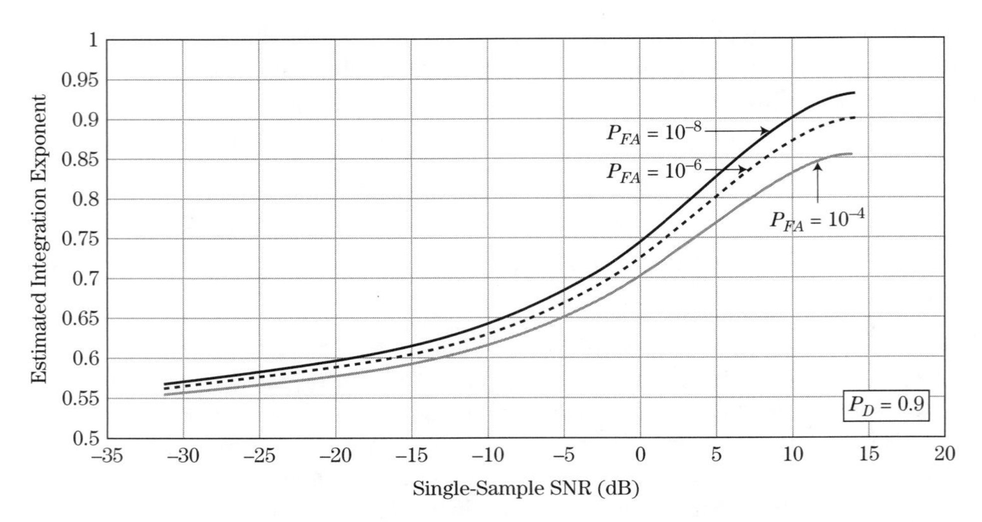

FIGURE 14-28 ■ Estimated noncoherent integration exponent for detection of a constant in complex Gaussian noise with a linear detector. Approximation based on Albersheim's equation [18].

the magnitude operation. The result depends on the particular statistics of the interference and signal and on the detector law. Noncoherent integration gain is usually quantified by computing the preintegration SNR required such that, after integration of N samples, the integrated sample x meets specified performance requirements such as particular probabilities of detection and false alarm, or angular tracking accuracy. The ratio of the predetection SNR required when there is no integration to that required when N samples are integrated is the integration gain. The result is usually expressed in the form of an integration exponent, that is, as N^{α} for some exponent α. For coherent integration, $\alpha = 1$.

Figure 14-28 shows an estimate of the noncoherent integration exponent as a function of SNR for the radar detection problem. Albersheim's equation [18], which provides an estimate of the single-sample SNR needed to achieve specified detection probabilities when N samples are noncoherently integrated with a linear detector, was adapted to estimate α for the case where a detection probability of 90% and several values of false alarm probability were specified.[16] The noncoherent integration is in the range of 0.85 to 0.95 for high single-sample SNRs but falls to nearly $\sqrt{N}$ for very low single-sample SNRs. Lower false alarm probabilities or higher detection probabilities produce still higher values of α at high SNR, but the low-SNR limit remains approximately 0.5. Thus, noncoherent integration is less efficient than coherent integration. This result should be expected because, in discarding the data phase, noncoherent integration does not take advantage of all of the information in the data.

14.9 | CORRELATION AS A SIGNAL PROCESSING OPERATION

The *deterministic cross-correlation* of two signals $x[n]$ and $y[n]$ is defined as

$$c_{xy}[m] = \sum_{n=-\infty}^{+\infty} x[n]\, y^*[n+m] \tag{14.63}$$

[16]See Chapters 3 and 15 for a discussion of Albersheim's equation.

Analogously to the statistical correlation, the index m is called the correlation *lag*. If $x[n] = y[n]$, $c_{xx}[m]$ is called the deterministic *autocorrelation* of $x[n]$. Equation (14.63) is the deterministic equivalent of the statistical concept of correlation for two stationary random processes discussed in Section 14.7.2. Deterministic correlation is clearly a linear operation in the sense that if $s[n] = x[n] + w[n]$, $c_{sy}[m] = c_{xy}[m] + c_{wy}[m]$. Note that

$$c_{xx}[0] = \sum_{n=-\infty}^{+\infty} x[n]\, x^*[n] = \sum_{n=-\infty}^{+\infty} |x[n]|^2 = E_x \tag{14.64}$$

where E_x is the energy in the signal $x[n]$. Furthermore, it is shown in many DSP and random process texts that the $m = 0$ lag is the peak of the autocorrelation function:

$$c_{xx}[m] \le c_{xx}[0] \tag{14.65}$$

This property does not hold for the cross-correlation between two different signals. Another important property of the cross-correlation function is the symmetry property

$$c_{xy}[m] = c_{yx}^*[-m] \tag{14.66}$$

which reduces to $c_{xx}[m] = c_{xx}^*[-m]$ in the autocorrelation case.

As with the statistical cross-correlation, an alternative definition of the deterministic cross-correlation is

$$\bar{c}_{xy}[m] = \sum_{n=-\infty}^{+\infty} x[n]\, y^*[n-m] \tag{14.67}$$

The two definitions are related by $\bar{c}_{xy}[m] = c_{xy}[-m]$. Again, readers must be alert to differences in definitions when comparing results from different sources.

Correlation is a measurement of similarity between two signals. Its primary use in signal processing is to detect the presence of a known signal in high noise and to estimate the known signal's location in the noisy data. Figure 14-29 illustrates this idea. Figure 14-29a shows a noisy signal $s[n]$ composed of zero mean white noise $w[n]$ and an 18-sample square pulse $x[n]$ of amplitude 2 located between $n = 10$ and $n = 27$. Figure 14-29b shows the cross-correlation $c_{sx}[m]$ of the noisy signal and a replica of the noise-free pulse. The peak at lag 10 indicates a high correlation between the reference pulse and the samples of $s[n]$ that begin at $n = 10$; this is exactly the portion of $s[n]$ where the square pulse is located in the noise.

Correlation is another signal processing operation that can be related to coherent integration. Let $x[n] = A[n]e^{j\phi[n]}$ be an arbitrary complex-valued signal. The autocorrelation peak occurs at lag $m = 0$. From (14.64),

$$\begin{aligned} c_{xx}[0] &= \sum_{n=-\infty}^{+\infty} x[n]\, x^*[n] = \sum_{n=-\infty}^{+\infty} \left\{A[n]e^{j\phi[n]}\right\}\left\{A^*[n]e^{-j\phi[n]}\right\} \\ &= \sum_{n=-\infty}^{+\infty} |A[n]|^2 = E_x \end{aligned} \tag{14.68}$$

Equation (14.68) shows that the autocorrelation peak occurs because the signal is multiplied by its own conjugate, forming a zero-phase product. Each term in the sum is thus real and adds in phase to produce the autocorrelation peak. In other words, the $m = 0$ lag coherently integrates the samples of $x[n]$.

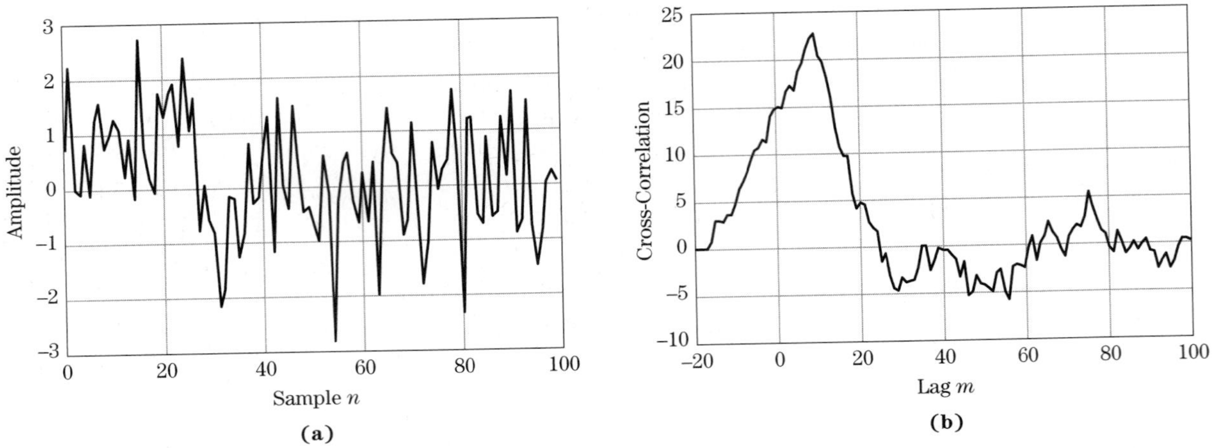

FIGURE 14-29 ■ Use of cross-correlation to locate a signal buried in noise. (a) Signal containing a square pulse in samples 10 through 17, plus noise in all samples. (b) Cross-correlation of the noisy data with a replica of the pulse.

14.10 MATCHED FILTERS

Chapters 15 and 18 show that the signal-to-noise ratio is the major determinant of performance in such applications as detection and tracking. In detection, for example, the probability of detection increases monotonically with SNR for a fixed probability of false alarm. In tracking, the error variance of track measurements (whether in range, angle, or Doppler) is inversely proportional to SNR; thus, as SNR increases, the error variance decreases. Consequently, maximizing the SNR of radar data is a fundamental goal of radar design, affecting everything from transmitter power and antenna size to the signal processing operations applied to the data.

The radar signal processor can contribute to the maximization of SNR by filtering or transforming the data in some way that removes the noise while reinforcing the desired signal component. This intuitive notion can be formalized by determining the LSI filter that will maximize the SNR. This is an optimization problem that can be stated and solved in many ways, all of which of course lead to the same answer. The approach here is based on the vector notation for signals and for convolution from Sections 14.2.3 and 14.6. Recall the vector expression for the output of an FIR filter in equation (14.38). The power in the output sample $y[n]$ is

$$|y|^2 = y^* y^T = \boldsymbol{H}^H \boldsymbol{X}^* \boldsymbol{X}^T \boldsymbol{H} \tag{14.69}$$

where the superscript H indicates a Hermitian (conjugate transpose) operation and the subscript n has been dropped for notational brevity. Now consider a finite-length signal consisting of signal and noise components: $\boldsymbol{X} = \boldsymbol{S} + \boldsymbol{W}$. The goal is to find the filter coefficient vector $\boldsymbol{H}$ that maximizes the SNR. Since the filter is linear, its effect on the signal power and noise power can be computed separately. The signal component power is, from (14.69), just $\boldsymbol{H}^H \boldsymbol{S}^* \boldsymbol{S}^T \boldsymbol{H}$. The noise power at the particular time index n is, similarly, $\boldsymbol{H}^H \boldsymbol{W}^* \boldsymbol{W}^T \boldsymbol{H}$. However, the average noise power is much more relevant. The expected value of the filtered noise sample is

$$|y|^2_{noise} = \boldsymbol{H}^H \mathbf{R_I} \boldsymbol{H} \tag{14.70}$$

where the linearity of the expected value operation has been used to bring it inside the vector products with $\boldsymbol{H}$ and the noise *covariance matrix* has been defined as

$$\mathbf{R_I} = \mathrm{E}\{\boldsymbol{W} \cdot \boldsymbol{W}^T\} \tag{14.71}$$

Covariance matrices have several special properties. For instance, they are Toeplitz matrices, meaning they are constant along each diagonal, they have Hermitian symmetry, meaning $\mathbf{R_I}^H = \mathbf{R_I}$, and they are positive definite [19].

The SNR is just the ratio of the signal and noise powers at the filter output:

$$SNR = \frac{\boldsymbol{H}^H \boldsymbol{S}^* \boldsymbol{S}^T \boldsymbol{H}}{\boldsymbol{H}^H \mathbf{R_I} \boldsymbol{H}} \tag{14.72}$$

To find the filter coefficient vector $\boldsymbol{H}$ that maximizes SNR, the Schwarz inequality is needed in the form

$$|\boldsymbol{P}^H \boldsymbol{Q}|^2 \leq \|\boldsymbol{P}\|^2 \|\boldsymbol{Q}\|^2 \tag{14.73}$$

where $\boldsymbol{P}$ and $\boldsymbol{Q}$ are two arbitrary vectors, and the norm of $\boldsymbol{P}$ is defined as $||\boldsymbol{P}|| = \boldsymbol{P}^H \boldsymbol{P}$. Because $\mathbf{R_I}$ is positive definite it can be factored as follows:

$$\mathbf{R_I} = \mathbf{A}^H \mathbf{A} \tag{14.74}$$

Contriving to choose $\boldsymbol{P} = \mathbf{A}\boldsymbol{H}$ and $\boldsymbol{Q} = (\mathbf{A}^H)^{-1}\boldsymbol{S}^*$ in the Schwarz inequality gives $\boldsymbol{P}^H \boldsymbol{Q} = \boldsymbol{H}^H \boldsymbol{S}^*$ and, with some basic matrix algebra, the Schwarz inequality gives

$$\begin{aligned} \boldsymbol{H}^H \boldsymbol{S}^* \boldsymbol{S}^T \boldsymbol{H} &\leq \|\mathbf{A}\boldsymbol{H}\|^2 \left\| \left(\mathbf{A}^H\right)^{-1} \boldsymbol{S}^* \right\|^2 \\ &= \left(\boldsymbol{H}^H \mathbf{R_I} \boldsymbol{H}\right) \left(\boldsymbol{S}^T \mathrm{R}_\mathbf{I}^{-1} \boldsymbol{S}^*\right) \end{aligned} \tag{14.75}$$

Dividing both sides by $\boldsymbol{H}^H \mathbf{R_I} \boldsymbol{H}$ gives

$$\frac{\boldsymbol{H}^H \boldsymbol{S}^* \boldsymbol{S}^T \boldsymbol{H}}{\boldsymbol{H}^H \mathbf{R_I} \boldsymbol{H}} = SNR \leq \boldsymbol{S}^T \mathbf{R_I}^{-1} \boldsymbol{S}^* \tag{14.76}$$

What choice of $\boldsymbol{H}$ will achieve this maximum SNR? The maximum is achieved when $\boldsymbol{P} = k\boldsymbol{Q}$ for some k. This condition becomes $\mathbf{A}\boldsymbol{H} = k\left(\mathbf{A}^H\right)^{-1} \boldsymbol{S}^*$, so that

$$\boldsymbol{H} = k\mathbf{R_I}^{-1} \boldsymbol{S}^* \tag{14.77}$$

The effect of $\mathbf{R_I}^{-1}$ is to whiten the interference prior to matched filtering with $\boldsymbol{S}^*$.

Equation (14.77) is of fundamental importance in radar signal processing; it is the basis not only of waveform matched filtering but also of many techniques in adaptive beamforming, ground moving target indication, and space-time adaptive processing.

While stationary interference has been assumed, the noise has not been assumed to be white. Thus, equation (14.77) is fairly general. If in fact the noise is white, an important special case, then $\mathbf{R_I} = \sigma_n^2 \mathbf{I}$ where $\mathbf{I}$ is an identity matrix. If k is chosen to equal $1/\sigma_n^2$, then

$$\boldsymbol{H} = \boldsymbol{S}^* \tag{14.78}$$

Thus, the optimum filter coefficients when the interference is white noise are just the conjugate of the samples of the desired signal $\boldsymbol{S}$. In this case, $\boldsymbol{H}$ is called the *matched filter*, because the coefficients are "matched" to the signal the filter is designed to detect. If that signal changes, then the filter coefficients must also be changed to maintain the maximum possible SNR.

The matched filter can also be derived for continuous-time signals using the appropriate version of the Schwarz inequality. The result is that the impulse response of the filter should satisfy

$$h(t) = ks^*(T_M - t) \tag{14.79}$$

where T_M is the time instant at which the SNR is maximized [3]. If the signal s is of finite length τ, then $T_M \geq \tau$ for causality. The derivation of equation (14.79) does not assume a finite impulse response; however, $h(t)$ will be infinite only if the desired signal s to which it is matched is infinite in duration.

Filtering a signal $s(t)$ with its matched filter from (14.79) corresponds to computing the continuous-time autocorrelation function $c_{ss}(l)$. To see this, write the convolution of $s(t)$ and its matched filter impulse response:

$$\begin{aligned} y(t) &= \int_{-\infty}^{\infty} h(u)s(t-u)du \\ &= \int_{-\infty}^{\infty} s^*(T_M - u)s(t-u)du \\ &= \int_{-\infty}^{\infty} s(v)s^*(v + T_M - t)dv \end{aligned} \tag{14.80}$$

The last step, obtained with the substitution $v = t - u$, is easily recognized as the autocorrelation of s evaluated at lag $T_M - t$. Thus, the signal component of the time waveform at the output of the matched filter is actually the autocorrelation function of that signal. The matched filter peak $y(T_M)$ is then $c_{ss}(0) = E_s$, where E_s is the total energy in the signal s. As previously discussed, the zero autocorrelation lag is equivalent to coherently integrating the signal s. The output of the vector matched filter $y = \boldsymbol{H}^T\boldsymbol{S}$ with the optimum coefficients $\boldsymbol{H} = \boldsymbol{S}^*$ is also easily seen to be equivalent to coherently integrating the signal vector samples, producing an output equal to the signal vector energy E_s. Once again, coherent integration is at the heart of a key signal processing operation.

Equation (14.76) shows that, in the case of white noise, the actual value of the peak SNR is

$$SNR_{\max} = \frac{E_s}{\sigma_n^2} \tag{14.81}$$

The same result is obtained in the continuous-time case. Equation (14.81) shows the important fact that, when the interference is white and a matched filter is used, the maximum SNR depends only on the signal energy, not on its detailed shape. This fact will be critical to the development of pulse compression and to its use in gaining independent control of detection performance and range resolution in Chapter 20.

There is a direct discrete-time equivalent to the analog matched filter of (14.79), which in turn gives a discrete correlation equivalent to (14.80):

$$\begin{aligned} h[n] &= ks^*[N_0 - n] \\ y[n] &= \sum_{m=-\infty}^{\infty} h[m]s[n-m] \\ &= \sum_{m=-\infty}^{\infty} s[m]s^*[m + N_0 - n] \end{aligned} \tag{14.82}$$

The discrete version exhibits the same properties (e.g., value of peak output, coherent integration effect) as the continuous version.

14.11 FURTHER READING

While there are many excellent texts devoted to digital signal processing fundamentals, few if any focus on aspects specific to radar and similar sensor systems. At this writing, the most recent text focused specifically on radar signal processing basics is *Fundamentals of Radar Signal Processing* by Richards [3]. Good background texts in general digital signal processing are too numerous to mention; two of the more successful texts are those by Mitra [1] and Oppenheim and Schafer [2]. Hayes has produced *Schaum's Outline of Digital Signal Processing* [20], which is an excellent concise reference and source of sample problems. Finally, Lyons's *Understanding Digital Signal Processing* [21], offers a very practical-minded review of DSP concepts a well as many "tricks of the trade."

14.12 REFERENCES

[1] Mitra, S.K., *Digital Signal Processing: A Computer-Based Approach*, 2d ed., McGraw-Hill, New York, 2001.

[2] Oppenheim, A.V., and Schafer, R.W., *Discrete-Time Signal Processing*, 3d ed., Prentice-Hall, Englewood Cliffs, NJ, 2009.

[3] Richards, M.A., *Fundamentals of Radar Signal Processing*, McGraw-Hill, New York, 2005.

[4] Bracewell, R.N., *The Fourier Transform and Its Applications*, 3d ed., McGraw-Hill, New York, 1999.

[5] Ercegovec, M., and Lang, T., *Digital Arithmetic*, Morgan Kauffman, San Francisco, CA, 2003.

[6] McClellan, J.H., and Purdy, R.J., "Applications of Digital Signal Processing to Radar," Chapter 5 in *Applications of Digital Signal Processing*, Ed. A.V. Oppenheim, Prentice-Hall, Englewood Cliffs, NJ, 1978.

[7] Papoulis, A., *The Fourier Integral and Its Applications*, McGraw-Hill, New York, 1987.

[8] Harris, F.J., "On the Use of Windows for Harmonic Analysis with the Discrete Fourier Transform," *Proceedings of the IEEE*, vol. 66, no. 1, pp. 51–83, January 1978.

[9] Johnson, D.H., and Dudgeon, D.E., *Array Signal Processing*, Prentice-Hall, Englewood Cliffs, NJ, 1993.

[10] Richards, M.A., "Relationship between Temporal and Spatial Frequency in Radar Imaging," Technical Memorandum. July 7, 2007. Available at http://www.radarsp.com.

[11] Jacobsen, E., and Kootsookos, P., "Fast Accurate Frequency Estimators," *IEEE Signal Processing Magazine*, pp. 123–125, May 2007.

[12] Cooley, J.W., and Tukey, J.W., "An Algorithm for the Machine Computation of Complex Fourier Series," *Mathematics of Computation*, vol. 19, pp. 297–301, April 1965.

[13] Burrus, C.S., and Parks, T.W., *DFT/FFT and Convolution Algorithms*, John Wiley & Sons, New York, 1985.

[14] "FFTW [Fastest Fourier Transform in the West]," n.d. Available at http://www.fftw.org.

[15] Shrader, W.W., and Gregers-Hansen, V., "MTI Radar," Chapter 2 in *Radar Handbook*, 3d ed., Ed. M.I. Skolnik, McGraw-Hill, New York, 2008.

[16] Rader, C.M., "A Simple Method for Sampling In-Phase and Quadrature Components," *IEEE Transactions Aerospace and Electronic Systems*, vol. AES-20, no. 6, pp. 821–824, November 1984.

[17] Kay, S., *Intuitive Probability and Random Processes Using MATLAB®*, Springer, New York, 2006.

[18] Albersheim, W.J., "Closed-Form Approximation to Robertson's Detection Characteristics," *Proceedings of the IEEE*, vol. 69, no. 7, p. 839, July 1981.

[19] Watkins, D.S., *Fundamentals of Matrix Computations*, Wiley, New York, 1991.

[20] Hayes, M.H., *Schaum's Outline of Digital Signal Processing*, McGraw-Hill, New York, 1998.

[21] Lyons, R.G., *Understanding Digital Signal Processing*, 2d ed., Prentice-Hall, New York, 2004.

14.13 PROBLEMS

1. Consider an electromagnetic wave with a temporal frequency of 1 GHz, propagating in the $+x$ direction. What is the numerical value of spatial frequency k_x of this wave, in radians per meter, and also in cycles/m?

2. How many pulses must be coherently integrated to produce an integration gain sufficient to increase the detection range for a given target from 5 to 50 miles?

3. In some cases, the spectrum replication property of sampling can be used as a substitute for demodulation. Given a signal $x_a(t)$ with the spectrum shown, what is the lowest sampling rate that will ensure that one of the spectrum replicas is centered at $f = 0$ with no aliasing?

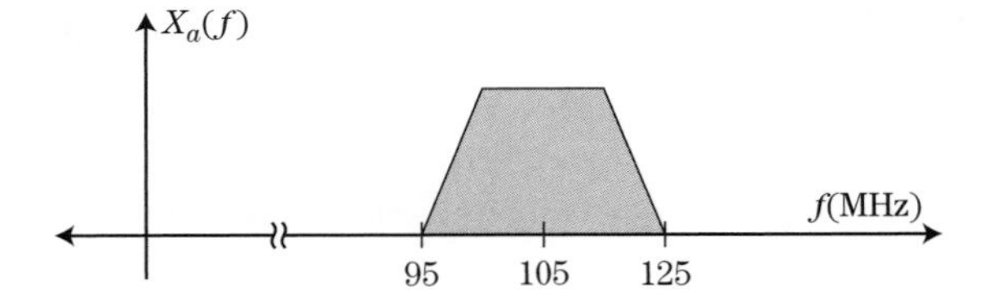

4. Consider a pulsed radar with a fast-time sampling rate equal to the pulse bandwidth $B = 10$ MHz, a PRF of 5 kHz, and a single receiver channel. If the range swath extends from 5 km to 50 km and the duration of the CPI is 4 ms, what is the total number of complex samples in the fast time–slow time data matrix?

5. The saturation voltage of an ADC is set at four times the input noise standard deviation, $A_{sat} = 4\sigma$. How many bits, b, are required in the ADC to provide an SQNR of at least 70 dB?

6. What is the 3 dB Doppler resolution of a system that collects 40 pulses of data at a PRF of 10 kHz and applies a Hamming window to the slow-time data before computing the Doppler spectrum?

7. Consider a sequence of 20 slow-time data samples collected at a PRF of 2 kHz. If a 1,000-point DFT of this sequence is computed, what is the spacing between DFT frequency samples in Hz?
8. Explicitly compute the SNR loss of equation (14.18) as a function of K for a triangular window of odd length $K + 1$ (so K is even) defined according to

$$w[k] = \begin{cases} 2k/K, & 0 \le k \le K/2 \\ 2 - 2k/K, & K/2 \le k \le K \\ 0. & \text{otherwise} \end{cases}$$

Numerically evaluate the result for $K = 4$ and $K = 20$, and give the results in dB. What is the asymptotic value of the loss in dB as $K \to \infty$ (give the result in both linear and dB units)? The following facts may be useful (careful about the limits!):

$$\sum_{k=1}^{n} k = n\,(n+1)/2, \qquad \sum_{k=1}^{n} k^2 = n\,(n+1)\,(2n+1)/6$$

Another hint: Sum just the first half of the triangle, and then use symmetry to get the sum of the whole function, being careful not to double-count any samples.
9. An X-band (10 GHz) pulse-Doppler radar collects a fast-time–slow-time matrix of 30 pulses by 200 range bins per pulse. This is converted to a range–Doppler matrix by applying a Hamming window and then a 64-point fast Fourier transform to each slow-time row. Suppose that there is a target with a constant radial velocity of 30 m/s approaching the radar at a range corresponding to range bin #100. The PRF is 6,000 samples/s. There is no ground clutter. Ignore thermal noise as well. For which FFT sample index k_0 is $|Y[k_0]|$ the largest? What velocity in m/s does this sample correspond to? What is the error between the apparent velocity based on the largest FFT sample, and the actual velocity?
10. Continuing Problem 9, in terms of the window function $w[m]$, what is the peak value of the DTFT (not DFT) of the windowed data in range bin #100, assuming that each slow-time sample has an amplitude of 1 before windowing? What is the numerical value of this peak? (Use a computing environment or spreadsheet program such as MATLAB® or Excel® to compute this value.) Now suppose the peak value of the magnitude of the FFT of the data $|Y[k_0]| = 15.45$. What is the straddle loss in dB?
11. Suppose also that $|Y[k_0 - 1]| = 11.61$ and $|Y[k_0 + 1]| = 14.61$. Use the quadratic interpolation technique of (14.29) and (14.30) to estimate the velocity of the target and the peak amplitude of the DTFT. Compute the loss in the new amplitude value relative to the true amplitude. Compare the new values of velocity error and straddle loss with those found in Problems 9 and 10.
12. One drawback in some situations of the K-point FFT is that it computes all of the DFT samples, even if only one or a few are required. The DFT of (14.25) can be used to compute individual frequency domain samples. In terms of K, what is the maximum value of L such that the number of complex multiplies to compute L samples of the K-point DFT using (14.25) is less than the number of complex multiplies to compute all K DFT samples using the FFT? What is the numerical value of L (must be an integer) for $K = 256$? Assume that the FFT requires $(K/2)\log_2 K$ complex multiplications, regardless of the value of K.
13. Fast convolution is to be used to convolve a 1,000-point input signal $x[n]$ with a filter impulse response $h[n]$ that is 100 samples long. What is the length of the filter output $y[n]$? Assuming the FFT size is restricted to be a power of 2, what is the minimum size FFTs that can be used?
14. Median filters can be used to filter "impulse noise," meaning isolated pulses of energy, from an otherwise relatively constant signal. What is the widest impulse in samples that will be removed by an 11-point median filter?

15. Show that the mean estimator of (14.55) is unbiased and consistent, that is, $E\{\hat{m}_x\} = m_x$ and $\lim_{L\to\infty}\{\mathrm{var}(\hat{m}_x)\} = 0$. Assume that $x[n]$ is stationary with mean m, variance σ_x^2, and that $x[n]$ is independent of $x[k]$ for $n \neq k$.

16. If 100 samples, each having an SNR of 0 dB, are coherently integrated, what will be the integrated SNR? What is the integration gain in dB? Assuming a receiver designed to achieve $P_D = 0.9$ and $P_{FA} = 10^{-8}$, use Figure 14-28 to estimate the number of 0 dB SNR samples that must be noncoherently integrated to achieve the same integration gain.

17. This problem applies the vector matched filter design equation (14.77) to the problem of detecting a sinusoid in zero mean complex white Gaussian noise with power σ_n^2. Let the filter order N be arbitrary. Show that the interference covariance matrix for this case is an identify matrix, $\mathbf{R_I} = \sigma_n^2\mathbf{I}$. Let the desired signal be a pure sinusoid of normalized radian frequency $\hat{\omega}_0$, so that $\boldsymbol{S} = \left[1\ \exp(j\hat{\omega}_0)\ \exp(j2\hat{\omega}_0)\ \cdots\ \exp(j(N-1)\hat{\omega}_0)\right]^T$. Find the matched filter coefficient vector $\boldsymbol{H}$. Show that when applied to a vector of data $\boldsymbol{X} = \boldsymbol{S} + \boldsymbol{W}$, the filter coherently integrates the signal components of the data.

18. A pulsed radar observes a target in a particular range bin that is approaching with a constant radial velocity of v m/s. In this situation, the slow-time signal observed in that range bin will be of the form $s[m] = A\exp(j\hat{\omega}_0 m),\ m = 0, 1, \ldots, M-1$ for some number of pulses M. What is the impulse response of the discrete-time matched filter for this signal, assuming it is designed to produce the peak SNR at time $m = M-1$? For a general input signal $x[m]$, show that the output of this matched filter at the peak is, to within a scale factor, the DTFT of the input evaluated at frequency $\hat{\omega}_0$.

CHAPTER 15

Threshold Detection of Radar Targets

Mark A. Richards

Chapter Outline

15.1 Introduction ... 547
15.2 Detection Strategies for Multiple Measurements ... 548
15.3 Introduction to Optimal Detection ... 552
15.4 Statistical Models for Noise and Target RCS in Radar ... 557
15.5 Threshold Detection of Radar Signals ... 560
15.6 Further Reading ... 584
15.7 References ... 584
15.8 Problems ... 585

15.1 INTRODUCTION

One of the most fundamental tasks of a radar is detection, the process of examining the radar data and determining if it represents interference only, or interference plus echoes from a target of interest. Once a target is detected, the system can turn its attention to processing the target information. Depending on the type of radar application, the system might be concerned with estimating the target radar cross section (RCS), measuring and tracking its position or velocity, imaging it, or providing fire control data to direct weapons to the target.

The basic concept of threshold detection was discussed in Chapter 3. Chapter 7 described common statistical models for the target echo power, including both probability distributions and pulse-to-pulse decorrelation models. Chapters 3 and 8 discussed coherent and noncoherent integration of data to improve the signal-to-noise ratio (SNR). In this chapter, these topics are brought together to provide a more detailed look at the optimal detection of fluctuating targets in noise. The analysis shows how the idea of threshold detection arises. The strategy for determining threshold levels and predicting detection and false alarm performance is demonstrated, and specific results are developed for the common Swerling target models. Also discussed are Albersheim's equation (first mentioned in Chapter 3) and Shnidman's equation, both very simple but useful analytical tools for estimating detection performance.

15.2 | DETECTION STRATEGIES FOR MULTIPLE MEASUREMENTS

15.2.1 Dwells and Coherent Processing Intervals

As discussed in Chapter 3, an individual target will generally be within the radar antenna mainbeam for a period of time, called the dwell time, that corresponds to a number of pulse repetition intervals (PRIs). That is, the radar may "hit" the target with multiple pulses during a dwell time. The data obtained during a dwell time might be organized into one or more coherent processing intervals (CPIs). Chapter 3 also showed that radar targets are generally detected using a threshold test procedure and that the best detection performance for a given P_{FA} is obtained if the signal-to-noise ratio of the data to be tested is maximized before the threshold test.

A CPI of coherent radar data and its organization into a fast-time/slow-time data matrix is shown in Figure 15-1. Each individual sample is a complex (in-phase [I] and quadrature [Q]) measurement of the amplitude and phase of the received echo signal. The gray shaded set of samples are the slow-time samples for a single range bin. Assuming no *range migration* (see Chapter 21), all of the echoes from a particular target in a CPI of data will be in a single range bin. If there are multiple CPIs in the dwell time, there will be multiple data matrices.

The term *coherent processing interval* suggests that the data within the CPI will be combined coherently, that is, using both amplitude and phase information. While this will usually be true, it is also possible to combine CPI samples noncoherently. Thus, the term CPI implies a block of coherent radar data but does not necessarily imply a specific means of combining that data.

At the point of threshold detection, a single value derived from the available data is compared with a threshold value and a decision made. The quantity to which the threshold test is applied is called the *detection statistic*.

15.2.2 Coherent, Noncoherent, and Binary Integration

The signal-to-noise ratio (SNR) of the detection statistic is often improved relative to that of a single target-plus-noise sample by *integrating* (adding) the multiple measured samples of the target and noise, motivated by the idea that the interference can be "averaged out" and the target echo reinforced by adding multiple samples. Thus, in general detection will be based on N samples of the combined target-plus-interference signal.

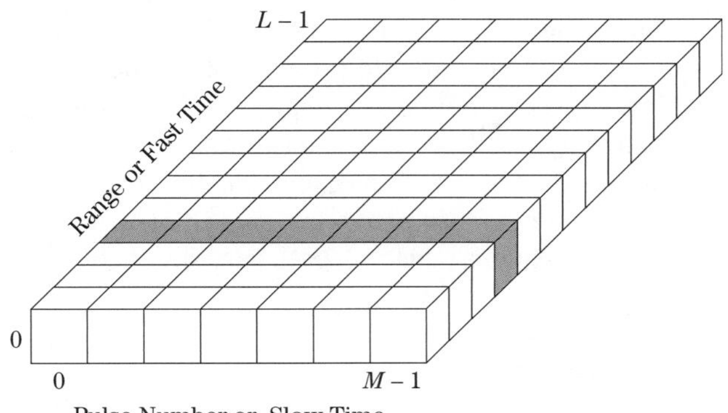

FIGURE 15-1 ■ One CPI of fast-time/slow-time data. The shaded region is the echo samples for the fourth range bin.

At least three types of integration may be applied to the data:

1. After coherent demodulation, to the baseband complex-valued (I and Q, or magnitude and phase) data. Combining complex data samples is referred to as *coherent integration*.
2. After envelope detection of the complex data, to the magnitude (or squared or log magnitude) data. Combining magnitude samples after the phase information is discarded is referred to as *noncoherent integration*.
3. After threshold testing, to the target present/target absent decisions. This technique is called *binary integration*, *m-of-n detection*, or *coincidence detection*.

In coherent integration, complex (magnitude and phase) data samples y_n are combined to form a new complex variable y:

$$y = \sum_{n=1}^{N} y_n e^{j\phi_n} \tag{15.1}$$

The phase weights $\exp(j\phi_n)$ are chosen to compensate the phase of the data samples y_n so that they add in phase with one another. If the SNR of a single complex sample y_n is χ_1 and the interference is uncorrelated from sample to sample, the coherently integrated complex data sample y has an SNR that is N times that of the single sample, that is, $\chi_N = N\chi_1$ (where χ_N and χ_1 are in linear units). That is, coherent integration attains an *integration gain* of N.

Coherent integration is effective only if the N data samples have a predictable phase relationship so that appropriate phase compensation can be applied in equation (15.1). Modern radar systems offer several opportunities for coherent integration. The most important ones were indicated in Figure 14-8, which showed the subsets of a coherent datacube used for various coherent operations. Depending on the particular system, the coherent operations used may include pulse compression of fast-time samples, pulse-Doppler processing of slow-time samples, and beamforming across phase centers. Synthetic aperture radar (SAR) imaging and space-time adaptive processing coherently integrate two-dimensional (2-D) subsets of the datacube.

For example, in pulse-Doppler processing, phase compensation of the data within a CPI for stationary and moving targets is typically accomplished with a K-point discrete Fourier transform (DFT) of the slow-time data, forming a range-Doppler map and effectively testing K different phase compensation functions corresponding to different Doppler shifts at once as discussed in Chapter 14. Applying the DFT is effective only if the pulses are collected at a constant pulse repetition interval (PRI) over a relatively short time interval. Doppler processing details are given in Chapter 17, and SAR is discussed in Chapter 21.

Noncoherent integration takes place after envelope detection, when phase information has been discarded. Instead, the magnitude or squared magnitude of the data samples is integrated. (Sometimes another function of the magnitude, such as the log magnitude, is used.) For example, noncoherent integration with a square law detector bases detection on the detection statistic

$$z = \sum_{n=1}^{N} |y_n|^2 \tag{15.2}$$

Both linear and square law detectors are considered in this chapter; the choice is primarily for convenience in discussing various topics. Noncoherent integration of N samples provides an integration gain less than N but greater than about $\sqrt{N}$.

Binary integration takes place after an initial detection decision has occurred on each of several CPIs of data. Because there are only two possible outputs of the detector each time a threshold test is made, the output is said to be binary. Multiple binary decisions can be combined in an "m-of-n" decision logic in an attempt to further improve performance. That is, a detection is not declared unless the target is detected in m of the n threshold tests. This type of integration was discussed in Chapter 3.

15.2.3 Data Combination Strategies

A system could elect to use none, one, or any combination of these integration techniques. Many systems use at least one integration technique, and a combination of either coherent or noncoherent with postdetection binary integration is common. The major costs of integration are the time and energy required to obtain the multiple datacube samples to be integrated and the computation required to combine those samples. The collection time is time that cannot be spent searching for targets elsewhere, or tracking already-known targets, or imaging other regions of interest. Modern systems vary as to whether the computational load is an issue: the required operations are simple but must be performed at a very high rate in many systems. Chapter 3 included examples of the trade-offs between different integration options within a given timeline in determining detection performance.

Suppose a radar collects two CPIs of data within a single dwell with two pulses per CPI and 5 range bins per pulse, giving a total of 20 data samples.[1] One processing approach would coherently integrate all 20 samples, giving a single integrated sample with a 13 dB integration gain compared with any one sample alone. There would be a single detection statistic (the magnitude or magnitude squared of the integrated sum) and a single threshold test. This strategy for detection testing of the available data is diagrammed in Figure 15-2a. In this figure, small white boxes indicate coherent (amplitude and phase) data samples, while gray boxes indicate noncoherent (magnitude or magnitude squared only, no phase) samples. The hexagons with an inscribed $\times$ are the detection statistic, the quantity to be tested against the threshold.

The phase relationship between the data in two different CPIs is generally not known, so the approach of Figure 15-2a is generally not practical. A very practical approach is to coherently integrate the data in fast time (pulse compression) and slow time (Doppler DFTs) within each CPI separately. Within a single CPI, the peak of the Doppler DFT in the range bin of interest would represent the coherently integrated target echo for the 10 samples of that CPI and would exhibit an integration gain of 10 dB. The values of the two resulting Doppler spectrum peaks could then be noncoherently integrated, producing a gain of at least 1.5 but less than 3 dB in the noncoherently integrated sample compared with the individual Doppler spectrum peaks. This again gives a single detection statistic that includes both coherently and noncoherently combined data and has an SNR gain greater than 10 dB but less than 13 dB, as compared to the SNR of a single sample. Figure 15-2b illustrates this approach.

[1] More realistic numbers would be perhaps three to eight CPIs per dwell, a few tens of pulses per CPI, and hundreds of range bins per pulse. The smaller numbers in this example are used to simplify Figure 15-2.

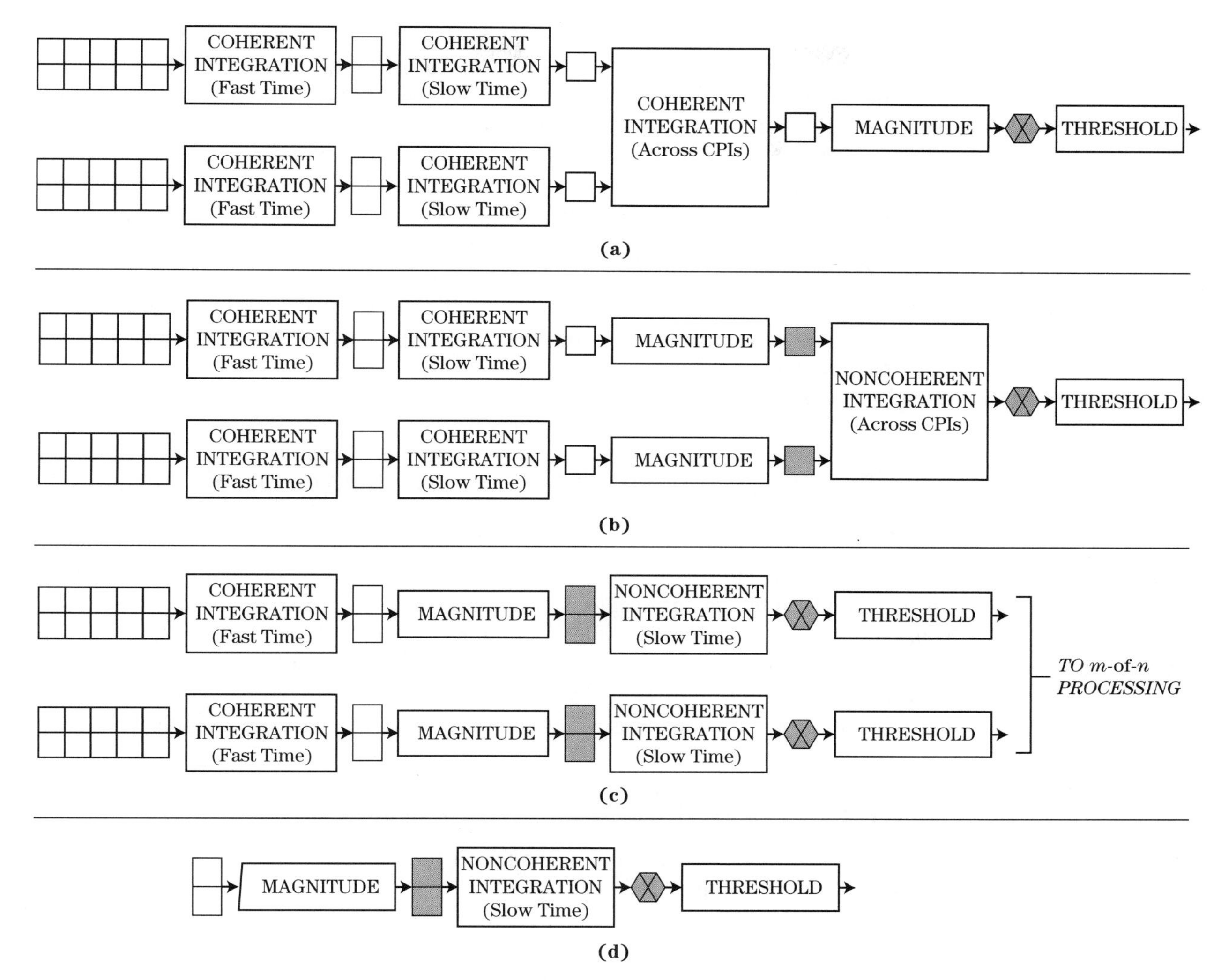

FIGURE 15-2 ■ Alternative detection strategies. (a) Coherent integration of all data. (b) Noncoherent integration of coherently integrated CPIs. (c) Mixed integration and threshold testing within each CPI followed by m-of-n detection. (d) Noncoherent integration only in slow time for a single range bin.

Figure 15-2c illustrates one version of the binary integration approach. Coherent integration in fast time and noncoherent integration in slow time are performed within a CPI to form a detection statistic, and then a threshold detection test is applied to each CPI separately. In this example, this would produce two separate target/no-target decisions. These could then be combined in a logic that declares a detection only if the target in that range bin was detected on at least one of the individual CPIs. (A more realistic example might require detection on perhaps two of five or three of eight CPIs.) Another version of this approach (not shown) would replace noncoherent integration in slow time with coherent integration (Doppler DFTs) prior to the threshold test and binary integration.

Figure 15-2d illustrates a simpler system that does not use pulse compression to effect coherent integration in fast time. In this case, samples from a given target are confined to a single range bin. Integration is in slow time only and can be coherent or noncoherent; the noncoherent case is diagrammed here.

Other combinations of coherent and noncoherent are also possible. Of the scenarios illustrated in Figure 15-2, cases (b) and (d) are the most common. Case (a) is unrealistic because of the lack of a known phase relationship between CPIs. Case (c) is not unreasonable, but it is more likely that coherent integration would be used in slow time if possible.

The purpose of the analysis in this chapter is to provide the tools to determine the probability of detection P_D for a given probability of false alarm P_{FA} for many of these scenarios so that the best processing strategy for the available data can be determined. It will be seen that the detection performance will depend on the number of data samples integrated before the threshold test, the combination of coherent and noncoherent integration strategies applied to those samples, the SNR of the samples, and the target fluctuation model that describes the target echo component of the samples.

15.3 INTRODUCTION TO OPTIMAL DETECTION

A single echo sample of radar data is composed of either interference alone or interference plus target echoes. The interference is, at a minimum, receiver noise, and might also include air or ground clutter echoes, electromagnetic interference (EMI) from other transmitting sources (e.g., radars, television stations, cellular telephones), and hostile jamming. The signals received from these interference sources are modeled as additive random processes, as discussed for clutter in Chapter 5. Thus, even if the target echo amplitude is entirely deterministic, the combined target-plus-interference signal is a random process.

15.3.1 Hypothesis Testing and the Neyman-Pearson Criterion

For any radar measurement that is to be tested for the presence of a target, one of two hypotheses can be assumed true:

1. The measurement is the result of interference only.
2. The measurement is the combined result of interference and echoes from a target.

The first hypothesis is denoted as the *null hypothesis,* H_0, and the second as H_1. A statistical description of the data under each hypothesis is needed. For simplicity, assume initially that this is a single echo sample (one fast-time sample from a single pulse) denoted y. Then two probability density functions (PDFs) $p_y(y|H_0)$ and $p_y(y|H_1)$ are required:

$p_y(y|H_0)$ = PDF of y given that a target was *not* present

$p_y(y|H_1)$ = PDF of y given that a target *was* present

Design of the detection algorithm and analysis of the resulting radar performance is dependent on developing models for these PDFs for the system and scenario at hand. Furthermore, a good deal of the radar system design problem is aimed at manipulating these two PDFs to obtain the most favorable detection performance.

The detection logic must examine each radar measurement to be tested and select one of the hypotheses as "best" accounting for that measurement. If H_0 best accounts for the data, then the system declares that a target was not present at the range, angle, or Doppler coordinates of that measurement; if H_1 best accounts for the data, then the system declares

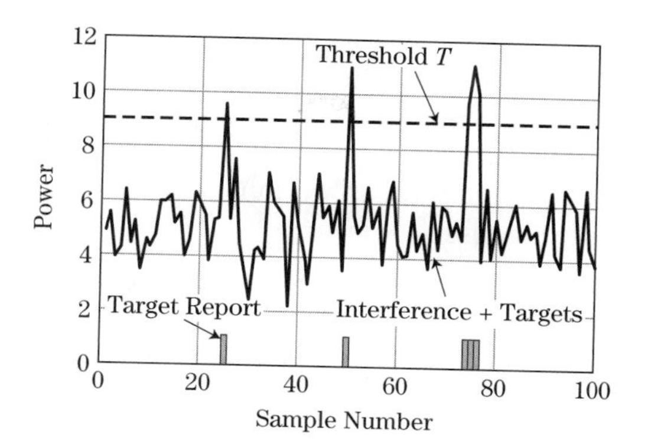

FIGURE 15-3 ■ The concept of threshold detection.

that a target was present.[2] The best procedure to use depends on the definition of "optimal" and the details of the random process models that describe the data.

Radar detection algorithms are usually designed according to the *Neyman-Pearson criterion*, a particular optimization strategy. This rule fixes the probability of false alarm, P_{FA}, that will be allowed by the detection processor and then maximizes the probability of detection, P_D, for a given SNR. Applying the Neyman-Pearson criterion to realistic radar detection problems leads to threshold detection using various detection statistics determined by the particular statistics of the data [1]. The threshold detection procedure is illustrated in Figure 15-3. The radar data shown could be received power versus range for a single pulse, or the Doppler power spectrum at a given range, or even a row of pixels in a SAR image. Whatever its source, a threshold value T is computed, and each data sample is compared with that threshold. If the sample is below the threshold, it is *assumed* to represent interference only. If it is above the threshold, it is similarly assumed to be too strong to represent interference only. In this case it must be interference plus a target echo, so a detection is declared at the range, velocity, or image location represented by that sample.

It is important to realize that these decisions can be wrong! A strong interference spike can cross the threshold, leading to a *false alarm*. Given a good model of the interference, the threshold can be selected to control the false alarm probability, P_{FA}. Similarly, a weak target echo might not add enough power to the interference to cause it to cross the threshold so that the target is not detected; this is called a *miss*, and its probability is $1 - P_D$.

The achievable combinations of P_D and P_{FA} are affected by the quality of the radar system and signal processor design. However, it will be seen that for a fixed system design, increasing P_D implies increasing P_{FA} as well. The radar system designer will generally decide what rate of false alarms can be tolerated based on the implications of acting on a false alarm, which may include overloading an operator monitoring a radar detection screen, using radar resources to start a track on a nonexistent target, or in extreme cases even firing a weapon!

Denote the number of detection decisions per unit time (usually 1 second) made by a particular radar as N_D. The *false alarm rate* (FAR), FAR, is the average number of false alarms per unit time. N_D, P_{FA}, and FAR are related according to

$$FAR = N_D P_{FA} \tag{15.3}$$

[2] In some detection problems, a third hypothesis is allowed: "don't know." Most radar systems, however, force a choice between "target present" and "target absent" on each detection test.

Since a radar may make tens or hundreds of thousands, even millions of detection decisions per second, values of P_{FA} must generally be quite low to maintain a tolerable FAR. Values in the range of 10^{-4} to 10^{-8} are common, and yet may still lead to false alarms every few seconds or minutes. Higher-level logic implemented in downstream data processing, for example in the tracking algorithms described in Chapter 19, is often used to reduce the number or impact of false alarms.

15.3.2 The Likelihood Ratio Test

It is shown in most detection texts that, given a particular data measurement y, the Neyman-Pearson criterion leads to the decision rule [2]

$$\frac{p_y(y|H_1)}{p_y(y|H_0)} \underset{H_0}{\overset{H_1}{\gtrless}} T_\Lambda \tag{15.4}$$

where T_Λ is an as yet unknown threshold value. Equation (15.4) is known as the *likelihood ratio test* (LRT). It states that the ratio of the two PDFs, each evaluated for the observed data y to produce a single numerical value (this ratio is called the *likelihood ratio* [LR]), should be compared with a threshold. If the likelihood ratio exceeds the threshold, choose H_1, that is, declare a target to be present. If it does not exceed the threshold, choose H_0 and declare that a target is not present. It will soon become clear that the LRT implicitly specifies the data processing operations to be carried out on the observed data y; what exactly those required operations are will depend on the particular PDFs. Example operations include taking the magnitude of the data or coherently or noncoherently integrating multiple samples.

The following notation is common shorthand for the LRT:

$$\Lambda(y) \underset{H_0}{\overset{H_1}{\gtrless}} T_\Lambda \tag{15.5}$$

where $\Lambda(y) = p_y(y|H_1)\,/\,p_y(y|H_0)$. Going a step further, because the decision depends only on whether the LRT exceeds the threshold, any monotone increasing operation can be performed on both sides of equation (15.5) without affecting which values of observed data y cause the threshold to be exceeded and therefore without affecting the performance (P_D and P_{FA}). Most common is to take the natural logarithm of both sides of equation (15.5) to obtain the *log-likelihood ratio test*:

$$\ln \Lambda(y) \underset{H_0}{\overset{H_1}{\gtrless}} \ln T_\Lambda \tag{15.6}$$

Some specific examples will be developed shortly to make clearer the use of the LRT and log-LRT.

The likelihood ratio Λ is the ratio of two random variables and so is also a random variable with its own probability density function, $p_\Lambda(\Lambda)$. If a specific model for $p_\Lambda(\Lambda)$ can be found (this is usually difficult), then P_{FA} can be expressed in terms of that PDF as

$$P_{FA} = \int_{T_\Lambda}^{+\infty} p_\Lambda(\Lambda)\,d\Lambda \tag{15.7}$$

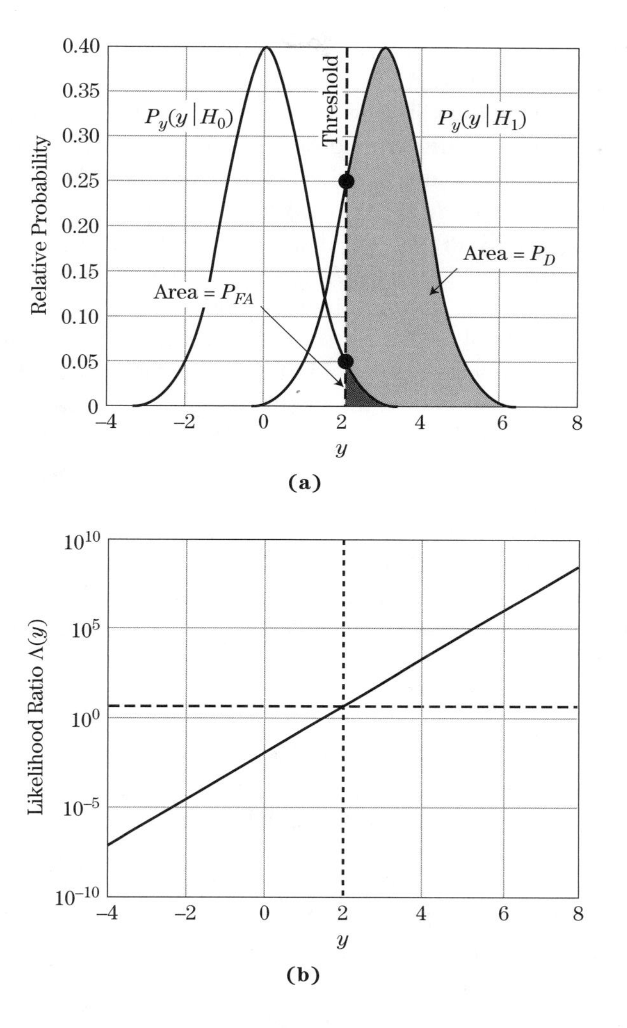

FIGURE 15-4 (a) Notional Gaussian probability density functions of the voltage y under H_0 (left) and H_1 (right). The black area represents P_{FA} and the gray + black area P_D. (b) Likelihood ratio for the PDFs of part (a). See text for discussion.

This equation can then be solved for T_Λ. More commonly, P_{FA} is found by determining a model for the PDF of the detection statistic z under H_0 (interference only) and then finding a threshold value T such that probability of y exceeding T is the desired P_{FA}:

$$P_{FA} = \int_{T}^{+\infty} p_z(z|H_0)dz \tag{15.8}$$

Figure 15-4 illustrates the relationship among the PDFs of the data, the likelihood ratio, and the thresholds T and T_Λ for a simple case. Suppose the noise is a single sample of a real-valued zero-mean Gaussian random process with variance σ_n^2, while the signal is simply a constant value of m.[3] Thus the target-plus-noise random process is Gaussian with a mean of m and a variance σ_n^2. The detection statistic is just the single data sample y. The PDFs $p_y(y|H_0)$ and $p_y(y|H_1)$ for the case $m = 3$ and $\sigma_n^2 = 1$ are shown in Figure 15-4a.

[3] These are not realistic models of the PDFs of radar data but are for illustration only. More realistic PDFs are considered in Section 15.4.

The signal-to-noise ratio is $\chi = m^2/\sigma_n^2 = 9$ (9.5 dB). A threshold T is shown as a vertical line at $y = 2$. P_D and P_{FA} are the areas under the right and left PDFs, respectively, from T to $+\infty$. T is found by adjusting the position of the threshold until the black area equals the acceptable false alarm probability. The detection probability is then the gray area (which includes the black area). In this example, $P_{FA} = 0.0228$ and $P_D = 0.841$.

The likelihood ratio $p_y(y|H_1)/p_y(y|H_0)$ for this example is shown in Figure 15-4b. The value of the noise-only PDF at the data threshold value, $p_y(2|H_0)$, is 0.054, while that of the target-plus-noise PDF, $p_y(2|H_1)$, is 0.252. These two values are indicated by the small circles on the dotted line in Figure 15-4a. The likelihood ratio at this point is therefore $\Lambda(2) = 0.252/0.04 = 4.48$, as shown by the intersection of the two dotted lines in Figure 15-4b. Thus, applying a threshold of $T = 2$ to the measured data y is equivalent to applying a threshold $T_\Lambda = 4.48$ to the likelihood ratio in this example.

Figure 15-4a makes it clear that P_D and P_{FA} both increase as the data threshold T moves left and decrease as it moves right. The achievable combinations of P_D and P_{FA} are determined by the degree to which the two distributions overlap. To achieve a high P_D and a low P_{FA} at the same time, the two PDFs must be well separated so that they overlap very little and the threshold value can be placed between them.

Figure 15-5 illustrates one form of the *receiver operating characteristic* (ROC) curve for this problem. A ROC curve is a plot of two of the three quantities P_D, P_{FA}, and SNR with the third as a parameter. Figure 15-5 plots P_D versus P_{FA} with the SNR χ as a parameter. For a given P_{FA}, P_D increases as SNR increases, a result which should also be intuitively satisfying.

If the achievable combinations of P_D and P_{FA} do not meet the performance specifications, what can be done? Consideration of Figure 15-4 suggests two answers. First, for a given P_{FA}, P_D can be increased by causing the two PDFs to move farther apart when a target is present. That is, the presence of a target must cause a larger shift in the mean, m, of the distribution of the detection statistic. Since the SNR is m^2/σ_n^2, this is equivalent to stating that one way to improve the detection/false alarm trade-off is to increase the SNR. The second way to improve the performance trade-off is to reduce the overlap of the PDFs by reducing their variance, that is, by reducing the noise power, σ_n^2. As with the first technique of increasing m, this is equivalent to increasing the signal-to-noise ratio. Thus, improving the trade-off between P_D and P_{FA} requires increasing the SNR χ. This is a fundamental result that will arise repeatedly.

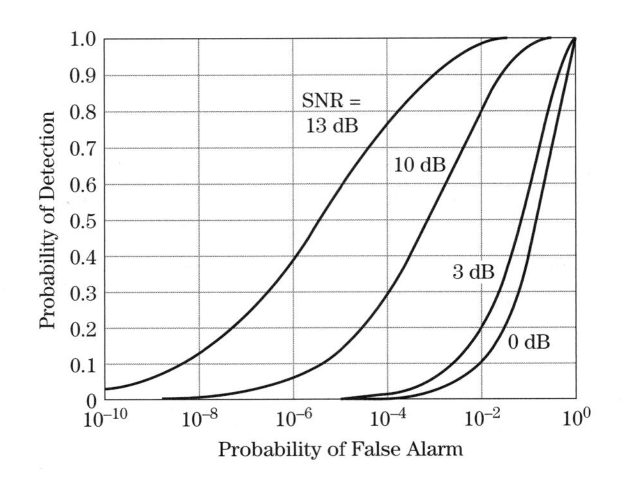

FIGURE 15-5 ■ Receiver operating characteristic for the Gaussian example for various values of the SNR.

Radar systems are designed to achieve specified values of P_D and P_{FA} subject to various conditions, such as specified ranges, target types, and interference environments. The designer can work with antenna design, transmitter power, waveform design, and signal processing techniques, all within cost and form factor constraints. The job of the designer is therefore to develop a radar system design that ultimately results in a pair of "target absent" and "target present" PDFs at the point of detection with a small enough overlap to allow the desired P_D and P_{FA} to be achieved. If the design does not do this, the designer must redesign one or more of the previously mentioned elements to reduce the variance of the PDFs, shift them farther apart, or both until the desired performance is obtained. Thus, a significant goal of radar system design is controlling the overlap of the noise-only and target-plus-noise PDFs, or equivalently, maximizing the SNR.

15.4 | STATISTICAL MODELS FOR NOISE AND TARGET RCS IN RADAR

The case of a real constant "target" in real Gaussian noise is not a realistic model for radar signals. In this section, more appropriate PDFs for describing radar interference and target statistics are reviewed. The concepts discussed in the preceding section can then be applied using these PDFs to estimate radar detection performance.

15.4.1 Statistical Model of Noise

The thermal noise at the output of each of the I and Q channels of a coherent radar receiver is typically modeled as a zero-mean Gaussian random process with variance $\sigma_n^2/2$. The I channel noise and the Q channel noise are assumed independent of each other. Under these circumstances, the complex noise signal $w[n] = I + jQ$ is a complex Gaussian noise process, often called a *circular* random process, with zero mean and a variance that is just the sum of the I and Q channel variances, namely, σ_n^2.

Define z as the magnitude of this noise (the signal at the output of a linear detector). It is a standard exercise in many random process textbooks to show that the PDF of z has a Rayleigh distribution [3],

$$p_z(z) = \begin{cases} \dfrac{2z}{\sigma_n^2} \exp\left(\dfrac{-z^2}{\sigma_n^2}\right), & z \geq 0 \\ 0, & z < 0 \end{cases} \quad \text{(linear detector)} \tag{15.9}$$

If z is instead the squared magnitude of the complex Gaussian noise (i.e., the signal at the output of a square law detector), the detected noise has an exponential PDF,

$$p_z(z) = \begin{cases} \dfrac{1}{\sigma_n^2} \exp\left(-\dfrac{z}{\sigma_n^2}\right), & z \geq 0 \\ 0, & z < 0 \end{cases} \quad \text{(square-law detector)} \tag{15.10}$$

For either detector type, the PDF of the noise phase θ is uniform,

$$p_\theta(\theta) = \begin{cases} \dfrac{1}{2\pi}, & -\pi \leq \theta < \pi \\ 0, & \text{otherwise} \end{cases} \tag{15.11}$$

15.4.2 Statistical Models of Radar Cross Section for Targets

A variety of probability density functions have been proposed for modeling target RCS fluctuations; the most common of these were discussed in Chapter 7. For each model, there is a corresponding PDF for the target reflectivity, which is proportional to the square root of the RCS. Received echo power is proportional to the target RCS, while echo voltage is proportional to the reflectivity. Tables 7-1 and 7-2 list some of the common PDFs used to model target RCS and reflectivity, respectively. These included the nonfluctuating, exponential, and fourth-degree chi-square. The choice of PDF to model the RCS fluctuations directly affects predicted detection performance, as will be seen in Section 15.5.

15.4.3 RCS Decorrelation Properties

Detection decisions are often based not on a single measurement but on a set of N measurements. For instance, the received complex voltage from a particular range bin might be measured on a series of N pulses. As suggested by the results in Section 15.3, combining all N measurements to improve the SNR before performing the threshold test will improve performance relative to using only one measurement.

When N target measurements are combined, the question arises as to whether their amplitudes or powers should be modeled as a single value selected from the target PDF repeated N times, N different values selected from the target PDF, or something in between. The first case is referred to as *scan-to-scan decorrelation* (because the target echoes do not decorrelate within a single "scan" or CPI), while the second is referred to as *pulse-to-pulse* decorrelation. In many situations, the actual decorrelation behavior may lie between these two extremes, but they make useful bounding cases for performance prediction.

The concept of pulse decorrelation can also be applied across CPIs as well as within them. Suppose the data within a CPI are coherently integrated and the magnitude of the result computed. This process can be repeated for each of N CPIs and the N results noncoherently integrated to form the final detection statistic. In this scenario, detection performance would depend on whether the target RCS decorrelated from CPI to CPI rather than from pulse to pulse.

RCS decorrelation is due to changes in radar-target aspect angle or radar frequency during the time interval over which the N data samples are collected. The choice of decorrelation model will be important in predicting detection performance in Section 15.5. As was shown in Chapter 7, the change in aspect angle required to decorrelate the echo amplitude of a complex target can be estimated as

$$\Delta\theta = \frac{c}{2L_w f} \quad \text{radians} \tag{15.12}$$

where L_w is the size of the target normal to the radar-target line of sight (LOS), that is, the width of the target as viewed from the radar.

It was also shown that successive target samples will be uncorrelated with one another if the radar frequency is changed. The frequency step required to decorrelate a complex target is approximately

$$\Delta f = \frac{c}{2L_d} \quad \text{Hz} \tag{15.13}$$

where L_d is the depth of the target projected along the radar boresight. Some systems deliberately change the radar frequency from pulse to pulse, a process called *frequency agility*, to ensure decorrelation of the target returns [4]. The reason for this will become apparent in Section 15.5.10. Frequency agility is not used within a CPI for the purpose of affecting the RCS decorrelation model if coherent integration is going to be applied to the CPI data.

As an example, consider a target the size of an automobile, about 2 m wide by 5 m long. At L-band (1 GHz), the target signature can be expected to decorrelate with between 30 and 75 mrad of aspect angle rotation (about 1.7° to 4.3°), depending on the target orientation. At W-band (95 GHz), this is reduced to only 0.316 to 0.79 mrad (0.018° to 0.0452°). The maximum frequency step required for decorrelation, which occurs when the target is viewed along its shortest axis, is 75 MHz. This value does not depend on the transmitted frequency.

The idea of target decorrelation (more generally, target fluctuations) is typically applied only in the context of noncoherent integration. Coherent integration usually occurs over relatively short time intervals, and the target is generally considered to exhibit constant RCS over the coherent integration time. Noncoherent integration often, though not always, occurs over longer time periods (e.g., multiple CPIs) that are more likely to result in observable RCS fluctuations. In some processing that requires long integration times, such as synthetic aperture imaging (see Chapter 21), target decorrelation can become a limiting factor.

15.4.4 Swerling Models

To model the performance of a radar detector, a model of the statistical behavior of the target RCS is required, including both a PDF and a decorrelation model. In combination with the noise model, the RCS model in turn determines the statistical model for the received voltage or power. As discussed in Chapter 7, *Swerling models* are one common set of target RCS models [5]. The Swerling models combine either the exponential or fourth-degree chi-square PDF with either the scan-to-scan or pulse-to-pulse decorrelation model. The exponential RCS model describes the behavior of a complex target consisting of many scatterers, none of which is dominant. The fourth-degree chi-square RCS model is an approximation to the *second-degree non-central chi square* (NCCS2) PDF that models targets having many scatterers of similar strength with one dominant scatterer. Table 15-1 defines the four cases. The terminology is sometimes stretched to include the nonfluctuating target case as the "Swerling 0" or "Swerling 5" model.

Choosing a Swerling model for analysis requires that a choice be made between the two PDFs and between the two decorrelation models. To choose the PDF, a judgment is needed as to whether the target RCS at the aspect angles of interest are likely to be dominated by one or two large scatterers or whether it is better described as the result of an ensemble of roughly equal scatterers. The choice of decorrelation model results from analyzing the geometry and motion of the radar and target over the CPI to determine whether the aspect

TABLE 15-1 ■ Swerling Models

Probability Density Function of RCS	Decorrelation	
	Scan-to-Scan	Pulse-to-Pulse
Exponential	Case 1	Case 2
Chi-square, degree 4	Case 3	Case 4

angle will change by an amount greater than $\Delta\theta$ during the CPI. If not, scan-to-scan decorrelation would be expected. If the aspect angle change between one target sample and the next is greater than $\Delta\theta$, then pulse-to-pulse decorrelation would be an appropriate model. If the time required to change the aspect angle by $\Delta\theta$ is greater than the time between successive samples, but less than the total CPI duration, the target decorrelation behavior will fall between the scan-to-scan and pulse-to-pulse limiting cases. In that case, calculations can be done for both cases and used to bound the range of detection results.

15.4.5 Extended Models of Target RCS Statistics

The concept of the Swerling models can easily be extended to other target models. For example, a model could be defined that uses a log-normal PDF for the target fluctuations with either pulse-to-pulse or scan-to-scan decorrelation to relate the N pulses in a block. Empirical observations have shown that such "long-tailed" distributions are often a better representation of observed radar data statistics than the traditional exponential and chi-square models, especially in high-resolution systems. Small-resolution cells are more likely to isolate one or a few scatterers, undermining the many-scatterer assumption of the traditional models and making large variations in observed RCS more likely. Detection probabilities for such models are sometimes difficult to develop, though a number of results are available in the literature (e.g., [6]). Other common models include the "expanded Swerling models" described in Chapter 7, which use the noncentral gamma density to make it possible to more closely match the density function to experimentally observed statistics for a wide range of systems while maintaining the (relatively) tractable computational results obtainable with target models based on chi-square and gamma densities [7,8].

15.4.6 Statistics of Targets in Interference and the Detection Statistic

The discussion of the aforementioned target models does not quite provide the information needed to carry out detection calculations. The data that will be subjected to the threshold detector will be the sum of the target echo and the interference, which is at least receiver noise and may also include clutter, jamming, and EMI. Thus, the probability density function of the target-plus-interference signal must be known. Furthermore, as shall be seen shortly, the optimum detector for N data samples will call for coherent or noncoherent integration of those samples followed by comparison of the resulting detection statistic to a threshold. Thus, the PDF of the sum of N outputs of the envelope detector when both target and interference are present will be needed.

15.5 THRESHOLD DETECTION OF RADAR SIGNALS

The models of the target and interference signals can now be used to develop equations for the performance of a radar detection scheme. Several basic choices distinguish common radar detection schemes:

- First, will detection be based on a single fast-time sample or multiple samples? In the latter case, are the data organized into one or several CPIs? Using multiple samples requires more radar resources to collect and process the data but obtains better results for a given single-sample SNR.

- What integration strategies will be used? There are three choices: coherent, noncoherent, and binary integration. Combinations of any two, or even all three, are also possible; four of these combinations were illustrated in Figure 15-2.
- If using noncoherent integration, what type of single-sample envelope detection law will be used? Common choices are linear and square law, as will be seen shortly; log detectors are also used.
- What target fluctuation model will be assumed? This requires specification of both a probability density function and a decorrelation model for the target samples. (Note that if detection is based on a single sample, decorrelation models are irrelevant.)
- Finally, will detection be based on a fixed threshold or on an adaptive threshold that responds to changes in the interference level? Adaptive threshold techniques are discussed in Chapter 16; here only fixed threshold detection will be considered.

In this section, the optimum detector under the Neyman-Pearson criterion for a nonfluctuating target, a single pulse, and a fixed threshold will first be presented. Next, the result for N samples of data with a predictable phase relationship will be developed, showing how coherent integration arises in detection. These results are then expanded to $N > 1$ samples with an unknown phase relationship for a nonfluctuating target, introducing noncoherent integration into the detection processing. Finally, the results are extended to include all four Swerling models of target RCS fluctuation.

15.5.1 Unknown Parameters

Realistic radar signals have unknown parameters. In most systems, the interference power is not known a priori. In this chapter it will be assumed that it is known, but Chapter 16 will show how to estimate it from the data and how that process affects detection performance. The range, amplitude, and Doppler shift of a target echo cannot be known before it has been detected. Even once detected, its exact phase will be unknown. Recall from Chapter 8 that the phase shift, θ, of the echo from a target at range R_0 is $-4\pi R_0/\lambda$ radians. Knowing the exact phase of the target echo implies knowing the range to the target very precisely, since a variation in one-way range of only $\lambda/4$ causes the received echo phase to change by $180°$. At microwave frequencies, this is typically only 15 to 30 cm (at L-band to ultra high frequency [UHF]) to a fraction of a centimeter (at millimeter wave frequencies). To complicate matters further, some parameters are linked. For example, the unknown echo amplitude varies with the unknown echo arrival time according to the appropriate version of the radar range equation. Thus, the LRT must be generalized to develop a technique that can work when some signal parameters are unknown.

15.5.2 The Optimum Detector for Nonfluctuating Radar Signals, One Sample

The PDFs used in Figure 15-4 were not realistic models of even the simplest radar detection scenario. Assume a coherent system, so that both I and Q channels are present and that the noise in each channel is independent and identically distributed (IID) zero-mean Gaussian noise with variance $\sigma_n^2/2$. The total noise power is then just the sum of the I and Q noise powers, which is σ_n^2.

Assuming a nonfluctuating target simply means that, when present, it adds a complex constant $m = \tilde{m}\exp(j\theta)$ to the noise sample y. "Nonfluctuating" means that the target

RCS is a constant so that the amplitude of the target component of the echo signal, $\tilde{m}$, is also constant. Because the target echo phase is extremely sensitive to small variations in range, the phase angle, θ, of the target component of the measured data is modeled as a random variable distributed uniformly over $(0,2\pi)$ and independent of the amplitude $\tilde{m}$.

To carry out the LRT, it is necessary to return to its basic definition of equation (15.4) and to determine $p_y(y|H_0)$ and $p_y(y|H_1)$ A detailed derivation of these PDFs is beyond the scope of this chapter; see [2] for details. The result for a single complex noise sample is

$$p_y(y|H_0) = \frac{1}{\pi\sigma_n^2} \exp\left[-|y|^2/\sigma_n^2\right] \tag{15.14}$$

As expected, the PDF depends only on $|y|$. Since the target echo is not present under H_0, neither the target amplitude nor phase appears in equation (15.14).

For the nonfluctuating target-plus-noise case, $p_y(y|H_1)$ must be determined, considering the effect of the random target phase, θ. The *Bayesian approach* for random parameters with known PDFs is applied to "average out" the θ dependence, resulting in a PDF that depend only on the magnitude of the data. The result is the Rician PDF [2]

$$p_y(y|H_1) = \frac{1}{\pi\sigma_n^2} \exp\left[-\frac{1}{\sigma_n^2}\left(|y|^2 + \tilde{m}^2\right)\right] I_0\left(\frac{2\tilde{m}|y|}{\sigma_n^2}\right) \tag{15.15}$$

where $I_0(\cdot)$ is the modified Bessel function of the first kind. Note that $p_y(y|H_1)$ uses only the magnitude $\tilde{m}$ of the complex signal sample but not the random phase of the target echo due to the averaging over θ.

The log-likelihood ratio is convenient in this case. It is straightforward to show that the log-LRT becomes

$$\ln\Lambda = \ln\left[\frac{p(y|H_1)}{p(y|H_0)}\right] = \ln\left[I_0\left(\frac{2\tilde{m}|y|}{\sigma_n^2}\right)\right] - \frac{\tilde{m}^2}{\sigma_n^2} \underset{H_0}{\overset{H_1}{\gtrless}} \ln T_\Lambda \tag{15.16}$$

Moving the term $\tilde{m}^2/\sigma_n^2$ to the right-hand side isolates the detection statistic on the left-hand side of the equation and gives a threshold test using a modified threshold T',

$$\ln\left[I_0\left(\frac{2\tilde{m}|y|}{\sigma_n^2}\right)\right] \underset{H_0}{\overset{H_1}{\gtrless}} \ln T_\Lambda + \frac{\tilde{m}^2}{\sigma_n^2} \equiv T' \tag{15.17}$$

Equation (15.17) defines the signal processing required for optimum detection in the presence of an unknown phase using a single complex data sample. It calls for taking the magnitude of the data sample, scaling it and passing it through the memoryless nonlinearity $\ln[I_0(\cdot)]$ to get the detection statistic, and comparing the result with a threshold.

As discussed previously, the phase is not the only unknown parameter of the target echo signal in practice. In this example, the amplitude $\tilde{m}$ of the echo depends on all of the factors in the radar range equation, including especially the unknown target radar cross section and, at least until it is successfully detected, the target's range. In addition, the target may be moving relative to the radar so that the echo is modified by a Doppler shift. It is straightforward to show that accounting for unknown target echo amplitude neither requires any change in the detector structure nor changes its performance [2]. The same is true of a Doppler shift in the single-sample case.

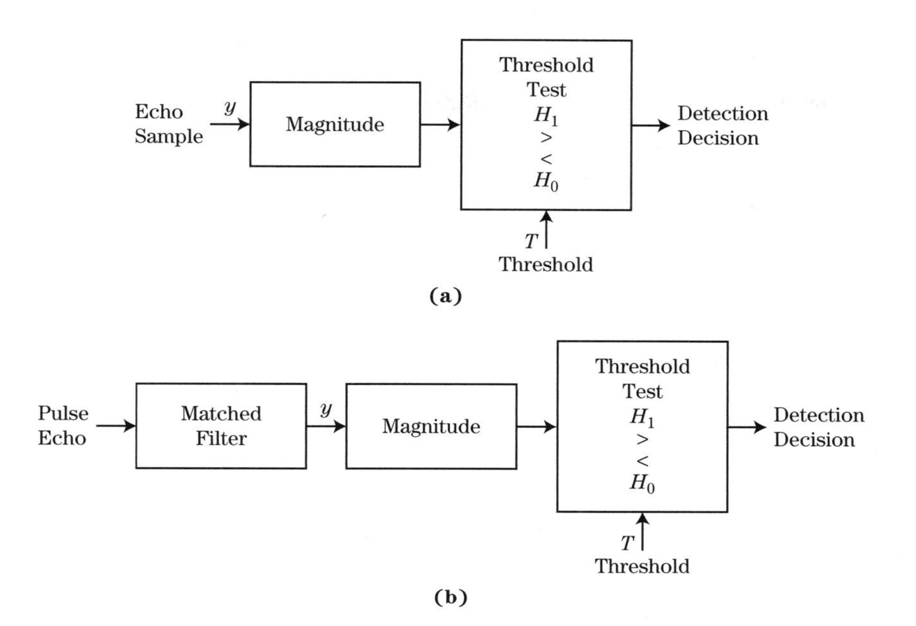

FIGURE 15-6 ■ Structure of optimal detector when the absolute signal phase is unknown.

15.5.3 Performance for the Nonfluctuating Signal in Gaussian Noise, $N = 1$

As a practical matter, it is desirable to avoid having to compute the natural logarithm and Bessel function in equation (15.17) for every threshold test, since these might occur millions of times per second in some systems. Because the function $\ln[I_0(\cdot)]$ is monotonic increasing, the same detection results can be obtained by simply comparing its argument $x = 2\tilde{m}|y|/\sigma_n^2$ with a modified threshold. After some further rearrangement, equation (15.17) then becomes simply

$$|y| \underset{H_0}{\overset{H_1}{\gtrless}} T \tag{15.18}$$

A detector that bases its decision on $|y|$ is called a *linear detector*, as opposed to a *square law detector* that would use $|y|^2$. Figure 15-6 illustrates the resulting optimal detector for the coherent receiver with an unknown phase. The matched filter maximizes the SNR of the complex data sample y; the linear detector and threshold test implement the LRT.

The performance of this detector will now be established. Let $z = |y|$. The detection test becomes simply $z \gtrless T$; thus, the distribution of z under each of the two hypotheses is needed. Under H_0 (target absent), the real and imaginary parts of y are IID Gaussian random processes with zero mean and variance $\sigma_n^2/2$, denoted as $\mathrm{N}\left(0, \sigma_n^2/2\right)$.[4] It follows from Section 2.2.6 of [2] that z is Rayleigh distributed:

$$p_z(z|H_0) = \begin{cases} \dfrac{2z}{\sigma_n^2} \exp\left(-\dfrac{z^2}{\sigma_n^2}\right), & z \geq 0 \\ 0, & z < 0 \end{cases} \tag{15.19}$$

[4]The notation $\mathrm{N}(u, v)$ denotes a normal (Gaussian) distribution of mean u and variance v.

The probability of false alarm is therefore

$$P_{FA} = \int_{T}^{+\infty} p_z(z|H_0)dz = \exp\left(-\frac{T^2}{\sigma_n^2}\right) \tag{15.20}$$

This equation can be inverted to obtain the threshold setting in terms of P_{FA}:

$$T = \sigma_n \sqrt{-\ln P_{FA}} \tag{15.21}$$

Equation (15.21) gives a rule for setting the threshold at the output of a linear detector to achieve the specified P_{FA}, assuming the noise power at the detector input is known.

Now consider H_1, (i.e., target present). The real part of y is distributed as $N(\tilde{m}\cos\theta, \sigma_n^2/2)$, while the imaginary part is distributed as $N(\tilde{m}\sin\theta, \sigma_n^2/2)$. It is shown in Section 2.2.7 of [2] that the PDF of z is

$$p_z(z|H_1) = \begin{cases} \dfrac{2z}{\sigma_n^2}\exp\left[-\dfrac{1}{\sigma_n^2}\left(z^2+\tilde{m}^2\right)\right] I_0\left(\dfrac{2\tilde{m}^2 z}{\sigma_n^2}\right), & z \geq 0 \\ 0, & z < 0 \end{cases} \tag{15.22}$$

Equation (15.22) is the Rician PDF. The probability of detection is obtained by integrating it from T to $+\infty$. It is convenient to make the substitutions $t = z\big/\sqrt{\sigma_n^2/2}$ and $\alpha = \sqrt{2\tilde{m}^2/\sigma_n^2}$ to put the integral in the more standard form

$$Q_M(\alpha, t) = \int_{T}^{+\infty} t\exp\left[-\frac{1}{2}(t^2+\alpha^2)\right] I_0(\alpha t)dt \tag{15.23}$$

The expression $Q_M(\alpha, t)$ is known as *Marcum's Q function*. It arises frequently in radar detection calculations. A closed form for this integral is not known. An example of a MATLAB program for evaluating it iteratively is given in [2]. A comparison of several numerical algorithms for its computation is given in [9].

In terms of Marcum's Q function, the probability of detection is

$$P_D = Q_M\left(\sqrt{\frac{2\tilde{m}^2}{\sigma_n^2}}, \sqrt{\frac{2T^2}{\sigma_n^2}}\right) \tag{15.24}$$

Finally, noting that $\tilde{m}^2/\sigma_n^2$ is the signal-to-noise ratio χ and expressing the threshold in terms of the false alarm probability using (15.21) gives

$$P_D = Q_M\left(\sqrt{2\chi}, \sqrt{-2\ln P_{FA}}\right) \tag{15.25}$$

The performance of this detector for a nonfluctuating target in Gaussian noise is given in Figure 15-7. The general behavior is very similar to the real-valued constant-in-Gaussian-noise case of Figure 15-5.

A more interesting comparison is to the optimum coherent detector that would have been obtained if the target echo phase was known exactly. Although the equations are not derived here (see [1]), the resulting ROC is shown in Figure 15-8a. Again, the general shape of the performance curves is very similar to that of the envelope detector in Figure 15-7. Close inspection, however, shows that for a given SNR χ and P_{FA}, the coherent detector

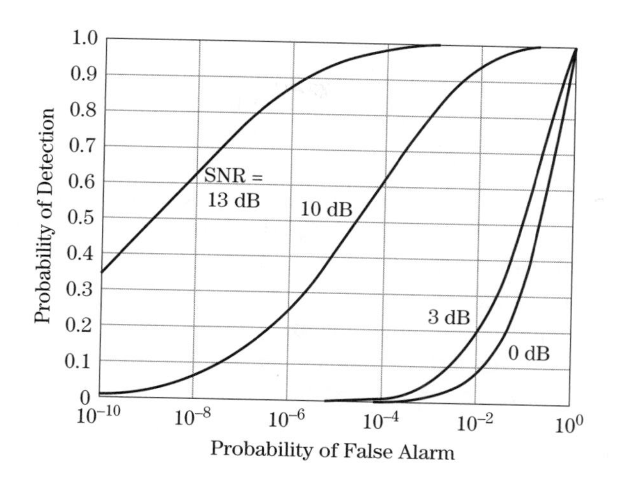

FIGURE 15-7 ■ Performance of the linear envelope detector for the Gaussian example with unknown phase.

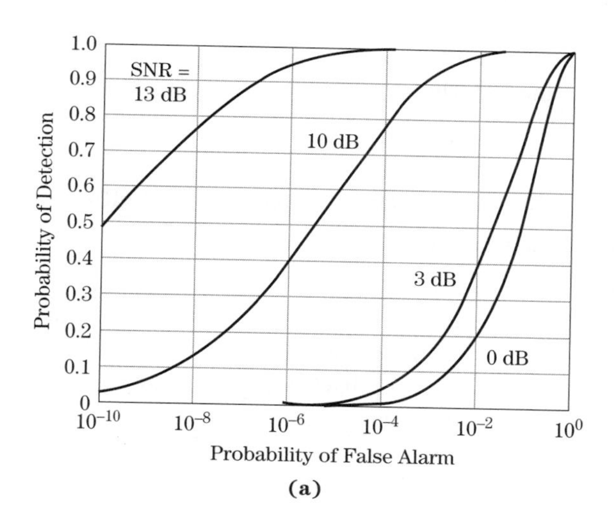

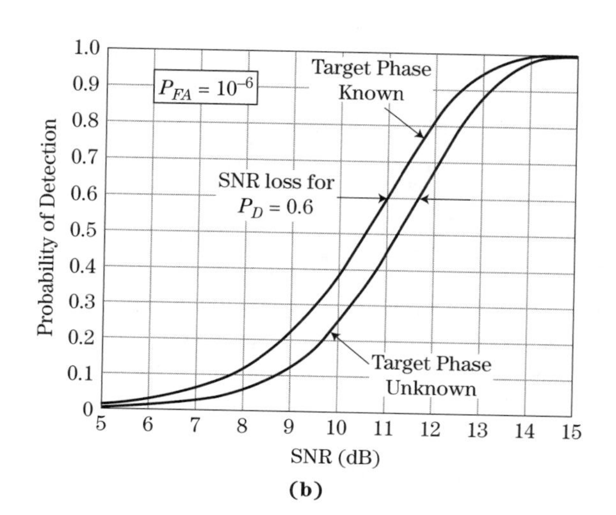

FIGURE 15-8 ■ (a) ROC for a coherent detector with known target phase (compare to Figure 15-7). (b) Detector loss due to unknown target phase.

obtains a higher P_D. This occurs because the coherent detector uses more information about the signal, namely, its phase.

To make this point clearer, Figure 15-8b plots the detection performance as a function of χ with P_{FA} fixed (at 10^{-6} in this example) for both the known and unknown phase cases (coherent and linear detectors). This figure shows that, to achieve the same probability of detection, the linear envelope detector requires about 0.6 dB higher SNR than the coherent detector at $P_D = 0.9$ and about 0.7 dB more at $P_D = 0.5$. The extra signal-to-noise required to maintain the detection performance of the envelope detector compared with the coherent case is called an *SNR loss*. SNR losses can result from many factors; this particular one is often called the *detector loss*. It represents extra SNR that must be obtained in some way if the performance of the envelope detector is to match that of the ideal coherent detector. Increasing the SNR in turn implies one or more of many radar system changes, such as greater transmitter power, a larger antenna gain, and reduced range coverage.

The phenomenon of detector loss illustrates a very important point in detection theory: the less is known about the signal to be detected, the higher must be the SNR to detect with a given combination of P_D and P_{FA}. In this case, not knowing the absolute phase of the signal has cost about 0.6 dB. Inconvenient though it may be, this result is intuitively satisfying: the worse the knowledge of the signal details, the worse the performance of the detector should be.

15.5.4 Optimum Detector for Nonfluctuating Radar Signals with Coherent Integration

In many cases the raw data contain more than one sample of data from the same target, and those data have a distinct phase relationship. For example, suppose the available data are N slow-time samples from a range bin containing a target moving at constant radial velocity with respect to the radar. As discussed in Chapter 8, a model for the slow-time target-only data $s[m]$ for the range bin of interest is

$$s[m] = Ae^{j(2\pi f_d mT+\theta)} = Ae^{j\theta}e^{j2\pi f_d mT}, \quad 0 \le m \le N-1 \tag{15.26}$$

where θ is the phase due to the target range on the first sample. These N samples of data can be compactly expressed as an $N \times 1$ column vector $\boldsymbol{S}$ of the form

$$\begin{aligned} \boldsymbol{S} &= Ae^{j\theta}\left[1 \; e^{j2\pi f_d T} \; \cdots \; e^{j2\pi f_d (N-1)T}\right]^T \\ &= Ae^{j\theta}\tilde{\boldsymbol{M}} \end{aligned} \tag{15.27}$$

where $\tilde{\boldsymbol{M}}$ is a vector representing the phase progression in the target signal due to the Doppler shift. Thus, to within a complex constant, $\tilde{\boldsymbol{M}}$ is a model of the expected signal structure for a target with a Doppler shift of f_d Hz. This vector is called a *steering vector* in Doppler. The total slow-time data $y[m]$ is $s[m]$ plus complex Gaussian noise $w[m]$:

$$y[m] = s[m] + w[m] \tag{15.28}$$

The SNR χ_1 of a single sample of $y[m]$ is A^2/σ_n^2, where σ_n^2 is the variance of the noise.

Again assuming the complex Gaussian noise case, the slow-time data vector $\boldsymbol{Y}$ is N independent samples of complex Gaussian noise under H_0. Their joint PDF is [2]

$$p_{\boldsymbol{Y}}(\boldsymbol{Y}|H_0) = \frac{1}{\pi^N \sigma_n^{2N}} \exp\left[-\frac{1}{\sigma_n^2}\boldsymbol{Y}^H\boldsymbol{Y}\right] \tag{15.29}$$

For the target-plus-noise case, the data are samples of complex Gaussian noise plus the target echo. Collecting the noise samples into a vector $\boldsymbol{W}$, the joint PDF of the data $\boldsymbol{Y} = \boldsymbol{S} + \boldsymbol{W}$ becomes

$$p_{\boldsymbol{Y}}(\boldsymbol{Y}|H_1, \theta) = \frac{1}{\pi^N \sigma_n^{2N}} \exp\left[-\frac{1}{\sigma_n^2}\left(\boldsymbol{Y} - Ae^{j\theta}\tilde{\boldsymbol{M}}\right)^H\left(\boldsymbol{Y} - Ae^{j\theta}\tilde{\boldsymbol{M}}\right)\right] \tag{15.30}$$

Expanding the exponent in (15.30) and defining the integrated signal energy $E = N \cdot A^2$ gives

$$\begin{aligned} p_{\boldsymbol{Y}}(\boldsymbol{Y}|H_1, \theta) &= \frac{1}{\pi^N \sigma_n^{2N}} \exp\left[-\frac{1}{\sigma_n^2}\left(\boldsymbol{Y}^H\boldsymbol{Y} - 2\mathrm{Re}\left\{Ae^{-j\theta}\tilde{\boldsymbol{M}}^H\boldsymbol{Y}\right\} + E\right)\right] \\ &= \frac{1}{\pi^N \sigma_n^{2N}} \exp\left[-\frac{1}{\sigma_n^2}\left(\boldsymbol{Y}^H\boldsymbol{Y} - 2A\left|\tilde{\boldsymbol{M}}^H\boldsymbol{Y}\right|\cos(\phi - \theta) + E\right)\right] \end{aligned} \tag{15.31}$$

where ϕ is the unknown, but fixed, phase of the inner product $\tilde{\boldsymbol{M}}^H\boldsymbol{Y}$ and the fact that $\tilde{\boldsymbol{M}}^H\tilde{\boldsymbol{M}} = N$ has been used.

It is important to note that the inner product $\tilde{\boldsymbol{M}}^H\boldsymbol{Y}$ represents *matched filtering* of the data samples $\boldsymbol{Y}$ with the target model represented by the steering vector $\tilde{\boldsymbol{M}}$. Writing this sum out explicitly and using equation (15.27) gives

$$\tilde{\boldsymbol{M}}^H\boldsymbol{Y} = \sum_{m=0}^{N-1} y[m]\,\tilde{M}^*[m] = \sum_{m=0}^{N-1} y[m]\, e^{-j2\pi f_d mT} \tag{15.32}$$

Equation (15.32) is the discrete-time Fourier transform (DTFT) of the slow-time data $y[m]$. If the value of f_d assumed in the steering vector matches the actual value of the Doppler shift present in the data, then the operation $\tilde{\boldsymbol{M}}^H\boldsymbol{Y}$ implements coherent integration of the data. Specifically, if the data are of the form $y[m] = s[m] + w[m]$, where $s[m]$ is of the form given in (15.26) and $w[m]$ is complex Gaussian noise, the result of (15.32) becomes

$$\begin{aligned} \tilde{\boldsymbol{M}}^H\boldsymbol{Y} &= \sum_{m=0}^{N-1}\left(Ae^{j\theta}e^{+j2\pi f_d mT} + w[m]\right) e^{-j2\pi f_d mT} \\ &= NAe^{j\theta} + \sum_{m=0}^{N-1} w[m]\, e^{-j2\pi f_d mT} \\ &\equiv NAe^{j\theta} + w_N \end{aligned} \tag{15.33}$$

where w_N is a weighted sum of N complex Gaussian noise samples. The SNR, χ_N, of the matched filter output $\tilde{\boldsymbol{M}}^H\boldsymbol{Y}$ is increased by a factor of N relative to the single-sample case, to $NA^2/\sigma_n^2 = N\chi_1$. Thus, when the target component of the data can be modeled as having a known relative phase from one sample to the next, the LRT dictates coherent integration of that data!

Notice that $p_{\boldsymbol{Y}}(\boldsymbol{Y}|H_1,\theta)$ does indeed display an explicit dependence on θ. This dependence is removed by averaging over the PDF of the phase:

$$p_{\boldsymbol{Y}}(\boldsymbol{Y}|H_1) = \int p_y(\boldsymbol{Y}|H_1,\theta)p_\theta(\theta)d\theta. \tag{15.34}$$

Assuming a uniform random PDF for θ, defining $\theta' = \phi - \theta$, and applying (15.30) to (15.34) gives, after minor rearrangement,

$$p_{\boldsymbol{Y}}(\boldsymbol{Y}|H_1) = \frac{1}{\pi^N\sigma_n^{2N}}e^{-(\boldsymbol{Y}^H\boldsymbol{Y}+E)/\sigma_n^2}\frac{1}{2\pi}\int_0^{2\pi}\exp\left[\frac{2}{\sigma_n^2}\left|\tilde{\boldsymbol{M}}^H\boldsymbol{Y}\right|\cos\theta'\right]d\theta' \tag{15.35}$$

Equation (15.35) is a standard integral. Specifically, integral 9.6.16 in [10] is

$$\frac{1}{\pi}\int_0^{\pi} e^{\pm z\cos\theta}d\theta = I_0(z) \tag{15.36}$$

where $I_0(z)$ is the modified Bessel function of the first kind. Using this result and properties of the cosine function, equation (15.34) becomes

$$p_{\boldsymbol{Y}}(\boldsymbol{Y}|H_1) = \frac{1}{\pi^N\sigma_n^{2N}}e^{-(\boldsymbol{Y}^H\boldsymbol{Y}+E)/\sigma_n^2}\,I_0\left(\frac{2\left|\tilde{\boldsymbol{M}}^H\boldsymbol{Y}\right|}{\sigma_n^2}\right) \tag{15.37}$$

The log LRT now becomes

$$\ln\Lambda = \ln\left[I_0\left(\frac{2\left|\tilde{\boldsymbol{M}}^H\boldsymbol{Y}\right|}{\sigma_n^2}\right)\right] - \frac{E}{\sigma_n^2} \underset{H_0}{\overset{H_1}{\gtrless}} \ln T_\Lambda \tag{15.38}$$

or, equivalently,

$$\ln\left[I_0\left(\frac{2\left|\tilde{\boldsymbol{M}}^H\boldsymbol{Y}\right|}{\sigma_n^2}\right)\right] \underset{H_0}{\overset{H_1}{\gtrless}} \ln T_\Lambda + \frac{E}{\sigma_n^2} \equiv T' \tag{15.39}$$

Equation (15.39) defines the signal processing required for optimum detection in the presence of an unknown phase. It calls for taking the magnitude of the matched filter output $\tilde{\boldsymbol{M}}^H\boldsymbol{Y}$, passing it through the memoryless nonlinearity $\ln[I_0()]$, and comparing the result with a threshold. This result is appealing in that the matched filter is still applied to utilize the *internal* phase structure of the known signal and get the maximum integration gain, but then a magnitude operation is applied because the absolute phase of the result can't be known. Again, the argument of the Bessel function is a signal-to-noise ratio.

As before, the same detection results can be obtained by simply comparing its argument $2\left|\tilde{\boldsymbol{M}}^H\boldsymbol{Y}\right|/\sigma_n^2$ with a modified threshold. Equation (15.39) then becomes simply

$$\left|\tilde{\boldsymbol{M}}^H\boldsymbol{Y}\right| \underset{H_0}{\overset{H_1}{\gtrless}} T \tag{15.40}$$

Figure 15-9 illustrates the optimal detector for the coherent detector with an unknown phase.

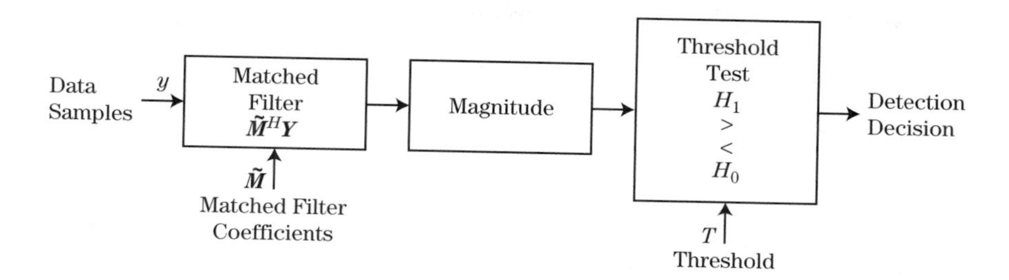

FIGURE 15-9 ■ Structure of optimal coherent detector.

Observe that equations (15.37) through (15.40) are essentially identical to equations (15.15) through (15.17) for the single-sample case but with $\left|\tilde{M}^H Y\right|$ substituted for $\tilde{m}|y|$ (and the scalar $\tilde{m}$ absorbed into the threshold in the single-sample case). This means that the performance of the optimal coherent detector will be the same as that of the optimal single-sample detector but with an SNR increased by the number of coherently integrated samples N. To confirm this, start by defining the detection statistic $z = |\tilde{M}^H Y|$. The detection test becomes simply $z \gtrless T$; thus the distribution of z under each of the two hypotheses is needed. As in the known phase case, under H_0 (target absent) $\tilde{M}^H Y \sim \mathrm{N}(0, E\sigma_n^2)$;[5] thus, the real and imaginary parts of $\tilde{M}^H Y$ are independent of one another and each is distributed as $\mathrm{N}(0, E\sigma_n^2/2)$. The detection statistic z is Rayleigh distributed:

$$p_z(z|H_0) = \begin{cases} \dfrac{2z}{E\sigma_n^2} \exp\left(-\dfrac{z^2}{E\sigma_n^2}\right), & z \geq 0 \\ 0, & z < 0 \end{cases} \tag{15.41}$$

The probability of false alarm is

$$P_{FA} = \int_T^{+\infty} p_z(z|H_0)dz = \exp\left(-\frac{T^2}{E\sigma_n^2}\right) \tag{15.42}$$

The threshold setting in terms of P_{FA} is

$$T = \sqrt{-E\sigma_n^2 \ln P_{FA}} \tag{15.43}$$

Now consider H_1, (i.e., target present). In this case $\tilde{M}^H Y \sim \mathrm{N}(E, E\sigma_n^2)$. Since E is real-valued, the real part $\tilde{M}^H Y$ is distributed as $\mathrm{N}(E, E\sigma_n^2/2)$ while the imaginary part is distributed as $\mathrm{N}(0, E\sigma_n^2/2)$. The PDF of z is

$$p_z(z|H_1) = \begin{cases} \dfrac{2z}{E\sigma_n^2} \exp\left[-\dfrac{1}{E\sigma_n^2}\left(z^2 + E^2\right)\right] I_0\left(\dfrac{2z}{\sigma_n^2}\right), & z \geq 0 \\ 0, & z < 0 \end{cases} \tag{15.44}$$

[5] The notation $x \sim \mathrm{N}(\mu, \sigma^2)$ means that a random variable x is distributed normally (Gaussian) with mean μ and variance σ^2.

The probability of detection is obtained by integrating this Rician PDF from T to $+\infty$. The result can be expressed in terms of Marcum's Q function as

$$P_D = Q_M\left(\sqrt{\frac{2E}{\sigma_n^2}}, \sqrt{\frac{2T^2}{E\sigma_n^2}}\right) \tag{15.45}$$

Finally, noting that $E/\sigma_n^2 = NA^2/\sigma_n^2$ is the signal-to-noise ratio χ_N and expressing the threshold in terms of the false alarm probability using (15.20) gives

$$P_D = Q_M\left(\sqrt{2\chi_N}, \sqrt{-2\ln P_{FA}}\right) \tag{15.46}$$

Equation (15.46) is identical to equation (15.25), but with the single-sample SNR χ replaced by χ_N. Thus, as expected the performance is identical to that shown in Figure 15-7, provided the SNR is interpreted as the integrated SNR χ_N. This gives the important conclusion that single-sample detection performance results can be applied to the coherent integration case by simply using the coherently integrated SNR.

It is usually the case that the energy E in $\boldsymbol{S}$ or $\tilde{\boldsymbol{M}}$ and the noise power σ_n^2 are not known. Fortunately, equation (15.46) does not depend on them individually but only on their ratio χ_N so that it is possible to generate the ROC without this information. However, actually implementing the detector requires a specific value of the threshold T as given in equation (15.43), and this does require knowledge of both E and σ_n^2. One way to avoid this problem is to replace the matched filter coefficients $\tilde{\boldsymbol{M}}$ with a normalized coefficient vector $\hat{\boldsymbol{M}} = \tilde{\boldsymbol{M}}/E_{\tilde{\mathbf{m}}}$, where $E_{\tilde{\mathbf{m}}}$ is the energy in $\tilde{\boldsymbol{M}}$ (N in this example). This choice simply normalizes the gain of the matched filter to 1. The energy in this modified sequence $\hat{\boldsymbol{M}}$ is $\hat{E} = 1$, leading to a modified threshold

$$\hat{T} = \sqrt{-\sigma_n^2 \ln P_{FA}} \tag{15.47}$$

The reduced matched filter gain, along with the reduced threshold, results in no change to the ROC, so equation (15.46) remains valid. Setting of the threshold $\hat{T}$ still requires knowledge of the noise power, σ_n^2; removal of this restriction is the subject of Chapter 16.

15.5.5 Optimum Detector for a Nonfluctuating Target with Noncoherent Integration

Now consider detection based on noncoherent integration of N samples of a nonfluctuating target in white Gaussian noise. This is called the "Swerling 0," "Swerling 5," or "Marcum" case. Recall that "nonfluctuating" refers to the amplitude of the samples but does not imply that their phases have any known relationship. This model arises in several different types of radar detection problems. In the first, the N samples could be slow-time samples in a single CPI, as suggested by Figure 15-2d. Alternatively, they could be a single sample from each of N scans of a target region. The SNR of a single sample is denoted χ_1. In the second situation, some number of data samples are coherently integrated in each of N CPIs, and then the coherently integrated result from each of the CPIs are combined for detection without assuming any known phase relationship between the data from different CPIs. For example, pulse compression and Doppler DFTs could be applied to the data matrix for each CPI, and the N samples from a given range-Doppler cell could then be used to test for the presence of a target at that range and Doppler shift. This is the case illustrated in Figure 15-2b. Another variation would be to coherently integrate in fast time only (pulse compression) and then to combine the results for a given range bin across slow time without assuming any particular slow-time phase relationship. This is the case illustrated for one of the two CPIs before threshold testing in Figure 15-2c. In the cases that include coherent integration, the SNR of the samples to be combined by noncoherent integration is $\chi_N = N\chi_1$.

Now consider detection based on the N available samples. The amplitude and absolute phase of the target component are unknown. Thus, an individual data sample y_n is again a complex constant $s = \tilde{m}\exp(j\theta)$ for some real amplitude $\tilde{m}$ and phase θ plus a white Gaussian noise sample of power $\sigma_n^2/2$ in each of the I and Q channels (total noise power σ_n^2). The PDFs of $z_n = |y_n|$ under H_0 (no target) and H_1 (target present) are again the Rayleigh and Rician densities of equations (15.19) and (15.22), respectively, with z replaced by z_n. For a vector $\mathbf{Z}$ of N such samples, the joint PDFs are, for each $z_n \geq 0$,

$$p_Z(\mathbf{Z}|H_0) = \prod_{n=1}^{N} \frac{2z_n}{\sigma_n^2}\exp\left(-z_n^2/\sigma_n^2\right), \text{ and} \tag{15.48}$$

$$p_Z(\mathbf{Z}|H_0) = \prod_{n=1}^{N} \frac{2z_n}{\sigma_n^2}\exp\left[-\left(z_n^2+\tilde{m}^2\right)/\sigma_n^2\right] I_0\left(\frac{2\tilde{m}z_n}{\sigma_n^2}\right) \tag{15.49}$$

The LRT and log-LRT become

$$\begin{aligned}\Lambda &= \prod_{n=1}^{N} \exp\left(-\tilde{m}^2/\sigma_n^2\right) I_0\left(\frac{2\tilde{m}z_n}{\sigma_n^2}\right) \\ &= \exp\left(-\tilde{m}^2/\sigma_n^2\right)\prod_{n=1}^{N} I_0\left(\frac{2\tilde{m}z_n}{\sigma_n^2}\right) \underset{H_0}{\overset{H_1}{\gtrless}} T_\Lambda \end{aligned} \tag{15.50}$$

$$\ln\Lambda = -\frac{\tilde{m}^2}{\sigma_n^2} + \sum_{n=1}^{N}\ln\left[I_0\left(\frac{2\tilde{m}z_n}{\sigma_n^2}\right)\right] \underset{H_0}{\overset{H_1}{\gtrless}} \ln T_\Lambda \tag{15.51}$$

Incorporating the term involving the ratio of signal power and noise power on the left-hand side into the threshold gives

$$\sum_{n=1}^{N}\ln\left[I_0\left(\frac{2\tilde{m}z_n}{\sigma_n^2}\right)\right] \underset{H_0}{\overset{H_1}{\gtrless}} \ln\left(T_\Lambda\right) + \frac{\tilde{m}^2}{\sigma_n^2} \equiv T' \tag{15.52}$$

Equation (15.52) shows that, given N noncoherent samples of a nonfluctuating target in white noise, the optimal Neyman-Pearson detection test scales each sample by the quantity $2\tilde{m}/\sigma_n^2$, passes it through the monotonic nonlinearity $\ln[I_0(\cdot)]$, and then noncoherently integrates the processed samples and performs a threshold test. This result is very similar to the $N = 1$ case of equation (15.17). The integration is considered noncoherent because only the magnitude of the original data is being used; the phase information was discarded by the envelope detector.

There are three practical problems with this equation. First, as noted earlier, it is desirable to avoid computing the function $\ln[I_0(\cdot)]$ possibly millions of times per second. Second, both the target amplitude $\tilde{m}$ and the noise power σ_n^2 must be known to perform the required scaling. Third, performance analysis of the resulting detector would require solving the difficult problem of finding the PDF of the sum of N samples of the transformed data $\ln\left[I_0\left(2\tilde{m}z_n/\sigma_n^2\right)\right]$, where z_n is the magnitude of a sample of either complex Gaussian noise, or complex Gaussian noise plus a constant target signal.

15.5.6 Linear and Square Law Detectors

It was noted in Section 15.5.3 that the $\ln[I_0(x)]$ function could be replaced by its argument x without altering the performance in the case of detection using a single sample ($N = 1$). When integrating $N > 1$ samples, a simpler detector characteristic is again desirable to simplify computation. While applying a monotonic increasing transformation to the right-hand side and to the entire summation on the left-hand side of (15.52) would not change the outcome of the comparison, this is not true in general if the transformation is applied instead to the individual $\ln[I_0(\cdot)]$ terms before the summation. However, it is possible to find a simpler alternate detector law by seeing what approximations can be made to the $\ln[I_0(\cdot)]$ function.

A standard series expansion for the Bessel function is $I_0(x) = 1 + x^2/4 + x^4/64 + \cdots$. Thus for small x, $I_0(x) \approx 1 + x^2/4$. Furthermore, one series expansion of the natural logarithm is $\ln(1+z) = z - z^2/2 + z^3/3 + \cdots$. Combining these gives

$$\ln[I_0(x)] \approx \frac{x^2}{4}, \qquad x \ll 1 \tag{15.53}$$

Equation (15.53) shows that if x is small, the optimal detector is well approximated by a square law detector.

For large values of x, $I_0(x) \approx e^x \big/ \sqrt{2\pi x}$, $x \gg 1$; then

$$\ln[I_0(x)] \approx x - \frac{1}{2}\ln(2\pi) - \frac{1}{2}\ln(x) \tag{15.54}$$

The constant term on the right of (15.54) can be incorporated into the threshold in equation (15.52), while the linear term in x quickly dominates the logarithmic term for $x \gg 1$. This leads to the linear detector approximation for large x:

$$\ln[I_0(x)] \approx x, \qquad x \gg 1 \tag{15.55}$$

Figure 15-10 illustrates the fit between the square law and linear approximations and the exact $\ln[I_0(x)]$ function. The square law detector is an excellent fit for $10 \cdot \log_{10}(x) < 5$, while the linear detector fits the $\ln[I_0(x)]$ very well for $10 \cdot \log_{10}(x) > 10$.

Equations (15.53) and (15.55) show that, despite the summation, a linear or square law detector is still a close approximation to the optimal detection rule of equation (15.52). This is convenient since computation of these detectors is much simpler. Whether x is "big" or "small," and thus the choice of a linear or square law detector, depends not only on the actual signal strength but also on how that signal is digitized and numerically represented.

Finally, note that it is easy to compute the squared magnitude of a complex-valued test sample as simply the sum of the squares of the real and imaginary parts. The linear magnitude requires a square root and is less computationally convenient. A family of computationally simple approximations to the magnitude function is presented in [11].

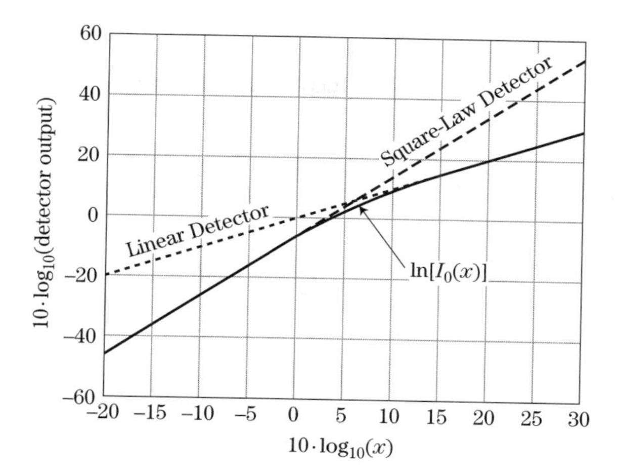

FIGURE 15-10 ■ Approximation of the $\ln[I_0(\cdot)]$ detector characteristic by the square law detector when its argument is small and the linear detector when its argument is large.

15.5.7 Square Law Detector Performance for a Nonfluctuating Target, $N > 1$

The test in equation (15.52) can now be simplified by applying the square law detector approximation of equation (15.53), giving the new test

$$\sum_{n=1}^{N} \frac{\tilde{m}^2 z_n^2}{\sigma_n^4} \underset{H_0}{\overset{H_1}{\gtrless}} T' \tag{15.56}$$

Combining the remaining constants into the threshold and defining the noncoherent sum z of the individual detected samples gives the final noncoherent integration detection rule:

$$z \equiv \sum_{n=1}^{N} z_n^2 \underset{H_0}{\overset{H_1}{\gtrless}} \frac{\sigma_n^4 T'}{\tilde{m}^2} \equiv T \tag{15.57}$$

Equation (15.57) states that the squared magnitudes of the data samples are simply integrated and the integrated sum compared with a threshold to decide whether a target is present.

The performance of the detector given in (15.57) must now be determined. It is convenient to scale the z_n, replacing them with the new variables $z'_n = z_n/\sigma_n$ and thus replacing z with $z' = \sum (z'_n)^2 = z/\sigma_n^2$; such a scaling does not change the performance but merely alters the threshold value that corresponds to a particular P_{FA}. The PDF of z'_n is still the Rayleigh voltage of equation (15.19) under H_0 or the Rician voltage of equation (15.22) under H_1, but now with unit noise variance. Since a square law detector is being used, the PDF of $y_n = (z'_n)^2$ is needed. This is exponential under H_0 and NCCS2 under H_1 (again with unit noise power) [2],

$$p_{y_n}(y_n|H_0) = \begin{cases} e^{-y_n}, & y_n \geq 0 \\ 0, & y_n < 0 \end{cases} \tag{15.58}$$

$$p_{y_n}(y_n|H_1) = \begin{cases} e^{-(y_n+\chi)} I_0\left(2\sqrt{\chi y_n}\right), & y_n \geq 0 \\ 0, & y_n < 0 \end{cases} \tag{15.59}$$

where $\chi = \tilde{m}^2/\sigma_n^2$ is the signal-to-noise ratio. Since z' is the sum of N scaled random variables $y_n = (z'_n)^2$, the PDF of z' is the N-fold convolution of the PDF given in equation (15.58) or (15.59).

In the H_0 case, the resulting PDF can be shown to be the Erlang density (a special case of the gamma density) [3]:

$$p_{z'}(z'|H_0) = \begin{cases} \dfrac{(z')^{N-1}}{(N-1)!}e^{-z'}, & z' \geq 0 \\ 0, & z' < 0 \end{cases} \tag{15.60}$$

Note that this reduces to the exponential PDF when $N = 1$, as would be expected since in that case z' is the magnitude squared of a single sample of complex Gaussian noise. The probability of false alarm is obtained by integrating equation (15.60) from some threshold value to $+\infty$. The result is [1]

$$P_{FA} = \int_T^{\infty} \frac{(z')^{N-1}}{(N-1)!}e^{-z'}dz' = 1 - I\left(\frac{T}{\sqrt{N}}, N-1\right) \tag{15.61}$$

where

$$I(u, M) = \int_0^{u\sqrt{M+1}} \frac{e^{-\tau}\tau^M}{M!}d\tau \tag{15.62}$$

is Pearson's form of the incomplete gamma function. For a single sample ($N = 1$), equation (15.61) reduces to the especially simple result

$$P_{FA} = e^{-T} \tag{15.63}$$

so that $T = -\ln(P_{FA})$. Note that this value of threshold is the square of that found in equation (15.21) (with $\sigma_n = 1$), because this threshold is applied to the squared magnitude of the data samples whereas the earlier threshold was applied to just the linear magnitude of the data samples. Equation (15.63), which was also seen in Chapter 3, can be used to determine the probability of false alarm P_{FA} for a given threshold T or, more likely, to determine the required value of T for a desired P_{FA}.

Now the probability of detection, P_D, corresponding to the same threshold must be determined. Start by finding the PDF of the normalized, integrated, and square law detected samples under H_1. The result is [1,5]

$$p_{z'}(z'|H_1) = \left(\frac{z'}{N\chi}\right)^{(N-1)/2} e^{-z-N\chi} I_{N-1}\left(2\sqrt{N\chi z'}\right) \tag{15.64}$$

P_D is found by integrating equation (15.64). One version of the result, given in [12], is

$$\begin{aligned} P_D &= \int_T^{\infty} \left(\frac{z'}{N\chi}\right)^{(N-1)/2} e^{-z'-N\chi} I_{N-1}\left(2\sqrt{N\chi z'}\right)dz' \\ &= Q_M\left(\sqrt{2N\chi}, \sqrt{2T}\right) + e^{-(T+N\chi)} \sum_{r=2}^{N} \left(\frac{T}{N\chi}\right)^{(r-1)/2} I_{r-1}\left(2\sqrt{N\chi T}\right) \end{aligned} \tag{15.65}$$

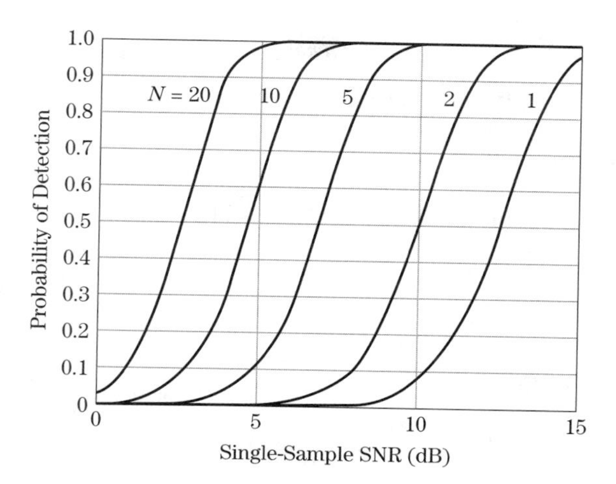

FIGURE 15-11 ■ Effect of noncoherent integration on detection performance for a nonfluctuating target in complex Gaussian noise.

The series term in this expression applies only when $N \geq 2$. Equations (15.61) and (15.65) define the performance achievable with noncoherent integration using a square law detector.

Figure 15-11 shows the effect of the number of samples noncoherently integrated, N, on the receiver operating characteristic. This figure shows that noncoherent integration reduces the single-sample SNR required to achieve a given P_D and P_{FA} but not by the factor N achieved with coherent integration. For example, consider the single-sample SNR required to achieve $P_D = 0.9$. For $N = 1$, this is 14.2 dB; for $N = 10$, it drops to 6.1 dB, a reduction of 8.1 dB, but less than the 10 dB that corresponds to the factor of 10 increase in the number of pulses integrated. In the next section a simple approximation for estimating this reduction in required single-sample SNR, called the *noncoherent integration gain*, will be developed.

It is important to realize that the noncoherent integration gain of 8.1 dB in the preceding example does not imply that the SNR of the square law detected and noncoherently integrated detection statistic z (or its linearly detected equivalent) is 8.1 dB higher than that of a single sample y. Because of the nonlinear magnitude-squared or magnitude operation, z cannot be expressed as the sum of separate noise and target contributions. Noncoherent integration gain represents the reduction in the SNR of the individual coherent data samples required to achieve certain values of P_D and P_{FA} when multiples samples are combined via noncoherent integration. Consequently, the noncoherent integration gain depends on the particular values of P_D and P_{FA} desired. It also depends on the SNR of the individual coherent samples.

15.5.8 Albersheim's Equation

The performance results for the case of N envelope-detected samples of a nonfluctuating target in complex Gaussian noise are given by equations (15.61) and (15.65). While relatively easy to implement in a modern software analysis system such as MATLAB, these equations do not lend themselves to hand calculation. Fortunately, there is a simple closed-form expression relating P_D, P_{FA}, and SNR that can be computed with simple scientific calculators. This expression is known as *Albersheim's equation* [13,14].

Albersheim's equation is an empirical approximation to the results in [5] for computing the single-sample SNR, χ_1, required to achieve a given P_D and P_{FA}. It applies under the following conditions:

- Nonfluctuating target in Gaussian (IID in I and Q) noise.
- Linear (not square law) detector.
- Noncoherent integration of N samples.

The estimate is given by the series of calculations [13]

$$
\begin{aligned}
A &= \ln\left(\frac{0.62}{P_{FA}}\right) \\
B &= \ln\left(\frac{P_D}{1-P_D}\right) \\
\chi_{1_{\mathrm{dB}}} &= -5\log_{10} N + \left(6.2 + \left(\frac{4.54}{\sqrt{N+0.44}}\right)\right)\cdot \log_{10}(A + 0.12AB + 1.7B)\ \mathrm{dB}
\end{aligned}
\tag{15.66}
$$

Note that $\chi_{1_{\mathrm{dB}}}$ is in decibels, not linear power units. The error in the estimate of $\chi_{1_{\mathrm{dB}}}$ is less than 0.2 dB for $10^{-7} \le P_{FA} \le 10^{-3}$, $0.1 \le P_D \le 0.9$, and $1 \le N \le 8096$ [13], a very useful range of parameters.

As a simple example of the use of Albersheim's equation, suppose $P_D = 0.9$ and $P_{FA} = 10^{-6}$ are required for a nonfluctuating target in a system using a linear detector. If detection is to be based on a single sample, what is the required SNR of that sample? Compute $A = \ln(0.62 \times 10^6) = 13.34$ and $B = \ln(0.9) = 2.197$. With $N = 1$, equation (15.66) then gives $\chi_{1_{\mathrm{dB}}} = 13.14$ dB; on a linear scale, this is $\chi_1 = 20.59$.

If $N = 100$ samples are noncoherently integrated, it should be possible to obtain the same P_D and P_{FA} with a lower single-sample SNR. To confirm this, again apply Albersheim's equation but now with $N = 100$. The intermediate parameters A and B are unchanged. $\chi_{1_{\mathrm{dB}}}$ is now reduced to -1.26 dB, a reduction of 14.4 dB. This noncoherent integration gain of 14.4 dB, a factor of 27.54 on a linear scale, is much better than the $\sqrt{N}$ rule of thumb sometimes given for noncoherent integration, which would give a gain factor of only 10 for $N = 100$ samples integrated. Rather, the gain is approximately $N^{0.7}$ in this example.

Albersheim's equation is useful because it requires no function more exotic than the natural logarithm and square root for its evaluation. It can thus be evaluated on virtually any scientific calculator. If a somewhat larger error can be tolerated, it can also be used for square law detector results for the nonfluctuating target, Gaussian noise case. Specifically, square law detector results are within 0.2 dB of linear detector results over a wide range of parameters [15,16]. Thus, the same equation can be used for calculations over the range of previously given parameters with errors not exceeding 0.4 dB.

Equation (15.66) provides for the calculation of χ_1 given P_D, P_{FA}, and N. It is possible, however, to solve (15.66) for either P_D or P_{FA} in terms of the other and χ_1 and N, extending further the usefulness of Albersheim's equation. For instance, the following calculations show how to estimate P_D given the other factors:

$$
A = \ln\left(\frac{0.62}{P_{FA}}\right), \quad Z = \frac{\chi_{1_{\mathrm{dB}}} + 5\log_{10} N}{6.2 + \dfrac{4.54}{\sqrt{N+0.44}}}, \quad B = \frac{10^Z - A}{1.7 + 0.12A}
$$
$$
P_D = \frac{1}{1+e^{-B}}
\tag{15.67}
$$

In equation (15.67), A and B are the same values as in (15.66), though B cannot be computed in terms of P_D, since P_D is now the unknown. A result similar to (15.67) can be derived for computing P_{FA} in terms of P_D and $\chi_{1_{\text{dB}}}$.

Albersheim's equation can also be used to estimate the signal-to-noise ratio gain for noncoherent integration of N samples of a nonfluctuating target. The noncoherent integration gain is the reduction in single-sample SNR required to achieve a specified P_D and P_{FA} when N samples are combined. Coherent integration provides an integration gain of a factor of N, while noncoherent integration is less efficient, providing a gain of a factor of N^{α} for some $\alpha < 1$ (e.g., $\alpha \approx 0.7$ in the earlier example).

Figure 15-12a plots the value of α for the nonfluctuating, linear detector case, estimated using for Albersheim's equation with $P_D = 0.9$. The abscissa is the single-sample SNR, that is, the SNR before noncoherent integration. This figure shows that when the data are very noisy to begin with (single-sample SNR $\ll$ 0 dB), the integration is inefficient, with α falling to about 0.55. (Note that $\alpha = 0.5$ would be a factor of $\sqrt{N}$.) If the data are very "clean" to begin with (single-sample SNR > 10 dB), α is in the neighborhood of 0.9. Since coherent integration corresponds to $\alpha = 1$, noncoherent integration is almost as efficient as coherent integration when the initial SNR is high. For lower values of P_D, the efficiency

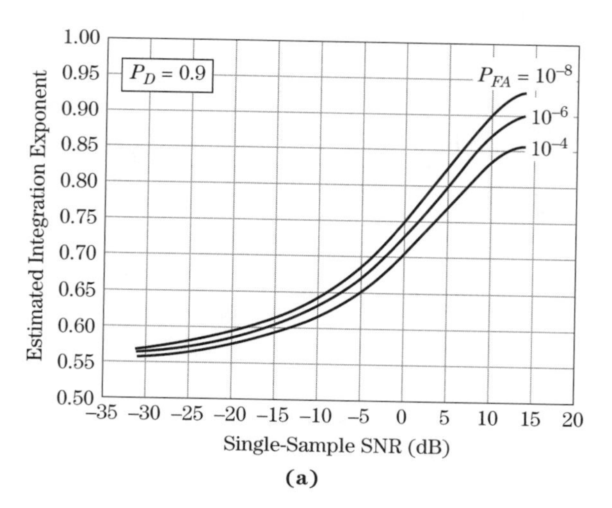

(a)

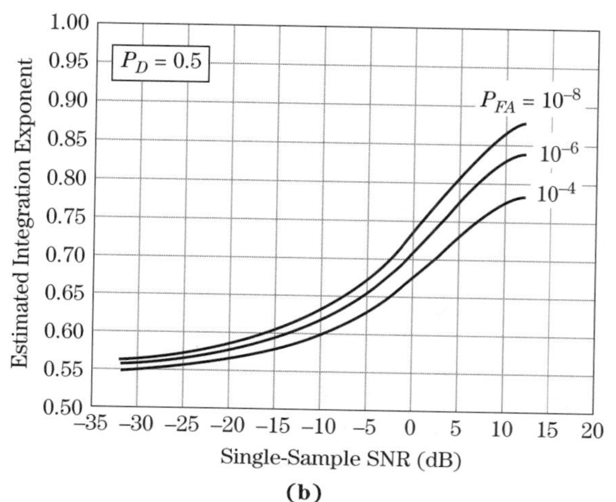

(b)

FIGURE 15-12 ■ Noncoherent integration gain exponent for a nonfluctuating target, estimated using Albersheim's equation. (a) $P_D = 0.9$. (b) $P_D = 0.5$.

falls somewhat for high-SNR data, as shown in Figure 15-12b, but never falls below $\sqrt{N}$ for low-SNR data.

15.5.9 Fluctuating Targets

The analysis in the preceding section considered only nonfluctuating targets, also called the "Swerling 0," "Swerling 5," or "Marcum" case. A more realistic model allows for target fluctuations, in which the target RCS is drawn from either the exponential or chi-square PDF, and the RCS of a group of N noncoherently integrated samples follows either the pulse-to-pulse or scan-to-scan decorrelation model, as described in Section 15.4.4. Note that representing the target by one of the Swerling models 1 through 4 has no effect on the probability of false alarm, since that is determined only by the PDF when no target is present; thus equation (15.61) still applies.

The strategy for determining probability of detection depends on the Swerling model used. Figure 15-13 illustrates the approach [5]. In all cases, the PDF of the magnitude of a single sample is still NCCS2, assuming a square law detector. However, the SNR χ is now a random variable because the target RCS is a random variable.

In the scan-to-scan decorrelation cases (Swerling models 1 and 3), the target RCS and thus SNR, while random, is the same value for all N pulses integrated to form z'. The PDF of a single sample is still given by equation (15.59). That result is then averaged over the SNR fluctuations to get an "average" PDF for the sum of N nonfluctuating target samples. For instance, in the Swerling 1 case, the PDF of the target RCS and therefore of the SNR is exponential:

$$p_\chi(\chi) = \frac{1}{\bar{\chi}} e^{-\chi/\bar{\chi}} \tag{15.68}$$

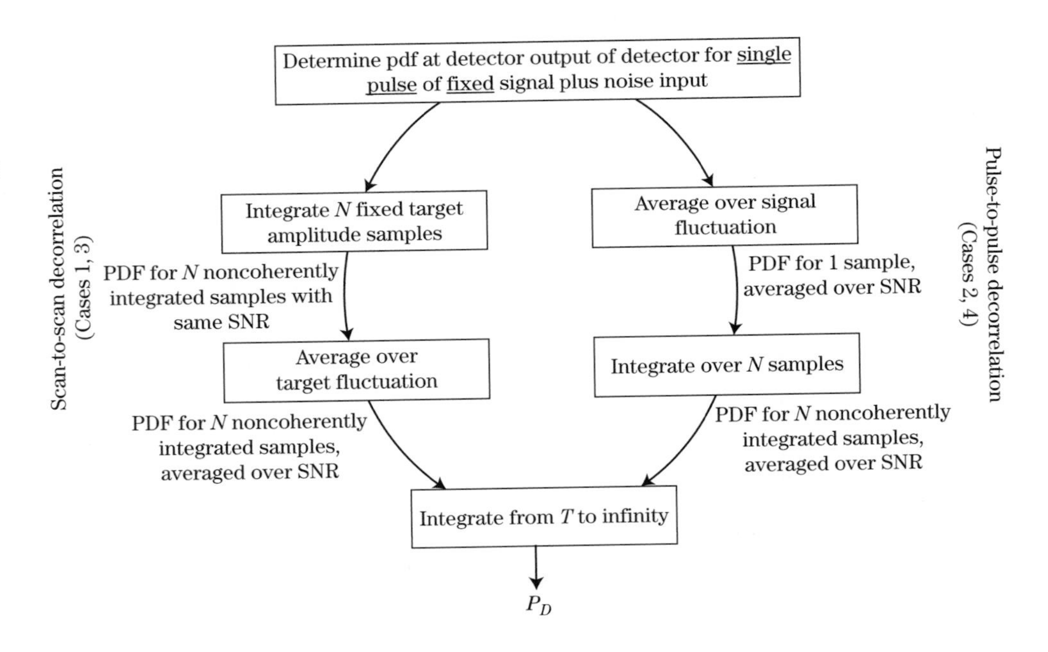

FIGURE 15-13 ■ The strategy for computing P_D for fluctuating targets depends on the RCS fluctuation model.

where $\bar{\chi}$ is the average SNR over the N samples. The detailed calculations are beyond the scope of this chapter, but the resulting PDF of z' under H_1 is [1,12]

$$p_{z'}(z'|H_1) = \frac{1}{N\bar{\chi}}\left(1 + \frac{1}{N\bar{\chi}}\right)^{N-2} I\left[\frac{z'}{(1 + 1/N\bar{\chi})\sqrt{N-1}}, N-2\right] e^{-z'/(1+N\bar{\chi})} \quad (15.69)$$

Integrating this PDF from the threshold T to $+\infty$ and assuming that $P_{FA} \ll 1$ and the integrated average SNR $N\bar{\chi} > 1$ (conditions that are almost always true in any scenario where target detection is likely to be successful) results in the following expression for the probability of detection in the Swerling 1 case [1,12]:

$$P_D \approx \left(1 + \frac{1}{N\bar{\chi}}\right)^{N-1} e^{-T/(1+N\bar{\chi})}, \quad P_{FA} << 1, \quad N\bar{\chi} > 1 \quad (15.70)$$

Equation (15.70) is exact when $N = 1$; in this case it reduces to

$$P_D = \exp\left(-\frac{T}{1+\chi_1}\right), \quad N = 1 \quad (15.71)$$

(When there is only one pulse, the average SNR $\bar{\chi}$ is just the single-pulse SNR χ_1.) For the $N = 1$ case, equation (15.63) can then be used in (15.71) to write a direct relationship between P_D and P_{FA}:

$$P_D = (P_{FA})^{1/(1+\chi_1)} \quad (15.72)$$

This equation was seen also in Chapter 3.

In the Swerling 2 or 4 cases, the samples exhibit pulse-to-pulse decorrelation, meaning they are in fact uncorrelated with one another. In this case, each of the N samples integrated has a different value of SNR. The PDF of z' is therefore averaged over the SNR fluctuations first to get an "average" PDF for a single sample. The PDF of a detection statistic formed as the sum of N samples having that "average" PDF is then computed. The result is [1,12]

$$p_{z'}(z'|H_1) = \frac{z'^{N-1}\exp\left[z'/(1+\bar{\chi})\right]}{(1+\bar{\chi})^N (N-1)!}. \quad (15.73)$$

Integrating (15.73) gives the probability of detection, which can be shown to be [12]

$$P_D = 1 - I\left[\frac{T}{(1+\bar{\chi})\sqrt{N}}, N-1\right] \quad (15.74)$$

Results for Swerling 3 and 4 targets can be obtained by repeating the previously given analyses for the Swerling 1 and 2 cases, but with a chi-square instead of exponential density function for the SNR:

$$p_\chi(\chi) = \frac{4\chi}{\bar{\chi}^2} e^{-2\chi/\bar{\chi}} \quad (15.75)$$

Derivations of the resulting expressions for P_D can be found in [12,17] and many other radar detection texts. Table 15-2 summarizes one form of the resulting expressions.

Figure 15-14 compares the detection performance of the four Swerling model fluctuating targets and the nonfluctuating target for $N = 10$ samples as a function of the average single-sample SNR for a fixed $P_{FA} = 10^{-8}$. Assuming that the primary interest is in relatively high (> 0.5) values of P_D, the upper half of the figure is of greatest interest. In this case, the nonfluctuating target is the most favorable, in the sense that it achieves a given probability of detection at the lowest SNR. The worst case (highest required SNR

TABLE 15-2 ■ Probability of Detection for Swerling Model Fluctuating Targets with a Square Law Detector

Case	P_D	Comments
0 or 5	$Q_M(\sqrt{2N\bar{\chi}}, \sqrt{2T}) + e^{-(T+N\bar{\chi})} \sum_{r=2}^{N} \left(\frac{T}{N\bar{\chi}}\right)^{\frac{r-1}{2}} I_{r-1}(2\sqrt{NT\bar{\chi}})$	Second term applies only for $N \geq 2$
1	$\left(1 + \frac{1}{N\bar{\chi}}\right)^{N-1} e^{-T/(1+N\bar{\chi})}$	Approximate for $P_{FA} \ll 1$ and $N\bar{\chi} > 1$; exact for $N = 1$
2	$1 - I\left[\frac{T}{(1+\bar{\chi})\sqrt{N}}, N-1\right]$	
3	$\left(1 + \frac{2}{N\bar{\chi}}\right)^{N-2} \left[1 + \frac{T}{1+(N\bar{\chi}/2)} - \frac{2(N-2)}{N\bar{\chi}}\right] e^{-T/(1+(N\bar{\chi}/2))}$	Approximate for $P_{FA} \ll 1$ and $N\bar{\chi}/2 > 1$; exact for $N = 1$ or 2
4	$\left\{c^N \sum_{k=0}^{N} \frac{N!}{k!\,(N-k)!} \left(\frac{1-c}{c}\right)^{N-k}\right\} \left\{\sum_{l=0}^{2N-1-k} \frac{e^{-cT}\,(cT)^l}{l!}\right\}, \quad T > N(2-c)$ $1 - \left\{c^N \sum_{k=0}^{N} \frac{N!}{k!\,(N-k)!} \left(\frac{1-c}{c}\right)^{N-k}\right\} \left\{\sum_{l=2N-k}^{\infty} \frac{e^{-cT}\,(cT)^l}{l!}\right\}, \quad T < N(2-c)$	$c \equiv \frac{1}{1+(\bar{\chi}/2)}$
	$P_{FA} = 1 - I\left(\frac{T}{\sqrt{N}}, N-1\right)$ in all cases	

$I(\cdot,\cdot)$ is Pearson's form of the incomplete Gamma function; $I_k(\cdot)$ is the modified Bessel function of the first kind and order k

Source: After Meyer and Mayer [12] (with permission).

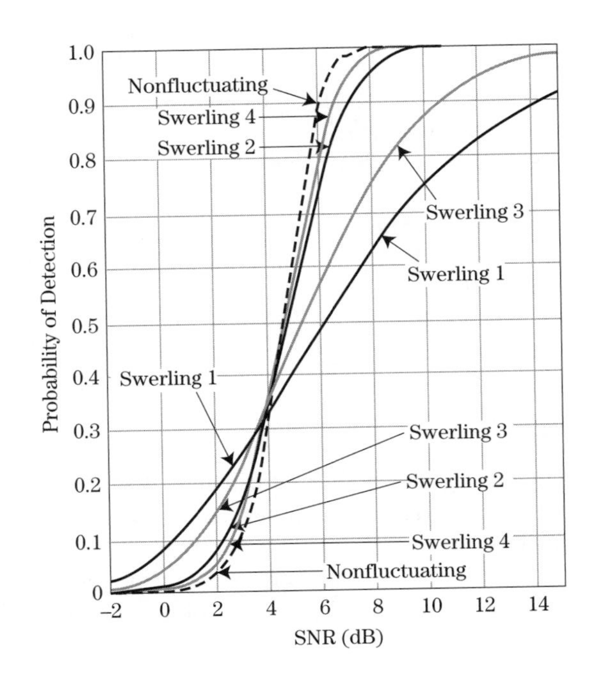

FIGURE 15-14 ■ Comparison of detection performance for fluctuating (Swerling) and nonfluctuating target models using noncoherent integration of $N = 10$ pulses and a fixed $P_{FA} = 10^{-8}$.

for a given P_D) is the Swerling case 1, which corresponds to scan-to-scan decorrelation and an exponential PDF for the target RCS. For instance, $P_D = 0.9$ requires $\bar{\chi} \approx 6$ dB for the nonfluctuating case, but $\bar{\chi} \approx 14.5$ dB for the Swerling 1 case, a difference of about 8.5 dB! This figure makes clear the fact that the assumed target model has a major impact on estimated detection performance.

At least two general conclusions can be drawn from this figure. First, for $P_D > 0.5$, nonfluctuating targets are easier to detect than any of the Swerling cases; target fluctuations make detection more difficult by requiring a higher SNR for a given P_D. Second, pulse-to-pulse fluctuations (Swerling 2 and 4) aid target detectability compared with scan-to-scan detectability. For instance, a Swerling 2 target is easier to detect than a Swerling 1 target that shares the same PDF for target fluctuations, and a Swerling 4 target is easier than a Swerling 3 target. Finally, note that the converse of these statements is true for detection probabilities less than about 0.35 in this case.

15.5.10 Frequency Agility

Figure 15-14 shows that, for reasonably high signal-to-noise ratios, the probability of detection is higher for Swerling 2 or 4 targets than it is for Swerling 1 or 3 targets. This suggests that if the target RCS is going to fluctuate, it is preferable to have pulse-to-pulse fluctuations. If the radar-to-target aspect angle does not change enough during the CPI to decorrelate the target echoes, the radar can take advantage of the fact that changing the radar frequency will decorrelate the measurements. As was discussed in Section 15.4.3, the technique of changing the radar frequency pulse-to-pulse to improve detection performance is called *frequency agility*.[6] The frequency step required to decorrelate a complex target was given in equation (15.13) as approximately $\Delta f = c/2L_d$, where L_d is the depth of the target projected along the radar boresight. The required frequency step size is typically a few tens of megahertz.

15.5.11 Shnidman's Equation

Albersheim's equation provided a simple way to compute the single-sample SNR required to achieve a specified P_D and P_{FA} for a nonfluctuating target and a linear detector or, with slightly greater error, a square law detector, over a useful range of parameters. Furthermore, it can be rearranged to solve for P_D (or P_{FA}) in terms of N, SNR_1, and P_{FA} (or P_D). While an extremely useful tool, Albersheim's equation has the serious limitation that it does not treat the case of fluctuating targets. As was seen in Figure 15-14, the nonfluctuating case is optimistic, especially compared with cases characterized by scan-to-scan decorrelation behavior.

Recently, a new empirical approximation has been developed that addresses this limitation. *Shnidman's equation*, like Albersheim's, is an analytically based but ultimately empirical approximation to compute the required single-sample SNR to achieve a specified P_D and P_{FA} when N pulses are noncoherently integrated [18]. However, it applies to all four Swerling models.

Though somewhat lengthier to express than Albersheim's equation, the actual calculation are equally simple. It begins by selecting values for the parameters K and α; the

[6]The term *frequency agility* generally implies changing the radar frequency for the purpose of improving detection. When this is done for electronic counter-countermeasure (antijamming) purposes, it is generally called *frequency diversity*.

choice of K determines the Swerling model to be represented. Two derived parameters η and X_∞ are then computed:[7]

$$K = \begin{cases} \infty, & \text{nonfluctuating target ("Swerling 0/5")} \\ 1, & \text{Swerling 1} \\ N, & \text{Swerling 2} \\ 2, & \text{Swerling 3} \\ 2N, & \text{Swerling 4} \end{cases}$$

$$\alpha = \begin{cases} 0, & N < 40 \\ \frac{1}{4}, & N \geq 40 \end{cases} \tag{15.76}$$

$$\eta = \sqrt{-0.8\ln\left(4P_{FA}\left(1 - P_{FA}\right)\right)} + \text{sign}\left(P_D - 0.5\right)\sqrt{-0.8\ln\left(4P_D\left(1 - P_D\right)\right)}$$

$$X_\infty = \eta\left[\eta + 2\sqrt{\frac{N}{2} + \left(\alpha - \frac{1}{4}\right)}\right]$$

Next, the series of constants C_1, C_2, C_{dB}, and C are computed:

$$C_1 = \left\{\left[\left(17.7006P_D - 18.4496\right)P_D + 14.5339\right]P_D - 3.525\right\}\Big/K$$

$$C_2 = \frac{1}{K}\left\{\exp\left(27.31P_D - 25.14\right) + \left(P_D - 0.8\right)\left[0.7\ln\left(\frac{10^{-5}}{P_{FA}}\right) + \frac{(2N - 20)}{80}\right]\right\}$$

$$C_{dB} = \begin{cases} C_1, & 0.1 \leq P_D \leq 0.872 \\ C_1 + C_2, & 0.872 < P_D \leq 0.99 \end{cases}$$

$$C = 10^{C_{dB}/10} \tag{15.77}$$

Notice that C_1 and C_2 will equal 0, and therefore C equals 1, for the nonfluctuating case. Finally,

$$\chi_1 = \frac{C \cdot X_\infty}{N}$$

$$\chi_{1_{dB}} = 10\log_{10}(SNR_1) \tag{15.78}$$

Equations (15.76) through (15.78) constitute Shnidman's equation.

The accuracy of these equations is better than 1 dB for the parameter range $0.1 \leq P_D \leq 0.99$, $10^{-9} \leq P_{FA} \leq 10^{-3}$, and $1 \leq N \leq 100$, with the error being less than 0.5 dB over nearly all of this parameter range. (A variant is given in [18] that maintains the same accuracy for values of P_D up to 0.9992 if N is restricted to 20 or less.) The greatest errors occur for the Swerling 1 case near the two extremes of P_D. Compared to Albersheim's equation, this is a more restricted but still very useful range of N, but a significantly better

[7]The function sign(x) equals 1 if the argument $x \geq 0$ and -1 if $x < 0$.

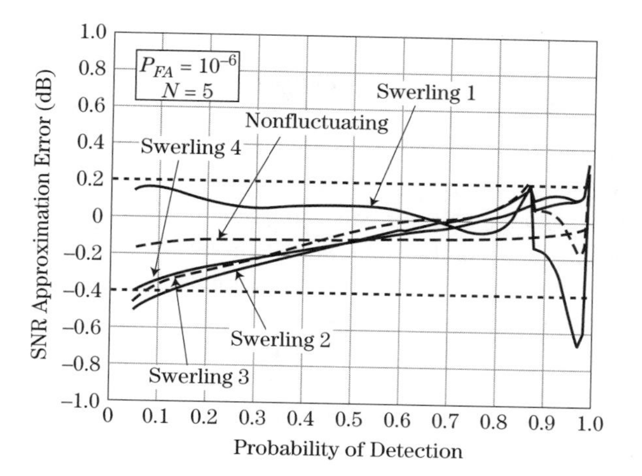

FIGURE 15-15 ■ Example of error in Shnidman's approximation.

upper limit on P_D (0.99 instead of 0.9), at the expense of a looser error bound (0.5 to 1 dB instead of 0.2 dB). Figure 15-15 plots a representative set of approximation error curves for the various Swerling cases for the particular case of $N = 5$ and $P_{FA} = 10^{-6}$. While Shnidman's equation applies directly only to a square law detector, the same argument used to extend Albersheim's equation to both detector types, at the cost of an increase in error of about 0.2 dB, can be applied to extend Shnidman's equation to linear detectors with a similar error increase.

15.5.12 Detection in Clutter

The specific results derived in the previous sections, for both nonfluctuating and fluctuating targets, all assumed that the interference was complex, white Gaussian noise. While appropriate for many radar systems and also for cases where noise jammers are the dominant interference, in many other systems the dominant interference is clutter, not noise. The probability density function of a single sample of envelope-detected clutter, discussed in Chapter 5, is often not well modeled by the Rayleigh voltage/exponential power function that describes noise. A very wide variety of models are described in the literature; see the book by Long [19] for an excellent introduction.

In addition to changes in the single-sample PDFs, the decorrelation models used to describe clutter are often complex, primarily because they are functions not only of radar-clutter geometry but, in many cases, of such factors as weather and the seasons. For example, tree clutter decorrelates faster when the wind speed increases due to increased movement of the branches and leaves. Furthermore, deciduous tree clutter statistics vary drastically between summer, when the branches are fully leafed out, and winter when they are bare. These changes affect the clutter reflectivity and PDFs as well as the decorrelation behavior.

Despite these difficulties, the basic procedure for determining detection performance in clutter is the same as described here; the difference is that the PDFs $p_y(y|H_0)$ and $p_y(y|H_1)$ will change, leading to new likelihood ratios and detection statistics, and new ROC equations. In practice, non-Gaussian PDFs usually lead to more difficult calculations, though results are available for many specific models in the literature. Good examples of studies of coherent detection in non-Gaussian clutter are available in [20–22].

15.5.13 Binary Integration

If the entire threshold detection process is repeated n times, n binary decisions will be available. Binary integration was discussed in some detail in Chapter 3. The detection probability of a properly implemented binary integration detector is higher than the P_D obtained on a single detection test. Conversely, the P_{FA} is lower than the single-test P_{FA}. Thus, the binary integration technique can improve both detection and false alarm performance. Alternatively, the single-dwell P_D and P_{FA} requirements can be relaxed while still meeting the overall performance specification. Examples are given in Chapter 3.

The analysis in Chapter 3 implicitly assumes a nonfluctuating target, since the single-trial P_D was assumed to be the same on each trial. The results can be extended to fluctuating targets [23,24]; these studies also indicate optimum choices for m as a function of the target model and n. Finally, other binary integration strategies exist with somewhat different properties; for instance, the decision logic can require that the m hits be contiguous rather than just any m hits out of n. See [25] for references to the literature for these and other variations on binary integration.

15.6 FURTHER READING

An excellent concise reference for modern detection theory is Chapter 5 of the text by Johnson and Dudgeon [26]. When greater depth is needed, another excellent modern reference with a digital signal processing point of view is volume II of Kay's text [2]. An important classical textbook in detection theory is Part I of Van Trees's series [27], while Meyer and Mayer [12] and DiFranco and Rubin [17] provide classical in-depth analyses and many detection curves specifically for radar applications. A good recent source on basic radar detection models and calculations is the series of papers by Shnidman [6,8,16,23].

15.7 REFERENCES

[1] Richards, M.A., *Fundamentals of Radar Signal Processing*, McGraw-Hill, New York, 2005.

[2] Kay, S.M., *Fundamentals of Statistical Signal Processing, Vol. II: Detection Theory,* Prentice-Hall, Upper Saddle River, NJ, 1998.

[3] Papoulis, A., and Pillai, S.U., *Probability, Random Variables, and Stochastic Processes*, 4th ed., McGraw-Hill, New York, 2002.

[4] Ray, H., "Improving Radar Range and Angle Detection with Frequency Agility," *Microwave Journal,* pp. 64ff, May 1966.

[5] Swerling, P., "Probability of Detection for Fluctuating Targets," *IRE Trans. Information Theory*, vol. IT-6, pp. 269–308, April 1960.

[6] Shnidman, D.A., "Calculation of Probability of Detection for Log-Normal Target Fluctuations," *IEEE Trans. Aerospace and Electronic Systems*, vol. AES-27, no. 1, pp. 172–174, January 1991.

[7] Swerling, P., "Radar Probability of Detection for Some Additional Fluctuating Target Cases," *IEEE Trans. Aerospace and Electronic Systems*, vol. AES-33, no. 2, pp. 698–708, April 1997.

[8] Shnidman, D.A., "Expanded Swerling Target Models," *IEEE Transactions on Aerospace and Electronic Systems*, vol. AES-39, no. 3, pp. 1059–1068, July 2003.

[9] Cantrell, P.E., and Ojha, A.K., "Comparison of Generalized Q-Function Algorithms," *IEEE Trans. Information Theory*, vol. IT-33, no. 4, pp. 591–596, July 1987.

[10] Abramowitz, M., and Stegun, I.A., *Handbook of Mathematical Functions: with Formulas, Graphs, and Mathematical Tables*, U.S. National Bureau of Standards, Applied Mathematics Series – 55, Washington, DC, 1964.

[11] Filip, A.E., "A Baker's Dozen Magnitude Approximations and Their Detection Statistics," *IEEE Trans. Aerospace & Electronic Systems*, pp. 86–89, January 1976.

[12] Meyer, D.P., and Mayer, H.A., *Radar Target Detection*, Academic Press, New York, 1973.

[13] Albersheim, W.J., "Closed-Form Approximation to Robertson's Detection Characteristics," *Proceedings IEEE*, vol. 69, no. 7, p. 839, July 1981.

[14] Tufts, D.W., and Cann, A.J., "On Albersheim's Detection Equation," *IEEE Trans. Aerospace and Electronic Systems*, vol. AES-19, no. 4, pp. 643–646, July 1983.

[15] Robertson, G.H., "Operating Characteristic for a Linear Detector of CW Signals in Narrow Band Gaussian Noise," *Bell System Technical Journal*, vol. 46, no. 4, pp. 755–774, April 1967.

[16] Shnidman, D.A., "Radar Detection Probabilities and Their Calculation," *IEEE Transactions on Aerospace and Electronic Systems*, vol. AES-31, no. 3, pp. 928–950, July 1995.

[17] DiFranco, J.V., and Rubin, W.L., *Radar Detection,* Artech House, Dedham, MA, 1980.

[18] Shnidman, D.A., "Determination of Required SNR Values," *IEEE Trans. Aerospace & Electronic Systems*, vol. AES-38, no. 3, pp. 1059–1064, July 2002.

[19] Long, M.W., *Radar Reflectivity of Land and Sea*, Artech House, Dedham, MA, 2001.

[20] Gini, F., "Sub-optimum Coherent Radar Detection in a Mixture of K-distributed and Gaussian Clutter," *Proc. IEE Radar, Sonar and Navigation,* vol. 144, no. 1, pp. 39–48, February 1997.

[21] Sangston, K.J., and Gerlach, K.R., "Coherent Detection of Radar Targets in a Non-Gaussian Background," *IEEE Trans. Aerospace & Electronic Systems*, vol. AES30, no. 2, pp. 330–340, April 1994.

[22] Sangston, K.J., Gini, F., Greco, M.V., and Farina, A., "Structures for Radar Detection in Compound Gaussian Clutter," *IEEE Trans. Aerospace & Electronic Systems*, vol. AES35, no. 2, pp. 445–458, April 1999.

[23] Shnidman, D.A., "Binary Integration for Swerling Target Fluctuations," *IEEE Trans. Aerospace & Electronic Systems*, vol. AES-34, no. 3, pp. 1043–1053, July 1998.

[24] Weiner, M.A., "Binary Integration of Fluctuating Targets," *IEEE Trans. Aerospace & Electronic Systems*, vol. AES-27, no. 1, pp. 11–17, July 1991.

[25] Skolnik, M.I., *Introduction to Radar Systems*, 3d ed., McGraw-Hill, New York, 2001.

[26] Johnson, D.H., and Dudgeon, D.E., *Array Signal Processing*, Prentice-Hall, Englewood Cliffs, NJ, 1993.

[27] Van Trees, H.L., *Detection, Estimation, and Modulation Theory, Part I: Detection, Estimation, and Linear Modulation Theory,* Wiley, New York, 1968.

15.8 | PROBLEMS

1. Consider detection of a real-valued constant in zero-mean real-valued Gaussian noise. Let the noise variance $\sigma_n^2 = 2$, the number of samples $N = 1$, and the constant $m = 4$. What is the SNR for this case? Sketch approximately the distributions $p(y|H_0)$ and $p(y|H_1)$; be sure to label appropriate numerical values on the axes.

2. Write the likelihood ratio and log-likelihood ratio that applies to problem 1. Simplify the resulting expressions.
3. Continuing with the same parameters given in problem 1, what is the required value of the threshold T to achieve $P_{FA} = 0.01$ (1%)? Look-up tables or MATLAB can be used to calculate the values of functions such as $\text{erf}(\cdot)$, $\text{erfc}(\cdot)$, $\text{erf}^{-1}(\cdot)$, or $\text{erfc}^{-1}(\cdot)$ that may be needed.
4. With the threshold selected in problem 3, what is the resulting value of P_D? Look-up tables or MATLAB can be used to calculate the values of functions such as $\text{erf}(\cdot)$, $\text{erfc}(\cdot)$, $\text{erf}^{-1}(\cdot)$, or $\text{erfc}^{-1}(\cdot)$ that may be needed.
5. Suppose m from problem 1 is increased to double the SNR on a linear (not dB) scale; $\sigma_n^2 = 2$ and $N = 1$ still. What is the new value of m? Sketch approximately the distributions $p(y|H_0)$ and $p(y|H_1)$ with this new value of m; be sure to label appropriate numerical values on your axes.
6. If the same value of the threshold T found in problem 3 is retained when m is increased to the value in problem 5, does the P_{FA} change, and, if so, what is the new value?
7. If the same value of the threshold T found in found in problem 3 is retained when m is increased to the value in problem 5, does the P_D change, and, if so, what is the new value?
8. Go back to the case of $m = 4$, but now reduce the noise variance to $\sigma_n^2 = 1$. $N = 1$ still. What is the SNR now? Again, sketch approximately the distributions $p(y|H_0)$ and $p(y|H_1)$ with this value of m and σ_n^2; be sure to label appropriate numerical values on the axes.
9. Assuming the same threshold T from problem 3 is maintained with the values of m and σ_n^2 from problem 8, does the P_{FA} change from the value in problem 3, and, if so, what is the new value?
10. If the same value of the threshold T is retained with the values of m and σ_n^2 from problem 8, does the P_D change from the value in problem 4, and, if so, what is the new value?
11. Compute the threshold T and probability of detection P_D for the case of a constant in zero-mean complex Gaussian noise, but now with *unknown phase*. Use $\tilde{m} = 4$, $\sigma_n^2 = 2$, $N = 1$, and $P_{FA} = 0.01$ again. It will be necessary to evaluate the Marcum Q function Q_M. The website for this book includes a MATLAB routine marcum that can be used for the numerical evaluation.
12. Use Albersheim's equation to estimate the single-sample SNR χ_1 required to achieve $P_{FA} = 0.01$ and P_D equal to the same value obtained in problem 11. How does the result compare with the actual SNR in problem 1?
13. Repeat problem 12 using Shnidman's equation in place of Albersheim's equation.
14. Using the PDFs for the case of a nonfluctuating target in complex Gaussian noise given by equations (15.14) and (15.15), derive the LRT of equation (15.17).
15. Coherently integrating N samples of signal-plus-noise produces an integration gain of N on a linear (not dB) scale; that is, if the SNR of a single sample y_i is χ, the SNR of $z = \sum_{i=1}^{N} y_i$ is $N\chi$. It is also often said that *non*coherent integration produces an integration gain of about $\sqrt{N}$. In problems 15 through 18, Albersheim's equation will be used to see if this is accurate for one example case. Throughout these problems, assume $P_D = 0.9$ and $P_{FA} = 10^{-6}$ is required and that a linear (not square law) detector is used. Start by considering detection based on a single sample, $N = 1$. Use Albersheim's equation to estimate the signal-to-noise ratio, χ_1, needed for this single sample to meet the previously given specifications. Give the answer in dB. Be careful about comparing or combining things on the same (linear or dB) scales throughout these four related problems.
16. Continuing, suppose the system now noncoherently integrates 100 samples to achieve the same P_D and P_{FA}. Each individual sample can then have a lower SNR. Use Albersheim's equation

again to estimate the signal-to-noise ratio χ_{nc} in dB of each sample needed to achieve the required detection performance.

17. Continuing, now consider *coherent* integration of 100 pulses. What is the signal-to-noise ratio χ_c required, in dB, for each sample such that the coherently integrated SNR will be equal to the value χ_1 found in problem 8?

18. Finally, the noncoherent integration gain is the ratio χ_1/χ_{nc}, where χ_1 and χ_{nc} are on a linear (not dB) scale. To compare it with the $\sqrt{N}$ estimate, find α such that $\chi_1/\chi_{nc} = N^{\alpha}$. Is the noncoherent integration gain better or worse than $\sqrt{N}$ in this case? Is it better or worse than coherent integration?

19. Rearrange Albersheim's equation to derive a set of equations for P_{FA} in terms of P_D, N, and single-pulse SNR in dB, χ_1.

20. Use Shnidman's equation to estimate the single-pulse SNR χ_1 in dB required to achieve $P_{FA} = 10^{-8}$ and $P_D = 0.9$ when noncoherently integrating $N = 10$ samples. Do this for all four Swerling cases and for the nonfluctuating case. (*Hint*: The answers should match the data in Figure 15-14.)

CHAPTER 16

Constant False Alarm Rate Detectors

Byron Murray Keel

Chapter Outline

16.1 Introduction 589
16.2 Overview of Detection Theory 590
16.3 False Alarm Impact and Sensitivity 592
16.4 CFAR Detectors 593
16.5 Cell Averaging CFAR 597
16.6 Robust CFARs 607
16.7 Algorithm Comparison 616
16.8 Adaptive CFARs 618
16.9 Additional Comments 619
16.10 Further Reading 620
16.11 References 620
16.12 Problems 622

16.1 INTRODUCTION

The process of detecting a target begins with comparing a radar measurement with a threshold. Measurements exceeding the threshold are associated with returns from a target, and measurements below the threshold are associated with thermal noise or other interference sources including intentional jamming and background returns from terrain and bodies of water. The detector threshold is selected to achieve the highest possible probability of detection for a given signal-to-noise ratio (SNR) and probability of false alarm. A false alarm occurs when, in the absence of a target, a source of interference produces a measured value that exceeds the detection threshold. A radar system is designed to achieve and maintain a specified probability of false alarm. False alarms drain radar resources by appearing as valid target detections requiring subsequent radar actions and thus degrade system performance.

If the statistics of the interference are known a priori, a threshold may be selected to achieve a specific probability of false alarm. In many cases, the form of the probability density function (PDF) associated with the interference is known, but the parameters of the distribution are either unknown or change temporally or spatially. Constant false alarm rate (CFAR) detectors are designed to track changes in the interference and to adjust the detection threshold to maintain a constant probability of false alarm.

16.2 OVERVIEW OF DETECTION THEORY

An introduction to constant false alarm rate detectors begins with a review of detection theory. Detection decisions are based on measurements of reflected signals received at the radar and thermal noise inherently present in the receiver. Samples or measurements may be collected in one or more dimensions, including range, cross-range, angle, and Doppler. In general, the received signals are sampled at an interval spacing equal to the radar system's resolution in the dimension in which they are collected. The radar detector is tasked with comparing the measurement with a threshold and choosing between two hypotheses. Measurements exceeding the threshold are declared to contain returns from targets as well as energy from interfering sources and are associated with the target-plus-interference hypothesis (commonly referred to as the H_1 hypothesis). Measurements below the threshold are declared to contain energy only from interfering sources and are associated with the null hypothesis, H_0. Interfering sources include receiver noise, intentional jamming, unintentional electromagnetic interference, and background returns (often referred to as "clutter"). Clutter returns are defined as reflections from objects in the scene that are not viewed as "targets" by the radar (e.g., reflections from precipitation or terrain).

A diagram of a radar's basic processing chain is provided in Figure 16-1. The analog-to-digital converter (ADC) converts the received analog signal into digital samples. Signal conditioning is then applied to these samples to maximize the signal-to-interference-plus-noise-ratio (SINR) prior to the detector. Signal conditioning algorithms include, but are not limited to, Doppler processing, pulse compression, array processing, and space-time adaptive processing (see Chapters 17 and 20). In general, these algorithms operate on complex samples. The real and imaginary parts of a complex sample correspond to the receiver's in-phase (I) and quadrature (Q) channels. Complex signals at the output of the conditioning process are then passed to a rectifier. A rectifier converts the complex samples into either their magnitude or magnitude-squared values. A linear rectifier outputs a complex sample's magnitude, and a square law rectifier outputs its magnitude squared. A detector employing a linear (or square law) rectifier is referred to as a linear (or square law) detector. A log rectifier may also be applied [1,2]. A log rectifier compresses the dynamic range of the measured values and is less susceptible to target masking. Target masking refers to missed detections caused by target returns that bias the CFAR detection threshold. Target masking and its impact on CFAR detection performance are examined in detail in subsequent sections.

The detector operates on the output of the rectifier and yields one of three possible outcomes: a correct decision, a missed detection, or a false alarm. A correct decision is one in which the detector correctly declares the presence or absence of a target. A missed detection is one in which the detector declares the absence of a target when in truth the measurement contains a target return. A false alarm occurs when the detector declares the presence of a target and in reality a target's return is not present in the measured data.

In radar, both interference and target returns are modeled as random processes characterized by PDFs and power spectral densities or, equivalently, autocorrelation functions.

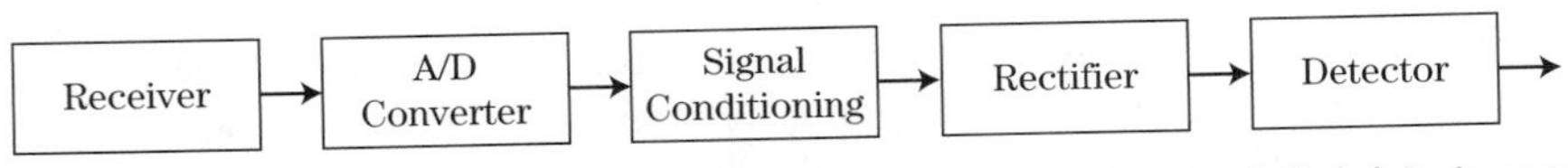

FIGURE 16-1 ■ In a modern radar, the processing chain operates on digital data to maximize SINR, detect targets, and control false alarms.

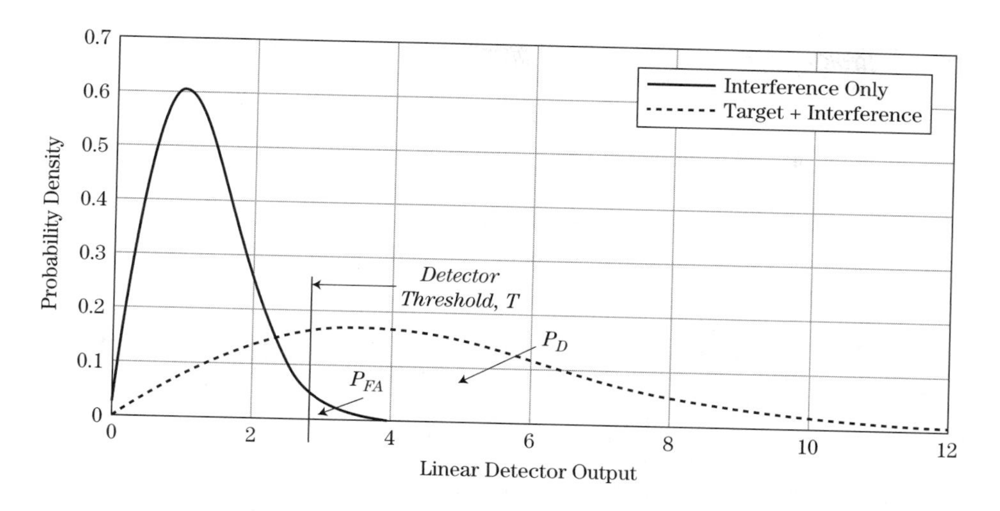

FIGURE 16-2 ■ Setting the detector threshold determines the P_{FA} of the null hypothesis and the P_D of the target present hypothesis.

In many cases, both the target and interference are Rayleigh distributed at the output of a linear rectifier [3]. Figure 16-2 contains a plot of the PDF associated with Rayleigh distributed interference (solid curve) and a plot of the PDF associated with a Swerling 1 or 2 [3] target combined with Rayleigh distributed interference (dashed curve). Section 15.5.9 showed that, in the absence of noncoherent integration (i.e., $N = 1$), both a Swerling 1 and 2 target produce a Rayleigh distributed voltage at the output of a linear rectifier.

Due to the stochastic nature of target returns and interference, a detector's performance is described in terms of probabilities. The likelihood of detecting a target is specified in terms of a probability of detection, P_D, and the likelihood of a false alarm is specified in terms of a probability of false alarm, P_{FA}. A detector may be designed to optimize performance based on a cost function that weights each decision (i.e., correct decision, missed detection, and false alarm).

The Neyman-Pearson (NP) detector, described in Chapter 15, employs a fixed threshold that maximizes P_D given a specific probability of false alarm. The NP detector is employed assuming that the interference is independent and identically distributed (IID) over all resolution cells to which the fixed threshold is to be applied and that the parameters of the interference distribution are known. In many cases, the assertion that the distribution parameters are known is not valid. The solution is to employ a detector that estimates the parameters of the distribution from the measured data and uses the estimate to set the detector threshold. This chapter focuses on methods for adaptively setting the detector threshold to maintain a constant false alarm rate.

Once a threshold, T, has been selected, performance metrics such as P_D and P_{FA} may be computed using the target-plus-interference and interference-only PDFs. P_D is calculated by computing the area under the target-plus-interference PDF to the right of the threshold, and P_{FA} is calculated by computing the area under the interference-only PDF to the right of the threshold. The area under each PDF associated with P_D and P_{FA} is illustrated in Figure 16-2. In general, threshold selection is a trade-off between P_D and P_{FA} for a given SINR and the cost associated with a correct and incorrect decision.

The occurrence of a false alarm represents a drain on limited radar resources. The frequency of false alarms has a direct impact on overall radar system performance and is quantified using a false alarm rate metric. *False alarm rate* is defined as the number of false alarms occurring within a given time interval. False alarm rate, *FAR*, is computed using the expression

$$FAR = \frac{P_{FA} M}{T_M} = N_D P_{FA} \tag{16.1}$$

where M represents the number of resolution cells collected over a specific time interval defined by T_M and $N_D = M/T_M$ is the number of detection decisions per unit time.

16.3 FALSE ALARM IMPACT AND SENSITIVITY

The initial task assigned to a radar system is to detect targets. Once a target is detected, radar resources are then allocated to other functions such as verification, track initiation, discrimination, or the measurement of some physical property associated with the target. False alarms cause the radar to invoke actions that consume finite resources. For example, verify pulses or track initiate pulses may be commanded in response to false alarms and as a result reduce the time available for other actions. A large number of false alarms may also overload the data bus and signal/data processor, which would result in dropped detections or tracks and a reduction in the timeline and processing capacity available for other radar modes (e.g., discrimination). Radars are often sized to accommodate a specific false alarm rate. A significant increase in the false alarm rate produces a substantial decrease in system performance.

In most cases, changes in the false alarm rate correspond to fluctuations in the interference power. For example, thermal noise power in the receiver may vary with time due to internal heating and cooling as well as to changes in ambient radiation impinging on the antenna. Interference power levels may also fluctuate due to intentional jamming or variations in terrain reflectivity associated with changes in terrain type or grazing angle.

A square law detector is commonly applied in radar systems and exhibits several desirable properties. In many instances, the PDFs at the output of a square law detector and the resulting mathematical analysis are relatively tractable. In addition, for Rayleigh distributed interference, the maximum likelihood (ML) estimator of the mean interference power takes the form of a square law detector. The mean power estimate is used as a statistic in a number of the CFAR algorithms examined in subsequent sections.

Interference that is Rayleigh distributed at the output of a linear detector will be exponentially distributed at the output of a square law detector, as shown in equation (15.10) and repeated here [3],

$$p_z(z) = \begin{cases} \dfrac{1}{\sigma_i^2} \exp\left(\dfrac{-z}{\sigma_i^2}\right), & z \geq 0 \\ 0, & z < 0 \end{cases} \tag{16.2}$$

where $z = |y|^2$ is the output of the detector when its input is the complex interference signal y, and σ_i^2 is the mean or expected value of z. The output of a square law detector is interpreted as a power measurement, whereas the output of a linear detector is interpreted as a voltage measurement. Thus, the quantity σ_i^2 represents the mean of the interference power. Given exponentially distributed interference, the probability of a false alarm is computed by integrating equation (16.2) from the threshold T to $+\infty$, or

$$P_{FA} = \int_{T}^{\infty} \frac{1}{\sigma_i^2} \exp\left(-\frac{z}{\sigma_i^2}\right) dz \tag{16.3}$$

which gives

$$P_{FA} = \exp\left(-\frac{T}{\sigma_i^2}\right) \tag{16.4}$$

For a fixed threshold, an increase in the interference power causes an increase in P_{FA} as evident in equation (16.4).

Suppose that the mean interference power is increased by a factor of κ and that the threshold is not adjusted. The threshold is initially set based on a desired probability of false alarm and an assumed interference power. Using equation (16.4), the probability of false alarm resulting from an increase in the interference power is

$$P_{FA_{final}} = (P_{FA_{initial}})^{1/\kappa} \tag{16.5}$$

where $P_{FA_{initial}}$ is the initial probability of false alarm, and $P_{FA_{final}}$ is the resultant probability of false alarm. Consider an increase in the interference power by a factor of 3 dB (i.e., $\kappa = 2$). In this case, $P_{FA_{final}} = \sqrt{P_{FA_{initial}}}$. For $P_{FA_{initial}} = 10^{-4}$, or 1 false alarm per 10,000 observations, the probability of false alarm increases to 10^{-2}, or 1 false alarm in 100 observations. This example illustrates a dramatic increase in probability of false alarm or false alarm rate for a relatively small increase in interference power. As expected, P_{FA} increases because the threshold is not adjusted to account for the increase in the interference power.

16.4 | CFAR DETECTORS

A desirable property of a detector is an ability to maintain a given probability of false alarm in the presence of heterogeneous or changing interference. A detector possessing this property is termed a constant false alarm rate detector. CFAR detectors estimate statistics of the interference from radar measurements and adjust the detector threshold to maintain a constant false alarm rate or, equivalently, a fixed P_{FA}.

In response to increasing interference, it can be argued that a higher threshold maintains the false alarm rate at a cost of degraded detection performance. This argument is valid, but there are other ways to increase the probability of detection once the threshold is adjusted. If the interference is associated with thermal noise or a noise jammer, one option is to increase the pulse length in order to place more energy on target and thus increase the SNR. As discussed in Chapters 3 and 15, P_D increases with increasing SNR.

An examination of CFAR algorithms begins with the cell-averaging (CA) CFAR [4]. The CA-CFAR exhibits optimum performance in a homogeneous interference environment. In many operational environments, heterogeneous conditions exist including spatial and temporal variations in the interference power and closely spaced target returns that may bias the threshold estimate. CFAR algorithms designed to operate in heterogeneous environments include greatest-of [5-7], smallest-of [6-8], censored [9,10], and order statistic [11,12] CFARs. These and other CFAR algorithms are examined in subsequent sections.

16.4.1 Review of the Neyman-Pearson Square Law Detector

Before describing the CA-CFAR algorithm in more detail, it is useful to examine the Neyman-Pearson detector and its properties [13]. An NP detector maximizes P_D given a desired P_{FA}. The threshold is fixed and is derived from a known interference PDF. In applying the NP detector, it is assumed that the interference environment is homogenous and that the parameters of the distribution are known.

Chapter 15 developed the NP detector for different combinations of linear and square law detectors; nonfluctuating and fluctuating targets; and coherent or noncoherent integration. In all cases, complex Gaussian interference was assumed. In this chapter, the focus is restricted to square law detectors and single-pulse or coherent integration processing. Noncoherent integration is not applied. Under these conditions, the likelihood ratio [13] reveals that the optimum form of the detector for a Swerling 1 or 2 target is square law. Swerling 1 and 2 targets exhibit exponentially distributed radar cross section fluctuations.

Now consider a Swerling 1 target embedded in Rayleigh distributed interference. Under the null hypothesis H_0, the in-phase and quadrature channels contain normally distributed interference and are modeled as

$$y = \mathrm{I}_i + j\mathrm{Q}_i \tag{16.6}$$

where $\mathrm{I}_i \sim \mathrm{N}\left(0, \sigma_i^2/2\right)$ and $\mathrm{Q}_i \sim \mathrm{N}\left(0, \sigma_i^2/2\right)$, respectively. Under the H_1 hypothesis, the receiver output is modeled as

$$y = (\mathrm{I}_i + \mathrm{I}_t) + j(\mathrm{Q}_i + \mathrm{Q}_t) \tag{16.7}$$

where $\mathrm{I}_t \sim \mathrm{N}\left(0, \sigma_t^2/2\right)$, and $\mathrm{Q}_t \sim \mathrm{N}\left(0, \sigma_t^2/2\right)$. The subscripts i and t denote interference and target, respectively. The output, z, of a square law detector under hypothesis H_0 (interference only) is exponentially distributed according to the PDF given in equation (16.2). The output for the target-plus-interference case is also exponentially distributed [3]

$$p_z(z) = \begin{cases} \dfrac{1}{\sigma_i^2 + \sigma_t^2} \exp\left(\dfrac{-z}{\sigma_i^2 + \sigma_t^2}\right), & z \geq 0 \\ 0, & z < 0 \end{cases} \tag{16.8}$$

It is common to write equation (16.8) in terms of SINR

$$p_z(z) = \begin{cases} \dfrac{1}{\sigma_i^2 (1 + SINR)} \exp\left(\dfrac{-z}{\sigma_i^2 (1 + SINR)}\right), & z \geq 0 \\ 0, & z < 0 \end{cases} \tag{16.9}$$

where $SINR = \sigma_t^2/\sigma_i^2$.

A Neyman-Pearson detector maximizes P_D under the constraint of achieving a specific probability of false alarm. The threshold is derived by solving equation (16.4) for the threshold T, yielding

$$T = -\ln(P_{FA})\sigma_i^2 \tag{16.10}$$

The threshold consists of the product of two terms. The first term is a function of the desired P_{FA}, and the second, σ_i^2, represents the interference power at the output of the square law detector.

Similarly, a CFAR detector's threshold is expressed, in general, as the product of two terms

$$T = \alpha \hat{g} \tag{16.11}$$

where $\hat{g}$ is a statistic associated with the interference and is estimated from the measured data, and α is the CFAR constant and is a function of the desired P_{FA}. These parameters will be examined in more detail in subsequent sections.

Computing the threshold in equation (16.10) for a specific P_{FA} requires knowledge of the interference power. As previously discussed, in many scenarios the interference power is not known a priori and may vary both temporally and spatially. In addition, small increases in interference power can produce significant increases in the false alarm rate. A detector designed to adapt its threshold to changing interference levels is required if a constant false alarm rate is to be achieved in practice.

16.4.2 Basic CFAR Architecture

In this section, the basic architecture and elements associated with a CFAR algorithm are examined. A generic CFAR detector is depicted in Figure 16-3. The samples at the output of the rectifier are stored in a data window for presentation to the detector. The CFAR window resides within the data window and is composed of leading and lagging reference windows, guard cells (Gs), and a cell under test (CUT).

Data and CFAR windows may be single- or multidimensional. An example of a one-dimensional (1-D) window is a range window consisting of N_R resolutions cells. A generic, 1-D data sequence and CFAR window are illustrated in Figure 16-4. An example

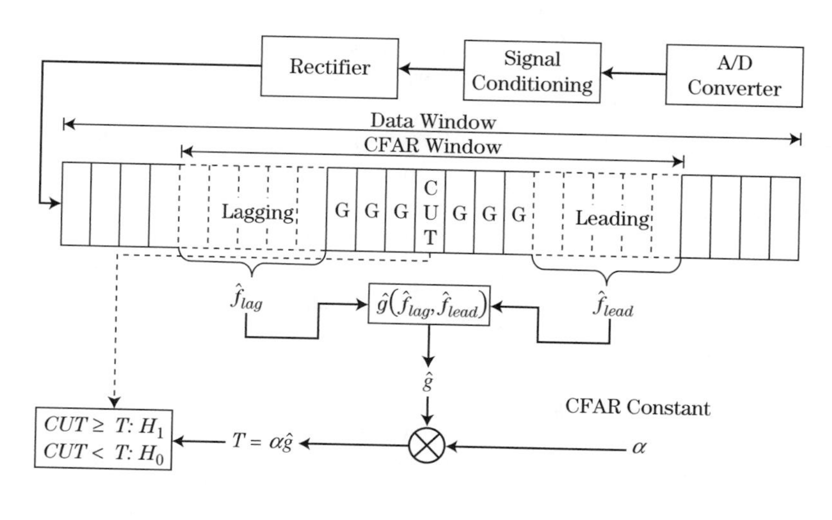

FIGURE 16-3 ■ The Constant-False-Alarm Rate (CFAR) detector uses measured samples to derive a threshold designed to maintain a constant P_{FA} in heterogeneous environments.

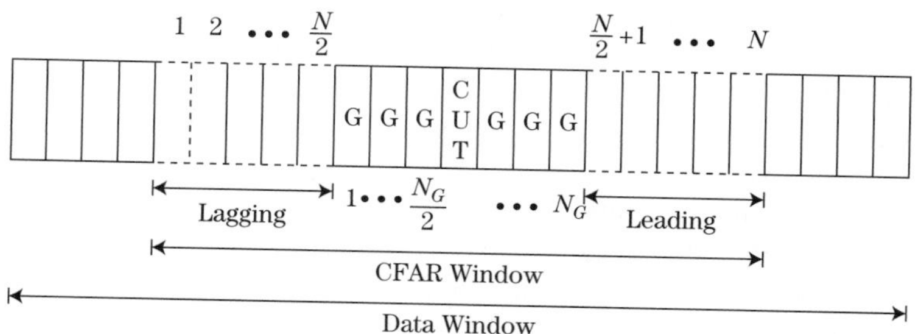

FIGURE 16-4 ■ The one-dimensional CFAR window is a subset of the data window.

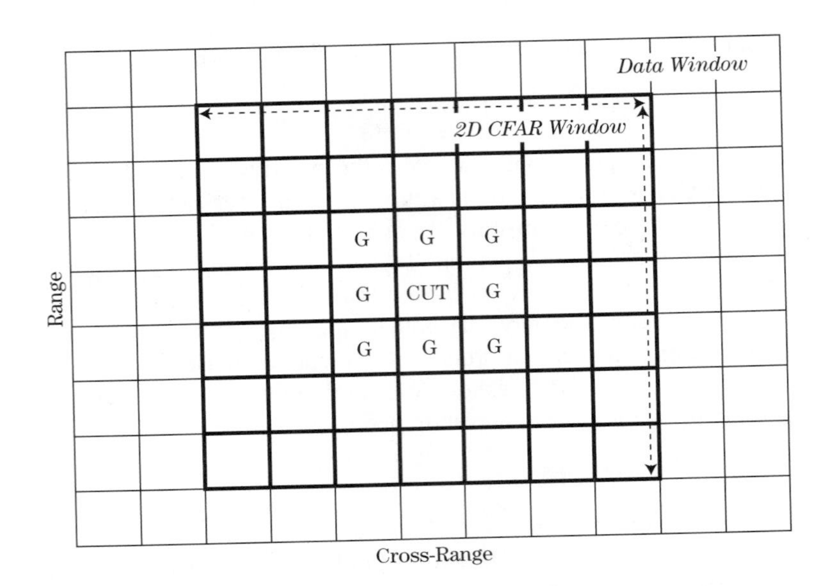

FIGURE 16-5 ■ An area CFAR uses two-dimensional data and CFAR windows.

of a two-dimensional (2-D) data set is a synthetic aperture radar (SAR) image consisting of N_R range and N_{CR} cross-range cells. A 2-D data set and CFAR window are depicted in Figure 16-5.

In Figure 16-4, the CUT is located in the center of the CFAR window. The term *cell under test* refers to the current cell to which the CFAR threshold is to be applied. In general, a fixed number of resolution cells on either side of the cell under test are denoted as *guard* or *gap cells* (indicated by a "G" in Figure 16-4). Measurements contained in the guard cells are not used to estimate the interference statistic, as they may contain returns associated with the target in the cell under test, which will bias the interference estimate. Reference windows are defined outside the guard cell region and are denoted in the 1-D case as the *leading* and *lagging windows*. The leading window is defined as the window in the direction of movement through the data window. Measurements contained in the reference windows are used to estimate the interference statistic.

The CFAR window is moved through the data window one sample or cell at a time. At each position, a detection decision is made regarding the measurement in the CUT. The detection threshold applied to the CUT is derived from measurements contained in the leading and lagging windows.

The question of how to apply the detector to the cells located within $N/2$ samples of the ends of the data window is often asked. For those cells, the CUT is not required to lie midway between the leading and lagging windows and may be shifted within the CFAR window, resulting in a nonsymmetric window. The CFAR algorithms examined in this chapter assume a symmetric CFAR window but can be modified to account for asymmetry. Note that some algorithms process the two reference windows separately and compute individual interference statistics. In these cases, extra care is needed when deriving thresholds and performance bounds for asymmetric windows. In general, it is desirable to maintain equal length reference windows (leading and lagging) but to allow the CUT and guard cells to be repositioned within the reference window if needed to test for targets at the beginning and end of the data window.

A CFAR window may be defined and applied in two dimensions (e.g., range and cross-range), as shown in Figure 16-5. For the 2-D case, the labels "leading" and "lagging" may be less intuitive, but this does not prevent the cells from being partitioned into two windows. A 2-D CFAR is commonly applied to a SAR image to detect targets or regions of interest. Detection is possible since the returns from targets are, in general, larger than returns from neighboring terrain. Over a large area, the returns from terrain may appear heterogeneous, but over localized regions, the terrain is often homogeneous, thus accommodating some form of CFAR processing.

In some systems, where measurements are collected over multiple dimensions (e.g., fast time and slow time), the CFAR detector may be applied in only a single dimension. For example, consider a moving target detector employing both range and Doppler processing. In many cases, the Doppler dimension is partitioned into *endo-* and *exo-clutter* regions. The endo-clutter region is heterogeneous in nature, and a clutter boundary exists between the two regions. In these cases, the Doppler dimension is less conducive to CFAR processing, and the CFAR is applied in only the range dimension.

Referring to the CFAR architecture in Figure 16-3, the interference statistics are computed independently for the leading and lagging windows and are denoted $\hat{f}_{lead}$ and $\hat{f}_{lag}$, respectively. The two statistics are combined using one of a number of possible operations, including mean, minimum, or maximum, to form a composite statistic, $\hat{g}$. A CFAR constant, α, is selected based on a desired P_{FA} and is a function of both the P_{FA} and specific CFAR parameters (e.g., reference window size). The threshold, T, is defined by the product of the CFAR constant, α, and the composite interference statistic, $\hat{g}$. Once the threshold has been estimated, it is applied to the CUT, and a detection decision is made. Next, the CFAR window is shifted by one resolution cell, and the process is repeated until a detection decision is made for each cell within the data window.

16.5 CELL AVERAGING CFAR

The cell averaging CFAR was first introduced by Finn and Johnson [4] in 1968. The algorithm is relatively simple in that it computes a threshold based on an estimate of the average interference power in the reference window. The CA-CFAR is designed to operate in a fairly benign interference/target environment. The following assertions are made when applying a CA-CFAR:

1. The interference in the leading and lagging windows and in the CUT is IID.
2. With a target return present in the CUT, the leading and lagging windows do not contain returns from other targets that bias the threshold estimate.

Environments for which these conditions hold are labeled "homogeneous." For many real-world target/clutter environments, these conditions are too restrictive, and a more robust CFAR algorithm is required. CFAR algorithms designed to operate in heterogeneous environments are examined in subsequent sections.

The CA-CFAR is derived to illustrate the steps involved in developing a CFAR algorithm and to identify the limitations and performance bounds associated with the detector. The derivation is based on several assertions:

1. The interference in the reference window and CUT is IID.
2. The interference is Rayleigh distributed in voltage.
3. The rectifier is square law, and thus the interference at the output is exponentially distributed.

4. The mean of the interference power at the output of the rectifier is unknown and must be estimated from the samples in the reference window.
5. The target is modeled as either a Swerling 1 or 2 (Rayleigh voltage).

The first step in the derivation is to compute an estimate of the average interference power using the samples in the reference window. The interference at the output of the square law detector is exponentially distributed

$$p_z(z) = \frac{1}{\sigma_i^2} \exp\left(\frac{-z}{\sigma_i^2}\right) \qquad z \geq 0 \tag{16.12}$$

where σ_i^2 is the mean of the interference power. For the samples in the reference window, the joint PDF is

$$p_z(z) = \frac{1}{\left(\sigma_i^2\right)^N} \exp\left(-\sum_{n=1}^{N} z_n \Big/ \sigma_i^2\right) \qquad z_n \geq 0 \tag{16.13}$$

where $z = \{z_1, z_2, \ldots, z_N\}$. Given the joint PDF, the maximum likelihood estimate [14] of the interference power is

$$\hat{\sigma}_i^2 = \frac{1}{N} \sum_{n=1}^{N} z_n \tag{16.14}$$

The CA-CFAR applies the maximum likelihood estimator in (16.14) to the samples in the leading and lagging windows to form an estimate of the interference power.

The CA-CFAR threshold, T_{CA}, is defined by the product of the power estimate in equation (16.14) and the CA-CFAR constant, α_{CA}, or

$$T_{CA} = \alpha_{CA} \hat{\sigma}_i^2 \tag{16.15}$$

The CFAR constant is a function of both the desired P_{FA} and the number of samples in the reference window. An expression relating the three parameters, α_{CA}, P_{FA}, and N, is defined in subsequent paragraphs.

The next step is to compute the average probability of detection as a function of the estimated threshold. The probability of detection is obtained by integrating the PDF in equation (16.9) from the estimated threshold to infinity. Given the CA-CFAR threshold in (16.15), the PDF in (16.9) may be integrated to obtain the probability of detection for a Swerling 1 or 2 target

$$P_D = \int_{\alpha_{CA}\hat{\sigma}_i^2}^{\infty} \frac{1}{\sigma_i^2(1 + SINR)} \exp\left(\frac{-z}{\sigma_i^2(1 + SINR)}\right) dz \tag{16.16}$$

or

$$P_D\left(\hat{\sigma}_i^2\right) = \exp\left(\frac{-\alpha_{CA}\hat{\sigma}_i^2}{\sigma_i^2\,(1 + SINR)}\right) \tag{16.17}$$

The probability of detection defined in (16.17) is a function of the interference power estimate, which is a random variable, and thus $P_D\left(\hat{\sigma}_i^2\right)$ is also a random variable. To compute the average probability of detection, the PDF associated with the interference power estimate is required and is defined by [3,15]

$$p_{\hat{\sigma}_i^2}\left(\hat{\sigma}_i^2\right) = \frac{N^N \left(\hat{\sigma}_i^2\right)^{N-1} \exp\left(\dfrac{-N\hat{\sigma}_i^2}{\sigma_i^2}\right)}{\left(\sigma_i^2\right)^N (N-1)!} \quad \hat{\sigma}_i^2 \geq 0 \tag{16.18}$$

The average or mean probability of detection, $\bar{P}_D$, is found by integrating (16.17) over all possible values of the interference statistic defined in (16.18), or

$$\bar{P}_D = \int_0^\infty p_{\hat{\sigma}_i^2}(\hat{\sigma}_i^2) \exp\left(\frac{-\alpha_{CA}\hat{\sigma}_i^2}{\sigma_i^2(1+SINR)}\right) d\hat{\sigma}_i^2 \tag{16.19}$$

Using the fact that [16]

$$\int_0^\infty x^n \exp(-ax)dx = \frac{n!}{a^{n+1}} \tag{16.20}$$

where x is a real valued variable, $a > 0$, and n is a positive integer, it can be shown [3] that (16.19) reduces to a simple closed-form expression

$$\bar{P}_D = \left[1 + \frac{\dfrac{\alpha_{CA}}{N}}{(1+SINR)}\right]^{-N} \tag{16.21}$$

The average probability of false alarm is found by setting $SINR$ equal to zero, corresponding to the interference-only condition. This gives

$$\bar{P}_{FA} = \left[1 + \frac{\alpha_{CA}}{N}\right]^{-N} \tag{16.22}$$

Note that the average probability of false alarm is independent of the interference power. The detector thus achieves a constant false alarm rate without a priori knowledge of the interference power. This property is used to define a "CFAR" detector.

As a notational convenience, the overbar on P_D and P_{FA} will not be used in the remainder of this chapter. The reader should remember that when referring to the performance of a CFAR detector, references to P_D and P_{FA} denote an average probability obtained by integrating over all possible values of the interference statistic.

The CA-CFAR constant is found by solving for α_{CA} in (16.22), giving

$$\alpha_{CA} = N\left[P_{FA}^{-1/N} - 1\right] \tag{16.23}$$

This expression for the CFAR constant applies only to the cell-averaging case and should not be associated with other CFAR algorithms.

In some cases, the number of operations is reduced by defining an interference statistic that is not normalized by $1/N$

$$\hat{g}'_{CA} = \sum_{n=1}^{N} z_n \tag{16.24}$$

and incorporating the $1/N$ factor into the CFAR constant. For this case, the resultant CFAR constant is defined by

$$\alpha'_{CA} = \left[P_{FA}^{-1/N} - 1\right] \tag{16.25}$$

and the CA-CFAR threshold is expressed as

$$T_{CA} = \alpha'_{CA}\hat{g}'_{CA} \tag{16.26}$$

16.5.1 CA-CFAR Performance

In the following sections, the performance of a CA-CFAR is examined in terms of CFAR loss, target masking, and clutter boundaries. In a homogeneous environment, the CA-CFAR is designed to achieve an average probability of false alarm. The uncertainty or variance in the CFAR statistic produces a threshold that is greater than the threshold associated with the NP detector for the same P_{FA}. The higher threshold implies that an SINR greater than that associated with an NP detector is required to achieve a given P_D. In a radar system, the CFAR detector's higher SINR requirement is interpreted as a "loss," since the lower SINR associated with the NP detector is not sufficient to achieve the desired P_D when applying the CFAR threshold. "CFAR loss" is defined as the ratio of the SINR required by a CFAR detector to that required by an NP detector for a given P_D and should be accounted for in the radar system design.

In a heterogeneous environment, multiple targets and changes in the interference power degrade CFAR performance. Target returns in the reference window bias the threshold estimate and may prevent the target in the cell under test from being detected. This condition is known as *target masking*. *Clutter boundaries* are defined by significant, localized changes in the interference power. The occurrence of a clutter boundary may lead to an increased number of false alarms and to masking of targets located near the boundary.

16.5.2 Homogeneous Performance

A numerical example is used to examine the performance of a CA-CFAR operating in a homogenous environment. Consider a single target embedded in IID interference. The measured returns and CA-CFAR threshold are plotted in Figure 16-6. The data window consists of 1,000 cells, and a Swerling 1 target is positioned in the 500th cell. The target is embedded in Rayleigh distributed interference with an average $SINR = 20$ dB. The

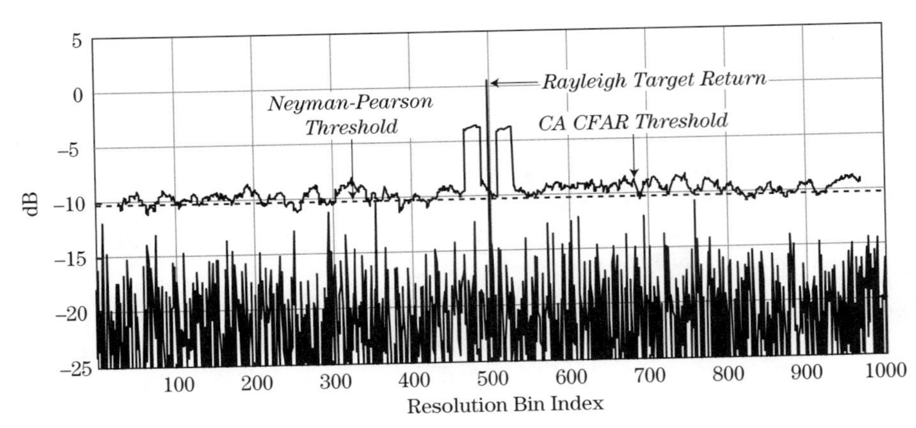

FIGURE 16-6 ■ The Cell-Averaging (CA)-CFAR ($N = 40$, $N_G = 20$, $P_{FA} = 10^{-4}$) varies the threshold over the data window containing a single target with $SINR = 20$ dB.

CFAR parameters are $N = 40$, $N_G = 20$, and $P_{FA} = 10^{-4}$. A large value of N_G is chosen to illustrate the influence of guard cells on the CFAR threshold.

The plot compares the CFAR-derived threshold (solid curve) with the Neyman-Pearson threshold (dashed curve), which requires knowledge of the interference power. The target's return exceeds both the CFAR and NP threshold. A number of the interference samples are close to the CFAR threshold, but in this realization none exceed the threshold. This is not surprising since the probability of a false alarm is 1 in 10,000, but only 1,000 cells were examined. The CA-CFAR exhibits a threshold that, in general, exceeds the threshold of the NP detector. The higher CA threshold translates into a larger SINR requirement, compared with the NP detector, to achieve a given P_D.

The impact of a target return located in either the leading or lagging reference window is illustrated in Figure 16-6. As the CFAR window moves to the right, the target initially falls in the leading window and subsequently appears in the lagging window. The presence of a target return in either reference window biases the threshold higher. Note that when the target return falls in either the CUT or the guard cell region, the return is not used in computing the interference statistic. A target present in the reference window imposes a bias on the threshold that may mask a target in the CUT. This condition is referred to as *mutual target masking.*

16.5.3 CFAR Loss

As observed in Figure 16-6, the CA-CFAR threshold appears, on average, to be higher than the NP threshold. The higher threshold leads to a reduction in P_D. To achieve a P_D equivalent to that of the NP detector requires a higher SINR. The ratio of the *SINR* required for a CA-CFAR detector to that required for an NP detector, for a given value of P_D and P_{FA}, is defined as the CFAR loss.

Expressions for P_D and P_{FA} are derived for the NP detector using equations (16.3) and (16.9), respectively. P_{FA} is defined by

$$P_{FA} = \exp\left(\frac{-T}{\sigma_i^2}\right) \tag{16.27}$$

and P_D is defined by

$$P_D = \exp\left(\frac{-T}{\sigma_i^2\,(1 + SINR)}\right) \tag{16.28}$$

Solving for *SINR* as a function of P_D and P_{FA} yields

$$SINR_{NP} = \frac{\ln\left(\frac{P_{FA}}{P_D}\right)}{\ln(P_D)} \tag{16.29}$$

where $SINR_{NP}$ is the SINR associated with the NP detector. Equation (16.29) defines the SINR required to achieve a specified value of P_D and P_{FA} when employing an NP detector.

For a CA-CFAR detector, the SINR required to achieve a specified value of P_D and P_{FA} is derived from equations (16.21) and (16.22) and is

$$SINR_{CA} = \frac{\left(\frac{P_D}{P_{FA}}\right)^{1/N} - 1}{1 - (P_D)^{1/N}} \tag{16.30}$$

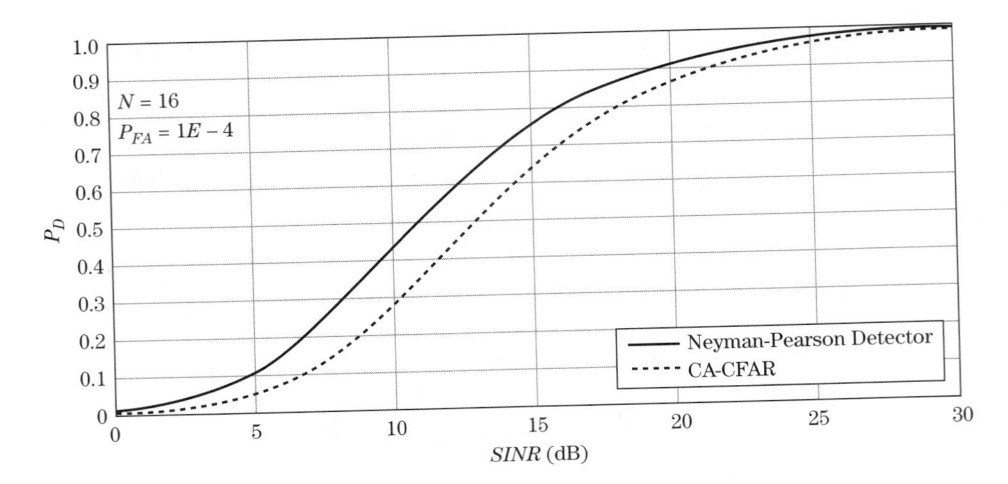

FIGURE 16-7 ■ ROC for a Neyman-Pearson detector and CA-CFAR. Additional *SINR* is required for a given P_D when employing CA-CFAR.

Note that the required *SINR* is a function of three parameters: P_D, P_{FA}, and N. Equation (16.30) is valid only for a Swerling 1 or 2 target and does not apply to noncoherently integrated returns. For a CA-CFAR, CFAR loss is defined as

$$L_{CA\text{-}CFAR} = \frac{SINR_{CA}}{SINR_{NP}} \tag{16.31}$$

In general, CFAR loss is defined as

$$L_{CFAR} = \frac{SINR_{CFAR}}{SINR_{NP}} \tag{16.32}$$

where $SINR_{CFAR}$ is the SINR associated with a given CFAR algorithm.

Receiver operating characteristic (ROC) curves are used in radar to relate SINR and detection performance. Receiver operating curves for the NP and CA-CFAR detectors are derived using (16.29) and (16.30), respectively, and are provided in Figure 16-7. The curves are based on $P_{FA} = 10^{-4}$ and $N = 16$. The location of the CA-CFAR ROC, which is to the right of the NP ROC, is indicative of a higher SINR requirement for a given P_D. CFAR loss is, therefore, the difference in SINR between the two ROCs for a given P_D. For 90% P_D, the CA-CFAR loss is ~2 dB in this example.

CFAR loss is a function of three parameters: P_D, P_{FA}, and N. For a given P_D, CFAR loss decreases with increasing P_{FA} and increasing N. The dependence on N is illustrated in Figure 16-8 for three values of P_{FA} (10^{-4}, 10^{-6}, and 10^{-8}) and 90% P_D. The decrease in CFAR loss corresponding to an increase in N is intuitive in that the variance in the interference power estimate decreases with increasing N, thus leading to a more accurate estimate of the interference power. As the CFAR loss decreases or equivalently as N increases, CFAR performance approaches that of the NP detector [6].

Based on the previous discussion, a large reference window could be used to minimize CFAR loss. A problem with this assertion is that as the size of the reference window is increased, the likelihood of encompassing multiple targets or a heterogeneous interference environment also increases. For example, a large reference window increases the likelihood that two or more targets will reside within the CFAR window and thus increases the potential for mutual target masking. Typically, the number of bins comprising the reference window ranges from 20 to 40, but this number varies depending on the resolution of the system and the target/interference environment.

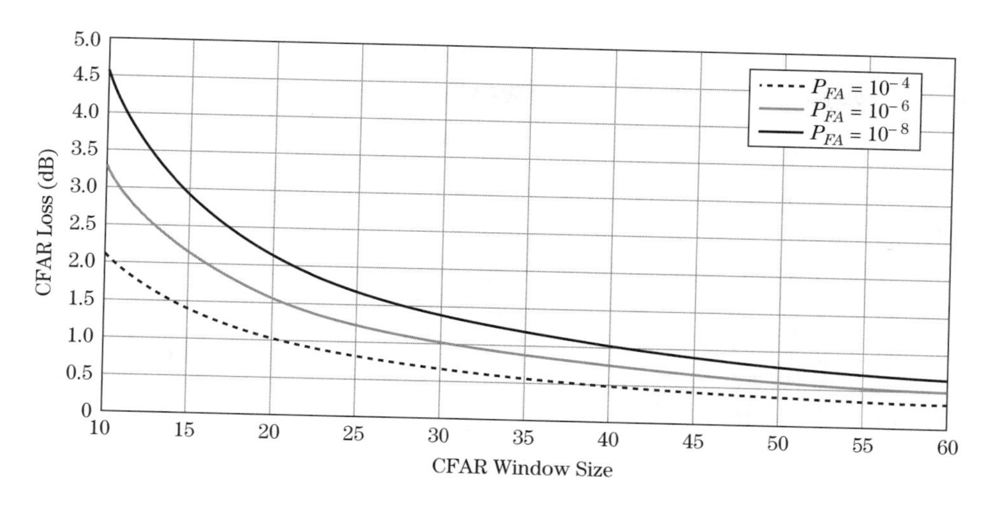

FIGURE 16-8 ■ CFAR loss as a function of CA-CFAR window size for three different values of P_{FA} and a 90% P_D.

Robust CFARs, examined in subsequent sections, are designed to operate in heterogeneous environments. The robustness of these algorithms is achieved at a price that includes greater CFAR loss and a higher degree of complexity and increased computational expense.

16.5.4 CA-CFAR Performance in Heterogeneous Environments

In this section, the performance of a CA-CFAR is examined under various heterogeneous conditions. Heterogeneous conditions exist when either of the following is true:

1. Target returns are present in either, or both, the leading or lagging windows and a target is simultaneously present in the CUT.
2. Interference sources are not identically distributed throughout the entire reference window.

Target returns in the reference window may arise from two sources:

1. A target located in the CUT whose physical extent occupies several resolution bins (e.g., a 3 m length target occupying three 1 m resolution bins).
2. Multiple targets.

In general, these heterogeneous conditions degrade CFAR performance resulting in either a loss in P_D or increase in P_{FA}. In the next two sections, the impact of targets and clutter boundaries on CA-CFAR performance is examined.

16.5.4.1 Masking

Target masking occurs when target returns located within the reference window bias the threshold above the return in the CUT. Target masking may be partitioned into two categories: self-masking and mutual target masking.

Self-masking is associated with an extended target. An extended target is defined as one whose physical extent causes it to occupy more than one resolution cell. With a sample of an extended target located in the CUT, the remaining samples associated with the target bias the threshold if one or more samples lie within the reference window. The biased threshold may mask the presence of the extended target resulting in *self-masking*.

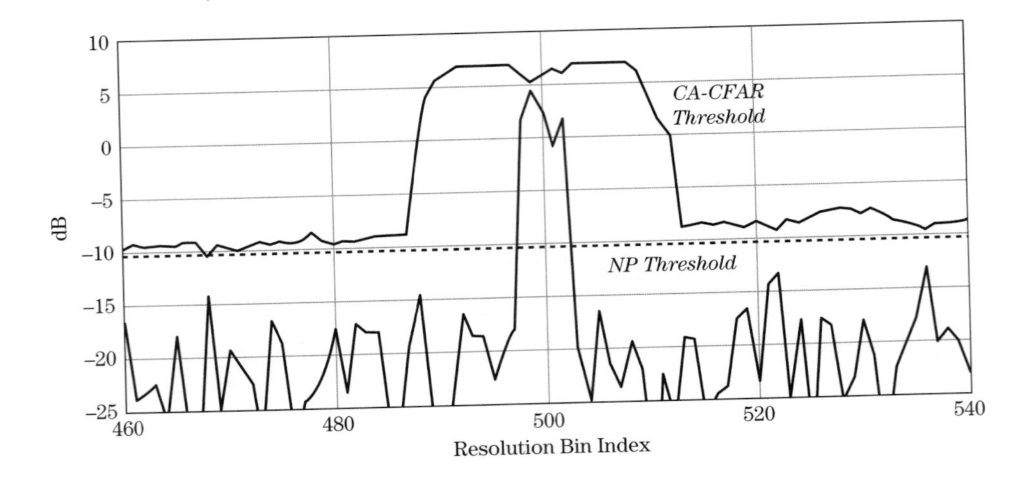

FIGURE 16-9 ■ A CA-CFAR ($N = 20$) with no guard cells exhibits self-masking when an extended target consisting of 5 Rayleigh distributed scatterers is encountered.

An example of self-masking is illustrated in Figure 16-9. The extended target is composed of five Rayleigh distributed scatterers, each occupying a different resolution cell. A CA-CFAR is applied with $N = 20$ and no guard cells for illustration. The CFAR threshold is biased above the target, and none of the target's samples are detected. The extended target experiences self-masking.

Guard or gap cells may be added to both sides of the CUT to suppress self-masking. Figure 16-10 contains a plot of the previous data set with a new threshold obtained using a CFAR window consisting of 16 guard cells ($N_G = 16$), eight on each side of the CUT, and $N = 20$. As is evident from the plot, the guard cells prevent self-masking. The number of guard cells employed is a function of the maximum target extent and the resolution of the system. The minimum number of guard cells to place on either side of the CUT is equal to the target's extent divided by the resolution cell size.

Mutual target masking occurs when target returns not associated with the target in the CUT fall within the reference window and bias the threshold. To illustrate mutual target masking, the returns from two Swerling 1 targets are plotted in Figure 16-11. The Swerling targets have an average SINR equal to 20 dB. The targets are point targets and are separated by 10 resolution cells. A CA-CFAR is applied with $N = 20$ and $N_G = 16$. In Figure 16-11, the target on the right masks the presence of the target on the left. The target

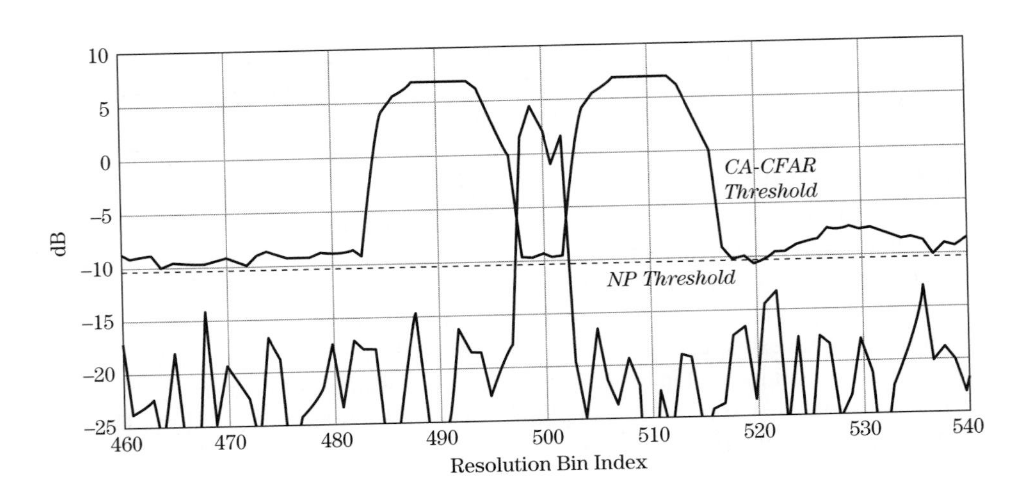

FIGURE 16-10 ■ Adding guard cells to the CA-CFAR ($N = 20$, $N_G = 16$) produces an adaptive threshold that detects the extended target.

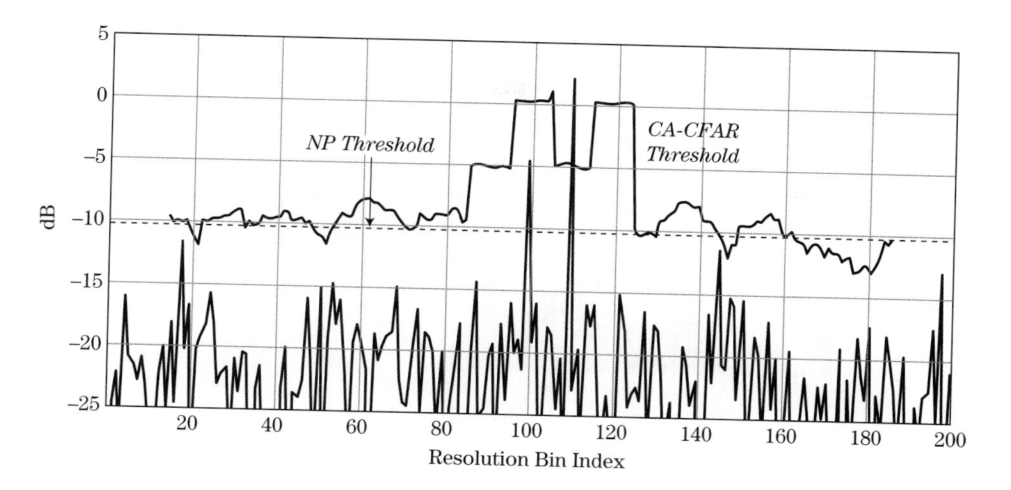

FIGURE 16-11 Mutual target masking can occur for closely-spaced targets: two Rayleigh distributed point targets separated by 10 resolution bins with average $SINR = 20$ dB. CA-CFAR has $N = 20$ and $N_G = 16$.

on the right is not masked. The bias associated with the smaller target was not sufficient to mask the presence of the larger target.

Target masking is statistical in nature as observed in the previous example, where one target masked the presence of the other but the converse did not occur. Masking is a function of the reference window size, the desired P_{FA}, the number of interfering targets, and the ratio of the interfering target's power to that of the target in the CUT.

Gandhi and Kassam [7] derive an expression for the P_D associated with a CA-CFAR with target returns present in the reference window. The target returns are assumed to be Rayleigh distributed and independent. Consider a target in the CUT with an SINR denoted $SINR_{CUT}$ and M interfering targets, each with SINR denoted $SINR_{IT}$. The P_D achieved with a CA-CFAR in the presence of M interfering targets is [7]

$$P_{D_M} = \left[1 + \frac{\alpha'_{CA}(1 + SINR_{IT})}{1 + SINR_{CUT}}\right]^{-M} \left[1 + \frac{\alpha'_{CA}}{1 + SINR_{CUT}}\right]^{M-N} \tag{16.33}$$

where α'_{CA} is defined in equation (16.25).

A plot of P_D as a function of M for a $P_{FA} = 10^{-6}$, $SINR_{CUT} = 20$ dB, and $SINR_{IT} = \{5, 10, 15, 20 \text{ dB}\}$ is provided in Figure 16-12. P_D is reduced from 82% for $M = 0$ to 42% for $M = 1$ when $SINR_{CUT} = SINR_{IT} = 20$ dB.

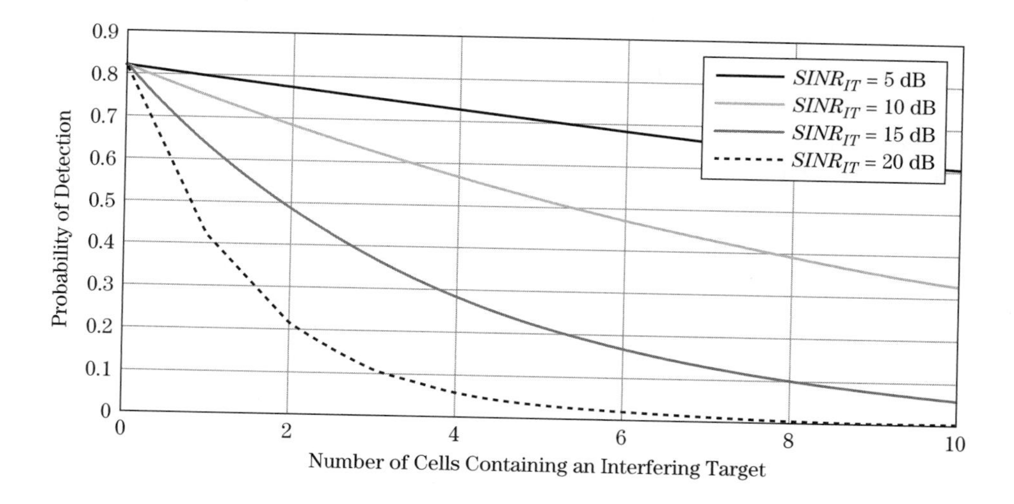

FIGURE 16-12 For a CA-CFAR, P_D as a function of the number of targets in the reference window. Simulation parameters include $SINR_{CUT} = 20$ dB, $N = 20$ cells, and $P_{FA} = 10^{-6}$.

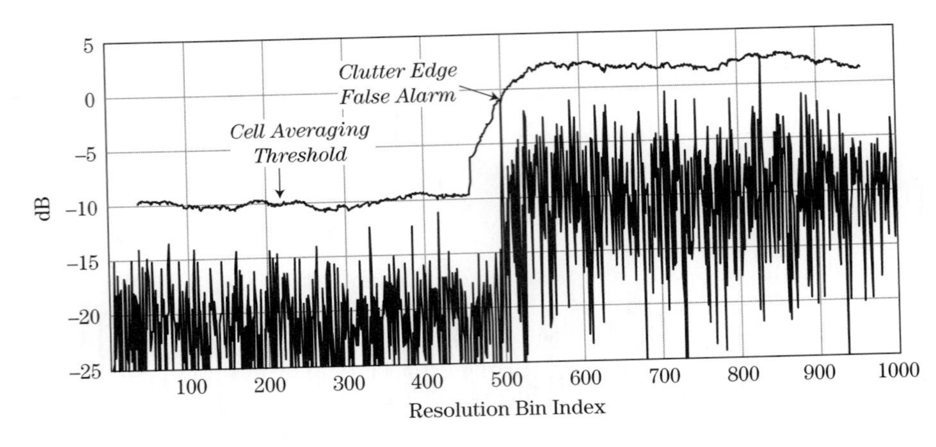

FIGURE 16-13 ■ The CA-CFAR threshold generates a clutter edge false alarm at the clutter boundary.

16.5.4.2 Clutter Boundaries

Significant and abrupt changes in terrain reflectivity impact both the false alarm rate and target masking. Abrupt changes in terrain reflectivity are termed *clutter boundaries* and are indicative of a change in terrain type or grazing angle. A data window containing a clutter boundary is illustrated in Figure 16-13. The presence of a clutter boundary has two primary effects on CFAR performance. The first is a reduction in P_D for targets positioned on the low reflectivity side of the clutter boundary. This condition is termed clutter edge masking. The second is an increase in the number of false alarms near the clutter boundary. A clutter edge false alarm is illustrated in Figure 16-13. The lower reflectivity region biases the threshold down as the CFAR window passes over the clutter boundary resulting in a false alarm.

Gandhi and Kassam [7] derive bounds on detection performance for a CA-CFAR in proximity to a clutter boundary. Consider a reference window that contains a clutter boundary. The clutter returns to the left of the clutter boundary are defined to have an average power level of $\sigma_{C_L}^2$, and clutter returns to the right of the clutter boundary are larger in terms of reflectivity and are defined to have an average power level of $\sigma_{C_H}^2$ where $\sigma_{C_L}^2 < \sigma_{C_H}^2$. Target and clutter returns are immersed in receiver noise with noise power defined by σ_n^2. For a target located in the lower clutter reflectivity region, P_D may be expressed as [7]

$$P_{D_{CB}} = \left[1 + \frac{\alpha'_{CA}\dfrac{\sigma_{C_H}^2 - \sigma_{C_L}^2}{\sigma_n^2 + \sigma_{C_L}^2}}{1 + \dfrac{\sigma_t^2}{\sigma_n^2 + \sigma_{C_L}^2}}\right]^{-M_H} \left[1 + \frac{\alpha'_{CA}}{1 + \dfrac{\sigma_t^2}{\sigma_n^2 + \sigma_{C_L}^2}}\right]^{M_H - N} \tag{16.34}$$

where M_H is the number of samples in the leading window that lie in the higher reflectivity region, and σ_t^2 is the power in the target return. The quantity $\sigma_t^2/(\sigma_n^2 + \sigma_{C_L}^2)$ represents the target's SINR in the lower reflectivity region, and $(\sigma_{C_H}^2 - \sigma_{C_L}^2)/(\sigma_n^2 + \sigma_{C_L}^2)$ represents the differential clutter reflectivity between the two regions normalized by the total interference in the lower reflectivity region. For purposes of notation, define $\Delta C = (\sigma_{C_H}^2 - \sigma_{C_L}^2)/(\sigma_n^2 + \sigma_{C_L}^2)$. Using (16.34), P_D is plotted in Figure 16-14 as a function of M_H for four values of ΔC (5, 10, 15, and 20 dB) and with $SINR = 20$ dB, $N = 20$, and $P_{FA} = 10^{-4}$. As observed, the potential for a clutter boundary to mask a nearby target is a function of both the number of higher reflectivity returns present in the reference window and the difference in reflectivity between the two regions.

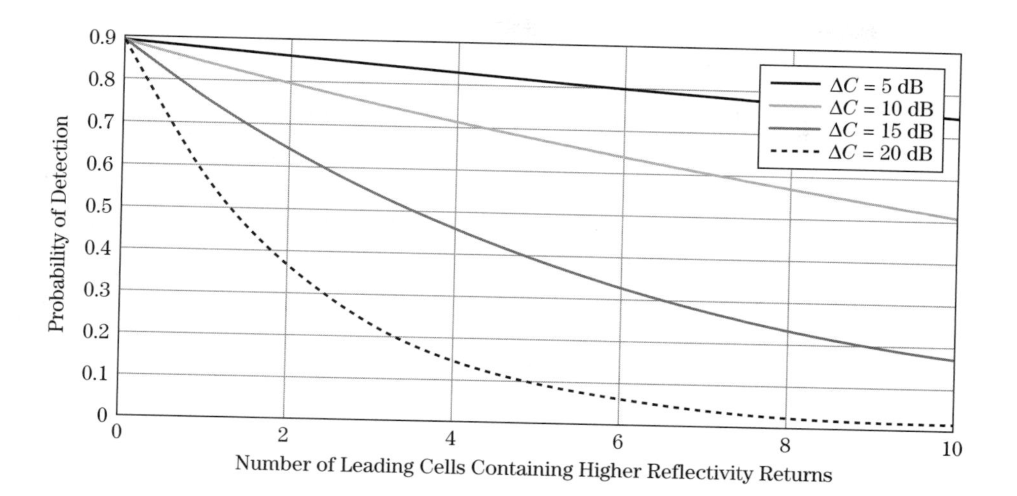

FIGURE 16-14 ■ For a CA-CFAR, P_D as a function of the number of returns associated with the higher reflectivity region. Simulation parameters include $SINR = 20$ dB, $N = 20$, and $P_{FA} = 10^{-4}$.

Clutter boundaries also influence the likelihood of a false alarm. Whether the boundary causes an increase or decrease in the P_{FA} depends on the location of the CUT. If the CUT is located in the lower reflectivity region, a reduction in the false alarm rate is observed as samples from the higher reflectivity region fall in the leading window and bias the threshold higher. An expression for P_{FA} when the CUT is positioned in the lower reflectivity region is [7]

$$P_{FA_{CB-low}} = \left[1 + \left(1 + \frac{\sigma_{C_H}^2 - \sigma_{C_L}^2}{\sigma_{C_L}^2 + \sigma_n^2}\right)\alpha'_{CA}\right]^{-M_H} [1 + \alpha'_{CA}]^{M_H - N} \qquad (16.35)$$

where M_H is the number of samples in the higher reflectivity region contained within the leading reference window. If the CUT is located in the higher reflectivity region, a higher false alarm rate is observed provided the lagging window contains some samples associated with the lower reflectivity region. The expression for P_{FA} when the CUT is positioned in the higher reflectivity region is [7]

$$P_{FA_{CB-high}} = [1 + \alpha'_{CA}]^{-M_H} \left[1 + \frac{\alpha'_{CA}}{\left(1 + \frac{\sigma_{C_H}^2 - \sigma_{C_L}^2}{\sigma_{C_L}^2 + \sigma_n^2}\right)}\right]^{M_H - N} \qquad (16.36)$$

The two expressions for P_{FA} in equations (16.35) and (16.36) are plotted in Figures 16-15 and 16-16, respectively, for $N = 20$ and $P_{FA} = 10^{-3}$. P_{FA} is maximum when the CUT lies in the higher reflectivity region, and the lagging reference window contains returns exclusive to the lower reflectivity region.

16.6 ROBUST CFARs

As previously discussed, a CA-CFAR's performance may degrade significantly in the presence of interfering targets and clutter boundaries. Since the introduction of the CA-CFAR in 1968, researchers have developed alternative CFAR algorithms that are designed to achieve good performance under specific heterogeneous conditions. A level of robustness is achieved at the expense of increased CFAR loss, additional complexity, higher

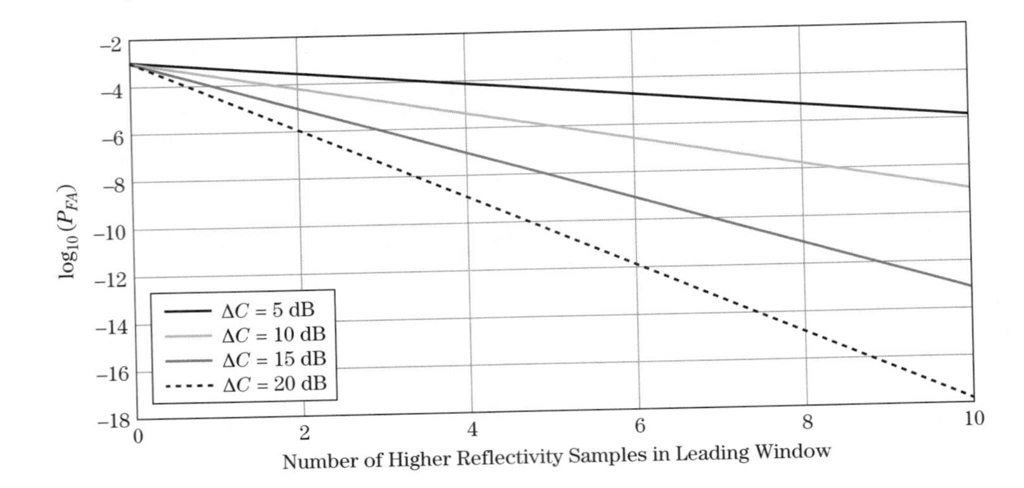

FIGURE 16-15 ■ For a CA-CFAR, P_{FA} as a function of the number of clutter cells associated with the higher reflectivity region when the cell-under-test (CUT) is located in the lower reflectivity region. Simulation parameters include $SINR = 20$ dB, $N = 20$, and $P_{FA} = 10^{-3}$.

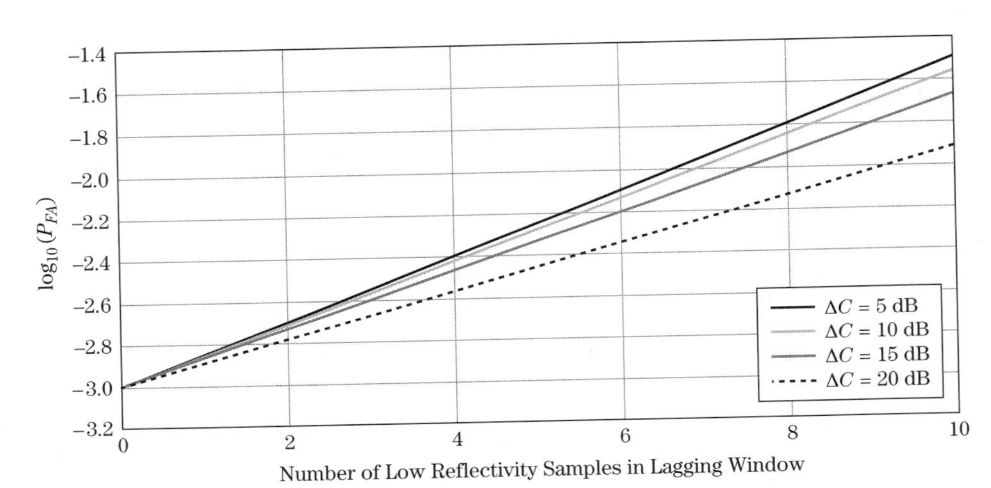

FIGURE 16-16 ■ For a CA-CFAR, P_{FA} as a function of the number of clutter cells associated with the lower reflectivity region when the CUT is located in the higher reflectivity region. Simulation parameters include $SINR = 20$ dB, $N = 20$, and $P_{FA} = 10^{-3}$.

computational cost, and CFAR constants that require iterative solutions. In general, robust CFAR algorithms require the user to possess some a priori knowledge about the target/clutter environment. For example, some algorithms require the user to define, a priori, the number of targets that may be present in the reference window and to use the number as an input to the algorithm to suppress mutual target masking.

Some of the more common CFARs include greatest-of CA-CFAR (GOCA-CFAR), smallest-of CA-CFAR (SOCA-CFAR), trimmed mean (TM) or censored (CS) CFAR, and order statistics (OS) CFAR. GOCA-CFAR is designed to minimize the number of clutter edge false alarms. SOCA-CFAR, TM-CFAR or CS-CFAR, and OS-CFARs are designed to suppress mutual target masking. The architecture and performance of each algorithm is examined in the following sections.

16.6.1 Greatest-of CA-CFAR

Hansen and Sawyers [5] developed the "greatest-of" CA-CFAR to reduce clutter edge false alarms. Clutter edge false alarms are suppressed by computing the average interference power in the lagging and leading windows separately and selecting the larger of the two sample means as the CFAR statistic. Mathematically, the GOCA-CFAR may be expressed as

$$\hat{g}_{GO} = \max(\hat{f}_{GO,\,lag}, \hat{f}_{GO,\,lead}) \tag{16.37}$$

where

$$\hat{f}_{GO,\,lag} = \sum_{i=1}^{N/2} z_i \tag{16.38}$$

and

$$\hat{f}_{GO,\,lead} = \sum_{i=N/2+1}^{N} z_i \tag{16.39}$$

Note that the $1/2N$ scale factor needed to compute the sample mean is accounted for in the CFAR constant. In a homogenous interference environment, the average P_D associated with a GOCA-CFAR [5–7] is defined by

$$P_{D_{GO}} = 2\left\{\left[1 + \frac{\alpha_{GO}}{1 + SINR}\right]^{-\frac{N}{2}} - \left[2 + \frac{\alpha_{GO}}{1 + SINR}\right]^{-\frac{N}{2}} \times \sum_{k=0}^{N/2-1} \binom{\frac{N}{2} - 1 + k}{k} \left[2 + \frac{\alpha_{GO}}{1 + SINR}\right]^{-k}\right\} \tag{16.40}$$

and the average P_{FA} is found by setting $SINR = 0$, yielding

$$P_{FA_{GO}} = 2\left\{[1 + \alpha_{GO}]^{-\frac{N}{2}} - [2 + \alpha_{GO}]^{-\frac{N}{2}} \sum_{k=0}^{N/2-1} \binom{\frac{N}{2} - 1 + k}{k} [2 + \alpha_{GO}]^{-k}\right\} \tag{16.41}$$

The binomial coefficients, denoted $\binom{m}{n}$ where m and n are integers, are defined by

$$\binom{m}{n} = \frac{m!}{(m-n)!n!} \tag{16.42}$$

and are found in many of the subsequent expressions for P_D and P_{FA}.

The GOCA-CFAR threshold is defined as the product of the statistic in equation (16.37) and the CFAR constant embedded in equation (16.41). Solving for the CFAR constant in (16.41) is more difficult than in the case of a CA-CFAR and requires an iterative solution. A complex relationship between P_{FA} and α is inherent in most robust CFAR algorithms.

CFAR loss is an important performance metric and is used in comparing the strength of different CFAR algorithms. Hansen and Sawyers [5] show that the additional CFAR loss associated with a GOCA-CFAR is ≤ 0.3 dB. This additional loss is defined relative to a CA-CFAR with an equivalent length reference window. The relationship holds even for small reference window sizes (e.g., $N = 4$). The additional CFAR loss typically ranges from 0.1 to 0.3 dB. Figure 16-17 contains a plot of the ROC associated with a CA-CFAR and a GOCA-CFAR with $N = 16$ and $P_{FA} = 10^{-6}$. As expected, the ROC associated with

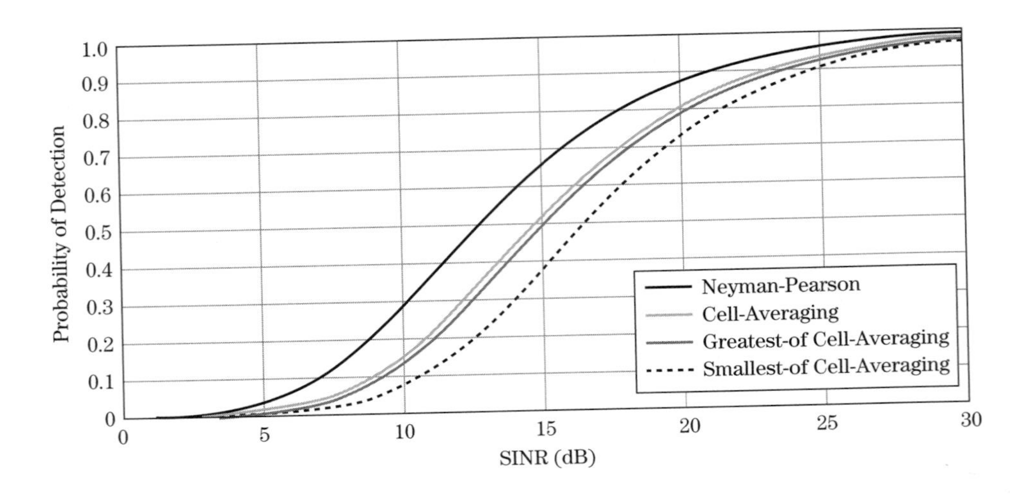

FIGURE 16-17 ■ ROC for an NP detector and three CFARs – CA, GO, and SO with $N = 16$ and $P_{FA} = 10^{-6}$.

the GOCA-CFAR appears to the right of the CA-CFAR. The larger CFAR loss is due to the fact that fewer samples are used in computing the GOCA-CFAR statistic. Table 16-1 contains a list of CFAR loss values recorded at the 90% P_D level for both the CA and GOCA-CFARs. The losses are defined for selected values of P_{FA} and N. Note that the difference between the CA and GOCA-CFAR is less than or equal to 0.3 dB.

A GOCA-CFAR reduces clutter edge false alarms by biasing the threshold above that of the CA-CFAR. The threshold tends to track the higher reflectivity region. Figure 16-18 contains simulated Rayleigh distributed interference for two clutter regions where the reflectivity between the two regions differs by 20 dB. The reference window is sized for $N = 40$, and the desired P_{FA} equals 10^{-3}. Superimposed on the plot are the CA and GOCA thresholds. The GOCA threshold tends to ride above the CA threshold as the CFAR window passes over the clutter boundary. In this particular data set, a false alarm is not observed; however, the GOCA-CFAR, with the higher threshold, is less susceptible to false alarms within the transition region. Gandhi and Kassam [7] provide an expression for a GOCA-CFAR's P_{FA} when a clutter boundary is present in the reference window.

GOCA-CFAR, with its ability to suppress clutter edge false alarms, exhibits degraded performance in the presence of interfering targets. Interfering targets capture the greatest-of logic and bias the CFAR threshold. Furthermore, the bias is greater relative to a

TABLE 16-1 ■ The CFAR Loss Associated with a CA-CFAR, GOCA-CFAR, and SOCA-CFAR

		CFAR Loss (dB)		
N	P_{FA}	CA-CFAR	GOCA-CFAR	SOCA-CFAR
8	10^{-4}	2.7	3.0	5.3
16	10^{-4}	1.3	1.5	2.3
24	10^{-4}	0.9	1.0	1.5
32	10^{-4}	0.6	0.8	1.0
8	10^{-6}	4.3	4.6	8.9
16	10^{-6}	2.0	2.2	3.8
24	10^{-6}	1.3	1.5	2.3
32	10^{-6}	1.0	1.1	1.7

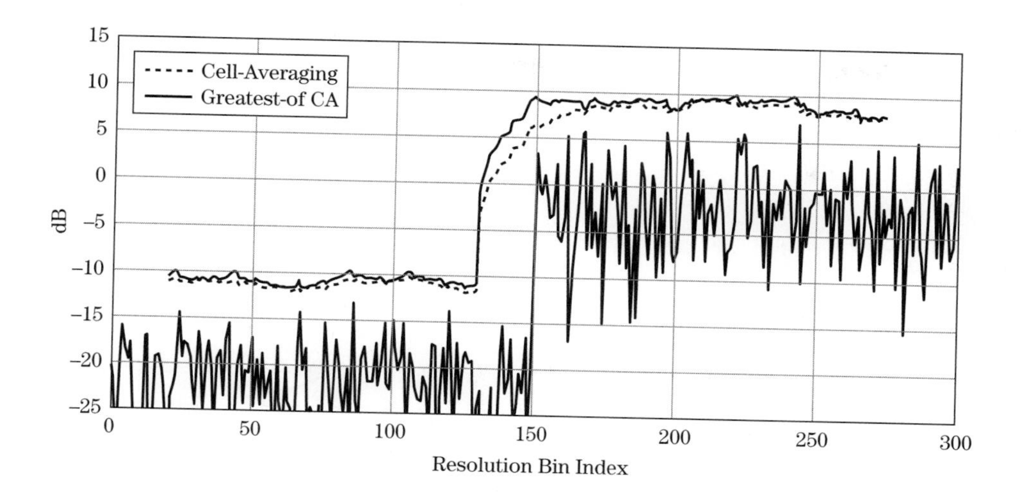

FIGURE 16-18 ■ The Greatest-Of CA-CFAR threshold is higher than a simple CA-CFAR threshold at a clutter boundary. The higher threshold reduces clutter edge false alarms.

CA-CFAR because the effective size of the reference window is reduced by a factor of 2. The reduction in the size of the reference window occurs because the estimate of the interference is computed separately for the leading and lagging windows and only one estimate is used to set the threshold. Weiss [6] has examined CFAR performance for the case of a single, Swerling 1 target located in either the leading or lagging window and has provided bounds on P_D for both CA and GOCA-CFARs. The bounds are defined for the limiting cases where $SINR_{CUT}$ and $SINR_{IT}$ tend toward infinity and the ratio $SINR_{IT}/SINR_{CUT}$ is a constant. The bound on P_D for GOCA is

$$\lim_{SINR_{CUT},\, SINR_{IT}\to\infty} P_{D_{GO}} = \left[1 + \frac{SINR_{IT}}{SINR_{CUT}}\alpha_{GO}\right]^{-1} \tag{16.43}$$

and for CA is

$$\lim_{SINR_{CUT},\, SINR_{IT}\to\infty} P_{D_{CA}} = \left[1 + \frac{SINR_{IT}}{SINR_{CUT}}\alpha'_{CA}\right]^{-1} \tag{16.44}$$

Note that the GOCA-CFAR constant, α_{GO}, in equation (16.43) is larger than the CA-CFAR constant, α'_{CA}, in equation (16.44) [6]. Given an interfering target, the larger GOCA-CFAR constant produces a lower P_D compared with a CA-CFAR.

Figure 16-19 contains a plot comparing the detection performance of a CA-CFAR and GOCA-CFAR, based on equations (16.43) and (16.44), when a single interferer lies within the reference window. For this example, the simulation parameters are $N = 20$ and $P_{FA} = 10^{-4}$. The abscissa in Figure 16-19 represents the ratio of the power in the interfering target to that in the target located in the CUT. With an interfering target 1/10th as large as the target in the CUT, the CA-CFAR achieves a $P_D \approx 95\%$, whereas the GOCA-CFAR achieves a $P_D \approx 90\%$.

As a final comment, when applying the bounds in equations (16.43) and (16.44), care must be taken to recognize the limiting conditions placed on SINR. Also, note that an exact expression for P_D for a CA-CFAR with interfering targets is provided in (16.33).

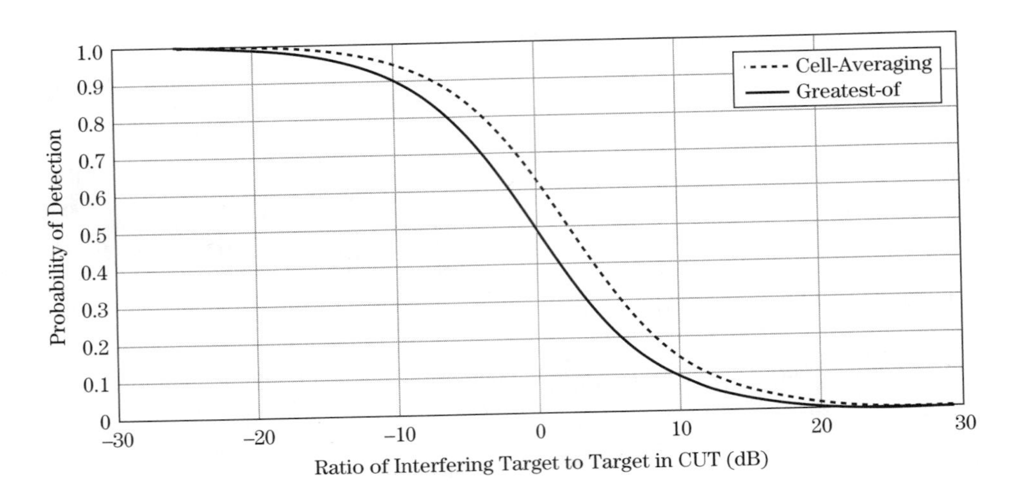

FIGURE 16-19 Weiss' performance bounds for a CA and GOCA-CFAR in the presence of an interfering target. The CFAR parameters are $N = 20$ and $P_{FA} = 10^{-4}$.

16.6.2 Suppression of Mutual Target Masking

Several CFAR algorithms have been designed to address mutual target masking: SOCA-CFAR, CS-CFAR or TM-CFAR, and OS-CFAR. These algorithms are described in the following sections.

16.6.2.1 Smallest-of CA-CFAR

Trunk [8] proposed a smallest-of CA-CFAR to address mutual target masking. The SOCA-CFAR estimates the interference power in the lagging and leading reference windows and selects the smaller of the two estimates as the CFAR statistic. The smaller of the two estimates is selected to suppress interfering targets that may reside in either the leading or lagging window but not targets present simultaneously in both windows. The SOCA-CFAR statistic is defined as

$$\hat{g}_{SO} = \min(\hat{f}_{SO,lag}, \hat{f}_{SO,lead}) \tag{16.45}$$

where

$$\hat{f}_{SO,lag} = \sum_{i=1}^{N/2} z_i \tag{16.46}$$

and

$$\hat{f}_{SO,lead} = \sum_{i=N/2+1}^{N} z_i \tag{16.47}$$

The SO threshold is defined as

$$T_{SO} = \alpha_{SO}\hat{g}_{SO} \tag{16.48}$$

In a homogenous interference environment, the average P_D for a SOCA-CFAR [6] is defined as

$$P_{D_{SO}} = 2\left[2 + \frac{\alpha_{SO}}{1 + SINR}\right]^{-\frac{N}{2}} \sum_{k=0}^{N/2-1} \binom{\frac{N}{2} - 1 + k}{k} \left[2 + \frac{\alpha_{SO}}{1 + SINR}\right]^{-k} \tag{16.49}$$

and the average P_{FA} is found by setting $SINR = 0$, giving

$$P_{FA_{SO}} = 2[2+\alpha_{SO}]^{-\frac{N}{2}} \sum_{k=0}^{N/2-1} \binom{\frac{N}{2}-1+k}{k} [2+\alpha_{SO}]^{-k} \tag{16.50}$$

SOCA-CFAR exhibits two properties that limit its practical application. First, a SOCA-CFAR's ability to suppress mutual target masking is limited to cases where the interfering targets are restricted to either the leading or lagging windows. Performance is severely degraded when interfering targets are present in both windows. Second, a SOCA-CFAR exhibits a relatively large CFAR loss when compared with other CFAR algorithms with equivalent size reference windows. The SOCA CFAR loss is observed in Figure 16-17. At 90% P_D, the CFAR loss associated with the SOCA-CFAR is 1.8 dB greater than a CA-CFAR and 1.6 dB greater than a GOCA-CFAR. The difference in CFAR loss between a CA-CFAR and a SOCA-CFAR varies depending on the value of N and P_{FA} as illustrated in Table 16-1. For these cases, the difference in CFAR loss ranges from 0.4 dB to 4.6 dB.

16.6.2.2 Trimmed Mean or Censored CFAR

Rickard and Dillard [9] and Ritcey [10] define a censored CFAR, which rank orders the measured samples in the reference window and discards the largest N_C samples prior to computing the CFAR statistic. The assertion is that the largest N_C samples may contain returns from interfering targets and therefore should not be used in estimating the CFAR statistic. The CFAR statistic is an estimate of the average power in the remaining reference cells. The CS-CFAR is capable of removing N_C interfering targets from the reference window. The user is required to assume, a priori, the maximum number of targets that may be present in the reference window and to use this number to set N_C. For a fixed reference window size, increasing N_C increases the CFAR loss as the number of samples used in estimating the CFAR statistic decreases; therefore, N_C should be selected taking into account the number of potential interfering targets and the CFAR loss incurred.

Gandhi and Kassam [7] present a more general form of the CS-CFAR algorithm termed a trimmed mean CFAR, which discards the N_{T_L} largest and N_{T_S} smallest samples. The TM-CFAR is a rank-ordered approach that computes the mean interference power from a subset of samples. A TM-CFAR may be tailored to address different interference environments by adjusting the two parameters, N_{T_S} and N_{T_L}, for specific conditions. It can shown that an OS-CFAR and CA-CFAR are special cases of the TM-CFAR with $(N_{T_S}, N_{T_L}) = (k-1, N-k)$ and $(0, 0)$, respectively. An OS-CFAR and the variable k are defined in the next section.

In general, N_{T_L} is selected to remove interfering targets, and N_{T_S} is selected to suppress clutter edge false alarms. As in the case of a CS-CFAR, N_{T_L} is selected to match an a priori estimate of the maximum number of interfering targets in the reference window. If a goal is to minimize clutter edge false alarms, then N_{T_S} should be selected as a significant percentage of N. To address both clutter edge false alarms and target masking, Gandhi [7] suggests that for $N = 24$, let $N_{T_S} = 18$ to 20 and $N_{T_L} = 1$ to 3.

Expressions for the average P_D and P_{FA} for the both the TM-CFAR and CS-CFARs are given by

$$P_{D_{TM}} = \prod_{i=1}^{N-N_{T_S}-N_{T_L}} \gamma_i(\nu)\bigg|_{\nu = \frac{\alpha_{TM}}{1+SINR}} \tag{16.51}$$

and

$$P_{FA_{TM}} = \prod_{i=1}^{N-N_{T_S}-N_{T_L}} \gamma_i(\nu)\bigg|_{\nu = \alpha_{TM}} \tag{16.52}$$

where

$$\gamma_1 = \frac{N!}{N_{T_S}!(N - N_{T_S} - 1)!(N - N_{T_S} - N_{T_L})} \sum_{k=0}^{N_L} \frac{\binom{N_{T_S}}{k}(-1)^{N_{T_S}-k}}{\frac{N-k}{N-N_{T_S}-N_{T_L}} + \nu} \tag{16.53}$$

and

$$\gamma_i = \frac{\frac{N-N_{T_S}-i+1}{N-N_{T_S}-N_{T_L}-i+1}}{\frac{N-N_{T_S}-i+1}{N-N_{T_S}-N_{T_L}-i+1} + \nu} \quad i = 2, \ldots, (N - N_{T_S} - N_{T_L}) \tag{16.54}$$

The CFAR constant in equation (16.52) contains the scale factor used in computing the sample mean. Note that the CS-CFAR is a special case of the TM-CFAR with $N_{T_S} = 0$ and $N_{T_L} = N_C$.

16.6.2.3 Order Statistics CFAR

Rohling [11,12] defines an order statistics CFAR that is designed to suppress target masking. The OS-CFAR rank orders the N samples in the CFAR reference window and selects the k-th sample as the CFAR statistic. The CFAR is thus capable of rejecting $N - k$ interfering targets. In addition, an OS-CFAR is capable of suppressing clutter edge false alarms provided $k > N/2$ [7,11].

Gandhi and Kassam [7] provide the following expressions for the average P_D and P_{FA}:

$$P_{D_{OS}} = \prod_{i=0}^{k-1} \frac{N - i}{N - i + \frac{\alpha_{OS}}{1+SINR}} \tag{16.55}$$

and

$$P_{FA_{OS}} = \prod_{i=0}^{k-1} \frac{N - i}{N - i + \alpha_{OS}} \tag{16.56}$$

where α_{OS} is the OS-CFAR constant. An equivalent expression for the average P_{FA} is [11]

$$P_{FA_{OS}} = k\binom{N}{k}\frac{(k - 1)!(\alpha_{OS} + N - k)!}{(\alpha_{OS} + N)!} \tag{16.57}$$

Rohling [11] examines the CFAR loss associated with an OS-CFAR and shows that it exhibits a relatively broad minimum as a function of k. To achieve a CFAR loss near the minimum, a reasonable value of k is $3N/4$. This value of k also supports the suppression of $N/4$ interfering targets.

For a given value of N, an OS-CFAR exhibits a CFAR loss greater than that exhibited by a CA-CFAR. With its ability to suppress mutual target masking, the length of an OS-CFAR's reference window may be extended to reduce the CFAR loss. The longer reference window may contain additional interfering targets, but the algorithm is designed to suppress them. From a CFAR loss perspective, similar detection performance may be achieved in a homogenous interference environment using a CA-CFAR with $N = 16$ or an OS-CFAR with $N = 24$ [11].

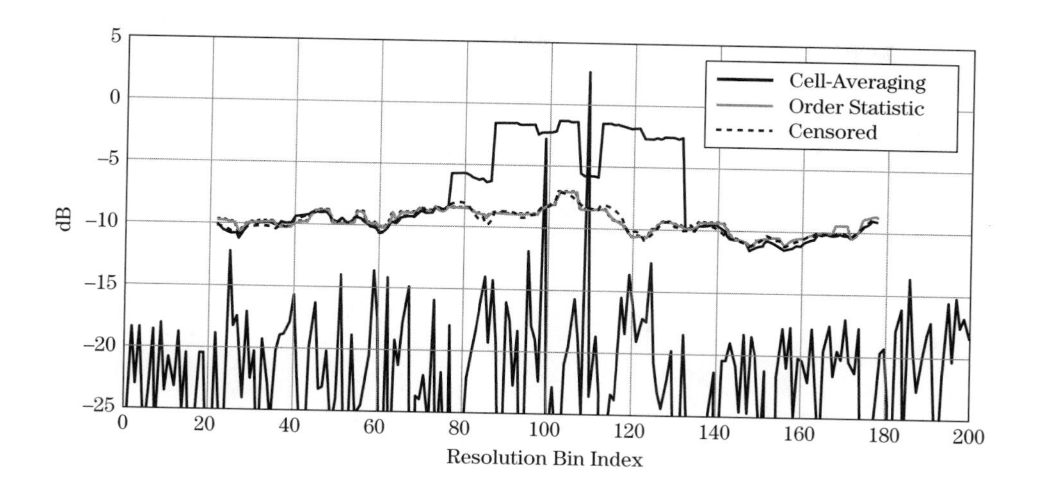

FIGURE 16-20 ■ The Order Statistic (OS), and Censored CS(2) CFARs mitigate mutual target masking while the CA-CFAR misses one target.

An OS-CFAR may also be used to address self-target masking. An OS-CFAR eliminates the need for guard cells (i.e., $N_G = 0$) provided the total number of reference cells containing target returns is less than $N - k$.

The discussion to this point has been limited to square law detectors. However, for the case of an OS-CFAR a simple expression does exists that relates the CFAR constants associated with a linear and square law detector. The two OS-CFAR constants are related via [11]

$$\alpha_{OS_{\text{linear}}} = \sqrt{\alpha_{OS_{\text{square law}}}} \tag{16.58}$$

This relationship does not apply to other CFAR algorithms (e.g., CA-CFAR).

16.6.2.4 CS- and OS-CFAR Numerical Example

The ability of a CS-CFAR and OS-CFAR to mitigate target masking is illustrated in the following example. Consider two Swerling 1 targets embedded in Rayleigh distributed interference with an average $SINR = 20$ dB. Figure 16-20 contains a plot of the CA-, CS-, and OS-CFAR thresholds derived from the simulated returns. The detectors are designed to achieve a $P_{FA} = 10^{-4}$. The reference window consists of 40 cells. The CS-CFAR discards the largest two samples (denoted CS(2)), and the OS-CFAR uses the 30th sample to compute the CFAR statistic. In this example, both the CS- and OS-CFARs detect the two targets, whereas the CA-CFAR detects only the larger target. The OS-CFAR is capable of suppressing 10 interfering targets, and the CS-CFAR is capable of suppressing two interfering targets.

16.6.3 Combining GO with CS or OS

Both order statistics and censored CFARs are designed to address mutual target masking. Each algorithm rank orders the samples in the reference window and excludes some number of the largest samples when computing the interference statistic. In suppressing some of the larger returns, these CFAR techniques are susceptible to clutter edge false alarms. Several authors [17–20] have examined combining greatest-of with OS- and CS-CFARs to provide robustness in the presence of multiple targets and clutter boundaries. The first step is to apply either an OS- or CS-CFAR to the leading and lagging reference

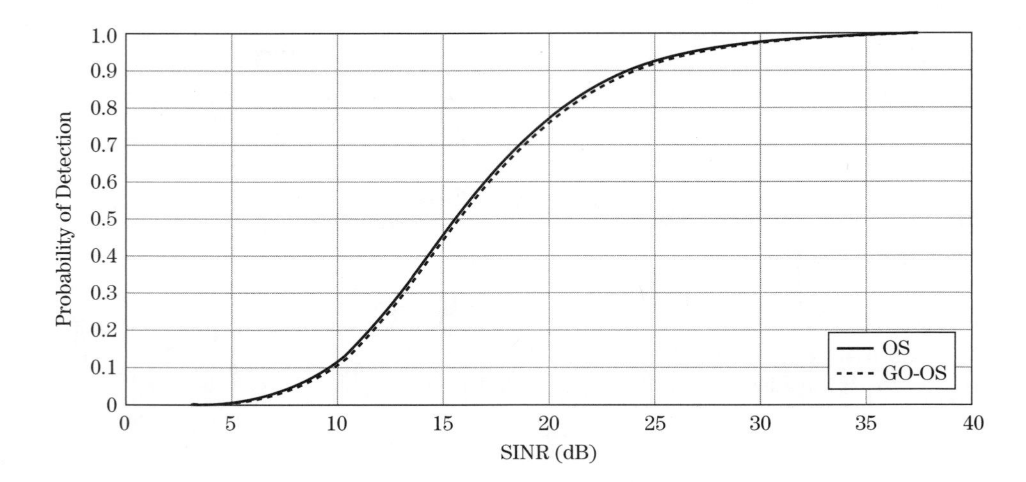

FIGURE 16-21 ■ ROC for OS and GO-OS CFARs with $N = 16$ and $P_{FA} = 10^{-6}$.

windows separately to address mutual target masking. Next, the larger of the two statistics is then selected to suppress clutter edge false alarms. The acronyms GO-OS and GO-CS CFAR are used to denote greatest-of order statistics and greatest-of censored CFAR, respectively.

Elias-Fuste [17] provides expressions for P_D and P_{FA} when applying GO-OS CFAR. The average P_D is defined by

$$P_{D_{GO-OS}} = 2k^2 \binom{N/2}{k}^2 \sum_{j=0}^{N/2-k} \sum_{i=0}^{N/2-k} \binom{N/2-k}{j} \binom{N/2-k}{i} \times \cdots$$
$$\cdots \frac{(-1)^{N-2k-j-i}}{N/2-i} \frac{\Gamma(N-j-i)\Gamma(\alpha_{GO-OS}/(1+SINR)+1)}{\Gamma(N-j-i+\alpha_{GO-OS}/(1+SINR)+N/2+1)} \tag{16.59}$$

and the average P_{FA} is defined by

$$P_{FA_{GO-OS}} = 2k^2 \binom{N/2}{k}^2 \sum_{j=0}^{N/2-k} \sum_{i=0}^{N/2-k} \binom{N/2-k}{j} \binom{N/2-k}{i} \times \cdots$$
$$\cdots \frac{(-1)^{N-2k-j-i}}{N/2-i} \frac{\Gamma(N-j-i)\Gamma(\alpha_{GO-OS}+1)}{\Gamma(N-j-i+\alpha_{GO-OS}+N/2+1)} \tag{16.60}$$

where $\Gamma()$ is the Gamma function. Figure 16-21 contains a plot comparing the ROCs for an OS- and GO-OS CFAR. The reference window is length 16, and detector is designed to exhibit a P_{FA} equal to 10^{-6}. The additional CFAR loss associated with combining GO and OS is only a few tenths of a dB.

16.7 ALGORITHM COMPARISON

The selection of a CFAR processor and its parameters is highly dependent on the heterogeneity of the interference environment. In a homogeneous environment, the cell averaging CFAR [6] maximizes P_D for a given reference window size, and its performance approaches that of the Neyman-Pearson detector as the window size is increased. However,

TABLE 16-2 ■ CFAR Algorithms and the Environments in which They Were Designed to Operate

	Environment			
CFAR	Homogeneous	Interfering Targets	Clutter Boundaries	Interfering Targets and Clutter Boundaries
CA	X			
GOCA			X	
SOCA		X		
CS		X		
TM		X	X	X
OS		X	X	X
GO-OS		X	X	X
GO-CS		X	X	X

as previously discussed, heterogeneous interference significantly degrades the P_D and P_{FA} performance of a CA-CFAR. To operate in heterogeneous environments, "robust" CFARs have been developed to address specific operational conditions. In many cases, a CFAR is designed to address either target masking or clutter edge false alarms. The burden is on the user to possess a priori knowledge of the interference environment and to select the CFAR best tailored to that environment. Table 16-2 contains a list of CFAR algorithms and the interference environments in which they are designed to operate. Note that the SOCA-CFAR is capable of suppressing target masking only when the interferers fall in one of the two reference windows and not both.

When selecting a CFAR algorithm, the impact of CFAR loss on detection performance must be considered. To illustrate the point, consider the ROCs for five CFAR algorithms: CA, GOCA, SOCA, CS, and OS. The ROCs are plotted in Figure 16-22. Each detector is designed to achieve a P_{FA} equal to 10^{-6}. All reference windows are length 16, and the CS-CFAR discards the largest four samples. For the OS-CFAR, $k = 12$. The CFAR loss associated with the CA-CFAR is 2 dB at the 90% P_D mark. In comparison, the GOCA-CFAR exhibits 2.25 dB of loss, and the CS- and OS-CFARs exhibit approximately 2.9 dB of loss. The CS-CFAR ROC appears to left of the OS-CFAR ROC, and the difference in CFAR loss between them is less than 0.05 dB. As with any of the CFAR algorithms,

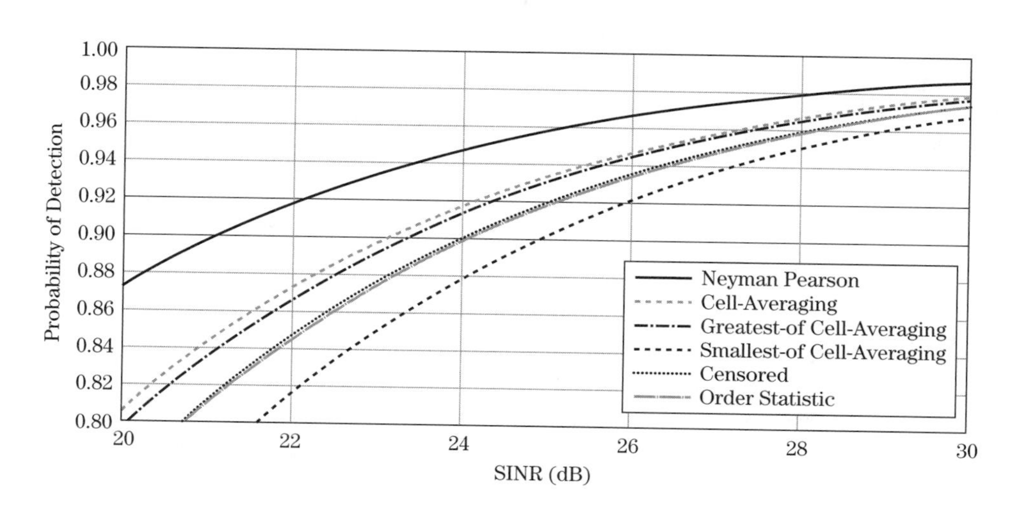

FIGURE 16-22 ■ ROCs for NP detector and CA, GOCA, SOCA, CS(4), and OS ($k = 3N/4$) CFARs with $N = 16$ and $P_{FA} = 10^{-6}$.

the size of the reference window can be increased to decrease the CFAR loss; however, the impact of heterogeneity and additional computations must also be considered when employing a larger window.

16.8 ADAPTIVE CFARs

The robust CFARs examined in the previous sections rely heavily on the user's a priori knowledge of the interference environment to select the most appropriate CFAR and its parameters. For example, when applying a censored CFAR, the number of cells to discard must be selected a priori. The number of cells is dependent on the user's knowledge or assumptions regarding the number of targets that might lie in the reference window. Since the mid 1980s, a number of algorithms have appeared in the literature that are designed to adaptively select CFAR parameters based on the measured data rather than on the user's assumptions regarding the environment. In this text, these algorithms are referred to as *adaptive* CFARs. One of the first adaptive algorithms to appear in the literature was the heterogeneous clutter estimating (HCE) CFAR developed by Finn [21]. The algorithm seeks to identify the location of the clutter boundary within the reference window and to use only samples associated with the same distribution as the CUT to compute the interference statistic. The HCE-CFAR is designed to address both clutter edge false alarms and target masking.

Adaptive CFARs offer significant P_D and P_{FA} performance advantages. The price for performance is a significant increase in the number of computations. Several adaptive CFARs appear in the literature [21–25]. The generalized censored mean level detector (GCMLD) [22] is an adaptive censored CFAR. The GCMLD estimates the number and location of interfering targets contained within the reference window and discards them prior to calculating the interference statistics. The greatest-of/smallest-of CFAR [25] is designed to estimate the location of a clutter boundary and is similar in its approach to the HCE-CFAR. Himonas and Barkat [25] have also combined the properties of the GO/SO and GCMLD into a single CFAR termed the generalized two-level censored mean level detector to address both target masking and clutter edge false alarms.

To illustrate the potential performance gains associated with an adaptive CFAR, consider the performance of a CA, CS, and GCMLD with two interfering targets located within the reference window. Figure 16-23 contains the ROC for a CA- and CS-CFAR

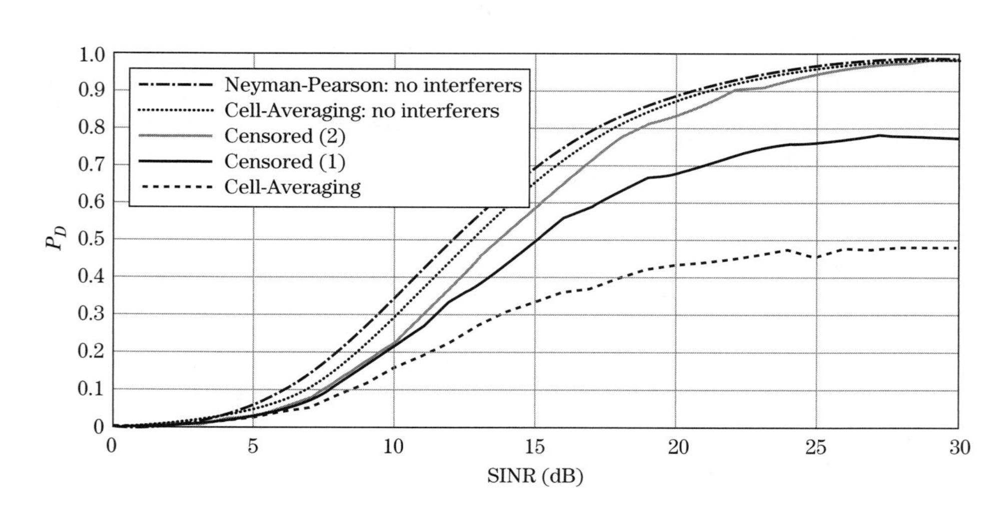

FIGURE 16-23 ■ ROC for CA, CS(1), and CS(2) CFARs with two interferers in the reference window and $SINR = SINR_{IT}$.

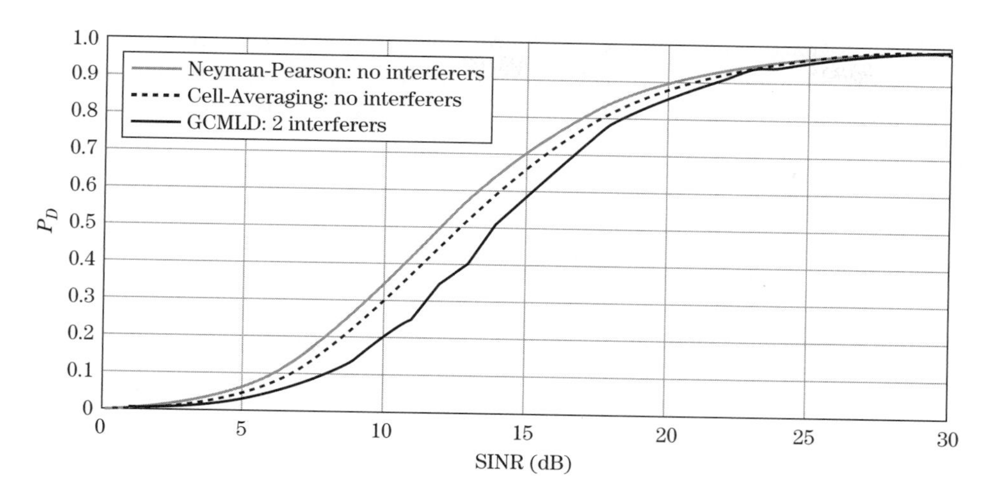

FIGURE 16-24 ■ ROC for an adaptive-censored (GCMLD) CFAR with two interferers in the reference window and $SINR = SINR_{IT}$ shows excellent performance without preselecting the number of samples to discard.

with two interfering targets and $SINR_{IT} = SINR_{CUT}$. The performance of an NP detector and CA-CFAR, with no interferers present, is also plotted as a reference. The CFARs are designed with $N = 30$ and $P_{FA} = 10^{-5}$. With two interferers, the CA-CFAR yields less than 50% P_D. The CS-CFAR, discarding the largest return (denoted CS(1)), yields less than 80% P_D. If the number of interfering targets was guessed correctly so that the largest two samples were discarded, then performance approaches that of a CA with no interferers. As expected, the CS(2)-CFAR exhibits some additional CFAR loss over the CA-CFAR with no interferers present.

The GCMLD operates on the measured data with no assumption regarding the number of targets in the reference window. It estimates the number of interferers from the measurements in the reference window. The performance of the GCMLD is illustrated in Figure 16-24 against the two interferers. The ROC for the GCMLD approaches that of the CS(2)-CFAR. The GCMLD and other adaptive CFARs hold promise for the future, provided that computational resources are available to support real-time implementation.

16.9 ADDITIONAL COMMENTS

16.9.1 Non-Rayleigh Backgrounds

The CFAR algorithms examined in this chapter are designed to operate in environments composed of Rayleigh distributed interference. However, some backgrounds (e.g., low grazing angle terrain, high-resolution imagery, reflections from the sea) are more accurately described using log normal, Weibull, or K distributions, which exhibit longer tails. These distributions are defined by both a scale and a shape parameter [26], which must be estimated to characterize the interference. CFAR algorithms have been designed for operation in these environments and are described in [26–31].

16.9.2 Clutter Map CFAR

The CFAR algorithms covered in this chapter use samples contained within a reference window to estimate the interference statistics. The reference window is moved in one or more dimensions, and the interference within the window is assumed to be relatively homogeneous. Clutter boundaries may exist, but large sections of the reference window contain homogeneous interference. In environments where the interference is heterogeneous from resolution cell to resolution cell, a clutter map CFAR may be employed. A clutter map CFAR computes the interference statistics within a given resolution cell using samples collected temporally (e.g., scan to scan). Nitzberg and Levanon [32,33] provide a discussion of the implementation and performance of clutter map CFARs.

16.10 FURTHER READING

Survey papers that describe and compare the performance of different CFAR algorithms are worthy of further examination. Two excellent survey papers are Weiss [6], which examines CA-, GO-, and SO-CFARs, and Gandhi and Kassam [7], which examines CA-, GO-, SO-, OS-, and TM-CFARs. Levanon's *Radar Principles* [3] contains an excellent chapter on CFAR. The chapter includes a derivation of the CA- and OS-CFARs and introduces the reader to the concept of CFAR loss. Belcher, in *Radar Design Principles* by Nathason [34], provides a brief but comprehensive overview and comparison of a number of CFAR algorithms without delving into the complex derivations and equations. Richards' [13] recent text *Fundamentals of Radar Signal Processing* also provides an excellent and comprehensive chapter on CFAR, which includes derivations and performance metrics.

16.11 REFERENCES

[1] Hansen, V.G., and Ward, H.R., "Detection Performance of the Cell Averaging Log/CFAR Receiver," *IEEE Transactions on Aerospace and Electronic Systems,* vol. AES-8, no. 5, pp. 648–652, September 1972.

[2] Novak, L.M., "Radar Target Detection and Map-Matching Algorithm Studies," *IEEE Transactions on Aerospace and Electronic Systems*, vol. AES-16, no. 5, pp. 620–625, September 1980.

[3] Levanon, N., *Radar Principles*, John Wiley and Sons, New York, 1988.

[4] Finn, H.M., and Johnson, R.S., "Adaptive Detection Mode with Threshold Control as a Function of Spatially Sampled Clutter-Level Estimates," *RCA Review*, pp. 414–464, September 1968.

[5] Hansen, V.G., and Sawyers, J.H., "Detectability Loss Due to "Greatest of" Selection in a Cell-Averaging CFAR," *IEEE Transactions on Aerospace and Electronic Systems*, vol. AES-16, no. 1, pp. 115–118, January 1980.

[6] Weiss, M., "Analysis of Some Modified Cell-Averaging CFAR Processors in Multiple-Target Situations," *IEEE Transactions on Aerospace and Electronic Systems*, vol. AES-18, no. 1, pp. 102–114, January 1982.

[7] Gandhi, P.P., and Kassam, S.A., "Analysis of CFAR Processors in Nonhomogeneous Background," *IEEE Transactions on AES*, vol. 24, no. 4, July 1988.

[8] Trunk, G.V., "Range Resolution of Targets Using Automatic Detectors," *IEEE Transactions on Aerospace and Electronic Systems*, vol. AES-14, no. 5, September 1978.

[9] Rickard, J.T., and Dillard, G.M., "Adaptive Detection Algorithms for Multiple-Target Situations," *IEEE Transactions on Aerospace and Electronic Systems*, vol. AES-13, no. 4, July 1977.

[10] Ritcey, J.A., "Performance Analysis of the Censored Mean-Level Detector," *IEEE Transactions on Aerospace and Electronic Systems*, vol. AES-22, no. 4, pp. 443–454, July 1986.

[11] Rohling, H., "Radar CFAR Thresholding in Clutter and Multiple Target Situations," *IEEE Transactions on Aerospace and Electronic Systems*, vol. AES-19, no. 4, pp. 608–621, July 1983.

[12] Rohling, H., "New CFAR-Processor Based on an Ordered Statistic," pp. 271–275 in *Proceedings of the IEEE International Radar Conference,* Arlington, VA, 1985.

[13] Richards, M.A., *Fundamentals of Radar Signal Processing*, McGraw-Hill, New York, 2005.

[14] Kay, S.M., *Fundamentals for Statistical Signal Processing, vol. 1: Estimation Theory*, Prentice Hall, Englewood Cliffs, NJ, 1993.

[15] DiFranco, J.V., and Rubin, W.L., *Radar Detection*, Prentice-Hall, Englewood Cliffs, NJ, 1968.

[16] Beyer, W.H., *CRC Standard Mathematical Tables,* 26th ed., CRC Press, Boca Raton, FL, 1981.

[17] Elias-Fuste, A.R., "Analysis of Some Modified Ordered Statistic CFAR: OSGO and OSSO CFAR," *IEEE Transactions on Aerospace and Electronic Systems*, vol. 26, no. 1, pp. 197–201, January 1990.

[18] Ritcey, J.A., and Hines, J.L., "Performance of Max-Mean Level Detector with and without Censoring," *IEEE Transactions on Aerospace and Electronic Systems*, vol. AES-25, no. 2, pp. 213–222, March 1989.

[19] Wilson, S.L., "Two CFAR Algorithms for Interfering Targets and Nonhomogeneous Clutter," *IEEE Transactions on Aerospace and Electronic Systems,* vol. 29, no. 1, pp. 57–72, January 1993.

[20] Ritcey, J.A., and Hines, J.L., "Performance of MAX Family of Order-Statistic CFAR Detectors," *IEEE Transactions on Aerospace and Electronic Systems*, vol. 27, no. 1, pp. 48–57, January 1991.

[21] Finn, H.M., "A CFAR Design for a Window Spanning Two Clutter Fields," *IEEE Transactions on Aerospace and Electronic Systems*, vol. AES-22, no. 2, pp. 155–169, March 1986.

[22] Himonas, S.D., and Barkat, M., "A Robust Radar CFAR Detector for Multiple Target Situations," *Proceedings of the IEEE National Radar Conference,* pp. 85–90, 1989.

[23] Gandhi, P.P., and Kassam, S.A., "An Adaptive Order Statistic Constant False Alarm Rate Detector," pp. 85–88 in *Proceedings of the IEEE National Radar Conference,* 1989.

[24] Himonas, S.D., "Adaptive Censored Greatest-of CFAR Detection," *IEE Proceedings-F*, vol. 139, no. 3, pp. 247–255, June 1992.

[25] Himonas, S.D., and Barkat, M., "Automatic Censored CFAR Detection for Nonhomogeneous Environments," *IEEE Transactions on Aerospace and Electronic Systems*, vol. 28, no. 1, January 1992.

[26] Guida, M., Longo, M., and Lops, M., "Biparametric CFAR Procedures for Lognormal Clutter," *IEEE Transactions on Aerospace and Electronic Systems,* vol. 29, no. 3, pp. 798–809, July 1993.

[27] Schleher, D.C., "Harbor Surveillance Radar Detection Performance," *IEEE Journal of Oceanic Engineering,* vol. 2, no. 4, pp. 318–325, October 1997.

[28] Gandhi, P.P., Cardona, E., and Baker, L., "CFAR Signal Detection in Nonhomogeneous Weibull Clutter and Interference," pp. 583–588 in *IEEE International Radar Conference*, 1995.

[29] Ravid, R., and Levanon, N., "Maximum-Likelihood CFAR for Weibull Background," *IEE Proceedings–F*, vol. 139, no. 3, pp. 256–264, June 1992.

[30] Levanon, N., and Shor, M., "Order Statistics CFAR for Weibull Background," *IEE Proceedings–F*, vol. 137, no. 3, pp. 157–162, June 1990.

[31] Rifkin, R., "Analysis of CFAR Performance in Weibull Clutter," *IEEE Transactions on Aerospace and Electronic Systems*, vol. 30, no. 2, pp. 315–329, April 1994.

[32] Nitzberg, R., "Clutter Map CFAR Analysis," *IEEE Transactions on Aerospace and Electronic Systems,* vol. 22, no. 4, pp. 419–421, July 1986.

[33] Levanon, N., "Numerically Efficient Calculations of Clutter Map CFAR Performance," *IEEE Transactions on Aerospace and Electronic Systems,* vol. 23, no. 6, pp. 813–814, November 1987.

[34] Nathanson, F.E., *Radar Design Principles*, 2d ed., McGraw-Hill, Inc., New York, pp. 129–142, 1991.

16.12 PROBLEMS

1. A CFAR is designed to maintain a probability of false alarm equal to 10^{-6}. Detections are performed on a per pulse basis. The pulse repetition frequency is 1 kHz. During a pulse repetition interval, 2000 samples are collected in range. What is the false alarm rate?

2. For Rayleigh distributed interference with an average interference power of 2 mwatts per receiver channel (assume both I and Q channels are present), what is the threshold required to achieve a $P_{FA} = 10^{-4}$? Assume a Neyman-Pearson detector.

3. The Neyman-Pearson threshold is set to achieve a $P_{FA} = 10^{-6}$. The interference power level changes by 6 dB. What is the new P_{FA} if the threshold remains unchanged?

4. Calculate the average P_D for a CA-CFAR with $N = 20$ and $P_{FA} = 10^{-4}$ in a homogenous environment. Assume the target in the CUT has $SINR = 22$ dB.

5. Calculate the CFAR loss associated with a CA-CFAR. Assume $P_D = 0.85$, $P_{FA} = 10^{-6}$, and $N = 20$.

6. For a CA-CFAR, calculate the SINR required to achieve a $P_D = 0.95$, with $N = 16$ and $P_{FA} = 10^{-4}$ in a homogeneous environment.

7. For a CA-CFAR, calculate P_D with one interfering target in the reference window. Assume $N = 20$, $SINR_{CUT} = SINR_{IT} = 20$ dB, and $P_{FA} = 10^{-4}$.

8. Given a clutter boundary and the potential for clutter edge masking, what is the P_D associated with a target in the lower reflectivity region with $SINR = 20$ and $\Delta C = 15$ dB. Assume $P_{FA} = 10^{-5}$, $N = 20$, and five samples of the leading window appear in the higher reflectivity region.

9. Given a clutter boundary and the potential for a clutter edge false alarm, what is the P_{FA} when the CUT of a CA-CFAR lies in the higher reflectivity region with $\Delta C = 20$ dB? Assume $N = 14$, $P_{FA} = 10^{-4}$, and five samples of the lagging window appear in the lower reflectivity region.

10. For a GO-CA CFAR, which statement is true:

a. Reduces mutual target masking. Increases clutter edge false alarms.

b. Reduces both mutual target masking and clutter edge false alarms.

c. Reduces clutter edge false alarms. Increases mutual target masking.

11. A censored CFAR has a reference window containing the following samples at the output of a square law detector. Compute the CFAR statistic with $N_C = 3$. How many interferers is the CFAR capable of rejecting? Leading = {1.1, 0.5, 2.8, 0.8, 0.99, 0.3}, Lagging = {0.65, 0.29, 0.87, 1.6, 0.3, 1.6}

12. Using the reference window data in problem 11, compute the CFAR statistic for an order statistic CFAR with $k = 0.75N$. How many interferers is the CFAR capable of rejecting?

13. Using the reference window data in problem 11, compute the CFAR statistic for a CA, GOCA, and SOCA-CFAR. Use the sample mean to compute the statistic in each case. For the CA-CFAR, compute the threshold given a $P_{FA} = 10^{-6}$.

14. Using the reference window data in problem 11, compute the CFAR statistic for a GO-OS CFAR with $k = 4$.

15. Given a scenario where at most three targets may appear in the reference window, which algorithm provides the most robustness to target masking? Assume $N = 16$.

a. Censored CFAR, $N_C = 2$.

b. Order Statistic CFAR, $k = 14$.

c. Greater-of CA-CFAR.

d. Smaller-of CA-CFAR.

e. Order Statistic CFAR, $k = 12$.

CHAPTER

17

Doppler Processing

Mark A. Richards

Chapter Outline

17.1	Introduction	625
17.2	Review of Doppler Shift and Pulsed Radar Data	626
17.3	Pulsed Radar Doppler Data Acquisition and Characteristics	627
17.4	Moving Target Indication	629
17.5	Pulse-Doppler Processing	644
17.6	Clutter Mapping and the Moving Target Detector	665
17.7	Pulse Pair Processing	668
17.8	Further Reading	673
17.9	References	673
17.10	Problems	674

17.1 INTRODUCTION

Doppler processing refers to the use of Doppler shift information to achieve one or both of two goals. The first is to enable detection of targets in environments where clutter is the dominant interference. The second is to measure Doppler shift, and thus radial velocity, of targets. In this chapter, two general classes of Doppler processing will be discussed: *moving target indication* (MTI) and *pulse-Doppler processing*. MTI processing addresses the first goal; pulse-Doppler processing addresses both.

Like target returns, clutter signals are echoes of external objects, so increasing radar power does not improve the signal-to-clutter ratio (SCR), as is seen in the signal-to-clutter form of the radar range equation in Chapter 2, equation (2.31). The SCR can be improved by decreasing the antenna azimuth beamwidth and, if the range cell extent is beam-limited, decreasing the elevation beamwidth. In a scenario where the ground clutter cell extent is pulse-limited or where volume clutter (weather, chaff) is a limiting factor, improving the range resolution with a shorter pulse or a higher time-bandwidth pulse will also help. Once the antenna and waveform parameters are set, however, Doppler processing is the principal means of improving SCR.

Chapter 8 described the Doppler effect in radar systems and typical components of the Doppler spectrum observed by a pulsed radar. Also discussed were the way a pulsed radar collects the coherent range/pulse number (or fast-time/slow-time) data matrix and measures the Doppler spectrum via the "spatial Doppler" pulse-to-pulse phase history.

This chapter focuses on processing the slow-time data in a given range bin to analyze the Doppler content of this signal, to reduce interference from clutter, and to enable detection and parameter estimation of moving targets. The emphasis is on basic concepts for detecting moving targets in clutter and on the generic slow-time filtering and spectral analysis techniques and measurements used to make this possible. Introductions to some more specialized considerations such as ambiguity resolution and blind zones are included.

17.2 | REVIEW OF DOPPLER SHIFT AND PULSED RADAR DATA

17.2.1 Doppler Shift

As discussed in Chapter 8, the Doppler shift (*change* in radar frequency) observed by a monochromatic or narrowband radar illuminating a target with a radial velocity component of v m/s is, to a very good approximation,

$$f_d = \frac{2v}{c} f = \frac{2v}{\lambda} \tag{17.1}$$

where f and λ are the transmitted frequency and wavelength, respectively, and c is the speed of light. A positive value of v indicates an approaching target and thus a positive Doppler shift.

The numerical values of Doppler shift are small compared with the radar frequencies because typical target velocities are small compared with the speed of electromagnetic wave propagation. Table 17-1 (a reprint of Table 8-1) gives the magnitude of the Doppler shift corresponding to a radial velocity of 1 meter per second, knot, or mile per hour at representative radar frequencies. A Mach 2 aircraft observed with an L-band radar would cause a Doppler shift of only 4.4 kHz in a 1 GHz carrier frequency.

TABLE 17-1 ■ Doppler Shift as a Function of Velocity and Frequency

Radar Frequency f		Doppler Shift f_d (Hz)		
Band	Frequency (GHz)	1 m/s	1 knot	1 mph
L	1	6.67	3.43	2.98
S	3	20.0	10.3	8.94
C	5	33.3	17.1	14.9
X	10	66.7	34.3	29.8
K_u	16	107	54.9	47.7
K_a	35	233	120	104
W	95	633	326	283

17.3 | PULSED RADAR DOPPLER DATA ACQUISITION AND CHARACTERISTICS

17.3.1 Review of Pulsed Radar Data Matrix and Doppler Signal Model

Measurement and processing of Doppler data in a pulsed radar begins with the fast time/slow time (range/pulse number) matrix $y[l, m]$ of coherent, complex baseband data shown in Figure 17-1. The sampling rate in the fast time or range dimension (vertical in this figure) is at least equal to the fast time signal bandwidth, which in most cases equals the transmitted pulse bandwidth, $f_s \geq B$. For a radar using a simple pulse of length τ, $B \approx 1/\tau$. If pulse compression waveforms such as linear frequency modulation (LFM) chirps are used, B is the bandwidth of the modulated pulse.

The slow-time or pulse number dimension (horizontal in the figure) is sampled at the pulse repetition interval (PRI) T of the radar. Thus, the sampling rate in this dimension is the pulse repetition frequency (PRF). Each row of the matrix represents a series of measurements from the same range bin over M successive pulses. The total amount of time MT represented by the data matrix is called the *coherent processing interval* (CPI). In the absence of windowing, the Rayleigh (peak-to-null) Doppler resolution is $1/MT$ Hz. There may be several CPIs in a *dwell*, which is the amount of time a given target is within the antenna mainbeam on a single scan.

Suppose a target is present in range bin l_0, approaching the radar with a radial velocity v. It was shown in Chapter 8 that the model of the baseband slow-time target signal (not including noise or clutter) is the *spatial Doppler* signal

$$y[l_0, m] = A \exp\left(-j\frac{4\pi}{\lambda}R_0\right) \exp\left[+j2\pi\left(\frac{2v}{\lambda}\right)mT\right], \quad m = 0, \ldots, M-1 \quad (17.2)$$

That is, the slow-time data sequence forms a complex sinusoid at the Doppler frequency $2v/\lambda$ Hz. Equation (17.2) is valid under the "stop-and-hop" assumption discussed in

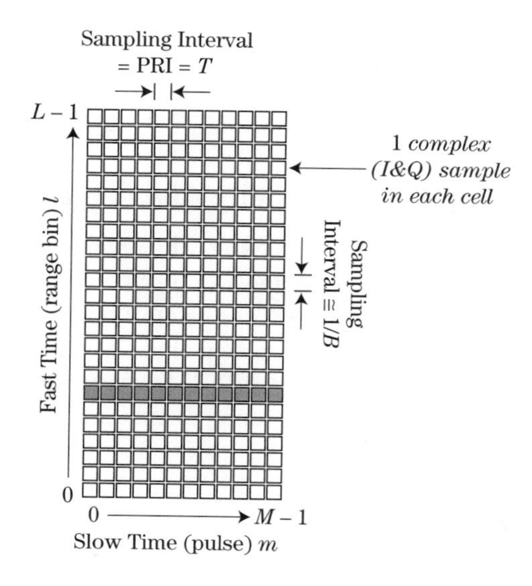

FIGURE 17-1 ■ Notional two-dimensional pulse-Doppler data matrix. The shaded samples are the slow-time signal for the seventh range bin, $y[6,m]$.

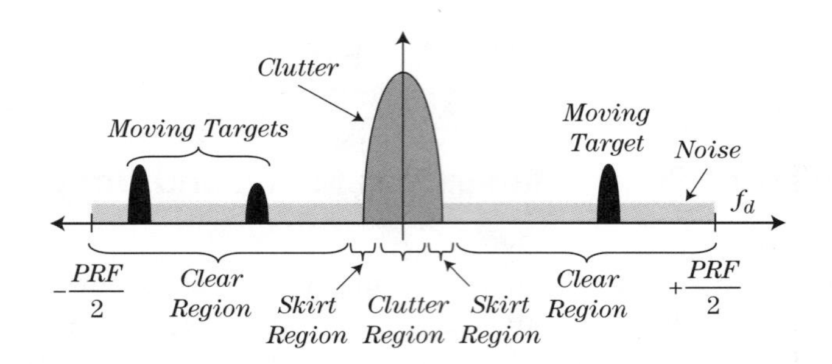

FIGURE 17-2 ■ The principal period of a notional generic Doppler spectrum containing noise, clutter, and target components. A stationary radar has been assumed.

Chapter 8 and [1]. For the velocities and CPI durations seen in conventional Doppler processing, it is usually safe to assume that range migration over the CPI is less than a range bin and can be ignored.

17.3.2 Generic Doppler Spectrum for a Single Range Bin

In general, the spectrum of the slow-time signal from a single range bin consists of noise, clutter, and one or more target signals. The distribution of these signals in range and velocity (or Doppler) was discussed in some detail in Chapter 8. Figure 17-2, which repeats Figure 8-16, shows a notional generic Doppler spectrum as observed from a stationary radar for a single range bin containing clutter, noise, and three moving targets.

In many situations, the relative amplitudes of the clutter, target, and noise signals are generally as shown: the target returns are above the noise floor (*signal-to-noise ratio* [SNR] > 1) but below the clutter (SCR < 1). In this case, targets cannot be detected reliably based on amplitude in the slow-time domain alone because the presence or absence of the target makes little difference to the overall power of the received signal, which is dominated by the clutter. Doppler processing, however, can separate moving target signals from the clutter signals in the frequency domain. The clutter can be explicitly filtered out, leaving the target returns as the strongest signal present, or the spectrum can be computed explicitly so that targets outside of the clutter region can be located by finding frequency components that significantly exceed the noise floor.

As shown in Chapter 8, the simple spectrum of Figure 17-2 is complicated by its distribution over range; moving platforms, which introduce sidelobe clutter and altitude lines; and ambiguities, which result in range and velocity foldovers. Nonetheless, it has all the features needed to introduce the basic concepts and algorithms of Doppler processing.

17.3.3 Review of Range and Velocity Aliasing and Coverage

What happens if the radar views a target having a Doppler shift magnitude greater than *PRF*/2? As was discussed in Chapter 14, the Doppler spectrum obtained by computing the discrete-time Fourier transform (DTFT), discrete Fourier transform (DFT), or fast Fourier transform (FFT) of a slow-time data sequence is periodic in frequency, with the principal period ranging from $-PRF/2$ to $+PRF/2$ Hz, corresponding to a velocity range of $\pm\lambda PRF/4$. Consequently, all frequency components are aliased, or replicated, every *PRF* Hz. Figure 17-3 illustrates this periodicity for a spectrum similar to Figure 17-2. Targets at a Doppler shift f_d outside of the $\pm PRF/2$ range will appear in the principal period at an apparent Doppler frequency

$$f_{d_a} = f_d - k_a PRF \tag{17.3}$$

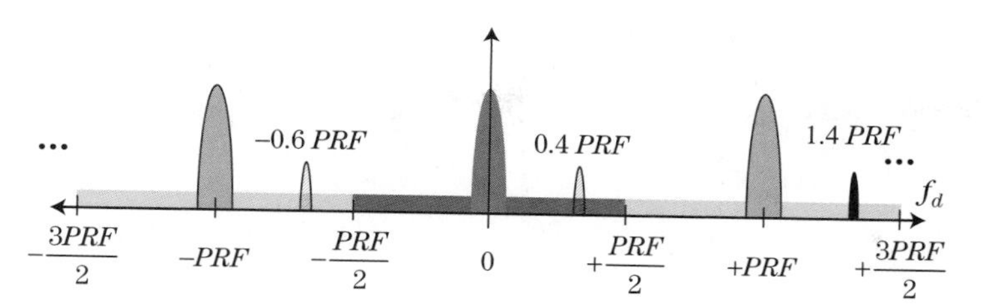

FIGURE 17-3 ■ Replication of Doppler spectrum of sampled slow-time data, showing aliasing of a high-speed target.

where k_a is an integer chosen such that f_{d_a} is between $-PRF/2$ and $+PRF/2$. For example, the target having a Doppler shift of $1.4PRF$ will have the aliases shown at $0.4PRF$ and $-0.6PRF$ as well as others at all frequencies $1.4PRF + k_a PRF$ for any integer k_a. The $k_a = -1$ alias at $0.4PRF$ will be the apparent Doppler shift for this target because it falls in the principal period of the spectrum.

The *unambiguous Doppler coverage* of the radar is the width of the range of Doppler shifts that can be measured without aliasing and is simply

$$f_{d_{ua}} = PRF \tag{17.4}$$

Caution must be used to avoid confusing the unambiguous Doppler coverage $f_{d_{ua}}$ with the limits $\pm f_{d_{ua}}/2$ of the principal period of the velocity spectrum. The range extent that can be represented unambiguously given T or PRF is

$$R_{ua} = \frac{c\,T}{2} = \frac{c}{2PRF} \tag{17.5}$$

This quantity is called the *unambiguous range* or the *range coverage*. Targets at range $R > R_{ua}$ will alias to the apparent range

$$R_a = R - k_a R_{ua} \tag{17.6}$$

where k_a is an integer chosen such that R_a is between zero and R_{ua}.

Combining equations (17.4) and (17.5) gives the combined range-Doppler coverage as

$$R_{ua} f_{d_{ua}} = \frac{c}{2}. \tag{17.7}$$

The equivalent range-velocity coverage is

$$R_{ua} v_{ua} = \frac{\lambda c}{4} \tag{17.8}$$

Equation (17.7) makes it clear that the total unambiguous range-Doppler coverage is independent of the PRF. Increasing the unambiguous range coverage by increasing the PRI reduces unambiguous Doppler coverage and vice versa. For a constant PRI system, the total unambiguous range-Doppler coverage can only be increased by increasing the wavelength—that is, using a lower radar frequency. Section 17.4.2 will show that the use of variable PRIs offers another way to improve total range-Doppler coverage.

17.4 MOVING TARGET INDICATION

Doppler processing is the term applied to filtering or spectral analysis of the slow-time signal $y[m]$ received from a fixed range bin over a period of time corresponding to several pulses. Doppler processing is applied independently to each range bin of interest. There are two major classes of Doppler processing: moving target indication and pulse-Doppler processing. In this chapter, MTI refers to the case where the slow-time signal is processed

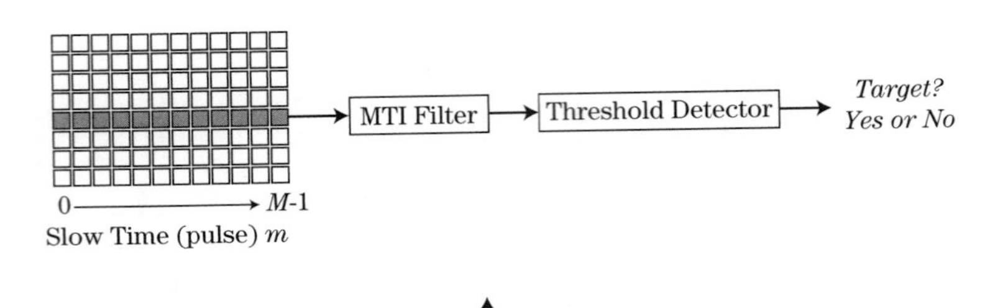

FIGURE 17-4 ■ MTI filtering and detection process.

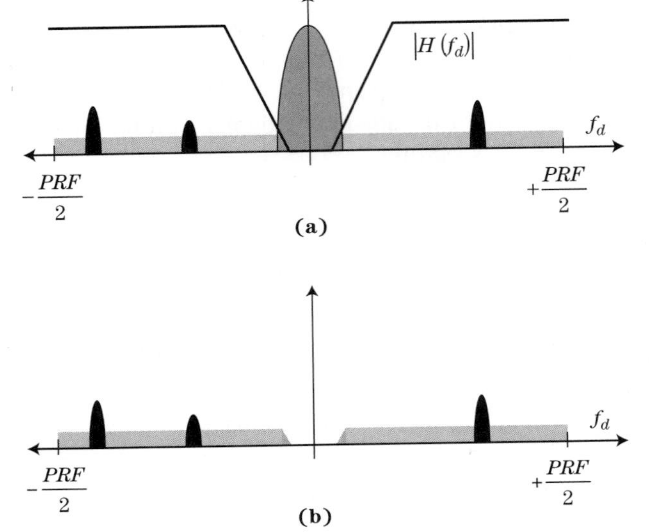

FIGURE 17-5 ■ Desired effect of the MTI filter. (a) Slow-time spectrum before MTI filtering and notional MTI filter frequency response $|H(f_d)|$. (b) After MTI filtering.

entirely in the time domain, usually using a single high-pass filter. Pulse-Doppler processing refers to the case where the signal is processed in the frequency domain, usually using an FFT.[1] As will be seen, MTI processing produces limited information at very low computational cost; pulse-Doppler processing requires more computation but produces more information and greater *signal-to-interference ratio* (SIR) improvement. Only coherent Doppler processing using digital implementations is considered, since this is the approach taken in most modern radars. Alternative systems using noncoherent Doppler processing and implementations based on analog technologies are described in [2–6], among many others.

MTI processing applies a linear filter to the slow-time data sequence to suppress the clutter component. Figure 17-4 illustrates the process. The type of filtering needed can be understood by considering the notional spectrum of the slow-time data shown in Figure 17-5. In this figure, it is assumed that knowledge of the platform motion and scenario geometry has been used to center the clutter spectrum at zero Doppler frequency. Clearly, some form of high-pass filter is needed to attenuate the clutter without filtering out moving targets in the clear portions of the Doppler spectrum.

The output of the high-pass MTI filter will be a modified slow-time signal containing components due to noise and, possibly, one or more targets. This signal is passed to a detector, typically based on the amplitude or squared amplitude of the data and possibly involving noncoherent detection as well (see Chapters 3 and 15). If the amplitude of the filtered signal exceeds the detector threshold (i.e., its energy is too great to likely be the

[1] Skolnik [2] distinguishes MTI and pulse-Doppler by defining pulse-Doppler as a system that uses a PRF high enough to avoid blind speeds. Here, the two cases are distinguished based on the type of processing operations used and the information obtained.

result of noise alone), a target will be declared; otherwise, the data are declared to represent interference only.

Note that in MTI processing, the presence or absence of a moving target is the only information obtained. The filtering process of Figures 17-4 and 17-5 does not provide any estimate of the Doppler frequency at which the target energy causing the detection occurred; thus, it "indicates" the presence of a moving target but does not determine whether the target is approaching or receding or at what radial velocity. Furthermore, it provides no indication of the number of moving targets present. If multiple moving targets are present in the slow-time signal from a particular range bin, the result will still be only a "target present" decision from the detector. On the other hand, MTI processing is very simple and computationally undemanding.

17.4.1 Pulse Cancellers

The major MTI design decision is the choice of the particular MTI filter to be used. MTI filters are typically low-order, simple *finite impulse response* (FIR; also called tapped delay line or nonrecursive) designs [7]. Indeed, some of the most common MTI filters are based on very simple heuristic design approaches. For example, suppose a stationary radar illuminates a stationary clutter scatterer. After demodulation, the measured sample of the received signal in the appropriate range bin will be of the form $Ae^{j\phi}$ for some amplitude A and phase ϕ. If the measurement is repeated, the same value will be measured again (ignoring noise). Subtracting the echoes from successive pairs of pulses would cancel the clutter return completely.

Now consider the same scenario, but with a moving target. While the amplitude of the successive echoes may be identical, the range to the target will change between pulses by an amount $\delta R = vT$ m, where v is the radial velocity of the target and T is the pulse repetition interval. Consequently, the phase of the echo will change by $(4\pi/\lambda)\delta R$ radians as can be seen in equation (17.2). Subtracting these two measurements will not result in a zero signal due to the different phases.

This reasoning motivates the *two-pulse MTI canceller*, also referred to as the *single canceller* or *first-order canceller*. Figure 17-6a illustrates the flowgraph of a two-pulse canceller, which is an especially simple FIR digital filter. The input data are a sequence of baseband complex (in-phase and quadrature, or I and Q) data samples from the same range bin over successive pulses, forming a discrete-time sequence $y[m]$ with a sampling

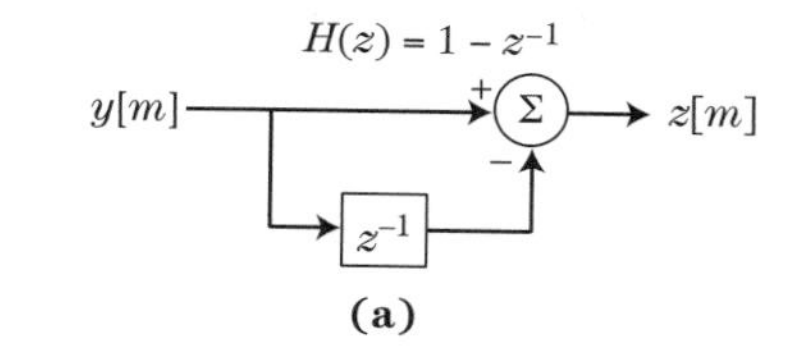

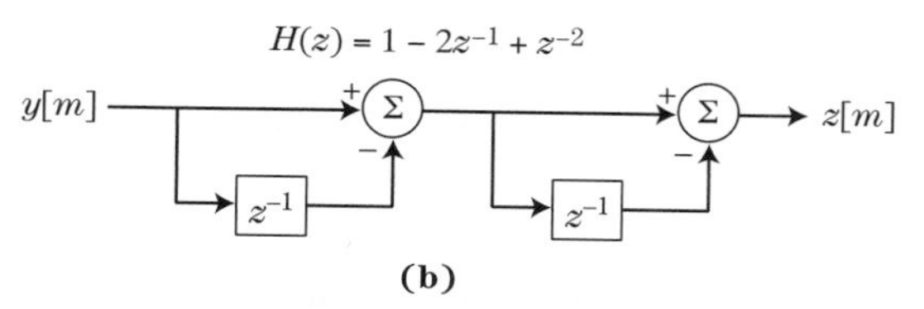

FIGURE 17-6 ■ Flowgraphs and transfer functions of basic MTI cancellers. (a) Two-pulse canceller. (b) Three-pulse canceller.

interval T equal to the pulse repetition interval. The discrete-time transfer function of this filter is simply $H(z) = 1 - z^{-1}$. The frequency response as a function of analog Doppler frequency f_d in hertz is obtained by setting $z = e^{j2\pi f_d T}$:

$$
\begin{aligned}
H(f_d) &= (1 - z^{-1})\big|_{z=e^{j2\pi f_D T}} = 1 - e^{-j2\pi f_d T} \\
&= e^{-j\pi f_d T}(e^{+j\pi f_d T} - e^{-j\pi f_d T}) \\
&= 2je^{-j\pi f_d T}\sin(\pi f_d T)
\end{aligned}
\tag{17.9}
$$

The frequency response may also be expressed in terms of normalized frequency $\hat{f} = f_d T$ cycles/sample or the radian equivalent, $\hat{\omega} = \omega T = 2\pi f_d T$ radians/sample. For example, in terms of normalized radian frequency, the frequency response of the two-pulse canceller is

$$
H(\hat{\omega}) = 2je^{-j\hat{\omega}/2}\sin(\hat{\omega}/2) \tag{17.10}
$$

Note that as f_d ranges from $-PRF/2$ to $+PRF/2$ ($-1/2T$ to $+1/2T$), $\hat{f}$ ranges from -0.5 to $+0.5$ and $\hat{\omega}$ from $-\pi$ to $+\pi$.

Figure 17-7a plots the magnitude of this frequency response. Note that the filter is high-pass in nature, with a null at zero frequency to suppress the clutter energy. Spectral components representing moving targets may either be partially attenuated or amplified,

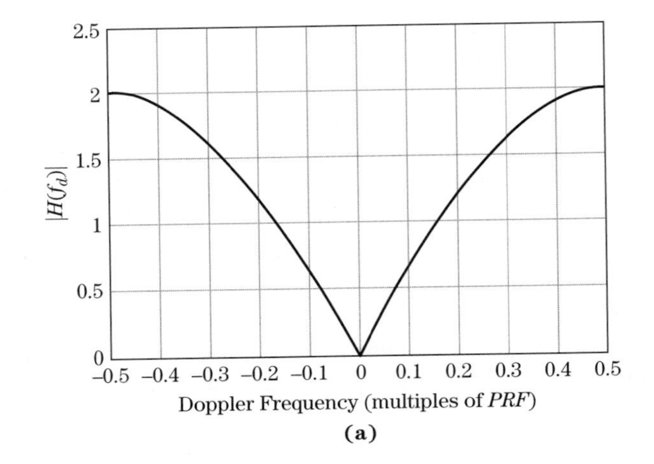

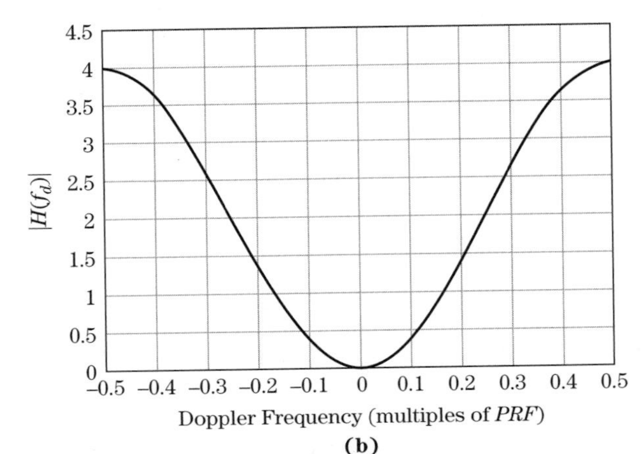

FIGURE 17-7 ■ Frequency response of basic MTI cancellers. (a) Two-pulse canceller. (b) Three-pulse canceller.

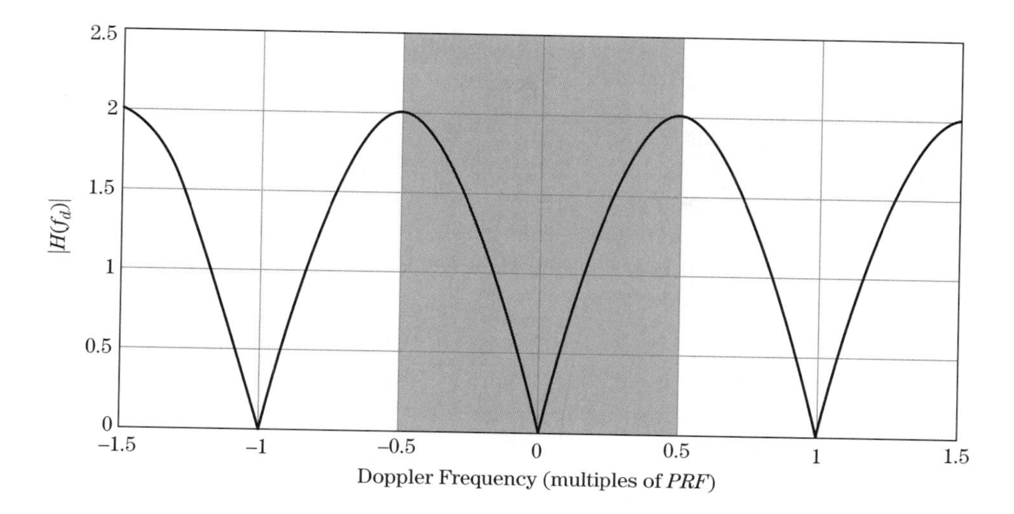

FIGURE 17-8 "Blind" Doppler frequencies.

depending on their precise location on the Doppler frequency axis. Also recall that, like all discrete-time filters, the frequency response is periodic with a period of 1 cycle/sample, corresponding to a period of 2π in normalized radian frequency $\hat{\omega}$ or $1/T = PRF$ in analog frequency in hertz. This is illustrated in Figure 17-8, which extends Figure 17-7a to three periods. The shaded area highlights the principal period from $-PRF/2$ to $+PRF/2$; this is all that is normally plotted. The implications of the periodicity will be considered in Section 17.4.2.

The two-pulse canceller imposes a very low computational load; Figure 17-6a shows that its implementation requires no multiplications and only one subtraction per output sample. As Figure 17-7a shows, however, it is a poor approximation to an ideal high-pass filter for clutter suppression. The next traditional step up in MTI filtering is the three-pulse (second-order or double) canceller, obtained by cascading two two-pulse cancellers. The flowgraph and frequency response are shown in Figures 17-6b and 17-7b. The three-pulse canceller improves the null depth and width in the vicinity of zero Doppler and requires only two subtractions per output sample, but there is still a large variation in the filter gain or attenuation for moving targets at various Doppler shifts away from zero Doppler.

Despite their simplicity, the two-and three-pulse cancellers can be very effective against clutter with moderate to high pulse-to-pulse correlations. This is because highly correlated clutter corresponds to a narrow power spectrum, so that a high fraction of the clutter energy falls within the filter notch at zero Doppler shift. Section 17.4.3 addresses the clutter suppression obtainable with pulse cancellers.

The idea of cascading two-pulse canceller sections to obtain higher-order filters can be extended to the N-pulse canceller, obtained by cascading $N - 1$ two-pulse canceller sections. The transfer function of the N-pulse canceller is therefore

$$H_N(z) = (1 - z^{-1})^{N-1} \tag{17.11}$$

Other types of digital high-pass filters could also be designed for MTI filtering. For example, an FIR high-pass filter could be designed using standard digital filter design techniques such as the window method or the Parks-McClellan algorithm [7]. Alternatively, *infinite impulse response* (IIR) high-pass filters could be designed. However, most MTI filters are low order, and, with only a few filter coefficients to optimize, more elaborate MTI filter designs provide only modest performance improvements over pulse cancellers. Two- or three-pulse cancellers are commonly used for initial MTI filtering due to their reasonable effectiveness and computational simplicity.

The N-pulse cancellers previously described are widely used. Nonetheless, they are motivated by heuristic ideas. Can a more effective pulse canceller be designed? Since the goal of MTI filtering is to maximize the signal-to-clutter ratio, it should be possible to apply the optimum filter concept, used previously in Chapter 14 to develop the matched filters, to this problem. The details are worked out in [1]. Assuming the clutter power is σ_c^2 and the noise power is σ_n^2, the resulting filter coefficients in the two-pulse case are

$$\boldsymbol{H} = \hat{k}\left[\sigma_c^2 + \sigma_n^2 - \rho^* \sigma_c^2\right]^T \tag{17.12}$$

where $\hat{k}$ absorbs all scale factors. In this equation, ρ is the first normalized autocorrelation lag of the clutter, and the superscript T denotes matrix transpose.

To interpret this result, consider the case where the clutter is the dominant interference and is highly correlated from one pulse to the next. Then σ_n^2 is negligible compared with σ_c^2, and ρ is close to 1. Absorbing σ_c^2 into $\hat{k}$, the matched filter coefficients are then 1 and approximately -1, that is, nearly the same as the two-pulse canceller. Despite its simplicity, the two-pulse canceller is therefore nearly a first-order matched filter for MTI processing when the clutter-to-noise ratio is high and the successive clutter pulses are highly correlated. In the limit of very high clutter-to-noise ratio and perfectly correlated clutter, the two-pulse canceller is exactly the first-order matched MTI filter.

17.4.2 Blind Speeds and Staggered PRFs

The frequency response of all discrete-time filters is periodic, repeating with a period of one in the normalized cyclical frequency, corresponding to a period of $PRF = 1/T$ Hz of Doppler shift. Since MTI filters are designed to have a null at zero frequency, they will also have nulls at Doppler frequencies that are multiples of the pulse repetition frequency. Consequently, a target moving with a radial velocity that results in a Doppler shift equal to a multiple of the PRF will be filtered out by the MTI filter. This is illustrated for a two-pulse canceller in Figure 17-8, which shows three periods of the frequency response. The first positive and negative blind Doppler frequencies are

$$f_b = \pm PRF \tag{17.13}$$

Velocities that result in these unfortunate Doppler shifts are called *blind speeds* because the target return will be suppressed by the MTI filter; the system is “blind” to such targets. From a digital signal processing point of view, blind speeds represent target velocities that will be aliased to zero frequency. The first blind speeds are obtained by scaling the blind Doppler frequencies:

$$v_{blind} = \frac{\lambda}{2} f_b = \pm \frac{\lambda}{2} PRF \tag{17.14}$$

As the PRF is increased for a given radio frequency (RF), the unambiguous range decreases, and the first blind Doppler frequency increases. Blind Doppler frequencies could be avoided by choosing the PRF high enough so that the first blind Doppler frequency exceeds any actual velocity likely to be observed for targets of interest. Unfortunately, sometimes no PRF will allow unambiguous coverage of both the range and Doppler intervals of interest. For example, suppose a designer requires an unambiguous range of $R_{ua} = 100$ km and $v_{ua} = 100$ m/s (± 50 m/s) of unambiguous velocity coverage. The maximum RF at which this is possible is 2.25 GHz, as can be seen from equation (17.8). If the radar is required to be at X-band (10 GHz), then the combination of 100 km unambiguous range coverage and 100 m/s unambiguous velocity coverage is not obtainable with any PRF, and some ambiguity must be accepted in range, Doppler, or both.

The use of *staggered PRFs* is a data collection and processing technique that raises the first blind speed significantly without significantly degrading unambiguous range [5,8]. PRF staggering can be performed on either a pulse-to-pulse or CPI-to-CPI basis; the latter is also called a block-to-block or dwell-to-dwell basis. The CPI-to-CPI case is common in airborne pulse-Doppler radars and is discussed in Section 17.5.8.

Pulse-to-pulse stagger varies the pulse repetition interval, or equivalently the PRF, from one pulse to the next within a single CPI. The resulting slow-time data sequence is then passed through a conventional MTI filter such as a two- or three-pulse canceller. As will be seen, this increases the Doppler coverage within a single CPI. One disadvantage is that the slow-time data sequence in a given range bin is now nonuniformly sampled in slow time, making it more difficult to apply Doppler filtering to the data and greatly complicating analysis. Another is that range-ambiguous mainlobe clutter, if any, can cause large pulse-to-pulse amplitude changes as the PRF varies, since the range of the second-time-around clutter that folds into each range cell will change as the PRF changes. Consequently, pulse-to-pulse PRF stagger is generally used only in low PRF modes in which no range ambiguities are expected.

Consider a system using a set of P staggered PRFs $\{PRF_p\} = \{PRF_0, PRF_1, \ldots, PRF_{P-1}\}$. The corresponding set of pulse repetition intervals is $\{T_p\} = \{1/PRF_p\}$. Assume that each of the PRFs can be expressed as an integer multiple of the *greatest common divisor* (gcd) of the set, f_g:

$$\begin{aligned} PRF_p &= k_p \,\mathrm{gcd}(PRF_0, \ldots, PRF_{P-1}) \\ &\equiv k_p f_g \end{aligned} \tag{17.15}$$

The set of integers $\{k_p\}$ is called the *staggers*,[2] and the ratio $k_m{:}k_p$ of any two of them is called a *stagger ratio*. A CPI of data is collected by transmitting the first pulse (pulse number 0) and sampling the desired range bins. The transmitter then waits T_0 seconds and transmits pulse number 1 to get the second measurement in each range bin. The transmitter then waits T_1 seconds before transmitting pulse number 2 and so forth. If the CPI contains more than P pulses, then after pulse number $P-1$ the transmitter waits T_{P-1} seconds, transmits the P-th pulse, and then cycles back to the first PRI, waiting T_0 seconds and transmitting pulse number $(P+1)$ and so on until the full CPI of M pulses has been collected.

For a fixed PRF, any MTI filter will exhibit blind Doppler frequencies at all integer multiples of the PRF. Similarly, the first true blind Doppler frequency of a system using staggered PRFs will be the lowest frequency that is blind at all of the individual PRFs—that is, the *least common multiple* (lcm) of the set [1]:

$$\begin{aligned} f_s &= \mathrm{lcm}\,(PRF_0, \ldots, PRF_{P-1}) \\ &= f_g \mathrm{lcm}\,(k_0, \ldots, k_{P-1}) \end{aligned} \tag{17.16}$$

[2] Some authors work in terms of the PRIs instead of the PRFs and use the term *staggers* to refer to the ratio of the $\{PRI_p\}$.

A measure of effectiveness of this technique is how much the blind speed of the staggered system is increased relative to that of an unstaggered system with the same average PRI. It is straightforward to show that the ratio of the first blind Doppler frequency f_s of the staggered PRF system and the blind Doppler frequency of an unstaggered system with a PRF f_{us} corresponding to the average PRI is [1]

$$\frac{f_s}{f_{us}} = \frac{1}{P}\operatorname{lcm}(k_0, \ldots, k_{P-1})\left(\sum_{p=0}^{P-1}\frac{1}{k_p}\right) \tag{17.17}$$

For example, a two-PRF system with a stagger ratio of 3:4 would have a first blind Doppler frequency 3.5 times that of a system using a fixed PRI equal to the average of the two individual PRIs. If a third PRF is added to give the set of staggers $\{3, 4, 5\}$, the first blind Doppler frequency will be 15.67 times that of the comparable unstaggered system.

The unambiguous range R_{us} of the unstaggered system is the range corresponding to the unstaggered PRF, f_{us}, which is just $c/2f_{us}$. The unambiguous range R_{min} of the staggered PRF system is the shortest of the unambiguous ranges corresponding to the individual PRFs. It is easy to show that the ratio of these two ranges is

$$\frac{R_{min}}{R_{us}} = \frac{P}{\max\{k_p\}\left(\sum_{p=0}^{P-1}\frac{1}{k_p}\right)} \tag{17.18}$$

For the two-PRF system with staggers $\{3, 4\}$, this ratio is 6/7, representing a reduction of unambiguous range of 14% in exchange for the increase in velocity coverage by a factor of 3.5. For the $\{3, 4, 5\}$ case, the reduction in unambiguous range is a factor of 45/47, or 8%, while the Doppler coverage is increased by a factor of 15.67.

The slow-time data in a staggered PRF MTI system has a nonuniform sampling interval from one sample to the next. Unfortunately, the response of a digital filter to such an input, while still linear, is not time-invariant, so the frequency response of a pulse-to-pulse staggered system cannot be determined using conventional Fourier analysis techniques. Instead, an approach based on first principles can be used to explicitly compute the frequency response of a specified MTI filter with staggered PRF data by determining the average amplitude of the filter output when the input is a pure complex sinusoid of arbitrary frequency and random initial phase [5,8,9]. The result for the squared magnitude of the frequency response of the two-pulse canceller with staggered PRFs is [1]

$$\left|H_{2,P}(f)\right|^2 = \frac{4}{P}\sum_{p=0}^{P-1}\sin^2(\pi f T_p) = \frac{4}{P}\sum_{p=0}^{P-1}\sin^2(\pi f/PRF_p) \tag{17.19}$$

where the notation $H_{N,P}(f)$ indicates the frequency response of an N-pulse canceller using P staggers. The response of more general MTI filters can be obtained using a similar technique. The actual frequency in hertz rather than normalized frequency $\hat{\omega}$ or $\hat{f}$ must be used in equation (17.19) because the nonuniform sampling rate invalidates the usual definition of normalized frequency.

Figure 17-9 compares the frequency response of a two-pulse canceller using two ($P = 2$) PRFs versus conventional single-PRF operation. The staggered case uses PRFs of 750 and 1,000 pulses per second; thus, $f_g = 250$ Hz and the set of staggers k_p is $\{3, 4\}$. The first blind Doppler frequency occurs at the least common multiple of 750 and 1,000 Hz,

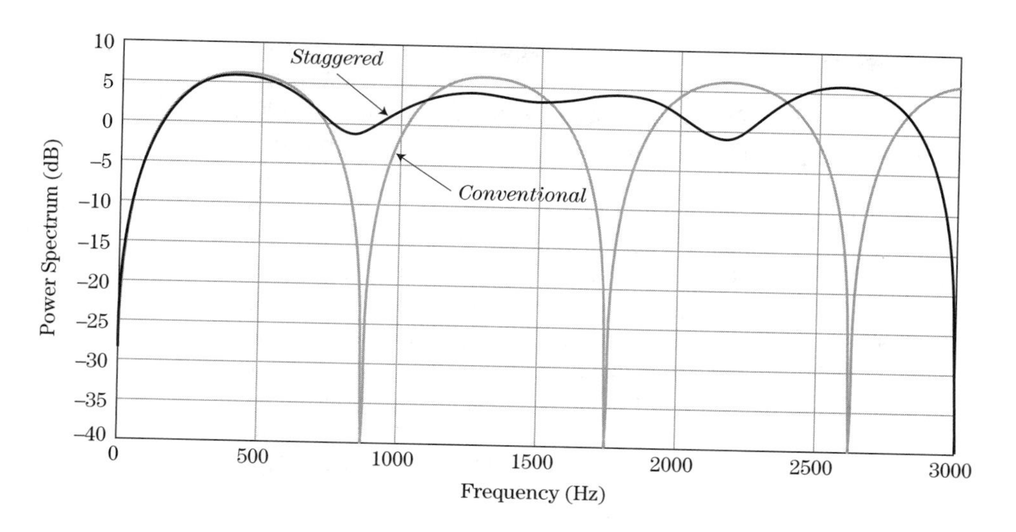

FIGURE 17-9 ■ Comparison of two-pulse canceller frequency response with conventional unstaggered waveform and 3:4 staggered waveform.

which is 3,000 Hz. f_{us}, which is the reciprocal of the average PRI $T_{avg} = 1.167$ ms, is 857.14 Hz. The conventional unstaggered response collected with the PRI $= T_{avg}$ shows blind Doppler shifts equal to integer multiples of 857.14 Hz. Thus, staggering the PRF has increased the blind Doppler frequency by a factor of 3.5 (= 3,000/857.14), consistent with equation (17.17). Note also that the unambiguous range corresponding to the highest PRF used (1,000 Hz) is $R_{min} = 150$ km, while the unambiguous range of the equivalent unstaggered system having the PRF 857.14 pulses/second is $R_{us} = 175$ km, a reduction by a factor of 6/7 (14%), as previously predicted.

The reader is cautioned that no effort has been made in this simple example to optimize the number or choice of PRFs. The "passband" region of the staggered response (about 300 to 2,700 Hz) shows frequency response variations of about 9 dB. Good staggered PRF systems spend considerable effort on PRF selection. Use of more than two PRIs in the stagger sequence and careful selection of their ratios can result in overall MTI frequency responses with much less variability than shown in this simple example [10,11]. Another design approach uses randomized PRIs rather than the fixed schedule described here to extend the blind speed [12].

17.4.3 MTI Figures of Merit

The goal of MTI filtering is to suppress clutter. In doing so, the MTI filter also attenuates or amplifies the target return, depending on the particular target Doppler shift. The change in signal and clutter power then affects the probabilities of detection and false alarm achievable in the system in a manner dependent on the particular design of the detection system.

There are three traditional MTI filtering figures of merit in wide use [13]. *Clutter attenuation* measures only the reduction in clutter power at the output of the MTI filter compared with the input but is simplest to compute. *Improvement factor* quantifies the increase in signal-to-clutter ratio due to MTI filtering; as such, it accounts for the effect of the filter on the target as well as on the clutter. *Subclutter visibility* is a more complex measure that also takes into account the detection and false alarm probabilities and the detector characteristic. Because of its complexity, it is less often used. In this chapter, attention is concentrated on clutter attenuation, *CA*, and improvement factor, *I*. Due to their common use in evaluating adaptive processing systems, two newer metrics, *minimum detectable velocity* (MDV) and *usable Doppler space fraction* (UDSF), are introduced later in the chapter in the context of pulse-Doppler processing [14].

Calculation of clutter attenuation and improvement factor can be approached in several ways: frequency domain approaches using clutter power spectra and MTI filter transfer functions; an approach using the autocorrelation functions of the input and output of the MTI filter; and a method based on vector analysis. Only the frequency domain approach, which is perhaps the most intuitive, is described here. A description and examples of the other two methods are available in [1].

Clutter attenuation directly evaluates the MTI filter's effectiveness at its main function of suppressing the clutter energy. It is simply the ratio of the clutter power at the input of the MTI filter to the clutter power at the output, which can be calculated by integrating the clutter power density spectrum before and after the MTI filter is applied:

$$CA = \frac{\sigma_{ci}^2}{\sigma_{co}^2} = \frac{\displaystyle\int_{-PRF/2}^{PRF/2} S_{cc}(f_d)df_d}{\displaystyle\int_{-PRF/2}^{PRF/2} S_{cc}(f_d)|H(f_d)|^2 df_d} \tag{17.20}$$

Here σ_{ci}^2 and σ_{co}^2 are the clutter power at the filter input and output, respectively; $S_c(f_d)$ is the sampled clutter power spectrum; and $H(f_d)$ is the discrete-time MTI filter frequency response. In equation (17.20) each term is expressed in terms of analog frequency, f_d, but could also be expressed in terms of normalized frequencies $\hat{\omega}$ or $\hat{f}$. Since the MTI filter presumably reduces the clutter power, CA will be greater than 1. In fact, clutter attenuation can be 20 dB or more in favorable conditions. However, it also depends on the clutter itself through $S_c(f_d)$. A change in clutter power spectrum due to changing terrain or weather conditions will alter the achieved clutter cancellation. The shape of the clutter power spectrum and its spread in meters per second are determined by the physical phenomenology (type of clutter, clutter motion, weather conditions) and are not under the radar engineer's control. However, the percentage of the discrete-time spectrum width in hertz to which a given clutter power spectrum is mapped depends on the PRF and wavelength and therefore is determined by the system design. Thus, the width of the clutter spectrum relative to the PRF is a combination of factors, some influenced by the radar design and some not.

Improvement factor I is defined formally as the SCR at the filter output divided by the SCR at the filter input, averaged over all target radial velocities of interest [13]. Considering for the moment only a specific target Doppler shift, the improvement factor can be factored into the form [8]

$$I = \frac{(SCR)_{out}}{(SCR)_{in}} = \left(\frac{S_{out}}{S_{in}}\right)\left(\frac{C_{in}}{C_{out}}\right) = G \cdot CA \tag{17.21}$$

where G is the MTI filter *gain* at the Doppler shift of interest. Figure 17-7 makes clear that the effect of the MTI filter on the target signal is a strong function of the target Doppler shift. Thus, G is a function of target velocity, whereas clutter attenuation CA is not. The value of CA in equation (17.21) is obtained from equation (17.20). The value of target power gain, $G(f_{d_0})$, for a target at some specific Doppler frequency, f_{d_0}, is determined by the frequency response of the MTI filter at that frequency:

$$G(f_{d_0}) = \left(\frac{S_{out}}{S_{in}}\right) = \left|H(f_{d_0})\right|^2 \tag{17.22}$$

With this definition of G, the improvement factor is a function of the target Doppler shift f_{d_0}.

To reduce I to a single number, the definition calls for averaging uniformly over all target Doppler shifts "of interest" [13]. It is more common to assume the target velocity is completely unknown a priori and use the average target power gain over a full period of the Doppler spectrum, which is just

$$G = \frac{1}{PRF} \int_{-PRF/2}^{PRF/2} |H(f_d)|^2 df_d \tag{17.23}$$

This equation gives an average value of G of 2 for a two-pulse canceller and 6 for a three-pulse canceller [4]. Combining (17.23) and (17.20) in (17.21) gives the improvement factor as

$$I = \frac{\left\{\int_{-PRF/2}^{PRF/2} |H(f_d)|^2 df_d\right\}\left\{\int_{-PRF/2}^{PRF/2} S_{cc}(f_d) df_d\right\}}{PRF\left\{\int_{-PRF/2}^{PRF/2} S_{cc}(f_d)|H(f_d)|^2 df_d\right\}} \tag{17.24}$$

Table 17-2 shows the improvement factor predicted by equation (17.24) for various clutter spectral widths, assuming the clutter spectrum is Gaussian in shape.[3] If the clutter spectrum is narrow compared with the PRF, then the improvement factor can be 20 dB or more even for the simple two-pulse canceller. If the clutter spectrum is wide, much of the clutter power will be in the passband of the MTI high-pass filter, and the improvement factor will be small.

TABLE 17-2 ■ Improvement Factor for Gaussian Clutter Power Spectrum

	Improvement Factor (dB)	
Standard Deviation of Clutter Power Spectrum (Hz)	Two-Pulse Canceller	Three-Pulse Canceller
PRF/6	3.7	5.7
PRF/10	7.5	12.5
PRF/20	13.2	21.7
PRF/100	24	51

[3]Clutter spectra having a standard deviation wider than $PRF/6$ do not decay to nearly zero at $f_d = \pm PRF/2$ and therefore will not be Gaussian in shape due to the replication of the spectrum with sampled data. The Gaussian spectrum assumed for this table is not valid for these very wide clutter spectra.

Additional MTI metrics can be defined. Improvement factor is the average of the improvement in signal-to-clutter ratio over one Doppler period. At some Doppler shifts, the target is above the clutter energy, while at others it is below the clutter and therefore not detectable. I does not indicate over what percentage of the Doppler spectrum a target can be detected. The concept of *MTI visibility factor* or *target visibility*, V, has been proposed to quantify this effect [15]. V is the percentage of the Doppler period over which the improvement factor for a target at a specific frequency is greater than or equal to the average improvement factor I.

17.4.4 Limitations to MTI Performance

The basic idea of MTI processing is that repeated measurements (pulses) of a stationary target yield the same echo amplitude and phase; thus, successive echo samples, when subtracted from one another, should cancel. Any effect internal or external to the radar that causes the received echo from a stationary target to vary will cause imperfect cancellation, limiting the improvement factor.

The simplest example is transmitter amplitude instability. If two transmitted pulses differ in amplitude by 10% (equivalent to $20\log_{10}(1.1/1) = 0.83$ dB), then the signal resulting from subtracting the two echoes from a perfectly stationary target will have an amplitude that is 10% that of the individual echoes. Consequently, clutter attenuation can be no better than $20\log_{10}(1/0.1) = 20$ dB. For a two-pulse canceller with an average signal gain G of 2 (3 dB), the maximum achievable improvement factor is 23 dB.

A more realistic analysis of the limitations due to pulse-to-pulse amplitude variations can be obtained by modeling the amplitude of the m-th transmitted pulse as $A[m] = k(1 + a[m])$, where $a[m]$ is a zero mean, white random process with variance σ_a^2 that represents the percentage variation in transmitted amplitude, and k is a constant. The received signal will have a complex amplitude of the form $k'(1+a[m])\exp(j\phi)$, where ϕ is the phase of the received slow-time sample and the constant k' absorbs all the radar range equation factors. The average power of this signal, which is the input to the pulse canceller, is

$$\mathrm{E}\{|y[m]|^2\} = k'^2\mathrm{E}\{1 + 2a[m] + a^2[m]\} = k'^2\left(1 + \sigma_a^2\right) \tag{17.25}$$

where $\mathrm{E}\{\cdot\}$ is the expected value operator. The expected value of the two-pulse canceller output power will be

$$\begin{aligned}\mathrm{E}\left\{|(y[m] - y[m-1])|^2\right\} &= \mathrm{E}\left\{\left|k'e^{j\phi}(a[m] - a[m-1])\right|^2\right\} \\ &= k'^2\mathrm{E}\{a^2[m]\} - 2\mathrm{E}\{a[m]a[m-1]\} + k'^2\mathrm{E}\left\{a^2[m-1]\right\} \\ &= 2k'^2\sigma_a^2\end{aligned} \tag{17.26}$$

The average clutter cancellation is thus

$$CA = \frac{\text{input power}}{\text{output power}} = \frac{k'^2\left(1 + \sigma_a^2\right)}{2k'^2\sigma_a^2} = \frac{1 + \sigma_a^2}{2\sigma_a^2} \quad \text{(amplitude jitter)} \tag{17.27}$$

For example, an amplitude variance of 1% ($\sigma_a^2 = 0.01$) limits two-pulse clutter cancellation to a factor of 50.5, or 17 dB. Because the average target gain G of the two-pulse canceller is $G = 2$ (3 dB), the limit to the improvement factor I is $17 + 3 = 20$ dB.[4]

[4]The effect of the jitter on the gain, G, should also be computed to be completely correct, but that effect is small and can usually be neglected.

Another limiting factor is phase drift in either the transmitter or receiver. This can occur, for example, due to instability in coherent local oscillators used either as part of the waveform generator on the transmit side or in the demodulation chains on the receiver side. A stationary radar viewing a stationary scatterer would expect the same phase as well as amplitude of two successive measurements. However, phase is measured by reference to an oscillator within the receiver. If the reference phase changes between measurements, the apparent measured phase will change, resulting again in imperfect cancellation of the two measurements when subtracted. This effect can be analyzed by modeling the error in the measured phase as a zero mean Gaussian random error. An analysis approach similar to that used for amplitude (see [16] for details) shows that the limitation on two-pulse cancellation due to the phase noise is

$$CA = \frac{\text{input power}}{\text{output power}} = \frac{1}{2\left(1 - e^{-\sigma_\phi^2}\right)} \text{ (phase jitter)} \tag{17.28}$$

Other sources of limitation due to radar system instabilities include instability in transmitter or oscillator frequencies; transmitter phase drift; coherent oscillator locking errors; PRI jitter; pulse width jitter; and quantization noise. Simple formulas to bound the achievable clutter attenuation due to each of these error sources are given in [2, 4]. These formulas can be used to construct an error budget and determine allowable tolerances on each error source. Another approach to improving canceller performance is to measure the actual errors when possible and compensate for them in the processing. For example, the actual power of each transmitted pulse could be measured by the radar and used to adjust the received signal voltages on a pulse-by-pulse basis to improve MTI cancellation.

External to the radar, the chief factor limiting MTI improvement factor is simply the width of the clutter spectrum itself. Wider spectra put more clutter energy outside of the MTI filter null so that less of the clutter energy is filtered out. This effect was illustrated numerically in Table 17-2. The clutter spectrum width is determined first and foremost by the inherent Doppler spread of the clutter scatterers. It can be increased by radar system effects and instabilities such as the amplitude and phase jitters previously discussed and by radar platform motion, as shown in the next subsection. For instance, a scanning antenna adds some amplitude modulation due to antenna pattern weighting to the clutter return that broadens the observed spectral width somewhat. In some cases, the clutter power spectrum may not be centered on zero Doppler shift. A good example is rain clutter: moving weather systems will have a nonzero average Doppler representing the rate at which the rain cell is approaching or receding from the radar system. Unless this average motion is detected and compensated, the MTI filter null will not be centered on the clutter spectrum, and cancellation will be poor.

17.4.5 MTI from a Moving Platform

The largest source of clutter offset and spreading is radar platform motion; formulas for these two effects were given in Chapter 8. The offset in center frequency of the clutter spectrum can be as much as a few kHz for fast aircraft, while the motion-induced spectral spread can be tens to a few hundreds of Hz. For space-based radars, both can be another order of magnitude larger. This clutter spreading adds to the intrinsic spread of the clutter spectrum due to internal motion and can often be the dominant effect determining the observed clutter spectral width and the resulting MTI performance limits.

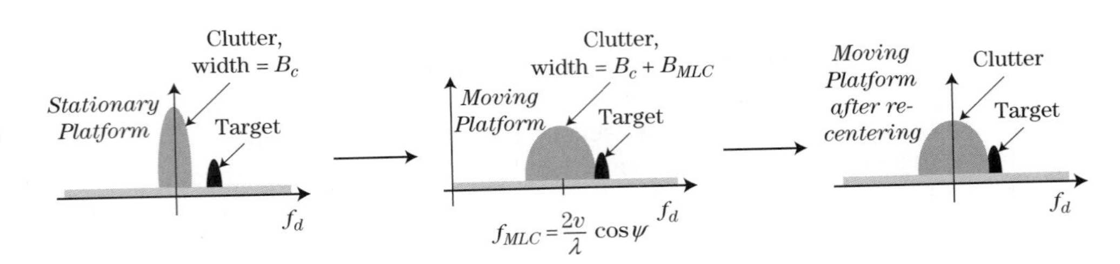

FIGURE 17-10 ■ Illustration of the effect of a moving radar platform on the Doppler spectrum and the detection of "slow movers."

This phenomenon is illustrated in Figure 17-10, which shows the shifting and spreading of the mainlobe clutter by the platform motion followed by the recentering of the spectrum at zero Doppler. The clutter spectrum width is B_c Hz when observed from a stationary radar, but when observed from the moving platform it increases to $B_c + B_{MLC}$ Hz, where $B_{MLC} = 2v\theta_3 \sin\psi/\lambda$ Hz. Because of this spreading, a relatively slow-moving target ("slow mover," typically a ground target) that was in the clear region (left portion of the figure) may now have to compete with clutter energy as well (center or right portion of the figure), making its detection more difficult.

Displaced phase center antenna (DPCA) processing is a technique for countering the effect of platform-induced clutter spectral spreading on MTI filtering. It is a special case of the more general *space-time adaptive processing* (STAP) [1,14].

In its simplest form, DPCA applies two-pulse MTI cancellation to data collected using two receive apertures on the side of the platform in a sidelooking configuration. The two apertures are denoted the "fore" and "aft" apertures. As the aircraft flies forward, the aft aperture passes through the same coordinates in space as the fore aperture, just delayed by $\Delta x/v$ seconds, where Δx is the spacing of the two receivers. If a data sample for a particular range bin collected on the aft aperture is subtracted from an earlier sample of the same range bin collected on the fore aperture when it was at the same location, the effect will be to subtract two samples of the same range bin taken from the same point in space. These two data samples will therefore be relatively well-correlated so that two-pulse cancellation is effective. Combining the data from two subapertures in this way implicitly avoids clutter spreading effects, improving clutter cancellation and slow target visibility. References for basic DPCA are [2,17,18].

Figure 17-11 illustrates the concept for a typical implementation using a sidelooking phased array antenna divided into two subapertures. Each half of the antenna has its own receiver, so there are in effect two receive apertures having respective phase centers R1 and R2, which are Δx m apart. Assume for the moment that the antenna also transmits using only one of the subapertures at a time.

Suppose the fore subaperture of the antenna is used to transmit and receive the first pulse and that its phase center R1 is centered at position x_0 at that time, as shown in Figure 17-11a. Now consider the motion of the platform over one PRI. The aft aperture will move vT m in T seconds and therefore be in the same position as the fore aperture one pulse earlier if the aperture spacing $\Delta x = vT$, as suggested by Figure 17-11b. More generally, the aft aperture will be in the same position as was the fore aperture M_s pulses earlier if

$$vM_sT = \Delta x \tag{17.29}$$

M_s is the "time slip" in pulses; M_sT is the time slip in seconds. Equation (17.29) is known as the *DPCA condition*.

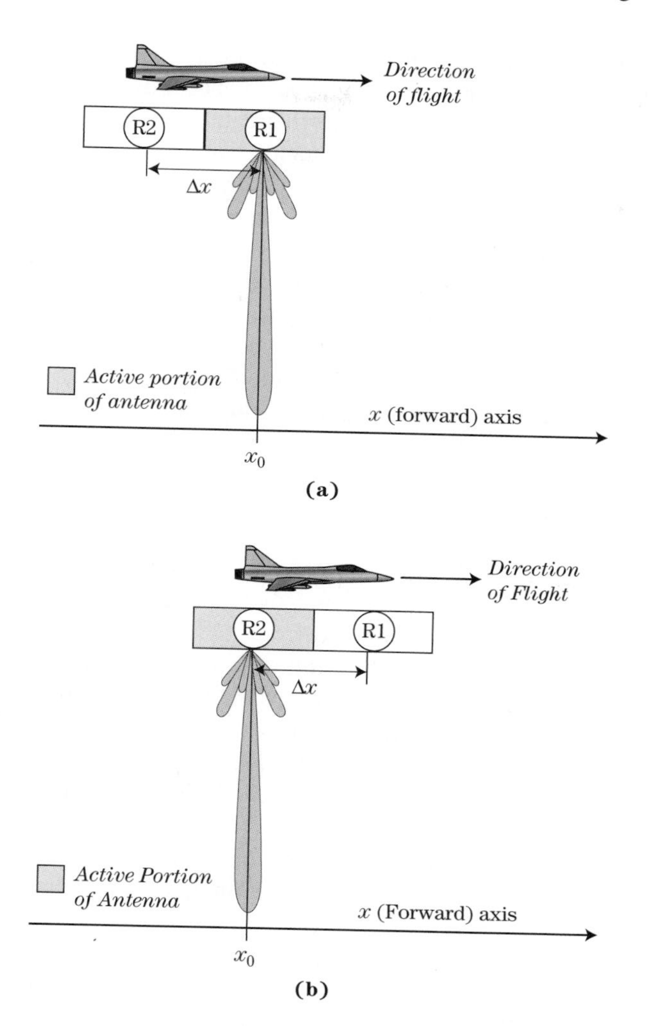

FIGURE 17-11 Relationship of transmit and receive aperture phase centers in DPCA processing.

The significance of the DPCA condition is that if it is satisfied, the data stream received on the aft receive aperture is geometrically equivalent to the data stream received on the forward receive aperture M_s pulses earlier. Consequently, two-pulse cancellation can be implemented by taking each sample in a given range bin from the R1 data stream and subtracting the sample from the same range bin in the R2 data stream taken M_s pulses later, as illustrated in Figure 17-12 for $M_s = 3$. Even though these data samples were collected

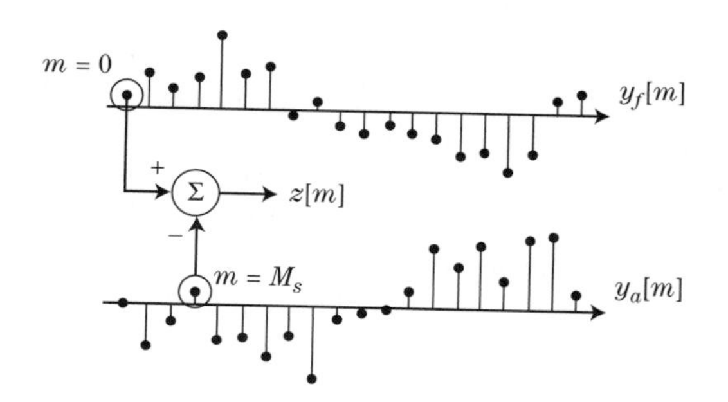

FIGURE 17-12 Illustration of combining of data to achieve two-pulse cancellation across two received data streams in DPCA with a time slip $M_s = 3$ pulses.

on different receive apertures and different pulses, their effective transmit/receive phase centers are the same, so they appear equivalent to successive pulses from a *stationary* antenna. This phase center stationarity results in greater correlation of the R2 and delayed R1 data streams, providing better two-pulse cancellation and improving the detection of slow-moving ground targets.

The radar and platform operators can vary the platform velocity, PRI, and time slip M_s to attempt to satisfy the DPCA condition. However, this may be difficult to do in general. The platform velocity, generally the cruise velocity of an aircraft or orbital speed of a spacecraft, may be reasonably variable over only a small range. The PRI is heavily constrained by range and Doppler ambiguity concerns. Thus, M_s will often not be integer. For example, if $\Delta x = 3$ m, $v = 225$ m/s, and $T = 2$ ms, then $M_s = 6.67$ pulses. While conventional band-limited interpolation could be used to implement fractional-PRI timing adjustments, in practice there will also be amplitude and gain mismatches between channels that will make it impossible to achieve high cancellation ratios even if the time alignment is perfect. A typical DPCA implementation will therefore round M_s to the nearest integer for coarse alignment of the two data streams and then will modify the basic two-pulse canceller weights to minimize the clutter residue at the processor output and therefore to maximize the improvement factor. The coefficients used to combine the fore and aft data streams are computed from the data itself, making the system an *adaptive* DPCA (A-DPCA) processor. In addition, it is common to perform pulse-Doppler processing on both the fore and aft slow-time signals first and then to apply A-DPCA separately to each DFT bin in the clutter region of the spectrum. (There is no advantage to applying it in the clear region, where noise is the dominant interference.) Separate adaptive weights can then be computed for each clutter-region DFT bin, further improving the clutter cancellation. Details are given in [1,18].

In practice, the full antenna is often used on transmit to maximize the gain and directivity. Consequently, the phase center of the antenna on transmit is in the middle of the full antenna, halfway between the two receive phase centers R1 and R2. The effective phase center for the transmission of a pulse and its reception at R1 is then halfway between the center of the full antenna and R1; a similar result holds for reception on the aft aperture. The effective spacing of the phase centers for the fore and aft data stream is then only $\Delta x/2$, making the DCPA condition

$$vM_sT = \frac{\Delta x}{2} \quad \Rightarrow \quad 2vM_sT = \Delta x \tag{17.30}$$

In the example just given, the required time slip would now be 3.33 pulses.

17.5 PULSE-DOPPLER PROCESSING

Pulse-Doppler processing is the second major class of Doppler processing. Recall that in MTI processing, the fast-time/slow-time data matrix is high-pass filtered in the slow-time dimension, yielding a new fast-time/slow-time data sequence in which the clutter components have been attenuated. Pulse-Doppler processing differs in that filtering in the slow-time domain is replaced by explicit spectral analysis of the slow-time data for each range bin. Thus, the result of pulse-Doppler processing is a range/Doppler data matrix in which the dimensions are fast time and Doppler frequency. In the range/Doppler data, the energy from a moving target is separated from that of the clutter and competes only with the noise in the target's Doppler bin. In addition, an estimate of the Doppler shift (and thus

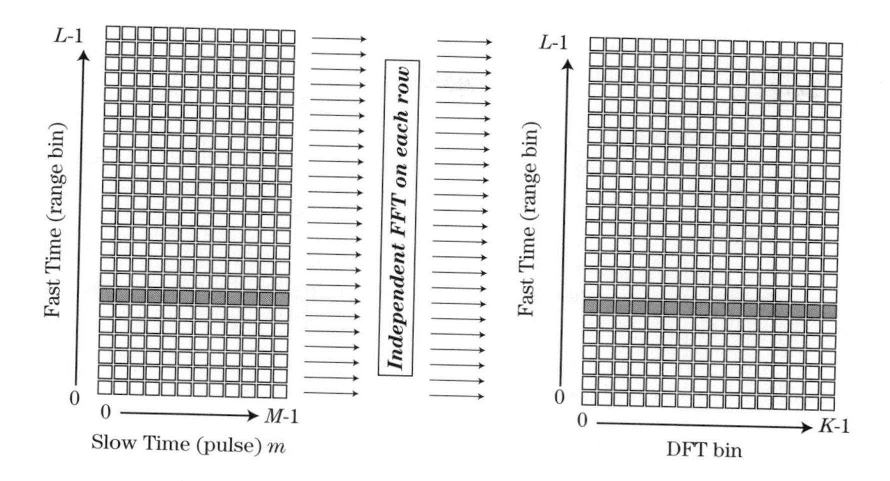

FIGURE 17-13 ■ Conversion of the fast-time/slow-time data matrix to a range-Doppler matrix by applying a DFT to each slow-time row.

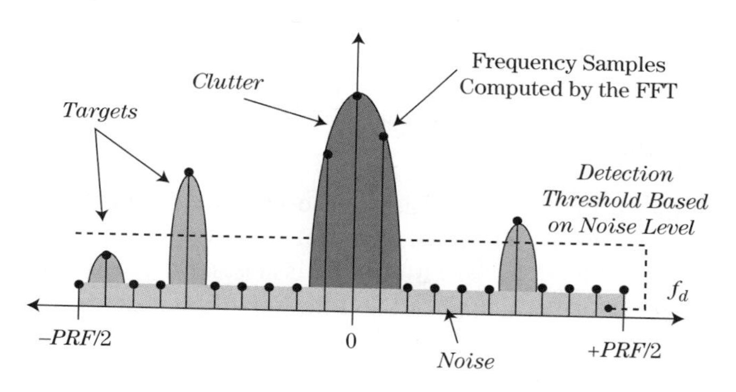

FIGURE 17-14 ■ The concept of pulse-Doppler processing for detection of moving targets.

radial velocity) of detected targets can be estimated based on the Doppler bin in which the detection occurs.

The spectral analysis is most commonly performed by computing the DFT of each slow-time row of the data using an FFT algorithm as shown in Figure 17-13, but other techniques can also be used [19]. The DFT size K can be equal to or greater than M.[5] Good general discussions of pulse-Doppler processing are contained in [20–22]. The DFT bins 0 through K-1 correspond to Doppler frequencies $-PRF/2$ to $+PRF/2$ as described in Chapter 14.

Figure 17-14 illustrates the concept of pulse-Doppler processing for moving target detection using a notional pulse-Doppler spectrum for a single range bin, similar to the one in Figure 17-2. The function in the background is the DTFT of the slow-time data, whereas the circles represent the K samples of the DTFT computed by the DFT. White thermal noise is present in every DFT sample at, on average, equal power. Assuming the clutter has been centered at zero Doppler, those spectral samples at or near 0 Hz will consist of both clutter and noise. In many cases of practical interest, the clutter dominates the noise in this clutter region. A target, if present, may appear anywhere in the spectrum, as appropriate to its Doppler shift. If the target is in the clear region, away from the clutter energy, only thermal noise will interfere with its detection. Each Doppler spectrum sample in the clear

[5] It is also possible to have $K < M$ in a sensible way, a process called *data turning*. This is used infrequently today because of increased computing power, but the technique is described in [1].

region is individually compared with a threshold based on the noise level to determine whether the signal at that frequency appears to be noise only or noise plus a target. In the figure, two of the three targets shown would be detected. The DFT samples that are clutter dominated are often simply discarded on the grounds that the signal-to-interference ratio will likely be too low for successful detection. However, some systems use a technique called clutter mapping, discussed in Section 17.6, to attempt detection of strong targets in this clutter region.

The advantages of pulse-Doppler processing are that it provides an estimate of the radial velocity component of a moving target (based on the DFT bin in which the detection occurred) and that it provides a way to detect multiple targets (based on detections in multiple DFT bins), provided they are separated enough in Doppler to be resolved. Pulse-Doppler processing also combines more data samples in a given range bin than does MTI processing, thus achieving a greater integration gain in SNR. The chief disadvantages are greater computational complexity of pulse-Doppler processing compared with MTI filtering and longer required dwell times due to the use of more pulses for the Doppler measurements.

17.5.1 The Discrete-Time Fourier Transform of a Constant-Velocity Target

To understand some of the details of basic pulse-Doppler measurements, it is useful to consider the Fourier spectrum of an ideal, constant radial velocity, moving point target and the effects of a sampled Doppler spectrum. The issues are the same as those considered when discussing the DTFT of a sinusoid and the sampling of the DTFT by the DFT in Chapter 14. Consider a radar illuminating a moving target in a particular range bin over a CPI of M pulses. If the target radial velocity is such that the Doppler shift is f_d Hz, then the slow-time received signal in that range bin after quadrature demodulation is

$$y[m] = Ae^{j2\pi f_d mT}, \quad m = 0, \ldots, M-1 \tag{17.31}$$

where T is the radar's pulse repetition interval. The signal of equation (17.31) is the same signal considered in equation (14.13), except for the change from normalized frequency $\hat{f}_0$ in cycles to analog frequency f_d in Hz; they are related according to $\hat{f}_0 = 2\pi f_d T$. Equation (14.16) gave the DTFT of this signal; expressing it in terms of analog frequency gives

$$Y(f) = A\frac{\sin[\pi(f - f_d)MT]}{\sin[\pi(f - f_d)T]} e^{-j\pi(M-1)(f-f_d)T} \tag{17.32}$$

The magnitude of this asinc function is illustrated in Figure 17-15a for the case where $f_d = PRF/4$ and $M = 20$ pulses. As would be expected, the mainlobe of the response is centered at $f = f_d$ Hz. So long as $M \geq 4$, the Rayleigh (peak-to-null) mainlobe bandwidth is $1/MT = PRF/M$ Hz. The width of the mainlobe at the -3 dB points is $0.89/MT = 0.89PRF/M$ Hz. These mainlobe width measures determine the Doppler resolution of the radar system. Note that both are inversely proportional to MT, which is the total elapsed time of the set of pulses used to make the spectral measurement. Thus, Doppler resolution is determined by the observation time of the measurement. Longer observation allows finer Doppler resolution.

The first sidelobe of the asinc function is 13.2 dB below the response peak. Because of these high sidelobes, it is common to use a data window to weight the slow-time data

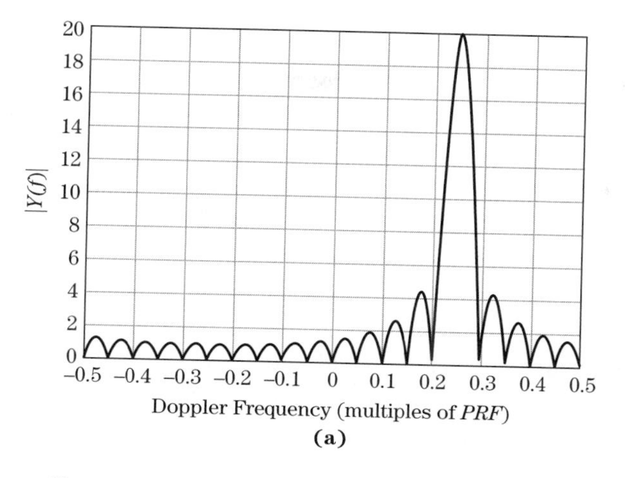

FIGURE 17-15 ■ Magnitude of the discrete-time Fourier transform of a pure complex sinusoid representing an ideal moving target slow-time data sequence with $f_d = PRF/4$ and $M = 20$ pulses. (a) No window. (b) Hamming window.

samples, $y[m]$, prior to computing the DTFT. To analyze this case, replace $y[m]$ by $w[m]y[m]$ in the computation of the DTFT. The result is

$$Y_w(f) = A\sum_{m=0}^{M-1} w[m]e^{-j2\pi(f-f_d)mT} = W(f - f_d) \tag{17.33}$$

where the notation $Y_w(f)$ is used to emphasize that the spectrum is computed with a nontrivial window applied to the data. This is simply the Fourier transform of the window function itself, shifted to be centered on the target Doppler frequency, f_d, rather than at zero. Figure 17-15b illustrates the effect of the window on the DTFT for the same data used in part a of the figure. Harris gives an extensive description of common window functions and their characteristics [23], while Nuttall provides some corrections and additional windows [24].

Note that the length M window should be applied to the data *before* it is zero padded. Applying a K-point window to the full length of a zero padded sequence has the effect of multiplying the data by a truncated, asymmetric window (the portion of the actual window that overlaps the M nonzero data points), resulting in greatly increased sidelobes.

In general, nonrectangular windows cause an increase in mainlobe width; a decrease in peak amplitude; and a decrease in signal-to-noise ratio in exchange for large reductions in peak sidelobe level. These effects were clearly visible in comparing the DTFTs of a rectangular window (no window) and a Hamming window on a decibel scale in Figure 14-14. It was shown there that the reduction in peak power and the SNR reduction depend only on the window and are

$$\text{DTFT peak power reduction} = \frac{1}{M^2}\left|\sum_{m=0}^{M-1} w[m]\right|^2 \tag{17.34}$$

$$\text{SNR loss} = \frac{\left|\sum_{m=0}^{M-1} w[m]\right|^2}{M\sum_{m=0}^{M-1} |w[m]|^2} \tag{17.35}$$

Both of these results are derived in [1]. Table 14-1 gave values of these metrics, along with resolution reduction and straddle loss, for several common windows. Values of 4 to 8 dB are common for the peak power reduction, while SNR losses are often in the range of 1 to 3 dB. The corresponding resolution loss (increase in 3 dB mainlobe width) is typically 35% to 85%. Note that as sidelobes are reduced within a given class of windows (e.g., Taylor windows [23]), losses in peak power, SNR, and resolution tend to increase.

17.5.2 Sampling the Doppler Spectrum and Straddle Loss

MTI processing operates in the slow-time domain, that is, directly on the time signal represented by a row of $y[l,m]$. In contrast, pulse-Doppler processing explicitly calculates the Doppler spectrum in each range bin and then operates in the frequency domain. In a digital processor, this must be done with a discrete Fourier transform or other discrete spectral analysis technique so that a sampled spectrum is obtained, as was implied in Figure 17-14. The question then arises as to how closely successive samples of the computed Doppler spectrum should be spaced; that is, what should the Doppler sampling interval be?

This question was also addressed in Chapter 14. Assuming a DFT is used to compute the Doppler spectrum, it was seen that the DFT size K is normally chosen so that $K \geq M$, the number of slow-time samples. This choice ensures that the slow-time sequence can be reconstructed from the Doppler DFT spectrum. If $K > M$ the data are extended to a K-point sequence by simply appending $K - M$ zeroes to the end of the sequence, an operation called *zero padding*. K might be chosen larger than M either to reduce straddle loss as described shortly, to improve the detail in the Doppler spectrum measurement, or to enable the use of an FFT algorithm that constrains K (e.g., to be a power of 2).

When actually computing the sampled spectrum, whether by the DFT or other means, one would like to be confident that the sampled spectrum captures all of the important features of the underlying DTFT. For example, if the DTFT exhibits significant peaks, the hope is that one of the spectral samples will fall on or very near that peak so that the sampled spectrum captures this feature. Figure 17-16 illustrates the DTFT of the same signal used in Figure 17-15a and the DFT spectral samples that result when $K = M$ ($= 20$ in this example). The k-th DFT sample corresponds to a normalized frequency of $\hat{f}_k = k/K$ cycles, or an analog frequency of $k \cdot PRF/K$ Hz. In Figure 17-16a, $\hat{f}_k = 0.25$, corresponding to $f_d = PRF/4$. One DFT sample (specifically, $k = 5$) falls on the peak of the asinc function while all of the others fall on its zeroes, so that the DFT becomes an impulse function. This could be viewed as an ideal measurement, since the discrete spectrum indicates a single sinusoid at the correct frequency and nothing else, but it does not reveal the mainlobe width or sidelobe structure of the underlying DTFT.

The good result of Figure 17-16a depends critically on the actual sinusoid frequency exactly matching one of the DFT sample frequencies. If this is not the case, then the

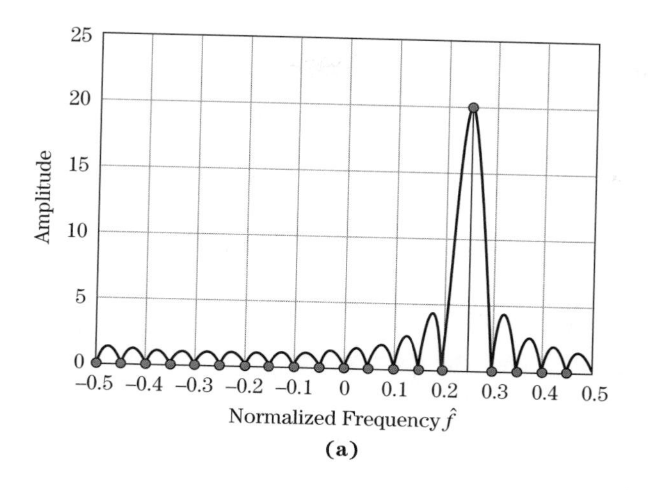

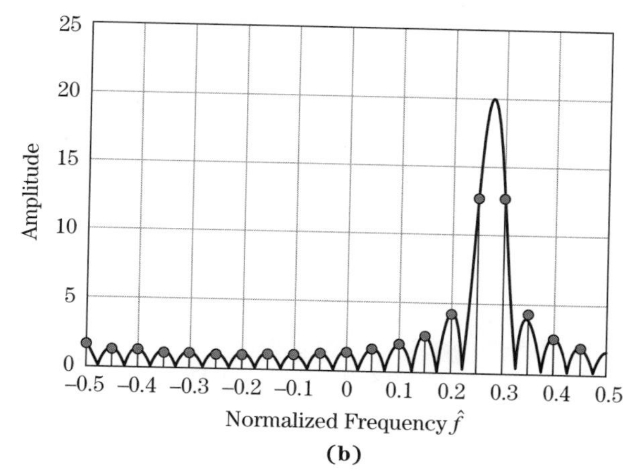

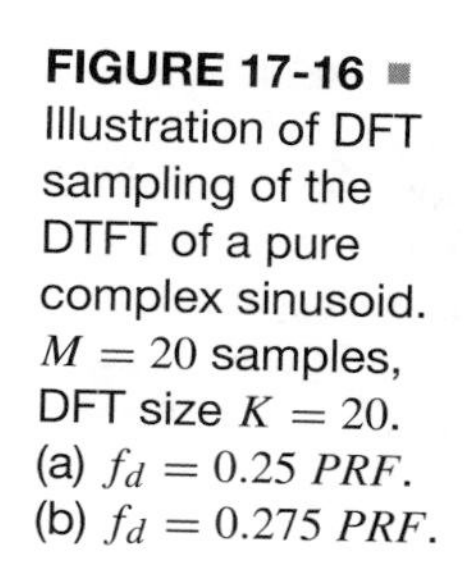
FIGURE 17-16 ■ Illustration of DFT sampling of the DTFT of a pure complex sinusoid. $M = 20$ samples, DFT size $K = 20$. (a) $f_d = 0.25\ PRF$. (b) $f_d = 0.275\ PRF$.

DFT samples will fall somewhere on the asinc function other than the peak and zeros. Figure 17-16b shows the result when the example is modified by changing the normalized frequency $\hat{f}_0$ from 0.25 to 0.275, exactly halfway between two DFT sample frequencies. Now a pair of DFT samples straddles the actual underlying peak of the asinc function, while the other samples fall near the sidelobe peaks. Even though the underlying asinc function is identical in shape in both cases, differing only by a half-bin shift, the effect on the apparent spectrum measured by the DFT is dramatic: the single peak splits into two spectral lines, and large sidelobes appear where before there apparently were none.

The peak amplitude of the underlying DTFT—and thus of the DFT in Figure 17-16a—is 20, whereas in Figure 17-16b the apparent peak amplitude of the spectrum is about 13. This reduction in measured amplitude from the true value is called a *straddle loss*.[6] For a given signal length M, the straddle loss is always greatest for signal frequencies exactly halfway between DFT sample frequencies, and, for a given frequency, the maximum straddle loss increases when the DFT size K is decreased (fewer frequency samples). Usually the smallest DFT size considered is $K = M$. If a window is used, straddle loss is measured relative to the peak of the DTFT of the windowed data.

[6]Straddle loss is also called *scallop loss* by some authors, such as [25].

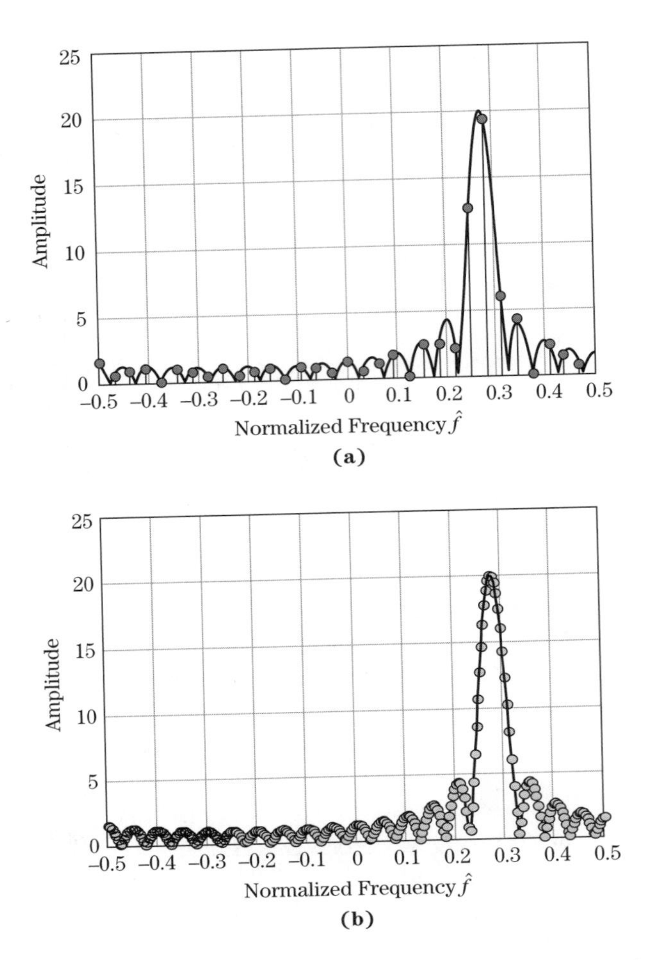

FIGURE 17-17 ■ DFT of a pure complex sinusoid when $M = 20$ and $K > M$. Compare with Figure 17-16. (a) $K = 32 = 1.6M$. (b) $K = 256 = 12.8M$.

The maximum straddle loss can be computed by repeating the two cases of Figure 17-16 using various windows and $K = M$. The worst-case loss is 3.92 dB for a rectangular window with $M = 20$. It is less for other windows because they have broader mainlobes, so the peak amplitude does not fall off as quickly as the frequency varies from the peak of the DTFT. The worst-case straddle loss for $M = 32$ is given for several example windows in Table 14-1. Thus, while any nonrectangular window causes a reduction in peak DTFT gain, the mostly unwanted mainlobe broadening that also accompanies windowing has the desirable property of reducing the *variability* in DTFT gain, and therefore the maximum straddle loss, as the Doppler shift of the target varies.

One obvious way to reduce straddle loss is to sample the Doppler frequency axis more densely—that is, to choose the number of spectrum samples $K > M$. The resulting samples are more closely spaced, and thus the maximum amount by which a sample frequency can miss the peak frequency of the DTFT is reduced. Figure 17-17 shows the result when the sinusoid frequency $\hat{f}_s = 0.275$ again, but the sampling density is increased to 1.6 M samples per Doppler spectrum period (32 samples in this case, the next power of 2 above 20) and then to 12.8 M samples per spectrum period (256 samples). Increasing the sample density causes the apparent spectrum measured by the DFT to increasingly resemble the underlying asinc of the DTFT.

The worst-case straddle loss can be limited to a specified value, at least for this idealized signal, by appropriate choice of the number of samples in the spectrum. Figure 17-18

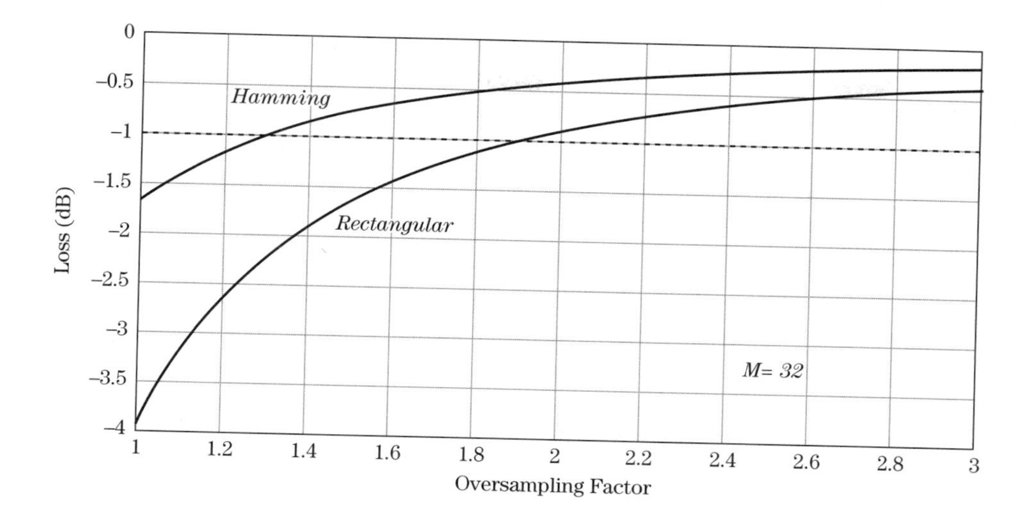

FIGURE 17-18 ■ Maximum straddle loss as a function of Doppler spectrum oversampling factor.

shows the worst-case straddle loss as a function of the degree of oversampling α, where the DFT size $K = \alpha M$. The loss can be limited to 1 dB if the DFT size is at least $1.92M$ for a rectangular window and $1.3M$ for a Hamming window.

17.5.3 Signal-to-Noise Ratio in the Doppler Spectrum

The DFT is a form of coherent integration. If the input signal is a sinusoid in white noise in the time domain, the white noise power will be spread uniformly (on average) in the frequency domain, while the energy in the sinusoid will be concentrated in a peak at the DFT bin closest to the sinusoid frequency. Specifically, if complex white noise with total variance σ_n^2 ($\sigma_n^2/2$ in each of the real and imaginary parts) is added to the sinusoidal data of equation (17.31), the SNR as measured in the time domain will be

$$SNR_t = \frac{\text{signal power}}{\text{noise power}} = \frac{|A|^2}{\sigma_n^2} \tag{17.36}$$

The DTFT of the signal component is given in equation (17.32); it has a peak amplitude of MA. The noise will contribute an equal average power to each DFT sample of $M\sigma_n^2$ [1]. The SNR measured in the frequency domain at the DTFT peak is therefore

$$SNR_f = \frac{|MA|^2}{M\sigma_n^2} = M \cdot SNR_t \tag{17.37}$$

Thus, the DTFT of an M-point sinusoid in white noise gives a coherent integration gain of a factor of M (13 dB for the $M = 20$ case in the preceding examples). Pulse-Doppler processing not only allows separation of the moving target signals from the clutter but also improves the SNR, as should be expected from combining multiple pulses of data.

In practice, the spectrum is computed using the DFT. The noise power in the DFT samples is the same no matter what the sample frequencies are, since white noise is distributed uniformly across all frequencies. Consequently, the DFT spectrum achieves the full integration gain of M only if there is a DFT sample at the peak of the underlying DTFT. That is, the integration gain will be reduced by the straddle loss, as illustrated in Figure 17-19. In this case, which used a rectangular window but suffers worst-case straddle loss, the actual integration gain will be 13 dB − 3.92 dB (from Table 14-1), or 9.08 dB.

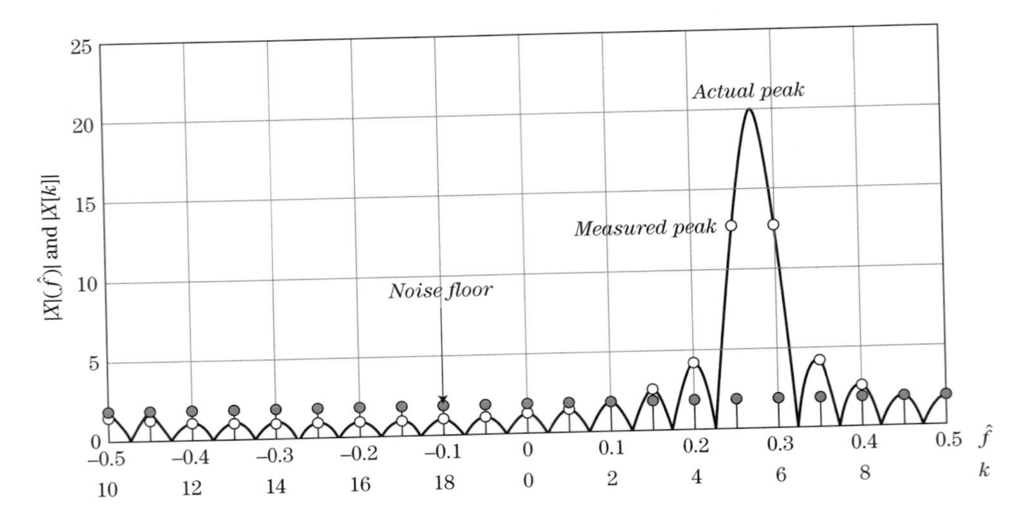

FIGURE 17-19 ■ SNR reduction due to straddle loss.

If a window is used on the data, the SNR loss of that window is incurred, but the worst-case straddle loss is reduced. For this example with a Hamming window, the integration gain would be 13 dB minus the 1.44 dB SNR loss and the 1.68 dB worst-case straddle loss, giving a net integration gain of 9.88 dB, which is 0.8 dB better than the no-window case.

The ideal integration gain and SNR loss due to the window do not depend on the DFT size, so zero padding to use a larger DFT will not increase the number of actual data samples and so will not improve the SNR. However, increasing the DFT size does limit the straddle loss, so a larger DFT may somewhat improve the SNR in the frequency domain.

17.5.4 Matched Filter and Filterbank Interpretations of Pulse-Doppler Processing with the DFT

As with MTI filters, it is interesting to see how DFT-based pulse-Doppler processing for detection compares with the matched filter. To derive the optimum MTI filter, it was assumed in [1] that the target Doppler shift is unknown, leading to optimum MTI filters for small order N that were nearly equivalent to simple pulse cancellers. In contrast, DFT-based pulse-Doppler processing attempts to separate target signals based on their particular Doppler shift. Assume that the baseband slow-time signal is a pure complex sinusoid (corresponding to a constant radial velocity moving target) at a Doppler shift of f_d Hz and that a K-point DFT is to be applied to the M-point slow-time data sequence. The data must be zero padded with $K-M$ zeroes. The model of the zero padded signal vector is

$$\boldsymbol{T} = \hat{A}\left[\underbrace{1 \quad e^{j2\pi f_d T} \; \ldots \; e^{j2\pi f_d (M-1)T}}_{M \text{ samples}} \quad \underbrace{0 \quad \cdots \quad 0}_{K-M \text{ samples}}\right]^T \tag{17.38}$$

If the interference consists only of white noise (no correlated clutter), then the interference covariance matrix $\mathbf{S_I}$ is just $\sigma_n^2 \mathbf{I}_K$, where $\mathbf{I}_K$ is the K-th order identity matrix. The slow-time data vector $\boldsymbol{Y}$ is also zero padded to produce the input to the matched filter

$$\boldsymbol{Y} \equiv \left[\underbrace{y[m] \; y[m+1] \cdots y[m+M-1]}_{M \text{ samples}} \quad \underbrace{0 \quad \cdots \quad 0}_{K-M \text{ samples}}\right]^T \tag{17.39}$$

The optimum filter coefficient vector is $\boldsymbol{H}_{\text{opt}} = \mathbf{S}_{\mathbf{I}}^{-1}\boldsymbol{T}^*$ (see Chapter 14 or [1]). Filtering the data $\boldsymbol{Y}$ with the filter $\boldsymbol{H}_{\text{opt}}$ gives the output

$$z = \boldsymbol{H}_{\text{opt}}^T \boldsymbol{Y} = \frac{\hat{A}}{\sigma_n^2} \sum_{m=0}^{M-1} y[m] e^{-j2\pi f_d m T} \tag{17.40}$$

which, to within a scale factor, is simply the DTFT of the slow-time data $y[m]$ evaluated at frequency f_d. When $f_d = k/KT = k \cdot PRF/K$ for some integer k, equation (17.40) is the K-point DFT of the data sequence $y[m]$. Consequently, the discrete Fourier transform is a matched filter to ideal, constant radial velocity, moving target signals provided that the Doppler shift equals one of the DFT sample frequencies and the interference is white.

The K-point DFT computes K different outputs from each input vector. It can be shown that the DFT effectively implements a bank of K matched filters at once, each tuned to a different Doppler frequency, and (by computing the DTFT of $\boldsymbol{H}_{\text{opt}}$) that the frequency response shape of each matched filter is an asinc function [1]. Thus, the k-th DFT sample corresponds to filtering the data with a band-pass filter having a frequency response with an asinc function shape centered at the frequency of the k-th DFT sample. If the data are windowed before processing with a window function $w[m]$, the DFT still implements a band-pass filter centered at each DFT frequency, but the filter frequency response shape now becomes that of the DTFT of the window function.

Of course, it is possible to build a literal bank of band-pass filters, each one perhaps individually designed, and some systems are constructed this way. For example, the zero-Doppler filter in the filterbank can be optimized to match the expected clutter spectrum or even made adaptive to account for changing clutter conditions. Most commonly, however, the DFT is used for Doppler spectrum analysis. This places several restrictions on the characteristics of the equivalent filterbank:

- There will be K filters in the bank, where K is the DFT size.
- The filter center frequencies will be equally spaced, equal to the DFT sample frequencies PRF/K.
- All the passband filter frequency response shapes will be identical and equal to the DTFT of the window used, differing only in center frequency.

The advantages to this approach are simplicity and speed with reasonable flexibility. The DFT provides a simple and computationally efficient (via the FFT) implementation of the filterbank; the number of filters can be changed by simply changing the DFT size; the filter shape can be changed by simply choosing a different window; and the filter maximizes the output signal-to-noise ratio for targets coinciding with a DFT filter center frequency in a noise-limited interference environment.

17.5.5 Fine Doppler Estimation

Peaks in the DFT output that are sufficiently above the noise level to cross an appropriate detection threshold are interpreted as signals from moving targets, that is, as samples of the peak of an asinc component of the form of equation (17.32) or its windowed equivalent, equation (17.33). As has been emphasized, there is no guarantee that a DFT sample will fall exactly on the asinc function peak. Consequently, the amplitude of the DFT sample giving rise to a detection and its frequency are only approximations to the actual amplitude and frequency of the asinc peak. In particular, the estimated Doppler frequency of the peak based on the DFT index of the detection can be off by as much as one-half Doppler bin, equal to *PRF*/2*K* Hz.

If the DFT size K is significantly larger than the number of pulses (data sequence length) M, then several DFT samples will be taken on the asinc mainlobe, and the largest may well be a good estimate of the amplitude and frequency of the asinc peak. However, if K is equal to or only slightly larger than M, the Doppler samples may be far apart, and a half-bin error may be intolerable due to excessive straddle loss and frequency error. One way to improve the estimate of the true Doppler frequency, f_d, is to interpolate the DFT in the vicinity of the detected peak.

The most obvious way to interpolate the DFT is to zero pad the data and compute a larger DFT. This approach is computationally expensive and interpolates the whole spectrum. If finer sampling is needed only over a small portion of the spectrum, the zero padding approach is inefficient.

A simple but very serviceable technique for interpolating local peaks is the local quadratic interpolation introduced in Chapter 14 and illustrated in Figure 17-20. Recall that if a peak in the DFT magnitude spectrum is observed at index $k = k_0$ with magnitude $|Y[k_0]|$, the frequency of an interpolated peak can be estimated using the formula (see Chapter 14 or [1])

$$k' = k_0 + \Delta k = k_0 + \frac{-\frac{1}{2}\{|Y[k_0+1]| - |Y[k_0-1]|\}}{|Y[k_0-1]| - 2|Y[k_0]| + |Y[k_0+1]|} \tag{17.41}$$

The frequency bin estimate k' is converted to frequency in hertz using $f'_d = k'PRF/K$. Once k' is computed, an improved amplitude estimate can also be computed. However, the techniques described here are more effective for frequency estimation than for amplitude estimation.

The accuracy of this technique improves when the spacing of DFT samples is smaller relative to the frequency resolution. Figure 17-21a shows the residual error after applying equation (17.41), as the difference between the actual input frequency and the nearest DFT bin varies between zero and one-half DFT bins. These data are for a 30-sample signal with no noise. When the DFT size is also 30, the reduction in error is modest. However, as the DFT size is increased, the worst-case residual errors shrink rapidly. For an oversampling factor of two or larger ($K \geq 60$), the worst-case error is less than 0.03 DFT bins.

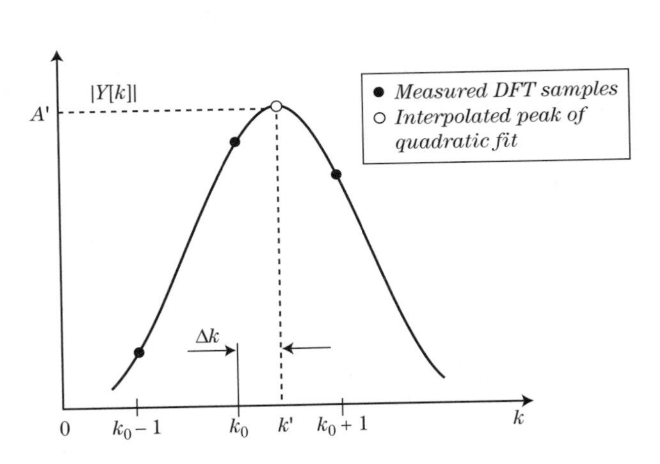

FIGURE 17-20 ■ Refining the estimated target amplitude and Doppler shift by interpolation around the DFT peak. k' is the estimated "bin" number of the peak, and A' is the estimated amplitude.

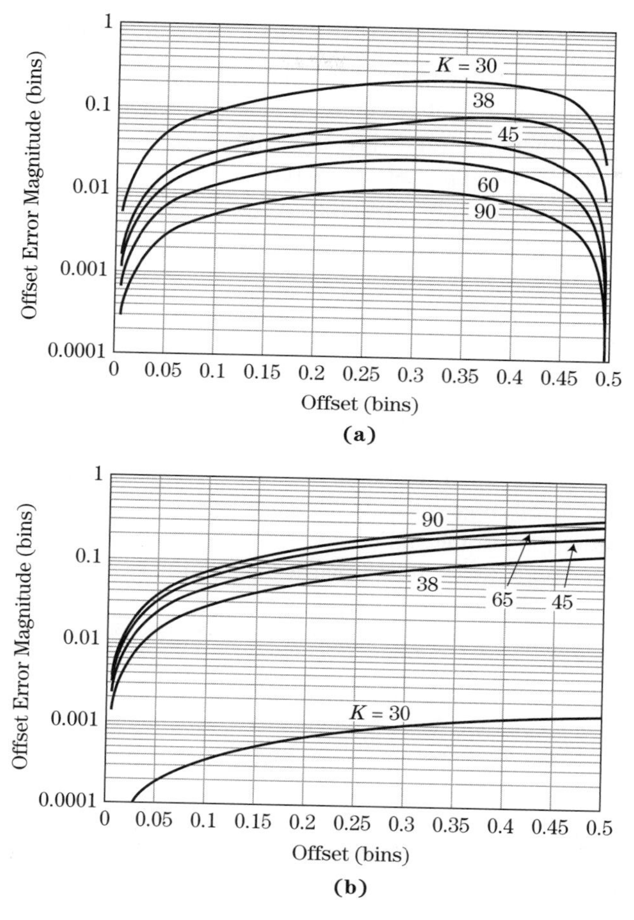

FIGURE 17-21 Accuracy of DFT frequency estimator as a function of DFT size K. Signal length $M = 30$ in all cases. (a) Magnitude-based estimator. (b) Complex estimator.

Another frequency estimator that achieves extremely good results when $K = M$ (minimum size DFT) uses the complex DFT samples

$$k' = k_0 + \Delta k = k_0 + Re\left\{\frac{Y[k_0+1] - Y[k_0-1]}{Y[k_0-1] - 2Y[k_0] + Y[k_0+1]}\right\} \tag{17.42}$$

Figure 17-21b illustrates the frequency estimation performance of this complex interpolator on the same example. The worst-case error when $K = M$ is only 0.0014 bins. However, when the DFT is oversampled ($K > M$), the performance of the complex estimator gets worse rather than better. Instead, the magnitude estimator is better for oversampling factors of about 1.25 or more.

Further complications arise when the data are windowed before the DFT. The magnitude estimator is slightly more accurate on Hamming windowed data, while in the complex estimator the error for all of the oversampling factor cases in the previous example tends to approach the $K = 90$ case. Specifically, the $K = 30$ case no longer exhibits extremely low errors. However, very similar estimators with slightly different weighting factors, dependent on the particular window used, are available (see [25] for details). In general, while the frequency interpolation technique is simple and computationally inexpensive, the particular estimator coefficients must be chosen to match the window and DFT oversampling factor. Additionally, these nonlinear frequency estimators generally perform well only when the post-DFT SNR exceed approximately 3 dB [25].

17.5.6 Modern Spectral Estimation in Pulse-Doppler Processing

So far, the discrete Fourier transform, implemented with the fast Fourier transform algorithm, has been used exclusively to compute the spectral estimates needed for pulse-Doppler processing. Other spectral estimators can be used. One that has been applied to radar is the autoregressive (AR) model, which represents the actual discrete-time spectrum $Y(\hat{\omega})$ of the slow-time signal with a spectrum of the form

$$\hat{Y}(\hat{\omega}) = \frac{\alpha}{1 + \sum_{p=1}^{P} a_p e^{-j\hat{\omega}p}} \tag{17.43}$$

The algorithm finds the set of model coefficients $\{a_p\}$ that optimally fits $\hat{Y}(\hat{\omega})$ to $Y(\hat{\omega})$ in the least-squares sense for a given model order P. These coefficients are found by solving a set of "normal equations" derived from the autocorrelation of the slow-time data $y[m]$ using one of a variety of related methods [26]. Finally, the $\{a_p\}$ are used to compute an estimated spectrum according to (17.43), which can then be analyzed for target detection and parameter estimation or other functions.

One advantage of AR modeling is that it has the potential to achieve finer spectral resolution than Fourier techniques for a given amount of data. This is illustrated in Figure 17-22, which compares the DTFT and the AR Doppler spectrum computed using the

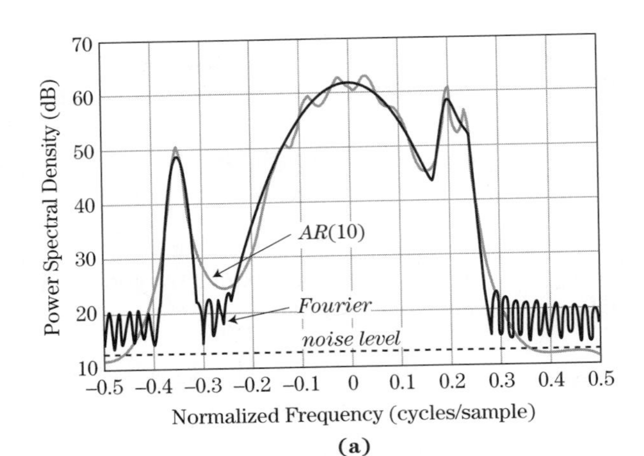

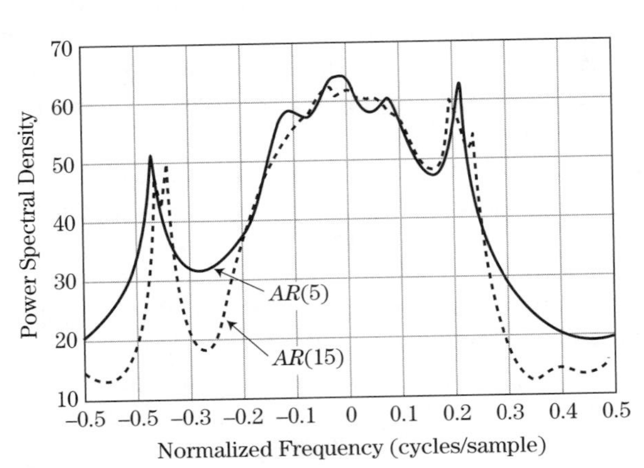

FIGURE 17-22 ■ Comparison of Fourier and covariance-method autoregressive spectrum estimators on simulated Doppler data. (a) Fourier and AR(10) estimates. (b) AR(5) and AR(15) estimates on same data.

covariance method [26] with the magnitude squared of the DFT on 50 samples of simulated slow-time radar data consisting of three sinusoids in white noise and log-normal clutter having a Gaussian power spectrum. The sinusoids are at normalized frequencies $\hat{f} = -0.35, 0.2$, and 0.23 cycles/sample. Figure 17-22a compares the average of 100 realizations of each of the AR and Fourier spectral estimates. The order $P = 10$ for the AR estimator. Note that the two sinusoids at $\hat{f} = 0.2$ and 0.23 are not well resolved in the Fourier estimate but are resolved at least weakly in the AR(10) estimate. Also note the lack of sidelobes and the rippled approximation to the clutter lobe in the AR estimate.

It is necessary to determine a model order P to use the AR method, and the results depend significantly on this choice. In general, using too low a value for P can cause the algorithm to fail to resolve spectral peaks, while too high an order may lead to spurious peaks. For example, Figure 17-22b shows the AR estimate on the same data for smaller ($P = 5$) and larger ($P = 15$) model orders. The low-order model is no longer able to resolve the two positive-Doppler sinusoids, and the quality of the clutter lobe estimate is degraded. The high-order AR estimator resolves the two positive-frequency sinusoids but splits the estimate of the negative-frequency sinusoids into two peaks, suggesting two sinusoids where there is actually only one. Algorithms exist to help estimate an appropriate order from the data [26]. Also, depending on details of the particular AR method used, the amount of computation required grows at a rate between P^2 and P^3 and so increases rapidly with the AR model order. Finally, AR and other modern spectral estimation methods often produce poor results, rather than degrading "gracefully," at low SNRs. While modern spectral estimation techniques have been explored for several aspects of radar signal processing, the vast majority of systems at the time of this writing in 2009 rely primarily on Fourier-based spectral estimation methods.

17.5.7 Metrics for Pulse-Doppler Detection of Moving Targets

In Section 17.4.3, improvement factor was presented as one metric for quantifying the effect of MTI filtering on detection performance. Improvement factor is independent of target Doppler shift. This is appropriate, since MTI filtering does not provide a measure of Doppler shift for detected targets. However, pulse-Doppler detection does indicate target Doppler, so a metric is desired that better reflects the variations in detectability at different Doppler shifts due to variations in total interference.

The minimum detectable velocity and usable Doppler space fraction are two related metrics that are increasingly common in pulse-Doppler and space-time adaptive processing. Consider the notional Doppler spectrum shown in Figure 17-23a. Targets anywhere in the spectrum compete with both noise and clutter because sidelobes of the clutter, even if below the noise, exist throughout the spectrum, so there is some clutter power at all Doppler shifts. The signal-to-interference ratio is the ratio of the target power to the combined noise and clutter power. In this example, the SIR approaches just the SNR for fast-moving targets because the clutter power is very small at high Doppler shifts. As the Doppler shift approaches zero, the SIR decreases because of the increased clutter power. The stationary target at $f_d = 0$ is unlikely to be detectable, while the fast-receding target at a large negative Doppler shift will be detectable provided it is far enough above the noise floor. The target at the edge of the clutter spectrum will have a higher SIR than a stationary target but a lower SIR than a fast target and thus will have an intermediate detection probability.

Figure 17-23b illustrates one way to quantify this observation. The solid curve is the SIR loss, which is the reduction in SIR relative to the noise-limited case (no clutter)—the

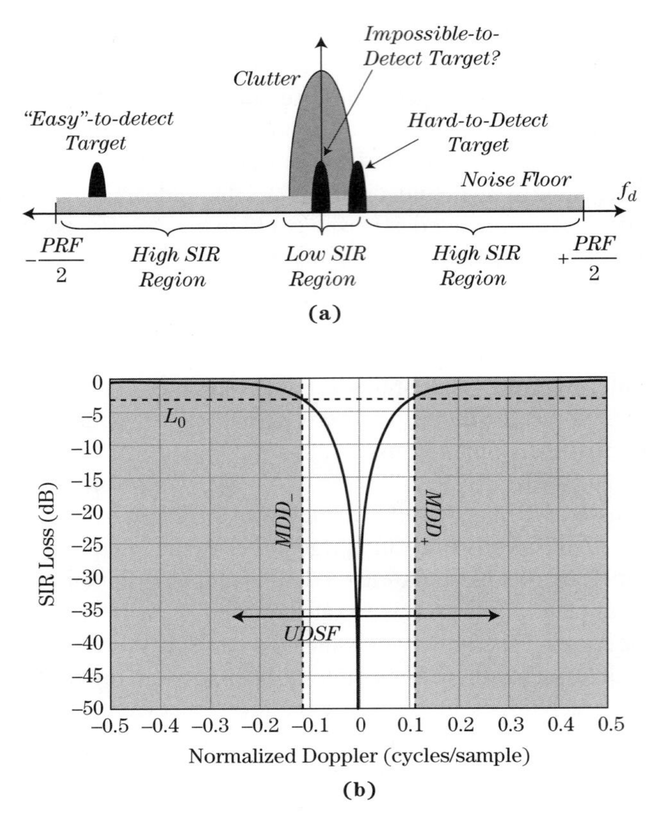

FIGURE 17-23 ■ Minimum detectable velocity and usable Doppler space fraction. (a) Notional spectrum showing three targets with varying signal-to-interference environments. (b) Typical SIR loss curve and resulting MDD and UDSF.

highest SIR possible. Thus, the SIR loss is very large for stationary targets but approaches zero dB for fast targets far removed from the clutter. Suppose systems analysis determines an "acceptable" value of the SIR loss, L_0, based on detection performance requirements. A typical value of L_0 might be -3 dB, as shown in the Figure 17-23b. The *minimum detectable Doppler* (MDD) is the Doppler shift above which the SIR loss is greater than L_0.[7] Because the SIR loss curve is often not symmetric about the clutter notch, positive and negative MDDs may be defined, and the single overall MDD is the average of their magnitudes

$$MDD = \frac{1}{2}(MDD_+ - MDD_-) \tag{17.44}$$

The minimum detectable velocity is simply the MDD converted to velocity units, $MDV = \lambda \cdot MDD/2$. The usable Doppler space fraction is the portion of the Doppler period for which the SIR loss is greater than L_0. In Figure 17-23b, this is the shaded area and equals approximately 78% of the spectrum width.

[7]Caution is needed with the terminology because SIR loss is expressed as a negative number. In this example, an SIR loss greater than L_0 would be, say, -1 or -2 dB, representing reduced clutter power and higher detection probability.

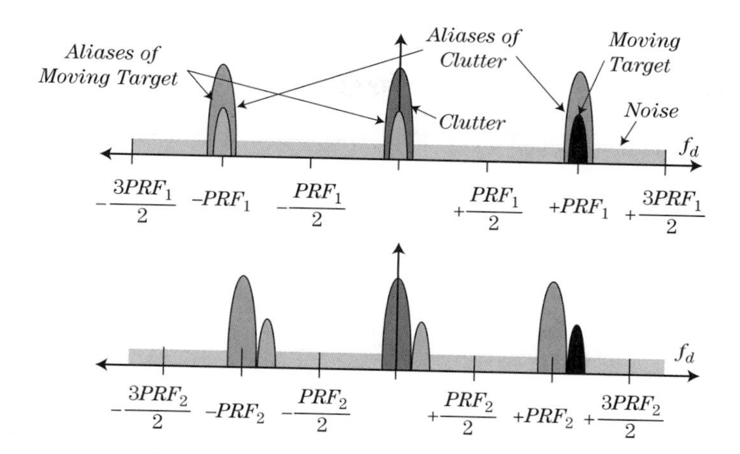

FIGURE 17-24 ■ Illustration of the use of two PRFs to avoid blind speeds in pulse-Doppler radar. The target Doppler shift equals the PRF in the upper plot.

17.5.8 Dwell-to-Dwell Stagger

Pulse-Doppler processing sometimes is combined with pulse cancellers. In this case, the applicability of the concept of blind speeds is clear. If a pulse canceller is not used, then there is no high-pass filter, and a target whose Doppler shift equals an integer multiple of the *PRF* will not be filtered out, as in MTI processing. However, the target energy will still be indistinguishable from clutter energy, since it will alias to the zero frequency portion of the spectrum and combine with the clutter energy. Thus, targets having Doppler shifts equal to a multiple of the PRF will still go undetected, and the corresponding target velocities are still blind speeds.

In dwell-to-dwell PRF stagger, a CPI of M pulses is transmitted at a fixed PRF. A second CPI is then transmitted at a different fixed PRF. This concept is illustrated in Figure 17-24, which shows a notional Doppler spectrum for two different PRFs. The plots are shown on the same analog frequency scale. First consider the spectrum plot in Figure 17-24a, which corresponds to data collected at PRF_1. A target whose Doppler equals PRF_1 will be aliased to zero Doppler shift, where it will be undetectable if clutter is present. Alternatively, the periodic repetitions of the clutter spectrum will coincide with the target. If the same target scenario is measured with a lower PRF_2, the spectrum shown in Figure 17-24b results. In this case, the Doppler shift of the target no longer matches the PRF. The target energy aliases to a nonzero Doppler, where it does not compete with the clutter and is still detectable.

In some systems, as many as eight PRFs may be used. The first velocity that is completely blind at all of the PRFs is the least common multiple of the individual blind speeds, which will be much higher than any one of them alone. Typically, target detections are accepted and passed to subsequent processing only if they occur in some minimum fraction of the PRFs used, for example one of two or three, or three of eight PRFs. In medium PRF systems, particularly those using a small number of PRFs, each of the "major" PRFs used for extending the unambiguous Doppler region may be accompanied by one or two additional "minor" PRFs to resolve range ambiguities as well (see Section 17.5.10).

The advantages of a dwell-to-dwell stagger system are that multiple-time-around clutter can be canceled using coherent MTI and that the radar system stability, particularly in the transmitter, is not as critical as with a pulse-to-pulse stagger system [5]. A major disadvantage is that the use of multiple CPIs consumes large amounts of the radar timeline.

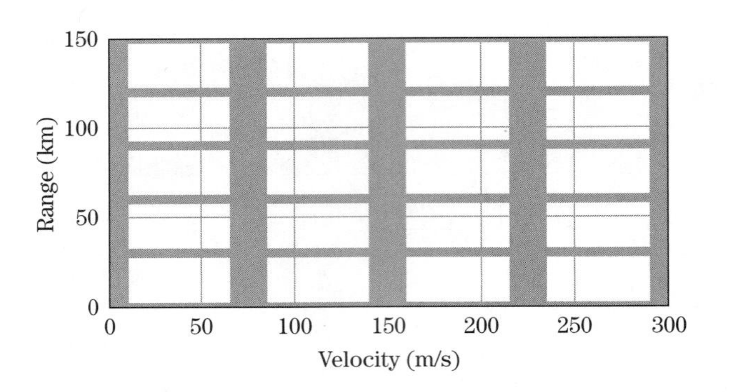

FIGURE 17-25 ■ Blind zone map. See text for parameters.

17.5.9 Blind Zones

For a given PRF, the radar is "blind" to targets having Doppler shifts at multiples of the PRF, as just discussed. The radar is also "blind" to targets at ranges that produce returns arriving at the radar while it is transmitting, regardless of the Doppler. For a given pulse length τ, this occurs for τ seconds at the beginning of every PRI. The radar is said to be *eclipsed* during these time intervals. A range-Doppler diagram illustrating the combined values of Doppler at which the radar is blind due to eclipsing, clutter, or both is called a *blind zone map*.

Figure 17-25 is an example of a simple blind zone map for a radar operating at 10 GHz ($\lambda = 3$ cm). The pulse length $\tau = 10\ \mu$s, the PRI is $T = 200\ \mu$s, and the clutter spectrum is assumed to have a width of 20 m/s. Thus, the unambiguous range is $cT/2 = 30$ km, while the unambiguous velocity is $\lambda/2T = 75$ m/s. Only the positive velocity axis is shown; the negative velocity region would be a mirror image.

Note that the range dimension is blind at any Doppler over intervals of 1.5 km (corresponding to the 10 μs pulse transmission time) starting at ranges of 0 km and integer multiples of the unambiguous range. These are the time intervals in which the radar is transmitting so a monostatic radar could not also be receiving. Similarly, the radar is assumed blind over all ranges at any Doppler frequency dominated by clutter, which occur in 20 m/s wide bands centered at multiples of the unambiguous velocity. A more realistic clutter map would extend the range blind zone extent to allow for transmit/receive switching time and strong close-in clutter echoes. Note that the blind zone map for a different PRI has the same basic shape but is rescaled in each dimension. For instance, increasing the PRI will increase the spacing between blind regions in range but will decrease the spacing between blind regions in velocity. Also, if an M-of-N detection rule is used, a particular range-Doppler resolution cell is considered blind only if it is not in the clear region on at least M of the N PRFs used.

17.5.10 PRF Regimes and Ambiguity Resolution

Measurements made with a pulse burst waveform can be ambiguous in range, Doppler, or both. Pulse-Doppler radars in particular frequently operate in scenarios that are ambiguous in one or both of the range and Doppler dimensions. Modern airborne pulse-Doppler radars operate in a dizzying variety of modes having various range and Doppler coverage and resolution requirements. Pulse burst waveforms using a variety of constituent pulses, including simple pulses, linear frequency modulation (LFM), and Barker phase codes at a minimum, are common. To meet the various mode requirements, PRFs ranging from several hundred Hz to 100 kHz or more are used.

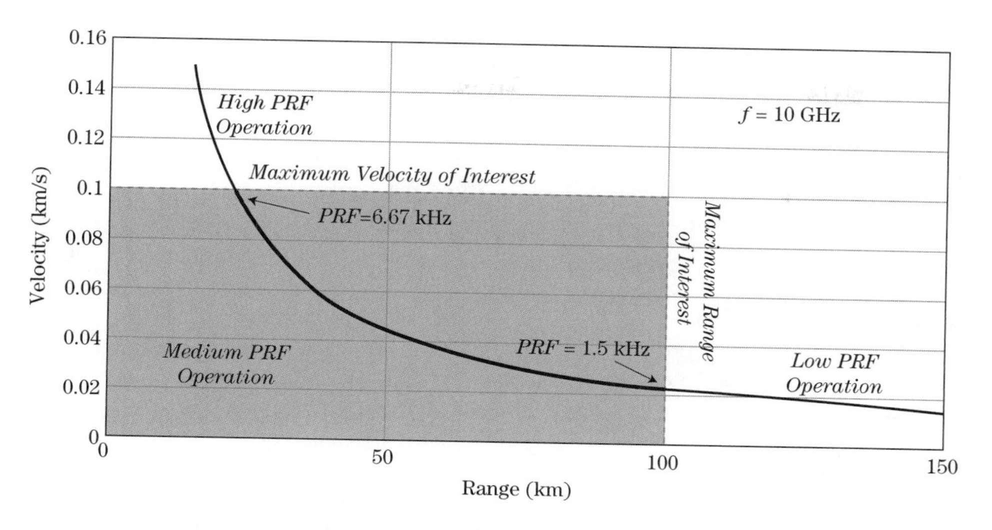

FIGURE 17-26 ■ Low, medium, and high PRF regimes for a notional X-band radar.

Pulse-Doppler radar operation is commonly divided into three "regimes" according to their ambiguity characteristics. Given an unambiguous range, R_{ua}, and unambiguous velocity, v_{ua}, of interest, the radar is considered to be in a *low PRF mode* if the PRF is sufficiently low to be unambiguous in range but is ambiguous in velocity. The *high PRF mode* is the opposite: the system is ambiguous in range but not in velocity. In a *medium PRF mode*, the radar is ambiguous in both. Since any finite PRF results in ambiguity at some range and velocity, the boundary between these regimes clearly depends on the radar scenario, which determines the maximum unambiguous range and velocity of interest.

The trade-off is summarized in Figure 17-26. The line plots the achievable combinations of R_{ua} and v_{ua} as the PRF varies at 10 GHz. If the desired range and velocity coverage are 100 km and 100 m/s, the shaded area indicates a range of PRFs (in this case, 1.5 kHz to 6.67 kHz) that will result in ambiguities in both dimensions. PRFs above 6.67 kHz will be ambiguous in range but not Doppler; those below 1.5 kHz will be ambiguous in Doppler but not range.

Each of the different PRF regimes has different advantages and disadvantages. For instance, high PRF operation provides a relatively large clutter-free region in the Doppler spectrum, while low PRF provides accurate ranging over a long unambiguous range interval. Table 17-3 summarizes these as well as some of the major applications of each mode in the context of airborne radar. A full discussion is beyond the scope of this chapter; the reader is referred to [20,21] for details.

Fortunately, techniques exist that can resolve ambiguities, although at the cost of extra measurement time and processing load. Consider range ambiguity resolution first. Once a PRF is selected, it establishes an unambiguous range $R_{ua} = c/2PRF = cT/2$. A target at an actual range $R_t > R_{ua}$ will be detected at an apparent range R_a that satisfies

$$R_t = R_a + kR_{ua} \tag{17.45}$$

for some integer k. That is, when the radar detects a target at apparent range R_a, the actual range could be R_a, or R_a plus any multiple of R_{ua}. It is convenient to express the apparent, true, and unambiguous ranges in terms of range bins, for example, $n_a = R_a/\Delta R$, where ΔR is the range bin spacing. Equation (17.45) then becomes

$$n_t = n_a + kN \tag{17.46}$$

TABLE 17-3 ■ Some Advantages and Disadvantages of the PRF Regimes in Airborne Radar

PRF Regime	Advantages	Disadvantages
Low	• Precise range measurement • Fine-range resolution possible • Sidelobe clutter rejection via range gating • Simple processing	• Highly ambiguous Doppler • Poor detection performance in look-down modes • High peak power or pulse compression required
Medium	• Good detection over wide range of target Dopplers • Good rejection of both mainlobe and sidelobe clutter • Accurate ranging • Reduced eclipsing compared with high PRF operation	• Sidelobe clutter at all velocities • Large number of PRF and pulse width combinations • Complex range and Doppler ambiguity resolution processing • Poor performance for large targets in sidelobes
High	• High average power • Unambiguous Doppler • Mainlobe clutter rejection without rejecting targets	• Highly ambiguous range • Increased eclipsing of targets • Complex, reduced-accuracy ranging • Reduced sensitivity to low-Doppler targets due to sidelobe clutter

The basic approach to resolving range ambiguities relies on multiple PRFs. Suppose that there are N_i range bins in the unambiguous range interval on PRF i; then $R_{ua_i} = N_i \Delta R$. The unambiguous range is different for each PRF. Assuming that the range bin spacing is the same for each PRF used, the true range bin must satisfy equation (17.46) for each of the PRFs used:

$$n_t = n_{a_0} + k_0 N_0 = n_{a_1} + k_1 N_1 = \ldots \tag{17.47}$$

The set of equations (17.47) can be solved using the Chinese remainder theorem (CRT; see [27] or [1] for details). However, a simpler and more easily understood approach is the *coincidence algorithm* for determining n_t. This technique is essentially a graphical implementation of the CRT [27,28]. The method is best illustrated with an example.

Presume that three PRFs are used. Suppose there are two targets, denoted a and b, with true ranges corresponding to range bins $n_a = 6$ and $n_b = 11$. Further, suppose the PRFs are such that the number of range bins in each unambiguous range interval are $N_0 = 7$, $N_1 = 8$, and $N_2 = 9$.[8] Because the true range bin of target a (range bin 6) is less than the number of range bins in the unambiguous range interval for each PRF (7, 8, or 9), target a is range-unambiguous at each PRF. Conversely, target b is range-ambiguous at each PRF. The measured data (apparent range bin of each target) will be

$$\begin{aligned} n_{a_0} &= n_{a_1} = n_{a_2} = 6 \\ n_{b_0} &= 4, \quad n_{b_1} = 3, \quad n_{b_2} = 2 \end{aligned} \tag{17.48}$$

This measurement scenario is illustrated in Figure 17-27. The light gray range bins denote the detections of target a in bin $n_a = 6$ at each PRF, while the dark gray bins show the detection of target b at bins 4, 3, and 2 in the three PRFs.

[8]The actual number of range bins in an unambiguous range interval is often much larger, but small values are convenient for illustration.

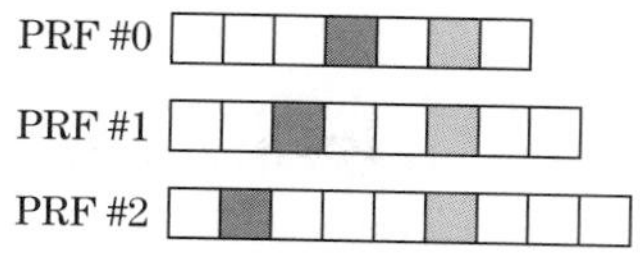

FIGURE 17-27 ■ Measured range data at each of three PRFs for coincidence algorithm for range ambiguity resolution.

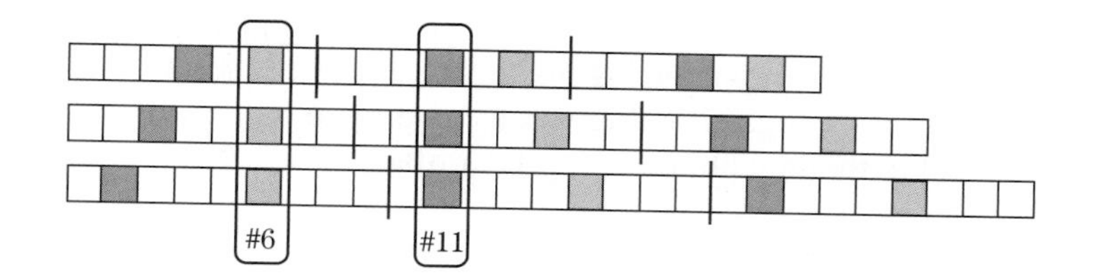

FIGURE 17-28 ■ Coincidence detection of target ranges in replicated range data.

The coincidence algorithm takes the pattern of detections at each PRF and replicates it, as shown in Figure 17-28. In essence, the replication implements equation (17.46), placing a tentative detection at each value of $n_a + kN_i$ and $n_b + kN_i$. These tentative detections represent the plausible ranges for each target at each PRF. The algorithm then searches for a range bin that exhibits a detection at all three PRFs, which indicates that the range bin is consistent with the measurements at all three PRFs. As shown in the figure, this process correctly detects the true range bins $n_a = 6$ and $n_b = 11$ in this example.

One practical difficulty with this approach is that there is no guarantee that the actual range, R_t, will be an integer multiple of the range bin spacing, ΔR, as previously assumed; the target may in fact straddle range bins. In addition, noise in the measurements may cause the target to be located in an incorrect range bin. The basic coincidence algorithm as previously described would produce a large error in the estimated true range in the presence of such measurement errors. However, various extensions have been developed to reduce the sensitivity to measurement errors. In one approach, exact coincidence is not required to declare a target. Instead, a tolerance of N_T range bins is established and a target is declared if a detection occurs in all three PRFs at some range bin $n_t \pm N_T$. Depending on the range bin size and SNR, N_T will typically be only 1 or 2 range bins. A more sophisticated version of this basic idea is described in [29].

In the example of Figure 17-28, three PRFs proved sufficient to resolve two different range-ambiguous targets. In general, N PRFs are required to successfully disambiguate $N - 1$ targets. If the number of targets exceeds $N - 1$, *ghosts* can appear [20]. Ghosts are false targets resulting from false coincidences of range-ambiguous data from different targets. The problem is illustrated in Figure 17-29, which repeats the example of Figure 17-28 using only the first and third of the previous three PRFs. While targets will still be detected at the correct bins $n_a = 6$ and $n_b = 11$, a third coincidence occurs between detections from targets 1 and 2 at range bin $n_c = 20$, representing an apparent third target.

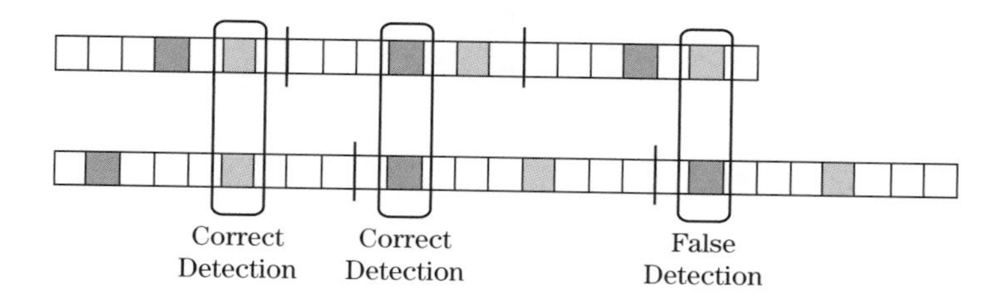

FIGURE 17-29 ■ Illustration of the formation of ghosts in range ambiguity resolution.

Unless additional data (e.g., tracking information) are available, the signal processor has no way of recognizing that the last coincidence is among detections from different targets. Thus, the processor will declare the presence of three targets in this example: the two correct targets and one "ghost." Use of a third PRF as in Figure 17-28 eliminates this ghost.

In a medium or low PRF mode, the radar will suffer velocity ambiguities. This problem is identical to that of range ambiguities: given an apparent Doppler shift, f_{d_a}, the actual Doppler shift, f_{d_t}, must be of the form $f_{d_a} + k \cdot PRF$ for some integer k. Use of the DFT for spectral estimation results in quantization of the Doppler spectrum into Doppler bins (equivalently, velocity bins), analogous to range bins in the range dimension. The same techniques used for range disambiguation can therefore be used to resolve velocity ambiguities, subject to the same limitations and artifacts.

17.5.11 Transient Effects

In range-ambiguous medium and high PRF modes, each transmitted pulse may result in significant target or clutter contributions from multiple ambiguous range intervals. All of the discussion in this chapter has assumed a steady-state scenario, in the sense that each slow-time sample in a given range bin is assumed to contain all clutter and target contributions that fold over into that range bin. However, this is not the case for the first few slow-time samples. For each new CPI, several pulses, known as *clutter fill* pulses, must be transmitted before a steady-state situation is achieved.

Figure 17-30 illustrates the need for clutter fill pulses. Part a shows 19 range bins of data that might be received for a single transmitted pulse. Targets are present in bins 6, 11, and 16. Part b shows the fast-time/slow-time data matrix that would be observed, assuming the PRF corresponds to seven range bins. The targets are thus distributed over three range ambiguity intervals. Because the second target is in bin 11, it does not appear until after the second pulse is transmitted, after which it appears as if it were in bin 4. In this example, it is not until the third pulse that the same pattern of target echoes is observed on each pulse. Data from the first two pulses should not be used in Doppler processing. In general, if significant target or clutter returns are received over R range-ambiguity intervals, $R-1$ clutter fill pulses will be required before steady-state conditions are achieved on the R-th pulse. Additional pulses may be used to set the automatic gain control of the receiver and are also not used for Doppler processing.

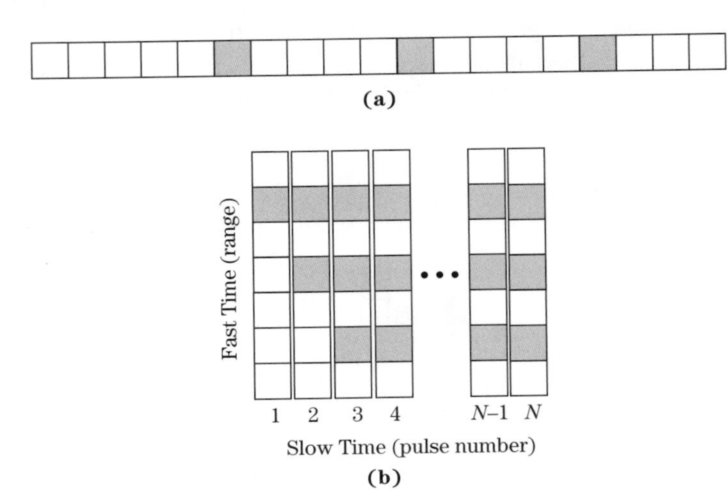

FIGURE 17-30 ■ Illustration of clutter fill pulses and MTI filter transients. (a) Single-pulse response. (b) Fast-time/slow-time data matrix if PRI is equivalent to 7 range bins.

Steady-state operation of the digital filters used for MTI processing occurs when the output value depends only on actual data input values rather than on any initial (typically zero-valued) samples used to initialize the processing. For FIR filters of order N (length $N + 1$), the first N outputs are transients and are discarded in some systems. For simple single or double cancellers, this is only one or two samples. For example, in Figure 17-30b, the slow-time data achieves steady state on the third pulse, but the output of a two-pulse canceller would not achieve steady state until the fourth pulse; for earlier pulses, at least one of the two input samples would not yet contain data from the further ambiguous ranges.

Now consider a combined MTI-pulse-Doppler system using an N-pulse canceller and a CPI of M pulses for the pulse-Doppler spectral analysis. If the radar operates in a range-ambiguous PRF with R range ambiguities, a total of $M + N + R - 2$ pulses will be required to obtain M steady-state pulses for the pulse-Doppler DFT: $R - 1$ clutter fill pulses to get to steady state input statistics; $N - 1$ more for the pulse canceller transient; and, finally, M for the actual CPI data.

17.6 | CLUTTER MAPPING AND THE MOVING TARGET DETECTOR

17.6.1 Clutter Mapping

All of the MTI and pulse-Doppler processing discussed so far has been focused on reducing the clutter power that interferes with the signature of a moving target, thus improving the signal-to-interference ratio and ultimately the probability of detection. These techniques are not effective for targets with little or no Doppler shift and therefore are not separable from the clutter based on Doppler shift. *Clutter mapping* is a technique for detection of moving targets with zero or low Doppler shift. It is typically used by ground-based scanning radars such as airport surveillance radars. It is intended for maintaining detection of targets on crossing paths—that is, passing orthogonal to the radar line of sight so that the radial velocity is zero. Such targets are discarded by MTI and conventional pulse-Doppler processing. Clutter mapping can be effective if the target radar cross section (RCS) is relatively large and the competing clutter is relatively weak, a situation depicted in Figure 17-31. This situation is most likely to arise in ground-to-air radar. In such situations the antenna is aimed somewhat upward, so the ground clutter is viewed primarily through the antenna sidelobes while the target is viewed through the mainlobe.

The concept of clutter mapping is shown in Figure 17-32, which presumes that conventional pulse-Doppler processing is applied to targets having Doppler shifts sufficient

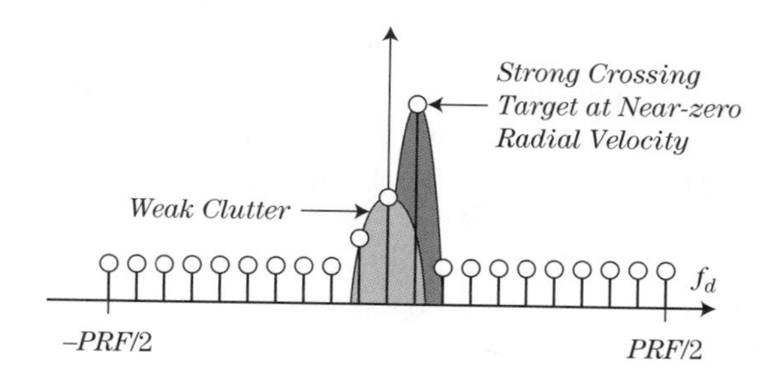

FIGURE 17-31 ■ Pulse-Doppler spectrum for the case of a large RCS crossing target in relatively weak clutter.

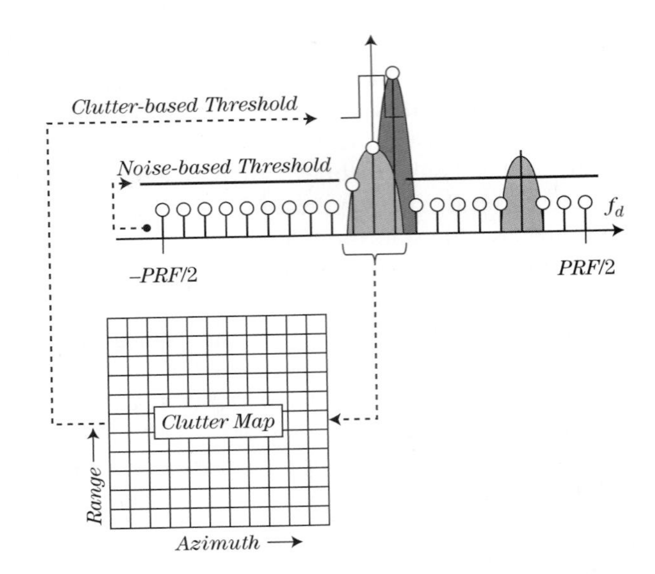

FIGURE 17-32 ■ The concept of clutter mapping for detection of strong targets in clutter.

to separate them from the ground clutter. The output of the zero-Doppler bin is used to create a stored map of the recent average clutter echo power for each range-azimuth cell in the radar's search area. Many clutter map systems combine multiple range cells at a given azimuth direction to form clutter map cells that are larger than the range-azimuth resolution cells.

This map is updated continuously to allow for clutter variations due to weather and other environmental changes. A typical approach updates the clutter measurement in each range-angle cell using a simple first-order recursive filter of the form

$$\hat{x}[n] = (1 - \gamma)\,\hat{x}[n-1] + \gamma x[n] \tag{17.49}$$

where $\hat{x}[n]$ is the estimate of the clutter reflectivity at time n (n usually indexes complete radar scans of an area), and $x[n]$ is the currently measured clutter sample at time n. The factor γ controls the relative weight of the current measurement versus the preceding measurements. Thus, the interference level is estimated by averaging over time in a fixed range-angle cell rather than averaging over spatial dimensions at a fixed time as is done in conventional cell-averaging constant false alarm rate (CFAR) detection processing (see Chapter 16). The averaging time constant is typically on the order of 0.5 to 1.0 minutes [2,4].

On each scan, the received power in each Doppler bin away from zero Doppler (clear region) is applied to a conventional threshold detector, using a threshold based on the noise that dominates the interference in those bins. The current zero-Doppler received power for each range-azimuth cell, instead of being discarded, is applied to a separate detector using a threshold based on the average clutter power level for that cell stored in the clutter map. The zero-Doppler threshold thus varies with range and azimuth direction. The details of threshold detection are discussed in Chapters 3 and 15; the clutter map procedure, which is a form of CFAR detection, is analyzed in [1].

Instead of using the zero-Doppler output of a pulse-Doppler processor (typically a fast Fourier transform of the slow-time data), many clutter maps systems pass the in-phase/quadrature (I/Q) slow-time data through a separate *zero-velocity filter,* as shown in

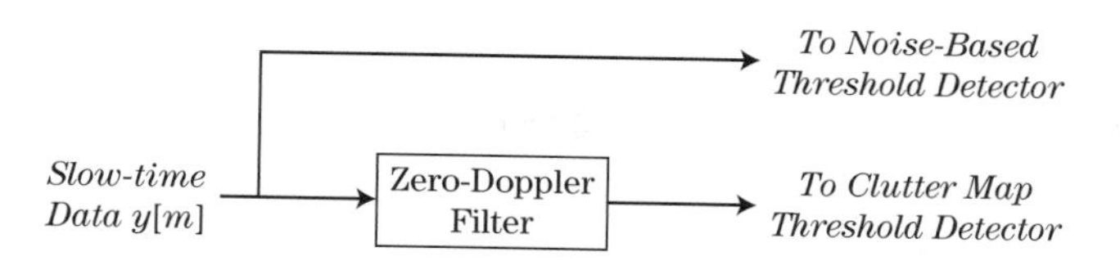

FIGURE 17-33 ■ Zero-Doppler filter used to isolate low-Doppler targets and ground clutter.

Figure 17-33. The zero-velocity filter serves the opposite purpose of an MTI filter. It is a low-pass design whose output consists primarily of ground clutter and crossing target returns. The design of the zero-Doppler filter can be optimized for the clutter environment at a specific radar site and can also be made adaptive to clutter changes, for instance, due to weather in the area.

The clutter map concept can be generalized to aid detection in rain clutter, which typically does not appear centered at zero Doppler due to overall storm motion. For example, clutter maps could be maintained for each Doppler region of the spectrum. The clutter map can then be used instead of the noise level to set the threshold for a particular range-azimuth cell if the clutter level for that cell exceeds the noise.

The clutter map is based on actual clutter returns from the individual radar and thus is site specific. Because of this, knowledge of certain local terrain features that cause consistent false alarms can be used to edit, or "censor," the detection map to eliminate likely false alarms. An example of such censoring could be removal of false alarms along the path of a local highway or of large discrete clutter returns that are otherwise difficult to do away with [30].

17.6.2 The Moving Target Detector

The *moving target detector* (MTD) is a term applied to the Doppler processing system used in many airport surveillance radars. The MTD combines all of the techniques previously discussed as well as others to achieve good moving target detection performance over a full range of target Doppler shifts. Several generations of the MTD have been fielded. Versions used in the ASR-9 and ASR-12 airport surveillance radars are described in [30,31], respectively.

A block diagram similar to the original MTD is shown in Figure 17-34 [4]. There are two major channels: the lower one for handling targets near zero-Doppler shift (and thus competing primarily with ground clutter); and the upper one for targets away from zero-Doppler shift. The upper channel begins with a standard three-pulse canceller. The data are weighted for frequency-domain sidelobe reduction and then are applied to an eight-point FFT for pulse-Doppler analysis. The zero-Doppler bin of the FFT output is not used; targets in this Doppler region are handled in the lower channel. Two PRFs are used in a block-to-block stagger to extend the unambiguous Doppler region. The small eight-point DFT size is probably a consequence in part of the limited digital processing power available at the time and in part of the need for multiple CPIs per antenna scan. The individual FFT samples are applied to a 16-range bin CFAR threshold detector, with thresholds selected separately for each frequency bin.

To provide some detection capability for crossing targets, the lower channel uses a site-specific zero-velocity filter to isolate the echo from clutter and low-Doppler targets. The output is again applied to a clutter map threshold detector. The original MTD updated the clutter map using an eight-scan moving average, corresponding to 32 seconds of data history [4].

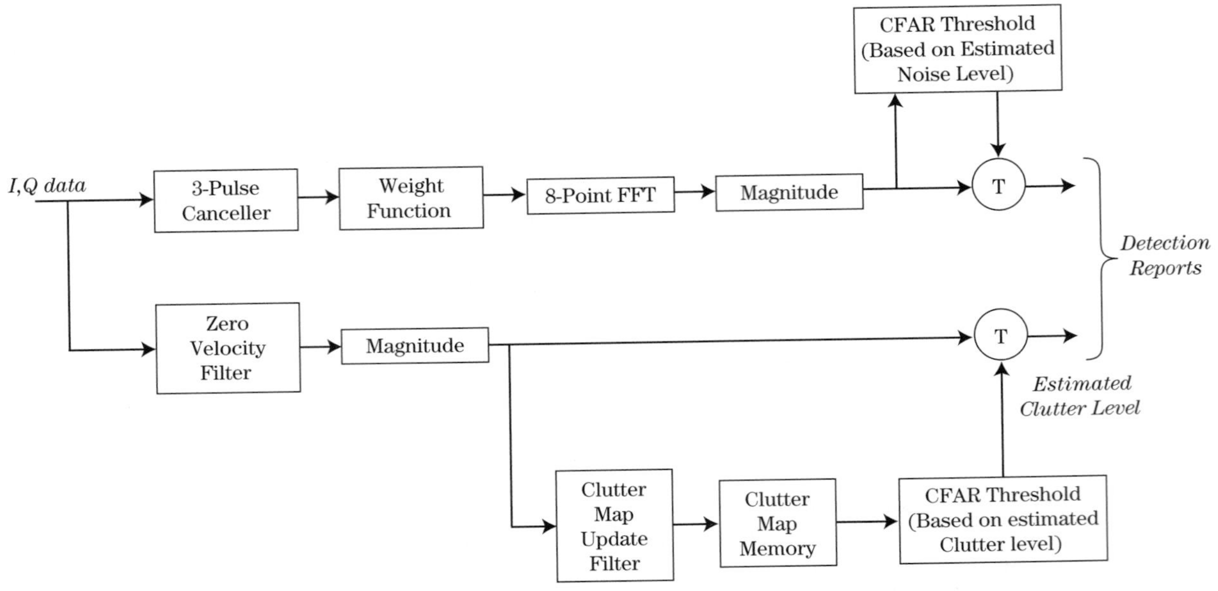

FIGURE 17-34 ■ Block diagram of a complete "moving target detector" system combining MTI, pulse-Doppler, and clutter mapping.

17.7 PULSE PAIR PROCESSING

Pulse pair processing (PPP) is a form of Doppler processing common in meteorological radar. Unlike the MTI and pulse-Doppler techniques discussed so far in this chapter, the goal of pulse pair processing is not clutter suppression to enable the detection of moving targets. Rather, it is to estimate three values for each radar resolution cell: (1) the echo power due to particulates (rain, snow, hail) in that cell; (2) the average radial velocity of those particulates; and (3) the variance of the velocity spectrum. Each of these can be estimated using either time- or frequency-domain algorithms, all of which are included under the PPP rubric. Only the time-domain case is considered here; the frequency-domain version is described in [1,32].

The power, velocity, and variance values provide information on the amount of precipitation, wind velocity, and turbulence in each cell. These measurements are useful in themselves but also provide the basic data upon which higher-level algorithms for a wide variety of meteorological products such as integrated rainfall, storm tops, severe storm and tornado warnings, and windshear detection are based. Pulse pair processing is used extensively in both ground-based and airborne weather radars for storm tracking and weather forecasting.

Pulse pair processing assumes the radar is looking generally upward if it is ground based or forward if it is airborne. Consequently, it is assumed that ground clutter competing with the weather signatures is small or negligible or has been removed by MTI filtering. Consider the slow-time data sequence $y[m]$ from a particular range bin. The PPP algorithm assumes that the Doppler spectrum $S_y(f_d)$ is of the form shown in Figure 17-35. It consists only of white noise and a single spectral peak due to backscatter from weather-related phenomena:

$$S_{yy}(f_d) = S_{ww}(f_d) + S_{nn}(f_d) \tag{17.50}$$

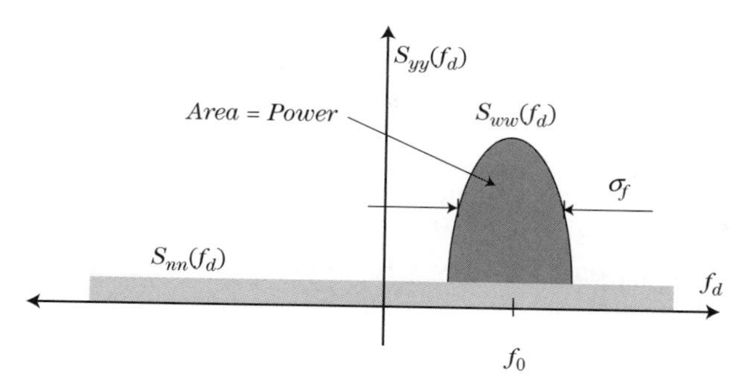

FIGURE 17-35 ■ Notional slow-time power spectrum assumed in pulse pair processing.

The weather peak $S_{ww}(f_d)$ is usually assumed to be approximately Gaussian shaped and is characterized by its amplitude, mean, and standard deviation. If the sampling interval, T (equal to the radar PRI), is chosen sufficiently small to guarantee that $S_{ww}(f_0 \pm 1/2T) \approx 0$, then the discrete-time spectrum and autocorrelation will also form a Gaussian pair to a very good approximation. If the Gaussian weather component has power σ_w^2 and a variance of σ_f^2 Hz in the frequency domain, its autocorrelation and power spectrum will be

$$\phi_{ww}[k] = \sigma_w^2 e^{-2\pi^2\sigma_f^2 k^2 T^2} e^{+j2\pi f_0 kT} \tag{17.51}$$

and

$$S_{ww}(f_d) = \frac{\sigma_w^2}{\sqrt{2\pi}\sigma_f T} e^{-(f_d - f_0)^2 / 2\sigma_f^2} \tag{17.52}$$

The three basic PPP measurements of power, mean Doppler (velocity), and variance are obtained from the total area of the $S_{ww}(f_d)$ power spectrum component, its mean Doppler shift, f_0, and its variance (commonly called the *spectral width*), σ_f^2.

If the white noise has power σ_n^2, its autocorrelation is $\phi_{nn}[k] = \sigma_n^2\delta[k]$, and the autocorrelation of the weather + noise data is

$$\phi_{yy}[k] = \phi_{nn}[k] + \phi_{ww}[k] = \sigma_n^2\delta[k] + \sigma_w^2 e^{-2\pi^2\sigma_f^2 k^2 T^2} e^{+j2\pi f_0 kT} \tag{17.53}$$

while the power spectrum is

$$S_{yy}(f_d) = S_{nn}(f_d) + S_{ww}(f) = \sigma_n^2 + \frac{\sigma_w^2}{\sqrt{2\pi}\sigma_f T} e^{-(f_d - f_0)^2 / 2\sigma_f^2} \tag{17.54}$$

The time-domain PPP method estimates the power, velocity, and spectral width from the autocorrelation function of the data. The echo power can be estimated from the zeroth autocorrelation lag:

$$\hat{P} = \phi_{yy}[0] = \sigma_n^2 + \sigma_w^2 = \left(1 + \frac{\sigma_n^2}{\sigma_w^2}\right)\sigma_w^2 \tag{17.55}$$

If the weather-to-noise power ratio $\sigma_w^2/\sigma_n^2 \gg 1$, then $\hat{P} \approx \sigma_w^2$.

The velocity can be estimated from the first autocorrelation lag,

$$\phi_{yy}[1] = \sigma_n^2(0) + \sigma_w^2 e^{-2\pi^2\sigma_f^2 T^2} e^{+j2\pi f_0 T} \approx \alpha e^{+j2\pi f_0 T} \tag{17.56}$$

where α absorbs all of the amplitude factors, again provided that $\sigma_w^2 \gg \sigma_n^2$. The argument of this complex value can be solved to estimate the Doppler center frequency of the weather spectrum:

$$\tilde{f}_0 = \frac{1}{2\pi T} \arg\{\phi_{yy}[1]\} \tag{17.57}$$

Multiplying $\tilde{f}_0$ by $\lambda/2$ converts the result into units of velocity. Note that the frequency estimate will be aliased if $\tilde{f}_0$ falls outside the range $(-PRF/2, +PRF/2)$. This is not uncommon in weather radars, which often required long unambiguous ranges. For instance, the WSR-88D Next-Generation Radar (NEXRAD) radar used by the U.S. National Weather Service operates at a frequency of about 3 GHz, giving a wavelength of 10 cm. A typical PRF of 650 pulses per second gives an unambiguous range of $R_{ua} = 231$ km but an unambiguous velocity range of only ± 16.25 m/s (about ± 36 miles per hour).

The name "pulse pair processing" derives from the use of the first autocorrelation lag. Given an M-pulse CPI, $\phi_{yy}[1]$ is estimated by the *deterministic* autocorrelation computation as

$$c_{yy}[1] = \sum_{m=0}^{M-1} y[m] \cdot y^*[m+1] \tag{17.58}$$

Thus, the first lag is computed by averaging the conjugate product of pairs of samples. In essence, equation (17.57) estimates the phase change from one pulse to the next and converts it to an equivalent frequency.

The spectral width can also be estimated from the first two autocorrelation lags. Using equations (17.55) and (17.56), the ratio of their magnitudes is

$$\frac{|\phi_{yy}[1]|}{|\phi_{yy}[0]|} = \frac{\sigma_w^2 e^{-2\pi^2\sigma_f^2 T^2}}{\left(1 + \dfrac{\sigma_n^2}{\sigma_w^2}\right)\sigma_w^2} = \frac{e^{-2\pi^2\sigma_f^2 T^2}}{\left(1 + \dfrac{\sigma_n^2}{\sigma_w^2}\right)} \approx e^{-2\pi^2\sigma_f^2 T^2} \tag{17.59}$$

The last approximation again assumes that $CNR >> 1$. The spectral width is then estimated as

$$\hat{\sigma}_f^2 = -\frac{1}{2\pi^2 T^2} \ln\left\{\frac{|\phi_{yy}[1]|}{|\phi_{yy}[0]|}\right\} \tag{17.60}$$

Equations (17.55), (17.57), and (17.60) are the time-domain pulse pair processing estimators. They can be computed from only two autocorrelation lags of the slow-time data. The autocorrelation lags are in turn estimated using the deterministic autocorrelation function described in Chapter 14. The $k = 0$ and $k = 1$ lags specifically become

$$\begin{aligned} c_{yy}[0] &= \sum_{m=0}^{M-1} y[m]\, y^*[m] = \sum_{m=0}^{M-1} |y[m]|^2 \\ c_{yy}[1] &= \sum_{m=0}^{M-2} y[m]\, y^*[m+1] \end{aligned} \tag{17.61}$$

Since the data used to compute $c_{yy}[k]$ are random variables, $\widehat{s_y[k]}$ is itself a random variable. It is easy to show that the expected value of $c_{yy}[k]$ is the true statistical autocorrelation $\phi_{yy}[k]$[7].

The basic PPP measurements of signal power, frequency, and spectral width can also be performed in the frequency domain (see [1,32] for details). Generally, the time-domain estimators are preferred if the signal-to-noise ratio is low or the spectral width is very narrow [32]. In the latter case, the signal is closer to the pure sinusoid assumption that motivated the time-domain estimator. In addition, the time-domain methods are more computationally efficient because no Fourier transform calculations are required. Conversely, the frequency-domain estimators tend to provide better estimators at high SNR and large spectral widths. An analysis of the errors in PPP is given in [33].

Figure 17-36 shows images of the three basic PPP data products from the U.S. National Weather Service's KLIX WSR-88D NEXRAD weather radar in New Orleans, Louisiana [34]. The images show Hurricane Katrina at 8:49 a.m. Central Daylight Time on August 29, 2005. About five minutes later the radar stopped transmitting. While shown here in grayscale, the images are more striking and easier to interpret in color.[9]

Figure 17-36a is the base (lowest elevation angle) reflectivity estimate, which is a version of the PPP power estimate scaled to units of dBz—a decibel-scale version of a common meteorological reflectivity measure z related to the water droplet size distribution and ultimately to the precipitation rate. Details are given in [1,34]. The counterclockwise rotational pattern of hurricanes in the northern hemisphere is clear in this image.

Figures 17-36b and 17-36c are shown on a magnified scale because the NEXRAD radars have less range in the velocity modes than in the reflectivity mode. Part b is the velocity map. The radar is at the center of the circular data region. Lighter areas represent negative velocities (away from the radar), while darker areas represent positive velocities. The two small regions where the measured velocity changes abruptly, at about "three o'clock" and "seven o'clock" relative to the radar location, are likely regions of aliased velocities. In velocity mode, the NEXRAD radar unambiguous velocity is typically 28 m/s or about 62 mph, too low for hurricane-force winds.

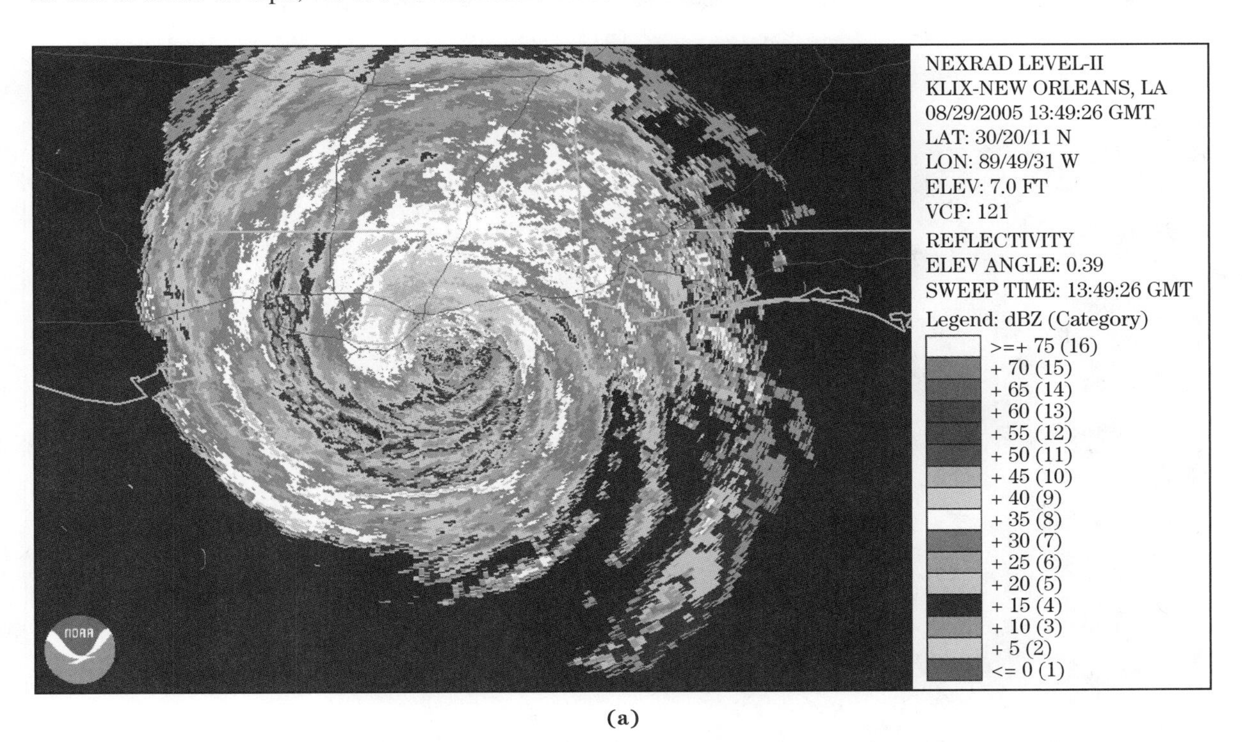

(a)

FIGURE 17-36 ■ Sample images of Hurricane Katrina illustrating pulse pair processing. (a) Reflectivity on dBz scale. (b) Velocity. (c) Spectral width. Data from U.S. National Climactic Data Center. (U.S. government work, not protected by copyright.)

[9] A variety of images of Hurricane Katrina data products in color are available at http://www.ncdc.noaa.gov/oa/radar/jnx/jnt-katrina.php.

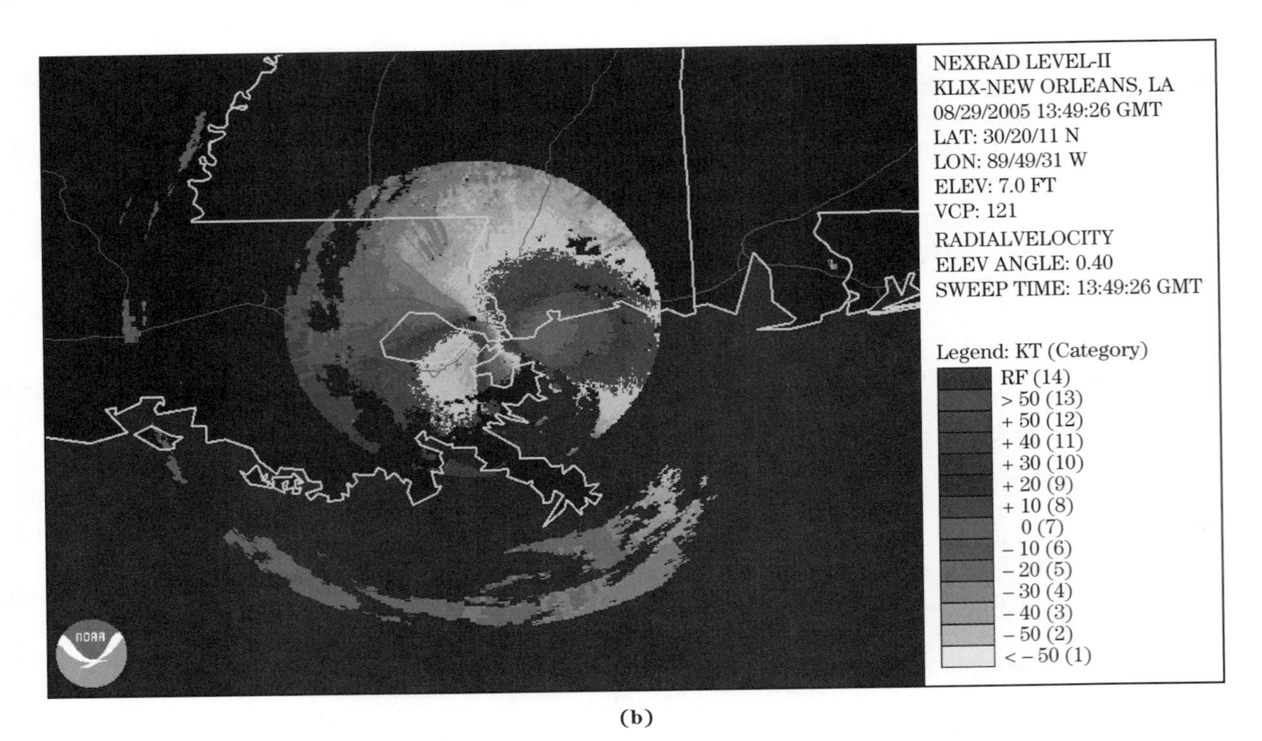

(b)

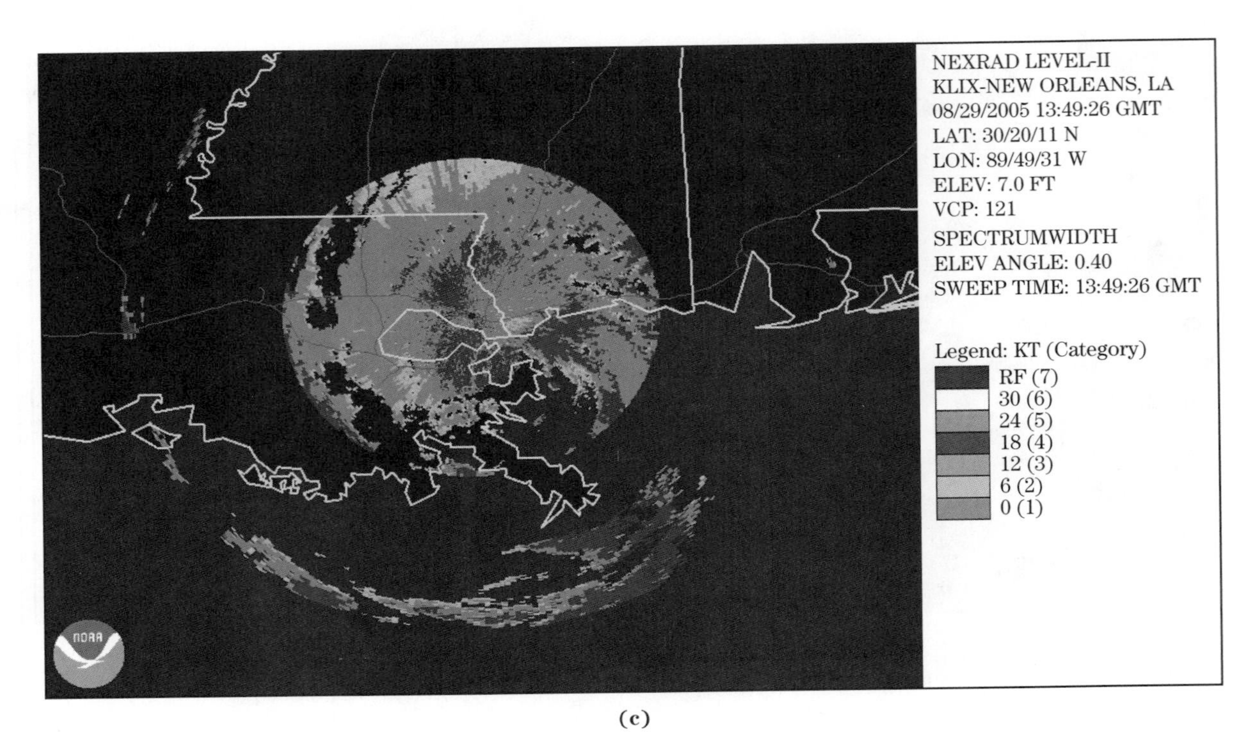

(c)

FIGURE 17-36 ■ *Continued*

Figure 17-36c is the spectral width, which is 9 knots (about 10 mph) or less. This indicates that the power spectrum, $S_{ww}(f)$, is a well-defined, fairly narrow spike and thus also that the velocity estimates are likely to be reliable.

17.8 | FURTHER READING

Aspects of Doppler processing are covered in virtually all books that deal with radar signal processing. Coverage similar in style and scope to this chapter, but in somewhat more depth, is available in Richards [1]. A step up in depth and scope is provided by Nathanson [4], while much more in-depth treatment is provided by Schleher [5]. Additional emphasis on issues particular to airborne MTI and pulse-Doppler radar is given in Morris and Harkness [20] and Stimson [21]. The particular techniques and considerations of weather radar are covered extensively by Doviak and Zrnic [32].

17.9 | REFERENCES

[1] Richards, M.A., *Fundamentals of Radar Signal Processing*, McGraw-Hill, New York, 2005.

[2] Skolnik, M.I., *Introduction to Radar Systems*, 3d ed., McGraw-Hill, New York, 2001.

[3] Skolnik, M.I. (Ed.), *Radar Handbook*, 2d ed., McGraw-Hill, New York, 1990.

[4] Nathanson, F.E., *Radar Design Principles,* 2d ed., McGraw-Hill, New York, 1991.

[5] Schleher, D.C., *MTI and Pulse Doppler Radar*, Artech House, Boston, 1991.

[6] Eaves, J.L., and Reedy, E.K. (Eds.), *Principles of Modern Radar*, Van Nostrand Reinhold, New York, 1988.

[7] Oppenheim, A.V., and Schafer, R.W., *Discrete-Time Signal Processing*, 3d ed., Prentice-Hall, Englewood Cliffs, NJ, 2009.

[8] Levanon, N., *Radar Principles*, John Wiley & Sons, New York, 1988.

[9] Roy, R., and Lowenschuss, O., "Design of MTI Detection Filters with Nonuniform Interpulse Periods," *IEEE Transactions on Circuit Theory*, vol. CT-17, no. 4, pp. 604–612, November 1970.

[10] Prinsen, P.J.A., "Elimination of Blind Velocities of MTI Radar by Modulating the Interpulse Period," *IEEE Transactions on Aerospace and Electronic Systems*, vol. AES-9, no. 5, pp. 714–724, September 1973.

[11] Hsiao, J.K., and Krestchmer Jr., F.F., "Design of a Staggered-P.R.F. Moving Target Indication Filter," *Radio and Electronic Engineer*, vol. 43, no. 11, pp. 689–694, November 1973.

[12] Vergara-Dominguez, L., "Analysis of the Digital MTI Filter with Random PRI," *IEE Proceedings*, part F, vol. 140, no. 2, pp. 129–137, April 1993.

[13] *IEEE Standard Radar Definitions*, IEEE Standard 686-2008, Institute of Electrical and Electronics Engineers, New York.

[14] Guerci, J.R., *Space-Time Adaptive Processing for Radar,* Artech House, Boston, 2003.

[15] Kretschmer Jr., F.F., "MTI Visibility Factor," *IEEE Transactions on Aerospace and Electronic Systems*, vol. AES-22, no. 2, pp. 216–218, March 1986.

[16] Richards, M.A., "Coherent Integration Loss due to White Gaussian Phase Noise," *IEEE Signal Processing Letters*, vol. 10, no. 7, pp. 208–210, July 2003.

[17] Staudaher, F.M., "Airborne MTI," Chapter 16 in *Radar Handbook*, 2d ed., Ed. M.I. Skolnik, McGraw-Hill, New York, 1990.

[18] Shaw, G.A., and McAulay, R.J., "The Application of Multichannel Signal Processing to Clutter Suppression for a Moving Platform Radar," IEEE Acoustics, Speech, and Signal Processing (ASSP) Spectrum Estimation Workshop II, Tampa, FL, November 10–11, 1983.

[19] Kay, S.M., *Modern Spectral Estimation*, Prentice-Hall, Englewood Cliffs, NJ, 1988.

[20] Morris, G.V., and Harkness, L. (Eds.), *Airborne Pulse Doppler Radar*, 2d ed., Artech House, Boston, 1996.

[21] Stimson, G.W., *Introduction to Airborne Radar*, 2d ed., SciTech Publishing, Raleigh, NC, 1998.

[22] Long, W.H., Mooney, D.H., and Skillman, W.A., "Pulse Doppler Radar," Chapter 17 in *Radar Handbook*, 2d ed., Ed. M.I. Skolnik, McGraw-Hill, New York, 1990.

[23] Harris, F.J., "On the Use of Windows for Harmonic Analysis with the Discrete Fourier Transform," *Proceedings of the IEEE*, vol. 66, no. 1, pp. 51–83, January 1978.

[24] Nuttal, A.H., "Some Windows with Very Good Sidelobe Behavior," *IEEE Transactions on Acoustics, Speech, and Signal Processing*, vol. 29, no. 1, pp. 84–91, February 1981.

[25] Jacobsen, E., and Kootsookos, P., "Fast, Accurate Frequency Estimators," *IEEE Signal Processing Magazine*, pp. 123–125, May 2007.

[26] Hayes, M.H., *Statistical Digital Signal Processing and Modeling*, Wiley, New York, 1996.

[27] Trunk, G., and Brockett, S., "Range and Velocity Ambiguity Resolution," *Record of the 1993 IEEE National Radar Conference,* pp. 146–149, April 20–22, 1993.

[28] Hovanessian, S.A., "An Algorithm for Calculation of Range in a Multiple PRF Radar," *IEEE Transactions on Aerospace and Electronic Systems*, vol. 12, no. 2, pp. 287–290, March 1976.

[29] Trunk, G., and Kim, M.W., "Ambiguity Resolution of Multiple Targets Using Pulse-Doppler Waveforms," *IEEE Transactions on Aerospace and Electronic Systems*, vol. 30, no. 4, pp. 1130–1137, October 1994.

[30] Taylor Jr., J.W., and Brunins, G., "Design of a New Airport Surveillance Radar (ASR-9), *Proceedings of the IEEE*, vol. 73, no. 2, pp. 284–289, February 1985.

[31] Cole, E.L., DeCesare, P.A., Martineau, M.J., Baker, R.S., and Buswell, S.M., "ASR-12: A Next Generation Solid State Air Traffic Control Radar," *Proceedings of the 1988 IEEE Radar Conference,* Dallas, TX, pp. 9–14, May 1998.

[32] Doviak, D.S., and Zrnic, R.J., *Doppler Radar and Weather Observations,* 2d ed., Academic Press, San Diego, CA, 1993.

[33] Abeysekera, S.S., "Performance of Pulse-Pair Method of Doppler Estimation," *IEEE Transactions on Aerospace and Electronic Systems*, vol. 34, no. 2, pp. 520–531, April 1998.

[34] NOAA Satellite and Information Service, "NCDC Radar Resources," 2009. Available at: http://www.ncdc.noaa.gov/oa/radar/radarresources.html.

17.10 PROBLEMS

1. Consider a stationary X-band (10 GHz) radar observing a moving target. The PRF is 6 kHz. What will be the apparent Doppler shift of the target if its radial velocity relative to the radar is $v = 30$ m/s? Repeat for $v = 120$ m/s and -150 m/s. Note: "apparent Doppler shift" is the frequency of the target signal contribution in the principal Doppler spectrum interval of $-PRF/2 \leq f_d \leq PRF/2$.

2. In terms of the radar wavelength λ, what is the two-way range change between pulses when the target Doppler shift equals the blind speed f_b?

3. Compute the clutter attenuation CA when using a two-pulse canceller, assuming the clutter Doppler power spectrum is of the form

$$S_c(f_d) = \begin{cases} \sigma_c^2, & |f_d| \leq f_{co} \\ 0, & f_{co} < |f_d| \leq PRF/2 \end{cases}$$

Assume $f_{co} \leq PRF/2$ and express your answer in terms of f_{co}.

4. Consider an $M = 32$ pulse sequence of slow-time data, collected with $PRF = 10$ kHz. A Cooley-Tukey radix 2 FFT algorithm is used to compute the Doppler spectrum of the data. If the Doppler frequency samples are to have a spacing of 100 Hz or less, what is the minimum FFT size K that should be used? What is the resulting spacing of the Doppler frequency samples in hertz?

5. Consider the same data and radar parameters used in problem 4, but suppose the spectrum is computed with a $K = 48$ point DFT. What is the oversampling factor α? Estimate the maximum straddle loss in dB for these parameters for both rectangular and Hamming windows using Figure 17-18.

6. Consider a C-band (5 GHz) radar using a pulse repetition frequency of $PRF = 3{,}500$ pulses per second. The radar collects 30 pulses of data. For a given range, the slow-time data sequence is zero padded and input to a 64-point DFT to compute the Doppler spectrum. What is the spacing of the DFT samples in normalized radian frequency (i.e., on the $-\pi$ to $+\pi$ scale)? What is the spacing in hertz? In meters per second? What is the Rayleigh resolution (peak-to-first null width) in Doppler, in hertz? In meters/second?

7. Consider a pulse-to-pulse staggered PRF system using a series of $P = 3$ PRFs, namely, {10 kHz, 12 kHz, 15 kHz}.

   a. What is the first blind Doppler frequency, f_{us}, of a constant PRF system having the same average PRI as the staggered system?

   b. What is the first blind Doppler frequency, f_s, of the staggered system?

   c. What is the ratio f_s/f_{us} of the first blind Doppler frequency of the staggered system to the first blind speed of a system having a PRF corresponding to the average PRI?

8. For the same staggered PRF system of problem 7:

   a. What is the maximum unambiguous range, R_{us}, corresponding to the average PRI from problem 7? (This would be the unambiguous range of a constant PRI system that used the same amount of time to collect N pulses as the staggered PRF system.)

   b. For the three PRFs in problem 7, what is the minimum unambiguous range R_{min}? Hint: this will be the shortest of the unambiguous ranges corresponding to the individual PRFs. (This will be the unambiguous range of the staggered PRF system.)

   c. What is the factor by which the range coverage (unambiguous range) is reduced in the staggered PRF system (part b) relative to the unstaggered system (part a)? Compare this with the factor predicted by equation (17.15).

   d. Compare the total unaliased range-Doppler coverage (product of unambiguous range and velocity) for the staggered and unstaggered cases.

9. Consider range-ambiguity resolution using three PRFs. Suppose the three PRFs correspond to $N_0 = 4$, $N_1 = 5$, and $N_2 = 7$ range cells. A single target is detected in the first range bin on the first PRF, the fourth range bin on PRF 2, and the second range bin on PRF 3—that is,

$n_{a_0} = 1, n_{a_1} = 4, n_{a_2} = 2$. Use the graphical technique to determine the true range bin number for this target.

10. Suppose a radar has a pulse length of $\tau = 10\ \mu s$ and a PRF of 10 kHz. Assume that the clutter observed by the radar has a two-sided spectral width of 1 kHz (i.e., the clutter spectrum occupies the range from −500 Hz to +500 Hz). Sketch the blind zone map for these operating conditions. For the vertical axis, use time in seconds from 0 to 400 μs; for the horizontal axis, use Doppler frequency in Hz from −20,000 to +20,000 Hz.

11. Find the "time constant" of the clutter map update filter of equation (17.49) as a function of γ. The time constant is the number of map updates (i.e., the value of the index n, which usually counts radar scans) are required before the contribution of the initial stored clutter value $\hat{x}[0]$ to the current value $\hat{x}[n]$ is reduced to 10% or less of its contribution at $n = 0$.

12. A weather radar has a PRF of 2 kHz. Using a series of 50 samples of data from a particular range bin and look direction, the following values of the deterministic autocorrelation function are computed: $\widehat{\phi_{yy}}[0] = 50$, $\widehat{\phi_{yy}}[1] = 30\exp(-j\pi/3)$. Use the PPP time-domain method to compute the estimated power, mean frequency, and spectral width of the echo in Hz.

CHAPTER 18

Radar Measurements

W. Dale Blair, Mark A. Richards, David G. Long

Chapter Outline

18.1 Introduction ... 677
18.2 Precision and Accuracy in Radar Measurements ... 678
18.3 Radar Signal Model ... 683
18.4 Parameter Estimation ... 685
18.5 Range Measurements ... 690
18.6 Phase Measurement ... 695
18.7 Doppler and Range Rate Measurmements ... 696
18.8 RCS Estimation ... 699
18.9 Angle Measurements ... 700
18.10 Coordinate Systems ... 709
18.11 Further Reading ... 710
18.12 References ... 710
18.13 Problems ... 711

18.1 INTRODUCTION

A radar is designed to transmit electromagnetic energy in a format that permits the extraction of information about the target from its echo. Once a target is detected, the next goal is often to precisely locate that target in three-dimensional space, which requires accurate measurements of the distance and angle (both azimuth and elevation) to the target. In addition, it is often desirable to estimate the radar cross section (RCS) and radial velocity of the target as well.

There are major differences in the way these quantities are measured. The range to the target is measured by estimating the two-way time delay of the transmitted signal. The radial velocity or *range rate* of the target is measured by estimating the Doppler shift of the echo signals. The angular position of the target is measured by comparing signal strength in multiple antenna beams offset in angle from one another, obtained either with an antenna structure that forms multiple offset beams (e.g., *monopulse*) or by scanning a single beam across or around the target (e.g., *conical scan*). A radar measures the RCS by using a propagation model (radar range equation) with the measured range and transmitted and received pulse powers to deduce the RCS.

The basic radar range, angle, and radial velocity measurements previously discussed are usually made repeatedly and then combined through kinematic state estimation or filtering of the measurements to produce improved three-dimensional position, velocity, and acceleration estimates, a process called *tracking* the target. The procedures that combine the individual measurements are called *track filtering* algorithms. For closely spaced objects, radar resolution and issues of data association play important roles in the kinematic state estimation, as detections may result from reflected energy from multiple targets and clutter, thermal noise, electromagnetic interference (EMI), and jamming signals from electronic attack sources. Chapter 19 considers how the measurements are combined in the track filtering process.

In this chapter, some of the common techniques for radar measurements are described, along with factors determining their accuracy and precision. Space limitations preclude considering additional techniques and error sources.

18.2 | PRECISION AND ACCURACY IN RADAR MEASUREMENTS

18.2.1 Precision and Accuracy

The quality of the measurement of a single quantity such as position in range or angle is characterized by its *precision* and *accuracy*. Accuracy measures the difference between the measured value and true value. Precision characterizes the repeatability of multiple measurements of the same quantity, even when the accuracy is poor. Thus, accuracy corresponds to the mean error while precision is quantified by the error standard deviation. Figure 18-1 illustrates the difference between these two concepts. Generally, the goal is to minimize both.

In radar, measurement errors are due to a combination of many factors, including interference such as noise and clutter; target phenomenology such as *glint* and *scintillation*; signal propagation characteristics such as multipath and turbulence; system measurement limitations such as signal quantization and signal sampling rate; and system uncertainties such as gain calibration, channel-to-channel phase calibration, and antenna pointing errors. Error sources can impact both the accuracy and precision of the measurements.

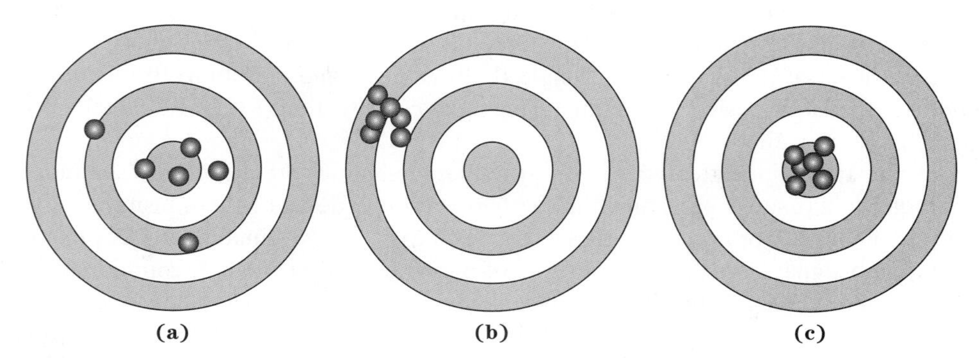

FIGURE 18-1 ■ Illustrating precision and accuracy in target shooting. (a) Accurate but imprecise (low error mean but high standard deviation). (b) Precise but inaccurate (low standard deviation but high mean error). (c) Precise and accurate (low standard deviation and low mean error).

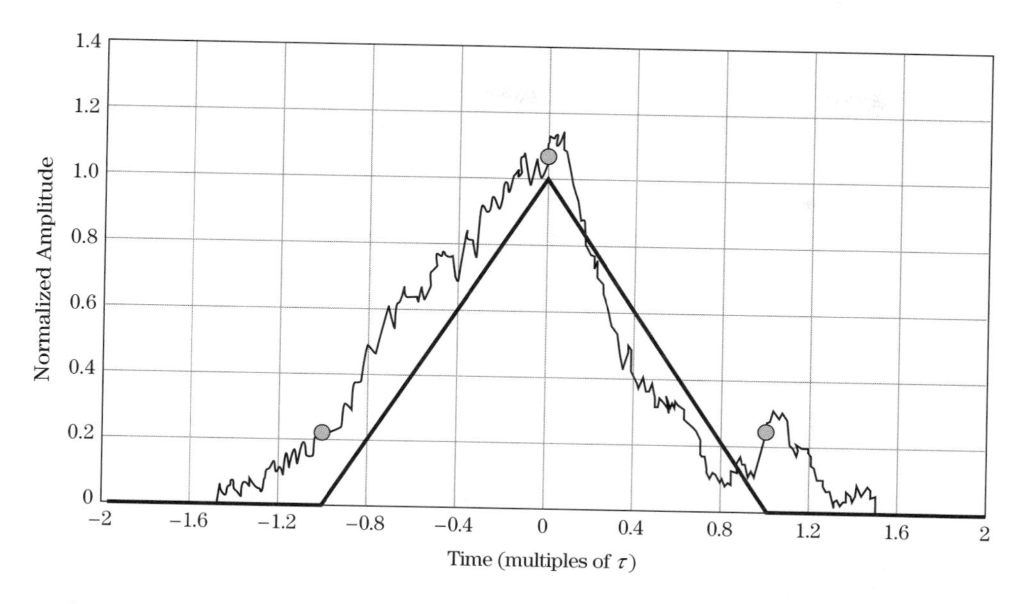

FIGURE 18-2 ■ Noise-free (bold line) and noisy (thin line) matched filter outputs for an ideal rectangular pulse of length τ seconds. The gray circles are samples taken at intervals of τ seconds.

As an example of the effects of noise on the measurement accuracy and precision, consider the output of a matched filter detector for a single simple pulse. In the absence of noise the output is a triangle function (see Chapter 20) as shown in Figure 18-2. Time $t = 0$ corresponds to the delay to the target. In the noise-free case, the time delay of the target echo and thus the range can be measured with no error by locating the peak of the response. A realistic radar, however, must account for the effects of noise and interference. When noise is added to the matched filter input signal, the output peak can be shifted by the noise, as shown in the example. (The gray circles represent samples taken at intervals of τ seconds; these will be discussed later in this chapter.)

Because the interference is random, the error in the peak location is a random variable (RV). If the peak measurement is repeated on noisy data many times, the probability density function (PDF) of that RV can be estimated using a histogram of the measurement data. For example, Figure 18-3a is the histogram of the error for 10,000 trials at an signal-to-noise ratio (SNR) of 20 dB. The mean of the error is the accuracy of the measurement, while the standard deviation is the precision. In this example, the mean of the error is near zero ($-0.0072\ \tau$), so the measurement is very accurate. In this case the accuracy is not dependent on the SNR. However, the range measurement precision does depend on the SNR. As shown in Figure 18-3b, when the SNR is increased, the precision (standard deviation of the error) decreases.

As another example, consider the accuracy and precision of a location angle estimate based on the notional output from the radar receiver for a fixed range bin as the radar system scans past a single, isolated point target. Assume a high pulse repetition frequency, relative to the antenna scan rate, so that the angle samples are closely spaced. In the absence of noise, the measured output voltage is proportional to the two-way antenna voltage pattern, as illustrated in Figure 18-4a for the case of a sinc-squared two-way voltage pattern. The angular position of the target can be accurately determined by simply finding the angle that gives us the peak output power. Thus, the target is located in angle to a precision much better than the angular resolution, which is typically considered to be either the 3 dB or Rayleigh (peak-to-first null) beamwidth of the antenna pattern (one Rayleigh width on the plot).

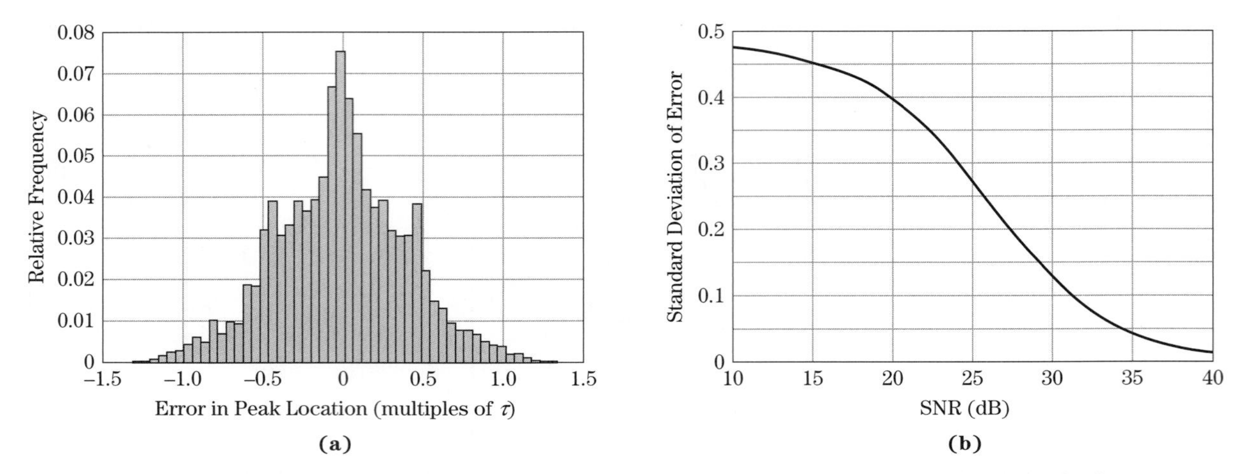

FIGURE 18-3 ■ Statistics of range estimation error using peak detection method. (a) Histogram of peak location measurement error for SNR = 20 dB and 10,000 trials. (b) Standard deviation of error versus SNR.

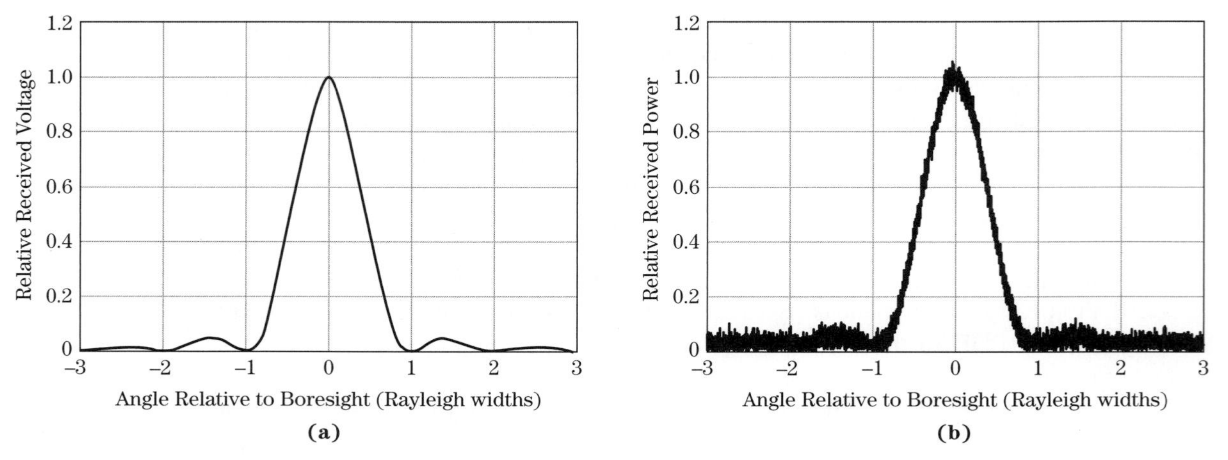

FIGURE 18-4 ■ Received voltage from an angle scan of a single-point target. (a) No noise. (b) 30 dB signal-to-noise ratio.

Now add noise or other interference to the problem. The receiver output consists of the sum of the target echo, weighted by the antenna pattern, and the noise. The noise may cause the observed peak to occur at an angle other than the true target location, as seen in Figure 18-4b; the actual peak in this sample is at -0.033 Rayleigh widths. As might be imagined, the larger the noise, the greater the likely deviation of the measurement from the noise-free case.

As before, the measured location of the peak receiver output, and thus the estimated angle to the target, is now a random variable. If the peak measurement is repeated on noisy data many times, the PDF of that RV can be estimated using a histogram of the measurement data. Figure 18-5 is an example of the histogram for the observed peak

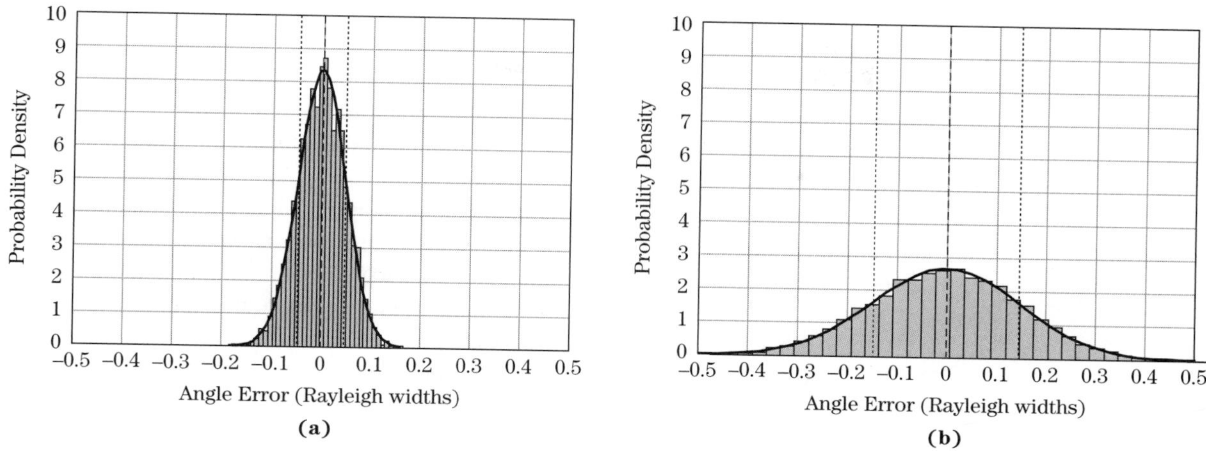

FIGURE 18-5 ■ Histograms of angle error using peak power method. The dark curve is a zero-mean Gaussian PDF with the same variance as the error data. The center dashed line indicates the data mean, with secondary dashed lines at one standard deviation from the mean. (a) 30 dB SNR at boresight. (b) 10 dB SNR at boresight. Note the wider spread (larger standard deviation) of errors at lower SNR.

location when additive complex Gaussian noise is included for two values of peak signal-to-noise ratio. The difference between the mean of the PDF of the peak location and the actual target location is the accuracy of the angle measurement, while the standard deviation of the PDF is the precision of the measurement. In Figure 18-5b, the SNR is 20 dB lower than in 18.5a, a factor of 100. Note that the angle error distribution is wider in the lower SNR case, that is, has a higher variance. In this example, in fact, the variance of the distribution in Figure 18-5b is 9.54 times that of the distribution in Figure 18-5a. This factor is approximately the square root of the 100× change in SNR, suggesting that the variance of the angle estimation error is inversely proportional to the square root of the SNR. The precision is therefore inversely proportional to the fourth root of SNR. On the other hand, the expected value of the error in the peak location is zero for both values of SNR, and hence the accuracy is high and independent of the SNR.

Noise is not the only limitation on measurement quality. Other factors such as resolution and sampling density also come into play. For example, in the angle measurement case, the output power is not measured on a continuous angle axis but only at discrete angles determined by the radar's PRF and the antenna scan rate, resulting in quantization of the angle estimates. Quantization also occurs due to discretized range bins (e.g., the gray circles in Figure 18-2) and arises in sampling and digital signal processing of the signal. Quantization effects can degrade both accuracy and precision; however, consideration of quantization effects is outside of the scope of this chapter. Resolution and sampling density are considered further in Section 18.5.

18.2.2 Accuracy and Performance Considerations

The emphasis in this chapter is on measurement precision. Nevertheless, accuracy considerations are important. Factors affecting measurement accuracy include not only noise and resolution but also signal and target characteristics and radar hardware considerations.

For example, any uncertainty in the antenna boresight angle, due perhaps to mounting or pattern calibration errors or uncertainty, will affect the accuracy of a location measurement. Radiofrequency (RF) hardware or antenna gain calibration errors (gain uncertainty) will affect the accuracy of target signal power measurements. Mismatched in-phase (I) and quadrature (Q) channels in the receiver can adversely affect both power and phase measurements [1] and may bias range measurements when range compression is used. System timing and frequency error and stability also degrade accuracy. Signal propagation effects are another important factor in overall measurement accuracy: atmospheric refraction of the radar signal and multipath can introduce pointing and range measurement error, as was discussed in Chapter 4. All of these effects shift the mean of the parameter measurement, thereby degrading the accuracy.

Because a key source of accuracy degradation is due to error in the values assumed and used for the hardware parameters such as receiver gain, care must be taken to measure or calibrate the system and antenna gain and the antenna pointing angle and beamwidth. The radar design usually includes mechanisms to minimize changes in these parameters. Periodic system calibration is common. One technique is to incorporate a calibration feedback loop wherein an attenuated transmit signal is fed into the antenna or receiver to measure the receiver power from which the system gain can then be inferred. While there is always some residual uncertainty, careful calibration can improve knowledge of the values used in computing signal power, timing, channel-to-channel phase and gain matching, and frequency measurements.

Since there is always residual uncertainty in calibration and knowledge of design parameters, how can the measurement accuracy be estimated? While a detailed treatment of this topic is outside of the scope of this chapter, some comments are provided. A simple, commonly used approach is based on *root sum of squares* (rss). The accuracy estimate is computed as the square root of the sum of the squared errors separately determined for each error source. Specifically, given errors $\{\varepsilon_1, \varepsilon_2, \ldots, \varepsilon_N\}$, the rss estimate of the total error ε is

$$\varepsilon = \sqrt{\varepsilon_1^2 + \varepsilon_1^2 + \ldots \varepsilon_N^2} \tag{18.1}$$

The rss error effectively treats each error source as independent and Gaussian.

To use rss, an estimate of the uncertainty of each error source is generated, usually by the system or subsystem designer. For example, suppose that for a particular radar the uncertainty in azimuth boresight pointing angle of the antenna is estimated in testing to be ± 0.03 degrees. This implies that the actual pointing angle is unknown but falls within this range. Alternately, this is the expected range of pointing angle errors for multiple antennas made for the project. Continuing the example, suppose the antenna mounting error is expected to fall in the range of ± 0.05 degrees. The rss of these values is $\sqrt{(0.03)^2 + (0.05)^2} = 0.058$ degrees.

The rss technique provides an imperfect, but often reasonably realistic, estimate of the expected pointing error for a particular unit. rss estimates are popular because they do not require the knowledge of the individual error PDFs needed for more sophisticated error estimation based on signal flow model approaches. Worst-case analysis can also be used and results in a more conservative estimate.

Errors in one system parameter also result in errors in other radar measurements. Continuing the example, antenna pointing errors alter the antenna gain in the direction of a target, in turn altering the received echo power. Typically, the effect of each source of pointing error on the power is determined by assuming the nominal values of the radar's other operational parameters (e.g., transmitted power, system losses) and varying one source of pointing error to estimate the resulting variation in received power. The power variances due to each identified error source are combined using the rss technique to provide an estimate of the range of the expected power error.

18.3 RADAR SIGNAL MODEL

The radar range equations presented in earlier chapters of this book are expressions of average power or energy and are often used for system-level trade studies in radar system design and analysis. However, radar measurements are typically formed with the voltage signals. Thus, the voltage form of the radar signals is used for the modeling and analysis of radar measurements for tracking studies. Since voltage is proportional to the square root of power, a general model of the RF echo signal received in the radar, for either a conventional antenna or the sum channel of a monopulse from a single target, can be written as

$$s(t) = 2\sqrt{\frac{P_t}{(4\pi)^3}\frac{\lambda}{R^2}}\xi V_s^2(\theta,\phi)\,p(t)\cos(\omega_c t + \omega_d t + \psi) + w_s(t) \tag{18.2}$$

where

P_t = transmitted power.
λ = wavelength.
R = range to target.
ξ = voltage reflectivity of the target.
$V_s(\theta,\phi)$ = voltage gain of the antenna at the angles (θ,ϕ).
(θ,ϕ) = angular location of the target relative to antenna boresight.
$p(t)$ = envelope of the matched filter output for the transmitted pulse.
ω_c = carrier frequency of the transmitted waveform.
ω_d = Doppler shift of the received waveform.
ψ = phase of the target echo.
$w_s(t)$ = receiver noise.

The subscripts s are in anticipation of the discussion of monopulse angle estimation in Section 18.9.2.

The "voltage reflectivity," ξ, of the target is related to its RCS σ according to

$$\sigma = \xi^2/2 \tag{18.3}$$

Note that in equation (18.2) the normalized antenna voltage gain pattern $V_s(\theta,\phi)$ is assumed to be the same on transmit and receive, so the two-way pattern is $V_s^2(\theta,\phi)$ as shown. The antenna gain pattern is assumed to be the product of two orthogonal voltage patterns,

$$V_s(\theta,\phi) = W(\theta)U(\phi) \tag{18.4}$$

where $W(\theta)$ and $U(\phi)$ are the elevation and azimuth voltage patterns, respectively.

Demodulation by coherent mixing with quadrature oscillators at the frequency ω_c removes the carrier term of equation (18.2), as described in Chapter 11. Two common sources of error frequently reduce the measured amplitude of the signal $s(t)$. First, if $s(t)$ is mixed with ω_c rather than $\omega_c + \omega_d$, a frequency mismatch occurs in the matched filter for $p(t)$, resulting in a loss in SNR referred to as *Doppler loss*, as discussed in [2] and in Chapter 20. When attempting to use a radar system designed for air targets to detect and track space targets traveling at significantly higher velocity, the Doppler loss can be sufficiently high to prevent detection. Second, the need to detect targets at a priori unknown ranges and to detect multiple, closely spaced objects in the same dwell dictates that the output of the matched filter be sampled periodically in fast time at the bandwidth of the signal over the range interval (*range window*) of interest. There is no guarantee that one of the samples will fall on the peak of the matched filter response, as seen in the example in Figure 18-2. Instead, the energy of a target echo may be captured in adjacent samples that straddle the peak, reducing the measured SNR. This reduction in SNR is often referred to as *straddle loss*. However, as will be seen, the signals in the adjacent cells can be used to improve the range estimate precision beyond the resolution and to reduce straddle loss.

Ignoring any Doppler and straddle losses, the I and Q components of the sampled output of the matched filter with gain p_0 are given by

$$s_I = \alpha \cos \psi + w_{sI} \quad s_Q = \alpha \sin \psi + w_{sQ} \tag{18.5}$$

where

$$\alpha = \frac{\sqrt{P_t}}{(4\pi)^{3/2}} \frac{\lambda}{R^2} \xi V_s^2(\theta, \phi) p_0, \quad w_{sI} \sim \mathrm{N}(0, \sigma_s^2/2), \quad w_{sQ} \sim \mathrm{N}(0, \sigma_s^2/2) \tag{18.6}$$

where

p_0 = pulse amplitude.
$\sigma_s^2 = kT_0 F B_{IF}$ = total noise power at the receiver output.
k = Boltzmann's constant.
T_0 = 290 K = standard noise temperature.
F = receiver noise figure.
B_{IF} = receiver intermediate frequency bandwidth.

The other variables are as given previously, and the notation $\sim\mathrm{N}(0, \sigma^2)$ indicates a normally distributed (Gaussian) random variable with zero mean and a variance of σ^2. Any gain due to coherent integration such as pulse compression, moving target indication (MTI), or pulse-Doppler processing with a discrete Fourier transform (DFT) is included in s_I and s_Q. The integration of s_I and s_Q from multiple pulses after pulse compression and Doppler processing is typically accomplished using only the measured amplitude of the pulses, ignoring the phase, and is referred to as *noncoherent integration*. Often, channel-dependent calibration corrections for gain, time delay, and phase shift are also applied to the measured values to improve their accuracy.

Letting Λ and φ denote the measured amplitude and phase of the signals in equation (18.5), including the calibration corrections and noise contributions, gives

$$s_I = \Lambda \cos \varphi, \quad s_Q = \Lambda \sin \varphi \tag{18.7}$$

It is useful to define the *observed SNR* as the ratio of total signal power to total noise power

$$SNR_{\text{obs}} = \frac{\Lambda^2}{\sigma_s^2} \tag{18.8}$$

Λ includes both signal and noise contributions, so SNR_{obs} is actually a signal-plus-noise-to-noise ratio. The SNR of a target can therefore be computed from the expected value of the observed SNR, that is,

$$SNR = \mathrm{E}\{SNR_{\text{obs}}\} - 1 \tag{18.9}$$

where $\mathrm{E}\{\cdot\}$ denotes the expected value.

18.4 PARAMETER ESTIMATION

The goal of the radar measurement process is to estimate the various parameters of the target reflected in the signal $s(t)$. The primary parameters of interest include the reflectivity amplitude, ξ, the Doppler shift, ω_d, the angular direction to the target, (θ, ϕ), and the time delay to the target, which is reflected in the sampling time at which the signal was measured and in the signal phase, ψ. Before addressing techniques for measuring each of these, it is useful to first discuss the general idea of an *estimator* and the achievable precision.

18.4.1 Estimators

Consider an observed signal $y(t)$ that is the sum of a target component $s(t)$ and a noise component $w(t)$:

$$y(t) = s(t) + w(t) \tag{18.10}$$

The signal $y(t)$ is a function of one or more parameters α_i. These might be, for example, the time delay, amplitude, Doppler shift, or angle of arrival (AOA) of the target component. The goal is to estimate the parameter values given a set of observations of $y(t)$. This is done using an *estimator*.

Suppose $y(t)$ is sampled multiple times (intrapulse or over multiple pulses) to produce a vector of N observations,

$$\boldsymbol{Y} = \{y_1, y_2, \ldots y_N\} \tag{18.11}$$

Because of the noise, the data $\boldsymbol{Y}$ is a random vector that depends on the parameter α. Thus, $\boldsymbol{Y}$ is described by a conditional PDF $p(\boldsymbol{Y}|\alpha)$. Now define an estimator f of a parameter α based on the data $\boldsymbol{Y}$,

$$\hat{\alpha} = f(\boldsymbol{Y}) \tag{18.12}$$

Because $\boldsymbol{Y}$ is random, the estimate $\hat{\alpha}$ is also a random variable and therefore has a probability density function with a mean and variance.

Two desirable properties of an estimator are that it be *unbiased* and *consistent*. These mean that the expected value of the estimate equals the actual value of the parameter, and that the variance of the estimate decreases to zero as more measurements become available:

$$\begin{aligned} &\mathrm{E}\{\hat{\alpha}\} = \alpha_i && \text{(unbiased)} \\ &\lim_{N\to\infty}\{\sigma_{\hat{\alpha}}^2\} \to 0 && \text{(consistent)} \end{aligned} \tag{18.13}$$

In other words, a desirable estimator produces estimates that are, on average, accurate and whose precision improves with more data.

A simple example of a good estimator is one that estimates the value of a constant signal A in the presence of white (and thus zero-mean) noise $w[n]$ of variance σ_w^2 by averaging N samples of the noisy signal $y[n] = A + w[n]$. In this case, the parameter α is the unknown amplitude A, the vector $\boldsymbol{Y}$ is composed of the N samples of $y[n]$, and the estimator is

$$\begin{aligned}\hat{A} &= f(\boldsymbol{Y}) = \frac{1}{N}\sum_{n=0}^{N-1} y[n] \\ &= \frac{1}{N}\sum_{n=0}^{N-1}(A + w[n]) = A + \frac{1}{N}\sum_{n=0}^{N-1} w[n]\end{aligned} \tag{18.14}$$

Note that the expected value of $\hat{A} = A$, so the estimator is unbiased. The variance of the second (noise) term is σ_w^2/N; this is also the variance of $\hat{A}$. Thus, the variance of the estimator tends to zero as the number of data samples increases, so it is also consistent. The expected value of the square root of the variance (the standard deviation) of the estimate is the measurement precision.

The absolute value of the variance of the estimate is less significant than its value relative to the value of A. Normalizing the estimator variance by the signal power A^2 gives the normalized measurement variance

$$\frac{\sigma_w^2}{NA^2} = \frac{1}{N \cdot SNR} \tag{18.15}$$

where the SNR for this problem is A^2/σ_w^2. The normalized estimate variance is thus an example of an estimator whose variance is inversely proportional to the SNR.

Many types of estimators exist. Two of the most commonly used are *minimum variance* (MV) estimators and *maximum likelihood* (ML) estimators. A minimum variance or minimum variance unbiased (MVU) estimator is one that is both unbiased and minimizes the mean square error between the actual value of the parameter being estimated and its estimate [3]. In the context of this chapter, it minimizes $(\hat{\alpha} - \alpha)^2$ under the condition that $\text{E}\{\hat{\alpha}\} = 0$.

The maximum likelihood estimator is one that chooses $\hat{\alpha}$ to maximize the likelihood of the specific observed data values $\boldsymbol{Y}$. For example, suppose an observed signal sample, s, is assumed to be the sum of a constant value, A, and zero-mean Gaussian noise of variance, σ^2. The observation s is then Gaussian with mean A and variance σ^2, $s \sim \text{N}(A, \sigma^2)$. The goal is to estimate the mean. For concreteness, suppose $A = 3$ and $\sigma^2 = 1$, and a single measurement results in the observed value $s = 2.8$. The ML estimate of A based on s is $\hat{A}_{ML} = 2.8$ because that is the value of the Gaussian mean that maximizes the chance that the measured value is 2.8. As will be seen later, if multiple measurements of s are available, the ML estimator of A is the sample mean.

The ML estimator is often a good practical choice because its form is often relatively easy to determine. In addition, in the case of Gaussian noise it is equivalent to the MV estimator and thus is the optimum estimator.

As previously noted the standard deviation of the estimate error describes the measurement precision. The standard deviation depends on the estimator chosen, and sometimes can be hard to compute for a particular estimator. However, as described in the following section, the error standard deviation can be bounded.

18.4.2 The Cramèr-Rao Lower Bound

In the angle measurement example in Section 18.2.1, it was seen that the variance of the angle estimate decreased with increasing SNR. Similar behavior is typically observed for range and Doppler estimates as well. Specifically, for a parameter α, the variance $\sigma_{\hat{\alpha}}^2$ of the estimated value $\hat{\alpha}$ often behaves as

$$\sigma_{\hat{\alpha}}^2 = \frac{k}{SNR} \tag{18.16}$$

for some constant k. Is this behavior predictable and, if so, what can be said about the constant k?

The *Cramèr-Rao lower bound* (CRLB) is a famous result that addresses these questions. The CRLB, $J(\alpha)$, establishes the minimum achievable variance (square of precision) of an unbiased estimator of the parameter α. The square root of the CRLB is thus the best achievable precision. Any particular unbiased estimator must have a variance equal to or greater than the CRLB, and the quality of a particular estimator can be judged by how close its actual variance comes to achieving the CRLB.

An important metric for describing an estimator's performance is the *root mean square* (rms) *error*. For zero-mean error the square root of the CRLB is the minimum achievable rms error. While the CRLB does not depend on the estimator, the rms error is a function of the estimator employed.

Derivation of the CRLB is beyond the scope of this chapter; good (and very similar) derivations are given in [3,4]. The bound states that the minimum variance in the estimate $\hat{\alpha}$ is

$$\sigma_{\hat{\alpha}}^2 \geq J(\alpha) = \frac{1}{\mathrm{E}\left[\{\partial \ln\{p(\boldsymbol{Y}|\alpha)\}/\partial\alpha\}^2\right]} \tag{18.17}$$

where $p(\boldsymbol{Y}|\alpha)$ is the conditional probability density function (PDF) of the data vector $\boldsymbol{Y}$ given some particular value of the parameter α. Under some mild conditions, the CRLB has an alternate form,

$$\sigma_{\hat{\alpha}}^2 \geq J(\alpha) = \frac{-1}{\mathrm{E}\left[\partial^2 \ln\{p(\boldsymbol{Y}|\alpha)\}/\partial\alpha^2\right]} \tag{18.18}$$

The choice between equation (18.17) or (18.18) is a matter of convenience, depending on the functional form of $\ln\{p(\boldsymbol{Y}|\alpha)\}$. More general forms of the CRLB that include bias are possible [3,4].

The CRLB can be further simplified for the special but very common and important case of a signal in additive Gaussian noise. Suppose the data $\boldsymbol{Y}$ is N observations of a real discrete signal s that is a function of some parameter α in real white Gaussian noise,

$$y[n] = s[n;\alpha] + w[n], \quad n = 0, \ldots, N-1, \quad w[n] \sim \mathrm{N}(0, \sigma_w^2) \tag{18.19}$$

Starting from equation (18.18), it can be shown that for this case the CRLB is [3]

$$J(\alpha) = \frac{\sigma_w^2}{\sum_{n=0}^{N-1}\left(\frac{\partial s[n;\alpha]}{\partial\alpha}\right)^2} \tag{18.20}$$

It is sometimes more convenient to deal with continuous time rather than sampled signals. The continuous time equivalents to equations (18.19) and (18.20) are

$$y(t) = s(t;\alpha) + w(t), \quad -T/2 \le t \le T/2, \quad w(t) \sim \mathrm{N}(0, \sigma_w^2) \tag{18.21}$$

$$J(\alpha) = \frac{\sigma_w^2}{\displaystyle\int_{t=-T/2}^{T/2} \left(\frac{\partial s(t;\alpha)}{\partial \alpha}\right)^2 dt} \tag{18.22}$$

where T is the duration of the signal of interest.

As a simple illustration of equation (18.20), consider estimating the value of a constant m in Gaussian noise. In this case, $s[n;\alpha] = m$, so $\partial s[n;\alpha]/\partial\alpha = \partial(m)/\partial m = 1$ and the CRLB for estimating m becomes

$$J(\alpha) = J(\hat{m}) = \frac{\sigma_w^2}{N}, \tag{18.23}$$

which is achieved for the mean estimator in equation (18.14) and shows that, as expected, the minimum achievable variance of the estimate decreases with the number of measurements available. A better metric than the variance in $\hat{m}$ is the variance normalized to the actual signal power m^2, *i.e.* the relative error. The CRLB for this quantity is

$$J\left(\frac{\hat{m}}{m}\right) = \frac{\sigma_w^2}{N \cdot m^2} = \frac{1}{N \cdot SNR} \tag{18.24}$$

Notice that the CRLB of the normalized error estimate varies as 1/*SNR*. This proves to be the case in many radar parameter estimation problems.

18.4.3 Precision and Resolution for the Gaussian Case

Equation (18.20) states that the minimum achievable precision of a measurement increases as the square of the derivative of the signal with respect to the parameter of interest increases. Loosely interpreted, the more rapidly the signal varies, the better the precision. For example, if the parameter of interest is range, then a matched filter output with a steep leading and trailing edge will allow better range measurement precision than a broad, slowly rising and falling output waveform. Similarly, a narrow antenna mainlobe should allow better angular precision than a wide one. This should not seem surprising. A waveform with sharp edges suggests a high bandwidth in the dimension of interest: temporal bandwidth in time or range or spatial bandwidth in angle. Thus, it might be anticipated that higher bandwidths lead to lower CRLBs, at least in the Gaussian noise case.

This is in fact the case. As an example, consider time-delay estimation. The parameter α in equation (18.20) is then the time delay, t_0, of the signal echo, and the signal itself is

$$s[n;\alpha] = s[nT_s - t_0] \tag{18.25}$$

where T_s is the sampling interval in fast time. Using equation (18.25) in (18.20), several references [3–5] derive the result that

$$\sigma_{t_0}^2 \ge J(t_0) = \frac{2E}{N_0} \cdot \frac{1}{B_{rms}^2} \tag{18.26}$$

where E is the energy in the signal s, N_0 is the noise power spectral density, and B_{rms} is the rms bandwidth of s, defined as

$$B_{rms} = \sqrt{\frac{\int_{-\infty}^{\infty} (2\pi f)^2 |S(f)|^2 \, df}{\int_{-\infty}^{\infty} |S(f)|^2 df}} \tag{18.27}$$

where $S(f)$ is the Fourier transform of $s(t)$.

B_{rms} is the square root of the normalized variance of $|S(f)|^2$. It is proportional to other more common measures of bandwidth, such as the 3 dB or Rayleigh bandwidths. For example, if $S(f)$ is a rectangular spectrum of width B Hz, then $B_{rms} = (\pi/\sqrt{3})B \approx 1.81\,B$.

Similarly, the *rms time duration*, τ_{rms}, is defined as

$$\tau_{rms} = \sqrt{\frac{\int_{-\infty}^{\infty} t^2 |s(t)|^2 \, dt}{\int_{-\infty}^{\infty} |s(t)|^2 \, dt}} \tag{18.28}$$

Following a procedure similar to that used to derive $J(t_0)$, it can be shown that the Doppler frequency estimate CRLB obeys [4]

$$\sigma_{\omega_d}^2 \geq J(\omega_d) = \frac{2E}{N_0} \cdot \frac{1}{\tau_{rms}^2} \tag{18.29}$$

The term $2E/N_0$ in equations (18.26) and (18.29) is recognized as the peak SNR at the output of a matched filter (Chapter 20 or [1]), so that those two equations can be rewritten as

$$\sigma_{t_0}^2 \geq \frac{1}{SNR \cdot B_{rms}^2} \tag{18.30}$$

$$\sigma_{\omega_d}^2 \geq \frac{1}{SNR \cdot \tau_{rms}^2} \tag{18.31}$$

Equation (18.30) shows that the time-delay estimation variance improves with both SNR and bandwidth. Since range is related to time delay by $R = ct/2$, this result can also be scaled and the square root taken to provide the range precision,

$$\sigma_R \geq \frac{c}{2\sqrt{SNR} \cdot B_{rms}} \tag{18.32}$$

Finally, recalling that range resolution is generally of the form $\Delta R = c/2B$, where B is an appropriate bandwidth (see Chapter 20), the range precision can be expressed as

$$\sigma_R \geq \frac{\Delta R}{\sqrt{SNR}} \tag{18.33}$$

Equation (18.33) shows that range estimation precision generally improves with finer-range resolution as well as with SNR. In fact, since $SNR > 1$ (and often $SNR \gg 1$) in any realistic detection scenario, this equation suggests that the lower bound on precision can be a small fraction of the resolution, as was seen in the first few examples in this chapter.

For instance, if $SNR = 10$ (10 dB), the lower bound on precision above is 32% of the rms range resolution; if it is 100 (20 dB), the bound is 10% of the rms resolution. In many practical systems, however, system limitations and signal variability create additional limits on estimation precision.

18.4.4 Signal and Target Variability

The derivations presented in the previous sections are based on the assumption that the target and signal characteristics remain stationary over the time period required to collect the samples. If the target characteristics or the signal propagation conditions vary over the sampling period, *signal variability* is introduced, which increases the estimate variation and thus degrades the measurement precision. A major source of target variability is changes in the target aspect angle that cause *scintillation* (exponentially distributed fluctuations of the target RCS as described in Chapter 7) and glint. *Glint* is a fluctuation of the apparent angle of the target relative to the radar boresight caused by variations in the orientation of the phase front of the echo signal at the antenna, possibly due to changes in the scattering location on the target. Multipath and diffraction in the signal path can also introduce signal fluctuation as well as possible multiple apparent targets.

If appropriate models for signal and target variability are available, these can be included in the probability density function used in deriving measurement estimators and computing the CRLB. In the interest of clarity and brevity, only a few specific Swerling cases for signal and target variability are considered in this chapter.

18.5 RANGE MEASUREMENTS

The previous discussion considered range measurement precision. Additional methods and accuracy analyses are provided in this section after a discussion on the effects of resolution and sampling. Note that although the discussion on resolution and sampling focuses on range and angle measurement, these issues also impact phase and Doppler measurement.

18.5.1 Resolution and Sampling

A typical radar collects target returns for each transmitted pulse over some finite time-delay interval or range window. The start of the range window may be positioned at any range beyond the blind range of the radar. The output of the matched filter is sampled periodically over the range window, usually at a rate approximately equal to the bandwidth of the waveform but sometimes at higher rates, for example, in a dedicated tracking mode. Thus, the time delay or range estimation is accomplished with a sequence of periodic samples of the matched filter output.

The range resolution of the radar can be generally expressed as

$$\Delta R = \alpha \frac{c}{2B} \tag{18.34}$$

where B is the waveform bandwidth, and α is a factor in the range $1 < \alpha < 2$ that represents the degradation in range resolution resulting from system errors or range sidelobe reduction techniques such as windowing. The bandwidth B/α corresponds to the bandwidth of the intermediate frequency (IF) filter and defines the Nyquist sampling rate at the output of the matched filter unless some technique such as stretch processing as discussed in [1] is employed.

Most radars sample the receiver output at a rate of approximately 1 to 1.5 samples per 3 dB resolution cell in range and angle. For a simple pulse waveform, $B \approx 1/\tau$ so that the matched filter output of Figure 18-2 would be sampled approximately every τ seconds. Figure 18-2 illustrated this sampling strategy. Depending on the alignment of the target's actual range with the sampling times, only two to three samples may be taken on the response to a single point target. This sparse set of measurements must be used to estimate the peak location. The coarsely time-quantized nature of the sampling contributes to variability in the resulting estimates. In contrast, radars designed primarily for precision tracking may provide many more samples per resolution cell, allowing the use of different techniques.

The resolution of a radar system limits its ability to distinguish and accurately locate closely spaced targets, as described in Chapter 1. Figure 18-6 illustrates radar angular and range resolution. The two solid lines in the figure denote the one-way 3 dB beamwidth of the antenna, denoted here as θ_3. While the 3 dB points on the antenna gain pattern are often considered to define the angular *field of view* (FOV) of a radar system, the antenna gain pattern changes smoothly from the peak of the mainlobe to the first null and includes sidelobes at various angles from the antenna boresight. Although attenuated relative to echoes in the mainlobe, the energy of a sufficiently strong echo from a target or clutter outside the 3 dB points will often be observed in the output of the matched filter and can result in a detection threshold crossing. No specific angle in the antenna gain pattern other than a null ensures that target echo energy is not present in the output of the matched filter.

The dashed lines in Figure 18-6 partition the observation volume into *range gates*. This figure illustrates some of the problems that can occur when estimating range using

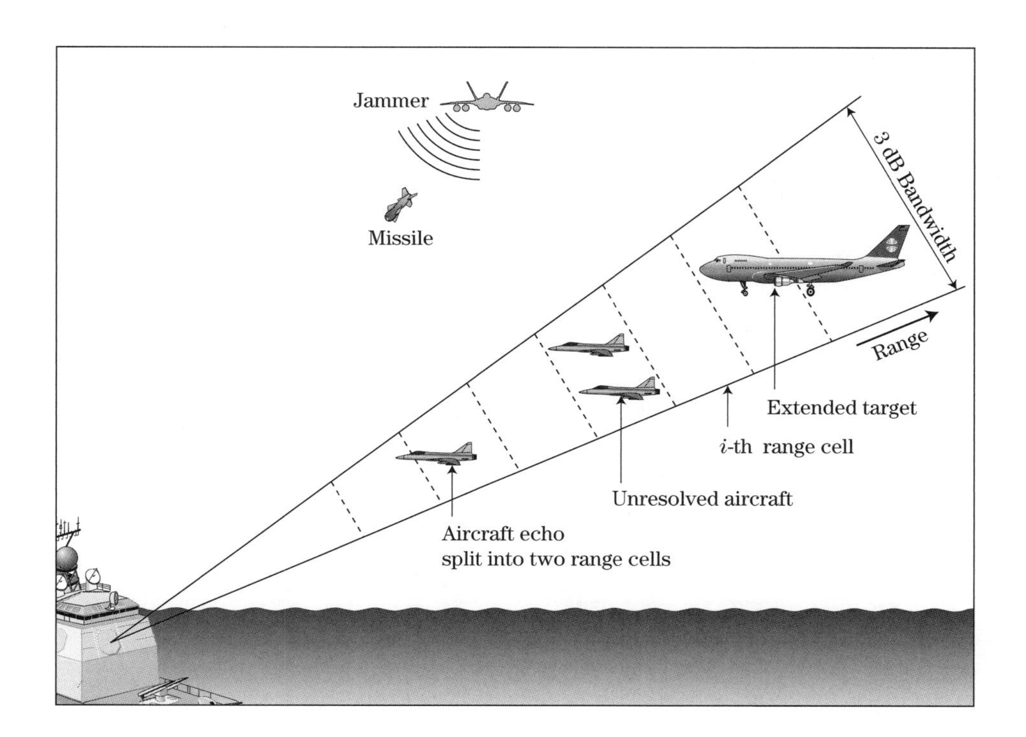

FIGURE 18-6 ■ Illustration of radar resolution with various targets. The solid lines show the two-way 3 dB beam pattern, while the dashed lines indicate range bins.

a sampling interval equal to the range resolution. Each range cell is associated with a sample of the output of the matched filter through which the received echo signal is passed. The center of each range cell corresponds to the sample time. A range and angle measurement is made whenever the output of the matched filter exceeds the detection threshold. The small single aircraft is smaller than the extent of the range cell, but its position straddles two range cells, causing the echoed energy from the aircraft to be split between two adjacent range cells and incurring a straddle loss. Typically, every measurement experiences some straddle loss. The two closely spaced aircraft are in the same range cell, and only one measurement with energy from both targets results. The large airliner extends over three range cells. Such targets are often referred to as *extended targets*. Significant echoes from the airliner are observed in three consecutive samples of the matched filter. Part of the challenge in radar tracking and data association is deciding whether a sequence of detections is one extended target or multiple closely spaced targets. Also, the signals from the jammer enter into the measurements of the missile between it and the radar and affect all of the range cells that are collected along the line of sight to the missile.

Continuing the discussion on range measurement, the various situations that can result are depicted in Figure 18-7, where the sequence of rectangles represent consecutive samples of the matched filter output and the fraction of the target energy captured in the sample is represented by the percentage fill of the rectangle. Some of these situations are analogous to those depicted in Figure 18-6. On the left side of the top sequence in Figure 18-7, the reflected energy is captured in a single sample of the matched filter output. However, this is typically not the case.

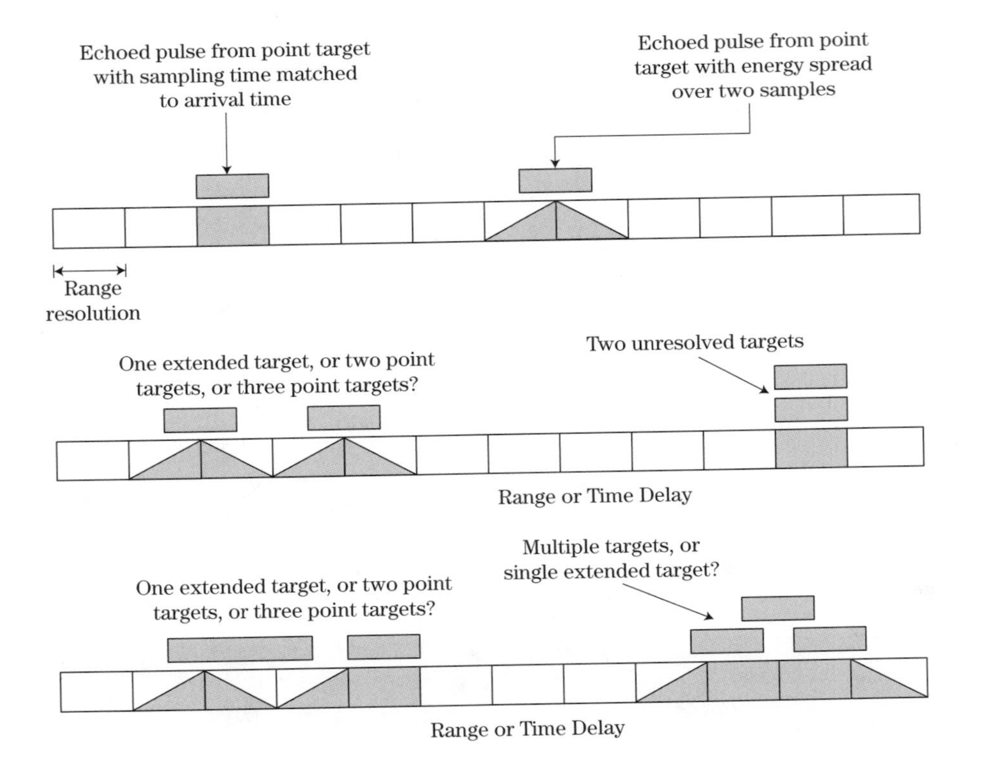

FIGURE 18-7 ■ Resolving closely spaced objects in range.

The more common case is illustrated by the sample of the echo in the top right of Figure 18-7, where the reflected energy is captured in two consecutive samples. In this case, the Cramèr-Rao lower bound on the variance of the range estimate is closely achieved by a power-weighted centroid of the ranges of the two consecutive samples. This technique, along with the split-gate centroid method of measuring range, are described and their precision analyzed in the next section. Examples of closely spaced single targets and multiple targets are shown in the other rows.

Sidelobes in range, angle, or Doppler complicate measurements further. Sidelobes are always present in angle due to the antenna sidelobes and in Doppler if the spectrum is computed using Fourier techniques, as will usually be the case. They will be present in range when pulse compression techniques are used. Sidelobes result in energy from the echo of a target being present in the data samples adjacent to those that correspond to the actual target location. For example, the echo of a small aircraft target with a relatively low RCS that is within a few of range cells of an airliner with a large RCS may be obscured by the range sidelobes from the airliner, a phenomenon known as *masking* (see Chapter 17 for an example of masking in the Doppler dimension). Waveform design methods and signal windowing techniques are often employed to reduce the impacts of range sidelobes. While special signal processing techniques can be employed to assist in the detection of unresolved or extended targets, multiple targets can appear as a single target measurement and one target can produce multiple measurements. Thus, close coordination of the signal processing and target tracking is needed.

18.5.2 Split-Gate and Centroid Range Measurement

In Sections 18.2.1 and 18.4.3 it was shown that target range measurement precision using the signal peak is dependent on the SNR. Rather than use the signal peak to estimate the range, a better approach is to use one of a number of techniques that combine multiple samples to provide some integration against noise. One such technique is an *early/late gate* or *split-gate* tracker. This technique is suitable for signals that produce outputs symmetric about the true location in the absence of noise. This is the case for both the angle and range measurement examples considered so far. The early/late gate tracker slides a two-part window across the data and integrates the energy within each of the two "gates" (halves of the window). When the energy in each gate is equal, the gate is approximately centered on the peak signal.

One way to implement the early/late gate tracker is to convolve the noisy output with the impulse response $h[l]$ shown in Figure 18-8b as the solid black line overlaid on the noisy data. The ideal (noise-free) matched filter output is shown in Figure 18-8a. Because of the difference in sign of the impulse response in the two gates, the early/late gate tracker produces an output near zero when centered on the signal peak. This is depicted in Figure 18-9, which shows the magnitude of the output of the early/late gate tracker for input SNRs of 20 dB and 5 dB. The peak output of the matched filter (not shown) occurs at samples 99 and 107 in these two examples, while the early/late gate tracker correctly estimates the target range to be at sample 100 for the 20 dB case and misses by only 1 sample, at range bin 101, in the noisier 5 dB case. An alternative implementation does not take the magnitude at the tracker output; in this case, the range estimate is the zero crossing in the tracker output.

The precision of an early/late gate tracker is limited to the sampling density of the signal to which it is applied. This may be quite adequate if the data are highly oversampled

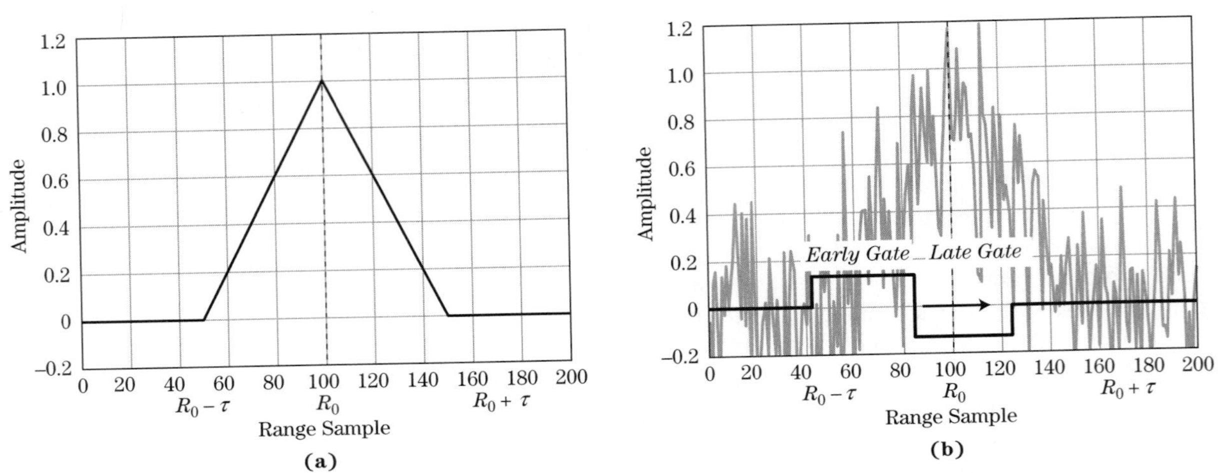

FIGURE 18-8 ■ (a) Triangular output of matched filter for an ideal rectangular pulse with no noise. (b) Output with 10 dB SNR. Also shown is an early/late gate tracker impulse response.

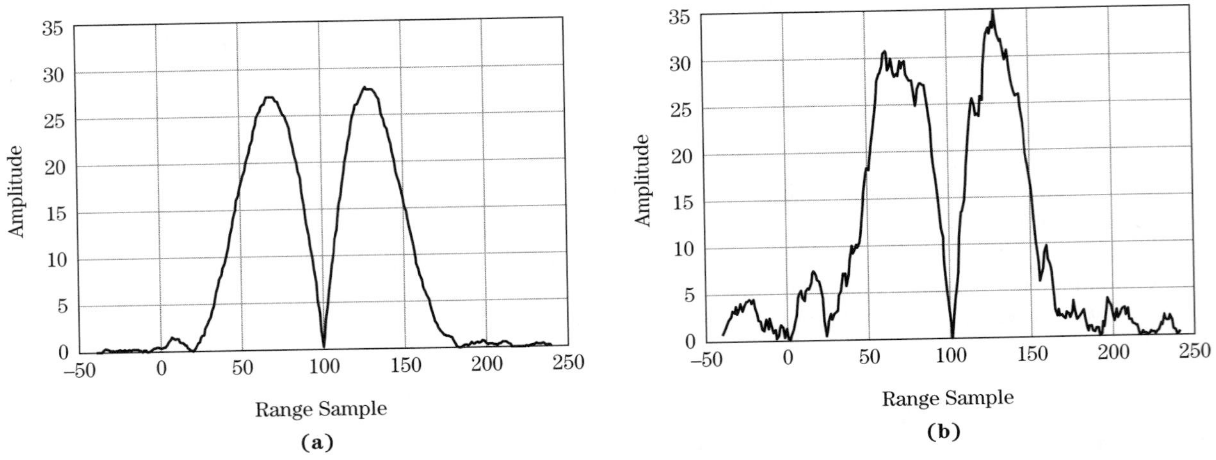

FIGURE 18-9 ■ (a) Magnitude of early/late gate tracker for simple pulse matched filter output with 20 dB peak SNR. (b) Output with 5 dB peak SNR.

(many samples per resolution cell), but, as discussed earlier, range is often estimated using data sampled at intervals approximately equal to a resolution cell. An alternative to the early/late gate tracker is the *centroid* tracker. The centroid, C_x, over a region $l \in [L_1, L_2]$ of a sequence $x[l]$ is defined as

$$C_x = \frac{\sum_{l=L_1}^{L_2} l \cdot x[l]}{\sum_{l=L_1}^{L_2} x[l]} \tag{18.35}$$

Note that the centroid can take on a noninteger value, so it inherently interpolates between samples.

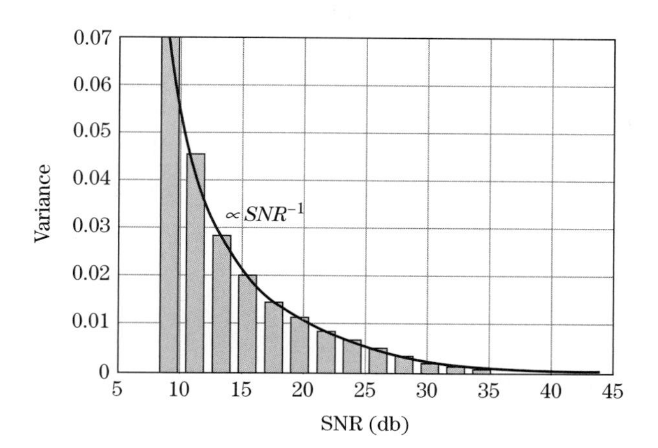

FIGURE 18-10 ■ Example of variance of power-weighted centroid range estimation for ideal simple pulse and matched filter. The variance is normalized to the range resolution.

To demonstrate the performance of the two-point power centroid, consider an ideal rectangular pulse of duration τ and a matched filter. The output of the matched filter is a triangular pulse of duration 2τ. A square law detector is assumed. The magnitude squared of the filter output is sampled at the nominal range resolution (i.e., every τ seconds). A two-point power centroid is applied and the range estimation error is calculated after removing the bias of the estimator for the noise-free case. The error depends on the location of the actual target peak relative to the sampling times, so it is averaged over sampling times that vary by $\pm\tau/2$ seconds relative to the peak. The resulting variance of the average estimation error, normalized to the range resolution, is shown in Figure 18-10 as a function of SNR. The error variance decreases approximately as $1/SNR$ for the range of SNR shown.

An approximation to the CRLB of the range measurement for a simple, uncoded, rectangular pulse is given in [6] as

$$\sigma_R^2 \geq \frac{c^2\tau}{8SNR \cdot B_{IF}} \tag{18.36}$$

where B_{IF} is the bandwidth of the IF filter, and τ is the pulse duration. A power-weighted centroid of the ranges of the two consecutive samples approaches this CRLB. More detail on measurement precision for closely spaced targets is given in [6]. Resolving two closely spaced targets in range is addressed in [7,8]. When range is measured with a linear frequency modulated (LFM) waveform, any uncompensated Doppler shift results in a shift in the apparent range of the target, a phenomenon called *range-Doppler coupling* (see Chapter 20). The CRLB for range measurement using an LFM waveform with a rectangular envelope and a known range rate is given in [9] as

$$\sigma_R^2 \geq J(R) = \frac{3c^2}{8\pi^2 SNR \cdot B^2} \tag{18.37}$$

where B is the swept bandwidth of the LFM waveform.

18.6 PHASE MEASUREMENT

Estimating the echo time of arrival using the methods previously described ignores the signal phase. However, signal phase can also be estimated if the Doppler shift is known. Combining terms to simplify the signal model in equation (18.2) when the frequency is known, the radar return can be written as

$$s(t) = p(t)\cos(\omega t + \varphi) \tag{18.38}$$

where ω is the frequency and φ is the carrier phase. It can then be shown that the optimal estimator for the phase is [4]

$$\hat{\varphi} = -\tan^{-1}\frac{P_s}{P_c} \tag{18.39}$$

$$P_s = \int p(t)\sin(\omega t)dt \qquad P_c = \int p(t)\cos(\omega t)dt$$

where the sine and cosine filter integrals are over the analysis time window. When the SNR is sufficiently large the rms error is [4]

$$\sigma_{\hat{\varphi}}^2 \geq \frac{1}{SNR \cdot \tau B} \tag{18.40}$$

where τB is the pulse time-bandwidth product. Note that, like range estimation, the phase measurement variance is inversely proportional to the measurement SNR.

18.7 | DOPPLER AND RANGE RATE MEASUREMENTS

In most modern tracking radars, range rate measurements are accomplished by pulsed Doppler waveforms that include a periodic sequence of pulses. In pulse-Doppler processing, the output of the matched filter is sampled throughout the range window for each pulse for time delay (i.e., range) estimation, and samples of the matched filter output from the multiple pulses at each range are processed with a DFT to estimate the Doppler frequency, f_d and the corresponding radial velocity $v = 2f_d/\lambda$. The negative of the radial velocity is the range rate, $\dot{R}$. The peak-to-null (Rayleigh) Doppler resolution is the inverse of the duration of the pulse-Doppler waveform dwell time, T_d (see Chapter 14),

$$\Delta f_d = \frac{1}{T_d} \tag{18.41}$$

Because velocity is related to Doppler shift according to $f_d = 2v/\lambda$, the resolution of velocity or range rate is

$$\Delta v = \Delta \dot{R} = \frac{\lambda}{2T_d} = \frac{c}{2T_d f_t} \tag{18.42}$$

where f_t is the carrier frequency of the transmitted waveform.

Pulse-Doppler waveforms suffer potential *ambiguities* in range and range rate. The minimum unambiguous range is

$$R_{ua} = \frac{cT}{2} \tag{18.43}$$

where T is the slow-time sampling interval (PRI). The ambiguous interval in Doppler frequency is $1/T$ Hz. Assuming both positive and negative frequencies (and thus range rates) are of interest, the maximum unambiguous Doppler frequency, velocity, and range rate are then given by

$$
\begin{aligned}
f_{d_{ua}} &= \pm\frac{1}{2T} = \pm\frac{PRF}{2} \\
v_{ua} &= \dot{R}_{ua} = \pm\frac{\lambda}{2}PRF = \pm\frac{c}{2f_t}PRF
\end{aligned}
\tag{18.44}
$$

Thus, the range rate resolution is specified by the dwell time of the waveform, while the maximum unambiguous range rate is specified by the PRF of the waveform.

The CRLB for measuring frequency in hertz for the signal model of equation (18.2) with M measurements, assuming the initial phase and amplitude are also unknown, is shown by several authors to be [3,4]

$$
\sigma_f^2 \geq J(f) = \frac{3f_s^2}{\pi^2 SNR \cdot M(M^2-1)} \approx \frac{3f_s^2}{\pi^2 SNR \cdot M^3} \quad \text{Hz}^2 \tag{18.45}
$$

where f_s is the sampling frequency in samples/second. Notice that the CRLB decreases as M^3. One factor of M comes from the increase in SNR when integrating multiple samples, while a factor of M^2 comes from the improved resolution of the frequency estimate. This can be seen by putting equation (18.45) into terms of resolution and SNR for the case of an estimator based on the discrete-time Fourier transform (DTFT), of which the DFT is a special case. The Rayleigh frequency resolution in hertz of an M-point DFT is f_s/M, and the SNR of a sinusoidal signal in the frequency domain is M times the time-domain SNR, assuming white noise (see Chapter 14). Denoting the frequency domain resolution and SNR as Δf and SNR_f, equation (18.45) becomes

$$
\sigma_f \geq \frac{\sqrt{3}}{\pi}\frac{\Delta f}{\sqrt{SNR_f}} \quad \text{Hz} \tag{18.46}
$$

Equation (18.46) states that the precision of the frequency estimate is proportional to the resolution divided by the square root of the applicable SNR, analogous to the range estimation case of equation (18.33).

A number of single and multi-pulse frequency estimation schemes exist. These can be divided into coherent and noncoherent techniques [4]. Several are considered in the following. Other general frequency estimation techniques are considered in [3].

18.7.1 DFT Methods

An obvious estimator of Doppler frequency is the discrete Fourier transform, usually implemented with the fast Fourier transform (FFT) algorithm. Given M samples of slow time data $y[m]$, the K-point DFT is

$$
Y[k] = \sum_{m=0}^{M-1} y[m]\exp(-j2\pi mk/K), \quad k = 0,\ldots,K-1 \tag{18.47}
$$

where $K \geq M$. The DFT and its properties are discussed extensively in Chapter 14.

The frequency of a signal is estimated by computing its DFT and then finding the value of k that gives the largest value of $Y[k]$, that is, by finding the peak of the DFT. The k-th index corresponds to a frequency of $k{\cdot}PRF/K$ Hz. Consider a signal

$$
y[m] = A\exp(j2\pi f_0 mT) + w[m], \quad m = 0,\ldots,M-1 \tag{18.48}
$$

where $w[m]$ is white Gaussian noise with power σ_w^2. The SNR of the individual data samples is A^2/σ_w^2. The DFT effectively integrates the M data samples, increasing the SNR to a maximum of MA^2/σ_w^2 if f_0 equals one of the DFT frequencies. If f_0 does not equal one of these frequencies, the integrated SNR is up to 3.92 dB lower (depending on the frequency difference and the data window used; see Chapter 14 for details). This reduction is another example of straddle loss.

If the integrated SNR is at least 10 dB, it is virtually certain that the DFT peak will occur at the index closest to the true frequency as desired. Then, for a sinusoidal signal at an arbitrary frequency f_0, the maximum error in the estimated frequency $\hat{f}_0$ is $PRF/2K$ Hz. Thus, increasing the DFT size reduces the maximum frequency estimation error, at the cost of increased computational load for the larger DFT. The same CRLB for frequency estimation given by equation (18.45) applies [3].

18.7.2 DFT Interpolation Methods

Frequency estimation precision can be improved by following the DFT calculation with one of a number of interpolation methods. The centroiding technique discussed earlier for range processing can be applied to the DFT also. Another common method is to fit a low-order polynomial through the DFT peak and its nearest neighbors. The estimated frequency is the location of the peak of the interpolated polynomial. A typical interpolator, using only the magnitude of the DFT samples, estimates the frequency as [10]

$$\Delta k = \frac{-\frac{1}{2}\{|Y[k_0+1]| - |Y[k_0-1]|\}}{|Y[k_0-1]| - 2|Y[k_0]| + |Y[k_0+1]|}$$
$$\hat{f}_0 = \frac{(k_0 + \Delta k)}{K} PRF \tag{18.49}$$

where k_0 is the index at which the DFT peak occurs. Figure 18-11a shows the error in this estimate in the absence of noise, $f_0 - \hat{f}_0$, as a function of the starting offset $f_0 - k_0 PRF/K$ for $M = 30$ data samples and various DFT sizes K. Note that the error is reduced as the DFT size increases.

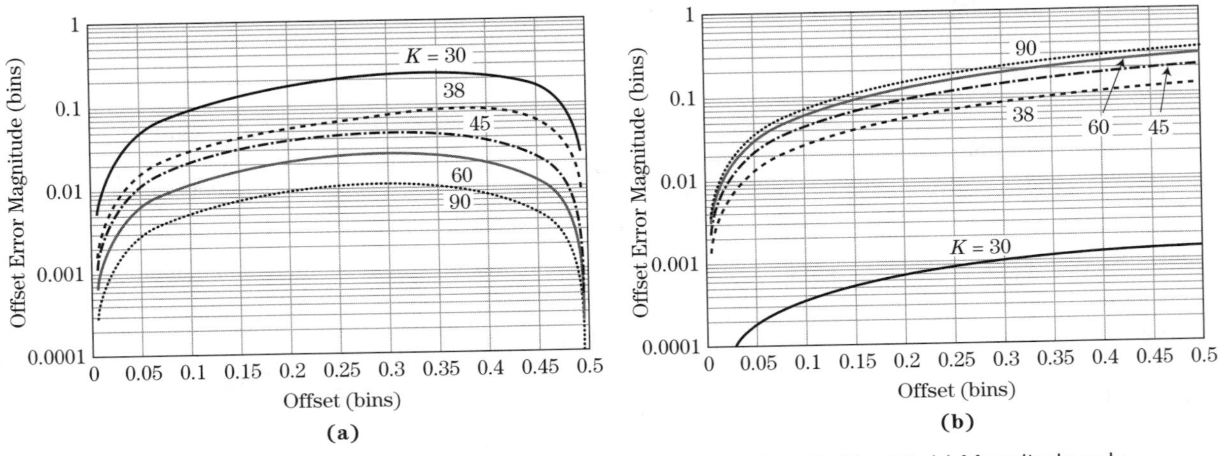

FIGURE 18-11 ■ Error in DFT interpolators. Data length $M = 30$. (a) Magnitude only. (b) Magnitude and phase.

Figure 18-11b shows the precision of a similar interpolator that uses the complex DFT data instead of just the magnitude. The estimate is given by [10]

$$\Delta k = -\text{Re}\left\{\frac{Y[k_0+1]-Y[k_0-1]}{2Y[k_0]-Y[k_0-1]-Y[k_0+1]}\right\}$$
$$\hat{f}_0 = \frac{(k_0+\Delta k)}{K}PRF \tag{18.50}$$

This estimator achieves much lower error than the estimator of equation (18.49) when $K = M$, but the error actually gets worse as K gets larger! Another complication with these interpolators is that they must be modified if a window is used on the data, and the modification required depends on the specific window being used. Examples and additional details are given in [10].

18.7.3 Pulse Pair Estimation

Pulse pair processing (PPP) is a specialized form of Doppler measurement common in meteorological radar. In PPP, it is assumed that the spectrum of the slow-time data consists of noise and a single Doppler peak, generally not located at zero Doppler (though it could be). The goal of PPP is to estimate the power, mean velocity, and *spectral width* (variance) of this peak. PPP is used extensively in both ground-based and airborne weather radars for storm tracking and weather forecasting. In airborne radars, it is also used for windshear detection. Pulse pair processing was discussed in Section 17.7, and estimators for the power, mean velocity, and spectral width were described there. Additional detail is available in [1,11].

18.8 RCS ESTIMATION

The radar cross section, σ, of a target is related to the target echo signal amplitude, ξ, through the radar range equation as shown in equation (18.6). The actual amplitude available for measurement is Λ of equation (18.7). Consider the measured power $P_s = \Lambda^2$ of the complex signal $s = s_I + js_Q$. P_s is the combination of the target echo power and the noise power and thus is a random variable. The expected value of P_s is

$$\overline{P_s} = \alpha^2 + \sigma_w^2 \tag{18.51}$$

If $\overline{P_s}$ can be estimated and an estimate of the noise power σ_w^2 is available, then equation (18.51) can be used to estimate α^2, which can be used in turn with the radar range equation and equation (18.3) to estimate the amplitude reflectivity, ξ, and the RCS, σ. In this section procedures for estimating $\overline{P_s}$ from N samples are considered.

The optimal estimator for the power depends on its PDF. Recall from Chapter 15 that various fluctuation models are used to model target echoes and that the nonfluctuating and Swerling models are among the most common. However, a simple suboptimal estimator usable with any PDF is the sample mean,

$$\widehat{\overline{P_s}} = \frac{1}{N}\sum_{n=0}^{N-1} P_s[n] \tag{18.52}$$

where $P_s[n]$ denotes the n-th sample of the measured power, typically the power measured on the n-th pulse of an N-pulse dwell. The CRLB for this case was given by equation (18.23).

Now consider the Swerling 2 target. In this case, the RCS is exponentially distributed and is uncorrelated pulse to pulse. The resulting measured power is also exponentially distributed (see Chapter 15 or [1]). Specifically,

$$p_{P_s}(P_s) = \frac{1}{\overline{P}_s} \exp\left(-P_s/\overline{P}_s\right) \tag{18.53}$$

where $\overline{P}_s$ is the expected value of the received power. Note that this PDF is fully specified by the single parameter, $\overline{P}_s$. It is easy to show that the maximum likelihood estimator of the parameter $\overline{P}_s$ is simply the sample mean, that is, equation (18.52) [1,12]. Consequently, the sample mean is an optimal estimator for this case. Problem 5 at the end of this chapter derives the CRLB for this case.

In the Swerling 4 and nonfluctuating cases, the PDF of P_s is significantly more complicated, and simple analytical expressions for the ML estimator do not exist. The CRLB also does not have a closed form. For some cases, it can be shown that for moderate to high SNRs, the ML estimator is approximately equal to the sample mean of equation (18.52) [12]. Thus in practice, the sample mean is usually used to estimate $\overline{P}_s$ for most common target models.

18.9 | ANGLE MEASUREMENTS

A target echo that is received through a standard antenna pattern gives no information about the angular location of the target other than it is most likely within the mainlobe of the beam. However, if a target has a large RCS, it may not even be within the 3 dB beamwidth of the antenna pattern. In most radars that rotate for scanning the field of regard, the amplitudes of the echoed signals are collected for multiple positions of the antenna boresight as it scans by the target, and centroiding is used to estimate the angular location of the target. However, for tracking radars that support control functions, radars that measure two angular coordinates, and electronically scanned radars that scan while tracking, centroiding the signals from multiple positions of the antenna pattern is not a viable option because many beam positions are needed to overcome the RCS fluctuations of the target for an accurate angle-of-arrival estimate.

One of the techniques used early on to improve the angle measurements of this type of radars was *sequential lobing*, which uses two consecutive dwells on the target to refine each angle measurement [13]. As illustrated in Figure 18-12, the first measurement is taken with the boresight of the antenna pointing slightly to one side of the predicted target position, while the second measurement is taken with the boresight of the antenna pointing slightly to the other side of the predicted position. The target is then declared to be closer to the angle of the measurement with the larger amplitude, and the predicted angle of the target is corrected. However, sequential lobing is very susceptible to pulse-to-pulse amplitude fluctuations of the target echoes, which are common in radar measurements due to scintillation in the RCS of the target. Furthermore, when tracking in azimuth and elevation, sequential lobing requires lobe switching between azimuth and elevation or conical scanning (see Figure 18-13) [13], both of which are inefficient with respect to radar time and energy and are easily jammed or deceived by the intended target.

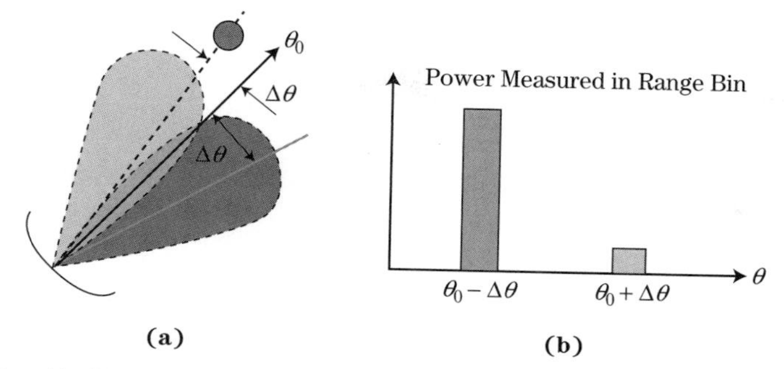

FIGURE 18-12 ■ Notional illustration of sequential lobing for direction determination. In (a) the target is observed first at one angle, then at a second. Due to the antenna gain pattern the differences in observation angle result in different return powers as illustrated in (b).

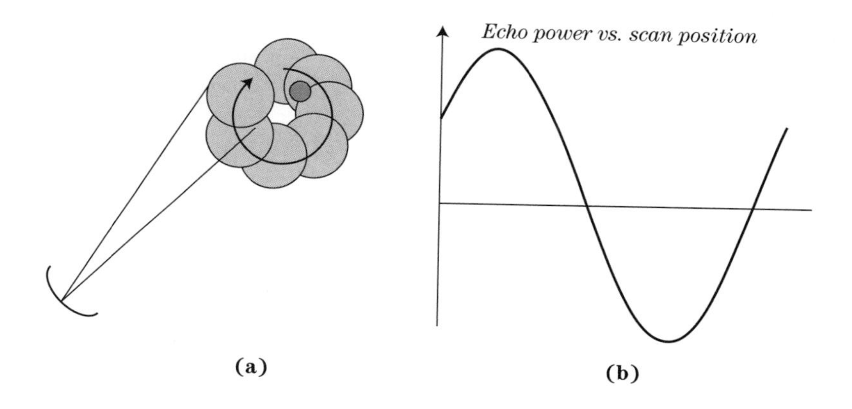

FIGURE 18-13 ■ Conical sequential lobing illustration.

Monopulse is a simultaneous lobing technique that was developed to overcome the shortcomings of sequential lobing and conical scanning [13]. Monopulse antennas were described in Chapter 9. In an *amplitude comparison monopulse* radar system, a pulse is transmitted directly at the predicted position of the target, and the target echo is received with two squinted beams ("split beams"). Figures 18-14 and 18-15 illustrate the transmit and difference antenna patterns for a single-dimensional monopulse. The angle of arrival of the target is typically estimated with the in-phase part (i.e., the real part) of the monopulse ratio, which is formed by dividing the difference of the two received signals by their sum (see Figure 18-15). When tracking in azimuth and elevation, four beams are used for receive, and two monopulse ratios are typically formed. The simultaneous lobing of monopulse allows the transmitted energy to be directed at the predicted position of the target and eliminates the errors due to amplitude fluctuations by forming a refinement of the angular precision with a single pulse.[1] Since the lobing is simultaneous rather than sequential, monopulse is very efficient with respect to radar time and energy and difficult

[1] The term *monopulse* originated with this idea of a single-pulse refinement of the angular accuracy. Some confusion exists concerning monopulse radars because many monopulse radars use multiple pulses to form an angular measurement. The monopulse ratios formed with each pulse are what distinguishes a radar as monopulse.

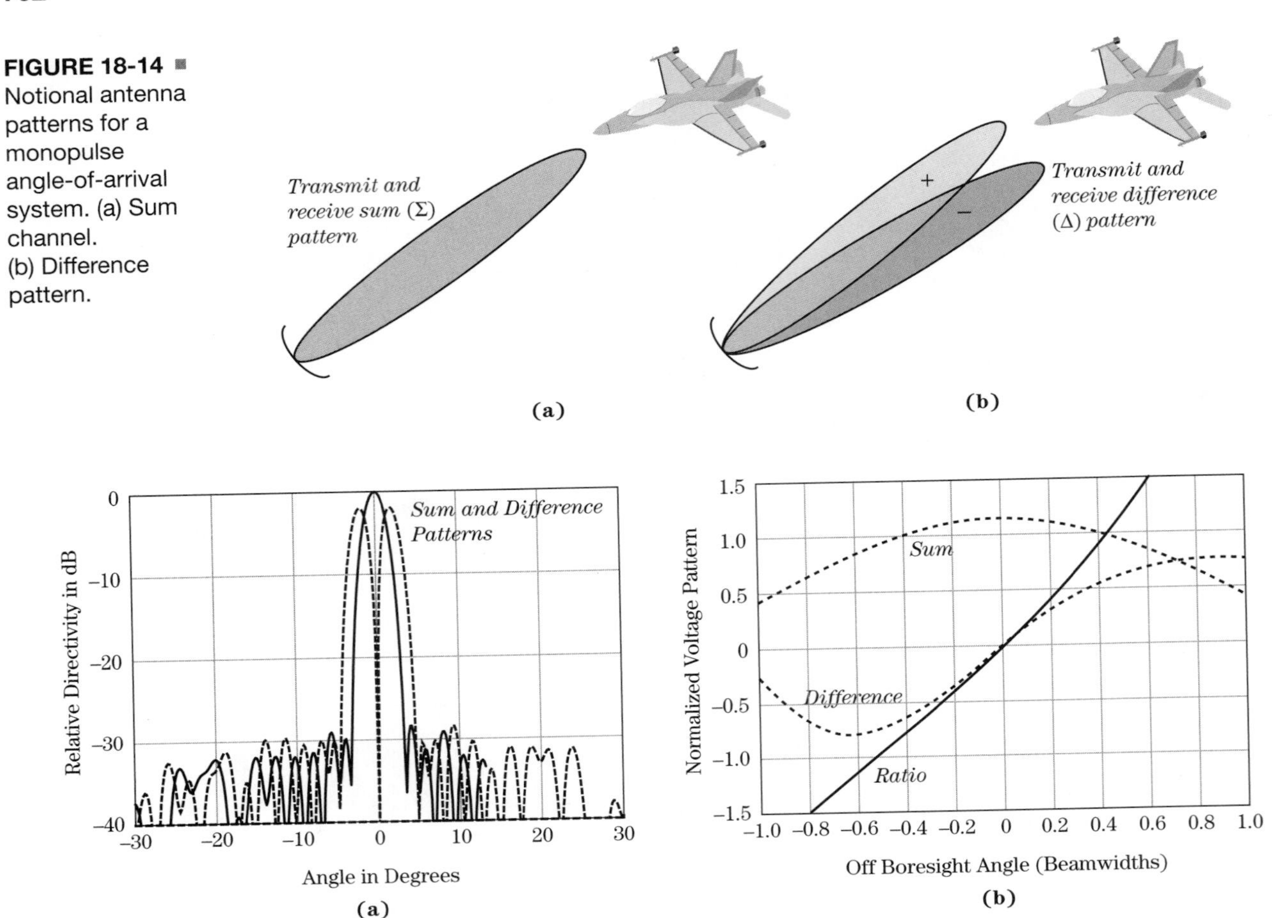

FIGURE 18-14 ■ Notional antenna patterns for a monopulse angle-of-arrival system. (a) Sum channel. (b) Difference pattern.

FIGURE 18-15 ■ Monopulse antenna gain patterns. (a) Solid line is the sum channel gain while dotted line is the difference channel gain. (b) Plots of the sum and difference channel gains and their ratio near the boresight.

to jam or deceive. Both of these benefits are particularly important to electronically steered radars that are required to maintain simultaneous tracks on many targets.

An alternative approach to amplitude comparison monopulse is *phase comparison monopulse* [4]. Phase comparison monopulse is a form of radar interferometry that relies on phase differences between antenna beams, often achieved by offsetting two antennas, to determine the angle to the target. As illustrated in Figure 18-16 a baseline d between the two beam phase centers creates a path length difference, $d\sin\theta$, to the target, which results in a phase difference in the signals. An important practical limitation in AOA measurement using phase comparison monopulse is that, to prevent ambiguity in the estimated direction, the target angle and antenna baseline separation must meet the requirement [13]

$$|\sin\theta| < \frac{\lambda}{8d} \tag{18.54}$$

Both monopulse and centroiding are used for AOA estimation in modern radar systems. Implementing a monopulse AOA system is more expensive than implementing a centroiding AOA system. However, the time occupancy requirements of electronically

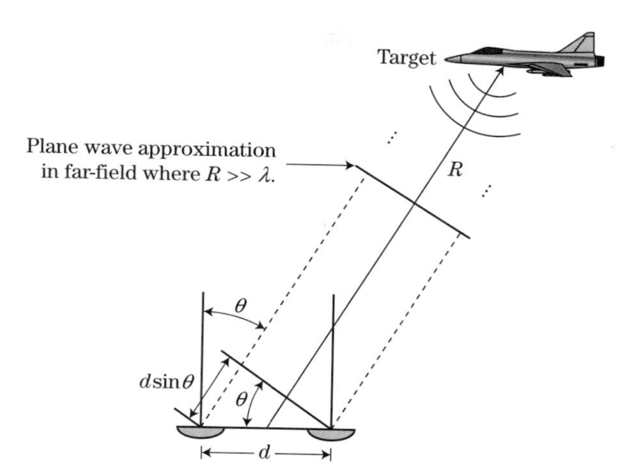

FIGURE 18-16 ■ Geometry of phase comparison monopulse. The displacement distance d between the antennas causes different path lengths to the target, which gives rise to a difference in phase of the return echo between the two beams. The target is assumed to be far enough away that the far-field approximation applies.

scanned radars for multiple target tracking dictate the use of monopulse AOA estimation. In the remainder of this section, centroiding and monopulse AOA estimation are discussed.

18.9.1 Angle Centroiding for Scanning Radars

When considering the AOA estimation method for a rotating radar, centroiding should be the first consideration. The estimation error of an angle centroider depends on the single-pulse SNR, the number of pulses transmitted within the antenna beamwidth, the antenna gain pattern, and target fluctuation model. Two basic approaches to AOA centroiding are the *binary integration* approach [14] and the ML approach [15]. The binary integration approach is motivated by the need to limit memory and processing in real-time radar systems, while the ML approach is motivated by the need for more accurate estimation. However, studies have found that the AOA estimation of binary integration can approach the CRLB for sufficiently high SNR and targets with rather stable radar cross sections. For Swerling targets, the CRLB for the AOA has the following form [14,16]

$$\sigma_{\hat{\theta}}^2 = \frac{\theta_3^2}{N \cdot SNR \cdot \alpha^2} \tag{18.55}$$

where θ is the AOA, $\hat{\theta}$ is the estimate of θ, θ_3 is the 3 dB beamwidth of the antenna pattern, N is the number of pulses in the azimuth interval, and α is a constant that depends on the ratio of the azimuth observational interval and the 3 dB antenna beamwidth. For a large azimuth observation interval, $\alpha \approx 0.339$. While the performance of a centroider depends on many parameters, an AOA centroider can be expected to achieve a precision on the order of $\sigma_{\hat{\theta}} \approx \theta_3/10$.

In the binary integration approach, the samples of the matched filter output are quantized to 0 or 1. For a given scan of the radar by the target, an AOA estimate can be produced by taking a simple average of the angle of the antenna boresight at the first detection of the target (i.e., the leading edge) and the angle of the antenna boresight at the final detection

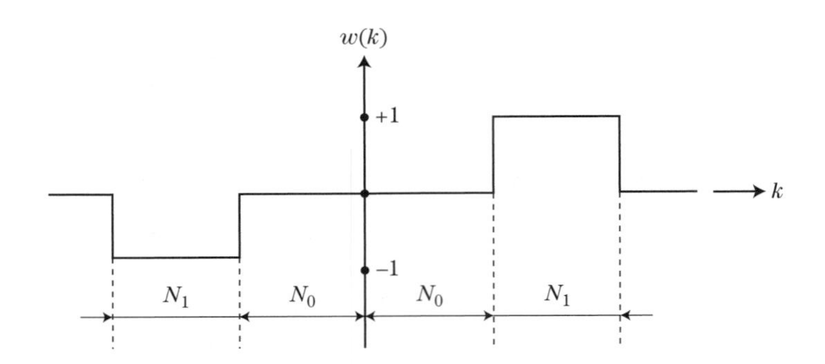

FIGURE 18-17 ■ Weights of a binary Bernstein estimator for angle-of-arrival centroiding.

of the target (i.e., the trailing edge). For targets with very stable echoes, this simple approach performs reasonably well. However, for targets with fluctuating amplitudes, a better method is needed. An improved approach is represented by the Bernstein estimator [14] in which the quantized detections are convolved with the weights shown in Figure 18-17. The Bernstein estimator uses N_1 detection opportunities that occur as the antenna boresight scans toward the target and N_1 detection opportunities that occur as the antenna boresight scans away from the target to reduce the sensitivity of the AOA estimate to amplitude fluctuations. The convolution of those detection opportunities with the Bernstein weights gives rise to a zero crossing that corresponds to the AOA estimate of the target. The $2N_0 - 1$ detection opportunities that occur when the antenna boresight is closely aligned with the target are ignored by the Bernstein estimator as those samples provide very little or no information concerning the AOA of the target. Typically, $2(N_0 + N_1) \approx N$. Optimal values for N_0 and N_1 are a function of the SNR of the target and the fluctuation model. Thus, the diversity of targets with respect to RCS and fluctuations make it impossible to achieve optimal AOA estimation without adaptation.

The ML approach to centroiding for AOA estimation involves formulating the measured amplitude of the target echoes in terms of the antenna gain pattern and the angle of the target and computing the ML estimate of the target AOA. The target amplitude is typically treated as a fixed, unknown quantity over the scan, which is a reasonable assumption for many targets if the pulses include at least 10 MHz of frequency agility.

18.9.2 Monopulse AOA Estimation

While most monopulse systems include AOA estimation to two angular coordinates, only monopulse processing in a single angular coordinate is considered in this section. The focus is on amplitude-comparison monopulse. Only the monopulse processing for Rayleigh targets is considered here. The monopulse signals are formulated, and conventional monopulse processing is discussed first.

18.9.2.1 Single Target

Using the notation in Section 18.3, equation (18.2) gives the sum signal for the monopulse processing, while the vertical difference signal is given by

$$d(t) = 2\sqrt{\frac{P_t}{(4\pi)^3}\frac{G_t\lambda}{R^2}} A V_s(\theta,\phi) V_{dV}(\theta,\phi) p(t) \cos(\omega_c t + \omega_d t + \psi) + w_d(t) \qquad (18.56)$$

where $V_s(\theta, \phi)$ denotes the voltage gain of the antenna in the sum channel at (θ, ϕ), $V_{dV}(\theta, \phi)$ denotes the voltage gain of the antenna in the vertical difference channel at

(θ, ϕ), (θ, ϕ) denotes the angular location of the target relative to antenna boresight, and $w_d(t)$ denotes the receiver noise in the vertical difference channel. A similar difference signal will also be received for the horizontal coordinate. Note that "vertical" and "horizontal" are used to denote the two orthogonal coordinates of the radar. These differ from the local horizontal and vertical if the antenna boresight is tilted out of the horizontal plane.

Monopulse is implemented slightly differently in phased array radars. In a phased array radar, two beamforming networks (three networks for monopulse in two angles) are used to simultaneously form sum and difference beams. The sum pattern as given in Section 18.3 may be used for transmit and receive in the sum channel, while the difference pattern may be the result of subtracting two beams, each formed with one-half of the array. Ignoring the Doppler and straddle losses, the in-phase and quadrature components of the sum (s_I and s_Q) signal and difference (d_I and d_Q) signal are given by

$$\begin{aligned} s_I &= \alpha \cos\phi + w_{sI} \qquad & s_Q &= \alpha \sin\phi + w_{sQ} \\ d_I &= \alpha\eta \cos\phi + w_{dI} \qquad & d_Q &= \alpha\eta \sin\phi + w_{dQ} \end{aligned} \tag{18.57}$$

where

$$\alpha = \kappa V_s(\theta, \phi), \qquad \eta = \frac{V_{dV}(\theta, \phi)}{V_s(\theta, \phi)}, \tag{18.58}$$

$$w_{sI} \sim \mathrm{N}(0, \sigma_s^2), \quad w_{sI} \sim \mathrm{N}\left(0, \sigma_s^2\right), \quad w_{dI} \sim \mathrm{N}\left(0, \sigma_d^2\right), \quad w_{dQ} \sim \mathrm{N}\left(0, \sigma_d^2\right)$$

and κ is a constant that depends on the target amplitude, system gain, and transmit power.

The receiver errors in the sum and difference channels are typically uncorrelated except for a possible real-valued correlation or $\mathrm{E}\{n_{sI}n_{dI}\} = \mathrm{E}\{n_{sQ}n_{dQ}\} = \rho\sigma_d\sigma_s$ [13]. Since the parameter η defines an off-boresight angle θ, η is referred to as the AOA parameter. Thus, monopulse processing involves the estimation of η and using the estimate of η to compute the AOA relative to the antenna boresight.

In a typical monopulse system, the AOA with respect to the antenna boresight is approximated by

$$\hat{\theta} \approx \frac{\theta_3}{k_m}\eta \tag{18.59}$$

where $1 < k_m < 2$ and θ_3 is the 3 dB beamwidth of the antenna pattern. The linear approximation to the monopulse error function is usually appropriate for $-\theta_3/2 \leq \theta \leq \theta_3/2$. Denoting $s = s_I + js_Q$ and $d = d_I + jd_Q$, the in-phase and quadrature parts of the monopulse ratio are given by

$$y_I = \mathrm{Re}\left(\frac{d}{s}\right) = \frac{d_I s_I + s_Q d_Q}{s_I^2 + s_Q^2}, \qquad y_Q = \mathrm{Im}\left(\frac{d}{s}\right) = \frac{d_Q s_I - d_I s_Q}{s_I^2 + s_Q^2} \tag{18.60}$$

Typically, y_I is taken as the estimate of the direction of arrival (DOA), which gives the AOA estimate as

$$\hat{\theta} = \frac{\theta_3}{k_m}\left(y_I - \rho\frac{\sigma_d}{\sigma_s}\right) \tag{18.61}$$

The variance of y_I is often reported in the literature [13] as

$$\sigma_{y_I}^2 \approx \frac{1}{2 \cdot SNR}\left[\frac{\sigma_d^2}{\sigma_s^2} + \eta^2 - 2\rho\frac{\sigma_d}{\sigma_s}\right], \qquad SNR > 13 \text{ dB} \tag{18.62}$$

where SNR is the signal-to-noise of the sum channel signal. Estimates of the variance of y_I are often computed by setting $\eta = y_I$. An estimate of the variance of the AOA is then given by

$$\sigma_{\hat{\theta}}^2 \approx \frac{\theta_3^2}{k_m^2}\sigma_{y_I}^2 \approx \frac{\theta_3^2}{2k_m^2 SNR}\left[\frac{\sigma_d^2}{\sigma_s^2} + y_I^2 - 2\rho\frac{\sigma_d}{\sigma_s}\right], \quad SNR > 13 \text{ dB} \tag{18.63}$$

Several authors have shown that y_I is a notably biased[2] estimate of the DOA at moderate and low SNR. The bias is often reported in the literature [13] as

$$E(y_I) - \eta = \left(\rho\frac{\sigma_d}{\sigma_s} - \eta\right)\exp(-SNR) \tag{18.64}$$

Seifer showed in [17] that this is an optimistic assessment of the bias because the measured amplitude of the sum signal is subjected to a threshold test prior to monopulse processing.

For a Rayleigh target with real uncorrelated receiver errors, the joint PDF of the in-phase and quadrature parts of the monopulse ratio is given by

$$p_{y_I,y_Q}\left(y_I, y_Q \middle| SNR_{\text{obs}}\right) = p_{y_I}\left(y_I | SNR_{\text{obs}}\right) p_{y_Q}\left(y_Q \middle| SNR_0\right) \tag{18.65}$$

where SNR_{obs} is the observed SNR $\alpha^2\eta^2/\sigma_d^2$ and

$$\begin{aligned} p_{y_I}(y_I|SNR_{\text{obs}}) &= \text{N}\left(\frac{SNR \cdot \eta + \rho\sigma_d\sigma_s^{-1}}{SNR + 1}, \frac{\gamma}{2SNR_{\text{obs}}}\right) \\ p_{y_Q}(y_Q|SNR_{\text{obs}}) &= \text{N}\left(0, \frac{\gamma}{2SNR_{\text{obs}}}\right) \\ \gamma &= \left[\frac{\sigma_d^2}{\sigma_s^2}\left(1 - \frac{\rho^2}{SNR + 1}\right) + \frac{SNR \cdot \eta\left(\eta - 2\rho\sigma_d\sigma_s^{-1}\right)}{SNR + 1}\right] \end{aligned} \tag{18.66}$$

The notation $\text{N}(\mu, \sigma^2)$ denotes a normal (Gaussian) distribution with mean μ and variance σ^2. Thus, y_I and y_Q are conditionally Gaussian, independent random variables. Note that equation (18.66) shows that even for $\rho = 0$, y_I is a notably biased observation of the DOA η for values of SNR less than about 13 dB.

Let $\bar{y}_I$ denote the mean of y_I given SNR_{obs}. Then, for N independent samples or pulses, the ML estimate of $\bar{y}_I$ is given by

$$\hat{y}_I = \frac{1}{NY_N}\sum_{k=1}^{N} SNR_{\text{obs}_k} y_{I_k}, \qquad Y_N = \frac{1}{N}\sum_{k=1}^{N} SNR_{\text{obs}_k} \tag{18.67}$$

where SNR_{obs_k} and y_{I_k} denote the observed SNR and in-phase monopulse ratio for pulse k, respectively. Thus, the estimate $\hat{y}_I$ is a "power" or "energy" weighted sum of the N monopulse ratios. Since the y_{Ik} are Gaussian random variables, $\hat{y}_I$ is the minimum variance

---

[2] *Notably biased estimate* is used here to denote an estimate with bias greater than 5% of the true value.

estimate of $\bar{y}_I$ and a Gaussian random variable with variance given by

$$\sigma^2_{\hat{y}_I} = \frac{\gamma}{2NY_N} \tag{18.68}$$

where γ was defined in equation (18.68). However, note that $\hat{y}_I$ being the minimum variance estimate of $\bar{y}_I$ does not imply that $\hat{y}_I$ is the minimum variance estimate of η [6]. For target tracking, an estimate of the variance $\sigma^2_{\hat{y}_I}$ is found by setting $\eta = y_I$ and $SNR = Y_N - 1$ in p, resulting in

$$\hat{\sigma}^2_{\hat{y}_I} = \frac{1}{2NY_N}\left[\frac{\sigma_d^2}{\sigma_S^2}\left(1 - \frac{\rho^2}{Y_N}\right) + \left(1 - \frac{1}{Y_N}\right)\hat{y}_I\left(\hat{y}_I - 2\rho\sigma_d\sigma_S^{-1}\right)\right] \tag{18.69}$$

Since the y_{Ik} are Gaussian, $\sigma^2_{\hat{y}_I}$ is the conditional CRLB for $\bar{y}_I$ given $\{SNR_{\text{obs}_k}\}_{k=1}^N$ The term *conditional* here is used to denote the fact that the CRLB is developed with the amplitude-conditioned PDF. While $\hat{\sigma}^2_{\hat{y}_I}$ provides the variance of $\hat{y}_I$ for real-time or actual tracking, $\sigma^2_{\hat{y}_I}$ cannot be used for performance prediction because it is a function of $\{SNR_{\text{obs}_k}\}_{k=1}^N$, which are measured quantities. The modified Cramèr-Rao lower bound (MCRLB) [18], while somewhat looser than the CRLB, can be used for performance prediction. The MCRLB of $\bar{y}_I$ is given by

$$\sigma^2_{\hat{y}_I} \geq J(\bar{y}_I) = \frac{p}{2N(SNR + 1)} \tag{18.70}$$

18.9.2.2 Multiple Unresolved Targets

Conventional monopulse processing is based on the assumption that each observation includes a single resolved object. When measurements of two closely spaced objects are made and conventional monopulse processing is employed, the resulting AOA measurement may not reflect the existence of two targets. Since the position and velocity estimates of a target determine the association of any subsequent measurements to the target, failure to detect the presence of the interference of a second target and address it in the DOA estimation can be catastrophic to the performance of the tracking algorithm. For two unresolved Rayleigh targets, the in-phase and quadrature parts of the sum (s_I and s_Q) and (d_I and d_Q) difference signals are given by [19]

$$\begin{aligned} s_I &= \alpha_1\cos\phi_1 + \alpha_2\cos\phi_2 + w_{sI}, \quad & s_Q &= \alpha_1\sin\phi_1 + \alpha_2\sin\phi_2 + w_{sQ} \\ d_I &= \alpha_1\eta_1\cos\phi_1 + \alpha_2\eta_2\cos\phi_2 + w_{dI}, \quad & d_Q &= \alpha_1\eta_1\sin\phi_1 + \alpha_2\eta_2\sin\phi_2 + w_{dQ} \end{aligned} \tag{18.71}$$

where η_i is the DOA parameter for target i. With SNR_i denoting the average SNR for target i, the joint PDF of the in-phase and quadrature parts of the monopulse ratio for uncorrelated receiver errors is again given by equation (18.65), but now with

$$\begin{aligned} p_{y_I}\left(y_I|SNR_{\text{obs}}\right) &= \mathrm{N}\left(\frac{SNR_1\eta_1 + SNR_2\eta_2}{SNR_1 + SNR_2 + 1}, \frac{\kappa}{2SNR_0}\right) \\ p_{y_Q}\left(y_Q|SNR_{\text{obs}}\right) &= \mathrm{N}\left(0, \frac{q\kappa}{2SNR_0}\right). \\ \kappa &= \left[\frac{\sigma_d^2}{\sigma_s^2} + \frac{SNR_1\eta_1^2 + SNR_2\eta_2^2 + SNR_1 SNR_2(\eta_1 - \eta_2)^2}{SNR_1 + SNR_2 + 1}\right] \end{aligned} \tag{18.72}$$

Thus, y_I and y_Q are conditionally Gaussian, independent random variables. Note that equation (18.72) shows that the mean of y_I is a power or energy weighted average of the DOAs of the two targets. With $\bar{y}_I$ denoting the mean of y_I given SNR_0, the ML estimate of $\bar{y}_I$ for N independent pulses is given by equation (18.67). Since the y_{Ik} are Gaussian random variables, $\hat{y}_I$ is the minimum variance estimate of $\bar{y}_I$ and a Gaussian random variable with variance given by

$$\sigma^2_{\hat{y}_I} = \frac{\kappa}{2NY_N} \tag{18.73}$$

The MCRLB of $\bar{y}_I$ is given by

$$\sigma^2_{\bar{y}_I} \geq J(\bar{y}_I) = \frac{\kappa}{2N(SNR_1 + SNR_2 + 1)} \tag{18.74}$$

Figure 18-18 gives a comparison of the MCRLBs of $\bar{y}_I$ for a single target and two unresolved, equal-amplitude targets versus the total SNR in a single frequency (i.e., $N = 1$). To obtain $\bar{y}_I = 0$ in both cases, the single target was set at the antenna boresight, and the two targets were situated symmetrically about the boresight (i.e., $\eta_1 = -\eta_2$) and separated by one-half of a beamwidth by setting $\eta_1 = 0.4$ for $k_m = 1.6$. Figure 18-18 shows that a total SNR of 14 dB gives a MCRLB of 0.02 for the single-target case and 0.095 for the two-target case. This figure also shows that doubling the energy in a single-frequency waveform to obtain a total SNR of 17 dB gives an MCRLB of 0.01 for the single-target case and 0.09 for the two-target case. Thus, doubling the energy in a single-frequency pulse gives only a small reduction in the MCRLB for the two-target case. However, if the energy in the pulse is doubled and a second frequency is added so that two independent observations of the unresolved targets are obtained, the MCRLB would be reduced by about 50 percent because the errors are Gaussian (not shown in figure). Therefore, frequency agility is critical to improving the monopulse angle estimation when two unresolved targets are present.

The failure to detect the presence of the interference of a second target can be catastrophic to the performance of a tracking algorithm. A *generalized likelihood ratio test* (GLRT) for detection of the presence of unresolved Rayleigh targets is developed in [3]. Waveform design and antenna boresight pointing for AOA estimation of two unresolved Rayleigh targets with a monopulse radar is addressed further in [19–21].

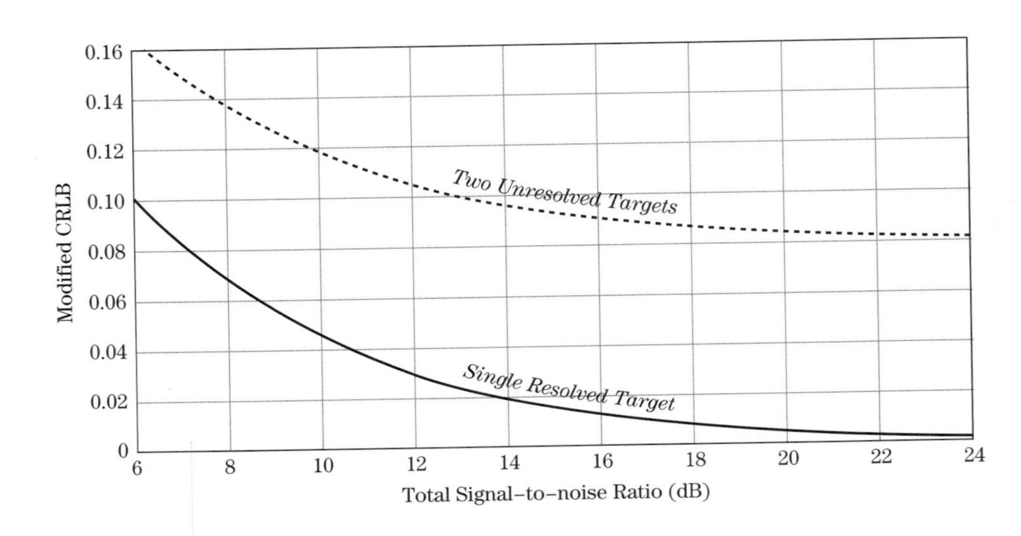

FIGURE 18-18 ■ MCRLBs of $\bar{y}_I$ for a single pulse and single target at the boresight and two targets separated by one-half beamwidth and symmetric about the boresight.

18.10 COORDINATE SYSTEMS

A radar system measures target position in spherical coordinates relative to the boresight of the radar antenna: range, R, azimuth angle, θ, and elevation angle, ϕ. These two angle coordinates are measured in orthogonal planes in a coordinate system centered on the antenna. If the x direction is the "horizontal" dimension of the antenna, the z direction is the "vertical" dimension of the antenna, and the y dimension is the normal to the antenna face (the boresight direction in a mechanically-scanned antenna), then elevation is measured from the horizontal x-y plane in the vertical plane containing the radar boresight, while azimuth is measured in the horizontal x-y plane containing the boresight. The measurement errors in these dimensions are generally independent of one another. Note that this is different from the usual definition of spherical coordinates in which the elevation angle is measured from vertical.

Frequently it is the case that it is desirable to track a target in a different coordinate system, typically a Cartesian (x-y-z) system centered at the radar platform or some other convenient reference point. The transformation from the azimuth-elevation spherical measurement coordinates back to antenna-centered Cartesian coordinates with the same origin is given by equation (18.75) and illustrated in Figure 18-19.

$$\begin{aligned} x &= R\cos\theta\cos\phi \\ y &= R\sin\theta\cos\phi \\ z &= R\sin\theta \end{aligned} \tag{18.75}$$

Equation (18.75) shows that measurement errors in Cartesian coordinates are a nonlinear combination of the spherical coordinate errors. Consequently, the errors in the three Cartesian coordinates are *coupled*: the error in x is not independent of the error in y and so forth. This complicates the design and analysis of optimal tracking algorithms [22]. In practice, the transformation of the uncertainty model in the measurement frame to the track filter reference frame is typically accomplished by linearizing the model in equation (18.75) at the current estimates of R, θ, and ϕ.

The antenna-centered coordinates can in turn be rotated and translated into alternative Cartesian systems, for instance into coordinates aligned with the platform flight path, or to East-North-Up coordinates. Coordinate systems and coordinate conversion are a major topic in track filtering and are addressed further in Chapter 19.

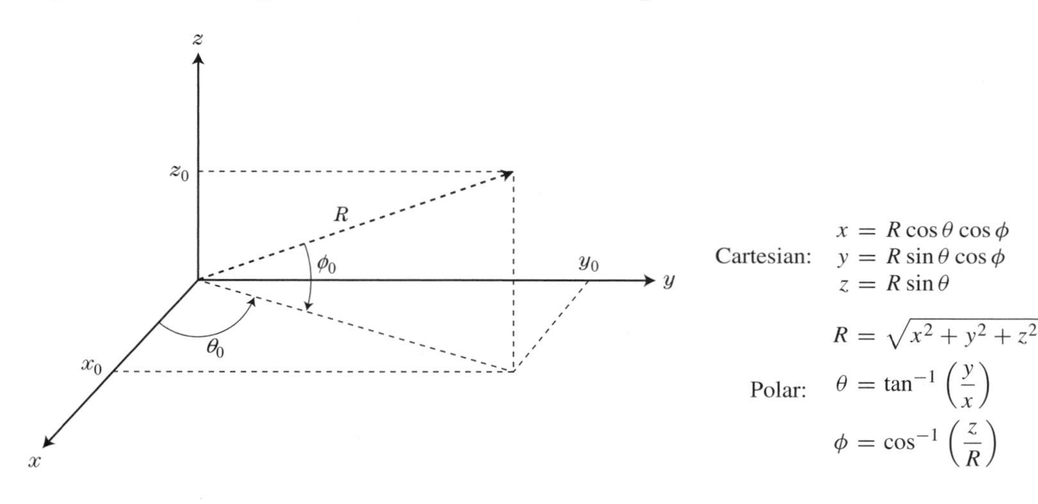

FIGURE 18-19 ■ Spherical and Cartesian representations of a three-dimensional vector.

18.11 FURTHER READING

Additional material and alternative presentations on radar measurement accuracy may be found in several sources. A description of number of practical approaches in radar measurement can be found in Barton and Ward's handbook [6]. Peebles [4] and Raemer [23] provide extensive theoretical results and derivations in radar measurement theory. A good introduction to radar system design and performance analysis is provided by the classic text [24]. Detailed background on the theory of measurement in the presence of noise is provided in the well-known book [25]. A popular forum for publication of current research in this area is the *IEEE Transactions on Aerospace Systems*.

18.12 REFERENCES

[1] Richards, M.A., *Fundamentals of Radar Signal Processing*, McGraw-Hill, New York, 2005.

[2] Blair, W.D., and Keel, B.M., "Radar Systems Modeling for Target Tracking," in *Multitarget-Multisensor Tracking: Advanced and Applications, Vol. III*, Ed. Y. Bar-Shalom and W.D. Blair, Artech House, Dedham, MA, 2000.

[3] Kay, S.M., *Fundamentals of Statistical Signal Processing, Vol. 1: Estimation Theory*, Prentice-Hall, Upper Saddle River, NJ, 1993.

[4] Peebles Jr., P.Z., *Radar Principles*, Wiley, New York, 1998.

[5] Levanon, N., *Radar Principles*, Wiley, New York, 1988.

[6] Barton, D.K., and Ward, H.R., *Handbook of Radar Measurement*, Artech House, Dedham, MA, 1984.

[7] Trunk, G.V., "Range Resolution of Targets Using Automatic Detectors," *IEEE Trans. Aerospace and Electronic Systems*, vol. AES-14, no. 5, pp. 750–755, Sept. 1978.

[8] Trunk, G.V., "Range Resolution of Targets," *IEEE Trans. Aerospace and Electronic Systems*, vol. AES-20, no. 6, pp. 789–797, Nov. 1984.

[9] Cook, C.E., and Bernfeld, M., *Radar Signals: An Introduction to Theory and Application*, Artech House, Norwood, MA, 1993.

[10] Jacobsen, E., and Kootsookos, P., "Fast Accurate Frequency Estimators," *IEEE Signal Processing Magazine*, pp. 123–125, May 2007.

[11] Doviak, R.J., and Zurnic, D.S., *Doppler Radar and Weather Observations*, 2d ed., Academic Press, San Diego, 1993.

[12] Blair, W.D., and Brandt-Pearce, M., "Estimation and Discrimination for Swerling Targets," *Proceedings Twenty-Eighth Southeastern Symposium on System Theory, 1996*, pp. 280–284, March 31–April 2, 1996.

[13] Sherman, S.M., *Monopulse Principles and Techniques*, Artech House, Dedham, MA, 1984.

[14] Galati, G., and Struder, F.A., "Angular Accuracy of the Binary Moving Window Radar Detector," *IEEE Trans. Aerospace and Electronic Systems*, vol. AES-18, no. 4, pp. 416–422, July 1982.

[15] Galati, G., and Struder, F.A., "Maximum Likelihood Azimuth Estimation Applied to SSR/IFF Systems," *IEEE Trans. Aerospace and Electronic Systems*, vol. AES-26, no. 1, pp. 27–43, Jan. 1990.

[16] Swerling, P., "Maximum Angular Accuracy of a Pulsed Search Radar," *Proc. of the IRE*, vol. 44, pp. 1145–1155, Sept. 1956.

[17] Seifer, A.D., "Monopulse-Radar Angle Measurement in Noise," *IEEE Trans. Aerospace and Electronic Systems*, vol. AES-30, pp. 950–957, July 1994.

[18] Gini, F., Reggiannini, R., and Mengali, U., "The Modified CramePr-Rao Bound in Vector Parameter Estimation," *IEEE Trans. on Communications*, vol. 46, no. 1, pp. 52–60, Jan. 1998.

[19] Blair, W.D., and Brandt-Pearce, M., "Monopulse DOA Estimation of Two Unresolved Rayleigh Targets," *IEEE Trans. Aerospace and Electronic Systems*, vol. AES-37, no. 2, pp. 452–469, Apr. 2001.

[20] Willett, P.D., Blair, W.D., and Zhang, X., "The Multitarget Monopulse CRLB for Matched Filter Samples," *IEEE Trans. Signal Processing*, vol. 55, no. 8, pp. 4183–4198, August 2007.

[21] Zhang, X., Willett, P., and Bar-Shalom, Y., "Monopulse Radar Detection and Localization of Multiple Unresolved Targets via Joint Bin Processing," *IEEE Trans. Signal Processing*, vol. 53, no. 4., pp. 122–1236., Apr. 2005

[22] Brookner, E., *Tracking and Kalman Filtering Made Easy*, John Wiley & Sons, Hoboken, NJ, 1998.

[23] Raemer, H.R., *Radar Systems Principles*, CRC Press, Boca Raton, FL, 1997.

[24] Barton, D.K., *Radar System Analysis and Modeling*, Artech, New York, 2005.

[25] Van Trees, H.L., *Detection, Estimation, and Modulation Theory*, John Wiley & Sons, Inc., New York, 1968.

18.13 | PROBLEMS

1. If every RF power measurement has a calibration uncertainty of 0.1 dB, discuss the relative accuracy of RF path calibration by separately calibrating individual components versus doing a single end-to-end calibration.
2. Show that the general centroid estimator in equation (18.35) is unbiased when estimating a fixed parameter in Gaussian noise. Is this result changed for a power law detector?
3. Show the steps in deriving the CRLB equation (18.23) for estimating a constant in independent Gaussian noise.
4. Find the joint probability density function $p(y_1, \ldots y_N|m)$ for N observations $y_k = m + w_k$ of a constant m in independent Gaussian noise given a particular value of m. Determine the maximum likelihood estimate by solving for the argument of $p(y_1, \ldots y_N|m)$ that maximizes the joint probability function. Determine if the maximum likelihood estimate is biased.
5. Derive the CRLB for estimating the RCS for Swerling 2 target in the presence of independent Gaussian noise using the distribution for the measured power given in equation (18.53).
6. For problem 5, show that the maximum likelihood estimator for the signal power is the sample mean.
7. Plot the MCRLB of the weight sums of N monopulse ratios for a single target versus the various parameters in equation (18.70). Comment on the effects of parameters. For an ideal monopulse system, what would these parameter values be and what would the MCRLB be?
8. What is the minimum SNR required in a 40 MHz bandwidth LFM radar to ensure a single-pulse range precision of 30 m?
9. Repeat problem 8 for an unmodulated CW pulse of duration 1 μs.
10. What is the minimum SNR required for an unbiased phase estimator that must produce a minimum error of 10 degrees for $N = 1$ and $N = 10$ pulses? The waveform is a simple pulse having a time-bandwidth product $\beta\tau = 1$.

11. What is the value of the CRLB for frequency estimation for a radar with an SNR of 20 dB, a pulse length of 10 μs, and a single ($N = 1$) pulse? What is the value for $N = 20$ pulses?

12. Compute the rms bandwidth of (a) a CW pulse with a rectangular envelope, and (b) an LFM chirp. For part (b), assume that the frequency spectrum of the chirp is rectangular. What are the implications of the signal modulation bandwidth on the bound on time-delay measurement precision?

13. Derive equations (18.26) and (18.27) assuming the signal *mean time duration* $\bar{\tau}$ and *mean bandwidth* $\bar{B}$ are zero, where

$$\bar{\tau} = \frac{\int_{-\infty}^{\infty} t|s(t)|^2 dt}{\int_{-\infty}^{\infty} |s(t)|^2 dt} \qquad \bar{B} = \frac{\int_{-\infty}^{\infty} 2\pi f |S(f)|^2 df}{\int_{-\infty}^{\infty} |S(f)|^2 df}$$

CHAPTER 19

Radar Tracking Algorithms

W. Dale Blair

Chapter Outline

19.1	Introduction	713
19.2	Basics of Track Filtering	719
19.3	Kinematic Motion Models	746
19.4	Measurement Models	751
19.5	Radar Track Filtering	757
19.6	Measurement-to-Track Data Association	760
19.7	Performance Assessment of Tracking Algorithms	766
19.8	Further Reading	767
19.9	References	768
19.10	Problems	770

19.1 INTRODUCTION

When radar systems are discussed in the literature or the remainder of this text, it is in the context of a sensor providing observations of the environment. While some of those measurements are responses from coherent waveforms of finite duration, the environment is treated as stationary with at most linear motion on the targets. Target tracking addresses the integration of measurements into a longer-term picture as illustrated in Figure 19-1. Target tracking is separated into two parts: *track filtering* and *measurement-to-track data association*.

Track filtering is the process of estimating the trajectory (i.e., position, velocity, and possibly acceleration) of a track from measurements (e.g., range, bearing, and elevation) that have been assigned to that track. As shown in Figure 19-1, a trajectory estimate always has an associated uncertainty that is characterized by the covariance of the estimate. The position and velocity estimates are used to predict the next measurement and, in the case of an electronically scanned (or phased array) radar, the position of the beam for the next measurement as illustrated in Figure 19-2.

Measurement-to-track data association (or data association) is the process of assigning a measurement to an existing track or as a detection of a newly found target or false signal. In Figure 19-2, three measurements (or threshold exceedances) occur in the dwell. Two measurements fall in the validation region (or gate) for the predicted measurement of the track, while the third measurement falls outside the validation region. Typically, the

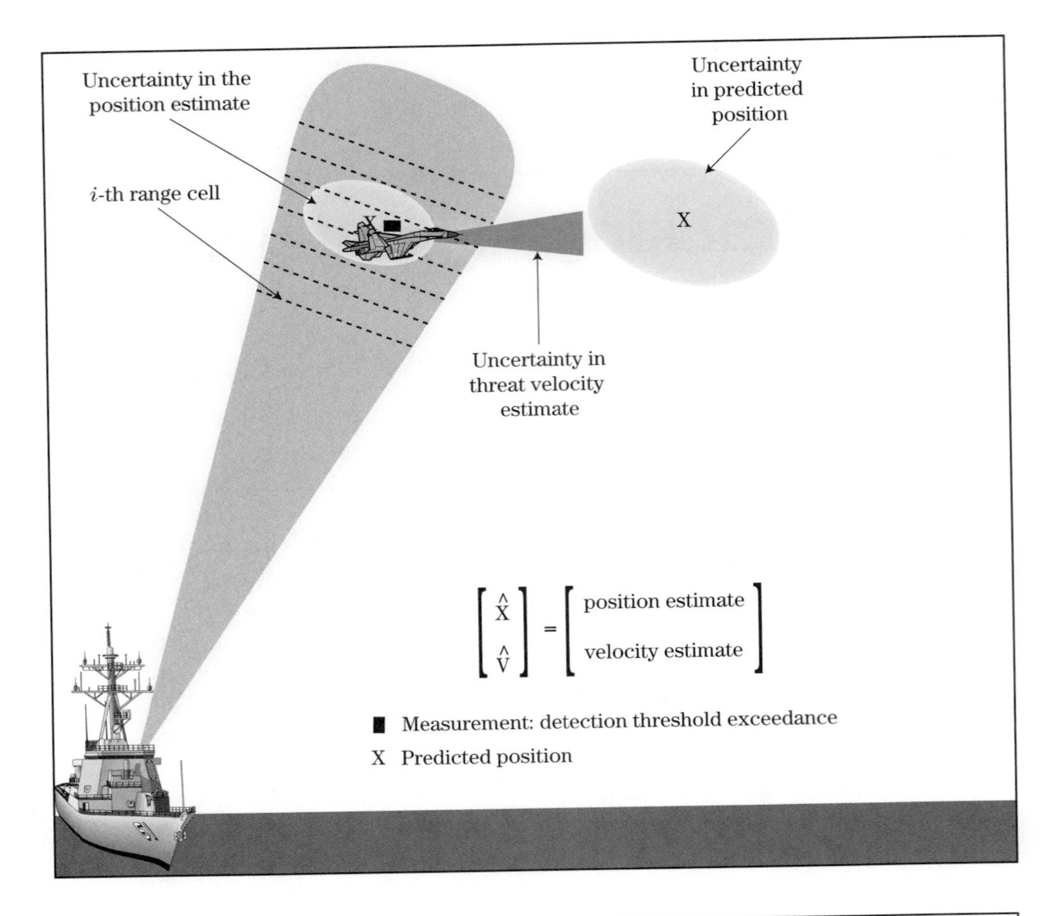

FIGURE 19-1 ■ Tracking and prediction for a phased array radar.

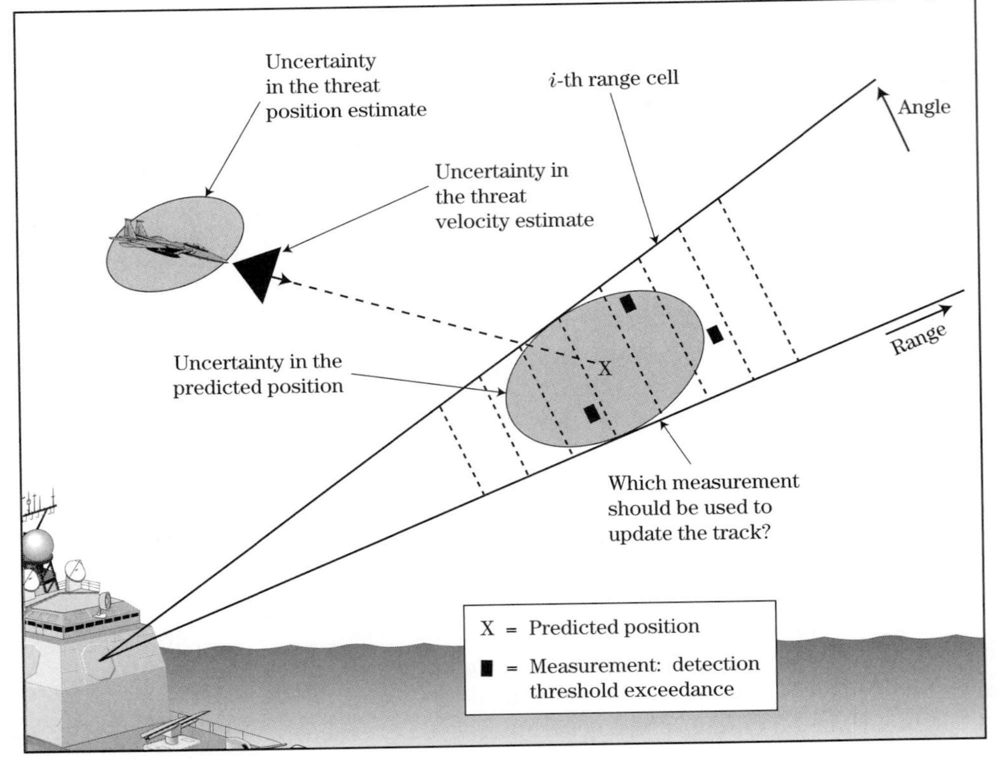

FIGURE 19-2 ■ Gating and assignment algorithms couple new measurements to a track.

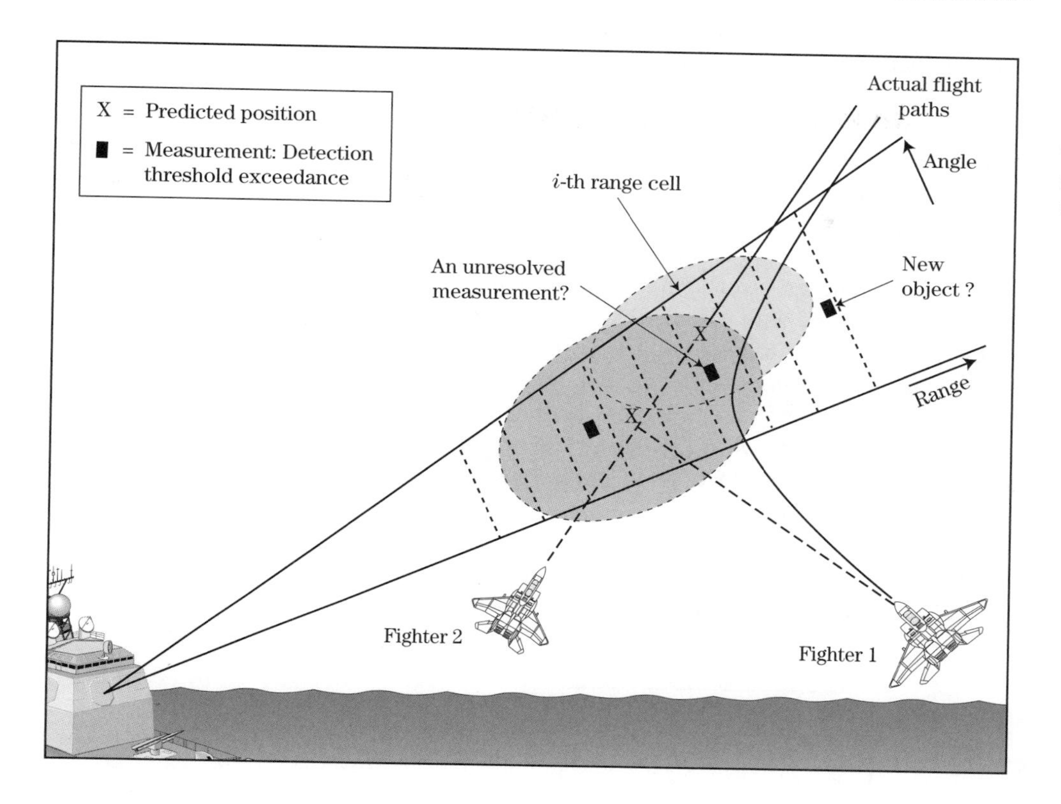

FIGURE 19-3 ■ Resolution and measurement-to-track association are critical for accurate tracking.

measurement in the validation region that is closest to the predicted measurement is used to update the estimates of position and velocity of the track, while the second measurement in the validation gate and the measurement outside the validation region are considered to be false measurement or used to initiate a new tentative track.

The track filtering and measurement-to-track association is further complicated by target maneuvers, closely spaced targets, and limited resolution of the radar. In Figure 19-2, a target maneuver could have easily switched the roles of the three measurements in the assignment. In this case, the measurement outside the validation gate is likely the target-originated measurement, while the measurements inside the gate are false alarms. Thus, validation gates that account for target maneuvers are critical. Figure 19-3 shows measurements that further illustrate the tracking and data association problems in a multitarget environment. Two fighters are shown to be entering into formation. The data association problem is complicated by the overlap in the uncertainty in the predicted positions of the two fighters at the measurement time. The track of Fighter 1 has two possible measurements, while the track of Fighter 2 has one possible measurement. Also, one of the measurements could represent a new target that is not currently under track. In the measurement-to-track assignment problem, the measurement in both gates of the predicted positions of the two targets could be a merged measurement of the two aircraft or a resolved measurement of one of the two aircraft. If this measurement is assigned as a resolved measurement to the track on Fighter 1, a misdetection of Fighter 2 will be declared. On the other hand, if the measurement is assigned as a resolved measurement to Fighter 2, the other measurement will be assigned to Fighter 1. From the actual trajectories in Figure 19-3, the correct measurement-to-track assignment is the assignment of the measurement in both validation gates to both aircraft as a merged measurement and the other possible measurement as a false alarm. If that merged measurement is assigned to Fighter 2 and the other measurement is assigned to Fighter 1, Fighter 1 will not be found near the predicted position in the following radar dwells and the track of Fighter 1 is likely to be lost. Various approaches to data association are given in [1,2].

In much of the literature on radar systems, target tracking is viewed from a single signal or single target perspective in that a single signal is selected by time of arrival or frequency, and an error signal is generated for tracking. For example, in [3] range tracking is achieved with early and late gates, and the difference in the signal amplitudes in the two gates are used to adjust the range estimate. For a radar system to perform *multiple target tracking* (MTT), the data processing must be treated differently. The role of the target tracking in a MTT radar is illustrated in Figure 19-4. The most significant differences with the view of [3] is the presence of the measurement-to-track assignment and track management functions in which multiple measurements are considered for the assignment to multiple tracks.

Targets, clutter, and electronic countermeasurements (ECMs) from the environment and false alarms from the receiver noise can cause tracks to be formed in the radar system. In target tracking, *track* denotes the position and velocity estimate that represents

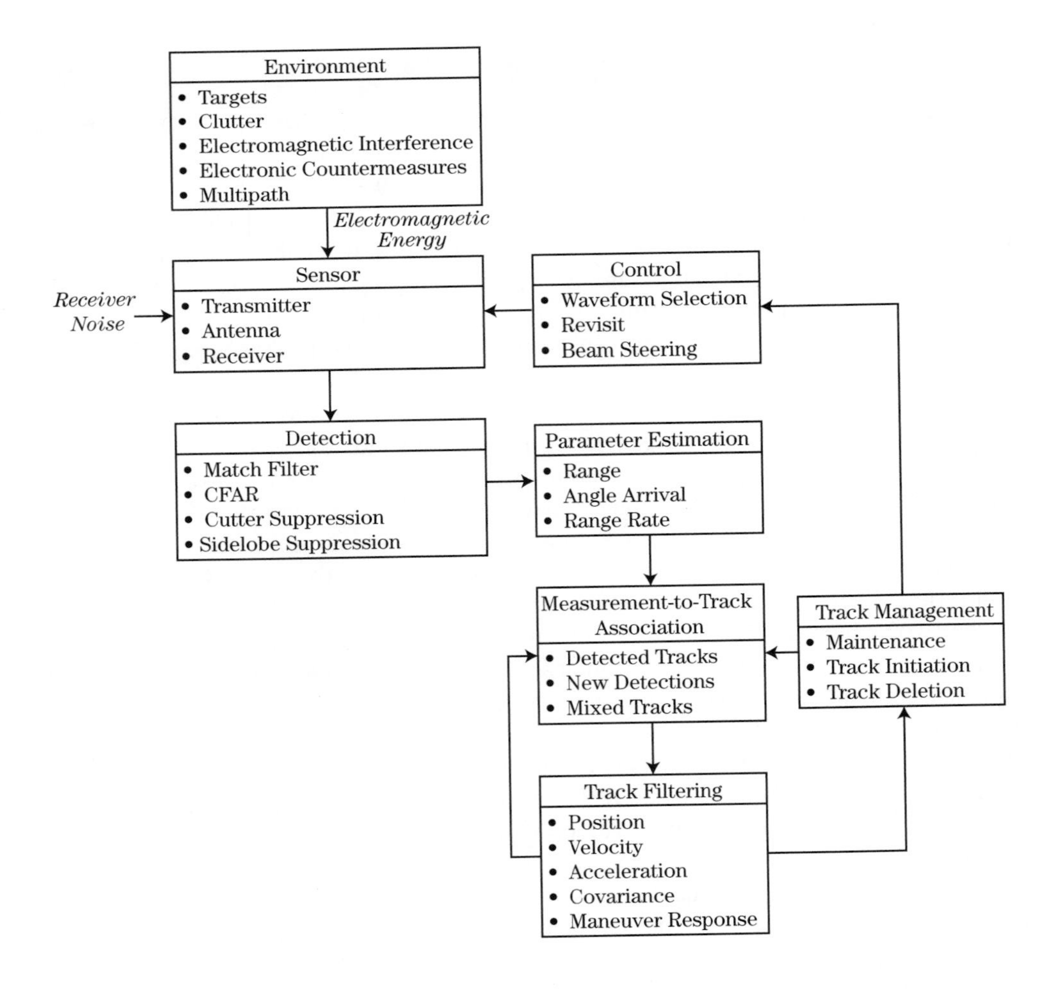

FIGURE 19-4 ■ Target tracking algorithms are highly sophisticated in modern radar systems.

something perceived to be in the environment, while *target* or *truth object* denotes something actually in the environment that should be tracked by the radar. Tracks are based solely on data collected by the radar, and the number of tracks held by a radar system can be larger (extra) or smaller (missed) than the number of targets. Ideally, the number of tracks will equal the number of targets. Clutter rejection techniques such as Doppler processing are used to reduce the clutter in a radar system. The radar returns are also monitored to detect the presence of ECM, and special processing is employed to minimize the number of false tracks introduced.

In the detection processing, signal amplitude and short time signal correlations are examined to reduce the number of false alarms and clutter detections reported to the tracker. Higher detection thresholds for a given signal-to-noise ratio (SNR) are used to the reduce the effects of false alarms. However, the higher detection thresholds also reduce the probability of detecting the true target. Constant false alarm processing (CFAR) is a classical approach to limiting the number of false detections. In CFAR processing, the detection threshold for a range cell or matched filter output is adjusted to reflect the noise level of the nearby range cells. Special processing can also be employed to prevent the declaration of threshold exceedances due to range sidelobes as detections of targets. Space-time adaptive processing (STAP) or Doppler processing can be employed to reduce the effects of clutter.

In the parameter estimation, estimates of the range, angle of arrival, and possibly range rate are produced for all of the detections and provided as measurements to the measurement-to-track association function and track filter. In the measurement-to-track association, measurements are assigned to existing tracks, and unassigned measurements are declared as a false alarms or new targets. Tracks that are not assigned a measurement are declared as missed. A track miss does not necessarily imply a missed detection, because a measurement from the target associated with the track could have been assigned incorrectly to a different track or the measurement could have failed the gating test with the correct track. Measurements that are assigned to existing tracks are provided to the track filter for updating the estimates of position and velocity. Unassigned measurements are considered with other unassigned measurements at different times for initiation of new tracks. Track misses are monitored for deleting a track or declaring a it as lost. The track management function monitors the number of measurements assigned to a track and declares the maturity of the track as a tentative or firm. It also monitors track misses and declares a track as lost for removal from the list of confirmed tracks. The control function monitors the list of tentative and firm tracks and clutter along with the track accuracies to schedules the timing and waveform for radar dwells.

As illustrated in Figure 19-5, tracking targets can be viewed as four functional areas: detection, resolution, association, and filtering. As shown in the upper left corner of Figure 19-5, detection is the first step in tracking a target. Detection of target echoes occurs in the presence of false alarms and clutter. The detection threshold that sets the false alarm rate and probability of detection also sets the performance limits of the tracker. Clutter is often countered with Doppler processing for clutter rejection, a clutter map, or tracking of moving clutter. As indicated in Figure 19-5, the second step of tracking targets is resolving the signals of closely spaced objects or nearby clutter into separate/isolated measurements. In radar systems, closely spaced targets are typically resolved in either range, angle, or range rate. Radars tend to be very effective in resolving closely spaced objects in range with waveforms of wider bandwidths and resolving targets from background clutter with Doppler processing. The third step of tracking targets is assigning measurements to tracks.

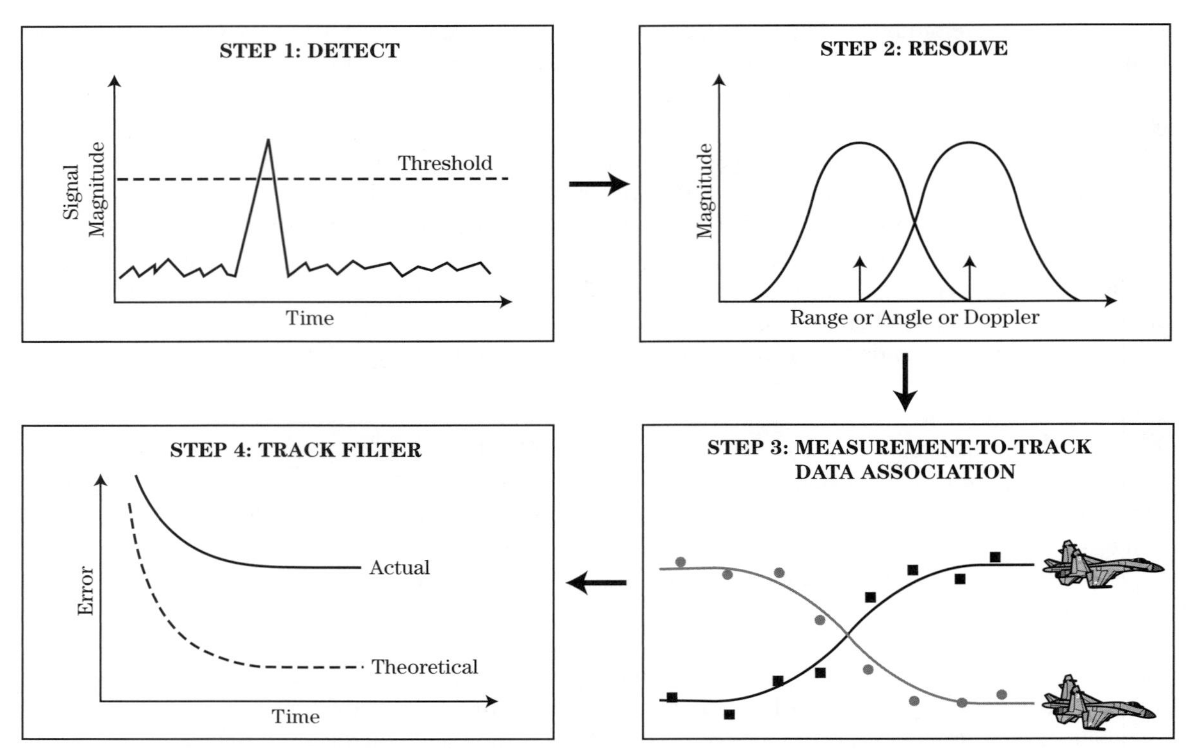

FIGURE 19-5 ■ Functional tracking components begin with detections and end with track filtering before the process repeats.

This involves assigning measurements to existing tracks, initiating new tracks, and deleting tracks that are not perceived to represent a truth object. As shown in Figure 19-5, the fourth step involves the track filtering to estimate the position, velocity, and possibly acceleration of the target. The performance of the track filter is often analyzed under ideal conditions and the performance is found to be below expectations when tested in the field.

This chapter addresses the fundamentals of tracking a single target with a radar, while the advanced topics associated with tracking multiple targets are addressed in [4]. The basics of *kinematic state estimation* (or track filtering) are discussed in Section 19.2. The section discusses the parametric and stochastic state estimation approaches to track filtering and addresses the use of least squares estimation, *Kalman filtering*, *alpha-beta filters*, and the *interacting multiple model* (IMM) estimator for tracking maneuvering targets. The kinematic models commonly used in track filtering are reviewed in Section 19.3, while the measurement models are discussed in Section 19.4. The application of the estimation methods discussed in Section 19.2 to the models presented in Sections 19.3 and 19.4 are discussed in Section 19.5. Measurement-to-track data association for a single target in the presence of false signals is discussed in Section 19.6, where the topics include measurement validation and gating, *strongest neighbor tracking*, *nearest neighbor tracking*, and the *probabilistic data association filter* (PDAF). In Section 19.7, performance assessment of track filtering algorithms is discussed. Key results and suggestions for further reading are provided in Section 19.8.

19.2 BASICS OF TRACK FILTERING

Given the decision to assign a measurement to a particular track, an update to the kinematic state estimate is performed for the track with that measurement. For most radar systems, the target motion is modeled in Cartesian coordinates, while radar measurements are typically in polar or spherical coordinates. For this basic treatment of track filtering, the kinematic state of the target will be represented in a single Cartesian coordinate, and the radar measurement will be treated as a linear observation of that kinematic state.

Let $\boldsymbol{X}_k$ denote the kinematic state vector at time t_k and $\boldsymbol{Z}_k$ denote the measurement at time t_k. The state estimate for t_k given measurements through t_j is denoted by $\boldsymbol{X}_{k|j}$. Thus, when the state vector has a single subscript, it denotes a truth or modeled value, and when it has two subscripts, it represents an estimate. The state estimate $\boldsymbol{X}_{k|k}$ is referred to as the filtered state estimate, while $\boldsymbol{X}_{k|k-1}$ is referred to as the one-step prediction of the state. The estimate $\boldsymbol{X}_{k|k+1}$ is referred to as the one-step smoothed estimate.

Track filtering algorithms typically fall into one of two groups. The first group uses a parametric estimation approach that presumes a perfect model for the target motion and the time period over which the model is applied is limited to prevent distortion of the data. In this approach, the covariance of the state estimate (or track) will approach zero as more data are processed. As the covariance of the track error approaches zero, the gain for processing new data will approach zero. When this processing gain reaches a very small number, all future data will be essentially ignored. Thus, since all motion models are imperfect in practice, the time period for which the perfect model is applied is limited to alleviate the distortion that results when new data are ignored. Least squares or maximum likelihood (ML) estimation are examples of the parametric approach to track filtering.

The second group uses a stochastic state estimation approach that presumes an imperfect model for the target motion. In this approach, the target motion model includes a random process, and a perfect estimate of the kinematic state is not possible. In other words, the covariance of the track does not approach zero as the window of data expands. As the model is applied over an expanding window of measurements, the covariance of the track settles to a stable, slowing changing value. If the measurement rate is fixed and the data quality is uniform, the filter will achieve "steady-state" conditions in which the covariance is the same value after each measurement update. The Kalman filter and alpha-beta filter are examples of the the stochastic state estimation approach.

One of the most critical items of a tracking system that supports any automatic decision system is track filter consistency. A track filter is considered to be *consistent* if the following three criteria are satisfied [5]:

1. The state errors should be acceptable as zero mean and have magnitudes commensurate with the state covariance as yielded by the filter.
2. The *innovations* (i.e., residuals or difference between the measurement and predicted measurement) should be acceptable as zero mean and have magnitudes commensurate with the innovation covariance as yielded by the filter.
3. The innovations should be acceptable as a white error process.

In other words, a track filter that is consistent produces a state error covariance that accurately represents the errors in the state estimate. Thus, track filter consistency is critical for effective fusion of data from multiple sensors with diverse accuracies. Maneuvering targets pose a particularly difficult challenge to achieving track filter consistency. In fact, [6] includes an illustrative example that shows that more data does not necessary mean better estimates when a Kalman filter is used to track a maneuvering target. Using a Kalman filter to track a maneuvering target will not provide reliable tracking performance because the loss of track filter consistency prevents reliable decision making for simultaneously adapting the filter parameters and performing data association.

In this section, parametric and stochastic state approaches are discussed in further detail.

19.2.1 Parametric Estimation

For a typical parametric approach, the dynamical motion and measurement equations for tracking a target are given by

$$\boldsymbol{X}_{k+1} = \mathbf{F}_k \boldsymbol{X}_k \tag{19.1}$$

$$\mathbf{Z}_k = \mathbf{H}_k \boldsymbol{X}_k + \boldsymbol{w}_k \tag{19.2}$$

where $\boldsymbol{X}_k$ is the state vector for the target at time t_k, $\mathbf{F}_k$ defines the kinematic motion constraint between between time t_k and time t_{k+1}, $\mathbf{Z}_k$ is the sensor measurement vector at time t_k, $\mathbf{H}_k$ is the output matrix that relates the kinematic state to the measurement, and $\boldsymbol{w}_k$ is the sensor measurement error that is considered to be a zero-mean, Gaussian random variable with covariance $\mathbf{R}_k$, denoted $\boldsymbol{w}_k \sim \mathrm{N}(0, \mathbf{R}_k)$.

Figure 19-6 gives the position of a target trajectory in a single coordinate versus time and measurements for times t_1 through t_9. Figure 19-6 also includes an illustration of a parametric estimate based on constant velocity motion for the target. Note that the slope of the line illustrating the velocity estimate is constant indicating a velocity estimate that is fixed for all time t_k. The constant velocity estimate of the trajectory poorly represents the true trajectory near t_5 and t_9. If these errors are acceptable for the application of interest, the parametric estimator based on constant velocity is an acceptable solution. However, if the time expands beyond t_9, it is likely to become unacceptable.

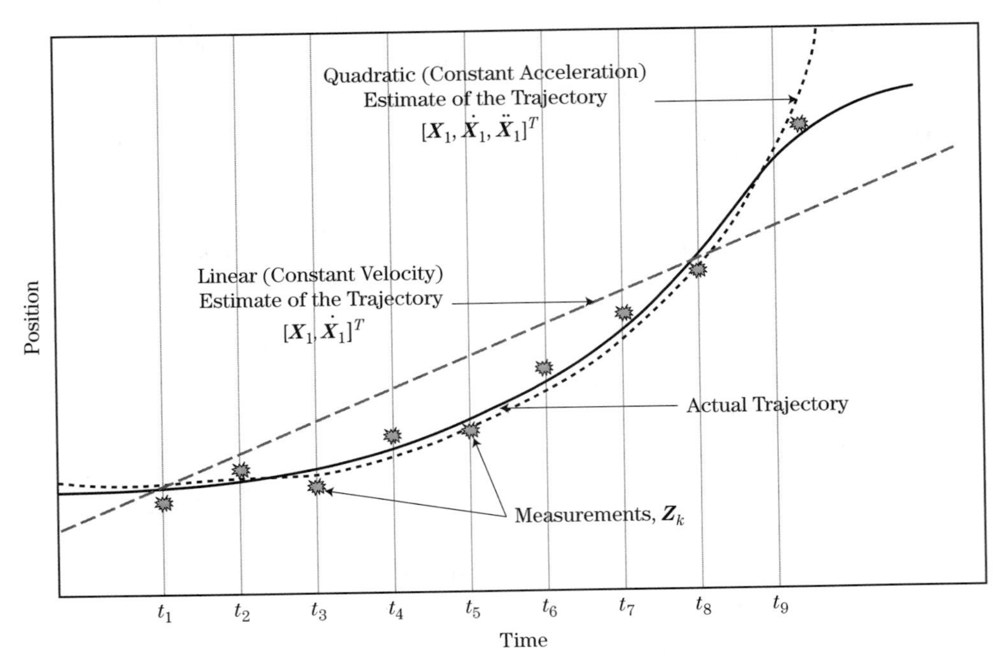

FIGURE 19-6 ■ Parametric approach to track filtering performs curve fitting to measurement data.

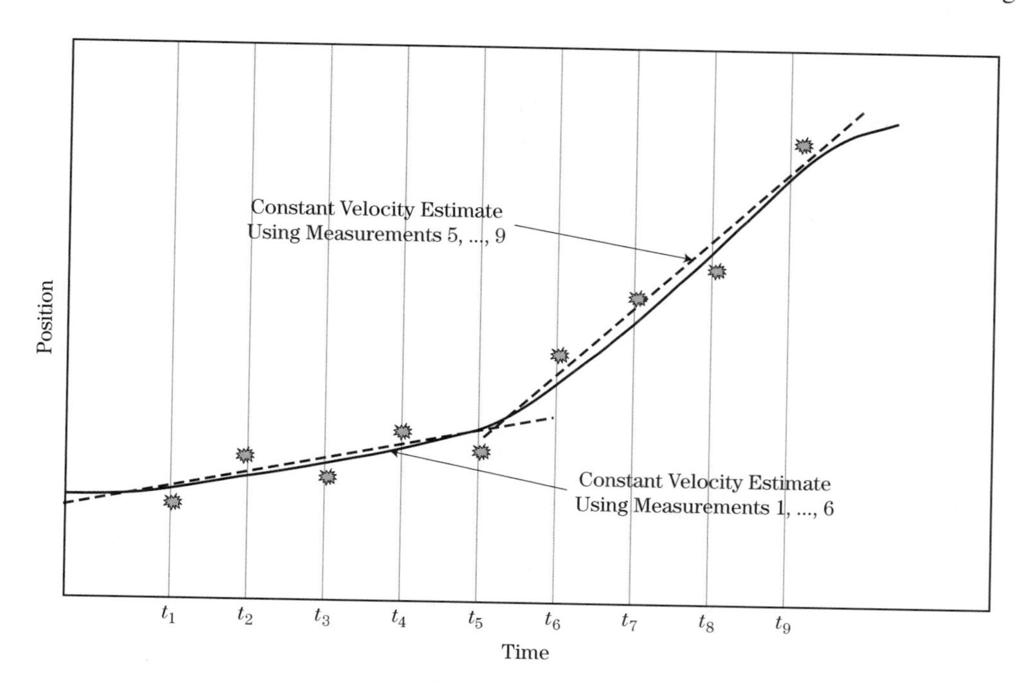

FIGURE 19-7 ■ Limiting memory in parametric approach to track filtering accommodates modeling errors.

Often, the first response to a constant velocity estimator that gives unacceptable performance is to use a higher order estimator (e.g., constant acceleration estimator). Figure 19-6 includes an illustration of a parametric estimate based on constant acceleration motion for the target. The constant acceleration estimator gives smaller errors than the constant velocity estimator near t_5 and t_9. However, estimating an additional parameter in this case results in estimates of position and velocity with higher variances. Furthermore, as the final time expands beyond t_9, the constant acceleration estimate is likely to become unacceptable.

In the application of parametric estimators, the period of time for application of the "perfect model" (i.e., the memory of the estimator) is limited to reduce the mismatch between the true trajectory and the assumed model. Figure 19-7 illustrates the application of the constant velocity estimator with a memory of five measurements. The state estimate for each time is computed from the data in a window of five measurements prior to and including the time of the estimate. This estimator is often referred to as a *sliding window estimator*. Note that the estimates based on fewer measurements will have a higher variance, but the distortions of the underlying true trajectory will be less.

Using (19.1) and (19.2), the measurement at t_N can be rewritten as

$$\boldsymbol{Z}_N = \mathbf{H}_N \boldsymbol{X}_N + \boldsymbol{w}_N = \mathbf{H}_k \mathbf{F}_{N-1} \mathbf{F}_{N-2} \ldots \mathbf{F}_1 \boldsymbol{X}_1 + \boldsymbol{w}_k \tag{19.3}$$

Given N measurements between t_1 and t_N, an augmented measurement equation can be written in terms of $\boldsymbol{X}_1$ as

$$\bar{\boldsymbol{Z}}_1^N = \begin{bmatrix} \boldsymbol{Z}_1 \\ \boldsymbol{Z}_2 \\ \vdots \\ \boldsymbol{Z}_{N-1} \\ \boldsymbol{Z}_N \end{bmatrix} = \begin{bmatrix} \mathbf{H}_1 \\ \mathbf{H}_2 \mathbf{F}_1 \\ \vdots \\ \mathbf{H}_{N-1} \mathbf{F}_{N-2} \mathbf{F}_{N-3} \ldots \mathbf{F}_1 \\ \mathbf{H}_N \mathbf{F}_{N-1} \mathbf{F}_{N-2} \ldots \mathbf{F}_1 \end{bmatrix} \boldsymbol{X}_1 + \begin{bmatrix} \boldsymbol{w}_1 \\ \boldsymbol{w}_2 \\ \vdots \\ \boldsymbol{w}_{N-1} \\ \boldsymbol{w}_N \end{bmatrix} = \bar{\mathbf{H}}_N \boldsymbol{X}_1 + \bar{\boldsymbol{W}}_N \tag{19.4}$$

In this case, $\mathrm{E}\{\bar{\boldsymbol{W}}_N\} = 0$, and

$$\bar{\mathbf{H}}_N = \begin{bmatrix} \mathbf{H}_1 \\ \mathbf{H}_2\mathbf{F}_1 \\ \vdots \\ \mathbf{H}_{N-1}\mathbf{F}_{N-2}\mathbf{F}_{N-3}\ldots\mathbf{F}_1 \\ \mathbf{H}_N\mathbf{F}_{N-1}\mathbf{F}_{N-2}\ldots\mathbf{F}_1 \end{bmatrix} \tag{19.5}$$

$$\bar{R}_N = \mathrm{E}\{\bar{\boldsymbol{W}}_N\bar{\boldsymbol{W}}_N^T\} = \begin{bmatrix} \mathbf{R}_1 & \mathbf{0} & \ldots & \mathbf{0} \\ \mathbf{0} & \mathbf{R}_2 & \ldots & \vdots \\ \vdots & & \ddots & \\ \mathbf{0} & \ldots & \mathbf{0} & \mathbf{R}_N \end{bmatrix} \tag{19.6}$$

In this case, $\bar{\mathbf{R}}_N$ is a block diagonal matrix because the measurement errors between any two times are assumed to be zero mean and independent. When the errors are correlated across time, the off-diagonal elements of the matrix are nonzero. The minimum mean squared error (MMSE) for Gaussian errors or weighted least squares estimate (LSE) [7] of $\boldsymbol{X}_1$ given measurements from t_1 to t_N is given by

$$\boldsymbol{X}_{1|N} = \left(\bar{\mathbf{H}}_N^T\bar{\mathbf{R}}_N^{-1}\bar{\mathbf{H}}_N\right)^{-1}\bar{\mathbf{H}}_N^T\bar{\mathbf{R}}_N^{-1}\bar{\mathbf{Z}}_1^N \tag{19.7}$$

and the covariance of the estimate is given by

$$\mathbf{P}_{1|N} = \left(\bar{\mathbf{H}}_N^T\bar{\mathbf{R}}_N^{-1}\bar{\mathbf{H}}_N\right)^{-1} \tag{19.8}$$

Note that the covariance will reflect only the actual errors in the state estimate to the degree that the kinematic model (19.1) is accurate and the measurement errors are zero-mean Gaussian with covariance $\bar{\mathbf{R}}_N$. The $\boldsymbol{X}_{1|N}$ and $\mathbf{P}_{1|N}$ are smoothed estimates. For a state estimate and covariance at time t_j, $t_1 < t_j \le t_N$,

$$\boldsymbol{X}_{j|N} = \mathbf{F}_{j-1}\ldots\mathbf{F}_1\boldsymbol{X}_{1|N} \tag{19.9}$$

$$\mathbf{P}_{j|N} = \mathbf{F}_{j-1}\ldots\mathbf{F}_1\mathbf{P}_{1|N}\mathbf{F}_1^T\ldots\mathbf{F}_{j-1}^T \tag{19.10}$$

In this case, $\boldsymbol{X}_{j|N}$ is an MMSE estimate of $\boldsymbol{X}_j$. For the Gaussian errors, $\boldsymbol{X}_{j|N}$ is also the ML estimate of $\boldsymbol{X}_j$.

For the parametric approach, constant velocity and constant acceleration are the two most common assumptions for the motion of the target. Thus, constant velocity filtering and constant acceleration filtering are discussed here.

19.2.1.1 Constant Velocity Filtering

Let x_k denote the position of the x coordinate of the target at time t_k and $\dot{x}_k$ denote the velocity of the x coordinate of the target at time t_k. For a target moving with a constant velocity, the kinematic state at time t_{k+1} in terms of the state at time t_k is given by

$$x_{k+1} = x_k + (t_{k+1} - t_k)\dot{x}_k \tag{19.11}$$

$$\dot{x}_{k+1} = \dot{x}_k \tag{19.12}$$

This evolution of the target state can be written as

$$\boldsymbol{X}_{k+1} = \begin{bmatrix} x_{k+1} \\ \dot{x}_{k+1} \end{bmatrix} = \begin{bmatrix} 1 & (t_{k+1} - t_k) \\ 0 & 1 \end{bmatrix} \begin{bmatrix} x_k \\ \dot{x}_k \end{bmatrix} = \mathbf{F}_k \boldsymbol{X}_k \tag{19.13}$$

For a measurement of the position of the target at t_k,

$$Z_k = x_k + w_k = \begin{bmatrix} 1 & 0 \end{bmatrix} \begin{bmatrix} x_k \\ \dot{x}_k \end{bmatrix} + w_k = \mathbf{H}_k \boldsymbol{X}_k + w_k \tag{19.14}$$

where w_k is a scalar Gaussian random variable with variance σ_{wk}^2 and $\mathbf{H}_k = \begin{bmatrix} 1 & 0 \end{bmatrix}^T$. In this case, (19.5) and (19.6) are given by

$$\bar{\mathbf{H}}_N = \begin{bmatrix} 1 & 0 \\ 1 & t_2 - t_1 \\ \vdots & \vdots \\ 1 & t_{N-1} - t_1 \\ 1 & t_N - t_1 \end{bmatrix} \tag{19.15}$$

$$\bar{\mathbf{R}}_N = \begin{bmatrix} \sigma_{w1}^2 & 0 & \dots & 0 \\ 0 & \sigma_{w2}^2 & \dots & 0 \\ \vdots & & \ddots & \vdots \\ 0 & 0 & \dots & \sigma_{wN}^2 \end{bmatrix} \tag{19.16}$$

Thus, $\bar{\mathbf{H}}_N$ and $\bar{\mathbf{R}}_N$ have rather simple forms that can be easily manipulated. For N measurements with uniform sampling in time with period T and equal variances $\sigma_{wk}^2 = \sigma_w^2$,

$$\bar{\mathbf{H}}_N = \begin{bmatrix} 1 & 0 \\ \vdots & \vdots \\ 1 & (N-2)T \\ 1 & (N-1)T \end{bmatrix} \tag{19.17}$$

$$\bar{\mathbf{R}}_N = \sigma_w^2 \mathbf{I}_N \tag{19.18}$$

where $\mathbf{I}_N$ is an N dimensional identity matrix. For this case, the smoothed state estimate and covariance for t_1 are given by

$$\boldsymbol{X}_{1|N} = \frac{2}{N(N+1)} \begin{bmatrix} \displaystyle\sum_{k=1}^{N} (2N + 2 - 3k) Z_k \\ \displaystyle\frac{3}{(N-1)T} \sum_{k=1}^{N} (2k - N - 1) Z_k \end{bmatrix} \tag{19.19}$$

$$\mathbf{P}_{1|N} = \frac{2\sigma_w^2}{N(N+1)} \begin{bmatrix} 2N - 1 & -\dfrac{3}{T} \\ -\dfrac{3}{T} & \dfrac{6}{(N-1)T^2} \end{bmatrix} \tag{19.20}$$

The filtered state estimate and covariance for t_N are given by

$$\mathbf{X}_{N|N} = \frac{2}{N(N+1)} \begin{bmatrix} \sum_{k=1}^{N} (3k - N - 1) Z_k \\ \frac{3}{(N-1)T} \sum_{k=1}^{N} (2k - N - 1) Z_k \end{bmatrix} \tag{19.21}$$

$$\mathbf{P}_{N|N} = \frac{2\sigma_w^2}{N(N+1)} \begin{bmatrix} 2N - 1 & \frac{3}{T} \\ \frac{3}{T} & \frac{6}{(N-1)T^2} \end{bmatrix} \tag{19.22}$$

Comparing the equations for the velocity estimates in (19.19) and (19.21) shows that they are identical for both times. However, the position estimates are different. A comparison of the expressions for the elements of the covariances in (19.20) and (19.22) shows that they are almost identical. The variances of the position and velocity estimates are identical. However, the covariance in the (1,2) or (2,1) elements of (19.20) for the smoothed position and velocity estimates in (19.19) is negative, while the covariance in the (1,2) or (2,1) elements of (19.22) for the filtered position and velocity estimates in (19.21) is positive.

A scalar coordinate of the trajectory of a maneuvering target is given in Figure 19-8 along with 120 measurements at integer times. The measurement errors have a standard deviation of 120 m. The trajectory starts at t_0 seconds with a initial position of $x_0 = 150$ m and initial velocity of $\dot{x}_0 = -20$ m/s and it moves with constant velocity until $t = 20$ s. From $t = 20$ until $t = 30$ s, the target maneuvers with acceleration of $\ddot{x} = 10$ m/s^2. From $t = 30$ until $t = 70$ s, the target moves with constant velocity motion. Then, the target maneuvers with acceleration of $\ddot{x} = -10$ m/s^2 from $t = 70$ until $t = 80$ s. Then, the target moves with constant velocity motion until $t = 120$ s. A constant velocity (CV) estimate based on all 120 measurements is also given in Figure 19-8. Note that the position estimates make a perfectly straight line and the velocity estimate (i.e., the slope of the line) is a constant for all time. The position estimates poorly match the true position near times of 0, 25, 75, and 120 s. On the other hand, the position estimates closely match the true position near times of 10, 50, and 90 s. The velocity estimates poorly match the true velocity

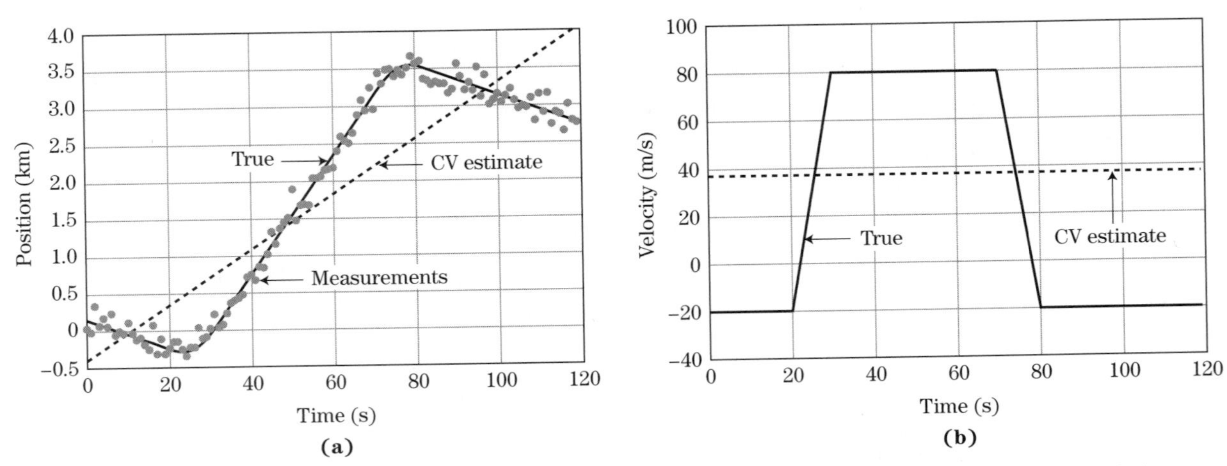

FIGURE 19-8 ■ Trajectory with measurements and the average constant velocity (CV) estimate. (a) Position. (b) Velocity.

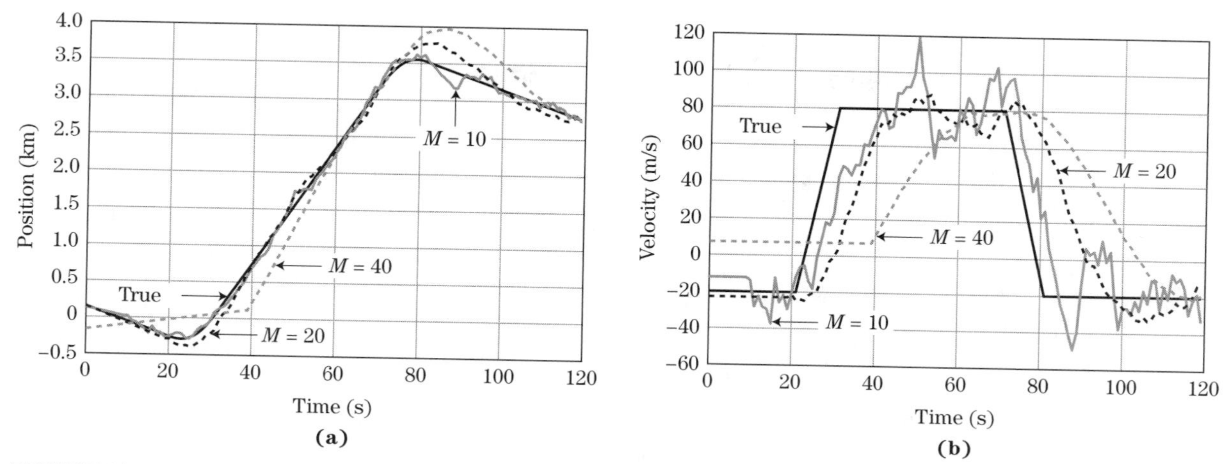

FIGURE 19-9 ■ Average CV position (a) and velocity (b) estimates with true values for memory of 10, 20, and 40 measurements.

at all times except those times near 25 and 75 s. The errors in the CV estimates will grow as the target maneuvers and continues to move beyond $t = 120$ s with a constant velocity. Thus, the number of measurements M (or memory) of the estimator must be limited.

If the distortions of the trajectory that are present in the estimates are unacceptable, the memory is limited to reduce the distortion or model mismatch. The averages of the CV estimates with memory of $M = 10, 20$, and 40 measurements over 2,000 Monte Carlo runs are shown in Figure 19-9. The first M measurements are used to initialize the tracking and the estimate at $t_k = t_{M-1}$ is a filtered estimate, while all of the estimates for $t_k < t_{M-1}$ are smoothed estimates. For $t_k \geq t_{M-1}$, the CV estimates are filtered estimates based on the current and $M-1$ previous measurements. In other words, the velocity estimate for the CV estimator with $M = 40$ is the same for $0 \leq t_k \leq 39$ s, and it will change value at $t_k = 40$ s. Note that the distortion or error in the position and velocity estimates as the maneuvers increase as M increases. For $M = 10$, distortion in the estimates is very little, while the distortion for $M = 40$ is large for the first 40 s and near 85 s. The *root mean squared error* (RMSE) of the CV estimator with $M = 10, 20$, and 40 measurements is given in Figure 19-10. Note that the smallest values of RMSE are achieved by the CV estimator with $M = 40$ after the maneuver has been expunged from the memory of the filter (i.e., $t = 70$ and 115 s). Also, note that the values of the RMSE of the CV estimator with $M = 10$ are the largest at those times. However, the values of RMSE of the CV estimator with $M = 10$ are the smallest during the maneuvers. Thus, the implementation or design of the CV estimator involves selecting the memory that best meets the requirements of the application.

A key aspect of a track filter is the accuracy of the covariance estimate. Generation of an accurate covariance is a challenge for maneuvering targets because most all estimation algorithms poorly model the acceleration and the uncertainty in the presence or absence of a maneuver. The accuracy of the covariance is measured through the normalized estimation error squared at time t_k given measurements through t_j by

$$C_{k|j} = (\boldsymbol{X}_{k|j} - \boldsymbol{X}_k)^T \mathbf{P}_{k|j}^{-1} (\boldsymbol{X}_{k|j} - \boldsymbol{X}_k) \tag{19.23}$$

For Gaussian errors in the state estimate with respect to the true state $\boldsymbol{X}_k$ and an accurate covariance $\mathbf{P}_{k|j}$, $C_{k|j}$ is chi-square distributed [7] with n degrees of freedom, where n is the size of the state vector. The average *normalized estimation error squared* (NEES)

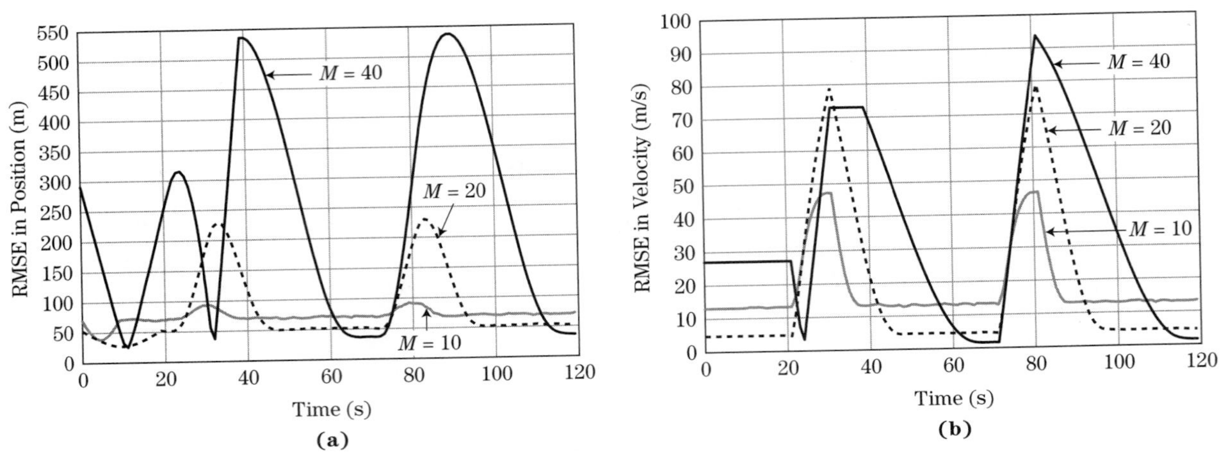

FIGURE 19-10 ■ RMSE of CV estimates with memory of 10, 20, and 40 measurements. (a) Position RMSE. (b) Velocity RMSE.

for the CV estimator with $M = 10, 20$, and 40 and 2,000 Monte Carlo runs is shown in Figure 19-11. For a single sample of the NEES for the CV estimate, it should be chi-square distributed with two degrees of freedom. The average value for $C_{k|k}$ should be two, and the chi-square tables in [7] give the 95% containment region for the NEES as $0.54 \leq C_{k|k} \leq 7.38$. According, in Figure 19-11, the average NEES of all three CV estimators closely matches two when the maneuver has been expunged from it memory. When more than a few seconds of target maneuver are included in the memory of the CV estimator, the average NEES is greater than 7.38 and falls outside the 95% confidence region for a two degree of freedom chi-square random variable. Thus, the presence of the maneuver significantly degrades the quality of the covariance estimate, and the quality degrades as M grows.

19.2.1.2 Constant Acceleration Filtering

Let $\ddot{x}_k$ denote the acceleration of the x coordinate of the target at time t_k. For a target moving with a constant acceleration, the kinematic state at time t_{k+1} given in terms of the state at time t_k is given by

$$x_{k+1} = x_k + (t_{k+1} - t_k)\dot{x}_k + \frac{1}{2}(t_{k+1} - t_k)^2\ddot{x}_k \tag{19.24}$$

$$\dot{x}_{k+1} = \dot{x}_k + (t_{k+1} - t_k)\ddot{x}_k \tag{19.25}$$

$$\ddot{x}_{k+1} = \ddot{x}_k \tag{19.26}$$

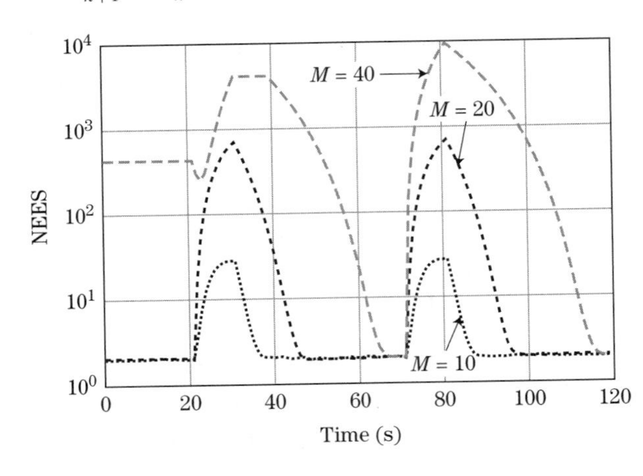

FIGURE 19-11 ■ Average NEES of CV estimates with memory of 10, 20, and 40 measurements.

This evolution of the target state can be written as

$$\boldsymbol{X}_{k+1} = \begin{bmatrix} x_{k+1} \\ \dot{x}_{k+1} \\ \ddot{x}_{k+1} \end{bmatrix} = \begin{bmatrix} 1 & (t_{k+1} - t_k) & \frac{1}{2}(t_{k+1} - t_k)^2 \\ 0 & 1 & (t_{k+1} - t_k) \\ 0 & 0 & 1 \end{bmatrix} \begin{bmatrix} x_k \\ \dot{x}_k \\ \ddot{x}_k \end{bmatrix} = \mathbf{F}_k \boldsymbol{X}_k \quad (19.27)$$

For measurements of the position of the target at t_k,

$$Z_k = x_k + w_k = \begin{bmatrix} 1 & 0 & 0 \end{bmatrix} \begin{bmatrix} x_k \\ \dot{x}_k \\ \ddot{x}_k \end{bmatrix} + w_k = \mathbf{H}_k \boldsymbol{X}_k + w_k \quad (19.28)$$

where w_k is a scalar Gaussian random variable with variance σ_{wk}^2. In this case, the smoothed estimate $\boldsymbol{X}_{1|N}$ is given by (19.7) and (19.8) with $\bar{\boldsymbol{Z}}_1^N$ given by (19.4), and $\mathbf{R}_N$ and $\bar{\mathbf{H}}_N$ are given by

$$\bar{\mathbf{H}}_N = \begin{bmatrix} 1 & 0 & 0 \\ 1 & t_2 - t_1 & \frac{1}{2}(t_2 - t_1)^2 \\ \vdots & \vdots & \vdots \\ 1 & t_{N-1} - t_1 & \frac{1}{2}(t_{N-1} - t_1)^2 \\ 1 & t_N - t_1 & \frac{1}{2}(t_N - t_1)^2 \end{bmatrix} \quad (19.29)$$

$$\bar{\mathbf{R}}_N = \begin{bmatrix} \sigma_{w1}^2 & 0 & \dots & 0 \\ 0 & \sigma_{w2}^2 & \dots & 0 \\ \vdots & \vdots & \ddots & \\ \vdots & 0 & \dots & \sigma_{wN}^2 \end{bmatrix} \quad (19.30)$$

For N measurements with uniform sampling with period T and equal variances $\sigma_{wk}^2 = \sigma_w^2$,

$$\bar{\mathbf{H}}_N = \begin{bmatrix} 1 & 0 & 0 \\ 1 & T & \frac{1}{2}T^2 \\ \vdots & \vdots & \vdots \\ 1 & (N-2)T & \frac{1}{2}(N-2)^2T^2 \\ 1 & (N-1)T & \frac{1}{2}(N-1)^2T^2 \end{bmatrix} \quad (19.31)$$

$$\bar{\mathbf{R}}_N = \sigma_w^2 \mathbf{I}_N \quad (19.32)$$

The estimate of the filtered state estimate and covariance are expressed in a form similar to (19.21) and (19.22) in [7]. However, the covariance is a 3×3 matrix and the variances of position and velocity are larger for a given N than in the case of the constant velocity estimator.

A constant acceleration estimate based on all 120 measurements could be envisioned for Figure 19-8. Note that the position estimates would form a perfectly quadratic line, the velocity would change linearly, and the acceleration estimate would be a constant for all time. The position and velocity estimates would poorly match the true values at one of the two ends of the trajectory. Thus, the number of measurements (or memory) of the estimator would have to be limited in order ensure reasonable performance.

19.2.2 Stochastic State Estimation

For stochastic state estimation, measurements are defined by (19.2) and the dynamical motion model is given by

$$\boldsymbol{X}_{k+1} = \mathbf{F}_k \boldsymbol{X}_k + \mathbf{G}_k \boldsymbol{v}_k \tag{19.33}$$

where

$\boldsymbol{v}_k$ = error in the system processes at time t_k with $v_k \sim \mathrm{N}(0, \mathbf{Q}_k)$.

$\mathbf{G}_k$ = relates the system errors to the target state at time t_k.

For a nearly constant velocity motion model, the state vector of the target in a scalar coordinate is given by

$$\boldsymbol{X}_k = \begin{bmatrix} x_k & \dot{x}_k \end{bmatrix}^T \tag{19.34}$$

where x_k represents the position of the target at time t_k, and $\dot{x}_k$ represents the velocity of the target. Process noise is included to account for the uncertainty associated with unknown maneuvers and unmodeled dynamics of the target under track. Since the evolution of the state includes a stochastic process, the state estimates are a stochastic process, and the covariance of the estimate will not achieve zero and grow in the absence of measurements.

The Kalman filter gives the MMSE and minimum variance estimate of the stochastic state $\boldsymbol{X}_k$. Since the random processes $\boldsymbol{v}_k$ and $\boldsymbol{w}_k$ are additive Gaussian and the Kalman filter is a linear filter, the state estimation error of the Kalman filter will be Gaussian. Thus, only the mean and covariance are needed to fully characterize the state estimation error. The Kalman filter is a predictor-corrector algorithm with the predictor accounting for changes in time and the corrector accounting for the measurement processing. The Kalman algorithm is defined by the following equations, where $\boldsymbol{X}_{k|j}$ denotes the state estimate at t_k given measurements through t_j, and $\mathbf{P}_{k|j}$ denotes the state error covariance at t_k given measurements through t_j:

Prediction of the state estimate and covariance to the next time:

$$\boldsymbol{X}_{k|k-1} = \mathbf{F}_{k-1} \boldsymbol{X}_{k-1|k-1} \tag{19.35}$$

$$\mathbf{P}_{k|k-1} = \mathbf{F}_{k-1} \mathbf{P}_{k-1|k-1} \mathbf{F}_{k-1}^T + \mathbf{G}_{k-1} \mathbf{Q}_{k-1} \mathbf{G}_{k-1}^T \tag{19.36}$$

Update of the state estimate and covariance with the measurement:

$$\boldsymbol{X}_{k|k} = \boldsymbol{X}_{k|k-1} + \mathbf{K}_k [\boldsymbol{Z}_k - \mathbf{H}_k \boldsymbol{X}_{k|k-1}] = \boldsymbol{X}_{k|k-1} + \mathbf{K}_k \widetilde{\boldsymbol{Z}}_k \tag{19.37}$$

$$\mathbf{P}_{k|k} = [\mathbf{I} - \mathbf{K}_k \mathbf{H}_k] \mathbf{P}_{k|k-1} \tag{19.38}$$

$$\mathbf{K}_k = \mathbf{P}_{k|k-1} \mathbf{H}_k^T \mathbf{S}_k^{-1} \tag{19.39}$$

$$\mathbf{S}_k = \mathbf{H}_k \mathbf{P}_{k|k-1} \mathbf{H}_k^T + \mathbf{R}_k \tag{19.40}$$

where $\mathbf{K}_k$ is referred to as the Kalman filter gain, $\widetilde{\boldsymbol{Z}}_k$ denotes the filter residual vector, and the $\mathbf{S}_k$ is the covariance of the measurement residual.

The processing of the Kalman filter for nearly constant velocity filtering is illustrated in Figure 19-12. The state estimate at time t_1 is predicted with constant velocity motion to t_2. The predicted state estimate $\boldsymbol{X}_{2|1}$ is then updated with measurement $\boldsymbol{Z}_2$ to produce the filtered state estimate $\boldsymbol{X}_{2|2}$. The filtered state estimate is an optimal weighting of the predicted measurement and the measurement. The filtered state estimate will tend to

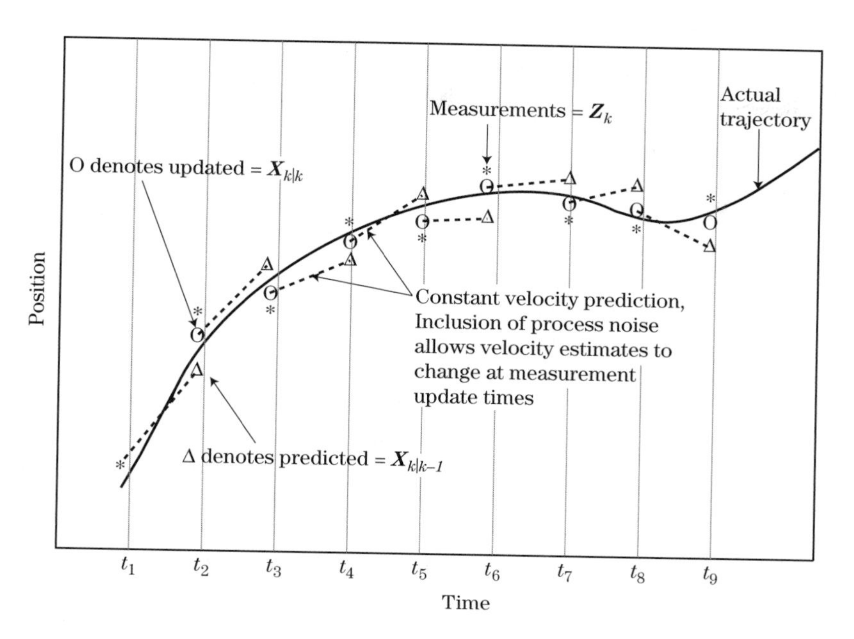

FIGURE 19-12 ■ Kalman filter process performs track filtering by correcting the predicted state with measurements.

"favor" the one with the smaller covariance. Note that the filtered state estimate lies on a line connecting the measurement and the position of the predicted measurement. This filter step is repeated for each measurement. Note that the velocity estimates (i.e., the slope of the prediction lines) are not constant over the window of data. In fact, it changes after every measurement update. The higher the process noise covariance is set, the more the velocity will change between measurement updates. For example, tracking a highly maneuvering air target will involve a large covariance for the process noise acceleration and the velocity estimate will make large changes between updates. For tracking ballistic targets above the atmosphere, the process noise covariance will be smaller and the velocity estimate will change much more slowly over time.

An alternate depiction of the processing of the Kalman filter is shown in Figure 19-13 as a predictor-corrector algorithm. The filter process is recursive. It starts with the initial

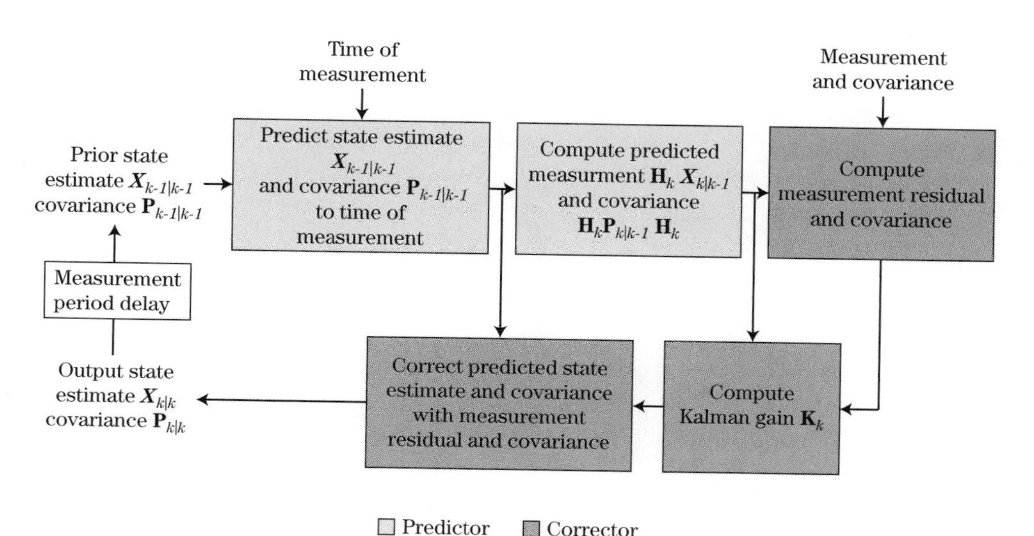

FIGURE 19-13 ■ Kalman filter is a predictor-corrector algorithm.

state estimate $\boldsymbol{X}_{k-1|k-1}$ and covariance $\mathbf{P}_{k-1|k-1}$ at time t_{k-1}. The filtered state estimate and covariance are predicted to the time of the measurement t_k to obtain $\boldsymbol{X}_{k|k-1}$ and $\mathbf{P}_{k|k-1}$. The predicted measurement and its covariance are then computed to complete the predictor. The measurement and its covariance are then used to compute the measurement residual and its covariance. Then the Kalman gain is computed and the predicted state is updated with the measurement residual and the Kalman gain to form the new filtered state estimate $\boldsymbol{X}_{k|k}$. The predicted covariance is updated with the Kalman gain to form the new filtered covariance $\mathbf{P}_{k|k}$. The new filtered state estimate $\boldsymbol{X}_{k|k}$ and covariance $\mathbf{P}_{k|k}$ are then retained until the next filter cycle in which they become the new initial state estimate and covariance.

Nearly constant velocity and *nearly constant acceleration* are the most commonly used motion models [2,7]. The choice of a model is governed by the maximum acceleration of the target, the quality of the sensor measurements, and the measurement rate. Since no algorithm exists for predicting the best model, experience in the design process and Monte Carlo simulations are typically used to assess the best model for a tracking system. If the quality of the measurements and the measurement rate are not sufficient to estimate acceleration during a maneuver, models that do not include acceleration in the target state will be the better models. For example, if only two or three measurements are taken during maneuvers, accurate estimation of the acceleration will not be possible and the acceleration should not be included in the target state. If the quality of the measurements and the measurement rate allow for estimation of the acceleration during a maneuver and the bias in the position estimates of the nearly constant velocity filter during the maximum acceleration of the target [8] is larger than the measurement errors, then acceleration should be included the target state.

Process noise is included to account for the uncertainty associated with the unknown maneuvers of the target. The process noise for nearly constant velocity motion is typically treated as *discrete white noise acceleration* (DWNA) errors or discretized *continuous-time white noise acceleration* (CWNA) [7]. Nearly constant velocity filtering with the two different models for process noise and two different models for measurements is discussed in the next three sections. The measurement models reflect standard radar measurements of position with linear frequency modulation (LFM) waveforms and fixed frequency waveforms. Nearly constant acceleration tracking and multiple model tracking for highly maneuvering targets are addressed to complete this section on stochastic state estimation.

19.2.2.1 Nearly Constant Velocity Filtering with DWNA

Nearly constant velocity motion with DWNA in a single coordinate is given by (19.33) with

$$\boldsymbol{X}_k = \begin{bmatrix} x_k & \dot{x}_k \end{bmatrix}^T \tag{19.41}$$

$$\mathbf{F}_k = \begin{bmatrix} 1 & \delta_k \\ 0 & 1 \end{bmatrix} \tag{19.42}$$

$$\boldsymbol{G}_k = \begin{bmatrix} \dfrac{\delta_k^2}{2} \\ \delta_k \end{bmatrix} \tag{19.43}$$

$$Q_k = \sigma_{vk}^2 \tag{19.44}$$

where $\delta_k = t_k - t_{k-1}$. The v_k is a white noise acceleration error that is constant or fixed between t_k and t_{k-1} in the state process with $v_k \sim \mathrm{N}(0, \sigma_{vk}^2)$. For the nearly constant velocity motion model, the process noise covariance matrix for DWNA is given by

$$\boldsymbol{G}_k Q_k \boldsymbol{G}_k^T = \sigma_{vk}^2 \begin{bmatrix} \frac{\delta_k^4}{4} & \frac{\delta_k^3}{2} \\ \frac{\delta_k^3}{2} & \delta_k^2 \end{bmatrix} \quad (19.45)$$

The σ_{vk} is the design parameter for the nearly constant velocity (NCV) filter with DWNA errors. Typically, the filter design process begins by setting σ_{vk} greater than one half of the maximum acceleration of the target and less than the maximum acceleration, and Monte Carlo simulations are conducted to further refine the selection of σ_{vk}. Further guidelines on the selection of σ_{vk} are developed in [9] and summarized next. Typically, the measurements are the position of the target. In this case, the measurement equation of (19.2) is defined by

$$\boldsymbol{H}_k = \begin{bmatrix} 1 & 0 \end{bmatrix} \quad (19.46)$$

When the error processes v_k and w_k are stationary (i.e., zero mean and constant variances) and the data rate is constant, the Kalman filter will achieve steady-state conditions in which the filtered state covariance and Kalman gain are constant. While these conditions are seldom satisfied in practice, the steady-state form of the filter can be used to predict average or expected tracking performance. The design and performance of the filter is characterized by the *maneuver index* or *random tracking index*

$$\Gamma_{DWNA} = \frac{\sigma_v T^2}{\sigma_w} = \frac{\beta}{\sqrt{1-\alpha}} \quad (19.47)$$

where $\sigma_v = \sigma_{vk}$, $\sigma_w = \sigma_{wk}$, and $T = t_k - t_{k-1}$ for all k. Under steady-state conditions, the NCV Kalman filter is equivalent to an alpha-beta filter. For the alpha-beta filter, the steady-state gains that occur after the transients associated with filter initialization diminish are given by

$$\boldsymbol{K}_k = \begin{bmatrix} \alpha & \frac{\beta}{T} \end{bmatrix}^T \quad (19.48)$$

where α and β are the optimal gains for DWNA given in [7,10]. The alpha-beta filter is rather simple and computationally efficient to implement because online real-time calculation of neither the state covariance nor gain is needed. The alpha-beta filter for a scalar coordinate is given by

$$x_{k|k-1} = x_{k-1|k-1} + T\dot{x}_{k-1|k-1} \quad (19.49)$$

$$\dot{x}_{k|k-1} = \dot{x}_{k-1|k-1} \quad (19.50)$$

$$x_{k|k} = x_{k|k-1} + \alpha[Z_k - x_{k|k-1}] \quad (19.51)$$

$$\dot{x}_{k|k} = \dot{x}_{k|k-1} + \frac{\beta}{T}[Z_k - x_{k|k-1}] \quad (19.52)$$

The steady-state error covariance of the alpha-beta filter is given by

$$\mathbf{P}_{\alpha\beta} = \sigma_w^2 \begin{bmatrix} \alpha & \frac{\beta}{T} \\ \frac{\beta}{T} & \frac{\beta(2\alpha-\beta)}{2(1-\alpha)T^2} \end{bmatrix} \quad (19.53)$$

Given the maneuver or tracking index, the steady-state filter gains are specified according to

$$\alpha = -\frac{1}{8}\Gamma_{DWNA}^2 - \Gamma_{DWNA} + \frac{1}{8}(\Gamma_{DWNA} + 4)\sqrt{\Gamma_{DWNA}^2 + 8\Gamma_{DWNA}} \tag{19.54}$$

$$\beta = \frac{1}{4}\Gamma_{DWNA}^2 + \Gamma_{DWNA} - \frac{1}{4}\Gamma_{DWNA}\sqrt{\Gamma_{DWNA}^2 + 8\Gamma_{DWNA}} \tag{19.55}$$

Figure 19-14 gives the gains versus the random tracking index, and also includes the corresponding CWNA case to be discussed in Section 19.2.2.3. Note that α approaches 1 and β approaches 2 for large values of the random tracking index. An additional relationship between α and β for DWNA [7,10] is given by

$$\beta = 2(2 - \alpha) - 4\sqrt{1 - \alpha} \tag{19.56}$$

While the alpha-beta filter can be implemented efficiently for steady-state conditions, poor tracking will be experienced if the steady-state gains are used for the settling of the alpha-beta filter from track initiation to steady state. Least squares techniques can be used to develop an effective gain scheduling method. Given $k = 0$ for the first measurement, the alpha and beta gains can be scheduled as a function of the measurement number according to

$$\alpha_k = \max\left\{\frac{2(2k+1)}{(k+1)(k+2)}, \alpha_{SS}\right\} \tag{19.57}$$

$$\beta_k = \max\left\{\frac{6}{(k+1)(k+2)}, \beta_{SS}\right\} \tag{19.58}$$

where $x_{0|-1} = 0$, and $\dot{x}_{0|-1} = 0$, and α_{SS} and β_{SS} are the steady-state values for α and β, respectively. Using these gains through N measurements gives a parametric least squares estimate of the position and velocity for constant velocity filtering for the first N measurements.

When a target undergoes a deterministic maneuver (i.e., a constant acceleration maneuver), the estimates are biased and the covariance matrix tends to be a biased estimate of track filter performance. When a target undergoes no maneuver (i.e., zero acceleration), the covariance matrix tends to also be a biased estimate of track filter performance, because

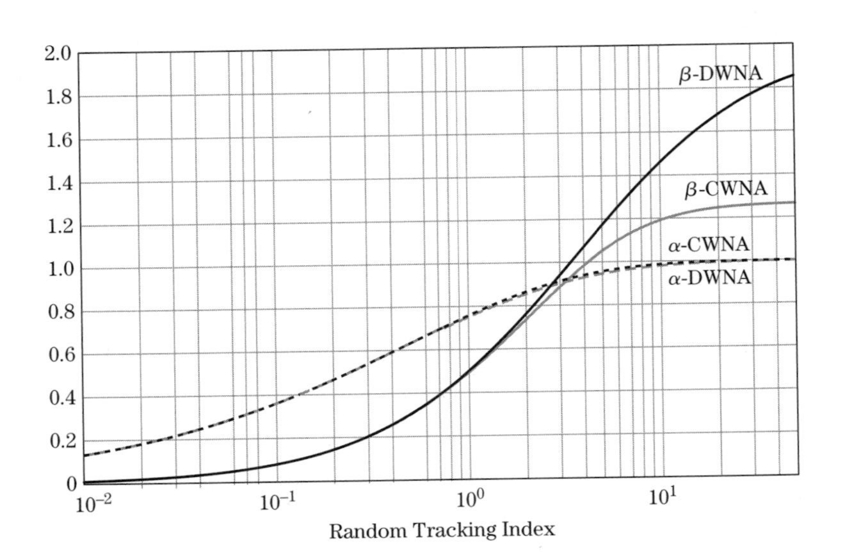

FIGURE 19-14 ■ α and β gains versus random tracking index for DWNA and CWNA models for maneuvers.

process noise is included in the filter for maneuver response, when there is no maneuver. Thus, to address both conditions of the performance prediction, the MSE can be written in terms of a sensor-noise only (SNO) covariance for no maneuver and a bias or maneuver lag for the constant acceleration maneuver. Let Γ_D denote the deterministic tracking index given by

$$\Gamma_D = \frac{A_{max} T^2}{\sigma_w} \tag{19.59}$$

where A_{max} denotes the maximum acceleration of the target. Then the maximum MSE (MMSE) in the filtered position estimates of the alpha-beta filter can be written as

$$MMSE^p = \sigma_w^2 \left[\frac{2\alpha^2 + \beta(2 - 3\alpha)}{\alpha(4 - 2\alpha - \beta)} + (1 - \alpha)^2 \frac{\Gamma_D^2}{\beta^2} \right] \tag{19.60}$$

The MMSE in the velocity estimates can be expressed as

$$MMSE^v = \frac{\sigma_w^2}{T^2} \left[\frac{2\beta^2}{\alpha(4 - 2\alpha - \beta)} + \left(\frac{\alpha - 0.5\beta}{\beta} \right)^2 \Gamma_D^2 \right] \tag{19.61}$$

For a given value of Γ_D, a unique α and β minimize (19.60) in conjunction with (19.56). These α and β define a unique Γ_{DWNA} given by (19.47) for the given Γ_D. Let the random tracking index be expressed in terms of the deterministic tracking index as

$$\Gamma_{DWNA} = \kappa_1(\Gamma_D)\Gamma_D \tag{19.62}$$

Using (19.47) and (19.59) in (19.62) gives

$$\sigma_v = \kappa_1(\Gamma_D) A_{max} \tag{19.63}$$

The $\kappa_1(\Gamma_D)$ that corresponds to minimizing (19.60) is given versus Γ_D in Figure 19-15 and denoted by the line labeled "MMSE". It is given approximately by

$$\kappa_1(\Gamma_D) = 1.67 - 0.74 \log(\Gamma_D) + 0.26[\log(\Gamma_D)]^2, \quad 0.01 \le \Gamma_D \le 10 \tag{19.64}$$

Thus, (19.64) along with (19.63) and the maximum acceleration of the target defines the process noise variance for the NCV Kalman filter so that the MMSE is minimized.

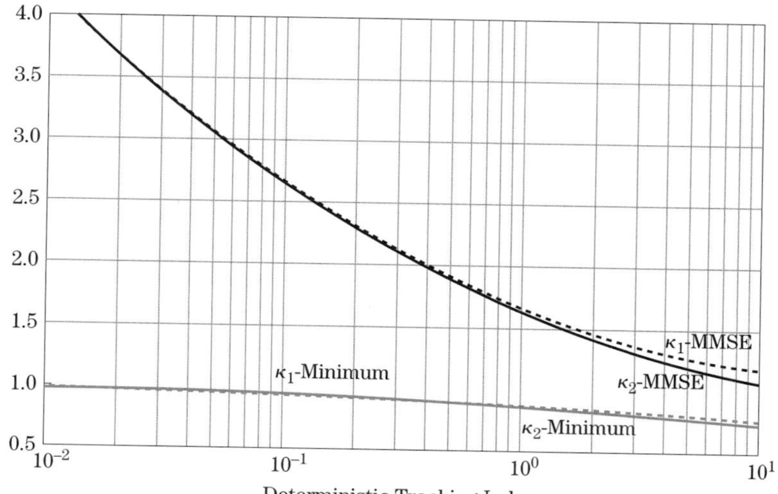

FIGURE 19-15 Values for $\kappa_1(\Gamma_D)$ and $\kappa_2(\Gamma_D)$ for defining the process noise variance in terms of the maximum acceleration.

Using the constraint that the MMSE in the filtered position estimate should not exceed the measurement error variance gives that $MMSE^p \leq \sigma_w^2$ or

$$\frac{2\alpha^2 + \beta(2 - 3\alpha)}{\alpha(4 - 2\alpha - \beta)} + \frac{(1 - \alpha)^2}{\beta^2}\Gamma_D^2 \leq 1 \tag{19.65}$$

For a given value of Γ_D, unique α and β satisfy (19.65) with equality in conjunction with (19.56). These α and β are the minimum gains and define the minimum Γ_{DWNA} (or σ_v) that is related to Γ_D by $\kappa_1(\Gamma_D)$. The κ_1 that defines the minimum σ_v is given versus Γ_D in Figure 19-15 and is denoted by the line labeled "MINIMUM". It is given approximately by

$$\kappa_1^{\min}(\Gamma_D) = 0.87 - 0.09\log(\Gamma_D) - 0.02[\log(\Gamma_D)]^2 \tag{19.66}$$

Thus, Figure 19-15 shows that the constraint is most always satisfied if the process noise variance is chosen to minimize the MMSE. For $\Gamma_D < 1.0$, Figure 19-15 shows that a wide range of values for κ_1 satisfy the constraint in (19.65) and the filter designer has freedom in their selection of σ_v.

When designing NCV filters for deterministic maneuvers, the question of the duration of a typical maneuver arises. Figure 19-16 gives the approximate time period in measurement samples for the maneuver lag to achieve 90% of the steady-state lag in position [11]. Thus, for $\Gamma_D \approx 1$, the method presented here for selecting the process noise variance is valid for maneuvers lasting three measurements or longer. For $\Gamma_D \approx 0.1$, this method for selecting the process noise variance is valid for maneuvers lasting six measurements or longer. For targets with maximum maneuvers that are not sustained so that the maximum bias in the estimate is achieved, an alternate design procedure should be considered.

For illustrating the filter design methodology, consider a target that maneuvers with 40 m/s^2 of acceleration from 40 to 60 s. The sensor measures the target position at a 1 Hz rate with errors defined by $\sigma_w = 120$ m. Thus, $\Gamma_D = 0.33$. Since the maneuver is sustained for more than four measurements, considering the maximum lag for design is acceptable. Then, (19.66) gives $\kappa_1^{\min} = 0.91$ and $\sigma_v = 36.4$ as the minimum acceptable

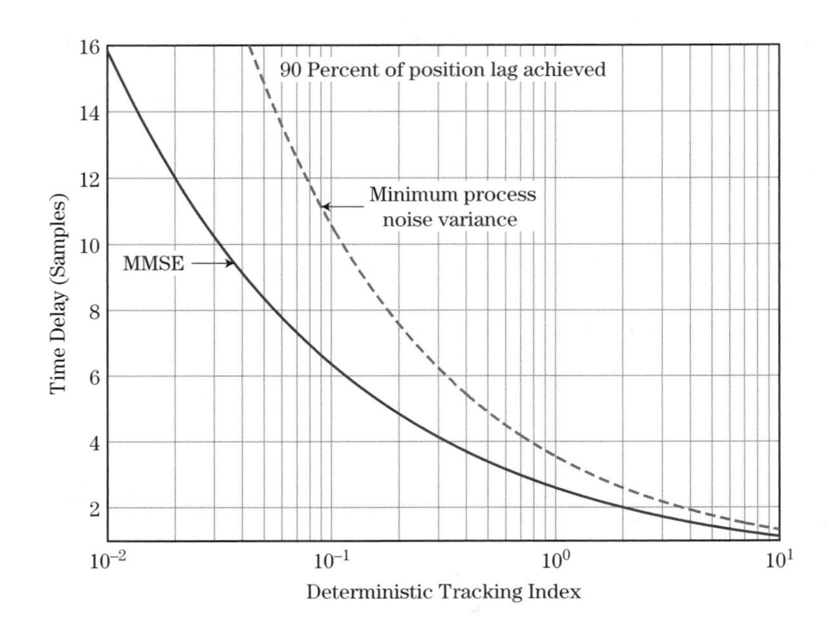

FIGURE 19-16 ■ Approximate time delay in filters achieving 90% of steady-state lag in position.

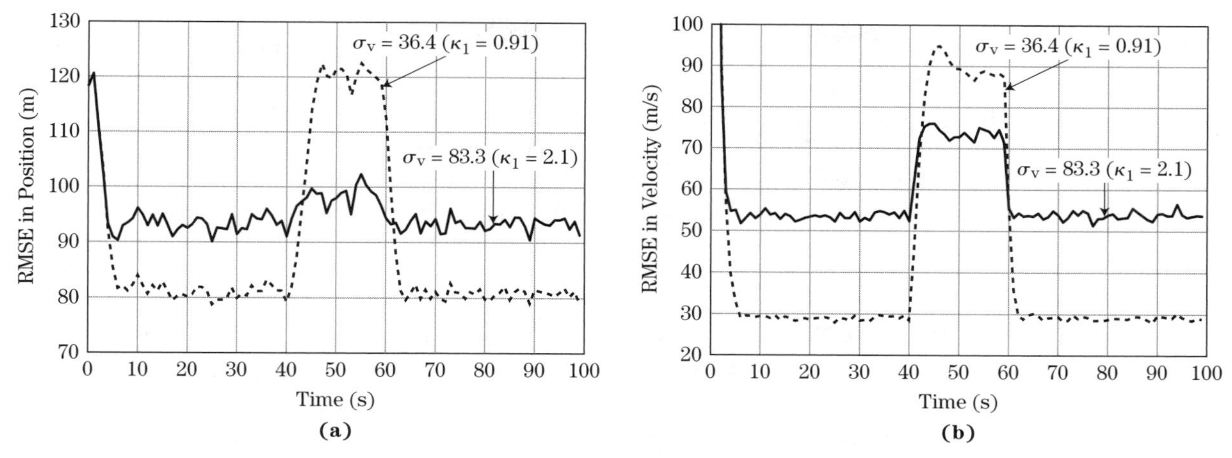

FIGURE 19-17 ■ RMSE in state estimates of two NCV filter designs for tracking maneuvering target. (a) RMSE in position. (b) RMSE in velocity.

value for σ_v. The design for minimizing MMSE gives $\kappa_1(0.33) = 2.1$ from (19.64) and $\sigma_v = 83.3$ as the process noise standard deviation that minimizes the MSE. Figure 19-17 shows the RMSE results of Monte Carlo simulations with 2,000 experiments for the two filter designs. Note that the RMSE in position during the maneuver for $\kappa_1 = 0.91$ and $\sigma_v = 36.4$ is closely matched to the standard deviation of the measurements of 120 m as anticipated. Also, note for $\kappa_1 = 2.1$ and $\sigma_v = 83.3$, the maximum RMSE in position is minimized and is only slightly larger than the RMSE in the absence of a maneuver.

Figure 19-18 gives the NEES as defined in (19.23) for the two filter designs. If the errors in the state estimates are zero-mean Guassian with covariance $\mathbf{P}_{k|k}$, then the NEES

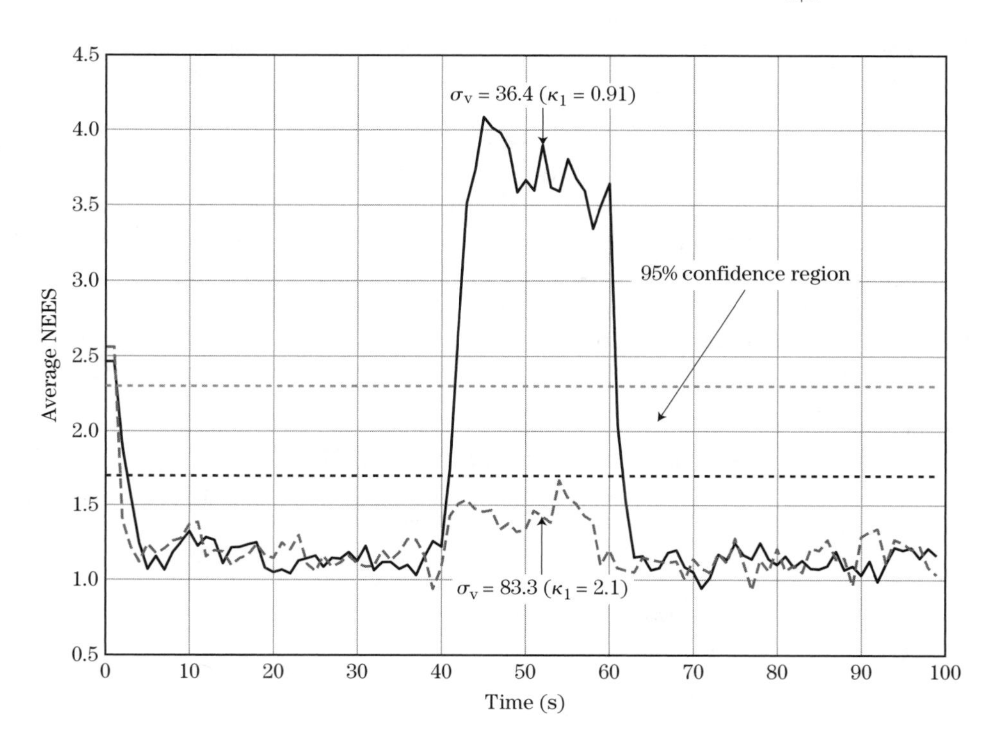

FIGURE 19-18 ■ Covariance consistency of state estimates of two NCV filter designs for tracking a maneuvering target.

is a chi-square random variable with two degrees of freedom. For a two degree of freedom chi-square random variable, the sample average should be two. In Figure 19-18, the two dotted lines denote the 95% confidence region for a 200 sample average of a two degree of freedom chi-square random variable. Thus, results indicate that the covariances for both filter designs are too large when the target is not maneuvering. For the design with $\kappa_1 = 2.1$, the covariance is also too large during the maneuver. For the design with $\kappa_1 = 0.91$, the covariance is too small during maneuver. Thus, Figure 19-18 illustrates the challenge with the error covariance that results from modeling deterministic-like maneuvers with a random acceleration model.

19.2.2.2 Nearly Constant Velocity Tracking with LFM Waveforms

When radars use an LFM waveform, the range measurement is coupled to the range rate-induced Doppler of the target [12]. For an LFM waveform, DWNA, and fixed track rate, the system for tracking in the range coordinate is given by (19.41) through (19.46) with the changes given by

$$\boldsymbol{X}_k = [\, r_k \quad \dot{r}_k \,]^T \tag{19.67}$$

$$\mathbf{H}_k = \mathbf{H} = [\, 1 \quad \Delta t \,] \tag{19.68}$$

where r_k and $\dot{r}_k$ are the range and range rate of the target, respectively, and Δt is the range-Doppler coupling coefficient. The range-Doppler coupling coefficient is defined by

$$\Delta t = \frac{f_1 \tau}{f_2 - f_1} \tag{19.69}$$

with f_1 denoting the initial frequency of the transmitted waveform, τ denoting the duration of the transmitted waveform, and f_2 denoting the final frequency in the waveform modulation.

As noted in [12] and shown analytically in [13], the range-Doppler coupling of the waveform affects the tracking accuracy. For example, an up-chirp waveform (i.e., $f_2 > f_1$ or $\Delta t > 0$) gives lower steady-state tracking errors than an equivalent down-chirp waveform (i.e., $f_2 < f_1$ or $\Delta t < 0$). The steady-state gains and covariance for tracking the range and range rate with LFM waveforms are derived in [13].

Figure 19-19 shows that the variances of range and range rate estimates decrease with increasing Δt for a given maneuver index Γ_{DWNA}. The correlation coefficients for the range and range rate estimates are between 0 and 1. As Δt becomes more positive, the correlation coefficient for range and range rate goes to zero at higher values of Γ_{DWNA}. This means that the range and range rate estimates are less correlated at higher values of Γ_{DWNA}. Thus, the accuracy of the predicted measurements will be improved at more positive Δt. Thus, α decreases with increasing Δt.

19.2.2.3 Nearly Constant Velocity Filtering with CWNA

Nearly constant velocity motion with continuous acceleration errors is described in a scalar coordinate by

$$\begin{bmatrix} \dot{x}(t) \\ \ddot{x}(t) \end{bmatrix} = \begin{bmatrix} 0 & 1 \\ 0 & 0 \end{bmatrix} \begin{bmatrix} x(t) \\ \dot{x}(t) \end{bmatrix} + \begin{bmatrix} 0 \\ 1 \end{bmatrix} \tilde{v}(t) \tag{19.70}$$

where $\tilde{v}(t)$ is a continuous time, zero-mean white noise process with $\mathrm{E}\{\tilde{v}(t)\tilde{v}(\tau)\} = \tilde{q}\delta_D(t-\tau)$, where $\mathrm{E}\{\cdot\}$ denotes the expected value, $\delta_D(\cdot)$ is the Dirac delta function.

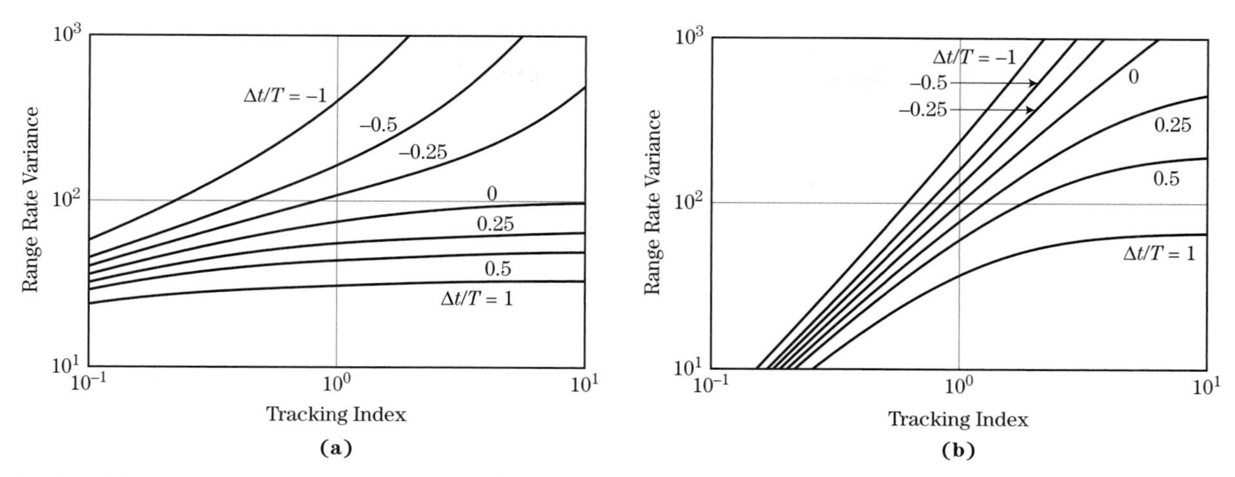

FIGURE 19-19 ■ Variances of the range and range rate estimates for tracking with LFM waveforms versus Γ_{DWNA}.

For time-invariant system, $\tilde{q}$ is the power spectral density. The discretized version of the motion model in a single coordinate is given by (19.33) with

$$\mathbf{X}_k = \begin{bmatrix} x_k & \dot{x}_k \end{bmatrix}^T \tag{19.71}$$

$$\mathbf{G}_k = \begin{bmatrix} \frac{1}{2\sqrt{3}}(\sqrt{\delta_k})^3 & \frac{1}{2}(\sqrt{\delta_k})^3 \\ 0 & \sqrt{\delta_k} \end{bmatrix} \tag{19.72}$$

$$\mathbf{Q}_k = \begin{bmatrix} \tilde{q} & 0 \\ 0 & \tilde{q} \end{bmatrix} \tag{19.73}$$

For the nearly constant velocity motion model, the process noise covariance matrix for CWNA is given by

$$\mathbf{G}_k\mathbf{Q}_k\mathbf{G}_k^T = \tilde{q} \begin{bmatrix} \frac{1}{3}\delta_k^3 & \frac{1}{2}\delta_k^2 \\ \frac{1}{2}\delta_k^2 & \delta_k \end{bmatrix} \tag{19.74}$$

where $\tilde{q}$ is the design parameter for the NCV Kalman filter. Typically, the filter design process begins by setting $\sqrt{\tilde{q}}/T$ greater than one half of the maximum acceleration of the target and less than the maximum acceleration, where T is a nominal sample period. Monte Carlo simulations are conducted to further refine the selection of $\tilde{q}$. Further guidelines on the selection of $\tilde{q}$ are developed in [9] and summarized below. Typically, the measurements are the position of the target and the measurement equation of (19.2) is defined by (19.46).

When the error processes $\tilde{v}(t)$ and w_k are stationary and the data rate is constant, the Kalman filter will achieve steady-state conditions in which the filtered state covariance and Kalman gain are constant. The design and performance of the filter is characterized by the maneuver index

$$\Gamma_{CWNA}^2 = \frac{\tilde{q}T^3}{\sigma_w^2} = \frac{\beta^2}{1-\alpha} \tag{19.75}$$

where $\sigma_w = \sigma_{wk}$, and $T = t_k - t_{k-1}$ for all k. Under steady-state conditions, the NCV filter is equivalent to an alpha-beta filter, where α and β are the optimal gains for CWNA model given in [7]. The steady-state error covariance of the alpha-beta filter is given as a function of α and β by (19.53). Given the random tracking index, the steady-filter gains are specified by

$$u = \frac{1}{3} + \sqrt{\frac{1}{12} + \frac{4}{\Gamma_{CWNA}^2}} \tag{19.76}$$

$$\alpha = \frac{12}{6(u + \sqrt{u}) + 1} \tag{19.77}$$

$$\beta = \frac{12\sqrt{u}}{6(u + \sqrt{u}) + 1} \tag{19.78}$$

Figure 19-14 gives the gains versus the random tracking index. Note that α approaches 1 and β approaches 1.3 for large values of the random tracking index. An additional relationship between α and β [7] is given by

$$\beta = 3(2 - \alpha) - \sqrt{3(\alpha^2 - 12\alpha + 12)} \tag{19.79}$$

While the alpha-beta filter can be implemented efficiently for steady-state conditions, poor tracking will be experienced if the steady-state gains are used for the settling of the alpha-beta filter from track initiation to steady-state. The filter initialization and settling can be accomplished by the gains in (19.57) and (19.58) for the first N measurements.

The MMSE in the filtered position and velocity estimates of the alpha-beta filter are given by (19.60) and (19.61). For a given value of Γ_D, a unique α and β minimize (19.60) in conjunction with (19.79). The α and β define a unique Γ_{CWNA} for the given Γ_D. Let the random tracking index be expressed in terms of the deterministic tracking index as

$$\Gamma_{CWNA} = \kappa_2(\Gamma_D)\Gamma_D \tag{19.80}$$

Using (19.75) and (19.59) in (19.80) gives

$$\tilde{q} = T\kappa_2^2(\Gamma_D)A_{max}^2 \tag{19.81}$$

where κ_2 that corresponds to minimizing (19.60) is given versus Γ_D in Figure 19-15 and denoted by the line labeled with MMSE. It is given approximately by

$$\kappa_2(\Gamma_D) = 1.62 - 0.79\log(\Gamma_D) + 0.24(\log(\Gamma_D))^2, \quad 0.01 \le \Gamma_D \le 10 \tag{19.82}$$

Thus, (19.82) along with (19.81) and the maximum acceleration of the target defines the process noise variance for the NCV filter. Using the constraint that the MMSE in the filtered position estimate should not exceed the measurement error variance gives that $MMSE^p \le \sigma_w^2$ or (19.65). For a given value of Γ_D, unique α and β satisfy (19.65) with equality in conjunction with (19.79). These values α and β are the minimum gains and define the minimum Γ_{CWNA} that is related to Γ_D by κ_2. The κ_2 that defines the minimum Γ_{CWNA} is given versus Γ_D in Figure 19-15, denoted by the line labeled "MINIMUM," and given approximately by

$$\kappa_2^{\min}(\Gamma_D) = 0.87 - 0.11\log(\Gamma_D) - 0.03[\log(\Gamma_D)]^2 \tag{19.83}$$

Thus, Figure 19-15 shows that the constraint is most always satisfied if the process noise variance is chosen to minimize the MMSE. For $\Gamma_D < 1.0$, Figure 19-15 shows that a wide range of values for κ_2 satisfy the constraint in (19.65) and the filter designer has freedom in the selection of σ_v. When designing NCV Kalman filters for deterministic maneuvers, the question of duration of a typical maneuver arises. Figure 19-16 gives the approximate time period in measurement samples for the maneuver lag to achieve 90% of the steady-state value. For targets with maximum maneuvers that are not sustained such that the maximum bias in the estimate is attained, an alternate design procedure should be considered.

19.2.2.4 Nearly Constant Acceleration Filtering

Nearly constant acceleration motion with piecewise constant acceleration errors in a scalar coordinate is defined by (19.33) with

$$\boldsymbol{X}_k = \begin{bmatrix} x_k & \dot{x}_k & \ddot{x}_k \end{bmatrix}^T \tag{19.84}$$

$$\mathbf{F}_k = \begin{bmatrix} 1 & \delta_k & \frac{1}{2}\delta_k^2 \\ 0 & 1 & \delta_k \\ 0 & 0 & 1 \end{bmatrix} \tag{19.85}$$

$$\boldsymbol{G}_k = \begin{bmatrix} \frac{1}{2}\delta_k^2 & \delta_k & 1 \end{bmatrix}^T \tag{19.86}$$

where v_k is a discrete Wiener process acceleration (DWPA) error that is constant between t_k and t_{k-1} in the state process with $v_k \sim \mathrm{N}(0, \sigma_{vk}^2)$ For the nearly constant acceleration motion model, the process noise covariance matrix for DWPA is given by

$$\boldsymbol{G}_k Q_k \boldsymbol{G}_k^T = \sigma_{vk}^2 \begin{bmatrix} \frac{1}{4}\delta_k^4 & \frac{1}{2}\delta_k^3 & \frac{1}{2}\delta_k^2 \\ \frac{1}{2}\delta_k^3 & \delta_k^2 & \delta_k \\ \frac{1}{2}\delta_k^2 & \delta_k & 1 \end{bmatrix} \tag{19.87}$$

where σ_{vk} is the design parameter for the nearly constant acceleration (NCA) filter with DWPA. Typically, the filter design process begins by setting σ_{vk} greater than one half of the maximum change in acceleration between t_{k-1} and t_k and less than the maximum change in acceleration. Monte Carlo simulations are conducted to further refine the selection of σ_{vk}. Typically, the measurements are of the position of the target. In this case, the measurement equation of (19.2) is defined by

$$\mathbf{H}_k = \begin{bmatrix} 1 & 0 & 0 \end{bmatrix} \tag{19.88}$$

When the error processes v_k and w_k are stationary (i.e., zero mean and constant variances) and the data rate is constant, the Kalman filter will achieve steady-state conditions in which the filtered state covariance and Kalman gain are constant. While these conditions are seldom satisfied in practice, the steady-state form of the filter can be used to predict average or expected tracking performance. The design and performance of the filter is characterized by the maneuver index or random tracking index

$$\Gamma_{DWPA} = \frac{\sigma_v T^2}{\sigma_w} \tag{19.89}$$

where $\sigma_v = \sigma_{vk}$, $\sigma_w = \sigma_{wk}$, and $T = t_k - t_{k-1}$ for all k. Under steady-state conditions, the NCA filter is equivalent to an alpha-beta-gamma filter. For the alpha-beta-gamma filter, the steady-state gains that occur after the transients associated with filter initialization diminish are given by

$$\boldsymbol{K}_k = \begin{bmatrix} \alpha & \frac{\beta}{T} & \frac{\gamma}{2T^2} \end{bmatrix}^T \tag{19.90}$$

where α, β, and γ are the optimal gains for DWPA given in [7,10]. The alpha-beta-gamma filter is rather simple and computationally efficient to implement because online real-time calculation of the filtered state covariance and gains is not needed. The alpha-beta-gamma filter for a scalar coordinate is given by

$$x_{k|k-1} = x_{k-1|k-1} + T\dot{x}_{k-1|k-1} + \frac{T^2}{2}\ddot{x}_{k-1|k-1} \tag{19.91}$$

$$\dot{x}_{k|k-1} = \dot{x}_{k-1|k-1} + T\ddot{x}_{k-1|k-1} \tag{19.92}$$

$$\ddot{x}_{k|k-1} = \ddot{x}_{k-1|k-1} \tag{19.93}$$

$$x_{k|k} = x_{k|k-1} + \alpha[Z_k - x_{k|k-1}] \tag{19.94}$$

$$\dot{x}_{k|k} = \dot{x}_{k|k-1} + \frac{\beta}{T}[Z_k - x_{k|k-1}] \tag{19.95}$$

$$\ddot{x}_{k|k} = \ddot{x}_{k|k-1} + \frac{\gamma}{2T^2}[Z_k - x_{k|k-1}] \tag{19.96}$$

where Z_k is the measured position of the target. The steady-state error covariance of the alpha-beta-gamma filter is given in [7,10] as

$$\mathbf{P}_{\alpha\beta} = \sigma_w^2 \begin{bmatrix} \alpha & \dfrac{\beta}{T} & \dfrac{\gamma}{2T^2} \\ \dfrac{\beta}{T} & \dfrac{8\alpha\beta + \gamma(\beta - 2\alpha - 4)}{8(1-\alpha)T^2} & \dfrac{\beta(2\beta - \gamma)}{4(1-\alpha)T^3} \\ \dfrac{\gamma}{2T^2} & \dfrac{\beta(2\beta - \gamma)}{4(1-\alpha)T^3} & \dfrac{\gamma(2\beta - \gamma)}{4(1-\alpha)T^4} \end{bmatrix} \tag{19.97}$$

Given the maneuver index, the steady-filter gains are found by solving three simultaneous equations given by

$$\Gamma_{DWPA} = \frac{\gamma^2}{4(1-\alpha)} \tag{19.98}$$

$$\beta = 2(2-\alpha) - 4\sqrt{1-\alpha} \tag{19.99}$$

$$\gamma = \frac{\beta^2}{\alpha} \tag{19.100}$$

For a given tracking index, the gains are found by the use of a table. The table is constructed by varying α from a small value to a value near 1 and for each value of α, values for β, γ, and Γ_{DWPA} are computed. These gains are plotted versus the tracking index in Figure 19-20. Note that for a given tracking index, alpha is larger for the alpha-beta-gamma filter than for the alpha-beta filter. Thus, according to (19.53) and (19.97), the position estimate of the alpha-beta filter will have a smaller variance than that of the alpha-beta-gamma filter for a given tracking index.

While the alpha-beta-gamma filter can be implemented efficiently for steady-state conditions, poor tracking will be experienced if the steady-state gains are used for the settling of the alpha-beta-gamma filter from track initiation to steady-state. Least squares techniques can be used to develop and effective gain scheduling method. Given $k = 0$ for the first measurement, the alpha, beta, and gamma gains can be scheduled as a function

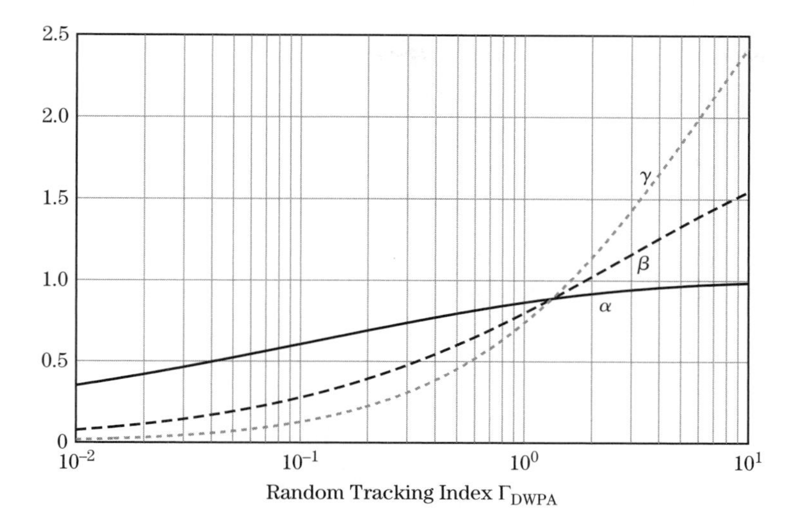

FIGURE 19-20 ■ Alpha, beta, and gamma gains versus random tracking index.

of the measurement number according to

$$\alpha_k = \max\left\{\frac{3(3k^2 + 3k + 2)}{(k+1)(k+2)(k+3)}, \alpha_{SS}\right\} \tag{19.101}$$

$$\beta_k = \max\left\{\frac{18(2k + 1)}{(k+1)(k+2)(k+3)}, \beta_{SS}\right\} \tag{19.102}$$

$$\gamma_k = \max\left\{\frac{60}{(k+1)(k+2)(k+3)}, \gamma_{SS}\right\} \tag{19.103}$$

where $x_{0|-1} = 0$, $\dot{x}_{0|-1} = 0$, and $\ddot{x}_{0|-1} = 0$, and α_{SS}, β_{SS}, and γ_{SS} are the steady-state values. Using these gains through N measurements gives a parametric least squares estimate of the position, velocity, and acceleration for constant acceleration filtering for the first N measurements.

To illustrate NCA filter tracking, consider the target trajectory and sensor used to generate the results in Figure 19-17. The target maneuvers with 40 m/s^2 of acceleration from 40 to 60 s, and the sensor measures the target position at a 1 Hz rate with errors defined by $\sigma_w = 120$ m. Let $\sigma_v = 7$ m/s^2 for the NCA filter design. Figure 19-21 shows the RMSE results of Monte Carlo simulations with 2000 runs. Note that the RMSE in both position and velocity have peaks at the beginning and end of the maneuver and this behavior is indicative of a second-order model with acceleration estimation. Also, note that the peak of the RMSE in position remains lower than the measurement error during the maneuver. However, the RMSE in position is significantly higher than that of the NCV filters in Figure 19-17, when the target is not maneuvering. Also, when the target is not maneuvering, the RMSE in velocity is higher than that of the NCV filter with $\sigma_v = 36.4$ and less than that for the NCV filter with $\sigma_v = 83.3$. Figure 19-22 gives the NEES as defined in (19.23) for the NCA filter. In Figure 19-22, the two dotted lines denote the 95% confidence region for a 200 sample average of a two degree of freedom chi-square random variable. Thus, results indicate that the covariance for NCA filter is too large when the target is not maneuvering. This is a result of modeling deterministic-like maneuvers with a random acceleration model.

19.2.2.5 Multiple Model Filtering for Highly Tracking Maneuvering Targets

Techniques for adapting the Kalman filter for target maneuvers have been studied for many years with very little success. Adaptive Kalman filtering is based on monitoring in real time

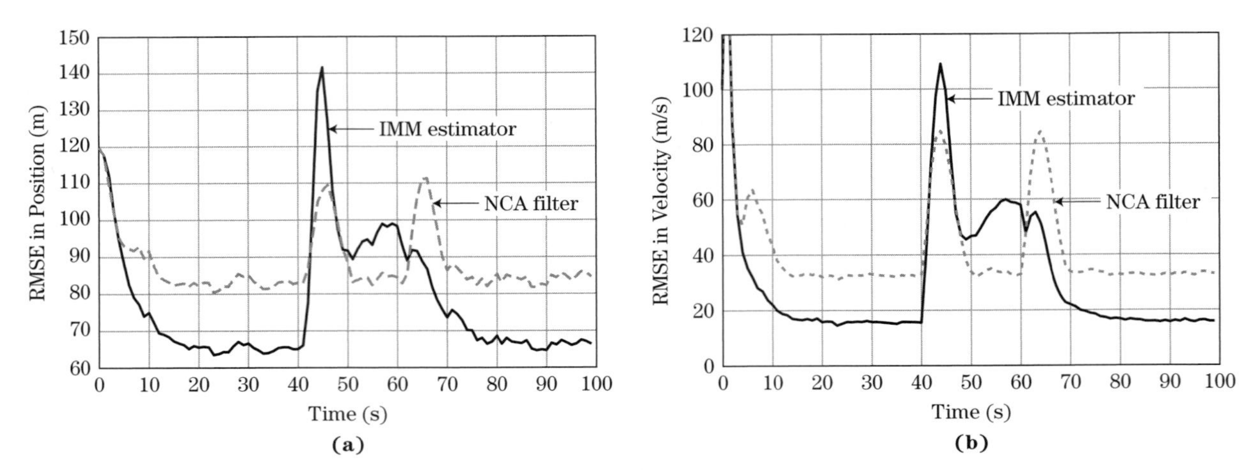

FIGURE 19-21 ■ RMSE in state estimates of NCA filter and IMM estimator for tracking a maneuvering target. (a) Position errors. (b) Velocity errors.

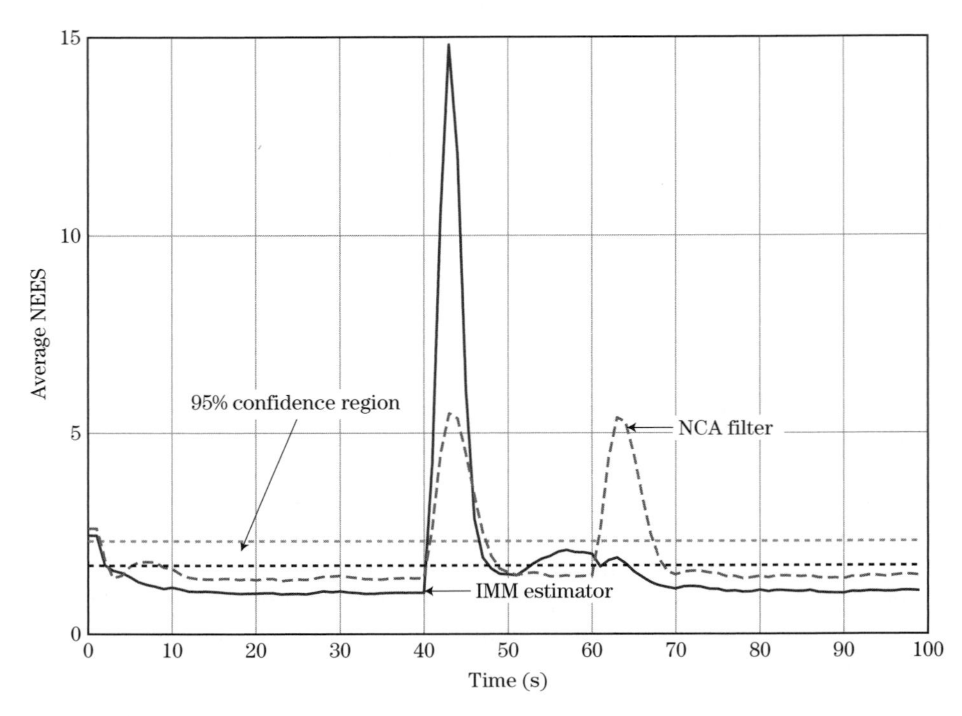

FIGURE 19-22 ■ Covariance consistency of state estimates of NCA filter and IMM estimator for tracking maneuvering target.

the whiteness of the residuals or the agreement of the magnitude of the innovations with their yielded error covariances. Thus, adaptive Kalman filtering for tracking maneuvering targets is based on detecting a breakdown in filter consistency and adjusting the filter parameters to regain filter consistency. Therefore, using an adaptive Kalman filter to track maneuvering targets will not provide reliable tracking performance because the loss of track filter consistency prevents reliable decision making for simultaneously adapting the filter parameters and performing data association.

As an alternative to adaptive Kalman filtering, the tracking of a maneuvering target can be treated as the state estimation of a system with Markovian switching coefficients.

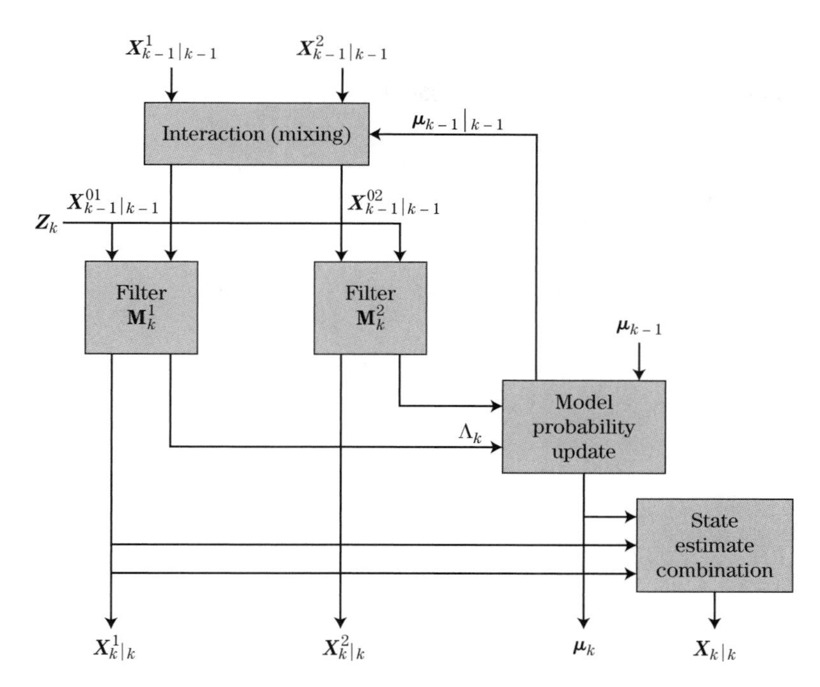

FIGURE 19-23 ■ Typical IMM estimator with two modes.

The dynamical motion and measurement equations for a system with Markovian switching coefficients are given by

$$X_{k+1} = \mathbf{F}_k(\theta_{k+1})X_k + G_k(\theta_k)v_k(\theta_k) \quad (19.104)$$

$$Z_k = \mathbf{H}(\theta_k)X_k + w_k \quad (19.105)$$

where θ_k is a finite-state Markov chain taking values in $\{1, \ldots, N\}$ according to the transition probabilities p_{ij} of switching from mode i to mode j. The interacting multiple model (IMM) estimator has become the well-accepted method for state and mode estimation of such multiple-mode systems with Markov switching between the modes [7]. Also, the IMM estimator is a nearly consistent estimator for the state of a maneuvering target.

The IMM estimator consists of a filter for each model, a model probability evaluator, an estimate mixer at the input of the filters, and an estimate combiner at the output of the filters. Figure 19-23 shows a flow diagram of the IMM estimator for two modes, where $X_{k|k}$ is the state estimate at time k based on both models, $X^j_{k|k}$ is the state estimate based on model j, $\mathbf{\Lambda}_k$ is the vector of mode likelihoods, and $\boldsymbol{\mu}_k$ is the vector of mode probabilities when all the likelihoods have been considered. With the assumption that the mode switching is governed by an underlying Markov chain, the mixing uses the mode probabilities and mode switching probabilities to compute a mixed state estimate for each filter. At the beginning of a filtering cycle, each filter uses a mixed estimate and a measurement to compute a new estimate and likelihood for the model within the filter. The likelihoods, the prior mode probabilities, and the mode switching probabilities are used to compute new mode probabilities. The overall state estimate is then computed with the new state estimates and their mode probabilities. In many applications, the IMM algorithm provides a 25% reduction of error in the position estimates and a 50% reduction of error in the velocity estimates that are produced by a Kalman filter. An illustrative example that compares the performances of the Kalman filter and the IMM estimator for tracking a maneuvering target is described in the introduction of [8]. The effectiveness of the IMM algorithm has been thoroughly documented in the literature and confirmed in a real-time tracking experiments [1,5,6,14–19].

Let $\mathbf{F}_k^i = \mathbf{F}_k(\theta_{k+1} = i)$, $\mathbf{G}_k^i = \mathbf{G}_k(\theta_{k+1} = i)$, and $\mathbf{H}_k^i = \mathbf{H}_k(\theta_k = i)$. Also, let $\boldsymbol{X}_{k|j}^i$ and $\mathbf{P}_{k|j}^i$ denote the state estimate and covariance at time t_k given data through time t_j under the condition that mode i, denoted as $\mathbf{M}_k^i$, is the true mode. The implementation of the IMM estimator involves the following five steps.

Step 1: State Mixing

Starting with $\boldsymbol{X}_{k-1|k-1}^i$ and the mode probability μ_{k-1}^i for each of r modes, compute the mixed estimate for each mode according

$$\boldsymbol{X}_{k-1|k-1}^{0i} = \sum_{j=1}^{r} \boldsymbol{X}_{k-1|k-1}^{j} \mu_{k-1|k-1}^{j|i} \tag{19.106}$$

where

$$\mu_{k-1|k-1}^{j|i} = \frac{\mu_{k-1}^{j}}{\mu_{k-1|k-1}^{i}} p_{ji}, \qquad \mu_{k-1|k-1}^{i} = \sum_{j=1}^{r} \mu_{k-1}^{j} p_{ji} \tag{19.107}$$

The covariance of the mixed estimate is given by

$$\mathbf{P}_{k-1|k-1}^{0i} = \sum_{j=1}^{r} \mu_{k-1|k-1}^{j|i} \left[\mathbf{P}_{k-1|k-1}^{j} + \left(\boldsymbol{X}_{k-1|k-1}^{j} - \boldsymbol{X}_{k-1|k-1}^{0i}\right)\left(\boldsymbol{X}_{k-1|k-1}^{j} - \boldsymbol{X}_{k-1|k-1}^{0i}\right)^T\right] \tag{19.108}$$

Step 2: Mode Conditioned Estimates

Starting with $\boldsymbol{X}_{k-1|k-1}^{0i}$ and $\mathbf{P}_{k-1|k-1}^{0i}$ for each of r modes, compute the mode-conditioned estimate for each mode according to the following.

Prediction of the mode-conditioned estimate to the next time:

$$\boldsymbol{X}_{k|k-1}^{i} = \mathbf{F}_{k-1}^{i} \boldsymbol{X}_{k-1|k-1}^{0i} \tag{19.109}$$

$$\mathbf{P}_{k|k-1}^{i} = \mathbf{F}_{k-1}^{i} \mathbf{P}_{k-1|k-1}^{0i} \left(\mathbf{F}_{k-1}^{i}\right)^T + \mathbf{G}_{k-1}^{i} Q_{k-1}^{i} \left(\mathbf{G}_{k-1}^{i}\right)^T \tag{19.110}$$

Update of the mode-conditioned estimate with the measurement:

$$\boldsymbol{X}_{k|k}^{i} = \boldsymbol{X}_{k|k-1}^{i} + \mathbf{K}_k^i \left[\mathbf{Z}_k - H_k^i \boldsymbol{X}_{k|k-1}^{i}\right] = \boldsymbol{X}_{k|k-1}^{i} + \mathbf{K}_k^i \widetilde{\mathbf{Z}}_k^i \tag{19.111}$$

$$\mathbf{P}_{k|k}^{i} = \left[\mathbf{I} - \mathbf{K}_k^i \mathbf{H}_k^i\right] \mathbf{P}_{k|k-1}^{i} \tag{19.112}$$

$$\mathbf{K}_k^i = \mathbf{P}_{k|k-1}^{i} \left(\mathbf{H}_k^i\right)^T \left(\mathbf{S}_k^i\right)^{-1} \tag{19.113}$$

$$\mathbf{S}_k^i = \mathbf{H}_k^i \mathbf{P}_{k|k-1}^{i} \left(\mathbf{H}_k^i\right)^T + \mathbf{R}_k \tag{19.114}$$

$\mathbf{K}_k^i$ is referred to as the Kalman filter gain for mode i, $\widetilde{\mathbf{Z}}_k^i$ denotes the filter residual vector for mode i, and $\mathbf{S}_k^i$ is the covariance of the measurement residual for mode i.

Step 3: Model Likelihood Computations

The likelihood of mode i at time t_k is computed according to

$$\Lambda_k^i = \frac{1}{\sqrt{|2\pi \mathbf{S}_k^i|}} \exp\left[-\frac{1}{2}\left(\widetilde{\mathbf{Z}}_k^i\right)^T \left[S_k^i\right]^{-1} \widetilde{\mathbf{Z}}_k^i\right] \tag{19.115}$$

where exp[·] denotes the exponential function and $|\cdot|$ denotes the matrix determinant.

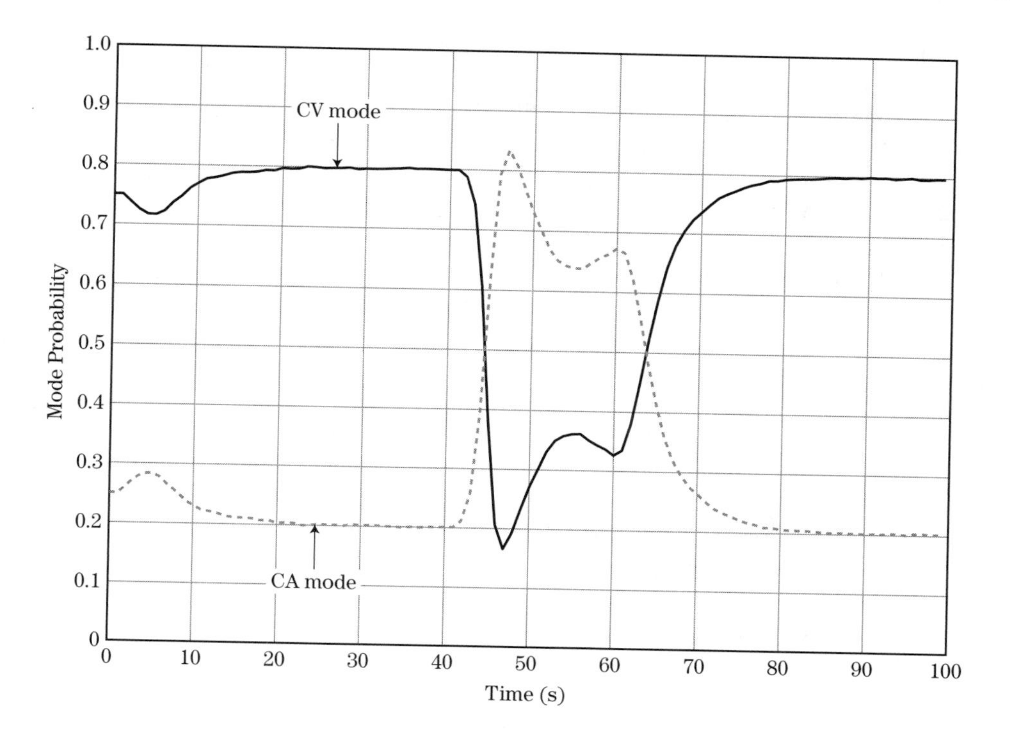

FIGURE 19-24 ■ Mode probabilities of two mode (i.e., NCV and NCA) IMM estimator for tracking a maneuvering target.

Step 4: Mode Probability Update

The probability of mode i at time t_k is computed according to

$$\mu_k^i = \frac{\Lambda_k^i}{\sum_{j=1}^r \mu_{k-1|k-1}^j} \mu_{k-1|k-1}^i \tag{19.116}$$

Step 5: Combination of State Estimates

The filtered state estimate at time t_k is computed for output according to

$$\boldsymbol{X}_{k|k} = \sum_{j=1}^{r} \mu_k^j \boldsymbol{X}_{k|k}^j \tag{19.117}$$

$$\mathbf{P}_{k|k} = \sum_{j=1}^{r} \mu_k^j \left[\mathbf{P}_{k|k}^j + \left(\boldsymbol{X}_{k|k}^j - \boldsymbol{X}_{k|k}\right)\left(\boldsymbol{X}_{k|k}^j - \boldsymbol{X}_{k|k}\right)^T\right] \tag{19.118}$$

To illustrate the tracking performance of the IMM estimator, consider the target trajectory and sensor used to generate the results in Figures 19-17 and 19-21. The target maneuvers with 40 m/s^2 of acceleration from 40 to 60 s, and the sensor measures the target position at a 1 Hz rate with errors defined by $\sigma_w = 120$ m. Consider an IMM estimator with NCV and NCA models. Let $\sigma_v^1 = 1$ m/s^2 for the NCV model and $\sigma_v^2 = 12$ m/s^2 for the NCA model. Let the mode switching probabilities be given by $p_{11} = 0.95$ with $p_{12} = 0.05$ and $p_{22} = 0.9$ with $p_{21} = 0.1$. Figure 19-24 illustrates the calculated probabilities of the CA and CV modes, showing that target acceleration causes the CA mode to be recognized as more likely, while the CV mode is deemed more likely when the target is not accelerating. Figure 19-21 shows the RMSE results of Monte Carlo simulations with 2,000 experiments. When the target is not maneuvering, the RMSE in position and velocity are significantly lower than that of the NCA filter and the NCV filters in Figure 19-17. Note that the RMSE in both position and velocity have peaks at the beginning of the maneuver. Also, note that the peak of the RMSE in position raises slightly above the measurement error at the beginning of the maneuver. During the maneuvering, the RMSE of the IMM estimator is similar to the that of the better design of the NCV filter and that of the NCA filter. In this simulation, the target acceleration moves instantly from 0 to 40 m/s^2 and this is almost unavoidable. In practice, targets require a system response time of a few seconds to achieve maximum acceleration from a nonmaneuver state. Figure 19-22 gives the NEES as defined in (19.23) for the IMM estimator. The results indicate that the covariance for IMM estimator is too large when the target is not maneuvering and adapts to the maneuvering after the initial peak. Figure 19-24 gives the mode probabilities of the IMM estimator for this example.

The IMM estimator can also play a key role in the allocation of radar resources of phased radars when tracking maneuvering targets. Instead of setting a quasiperiodic data radar for surveillance tracking [20], the time at which the next measurement is required to maintain the track is computed after every measurement, and then a radar dwell is scheduled for the required time. Thus, the sample period between consecutive measurements is adapted giving rise to aperiodic data. The near consistency of the IMM estimator allows for reliable computation of the next required revisit time for the radar [15,21]. The next required revisit time is computed as the latest time at which the target can be expected to be in the predicted track gate, which is defined by the radar beamwidth, the length of the range window used in track mode, and maximum acceleration of the target. The revist time is determined by selecting the minimum of the required revisit times for range, bearing, and elevation and applying an upper limit on the revisit period that is based on the maximum acceleration of the target. For the angles, the revisit times are computed to be the times at which the standard deviations of the predicted angles exceeds a fraction of the beamwidth. For the range, the revisit time is computed to be the time at which the standard deviation of the predicted range exceeds a fraction of the range window. Using the output of the IMM estimator to compute the next required revisit time gives revisit intervals that automatically reflect target maneuvers, target range, missed detections, fluctuating signal amplitudes, the length of the range window, and the radar beamwidth at the predicted location of the target. Analysis of the results from simulation studies and real-time experiments [15] indicate that the IMM estimator with adaptive revisit times provides a 50% reduction in radar time and energy required for surveillance tracking by a conventional approach.

19.3 KINEMATIC MOTION MODELS

For most tracking systems, the target state is modeled in Cartesian coordinates and maintained in a reference frame that is stabilized relative to the location of the platform. The dynamical equation that is commonly used to represent the motion of the target relative to the platform is given by

$$\boldsymbol{X}_{k+1} = \mathbf{F}_k \boldsymbol{X}_k + \boldsymbol{G}_k \boldsymbol{v}_k \tag{19.119}$$

The state vector for a NCV motion model is given by

$$\boldsymbol{X}_k = \begin{bmatrix} x_k & \dot{x}_k & y_k & \dot{y}_k & z_k & \dot{z}_k \end{bmatrix}^T \tag{19.120}$$

where (x_k, y_k, z_k) represents the position of the target in Cartesian coordinates at time t_k and $(\dot{x}_k, \dot{y}_k, \dot{z}_k)$ represents the velocity of the target. When tracking with an early warning surveillance radar that measures only the range and bearing of the target, the state vector may be of a reduced Cartesian space and include only the four elements associated with x_k and y_k. Note that in this case, x_k and y_k are in a tilted frame relative to the horizontal plane and that tilt angle changes as the target moves. Thus, the x_k and y_k in the tilted frame is certainly different from those in the stabilized frame. The state vector for an NCA motion model in full Cartesian space is given by

$$\boldsymbol{X}_k = \begin{bmatrix} x_k & \dot{x}_k & \ddot{x}_k & y_k & \dot{y}_k & \ddot{y}_k & z_k & \dot{z}_k & \ddot{z}_k \end{bmatrix}^T \tag{19.121}$$

The dynamical constraint for NCV or NCA motion in full Cartesian space is given by

$$\mathbf{F}_k = \begin{bmatrix} \mathbf{A}_k & \mathbf{0}_{m\times m} & \mathbf{0}_{m\times m} \\ \mathbf{0}_{m\times m} & \mathbf{A}_k & \mathbf{0}_{m\times m} \\ \mathbf{0}_{m\times m} & \mathbf{0}_{m\times m} & \mathbf{A}_k \end{bmatrix} \tag{19.122}$$

where $\mathbf{0}_{m\times m}$ denotes an m by m matrix of zeros. For NCV motion, $m = 2$, while $m = 3$ for NCA motion. The process noise is included to account for the uncertainty associated with the unknown maneuvers of the targets. The input process noise is characterized by $\boldsymbol{G}_k \mathbf{Q}_k \boldsymbol{G}_k^T$ that is given by

$$\boldsymbol{G}_k \mathbf{Q}_k \boldsymbol{G}_k^T = \begin{bmatrix} q_k^x \boldsymbol{B}_k \boldsymbol{B}_k^T & \mathbf{0}_{m\times m} & \mathbf{0}_{m\times m} \\ \mathbf{0}_{m\times m} & q_k^y \boldsymbol{B}_k \boldsymbol{B}_k^T & \mathbf{0}_{m\times m} \\ \mathbf{0}_{m\times m} & \mathbf{0}_{m\times m} & q_k^z \boldsymbol{B}_k \boldsymbol{B}_k^T \end{bmatrix} \tag{19.123}$$

The choice of a motion model is governed by the maximum acceleration of the target, the quality of the sensor measurements, and the measurement rate. Since no algorithm exists for predicting the best model, Monte Carlo simulations are typically used to assess the best model for a tracking system. If the quality of the measurements and the measurement rate are not sufficient to estimate acceleration, models that do not include acceleration in the target state will provide better tracking performance. If only two or three measurements are taken during maneuvers, accurate estimation of the acceleration will not be possible, and the acceleration should not be included in the target state. If the quality of the measurements and the measurement rate allow for estimation of the acceleration, and the bias in the position estimates of the nearly constant velocity filter during the maximum acceleration of the target [22] is larger than the measurement errors, then acceleration should be included the target state. In this section, the motion models for NCV, NCA, Singer correlated acceleration, and constant speed turns are summarized [2,5,7,17,23].

19.3.1 Nearly Constant Velocity Motion Model

The dynamical constraint for NCV motion is given by (19.122) with

$$\mathbf{A}_k = \begin{bmatrix} 1 & \delta_k \\ 0 & 1 \end{bmatrix} \tag{19.124}$$

where $\delta_k = t_k - t_{k-1}$ denotes the sample period and t_k denotes the time of measurement k. The process noise is typically treated as DWNA or CWNA [7].

For the NCV motion model, the process noise covariance matrix for DWNA is given by (19.123) with

$$\boldsymbol{B}_k \boldsymbol{B}_k^T = \begin{bmatrix} \frac{\delta_k^4}{4} & \frac{\delta_k^3}{2} \\ \frac{\delta_k^3}{2} & \delta_k^2 \end{bmatrix} \tag{19.125}$$

The random tracking index is often used to characterize the filter design. While expressing the process noise variances in the Cartesian coordinates is straightforward, expressing the measurement variances in Cartesian coordinates is more difficult because the measurement errors are typically characterized in spherical or polar coordinates. The maneuver index for the NCV filter with DWNA [7] is given by

$$\Gamma_{DWNA} = \frac{\sigma_v T^2}{\sigma_w} \tag{19.126}$$

where σ_v^2 denotes the variance of the acceleration errors, T is the nominal measurement period, and σ_w^2 is the variance of the measurement errors in the corresponding coordinate. The σ_w^2 in each coordinate will vary with the target location. An appropriate σ_w is identified for a radar tracking problem by picking the maximum or average over the three coordinates or computing it with target location. Once an appropriate σ_w^2 and the maximum acceleration of the target have been identified, the methods in Section 19.2.2.1 can be used for the filter design. The key item of that design is to select the appropriate scaling $\kappa_1(\Gamma_D)$ between maximum acceleration of the target and σ_v.

For the NCV motion model with CWNA, the process noise covariance matrix is given by (19.123) with

$$\boldsymbol{B}_k \boldsymbol{B}_k^T q_k = \tilde{q} \begin{bmatrix} \frac{\delta_k^3}{3} & \frac{\delta_k^2}{2} \\ \frac{\delta_k^2}{2} & \delta_k \end{bmatrix} \tag{19.127}$$

The square of the random tracking index for the NCV filter with CWNA [7] is given by

$$\Gamma_{CWNA}^2 = \frac{\tilde{q} T^3}{\sigma_w^2} \tag{19.128}$$

where $\tilde{q}$ denotes the power spectral density of the CWNA. Once an appropriate σ_w^2 and the maximum acceleration of the target have been identified, the methods in Section 19.2.2.3 can be used to design the filter. The key step in the design is to select the appropriate scaling $\kappa_2(\Gamma_D)$ between maximum acceleration of the target and $\tilde{q}$.

In the case of DWNA, the acceleration errors are assumed to be fixed during each sample period and independent between any two sample periods. Thus, if the time periods between measurements vary widely, the CWNA is more appropriate, because the basic target motion model is independent of the sampling interval. This is supported by the fact that the maneuver index of DWNA is a higher order function of sample period T than the maneuver index for CWNA. Also, if the process noise for CWNA is used for the illustrative example in [6], better tracking is achieved by doubling the data rate. Thus, the process noise based on CWNA is more appropriate for tracking with multiple radars because the sample periods between measurements tend to vary significantly. However, the DWNA model leads to simpler equations and it is most often used for illustration and publications.

19.3.2 Nearly Constant Acceleration Motion Model

The dynamical constraint for NCA motion is given by (19.122) with

$$\mathbf{A}_k = \begin{bmatrix} 1 & \delta_k & \frac{1}{2}\delta_k^2 \\ 0 & 1 & \delta_k \\ 0 & 0 & 1 \end{bmatrix} \tag{19.129}$$

For the NCA model, the process noise covariance matrix for DWPA is given by (19.123) with

$$\boldsymbol{B}_k \boldsymbol{B}_k^T = \begin{bmatrix} \frac{1}{4}\delta_k^4 & \frac{1}{2}\delta_k^3 & \frac{1}{2}\delta_k^2 \\ \frac{1}{2}\delta_k^3 & \delta_k^2 & \delta_k \\ \frac{1}{2}\delta_k^2 & \delta_k & 1 \end{bmatrix} \tag{19.130}$$

The maneuver index for the NCA filter with DWPA [7] is given by

$$\Gamma_{DWPA} = \frac{\sigma_v T^2}{\sigma_w} \tag{19.131}$$

where σ_v^2 denotes the variance of the acceleration errors. The σ_w^2 in each coordinate will vary with the target location. An appropriate value of σ_w is identified for a radar tracking problem by picking the maximum or average over the three coordinates or computing it with target location. The maneuver index is often used to characterize the filter design. However, selection of σ_v^2 presents a challenge. One design method begins by setting σ_v equal to the maximum incremental change in acceleration of the target between any two measurements. Then, Monte Carlo simulations are conducted to characterize performance and refine the selection of σ_v.

19.3.3 Singer Motion Model

The Singer motion model represents the acceleration errors as an auto-correlated sequence of accelerations. The autocorrelation decays exponentially with the time period between the measurements. The dynamical constraint for the Singer model is given by (19.122) with

$$\mathbf{A}_k = \begin{bmatrix} 1 & \delta_k & \tau_m^2\left[\frac{\delta_k}{\tau_m} + \exp\left[-\frac{\delta_k}{\tau_m}\right] - 1\right] \\ 0 & 1 & \tau_m\left[1 - \exp\left[-\frac{\delta_k}{\tau_m}\right]\right] \\ 0 & 0 & \exp\left[-\frac{\delta_k}{\tau_m}\right] \end{bmatrix} \tag{19.132}$$

where τ_m is the time constant for the maneuver. For the Singer model, the process noise covariance matrix is given in [24] by (19.123) with

$$\boldsymbol{B}_k \boldsymbol{B}_k^T = \sigma_v^2 \begin{bmatrix} b_{11} & b_{12} & b_{13} \\ b_{12} & b_{22} & b_{23} \\ b_{13} & b_{23} & b_{33} \end{bmatrix} \tag{19.133}$$

where

$$b_{11} = \tau_m^4 \left[1 + \frac{2\delta_k}{\tau_m} - \exp\left[-2\frac{\delta_k}{\tau_m}\right] - 2\frac{\delta_k^2}{\tau_m^2} + \frac{2\delta_k^3}{3\tau_m^3} - 4\frac{\delta_k}{\tau_m}\exp\left[-\frac{\delta_k}{\tau_m}\right] \right] \tag{19.134}$$

$$b_{12} = \tau_m^3 \left[1 - \frac{2\delta_k}{\tau_m} + \exp\left[-2\frac{\delta_k}{\tau_m}\right] - 2\left(1 - \frac{\delta_k}{\tau_m}\right)\exp\left[-\frac{\delta_k}{\tau_m}\right] + \frac{\delta_k^2}{\tau_m^2} \right] \tag{19.135}$$

$$b_{13} = \tau_m^2 \left[1 - \exp\left[-2\frac{\delta_k}{\tau_m}\right] - 2\frac{\delta_k}{\tau_m}\exp\left[-\frac{\delta_k}{\tau_m}\right] \right] \tag{19.136}$$

$$b_{22} = \tau_m^2 \left[\frac{2\delta_k}{\tau_m} - 3 - \exp\left[-2\frac{\delta_k}{\tau_m}\right] + 4\exp\left[-\frac{\delta_k}{\tau_m}\right] \right] \tag{19.137}$$

$$b_{23} = \tau_m \left[1 + \exp\left[-2\frac{\delta_k}{\tau_m}\right] - 2\exp\left[-\frac{\delta_k}{\tau_m}\right] \right] \tag{19.138}$$

$$b_{33} = 1 - \exp\left[-2\frac{\delta_k}{\tau_m}\right] \tag{19.139}$$

and σ_v^2 denotes the variance of the acceleration errors in the model. When $\delta_k \tau_m^{-1} \leq 0.1$, the state transition matrix approaches that of NCA and [24]

$$\boldsymbol{B}_k \boldsymbol{B}_k^T \approx \frac{2\sigma_v^2}{\tau_m} \begin{bmatrix} \frac{\delta_k^5}{20} & \frac{\delta_k^4}{8} & \frac{\delta_k^3}{6} \\ \frac{\delta_k^4}{8} & \frac{\delta_k^3}{3} & \frac{\delta_k^2}{2} \\ \frac{\delta_k^3}{6} & \frac{\delta_k^2}{2} & \delta_k \end{bmatrix} \tag{19.140}$$

When $\delta_k \tau_m^{-1} \geq 0.9$, the state transition matrix approaches that of NCV and [24]

$$\boldsymbol{B}_k \boldsymbol{B}_k^T \approx \sigma_v^2 \begin{bmatrix} \frac{2\delta_k^3 \tau_m}{3} & \delta_k^2 \tau_m & \tau_m^2 \\ \delta_k^2 \tau_m & 2\delta_k \tau_m & \tau_m \\ \tau_m^2 & \tau_m & 1 \end{bmatrix} \tag{19.141}$$

The selection of σ_v^2 presents a challenge. One design method begins by setting σ_v equal to the maximum incremental change in acceleration of the target between any two measurements. Then, Monte Carlo simulations are conducted to characterize performance and refine the selection of σ_v.

19.3.4 Nearly Constant Speed Motion Model

Nearly constant speed motion is very important because it tends to characterize the motion of targets performing maneuvers with control surfaces. The high acceleration maneuvers of many targets are achieved with control surfaces and the speed remains nearly constant during maneuvers. A common approach to modeling constant speed maneuvers is to assume a constant turn rate in the horizontal plane [7] and estimate the turn rate Ω. The limitations of this approach include the requirements that the turn rate remain nearly constant and the maneuver be performed in the horizontal plane. These are two serious limitations. This model has been extended to include both vertical and horizontal maneuvers, but the approach is rather complex and involves estimation of two parameters for the turn. An alternate approach is the use of a piecewise constant speed motion model in which the motion is characterized by incremental constant speed turns in an arbitrary plane. This approach is generally preferable. The dynamical constraint for the piecewise constant speed motion is given by (19.122) with

$$\mathbf{A}_k = \begin{bmatrix} 1 & \dfrac{\sin(\Omega_k\delta_k)}{\Omega_k} & \dfrac{1-\cos(\Omega_k\delta_k)}{\Omega_k^2} \\ 0 & \cos(\Omega_k\delta_k) & \dfrac{\sin(\Omega_k\delta_k)}{\Omega_k} \\ 0 & -\Omega_k\sin(\Omega_k\delta_k) & \cos(\Omega_k\delta_k) \end{bmatrix} \tag{19.142}$$

where

$$\Omega_k = \frac{\sqrt{\ddot{x}_k^2 + \ddot{y}_k^2 + \ddot{z}_k^2}}{\sqrt{\dot{x}_k^2 + \dot{y}_k^2 + \dot{z}_k^2}} \tag{19.143}$$

The process noise model for NCA motion is appropriate for the piecewise constant speed motion model. However, the selection of σ_v^2 continues to present a challenge. Again, σ_v can be set equal to the maximum incremental change in acceleration of the target between any two measurements. As before, Monte Carlo simulations are conducted to characterize performance and refine the selection of σ_v.

Targets that maneuver with control surfaces tend to retain a nearly constant speed through the maneuver. The use of this kinematic constraint has been investigated as a means to improve the tracking of highly maneuvering targets. The application of a kinematic constraint as a pseudo-measurement has been demonstrated to improve the tracking of constant speed, maneuvering targets [17,25].

19.4 MEASUREMENT MODELS

Radar measurements are typically in polar or spherical coordinates. For a radar measuring the target location in full Cartesian coordinates, the two angles are actually measured in two orthogonal planes, say U and V, in the aperture of the antenna [26]. In some cases, the radar measures the velocity of the target along the range vector between the target and the antenna. The general measurement equation is given by

$$\boldsymbol{Z}_k = h_k(\boldsymbol{X}_k) + \boldsymbol{w}_k \tag{19.144}$$

where

$\boldsymbol{Z}_k$ = measurement vector at time t_k.

$\boldsymbol{w}_k$ = measurement error at time t_k with $\boldsymbol{w}_k \sim \mathrm{N}(0, \mathbf{R}_k)$.

Typically, the target state estimate is maintained in a Cartesian reference frame that is stabilized (often in an inertial coordinate frame) relative to any motion of the radar antenna. In this case, an affine transform is typically used to define the relationship between the two coordinate systems and the measurement equation is given by

$$\boldsymbol{Z}_k = h_k(\mathbf{M}_k \boldsymbol{X}_k + \boldsymbol{L}_k) + \boldsymbol{w}_k \tag{19.145}$$

where

$\mathbf{M}_k$ = matrix that rotates the target state vector into the frame of the antenna at time t_k.
$\boldsymbol{L}_k$ = vector that translates the target state vector into the frame of the antenna at time t_k.

The typical formulation of the measurements as range, bearing, elevation, and range rate is presented first. Then, the sine space formulation of the measurements in the antenna frame is presented. The issue of unbiased converted measurements will be addressed in both sections. The measurement equation for LFM waveforms is then addressed. Finally, tracking with measurements of reduced dimension is discussed.

19.4.1 Measurements in Stabilized Coordinates

Typically, radar measurements are modeled in spherical or polar coordinates in which measurements as a function of $\boldsymbol{X}_k$ are given by

$$\boldsymbol{Z}_k = \begin{bmatrix} r_k \\ b_k \\ e_k \\ \dot{r}_k \end{bmatrix} = h_k(\boldsymbol{X}_k) + \boldsymbol{w}_k = \begin{bmatrix} h_r(\boldsymbol{X}_k) \\ h_b(\boldsymbol{X}_k) \\ h_e(\boldsymbol{X}_k) \\ h_d(\boldsymbol{X}_k) \end{bmatrix} + \begin{bmatrix} w_{rk} \\ w_{bk} \\ w_{ek} \\ w_{dk} \end{bmatrix} \tag{19.146}$$

where

r_k = measured range at time t_k.
b_k = measured bearing at time t_k.
e_k = measured elevation at time t_k.
$\dot{r}_k$ = Doppler-derived measurement of range rate at time t_k.
w_{rk} = error in range measurement with $w_{rk} \sim \mathrm{N}(0, \sigma_{rk}^2)$.
w_{bk} = error in bearing measurement with $w_{bk} \sim \mathrm{N}(0, \sigma_{bk}^2)$.
w_{ek} = error in elevation measurement with $w_{ek} \sim \mathrm{N}(0, \sigma_{ek}^2)$.
w_{dk} = error in Doppler range rate measurement with $w_{dk} \sim \mathrm{N}(0, \sigma_{dk}^2)$.

Also

$$h_r(\boldsymbol{X}_k) = \sqrt{x_k^2 + y_k^2 + z_k^2} \tag{19.147}$$

$$h_b(\boldsymbol{X}_k) = \begin{cases} \tan^{-1}\left(\dfrac{x_k}{y_k}\right), & |y_k| > |x_k| \\ \dfrac{\pi}{2} - \tan^{-1}\left(\dfrac{y_k}{x_k}\right), & |x_k| > |y_k| \end{cases} \tag{19.148}$$

$$h_e(\boldsymbol{X}_k) = \begin{cases} \tan^{-1}\left(\dfrac{z_k}{\sqrt{x_k^2 + y_k^2}}\right), & |\sqrt{x_k^2 + y_k^2}| > |z_k| \\ \dfrac{\pi}{2} - \tan^{-1}\left(\dfrac{\sqrt{x_k^2 + y_k^2}}{z_k}\right), & |z_k| > |\sqrt{x_k^2 + y_k^2}| \end{cases} \tag{19.149}$$

$$h_d(\boldsymbol{X}_k) = \frac{x_k \dot{x}_k + y_k \dot{y}_k + z_k \dot{z}_k}{\sqrt{x_k^2 + y_k^2 + z_k^2}} \tag{19.150}$$

where bearing b_k and elevation e_k measurements are two angles in a radar reference frame. The two cases for the bearing and elevation measurements are included to prevent singularities in the measurement functions. Note that a four-quadrant arctangent function or some other logic must be employed to ensure the bearing measurement is in the proper quadrant. While (19.147) through (19.150) represent the different kinematic measurements of a radar, many radars do not measure all four kinematic parameters. For example, most long-range surveillance radars measure only range and bearing (and possibly range rate). The kinematic state of the target is modeled in a plane that is tilted relative to the local horizontal plane and the angle of the tilt changes as the target range or altitude change.

The extended Kalman filter (EKF) for tracking with radar measurements uses a linearized model for the measurements. Thus,

$$\mathbf{H}_k = \begin{bmatrix} \dfrac{\partial h_r(\boldsymbol{X}_k)}{\partial \boldsymbol{X}_k} \\ \dfrac{\partial h_b(\boldsymbol{X}_k)}{\partial \boldsymbol{X}_k} \\ \dfrac{\partial h_e(\boldsymbol{X}_k)}{\partial \boldsymbol{X}_k} \\ \dfrac{\partial h_d(\boldsymbol{X}_k)}{\partial \boldsymbol{X}_k} \end{bmatrix}_{\boldsymbol{X}_k = \boldsymbol{X}_{k|k-1}} = \begin{bmatrix} \boldsymbol{H}_{rk} \\ \boldsymbol{H}_{bk} \\ \boldsymbol{H}_{ek} \\ \boldsymbol{H}_{dk} \end{bmatrix} \tag{19.151}$$

where the one-step ahead predicted state is given by

$$\boldsymbol{X}_{k|k-1} = \begin{bmatrix} x_{k|k-1} & \dot{x}_{k|k-1} & y_{k|k-1} & \dot{y}_{k|k-1} & z_{k|k-1} & \dot{z}_{k|k-1} \end{bmatrix}^T \tag{19.152}$$

The rows of H_k are given by

$$\boldsymbol{H}_{rk} = \begin{bmatrix} \dfrac{x_{k|k-1}}{r_{k|k-1}} & 0 & \dfrac{y_{k|k-1}}{r_{k|k-1}} & 0 & \dfrac{z_{k|k-1}}{r_{k|k-1}} & 0 \end{bmatrix} \tag{19.153}$$

$$\boldsymbol{H}_{bk} = \begin{bmatrix} \dfrac{y_{k|k-1}}{rh_{k|k-1}^2} & 0 & -\dfrac{x_{k|k-1}}{rh_{k|k-1}^2} & 0 & 0 & 0 \end{bmatrix} \tag{19.154}$$

$$\boldsymbol{H}_{ek} = \begin{bmatrix} -\dfrac{x_{k|k-1} z_{k|k-1}}{r_{k|k-1}^2 rh_{k|k-1}} & 0 & -\dfrac{y_{k|k-1} z_{k|k-1}}{r_{k|k-1}^2 rh_{k|k-1}} & 0 & \dfrac{rh_{k|k-1}}{r_{k|k-1}^2} & 0 \end{bmatrix} \tag{19.155}$$

$$\boldsymbol{H}_{dk} = \begin{bmatrix} \boldsymbol{H}_{dk}(1) & \dfrac{x_{k|k-1}}{r_{k|k-1}} & \boldsymbol{H}_{dk}(3) & \dfrac{y_{k|k-1}}{r_{k|k-1}} & \boldsymbol{H}_{dk}(5) & \dfrac{z_{k|k-1}}{r_{k|k-1}} \end{bmatrix} \tag{19.156}$$

where

$$H_{dk}(1) = \frac{\left(y_{k|k-1}^2 + z_{k|k-1}^2\right)\dot{x}_{k|k-1} - (y_{k|k-1}\dot{y}_{k|k-1} + z_{k|k-1}\dot{z}_{k|k-1})x_{k|k-1}}{r_{k|k-1}^3} \tag{19.157}$$

$$H_{dk}(3) = \frac{\left(x_{k|k-1}^2 + z_{k|k-1}^2\right)\dot{y}_{k|k-1} - (x_{k|k-1}\dot{x}_{k|k-1} + z_{k|k-1}\dot{z}_{k|k-1})y_{k|k-1}}{r_{k|k-1}^3} \tag{19.158}$$

$$H_{dk}(5) = \frac{rh_{k|k-1}^2\dot{z}_{k|k-1} - (x_{k|k-1}\dot{x}_{k|k-1} + y_{k|k-1}\dot{y}_{k|k-1})z_{k|k-1}}{r_{k|k-1}^3} \tag{19.159}$$

$$r_{k|k-1} = \sqrt{x_{k|k-1}^2 + y_{k|k-1}^2 + z_{k|k-1}^2} \tag{19.160}$$

$$rh_{k|k-1} = \sqrt{x_{k|k-1}^2 + y_{k|k-1}^2} \tag{19.161}$$

Since the EKF uses a linearized output matrix for $h(\boldsymbol{X}_k)$ in the covariance update, the EKF does not provide an optimal estimate of the target state. The results of [7,27] suggest that the performance of the EKF is degraded significantly when

$$\frac{r_k \max\{\sigma_{bk}^2, \sigma_{ek}^2\}}{\sigma_{rk}} \geq 0.4 \tag{19.162}$$

Performing a debiased coordinate conversion of the spherical measurements to Cartesian measurements was proposed in [27] and refined in [28] to reduce the impacts of this limitation of the EKF. In this case, the measurements of position become direct observations of the Cartesian coordinates of the target position and the measurement equation is a simple linear function of the state. However, the EKF will be required to process the Doppler range rate measurement because it is not easily be converted to a Cartesian measurement. Other issues associated with the processing of spherical radar measurements as Cartesian observations are discussed in [29].

19.4.2 Measurements in Sine Space

The angles of a target are actually measured in two orthogonal planes, say U and V, at the aperture of the radar antenna [26]. When considering the measurements in the U-plane and the V-plane, the radar measurements (direction cosines) as a function of $\boldsymbol{X}_k$ in a Cartesian reference frame at the antenna are given by

$$\boldsymbol{Z}_k = \begin{bmatrix} r_k \\ U_k \\ V_k \\ \dot{r}_k \end{bmatrix} = \begin{bmatrix} h_r(\boldsymbol{X}_k) \\ h_U(\boldsymbol{X}_k) \\ h_V(\boldsymbol{X}_k) \\ h_d(\boldsymbol{X}_k) \end{bmatrix} + \begin{bmatrix} w_{rk} \\ w_{Uk} \\ w_{Vk} \\ w_{dk} \end{bmatrix} \tag{19.163}$$

where

U_k = measured sine of the angle of the target in the U-plane at time t_k.
V_k = measured sine of the angle of the target in the V-plane at time t_k.
w_{Uk} = error in the U_k measurement with $w_{Uk} \sim \mathrm{N}(0, \sigma_{Uk}^2)$.
w_{Vk} = error in the V_k measurement with $w_{Vk} \sim \mathrm{N}(0, \sigma_{Vk}^2)$.

Also

$$h_U(\boldsymbol{X}_k) = \frac{x_k}{\sqrt{x_k^2 + y_k^2 + z_k^2}} \tag{19.164}$$

$$h_V(\boldsymbol{X}_k) = \frac{y_k}{\sqrt{x_k^2 + y_k^2 + z_k^2}} \tag{19.165}$$

The radar measurements as a function of $\boldsymbol{X}_k$ in a radar reference frame displaced from the antenna are given by

$$\boldsymbol{Z}_k = \begin{bmatrix} r_k \\ U_k \\ V_k \\ \dot{r}_k \end{bmatrix} = h_k(M_k\boldsymbol{X}_k + \boldsymbol{L}_k) + \boldsymbol{w}_k = \begin{bmatrix} h_r(M_k\boldsymbol{X}_k + \boldsymbol{L}_k) \\ h_U(M_k\boldsymbol{X}_k + \boldsymbol{L}_k) \\ h_V(M_k\boldsymbol{X}_k + \boldsymbol{L}_k) \\ h_d(M_k\boldsymbol{X}_k + \boldsymbol{L}_k) \end{bmatrix} + \begin{bmatrix} w_{rk} \\ w_{Uk} \\ w_{Vk} \\ w_{dk} \end{bmatrix} \tag{19.166}$$

The EKF or converted measurements is commonly used for tracking the target with measurements in sine space. Implementation of the EKF requires the following

$$\mathbf{H}_k = \begin{bmatrix} \dfrac{\partial h_r(\boldsymbol{X}_k)}{\partial \boldsymbol{X}_k} \\ \dfrac{\partial h_U(\boldsymbol{X}_k)}{\partial \boldsymbol{X}_k} \\ \dfrac{\partial h_V(\boldsymbol{X}_k)}{\partial \boldsymbol{X}_k} \\ \dfrac{\partial h_d(\boldsymbol{X}_k)}{\partial \boldsymbol{X}_k} \end{bmatrix}_{\boldsymbol{X}_k = M_k\boldsymbol{X}_{k|k-1} + \boldsymbol{L}_k} = \begin{bmatrix} \boldsymbol{H}_{rk} \\ \boldsymbol{H}_{Uk} \\ \boldsymbol{H}_{Vk} \\ \boldsymbol{H}_{dk} \end{bmatrix} \tag{19.167}$$

Let

$$\boldsymbol{X}'_{k|k-1} = M_k\boldsymbol{X}_{k|k-1} + \boldsymbol{L}_k \tag{19.168}$$

and

$$\boldsymbol{X}'_{k|k-1} = \begin{bmatrix} x'_{k|k-1} & \dot{x}'_{k|k-1} & y'_{k|k-1} & \dot{y}'_{k|k-1} & z'_{k|k-1} & \dot{z}'_{k|k-1} \end{bmatrix}^T \tag{19.169}$$

The elements of $\mathbf{H}_k$ are then given by (19.153) and (19.156) with $\boldsymbol{X}_{k|k-1} = \boldsymbol{X}'_{k|k-1}$ and

$$\boldsymbol{H}_{Uk} = \begin{bmatrix} \dfrac{\left(r'_{k|k-1}\right)^2 - \left(x'_{k|k-1}\right)^2}{\left(r'_{k|k-1}\right)^3} & 0 & -\dfrac{x'_{k|k-1}y'_{k|k-1}}{\left(r'_{k|k-1}\right)^3} & 0 & -\dfrac{x'_{k|k-1}z'_{k|k-1}}{\left(r'_{k|k-1}\right)^3} & 0 \end{bmatrix} \tag{19.170}$$

$$\boldsymbol{H}_{Vk} = \begin{bmatrix} -\dfrac{x'_{k|k-1}y'_{k|k-1}}{\left(r'_{k|k-1}\right)^3} & 0 & \dfrac{\left(r'_{k|k-1}\right)^2 - \left(y'_{k|k-1}\right)^2}{\left(r'_{k|k-1}\right)^3} & 0 & -\dfrac{y'_{k|k-1}z'_{k|k-1}}{\left(r'_{k|k-1}\right)^3} & 0 \end{bmatrix} \tag{19.171}$$

where

$$r'_{k|k-1} = \sqrt{\left(x'_{k|k-1}\right)^2 + \left(y'_{k|k-1}\right)^2 + \left(z'_{k|k-1}\right)^2} \tag{19.172}$$

$$rh'_{k|k-1} = \sqrt{\left(x'_{k|k-1}\right)^2 + \left(y'_{k|k-1}\right)^2} \tag{19.173}$$

Since the EKF uses a linearized output matrix for $h(\boldsymbol{X}_k)$ in the covariance update, the EKF does not provide an optimal estimate of the target state. The results of [30] suggest that the performance of the EKF is degraded significantly when

$$r_k \frac{\max\{\sigma_{Uk}^2, \sigma_{Vk}^2\}}{\sigma_{rk}} \sqrt{\frac{3 - 3\bar{U}_{0k}^2 + \bar{U}_{0k}^4}{1 - 2\bar{U}_{0k}^2 + \bar{U}_{0k}^4}} \geq \frac{2}{3} \tag{19.174}$$

where $\bar{U}_{0k} = \sqrt{U_k^2 + V_k^2}$. Performing a nearly unbiased coordinate conversion of the measurements in sine space to Cartesian measurements is proposed in [30] to reduce the impact on this limitation of the EKF. In this case, the measurements become direct observations of the Cartesian coordinates of the target position and the measurement equation is a simple linear function of the state.

19.4.3 Measurements with LFM Waveforms

When a radar use an LFM waveform, the range measurement is coupled to the range rate induced Doppler of the target [12]. For an LFM waveform, the output function for the range measurement is given by

$$h_{rd}(\boldsymbol{X}_k) = h_r(\boldsymbol{X}_k) + \Delta t \; h_d(\boldsymbol{X}_k) \tag{19.175}$$

where

$$\Delta t = \frac{f_t \tau}{f_2 - f_1} \tag{19.176}$$

with f_t denoting the frequency of the transmitted waveform, τ denoting the duration of the transmitted pulse, f_1 denoting the initial frequency in the waveform modulation, and f_2 denoting the final frequency in the waveform modulation. The linearized output matrix is given by

$$\mathbf{H}_{rdk} = \mathbf{H}_{rk} + \Delta t \; \mathbf{H}_{dk} \tag{19.177}$$

As noted in [12] and shown analytically in [13], the range-Doppler coupling of the waveform affects the tracking accuracy. For example, an up-chirp waveform (i.e., $f_2 > f_1$ or $\Delta t > 0$) gives lower steady-state tracking errors than an equivalent down-chirp waveform (i.e., $f_2 < f_1$ or $\Delta t < 0$).

19.4.4 Surveillance Radars with Measurements of Reduced Dimension

Early warning or long range surveillance radars typically measure only the range and bearing (and possibly range rate) of the target. In this case, the radar measurements are given as a function of $\boldsymbol{X}_k$ by

$$\mathbf{Z}_k = \begin{bmatrix} r_k \\ b_k \\ \dot{r}_k \end{bmatrix} = h_k(\boldsymbol{X}_k) + \boldsymbol{w}_k = \begin{bmatrix} h_r(\boldsymbol{X}_k) \\ h_b(\boldsymbol{X}_k) \\ h_d(\boldsymbol{X}_k) \end{bmatrix} + \begin{bmatrix} w_{rk} \\ w_{bk} \\ w_{dk} \end{bmatrix} \tag{19.178}$$

where measured range is defined in a plane that is tilted relative to the horizontal plane. Full Cartesian estimate of the target state is not achievable under most circumstances. For tracking, the measurements are usually defined as a function in a two coordinate Cartesian system. For example, the measured range is given by $r_k = \sqrt{\bar{x}_k^2 + \bar{y}_k^2}$, where $\bar{x}_k$ and $\bar{y}_k$ are the coordinates of the target in the tilted plane. Note that the tilt angle changes as the target moves toward the radar or changes altitude. Also, note that the measured range rate is not exactly equal to the derivative to the range r_k. The measured bearing will have the same functional form of $\bar{x}_k$ and $\bar{y}_k$, while the Doppler measurement cannot be expressed solely as a function of $\bar{x}_k$ and $\bar{y}_k$ and it remain a function of z_k and $\dot{z}_k$. Thus, the processing of Doppler measurements while tracking $\bar{x}_k$ and $\bar{y}_k$ is problematic. The range and range rate are often tracked in a separate filter.

19.5 RADAR TRACK FILTERING

After the data association has been accomplished and a measurement has been assigned to a track, kinematic state estimation is performed for the track with that measurement. For most radar systems, the target motion is modeled in Cartesian coordinates, while radar measurements are typically in polar or spherical coordinates. Since the measurements of (19.144) are a nonlinear function of the state, the EKF is often employed to estimate the state [7]. The EKF is a stochastic state estimation approach that presumes an imperfect model for the target motion. In this approach, the target motion model includes a random process and a perfect estimate of the kinematic state is not possible. However, in some applications, a parametric estimation approach is taken. In this approach, a perfect model for the target motion is assumed, and the time period over which the model is applied is limited to prevent distortion of the data by violation of the model. In this approach, the covariance of the state estimate (or track) will approach zero as more data are processed. As the covariance of the track approaches zero, the gain for processing new data will likewise approach zero. Nonlinear least squares or maximum likelihood estimation are the typical methods employed for a parametric approach to track filtering, and since a window or segment of the measurements is processed as a batch, the parametric estimator is often referred to as a batch estimator.

In this section, nonlinear least squares estimation and the EKF are presented for the radar tracking problem. A few comments on the converted measurement filter are also given.

19.5.1 Nonlinear Least Squares Estimation

For a parametric or batch estimator approach to radar tracking of a target with nonlinear measurements, the measurement equation is given by (19.144) and the dynamical motion model is given by

$$\boldsymbol{X}_{k+1} = \mathbf{F}_k \boldsymbol{X}_k \tag{19.179}$$

where $\boldsymbol{X}_k$ is the state vector for the target at time t_k. This estimator is often referred to as a *sliding window estimator* or *batch estimator*. The name “sliding window estimator” is derived from the idea that the estimate is based on a window of the N most recent measurements and that window slides along with time. Note that estimates based on fewer measurements will have a higher variance, but the distortions of the underlying true trajectory will be smaller.

Using (19.179) and (19.144), the measurement at t_N can be rewritten as

$$Z_N = h(X_N) + w_N = h(\mathbf{F}_{N-1}\mathbf{F}_{N-2}\ldots\mathbf{F}_1 X_1) + w_N \tag{19.180}$$

Given N measurements between t_1 and t_N, an augmented measurement equation can be written in terms of X_1 as

$$\bar{Z}_1^N = \begin{bmatrix} Z_1 \\ Z_2 \\ \vdots \\ Z_{N-1} \\ Z_N \end{bmatrix} = \begin{bmatrix} h_1(X_1) \\ h_2(\mathbf{F}_1 X_1) \\ \vdots \\ h_{N-1}(\mathbf{F}_{N-2}\mathbf{F}_{N-3}\ldots\mathbf{F}_1 X_1) \\ h_N(\mathbf{F}_{N-1}\mathbf{F}_{N-2}\ldots\mathbf{F}_1 X_1) \end{bmatrix} + \begin{bmatrix} w_1 \\ w_2 \\ \vdots \\ w_{N-1} \\ w_N \end{bmatrix} = \bar{h}_N(X_1) + W_N \tag{19.181}$$

where

$$\mathrm{E}\{W_N\} = 0 \tag{19.182}$$

$$\bar{\mathbf{R}}_N = \mathrm{E}\{W_N W_N^T\} = \begin{bmatrix} \mathbf{R}_1 & \mathbf{0} & \ldots & \mathbf{0} \\ \mathbf{0} & \mathbf{R}_2 & \ldots & \vdots \\ \vdots & & \ddots & \\ \mathbf{0} & \ldots & \mathbf{0} & \mathbf{R}_N \end{bmatrix} \tag{19.183}$$

In this case, $\bar{\mathbf{R}}_N$ is a block diagonal matrix because the measurement errors between any two times are assumed to be zero mean and independent. Had the errors included correlation across time, off-diagonal elements of the matrix would be nonzero. One of the advantages of parametric estimation is the ease of including correlation of the measurement errors at different times. The weighted nonlinear least-squares estimate (NLSE) [7] of X_1 given measurements from t_1 to t_N is found through an iterative process given by

$$X_{1|N}^{(i+1)} = X_{1|N}^{(i)} + \left((\bar{\mathbf{H}}_N^{(i)})^T \bar{\mathbf{R}}_N^{-1} \bar{\mathbf{H}}_N^{(i)}\right)^{-1} (\bar{\mathbf{H}}_N^{(i)})^T \bar{\mathbf{R}}_N^{-1} \left[\bar{Z}_1^N - \bar{h}_N(X_{1|N}^{(i)})\right] \tag{19.184}$$

where

$$\mathbf{H}_N^{(i)} = \left.\frac{\partial \bar{h}_N(X_1)}{\partial X_1}\right|_{X_1 = X_{1|N}^{(i)}} \tag{19.185}$$

The initialization of the iterative process can be achieved by using converted measurements and a linear least squares estimate with the first M measurements. The covariance of the estimate is given by

$$P_{1|N} = \left(\bar{\mathbf{H}}_N^T \bar{\mathbf{R}}_N^{-1} \bar{\mathbf{H}}_N\right)^{-1} \tag{19.186}$$

where $\bar{\mathbf{H}}_N$ is the partial derivative of $\bar{h}_N(X_1)$ evaluated at the final estimate. Note that the covariance will reflect only the actual errors in the state estimate to the degree that the kinematic model (19.179) is accurate and the measurements errors are zero-mean Gaussian with covariance $\bar{\mathbf{R}}_N$. The $X_{1|N}$ and $\mathbf{P}_{1|N}$ are smoothed estimates. For a state estimate and covariance at time t_j, $t_1 < t_j \leq t_N$,

$$\boldsymbol{X}_{j|N} = \mathbf{F}_{j-1} \ldots \mathbf{F}_1 \boldsymbol{X}_{1|N} \tag{19.187}$$

$$\mathbf{P}_{j|N} = \mathbf{F}_{j-1} \ldots \mathbf{F}_1 \mathbf{P}_{1|N} \mathbf{F}_1^T \ldots \mathbf{F}_{j-1}^T \tag{19.188}$$

For the Gaussian errors, $\boldsymbol{X}_{j|N}$ is also the ML estimate of $\boldsymbol{X}_j$. For the parametric approach, constant velocity and constant acceleration motion are the two most common assumptions for the motion of the target. However, more complicated models are typically used for applications like ballistic missile defense.

19.5.2 Extended Kalman Filter

For a stochastic state estimation for radar tracking with nonlinear measurements, the measurement equation is given by (19.144) and the dynamical motion model is given by

$$\boldsymbol{X}_{k+1} = \mathbf{F}_k \boldsymbol{X}_k + \mathbf{G}_k \boldsymbol{v}_k \tag{19.189}$$

where $\mathbf{G}_k$ is the input matrix for the target motion at time t_k, and $\boldsymbol{v}_k$ is the white noise error in the state process with $v_k \sim \mathrm{N}(0, \mathbf{Q}_k)$. The EKF for target state estimation with nonlinear measurements is given by the following equations:

State prediction:

$$\boldsymbol{X}_{k|k-1} = \mathbf{F}_{k-1} \boldsymbol{X}_{k-1|k-1} \tag{19.190}$$

$$\mathbf{P}_{k|k-1} = \mathbf{F}_{k-1} \mathbf{P}_{k-1|k-1} \mathbf{F}_{k-1}^T + \mathbf{G}_{k-1} \mathbf{Q}_{k-1} \mathbf{G}_{k-1}^T \tag{19.191}$$

State update with the measurement:

$$\boldsymbol{X}_{k|k} = \boldsymbol{X}_{k|k-1} + \mathbf{K}_k \widetilde{\boldsymbol{Z}}_k \tag{19.192}$$

$$\widetilde{\boldsymbol{Z}}_k = \boldsymbol{Z}_k - h_k(\boldsymbol{X}_{k|k-1}) \tag{19.193}$$

$$\mathbf{P}_{k|k} = [\mathbf{I} - \mathbf{K}_k \mathbf{H}_k] \mathbf{P}_{k|k-1} \tag{19.194}$$

$$\mathbf{K}_k = \mathbf{P}_{k|k-1} \mathbf{H}_k^T \mathbf{S}_k^{-1} \tag{19.195}$$

$$\mathbf{S}_k = \mathbf{H}_k \mathbf{P}_{k|k-1} \mathbf{H}_k^T + \mathbf{R}_k \tag{19.196}$$

In these equations, $\widetilde{\boldsymbol{Z}}_k$ is the filter residual at time t_k and

$$\mathbf{H}_k = \left[\frac{\partial h_k(\boldsymbol{X}_k)}{\partial \boldsymbol{X}_k} \right]_{\boldsymbol{X}_k = \boldsymbol{X}_{k|k-1}} \tag{19.197}$$

Since the measurements are a nonlinear function of the state and the EKF uses a linearized output matrix for $h_k(\boldsymbol{X}_k)$ in the covariance update, the EKF does not provide an optimal estimate of the target state. When the criteria of (19.162) are satisfied for radar measurements in spherical coordinates or the criteria of (19.174) is satisfied for measurements in sine space, the performance of the EKF is expected to be poor, and either converted measurements [28,30] or the measurement covariance adaptive extended Kalman filter (MCAEKF) [30] should be employed to reduce the impacts of the nonlinearities in the measurements on tracking performance.

Typically, the target state estimate is maintained in a Cartesian reference frame that is stabilized relative to any motion of the radar antenna. In this case, an affine transform is typically used to define the relationship between the two coordinate systems and the measurement equation is given in (19.145). The measurement update portion of the EKF for target state estimation with nonlinear measurements at a remote reference frame is given by the following:

Update of the state estimate with the measurement:

$$\boldsymbol{X}_{k|k} = \boldsymbol{X}_{k|k-1} + \mathbf{K}_k[\boldsymbol{Z}_k - h_k(\mathbf{M}_k\boldsymbol{X}_{k|k-1} + \boldsymbol{L}_k)] \tag{19.198}$$

$$\mathbf{P}_{k|k} = [\mathbf{I} - \mathbf{K}_k\mathbf{H}_k\mathbf{M}_k]\mathbf{P}_{k|k-1} \tag{19.199}$$

$$\mathbf{K}_k = \mathbf{P}_{k|k-1}\mathbf{M}_k^T\mathbf{H}_k^T\mathbf{S}_k^{-1} \tag{19.200}$$

$$\mathbf{S}_k = \mathbf{H}_k\mathbf{M}_k\mathbf{P}_{k|k-1}\mathbf{M}_k^T\mathbf{H}_k^T + \mathbf{R}_k \tag{19.201}$$

where

$$\mathbf{H}_k = \left[\frac{\partial h_k(\boldsymbol{X}_k)}{\partial \boldsymbol{X}_k}\right]_{\boldsymbol{X}_k=\mathbf{M}_k\boldsymbol{X}_{k|k-1}+\boldsymbol{L}_k} \tag{19.202}$$

19.5.3 Converted Measurement Filter

One of the attractions to the converted measurement filter is the nice linear relationship between the state and the measurements. For the converted measurement filter in Cartesian space, the measurement equation is given as a function of $\boldsymbol{X}_k$ by

$$\boldsymbol{Z}_k = \begin{bmatrix} x_k^m \\ y_k^m \\ z_k^m \end{bmatrix} = \begin{bmatrix} 1 & 0 & 0 & 0 & 0 & 0 \\ 0 & 0 & 1 & 0 & 0 & 0 \\ 0 & 0 & 0 & 0 & 1 & 0 \end{bmatrix} \boldsymbol{X}_k + \begin{bmatrix} w_{xk} \\ w_{yk} \\ w_{zk} \end{bmatrix} = \mathbf{H}\boldsymbol{X}_k + \boldsymbol{W}_k \tag{19.203}$$

where

x_k^m = measured x coordinate at time t_k.
y_k^m = measured y coordinate at time t_k.
z_k^m = measured z coordinate at time t_k.
w_{xk} = error in x-coordinate measurement at time t_k.
w_{yk} = error in y-coordinate measurement at time t_k.
w_{zk} = error in z-coordinate measurement at time t_k.

The converted measurements (x_k^m, y_k^m, z_k^m) are computed from the spherical or sine space measurements, and one should take care to use the unbiased transform as needed [28,30]. The measurement errors (w_{xk}, w_{yk}, w_{zk}) are cross-correlated and non-Gaussian, and one must be careful to use the covariance from the unbiased transform as needed. Since the measurement errors are non-Gaussian, the Kalman filter will not be optimal in a general sense. The Kalman filter will be the best linear estimator given that the measurements are unbiased and the measurement covariance is correct.

19.6 | MEASUREMENT-TO-TRACK DATA ASSOCIATION

All detections, or threshold exceedances, are processed to produce measurements that include an estimate of range and two angles that correspond to the location of a potential target. That measurement could be the result of a false alarm due to receiver noise or an echo from clutter. If the measurement originated from a target, it could be from a target currently under track or a newly found target. When tracking multiple closely spaced targets, the measurement could have originated with any of the targets. Furthermore, if any of the targets are employing countermeasures to electronic attack, the measurement may not correspond directly to the position of any target. Thus, the problem of accurate and reliable measurement-to-track association is very challenging, and it is the crux of the multitarget tracking problem. However, in this section, the measurement-to-track association problem is restricted to the problem of tracking a single target to limit the scope and length of the material to an appropriate level for this text.

The first step in the measurement-to-track association is validation of the candidate measurements. The measurements are compared with the predicted measurement based on the state estimate of the track. The comparison is usually achieved by computing the difference between the measurements and the predicted measurement and comparing that difference with the sum of the covariances of the measurement and the predicted measurement. Let the set of validated measurements for the target under track be denoted by

$$\mathbf{Z}_k^{1,m_k} = \left\{\mathbf{Z}_k^i, \mathbf{R}_k^i, \Re_{ok}^i\right\}_{i=1}^{m_k} \tag{19.204}$$

where $\mathbf{Z}_k^i$ is the i-th validated measurement, $\mathbf{R}_k^i$ is the covariance of the i-th measurement, and $\Re_{ok}^i$ is the observed signal-to-noise (i.e., signal-plus-noise-to-noise) ratio.

The second step in the measurement-to-track association is the processing of the validated measurements to update the track state estimate. Some techniques such as *statistical nearest neighbor* (NN) or *strongest neighbor* (SN) select one of the validated measurements to update the track and the standard Kalman filter equations are used to update the state estimate and covariance as if the selection is correct. Other techniques such as the *probabilistic data association filter* (PDAF) use all of the measurements in the validation gate to update the state estimate and covariance. In the case of the PDAF, the uncertainty in the origin of the measurements is captured in the filtering process and the covariance reflects this uncertainty.

The gating and measurement update process is illustrated in Figure 19-25. Note that in this example one measurement has been determined to be invalid with the use of a rectangular gate. Rectangular gates are computationally efficient to implement and computational efficient methods for classifying measurements as valid or invalid are critical to the successful implementation of real-time tracking algorithms. Note in Figure 19-25 that two measurements have been validated by the ellipsoidal gate that is the result of a likelihood test. The border of the ellipse represents a constant contour of the likelihood. Thus, any two measurements on the border of the ellipse will have the same likelihood. The statistical NN measurement is found by drawing a line from the center of the ellipse through each measurement to the border. The measurement with the smallest fraction of the distance to the border is the statistical NN. On the other hand, the PDAF processes both validated measurements with the additional hypothesis that the target was not detected and both measurements are false alarms. The PDAF can be thought to be processing a pseudo-measurement that is a result of blending the two validated measurements and the predicted measurement.

In the remainder of this section, the process of taking an initial, isolated detection to a firm track is discussed, and that is followed by a discussion of measurement validation and gating. Then, the equations for the NN filter, the SN filter, and PDAF are presented.

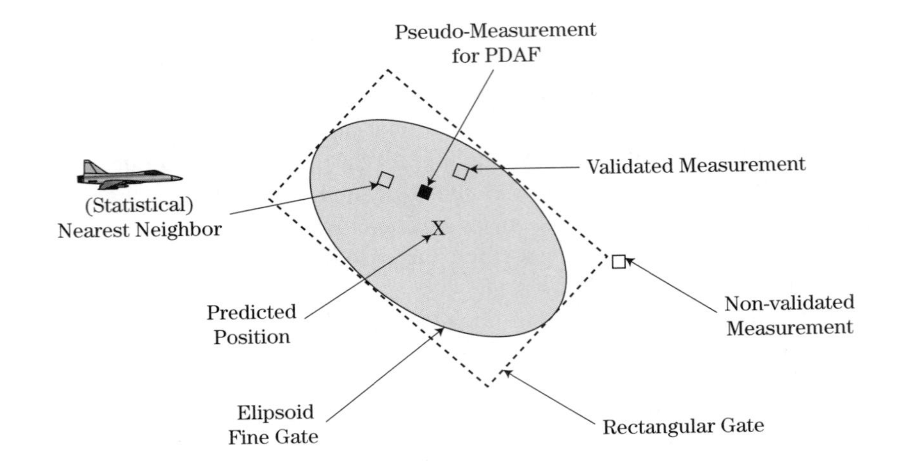

FIGURE 19-25 ■ Measurement validation, gating, and processing for multiple measurements.

- Conventional approach uses statistical nearest neighbor measurement for filtering.
- PDAF uses all validated measurements for filtering, which is conceptually equivalent to using a single pseudo-measurement with inflated covariance.
- PDAF does not directly support threat discriminqation in ballistic missile defense or combat identification in air defense.

19.6.1 Formation of a New Track

When a measurement is formed and it does not associate with any existing track, the measurement is first considered for association to any of the unassociated measurements from the previous scan. The measurements are usually converted to Cartesian coordinates for this assessment. If the distance between the measurement and an unassociated measurement from the previous scan is less than the maximum speed of the target times the scan period, then a tentative track is formed from the two measurements by setting the position estimate to the position of the latest measurement and the velocity estimate to the vector difference between the two measurements divided by the scan period. For rotating targets, the scan period is usually on the order of a few seconds, and the entire surveillance region is observed between the two measurements. The long scan period and large region observed between the two measurements can result in a rather challenging data sorting and management problem. If M measurements of the next the N scans (e.g., 5 of 7) associate to the track, the newly formed tentative track is promoted to a confirmed track.

For phased array radars with electronic scanning, an unassociated detection is usually followed by a confirmation dwell within a fraction of a second after the original detection. Since the scan period is small and the search region covered by a single dwell is also small, the data sorting and management is rather straightforward. Typically, if the confirmation dwell does not find a candidate measurement, the candidate track is dropped. If a measurement is found on the confirmation dwell and it associates to the measurement from the previous scan, a sequence of dwells is scheduled and if M measurements of the next N dwells are associated with the track, then the tentative track is promoted to a firm track.

19.6.2 Measurement Validation and Gating

Any real-time algorithm should include code that verifies the inputs as reasonable and statistically feasible. A computer program will process unreasonable data and produce unreasonable results, and those unreasonable results may drive an operator display, combat identification system, or a weapon system. For tracking, these functions are referred to as *measurement validation* and *gating*, and these functions are critical elements of any real-time tracker. The developers of advanced MTT algorithms such as *multiple hypothesis tracking* (MHT) or *multiple frame assignment* (MFA) often cite efficient and effective gating as critical to a successful implementation of a real-time tracker [2].

Consider a radar system with 1,000 active tracks and a radar dwell that results in five measurements. A brute force approach to the gating would require 5,000 tests or decisions. Since a radar performs a dwell every few milliseconds, a brute force approach is not a reasonable use of computational resources. Measurement gating is typically implemented as a sequence of tests of increasing computational complexity. Coarse gating is typically performed first and it should be a computationally efficient method for eliminating nearly all of the tracks from consideration for association to the measurement. Coarse gating techniques include segmentation of the surveillance space and track partitioning. *Segmentation* involves dividing the surveillance space into contiguous regions and mapping each new measurement into a region or regions. The measurement is then gated with only the tracks identified within that region. *Track partitioning* involves separating the tracks into groups of tracks that can be isolated from all other groups of tracks. The coarse gating is then performed with the centriod of the partition instead of the individual tracks.

Rectangular gating usually follows the coarse gating. Rectangular gating is simple in that it is usually performed in measurement space and involves a series of inequality tests on the individual coordinates of the difference between the measurement and track. Once an inequality test is failed, the track is rejected and no further testing is needed. For radar tracking, the range coordinate eliminates most of the measurement-to-track hypotheses and is often performed first.

The fine or ellipsoidal gating is the final test and performed only for measurement-to-track hypotheses that pass the rectangular gating process. The border of the ellipse as shown in Figure 19-25 represents the location of measurements with equal likelihood, and ellipsoidal gating is a likelihood test of the measurement-to-track hypothesis. Moving the focus back to the single target tracking problem of this section, the likelihood test is evaluated with a *Mahalanobis distance* between measurement i, and it is given by

$$d_k^i = \left(\widetilde{\mathbf{Z}}_k^i\right)^T \left(\mathbf{S}_k^i\right)^{-1} \widetilde{\mathbf{Z}}_k^i \tag{19.205}$$

where $\widetilde{\mathbf{Z}}_k^i$ is the residual of (19.193) between measurement i and the track, and $\mathbf{S}_k^i$ is the covariance of the residual in (19.196) between measurement i and the track and given by

$$\mathbf{S}_k^i = \mathbf{H}_k \mathbf{P}_{k|k-1} \mathbf{H}_k^T + \mathbf{R}_k^i \tag{19.206}$$

If measurement i originated from the target corresponding to the track, then d_k^i is chi-square distributed with n_Z degrees of freedom, where n_Z is the dimension of the measurement. Tables such as Table 1.5.4-1 in [7] for the chi-square distribution are used to select a gating threshold d_{th} for d_k^i with a given probability of gating P_G. Thus, $1 - P_G$ is the probability that measurement i is rejected for the track, given that measurement i originated from the target. That is, $1 - P_G$ is the probability of discarding the correct measurement due

TABLE 19-1 ■ Typical Threshold Values d_{th} for Gating

n_Z	$P_G = 0.995$	$P_G = 0.999$
1	7.88	10.8
2	10.6	13.8
3	12.8	16.3

to failure of the gating test. Some typical gating thresholds d_{th} are given in Table 19-1 for measurement dimensions of one, two, and three. If $d_k^i < d_{th}$, the measurement i is accepted as a validated measurement.

19.6.3 Nearest Neighbor Filter

The NN filter selects the most likely measurement from a group of validated measurements and performs the track filtering as if no uncertainty exists in the selection of the correct measurement. Thus, the state error covariance produced by the NN filter is often optimistic in that it does not reflect the data association errors in the state estimate. The measurement selection of the NN filter is given by

$$NN_k = \underset{1 \le i \le m_k}{\text{argmin}} \left\{ d_k^i \right\} \tag{19.207}$$

where d_k^i is given by (19.205) and the NN_k represents the index of the statistically nearest neighbor measurement at time t_k and argmin{} denotes the argument (i.e., i in this case) that provides the minimum. Then, the NN filter is given by standard equations for prediction or time update and the following equations for the measurement update:

$$\boldsymbol{X}_{k|k} = \boldsymbol{X}_{k|k-1} + \mathbf{K}_k \left[\mathbf{Z}_k^{NN_k} - h_k(\boldsymbol{X}_{k|k-1}) \right] \tag{19.208}$$

$$\mathbf{P}_{k|k} = [\mathbf{I} - \mathbf{K}_k \mathbf{H}_k] \mathbf{P}_{k|k-1} \tag{19.209}$$

$$\mathbf{K}_k = \mathbf{P}_{k|k-1} \mathbf{H}_k^T \left(\mathbf{S}_k^{NN_k} \right)^{-1} \tag{19.210}$$

where $\mathbf{S}_k^{NN_k}$ is the residual measurement covariance given by (19.206) for measurement NN_k and $\mathbf{H}_k$ is the linearized output matrix given in (19.197). Note that the measurement update equations do not reflect any uncertainty in the selection of the measurement from a group of m_k validated measurements.

19.6.4 Strongest Neighbor Filter

The SN filter selects the measurement with the highest signal-plus-noise-to-noise ratio (i.e., $\Re_{oi}$) from a group of validated measurements and performs the track filtering as if no uncertainty exists in the selection of the correct measurement. Thus, the state error covariance produced by the SN filter is often optimistic in that it does not reflect the data association errors in the state estimate. The measurement selection of the SN filter is given by

$$SN_k = \underset{1 \le i \le m_k}{\text{argmax}} \left\{ \Re_{ok}^i \right\} \tag{19.211}$$

where $\Re_{ok}^i$ is the observed SNR for measurement i, and SN_k represents the index of the strongest neighbor at time t_k and argmax{} denotes the argument (i.e., i in this case) that provides the maximum. Then, the SN filter is given by standard equations for prediction or time update and the following equations for the measurement update:

$$\boldsymbol{X}_{k|k} = \boldsymbol{X}_{k|k-1} + \mathbf{K}_k\left[\mathbf{Z}_k^{SN_k} - h_k(\boldsymbol{X}_{k|k-1})\right] \tag{19.212}$$

$$\mathbf{P}_{k|k} = [\mathbf{I} - \mathbf{K}_k\mathbf{H}_k]\mathbf{P}_{k|k-1} \tag{19.213}$$

$$\mathbf{K}_k = \mathbf{P}_{k|k-1}\mathbf{H}_k^T\left(\mathbf{S}_k^{SN_k}\right)^{-1} \tag{19.214}$$

where $\mathbf{S}_k^{SN_k}$ is the residual measurement covariance given by (19.206) for measurement SN_k. Note that the measurement update equations do not reflect any uncertainty in the selection of the measurement from a group of m_k validated measurements.

19.6.5 Probabilistic Data Association Filter (PDAF)

The PDAF uses all validated measurements and performs the track filtering while accounting for the uncertainty in the origin of the validated measurements. Thus, one of the most beneficial characteristics of the PDAF is that its state error covariance is nearly consistent and reflects the uncertainty in the measurement-to-track association when tracking a single target. In the PDAF, a measurement update from the Kalman filter is performed for each validated measurement, and the resulting hypothesized tracks are blended with the probability that each is true. Let β_k^i, $1 \le i \le m_k$, denote the probability that measurement i originated from the target. Also, let β_k^0 denote the probability that none of the validated measurements originated from the target. Then

$$\beta_k^i = \begin{cases} \dfrac{e_k^i}{b_0 + \sum_{i=1}^{m_k} e_k^i}, & 1 \le i \le m_k \\ \dfrac{b_0}{b_0 + \sum_{i=1}^{m_k} e_k^i}, & i = 0 \end{cases} \tag{19.215}$$

where

$$e_k^i = \frac{P_D V_k}{m_k\sqrt{|2\pi S_k^i|}}\exp\left[-\frac{1}{2}d_k^i\right] \tag{19.216}$$

$$b_0 = 1 - P_D P_G \tag{19.217}$$

d_k^i is given by (19.205), P_G is the probability of gating the target-originated measurement with the track, P_D is the probability of detection of the target under track, and V_k is the volume of the validation gate.

The PDAF is given by the standard equations for prediction or time update and the following equations for the measurement update.

$$\boldsymbol{X}_{k|k} = \boldsymbol{X}_{k|k-1} + \boldsymbol{U}_k \tag{19.218}$$

$$\mathbf{P}_{k|k} = [\mathbf{I} - \bar{\mathbf{K}}_k H_k]\mathbf{P}_{k|k-1} + \sum_{i=1}^{m_k}\beta_k^i\left[\mathbf{K}_k^i\boldsymbol{\varepsilon}_k^i\right]\left[\mathbf{K}_k^i\boldsymbol{\varepsilon}_k^i\right]^T - \boldsymbol{U}_k\boldsymbol{U}_k^T \tag{19.219}$$

$$\mathbf{K}_k^i = \mathbf{P}_{k|k-1}\mathbf{H}_k^T\left(\mathbf{S}_k^i\right)^{-1} \tag{19.220}$$

where

$$\boldsymbol{U}_k = \mathbf{P}_{k|k-1}\mathbf{H}_k^T\sum_{i=1}^{m_k}\beta_k^i\left(\mathbf{S}_k^i\right)^{-1}\boldsymbol{\varepsilon}_k^i \tag{19.221}$$

$$\boldsymbol{\varepsilon}_k^i = \mathbf{Z}_k^i - h_k(\boldsymbol{X}_{k|k-1}) \tag{19.222}$$

$$\bar{\mathbf{K}}_k = \mathbf{P}_{k|k-1}\mathbf{H}_k^T\sum_{i=1}^{m_k}\beta_k^i\left(\mathbf{S}_k^i\right)^{-1} \tag{19.223}$$

The PDAF is often presented in the literature for measurements with covariance that is uniform across all of the validated measurements. Thus, $\mathbf{R}_k^i = \mathbf{R}_k,\ 1 \le i \le m_k$, which implies for radar measurements that $\Re_{ok}^i = \Re_{ok},\ 1 \le i \le m_k$. While this assumption is typically invalid for a set of radar measurements, it does greatly simplify the PDAF equations when made. The measurement update equations for the PDAF with uniform measurement covariances are given by the following equations:

$$\boldsymbol{X}_{k|k} = \boldsymbol{X}_{k|k-1} + \mathbf{K}_k[\bar{\mathbf{Z}}_k - h_k(\boldsymbol{X}_{k|k-1}] \tag{19.224}$$

$$\mathbf{P}_{k|k} = \left[\mathbf{I} - (1-\beta_k^0)\mathbf{K}_k\mathbf{H}_k\right]\mathbf{P}_{k|k-1} + \mathbf{K}_k\left[\sum_{i=1}^{m_k}\beta_k^i\left[\boldsymbol{\varepsilon}_k^i\right]\left[\boldsymbol{\varepsilon}_k^i\right]^T - \bar{\boldsymbol{\varepsilon}}_k\bar{\boldsymbol{\varepsilon}}_k^T\right]\mathbf{K}_k^T \tag{19.225}$$

$$\mathbf{K}_k = \mathbf{P}_{k|k-1}\mathbf{H}_k^T(\mathbf{S}_k)^{-1} \tag{19.226}$$

where

$$\bar{\mathbf{Z}}_k = \sum_{i=0}^{m_k}\beta_k^i\mathbf{Z}_k^i \tag{19.227}$$

$$\mathbf{Z}_k^0 = h_k(\boldsymbol{X}_{k|k-1}) \tag{19.228}$$

$$\boldsymbol{\varepsilon}_k^i = \mathbf{Z}_k^i - h_k(\boldsymbol{X}_{k|k-1}) \tag{19.229}$$

$$\bar{\boldsymbol{\varepsilon}}_k = \sum_{i=1}^{m_k}\beta_k^i\boldsymbol{\varepsilon}_k^i \tag{19.230}$$

19.7 | PERFORMANCE ASSESSMENT OF TRACKING ALGORITHMS

Performance assessment of MTT algorithms is very much an outstanding research problem. The metrics are rather complex and involve such items as track completeness, track switches, track breaks, spurious tracks, and redundant tracks as well as the standard metrics of track accuracy and covariance consistency used in this chapter [31]. One of the major challenges in the assessment of MTT algorithms is the track-to-truth assignment problem [32] that is required to compute any of these metrics. However, performance metrics for tracking a single target are much simpler and track-to-truth assignment is not required, because only one target is present and only one track is generated by the tracker. The discussion of this section is restricted to the accuracy metrics for single target tracking.

Typically, for a Monte Carlo simulation, the kinematic trajectory for the truth object is fixed and the sensor measurement errors are randomized between experiments. The first step for computing the performance metrics involves setting the scoring times (i.e., identifying the times of the scenario when the track filter will be requested to report its best estimate of the kinematic state) throughout the entire scenario or truth trajectory. For each experiment, the squared errors for position and velocity are computed at each scoring time. These errors at each scoring time are averaged across the Monte Carlo experiments giving the RMSEs.

Let the error in the reported track state on the ith Monte Carlo experiment at time t_k be given by

$$\boldsymbol{\varepsilon}_k^{(i)} = \boldsymbol{X}_{k|j}^{(i)} - \boldsymbol{X}_k \tag{19.231}$$

where $\boldsymbol{X}_{k|j}^{(i)}$ is the state estimate at time t_k for experiment i given the most recent measurements at time t_j, $t_j \leq t_k$, and $\boldsymbol{X}_k$ is the true kinematic state at time t_k. Let $\boldsymbol{\varepsilon}_k^{(i)}(n)$ denote the n-th component of $\boldsymbol{\varepsilon}_k^{(i)}$. Then, $[\boldsymbol{\varepsilon}_k^{(i)}(1), \boldsymbol{\varepsilon}_k^{(i)}(3), \boldsymbol{\varepsilon}_k^{(i)}(5)]$ denotes the position vector, and $[\boldsymbol{\varepsilon}_k^{(i)}(2), \boldsymbol{\varepsilon}_k^{(i)}(4), \boldsymbol{\varepsilon}_k^{(i)}(6)]$ denotes the velocity vector. The RMSEs in position and velocity at t_k for M experiments are given by

$$RMSE_k^{pos} = \sqrt{\frac{1}{M}\sum_{i=1}^{M}\left[\left(\boldsymbol{\varepsilon}_k^{(i)}(1)\right)^2 + \left(\boldsymbol{\varepsilon}_k^{(i)}(3)\right)^2 + \left(\boldsymbol{\varepsilon}_k^{(i)}(5)\right)^2\right]} \qquad (19.232)$$

$$RMSE_k^{vel} = \sqrt{\frac{1}{M}\sum_{i=1}^{M}\left[\left(\boldsymbol{\varepsilon}_k^{(i)}(2)\right)^2 + \left(\boldsymbol{\varepsilon}_k^{(i)}(4)\right)^2 + \left(\boldsymbol{\varepsilon}_k^{(i)}(6)\right)^2\right]} \qquad (19.233)$$

The NEES is used the to measure the quality or consistency of the covariance produced by the estimator. The NEES is a Mahalanobis distance between the estimated state and the true kinematic state and it should reflect the true errors in the estimate relative to the covariance produced by the estimator. A chi-square test is performed to assess the correctness of the covariance matrices. For $\boldsymbol{X}_{k|j}$ of dimension N and M experiments, the NEES is given by

$$C_k = \frac{1}{NM}\sum_{i=1}^{M}\left(\boldsymbol{\varepsilon}_k^{(i)}\right)^T \left[\mathbf{P}_{k|j}^{(i)}\right]^{-1} \boldsymbol{\varepsilon}_k^{(i)} \qquad (19.234)$$

where $\mathbf{P}_{k|j}^{(i)}$ is the state covariance for experiment i at time t_k given the most recent measurements at time t_j. Under the case that the errors in the state estimate are zero-mean Gaussian with covariance $\mathbf{P}_{k|j}^{(i)}$, C_k is chi-square distributed with NM degrees of freedom. For example, consider a track filter the provides state estimates with $N = 6$ and a Monte Carlo simulation with $M = 50$ experiments. Thus, if errors in that state estimate are zero-mean Gaussian and the covariance is accurate, then C_k will be chi-square distributed with 300 degrees of freedom at each time t_k. From Table 1.5.4-1 of [7], C_k should be between 0.87 (261/300) and 1.14 (341/300) 90% of the time. If C_k is greater than 1.14, then the covariance underrepresents the errors in the state estimate, and if C_k is less than 0.87, then the covariance overrepresents the errors.

19.8 FURTHER READING

Further applications of the track filtering concepts of this chapter are discussed in [8]. Some topics include multiplatform-multisensor track filtering, tracking in the presence of electronic attack and electronic protection, modeling radars for assessing tracking performance, and an engineering guide to the IMM estimator. Another topic of interest in radar is tracking with monopulse measurements in the presence of sea-surface induced multipath [33–35]. Advanced track filtering techniques such as particle filters are addressed nicely in [36]. The book includes an introduction to nonlinear filters and particle filters and gives numerous applications. A good introduction to multitarget tracking and measurement-to-track association is given in [5]. The book includes much of the mathematical details that are needed by a novice. Practical insights into the application of tracking techniques without the mathematical details are presented in [2]. Thus, this book is a good source for those who can add the mathematics on their own. Since all radar tracking problems are actually MTT problems, additional reading about the metrics for performance assessment of MTT algorithms is recommended. One of the greatest challenges to performance assessment of MTT algorithms is the track-to-truth assignment. In other words, in MTT, the tracks in a good simulation do not necessarily map uniquely to the truth objects. The methodologies of assigning tracks-to-truth objects are addressed in [32]. Many of the metrics associated with MTT are presented in [31].

19.9 REFERENCES

[1] Bar-Shalom, Y., and Li, X.R., *Multitarget-Multisensor Tracking: Principles and Techniques*, YBS Publishing, Storrs, CT, 1995.

[2] Blackman, S.S., and Popoli, R., *Design and Analysis of Modern Tracking Systems,* Artech House, Norwood, MA, 1999.

[3] Morris, G., and Harkness, L., (Eds.), *Airborne Pulsed Doppler Radar,* 2d ed., Artech House, Norwood, MA, 1996.

[4] Blair, W.D., Register, A.H.,and West, P.D., "Multiple Target Tracking," W. L. Melvin, Ed., *Principles of Modern Radar Systems*, vol. II, forthcoming.

[5] Bar-Shalom, Y., and Li, X.R., *Estimation and Tracking: Principles, Techniques and Software*, Artech House, Dedham, MA, 1993. (reprinted by YBS Publishing, 1998).

[6] Blair, W.D., and Bar-Shalom, Y., "Tracking Maneuvering Targets with Multiple Sensors: Does More Data Always Mean Better Estimates?" *IEEE Trans. Aerospace and Electronic Systems*, vol. AES-32, no. 1, pp. 822–825, January 1996.

[7] Bar-Shalom, Y., Li, X.R., and Kirubarajan, T., *Estimation with Applications to Tracking and Navigation: Theory, Algorithms, and Software*, John Wiley & Sons, Inc., New York, 2001.

[8] Bar-Shalom, Y., and Blair, W.D. (Eds.), *Multitarget-Multisensor Tracking: Applications and Advances,* vol. 3, Artech House, Dedham, MA, 2000.

[9] Blair, W.D., "Design of Nearly Constant Velocity Filters for Tracking Maneuvering Targets," *Proceedings of the 11th International Conference on Information Fusion*, Cologne, Germany, July 2008.

[10] Kalata, P.R., "The Tracking Index: A Generalized Parameter for Alpha-Beta-Gamma Target Trackers," *IEEE Trans. on Aerospace and Electronic Systems*, vol. AES-20, no. 2, pp. 174–182, March 1984.

[11] Gray, J.E., and Murray, W., "The Response of the Transfer Function of an Alpha-Beta Filter to Various Measurement Models," in *Proc. of 23th IEEE Southeastern Symposium on System Theory,* pp. 389–393, March 1991.

[12] Fitzgerald, R.J., "Effects of Range-Doppler Coupling on Chirp Radar Tracking Accuracy," *IEEE Trans. Aerospace and Electronic Systems*, vol. AES-10, No. 3, pp. 528–532, July 1974.

[13] Wong, W., and Blair, W.D., "Steady-State Tracking with LFM Waveforms," *IEEE Trans. on Aerospace and Electronic Systems*, vol. AES-36, no. 1, pp. 701–709, April 2000.

[14] Bar-Shalom, Y., Chang, K.C., and Blom, H.A.P., "Tracking Maneuvering Targets Using Input Estimation Versus the Interacting Multiple Model Algorithm," *IEEE Trans. Aerospace and Electronic Systems*, vol. AES-25, no. 2, pp. 296–300, March 1989.

[15] Blair, W.D., Watson, G.A., et al., *Information-Based Radar Resource Allocation: FY96 Test-of-Concept Experiment (TOCE),* Technical Report No. NSWCDD/TR-97/22, Naval Surface Warfare Center, Dahlgren Division, Dahlgren, VA, February 1997.

[16] Blom, H.A.P., and Bar-Shalom Y., "The Interacting Multiple Model Algorithm for Systems with Markovian Switching Systems," *IEEE Trans. on Automatic Control Systems*, vol. AC-33, no. 8, pp. 780–783, August 1988.

[17] Watson, G.A., and Blair, W.D., "IMM Algorithm for Tracking Targets That Maneuver Through Coordinate Turns," pp. 236–247 in *Signal and Data Processing of Small Targets 1992* SPIE vol. 1698, ed. O.E. Drummond, 1992.

[18] Watson, G.A., and Blair, W.D., "Multiple Model Estimation for Control of a Phased Array Radar," pp. 275–286 in *Signal and Data Processing of Small Targets 1993,* SPIE vol. 1954, ed. E.O. Drummond, 1993.

[19] Kirubarajan, T., and Bar-Shalom, Y., "Kalman Filter Versus IMM Estimator: When Do We Need the Latter?" *IEEE Trans. on Aerospace and Electronic Systems*, vol. AES-39, no. 4, pp. 1452–1457, October 2003.

[20] Stromberg, D., "Scheduling of Track Updates in Phased-Array Radars," *Proc. of the 1996 IEEE National Radar Conference,* Ann Arbor, MI, pp. 214–219, May 13–16, 1996.

[21] Kirubarajan, T., Bar-Shalom, Y., Blair, W.D., and Watson, G.A., "IMMPDA Solution to Benchmark for Radar Resource Allocation and Tracking in the Presence of ECM," *IEEE Trans. on Aerospace and Electronic Systems*, vol. AES-34, no. 4, pp. 1115–1133, October 1998.

[22] Blair, W.D., *Fixed-Gain, Two-Stage Estimators for Tracking Maneuvering Targets,* Technical Report No. NSWCDD/TR-92/297, Naval Surface Warfare Center Dahlgren Division, Dahlgren, VA, 1992.

[23] Li, X.R., and Jilkov, V.P., "Survey of Maneuvering Target Tracking - Part I: Dynamic Models," *IEEE Trans. Aerospace and Electronic Systems*, vol. AES-39, no. 4, pp. 1333–1364, October 2003.

[24] Blackman, S.S., *Multiple Target-Target Tracking with Radar Applications,* Artech House, Dedham, MA, 1986.

[25] Alouni, A.T., and Blair, W.D., "Use of Kinematic Constraint in Tracking Constant Speed, Maneuvering Targets," *IEEE Trans. Automatic Control,* pp. 1107–1111, July 1993.

[26] Mehra, R.K., "A Comparison of Several Nonlinear Filters for Reentry Vehicle Tracking," *IEEE Trans. on Automatic Control*, vol. 16, no. 3, pp. 307–319, March 1971.

[27] Lerro, D., and Bar-Shalom, Y., "Tracking With Debiased Consistent Converted Measurements Versus the EKF," *IEEE Transactions on Aerospace and Electronic Systems*, vol. AES-29, no. 3, pp. 1015–1022, July 1993.

[28] Mo, L., Song, X., Zhou, Y., Sun, Z.K., and Bar-Shalom, Y., "Unbiased Converted Measurements for Tracking," *IEEE Transactions on Aerospace and Electronic Systems*, vol. AES-34, no. 3, pp. 1023–1027, July 1998.

[29] Daum, F.E., and Fitzgerald, R.J., "Decoupled Kalman Filters for Phased Array Radar Tracking," *IEEE Trans. Automatic Control*, vol. AC-28, no. 3, pp. 269–283, March 1983.

[30] Tian, X., and Bar-Shalom, Y., "Coordinate Convertion and Tracking for Very Long Range Radars," *IEEE Trans. on Aerospace and Electronic Systems*, vol. AES-45, no. 3, pp. 1073–1088, July 2009.

[31] Rothrock, R.L., and Drummond, O., "Performance Metrics for Mulitple-Sensor Multiple-Target Tracking," pp. 521–531 in *Signal and Data Processing of Small Targets 1999,* SPIE vol. 4048, ed. Drummond, 2000.

[32] Drummond, O., "Methodology for Performance Evaluation of Multitarget Multisensor Tracking," pp. 355–369 in *Signal and Data Processing of Small Targets 1999,* SPIE vol. 3809, ed. O.E. Drummond, 1999.

[33] Bruder, J.A., and Saffold, J.A., "Multipath Effects on Low-angle Tracking at Millimetre-wave Frequencies," *IEE Proceedings F,* vol. 138, no. 2, pp. 172–184, April 1991.

[34] Blair, W.D., and Keel, B.M., "Radar Systems Modeling for Target Tracking," in *Multitarget-Multisensor Tracking: Advanced and Applications,* vol. 3, ed. Y. Bar-Shalom and W.D. Blair, Artech House, Dedham, MA, 2000.

[35] Blair, W.D., and Brandt-Pearce, M., "Statistics of Monopulse Measurements of Rayleigh Targets in the Presence of Specular and Diffuse Multipath," *Proc. of the 2001 IEEE Radar Conference*, Atlanta, GA, pp. 369–375, May 2001.

[36] Ristic, B., Arulampalam, S., and Gordon, N., *Beyond the Kalman Filter: Particle Filters for Tracking Applications,* Artech House, Boston, MA, 2004.

19.10 PROBLEMS

1. *Process observations of a stationary object.* Let the state equation be given by $x_{k+1} = x_k$ and the measurement equation be given by $y_k = x_k + w_k$, with w_k being a Gaussian random process with $E\{w_k\} = 0$ and $E\{w_k^2\} = \sigma_w^2$. Derive the formulas for the LSE and covariance for the state of a stationary object as function of N measurements.

2. *Derive the alpha filter gain for tracking a stationary random process.* Let the state equation be given by $x_{k+1} = x_k + \frac{T^2}{2} v_k$, with v_k being a Gaussian random process with $\mathrm{E}\{v_k\} = 0$ and $\mathrm{E}\{v_k^2\} = q$. Let the measurement equation be given by $y_k = x_k + w_k$, with w_k being a Gaussian random process with $\mathrm{E}\{w_k\} = 0$ and $\mathrm{E}\{w_k^2\} = \sigma_w^2$. Derive expressions for steady-state gain α and covariance for the alpha filter for tracking this stationary random process.

3. *Develop a gain schedule for track initiation.* Use $\mathbf{K}_k = \mathbf{P}_{k|k}\mathbf{H}_k^T\mathbf{R}_k^{-1}$ and the covariance of the LSE from Problem 1 to develop the Kalman gain for the alpha filter of Problem 2 for processing the k-th measurement. The result of this processing is a LSE through the kth measurement.

4. *Derive the sensor noise only (SNO) variance.* Given the measurement noise variance σ_w^2, derive the SNO variance for the alpha filter derived in Problem 2. Hint: Write the current filtered state estimate $(k|k)$ of the alpha filter as a linear difference equation in terms of the previous filtered state estimate $(k-1|k-1)$ and the measurement. Then let the input be zero-mean white noise and compute the variance of the output in terms of the variance of the input.

5. *Derive the filter bias for a moving target.* Given a constant velocity target of velocity V_0, derive the lag or bias in the position estimates relative to the true values for the alpha filter derived in Problem 2. Hint: Write the alpha filter as a linear difference equation in terms of the previous filtered state and the measurement as a kTV_0 and use a z transform.

6. *Derive the bounds for the process noise variance.* Given targets that maneuver as much as 30 m/s^2, a sensor measurement rate of 2 s, and measurement variance of 1600 m^2 for the nearly constant velocity filter with discrete white noise acceleration acceleration (DWNA), find the minimum acceptable process noise variance and the process noise variance that minimizes the maximum mean squared error (MMSE).

7. *Compute the MMSE.* Compute the MMSE for the position and velocity estimates for the maximum acceleration of target and both values of the process noise variance found in Problem 6.

8. *Find the number of measurements required.* Given a sensor measurement rate of 1 s and measurement variance of 625 m^2, find the minimum number of measurements required to achieve a variance of the velocity estimate that is less than than 100 m^2/s^4.

9. *Derive the LSE and covariance for tracking with LFM waveforms.* Derive the formulas for the LSE and covariance for the state of a constant velocity object as function of N measurements. Let the state and measurement equations be defined by

$$\mathbf{X}_k = [r_k \quad \dot{r}_k]^T$$
$$\mathbf{F}_k = \mathbf{F} = \begin{bmatrix} 1 & T \\ 0 & 1 \end{bmatrix}$$
$$\mathbf{H}_k = \mathbf{H} = [1 \quad \Delta t]$$
$$R_k = R = \sigma_r^2$$

where r_k and $\dot{r}_k$ are the range and range rate of the target, respectively, σ_w^2 is the variance of the measurement errors, T is the sample period between measurements, and Δt is the range-Doppler coupling coefficient.

10. *Develop a gain schedule for track initiation.* Use the covariance of the LSE from Problem 9 to develop a gain schedule for processing the k-th measurement with the alpha-beta filter for tracking with LFM waveforms. The result of this processing is a LSE through the k-th measurement.

11. *Investigate the need for an unbiased transform.* Consider a target at a range of 1,000 km and bearing of zero. Thus, the Cartesian position of the target is $(x, y) = (1000, 0)$ km. Consider a radar that measures the bearing with errors of standard deviation of 3 mrad and range with errors of standard deviation of 25 m. For this radar-target scenario, use the criteria in (19.162) to assess the need for an unbiased transform of spherical to Cartesian coordinates in this case. Simulate 2,000 measurements of range and bearing by adding Gaussian errors to the true range and bearing of the target. Convert the measurements to Cartesian x and y, and plot the converted measurements. Compute the sample mean and sample covariance of the measurements in Cartesian space. The sample covariance is computed by subtracting the sample mean from each sample and forming the outer product of it with itself and then computing a sample average. Next, compute the measurements directly in Cartesian space as Gaussian errors in the x coordinate with standard deviation of 25 m and Gaussian errors in the y coordinate with standard deviation of 3,000 m. Plot these measurements. Do the two distributions of the measurements appear similar in Cartesian space? Repeat the above for a radar measuring range with a standard deviation of 1 m.

12. *Investigate the PDAF for a single validated measurement.* Consider the case of processing a single measurement with the PDAF. Consider a radar that measures the bearing with errors of standard deviation of 2 mrad and range with errors of standard deviation of 10 m. Consider a target at a range of 100 km and bearing of zero. Let the covariance of the predicted position in Cartesian coordinates be a diagonal matrix with standard deviations of 20 m in the x coordinate and 200 m in the the y coordinate. From the chi-squared tables, what is the gating threshold for a probability of gating of 0.999? For probabilities of detection of 0.5, 0.75, and 0.95, compute and plot β_k^1 versus the Mahalanobis distance (i.e., d_k^1) from 0 to the gating threshold. For a validated measurement near the edge of the gate and $P_D = 0.5, 0.75$, and 0.95, how much will the effect of the Kalman gain be reduced by the consideration that a false alarm may have been detected in place of the actual target?

13. *Design and analyze a radar tracking algorithm.* Consider a surveillance radar that measures range and bearing and scans with a period of 4 s. The radar measures range with standard deviation of 25 m and bearing with standard deviation of 0.02 degree. Targets that perform maneuvers with acceleration as much as 30 m/s^2 are expected. Use the design techniques of Section 19.2.2.1 to find process noise covariances for tracking in range and cross-range. Make a small angle assumption to approximate the cross-range errors in Cartesian space. The process noise covariance for tracking in cross-range will be a function of the range of the target. The radar is to track targets at ranges from 5 km to 400 km. (a) Recommend a single process noise covariance tracking targets at all ranges in the field of regard. (b) Recommend a range dependent function for selecting the process noise covariance for both coordinates for tracking.

(c) Develop a formulation of the process noise covariance that accounts for a different process noise variances in the range and cross range coordinates. It will be a function of the bearing of the target. (d) Assume target maneuvers are restricted to 30 m/s^2 perpendicular to the velocity vector of the target and 10 m/s^2 along the velocity vector. Extend the results of (c) to include these limitations on the target acceleration in the selection of the process noise covariance. (e) Consider a target starting at (250, 250) km and moving with a velocity of (−150, −150) m/s. After 100 s, the target maneuvers with acceleration of (−7, −7) for 15 s. At 200 s, the target starts maneuvering with acceleration of (20, −20) m/s^2 and the acceleration is orthogonal to the velocity for 16 s. At 300 s, the target starts maneuvering with acceleration of 24 m/s^2 orthogonal to the velocity for a right turn for 16 s. At 400 s, the target slows down at 10 m/s^2 for 15 s. The scenario ends at 500 s. Simulate the target and radar and compare the performances of the four track filter designs with 200 Monte Carlo runs. (f) Implement an IMM estimator with two NCV filters and compare its performance with that of the NCV filters. Use the NCV filter formulation of (d) and use the process noise covariance the minimizes the MMSE as the maneuver model.

CHAPTER 20

Fundamentals of Pulse Compression Waveforms

Byron Murray Keel

Chapter Outline

20.1 Introduction 773
20.2 Matched Filters 774
20.3 Range Resolution 782
20.4 Straddle Loss 786
20.5 Pulse Compression Waveforms 787
20.6 Pulse Compression Gain 788
20.7 Linear Frequency Modulated Waveforms 789
20.8 Matched Filter Implementations 794
20.9 Sidelobe Reduction in an LFM Waveform 797
20.10 Ambiguity Functions 800
20.11 LFM Summary 808
20.12 Phase-Coded Waveforms 808
20.13 Biphase Codes 817
20.14 Polyphase Codes 824
20.15 Phase-Code Summary 829
20.16 Further Reading 830
20.17 References 830
20.18 Problems 833

20.1 INTRODUCTION

A radar system employing a continuous wave (CW) pulse exhibits a range resolution and signal-to-noise ratio (SNR) that are both proportional to pulse width. SNR drives detection performance and measurement accuracy and is a function of the energy in the pulse. Energy and SNR are increased by lengthening the pulse. Range resolution defines a radar's ability to separate returns in range and is improved by decreasing the pulse width. An undesired relationship, coupled through the pulse width, exists between the energy in a CW pulse and the pulse's range resolution. Near the end of World War II, radar engineers applied intrapulse modulation to decouple the two quantities. Range resolution was shown to be inversely proportional to bandwidth. A waveform's bandwidth could be increased via modulation, achieving finer range resolution without shortening the pulse. Waveforms that decouple resolution and energy via intrapulse or interpulse modulation are termed *pulse compression* waveforms.

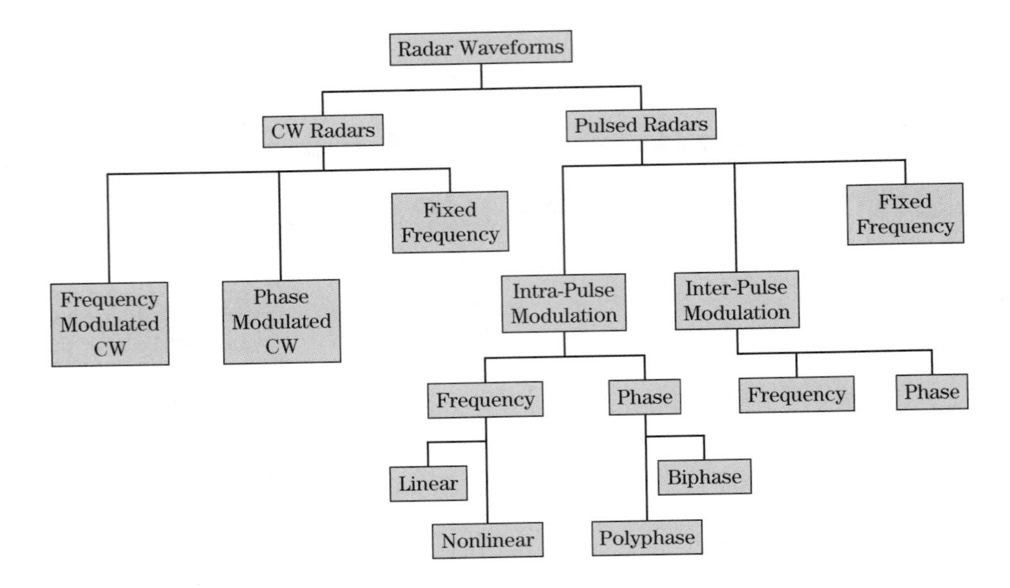

FIGURE 20-1 ■ Modern radars select from and employ many waveform modulations.

Modern radar systems employ both phase and frequency modulated waveforms. Waveform design takes into account a number of system requirements and constraints including percent bandwidth, sampling rate, dispersion, blind range, duty cycle, power, Doppler tolerance, and sidelobes as well as range resolution and SNR. A variety of waveform modulations have been developed since the 1950s to address these and other system requirements. Figure 20-1 contains a chart summarizing the various waveform modulations used in modern CW and pulsed systems.

This introductory chapter on pulse compression focuses on intrapulse linear frequency modulation and phase coding. The matched filter and resolution metrics are developed first. Pulse compression waveforms are then defined, and an overview of amplitude, phase, and frequency modulation is presented. The linear frequency modulated (LFM) waveform is explored in depth and serves as a basis for introducing a number of general concepts and properties including resolution, sidelobes, ambiguity surfaces, and processing gain. The properties of a CW pulse are examined and contrasted with the performance of pulse compression waveforms. In the latter sections, both biphase and polyphase codes are investigated. Special attention is paid to Barker codes, minimum peak sidelobe (MPS) codes, maximal length sequences, and Frank, P1, P2, P3, and P4 codes. The "Further Reading" section provides references to other frequency modulated (FM) waveforms and techniques (e.g., nonlinear FM, stepped frequency, stepped chirp, stretch processing) and phase codes (e.g., quadriphase codes).

20.2 MATCHED FILTERS

In a radar system, a filter is applied to the received signal to maximize SNR at a point in time corresponding to the delay to the target. The filter maximizing SNR is derived from the transmit waveform and is termed a *matched filter.* A number of radar performance metrics, including range resolution, are defined in terms of the waveform's filtered response. The matched filter also plays an important role in processing pulse compression waveforms. The matched filter and its properties are examined in the following sections.

20.2.1 Relevance of SNR to Radar Performance

The probability of detecting a target and the accuracy of a measurement are both functions of signal-to-noise ratio. At the output of a square law detector, the probability of detecting a Swerling I or II target employing a single or multiple, coherently combined, pulses in the presence of additive, Gaussian distributed noise is [1]

$$P_D = (P_{FA})^{1/(1+SNR)} \tag{20.1}$$

where P_D is the probability of detection and P_{FA} is the probability of false alarm (see Chapters 2 and 15). For a fixed P_{FA}, detection performance improves with increasing SNR. The Cramèr-Rao bound defines the lower limit on accuracy with which the range to a target with known velocity may be measured and is given by [2]

$$\sigma_R = \frac{c}{4\pi B_{rms}} \frac{1}{\sqrt{SNR}} \tag{20.2}$$

where σ_R is the standard deviation in the range measurement, c is the speed of light ($\approx 3 \times 10^8$ m/sec), and B_{rms} is the root mean square (rms) waveform bandwidth with units of hertz. Cramèr-Rao lower bounds also exist for amplitude, Doppler, and angle measurements [2] (see Chapter 18). In general, the accuracy of a measurement is inversely proportional to the square root of SNR. Given the dependence of measurement accuracy and detection performance on SNR, radar systems are designed to maximize this quantity.

20.2.2 Energy Form of the Radar Range Equation

The radar range equation relates SNR to system and target parameters. A common form of the radar range equation, similar to the form in equation (2.11), is

$$SNR = \frac{P_t G_t G_r \lambda^2 \sigma}{(4\pi)^3 R^4 k T_s B L_s} \tag{20.3}$$

where P_t is the peak transmit power, G_t and G_r are the transmit and receive antenna gains, respectively, λ is the transmit wavelength, σ is the target radar cross-section (RCS), R is the one-way radial range to the target, k is Boltzman's constant ($k \approx 1.38 \times 10^{-23}$ W/Hz/K), T_s is the system noise temperature, B is the receiver noise bandwidth in hertz, and L_s is the aggregate system loss. In general, the receiver bandwidth is matched to the pulse bandwidth.

The bandwidth in hertz, B, of an unmodulated pulse of duration τ is commonly defined as the reciprocal of the pulse width. The terms *unmodulated* or *simple* pulse in this text refer to a real, rectangular-shaped pulse whose symmetric spectrum is centered at baseband (i.e., zero hertz). The spectrum $X(\omega)$ of a unit amplitude, unmodulated pulse with duration τ is a sinc function defined by

$$X(\omega) = \frac{\tau \sin\left(\frac{\omega\tau}{2}\right)}{\frac{\omega\tau}{2}} \tag{20.4}$$

where ω represents frequency in units of radians per second. The spectrum bandwidth in hertz, defined at the -4 dB width, equals the reciprocal of the pulse width.

In a radar system, the receiver bandwidth is matched to the waveform bandwidth to maximize SNR. Substituting the reciprocal of the pulse width for the receiver bandwidth, the radar range equation in equation (20.3) takes the form

$$SNR = \frac{P_t \tau G_t G_r \lambda^2 \sigma}{(4\pi)^3 R^4 k T_s L_s} \tag{20.5}$$

The product of peak power and pulse width defines the energy, E, in a pulse or

$$E = P_t \tau \tag{20.6}$$

and is the first term in the numerator of equation (20.5). The radar range equation in (20.5) is known as the energy form of the equation.

In a radar system, peak power is limited, and the transmitter may be operated in saturation to maximize energy on target. Lengthening a pulse increases the energy in the pulse and is a simple and cost-effective method for improving SNR. Modern radars employ different pulse widths chosen by the radar designer to support various operating modes.

20.2.3 The Form of the Matched Filter

In a radar system, the received waveform is filtered to maximize SNR at a time delay corresponding to the target's range. For an arbitrary waveform, $x(t)$, defined over the time interval $0 \le t \le \tau$, and embedded in additive white noise, the filter that maximizes SNR takes the form [3–7]

$$h(t) = a x^*(-t) \qquad -\tau \le t \le 0 \tag{20.7}$$

and is referred to as the *matched filter*. The impulse response of the matched filter is a time-reversed and complex conjugated form of the transmit waveform, scaled by an arbitrary constant a, which is commonly set to 1. The form of the matched filter is advantageous since in a radar system the transmit waveform is known a priori, and thus the filter is known.

On receive, the matched filter $h(t)$ is convolved with the received waveform $x_r(t)$ to yield the output $y(t)$, or

$$y(t) = \int x_r(\alpha) h(t - \alpha) d\alpha \tag{20.8}$$

where α is a dummy variable of integration. Substituting (20.7) into (20.8) yields

$$y(t) = \int x_r(\alpha) x^*(\alpha - t) d\alpha \tag{20.9}$$

Applying the matched filter in (20.8) is equivalent to correlating the received signal with a copy of the transmit waveform as shown in (20.9). In most modern radars, the correlation is performed at baseband, after removal of the transmit radio frequency (RF) and receiver intermediate frequencies (IFs).

20.2.4 Point Target Model

A point target is defined as a scatterer with infinitesimal spatial extent. Some reflectors, such as a flat plate, sphere, dihedral, or trihedral, exhibit a response in range similar to a point target. The mathematical model for the reflectivity of a point target with unit RCS, located at a slant range R from the radar, is the Dirac delta function, $\delta_D(t - 2R/c)$.

The waveform reflected off a point target and received at the radar at time delay t_d is modeled as

$$x_r(t) = b\exp(j\phi)x(t - t_d) \quad t_d \le t \le (t_d + \tau) \tag{20.10}$$

where b is a constant proportional to the received voltage, and ϕ is the phase measured by the coherent detector. The measured phase is a function of the transmit frequency and the slant range to the scatterer, or

$$\phi = -2\pi f\frac{2R}{c} = -\frac{4\pi R}{\lambda} \tag{20.11}$$

where f is the transmit center frequency, and $R = ct_d/2$. The received waveform in equation (20.10) is an amplitude-scaled and time-delayed version of the transmit waveform. The model of the received waveform in (20.10) does not account for dispersive (i.e., frequency-dependent) distortions or Doppler shifts due to relative motion. The impact of Doppler is addressed in subsequent sections. A point target is used throughout the chapter to examine properties of the matched filter and range resolution.

20.2.5 Match Filtered Response Proportional to Waveform Energy

The radar range equation defines SNR at the output of a matched filter, and the form of equations (20.5) and (20.6) states that SNR is proportional to the waveform's energy. Thus, the energy in the transmit pulse should appear as a scale factor at the output of the matched filter. This relationship is now examined.

Applying the matched filter to the target return in equation (20.10) yields

$$y(t) = \int be^{j\phi}x(\alpha - t_d)x^*(\alpha - t)d\alpha \tag{20.12}$$

By design, the output SNR is maximized at $t = t_d$. The output of the filter at time delay t_d is

$$y(t_d) = be^{j\phi}\int_{t_d}^{t_d+\tau}|x(\alpha - t_d)|^2 d\alpha \tag{20.13}$$

The output at $t = t_d$ is proportional to the energy in the transmit pulse, which is defined as

$$E = \int_0^{\tau}|x(t)|^2 dt \tag{20.14}$$

Substituting (20.14) into (20.13) yields

$$y(t_d) = be^{j\phi}E \tag{20.15}$$

The preceding equations (20.12) through (20.15) show that the matched filter takes a scalar multiple of the energy in the waveform and positions it at a time delay associated with the point target. The relationship between the matched filter and the radar range equation is examined further in Section 20.2.8.

20.2.6 Fourier Relationships and the Matched Filter

The shape of the waveform spectrum, having applied the matched filter, establishes the shape of the time-domain response, and the duality between the two domains may be exploited to ascertain or influence the response. Given $x(t)$, the spectrum of the waveform is defined via the Fourier transform as

$$X(\omega) = \int x(t)\exp(-j\omega t)dt \equiv \Im\{x(t)\} \tag{20.16}$$

It is easy to show that the spectrum $H(\omega)$ of the matched filter [6,7] in equation (20.7) is

$$H(\omega) = X^*(\omega) \tag{20.17}$$

The filter spectrum in (20.17) is equal to the complex conjugate of the waveform's spectrum; thus, the filter is viewed as being "matched" in both the time and frequency domains.

Exploiting the duality between the time and frequency domains, the filtered output in (20.12) may be expressed in terms of its spectral components as

$$y(t) = \frac{1}{2\pi}\int b\exp(j\phi)X(\omega)\exp(-j\omega t_d)X^*(\omega)\exp(j\omega t)d\omega \tag{20.18}$$

An examination of the terms comprising equation (20.18) is instructive. The Fourier transform of the time-delayed and amplitude-scaled received signal in (20.10) is

$$\Im\{b\exp(j\phi)x(t-t_d)\} = b\exp(j\phi)X(\omega)\exp(-j\omega t_d) \tag{20.19}$$

The right side of (20.19) contains the first four terms inside the integral in equation (20.18). Time delay produces a linear phase ramp across the spectrum with a slope determined by the delay. The other terms in (20.18) include the waveform's spectrum $X(\omega)$, the matched filter's spectrum $X^*(\omega)$, and the Fourier kernel $\exp(j\omega t)$ associated with the inverse transform.

Grouping terms, (20.18) may be written as

$$y(t) = \frac{b\exp(j\phi)}{2\pi}\int |X(\omega)|^2\exp(-j\omega t_d)\exp(j\omega t)d\omega \tag{20.20}$$

The factors in (20.20) determine both the shape and location of the filtered response in the time domain. The product of the signal and matched filter spectra produces a squared magnitude response. The waveform's phase in the frequency domain has been removed by the filter. The shape of the squared magnitude response defines the time-domain response via Fourier transform pairs and may be intentionally modified or chosen to achieve a desired response. For example, spectral shaping is exploited in both linear and nonlinear frequency modulated waveforms to achieve low-range sidelobes [2,4,7]. The linear phase term contains the time-delay information and is responsible for positioning the filtered response in the time domain.

The response associated with N_{pt} point targets or scatterers may be modeled as

$$y(t) = \frac{1}{2\pi}\int |X(\omega)|^2 \left[\sum_{i=1}^{N_{pt}} b_i \exp(j\phi_i)\exp(-j\omega t_{d_i})\right]\exp(j\omega t)d\omega \tag{20.21}$$

where b_i, t_{d_i}, and ϕ_i are the amplitude, time delay, and phase associated with the i-th point target, respectively.

20.2.7 Derivation of the Matched Filter

Having examined Fourier relationships between the time and frequency domains, the derivation of the matched filter is relatively straightforward. The approach taken is similar to that found in [8]. Consider applying an arbitrary filter $H(\omega)$ to the return from a point target. The filtered signal is

$$y(t) = \frac{1}{2\pi}\int b\exp(j\phi)X(\omega)\exp(-j\omega t_d)H(\omega)\exp(j\omega t)d\omega \tag{20.22}$$

where $b\exp(j\phi)X(\omega)\exp(-j\omega t_d)$ is the spectrum of the return from a point target located at time delay t_d, b is the amplitude of the return, and ϕ is the measured phase.

The received signal is competing with thermal noise in the receiver. The two-sided power spectrum associated with white noise is defined as

$$N(\omega) = N_0 \tag{20.23}$$

where N_0 has units of watts per hertz. The term *white noise* means that the noise is uncorrelated, yielding a power spectrum that is constant over frequency. Applying an arbitrary filter $H(\omega)$ to white noise yields an expected output power given by

$$\overline{n^2(t)} = \frac{N_0}{2\pi}\int |H(\omega)|^2\, d\omega \tag{20.24}$$

where $n(t)$ is a realization of the noise as a function of time. $n(t)$ is a voltage, so squaring the realization yields the instantaneous noise power. The overbar denotes the expected value, $\overline{n^2(t)} = \mathrm{E}\{n^2(t)\}$ where E{ } is the expectation operator. The noise in a receiver channel is assumed to be Gaussian distributed with zero mean. Equation (20.24) thus represents the average noise power (or equivalently the variance of the noise) at the output of the filter.

For a target, the output of the filter in equation (20.22) at time delay t_d is

$$y(t_d) = \frac{b\exp(j\phi)}{2\pi}\int X(\omega)H(\omega)\, d\omega \tag{20.25}$$

and represents a voltage. The squared magnitude defines the power or

$$|y(t_d)|^2 = \left|\frac{b}{2\pi}\int X(\omega)H(\omega)\, d\omega\right|^2 \tag{20.26}$$

The signal-to-noise ratio at the output of the filter at time delay t_d is

$$SNR = \frac{|y(t_d)|^2}{\overline{n^2(t)}} \tag{20.27}$$

or

$$SNR = \frac{b^2 \left| \int X(\omega) H(\omega) d\omega \right|^2}{2\pi N_0 \int |H(\omega)|^2 d\omega} \tag{20.28}$$

The objective is to define a filter $H(\omega)$ that maximizes SNR. The Schwartz inequality may be applied to the numerator in equation (20.28). The Schwartz inequality states that, for any $X(\omega)$ and $H(\omega)$,

$$\left| \int X(\omega) H(\omega) d\omega \right|^2 \leq \int |X(\omega)|^2 d\omega \int |H(\omega)|^2 d\omega \tag{20.29}$$

For the equality to hold, the filter must be of the form

$$H(\omega) = a X^*(\omega) \tag{20.30}$$

where a is an arbitrary constant. Equation (20.30) defines the spectrum of the filter that maximizes the SNR at a time delay corresponding to the target's range. The impulse response in the time domain is obtained via the inverse Fourier transform applied to equation (20.30) and is

$$h(t) = a x^* (-t) \tag{20.31}$$

which is a time-reversed and complex conjugated copy of the transmit waveform and is equivalent to the filter previously defined in equation (20.7).

20.2.8 The Radar Range Equation and Matched Filter Relationship

The relationship between the radar range equation and the matched filter is shown by substituting the matched filter in (20.30) into (20.28), yielding

$$SNR = \frac{b^2 \frac{1}{2\pi} \int |X(\omega)|^2 d\omega}{N_0} \tag{20.32}$$

Now, Parseval's theorem [9] states that energy in the time-domain signal must equal the energy in the frequency domain, or

$$\int_{-\infty}^{\infty} |x(t)|^2 dt = \frac{1}{2\pi} \int_{-\infty}^{\infty} |X(\omega)|^2 d\omega \tag{20.33}$$

Recognizing that the energy in the transmit waveform is defined by

$$E = \int_{-\infty}^{\infty} |x(t)|^2 dt \tag{20.34}$$

equation (20.32) may be written as

$$SNR = \frac{E}{N_0} b^2 \tag{20.35}$$

By selecting $b^2 = G_t G_r \lambda^2 \sigma / (4\pi)^3 R^4 L_s$ and noting that $E = P_t \tau$ and $N_0 = kT_s$, it is evident that (20.35) is equivalent to the radar range equation (20.5); that is, both are of the form $SNR \propto E/N_0$.

20.2.9 The Match Filtered Response for a Simple Pulse

A simple (or CW) pulse is applied in many radar systems, and therefore an examination of its properties and its match filtered response is worthwhile and will provide the motivation for pulse compression waveforms. Consider a simple pulse with constant amplitude, A, and pulse width, τ, defined by

$$x(t) = A, \quad -\frac{\tau}{2} \leq t \leq \frac{\tau}{2} \tag{20.36}$$

and the corresponding matched filter, obtained from equation (20.7) with $a = 1$,

$$h(t) = A, \quad -\frac{\tau}{2} \leq t \leq \frac{\tau}{2} \tag{20.37}$$

The filtered response is a triangle defined by

$$y(t) = \begin{cases} A^2(t+\tau), & -\tau \leq t \leq 0 \\ A^2(\tau - t), & 0 < t \leq \tau \end{cases} \tag{20.38}$$

and is depicted in Figure 20-2. Note that the length of the filtered response is equal to twice the original pulse width. The expansion is a property of the filtering operation. For any finite duration waveform, the match filtered output exhibits a response whose duration is equal to twice the original waveform extent. The filtered signal represents a voltage, and the instantaneous power is defined as the square of the voltage. The peak voltage is $A^2\tau$, which equals the energy in $x(t)$, consistent with equations (20.14) and (20.15).

In many instances, the rectangular pulse is a good first-order model that facilitates analysis. In an actual system, the ideal pulse is not realizable due to bandwidth limits imposed by the hardware. Band-limiting or shaping of the waveform's spectrum increases the pulse duration and prevents instantaneous rise and fall times at the pulse boundaries.

When employing a simple pulse, the low-pass filter on receive serves as an approximation to the matched filter. The bandwidth of the filter is matched to the pulse bandwidth, or, equivalently, the reciprocal of the pulse width. A loss in SNR occurs if the spectrum of the waveform and the filter are not matched exactly except for a linear phase term.

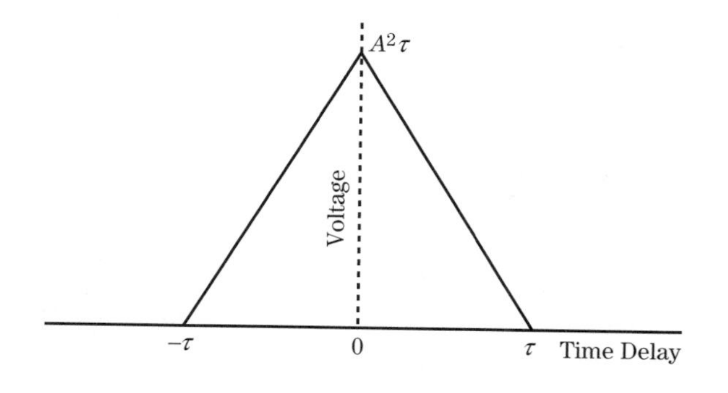

FIGURE 20-2 ■ The simple pulse of duration τ has a match filtered response of duration 2τ.

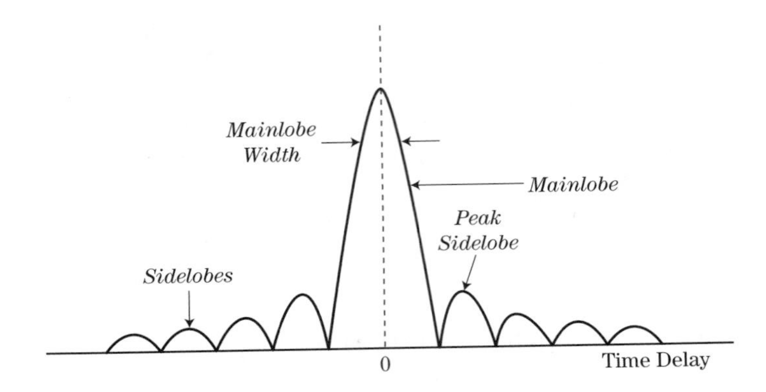

FIGURE 20-3 ■ A generic match filtered response includes the mainlobe and sidelobes.

20.2.10 Properties of the Match Filtered Response

Many waveform properties, including SNR, range resolution, and Doppler tolerance, are defined in terms of the match filtered response. In general, a waveform's filtered response exhibits both a mainlobe and sidelobe structure, as illustrated in Figure 20-3. The mainlobe is defined as the portion of the response positioned between the nulls that lie adjacent to the peak of the response, and the sidelobes are defined as the portion of the response outside the mainlobe. A simple, unmodulated pulse exhibits only a mainlobe.

With any waveform, SNR is maximized at the peak of the mainlobe. Points on the response that are located below the peak achieve a SNR that is less than the maximum. In a radar system, samples of the response do not always include the peak, and thus, the potential exists for a loss in SNR that must be accounted for in the design. The loss associated with not sampling the peak of the response is termed *straddle loss* (see Section 20.4).

Range resolution is a measure of the ability of a radar to distinguish between objects closely spaced in range and is defined in terms of the mainlobe width. A more formal definition and detailed examination of resolution is provided in subsequent sections.

Sidelobes are an undesired by-product of applying the matched filter to a modulated waveform. In the range dimension, sidelobes are interchangeably referred to as range or time sidelobes. Sidelobes are problematic because those associated with a large RCS target may be higher in amplitude than the mainlobe response of a weaker target and thus may mask the presence of the smaller target even when the two are well resolved in range. Sidelobe levels are commonly referenced relative to the peak of the mainlobe. Both peak and integrated sidelobe ratios are important performance metrics. Integrated sidelobe levels are relevant when operating in a distributed clutter environment, as the cumulative sidelobe contributions associated with clutter may degrade the quality of a target measurement. For example, integrated sidelobes contribute to the multiplicative noise in a synthetic aperture radar (SAR) image [10].

20.3 RANGE RESOLUTION

In many systems, an ability to resolve objects in range is required and is defined in terms of the radar's range resolution. Range resolution requirements vary depending on the application or radar mode. For example, search functions may employ a relatively coarse range resolution on the order of or greater than the target's physical extent. In contrast, target recognition (i.e., classification or identification) requires a resolution sufficient to resolve individual scattering centers located along the target. Resolution requirements on the order of 0.5 to 1 foot may be required in some systems [11]. In track, resolution requirements may be finer than those used during search.

Ultimately, a radar's ability to resolve closely spaced objects is governed by the shape and width of the waveform's mainlobe response. Two of the more common width metrics [2,4,6] used to define resolution are:

1. The *Rayleigh criterion*, which defines resolution as the separation between the peak and the first null.
2. The mainlobe width at a specific point below the peak of the response, such as the -3 dB point.

Each metric is described in more detail in the following sections.

20.3.1 Resolution as Defined by the Rayleigh Criterion

The Rayleigh criterion states that two point targets are resolved when the targets are separated in range such that the peak of the match filtered response of one target falls on the first null of the second target. The targets are considered to be resolved since no energy from one target is present at, or is competing with, the peak of the second target's return.

The Rayleigh criterion may be applied to any waveform, including pulse compression waveforms exhibiting both a mainlobe and sidelobe response. The Rayleigh separation is illustrated in Figure 20-4 using a simple pulse. The peak of the response and the first null are separated by τ seconds. For a simple pulse, the null occurs when the filtered response goes to zero. Given a monostatic radar, time delay represents the total round-trip delay; thus, two point target returns separated by a τ second delay are separated in range by

$$\delta R = \frac{c\tau}{2} \tag{20.39}$$

where c is the speed light ($\approx 3 \times 10^8$ m/sec). Equation (20.39) represents the Rayleigh resolution associated with a simple, ideal pulse of duration τ. Given that the match filtered response for the simple pulse is a triangle (Figure 20-2), the amplitude is one-half the peak amplitude (-6 dB) at a delay of $\pm\tau/2$ seconds from the peak. Thus, for a simple pulse the Rayleigh resolution is equivalent to the -6 dB mainlobe width.

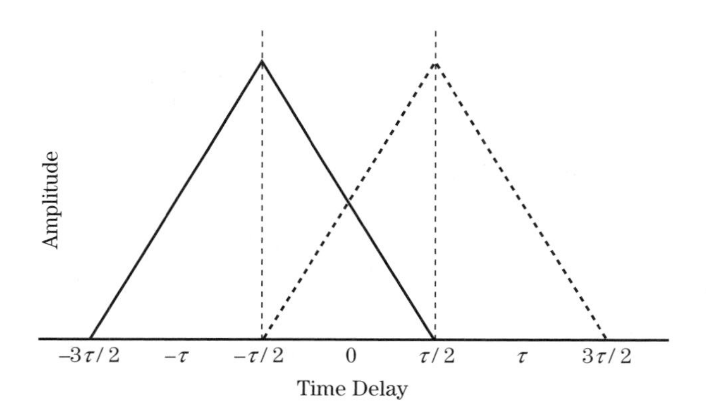

FIGURE 20-4 ■ Individual responses from two point targets separated by the Rayleigh resolution.

Some explanation is needed with regard to the terminology used to quantify resolution. The terms *improved* or *enhanced* resolution refer to a decrease in δR, while the terms *degraded* or *reduced* resolution refer to an increase in δR. Fine resolution implies relatively small values of δR, whereas coarse resolution implies relatively large values of δR. For a simple pulse, enhanced resolution is achieved by decreasing the pulse width.

20.3.2 Resolution Defined in Terms of Mainlobe Width

Range resolution may be defined with respect to the mainlobe width measured at a specified point below the peak. The -3 dB width is commonly used to define resolution and corresponds to the half-power point. In general, the physical separation associated with the -3 dB width is less than the separation associated with the Rayleigh resolution.

In some cases, several width metrics (e.g., -3, -6, -10, and -20 dB) may be employed to characterize the mainlobe width and roll-off. The shape of the mainlobe varies depending on the modulation employed and any weighting applied to suppress the range sidelobes. A single number is often used to quantify resolution, but depending on the application, a more detailed characterization of the mainlobe response may be required to fully assess performance.

20.3.3 Woodward's Range Resolution Constant

Woodward [2,4,5] describes a time resolution constant used to compare the range resolution of different waveforms. The metric defines a rectangle with peak amplitude and energy equal to that of the waveform's match filtered response. The width of such a rectangle is

$$T_{res} = \frac{\int_{-\infty}^{\infty} |y(t)|^2 dt}{|y(0)|^2} \tag{20.40}$$

Waveforms producing small values to T_{res} exhibit fine range resolution.

20.3.4 The Relationship between Bandwidth and Range Resolution

In Section 20.3.1, it was shown that for a simple pulse, range resolution is proportional to pulse width. This turns out to be a special case where pulse width and bandwidth are inversely related. In general, range resolution is inversely proportional to waveform bandwidth. This relationship is less intuitive but is demonstrated using the Fourier uncertainty principle.

The Fourier uncertainty principle [9] states that a signal's "width" in one domain is inversely proportional to the signal's "width" in the transform domain. The uncertainty principal uses second-order moments to define the width of a signal in the time domain as

$$D_t = \sqrt{\int_{-\infty}^{\infty} t^2 |y(t)|^2 dt} \tag{20.41}$$

and the signal's width in the frequency domain as

$$D_\omega = \sqrt{\int_{-\infty}^{\infty} \omega^2 |Y(\omega)|^2 d\omega} \tag{20.42}$$

where $y(t) \overset{\Im}{\longleftrightarrow} Y(\omega)$. Given that $y(t)$ is the output of the matched filter and $Y(\omega) = |X(\omega)|^2$, D_t is a measure of the width of the filtered time-domain response and D_ω is a measure of the power spectrum's width (or bandwidth). The Fourier uncertainty principle states that

$$D_t D_\omega \geq \sqrt{\frac{\pi}{2}} \tag{20.43}$$

which implies that the width of the time-domain response is inversely proportional to a measure of the waveform's bandwidth. The equality holds for Gaussian-shaped signals.

The inverse relationship between bandwidth and resolution applies to a simple pulse. As noted previously, the spectrum of a simple pulse is a sinc function with a -4 dB bandwidth defined by the inverse of the pulse width (i.e., $B = 1/\tau$). Substituting bandwidth for pulse width into equation (20.39) yields

$$\delta R = \frac{c}{2B} \tag{20.44}$$

For a simple pulse, equations (20.39) and (20.44) are equivalent definitions of resolution and correspond to the Rayleigh resolution or the -6 dB width.

Range resolution is commonly computed as

$$\delta R = \kappa \frac{c}{2B} \tag{20.45}$$

where B is the waveform bandwidth in hertz, and κ is scale factor used to account for intentional or unintentional factors that degrade resolution. The expression in (20.45) is not tied to a particular definition or measure of resolution (e.g., Rayleigh) or bandwidth. Instead, definitions of resolution and bandwidth and the scale factor κ are often chosen such that (20.45) holds. To fully assess "resolution," the entire mainlobe in terms of both width and roll-off must be characterized. In addition, the relative amplitude and phase difference between two scatterers also affects the shape of the combined response.

20.3.5 An Examination of Resolution Using Two Point Targets

For two closely spaced point targets, the shape of the composite response is a function of their separations, amplitudes, and phases. It is common to assume that if two scatterers are separated by the radar's "range resolution," then they are both visually distinguishable in range. This is not always the case. The metrics used to define resolution are not formulated to necessarily achieve a visually pleasing composite response in which the two scatterer returns are easily recognized.

Consider a simple pulse and two equal amplitude point targets separated in range by the Rayleigh resolution of $c\tau/2$ meters (τ seconds in time delay). The individual match filtered responses are depicted in Figure 20-4. The return from a point target produces a constant phase shift $\phi = -4\pi fR/c$ that is proportional to the range to the target and the radar's transmit center frequency. Two point scatterers separated in range by integer multiples of $\lambda/2$ exhibit a phase difference that is a multiple of 2π radians, equivalent to $0°$,

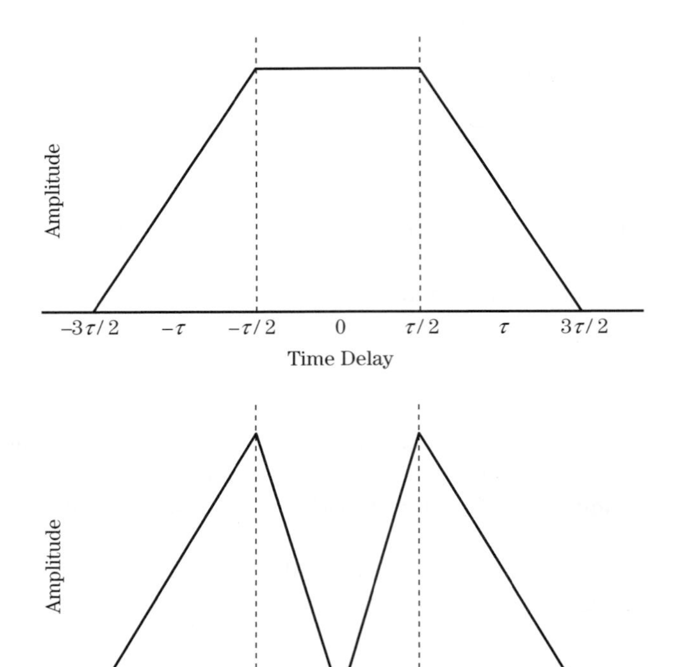

FIGURE 20-5 ■ Combined response for two point targets with phase difference equal to 0°.

FIGURE 20-6 ■ Combined response for two point targets with phase difference equal to 180°.

while two scatterers separated in range by odd multiples of $\lambda/4$ exhibit a phase difference equivalent to 180°. If the target amplitudes are equal and the phase difference is 0°, the combined response is that shown in Figure 20-5. The composite response does not provide a visual indication that two targets are present, even though the targets are separated by the Rayleigh resolution. Next assume the same scatterer separation and a different transmit frequency. If the phase difference between the two scatterers is 180°, then the combined response is that shown in Figure 20-6. Visually, the response suggests the presence of two targets. In both instances, the targets are resolved based on the Rayleigh criterion. In general, commonly applied definitions of resolution do not take into account amplitude and phase differences or the shape of the composite response.

20.4 STRADDLE LOSS

In a modern radar, the received signal is sampled in time and processed digitally. A sampled, match filtered response exhibits a loss in SNR if the samples do not include the theoretical peak value. The loss in SNR is defined as the ratio of the square of the largest sampled value to the square of the theoretical peak value and is commonly referred to as *straddle loss*, indicative of the fact that the peak falls between two sample bins. The largest or peak straddle loss occurs when the peak of the response falls mid-way between 2 samples and is calculated as

$$L_{\text{st\_pk}} = 20\log_{10}\left|\frac{y(t)}{y(0)}\right|_{t=\Delta t/2} \tag{20.46}$$

where Δt is the time between samples, and the peak of the response occurs at $t = 0$. For a given waveform, straddle loss decreases with increasing sample rate. The peak straddle loss associated with a simple pulse is -6 dB given a sample spacing equal to the pulse width (i.e., $\Delta t = \tau$).

For a point scatterer, the actual loss generally lies somewhere between the peak value and no loss. A more reasonable measure of performance is the average straddle loss. Average straddle loss is defined as

$$L_{\text{st\_avg}} = 10 \log_{10} \left(\frac{1}{\Delta t} \int_{-\Delta t/2}^{\Delta t/2} \left| \frac{y(t)}{y(0)} \right|^2 dt \right) \tag{20.47}$$

Peak and average straddle losses are a function of the shape of the match filtered response. The sharper the roll-off in amplitude the greater the straddle loss for a given sample spacing.

20.5 PULSE COMPRESSION WAVEFORMS

A simple, unmodulated pulse exhibits a coupling between energy and range resolution. Lengthening the pulse to increase the waveform's transmit energy degrades range resolution, and decreasing the pulse width to achieve finer range resolution reduces the energy. Pulse compression waveforms decouple energy and resolution by exploiting amplitude, phase, or frequency modulation to increase the waveform bandwidth while maintaining the pulse length, with the result that $B \gg 1/\tau$. The application of pulse compression waveforms originated near the end of World War II with the development of the linear frequency modulated waveform [4,12]. Since that time, numerous waveform modulations and processing techniques have been developed and employed. The breadth of modulations reflects the challenge of operating within the constraints of the radar hardware and requirements imposed by targets and interference.

20.5.1 Amplitude Modulation

Intrapulse amplitude modulation could be used to increase a waveform's bandwidth but at a cost of reduced efficiency. Traditionally, intrapulse amplitude modulation has not been employed in radar; however, amplitude modulated waveforms are described in the literature. An example is the Huffman coded waveform [13,14]. Huffman codes employ intrapulse amplitude and phase modulation to tailor the range sidelobes.

20.5.2 Frequency Modulation

In modern systems, both intra- and interpulse frequency modulations are employed. Intrapulse frequency modulated waveforms include both linear and nonlinear modulations [2,4,7]. The LFM waveform is employed in a large number of modern systems and exhibits some unique properties. The LFM waveform has also been combined with *stretch processing* [15] to achieve a reduction in processing bandwidth while preserving the resolution afforded by the transmit bandwidth. Stretch processing is employed in many high-range resolution (HRR) systems including SARs. *Nonlinear frequency modulated* (NLFM) waveforms achieve low-range sidelobes through modulation and circumvent the need to employ an amplitude weighting.

Interpulse modulation is applied in systems where some component of the hardware limits the instantaneous (i.e., intrapulse) bandwidth. A stepped frequency waveform

[16,17] employs interpulse modulation. The waveform consists of a series of narrow band pulses that are transmitted at different frequencies to create a large composite bandwidth. On receive, the pulses are combined coherently to achieve fine range resolution. Stepped frequency waveforms are also known as synthetic wideband waveforms. A stepped chirp [18] is another example of an interpulse modulated waveform. The individual pulses comprising a stepped chirp waveform are linear frequency modulated and separated in frequency. The returns from successive pulses are stitched together in the signal processor to create the return from a wider bandwidth LFM waveform.

20.5.3 Phase-Coded Waveforms

Phase-coded waveforms consists of N concatenated subpulses (or chips) where the phase is intentionally varied subpulse to subpulse to achieve a desired mainlobe and sidelobe response. In general, the length of an individual subpulse defines the range resolution of the waveform. Phase-coded waveforms are grouped into two categories: biphase and polyphase. Biphase-coded waveforms exhibit two possible phase states, typically 0 and 180 degrees, while polyphase codes exhibit more than two phase states. In general, the sidelobe levels of a phase-coded waveform decrease with increasing code length (number of subpulses). Phase-coded waveforms have been identified that yield the minimum peak sidelobe level [19–24] for a given code length. In addition, phase codes have been identified that yield low, predictable sidelobe levels and that are easy to synthesize [2,6,7,25,26]. Phase codes may also be designed address to Doppler tolerance [27] and electromagnetic interference [7,28].

20.6 PULSE COMPRESSION GAIN

Pulse compression gain is defined as the ratio of the SNR at the output of the matched filter to that prior to the filter. In a radar system, an anti-aliasing filter precedes the analog-to-digital converter (ADC). The anti-aliasing filter is generally a linear phase filter with a bandwidth matched to the waveform bandwidth, B. The filter limits the noise bandwidth to that of the waveform and supports Nyquist sampling by rejecting out-of-band signals. The SNR at the output of the anti-aliasing filter is defined by the radar range equation in equation (20.3).

When employing a pulse compression waveform, the matched filter follows the anti-aliasing filter in a digital implementation. The matched filter accounts for the waveform modulation and thus coherently integrates the waveform samples. In contrast, noise samples are noncoherently integrated. The net result is a gain in SNR at the output of the matched filter. The gain is also present with an analog implementation of the matched filter.

Pulse compression gain is defined by the waveform's *time-bandwidth* (TB) *product* τB. The resultant SNR is

$$SNR = \frac{P_t G_t G_r \lambda^2 \sigma}{(4\pi)^3 R^4 k T_s B L_s} \tau B = \frac{P_t \tau G_t G_r \lambda^2 \sigma}{(4\pi)^3 R^4 k T_s L_s} \tag{20.48}$$

and reduces to the energy form of the radar range equation. In many cases, the SNR at the output of the ADC is less than 0 dB. It is only after applying the matched filter that the signal appears above the noise floor. In modern systems, time-bandwidth products can range from 1 (for a simple pulse) to 10^6 or greater.

20.7 | LINEAR FREQUENCY MODULATED WAVEFORMS

During World War II, the need to detect targets at extended ranges and limits on transmit power forced radar engineers to employ long pulses to improve detection at the expense of degraded range resolution. However, near the end of the war, American, British, and German scientists were experimenting with intrapulse modulated waveforms and dispersive filters that, when combined, decoupled waveform energy and resolution. The war ended in 1945, and by the 1950s the once-classified work appeared in patents and papers. The documents describe an LFM waveform and a technique for synthesizing and compressing it [4,12]. The LFM waveform may be described at a high level as a sinusoid whose frequency changes linearly with time. The pulse is compressed by taking advantage of the fact that the propagation delay through a dispersive filter is frequency dependent. The filter time-aligns the frequencies, resulting in a compression of the pulse. The filter may also be used to synthesize the waveform by exciting it with a short CW pulse possessing the desired bandwidth. The frequency-dependent propagation delay through the filter creates an extended length pulse exhibiting a linear time versus frequency relationship.

The LFM waveform has some unique properties and is employed in many modern radar systems supporting search, track, and high-resolution modes. The waveform is considered to be Doppler tolerant and exhibits a range-Doppler coupled ambiguity surface [2,4,6,7]. The waveform also enables stretch processing [15], which reduces the required processing bandwidth in high-resolution systems.

20.7.1 Time-Domain Description of an LFM Waveform

A baseband LFM pulse is defined as

$$x(t) = A\cos\left(\pi\frac{B}{\tau}t^2\right) \qquad -\frac{\tau}{2} \leq t \leq \frac{\tau}{2} \tag{20.49}$$

where A is the waveform amplitude, B is the waveform bandwidth, and τ is the pulse duration. The time-domain response of an LFM pulse is plotted in Figure 20-7.

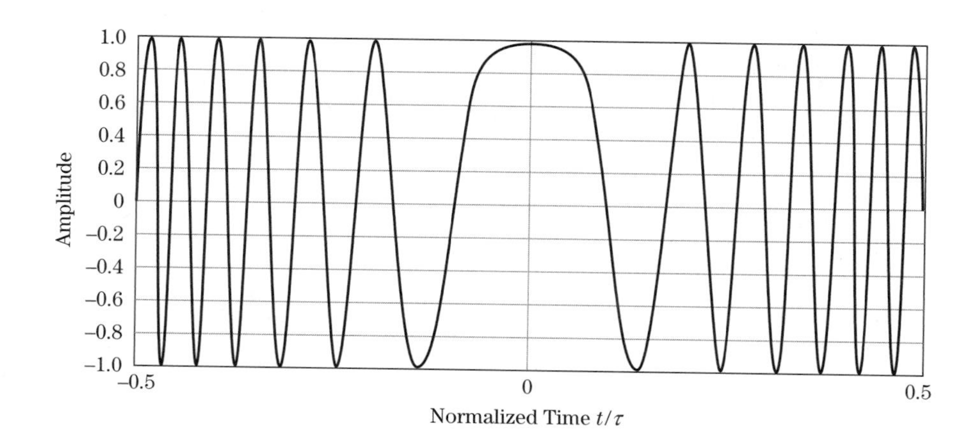

FIGURE 20-7 ■ Time-domain response, within the pulse, of a linear frequency modulated (LFM) waveform with a time-bandwidth product equal to 50.

On transmit, the LFM pulse is centered at an RF f_0 and is expressed as

$$x_{RF}(t) = A\cos\left(2\pi f_0 t + \pi\frac{B}{\tau}t^2\right) \quad -\frac{\tau}{2} \le t \le \frac{\tau}{2} \tag{20.50}$$

In most systems, the RF signal is mixed to baseband prior to compression, and a coherent detector is used in the downconversion process to form in-phase (I) and quadrature (Q) receive channels. The resultant complex, baseband signal is

$$x(t) = A\exp\left(j\pi\frac{B}{\tau}t^2\right) \quad -\frac{\tau}{2} \le t \le \frac{\tau}{2} \tag{20.51}$$

The time-varying phase $\phi(t)$ of an LFM waveform is quadratic

$$\phi(t) = \pi\frac{B}{\tau}t^2 \quad -\frac{\tau}{2} \le t \le \frac{\tau}{2} \tag{20.52}$$

and the instantaneous frequency in radians per second, defined as the derivative of the phase in (20.52), is

$$\frac{d\phi(t)}{dt} = 2\pi\frac{B}{\tau}t \quad -\frac{\tau}{2} \le t \le \frac{\tau}{2} \tag{20.53}$$

The instantaneous frequency in hertz is

$$f(t) = \frac{B}{\tau}t \quad -\frac{\tau}{2} \le t \le \frac{\tau}{2} \tag{20.54}$$

and is plotted in Figure 20-8. Note that the instantaneous frequency is linear with time—thus, the label *linear frequency modulation.* The waveform is commonly referred to as a "chirp" waveform because a similar pulse in the audible frequency range produces a chirping sound.

The plot of instantaneous frequency versus time is often used to depict an LFM waveform. An LFM waveform sweeps through B hertz in τ seconds. The ratio B/τ in (20.54) is the slope of the instantaneous frequency and is termed the *ramp rate* or *sweep rate*.

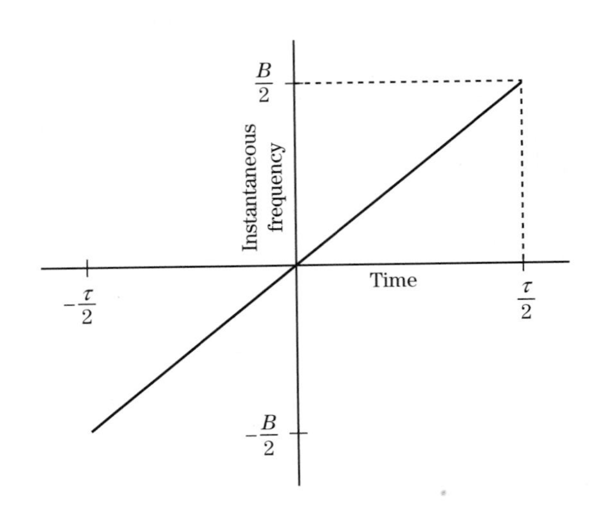

FIGURE 20-8 ■ Instantaneous frequency versus time for an LFM waveform.

20.7.2 Waveform Spectrum

Klauder [12] and Cook [4] provide closed-form expressions for the spectrum of an LFM waveform. The expressions contain Fresnel sine and cosine integrals. However, for reasonable time-bandwidth products (e.g., $\tau B > 10$), the LFM spectrum may be approximated by a much simpler expression

$$X(\omega) \approx |X(\omega)| \exp\left(-j\frac{1}{4\pi}\frac{\tau}{B}\omega^2\right) \exp\left(j\frac{\pi}{4}\right) \tag{20.55}$$

where $|X(\omega)| \approx 1$, $-\pi B \leq \omega \leq \pi B$, and is zero elsewhere. The spectrum in (20.55) consists of rectangle-shaped magnitude response, defined over the swept bandwidth, and a quadratic phase response.

The adequacy of the approximation in equation (20.55) is a function of the waveform TB product. To illustrate the dependence on TB, the spectra associated with two LFM waveforms are presented in Figure 20-9 with time-bandwidth products of 20 (light curve) and 100 (dark curve). As the time-bandwidth product increases, the spectrum becomes more rectangular in shape with a sharper transition region and with a larger percentage of the waveform energy contained within the nominal range of frequencies $-B/2 \leq f \leq B/2$. For $\tau B \geq 100$, approximately 98% to 99% of the waveform energy is contained in this region [12]. Note that neither the peak of the spectral response nor the -3 dB point occurs at $B/2$. For both waveforms, the -6 dB point occurs at approximately $\pm B/2$.

The quadratic phase term in equation (20.55) plays an important role in defining the response in the presence of uncompensated Doppler. The quadratic phase is unique to an LFM waveform and provides it with a degree of Doppler tolerance not found in other waveforms. The contribution of the quadratic phase is examined in Section 20.10.4.

20.7.3 Compressed Response

The compressed response is computed in the time domain by convolving the LFM waveform with its matched filter or

$$y(t) = \frac{1}{\tau}\int \exp\left(j\pi\frac{B}{\tau}\alpha^2\right) \exp\left(-j\pi\frac{B}{\tau}(\alpha - t)^2\right) d\alpha, \quad -\tau \leq t \leq \tau \tag{20.56}$$

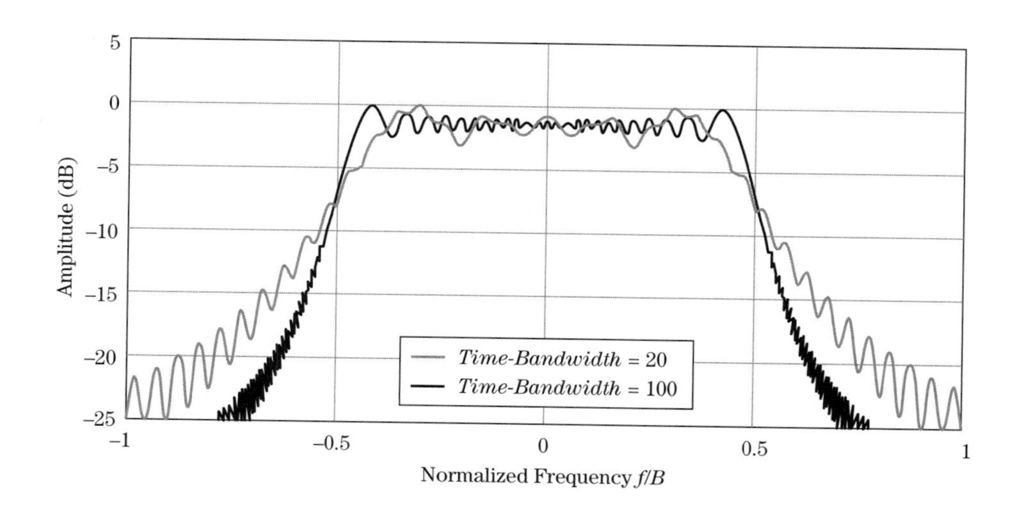

FIGURE 20-9 ■ Comparison of the spectra of LFM waveforms with time-bandwidth products of 20 (light curve) and 100 (dark curve).

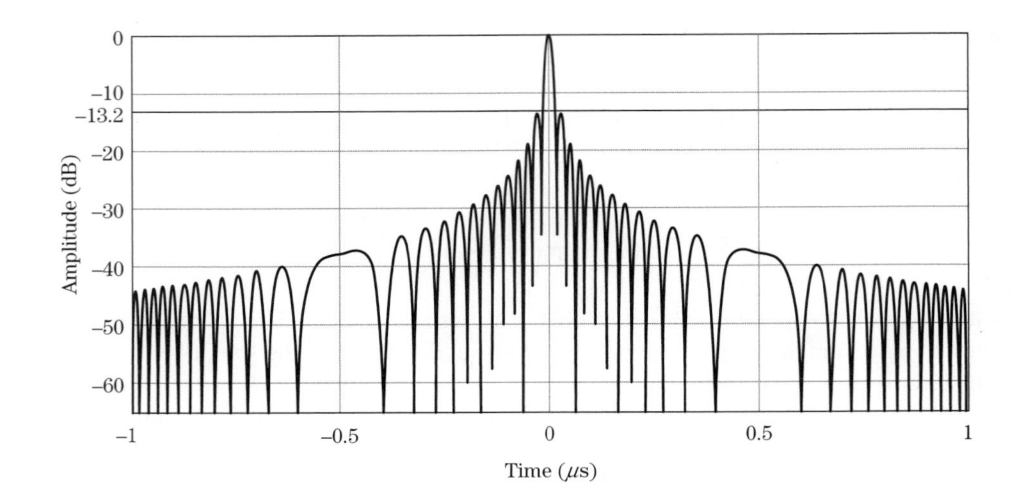

FIGURE 20-10 ■ Match filtered response for a 50 MHz, 1 μsec LFM waveform.

where α is a dummy variable of integration. The waveform and matched filter are both normalized to unity energy. The compressed response obtained by evaluating (2.56) is

$$y(t) = \left(1 - \frac{|t|}{\tau}\right) \frac{\sin\left[\left(1 - \frac{|t|}{\tau}\right)\pi Bt\right]}{\left(1 - \frac{|t|}{\tau}\right)\pi Bt}, \quad |t| \leq \tau \tag{20.57}$$

The response in (20.57) consists of the product of a term resembling a sinc function and a triangle function defined over the time interval $-\tau \leq t \leq \tau$ and zero elsewhere. Figure 20-10 contains a plot of the compressed response for an LFM waveform with a 50 MHz bandwidth and a 1 μsec pulse width ($\tau B = 50$). The peak sidelobes are close to the nominal -13.2 dB peak sidelobes associated with a sinc function.

The argument of the sine function in equation (20.57) may be written as

$$\left(1 - \frac{|t|}{\tau}\right)\pi Bt = \pi Bt - \frac{\pi Bt^2}{\tau}, \quad t \geq 0 \tag{20.58}$$

revealing both a linear and quadratic term. For $t \ll \tau$, the linear term dominates, and the argument is approximately equal to πBt. Thus, in the vicinity of the peak, the response approaches a true sinc function weighted by a triangle. Cook [4] states that for TB products greater than 20 the match filtered response resembles a sinc.

20.7.4 Rayleigh Resolution

Rayleigh resolution is defined as the separation between the peak and the first null of the match filtered response. The first null in (20.57) occurs when the argument of the sine function equals π. For TB products great than 10, the null occurs at

$$t \approx \pm\frac{1}{B} \tag{20.59}$$

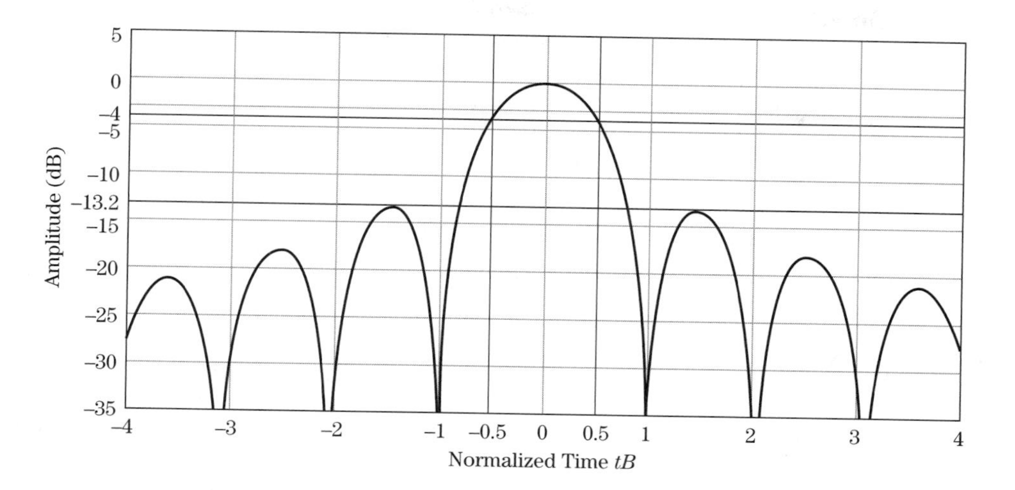

FIGURE 20-11 ■ Mainlobe and first 3 sidelobes for the LFM waveform match filtered response with a time-bandwidth product equal to 100.

Accounting for two-way propagation and the speed of light, the Rayleigh range resolution is

$$\delta R = \frac{c}{2B} \tag{20.60}$$

A portion of the compressed response associated with an LFM with $\tau B = 100$ is provided in Figure 20-11. The Rayleigh time resolution of $1/B$ is also the -4 dB width (provided no amplitude weighting is employed to suppress the range sidelobes). Note that the -3 dB width of the response, also shown in the figure, is less than the -4 dB width or Rayleigh resolution and thus represents a closer spacing between "resolved" scatterers.

20.7.5 The Nominal Sidelobe Response

Pulse compression waveforms exhibit range sidelobes that extend over a time interval equal to twice the pulse width and that are a function of the modulation employed. Sidelobes degrade radar performance by placing energy up and down range from its source. For example, range sidelobes associated with a large RCS target may mask the presence of a smaller target located within a pulse width of the larger target; thus, "low" sidelobes are a desirable property. Sidelobe performance is quantified in terms of both peak and integrated ratios. The *peak sidelobe ratio* (PSR) is defined as the ratio of the peak sidelobe to the peak of the mainlobe.

For time-bandwidth products greater than 20, the peak sidelobes of an LFM waveform are approximately -13.2 dB and occur adjacent to the mainlobe [4]. In general, frequency modulated waveforms exhibit high sidelobes adjacent to or near the mainlobe, and the sidelobes decrease with distance from the mainlobe. For an LFM waveform, the peaks of the sidelobes roll-off as $-20\log_{10}(\pi Bt)$. The minimum sidelobe peak is approximately $-20\log_{10}(\pi B\tau)$. This is illustrated in Figure 20-10 where the minimum sidelobe peak is approximately -44 dB for a time-bandwidth product of 50.

In a distributed clutter or multiple target/scatterer environment, the integrated sidelobe ratio is an important metric. For example, an HRR profile of a vehicle contains returns from scatterers located on the vehicle and returns from neighboring terrain projected onto the vehicle through the range sidelobes. Patches of resolved terrain contribute sidelobe energy onto the vehicle's range profile. The sidelobe energy, originating from resolved patches distributed in range, adds noncoherently at ranges associated with the vehicle. The cumulative, noncoherent contribution may be sufficient to degrade the HRR profile.

The integrated sidelobe ratio is defined as the ratio of the energy in the sidelobes to the energy in the mainlobe. Integrated sidelobes are a source of multiplicative noise [10] and must be weighted by the average power in the range cells containing the interference to determine their impact. The integrated sidelobe ratio for an unweighted LFM waveform is approximately −9.6 dB [29]. Amplitude weighting lowers the peak sidelobe and reduces the integrated sidelobe ratio (ISR; see Section 20.9).

20.8 MATCHED FILTER IMPLEMENTATIONS

The matched filter or compression operation may be implemented in analog hardware or performed digitally. In most modern systems, compression is performed digitally, but analog implementations still exist. The digital approach overcomes some of the disadvantages associated with analog compression including insertion loss, device dependence on waveform parameters, and pulse width and bandwidth limitations [30]. In addition, digital compression offers advantages including selectable sidelobe control and error compensation. A brief discussion of analog approaches is presented, followed by a more in-depth discussion of digital compression.

20.8.1 Dispersive Filters

Dispersive analog filters may be used to synthesize and compress an LFM pulse. Dispersive filters exhibit a time delay through the filter that is a function of frequency. A filter's dispersion is characterized by its *group delay*, t_{gd}, which is defined as the negative of the derivative of the filter's frequency-domain phase response $\Phi(\omega)$ or

$$t_{gd} = -\frac{d\Phi(\omega)}{d\omega} \tag{20.61}$$

The LFM waveform's phase response in (20.55) yields a group delay of

$$t_{gd\_LFM} = \frac{\tau}{2\pi B}\omega \tag{20.62}$$

The matched filter's group delay is

$$t_{gd\_LFM\_MF} = -\frac{\tau}{2\pi B}\omega \tag{20.63}$$

given the conjugate relationship between the waveform and matched filter spectrum.

A dispersive filter may be used to both expand and compress a pulse. A CW pulse with duration $1/B$ passing through a dispersive filter with group delay defined in (20.62) experiences a time expansion, producing a pulse of duration τ. The expansion occurs as each frequency component of the signal is delayed by a variable amount of time. The result is a linear frequency modulated pulse.

The pulse is compressed using a filter with the opposite sense phase, or negative group delay. The frequency-dependent delay aligns the various frequency components in time. The filter coherently integrates the response, producing a compressed pulse.

Bulk acoustic wave (BAW) and surface acoustic wave (SAW) devices are dispersive filters used in some radar systems. Farnett and Stevens [30] describe the performance of BAW and SAW devices. These devices exhibit *insertion loss* and are limited in the pulse lengths and bandwidths they support.

20.8.2 Digital Filters

Modern radars may employ either voltage-controlled oscillators or digital waveform generators to precisely modulate the transmit waveform. On receive, digital compressors have replaced their analog counterparts. A digital compressor correlates the digitized signal with a stored copy of the matched filter. The correlation operation may be implemented using fast Fourier transforms (FFTs) by exploiting the duality between correlation in the time-domain and multiplication in the frequency domain [31]. FFTs are efficient implementations of the discrete Fourier transform (DFT), which transforms discrete time samples into samples of the discrete-time Fourier transform (DTFT; see Chapter 14). The correlation operation, implemented using FFTs, is commonly referred to as *fast convolution* [32].

An illustration of a coherent receiver followed by a digital compressor is shown in Figure 20-12. In general, the received RF signal is mixed to some IF prior to mixing the signal to baseband. The analog signal at IF is denoted as $x_{IF}(t)$ in the figure. The IF signal is fed into a quadrature detector where in-phase $x_I(t)$ and quadrature phase $x_Q(t)$ signals are generated. The analog in-phase and quadrature signals are then fed into separate ADCs. The Nyquist sampling theorem requires that each signal (or channel) be sampled at a rate equal to or greater than twice the highest frequency component. For a baseband-centered, band-limited signal, the highest frequency is equal to one-half the bandwidth. The ADC is therefore required to sample at a rate equal to or greater than the waveform bandwidth. The baseband filter is designed with a bandwidth equal to the waveform bandwidth and suppresses out-of-band signals that alias. The in-phase and quadrature signals may be interpreted as the real and imaginary parts, respectively, of a complex signal. The digitized, complex signal is

$$x[n] = x_I[n] + jx_Q[n] \tag{20.64}$$

In some modern systems, sampling is performed at IF and the in-phase and quadrature channels are created digitally. In this case, the required sampling rate is greater than twice the waveform bandwidth (2.5 to 4 times the bandwidth depending on the implementation), and controlled aliasing is used to create a real, sampled signal centered at a frequency

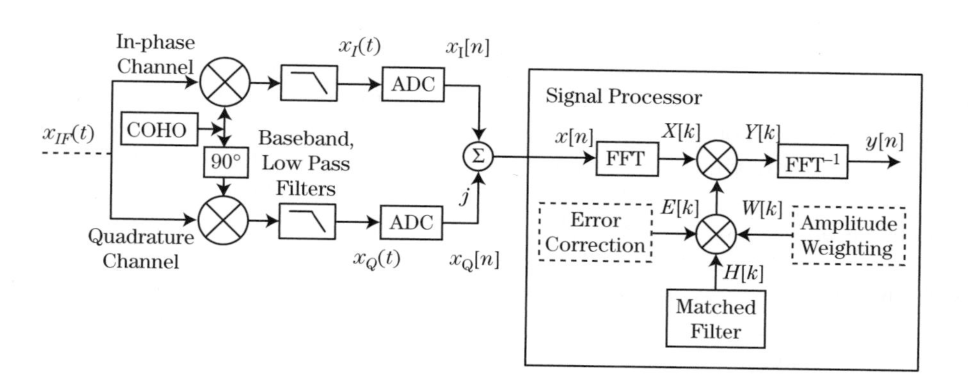

FIGURE 20-12 ■ Fast convolution is one way to implement digital pulse compression.

offset from baseband [6]. IF sampling circumvents the requirement to balance the analog in-phase and quadrature channels in both amplitude and phase.

In general, radar systems are designed to collect returns over a specified *range window*. A range window is defined in terms of a start range and an end range. The receiver is activated at a time delay corresponding to the start range and deactivated at a time delay corresponding to the completion of the collection interval that includes the end range. The time delay or extent associated with a range window is

$$T_{R_w} = \frac{2R_w}{c} \tag{20.65}$$

where R_w is the range extent of the window. To prevent eclipsing, a full pulse is collected from the beginning and end of the range window. The minimum collection time is therefore

$$T_C = \tau + \frac{2R_w}{c} \tag{20.66}$$

The number of complex samples collected over a range window is equal to the product of the ADC sampling rate and the collection time, or

$$N_{RW} = F_s\left(\tau + \frac{2R_w}{c}\right) \tag{20.67}$$

The digital compressor, employing fast convolution, receives the digitized signal $x[n]$ and passes it through the first FFT (or DFT), as shown in Figure 20-12. The DFT yields

$$x[n] \overset{DFT}{\longleftrightarrow} X[k] \tag{20.68}$$

where $X[k]$ is the DFT spectrum of the received signal. The matched filter's spectrum is defined by $H[k]$ and is generally precomputed and stored in memory. The compressor forms the product

$$Y[k] = X[k]H[k] \tag{20.69}$$

and the inverse transform yields

$$Y[k] \overset{DFT^{-1}}{\longleftrightarrow} y[n] \tag{20.70}$$

where $y[n]$ is the output of the filter. To ensure linear convolution over the entire range window, it is necessary to use the collection time in equation (20.66) as well as an appropriate DFT size (see Chapter 14).

As a waveform passes through the transmitter and receiver, modulations are induced that create a mismatch between the ideal matched filter and the received waveform. Extraneous modulations may cause an increase in the range sidelobes, a degradation in resolution, and a loss in processing gain. An error correction filter may be used to compensate for repeatable modulation errors. The error correction filter is formed using a *pilot pulse*. A pilot pulse consists of a copy of the transmit waveform that is propagated through the transmit/receive chain and, if radiated externally, may be reflected from a preselected point source (e.g., a corner reflector). The pilot pulse is modulated by error sources and is digitized on receive. The digitized signal is used to form an error compensation filter $E[k]$, as depicted in Figure 20-12. The error filter takes the place of an ideal matched filter. In general, phase modulations are the easiest to compensate for and are removed by applying the conjugate phase. When compensating for amplitude modulation, care must be taken to avoid division by zero.

20.9 | SIDELOBE REDUCTION IN AN LFM WAVEFORM

An LFM waveform's peak sidelobe ratio is −13.2 dB. The high peak sidelobes are a result of the spectrum's sharp transition regions, as illustrated in Figure 20-9. To reduce the nominal sidelobes, an amplitude weighting may be applied to the spectrum. The weighting tapers the spectrum's band edges or transition regions and produces a reduction in the peak sidelobe level. Amplitude weighting represents an intentional mismatch between the filter and the transmit waveform, which lowers the sidelobes but at a cost of degraded resolution and loss in processing gain.

A *Taylor weighting* is often applied in radar and achieves the minimum mainlobe width for a given peak sidelobe level conditioned on the fact that sidelobes decrease with distance from the mainlobe. The Taylor weighting is defined by two parameters: the peak sidelobe ratio, *PSR*, and the total number of sidelobes, $\bar{n}$, adjacent to the mainlobe beyond which the sidelobes start to decrease.

The Taylor weighting function, $W_{Taylor}(\omega)$, is defined by [2,4]

$$W_{Taylor}(\omega) = K\left\{1 + 2\sum_{m=1}^{\bar{n}-1} F_m \cos\left(\frac{m\omega}{B}\right)\right\} \quad -\pi B \le \omega \le \pi B \tag{20.71}$$

where

$$F_m = \begin{cases} \dfrac{(-1)^{m+1}\prod_{n=1}^{\bar{n}-1}\left[1 - \dfrac{m^2}{S(D^2 + (n-0.5)^2)}\right]}{2\prod_{\substack{n=1\\ n\ne m}}^{\bar{n}-1}\left(1 - \dfrac{m^2}{n^2}\right)}, & m = 1, 2, \ldots, (\bar{n}-1) \\ 0, & m \ge \bar{n} \end{cases} \tag{20.72}$$

$$D = \frac{1}{\pi}\cosh^{-1}\left[10^{-PSR/20}\right] \tag{20.73}$$

$$S = \frac{\bar{n}^2}{D^2 + (\bar{n}-0.5)^2} \tag{20.74}$$

and

$$K = \frac{1}{1 + 2\sum_{m=1}^{\bar{n}-1} F_m} \tag{20.75}$$

In equation (20.71), the spectral extent of the Taylor weighting is matched to the LFM waveform's bandwidth. The continuous weighting function in (20.71) is sampled at the DFT frequency bin spacing to produce a discrete weighting function, or

$$W_{Taylor}[k] = K\left\{1 + 2\sum_{m=1}^{\bar{n}-1} F_m \cos\left(\frac{m2\pi k}{M}\right)\right\}, \quad k = \left(-\frac{M}{2}+1\right), \ldots, 0, \ldots, \frac{M}{2} \tag{20.76}$$

where M is the DFT size and is assumed to be even in this case. Nyquist sampling is assumed with $F_s = B$.

TABLE 20-1 ■ SNR Loss Associated with a Taylor Weighting Function

	Peak Sidelobe Ratio (dB)								
	−20	−25	−30	−35	−40	−45	−50	−55	−60
$\bar{n}$	SNR Loss (dB)								
2	−0.21	−0.38	−0.51						
3	−0.21	−0.45	−0.67	−0.85					
4	−0.18	−0.43	−0.69	−0.91	−1.11	−1.27			
5	−0.16	−0.41	−0.68	−0.93	−1.14	−1.33	−1.49		
6	−0.15	−0.39	−0.66	−0.92	−1.15	−1.35	−1.53	−1.68	
7	−0.15	−0.37	−0.65	−0.91	−1.15	−1.36	−1.54	−1.71	−1.85
8	−0.16	−0.36	−0.63	−0.90	−1.14	−1.36	−1.55	−1.72	−1.87

The loss in SNR due to weighting is [6]

$$SNR_{loss} = \frac{\left[\sum\limits_{k=-M/2+1}^{M/2} W[k]\right]^2}{M \sum\limits_{k=-M/2+1}^{M/2} W^2[k]} \tag{20.77}$$

where $W[k]$ are the weighting coefficients indexed by k, and M is the number of coefficients. Table 20-1 contains processing losses for a Taylor window as a function of PSR and $\bar{n}$. Note that the loss increases with decreasing sidelobe level.

Weighting inherently reduces the waveform bandwidth resulting in degraded resolution. For a Taylor weighting, the mainlobe width, measured at the −4 dB point, is recorded in Table 20-2 for different values of PSR and $\bar{n}$. The −4 dB point was chosen since it corresponds to the Rayleigh resolution associated with an unweighted LFM waveform. The resolutions reported in Table 20-2 are normalized by $c/2B$ (i.e., the nominal Rayleigh resolution).

Figure 20-13 contains a portion of the response associated with an LFM waveform with (solid curve) and without (dashed curve) weighting. For the weighted case, a −40 dB, $\bar{n} = 4$, Taylor weighting has been applied. The waveform's time-bandwidth product is 500. The reduced sidelobes and mainlobe broadening are apparent. The peak sidelobe ratio is approximately −38 dB (2 dB above the design level of −40 dB). The elevated

TABLE 20-2 ■ 4 dB Resolution Associated with a Taylor Weighting Function

	Peak Sidelobe Ratio (dB)								
	−20	−25	−30	−35	−40	−45	−50	−55	−60
$\bar{n}$	4 dB Resolution Normalized by $c/2B$								
2	1.15	1.19	1.21						
3	1.14	1.22	1.28	1.33					
4	1.12	1.22	1.29	1.36	1.42	1.46			
5	1.11	1.20	1.29	1.36	1.43	1.49	1.54		
6	1.10	1.19	1.28	1.36	1.43	1.50	1.56	1.61	
7	1.09	1.19	1.28	1.36	1.43	1.50	1.56	1.62	1.67
8	1.08	1.18	1.27	1.35	1.43	1.50	1.57	1.63	1.68

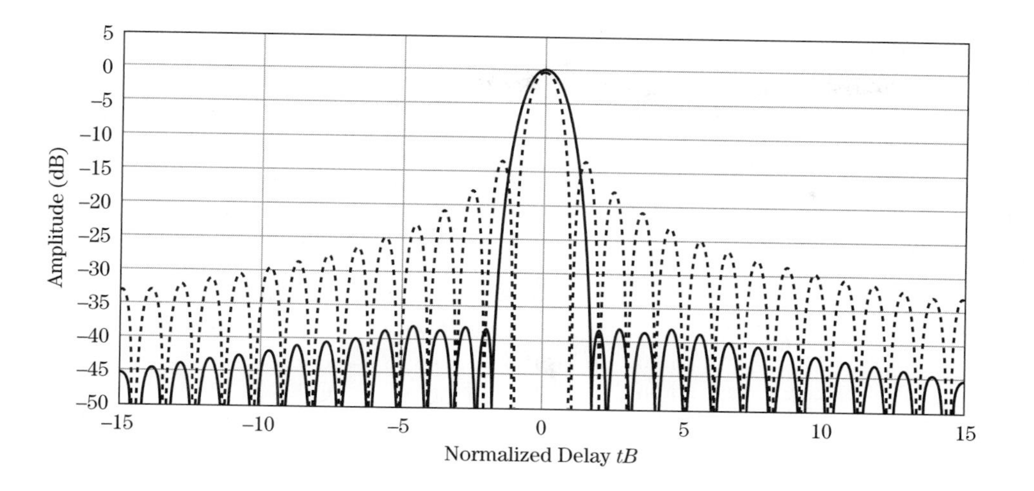

FIGURE 20-13 ■ A −40 dB, $\overline{n} = 4$, Taylor-weighted LFM waveform compressed response (solid curve) has significantly reduced sidelobes versus an unweighted LFM waveform response (dashed curve).

sidelobes are explained in the next section. Both responses have been normalized to their respective peak values so the loss in energy due to weighting is not apparent. Note that when applying a weighting it is the loss in SNR and not the loss in the waveform's energy that dictates performance.

20.9.1 Weighting and Time-Bandwidth Requirements

For a given taper, the expected sidelobe performance is achieved by applying the weighting to a signal with a rectangular envelope. The LFM waveform's spectrum is not a perfect rectangle, and therefore, some degradation in sidelobe performance is expected. As discussed in Section 20.7.2, the transition region sharpens with increasing time-bandwidth product, producing a more compact spectrum. As a result, sidelobes approach the design level as the time-bandwidth product is increased. Cook [4] states that a time-bandwidth product greater than 100 is required to achieve a peak sidelobe close to the taper's expected level.

A comparison of compressed responses for TB products of 20 and 100 is shown in Figure 20-14. A −40 dB, $\overline{n} = 4$, Taylor weighting is applied. The peak sidelobe associated with $\tau B = 20$ is −23.3 dB, and the peak sidelobe associated with $\tau B = 100$ is −35 dB.

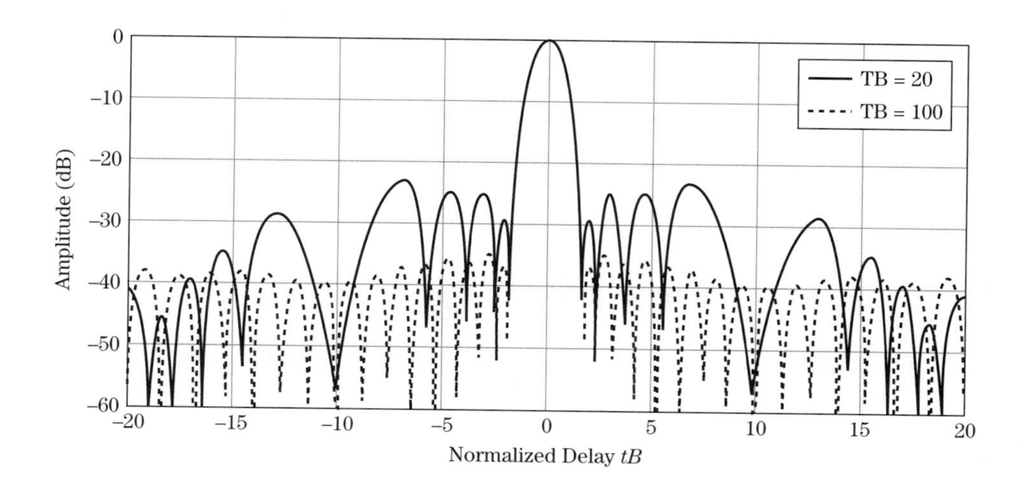

FIGURE 20-14 ■ A comparison of time-sidelobe responses for time-bandwidth products of 20 (solid curve) and 100 (dashed curve) when applying a −40 dB Taylor weighting.

TABLE 20-3 ■ Peak and Average Straddle Loss Associated with a Taylor Weighted LFM with $\tau B = 500$

	Peak Sidelobe Ratio (dB)									
	−20		−30		−40		−50		No Weighting	
	Straddle Loss									
$\bar{n}$	Peak	Average	Peak	Average	Peak	Average	Peak	Average	Peak	Average
3	−2.99	−0.89	−2.35	−0.72	−2.02	−0.63	−1.83	−0.57		
4	−3.07	−0.91	−2.32	−0.71	−1.93	−0.60	−1.69	−0.53		
5	−3.15	−0.93	−2.33	−0.71	−1.90	−0.59	−1.63	−0.51		
6	−3.21	−0.94	−2.35	−0.72	−1.89	−0.59	−1.61	−0.51	−3.92	−1.12
7	−3.25	−0.96	−2.37	−0.72	−1.89	−0.59	−1.59	−0.50		
8	−3.28	−0.96	−2.38	−0.73	−1.90	−0.59	−1.59	−0.50		

Note: The sampling rate is equal to the nominal waveform bandwidth, B.

Time sidelobes are shown to approach the design level as the time-bandwidth product is increased. For the case of a low time-bandwidth product, a heavier weighting may be applied to further reduce the sidelobes at the cost of additional SNR loss and degraded range resolution.

Phase codes, to be discussed in Section 20.12, are often used in lieu of frequency modulated pulses when low time-bandwidth products are driven by system constraints. For example, a 13-bit biphase Barker code achieves a −22.3 dB peak sidelobe with a time-bandwidth product of 13 and with no amplitude taper applied.

20.9.2 Straddle Loss Reduction

In addition to lowering the sidelobes, an amplitude taper reduces straddle loss. As discussed in Section 20.4, straddle loss is the loss in SNR that occurs when the peak of the match filtered response is not sampled. In most systems, the ADC sampling rate is proportional to the receive waveform bandwidth. For a fixed sampling rate, a broadening of the mainlobe reduces the straddle loss by decreasing the slope of the response near the peak and, as a result, diminishes the amplitude difference between the peak and the nearest sample. Table 20-3 contains peak and average straddle losses for a Taylor weighted LFM with a time-bandwidth product of 500. The waveform is sampled at a rate equal to the waveform bandwidth prior to weighting. Included in the table is the straddle loss associated with no weighting. Without weighting, the peak straddle loss is approximately −4 dB, and the average loss is about −1 dB. For a 40 dB Taylor, the peak straddle loss is approximately −1.9 dB, and the average loss is about −0.6 dB. This represents a reduction (improvement) in peak straddle loss of approximately 2 dB and a reduction in average loss of 0.4 dB.

20.10 AMBIGUITY FUNCTIONS

The matched filter defined in equation (20.31) is predicated on the assumption that the received waveform is an amplitude-scaled and time-delayed version of the transmit waveform. When a radial velocity component exists between the radar and a target, a Doppler shift is imparted to the waveform. A Doppler shift [33] is defined as a shift in frequency due to relative motion. The shift is computed as

$$f_d = \frac{2v_r}{\lambda} \tag{20.78}$$

where v_r is the radial component of velocity. For closing (approaching) targets, the sign of the radial velocity is positive so that the Doppler shift is positive, representing an increase in frequency. For opening targets, the sign of the radial velocity is negative and the Doppler shift is negative, indicating a decrease in frequency.

Without a priori knowledge, the Doppler shift represents an unintentional mismatch between the received waveform and the matched filter. In some cases, the Doppler shift is estimated and removed prior to applying the matched filter. However, some residual or uncompensated Doppler typically remains.

Ambiguity functions are used to characterize the response of the matched filter in the presence of uncompensated Doppler. The ambiguity function is defined as

$$A(t, f_d) = \left| \int x(\alpha) \exp\left(j2\pi f_d \alpha\right) x^* \left(\alpha - t\right) d\alpha \right| \tag{20.79}$$

where $x(\alpha)\exp(j2\pi f_d \alpha)$ is the Doppler-shifted waveform, and α is a dummy variable of integration. For $f_d = 0$, the ambiguity response is simply the magnitude of the match filtered response.

A three-dimensional plot of the ambiguity function is termed an *ambiguity surface*. To facilitate comparison of ambiguity surfaces, the waveform and matched filter are normalized to unit energy. A unit energy $x_u(t)$ instantiation of the waveform $x(t)$ is obtained via the normalization

$$x_u(t) = \frac{x(t)}{\sqrt{\int |x(t)|^2 dt}} \tag{20.80}$$

Several properties associated with an ambiguity function or surface are worthy of note. An ambiguity surface achieves its maximum value at zero delay and zero Doppler, or at $A(0, 0)$, and, if the waveform is normalized to unit energy, $A(0, 0) = 1$. The volume under the square of the ambiguity surface is unity

$$\int_{-\infty}^{\infty} \int_{-\infty}^{\infty} |A(t, f_d)|^2 dt \, df_d = 1 \tag{20.81}$$

given a unit energy waveform. The relationship in (20.81) implies that a waveform designed to lower the response in a given region of the ambiguity surface will produce an increase in another region because the total volume must remain constant. Additional information on ambiguity surfaces is provided in [4,7,34].

In designing a waveform to support a given mode of operation, a number of factors including energy, resolution, blind range extent, duty cycle, and Doppler tolerance must be considered. *Doppler tolerance* refers to the response of a waveform in the presence of uncompensated Doppler and may be assessed using an ambiguity surface.

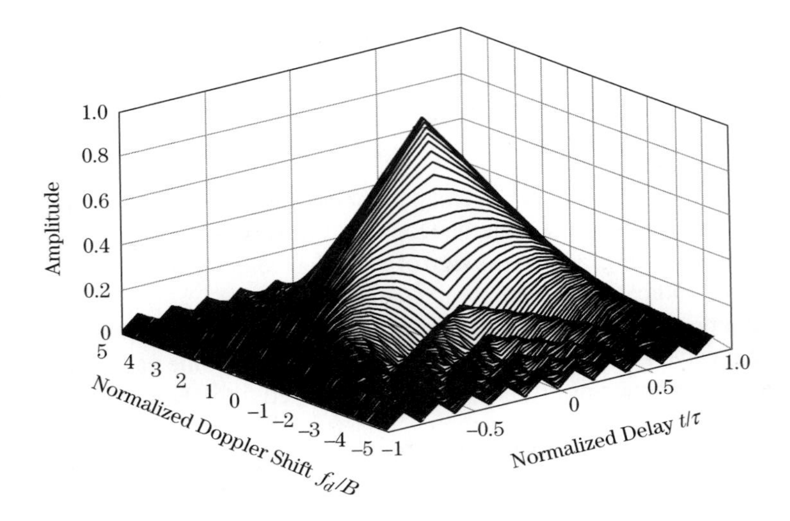

FIGURE 20-15 ■ The ambiguity surface for a simple pulse.

20.10.1 Ambiguity Function for a Simple Pulse

The ambiguity surface for a simple pulse is examined to contrast it with the surface associated with an LFM waveform. For an unmodulated pulse

$$x(t) = \frac{1}{\sqrt{\tau}} \qquad -\frac{\tau}{2} \le t \le \frac{\tau}{2} \tag{20.82}$$

with unit energy and pulse width τ, the ambiguity function [8] is

$$A(t, f_d) = \left| \left(1 - \frac{|t|}{\tau}\right) \frac{\sin\left(\pi f_d \tau \left(1 - \frac{|t|}{\tau}\right)\right)}{\pi f_d \tau \left(1 - \frac{|t|}{\tau}\right)} \right| \qquad |t| \le \tau \tag{20.83}$$

The ambiguity function in (20.83) is plotted in Figure 20-15 and is termed a *ridged* ambiguity surface [8,34]. In the figure, the time-delay axis is normalized by the pulse width, and the Doppler axis is normalized by the waveform bandwidth.

The surface exhibits a peak value at $A(0, 0)$. For $f_d = 0$ the ambiguity function reduces to the magnitude of the match filtered response for a simple pulse, or

$$A(t, 0) = \left| 1 - \frac{|t|}{\tau} \right| \qquad |t| \le \tau \tag{20.84}$$

A cut through the ambiguity surface at zero time delay characterizes the decrease in the peak value as a function of Doppler shift. The zero delay cut is a sinc function

$$A(0, f_d) = \left| \frac{\sin(\pi f_d \tau)}{\pi f_d \tau} \right| \qquad |t| \le \tau \tag{20.85}$$

The response is maximum at zero Doppler and is zero at multiples of $1/\tau$. A Doppler shift equal to $1/\tau$ represents a full cycle of Doppler across the pulse and annihilates the peak at zero delay. With one-quarter cycle of Doppler, the peak value is reduced by 1 dB.

20.10.2 Ambiguity Function for an LFM Waveform

A closed-form expression of the LFM waveform's ambiguity surface exists. For a unit energy LFM waveform

$$x(t) = \frac{1}{\sqrt{\tau}} \exp\left(j\pi \frac{B}{\tau} t^2\right) \quad |t| \leq \frac{\tau}{2} \tag{20.86}$$

the ambiguity function is [2,4,6–8]

$$A(t, f_d) = \left| \left(\left(1 - \frac{|t|}{\tau}\right) \frac{\sin\left[\pi\tau\left(1 - \frac{|t|}{\tau}\right)\left(f_d + \frac{B}{\tau}t\right)\right]}{\pi\tau\left(1 - \frac{|t|}{\tau}\right)\left(f_d + \frac{B}{\tau}t\right)} \right) \right| \quad |t| \leq \tau \tag{20.87}$$

As expected, a cut through the ambiguity response along the time-delay axis at zero Doppler is simply the magnitude of the match filtered response in equation (20.57), or

$$A(t, 0) = \left| \left(1 - \frac{|t|}{\tau}\right) \frac{\sin\left[\left(1 - \frac{|t|}{\tau}\right)\pi Bt\right]}{\left(1 - \frac{|t|}{\tau}\right)\pi Bt} \right| \quad |t| \leq \tau \tag{20.88}$$

20.10.3 Range-Doppler Coupling

An LFM waveform exhibits a coupling between range and Doppler. A Doppler shift translates the peak of the response along the time delay axis by an amount equal to $-(f_d/B)\tau$. The translation can be readily seen by comparing the arguments of the sine functions in equation (20.87) and equation (20.88). Note that the triangle $(1 - t/|\tau|)$ is not shifted and serves to reduce the amplitude of the time-shifted response. Figure 20-16 contains a plot of the ambiguity surface for an LFM waveform with a time-bandwidth

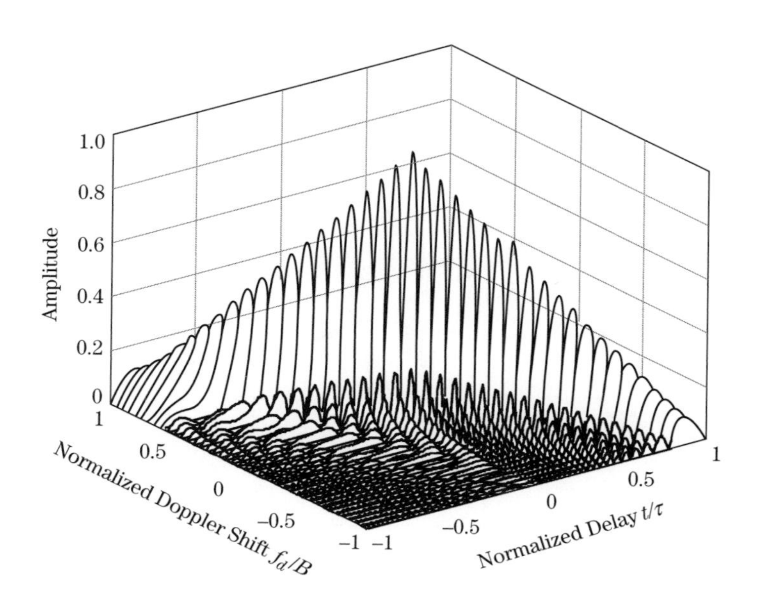

FIGURE 20-16 ■ The ambiguity surface for an LFM waveform with $\tau B = 20$.

product of 20. The surface is coarsely sampled in Doppler to aid the reader in visualizing the affects of range-Doppler coupling.

The LFM waveform exhibits a diagonal ridge in the (t, f_d) plane. For a Doppler shift equal to the inverse of the pulse width (equivalently, one cycle of Doppler across the uncompressed pulse), the peak of the response is shifted by a time delay equal to the Rayleigh resolution of an unweighted LFM waveform, and the peak is reduced by $(1 - 1/B\tau)$ (in voltage). In contrast, a single cycle of Doppler across a simple pulse drives the peak of the response to zero without the reemergence of a discernable mainlobe located at another time delay. The ambiguity surface for an LFM waveform is termed a *sheared ridge*.

The ratio of Doppler shift to waveform bandwidth is termed the *fractional Doppler shift* (FDS). An LFM waveform is considered *Doppler tolerant* given the preservation of the mainlobe and sidelobe structure in the presence of large fractional Doppler shifts (up to 50% or more).

Figures 20-17, 20-18, and 20-19 contain plots of the shift in the peak location, the loss in peak amplitude, and the degradation in resolution, respectively, for time-bandwidth products of 10, 100, and 1,000. The degradation in each metric is gradual and approximately linear for fractional Doppler shifts up to 50%.

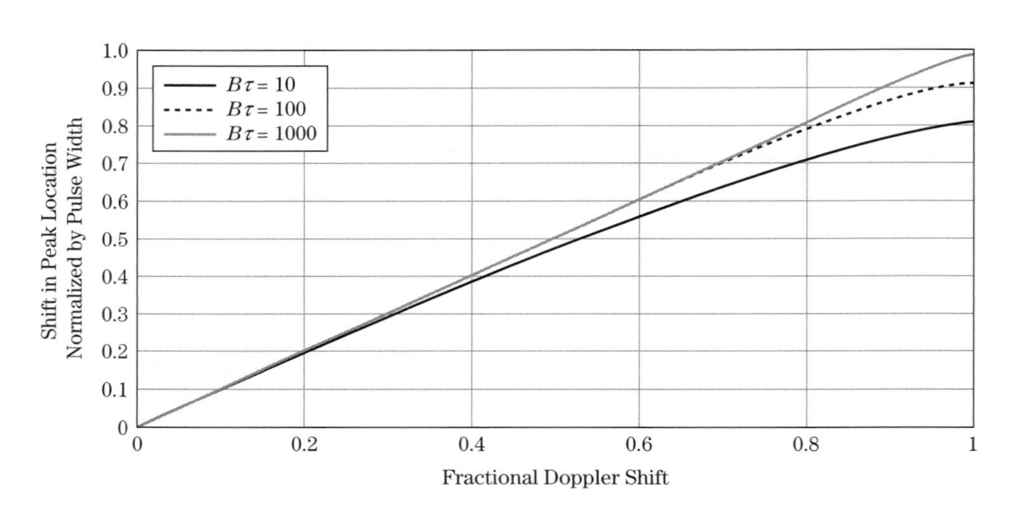

FIGURE 20-17 ■ Time shift in the peak of an LFM waveform's match filtered response as a function of Doppler shift.

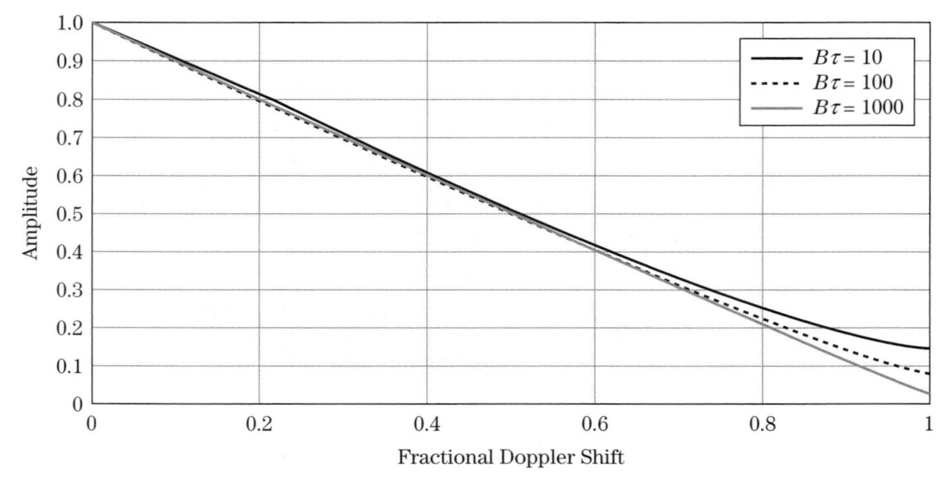

FIGURE 20-18 ■ Reduction in peak amplitude as a function of Doppler shift.

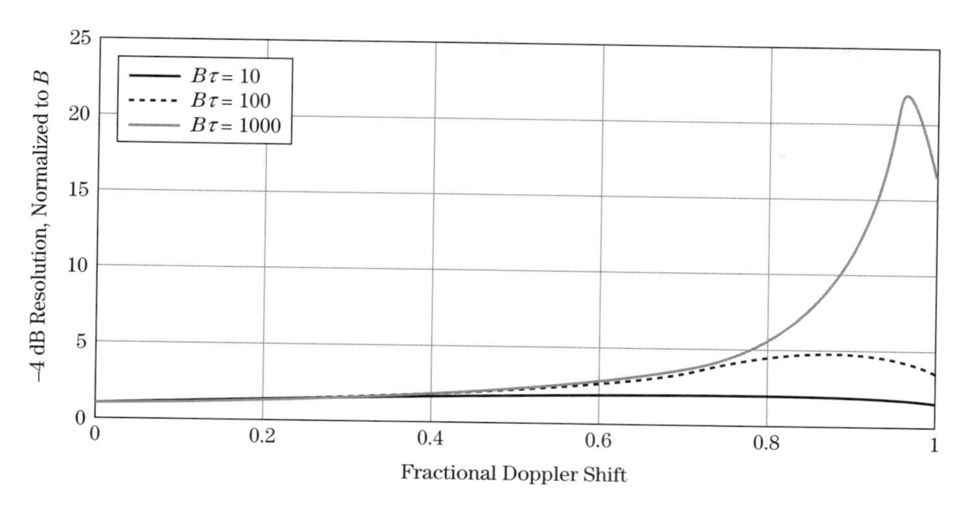

FIGURE 20-19 ■ Increase in −4 dB mainlobe width as a function of Doppler shift.

The sidelobe structure of the LFM waveform is also preserved in the presence of large fractional Doppler shifts. Figure 20-20 contains the compressed range response for an LFM waveform with a time-bandwidth product of 100 at fractional Doppler shifts of 0%, 25%, 50%, and 75%. The sidelobe structure is well behaved, and the peak sidelobe remains 12 to 13 dB below the peak of the mainlobe.

The sidelobe structure in the presence of uncompensated Doppler is not as well behaved when amplitude weighting is applied. The misalignment of the received waveform and the matched filter spectra, combined with amplitude weighting, produces a nonsymmetric weighting that degrades the sidelobe response. Figure 20-21 contains a plot of the compressed response for a −40 dB Taylor-weighted LFM waveform having a time-bandwidth product of 200. The fractional Doppler shifts are 0% (solid curve) and 15% (dotted curve), and the individual curves have been normalized to the peak of their responses. The response with no Doppler shift exhibits peak sidelobes of −38 dB, whereas the response experiencing a 15% fractional Doppler shift exhibits peak sidelobes of −25 dB, well short of the intended −40 dB.

The impact of Doppler should be considered when applying a weighting function to reduce the sidelobes. In general, the heavier the weighting the more the sidelobes degrade as a function uncompensated Doppler. Low sidelobes are often associated with tracking,

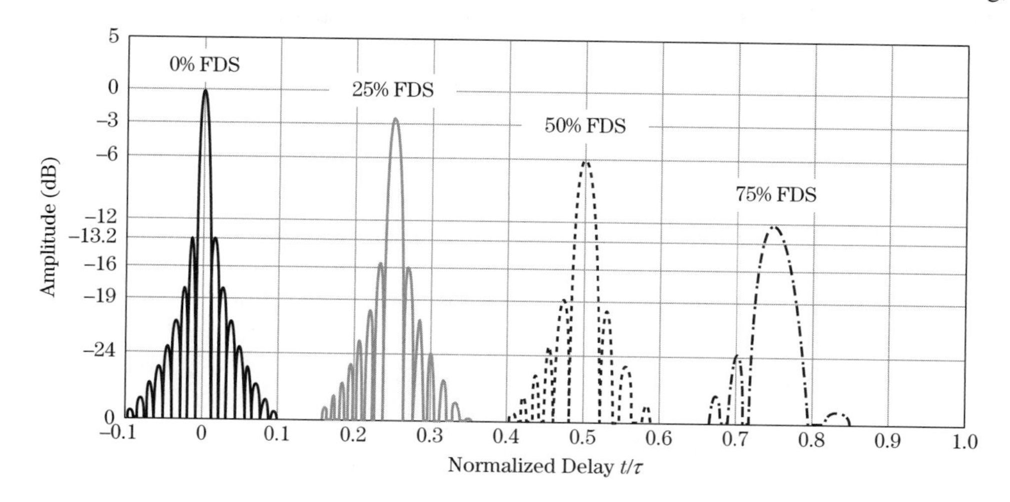

FIGURE 20-20 ■ Individual LFM match filtered responses for fractional Doppler shifts of 0%, 25%, 50%, and 75% illustrate both reduction in peak levels and broadening of the mainlobe.

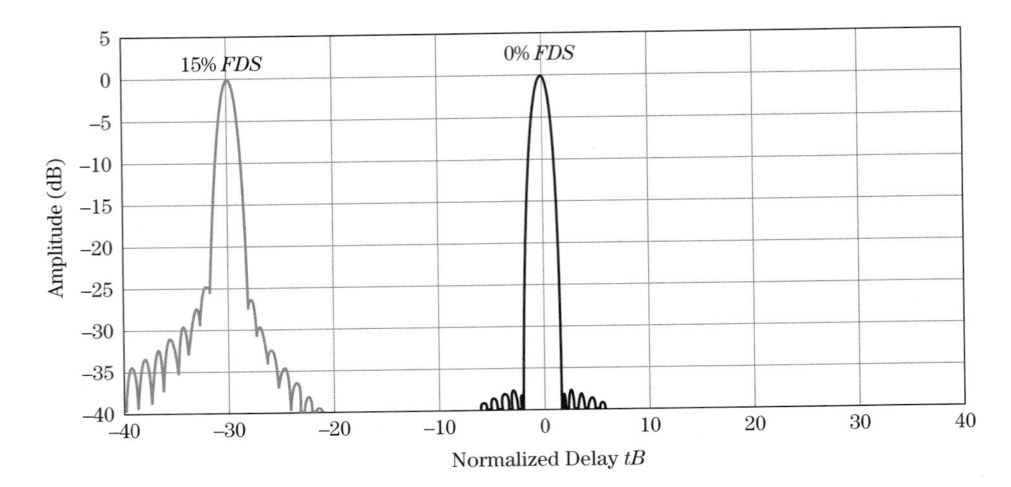

FIGURE 20-21 ■ Match filtered response for a −40 dB Taylor weighted LFM waveform with a time-bandwidth product of 200 and a fractional Doppler shift of 0% (dark curve) and 15% (light curve). Both curves have been normalized to the peak of their responses.

imaging, and discrimination. These radar modes may be supported by Doppler estimation techniques, which compensate for Doppler and thus reduce the impact on the compressed response.

In these examples, a large fractional Doppler shift has been used to illustrate the Doppler tolerance of an LFM waveform; however, fractional Doppler shifts less than 5%, and in most cases less than 1%, are typical. For example, a Mach 1 target (approximately 330 m/sec) at 10 GHz imparts a 22,000 Hz Doppler shift. For a 1 MHz bandwidth waveform, the fractional Doppler shift is 2.2%.

20.10.4 Spectral Interpretation of Range-Doppler Coupling

An examination of the LFM waveform's spectrum provides insight into the range-Doppler coupling and Doppler tolerance observed in the previous section. Consider the LFM spectrum defined in equation (20.55) and repeated here for convenience

$$X(\omega) \approx |X(\omega)| \exp\left(-j\frac{1}{4\pi}\frac{\tau}{B}\omega^2\right)\exp\left(j\frac{\pi}{4}\right) \tag{20.89}$$

where $|X(\omega)| \approx 1, -\pi B \leq \omega \leq \pi B$. For a waveform $x(t)$, the Doppler-shifted signal may be modeled as $x(t)\exp(j\omega_d t)$ where $\omega_d = 2\pi f_d$. The Fourier transform of $x(t)\exp(j\omega_d t)$ is

$$x(t)\exp(j\omega_d t) \stackrel{\Im}{\longleftrightarrow} X(\omega - \omega_d) \tag{20.90}$$

The spectrum of a Doppler-shifted LFM waveform is then

$$X(\omega - \omega_d) \approx |X(\omega - \omega_d)| \exp\left(-j\frac{1}{4\pi}\frac{\tau}{B}(\omega - \omega_d)^2\right)\exp\left(j\frac{\pi}{4}\right) \tag{20.91}$$

Applying a filter matched at zero Doppler yields the output spectrum $X(\omega - \omega_d)H(\omega)$:

$$\begin{aligned} Y(\omega) = {} & |X(\omega - \omega_d)| \exp\left(-j\frac{1}{4\pi}\frac{\tau}{B}(\omega - \omega_d)^2\right)\exp\left(j\frac{\pi}{4}\right) \\ & \times |X(\omega)| \exp\left(j\frac{1}{4\pi}\frac{\tau}{B}\omega^2\right)\exp\left(-j\frac{\pi}{4}\right) \end{aligned} \tag{20.92}$$

or

$$Y(\omega) \approx |X(\omega - \omega_d)||X(\omega)| \exp\left(j\frac{1}{4\pi}\frac{\tau}{B}2\omega \cdot \omega_d\right) \exp\left(j\frac{1}{4\pi}\frac{\tau}{B}\omega_d^2\right) \tag{20.93}$$

Substituting $\omega_d = 2\pi f_d$ yields

$$Y(\omega) \approx |X(\omega - 2\pi f_d)||X(\omega)| \exp\left(j\frac{f_d}{B}\tau\omega\right) \exp\left(j\frac{1}{4\pi}\frac{\tau}{B}(2\pi f_d)^2\right) \tag{20.94}$$

The coupling between range and Doppler is a result of the linear phase term in equation (20.94), which produces a time delay at the output of the filter equal to $-f_d\tau/B$.

Given that $|X(\omega)| \approx 1, -\pi B \leq \omega \leq \pi B$ for large time-bandwidth products, the magnitude response may be approximated as

$$|Y(\omega)| \approx |X(\omega - \omega_d)||X(\omega)| \approx \begin{cases} 1, & -\pi(B - f_d) \leq \omega \leq \pi B,\ f_d > 0 \\ 1, & -\pi B \leq \omega \leq \pi\,(B + f_d)\,,\ f_d < 0 \\ 0, & \text{otherwise} \end{cases} \tag{20.95}$$

The mismatch between the filter and Doppler shifted waveform yields a composite spectrum with reduced bandwidth approximately equal to $B - |f_d|$, which translates into degraded range resolution. The composite spectrum also contains less energy, which translates into a loss in peak amplitude.

In the frequency domain, the LFM waveform's quadratic phase produces a linear phase term in (20.94) that governs range-Doppler coupling. A waveform possessing a higher-order phase response will produce in the presence of uncompensated Doppler additional terms that degrade the compressed response. Thus, the LFM is unique in terms of its Doppler tolerance.

20.10.5 Dealing with Doppler Modulation in a Pulse Compression Waveform

In a radar system, Doppler shift and its impact on a waveform may be dealt with in several ways:

1. With a priori knowledge, waveform modulations and parameters may be selected to achieve a Doppler-tolerant waveform for the anticipated range of Doppler shifts. The ambiguity function is an essential tool supporting this type of analysis and design.
2. In some cases, a bank of matched filters may be employed—each one tuned to a different Doppler shift. The Doppler spacing between filters is based on the tolerance of the waveform. The received signal is passed through each filter, and the filter yielding the largest response is the one that most closely matches the received waveform in both time delay and Doppler.
3. A target's radial velocity may be estimated during track or via some other means. The velocity estimate may be used to center the received signal at baseband by applying a frequency shift to the transmit or receive waveform. Only a single Doppler shift (or target) may be compensated for at a time. Differences between the actual Doppler shift and the estimated frequency shift represent an uncompensated Doppler component. The impact of uncompensated Doppler must be considered in the design.

20.10.6 The V-LFM

The accuracy with which a radar is required to measure range is dependent on the function being performed. For example, in some cases range accuracy is less important in search than in track. In search, a higher degree of range-Doppler coupling may be acceptable. The transition from search to track requires one to consider techniques for estimating and compensating for Doppler. Rihaczek [34] describes a V-LFM waveform that consists of both an up-chirp and down-chirp segment. The up-chirp gives a time displacement of the target response equal to $-(f_d/B)\tau$, whereas the down-chirp gives a displacement in the opposite direction equal to $+(f_d/B)\tau$. The ambiguity surface consists of two sheared ridges with opposite slopes. A variant on the V-LFM uses two pulses closely spaced in time (one up-chirp and one down-chirp) to resolve the range-Doppler ambiguity and to estimate the observed Doppler shift. The estimated Doppler shift may then be applied as one transitions to track to compensate for Doppler. In track, the range-rate estimate is continually updated and may be applied to the next pulse to compensate for Doppler.

20.11 LFM SUMMARY

The LFM waveform has been in use since the 1950s and possesses some unique properties. The waveform exhibits a linear relationship between time and frequency and is known for its Doppler tolerance and sheared ridge ambiguity surface. In high-resolution systems, stretch processing [15] is often applied to an LFM waveform to reduce the processing bandwidth while maintaining the resolution afforded by the transmit bandwidth. The LFM waveform may also be applied in search and track. In this text, the waveform was used to introduce a number of general concepts including matched filter implementations, ambiguity surfaces, processing gain, sidelobe suppression, and Doppler tolerance.

20.12 PHASE-CODED WAVEFORMS

Phase codes form a second broad class of pulse compression waveforms. Phase-coded waveforms are composed of concatenated subpulses (or *chips*) where the phase sequencing or coding from subpulse to subpulse is chosen to elicit a desired mainlobe and sidelobe response. Waveform properties are largely dependent on the coding sequence employed, and both biphase- and polyphase-coded waveforms are used in modern radar systems.

20.12.1 The Structure and General Properties of Phase-Coded Waveforms

Phase-coded waveforms are constructed of concatenated subpulses or "chips." A chip of duration τ_{chip} is illustrated in Figure 20-22. A rectangular-shaped chip has a Rayleigh resolution defined by

$$\delta R = \frac{c\tau_{chip}}{2} \tag{20.96}$$

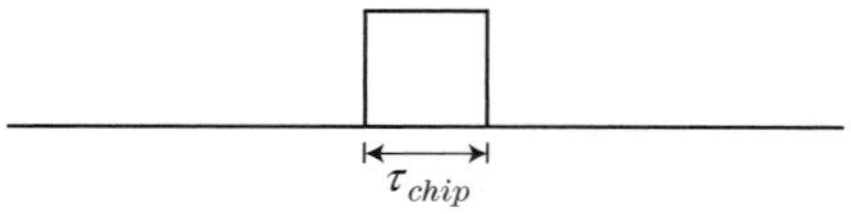

τ_{chip}

FIGURE 20-22 ■ A phase coded waveform consists of a set of concatenated sub-pulses or chips each of duration τ_{chip}.

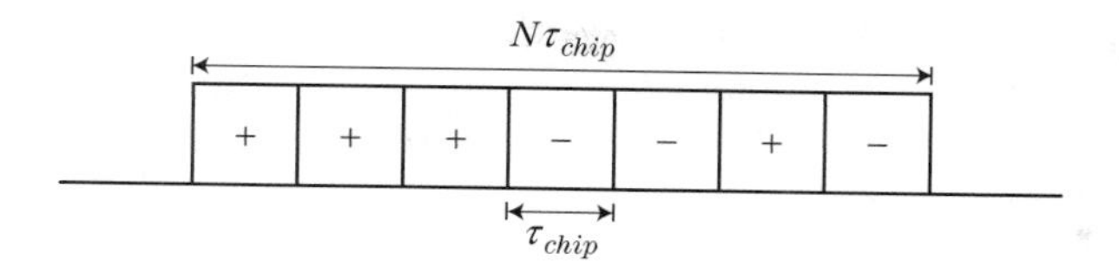

FIGURE 20-23 ■ Biphase coded waveforms consist of chips exhibiting 2 possible phase states.

and exhibits a sinc-shaped spectrum with a −4 dB bandwidth given by

$$B_{chip} = \frac{1}{\tau_{chip}} \tag{20.97}$$

Within a single chip, range resolution and waveform energy are inversely coupled. Increasing the chip width degrades resolution and increases the energy in the chip, while the converse occurs if the chip width is decreased.

Phase-coded waveforms are formed by concatenating N chips and selecting the phase of each chip to achieve a desired mainlobe and sidelobe response at the output of the corresponding matched filter. A baseband, biphase-coded waveform is illustrated in Figure 20-23. The allowable phase states, ϕ, are 0 and 180 degrees, and the chip amplitudes are defined by evaluating $\exp(j\phi)$ yielding either 1 or -1, respectively. The energy in a phase-coded waveform is proportional to the total pulse duration, τ, where

$$\tau = N\tau_{chip} \tag{20.98}$$

Appropriately chosen phase codes yield a Rayleigh range resolution defined by equation (20.96). Figure 20-24 contains a plot of the magnitude of the match filtered response associated with the phase code in Figure 20-23. The response has been normalized by the sequence length resulting a peak value of 1 and peak sidelobes of 1/7 (approximately 0.143). Using the Rayleigh criterion to define resolution, the distance between the mainlobe peak and first null is equal to the chip width.

The sidelobes in Figure 20-24 were achieved without amplitude weighting. The selection and ordering of the phase sequence determines the shape of the sidelobe response. This is in contrast to an LFM waveform where amplitude weighting facilitates sidelobe suppression but at a cost of degraded resolution and a loss in SNR.

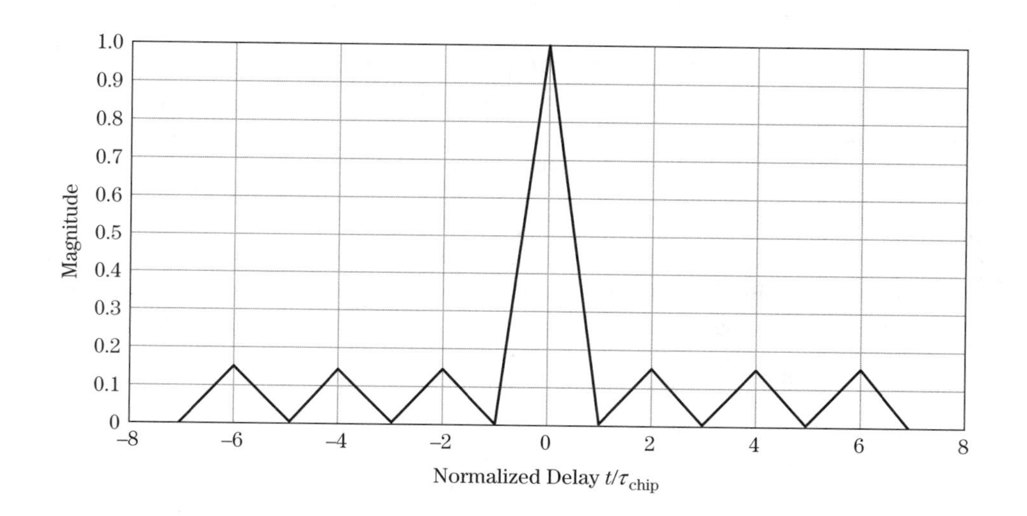

FIGURE 20-24 ■ Match filtered response for the Barker phase coded waveform (Figure 20-23) maintains equal peak sidelobes of level 1/N.

Phase-coded waveforms are compressed by applying a matched filter to the receive waveform. The filter may be implemented in the digital domain using fast convolution (see Section 20.8). The ADCs in Figure 20-12 are required to sample at a rate equal to or greater than the chip rate.

The time-bandwidth product associated with a phase-coded waveform is given by

$$\tau B = \tau B_{chip} = \frac{\tau}{\tau_{chip}} = N \tag{20.99}$$

and is equal to the number chips in the sequence. As with a frequency modulated waveform, the time-bandwidth product defines the gain in SNR at the output of the matched filter.

Phase codes are desired that exhibit low time sidelobes to meet both peak and integrated requirements. In general, sidelobe levels are inversely proportional to the sequence length. Long sequences may also be needed to satisfy SNR requirements, since the energy in a phase-coded waveform is proportional to the waveform duration $N\tau_{chip}$. As the chip duration is reduced to improve range resolution, energy in the waveform is decreased for a fixed length code. Thus, long codes may be needed to meet both the sidelobe and SNR requirements of a waveform.

20.12.2 Phase Codes Used in Radar

Selection of a phase code is based on a number of factors including the sidelobe response and sequence length. Optimal codes providing the minimum peak sidelobe (MPS) for a given sequence length have been identified for both biphase and polyphase codes [2,7,19–24,35–41]. These include the Barker codes, which achieve a 1:N peak sidelobe to mainlobe ratio. Biphase MPS codes have been identified through length 105 [22]. However, a requirement exists, in some cases, for longer sequences with predictable sidelobe levels. Nested codes are constructed by modulating one biphase code with another (e.g., Barker codes) to produce a code whose length is equal to the product of the two code lengths and whose peak sidelobe is defined by the longer of the two sequence lengths [7]. A *maximum length sequence* (MLS) is a biphase code of length $2^n - 1$ where n is an integer. The peak sidelobes of an MLS are inversely proportional to the square root of the sequence length [25,26].

Polyphase codes have the potential to exhibit lower sidelobes than a biphase code of equal length. Examples of polyphase codes with predictable sidelobe levels are the Frank, P1, and P2 codes [42,43]. Doppler tolerant polyphase codes include the P3 and P4 codes [27]. Quadriphase codes are polyphase codes designed to reduce the spectral energy located outside the nominal waveform bandwidth [28]. Spectral leakage represents a potential source of electromagnetic interference.

As with frequency modulated waveforms, mismatched filters may also be applied to a phase code to shape the sidelobe response [44–48]. Mismatched filters are used to reduce both the integrated and peak sidelobes as well as to tailor the sidelobe response in a given region.

The previous list is not meant to be comprehensive but includes some of the more common waveforms. Other phase codes are described in [2,4,6–8]. This chapter focuses on the MPS, MLS, Frank, P1, P2, P3, and P4 codes.

20.12.3 Phase Modulation

Phase-coded waveforms are partitioned into two categories: biphase and polyphase. Biphase-coded waveforms restrict their phase states to two values (e.g., 0° and 180°). Polyphase-coded waveforms exhibit more than two phase states (e.g., 0°, 90°, 180°, and 270°). The code sequence a_n is obtained by evaluating the complex exponential

$$a_n = \exp(j\phi_n) \quad n = 0, \ldots, N-1 \tag{20.100}$$

where ϕ_n is the phase (in radians) applied to the n-th chip. A biphase code consisting of phase states {0°, 180°} yields a code sequence consisting of elements $\{1, -1\}$. A phase-coded waveform is represented either in terms of its phase sequence expressed in degrees or radians or in terms of the complex or real sequence defined by evaluating equation (20.100). In some cases, biphase codes are expressed in terms of 0's and 1's instead of −1's and 1's, respectively. The conversion from $\{0, 1\}$ to $\{-1, 1\}$ is performed by multiplying the sequence by 2 and subtracting 1.

On transmit, a phase-coded waveform is mixed to an RF. An expression for the RF signal is

$$\begin{gathered} x_{RF}(t) = \cos\left\{2\pi f_c t + \phi_n\left[u\left(t-(n)\,\tau_{chip}\right) - u\left(t-(n+1)\,\tau_{chip}\right)\right]\right\} \\ 0 \le t \le \tau, 0 \le n \le N-1 \end{gathered} \tag{20.101}$$

where $u(t)$ is a unit step function. For example, given the sequence $\{1\ \ 1\ \ 1\ -1\ -1\ -1\ \ 1\ -1\ -1\ \ 1\ -1\}$, the baseband and RF modulated signals are shown in Figure 20-25. Note that the RF phase transitions occur at chip boundaries (denoted by asterisks) corresponding to a change or transition in the code sequence. A transition in the code sequence from 1 to −1 or from −1 to 1 corresponds to a 180° phase change.

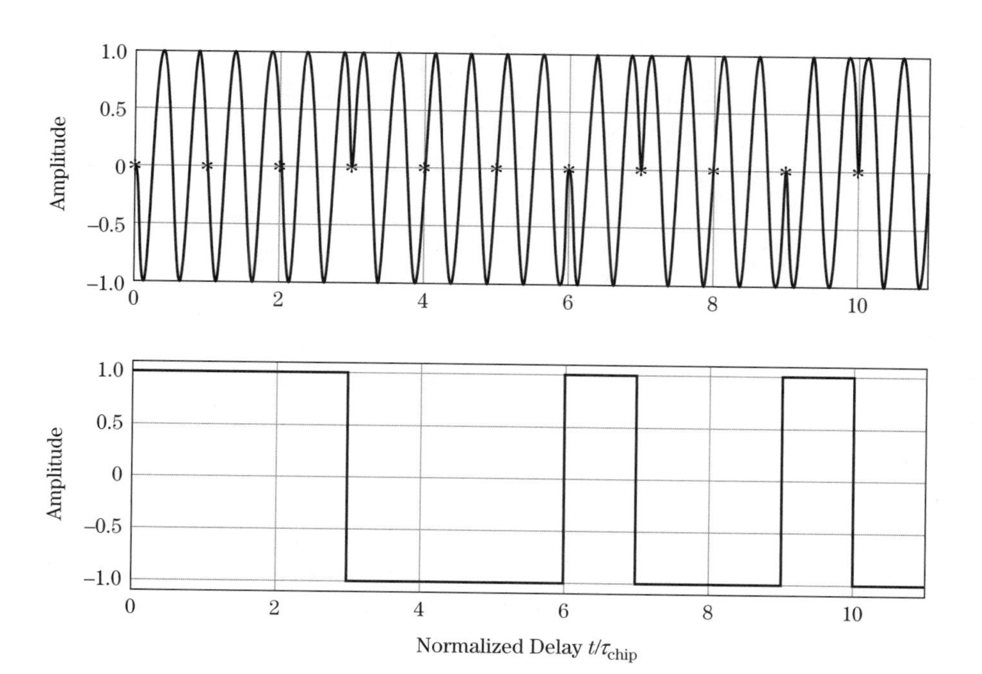

FIGURE 20-25 ■ Baseband (bottom) and RF modulated (top) phase coded waveform of length $N = 11$.

From a hardware perspective, biphase codes may be easier to implement than polyphase codes. However, with the advent of digital waveform generators, polyphase codes are just as likely to be used in a modern radar system as biphase codes. In general, polyphase-coded waveforms may be designed to achieve lower sidelobe levels than a biphase code of the same length. A polyphase code possesses more degrees of freedom (i.e., possible phase states), and these additional degrees of freedom may be exploited to achieve lower sidelobes. In addition, some polyphase codes are Doppler tolerant.

20.12.4 Equivalence Operations

Phase codes exhibiting identical match filtered magnitude responses are defined to be *equivalent*. Four operations may be applied to a code to generate an equivalent code. The magnitude response is preserved under the following operations [7]:

$$\hat{a}_n = a_{N-n}$$

$$\hat{a}_n = a_N^*$$

$$\hat{a}_n = \rho a_n, \text{ where } |\rho| = 1$$

$$\hat{a}_n = \rho^n a_n, \text{ where } |\rho| = 1$$

where $\hat{a}_n$ is an equivalent code and ρ is a unit-amplitude complex number.

20.12.5 Match Filtered Response of a Phase Code

A matched filter is applied to a phase-coded waveform to compress the waveform in range and maximize SNR. Consider the three chip sequence depicted in Figure 20-26 and the corresponding match filtered response in Figure 20-27. The response is obtained

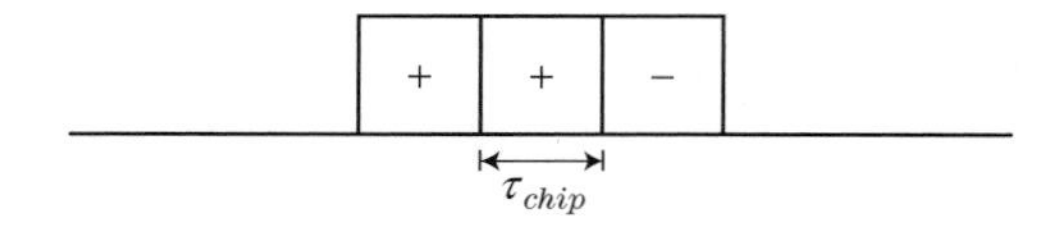

FIGURE 20-26 ■ Three chip Barker biphase code.

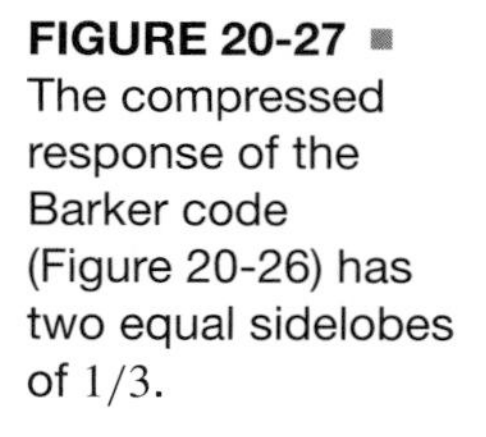

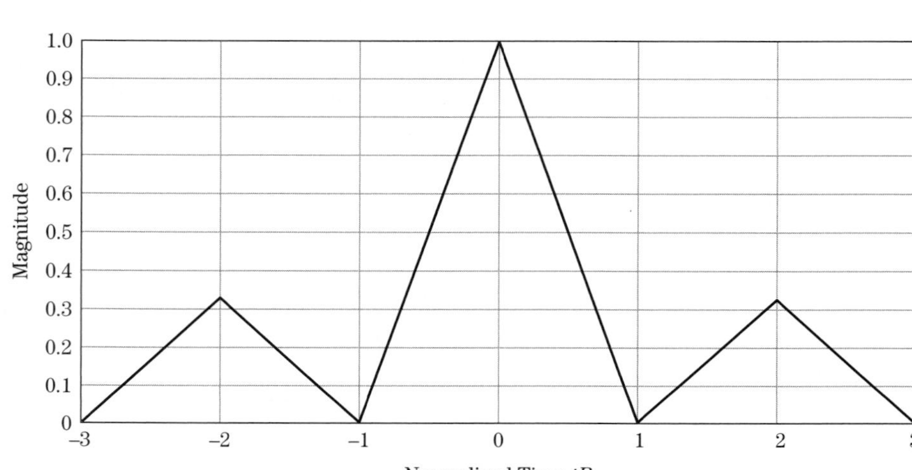

FIGURE 20-27 ■ The compressed response of the Barker code (Figure 20-26) has two equal sidelobes of 1/3.

by correlating the sequence $\{1, 1, -1\}$ with itself. Normalizing the sequence to unit energy (i.e., $\{1/\sqrt{3}, 1/\sqrt{3}, -1/\sqrt{3}\}$), the peak of the response is equal to 1, and the peak sidelobe has a value of $1/3$. Note that the output of the filter is interpreted as a voltage. Thus, the sidelobe-to-peak ratio is $1{:}N$ in voltage and $1{:}N^2$ in power. The abscissa in Figure 20-27 represents time delay and is normalized by the chip width. The Rayleigh resolution occurs at a time delay corresponding to the chip width. For a rectangular-shaped chip, the Rayleigh resolution also corresponds to the -6 dB width.

As noted previously, both the peak and integrated sidelobes are important performance metrics. Ignoring the affects of band-limiting, the peak and integrated sidelobes may be examined using the code sequence and ignoring the chip response. The phase code's autocorrelation sequence is defined as

$$y[k] = \sum_{n=0}^{N-1} a_n a_{n-k}^* \qquad k = -(N-1), \ldots, 0, \ldots (N-1) \tag{20.102}$$

and is equivalently obtained by sampling the waveform's match filtered response at multiples of the chip width. The ISR may be computed from the autocorrelation response and is defined as

$$ISR = \frac{2\sum_{k=1}^{N-1} |y[k]|^2}{N^2} \tag{20.103}$$

For the response in Figure 20-27, the ISR is 2/9, or -6.5 dB.

20.12.6 Spectrum of a Phase-Coded Waveform

The spectrum of a phase-coded waveform differs significantly from that of an LFM or NLFM waveform; however, properties associated with a phase-coded waveform may also be inferred from its spectrum. A derivation and examination of the spectrum of a phase-coded waveform is presented.

A phase-coded waveform $x(t)$ may be modeled in the time domain as

$$x(t) = \sum_{n=0}^{N-1} a_n p(t - n\tau_{chip}) \tag{20.104}$$

where an individual chip $p(t)$ is a unit amplitude rectangle defined by

$$p(t) = 1 \qquad -\frac{\tau_{chip}}{2} \leq t \leq \frac{\tau_{chip}}{2} \tag{20.105}$$

and $\{a_n\}$ is the code sequence. For biphase codes, $a_n \in \{1, -1\}$, and for polyphase codes $a_n \in \mathbb{C}$ where $\mathbb{C}$ represents the set of unit-amplitude complex numbers. For the waveform in equation (20.104), $t = 0$ is referenced to the center of the first chip. The Fourier transform of the waveform is

$$X(\omega) = P(\omega) \sum_{n=0}^{N-1} a_n \exp(-j\omega n\tau_{chip}) \tag{20.106}$$

where $P(\omega)$ is the Fourier transform of a single chip centered at $t = 0$. The time offset to each chip, $n\tau_{chip}$, appears in the frequency domain as the argument of a complex exponential. The spectrum consists of the product of the chip spectrum $P(\omega)$ and the DTFT of the phase sequence.

To compress the chip sequence, a matched filter is applied to the waveform spectrum. The matched filter was defined in Section 20.2.3 as $x^*(-t)$ in the time domain or $X^*(\omega)$ in the frequency domain. Applying $X^*(\omega)$ to $X(\omega)$ yields

$$Y(\omega) = |P(\omega)|^2 \sum_{n=0}^{N-1} a_n \exp(-j\omega n\tau_{chip}) \sum_{m=0}^{N-1} a_m^* \exp(j\omega m\tau_{chip}) \tag{20.107}$$

or

$$Y(\omega) = |P(\omega)|^2 C(\omega) \tag{20.108}$$

where

$$C(\omega) = \sum_{m=0}^{N-1} \sum_{n=0}^{N-1} a_n a_m^* \exp(-j\omega\tau_{chip}(n-m)) \tag{20.109}$$

The composite spectrum consists of the product of the magnitude-squared of the chip's spectrum and the double summation in equation (20.109), which is a function of the code sequence. The double summation may be expressed as

$$C(\omega) = \sum_{k=-(N-1)}^{N-1} c_k \exp(-j\omega k\tau_{chip}) \tag{20.110}$$

where c_k is defined as

$$c_k = \sum_{m=k}^{N-1-k} a_{m-k} a_m^* \qquad 0 \le k \le (N-1) \tag{20.111}$$

and

$$c_{-k} = c_k^* \tag{20.112}$$

The sequence c_k, in (20.111), represents the discrete autocorrelation of the code sequence, and $C(\omega)$ represents its DTFT.

The spectral components $C(\omega)$ and $P(\omega)$ are plotted in Figure 20-28 for the sequence $\{1, 1, -1\}$ and a rectangular chip. The spectrum of the chip is a sinc function. In the composite spectrum, $P(\omega)$ is modulated by $C(\omega)$, yielding $Y(\omega)$. A plot of the composite power spectrum is provided in the lower subplot of Figure 20-28. In general, the spectrum of a phase-coded waveform resembles a "noisy" or amplitude modulated sinc.

The shape of the compressed response is a function of both the coding sequence and the compressed chip response. The compressed response $y(t)$ may be written in terms of delayed and weighted copies of the filtered chip response $y_{chip}(t)$ where

$$y(t) = \sum_{k=-(N-1)}^{N-1} c_k y_{chip}(t - k\tau_{chip}) \tag{20.113}$$

and the chip response is obtained by taking the inverse Fourier transform of the chip spectrum

$$y_{chip}(t) = \frac{1}{2\pi} \int_{-\infty}^{\infty} |P(\omega)|^2 \exp(j\omega t) d\omega \tag{20.114}$$

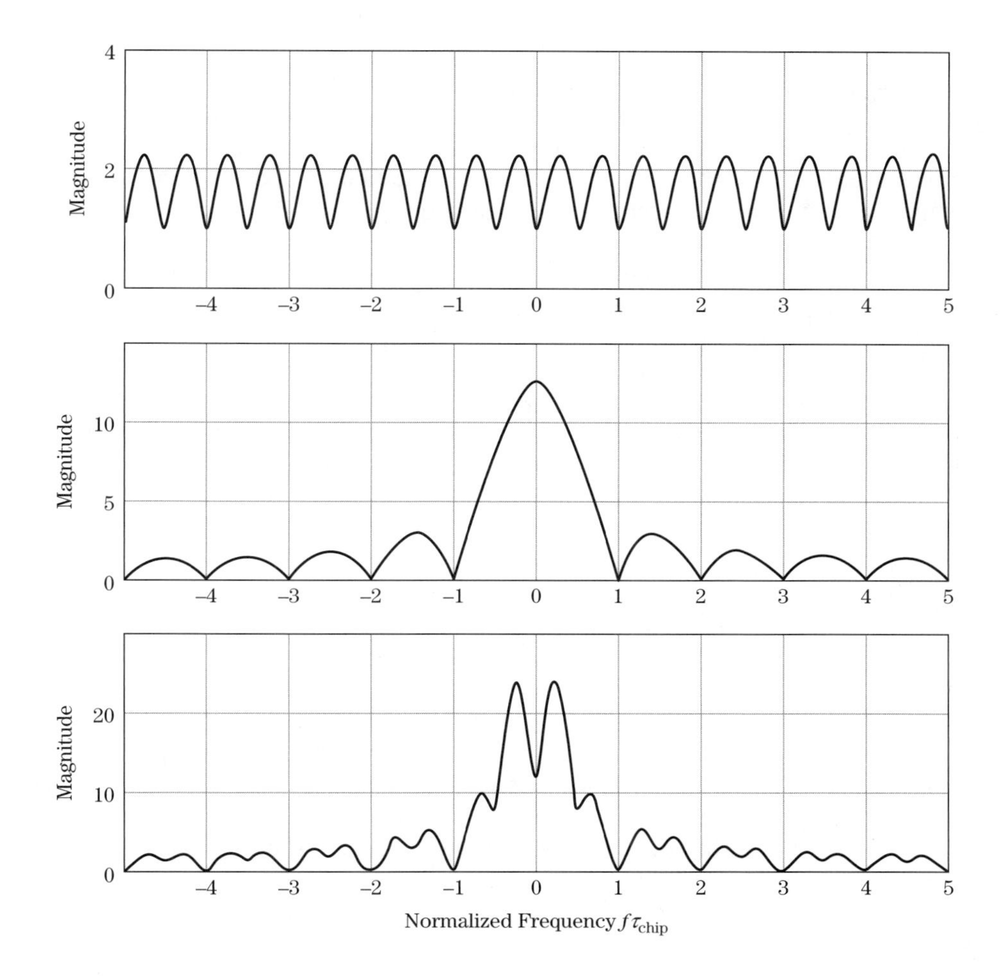

FIGURE 20-28 Top: Spectrum of the 3 bit phase code $C(\omega)$. Middle: Chip spectrum $P(\omega)$. Bottom: Composite spectrum $Y(\omega)$. See text for details.

The match filtered response is encoded in both the time and frequency domains. As an example, consider the three-chip sequence in Figure 20-26. For this sequence, the double summation in equation (20.110) becomes

$$\sum_{k=-(N-1)}^{N-1} c_k \exp(-j\omega k\tau_{chip}) = -e^{j\omega 2\tau_{chip}} + 3 - e^{-j\omega 2\tau_{chip}} \tag{20.115}$$

The filtered spectrum is then given by

$$Y(\omega) = -|P(\omega)|^2 e^{j\omega 2\tau_{chip}} + 3|P(\omega)|^2 - |P(\omega)|^2 e^{-j\omega 2\tau_{chip}} \tag{20.116}$$

The spectrum of a rectangular chip is a sinc, and the inverse transform of a squared sinc is a triangle. The compressed response therefore consists of three weighted triangles centered at time delays $t = \{-2\tau_{chip}, 0, 2\tau_{chip}\}$ as illustrated in Figure 20-27. Note that the match filtered spectrum $Y(\omega) = X(\omega)X^*(\omega) = |X(\omega)|^2$ is real and that $Y(\omega) \geq 0$. The spectrum in (20.116) is thus real and may be written as

$$Y(\omega) = -2|P(\omega)|^2 \cos(\omega 2\tau_{chip}) + 3|P(\omega)|^2 \tag{20.117}$$

The spectra associated with frequency modulated waveforms often have relatively simple shapes, for example, the approximate rectangular shape of an LFM waveform's spectrum. As a result, it is often easy to predict the match filtered time-domain response from the spectral shape. In contrast, the spectra of phase-coded waveforms have more complex shapes, making it difficult if not impossible to predict the match filtered time-domain response from the spectrum. For phase-coded waveforms, an autocorrelation is required to determine the time-domain response.

Band-limiting modifies the chip's spectrum and thus the composite spectrum. Filters in the transmitter and receiver increase the duration of a chip resulting in cross-talk, which degrades the sidelobe response and reduces bandwidth resulting in degraded resolution. The level of degradation depends on the degree of band-limiting imposed. The impact of band-limiting is illustrated in Section 20.14.2 using a Frank and P1 code.

In most instances, phase-code waveform properties are assessed assuming no band-limiting (i.e., no filtering of the ideal chip response). In these cases, an assessment of sidelobe performance may be performed using the code sequence and ignoring the chip response.

20.12.7 Doppler Tolerance

Doppler tolerance refers to the degree of degradation in the compressed response due to uncompensated Doppler. As discussed in Section 20.10.3, an LFM waveform is characterized as Doppler tolerant, exhibiting range-Doppler coupling and a preservation of the mainlobe and sidelobe response over large fractional Doppler shifts. Biphase-coded waveforms are considered to be *Doppler intolerant* when the residual Doppler exceeds one-quarter cycle over the uncompressed pulse length.

In the absence of a Doppler shift, the mainlobe of a phase code achieves its maximum at $k = 0$ in equation (20.111). In the presence of a Doppler shift f_d, the response at $k = 0$ is

$$c_0 = \frac{1}{N}\sum_{m=0}^{N-1} \exp(j2\pi f_d m \tau_{chip}) \tag{20.118}$$

for a unit energy phase code. Applying the finite sum formula, (20.118) reduces to

$$|c_0| = \frac{1}{N}\left|\frac{\sin(N\pi f_d \tau_{chip})}{\sin(\pi f_d \tau_{chip})}\right| \tag{20.119}$$

This quantity is zero when $f_d = 1/N\tau_{chip}$, which is equivalent to one full cycle of Doppler over the uncompressed pulse. One-quarter cycle of Doppler produces 1 dB of peak loss, and one-half cycle produces 4 dB of loss.

Doppler also causes each chip's phase to rotate, degrading the sidelobe response. A half-cycle of Doppler causes a phase change of 180° between the first and last chip. In general, biphase codes are designed to experience no more than a quarter-cycle of uncompensated Doppler. A quarter-cycle limits the degradation in the sidelobe response and corresponds to a 1 dB loss in SNR.

Some polyphase codes exhibit a Doppler tolerance similar to that of an LFM waveform, but not all polyphase codes are Doppler tolerant. In most cases, Doppler tolerance is an issue only when employing long pulses or in the presence of large Doppler shifts. If the Doppler shift is known or has been estimated, it may be removed through processing

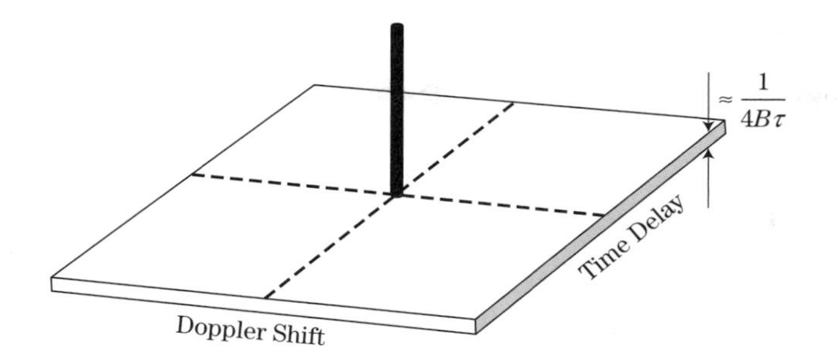

FIGURE 20-29 ■ The ambiguity surface associated with some phase coded waveforms is a thumb tack.

or modulation prior to the matched filter. In addition, a bank of matched filters may be employed, as described in Section 20.10.5.

In general, biphase codes do not exhibit range-Doppler coupling. In some cases, the ambiguity surface resembles a thumbtack [30,34]. A thumbtack ambiguity surface exhibits a narrow impulse-like response centered at zero Doppler and delay and a low plateau response extending over the waveform bandwidth in Doppler and pulse width in delay, as illustrated in Figure 20-29. Some biphase-coded waveforms exhibit large responses in the plateau region that are above the nominal sidelobe levels [30] and thus diverge from a thumbtack response.

A waveform possessing a thumbtack-like ambiguity surface may be used to simultaneously estimate a target's range and Doppler shift by employing a bank of filters, each matched to a different Doppler shift. The filter exhibiting the largest response is associated with the target's true range and Doppler.

The ambiguity surface associated with a polyphase code may resemble a thumbtack, a sheared ridge, or some other shape. The P3 and P4 polyphase codes [27] examined in Section 20.14.3 exhibit range-Doppler coupling and a Doppler tolerance similar to an LFM waveform.

20.13 BIPHASE CODES

Biphase codes have been identified that exhibit low, predictable sidelobe levels and in some cases achieve the minimum peak sidelobe for a given code length. Codes commonly applied in radar systems include Barker codes, MPS codes, and maximal length sequences [2,7,19–24,25–26].

20.13.1 Biphase Barker Codes

Biphase Barker codes exhibit a peak sidelobe to mainlobe voltage ratio of $1{:}N$, where N is the code length. The peak sidelobe ratio is defined in terms of power and is expressed as $20\log_{10}(1/N)$. The Barker codes achieve the lowest possible PSR for an aperiodic code. The known biphase Barker codes are listed in Table 20-4 with their corresponding peak sidelobe ratios. Equivalent codes are obtained using the transforms defined in Section 20.12.4.

The longest known Barker code is length 13 with a corresponding peak sidelobe ratio of -22.3 dB. Figure 20-30 contains a plot of the compressed response for a 13-bit Barker with unit energy. The peak sidelobe is $1/13 \approx 0.077$. It has been shown [4] that no odd-length biphase Barker sequence exists longer than length 13. No even-length sequences

TABLE 20-4 ■ A List of the Known Biphase Barker Codes

Code Length	Code Sequence	Peak Sidelobe Level, dB
2	+−, ++	−6.0
3	++−	−9.5
4	++−+, +++−	−12.0
5	+++−+	−14.0
7	+++−−+−	−16.9
11	+++−−−+−−+−	−20.8
13	+++++−−++−+−+	−22.3

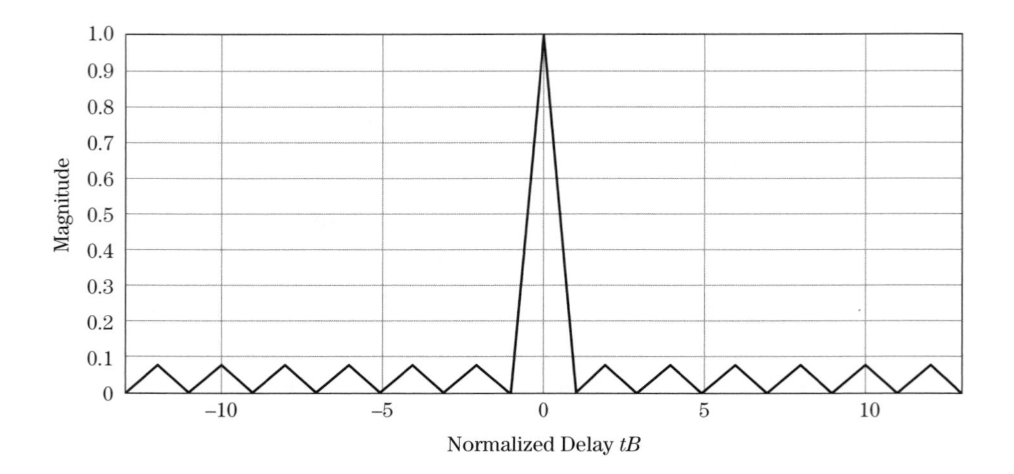

FIGURE 20-30 ■ Compressed response for the longest Barker biphase code: 13 chips.

have been found above length 4, and it can be shown that if a longer even-length sequence does exist, the sequence length must be a perfect square [4].

20.13.2 MPS codes

Given the length limitations of Barker codes, longer sequences with low peak sidelobe ratios are desired. Searches have been performed to identify biphase codes that yield MPS ratios for a given sequence length. Lindner [21] identified MPS codes up to length 40. Excluding the Barker codes, these MPS codes exhibit peak sidelobe ratios of either $2{:}N$ or $3{:}N$ in voltage. Codes with an MPS ratio equal to $2{:}N$ exist for lengths 6, 8–10, 12, 14–21, 25, and 28, while codes with an MPS ratio equal to $3{:}N$ exist for lengths 22–24, 26, 27, and 29–40. The number of MPS codes per sequence length varies significantly from 1 code for a 13-bit Barker code to 858 MPS codes for a 24-length sequence. Cohen et al. [19] developed techniques for judiciously searching for MPS codes and extended the known MPS codes to lengths 41 through 48. The MPS codes of length 41–48 exhibit peak sidelobe ratios of $3{:}N$. Figure 20-31 contains the compressed response for a 48-length biphase MPS code with a −24.08 dB peak sidelobe.

Kerdock et al. [20] identified an MPS code of length 51 with a peak sidelobe ratio of $3{:}N$, and Coxson and Russo [22] and Coxson et al. [24] extended the list of known MPS codes to include lengths 49, 50, and 52–70 with a peak sidelobe ratio of $4{:}N$. Nunn and Coxson [23] conjectured, given the current body of evidence, that no sequence greater than length 51 exhibits a peak sidelobe ratio of $3{:}N$ or less. They added to the list of codes exhibiting a peak sidelobe ratio of $4{:}N$ to lengths of 71–82 and the list of those exhibiting a

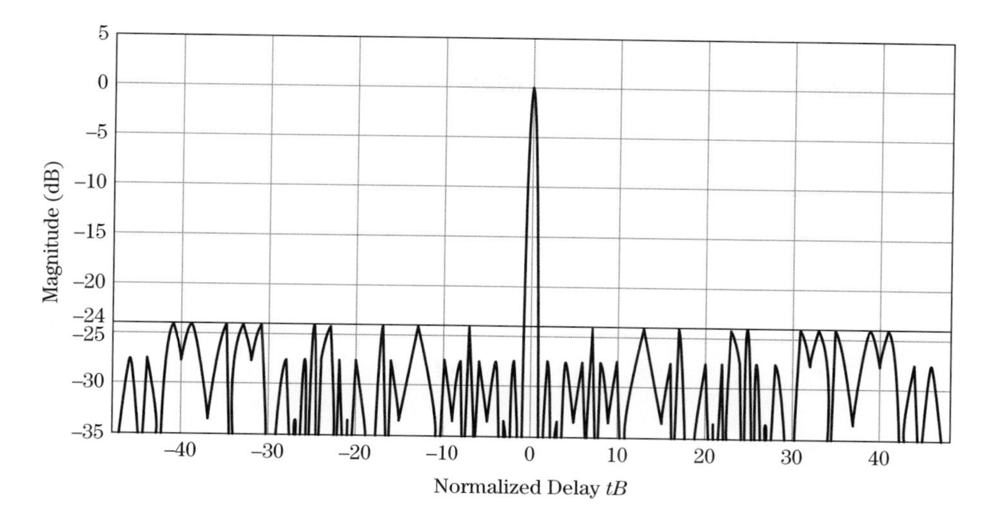

FIGURE 20-31 ■ Compressed response for a 48-length Minimal Peak Sidelobe (MPS) code achieves a 3:48 peak sidelobe ratio.

TABLE 20-5 ■ Summary of MPS Code Lengths and Corresponding Peak Sidelobe Levels

Peak Sidelobe Level (Voltage)	Code Length
1	2–5, 7, 11, 13
2	6, 8–10, 12, 14–21, 25, 28
3	22–24, 26, 27, 29–40, 41–48, 51
4	49, 50, 52–70, 71–82
5	83–105

peak sidelobe ratio of 5:N to lengths of 83–105. The searches were not exhaustive, but the codes achieve the best MPS ratio known to date for these code lengths. Table 20-5 contains a summary of MPS code lengths and their corresponding peak sidelobe levels. Cohen et al. [19] provides example MPS codes that also exhibit the lowest integrated sidelobe level. Cohen et al.'s list and other example MPS codes through length 105 [19–24] are provided in Table 20-6.[1] The number of codes identified for a given code length is also provided in Table 20-6. The count does not include equivalent codes obtained using the operations in Section 20.12.5.

20.13.3 Maximal Length Sequences

The biphase MPS codes, which include the Barker codes, achieve the lowest peak sidelobe for a given sequence length; however, in many radar applications an optimum code is not required given that a "good" code with relatively low sidelobe levels is available. Biphase maximum length sequences are used in radar applications and provide predictable peak sidelobe ratios that approach $10\log_{10}(1/N)$. Maximum length sequences are binary sequences whose pattern of 1's and -1's does not repeat within a period equal to $2^n - 1$, where n is an integer. MLSs are generated using a shift register

$$s_i = a_1 s_{i-1} \oplus a_2 s_{i-2} \oplus \ldots \oplus a_n s_{i-n} \qquad i = 1, \ldots, 2^n - 1 \tag{20.120}$$

[1]Note that the 103-length code is incorrect in [23] due to a typographical error. The corrected code was supplied by Carroll Nunn and is provided in Table 20-6.

TABLE 20-6 ■ Example Biphase MPS Codes through Length 105

Code Length	Peak Sidelobe	Number of Codes	Example Code (Hexadecimal)
2	1	1	2
3	1	1	6
4	1	1	D
5	1	1	1D
6	2	4	0B
7	1	1	27
8	2	8	97
9	2	10	D7
10	2	5	167
11	1	1	247
12	2	16	9AF
13	1	1	159F
14	2	9	1483
15	2	13	182B
16	2	10	6877
17	2	4	774B
18	2	2	190F5
19	2	1	5BB8F
20	2	3	5181B
21	2	3	16BB83
22	3	378	E6D5F
23	3	515	38FD49
24	3	858	64AFE3
25	2	1	12540E7
26	3	242	2380AD9
27	3	388	25BBB87
28	2	2	8F1112D
29	3	283	164A80E7
30	3	86	2315240F
31	3	251	2A498C0F
32	3	422	3355A780
33	3	139	CCAA587F
34	3	51	333FE1A55
35	3	111	796AB33
36	3	161	3314A083E
37	3	52	574276F9E
38	3	17	3C34AA66
39	3	30	13350BEF3C
40	3	57	2223DC3A5A
41	3	15	38EA520364
42	3	4	4447B874B4
43	3	12	5B2ACCE1C
44	3	15	FECECB2AD7
45	3	4	2AF0CC6DBF6
46	3	1	3C0CF7B6556
47	3	1	69A7E851988
48	3	4	156B61E64FF3
49	4	Not Reported	012ABEC79E46F
50	4	Not Reported	025863ABC266F
51	3	Not Reported	71C077376ADB4
52	4	Not Reported	0945AE0F3246F
53	4	Not Reported	0132AA7F8D2C6F

TABLE 20-6 ■ *(Continued)*

Code Length	Peak Sidelobe	Number of Codes	Example Code (Hexadecimal)
54	4	Not Reported	0266A2814B3C6F
55	4	Not Reported	04C26AA1E3246F
56	4	Not Reported	099BAACB47BC6F
57	4	Not Reported	01268A8ED623C6F
58	4	Not Reported	023CE545C9ED66F
59	4	Not Reported	049D38128A1DC6F
60	4	Not Reported	0AB8DF0C973252F
61	4	Not Reported	005B44C4C79EA350
62	4	Not Reported	002D66634CB07450
63	4	Not Reported	04CF5A2471657C6F
64	4	1859	55FF84B069386665
65	4	Not Reported	002DC0B0D9BCE5450
66	4	Not Reported	0069B454739F12B42
67	4	Not Reported	007F1D164C62A5242
68	4	Not Reported	009E49E3662A8EA50
69	4	Not Reported	0231C08FDA5A0D9355
70	4	Not Reported	1A133B4E3093EDD57E
71	4	Not Reported	63383AB6B452ED93FE
72	4	Not Reported	E4CD5AF0D054433D82
73	4	Not Reported	1B66B26359C3E2BC00A
74	4	Not Reported	36DDBED681F98C70EAE
75	4	Not Reported	6399C983D03EFDB556D
76	4	Not Reported	DB69891118E2C2A1FA0
77	4	Not Reported	1961AE251DC950FDDBF4
78	4	Not Reported	328B457F0461E4ED7B73
79	4	Not Reported	76CF68F327438AC6FA80
80	4	Not Reported	CE43C8D986ED429F7D75
81	4	Not Reported	0E3C32FA1FEFD2519AB32
82	4	Not Reported	3CB25D380CE3B7765695F
83	5	Not Reported	711763AE7DBB8482D3A5A
84	5	Not Reported	CE79CCCDB6003C1E95AAA
85	5	Not Reported	19900199463E51E8B4B574
86	5	Not Reported	3603FB659181A2A52A38C7
87	5	Not Reported	7F7184F04F4E5E4D9B56AA
88	5	Not Reported	D54A9326C2C686F86F3880
89	5	Not Reported	180E09434E1BBC44ACDAC8A
90	5	Not Reported	3326D87C3A91DA8AFA84211
91	5	Not Reported	77F80E632661C3459492A55
92	5	Not Reported	CC6181859D9244A5EAA87F0
93	5	Not Reported	187B2ECB802FB4F56BCCECE5
94	5	Not Reported	319D9676CAFEADD68825F878
95	5	Not Reported	69566B2ACCC8BC3CE0DE0005
96	5	Not Reported	CF963FD09B1381657A8A098E
97	5	Not Reported	1A843DC410898B2D3AE8FC362
98	5	Not Reported	30E05C18A1525596DCCE600DF
99	5	Not Reported	72E6DB6A75E6A9E81F0846777
100	5	Not Reported	DF490FFB1F8390A54E3CD9AAE
101	5	Not Reported	1A5048216CCF18F83E910DD4C5
102	5	Not Reported	2945A4F11CE44FF664850D182A
103	5	Not Reported	77FAA82C6E065AC4BE18F274CB
104	5	Not Reported	E568ED4982F9660EBA2F611184
105	5	Not Reported	1C6387FF5DA4FA325C895958DC5

Sources: Compiled from [19–24].

TABLE 20-7 ■ Coefficients for an Irreducible, Primitive Polynomial through Order 13

Degree of Polynomial	Coefficients (Octal)	Coefficients (Binary)
2	7	111
3	13	1011
4	23	10011
5	45	100101
6	103	1000011
7	211	10001001
8	435	100011101
9	1021	1000010001
10	2011	10000001001
11	4005	100000000101
12	10123	1000001010011
13	20033	10000000011011

employing feedback where $\{s_i, s_{i-1}, \ldots, s_{i-n}\}$ are the individual shift registers, each containing either a 0 or a 1. The operator $\oplus$ denotes modulo 2 addition. The feedback coefficients d_k, $k = 1, \ldots, n$, are obtained from the coefficients of an irreducible, primitive polynomial

$$h(x) = d_n x^n + d_{n-1} x^{n-1} + \cdots + d_1 x + 1 \tag{20.121}$$

where the coefficients d_k are either 0 or 1 and are equal to the shift register coefficients of (20.120), $d_n = a_n$. Cook [4] defines a primitive polynomial as one that divides evenly into $x^m + 1$ where $m > 2^n - 1$. Irreducible, primitive polynomials of degree n exists for all values of n [25], and Peterson [49] provides tables of primitive irreducible polynomials up to order 34. More than one irreducible polynomial may exist for a given order. Table 20-7 contains the coefficients for a single, primitive irreducible polynomial through order 13.

Figure 20-32 contains a shift register configured to generate a 31-length MLS. The sequence is obtained by recording the output of any one of the shift registers. The shift registers are initialized with a sequence of 1's and 0's, excluding the all-zero condition. The compressed response for a 127-length MLS is provided in Figure 20-33 and exhibits a peak sidelobe of approximately −21 dB. Note that the sidelobes are very low near the mainlobe and increase away from the mainlobe approaching the predicted peak sidelobe level before rolling off at the ends. This shape is characteristic of the compressed response associated with an MLS.

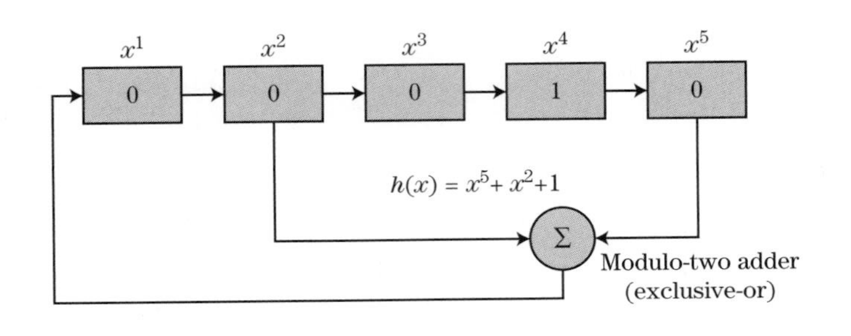

FIGURE 20-32 ■ A shift register employing feedback generates a 31-length Maximal Length Sequence.

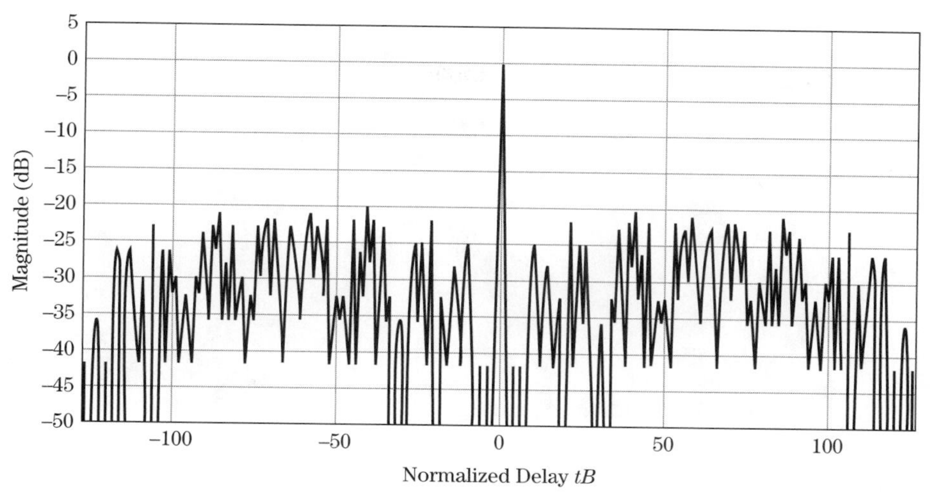

FIGURE 20-33 ■ The compressed response for a 127-length MLS.

20.13.4 Comparison of MLS and LFM Waveform Responses and Spectra

In general, the sidelobe structure of the match filtered response for a biphase-coded waveform is relatively uniform, whereas the sidelobes of a frequency modulated waveform roll-off with range. Figure 20-34 contains the compressed response for an MLS with a time-bandwidth product of 1,023 (gray curve) and an unweighted LFM waveform with a time-bandwidth product of 1,000 (black curve). The peak sidelobe ratio for the LFM waveform and MLS are −13.2 dB and −28.6 dB, respectively. The sidelobe roll-off associated with the LFM waveform and the more constant response associated with the MLS are evident. The ISR for the LFM waveform and MLS were calculated to be −9.7 dB and −4.8 dB, respectively. The relatively flat sidelobe response associated with the phase-coded waveform yields a higher ISR even though the peak sidelobe is much lower. An LFM weighted with a −30 dB Taylor and a TB of 1,000 yields an ISR of −22 dB. The roll-off of the sidelobes associated with a frequency modulated waveform aids in reducing the integrated sidelobe ratio.

Frequency modulated waveforms exhibit a compact spectral support in comparison with most phase-coded waveforms. Figure 20-35 contains a plot of the spectrum of a

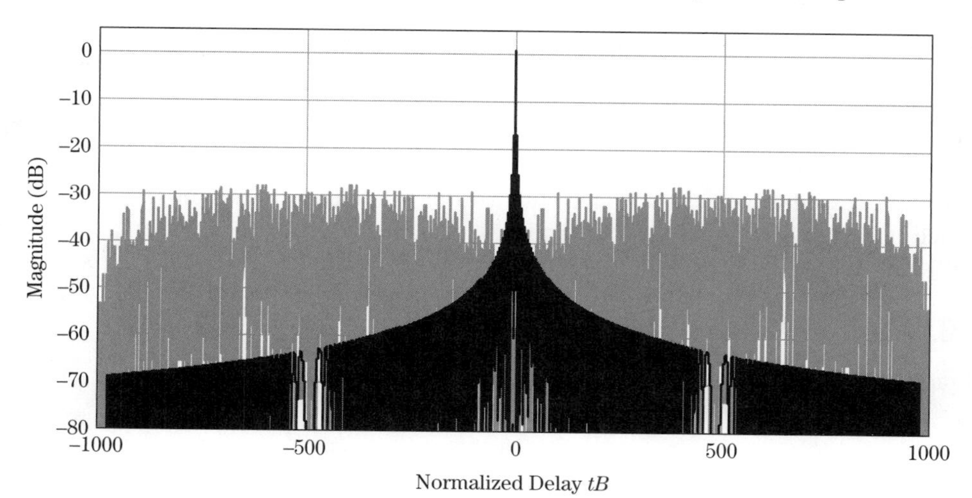

FIGURE 20-34 ■ Comparison of a compressed LFM waveform (black curve) (TB = 1000) with a compressed biphase MLS coded waveform (gray curve) (TB = 1023).

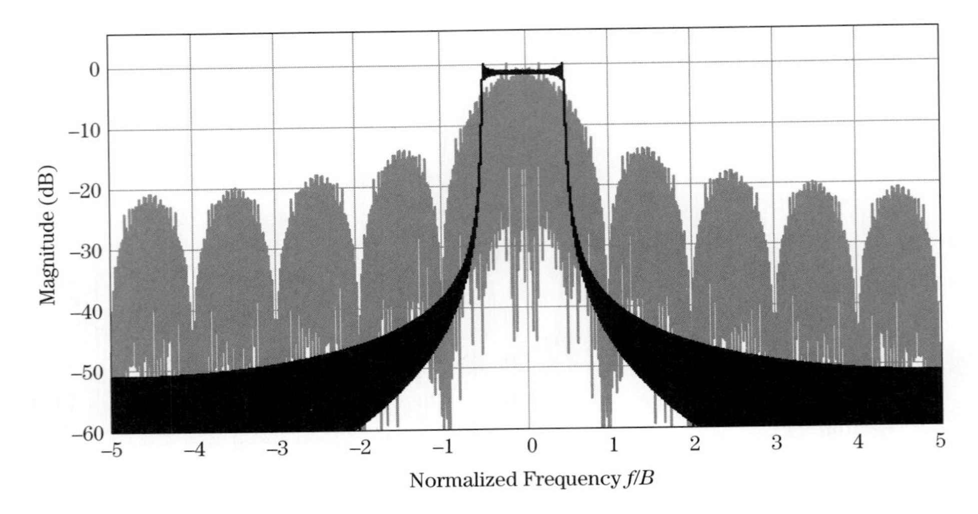

FIGURE 20-35 ■ Comparison of the spectra of an LFM waveform (black curve) with a 1023-length MLS coded waveform (gray curve).

biphase-coded waveform derived from a 1,023-length MLS with a 10 nanosecond chip width. The spectrum of the phase-coded waveform resembles a "noisy" sinc function. Overlaid on the MLS is the spectrum of an LFM waveform with a 100 MHz bandwidth. The phase-coded waveform exhibits a significant amount of energy outside the nominal 100 MHz bandwidth. This out-of-band energy is a potential source of *electromagnetic interference* (EMI) for a system operating nearby in frequency and in close physical proximity. Quadriphase codes and chip shaping have been used to reduce the energy located outside the nominal waveform bandwidth [7,28].

20.14 POLYPHASE CODES

Polyphase-coded waveforms possess more than two phase states and offer the potential for Doppler tolerance similar to that of an LFM waveform as well as lower peak sidelobes for a given length code. Polyphase Barker, Frank, P1, P2, P3, and P4 codes are examined in the following sections.

20.14.1 Polyphase Barker Codes

Polyphase Barker codes [35–41] exhibiting a peak sidelobe ratio of $1{:}N$ have been identified employing various search techniques. The phases are either unrestricted or restricted to a P-th root of unity such that the allowable phase increment is $\Delta\phi = 2\pi/P$ and P is an integer greater than 2. Borweinan and Ferguson [36] have identified polyphase Barker codes up to length 63. Several of the codes extracted from [36] are provided in Table 20-8. The compressed response associated with a 45-length polyphase Barker sequence is provided in Figure 20-36 with a theoretical peak sidelobe of −33.06 dB. Nunn and Coxson [39] have recently identified polyphase Barker codes of length 64–70 and 72, 76, and 77.

20.14.2 Frank, P1, and P2 Codes

Frank codes [42,43] are polyphase codes with a phase sequence defined by

$$\phi_{n,k} = \frac{2\pi}{M} nk \qquad n, k = 0, \ldots, (M-1) \tag{20.122}$$

TABLE 20-8 ■ Example Polyphase Barker Codes Where $\phi_n = 2\pi p_n/P$

Code Length, N	P	$p_n, n = 1, \ldots, N$
63	2,000	0, 0, 88, 200, 250, 89, 1832, 1668, 1792, 145, 308, 290, 528, 819, 1357, 1558, 1407, 1165, 930, 869, 274, 97, 10, 1857, 731, 789, 1736, 150, 1332, 1229, 390, 944, 1522, 1913, 648, 239, 1114, 1708, 200, 666, 1870, 1124, 1464, 265, 845, 1751, 1039, 53, 737, 1760, 798, 1880, 851, 1838, 1103, 419, 1711, 1155, 546, 1985, 1325, 754, 44
60	210	0, 0, 16, 208, 180, 153, 126, 161, 135, 78, 83, 98, 143, 127, 162, 153, 183, 141, 72, 207, 149, 167, 15, 13, 146, 58, 23, 109, 169, 208, 75, 143, 173, 199, 51, 50, 31, 142, 152, 84, 74, 6, 147, 205, 151, 66, 51, 151, 27, 101, 170, 75, 172, 91, 20, 131, 1, 78, 166, 68
57	240	0, 0, 18, 51, 31, 37, 6, 39, 43, 64, 128, 167, 187, 19, 22, 226, 163, 103, 97, 238, 200, 172, 111, 201, 72, 95, 75, 172, 2, 91, 49, 220, 209, 57, 212, 168, 116, 206, 110, 102, 25, 131, 2, 30, 143, 182, 42, 107, 216, 89, 10, 161, 29, 170, 106, 205, 86
54	200	0, 0, 23, 43, 16, 9, 40, 51, 20, 7, 67, 126, 178, 180, 71, 120, 144, 151, 61, 25, 45, 100, 86, 9, 172, 161, 142, 22, 85, 8, 96, 128, 81, 1, 18, 137, 0, 95, 132, 59, 44, 155, 16, 129, 157, 98, 47, 174, 73, 18, 145, 65, 170, 100
51	50	0, 0, 4, 4, 18, 20, 27, 25, 25, 26, 24, 15, 15, 14, 9, 32, 36, 2, 21, 17, 9, 27, 46, 49, 19, 29, 9, 32, 7, 45, 21, 46, 22, 47, 18, 35, 0, 22, 9, 31, 44, 5, 29, 21, 4, 49, 33, 24, 9, 49, 29

Source: Extracted from [36].

where the phase, $\phi_{n,k}$, is indexed over n for each value of k. The length of the code, N, is required to be a perfect square, $N = M^2$. The peak sidelobe ratio in dB [50] is $20\log_{10}\{1/[N\sin(\pi/M)]\}$ for N even and $20\log_{10}\{1/[2N\sin(\pi/2M)]\}$ for N odd. The upper plot in Figure 20-37 contains the compressed response for a 100-length Frank code. The peak sidelobe ratio is approximately -30 dB. Note that a biphase 127-length MLS would yield a PSR of approximately -21 dB. The additional degrees of freedom in the polyphase code support a lower PSR for nearly the same length code.

Frank codes exhibit range-Doppler coupling similar to an LFM waveform but are Doppler intolerant, exhibiting high near-in sidelobes in the presence of small fractional

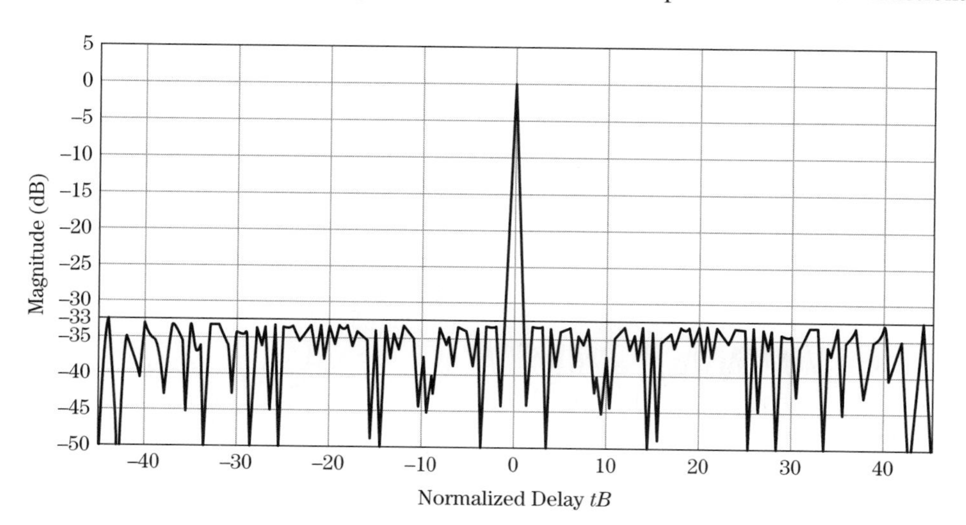

FIGURE 20-36 ■ The compressed response for a 45-length polyphase Barker code achieves a 1:45 peak sidelobe ratio.

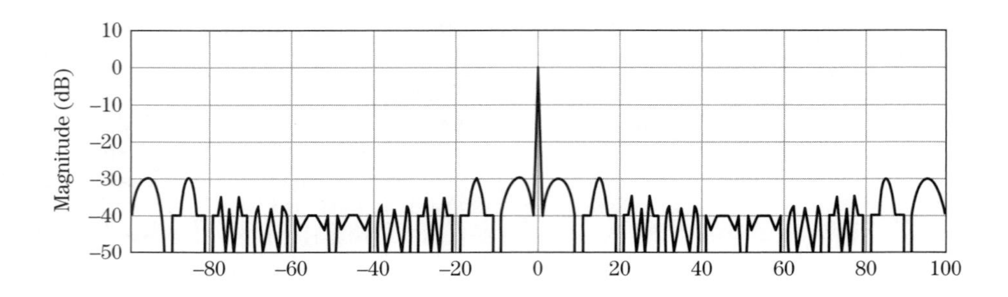

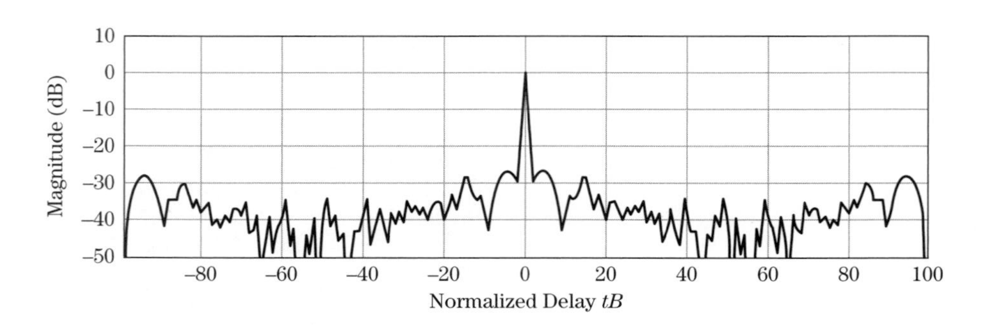

FIGURE 20-37 ■ The compressed response for a 100-length Frank code with no bandlimting (upper plot) and with bandlimiting (lower plot).

Doppler shifts. Figure 20-38 contains the compressed response associated with a 100-length Frank code with one cycle of Doppler across the uncompressed pulse. The mainlobe is displaced by one resolution cell, and the peak sidelobe has increased by approximately 11 dB.

Band-limiting causes the peak sidelobes of a Frank code to increase above their nominal level. Figure 20-39 contains a plot of the spectrum of a 100-length Frank code and the frequency response of a 20th-order finite impulse response filter. In this example, the filter's −6 dB width is equal to the waveform bandwidth ($B = 1/\tau_{chip}$). Figure 20-37 contains a plot of the autocorrelation with and without filtering. Careful examination reveals that the peak sidelobes have increased from −29.8 dB to −28.2 dB, an increase of 1.6 dB. Lewis and Kretschmer [51] describe the Frank code as exhibiting a slow phase

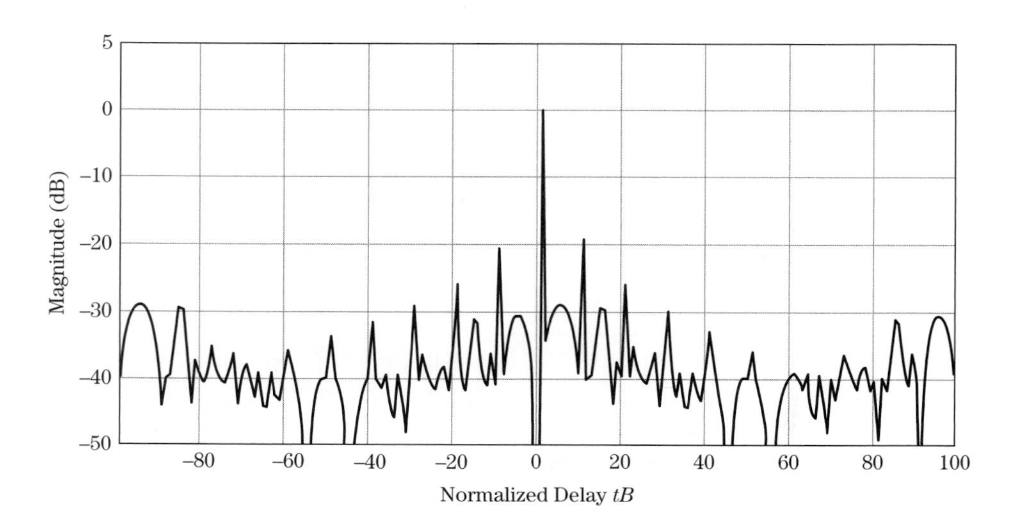

FIGURE 20-38 ■ The matched filter response for a 100-length Frank code with one cycle of Doppler over the uncompressed pulse.

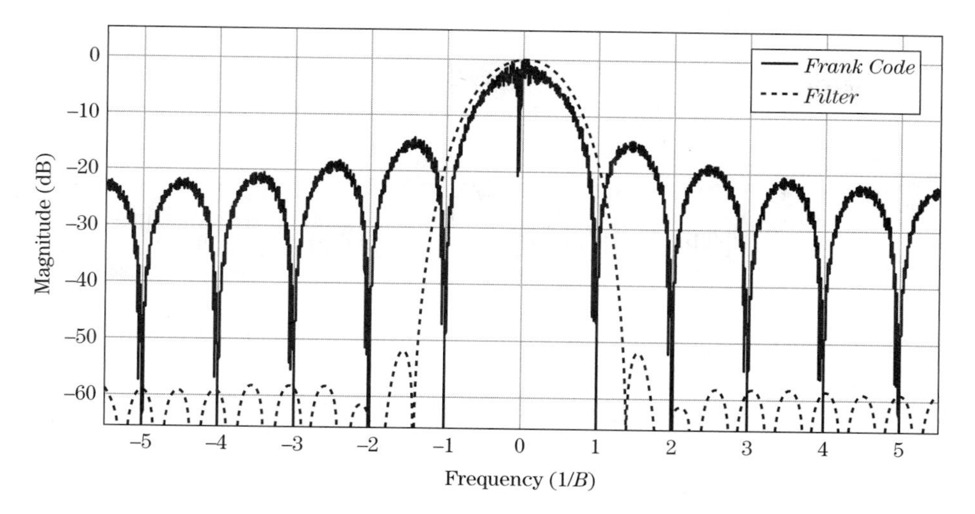

FIGURE 20-39 ■ Spectrum of a 100-length Frank code and a 20th order finite impulse response filter.

variation at the ends of the code and a faster phase variation near the middle of the code. The filter performs a smoothing operation [7] that integrates the slower varying phase terms producing the increased sidelobes near the end of the code. Band-limiting also broadens the mainlobe, which degrades resolution.

To address the impact of band-limiting, Lewis and Kretschmer [51] developed the P1 and P2 codes. The codes are designed to exhibit better sidelobe performance in the presence of band-limiting. The phase sequence defining a P1 code is

$$\phi_{n,k} = -\frac{\pi}{M}(M - 2k - 1)(Mk + n), \quad n = 0, \ldots, (M-1), k = 0, \ldots, (M-1) \tag{20.123}$$

and the phase sequence defining a P2 code is

$$\phi_{n,k} = \frac{\pi}{2M}(M - (2n + 1))(M - 2k), \quad n = 0, \ldots, (M-1), k = 0, \ldots, (M-1) \tag{20.124}$$

where the phase, $\phi_{n,k}$, is indexed over n for each value of k. Either even or odd values of M may be used for the P1 code. For the P2 code, M is required to be even to achieve low sidelobes. In contrast to the Frank code, the P1 and P2 codes exhibit a slow phase variation in the center of the code and a faster phase variation at the ends. The modification is intended to reduce the effects of band-limiting on the sidelobe structure. Figure 20-40 contains the autocorrelation response for a P1 code with and without filtering. The sidelobes associated with the filtered sequence in Figure 20-40 are lower than those depicted in Figure 20-37. This is particularly evident near the ends of the autocorrelation response.

The magnitude of the P1 code's autocorrelation response is identical to that of the Frank code. The peak sidelobes of the P2 match the Frank code, and the magnitude of the autocorrelation response approaches that of the Frank code as the time-bandwidth product is increased. For M odd, the ambiguity surface of a P1 code is identical to that of Frank code. For M even, the P1 and P2 ambiguity surfaces are similar to the ambiguity surface associated with a Frank code [51].

20.14.3 P3 and P4 Codes

The Doppler intolerance of the Frank, P1, and P2 codes led Lewis and Kretschmer [27] to develop the P3 and P4 codes, which exhibit a Doppler tolerance similar to that of an

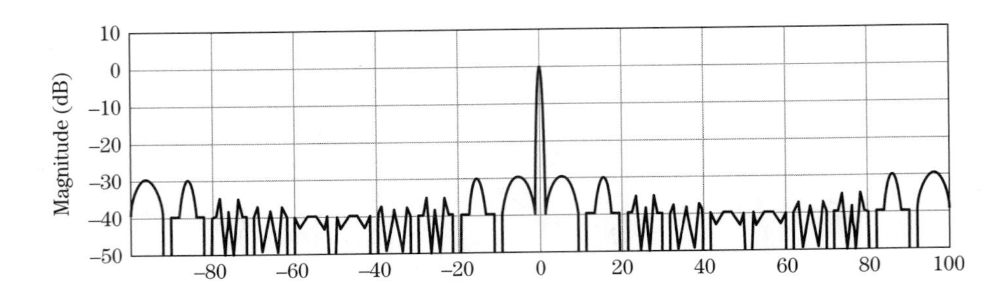

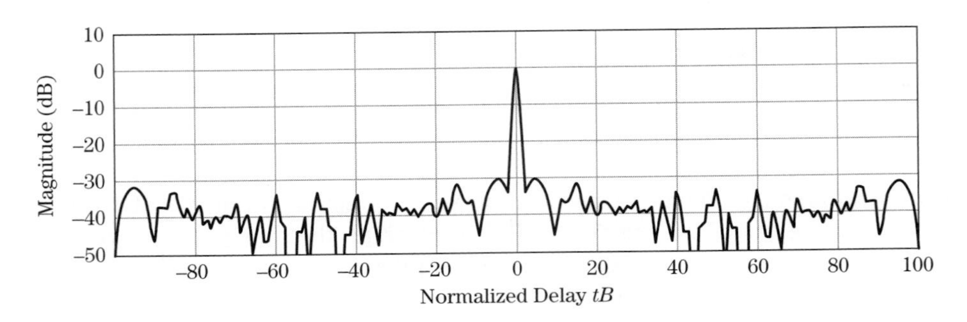

FIGURE 20-40 ■ The compressed response for a 100-length P1 code with no bandlimting (upper plot) and with bandlimiting (lower plot).

LFM waveform. The P3 and P4 codes are derived by sampling the quadratic phase of an LFM waveform. Consider an LFM waveform

$$x(t) = \exp\left(j\pi \frac{B}{\tau} t^2\right), \quad 0 \le t \le \tau \tag{20.125}$$

Sampling the phase in (20.125) at the chip interval τ_{chip} yields the phase sequence

$$\phi_n = \frac{\pi B}{\tau}(n\tau_{chip})^2 \tag{20.126}$$

For a rectangular chip, the chip width is the reciprocal of the bandwidth, $\tau_{chip} = 1/B$. Making the appropriate substitutions, the phase code in (20.126) reduces to

$$\phi_n = \frac{\pi}{N} n^2 \quad n = 0, \ldots, (N-1) \tag{20.127}$$

where N is the sequence length. The phase sequence in (20.127) defines a P3 code. The ambiguity surface for a P3 code is similar to that of an LFM waveform and does not exhibit the higher near-in sidelobes associated with Frank, P1, and P2 codes in the presence of small fractional Doppler shifts.

The P4 code is designed to be both Doppler tolerant and to perform well in the presence of precompression band-limiting. The P4 code's phase sequence is

$$\phi_n = \frac{\pi}{N} n^2 - \pi n \quad n = 0, \ldots, (N-1) \tag{20.128}$$

For large time-bandwidth products, the peak sidelobe ratio of the P3 and P4 codes [7] is approximately $10\log_{10}(2.5/\pi^2 N)$, which is about 4 dB higher than those associated with the same length Frank code. Figure 20-41 contains a plot of the match filtered response for a 100-length P4 code. Note that the peak sidelobes are approximately 4 dB higher than the Frank code response in Figure 20-37 (upper plot).

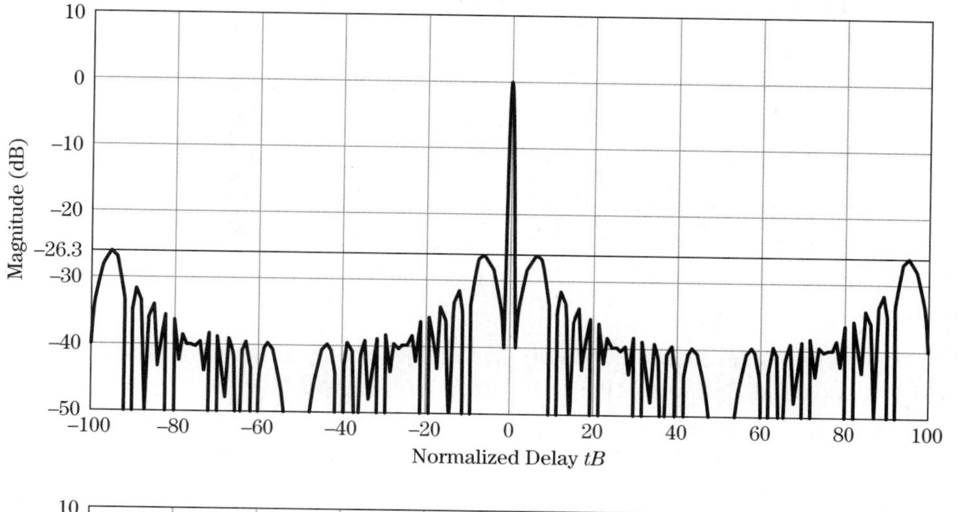

FIGURE 20-41 ■ The match filtered response for a 100-length P4 code.

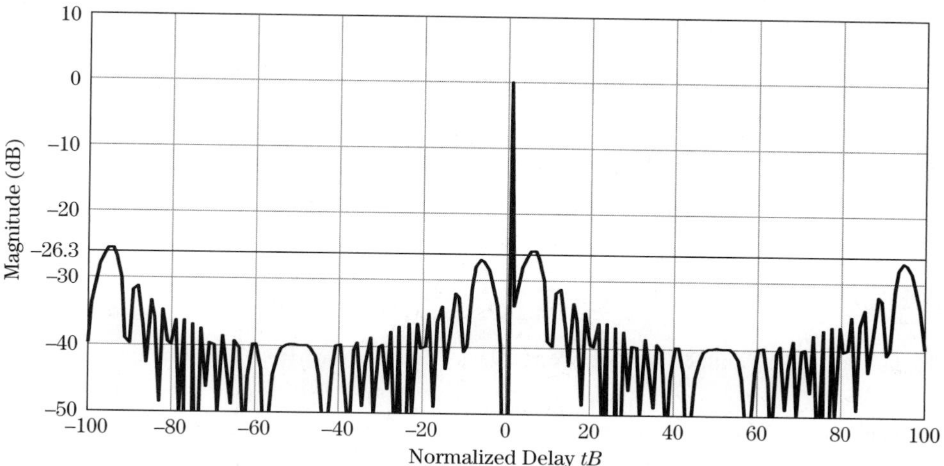

FIGURE 20-42 ■ The matched filter response for a 100-length P4 code with one cycle of Doppler over the uncompressed pulse length.

The Doppler tolerance of the P4 code relative to that of the Frank code is shown by comparing the responses in Figure 20-42 and Figure 20-38. Figure 20-42 contains the match filtered response for a P4 with a single cycle of Doppler across the uncompressed pulse. The effect of the Doppler shift is a displacement of the mainlobe by one resolution cell and a slight increase in the peak sidelobes. Figure 20-38 contains the match filtered response for a Frank of the same length with a single cycle of Doppler across the uncompressed pulse. The Frank code exhibits range-Doppler coupling similar to a P4 code, but the near-in sidelobes increase dramatically. The peak sidelobe has increased by ≈ 11 dB compared with ≈ 1 dB for the P4 code.

20.15 | PHASE-CODE SUMMARY

Phase codes are applied in many modern radar systems and decouple energy and resolution via chip-to-chip phase modulation. Range resolution is proportional to the chip width, and peak sidelobes are inversely related to the sequence length. Minimum peak sidelobe codes achieve the lowest sidelobe level for a given code length, but the longest known biphase MPS codes are on the order of length 105. Longer codes with predictable, but not necessarily minimum, sidelobe ratios are available including the maximal length sequences. Biphase coded waveforms are designed to constrain the uncompensated Doppler to one-quarter cycle over of the pulse. This constraint limits the SNR loss and preserves the sidelobe structure. Polyphase codes, in many cases, achieve a lower sidelobe level for a given sequence length and in some cases may be designed to achieve a Doppler tolerance similar to an LFM waveform.

20.16 FURTHER READING

Books principally devoted to the theory and application of radar waveforms are few in a number. Cook and Bernfeld [4] and Rihaczek [34] are two older texts still in print, and, more recently, Levanon and Mozeson [7] have written an excellent text on the subject. Many modern radar texts [2, 6, 8, 52, 53] also provide a thorough discussion of pulse compression waveforms.

A number of phase and frequency modulated waveforms and techniques are covered in the literature and are worthy of further investigation including nonlinear frequency modulation [54–60], stretch processing [15], stepped frequency [16,17], stepped chirp [18,61–63], Costas frequency arrays [2,7], mismatched filters [44–48], complementary codes [64], nested codes [7], P(n,k) codes [65], and quadriphase codes [28]. In addition, Sarwate and Pursely [26] provide bounds on autocorrelation and cross-correlation performance for periodic and aperiodic codes.

20.17 REFERENCES

[1] DiFranco, J.V., and Rubin, W.L., *Radar Detection*, Prentice-Hall, Inc., Englewood Cliffs, NJ, 1968.

[2] Peebles Jr., P.Z., *Radar Principles*, John Wiley & Sons, Inc., New York, 1998.

[3] Turin, G.L., "An Introduction to Matched Filters," *IRE Transactions on Information Theory*, pp. 311–319, June 1960.

[4] Cook, C.E., and Bernfeld, M., *Radar Signals, An Introduction to Theory and Application*, Artech House, Boston, 1993.

[5] Woodward, P.M., *Probability and Information Theory, with Applications to Radar*, Artech House, Dedham, MA, 1980.

[6] Richards, M.A., *Fundamentals of Radar Signal Processing*, McGraw-Hill, New York, 2005.

[7] Levanon, N., and Mozeson, E., *Radar Signals*, Wiley, New York, 2004.

[8] Levanon, N., *Radar Principles*, John Wiley & Sons, New York, 1988.

[9] Papoulis, A., *The Fourier Integral and Its Applications*, McGraw-Hill, New York, 1987.

[10] Carrara, W.G., Goodman, R.S., and Majewski, R.M., *Spotlight Synthetic Aperture Radar, Signal Processing Algorithm*, Artech House, Boston, 1995.

[11] Novak, L.M., Halversen, S.D., and Owirka, G.J., "Effects of Polarization and Resolution on SAR ATR," *IEEE Transactions on Aerospace and Electronics Systems*, vol. 33, pp. 102–116, January 1997.

[12] Klauder, J.R., Price, A.C., Darlington, S., and Albersheim, W.J., "The Theory and Design of Chirp Radars," *Bell System Technical Journal,* vol. 34, pp. 745–808, July 1960.

[13] Kretschmer Jr., F.F., "Pulse Compression Waveforms Using Huffman Codes," *Proceedings of the 1986 IEEE National Radar Conference,* pp. 59–64, 1986.

[14] Huffman, D.A., "The Generation of Impulse-Equivalent Pulse Trains," *IRE Transactions on Information Theory*, vol. IT-8, pp. S10–S16, September 1962.

[15] Caputi Jr., W.J., "Stretch: A Time-Transformation Technique," *IEEE Transactions on Aerospace and Electronics Systems*, vol. AES-7, no. 2, pp. 269–278, March 1971.

[16] Wehner, D.R., *High Resolution Radar*, 2d ed., Artech House, Boston, 1995.

[17] Scheer, J.A. and Kurtz, J.L., *Coherent Radar Performance Estimation*, Artech House, Boston, 1993.

[18] McGroary, F., and Lindell, K., "A Stepped Chirp Technique for Range Resolution Enhancement," *IEE National Telesystems Conference*, pp. 121–126, 1991.

[19] Cohen, M.N., Fox, M.R., and Baden, J.M., "Minimum Peak Sidelobe Pulse Compression Codes," *Proceedings of the IEEE International Radar Conference,* Arlington, VA, pp. 633–638, May 7–10, 1990.

[20] Kerdock, A.M., Mayer, R., and Bass, D., "Longest Binary Pulse Compression Codes with Given Peak Sidelobe Levels," *Proceedings of the IEEE,* vol. 74, no. 2, p. 366, 1986.

[21] Lindner, J., "Binary Sequences Up to Length 40 with Best Possible Autocorrelation Function," *Electronic Letters,* vol. 11, no. 21, p. 507, October 1975.

[22] Coxson, G., and Russo, J., "Efficient Exhaustive Search for Optimal-Peak-Sidelobe Binary Codes," *IEEE Transactions on Aerospace and Electronic Systems,* vol. 41, no. 1, pp. 302–308, January 2005.

[23] Nunn, C.J., and Coxson, G.E., "Best-Known Autocorrelation Peak Sidelobe Levels for Binary Codes of Length 71-105," *IEEE Transactions on Aerospace and Electronic Systems,* vol. 44, no. 1, pp. 392–395, January 2008.

[24] Coxson, G.E., Hirschel, A., Cohen, M.N., "New Results on Minimum-PSL Binary Codes," *Proceedings of the 2001 IEEE Radar Conference,* pp. 153–156, 2001.

[25] MacWilliams, F.J., and Sloane, N.J.A., "Pseudo-Random Sequences and Arrays," *Proceedings of the IEEE,* vol. 64, no. 12, pp. 1715–1729, December 1976.

[26] Sarwate, D.V., and Pursley, M.B., "Crosscorrelation Properties of Pseudorandom and Related Sequences," *Proceedings of the IEEE,* vol. 68, no. 5, pp. 593–619, 1980.

[27] Lewis, B.L., and Kretschmer Jr., F.F., "Linear Frequency Modulation Derived Polyphase Pulse Compression Codes," *IEEE Transactions on Aerospace and Electronic Systems,* vol. 18, no. 5, pp. 637–641, September 1982.

[28] Taylor, J., and Blinchikoff, H., "Quadriphase Code—A Radar Pulse Compression Signal with Unique Characteristics," *IEEE Transactions on Aerospace and Electronic Systems,* vol. 24, no. 2, pp. 156–170, March 1988.

[29] Keel, B.M., and Heath, T.H., "A Comprehensive Review of Quasi-Orthogonal Waveforms," *2007 IEEE Radar Conference,* pp. 122–127, April 2007.

[30] Farnett, E.C., and Stevens, G.H., "Pulse Compression Radar," in *Radar Handbook,* 2d ed., Ed. M. Skolnik, McGraw-Hill, New York, 1990.

[31] Ludeman, L.C., *Fundamentals for Digital Signal Processing*, Harper & Row, New York, 1986.

[32] Blankenship, P.E., and Hofstetter, E., "Digital Pulse Compression via Fast Convolution," *IEEE Transactions on Acoustics, Speech, and Signal Processing*, vol. ASSP-23, no. 2, pp. 189–201, April 1975.

[33] Morris, G., and Harkness, L., *Airborne Pulsed Doppler Radar*, 2d ed., Artech House, Boston, 1996.

[34] Rihaczek, A.W., *Principles of High-Resolution Radar*, Artech House, Boston, 1996.

[35] Friese, M., and Zotman, H., "Polyphase Barker Sequences up to Length 31," *Electronic Letters,* vol. 30, no. 23, pp. 1930–1931, November 1994.

[36] Borwein, P., and Ferguson, R., "Polyphase Sequences with Low Autocorrelation," *IEEE Transactions on Information Theory,* vol. 51, no. 4, pp. 1564–1567, April 2005.

[37] Brenner, A.R., "Polyphase Barker Sequences up to Length 45 with Small Alphabets," *Electronic Letters,* vol. 34, no. 16, pp. 1576–1577, August 1998.

[38] Friese, M., "Polyphase Barker Sequences up to Length 36," *IEEE Transactions on Information Theory*, vol. 42, no. 4, pp. 1248–1250, July 1996.

[39] Nunn, C.J., and Coxson, G.E., "Polyphase Pulse Compression Codes with Optimal Peak and Integrated Sidelobes," *IEEE Transactions on Aerospace and Electronic Systems,* vol. 45, no. 2, pp. 775–781, 2009.

[40] Gartz, K.J., "Generation of Uniform Amplitude Complex Code Sets with Low Correlation Sidelobes," *IEEE Transactions on Signal Processing,* vol. 40, no. 2, pp. 343–351, February 1992.

[41] Bomer, L., and Antweiler, M., "Polyphase Barker Sequences," *Electronics Letters,* vol. 25, no. 23, pp. 1577–1579, 1989.

[42] Frank, R.L., "Polyphase Codes with Good Nonperiodic Correlation Properties," *IEEE Transactions on Information Theory,* vol. 9, pp. 43–45, January 1963.

[43] Lewis, B.L., Kretschmer Jr., F.F., and Shelton, W.W., *Aspects of Radar Signal Processing*, Artech House, Norwood, MA, 1986.

[44] Baden, J.M., and Cohen, M.N., "Optimal Sidelobe Suppression for Biphase Codes," *1991 IEEE National Telesystems Conference,* vol. 1, pp. 127–131, 1991.

[45] Baden, J.M., and Cohen, M.N., "Optimal Peak Sidelobe Filters for Biphase Pulse Compression," *Record of the IEEE 1990 International Radar Conference,* pp. 249–252, May 1990.

[46] Griep, K.R., Ritcey, J.A., and Burlingame, J.J., "Polyphase Codes and Optimal Filters for Multiple User Ranging," *IEEE Transactions on Aerospace and Electronics Systems*, vol. 31, pp. 752–767, April 1995.

[47] Zoraster, S., "Minimum Peak Range Sidelobe Filters for Binary Phase-Coded Waveforms," *IEEE Transactions on Aerospace and Electronic Systems,* vol. AES-16, no. 1, pp. 112–115, January 1980.

[48] Griep, K.R., Ritcey, J.A., and Burlingame, J.J., "Poly-Phase Codes and Optimal Filters for Multiple User Ranging," *IEEE Transactions on Aerospace and Electronic Systems,* vol. 31, no. 2, pp. 752–767, April 1995.

[49] Peterson, W.W., and Weldon Jr., E.J., *Error-Correcting Codes*, MIT Press, Cambridge, MA, 1972.

[50] Mow, W.H., and Li, S.R., "Aperiodic Autocorrelation and Crosscorrelation of Polyphase Sequences," *IEEE Transactions on Information Theory,* vol. 43, no. 3, pp. 1000–1007, May 1997.

[51] Lewis, B.L., and Kretschmer Jr., F.F., "A New Class of Polyphase Pulse Compression Codes and Techniques," *IEEE Transactions on Aerospace and Electronic Systems,* vol. 17, no. 3, pp. 364–372, May 1981.

[52] Nathanson, F.E., *Radar Design Principles*, 2d ed., McGraw-Hill, New York, 1991.

[53] Skolnik, M.I., *Introduction to Radar Systems*, 3d ed., McGraw Hill, New York, 2001.

[54] Johnston, J.A., and Fairhead, A.C., "Waveform Design and Doppler Sensitivity Analysis for Nonlinear FM Chirp Pulses," *IEE Proceedings–F,* vol. 133, no. 2, pp. 163–175, April 1986.

[55] Newton, C.O., "Nonlinear Chirp Radar Signal Waveforms for Surface Acoustic Wave Pulse Compression Filters," *Wave Electronics*, pp. 387–401, 1974.

[56] Johnston, J.A., "Analysis of the Pulse Compression of Doppler Shifted Nonlinear Frequency Modulated Signals," *Electronic Letters,* vol. 20, pp. 1054–1055, December 1984.

[57] Fowle, E.N., "The Design of FM Pulse Compression Signals," *IEEE Transactions on Information Theory,* pp. 61–67, January 1964.

[58] Brandon, M.A., "The Design of a Non-linear Pulse Compression System to Give a Low Loss High Resolution Radar Performance," *Marconi Review,* Vol. 36, no. 188, pp. 1–45, 1973.

[59] Brandon, M.A., "The Spectra of Linear and Non-linear F.M. Used in Pulse Compression, and the Effects on the Resultant Compressed Pulse," *Marconi Review,* Vol. 36, no. 188, pp. 69–92, 1973.

[60] Key, E.L., Fowle, E.N., Haggarty, R.D., "A Method of Designing Signals of Large Time-Bandwidth Product," *IRE International Convention Record,* vol. 9, pt. 4, pp. 146–154, 1961.

[61] Wilkinson, A.J., Lord, R.T., and Inggs, M.R., "Stepped-Frequency Processing by Reconstruction of Target Reflectivity Spectrum," *Proceedings of the 1998 South African Symposium on Communications and Signal Processing,* pp. 101–104, 1998.

[62] Lord, R.T., and Inggs, M.R., "High Range Resolution Radar using Narrowband Linear Chirps Offset in Frequency," *Proceedings of the 1997 South African Symposium on Communications and Signal Processing*, pp. 9–12, September 1997.

[63] Lord, R.T., and Inggs, M.R., "High Resolution SAR Processing Using Stepped-Frequencies," *1997 IEEE International Geoscience and Remote Sensing Conference,* pp. 490–492, August 1997.

[64] Golay, M., "Complementary Series," *IRE Transactions on Information Theory,* pp. 82–87, April 1960.

[65] Felhauer, T., "Design and Analysis of New P(n,k) Polyphase Pulse Compression Codes," *IEEE Transactions on Aerospace and Electronic Systems*, vol. 30, no. 3, pp. 865–874, July 1994.

20.18 PROBLEMS

1. Compute the Rayleigh range resolution associated with a simple, rectangular pulse with $\tau = 2$ μsec. Assume the speed of light $c = 3 \times 10^8$ m/sec and the amplitude of the pulse is unity. Sketch the ideal matched filtered response and label the time delay axis at the points $\{-\tau, -\tau/2, 0, \tau/2, \tau\}$ and the amplitude axis at the points $\{0, 0.5, 1\}$. Note that the filtered response represents a voltage. What dB value does the 0.5 amplitude value represent? What is the range resolution associated with the -3 dB width of the response? How does it compare with the Rayleigh resolution?

2. Given a waveform $x(t)$ and its Fourier transform $X(\omega)$, the waveform's matched filter $h(t)$ is defined as $h(t) = x^*(-t)$. Using the Fourier transform, derive the spectrum of the matched filter $H(\omega)$ in terms of $X(\omega)$.

3. Derive the matched filter for a waveform $x(t)$ embedded in white noise with power spectral density N_0. Show all of your work.

4. Compute the Fourier transform for a simple pulse defined by $x(t) = 1, -\tau/2 \le t \le \tau/2$. Show all of your work, and put the result in terms of a sinc response.
5. Given an LFM waveform with $B = 250$ MHz and $\tau = 2\ \mu$sec, perform the following:
    a. Compute the waveform time-bandwidth product.
    b. In a baseband coherent detector, compute the minimum ADC rate required to support the waveform.
    c. Given a 1 km range window, compute the minimum collection time to avoid eclipsing returns within the window.
    d. Compute the number of complex samples obtained over the collection time in (c) given the sampling rate in (b).
    e. Compute the pulse compression integration gain in dB. If the SNR prior to the matched filter is –3 dB, compute the SNR at the output of the matched filter.
6. Derive the matched filter response for an LFM waveform defined by $x(t) = \exp(j\pi(B/\tau)t^2)$ $-\tau/2 \le t \le \tau/2$. Show all of your work.
7. Derive the ambiguity function for the LFM waveform defined in problem 6.
8. Compute the Doppler shift required to displace an LFM with $B = 50$ MHz and $\tau = 1$ msec by 3 Rayleigh range resolution cells. At 10 GHz, compute the radial velocity associated with the Doppler shift.
9. Compute the integrated sidelobe ratios for the Barker codes in Table 20-4.
10. For the biphase code $\{1\ 1\ -1\ 1\}$, apply the four equivalence operations defined in Section 20.12.5 to the code. Assume $\rho = \exp(j\pi/4)$. In addition, apply the conjugate equivalence operation (third property) to the code generated by the fourth property. Show that each code possesses the same autocorrelation magnitude response.
11. Using the equivalence properties defined in Section 20.12.5, prove that any polyphase code may be written with the first two phase values equal to zero radians.
12. Given a 127 length MLS sequence and a pulse width equal to 2.54 μsec, compute the chip width, the waveform bandwidth, the Rayleigh range resolution, waveform time-bandwidth product, and the pulse compression gain (in dB). Compute the approximate peak sidelobe ratio associated with the phase-coded waveform. Compute the minimum Doppler shift required to annihilate the peak of the match filtered response.
13. Draw the shift register with feedback required to implement a 511 maximal length sequence. Use the polynomial coefficients defined in Table 20-7. Initialize the shift register, avoiding the all-zeros condition. Select an output register, and compute the first four values generated by the shift register. Convert this sequence of 0's and 1's to a sequence of 1's and -1's.
14. Compute the peak sidelobe ratio associated with the following waveforms: 13 chip biphase Barker code, 69 chip poly-phase Barker code, 144 and 169 length P1 codes, a 1,023 length MLS sequence, and a 102 length MPS code.
15. Design a phase-coded waveform for a pulsed radar to meet all of the following requirements:
    a. Rayleigh range resolution $= 0.3$ meters.
    b. Pulse compression gain > 15 dB.
    c. Peak sidelobe ≤ -20 dB.
    d. Maximum allowable blind range within the first range ambiguity $= 50$ meters.
    e. Biphase-coded waveform.

CHAPTER 21

An Overview of Radar Imaging

Gregory A. Showman

Chapter Outline

21.1 Introduction 835
21.2 General Imaging Considerations 837
21.3 Resolution Relationships and Sampling Requirements 843
21.4 Data Collection 852
21.5 Image Formation 856
21.6 Image Phenomenology 875
21.7 Summary 888
21.8 Further Reading 888
21.9 References 889
21.10 Problems 890

21.1 INTRODUCTION

Synthetic aperture radar (SAR) is a combination of radar hardware, waveforms, signal processing, and relative motion that creates photograph-like renderings of stationary targets and scenes of interest. The principal product of any basic SAR implementation is a fine-resolution two-dimensional intensity image of the illuminated scene. SAR is widely employed by the remote sensing community for mapping and land-use surveying and by the military for detection, location, identification, and assessment of fixed targets.

To motivate this discussion, Figure 21-1 shows two pictures of a portion of the Albuquerque Airport: one a traditional optical photograph (Figure 21-1b) and the other a SAR image (Figure 21-1a). A cursory comparison of these images suggests the phenomenologies in optical and radar imaging are similar; indeed, SAR can be thought of as a kind of "radar photography." However, a closer examination reveals significant differences.

For example, the optical photograph, recorded in daytime, is dominated by returns with significant reflectivity in the visible spectrum, so that white concrete tarmac appears bright while road and runway blacktop render dark. The SAR image, in contrast, highlights objects that reflect radar energy back to the radar. Smooth, flat, horizontal surfaces like roads and runways tend to reflect energy away from the radar, so these objects are dark in the SAR imagery. Buildings and vehicles are often constructed with metal components arranged in strongly retro-reflective geometries, causing these man-made objects to produce high intensities in the SAR image, thereby increasing their visibility in the scene compared with optical photographs.

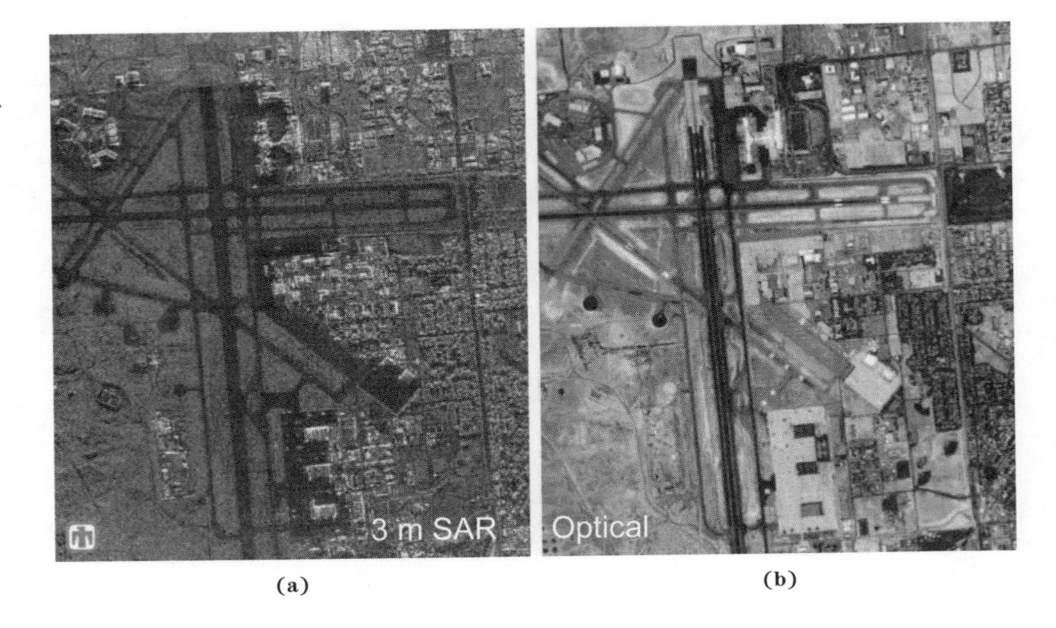

FIGURE 21-1 ■ Two renderings of Albuquerque Airport. (a) Synthetic Aperture Radar image, and (b) optical overhead photograph. (Images courtesy of Sandia National Laboratories.)

As the reader is likely familiar with the process and product of optical photography, the following is a list of some of the advantages of SAR over this more traditional imaging method:

- SAR, by definition, is built around radar returns, which allows quality imagery to be generated at night and in inclement weather, conditions that thoroughly thwart optical imaging.
- SAR emphasizes man-made returns, an important consideration for many military and remote sensing applications.
- It will be seen that SAR systems have the potential for very fine resolutions, down to a few inches. Moreover, these resolutions can be maintained at long ranges without the need to dramatically expand or augment existing radar hardware.
- Radar operation at low center frequencies permits penetration of cover and concealment. For example, *foliage penetration* (FOPEN) SAR systems can image vehicles hidden beneath treetop canopies. *Ground penetration* (GPEN) SAR is capable of revealing the presence of objects buried underground.
- SAR requires coherent operation; that is, the radar system must be capable of measuring both amplitude and phase and must use waveforms with well-controlled phase characteristics. One consequence of coherent operation and processing is that each pixel in a SAR image is complex, possessing both amplitude and phase information. Specialized modes have been developed that exploit the phase information to estimate terrain and target heights or to detect subtle changes that have occurred in the imaged scene over time.
- Most radar systems transmit and receive radiofrequency (RF) energy at a single polarization. Some SAR systems are capable of transmitting and receiving multiple polarizations, resulting in multipolarimetric SAR imagery. The additional information provided by multiple polarizations has proven useful for land and foliage type classification and ground vehicle identification.

All of these benefits come at some price, of course. While not dependent on ambient light, the radar must provide the hardware to illuminate the scene of interest. Optical systems can snap a picture in a fraction of a second, whereas fine-resolution SAR systems can require several seconds, or even minutes, to collect enough data to form an image. Precise timing and phase control must be built into the radar hardware to ensure coherent operation over dwells that may last tens of seconds. Modern SAR image formation places challenging demands on the speeds and memories of digital signal processing hardware, while the associated software or firmware must implement, efficiently and with tightly managed approximations, sophisticated filters and transformations.

The goal of this chapter is to familiarize the reader with the operation and implementation of radar imaging in general and synthetic imaging in particular. Topics related to coherent imaging methods such as SAR include resolution relationships, sampling requirements, data collection considerations, and approaches to image formation. A complete mathematical development of the SAR system impulse response is used to derive both a simple but crude Doppler-based imaging method as well as a very general, high-fidelity, matched filtering technique. Next, phenomenologies and artifacts particular to radar images are discussed. Finally, several elementary SAR image quality metrics are introduced.

One last clarification on the scope of this chapter is provided here. Recall the fundamental product of any SAR data acquisition and processing operation is a two-dimensional image. Each dimension entails a distinct method of measurement and coherent integration. The first of these two dimensions, termed *down-range*, is plumbed via wideband waveforms and focused using the pulse compression and matched filtering techniques presented in detail in Chapter 20. Conversely, the emphasis of this treatment is on the orthogonal, *cross-range* dimension. It will be shown that cross-range information for a scene of interest is gathered through the motion of the SAR platform, while focusing of cross-range returns is achieved by applying measurements to a suitable image formation algorithm.

21.2 | GENERAL IMAGING CONSIDERATIONS

SAR is typically presented using one of three paradigms:

1. SAR as a large synthetic antenna aperture.
2. SAR as range-Doppler imaging.
3. SAR as a signal processing exercise.

In this chapter, the first two paradigms are generally eschewed in favor of the signal processing view of SAR. However, for completeness, in this section all three approaches are reviewed.

21.2.1 SAR as a Large Synthetic Antenna Aperture

The concept of SAR as a means of forming a large synthetic antenna aperture is often found in traditional radar texts [e.g., 1,2]. This development begins by reviewing the noncoherent radar mapping mode known as *real-beam ground mapping* (RBGM). In RBGM a two-dimensional intensity map is made of a ground scene by first generating a one-dimensional

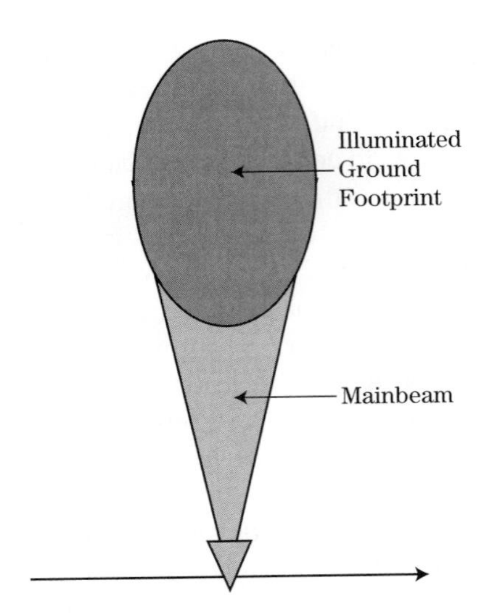

FIGURE 21-2 ■ Overhead view illustrates radar platform moving from left to right with side-looking antenna having main beam and ground footprint pointed towards the top.

range profile on a single pulse. Narrow range bins can be achieved by using short pulses or any of the wideband pulse compression techniques presented in Chapter 20. The return in any given bin is the coherent sum of the returns from all scatterers across the mainbeam extent at that particular range.

A two-dimensional map is produced by generating range profiles over a series of pulses and then placing the profiles next to one another in the order they were collected. The radar is operated so that the antenna illuminates a slightly different portion of the scene from pulse to pulse. For example, a stationary radar with a rotating antenna has a beam that sweeps through angle over time, and RBGM produces a two-dimensional intensity image as a function of slant range and azimuth angle. This polar-formatted data is often rendered using a plan position indicator (PPI) display.

When the goal is to survey large areas, a radar can be mounted on a vehicle with the antenna pointed to one side and oriented normal to the direction of travel, as seen in Figure 21-2. When the radar platform is an aircraft this arrangement is known as *side-looking airborne radar* (SLAR). In SLAR the antenna is not rotated; rather, the motion of the vehicle ensures a different portion of the earth's surface is illuminated from pulse to pulse. Scatterers in the scene drift into and out of the mainbeam as the aircraft moves along. The RBGM mode results in a map whose axes are the cross-track range to the scene and the along-track position of the aircraft. Cross-track range corresponds to the slant range to a bin in the range profiles and in radar imaging is sometimes known as "down-range." Displacement in the orthogonal direction, azimuthal for the stationary rotating radar, and along-track for the SLAR, is known generally as "cross-range."

The SLAR antenna beam pattern acts as a kind of low-pass filter, blurring scatterer returns in the cross-range direction of the image. The azimuth beamwidth of the antenna, then, sets a limit on the achievable image resolution. Note that the azimuthal extent of the beam on the ground is proportional to range, so that cross-range resolution gets worse as down-range increases. For a range R and a 3 dB azimuth beamwidth θ_3 (in radians), the cross-range resolution ΔCR is approximately

$$\Delta CR \approx R\theta_3 \tag{21.1}$$

Narrower beamwidths mean less blurring and finer resolutions. However, as discussed in Chapter 9, for a fixed operating wavelength, narrow beamwidths require large antennas. For an antenna length D and an RF wavelength λ, an approximate expression for the 3 dB beamwidth is

$$\theta_3 \approx \frac{\lambda}{D} \tag{21.2}$$

Combining (21.1) and (21.2) gives

$$\Delta CR \approx R\frac{\lambda}{D} \tag{21.3}$$

Equation (21.3) states that fine cross-range resolutions at long ranges require large antennas. Even moderate resolutions and ranges can quickly lead to antennas much too long to practically mount on a vehicle.

SAR can be thought of as an alternative solution to the "large antenna" problem. Instead of flying with a very large antenna, SAR uses a small antenna with a wide beam to collect radar returns from the scene as the aircraft flies along a straight line distance D_{SAR}, the SAR baseline. Once the measurements are recorded they are coherently combined to realize a synthetic aperture radar beamwidth θ_{SAR} consistent with the collection distance

$$\theta_{SAR} \approx \frac{\lambda}{2D_{SAR}} \tag{21.4}$$

Comparing equations (21.2) and (21.4), it can be seen that D_{SAR} is the length of a synthetic antenna aperture. (The factor of two appears in equation (21.4) because the transmit location moves with the platform, whereas the entire aperture is used on transmit in a real antenna. In essence, equation (21.4) is the beamwidth for a two-way pattern.) Substituting for the azimuth beamwidth in equation (21.1) with the SAR beamwidth in (21.4) yields the SAR version of (21.3),

$$\Delta CR \approx R\frac{\lambda}{2D_{SAR}} \tag{21.5}$$

By collecting data over longer distances, narrower SAR beamdwidths can be realized and fine cross-range resolutions achieved at long ranges.

SAR typically employs pulsed waveforms, so that the data recorded over D_{SAR} is collected at discrete, uniformly spaced locations determined by the platform velocity and radar pulse repetition frequency (PRF). This discretization along the synthetic aperture means SAR image formation can be roughly understood by resorting to the beam-steering developments for phased arrays or, more appropriately, digital arrays.

The large-aperture model becomes awkward when imaging at fine resolution or low frequencies. Equation (21.5) states that fine cross-range resolutions and long wavelengths require large synthetic arrays. When D_{SAR} is comparable to the range R, the scene of interest is in the near-field with respect to the synthetic array (Chapter 9), giving rise to spherical wavefronts at the array and similarly challenging artifacts.

21.2.2 SAR as Range-Doppler Imaging

SAR is sometimes described in the context of range-Doppler imaging, often in manuscripts specializing in pulse-Doppler processing [e.g., 3]. As discussed in Chapter 17, the Doppler

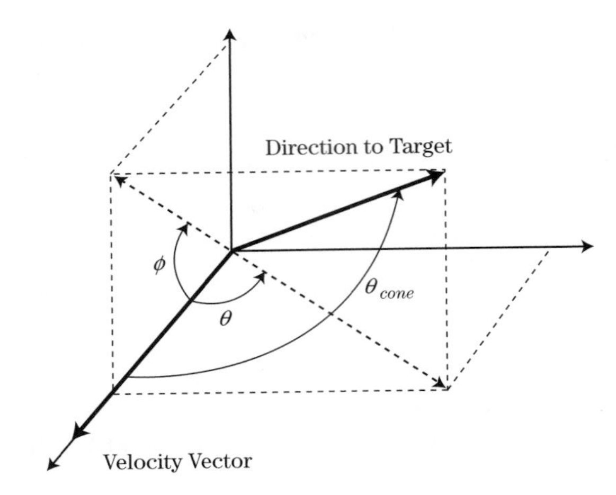

FIGURE 21-3 ■ Coordinate system showing azimuth, elevation, and cone angles, and vectors for platform velocity and direction to target.

frequency shift f_d generated by a stationary target at a radar moving at a speed v is

$$f_d = \frac{2v}{\lambda} \cos \theta_{cone} \tag{21.6}$$

where θ_{cone} is the *cone angle*, the angle between the radar velocity vector and the line-of-sight (LOS) vector from the radar to the target. The cone angle can be decomposed into azimuth θ and elevation ϕ angles defined with respect to the velocity vector

$$\cos \theta_{cone} = \cos \theta \cos \phi \tag{21.7}$$

Figure 21-3 contains a diagram of these angles.

Referencing azimuth and elevation to the velocity vector is consistent with nose-mounted forward-looking radar systems. As discussed in Section 21.2.1, imaging radars are often side-mounted and view a scene perpendicular to the velocity vector. Under these circumstances, azimuth is defined with respect to the side-looking direction, introducing a 90° shift in the azimuth coordinate system and thus causing the azimuth cosine in equation (21.7) to become a sine:

$$\cos \theta_{cone} = \sin \theta \cos \phi \tag{21.8}$$

Substituting (21.8) into (21.6) yields

$$f_d = \frac{2v}{\lambda} \sin \theta \cos \phi \tag{21.9}$$

Assuming a shallow elevation angle, (21.9) can be simplified to

$$f_d \approx \frac{2v}{\lambda} \sin \theta \tag{21.10}$$

Taking the partial derivative of (21.10) with respect to azimuth angle yields

$$\frac{\partial f_d}{\partial \theta} \approx \frac{2v}{\lambda} \cos \theta \tag{21.11}$$

and approximating ∂f_d and $\partial \theta$ with finite Doppler and azimuth extents Δf_d and $\Delta\theta$, respectively, gives

$$\Delta f_d \approx \left(\frac{2v}{\lambda}\cos\theta\right)\Delta\theta \tag{21.12}$$

Equation (21.12) relates azimuth extent to Doppler extent. In range-Doppler imaging the Doppler extent is set by Doppler filtering; specifically, the filter bandwidth B_d can be substituted for the Doppler width Δf_d in (21.12). Finally, equation (21.1) showed that an azimuth extent is related to a cross-range extent by the slant range. Making these substitutions into (21.12) gives

$$B_d \approx \left(\frac{2v}{\lambda}\cos\theta\right)\frac{\Delta CR}{R} \tag{21.13}$$

and rearranging the terms in (21.13) produces

$$\Delta CR \approx \frac{R\lambda}{2v\cos\theta}B_d \tag{21.14}$$

Equation (21.14) indicates that narrower Doppler filter bandwidths provide finer cross-range resolution. A simple approach to SAR imaging, then, is to apply the pulse history in each range bin to a Doppler filter bank, generated by a fast Fourier transform (FFT) for example, and to assign the output in each filter to a cross-range location in a manner consistent with equation (21.10).

As discussed in Chapter 17, the Doppler filter bandwidth B_d is inversely proportional to the data collection time, or *dwell time*, T_d. Applying this relationship to equation (21.14) yields

$$\Delta CR \approx \frac{R\lambda}{2vT_d\cos\theta} \tag{21.15}$$

The product of the platform velocity v and dwell time T_d is the data collection distance, equivalent to the SAR baseline D_{SAR} discussed in Section 21.2. Using $D_{SAR} = vT_d$, equation (21.15) becomes

$$\Delta CR \approx \frac{R\lambda}{2D_{SAR}\cos\theta} \tag{21.16}$$

Note that equation (21.16), developed using the Doppler framework, is equivalent to equation (21.5), which was formulated with the large antenna paradigm, when imaging broadside to the velocity vector ($\theta = 0$).

Like the large-aperture model, the Doppler model becomes cumbersome at low frequencies and fine resolutions. From equations (21.14), small cross-range resolutions at long wavelengths require narrow Doppler bandwidths, which in turn require long dwell times. Over long dwells platform motion causes the apparent scatterer locations to drift, so that returns move through multiple range and Doppler resolution cells. Severe migration over range and Doppler produces a badly blurred range-Doppler rendering of the scene. Consequently, range-Doppler imaging is appropriate for coarse resolutions at high carrier frequencies only.

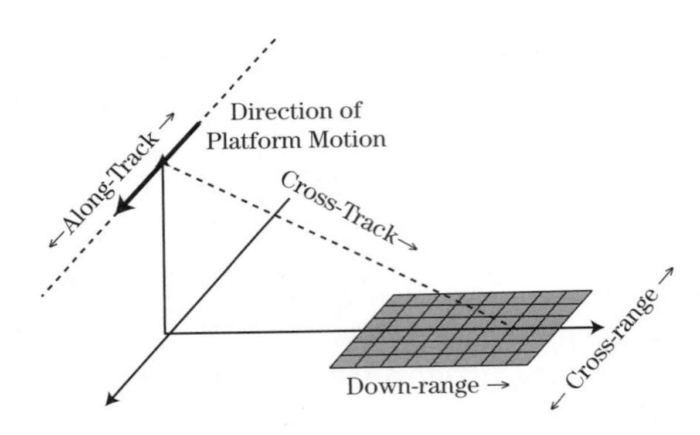

FIGURE 21-4 ■ Location in a SAR image is usually defined as a function of along-track and cross-track coordinates, or down-range and cross-range coordinates. For side-looking collections these coordinate systems are nearly equivalent.

21.2.3 SAR as a Signal Processing Exercise

Modern signal processing provides a very general and versatile framework for understanding SAR concepts and operation [4–6]. Indeed, the SAR discipline has benefited immensely from algorithms developed in related synthetic imaging disciplines. For example, two widely employed SAR image formation techniques were derived from principles first developed in the medical [7] and seismic imaging fields [8]. Moreover, signal processing accommodates SAR models and image formation methods developed using "inverse problem" approaches and constrained optimization algorithms [9].

From the signal processing point of view, SAR is just one example of a family of procedures that exploit sound, x-rays, positron emission, and other propagating energy sources for synthetic imaging. In general, image synthesis is a two-step procedure moving among three domains:

1. The object space, defined by the properties of the scene of interest.
2. The data space, comprising the raw radar measurements.
3. The image space, the final synthetic image.

In this development, all three spaces will be restricted to just two dimensions for simplicity. Hence, the SAR object space is the complex reflectivity of the scene versus down-range and cross-range location, the data are the complex returns recorded as a function of platform along-track location and "fast time" (range-delay time), and the final image is made up of complex pixels arrayed over down-range and cross-range.

The geometry shown in Figure 21-4 highlights the along-track, down-range, and cross-range dimensions. Note the additional term "cross-track," which is defined as range perpendicular to the platform direction of travel. For clarification, data "fast time" refers to the time delay into the measured record on each transmitted pulse and is equal to slant range divided by one-half the propagation speed. Analogously, the term "slow time" is sometimes used to refer to time over multiple pulses.[1]

[1]Of course, there is only one universal "time" for SAR data acquisition and processing. "Fast time" is reserved for rapid range-varying responses within the current time between transmitted pulses, so the start time of the transmission of each pulse is usually set to zero. "Slow time" is meant to describe changes over the dwell time, over many pulses, with a slow time of zero referenced to either the start or midpoint of the collection.

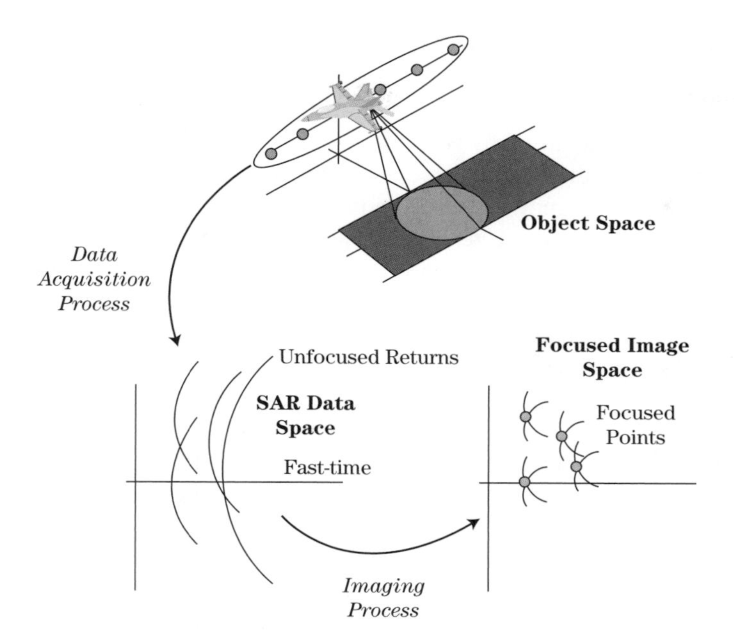

FIGURE 21-5 ■ The SAR process involves three spaces and two transformations. Data acquisition provides the transformation from scene reflectivity to raw data, while image formation operates on the raw data to yield the final SAR image.

From the point of view of the signal processing paradigm, the two steps of SAR imaging are as follows:

1. **Data acquisition:** Measurement of the response of the scene of interest over a range of frequencies and angles. The resolution obtainable in the final image depends on the frequency and angle parameters. This step is the transformation of the scene from object space to data space.
2. **Image formation:** Matched filtering of the measured data to the predicted responses from the scene of interest. This step is the transformation from the measured data to the final synthetic image. Image formation can be accomplished with any of a large number of algorithms, all of them exact implementations of, or approximations to, filtering matched to point target responses.

Figure 21-5 depicts these domains and transformations for SAR.

21.3 | RESOLUTION RELATIONSHIPS AND SAMPLING REQUIREMENTS

While many metrics are available to describe SAR image quality, the most important figure of merit in many applications is "resolution." *Resolution*, specifically spatial resolution, has units of distance (e.g., meters, inches) and is the limit on the ability to resolve, or to distinguish between, two closely spaced point scatterers. In essence, fine ("high") resolution means scatterers placed in proximity will appear as distinct returns in the final image. As SAR images are two-dimensional, a resolution value applies to each dimension.

In SAR imagery having coarse resolution, say tens or hundreds of meters, first-order terrain types (e.g., water, forest, urban) may be discerned by measuring pixel intensity and image texture. At finer resolutions, on the order of several meters, stands of trees, individual vehicles and buildings, and cultural features like roads and railways can be perceived. At 1 meter of resolution, individual trees are visible, and the size, shape, and orientation of buildings and vehicles are apparent. At a small fraction of a meter, SAR images render much like photographs; small objects like signs and lampposts can be detected and vehicles identified by type.

The wealth of information available in a submeter resolution image makes fine resolutions highly attractive, ignoring for the moment demands on hardware, processing, and data storage. But how are these resolutions achieved? Fundamentally, they require the scene of interest be measured over a wide range of angles and frequencies. To justify this assertion, consider the following thought experiment.

Imagine a stationary radar recording the return from a scene using a single-frequency waveform, that is, a pure continuous wave (CW) tone. All of the scatterers in the scene will interfere with one another and produce a single complex return at the radar output, one in-phase (I) and quadrature (Q) (or, equivalently, one amplitude and phase) value. The "resolution" of this system is very poor, equal to the size of the illuminated scene.

If the radar changes frequency, the scene scatterers will constructively and destructively interfere in a different way, yielding a different I and Q value. By measuring the scene over several discrete frequencies, a number of different complex measurements are produced. Chapter 14 discusses the Fourier relationship between frequency and time, which allows measurements collected over a frequency bandwidth to be used to resolve returns into range bins. In summary, measurements in frequency provide down-range resolution.

Similarly, if the radar is moved to a new location, the differential ranges from the radar to all the scatterers change, causing their returns to constructively and destructively interfere in an altered way, thus producing a modified I and Q value. Differential range changes most rapidly when the radar is moved not toward or away from the illuminated scene but rather over angle across the LOS to the scene. Movement over angle generates variations in the complex measurements even when the CW frequency is fixed. Chapter 17 notes the Fourier relationship between angular motion and azimuth location, which allows measurements over angle to be used to resolve and bin returns into cross-range. In summary, measurements over angle provide cross-range resolution.

21.3.1 Resolution Relationships

The range resolution Δ of any imaging system is equal to the ratio of the propagation speed c of the imaging energy through the supporting medium and the bandwidth B of the measurements

$$\Delta = \frac{c}{B} \tag{21.17}$$

where resolution is in units of distance. While there exist many formulae for resolution, (21.17) is consistent with the classical Rayleigh criterion for resolution, which states that two scatterers are resolved when the peak response from one falls at or outside the first null in the response from the other.

Imaging methods that employ two-way propagation, such as radar, realize a factor-of-two gain in resolution:

$$\Delta = \frac{c}{2B} \tag{21.18}$$

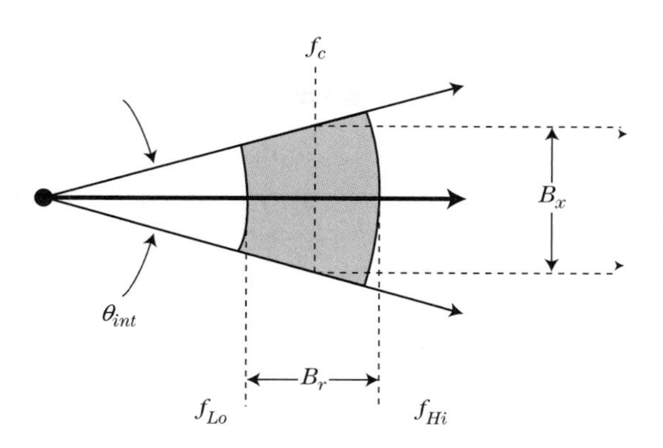

FIGURE 21-6 ■ Data support for a SAR collection is two dimensional and is a function of the waveform frequencies and integration angle. These quantities may be depicted in a polar coordinate system.

In SAR, the down-range resolution Δ_{DR} is given by equation (21.18) with RF bandwidth B_r provided by the frequency support of the radar waveform

$$\Delta R = \frac{c}{2B_r} \tag{21.19}$$

The cross-range resolution expression for SAR can be derived from (21.18) by referring to Figure 21-6, which depicts the angle-frequency support of measurements made with respect to the scene of interest. This information appears in a polar plot with waveform frequency and bandwidth along the radials and measurement azimuth over angle. Measurements are collected over an RF bandwidth B_r centered about a carrier frequency f_c, yielding low and high frequencies f_{Lo} and f_{Hi}. Measurements are also collected over a so-called *integration angle* θ_{int} defined by the upper and lower radials. Integration angle is the angular extent over which the SAR system collects returns from the scene of interest, as defined by the bearing angle from scene center to the radar.

Define a "cross-range bandwidth" B_x, perpendicular to the RF bandwidth B_r, as shown in the figure. This bandwidth is, for small integration angles, equal to the product of the carrier frequency f_c and integration angle θ_{int}

$$B_x \approx f_c\theta_{int} \tag{21.20}$$

Substituting for bandwidth in the resolution expression (21.18) with the cross-range bandwidth gives a formula for cross-range resolution ΔCR

$$\Delta CR = \frac{c}{2B_x} = \frac{c}{2f_c\theta_{int}} \tag{21.21}$$

Rewriting (21.21) in terms of the wavelength of the RF carrier λ_c gives

$$\Delta CR = \frac{\lambda_c}{2\theta_{int}} \tag{21.22}$$

Equation (21.22) is consistent with the expressions for cross-range resolution derived under the antenna (21.5) and Doppler (21.16) SAR paradigms, as integration angle is approximately equal to the ratio of the SAR aperture length D_{SAR} and the range to the center of the scene R:

$$\theta_{int} \approx \frac{D_{SAR}}{R} \tag{21.23}$$

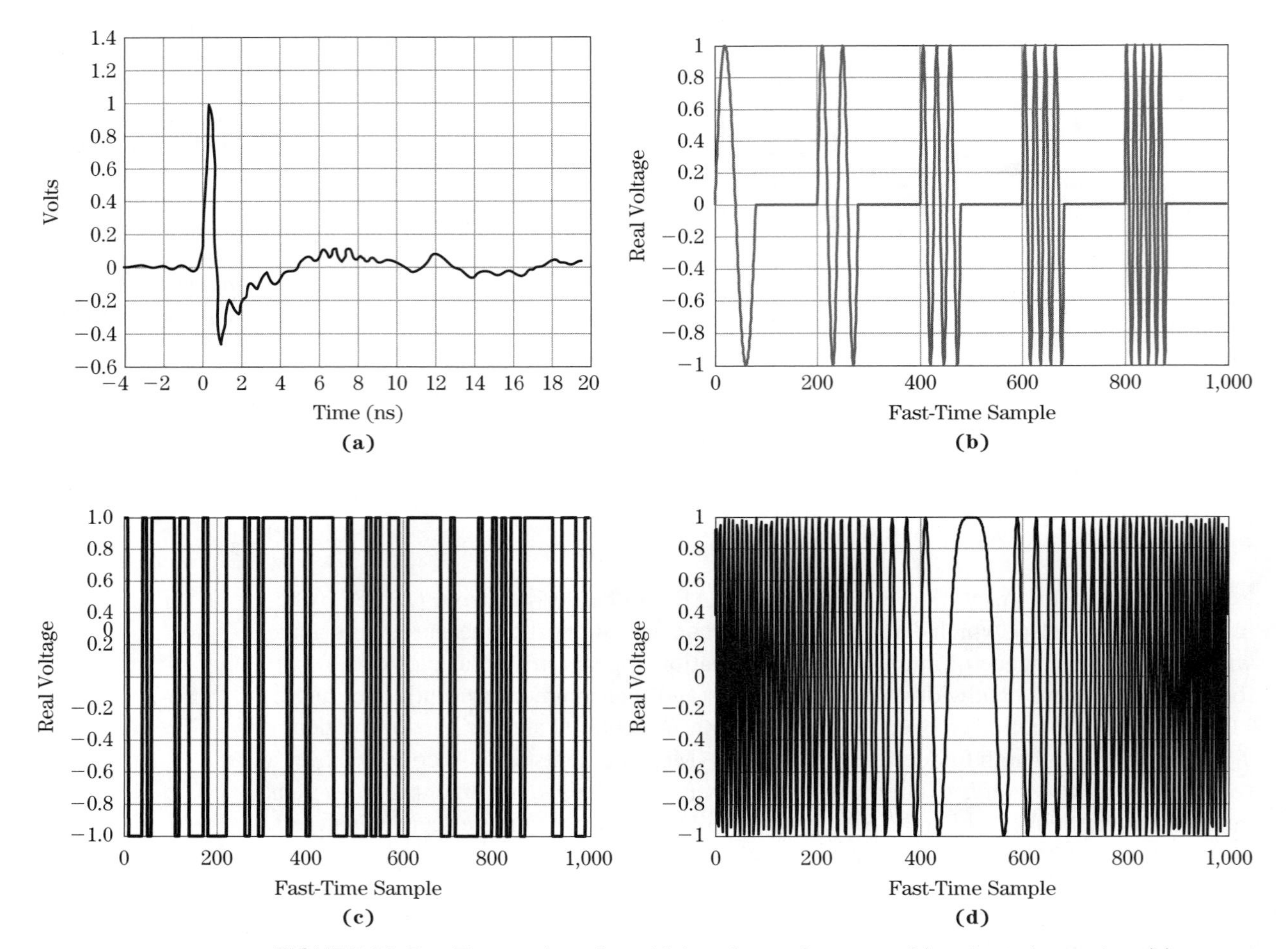

FIGURE 21-7 ■ Four options for wideband waveforms are (a) a short-time pulse, (b) a stepped-frequency series of pulses, (c) a bi-phase modulated pulse, or (d) a linear FM pulse.

Together, equations (21.19) and (21.22) support the earlier assertion that fine resolutions require measurement over frequency (wide waveform bandwidths) and angle (large integration angles). Additionally, equation (21.22) suggests that fine cross-range resolutions are easier to achieve at short wavelengths.

Figure 21-7 depicts the various waveforms a radar may use to collect measurements over a wide range of frequencies; these are described in more detail in Chapter 20. A short-duration transmit pulse, shown in Figure 21-7a, is arguably the clearest path to generating wide bandwidths and fine range resolutions. A narrow pulse obviously permits scene returns to be partitioned into very fine range bins. Due to the Fourier relationship between time and frequency, the record from a short-time pulse also provides measurements of the scene over a wide bandwidth. Thus, the fine resolution of the short pulse is consistent with the large bandwidth needed for fine resolution as seen in equation (21.19). However, short pulses tend to be energy limited and are therefore appropriate for short-range applications only.

Figure 21-7b shows a stylized representation of one solution to the energy limitations in a short-time pulse. The stepped frequency waveform consists of a series of pulses, each at a slightly different narrowband frequency. Every pulse is of long duration, sufficient to effectively immerse the entire scene of interest, thereby measuring its response at one pure frequency. The stepped-frequency radar transmits and receives one pulse at a time, collecting a single I and Q value at each frequency. The stepped-frequency waveform, then, explicitly measures the response of a scene over a wide bandwidth, one discrete frequency at a time. By Fourier transforming over the frequency measurements the radar data processor can synthesize the same fine-resolution range profile provided by a short-duration pulse of equivalent bandwidth. Moreover, the use of multiple long pulses means the stepped frequency waveform is not subject to energy limitations. However, it takes significant time to collect stepped-frequency measurements over a wide bandwidth, so this waveform is not appropriate for dynamic environments. While stepped frequency waveforms are common in carefully controlled antenna and radar cross section (RCS) measurement activities, they are rarely employed by moving platforms.

Phase or frequency modulation of a single long-duration pulse is a common solution to the competing requirements of wide bandwidths, high energy, and operation in dynamic environments. Figure 21-7c shows an example of a biphase modulated pulse. Phase modulation is commonly used in radars having moderate bandwidths, though application in wideband systems has been limited until recently due to the requirement for high-speed analog-to-digital conversion (ADC) and computationally intensive matched filtering. Existing SAR systems use frequency modulation, specifically linear frequency modulation (LFM), also known as a "chirp" waveform, almost exclusively (see Chapter 20). An example of the baseband representation of an LFM pulse appears in Figure 21-7d. The linear chirp waveform provides wide bandwidth over a long pulse having appreciable energy. In addition, LFM is the only modulation appropriate for so-called *dechirp-on-receive* or *stretch processing*, a computationally efficient matched filtering procedure for pulse compression that also tolerates low to moderate ADC rates. Finally, LFM waveforms are Doppler-tolerant, permitting high-speed platforms like satellites in low Earth orbit to accomplish SAR imaging without the added complication of a bank of Doppler-dependent matched filters.

In SAR, integration angle is provided by the relative motion between the radar platform and the scene of interest. Examples appear in Figure 21-8. In a *stripmap* SAR collection, shown in Figure 21-8a, the platform moves along a nominal straight-line path, with the radar antenna oriented off to one side and perpendicular to the flight path. As the platform moves, targets, clutter, and other scatterers move through the antenna mainbeam, and the effective bearing angle of the radar to these returns varies. The radar collects along a contiguous swath parallel to the radar direction of motion.

In a *spotlight* collection, shown in Figure 21-8b, the platform maintains a straight-line path, but the radar antenna is pointed toward a scene of interest. Beam-pointing may be accomplished by mechanical panning of a gimbaled antenna or, if the antenna is an

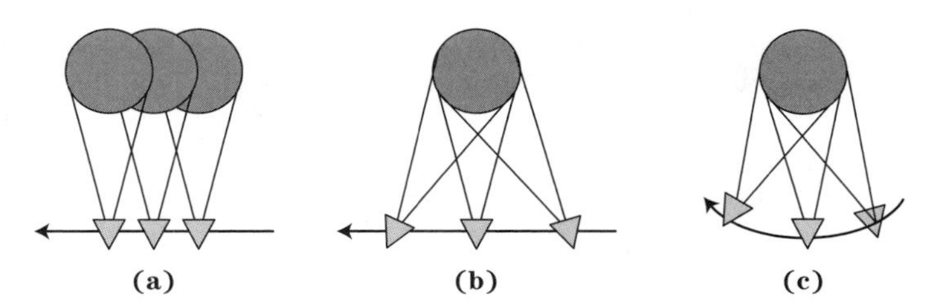

FIGURE 21-8 ■ Three collection options for achieving integration angle are (a) stripmap, (b) spotlight with mechanical or electronic steering of the real antenna beam onto a scene of interest, or (c) spotlight realized by flying a circle about the scene of interest.

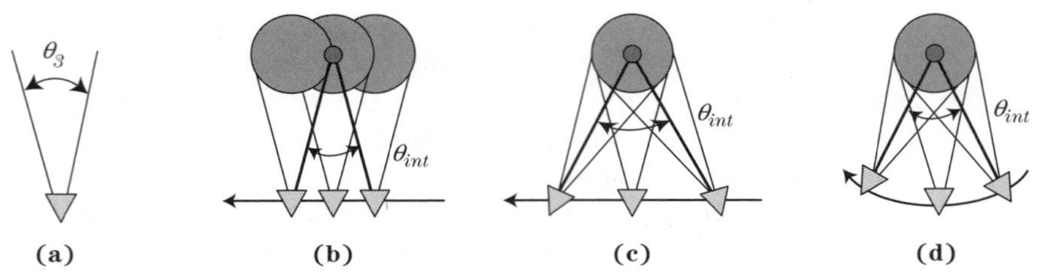

FIGURE 21-9 ■ Given the beamwidth of a real antenna (a), the stripmap integration angle (b) is limited by the real beamwidth, while spotlighting with linear (c) or circular (d) flight paths can realize integration angles greatly exceeding the real antenna beamwidth.

electronically scanned array (ESA), via analog beam steering. By aiming the beam at particular point as the platform flies past, the radar can collect data for that point over a longer time than in the fixed-beam stripmap mode. The extended dwell time afforded by beam-pointing increases the integration angle for scatterers in the scene of interest and so improves cross-range resolution. A drawback of spotlight operation is that longer acquisitions on scenes of interest come at the expense of returns from other regions along the flight path. That is, stripmap yields a moderate resolution image that is continuous along the flight path, whereas spotlight generates fine-resolution images intermittently, with gaps in coverage along the platform's ground track. Finally, a spotlight collection can be achieved by flying a curved path about a scene of interest, as shown in Figure 21-8c, so that beam-pointing is accomplished by the motion of the platform. If the path is circular, a full 360° of integration angle can be garnered against scatterers in the scene.

Figure 21-9 summarizes the integration angles provided by the various collection schemes. Figure 21-9a shows a physical antenna with a real beamwidth θ_3. In the stripmap collection shown in Figure 21-9b, measurements are collected against scatterers within the imaged swath only while they fall within the beamwidth θ_3 of the transmit/receive antenna. Simple geometry indicates the maximum stripmap integration angle is equal to the antenna beamwidth. From the resolution relation in equation (21.22) and setting $\theta_{int} \leq \theta_3$,

$$\Delta CR \text{ (stripmap)} \geq \frac{\lambda_c}{2\theta_3} \tag{21.24}$$

Achievable stripmap cross-range resolution is bound by the real antenna beamwidth: wider beamwidths provide increased integration angles and finer resolutions. On the other hand, the spotlight modes in Figures 21-9c and 21-9d clearly allow the integration angle to far exceed the antenna beamwidth and are therefore capable of much finer cross-range resolutions.

At this point it is convenient to highlight a nuance in the stripmap and spotlight terminology. Note that linear motion is common to the stripmap and the first spotlight methods shown in Figure 21-8. A linear collection geometry results in a specific scatterer response in the raw radar data, regardless of whether the real beam was steered. SAR image formation algorithms are designed based on the expected response of scatterers in the raw data. Thus, algorithms geared toward linear collections, sometimes called stripmap techniques, are equally applicable to both stripmap and linear spotlight collections. In contrast, the circular spotlight collection shown in Figure 21-8c gives rise to a very different scatterer signal history. Several imaging methods, sometimes referred to as spotlight techniques, are tailored toward data collected in a circular fashion (or data preprocessed, through digital delays and interpolation, to appear as though they were collected from a circular path). Thus, when employing the terms stripmap and spotlight, care must be taken in stating whether the appellation applies to the means of generating integration angle or to the image formation technique.

21.3.2 Synthetic Aperture Sampling Requirements

In the course of collecting SAR data a number of sampling requirements must be met lest aliased returns degrade the final SAR image. Time and frequency sampling is covered in Chapter 14; here the discussion will be restricted to requirements for sampling along the SAR collection path.

The diagram in Figure 21-10a contains an antenna with beamwidth θ_3 looking off to the right. The antenna platform is moving from bottom to top; equivalently, scatterers in the imaged swath move from top to bottom through the antenna mainbeam. Two scatterers are shown: the upper one moving into the top edge of the mainbeam, and the lower one moving out of the mainbeam.

As discussed in the previous section, SAR waveforms are typically pulsed; CW SAR systems are quite rare. Pulsed waveforms are required for monostatic antenna implementations and, through judicious time gating, allow returns to be constrained to the down-range swath of interest. Transmission and reception of pulses means the response of the illuminated scene is recorded at discrete, uniformly spaced locations along the radar flight path. Radar movement between two pulses causes a range decrease to scatterers along the leading edge of the antenna beam, as shown in Figure 21-10b. This range decrease is a function of the distance between sample points d and the angle to the scatter, which is equal to half the beamwidth. From the right triangle construction in Figure 21-10c, the far-field range decrease dR is

$$dR = d \sin\left(\frac{\theta_3}{2}\right) \tag{21.25}$$

Due to symmetry about the mainbeam direction, the distance to the receding scatterers increases by the same amount.

It has already been noted that SAR is a coherent collection and imaging process, requiring measurement of RF phase. The phase change on the leading edge scatter due to the motion of the platform in Figure 21-10b depends on the range change normalized by

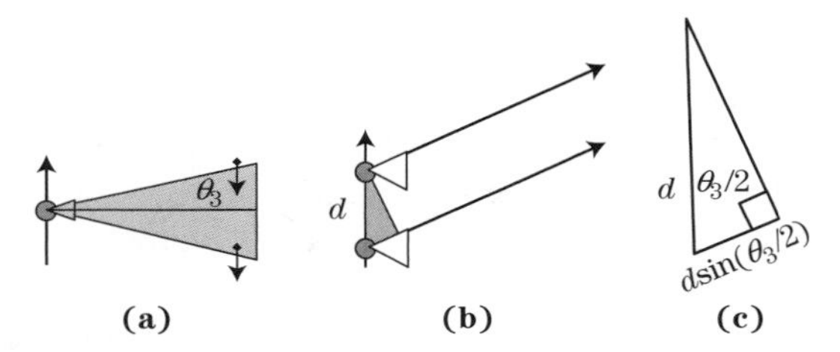

FIGURE 21-10 ■ Geometry for deriving the along-track sampling requirements for stripmap mode, (a) platform moving from bottom to top with antenna pointed to the right, (b) distance between along-track sample points d, and (c) differential range change to a scatterer on the leading edge of the mainbeam.

the RF wavelength. This phase decrease $\Delta\psi$ in radians is

$$\Delta\psi = 2\pi\left(2\frac{dR}{\lambda}\right) \tag{21.26}$$

The factor of 2 inside the parentheses appears because both the transmit and receive positions move with the radar, so the effective two-way range change is twice dR.

Combining (21.25) and (21.26) yields

$$\Delta\psi = 2\pi\left[2\frac{d}{\lambda}\sin\left(\frac{\theta_3}{2}\right)\right] \tag{21.27}$$

A coherent radar system is incapable of unambiguously measuring phase changes greater than 180°. A phase decrease of 179° is unique, but a decrease of 181° looks the same as an increase of 179°. Because the scatterers on the leading and trailing edges of the beam give rise to equal but opposite phase changes, the phase change to either side must not exceed 180°, or π radians:

$$|\Delta\psi| = \left|2\pi\left[2\frac{d}{\lambda}\sin\left(\frac{\theta_3}{2}\right)\right]\right| < \pi \tag{21.28}$$

Phase ambiguity is avoided by sampling along the flight path at sufficiently frequent intervals. From (21.28),

$$d < \frac{\lambda}{4\sin\left(\frac{\theta_3}{2}\right)} \tag{21.29}$$

As the value of an angle is always larger than the sine of that angle ($\sin\theta < \theta$), the bound in equation (21.29) can be simplified to

$$d < \frac{\lambda}{2\theta_3} \tag{21.30}$$

Equation (21.30) states that wider real beamwidths require finer sampling along the collection track. The sampling interval in (21.30) is set not by the wavelength at the carrier frequency but by the shortest wavelength λ_{min} of the waveform bandwidth

$$d < \frac{\lambda_{min}}{2\theta_3} \tag{21.31}$$

The consequences of violating the along-track sampling limit in equation (21.31) depend on the image formation technique and final image resolution. At coarse resolutions, undersampling is manifested primarily as a wrapping of returns in cross-range; that is, returns from a scatterer at a positive cross-range location beyond the limit of the final image will be aliased back into the image at a negative cross-range. At fine resolutions aliased returns tend to defocus and increase the effective noise level across the entire image. To preclude these degradations entirely it is prudent to use a more pessimistic value for beamwidth in equation (21.31), such as the null-to-null beamwidth $2\theta_R$, where θ_R is the peak-to-null (Rayleigh) beamwidth. Then, (21.31) becomes

$$d < \frac{\lambda_{min}}{2\,(2\theta_R)} \tag{21.32}$$

Antenna null-to-null beamwidths are typically about twice as wide as the 3 dB beamwidths θ_3 commonly used in resolution calculations.

The relationship in equations (21.32) was derived from a stripmap point of view as shown in Figure 21-10 but tends to hold for the circular spotlight collection geometry in Figure 21-8c as well. Sampling requirements for linear spotlight and squinted stripmap (with the antenna beam fixed away from broadside) are more complicated and depend strongly on the particular image formation technique applied to the data.

21.3.3 Miscellaneous Relationships

The formulae for resolution and sampling can be combined in various ways to derive constraints on SAR operations. For example, for square resolution cells the down-range and cross-range resolutions in equations (21.19) and (21.22) are equal, yielding

$$\frac{c}{2B_r} = \frac{\lambda_c}{2\theta_{int}} \tag{21.33}$$

Rearranging and replacing carrier wavelength with carrier frequency yields

$$B_r = f_c\theta_{int} \tag{21.34}$$

The right side of equation (21.34) is the "cross-range bandwidth" B_x in (21.20), so, not surprisingly, the RF and cross-range bandwidths must be equal for square resolution cells.

The null-to-null beamwidth of an antenna is a function of the physical size (length or diameter) D of the antenna and operating wavelength. A good rule of thumb for the peak-to-null beamwidth is

$$\theta_R \approx \frac{\lambda}{D} \tag{21.35}$$

when illumination is accomplished without amplitude tapering. (See Chapter 9 for more exact expressions for beamwidths and a discussion of the effects of tapering.) Using the minimum wavelength in equation (21.35) and inserting it into (21.32) gives an interesting relationship between along-track sampling and antenna size:

$$d < \frac{D}{4} \tag{21.36}$$

Equation (21.36) states that the radar must transmit and receive frequently enough to collect four pulses as the platform moves a distance equal to the antenna diameter.[2] Note the lack of dependence of the sampling requirement on wavelength/frequency: shorter wavelengths increase the phase sensitivity to range (21.26) but also decrease the antenna beamwidth (21.27) (for a fixed antenna size) at the same rate, producing a wash in the along-track sampling requirement.

A similar relationship for stripmap resolution can be developed with an approximate expression for antenna 3 dB beamwidth at the carrier frequency:

$$\theta_3 \approx \frac{\lambda_c}{D} \tag{21.37}$$

[2] The null-to-null beamwidth results in the factor of four. Some texts report a factor of two because they employ a less stringent beamwidth.

Combining (21.37) with (21.24) gives

$$\Delta CR \text{ (stripmap)} \approx \frac{D}{2} \tag{21.38}$$

Again, wavelength or frequency is absent this expression: higher frequencies result in narrower beamwidths for a given antenna size and therefore decreased integration angle, but wavelength decreases at the same rate for a wash in cross-range resolution. Note that equation (21.38) provides a lower bound on resolution; images having spoiled (coarser) resolution may be generated from full-resolution data.

Finally, equations (21.36) and (21.38) can be combined to relate the resolution limit to sampling requirements:

$$d < \frac{\Delta CR}{2} \text{(stripmap)} \tag{21.39}$$

Equation (21.39) is a handy rule of thumb for the required sampling interval when the stripmap resolution capability of a SAR system is documented but the operating frequency or antenna size is unknown.

21.4 DATA COLLECTION

As previously discussed, SAR is a two-step process: (1) data collection; and (2) image formation. In this section the data collection step is described in more detail.

This chapter began by noting the similarities between SAR and optical images, and now it is instructive to explore the differences in SAR and optical data collection. For example, to photograph a large region of terrain, a camera might be placed high overhead, as depicted in Figure 21-11. This geometry is common for aerial photogrammetry (map-making) and satellite photo-reconnaissance purposes and serves to minimize distortion of the scene in the final image. However, this geometry is completely unsuitable for radar imaging, as returns from the entire illuminated ground region would fall into a handful of range bins. By placing the radar off to the side, as shown in Figure 21-11, returns across the illuminated area can be subresolved into a large number of range bins. Imagining

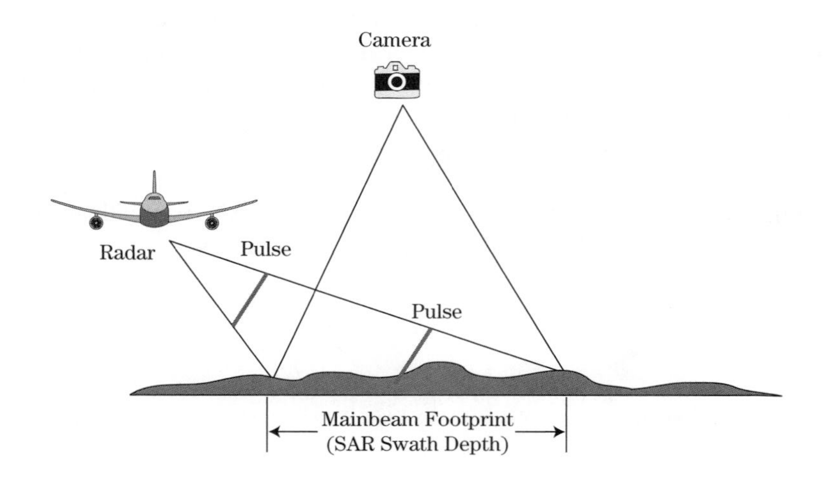

FIGURE 21-11 ■ Geometry of overhead photography of a scene of interest as compared to a typical SAR collection.

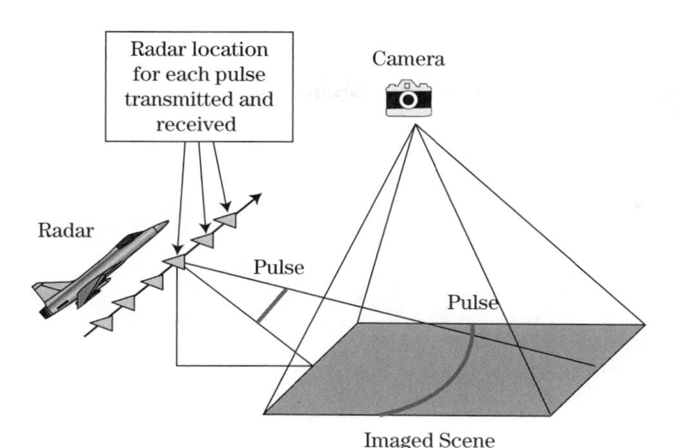

FIGURE 21-12 ■ The camera generates a 2-D image in one take, whereas the radar must collect data pulse-by-pulse over a long dwell.

for a moment the radar transmits a short-duration pulse, it can be seen that the outbound pulse will sweep across the SAR swath, allowing returns to be binned in time to form a high-resolution range profile.

In summary, optical imaging calls for steep look-down angles, generally 70° to 90°, whereas SAR imaging is typically accomplished at grazing angles of 10° to 45°. Images formed at steeper look-down angles suffer from dilation of the effective range resolution on the ground. (Resolution on the ground equals the resolution of the waveform divided by the cosine of the grazing angle; see Chapter 2.) Grazing angles shallower than 10° are theoretically acceptable. However, the radar not only receives returns but also provides the scene illumination. Real terrain tends to undulate mildly or is covered by high vegetation such as trees or bushes. Both phenomena cause large portions of the swath to be shadowed at extremely low grazing angles. (See Chapter 5 for a discussion of clutter shadowing effects.) For this reason SAR collection geometries are usually set to ensure grazing angles of at least 10°.

Figure 21-12 helps to clarify another important difference between optical and SAR imaging. A camera observes the entire scene through a two-dimensional recording medium (a chemical film emulsion or an electronic focal plane array) at the image focal plane and therefore requires just a fraction of a second to collect a data "snapshot" and form an image. At any given location the radar also views the entire scene; however, scatterers at the same slant range fall into a single range bin, so the radar's "snapshot" is a one-dimensional range profile. SAR image formation requires data over many pulses, each collected from a different location. These changing locations provide the integration angle required to achieve cross-range resolution. However, it does take some time for the platform to move and collect all of the measured data. For coarse resolutions this may be only a fraction of a second. Fine resolutions require large integration angles that, at long slant ranges, entail long SAR data collection distances. A slow-moving platform may require many seconds, even several minutes, to traverse the synthetic aperture and collect sufficient data for image formation.

Figure 21-13 further expands on the details behind a SAR data collection. In contrast to real-beam mapping systems such as SLAR wherein a very narrow antenna beam is used to resolve scatterers in cross-range, the SAR beamwidth is wide enough to illuminate the entire scene of interest and does so over many pulses. Each pulse transmission and reception position produces a one-dimensional stream of I and Q values.

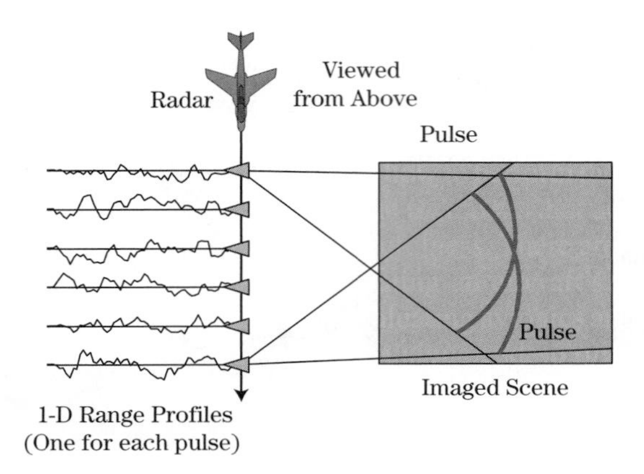

FIGURE 21-13 ■ Every pulse illuminates the entire scene of interest and produces a 1-D, complex range profile.

Figure 21-13 is also convenient to introduce a common simplification in the development of SAR collection geometries, the so-called *stop-and-hop model*. In reality the radar is moving continuously, and the radar location changes slightly in the time it takes for a pulse to propagate out to the scene and back. Hence, transmit and receive locations are not quite the same; the collection is bistatic, in the strictest sense. However, this displacement between transmit and receive location is usually very small compared with the range to the image scene and so may be neglected. From the signal processing point of view (e.g., sampling requirements, generation of matched filters for imaging), it can be assumed that the transmit and receive positions are colocated. In this model, the radar stops at one location, transmits and receives a pulse, then hops to the next location. The "virtual" collection point is usually chosen to be at the bisector between the transmit and receive positions.

The stop-and-hop model allows complicated two-way geometric calculations to be simplified to a one-way expression and a multiplicative factor of two for two-way propagation. It is appropriate when the transmit/receive displacement is very small compared with the imaging range, which is equivalent to restricting the platform speed to be much slower than the pulse propagation speed. This is always the case in radar, even for space-based collections where the platform speed can be several kilometers per second, but the stop-and-hop model may not be appropriate for some kinds of acoustic imaging, like ship-based sonar.

Finally, it is useful to justify the importance of phase information to SAR imaging. SAR is a coherent process and requires recording of both amplitude and phase. Recall that the entire scene is illuminated over the course of the SAR collection. For that reason the complex value recorded in any given range bin reflects the coherent accumulation of the returns from many scatterers distributed across the swath at that particular range. As the platform moves, the scatterers interfere constructively and destructively in different ways, causing amplitude and phase fluctuations over time. Therefore, the signal history bears no resemblance to the actual distribution of scatterers in the bin but contains the information the SAR image processor needs to separate those returns in cross-range in the final image.

In a way, coherent radar has two means to measure the range to a target. The first is relatively coarse and involves either measuring time delays on pulse returns or forming range responses using the frequency content in a modulated pulse. The fundamental product is the same: A one-dimensional profile consisting of range bins possessing a resolution on the order of many meters, several feet, or just a few inches. The profile allows the absolute range to a target to be measured to within a fraction of the range resolution. Thus, pulse timing and waveform modulation provides gross resolution and location of targets in down-range.

The second range estimate originates from phase information, which provides a very fine but highly ambiguous measure of range. Interestingly, these properties are exploited to isolate returns not in down-range but rather cross-range. Recall that one cycle of phase shift corresponds to a change in two-way distance equal to half the wavelength (for two-way collections). Radar wavelengths are almost always much smaller than range bin size, so phase provides a very accurate measure of range. Higher carrier frequencies afford more accuracy due to the smaller value of wavelength λ. For example, W-band systems operating at 94 GHz can use phase to measure range to within 3 mm. Unfortunately, this measurement is highly ambiguous, having an ambiguity interval equal to half a wavelength. For example, if an X-band radar having a wavelength of 3 cm measured a phase angle of 180°, the distance to the source scatterer would be $(1.5 \text{ cm}) \times (180°/360°) = 0.75$ cm plus an unknown integer number of half-wavelengths. The scatterer might be only 0.75 cm from the phase center of the radar antenna, or it could be 0.75 cm plus 1 million intervening half-wavelengths (1,500,000.75 cm = 15.0000075 km) from the phase center.

The ambiguous nature of the phase measurement makes absolute phase values meaningless in all but the most near-field synthetic imaging applications. Instead, the progression of phase as the location of the radar varies is the principal concern. The phase trend observed from pulse to pulse is a function of the cross-range location of a target, as seen in Figure 21-14. In the diagram at the top, a single scatterer is located broadside to the SAR collection points. If the target is very far from the synthetic array, its return will fall into the same range bin for all pulses. Indeed, in the far field, the slant range to the scatterer changes so little that the radar will record about the same phase angle, 10° in the example. The lower diagram shows a scatterer just off the centerline. This displacement away from broadside cause the slant range from the radar to slightly decrease as it moves from top to bottom. The range variation is not sufficient to cause the scatterer's return to drift through

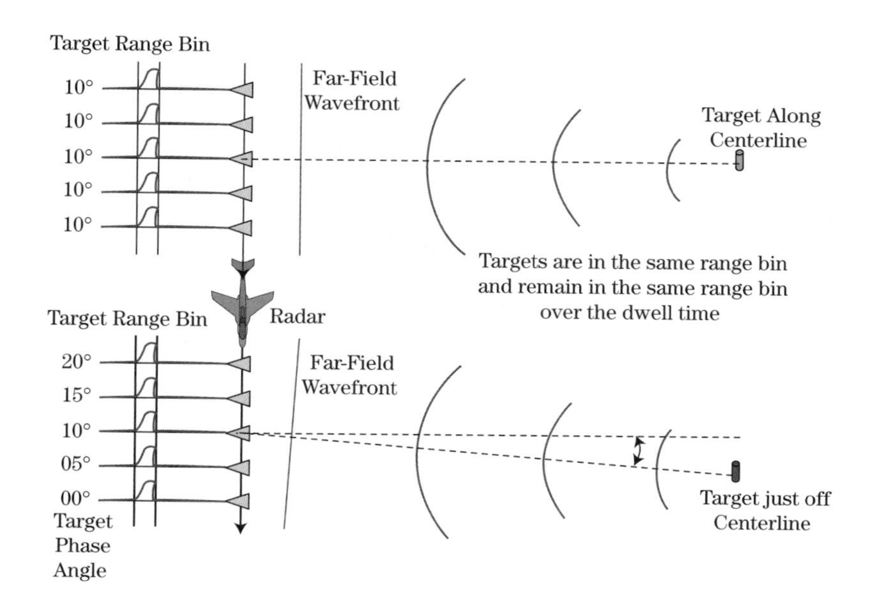

FIGURE 21-14 ■ Small changes in the cross-range location of a scatter cause a negligible change in the apparent range bin for that return but a significant change in the induced pulse-to-pulse phase progression for that return.

range bins (often referred to as *range migration*), but it is enough to generate a phase angle progression, 5° from pulse to pulse in this example. A scatterer removed still farther from broadside generates a larger phase change between radar positions, and a scatterer placed on the opposite side of the centerline causes a decreasing phase progression. Thus, phase history can be used to map returns to their appropriate cross-range locations. In summary, phase provides fine but highly ambiguous resolution and location of targets in range, but pulse-to-pulse phase progressions yield resolution and location in cross-range.

21.5 IMAGE FORMATION

Once raw radar data are acquired, image formation (i.e., the transformation from I and Q samples to the final SAR image) can be performed. There are a wide variety of image formation algorithms available that presume different collection geometries and make certain mathematical approximations, some very simple and suitable only for coarse resolution imaging, others more sophisticated and tailored to fine-resolution imaging, and some more computationally or memory efficient than others. All exploit the predicted return from a point source in the imaged scene. This *point spread response* (PSR) is the manifestation of an isotropic point scatterer in the raw data. In signal processing terminology, the PSR is the raw data "impulse response" of the SAR data acquisition system. Knowledge of the PSR is the key to designing an image formation algorithm to gather up the energy from a scatterer distributed throughout the raw data and focus it into a concentrated point-like return in the final SAR image.

In this section the PSR for a linear (stripmap) SAR collection geometry is derived and used to construct several image formation techniques.

21.5.1 SAR Coordinate Systems

Figure 21-15 revisits the domains and transformations introduced in Figure 21-5 but adds coordinate definitions. The top diagram shows the collection geometry; the location of

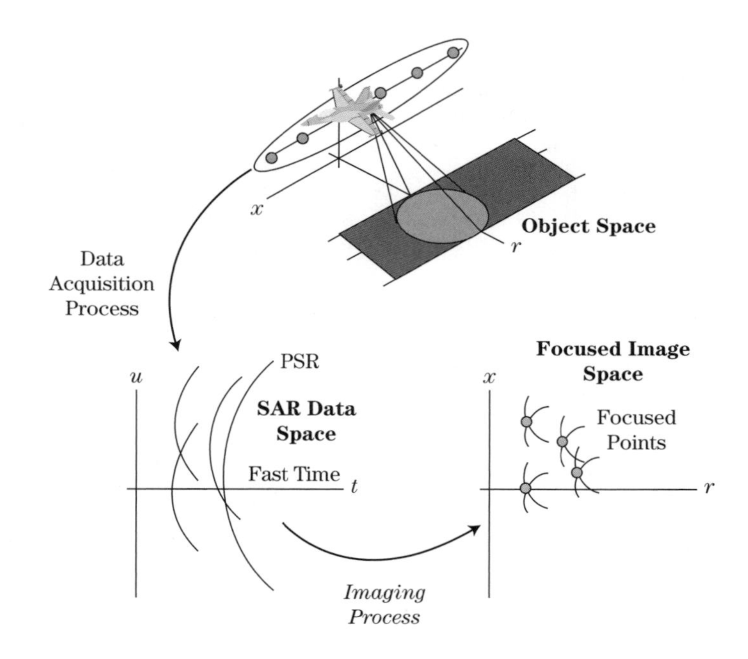

FIGURE 21-15 ■ The SAR process as depicted in Figure 21-5 with axes defined.

scatterers on the ground are determined by an along-track position x and cross-track range (distance from the platform flight line) r. The function $g(x, r)$ is the complex reflectivity of the imaged scene, including distributed natural clutter and man-made targets, as a function of along-track and cross-track location. Here $g(x, r)$ is modeled as a collection of isotropic point scatterers.

The second diagram in Figure 21-15 depicts the complex raw SAR data $d(u, t)$. This data is two-dimensional, with the vertical dimension showing the along-track location of the platform u where a given pulse was transmitted and received. In this development the scene coordinate x and data collection coordinate u are measurements of the same along-track position, but the former is dedicated to scatterer locations and the latter reserved for platform locations. The other dimension is the time delay t into the data on any given pulse. Here it is assumed that pulse compression has already been performed (see Chapter 20) if a long phase or frequency modulated pulse is used, so that t represents time into compressed range profiles.

The third diagram in Figure 21-15 shows the SAR image $f(x, r)$. The final image is formed to depict as accurately as possible the scene reflectivity, so the collection and image spaces share the same (x, r) coordinate system.

21.5.2 Linear Collection PSR

The PSR in the data $d(u, t)$ of a scatter located at (x, r) with reflectivity $g(x, r)$ is determined entirely by the slant range from the platform to the scatterer over the SAR data collection. The evolving slant range over the dwell allows the prediction of (1) amplitude variations due to geometric (R^4) loss, (2) any gross migration through compressed range bins, and (3) the fine pulse-to-pulse progression of carrier phase.

The platform is confined to the along-track axis and thus always has a zero cross-track component so that its location is $(u, 0)$. Using Figure 21-16 and the Pythagorean theorem, the slant range R to a scatterer at (x, r) from the platform at $(u, 0)$ is given by

$$R^2 = (u - x)^2 + r^2 \tag{21.40}$$

The slant range squared is simply the sum of the squared along-track range difference and the cross-track range to the scatterer. The hyperbolic form of equation (21.40) is commonly

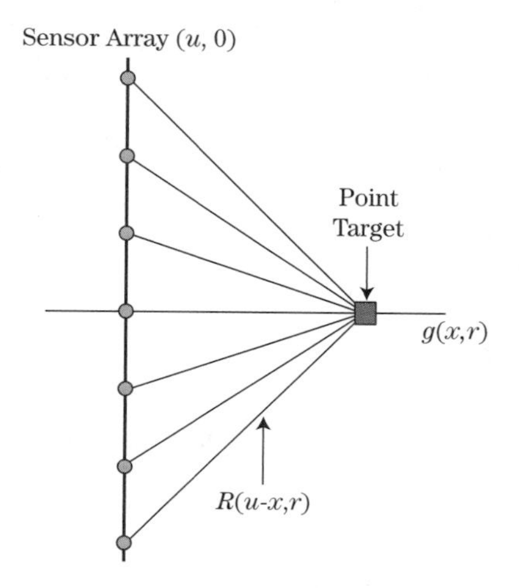

FIGURE 21-16 ■ Variation in slant range to a scatter at (x, r) due to platform motion.

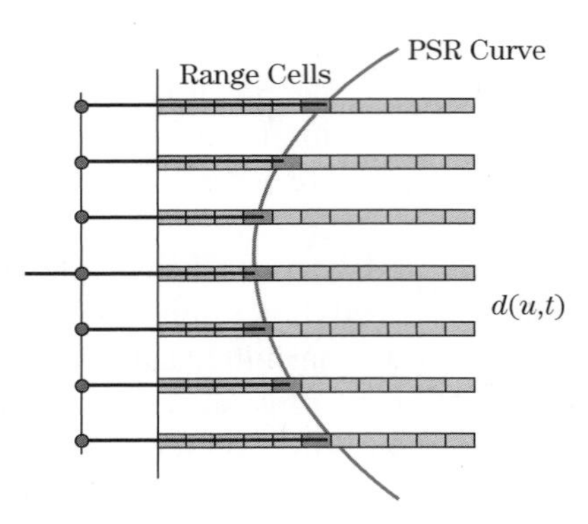

FIGURE 21-17 ■ The range variation to the scatterer over the dwell is described by the PSR. The PSR is manifested in the raw data as a hyperbolic migration of the return from a scatterer across compressed range bins.

associated with linear data collections. Writing the slant range explicitly as a function of the along-track difference and cross-track range gives

$$R(u; x, r) = \sqrt{(u - x)^2 + r^2} \tag{21.41}$$

The time delay t into the data for the return from a scatterer is then

$$\begin{aligned} t(u; x, r) &= \frac{2}{c} R(u; x, r) \\ &= \frac{2}{c} \sqrt{(u - x)^2 + r^2} \end{aligned} \tag{21.42}$$

The return from a point scatterer is concentrated in time at a delay given by equation (21.42). Figure 21-17 depicts the return from a point scatterer in the raw data, the PSR, over several pulse positions. Each pulse results in a range profile. The scatterer is manifested as a single strong return, highly concentrated in range, on any given pulse. Platform motion causes the range to the scatterer to change over the dwell yielding a time delay that takes the form of a hyperbola. The varying delay causes the return to migrate over time through range bins, first decreasing in time/range and then increasing. Because slant range determines measured phase angle, the hyperbolic time function also generates significant phase modulations (not shown in the figure) over the dwell.

It is useful to examine in more detail some properties of the PSR. From equation (21.42), the PSR of a scatterer at location (x_1, r) is nonzero at a time delay

$$t_1(u; x_1, r) = \frac{2}{c} \sqrt{(u - x_1)^2 + r^2} \tag{21.43}$$

The delay to the response from a second scatterer at the same cross-track range but displaced along-track by Δx to location $(x_1 + \Delta x, r)$ is

$$t_2(u; x_1 + \Delta x, r) = \frac{2}{c} \sqrt{(u - (x_1 + \Delta x))^2 + r^2} \tag{21.44}$$

Equation (21.44) can be rewritten as

$$t_2(u - \Delta x; x_1, r) = \frac{2}{c} \sqrt{((u - \Delta x) - x_1)^2 + r^2} \tag{21.45}$$

Defining a modified platform location $u' \equiv u - \Delta x$ for (21.45) yields

$$t_2(u'; x_1, r) = \frac{2}{c}\sqrt{(u' - x_1)^2 + r^2} \tag{21.46}$$

The PSRs for these two scatterers, given by equations (21.43) and (21.46), have the same hyperbolic form and differ only by a position shift in the u dimension. In general, all scatterers laying at the same cross-track range r give rise to PSRs with the same hyperbolic form but simply shifted in along-track location so that the PSR possesses *along-track invariance*. Therefore, a reference form for a given cross-track range can be derived from the PSR of a scatterer at an along-track location of zero ($x = 0$):

$$t_{REF}(u; r) = \frac{2}{c}\sqrt{u^2 + r^2} \tag{21.47}$$

Now perform a similar development for cross-track displacements. The PSR for a scatterer at $x = 0$ and r_1 is

$$t_1(u; r_1) = \frac{2}{c}\sqrt{u^2 + r_1^2} \tag{21.48}$$

and for a second scatterer displaced in cross-track range by Δr is

$$t_2(u; r_1 + \Delta r) = \frac{2}{c}\sqrt{u^2 + (r_1 + \Delta r)^2} \tag{21.49}$$

Unfortunately, the radical prevents expressing the cross-track range change as simply a time delay:

$$t_2(u; r_1 + \Delta r) \neq \frac{2}{c}\left(\sqrt{u^2 + r_1^2} + \Delta r\right) \tag{21.50}$$

From (21.50), the second PSR (21.49) is not simply the first PSR (21.48) delayed by a time consistent with the range change. The displacement in cross-track location actually changes the shape of the hyperbola, so the PSR exhibits *cross-track variance*. This effect is depicted in Figure 21-18. Four scatterers at the same along-track position and distributed in cross-track range produce hyperbolic PSRs centered at the same along-track position but delayed in time and having different shapes. The "bowing" of the hyperbola is more pronounced for scatterers close to the platform, and becomes shallower as cross-track

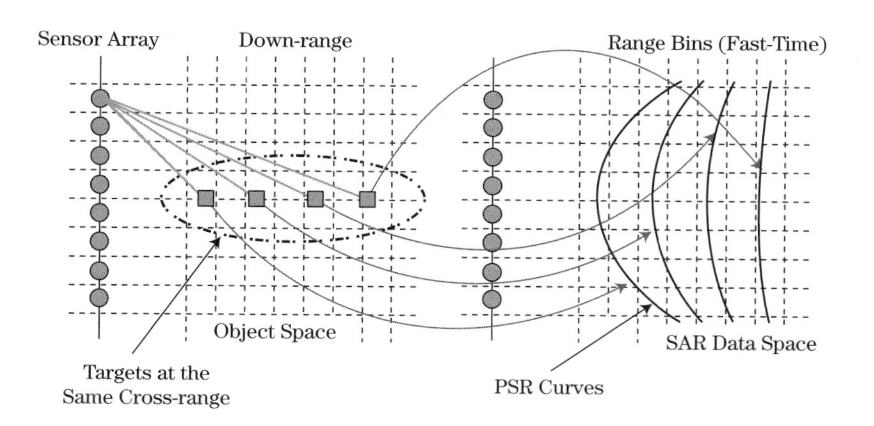

FIGURE 21-18 ■ Scatterers at different cross-track (down-range) locations have different PSR forms. Curvature decreases as range increases.

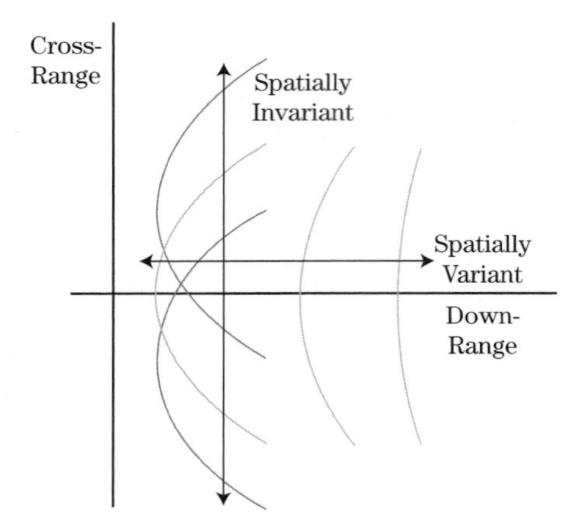

FIGURE 21-19 ■ Scatterers at different along-track (cross-range) location have the same PSR form, but with an along-track offset.

range increases. Mathematically, the PSR is spatially invariant along-track but spatially variant in the cross-track dimension.

Referring to Figure 21-19, the properties of the linear collection PSR can be summarized as follows:

1. The PSR is, in general, a two-dimensional (spatial and temporal) function in u and t.
2. The PSR is not separable in u and t (it cannot be constructed as the product of functions in u only and t only).
3. The PSR has a hyperbolic form that causes slant range to be dependent on the square root of platform and scatterer locations.
4. The PSR form varies with scatterer cross-track range.
5. The PSR form is the same for scatterers at the same cross-track range.

The first four properties complicate the image formation process, while the fifth provides structure an imaging algorithm can exploit, as shall be seen.

Knowledge of the PSR will now be used to derive *Doppler beam sharpening* (DBS), the oldest and most fundamental SAR image formation technique [10].

21.5.3 DBS Derivation

It is useful to begin with a qualitative presentation of DBS as a pulsed-Doppler imaging mode. Figure 21-20 depicts the platform moving from left to right with a side-looking radar pointed toward the top of the diagram. Range bins subdivide the illuminated scene in slant range. The Doppler frequency f_d in hertz from a scatterer is given by equation (21.6),

$$f_d = \frac{2v}{\lambda} \cos \theta_{cone}$$

where v is the platform ground speed, and θ_{cone} is the cone angle between the scatterer and the platform velocity vector. Azimuth angle θ is defined relative to the radar look direction. From Eq. (21.10)

$$f_d \approx \frac{2v}{\lambda} \sin \theta$$

Returns at different azimuth angles have different Doppler shifts. Because cross-range is a function of azimuth angle, Doppler filtering allows partitioning of the returns across the

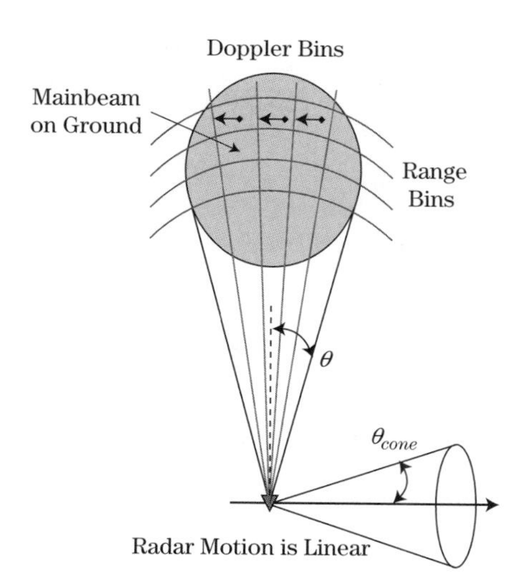

FIGURE 21-20 ■ Arranging ground returns as a function of slant range and Doppler yields a 2-D rendering of the scene of interest.

illuminated scene. DBS, then, uses range binning and Doppler filtering to create a range-Doppler map of the scene. The DBS image maps to slant range and cone angle but at long ranges and for small scene sizes closely approximates a rectilinear down-range-cross-range grid.

DBS is now examined from the PSR point of view. Figure 21-21 shows the hyperbolic PSRs of various scatterers in the range-compressed radar data. DBS assumes a relatively short collection time and coarse range resolution, so that any migration of the returns across range bins can be neglected. As shown in the figure, scatterer returns are then confined to a single range bin. This DBS approximation simplifies the first and second PSR properties from the previous section: the PSR is now a one-dimensional function confined to a single range bin. This greatly simplifies the task of the image former, which may now focus returns by operating on one range bin at a time.

The PSR in equation (21.42) can be rewritten as

$$
\begin{aligned}
t\,(u; x, r) &= \frac{2}{c}\sqrt{(u-x)^2 + r^2} \\
&= \frac{2}{c} r \sqrt{1 + \frac{(u-x)^2}{r^2}}
\end{aligned}
\tag{21.51}
$$

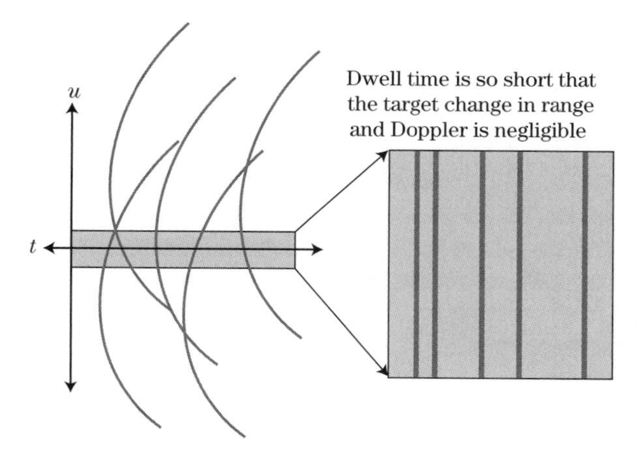

FIGURE 21-21 ■ For short collection times and coarse range bins each hyperbolic PSR is constrained to one range bin over the dwell.

The radical can be eliminated by performing a binomial expansion on its argument. The binomial expansion is simply the first two terms in a Taylor series expansion and is given as

$$(1+w)^p \approx 1 + pw, \quad w \ll 1 \tag{21.52}$$

Applying (21.52) with $p = 1/2$ to (21.51) gives

$$t(u;x,r) \approx \frac{2}{c} r \cdot \left(1 + \frac{(u-x)^2}{2r^2}\right), \quad (u-x) \ll r \tag{21.53}$$

In equation (21.53) it has been assumed that the along-track offset between the platform and the scatterer is much less than the cross-track range to the scatterer so that the condition of (21.52) is satisfied. Simplifying equation (21.53) yields

$$t(u;x,r) \approx \frac{2}{c}\left(r + \frac{(u-x)^2}{2r}\right) \tag{21.54}$$

Equation (21.54) is a quadratic approximation to the exact hyperbolic form of the PSR. The quadratic form changes the third PSR property to one more convenient for image formation.

Recall that DBS assumes that the slant range and time-delay variations are small enough that amplitude changes due to R^4 losses and migration through range bins can be neglected. Consequently, the primary manifestation of these range changes in the data is on the measured phase. Recall also that imaging relies not on absolute phase but on the phase progression as a function of platform location. Expanding equation (21.54),

$$t(u;x,r) \approx \frac{2}{c}\left(r + \frac{u^2}{2r} - \frac{ux}{r} + \frac{x^2}{2r}\right) \tag{21.55}$$

it is seen that the first and last term are functions of just the scatterer location (x, r). Only the second and third terms depend on the platform location u and so capture the time-delay changes over the collection. Reserving these terms to define a time-delay variation Δt as a function of u gives

$$\Delta t(u;x,r) \approx \frac{2}{c}\left(\frac{u^2}{2r} - \frac{ux}{r}\right) \tag{21.56}$$

The phase progression $\Delta\psi$ due to the time-delay variation of equation (21.56) is (in radians)

$$\begin{aligned}\Delta\psi &= 2\pi f_c \Delta t(u;x,r) \\ &\approx \frac{4\pi}{\lambda_c}\left(\frac{u^2}{2r} - \frac{ux}{r}\right)\end{aligned} \tag{21.57}$$

The first term does not depend on the along-track position of the scatterer. It is the *same* for all scatterers at a given cross-track range and serves to simply modulate the data by a quadratic phase function. If the center of a collection of length D_{SAR} is set to $u = 0$, then the quadratic term

$$\Delta\psi_{QT} \equiv \frac{4\pi}{\lambda_c}\left(\frac{u^2}{2r}\right)$$

has a maximum value $\Delta\psi_{QT,MAX}$ at $\pm D_{SAR}/2$ equal to

$$\Delta\psi_{QT,MAX} = \frac{\pi D_{SAR}^2}{2r\lambda_c} \tag{21.58}$$

Quadratic phase error (QPE) effects on imagery are generally negligible if the maximum angular deviation is much less a than a phase cycle. A commonly enforced limit is $\pi/2$:

$$\Delta\psi_{QT,MAX} < \frac{\pi}{2}$$

When combined with equation (21.58), this gives a synthetic aperture limit of

$$D_{SAR} < \sqrt{r\lambda_c} \tag{21.59}$$

Baseline DBS imaging employs geometries that meet this constraint and so allow the first term in equation (21.57) to be omitted, so that

$$\Delta\psi \approx -\frac{4\pi}{\lambda_c}u\cdot\left(\frac{x}{r}\right) \tag{21.60}$$

The phase progression of equation (21.60) is linearly proportional to platform location u and scatterer along-track location x. Figure 21-22a shows the form of equation (21.60) for three scatterers. The scatterer at broadside ($x = 0$) has no phase change as the platform position changes, so it has a phase progression slope of zero; scatterers on either side of broadside, with positive and negative x values, have positive and negative slopes.

The rate of change in phase as the platform moves can be examined by taking the derivative of the phase progression function with respect to platform position:

$$k_u \equiv \frac{\partial}{\partial u}\Delta\psi(u) \tag{21.61}$$

The term k_u in equation (21.61) is *spatial frequency*, the rate of change in phase angle in radians for a change in platform along-track position in meters with units of radians per meter. Spatial frequency is also sometimes known as *wavenumber*.

Spatial frequency is closely related to the more widely referenced temporal frequency, also known as Doppler, which has units of cycles per second (Hz) or radians per second.

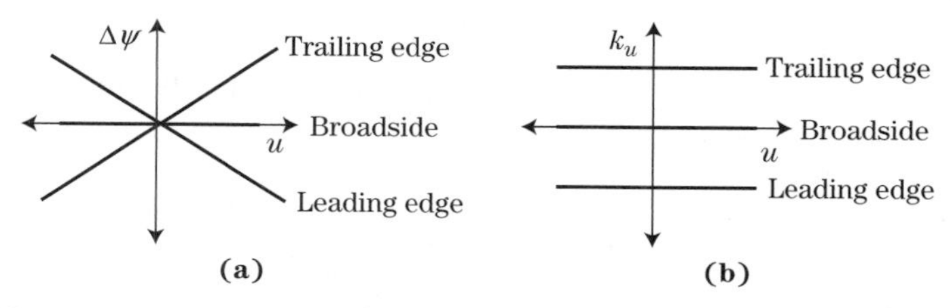

FIGURE 21-22 ■ In the DBS model, (a) phase progressions are approximately linear over the dwell, with slope mapping to cross-range location, and (b) spatial frequencies (Doppler) are approximately constant over the dwell, with frequency mapping to cross-range location.

When collecting SAR data against stationary scatterers, any change in phase from a scatterer from pulse to pulse is due solely to the change in the location of the platform. Therefore, there is a unique mapping of spatial frequency (wavenumber) to temporal frequency (Doppler) that is determined by the along-track sample interval and radar PRF, which in turn are related by the platform speed. Specifically, Doppler frequency is proportional to spatial frequency times the along-track speed of the platform v

$$f_d = -\frac{1}{2\pi} k_u v \tag{21.62}$$

where the factor of $1/2\pi$ puts the result in units of hertz.

The SAR community prefers spatial over temporal frequency because the former provides a more convenient measure of data and image properties. The amplitude and phase of the return from a scatterer are determined by the slant range from the platform to the scatterer. Slant ranges are a function of platform location, as indicated by the form of the PSR. In short, data records are a function of not *when* the radar measurement was made but from *where* the measurement was made. This dependency on position, not time, encourages the use of spatial frequencies.

Using the expression for phase progression in equation (21.60) in the definition of spatial frequency of equation (21.61) gives

$$k_u = -\frac{4\pi}{\lambda_c} \left(\frac{x}{r} \right) \tag{21.63}$$

For a fixed cross-track range r a scatterer has a constant spatial frequency over the dwell (i.e., for all u), as depicted in Figure 21-22b. The precise value depends on the cross-range location x. The frequency is highest for scatterers on the trailing edge of the antenna mainbeam (negative x), is zero for objects at broadside ($x = 0$) and is lowest for returns on the leading edge of the beam (positive x).

21.5.4 DBS Image Formation

The property that scatterers have constant spatial frequency over the dwell makes DBS image formation straightforward. As discussed in detail in Chapter 17, the Fourier transform is a convenient tool for analyzing the frequency content of a signal. The Fourier transform of a spatially sampled signal $s(u)$ to spatial frequency $S(k_u)$ takes the form

$$S(k_u) \equiv \int_{-\infty}^{\infty} s(u) e^{-jk_u u} du \tag{21.64}$$

For DBS this one-dimensional transform is applied to the two-dimensional raw SAR data by operating over the pulse history in each range bin, on a range bin by range bin basis. Mathematically, this can be written as

$$D(k_u, t) \equiv \int_{-\infty}^{\infty} d(u, t) e^{-jk_u u} du \tag{21.65}$$

so that range (fast time t) is transparent to the function. The output D is a two-dimensional complex function of time and spatial frequency. These dimensions map to scene position via the following relationships:

$$t = \frac{2R}{c} = \frac{2\sqrt{x^2 + r^2}}{c} \tag{21.66}$$

$$k_u = -\frac{4\pi}{\lambda_c}\left(\frac{x}{r}\right) \tag{21.67}$$

Now assume that the scene center is far from the platform and the cross-range extent is limited so that $x \ll r$, and the down-range scene extent is small enough that its influence on spatial frequency can be ignored. Using r_0 for the cross-track range to scene center, equations (21.66) and (21.67) can be approximated as

$$t \approx \frac{2}{c}r \tag{21.68}$$

$$k_u \approx -\frac{4\pi}{\lambda_c}\left(\frac{x}{r_0}\right) \tag{21.69}$$

With these approximations, the final SAR image $f(x, r)$ can be formed via a simple linear rescaling of the axes of the fast-time/spatial-frequency data:

$$f(x, r) = D\left[k_u\left(-\frac{\lambda_c r_0}{4\pi}\right) \to x,\ t\left(\frac{c}{2}\right) \to r\right] \tag{21.70}$$

The DBS image formation procedure, a spatial Fourier transform over the pulse history followed by mapping of the data into the scene coordinates, is summarized in Figure 21-23.

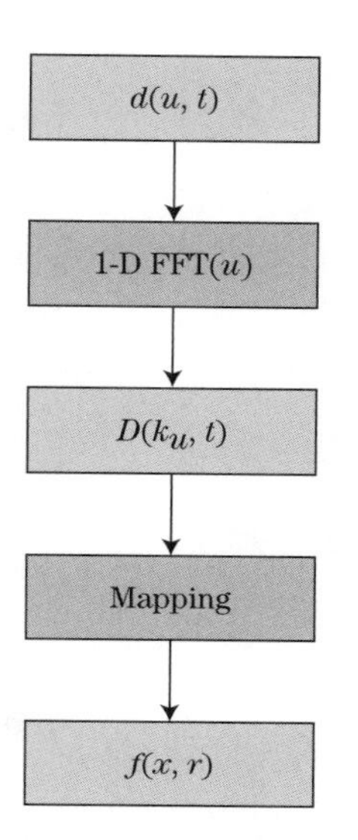

FIGURE 21-23 ■ DBS processing performs a spatial Fourier transform over the pulse history followed by mapping the data into scene coordinates.

21.5.5 DBS, Doppler, and the PSR

In the preceding two sections the PSR and a lengthy mathematical development were used to derive something that was already understood from consideration of the Doppler cone angle (θ_{cone}): a coarse SAR image can be formed by a one-dimensional Fourier transform of the data in each range bin. Indeed, the PSR derivation can be taken further to reproduce the Doppler relationship. Starting with equation (21.63),

$$k_u = -\frac{4\pi}{\lambda_c}\left(\frac{x}{r}\right)$$

and noting that the ratio (x/r) equals the tangent of the azimuth angle, which is approximately equal to the sine of the azimuth angle for small angles, gives

$$\frac{x}{r} = \tan\theta \approx \sin\theta$$

Spatial frequency then becomes

$$k_u \approx -\frac{4\pi}{\lambda}\sin\theta \tag{21.71}$$

The mapping between spatial frequency Doppler frequency is, from equation (21.62),

$$k_u = -\frac{2\pi}{v}f_d$$

Substituting this expression into equation (21.71) reproduces equation (21.10),

$$f_d \approx \frac{2v}{\lambda}\sin\theta \tag{21.72}$$

the relationship that is at the heart of the SAR Doppler paradigm. This development shows that the PSR can be used to develop and justify the Doppler filtering approach to SAR imaging.

21.5.6 DBS Example

Simulated data are now used to demonstrate the DBS image formation procedure. In this example the modeled radar and collection geometry have the following characteristics:

- Carrier frequency = 9.6 GHz.
- RF bandwidth = 3 MHz.
- Integration angle = 0.02°.
- Resolution = 50 meters (both dimensions).
- Number of along-track sample points (pulses) = 80.
- Range to scene center = 50 km.

Figure 21-24 contains diagrams of the collection geometry; note that the coordinate system origin is assigned to the center of the scene and not the platform location in these figures. The scene of interest, approximately 4 km × 4 km, appears to the right of Figure 21-24a, with the data acquisition locations at the far left. The axes ratio is distorted in Figure 21-24a to show both the scene and the platform positions. The synthetic aperture is relatively short so that the collection points all fall on top of one another in the diagram. The scene to be imaged is made up of five point scatterers whose locations are more easily discerned in the undistorted diagram in Figure 21-24b. One target is at the scene center, with the remaining targets distributed over down-range and cross-range.

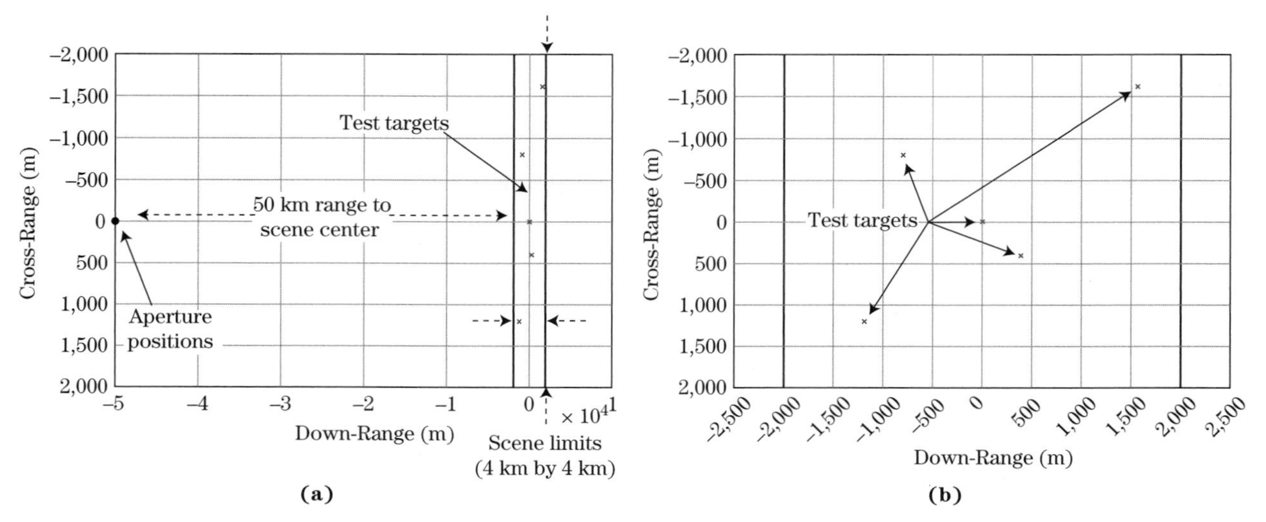

FIGURE 21-24 ■ Simulated DBS geometry emphasizes (a) the overall collection geometry, and (b) point scatterer locations in the scene of interest.

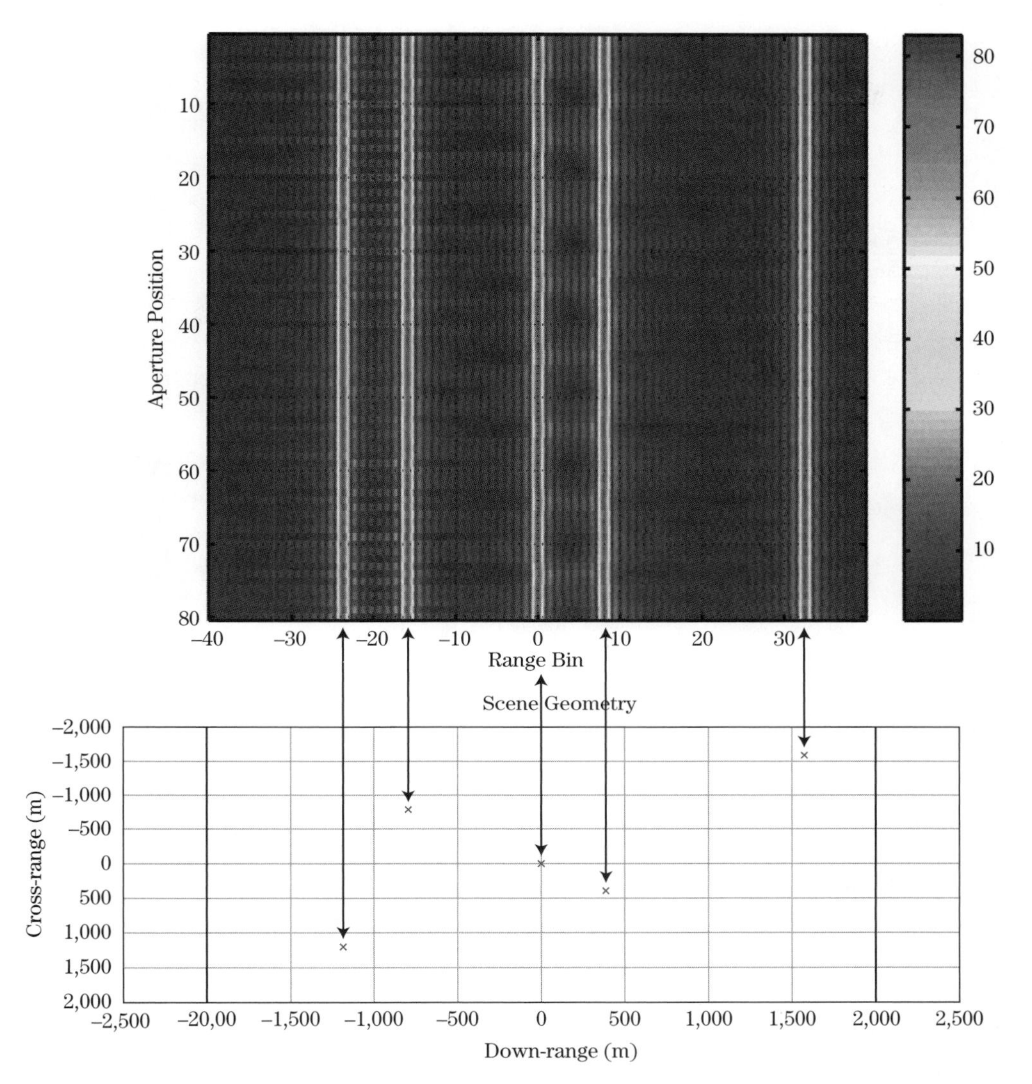

FIGURE 21-25 ■ Mapping of HRR profiles in the raw data to scatterer locations in the scene of interest.

An image of the magnitude of the recorded returns appears in Figure 21-25. The image shows the range-compressed raw data $d(u, t)$ with u given by aperture position number along the axis and t in the form of range bins numbered with respect to scene center. There are five point scatterer returns, one for each of the targets in the scene. For example, the return from the target at scene center appears at range bin number zero in Figure 21-25. Consistent with the assumption made under DBS imaging, the PSRs are confined to the same range bin over the collection.

Not shown in Figure 21-25 is the phase information. Figure 21-26 corresponds to the magnitude image in Figure 21-25 but indicates phase contribution by displaying the real part of the complex data $d(u, t)$. For example, the return in range bin zero displays no bipolar (positive and negative) fluctuations over the dwell, implying a constant phase and therefore zero Doppler and a cross-range location of zero. The other scatterers, in contrast, show some bipolar fluctuations. Low-frequency fluctuations are evidence of small Doppler (spatial frequency) and a location near the scene centerline, while higher frequencies reflect larger displacements. (The imaginary part of $d(u, t)$, not shown in Figure 21-26, has the

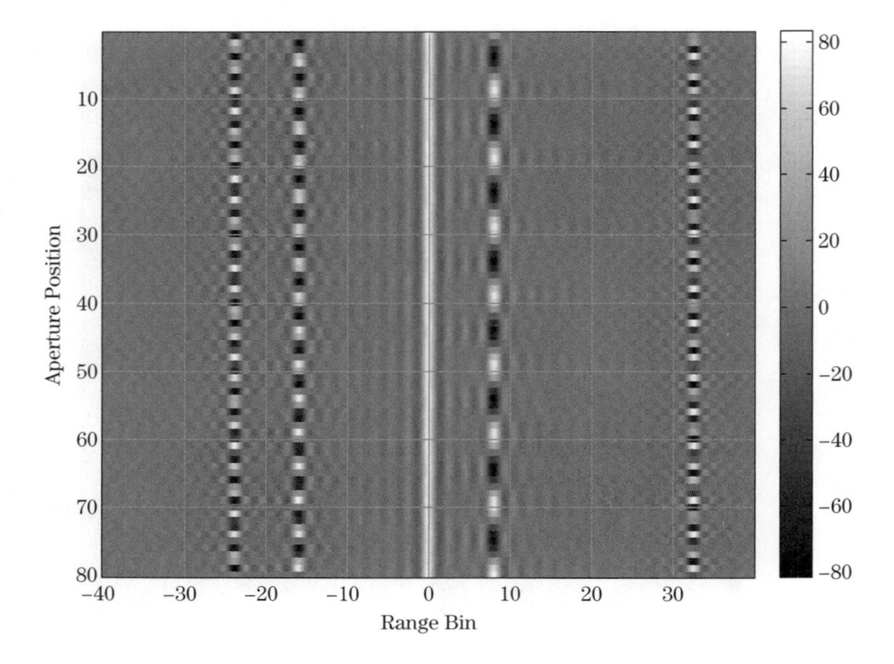

FIGURE 21-26 ■ Real part of the HRR profiles, with positive voltages in white and negative voltages in black, indicates phase changes over the dwell and, therefore, Doppler frequency.

same fluctuations frequencies as the real part, but the relative phasing between the real and imaginary components determines to which side of the centerline a scatterer falls.)

By Fourier transforming over the pulse history in each range bin (the columns in Figures 21-25 and 21-26), the linear phase signal of each scatterer is concentrated into a single frequency bin, which is then mapped to cross-range. The resulting DBS intensity image appears in Figure 21-27 on a decibel scale. The point responses in the image faithfully reproduce the true scatterer locations in Figure 21-24b. The target range Doppler functions show the two-dimensional "sinc" ($\sin x/x$) response commonly observed in Fourier analysis.

21.5.7 DBS Limitations

The DBS approach to image formation is easy to implement, consisting primarily of a Fourier transformation over each compressed range bin, which can be performed in a computationally efficient manner through application of the fast Fourier transform (FFT) algorithm. (See Chapter 17 for details on the FFT.) However, DBS is a suitable image formation algorithm only under tightly restrictive conditions. Recall the series of assumptions and approximations made to derive DBS from the PSR:

- Scatterer returns have the same hyperbolic form such that all returns are confined to a single range bin over the dwell.
- The hyperbolic form of the PSR is replaced with a quadratic form via the binomial approximation.
- A quadratic term in platform position u was neglected.
- The mapping from time and spatial frequency to along-track and cross-track location was simplified.

Figure 21-28 depicts an imaging geometry much more demanding than that used to demonstrate the DBS technique (Figure 21-24). Five scatterers are clustered into the scene

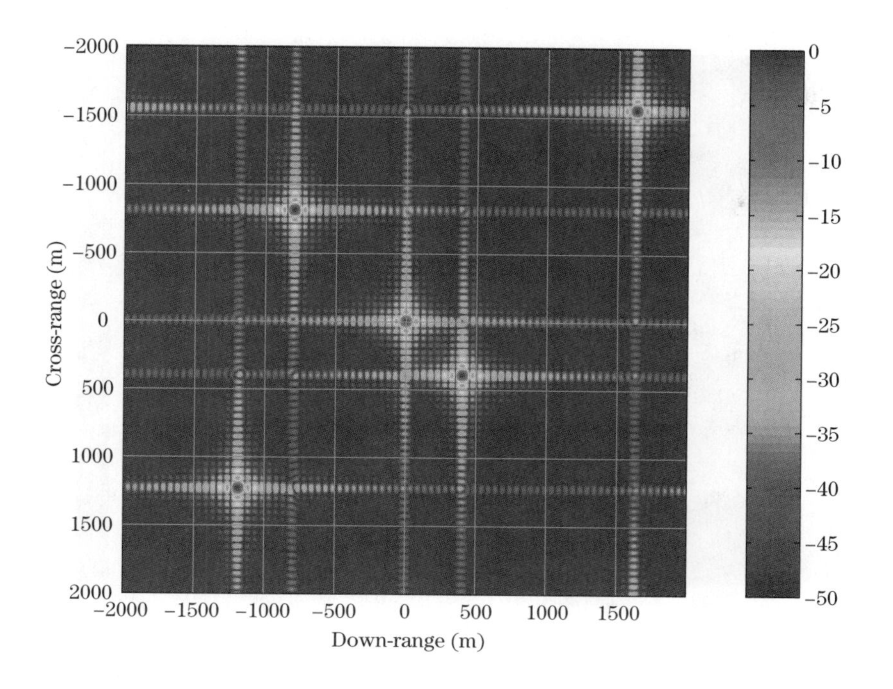

FIGURE 21-27 ■ DBS image of the simulated point targets.

of interest. For this simulation the radar and geometry parameters are

- Carrier frequency = 600 GHz.
- RF bandwidth = 750 MHz.
- Integration angle = 51°.
- Resolution = 0.2 meters (both dimensions).
- Number of along-track sample points (pulses) = 275.
- Range to scene center = 30 meters.

This example constitutes near-range imaging of a relatively large scene (compared with the range to the scene) with a low-frequency system employing a large RF bandwidth and wide integration angles to achieve fine resolution in down-range and cross-range. Figure 21-29 shows the real part of the resulting raw range-compressed data. The full hyperbolic form of the target PSRs is readily apparent. Target returns are not constrained to one range but

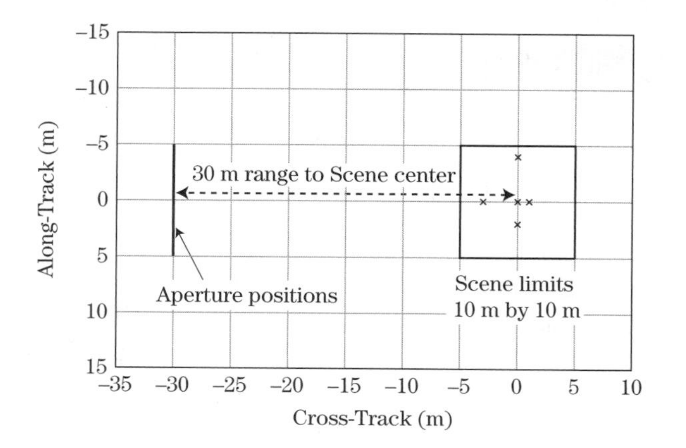

FIGURE 21-28 ■ A challenging, near-range collection geometry.

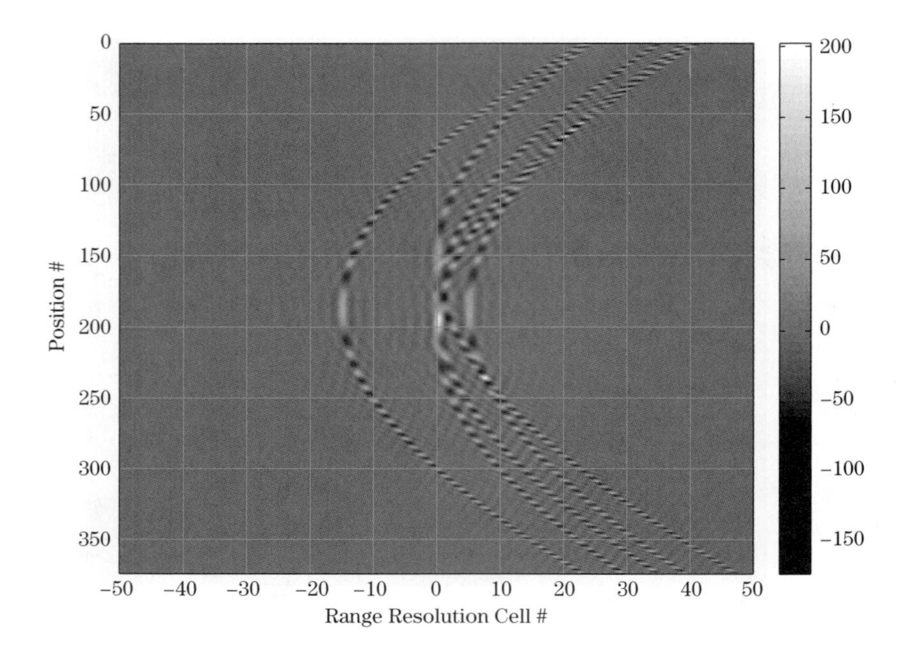

FIGURE 21-29 ■ Real part of the HRR profiles for the near-range collection. Tremendous range migration and higher-order phase modulations are apparent in the raw data.

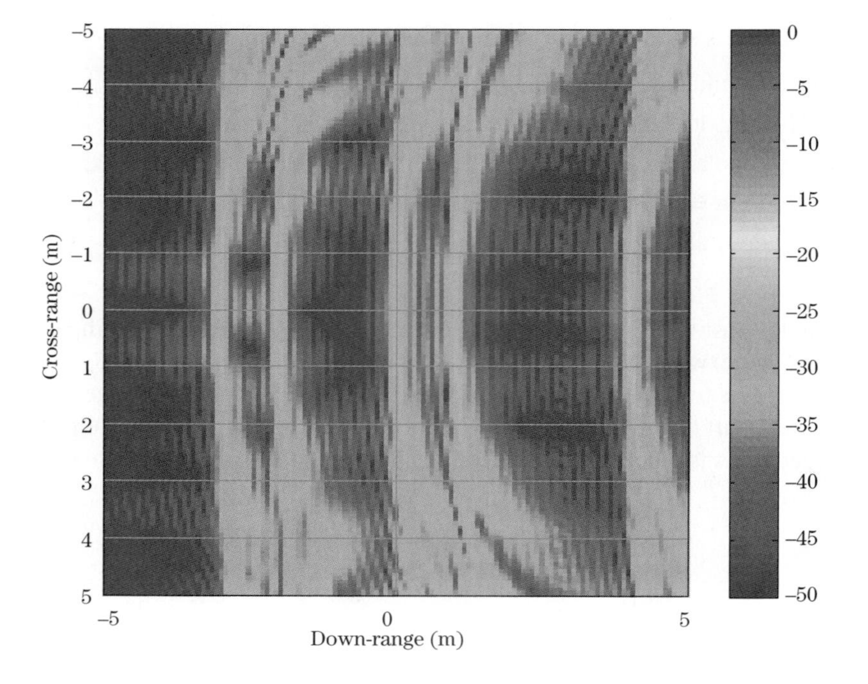

FIGURE 21-30 ■ The DBS image of the near-range point targets is distorted and unusable.

instead migrate over many range bins. A careful comparison of the returns reveals slightly different hyperbolic forms for scatterers at different down-range locations. The bipolar fluctuations of the PSRs indicate not linear phase (constant Doppler) or even quadratic phase but a complicated history consistent with the full hyperbolic PSR. Applying this data to DBS results in an image, provided in Figure 21-30, so terribly distorted and defocused as to be unusable.

21.5.8 Matched Filter Imaging

While DBS is simple to implement, its intrinsic limitations have encouraged the development of sophisticated image formation algorithms that exploit knowledge of the PSR more completely. At the other extreme from DBS, one very general SAR image formation algorithm applies the PSR to the measured data as a matched filter [11]. Recall (Chapters 14 and 20) the matched filter is designed to maximize the signal-to-noise ratio (SNR) at the output of the filter when the signal is deterministic and known. In SAR the signal is the return from a scatterer at a specific location of interest as described by the PSR. When the noise is white and Gaussian the matched filter impulse response is equal to the conjugate of the time-reversed signal of interest. Equivalently, the output of the matched filter equals the cross-correlation of the data with the conjugate of the PSR.

With these concepts in mind, matched filter image formation may be developed through the following procedure. First, construct an output image with all pixels set to zero. Next, for a particular output pixel location (x, r), compute the time delay to a scatterer at this location from a radar at $(u, 0)$ using (21.42)

$$t(u; x, r) = \frac{2}{c}\sqrt{(u - x)^2 + r^2}$$

This time delay is used to construct the expected, or reference, response for a unit-amplitude point target at that location. This reference signal $d_{REF}(u, t; x, r)$ has the form

$$d_{REF}(u, t; x, r) = \delta_D\left(t - t(u; x, r)\right) \tag{21.73}$$

where δ_D is the Dirac delta function, and $t(u; x, r)$ is given by (21.42). The Dirac delta function $\delta_D(t - t_0)$ is a continuous-time impulse function equal to zero when $t \neq t_0$ and defined by the “sifting” property that

$$\int_{-\infty}^{\infty} \delta_D(t - t_0) d(t) dt = d(t_0)$$

It can be inaccurately, but usefully, thought of as having a “large” value at $t = t_0$ and zero elsewhere. The reference response in (21.73) has the dimensions of the recorded data $d(u, t)$ but is the deterministic prediction of the return from a scatterer at (x, r) having an RCS equal to 1 and a phase angle equal to 0.

Finally, the correlator matched filter is realized by conjugating the reference signal, multiplying by the data, and coherently summing over the two-dimensional product,

$$f(x, r) = \int_{-\infty}^{\infty}\int_{-\infty}^{\infty} d(u, t) d^*_{REF}(u, t; x, r) du\, dt \tag{21.74}$$

to yield the amplitude and phase for the pixel at (x, r). While (21.74) has the form of infinite integrals over continuous time and location, the matched filter is implemented in real systems as a finite summation on digital data having discretized times and locations.

The operations represented by (21.42), (21.73), and (21.74) are straightforward and can be captured with just a few lines of computer code. Furthermore, image formation via the matched filter is not constrained to the linear collection shown in Figure 21-16. Indeed, data acquisition against a scene may be performed over an arbitrary set of locations in three-dimensional space; so long as the platform position is known at all times, the time delays

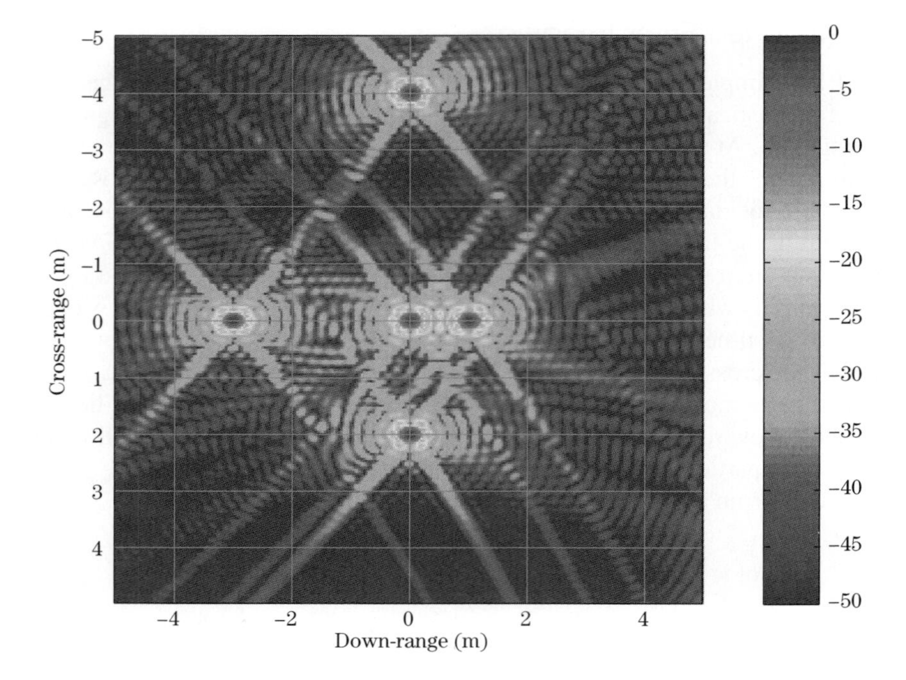

FIGURE 21-31 ■ Matched filter image of the near-range point targets is accurate.

to a down-range–cross-range location of interest can be calculated and the corresponding reference response d_{REF} generated and applied to the data as in (21.74).

Figure 21-31 contains a matched filter image formed on the wideband and wide-angle data collected as depicted in Figure 21-28. In contrast to the DBS image in Figure 21-30, the point target returns in Figure 21-31 are well focused and properly located and have responses consistent with the resolution provided by the bandwidth and integration angle. The unusual sidelobe patterns are not particular to the matched filter method but rather arise from the wide integration angle. Referring back to Figure 21-6, it can be seen that data collected over a small integration angle and a narrow bandwidth at a high carrier frequency have a polar region of support in the frequency domain that is nearly rectangular, which gives rise to a two-dimensional sinc response in the final image; see Figure 21-27 for an example. In contrast, data collected at low frequencies over a wide angle have a frequency domain region of support that is not rectangular but more like a pie-shaped wedge or an annulus with a limited angular extent. These non-rectangular regions of support produce impulse responses that are not sinc-like, as seen in Figure 21-31.

A downside to the matched filter is the need for significant processing power when imaging extended scenes at fine resolutions. The operation in (21.74) produces the complex value for just one image pixel and must be repeated for all pixels. Large images may contain tens or hundreds of millions of pixels. Furthermore, data sets and reference functions may be very large, often on the order of tens of thousands of fast-time samples and platform locations, so that the multiplication at the heart of (21.74) becomes demanding. Finally, a different reference response is required for every output pixel. Again, the reference response can be very large, and its core calculation involves a computationally nontrivial square root operation; again, see (21.42).

Matched filter processing demands may be alleviated somewhat by examining the reference response and noting that, for any given along-track location u, most of the samples over fast time t have zero value except at the time delay corresponding to the pixel location; see (21.73) and recall the Dirac delta function is always zero except when the argument is exactly zero. Therefore, most of the two-dimensional multiplication in (21.74) is a wasted scaling by zero. An alternative, and mathematically equivalent, approach uses (21.42) to calculate a set of time delays. These delays are used to index into the data and extract the data value at the time delay to the pixel. Therefore, one data value is extracted for each platform location. The resulting one-dimensional data subset is then multiplied by the corresponding one-dimensional reference response and integrated. This approach is justified mathematically by substituting (21.42) and (21.73) into (21.74), yielding

$$f(x,r) = \int_{-\infty}^{\infty} d\left(u, \frac{2}{c}\sqrt{(u-x)^2 + r^2}\right) du$$

Moving from two-dimensional to one-dimensional data manipulation and integration yields a significant computational savings. As the core operation now centers on applying time delays to, or projecting back into, the data, this image formation method is commonly referred to as *time-domain backprojection*. As the "time-domain" qualifier is rather redundant it is sometimes identified more succinctly as simply *backprojection*.

21.5.9 Image Formation Survey

Backprojection is just one of many image formation alternatives to DBS. Table 21-1 provides a survey of these methods with some salient characteristics. The table partitions the options into four categories:

1. Doppler methods, including DBS [12,13].
2. Fourier matched filter methods, which attempt to realize the matched filter as a computationally efficient multiplication in the frequency domain [14,15].
3. Tomographic methods, which view the SAR data acquisition process not as a PSR but rather as a set of one-dimensional projections of the scene of interest collected over a variety of azimuth angles [16,17].
4. Inverse methods, including matched filtering and backprojection [18–21].

Two of the most popular algorithms, the range migration algorithm (RMA) and the polar formatting algorithm (PFA), deserve further comment. In contrast to DBS, RMA not only assumes a hyperbolic PSR but also accommodates the down-range dependence in the hyperbolic form. RMA is often applied to stripmap acquisitions collected at low frequencies, wide bandwidths, and wide integration angles. PFA is commonly employed for fine-resolution imaging on spotlight collections against small (relative to stripmap) scene sizes. PFA is designed to be a computationally efficient inversion to the tomographic model of data acquisition. RMA and PFA are capable of generating high-quality SAR imagery possessing fine resolutions over a wide range of conditions. Both realize efficient matched filter imaging by operating in the frequency domain, and both require intensive interpolation of the data at some point in the image formation chain.

TABLE 21-1 ■ Survey of SAR Image Formation Algorithms

Category	Features	Limitations	Algorithms	Notes
Doppler	Requires only a 1-D FFT after pulse compression	No accounting for migration of scatterers through range bins or phase orders above linear	Doppler beam sharpening	Fourier transform provides the "matched filter"
			DBS with azimuth dechirp	Accounts for QPE by applying a conjugate quadratic modulation to the data prior to DBS
Fourier Matched Filter	Match filtering with the PSR implemented efficiently by multiplication in the frequency domain	Data acquisition must be linear with uniform spacing between along-track samples	Range-Doppler algorithm	Assumes a single PSR is good for the entire scene, so focused swath depth is limited
			Range migration algorithm	Range Doppler algorithm combined with Stolt data interpolation provides error-free imaging over deep swaths
			Chirp scaling algorithm	Exploits direct sampling of linear FM waveforms to realize a computationally efficient approximation to the Stolt interpolation
			Range stacking algorithm	Like RMA, but less computationally efficient
Tomographic	Projection-slice view of data acquisition; efficient image formation follows	Scene size; data must be time delayed and phase corrected prior to image formation	Rectangular formatting algorithm	Analogous to DBS in simplicity; capable of fine resolution, but scene size severely restricted
			Polar formatting algorithm	2-D data interpolation provides quality imaging at fine resolutions over reasonable scene sizes; scene extents ultimately limited by geometric considerations
Inverse	A more holistic, mathematical view of data acquisition and image formation	Demanding processing requirements, and sometimes memory requirements as well	Matched filtering	Employs a product of the data and a reference response; implementable in either time-distance or frequency domains
			Constant aperture backprojection	Delay and sum imaging; same data records used for all pixels
			Constant integration angle backprojection	Like constant aperture backprojection, but different subset of the overall data applied to each pixel to realize a constant integration angle
			Filtered backprojection	Wideband/wide-angle data is preequalized to improve impulse response in the final image
			Fast backprojection	Family of algorithms that garner computational efficiencies by making minor approximations to basic backprojection
			Quad-tree	Fast backprojection using multiresolution topologies
			Constrained inversion	SAR imaging treated as a mathematical inverse problem

21.6 IMAGE PHENOMENOLOGY

This section discusses a number of interesting effects and artifacts that are often observed in SAR imagery.

21.6.1 No Return Areas

Regions having low-pixel intensities in SAR images are known collectively as *no return areas* (NRAs). NRA sources include the following:

- Shadows, discussed in detail in the next section.
- Paved areas, including roads, parking lots, and airport runways and tarmacs. At typical radar wavelengths these surfaces are typically smooth, according to the Rayleigh criterion discussed in Chapter 5, and so specularly reflect most incident RF energy away from the radar receive antenna. However, isolated returns due to reflectors, signs, and lighting fixtures may still be present in these NRAs.
- Dry lake beds and planed soil and sandy surfaces, for the same reason as paved areas.
- Water surfaces, ranging from small ponds and rivers to lakes, waterways, and finally the open ocean. Still water is specularly reflective, providing very little return.

Figure 21-32 consists of three image chips extracted from larger SAR images. Each chip is dominated by a military vehicle in an open, grassy field; the particular vehicle and pose angle with respect to the collecting platform varies from chip to chip. Shadows cast by the vehicles are manifested as NRAs above (down-range from) the strong, direct vehicles returns. Figure 21-33 contains a medium-resolution SAR image of an airport in the California high desert. Runways and paved tarmac are clearly rendered as large, two-dimensional NRAs, while taxiways and roads appear as long, thin NRAs. Figure 21-34 shows a coarse-resolution range Doppler map of land and water returns near the intersection of the states of Delaware, Maryland, and Virginia. Regions of water are clearly visible as NRAs in contrast to the strong returns from dry terrain.

It should be noted that radar in general, and SAR in particular, has been used to image wave structure and measure ocean currents with great success. While returns from water in motion can be significant, they are usually far weaker than scattering from nearby terrain, so that even a choppy lake appears much darker than adjacent land in a SAR image.

NRAs in SAR imagery are not completely black (zero return), due to ubiquitous thermal noise, the range and Doppler sidelobes returns generated by nearby targets, and returns from terrain in the antenna azimuth and elevation sidelobes, aliased into the image via Doppler and range ambiguities, respectively. Nonetheless, large NRAs free of man-made objects, particularly dry lake beds, offer a convenient approximation to a free-space environment and are often used for SAR system quality assurance, system diagnostics, calibration, and performance assessment.

21.6.2 Shadowing

At the beginning of this chapter it was noted that the radar provides the illumination required by SAR imaging, in contrast to sensors like optical cameras and infrared imagers that rely on ambient or emissive energy. The radar transmit antenna acts as a spotlight, a point source (in typical geometries) for directing RF energy onto the scene of interest. Objects in

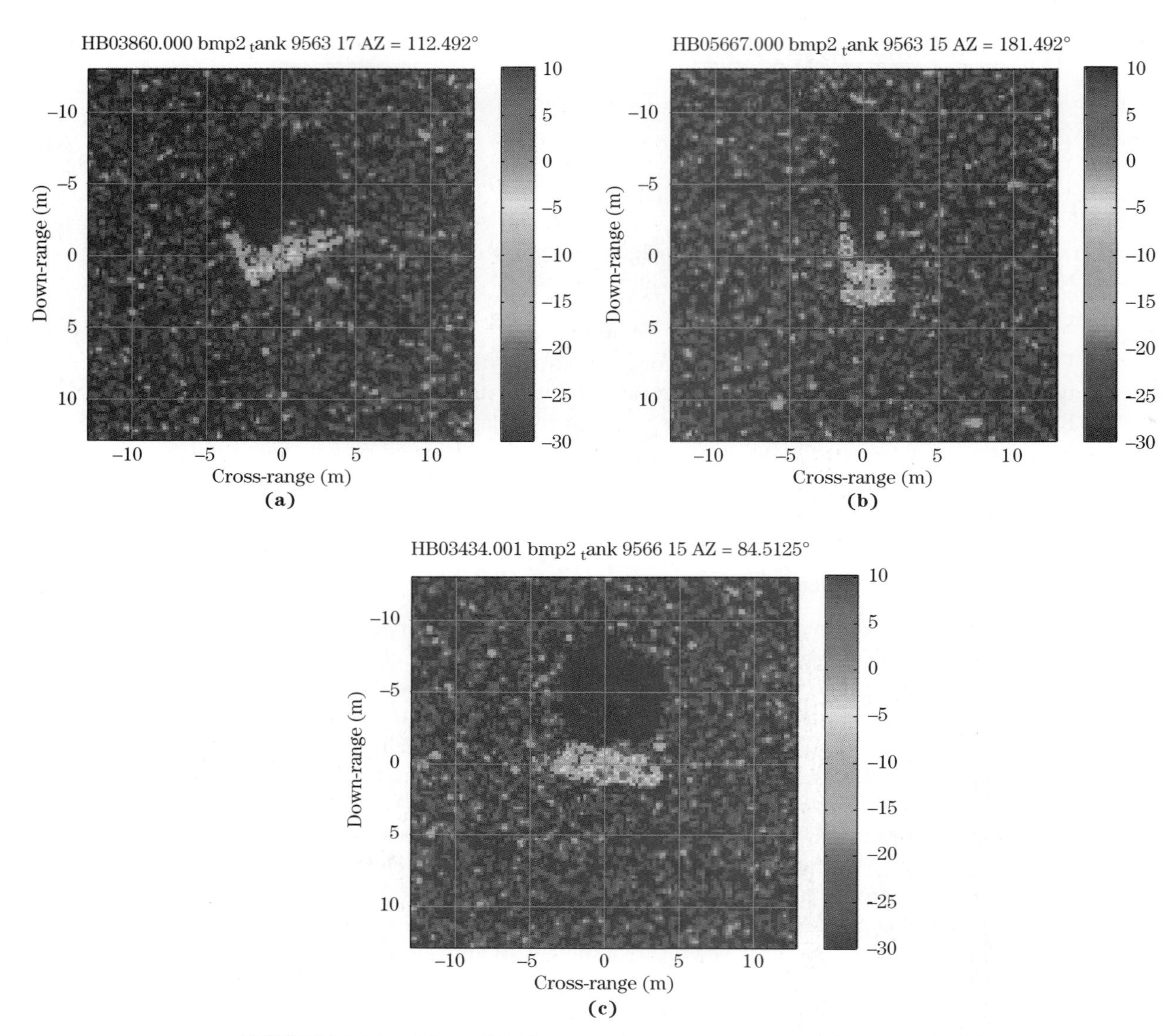

FIGURE 21-32 ■ Three SAR images of an armored vehicle with radar at the bottom of each image and the vehicle (a) at a non-cardinal pose angle, (b) broadside to the radar, and (c) end-on to the radar.

the scene scatterer some of this energy (a small fraction of the scattered energy is reflected back to the radar and recorded as raw SAR data) and absorb some of this energy. Many objects of interest reflect or absorb practically all of the incident RF energy, passing little or no energy for illumination of other scatterers down-range. These RF opaque objects thus cast radar shadows away from the radar illuminator. Shadows are apparent in formed SAR images as regions of low intensity down-range from the intervening objects. In Figure 21-32 the radar was located beyond the bottom of the images, and the vehicles cast shadows onto the distributed clutter returns from the surrounding field toward the top of the chips.

Large, solid objects having significant height tend to cast noticeable shadows. Mountains, trees, houses, and vehicles are all common sources of shadowing in SAR imagery.

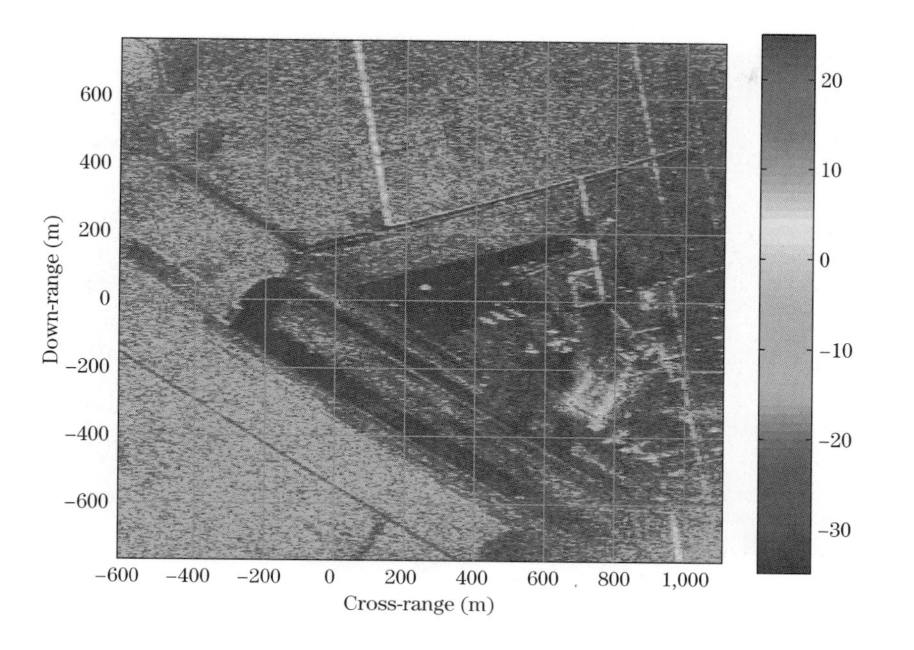

FIGURE 21-33 ■ SAR image of the Mojave Airport showing low-power areas corresponding to runways and tarmac, moderate-power areas off the runways due to natural clutter return, and strong man-made returns near the tarmac. (Data courtesy of Raytheon Corporation.)

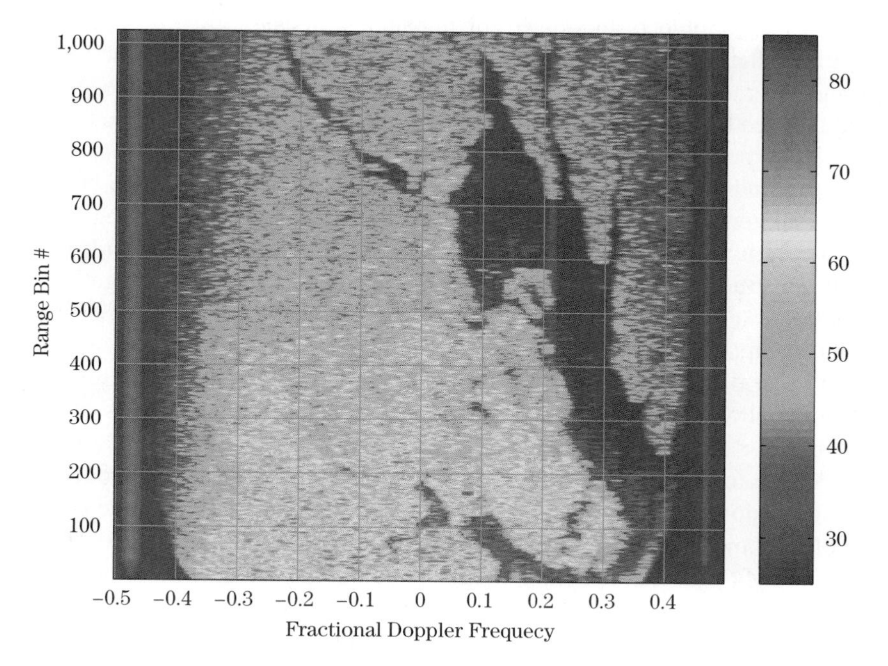

FIGURE 21-34 ■ Coarse-resolution DBS image emphasizing the contrast in power between land and water regions. (Data courtesy of Northrop Grumman Corporation.)

Figure 21-35 shows the shadow geometry for an isolated hill; the radar is above and to the left in the far field (outside the boundary of the image). Returns on the down-range side of the hill fall into shadow and would appear as an NRA in the final SAR image.

Shadow width is consistent with the cross-range extent of the object. In Figure 21-32, for example, the cross-range extents of the shadows and the vehicles returns are comparable over a wide range of vehicle pose angles. Shadow length on the ground is equal to the height of the object divided by the tangent of the local grazing angle. Obviously, shadows

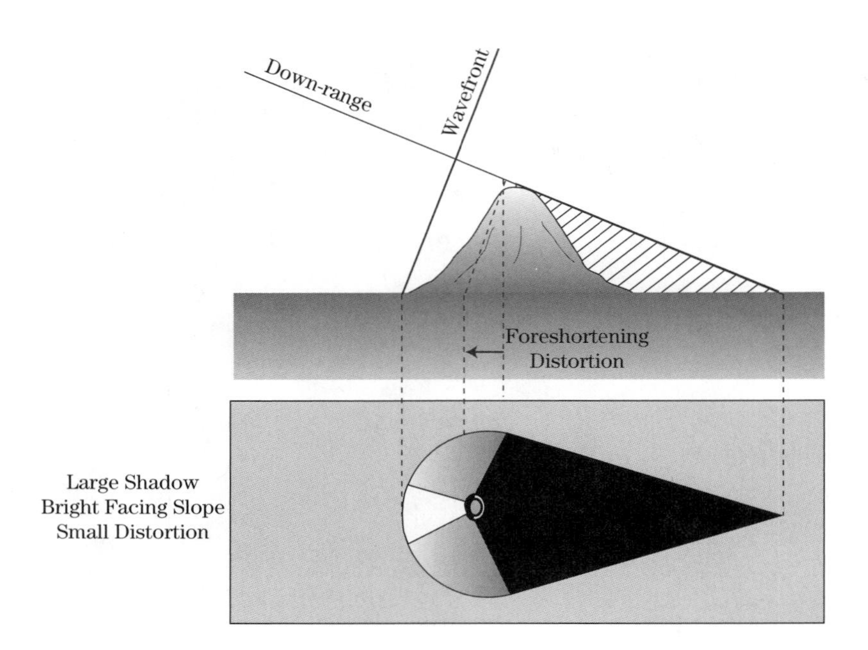

FIGURE 21-35 ■ Foreshortening: Targets having great height will appear collapsed in a SAR imaged collected with a significant grazing angle.

are longer for taller objects and lower grazing angles. Because grazing angle drops as the SAR platform height lessens or slant range grows, shadowing can become severe at long ranges and low radar altitudes. Even mild terrain undulations, for example, gently rolling hills and isolated tree stands, will cast very long shadows, plunging a large fraction of the image into darkness.

The size and shape of a shadow, when falling onto a flat region of homogeneous, distributed clutter, provides important information regarding the occluding object. Often the direct returns from buildings, ground vehicles, and trees are difficult to interpret. Coherent constructive and destructive interference can render these targets in the final SAR image as a cloud of noise-like returns or an amorphous collection of bright "blobs." Large, flat components angled away from the radar will reflect the bulk of the scattered energy in other directions, so that these items are not visible in a SAR image. Shadow borders, in contrast, faithfully represent the size, orientation, and shapes of intervening articles.

In addition to casting shadows, objects with nonzero height also produce a second interesting phenomenon. Because the object possesses finite height, some portion of the surface of the object facing the radar must have nonzero slope. In Figure 21-35, for example, the height of the hills casts a shadow, and the hill has height because the surface illuminated by the radar is not horizontal. A brief examination of the geometry indicates the local grazing angle on the illuminated side of the hill is higher than that of the surrounding level ground. For most surfaces, radar reflectivity increases with grazing angle (see Chapter 5). Thus, the return from the near side of the hill will be stronger than the scattering from flat Earth. Putting these two effects together, a tall object is often manifested in radar imagery through two phenomena: a localized bright patch next to a dark shadow directed down-range. The vehicles in the image chips in Figure 21-32 exhibit this feature.

Brightening and shadowing are convenient artifacts for image interpretation. For example, the fact that shadows are always cast down-range from the radar allows the viewer to deduce the direction of the radar with respect to the imaged scene.[3] If the collection geometry is unknown, shadow lengths from known objects can be used to estimate the grazing angle. Conversely, if the grazing angle is known, shadow lengths can be used to estimate object heights. For example, the grazing angles in Figure 21-32 are known to be 15° and 17° and shadow lengths appear to be between 7 and 10 meters, yielding an estimated vehicle height of 2 to 3 meters. This estimate is consistent with the reported 2.5 meter height for the BPM-2 armored fighting vehicles in the chips.

21.6.3 Foreshortening and Layover

The raw SAR data from a single aperture is two-dimensional, for example, fast-time delay t (proportional to slant range R) and platform along-track position u. However, objects and terrain in the real world exist in three dimensions. For example, scatterers may be located by their along-track (cross-range) location x, horizontal cross-track (down-range) position on the ground y, and height above a reference level z. The projection of three-dimensional scattering onto a two-dimension data set causes a loss of information that ultimately leads to an ambiguity in the actual location of a scatterer observed in a SAR image.

This ambiguity can be quantified mathematically. In the development of the SAR collection geometry, the scatterer height z and the platform height H were collapsed into a down-range distance r:

$$r = \sqrt{(H - z)^2 + y^2} \tag{21.75}$$

There is no error in doing this; the hyperbolic PSR is a function of r, and the final SAR image positions returns as a function of r. However, if the platform altitude and the terrain height over the scene are known, then equation (21.75) makes it possible to map down-range distance r to ground range y:

$$y = \sqrt{r^2 - (H - z)^2} \tag{21.76}$$

A special case of equation (21.76) occurs when the terrain is understood to be flat with height z for all scatterers in the scene of interest. Then equation (21.76) can be applied to a one-dimensional resampling of the initial image in down-range to produce a final image having a constant ground range interval between pixels. This process is sometimes referred to as the projection or mapping from the *slant plane* (the initial image in r) to the *ground plane* (the final image in y).

Notice that the mapping provided by equation (21.76) is not available if the scene height z is unknown. The situation is even worse when z is unknown and varies over the scene.

The ambiguity represented by equation (21.75) makes projecting from the slant plane to the ground plane a challenge, but that is not the only difficulty. By selecting a down-range distance of interest r_0 and rewriting equation (21.75),

$$(H - z)^2 + y^2 = r_0^2 \tag{21.77}$$

[3]There is no universal convention for orienting SAR imagery. The author has seen examples of SAR images oriented so that the radar was at the top, bottom, left, and right side of the image. Squinted collections and geo-referenced imagery can leave the radar at any arbitrary angle.

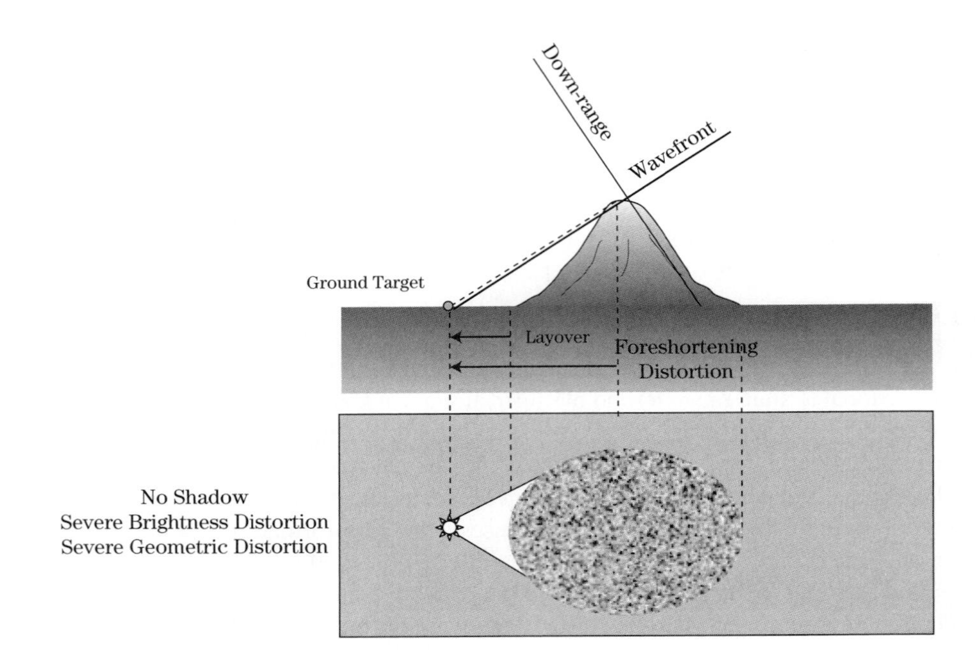

FIGURE 21-36 ■ Layover: In collections having steep lookdown angles and significant height variations, scatterer order may appear reversed in the down-range direction.

it can be seen that there is a loci of ground range and height positions (y,z) whose returns all fall into the same down-range r_0. These positions take the form of a circle with radius r_0 centered on the platform flight path.

The diagram in Figure 21-36 demonstrates this effect using a SAR collection having a very steep grazing angle. The "Wavefront" line denotes a surface having constant down-range r. The look-down geometry is so steep that the returns from the hilltop and a strong scatterer on the ground have the same down-range distance. Even though the hilltop is at a much greater ground range than the ground scatterer, the final SAR image renders both returns in the same set of pixels.

In general, terrain height variations cause distortions in the apparent location of returns in down-range. *Layover* is distortion so severe that the order of returns in down-range is reversed (they "lay over") with respect to their order on the ground. For example, in Figure 21-36 the top of the hill appears first in down-range; by sliding the wavefront line in slant range it can be seen that returns from the side of the hill appear next, followed by the base of the hill. This is in reverse order with respect to ground range: first the base, then the hillside, and finally the hilltop. These returns will appear in reverse order in the SAR image. Layover is common in fine-resolution imagery of buildings. Because most buildings have vertical walls, returns for the top of the building appear up-range (nearer) compared with returns from the base.

Milder distortion does not suffer from a change in order but only a change in scaling. *Foreshortening* is the term for an apparent compression of the range distribution of returns on a slope compared with returns from a level surface. Foreshortening occurs on the illuminated side of the hill in Figure 21-35. The moderate, but still significant, grazing angle in this geometry causes the ground range extent of the near-side slope to stay in order but to be compressed into a few down-range bins. Note that as the orientation of the slope approaches the angle of the wavefront all scatterers on that part of the slope compress down to just one range bin.

A priori knowledge of the terrain height over a scene, for example, from a digital elevation model data base, can be used in conjunction with equation (21.76) to resample from the slant plane to the ground plane and correct foreshortening. Theoretically, this approach can be used to correct layover as well, except for one complication: layover is almost always accompanied by the superposition of returns in down-range from scatterers having very different ground ranges. For example, in Figure 21-36 the pixels that contain the layover returns from the hillside also carry the returns from the ground extent between the strong ground target and the base of the hill. Multichannel SAR imaging methods are required to partition these contributions.

A couple of observations conclude this section. Notice the absence of shadowing on the far side of the hill in Figure 21-36 compared with Figure 21-35. As would be expected, steep look-down angles tend to reduce shadowing. However, they also result in degraded ground range resolutions (a consequence of the slant plane to ground plane projection), and an increase in the frequency and severity of foreshortening and layover. Finally, note that the slope of the back side of the hill aligns closely with the look direction, so that the returns from this slope are distributed over a long down-range extent, even though the slope has a moderate ground range extent. This rare phenomenon has no common term to describe it; as it is the opposite of foreshortening, it might be called *aft-lengthening*. Aft-lengthening requires that the terrain height decrease as range increases, but not so rapidly as to put the terrain into shadow.

21.6.4 Speckle

Speckle is an often misunderstood artifact of coherent imaging of distributed targets. Figure 21-37 shows an overhead view of a natural scene consisting of a large number of small scatterers. Examples of real terrain that might be modeled in this fashion include a sandy beach, a grassy field, or thick treetop canopies. The scene in Figure 21-37 has been subdivided into resolution cells, the size of each cell consistent with the down-range and cross-range resolution limits of the SAR system. Each resolution cell is composed of a large number of small scatterers. For example, a 1 m × 1 m cell on a grassy field may contain on the order of a million blades of grass. The exact number of blades varies from cell to cell. Furthermore, the blades in any given cell are randomly distributed and have various lengths and orientations.

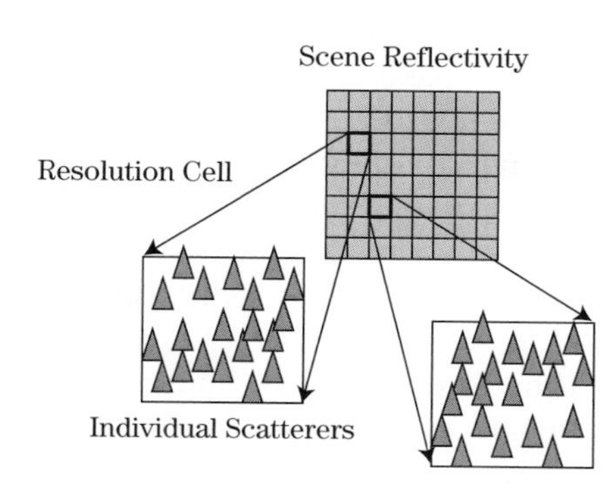

FIGURE 21-37 ■ Speckle is the natural consequence of coherent imaging on a scene having a large number of small, unresolved, randomly placed scatterers.

The complex return from such a resolution cell observed at the radar is the coherent superposition of the individual returns from each blade of grass. Sometimes the total number, distribution, orientations, and sizes of the blades conspire to coherently reinforce, causing a relatively high power level to be recorded by the radar. In other cells the blade returns may destructively interfere, producing a weak intensity value. Over the scene, the measured amplitude therefore fluctuates randomly from cell to cell. The pixel levels in the final SAR image reflect these intensity variations, as shown in Figure 21-37. The noise-like light and dark changes over the scene take on a "salt-and-pepper" appearance, as though these condiments had been randomly sprinkled over an otherwise constant-intensity image.

Speckle is apparent as noise-like fluctuations in amplitude from pixel to pixel over regions of distributed clutter in Figures 21-32, 21-33, and 21-34. A grassy field provides the speckled returns away from the vehicle and its shadow in Figure 21-32. In Figure 21-33, regions off the runways, roads, and tarmac are unprepared, consisting of moderately rough soil, sand, small rocks, and sparse scrub brush. Finally, the land terrain in Figure 21-34 is dominated by areas of trees whose random collections of leaves and branches generate speckle.

It should be understood that speckle is not a SAR model or system error, but a natural consequence of coherent imaging with finite resolution on a large number of randomly distributed, unresolved scatterers. While the appearance of speckle is noise-like, it is not additive interference, like thermal noise. Rather, it is accurately represented as a multiplicative artifact, a random modulation on an otherwise constant amplitude background level.

The coherent amplitude measured on a cell is due to the reflectivity of the scatterers in that cell, the path length between the radar and each scatterer, and the operating wavelength, which maps path length into phase. If the scatterers in the cells remain unchanged, the SAR platform can form an image of the scene, come back some time later, fly the exact same path with the same frequency and resolution, and form a second image that will be identical to the first, down to the cell-to-cell speckle amplitudes and phases. However, if on the second pass the radar operates in a different RF band, or if it collects data over a different angular extent, or if the scene has been disturbed (e.g., a grassy lawn was mowed), the path lengths and phase contributions within a given resolution cell will be different, and the speckle will be uncorrelated with the realization from the first pass.

The variation in observed speckle with angle and frequency presents an opportunity to smooth the speckle noise by averaging. One approach to *speckle reduction* [22] begins by collecting data over an integration angle or bandwidth much wider than that required for the desired resolution. The collection angles or bandwidths are chosen to be wide enough to support the generation of several images, or "looks," each made from disjoint subbands or subangles. Because the frequency domain regions of support for the images do not overlap, each look contains a different speckle realization. Assuming the images are coregistered, the intensity (energy) in each pixel can be averaged across the multiple looks, which acts to reduce speckle-induced fluctuations. This speckle-reduction procedure is known as *multilook processing*.

A similar benefit can be achieved by forming a single full-resolution image using the entire available bandwidth and integration angle. Next, the energy in each pixel is found; that is, phase is discarded. Finally, a low-order, two-dimensional, low-pass filter is applied to the intensity image. Typical filter impulse response sizes are 3×3, 5×5, and 7×7 pixels, with all impulse response coefficients given equal value, so that every contributing pixel is weighted the same. This moving average process tends to smooth

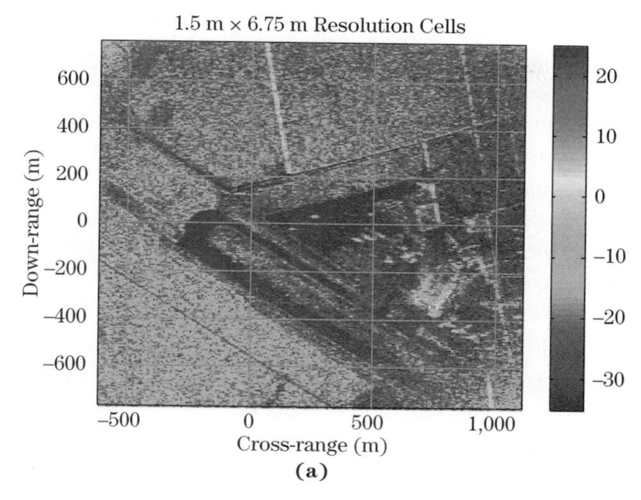

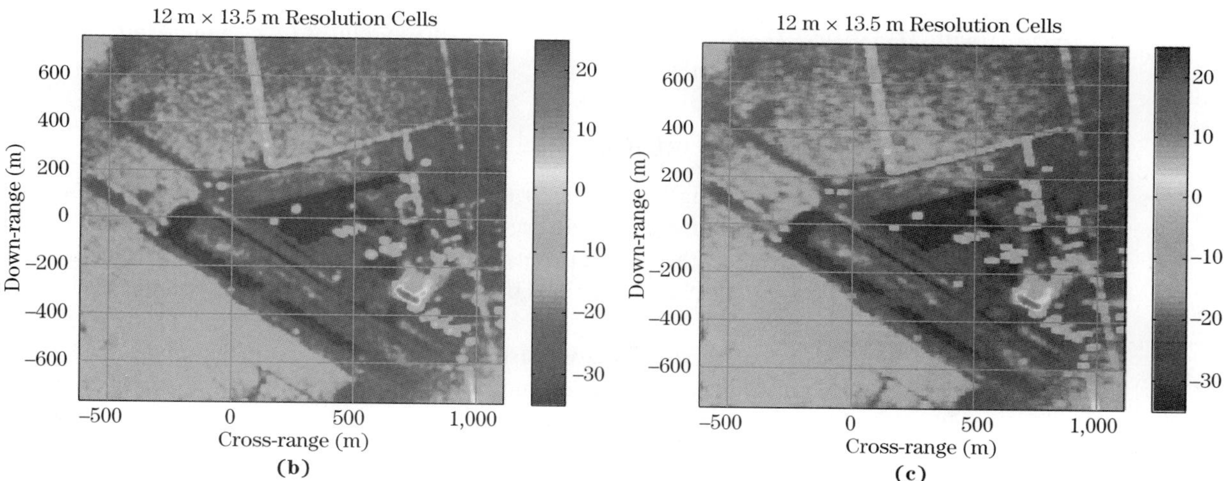

FIGURE 21-38 ■ Speckle reduction on (a) the original image using (b) multi-look processing and (c) image-domain filtering. (Data courtesy of Raytheon Corporation.)

out the speckle energy variations on distributed clutter. Mathematically, given an initial complex image $f(x, r)$, the final smoothed image $P(x, r)$ is found by

$$P(x, r) = |f(x, r)|^2 *_{x,r} h(x, r)$$

where $h(x, r)$ is the impulse response of a real low-order speckle-reducing filter, and $*_{x,r}$ denotes two-dimensional convolution.

Figure 21-38 provides examples of speckle-reduction processing. The original full-resolution image appears in Figure 21-38a; this is the same image from Figure 21-33. This particular collection featured a down-range resolution of 1.5 meters and a cross-range resolution of 6.75 meters, so that the down-range resolution is finer by more than a factor of four. (The frequency domain data was windowed prior to formation of the final image so that the actual resolutions in the image are coarser than 1.5 meters and 6.75 meters.) The power-average of several multilook images is given in Figure 21-38b. Each of the contributing images was formed using one-eighth of the frequency extent and one-half of the azimuth extent so that the final image resolution is 12.0 meters by 13.5 meters. (Again, frequency domain windowing further degrades the resolution in the final image.) An overlap of 50% between the subband/subangle data sets yielded 15 separate multilook realizations over frequency and 3 over angle for a total of 45 looks. Speckle reduction by two-dimensional filtering of the full-resolution power image is shown in Figure 21-38c. A 15×5 speckle reduction filter was applied to the original power image in Figure 21-38a. Because the original image was windowed the filter does not degrade intrinsic resolution by the full factor of 15 and 5, so that the final resolution is comparable to the multilook image in Figure 21-38b. Both speckle-reduction approaches reduced the intensity fluctuations on distributed clutter, for example, the desert terrain away from the runway.

Both speckle-reduction methods have two disadvantages. First, the final image contains no phase information, so speckle reduction is not an option for interferometric processing, where phase is critical. Second, the resolution in the final image is coarser than the potential resolution achievable using the full bandwidth and integration angle in the first technique and in the original image in the second technique. Comparing Figure 21-38a with 21-38b and 21-38c, the resolution on man-made returns (e.g., hangar, aircraft on the tarmac) is noticeably spoiled.

21.6.5 Man-Made Returns

SAR imagery tends to highlight returns from man-made objects such as buildings and vehicles. This feature is useful to both the military image analyst and the remote sensing surveyor. SAR is used by the former to detect and identify adversary facilities and assets of interest, and the latter to classify regions by use and to track urban and rural development over time.

A number of characteristics distinguish natural and man-made features in SAR. Natural returns like grassy areas (e.g., lawns, meadows) and trees tend to be spatially distributed with noise-like fluctuations in pixel intensity. Man-made returns, on the other hand, usually have significant structure and limited extent. For example, power lines and fences show up in SAR imagery as long, linear features with periodic variations in power. Vehicles and buildings are concentrated in down-range and cross-range consistent with their physical sizes. Man-made objects often contain electrically conductive materials such as steel or aluminum, making them highly radar reflective. Rocks and soil, in contrast, are primarily ceramic in makeup, while vegetation is organic by definition. Ceramics and organic chemicals are dielectrics with greatly reduced radar reflectivity compared with the metals often found in manufactured items. Finally, buildings, cars, trains, and ships are often constructed of large, flat panels. Such panels produce large returns when a SAR sensor is collecting at an angle normal to the panel surface. Moreover, collections of flat panels forming right angle junctions ("corners"), a common construction product, possess retro-reflective properties. Retro-reflective objects product strong radar returns over wide angular regions. A common retro-reflective interface is found where a vertical wall and the horizontal ground come together as shown in Figure 21-39. The intersection between a vertical and horizontal surface is one example of a *dihedral*, an object that effects two successive reflections, not just a single reflection, onto all incident radiation. When the surfaces are normal to one another, as in a dihedral, the incident and scattered angles are equal in the plane normal to the interface line. Therefore, the wall–ground combination retro-reflects energy back to the radar regardless of the platform elevation angle so long as the azimuth angle is normal to the seam between the ground and wall.

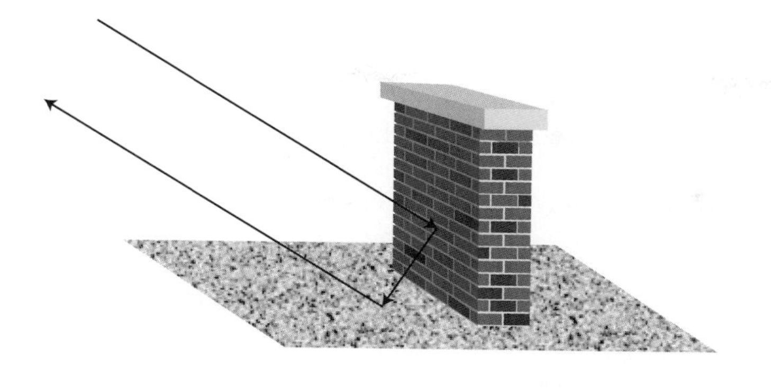

FIGURE 21-39 ■ Geometry for two-plane retro-reflection.

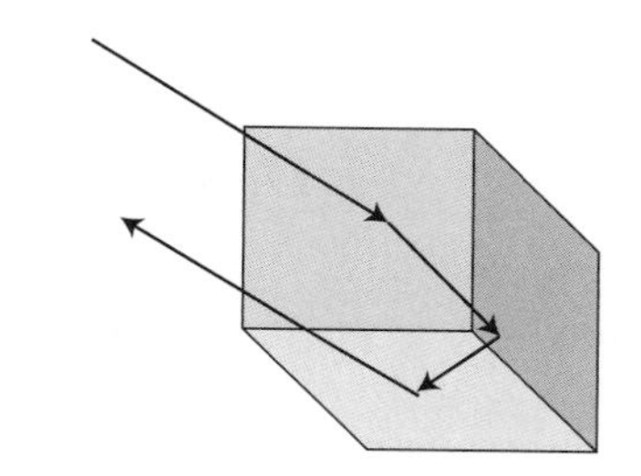

FIGURE 21-40 ■ Geometry for three-plane retro-reflection.

FIGURE 21-41 ■ Armored vehicle presents a complicated and faceted surface (see images in Figure 21-32).

Three surfaces normal to one another, as shown in Figure 21-40, form a *trihedral*, an object that imparts three reflections onto incident radiation and sends scattered energy back along the direction to the source over a wide range of azimuth and elevation angles. Trihedrals generate strong, persistent, and coherent returns over long SAR collections, yielding intense, point-like artifacts in the final image.

Figure 21-40 is a photograph of a BMP-2, the vehicle imaged in the chips in Figure 21-32. The vehicle construction is dominated by a number of large, flat surfaces. For example, the long panel along the side of the vehicles is responsible for much of the energy in Figure 21-32c. It is likely that the returns in this chip were augmented by dihedral returns created by interactions between the nearly vertical side of the BMP-2 and the ground. While the BMP-2's surfaces are flat at the macro level, a close examination of Figure 21-41 reveals a large number of panel breaks, added fixtures, and a complicated track-wheel-bogie arrangement along the side. These features present ample opportunity for retro-reflection. Evidence for retro-reflection is provided by the appearance of the vehicles in Figure 21-32 as collections of strong, point-like targets.

A number of man-made returns are rendered in Figure 21-33. Strong, isolated returns near the middle of the image are parked aircraft. Retro-reflections generated by hangar doors and structure appear as very intense pixels in the lower-right portion of the image. Finally, linear artifacts running through various portions of the image are almost certainly from chain-link fencing.

21.6.6 Signal-to-Noise and Clutter-to-Noise Ratios

The usual radar range equation can be used to predict the signal-to-noise ratio for a point target and the clutter-to-noise ratio (CNR) for a region of distributed clutter in a SAR image. Because SAR data acquisition and image formation are coherent processes, the pulse-Doppler (PD) radar range equation for SNR is preferred as a starting point,

$$SNR = \frac{P_t d G_t A_e \sigma}{(4\pi)^2 R^4 k T_0 F L_s B_d}$$

where P_t is the peak transmitted power, d is the transmit duty factor, G_t is the antenna gain on transmit, A_e is the effective aperture area on receive, σ is the target RCS, R is the slant range to the target, k is Boltzman's constant, T_0 is standard temperature, F is the system noise figure, L_s is system losses, and B_d is Doppler bandwidth.

To simplify the PD equation, make the following substitutions:

- Substitute in the average transmitted power P_{avg} using $P_{avg} = P_t d$.
- Substitute in the aperture gain on receive G_r using $G_r = 4\pi A_e/\lambda^2$.
- Substitute in the thermal noise power spectral density N_0 using $N_0 = kT_0F$.
- Substitute the coherent integration (dwell) time T_d for Doppler bandwidth, $T_d = 1/B_d$.

The PD radar range equation becomes

$$SNR = \frac{P_{avg} G_t G_r \lambda^2 \sigma T_d}{(4\pi)^3 R^4 N_0 L_s}$$

The following progression expands dwell time into an expression relevant to SAR operation

$$\begin{aligned} T_d &= \frac{D_{SAR}}{v} = \frac{R\theta_{int}}{v} = \frac{R\frac{\lambda}{2\Delta CR}}{v} \\ &= \frac{R\lambda}{2v\Delta CR} \end{aligned}$$

where v is platform velocity, D_{sar} is the SAR baseline length, θ_{int} is the integration angle, and ΔCR is cross-range resolution. This expression is accurate for SAR data acquisitions that are unsquinted, that is, collected at broadside. For squinted geometries more time is required to move the platform through the same integration angle. This effect can be represented by modifying the broadside dwell time by the sine of the cone angle

$$T_d = \frac{R\lambda}{2v \sin\theta_{cone} \Delta CR}$$

At broadside the cone angle is 90°, and dwell time simplifies to the first expression.

Substituting the new expression for dwell time into the PD equation yields

$$SNR = \frac{P_{avg} G_t G_r \lambda^3 \sigma}{2v \sin\theta_{cone} (4\pi)^3 R^3 N_0 L_s \Delta CR} \tag{21.78}$$

Interestingly, SNR is inversely proportional to platform speed and cross-range resolution: longer collection times, whether a result of slower speeds or finer resolutions, increase SNR.

CNR is easily computed from SNR by replacing the point-target RCS with the RCS of a clutter patch having dimensions established by the down-range and cross-range resolutions of the SAR image. The clutter patch area A_c is

$$A_c = \Delta CR \left(\frac{\Delta R}{\cos\delta} \right)$$

where δ is the grazing angle at the clutter patch. Here, down-range resolution is defined in the slant plane, so the resolution in the ground plane is dilated by the cosine of the grazing angle. The RCS of the clutter patch σ_c is the product of the clutter backscatter coefficient σ^0 and the area

$$\sigma_c = \sigma^o A_c$$

With these relationships, (21.78) yields CNR as

$$CNR = \frac{P_{avg} G_t G_r \lambda^3 \sigma^0 \Delta R}{2v \sin\theta_{cone} \cos\delta (4\pi)^3 R^3 N_0 L_s} \tag{21.79}$$

CNR is independent of cross-range resolution but decreases as down-range resolution grows finer.

The backscatter coefficient that provides a clutter power equal to the thermal noise power, σ_n, can be found from (21.79) by setting CNR to one and solving for σ^0:

$$\sigma_n = \frac{2v \sin\theta_{cone} \cos\delta (4\pi)^3 R^3 N_0 L_s}{P_{avg} G_t G_r \lambda^3 \Delta R} \tag{21.80}$$

The coefficient in (21.80) is termed the *noise-equivalent backscatter coefficient*; it is the clutter reflectivity that generates a power level in the SAR image equal to that of thermal noise. More to the point, σ_n is the equivalent terrain reflectivity presented by thermal noise. Because thermal noise is present throughout the image, this "noise-induced terrain" is always in the background. The noise-equivalent backscatter coefficient sets an upper limit on the image contrast between terrain types.

While SNR, CNR, and σ_n are all fundamental measures of image quality and interpretability, they are not a complete set of descriptors. For example, the background thermal noise level represented by σ_n is often greatly exceeded by other sources of interference including random digital quantization errors, down-range and cross-range sidelobe levels in the SAR image, and range and Doppler ambiguous clutter return originating from antenna azimuth and elevation sidelobes and aliasing back into the imaged area. These sources are all directly proportional to the quiescent distributed clutter power and thus are typically quantified through a *multiplicative noise ratio* (MNR) metric.

21.7 SUMMARY

This chapter began by presenting the primary motivation for SAR as the desire to generate photographic-quality images of large scenes from great distances at night and under all weather conditions. As the history of SAR development has been dominated by the quest for ever improved resolution, resolution relationships were established for both down- and cross-range image dimensions. Resolution along the former is simply a function of waveform bandwidth, but resolution in cross-range is the principal benefit of SAR processing and is provided by short wavelengths and large integration angles on the scene of interest. SAR achieves integration angle via motion of the radar platform. Because monostatic radars sample along the platform track at discrete locations, a limit was derived on the distance between these collection points.

Platform motion generates complex scatterer range and phase histories, complicating image formation. The form of these histories was derived for a linear, uniformly sampled SAR acquisition, yielding the hyperbolic PSR as the manifestation of a point target in the raw data. By making a number of approximations to the PSR the DBS imager was developed, an image formation technique historically justified using Doppler histories. While appropriate for coarse-resolution imaging, DBS is inappropriate for fine-resolution imaging over wide areas or at low center frequencies. Sophisticated algorithms like RMA and PFA account for the PSR more completely and are commonly employed under challenging image formation circumstances. Backprojection is a simple alternative to DBS and yet makes full use of the PSR. Built upon the matched filter, backprojection was developed and demonstrated on a very demanding simulated data set.

SAR is, by definition, coherent radar imaging. Coherent RF operation gives rise to a number of interesting artifacts in SAR imagery. NRAs appear as dark regions in the image and indicate terrain that reflects energy away from the radar, like water and paved surfaces and areas in shadow. Objects with finite height cast shadows down-range in the SAR image because the collection is made not directly overhead but away from the scene with a grazing angle between typically between 10° and 45° and with the collecting radar providing the RF illumination. The combination of grazing angle and nonzero terrain or object slope angle can lead to foreshortening, a down-range compression of returns in the SAR image, or even layover, a reversal in the order of the returns down-range. Speckle, an apparent noise-like modulation on distributed clutter returns in the final image, is the result of the coherent summation of many random, unresolved scatterers in each resolution cell. Pixel power variations due to speckle can be reduced by multilook averaging but with a sacrifice in final image resolution. Man-made objects are typically very apparent in SAR images because they are spatially concentrated along lines or in confined areas. In addition, man-made objects tend to generate much stronger returns compared with distributed clutter because these objects often contain metallic components or as a result of retro-reflective scattering from dihedral and trihedral shapes and structures.

21.8 FURTHER READING

Most radar textbooks contain a section or chapter dedicated to SAR imaging and its applications. Skolnik's [1] development is solid, although he adheres closely to the antenna array model. Curiously, the discussion of SAR is more limited in the most recent edition of his systems book [23]. The SAR chapter in Skolnik's current handbook [2], written by Sullivan, is more complete and draws on both the antenna array and range-Doppler SAR models. An expanded adaptation of Sullivan's development is available in [24]. Stimson [3] also provides a fine overview of the topic; while the array representation is discussed briefly, it is only natural that, in a book stressing airborne pulse-Doppler radar, the principal SAR model emphasizes Doppler histories and relationships. Richards' chapter in [4] begins with the array and Doppler viewpoints but goes on to formulate the phase history from a point scatterer. The resulting signal model is used to derive several image formation algorithms and establish their strengths and limitations. Finally, interferometric applications and postprocessing techniques like autofocus and speckle-reduction are covered.

A great number of textbooks entirely devoted to SAR have been published. The books by Carrara et al. [5] and Jakowatz et al. [6] are commonly employed within the SAR community and extensively cited in conference and journal papers. More importantly, these two texts are widely acknowledged to be accessible to the eager SAR neophyte.

21.9 REFERENCES

[1] Skolnik, M.I., *Introduction to Radar Systems,* 2d ed., McGraw-Hill Book Company, New York, 1980.

[2] Skolnik, M.I., *Radar Handbook,* 3d ed., McGraw-Hill Book Company, New York, 2008, pp. 17.1–17.37.

[3] Stimson, G.W., *Introduction to Airborne Radar*, Hughes Aircraft Company, El Segundo, CA, 1983.

[4] Richards, M.A., *Fundamentals of Radar Signal Processing*, McGraw-Hill Book Company, New York, 2005.

[5] Carrara, W.G., Goodman, R.S., and Majewski, R.M., *Spotlight Synthetic Aperture Radar*, Artech House, Boston, 1995.

[6] Jakowatz Jr., C.V., Wahl, D.E., Eichel, P.H., Ghiglia, D.C., and Thompson, P.A., *Spotlight-Mode Synthetic Aperture Radar: A Signal Processing Approach*, Kluwer Academic Publishers, Boston, 1996.

[7] Munson Jr., D.C., O'Brien, J.D., and Jenkins, W.K., "A Tomographic Formulation of Spotlight Mode Synthetic Aperture Radar," *Proceedings of the IEEE*, vol. 71, no. 8, pp. 917–925, August 8, 1983.

[8] Cafforio, C., Prati, C., and Rocca, E., "SAR Data Focusing Using Seismic Migration Techniques," *IEEE Transactions on Aerospace and Electronic Systems*, vol. 27, no. 2, pp. 194–207, March 1991.

[9] Çetin, M., and Karl, W.C., "Feature-Enhanced Synthetic Aperture Radar Image Formation Based on Nonquadratic Regularization," *IEEE Transaction on Image Processing,* vol. 10, no. 4, pp. 623–631, April 2001.

[10] Wiley, C.A., "Pulsed Doppler Radar Methods and Apparatus," U.S. patent #3,196,436, 1965 (originally filed 1951).

[11] McCorkle, J.W., "Focusing of Synthetic Aperture Wideband Data," *Proceedings of the 1991 IEEE International Conference on Systems Engineer*, pp. 1–5, August 1–3, 1991.

[12] Wiley, C.A., "Synthetic Aperture Radars—A Paradigm for Technology Evolution," *IEEE Transactions on Aerospace and Electronic Systems*, vol. AES-21, no. 3, pp. 440–443, May 1985.

[13] Sack, M., Ito, M.R., and Cumming, I.G., "Application of Efficient Linear FM Matched Filtering Algorithms to Synthetic Aperture Radar Processing," *IEE Proceedings,* vol. 132, pt. F, no. 1, pp. 45–57, February 1985.

[14] Bamler, R., "A Comparison of Range-Doppler and Wavenumber Domain SAR Focusing Algorithms," *IEEE Transactions on Geoscience and Remote Sensing*, vol. 30, no. 4, pp. 706–713, July 1992.

[15] Munson, D.C., and Visentin, R.L., "A Signal Processing View of Strip-Mapping Synthetic Aperture Radar," *IEEE Transactions on Acoustics, Speech, and Signal Processing,* vol. 37, no. 12, pp. 2131–2147, December 1989.

[16] Walker, J.L., and Carrara, W.G., "Method of Processing Radar Data from a Rotating Scene Using a Polar Recording Format," U.S. Patent #4,191,957, March 4, 1980.

[17] Walker, J., "Range-Doppler Imaging of Rotating Objects," *IEEE Transactions on Aerospace and Electronic Systems*, vol. AES-16, no. 1, pp. 23–51, January 1980.

[18] Yegulalp, A., "Fast Backprojection Algorithm for Synthetic Aperture Radar," *The Record of the 1999 IEEE Radar Conference*, pp. 542–546, April 20–22, 1999.

[19] McCorkle, J., and Rofheart, M., "An Order $N^2\log(N)$ Backprojector Algorithm for Focusing Wide-Angle Wide-Bandwidth Arbitrary Motion Synthetic Aperture Radar," *SPIE Radar Sensor Technology Conference Proceedings*, SPIE vol. 2747, pp. 25–34, April 1996.

[20] Oh, S.M., and McClellan, J.H., "Multiresolution Imaging with Quadtree Backprojection," *Conference Record of the Thirty-Fifth Asilomar Conference on Signals, Systems, and Computers, 2001*, pp. 105–109, November 4–7, 2001.

[21] Ulander, L.M., Hellsten, H., and Stentstrom, G., "Synthetic-Aperture Radar Processing Using Fast Factorized Back-Projection," *IEEE Transactions on Aerospace and Electronic Systems*, vol. 39, no. 3, pp. 760–776, July 2003.

[22] Oliver, C., and Quegan, S., *Understanding Synthetic Aperture Radar Images*, SciTech Publishing, Raleigh, NC, 2004.

[23] Skolnik, M.I., *Introduction to Radar Systems, Third Edition*, McGraw-Hill Book Company, New York, 2001.

[24] Sullivan, R.J., *Radar Foundations for Imaging and Advanced Topics*, SciTech Publishing, Raleigh, NC, 2004.

21.10 | PROBLEMS

1. Calculate the waveform bandwidths required for down-range resolutions (in the slant plane) of 150 meters, 15 meters, 1.5 meters, and 0.15 meters.
2. For a SAR system operating at X-band with a center frequency of 10 GHz:
   a. Determine the integration angles required to achieve cross-range resolutions of 150 meters, 15 meters, 1.5 meters, and 0.15 meters.
   b. Find the length of the SAR baseline on the scatterer at 100 km, given the four integration angles found in (a).
   c. If the platform is flying at 100 meters per second, find the coherent processing times for the four SAR baselines in (b).

3. For a SAR system with a center frequency of 10 GHz, a null-to-null beamwidth of 6°, and performing a stripmap collection using non-squinted geometry (antenna pointed broadside to the velocity vector):

   a. Find the minimum along-track sampling distance need to avoid aliasing.

   b. If the platform is flying at 100 meters per second, find the minimum PRF required to meet the along-track sampling limit in (a).

4. For an airborne SAR system with a center frequency of 10 GHz, and 3 dB beamwidth of 3°:

   a. Calculate the minimum (finest) cross-range resolution achievable for this system in a stripmap mode.

   b. Find the bandwidth required to achieve the same resolution down-range.

5. For an airborne SAR system performing a stripmap collection against a scene from 100 km slant range and having a 3 dB beamwidth of 6°:

   a. Using the PSR, calculate the maximum slant range change to a scatterer moving through the 3 dB beamwidth.

   b. If the radar center frequency is 10 GHz, find the maximum quadratic phase error (QPE) for a scatterer moving through the 3 dB antenna beamwidth.

6. For a stripmap SAR system employing an antenna 10 meters in length:

   a. Estimate the 3 dB and null-to-null beamwidth of the antenna.

   b. Calculate the finest achievable stripmap cross-range resolution and the required along-track sampling interval using the miscellaneous relationships as provided by equations (21.33–21.38).

7. For a space-based SAR system operating in a stripmap mode with a center frequency of 10 GHz, moving at 7 km/s, and having a null-to-null beamwidth of 0.35°:

   a. Find the maximum and minimum clutter Doppler shifts (in Hz).

   b. Determine the range of spatial frequencies in rad/m between the Doppler extrema in (a).

8. For a space-based SAR system operating in a stripmap mode with a center frequency of 10 GHz, moving at 7 km/s, having a 3 dB beamwidth of 0.175°, and collecting against a scene at a slant range of 1,000 km:

   a. Find the length of the SAR baseline for the 3 dB beamwidth and the coherent processing time.

   b. Using the PSR, calculate the maximum slant range change to a scatterer moving through the 3 dB beamwidth.

   c. What is the maximum quadratic phase deviation when using the 3 dB beamwidth?

Appendix A: Maxwell's Equations and Decibel Notation

A.1 MAXWELL'S EQUATIONS

The behavior of electromagnetic (EM) waves is governed by the four laws, or equations, of electromagnetism known collectively as *Maxwell's equations*. Before stating Maxwell's equations mathematically, it is useful to describe them more heuristically. Maxwell's equations, simply stated, in words are given in Table A-1.

Maxwell's equations state that charged particles (e.g., electrons) produce electric fields (Gauss's law for electricity) and moving charged particles (currents) produce magnetic fields (Ampere's law). Electric and magnetic fields can be visualized by the forces they exert on charged particles. A charged particle in the presence of an electric field will experience an *electric force*, while a moving charged particle in the presence of a magnetic field will experience a *magnetic force*. The sum of these two forces are given by the succinct equation [1]

$$\boldsymbol{F} = q(\boldsymbol{E} + \boldsymbol{v} \times \boldsymbol{B}) \tag{A.1}$$

where $\boldsymbol{F}$ is the EM force vector, q is the charge of the charged particle, $\boldsymbol{E}$ and $\boldsymbol{B}$ are, respectively, the vector electric and magnetic fields in which the charged particle is moving, $\boldsymbol{v}$ is the velocity vector of the charged particle, and the symbol denotes the vector cross-product.

Maxwell's equations also state that electric fields are produced by time-varying magnetic fields (Faraday's law) and vice versa, while magnetic fields are produced by time-varying electric fields (Maxwell's contribution of the Ampere-Maxwell law). This interaction gives rise to a electromagnetic waves that can propagate over long distances. Consider a charge that is oscillating back and forth (e.g., an electron in an alternating current [AC] circuit). Since this charge has a time-varying velocity (acceleration), it produces both electric and magnetic fields that change in time in accordance with Gauss's and Ampere's laws. It can be shown that if these two laws were the only laws of electromagnetism, then the amplitude of these electric and magnetic fields would be proportional to $1/R^2$, where R is the range from the charge, so that the fields would decay over a relatively short

TABLE A-1 ■ Maxwell's Equations in Words

Gauss's law for electricity	Electric charge produces electric fields.
Gauss's law for magnetism	There is no magnetic charge.
Faraday's law	Time-varying magnetic fields produce electric fields.
Ampere-Maxwell law	Moving charge (i.e., current, or Ampere's law) and time-varying electric fields (Maxwell's contribution) produce magnetic fields.

TABLE A-2 ■ Integral Form of Maxwell's Equations

Name of Law	Equation	Law in Words
Gauss's law for electricity	$\oint_{surface} \boldsymbol{E} \cdot d\boldsymbol{A} = \frac{q}{\varepsilon_0}$	The flux of $\boldsymbol{E}$ through any closed surface equals the net charge inside that surface, q, divided by ε_0.
Gauss's law for magnetism	$\oint_{surface} \boldsymbol{B} \cdot d\boldsymbol{A} = 0$	The flux of $\boldsymbol{B}$ through any closed surface is zero.
Faraday's Law	$\oint_{loop} \boldsymbol{E} \cdot d\boldsymbol{s} = -\int_{area} \frac{\partial \boldsymbol{B}}{\partial t} \cdot d\boldsymbol{A}$	The circulation of $\boldsymbol{E}$ around any closed loop equals the negative time derivative of the flux of $\boldsymbol{B}$ through any area bounded by that loop.
Ampere-Maxwell law	$\oint_{loop} \boldsymbol{B} \cdot d\boldsymbol{s} = \mu_0 i + \mu_0 \varepsilon_0 \int_{area} \frac{\partial \boldsymbol{E}}{\partial t} \cdot d\boldsymbol{A}$	The circulation of $\boldsymbol{B}$ around any closed loop equals μ_0 times the electric current, i, flowing through any area bounded by that loop plus $\mu_0\varepsilon_0$ times the time derivative of the flux of $\boldsymbol{E}$ through that area.

distance. However, Faraday's law states that the changing magnetic field also produces an electric field, while the Ampere-Maxwell law states that the changing electric field produces a magnetic field. These changing electric and magnetic fields reinforce each other far out into the space from the oscillating charge. The result is that the amplitudes are proportional to $1/R$ and therefore decay more slowly.

Mathematically, Maxwell's equations can be expressed in either differential form or integral form. Table A-2 gives the integral form as well as a more precise verbal description than given in Table A-1. The constants ε_0, μ_0 are, respectively, the *permittivity* and *permeability* of free space. Their numerical values are such that

$$1/\sqrt{\mu_0\varepsilon_0} = c \tag{A.2}$$

In free space (thus in the absence of charges and currents), Maxwell's equations can be reduced to the following vector differential equations:

$$\boldsymbol{\nabla}^2 \boldsymbol{E} = \mu_0\varepsilon_0 \frac{\partial^2 \boldsymbol{E}}{\partial t^2} \quad \text{and} \quad \boldsymbol{\nabla}^2 \boldsymbol{B} = \mu_0\varepsilon_0 \frac{\partial^2 \boldsymbol{B}}{\partial t^2} \tag{A.3}$$

where ($\boldsymbol{\nabla}$) is the *gradient* of a vector. Equations (A.3) are the so-called *wave equations*, the solutions of which are, respectively, electric and magnetic field waves (i.e., electromagnetic waves) propagating at the speed $1/\sqrt{\mu_0\varepsilon_0} = c$. A solution to the first wave equation of particular interest to radar is a traveling sinusoidal electric field wave, with the amplitude of each directional component (e.g., horizontal and vertical) having the form

$$E = E_0 \cos(kz - \omega t + \phi) \tag{A.4}$$

where k is the *wavenumber* $(2\pi/\lambda)$, ω is the radian *frequency*, and ϕ is the *phase* of the wave. These parameters and their significance are described in Section 1.3.1.

In free space, the oscillating electric, $\boldsymbol{E}$, and magnetic, $\boldsymbol{B}$, fields of the propagating EM wave are orthogonal and the direction of the propagating wave is orthogonal to both $\boldsymbol{E}$ and $\boldsymbol{B}$ (i.e., the EM wave is *transverse*). The direction of propagation is given by the direction of $\boldsymbol{S} = \boldsymbol{E} \times \boldsymbol{B}/\mu_0$ (the *Poynting vector*) and the intensity (power/area) of the wave is given by the magnitude of $\boldsymbol{S}$, that is, $|\boldsymbol{S}|$. The ratio $|\boldsymbol{E}|/|\boldsymbol{B}|$ is c, the speed of the propagation of the EM wave (speed of light).[1] Therefore, if the direction of propagation and the properties of either vector field ($\boldsymbol{E}$ or $\boldsymbol{B}$) is known, then the properties of the other vector field can be determined. Consequently, the EM wave is typically described in terms of the electric field.

A.2 THE UBIQUITOUS dB

The decibel (dB) is widely used in radar technology. Many radar parameters are expressed in units of dB due to the large dynamic range of these parameters. Radar cross section (RCS) is a good example. RCS values can range from 10^{-5} m^2 (insects) to over 10^6 m^2 (aircraft carriers). This represents 11 orders of magnitude, a range of a 100 billion to one (10^{11}). In dB units, these RCS values become –50 dB and 60 dB, respectively, a range of only 110. Thus, in dB, the scale becomes significantly compressed and easier to deal with mathematically.

The value of a quantity in dB is always computed relative to some reference value. The first step in converting a value to dB is therefore to divide it by the reference value. For example, consider power, P. Before taking the logarithm of P, it is divided by a reference power P_0, say, $P_0 = 1$ watt. Now the logarithm is taken:

$$\log_{10}\left(\frac{P}{P_0}\right) \tag{A.5}$$

This is the power in "bels." Multiplying this by 10 yields the power in decibels,

$$P \text{ in dB} = 10\log_{10}\left(\frac{P}{P_0}\right) \tag{A.6}$$

To convert a value x from dB units to linear units the inverse operation is performed:

$$x_{linear} = 10^{x_{\text{dB}}/10} \tag{A.7}$$

Shown in Figure A-1 is a scale comparing linear and dB values.

There is only one problem. This may be a convenient way for a transmitter engineer to express power, but a receiver engineer would have probably chosen the reference power to be $P_0 = 10^{-3}$ watt $= 1$ milliwatt. Somehow the choice of the reference parameter must be conveyed to the user. This is done by modifying the way the unit "dB" is written. In the case of power, for example, the unit is expressed as "dBW"(dB relative to 1 watt) if $P_0 = 1$ watt and "dBm" (dB relative to 1 milliwatt) if $P_0 = 1$ milliwatt.

---

[1] $\boldsymbol{H} = \boldsymbol{B}/\mu_0$ is also referred to as the "magnetic field" and $|\boldsymbol{E}|/|\boldsymbol{H}| \equiv Z_0 = 377$ ohms is the impedance of free space. In this case, $\boldsymbol{B}$ is sometimes referred to as the *magnetic induction* or *magnetic flux density*.

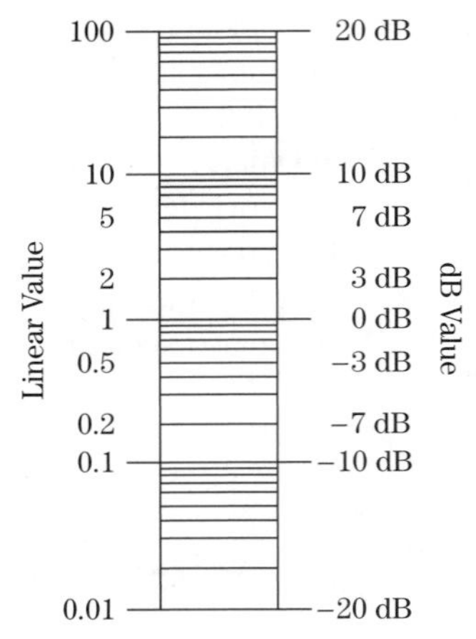

FIGURE A-1 ■ Linear and dB values.

Some features of measurements in dB to note are (1) values in dB can be determined only for positive parameters (the dB value of a negative parameter is not allowed), and (2) a negative dB value means that the linear value of the parameter is less than the reference value, that is, the ratio P/P_0 is less than 1.

Manipulating parameters expressed in dB simplifies some arithmetic operations; multiplication becomes addition and division becomes subtraction. This is due to the mathematical properties of the logarithm. For example, the equation $x = yz$, where x, y and z are linear values, becomes $x = y + z$ if x, y and z are expressed in dB. Similarly, $x = y/z$ becomes $x = y - z$ if x, y and z are expressed in dB. Also, the equation $z = y^a$ becomes $x = ay$ if x and y are expressed in dB (a is not in dB). This was of great utility before the age of hand-held scientific calculators and high-speed computers, but is of little use today.

The one exception is in the determination the linear value of a parameter given in dB when a calculator is not available. Table A-3 lists several "dBs to remember." With this table and the arithmetic properties of dBs, one can determine the approximate linear value of any parameter given in dB without resorting to a scientific calculator. For example, the linear equivalent of 7 dB can be determine by noting

$$7\ \text{dB} = 3\ \text{dB} + 3\ \text{dB} + 1\ \text{dB} \tag{A.8}$$

TABLE A-3 ■ dBs to Remember

dB	Linear
0	1
1	$1\frac{1}{4}$
3	2
10	10
$10x$	10^x
$-10x$	10^{-x}

TABLE A-4 ■ Radar Parameters Commonly Expressed in dB

Radar Parameter	dB Expression
Antenna gain	dBi (gain relative to isotropic)
Power gain	dB (power out/power in)
Power loss	dB (power in/power out)
Power	dBW (power relative to 1 watt)
	dBm (power relative to 1 milliwatt)
RCS	dBsm (RCS relative to 1 square meter)

and, from the table, that 3 dB corresponds to a linear value of 2, while 1 dB corresponds to a linear value of $1\frac{1}{4}$. From this and the arithmetic properties of dB, equation (A.8) becomes

$$7 \text{ dB} = 2 \times 2 \times 1\frac{1}{4} = 5 \tag{A.9}$$

An alternative way to reach the same conclusion is to note that

$$7 \text{ dB} = 10 \text{ dB} - 3 \text{ dB} \tag{A.10}$$

From the table, 10 dB = 10 and 3 dB = 2. Thus, equation (A.8) becomes

$$7 \text{ dB} = 10 \div 2 = 5 \tag{A.11}$$

Some radar parameters commonly expressed in dB are shown in Table A-4.

A.3 REFERENCE

[1] Lonngren, K.E., Savov, S.V., and Jost, R.J., *Fundamentals of Electromagnetics with MATLAB*, 2d ed., SciTech Publishing, Raleigh, NC, 2007.

Appendix B: Answers to Selected Problems

Chapter 1

1. $0.15x$ km.

3. 10.73 μs, 6.67 μs, 666.67 μs, 2.68 ms, 40.64 ns.

5. (a) 11.81 GHz (b) 85.714 GHz. (c) 34.88 GHz. (d) 333.33 MHz. (e) 3.33 GHz. (f) 984.25 MHz.

7. (a) 66.67 m. (b) 20 m. (c) 0.67 m. (d) 20 m. (e) 1125 m.

10. 1 kW = 30 dBW.

11. $150/x$ km.

13. (a) 2980 Hz (b) 22.0 kHz (c) 1.00 kHz (d) 707.1 Hz (e) 212.1 Hz (f) 500 Hz.

15. (a) 150 m (b) 150 m (c) 15 m (d) 1.5 m.

17. (a) 15.7 degrees (b) 4.25 degrees (c) 361.2 feet (d) 97.9 feet

Chapter 2

1. 2.06×10^{-14} W $= 2.06 \times 10^{-11}$ mW, -106.9 dBm.

3. 2.765, or 4.42 dB.

5. (a) 1.65 dB (b) 1.65 dB (c) 12.6 dB (d) 11.65 dB.

7. (a) -9.07 dB (b) -9.07 dB (c) $+1.88$ dB (d) $+0.93$ dB.

9. (a) 42.05 km (b) 28.12 km.

11. 17.5 ms.

13. 90 m^2, and -19.54 dB.

15. 62.3 dB.

17. 32.3 dB.

19. 158 kW.

Chapter 3

1. 90.

3. 1.0 seconds.

5. 1.11×10^{-2}.

7. 3 verifications.

9. 912 ms.

11. 1.75 seconds.

13. 2-of-3 detection: $P_D = 0.5$, $P_{FA} = 7.475 \times 10^{-5}$. 3-of-4 detection: $P_D = 0.6875$, $P_{FA} = 1.49 \times 10^{-4}$.

15. 9.9 dB.

Chapter 4

1. Atmospheric refraction, surface multipath.

3. Distance = 3000 m. Phase change = 62,832 radians.

5. Approximately 0.002 dB/km.

7. Radius of curvature $< \lambda/50$ for knife edge.

9. Approximately 0.126.

11. 114.6 m.

13. Specular reflection angle = 3.43°. Range extension = 0.16 m.

15. 12 dB.

17. Approximately 6 dB.

Chapter 5

1. 19.25° at $R = 10$ km, 60.2° at $R = 50$ km.

3. 100 km.

5. At L band, $\delta_C = 0.38$ rad (22°) for $\sigma_h = 10$ cm. Surface is "smooth" for any grazing angle when $\sigma_h = 1$ cm.

6. Rayleigh-resonance boundary: $a = 9.55 \times 10^{-3}$ m = 0.955 cm. Resonance-optics boundary: $a = 9.55$ cm.

8. 30 samples (1 kHz PRF), 6 (5 kHz), 1 (40 kHz).

Chapter 6

1. The angle of incidence from the transmitter with respect to the target surface normal is equal to the angle of reflection.

3. In a plane perpendicular to the edge.

4. For specular backscatter, surface and edge normals must point back toward the radar.

7. Edge diffraction and surface waves.

8. In free space there is equal energy in the E and H field components.

10. The permeability and permittivity, through the index of refraction.

11. Wavelength in the material is less than free space.

16. When the wave front curvature is so large that the phase change over the target dimensions is insignificant, typically less than 22.5°.

17. Zero.

20. An observer looking in the direction of propagation.

Chapter 7

1. $\sigma = \pi$ m^2. $a_t = 0.093$ m. $a_s/a_t = 10.75$.
3. $8(D/\lambda) = 40$ when $(D/\lambda) = 5$.
5. $\frac{1+2\left(1+\sqrt{2}\right)}{\left[1+\left(1+\sqrt{2}\right)\right]^2}\bar{\sigma}^2 = \frac{1}{2}\bar{\sigma}^2$
7. Exponential pdf: $P\{\sigma > 2\} = e^{-2} = 0.135$. 4-th degree chi-square: $P\{\sigma > 2\} = 5e^{-4} = 0.0916$.
9. $\Delta f = 50$ MHz.
11. $\Delta\theta = 1.14$ mrad $= 0.065°$. Choose scan-to-scan decorrelation.

Chapter 8

1. Error $\approx 2.2236\times10^{-4}$ Hz.
3. The radial velocity is initially close to 200 mph (89.4 m/s), decreases to zero when the aircraft is directly overhead, and decreases asymptotically to −200 mph (−89.4 m/s) as the aircraft flies away.
5. 1 MHz.
7. $T_d = 20$ ms. Rayleigh resolution = 50 Hz.
10. $\Delta f_d = 40$ Hz. $M = 25$.
11. Range change = 0.025 m. Phase change = $2\pi/3$ radians = 120°.
13. 329 Hz.
15. 65.1%.
17. 3 pulses. Apparent range = 20.4 km.

Chapter 9

1. $Q_t = 7.96 \times 10^{-7}$ W/m at target, same at clutter.
3. 10 GHz: $\theta_3 = 1.72°$, $D_{\max} = 10{,}970 = 40.4$ dBi. 32 GHz: $\theta_3 = 0.54°$, $D_{\max} = 112{,}460 = 50.5$ dBi.
5. $D_{\max} = 10{,}440 = 39.9$ dBi. Maximum gain = 37.9 dBi.
7. Feed diameter = 5.5 cm.
9. $\Delta x = 1.574$ cm. Number of elements = 97.
11. $\Delta\theta = 1.707°$. Pointing error = 74.5 km.
13. $\Delta\theta = 6.14° = 1.42$ beamwidths (1 m antenna), 14.2 beamwidths (10 m antenna).
15. 1600 elements. Additional cost = $487,500.

Chapter 10

1. 12.6 MHz of FM. Maximum ripple for 1 MHz of FM = 5 volts.
2. Noise density = −24 dBm/MHz. Total noise power = 3 dBm.
5. Drain voltage ripple = 570 microvolts. Gate voltage ripple = 5.7 microvolts.
7. −75 dBc.

9. Non-depressed collector mode: prime power = 3,750 W, combined efficiency = 13.3%. Depressed collector mode: prime power = 1,500 W, combined efficiency = 33.3%.
12. Answer (e).
14. Module power = 10 W. Array area = 0.56 m^2. Number of elements = 2,240.
16. Answers (a), (b), and (c).

Chapter 11

2. 50×, or 17 dB.
3. LO = 7.2 GHz. The $2f_1 - 3f_2$ intermodulation product falls within the IF band.
5. 1.75 volts.
7. $R_{\max} = -158.5$ km.
9. Noise factor = 3.05.
11. $T_S = 835$ Kelvin.
13. 59 dB.
14. Spurious free dynamic range = 45 dB.

Chapter 12

1. (a) −94 dBc/Hz (b) −6.07 dB
3. Yes.
5. (a) 450×10^3 m^2 (b) 28,125 m^2 (c) yes, two others between the first and last.
7. (a) −20.5 dB (b) 33.5 dB
9. 3,333.33 Hz, −11.11 dB
11. The STALO
13. 8 sections, 6 sections.

Chapter 13

2. Filter output length = 1,039 points. Number of multiplications = 41,560.
3. Number of complex multiplications = 24,576.
5. Input rate = 120 Mbps, Output rate = 80 Mbps (16 bits), 160 Mbps (single precision), 320 Mbps (double precision)
6. $N_p \leq \dfrac{B}{2 \cdot PRF}$
8. June 2019.

Chapter 14

1. $k_x = 20.94$ rads/m. $p = 3.33$ cycles/m.
3. $f_s = 35$ MHz.
7. 2 Hz.
9. $k_0 = 21$. $v_0 = 29.53$ m/s. Velocity error = 0.469 m/s.

12. $L_{max} = \lfloor \frac{1}{2} \log_2 K \rfloor$. For $K = 256$, $L_{max} = 4$.

13. Length of $y[n] = 1{,}099$ samples. FFT size $= 2{,}048$.

14. 5 samples.

16. Integrated SNR = integration gain = 20 dB. 464 samples integrated noncoherently.

Chapter 15

1. SNR = 8.

3. $T = 3.29$.

4. $P_D = 0.6922$.

6. P_{FA} does not change.

7. $P_D = 0.953$.

9. $P_{FA} = 5 \times 10^{-4}$.

11. $T = 3.035. P_D = 0.798$.

13. $\chi_1 = 8.14 = 9.1$ dB.

15. $\chi_1 = 13.14$ dB.

17. $\chi_c = -6.86$ dB.

18. $\alpha = 0.72$. Better than $\sqrt{N}$, worse than coherent integration (factor of N).

Chapter 16

1. $FAR = 2$ per second.

3. $P_{FA} = 0.0316$.

5. CFAR loss = 0.77 dB.

7. $P_D = 0.56$.

9. $P_{FA} = 0.029$.

11. $\hat{g} = 0.644$. The CFAR can reject up to 3 interfering targets.

13. $\hat{g}_{CA-CFAR} = 0.98$, $\hat{g}_{GOCA-CFAR} = 1.08$, $\hat{g}_{SOCA-CFAR} = 0.89$. $T_{CA} = 25.43$.

15. Answer (e).

Chapter 17

1. 2 kHz in all three cases.

2. Two-way range change $= \lambda$ m.

4. $K = 128$. Doppler frequency sample spacing = 78.125 Hz.

6. DFT sample spacing $= \pi/32 = 0.0982$ rad/sample, 54.6875 Hz, or 1.641 m/s. Rayleigh resolution = 116.67 Hz or 3.5 m/s.

7. (a) 12 kHz. (b) 60 kHz. (c) 5.

9. True range bin = 9.

12. $\hat{P} = 50$, $\tilde{f}_0 = -333.33$ Hz, $\hat{\sigma}_f^2 = 103{,}515$ Hz^2.

Chapter 18

1. Separate calibration: worst case uncertainty = 0.6 dB, rms = 0.26 dB. End-to-end: 0.1 dB uncertainty.

5. $J(\hat{m}) = \sigma_w^2/N$.

8. 3.3 dB.

9. −0.28 dB.

10. 32.8.

Chapter 19

1. $X_{N|N} = X_{1|N} = \frac{1}{N}\sum_{i=1}^{N} y_i,\ P_{N|N} = P_{1|N} = \frac{\sigma_w^2}{N}$.

3. $\alpha_k = \max\left\{\frac{1}{k}, \alpha_{SS}\right\}$.

5. Position bias or lag $= L_p = \left(\frac{1}{\alpha} - 1\right) V_0$.

7. Minimum MMSE: $MMSE^p = 1{,}535$, $MMSE^v = 1{,}700$. Minimum process noise covariance: $MMSE^p = 1{,}502$, $MMSE^v = 1{,}404$.

13. (a) Crossrange: $0.9 \le \kappa_1\,(0.34) \le 2.1$ at $R = 400$ km and $0.7 \le \kappa_1\,(28) \le 1.1$ at $R = 5$ km. Range: $0.7 \le \kappa_1\,(19) \le 1.1$. Choose $\kappa_1 = 1.1$ to achieve low peak errors at long ranges and minimum MMSE at short ranges in the cross range. The filter design is defined by $\sigma_v = 33\ (\text{m/s})^2$. (b) $k_1 = 0{:}9 + 0{:}7R/400$.

Chapter 20

1. Rayleigh resolution = 300 m. 0.5 amplitude corresponds to −6 dB. −3 dB range resolution = 175.7 m.

4. $X(\omega) = \tau\dfrac{\sin(\omega\tau/2)}{\omega\tau/2}$.

5. (a) Time-bandwidth product = 500. (b) 250 MHz. (c) 8.67 μs. (d) 2167 samples. (e) Gain = 27 dB, SNR = 24 dB.

8. Doppler shift = 3 kHz; radial velocity = 45 m/s.

9. Length 2, ISR = −3 dB; Length 3, ISR = −6.5 dB; Length 4, ISR = −6 dB; Length 5, ISR = −8 dB; Length 7, ISR = −9.1 dB; Length 11, ISR = −10.8 dB; Length 13, ISR = −11.5 dB.

12. Chip width = 20 ns; bandwidth = 50 MHz; Rayleigh resolution = 3 m; time-bandwidth product = 127; pulse compression gain = 21 dB; Approximate PSR = −21 dB; Doppler shift = 393.7 kHz.

14. 13 length bi-phase Barker, PSR = −22.3 dB; 69 length poly-phase Barker, PSR = −36.8 dB; 144 length P1, PSR = −31.4 dB; 169 length P1, PSR = −32.2 dB; 1023 length MLS, PSR = −30.1 dB; 102 length MPS, PSR = −26.2 dB.

Chapter 21

1. $B_r = 1$ MHz, 10 MHz, 100 MHz, and 1 GHz, respectively.

3. (a) $d = 0.143$ m (b) $PRF_{min} \approx 700$ Hz

5. (a) Change in slant range ≈ 137 m. (b) $\Delta\psi_{QT,MAX} \approx 57{,}400$ radians ≈ 9,130 cycles.

7. (a) $f_{d,min} \approx -1{,}400$ Hz, $f_{d,max} \approx +1{,}400$ Hz. (b) -1.28 rad/m $\le k_u \le +1.28$ rad/m.

Index

A

AAW. *See* Antiair warfare (AAW)
Absorption
  atmospheric, 117, 121
  bands, 125
A7C Corsair aircraft, 236, 238
Accuracy, in radar measurements. *See* Precision and accuracy, in radar measurements
ACF. *See* Autocorrelation function (ACF)
Active-aperture array, 353
Active array architectures, 339–341
  advantages of, 340–341
  vs. passive array architectures, 340–341
Active electronically scanned array (AESA), 339, 340, 371, 374, 375
Active phased array, 350
Adaptive DPCA (A-DPCA) processor, 644
Adaptive Kalman filtering, 740–741
ADC. *See* Analog-to-digital converter (ADC)
A-DPCA. *See* Adaptive DPCA (A-DPCA) processor
Aegis system, 48
Aerosols, smoke and, 130
AESA. *See* Active electronically scanned array (AESA)
AF. *See* Array factor (AF)
AGC. *See* Automated gain control (AGC)
AGL. *See* Aircraft above ground level (AGL)
AIM-7 missile, 44
Airborne interceptor, 46
Airborne moving target indication (AMTI), 465
Aircraft above ground level (AGL), 53
Air defense systems, 40
Air Force E3A Sentry (AWACS), 46
Airport surveillance radar, 49
Albersheim's equation, 106, 575–578, 581–583
  conditions for, 576
  vs. Shnidman's equation, 581–583
Algorithm metrics, of radar signal processor, 464
Aliased sinc, 508
Along-track position, 838
Alpha-beta filters, 34, 718, 719, 731–732, 738
Alpha-beta-gamma filter, 739–741
  for scalar coordinate, 740
Altera, 474
Altimeters, 361
AM. *See* Amplitude modulation (AM)
Ambiguity functions, 800–801
  for LFM waveform, 803
  range-Doppler coupling, 803–805
    spectral interpretation of, 806–807
  V-LFM, 808
Ambiguity surface, 801
  ridged, 802
  for simple pulse, 802
Amdahl's Law, 483
Amplifiers, for exciters, 451–452
Amplitude, 29
  threshold, 26
Amplitude comparison monopulse radar system, 701–702
Amplitude fluctuations, millimeter wave system and, 138
Amplitude modulation (AM), 379
AMTI. *See* Airborne moving target indication (AMTI)
Analog Devices, Inc., 474
Analog-to-digital converter (ADC), 4, 64, 392, 395, 409–414, 504, 590, 788, 795, 847
  configuration, 410
  radar signal processor and, 472–473
  sample of, 411
AN/FPS-115, 42
Angle measurements
  AOA estimation method. *See* AOA estimation method
  monopulse technique, 701–703
  sequential lobing, 700–701
Angle of arrival (AOA), 685
  estimation method. *See* AOA estimation method
  fluctuations, 138
Angular frequency, 6
Angular selectivity, 309, 310
AN/MPQ-39 C-band phased array radar, 45
AN/MPQ-64 sentinel, 40
Annapolis Microsystems, 474
AN nomenclature, 36–37
AN/SPQ-9A, 46
AN/SPQ-9B ship-based TWS radar, 47
AN/SPS-48 radar, 39
AN/SPS-49(V) radar, 38
AN/SPY-1 radar, 47–48
Antenna, 4. *See also* Radar antennas
  bistatic configurations, 18–20
  dwell time, 88
  isotropic, 310–311
  monostatic configurations, 18–20
  mounting error, 682
  one-dimensional pattern, 14
  pattern of, 14
  phased array. *See* Phased array antenna
  reflector, 322–326
  sidelobes, 14
Antiair warfare (AAW), 38
AN/TPQ-36 firefinder radar, 48
AN/TPQ-37 firefinder radar, 48
AN/TPS-75 air defense system, 40
AN/TPY-2 systems, 42
AOA. *See* Angle of arrival (AOA)
AOA estimation method, 702–704
  centroiding
    binary integration approach, 703–704
    ML approach, 703, 704
  monopulse
    multiple unresolved targets, 707–708
    single target, 704–707
AOA fluctuations. *See* Angle of arrival (AOA), fluctuations

Aperture, 312, 314
  defined, 314
  efficiency, 316, 323
  taper, 314
    errors, 316
API. *See* Application programming interface (API)
Application programming interface (API), 477
Applicationspecific integrated circuits (ASIC), 462, 474, 475, 480
Approximation, 82
AR model. *See* Autoregressive (AR) model
Array factor (AF), 327, 328, 330, 338
Array(s)
  antenna technology, 90
  defined, 326
  elements, 333–335
Artillery locating radars, 48
ASIC. *See* Applicationspecific integrated circuits (ASIC)
Aspect angle, 239
  of RCS, 249–250, 252, 253, 257, 261, 264, 266
ASR-9, 667
ASR-9 airport surveillance radar, 49
ASR-12 airport surveillance radars, 667
Atmosphere
  anomalous, refraction and, 130, 134–137
    ducting/trapping, 135–136
    subrefraction, 135
    superrefraction, 135
  attenuation of radar signals in, 121–123
  gases, attenuation and, 124–125
  layers, and radar propagation analysis, 120–121
  standard, refraction and, 130, 131–134
    effective earth radius and, 134
Atmospheric absorption, 117, 121
Atmospheric clutter
  frozen precipitation, 199–200
  models, 205
  rain reflectivity, average, 196–199
    versus frequency band, 197
  temporal spectra, 200
Atmospheric loss, 68–69
Atmospheric refraction, of electromagnetic waves, 15–17. *See also* Refraction
  cloud attenuation as, 16
  radar technology in, 16
Atmospherics, 118
Atmospheric turbulence, 118, 137–138
Atmospheric volumetric scattering, 118, 121–122, 168
ATR. *See* Automatic target recognition (ATR)
Attenuation
  of electromagnetic waves, 121–123
    circular polarization, 126
    dust and, 129
    fog and, 127–128
    hail and, 128–129
    one-way, 122
    rain rate and, 125–126, 127
    smoke and, 130
    snow and, 128–129
    water vapor and, 124–125
  frequency at sea level, 15
Autocorrelation function (ACF), 166, 187, 533
  of sea return, 194, 195
Autocorrelation sequence, of phase code, 812
Automated gain control (AGC), 408
Automatic target recognition (ATR), 36
Automotive collision avoidance radars, 53
Autoregressive (AR) model, 656
AWACS. *See* Air Force E3A Sentry (AWACS)
Azimuth angle, 27

B

Backprojection, of image formation, 873
Backscatter coefficient, 168
Backscatter cross section, 221
Backscatter reflectivity, 177–178
Ballistic missile defense (BMD) radars
  applications of, 41–43
Ballistic missile early warning system (BMEWS), 382
Ballistic Research Laboratory (BRL), 177
Band-pass filter, 391, 395–396
Bandwidth, of transmit signal, 431
Barker codes
  biphase, 817–818
    longest, compressed response for, 818
  polyphase, 824, 825
Barker phase coded waveform, 809
Baseband filter, 795
Batch estimator, 757
BAW devices. *See* Bulk acoustic wave (BAW) devices
Bayesian approach, 562
Bayliss function, 321
BBS synthesizer. *See* Broad-band synthesizer (BBS)
BDTI Benchmark Suites, 484
Beam-limited case, scattering coefficients, 169–170
Beam shape loss, 91
Beamwidth, 13, 236
  defined, 312
  factor, 316
Benchmarks, signal processor sizing and, 483–485
Bernstein estimator, 704
Bilinear transformation, 526
Binary integration, 550, 584, 703–704
Biphase-coded pulse, 30
Biphase-coded waveforms, 788, 809, 811–812, 816
Biphase codes
  Barker, 817–818
  MPS codes, 818–819, 820–821
Bisection bandwidth, 464
Bistatic antenna configurations, 18–20
Bistatic propagation, 119
Bistatic radar cross section, 221
BIT functions. *See* Built-in-test (BIT) functions
BLAS – LINPACK – LAPACK – ScaLAPACK series, 485
Blind Doppler frequencies, 634
Blind speeds, and staggered PRF, 634–637
Blind zone
  map, 660
  pulse-Doppler processing, 660
BMD. *See* Ballistic missile defense (BMD) radars
BMEWS. *See* Ballistic missile early warning system (BMEWS)
Boltzmann's constant, 404
Boundary admittance, 148
Bragg scattering, 194
Brewster's angle, 149, 150
BRL. *See* Ballistic Research Laboratory (BRL)
B_{rms}. *See* Rms bandwidth (B_{rms})
Broad-band synthesizer (BBS), 444, 445
Brute-force algorithm, 471
Built-in-test (BIT) functions, 392
Bulk acoustic wave (BAW) devices, 795

C

CA-CFAR. *See* Cell-averaging (CA)-CFAR
Canadian RADARSAT 2 system, 52
Cartesian coordinates, 10, 87, 709, 719
Cartesian space, 748
Cassegrain configuration, 325
Cathode-to-body voltage, 363
Causal, 524
C-band mixer, 430
CBE. *See* Cell Broadband Engine (CBE)
CDF. *See* Cumulative distribution function (CDF)

Cell-averaging (CA)-CFAR, 593, 597–599
CFAR loss, 601–603, 610
performance, 600
in heterogeneous environments, 603–607
in homogenous environment, 600–601
ROC for, 616–617
threshold, 598, 600
Cell Broadband Engine (CBE), 475
Cell under test (CUT), 595, 596
clutter boundaries and, 607
Censored (CS)-CFAR, 613–614, 615
combining greatest-of with, and ordered statistics CFAR, 615–616
ROC for, 616–617
Centroid tracker, 694
CFA. *See* Crossed-field amplifiers (CFA)
CFAR. *See* Constant false alarm rate (CFAR)
CFAR loss, 69
CFLOP. *See* Complex floating-point operations (CFLOP)
Channelized receivers, 392, 395–396
Chip, 438
Chirp pulse, 30
Chirp waveform. *See* Linear frequency modulation (LFM)
Chi-square, 558, 559, 560, 578, 579
Chi-square of degree 2, 254, 256
Circularly polarized, EM wave, 11
Circular polarization, 214
attenuation, 126
scattering matrix for, 223–224
Circular random process, 557
Circulator, 394
Clear region, 295
Closely spaced targets, 715
Clutter, 4, 590
atmospheric. *See* Atmospheric clutter
attenuation, 637, 638
backscatter coefficient of, 168
boundaries, 600
CA-CFAR performance and, 606–607
GOCA-CFAR and, 618
SOCA-CFAR and, 618
cancellation, 421
characteristics of
atmospheric clutter. *See* Atmospheric clutter
millimeter wave clutter, 200–202
overview of, 172
surface clutter. *See* Surface clutter
defined, 165
detect discrete target, 76
fill pulses, 664
interference, 105
mapping, 665–667
millimeter wave, 200–202
models, 202–205
versus noise signals, 165–169
polarization scattering matrix, 171
power spectra, 190–192
radar performance in, 202
reduction, effect of phase noise on, 424–425
region, 295
scattering coefficients, 169–171
statistical behavior of, 76
surface. *See* Surface clutter
volume, 77–78
Clutter, moving radar
altitude line (AL), 298–299
distribution, 299–300
hyperbolic isovelocity contours, 299
range-velocity distribution, 299, 300, 301
Doppler bandwidth, 296, 297
Doppler shift of, 296–297
ambiguities in, 301
radar-to-ground, 298
main lobe clutter (MLC), 298–299
range ambiguity, 300–301
sidelobe clutter (SLC), 298
specturm elements, 297–299
velocity ambiguity, 301–302
Clutter-to-noise ratio (CNR), 295, 887
CMOS. *See* Complementary metal oxide semiconductor (CMOS)
CNR. *See* Clutter-to-noise ratio (CNR)
Coarse resolutions, 841
Cobra Judy system, 42
Cognitive radar, 378
Coherent demodulation, 402–403
Coherent detector, 564, 565, 566, 568
Coherent integration, 67, 189, 536–537, 549, 559
Coherent-on-receive system, 353
Coherent oscillator (COHO), 431, 433
Coherent processing interval (CPI), 69, 88, 264, 419, 502, 548, 550, 551, 552, 627
Coherent system, 23, 419
local oscillator (LO) signal, 24
COHO. *See* Coherent oscillator (COHO)
Coincidence algorithm, 662–663
Coincidence detection. *See* Binary integration
COMBIC. *See* Combined Obscuration Model for Battlefield-Induced Contamination (COMBIC)
Combined Obscuration Model for Battlefield-Induced Contamination (COMBIC), 158
Commercial off-the-shelf (COTS) programmable processors, 462
open modular architectures of, 476–478
Complementary metal oxide semiconductor (CMOS), 410
Complex floating-point operations (CFLOP), 465
Computer simulation in error analysis, 316–317
Conditional CRLB, 707
Conical scan, 320, 677
Connectorized components, 446
Constant acceleration
estimator, 721
filtering, 726–727
Constant false alarm rate (CFAR), 69, 96, 460, 532, 593, 717
adaptive, 618–619
algorithms, selection of, 616–618
architecture of, 595–597
cell averaging. *See* Cell-averaging (CA)-CFAR
detection
processing, 666
theory, overview of, 590–592
loss, 69, 601–603
robust, 607
censored CFAR, 613–614
greatest-of CA-CFAR, 608–612
ordered statistics CFAR, 614–615
smallest-of CA-CFAR, 608, 612–613
trimmed mean-CFAR, 613–614
target masking and, 590
threshold, 595
vs. Neyman-Pearson threshold, 601
Constant gamma model, 174
Constant phase area, 230
Constant velocity (CV), 724
estimate, 720
filtering, 722–726
target, 646–648
Constructive interference, 9, 13, 51. *See also* Interference
Continuoustime white noise acceleration (CWNA), 730
NCV filtering with, 736–739
NCV motion model with, 748
Continuous wave (CW), 20, 222, 352, 844
pulse, 775
match filtered response for, 781
radars, 394
Converted measurement filter, 760
Convolution sum, 524

Cooley-Tukey FFT algorithm, 521
Coordinate systems, 709
Corner reflector, simple target, 248, 249
Corner reflectors, 226, 235
"Corner-turn" memories, 476
Corporate feeds, 354, 355
  vs. distributed feeds, 354, 355
Correlation lag, 534
Correlation length, 258, 259
Cosine, 92
COTS. *See* Commercial off-the-shelf (COTS) programmable processors
COTS 6U VME boards, 476
COTS 6U VME signal processor, 477–478
Covariance matrix, 541
Covariance of estimate, 722
CPI. *See* Coherent processing interval (CPI)
Cramèr-Rao lower bound (CRLB), 687–688, 775
Creeping waves, 226
Critical angle, 149
CRLB. *See* Cramèr-Rao lower bound (CRLB)
Cross-correlation function, 533
Crossed-field amplifiers (CFA), 353, 364
Crossed-field tubes *vs.* linear beam tubes, 357–359, 358–359
Cross range resolution, SAR, 838, 839, 841, 844, 845–846
  stripmap, 848
Cross-track range, 838, 840, 842, 857–860, 862, 864
Crowbar, 374
Crystal reference oscillator, 440–441
Crystal video receiver, 393
CS-CFAR. *See* Censored (CS)-CFAR
CTT ASM/050-1636-123A amplifier, 452
CTT RF amplifiers, 452
Cumulative distribution function (CDF), 185–186
CUT. *See* Cell under test (CUT)
CV. *See* Constant velocity (CV)
CW. *See* Continuous wave (CW)
CWNA. *See* Continuoustime white noise acceleration (CWNA)

D

1-D, CFAR window. *See* One-dimensional (1-D), CFAR window
2-D. *See* Two-dimensional (2-D) system
D/A converter. *See* Digital-to-analog (D/A) converter
Data association, 713–716, 717
Data collection, SAR, 852–856
  phase information, 854–855
  range variation, 855–856
  stop-and-hop model, 854
Datacube, radar, 502–503
  algorithms act in, 503
Data processing, 461
Data rates, signal processor sizing and, 485–487
DC. *See* Direct current (DC)
DCT. *See* Discrete cosine transforms (DCT)
2-D data set. *See* Two-dimensional (2-D) data set
DDS. *See* Direct digital synthesizer (DDS)
Dechirp-on-receive, LFM, 847
Decibels (dBi), 313
Decorrelation models, 583
Decorrelation time, 187–189
  for rain backscatter, 200
Delta beam, 320
Demodulation
  analog coherent detection implementation and mismatch errors, 403–404
  coherent, 402–403
  noncoherent, 400–402
Depressed collector, 374
Depression angle, 173
Despeckling, 168
Destructive interference, 9, 12, 13. *See also* Interference
Detection, 95, 547
  criterion, 109–111
  fundamentals
    closed-form solutions, 106
    concept of threshold, 95–96
    detection for nonfluctuating target, 106–107
    probabilities of, 96–97
    swerling 1 target model, 107–108
  statistics, 548, 550, 560
  theory, 590–592
Detector, 4
  loss, 566
DFT. *See* Discrete Fourier transform (DFT)
Dielectric permittivity, 213
Dielectric resonant oscillator (DRO), 441
Differential path length, 327
Diffraction, 11–15, 140–142, 226. *See also* Interference
  coefficient, 141
  extreme cases of, 12
  in multi-element antenna, 13
Diffuse scattering, 17–18, 153–154
Digital compressor, 795–796
Digital filter, 523–532, 795–796
  characteristics of, 525–528
  design of, 525–528
  FIR filters, 525–526
  implementation of, 528–530
  of LSI, 524
Digital receivers, 395
Digital signal processing (DSP), 392, 459–460
  correlation as, 538–540
  data collection, 502–503
  digital filtering, 523–532
  Fourier analysis, 506–522
  integration of, 536–538
  matched filters, 540–543
  overview, 495–496
  quantization, 504–506
  random signals, 532–535
  sampling, 496–503
  z transform, 522–523
Digital-to-analog (D/A) converter, 438
Digital up/down frequency conversion, 414
Dihedrals, 235, 884, 885
Diode detector, 400, 401
Diode limiters, 396
Dipole moments, 225
Dirac delta function, 498, 777, 871, 873
Direct current (DC), 350
Direct digital in-phase/quadrature (I/Q) sampling, 412–414
Direct digital synthesizer (DDS), 438–439, 444
Directivity, 62, 312–313, 316
  beamwidth and, 313
  gain and, 313
  sidelobe levels and, 314
Direct synthesizer (DSX), 444–445
Dirichlet function, 508
Discrete cosine transforms (DCT), 485
Discrete Fourier transform (DFT), 28, 464, 470, 499, 514–516, 549, 628, 643, 795
  matched filter and filterbank interpretations of pulse-Doppler processing with, 652–653
  methods
    for doppler and range rate measurements, 697–698
    interpolation methods, 698–699
Discrete ray tracing, 147
Discrete-time Fourier transform (DTFT), 294, 506–509, 567, 628, 697, 795
  of constant-velocity target, 646–648
Discrete white noise acceleration (DWNA)
  NCA model and, 749
  NCV filtering with, 730–736
  NCV motion model and, 748

Dispersive analog filters, 794–795
Displaced phase center antenna (DPCA) processing, 642–644
Distributed feeds, 354, 355
  vs. corporate feeds, 354, 355
Divergence factor, multipath, 151–152
D layer, ionosphere, 138
Doppler beam sharpening (DBS), for SAR imaging, 860–870
  derivation, 860–864
  example, 866–868
  image formation, 864–865
  limitations, 868–870
  spatial frequency, 863–864
Doppler bins, 82, 100
Doppler cell, 98
Doppler effect, 23
Doppler filter bandwidth, 841
Doppler filtering, 841
Doppler frequency, 195
Doppler loss, 684
Doppler processing, 66, 402, 717
  clutter mapping, 665–667
  Doppler shift, 626
  moving target detector (MTD), 667–668
  moving target indication (MTI), 629–631
    blind speeds and staggered PRFs, 634–637
    figures of merit, 637–640
    from moving platform, 641–644
    performance, limitations to, 640–641
    pulse cancellers, 631–634
  overview, 625–626
  pulse-Doppler processing. *See* Pulse-Doppler processing
  pulsed radar Doppler data acquisition, 627–629
  pulse pair processing (PPP), 668–673
Doppler shift, 23–24, 274–276, 294–295, 626, 805, 808, 816–817
  of clutter (moving radar), 296–297
    ambiguities in, 301
    radar-to-ground, 298
  as function of velocity and frequency, 626
  impacts of, on pulse compression waveform, 807
  measurement of radar, 28
  ratio of, 804
  unambiguous measurement of, 24–25
Doppler spectrum, 30
  clear region, 295
  DTFT, 294
  elements of, 293–294
  of fluctuating targets, 267–269
  for range bin, 294
  SNR in, 651–652
  and straddle loss, sampling, 648–651
  transition region, 295
Doppler tolerance, 788, 791, 816–817, 827–828
  defined, 801
  of P4 code, 829
Double-balanced mixer, 430
Double-delay filter, 424
Douglas sea number, 181
Down-range resolution, 845
DPCA condition, 642–643
DPCA processing. *See* Displaced phase center antenna (DPCA) processing
DRO. *See* Dielectric resonant oscillator (DRO)
Drop-size distribution model, 125
DSP. *See* Digital signal processing (DSP)
DSX. *See* Direct synthesizer (DSX)
DTFT. *See* Discrete-time Fourier transform (DTFT)
Dual polarized radars, 409
Ducting, 16
  anomalous atmosphere, 135–136
  sea reflectivity and, 184–185
Duplexer, function of, 391
Dust, and attenuation, 129
Duty cycle, 350–351
Dwell time, 31, 33, 88, 502, 548, 627
Dwell-to-dwell frequency change, 439–440
Dwell-to-dwell stagger
  pulse-Doppler processing, 659
DWNA. *See* Discrete white noise acceleration (DWNA)
Dynamical motion model, 728
Dynamic range, 496, 504
  receiver, 406
    coupling issues, 409
    gain control, 408–409
    sensitivity time control, 407–408
  SFDR, 411, 412, 413
  of signal-following ADC, 411–412

E

Early/late gate tracker, 693–694
East-North-Up coordinates, 709
Echo amplitude, 251, 259, 261
Echo area, 219
Eclipsed, 660
ECM. *See* Electronic countermeasures (ECM) system
Edge diffraction, 231, 232–234
Edge waves, 226
EED. *See* Electro-explosive devices (EED)
EEMBC, 484
Effective earth radius, 134
Effective number of bits (ENOB), 473
EKF. *See* Extended Kalman filter (EKF)
E layer, ionosphere, 139
Electric field, 5, 213
Electric permittivity, 212–213
Electro-explosive devices (EED), 384
Electromagnetic (EM) waves, 4, 59, 211
  atmospheric attenuation, 15
  atmospheric refraction, 15–17
  attenuation of, 121–123
  characteristics of, 212
  diffraction, 11–15
  electric, *E*, field, 5
  frequency of, 6–7
  fundamentals, 212–214
  intensity of, 10
  interaction with matter, 11–18
  magnetic, ***B***, field, 5
  phase of, 7–9
  physics of, 5
  polarization of, 10–11, 214–215
  propagation of, 4
    factor, 118–119
    paths and regions, 119–120
  reflection, 17–18, 215–219
  superposition of, 9–10
  transmission of, 156–158
  types of, 8
  wavelength of, 6
Electromagnetic fields, 6
Electromagnetic interference (EMI), 5, 294, 824
Electromagnetic velocity vector, 6
Electronically scanned array (ESA), 14, 80, 88, 848
Electronic countermeasures (ECM) system, 5, 362, 716, 717
Electronic warfare (EW) systems, 7, 370
  receivers, 392
Elevation angle, 27
Elliptic IIR filter, 527
EMI. *See* Electromagnetic interference (EMI)
EM waves. *See* Electromagnetic (EM) waves
End-region scattering, 226, 231–232
Energy conservation equation, 314
Engineering Refractive Effects Prediction (EREPS), 159
ENOB. *See* Effective number of bits (ENOB)
Envelope detector, 400

Equivalence operations, phase-coded waveforms, 812
EREPS. *See* Engineering Refractive Effects Prediction (EREPS)
Erlang density, 574
Error correction filter, 796
ESA. *See* Electronically scanned array (ESA)
Estimators, 685–686
  types of, 686
European Communications Office, 159
Evaporation duct, 136
EW. *See* Electronic warfare (EW) systems
EW bands, 8
Exciters
  components
    amplifiers, 451–452
    assembly approaches, 446–448
    frequency dividers, 450–451
    frequency multipliers, 450
    mixers, 448–449
    stable oscillators, 440–444
    switches, 449–450
    synthesizers, 444–446
  design considerations
    transmit signal, 429–437
    waveform generation, 437–440
  internal components of, 434
  overview, 417–418
  performance issues, 418–429
  phase noise, 419–421
    clutter reduction, effect on, 424–425
    imaging, effect on, 428–429
    pulse-Doppler processing, effect on, 425–428
    self-coherence effect and, 422–424
    spectral folding effects and, 421–422
  purpose of, 419
  timing and control circuits, 452–454
Exponential PDF, 253, 254, 255
Exponential PDF, 557
Extended Kalman filter (EKF), 753, 755–756, 759–760
Extended targets, defined, 692
Extinction efficiency, 129

F

False alarm rate (FAR), 553–554, 591–592. *See also* Constant false alarm rate (CFAR)
False alarms, 95, 552, 553–554
  average probability of, 599
  clutter boundaries and, 607
  defined, 97–98
  impact of, 592–593
  noise PDF and, 97–100
  probability of, 96–111, 592–593
Fan beam, 38
FAR. *See* False alarm rate (FAR)
Faraday shield, 215
Fast convolution, 528, 795
  approach, 471
Fastest Fourier Transform in the West (FFTW) library, 483, 485
Fast Fourier transform (FFT), 28, 67, 88, 460, 485, 496, 521, 628, 643, 795, 841
  algorithm, 470, 656, 697
Fast switching synthesizer (FSS), 444, 445
Fast time, 502
  SAR data acquisition, 842
FCC. *See* Federal Communication Commission (FCC)
FCR. *See* Fire control radar (FCR)
FDS. *See* Fractional Doppler shift (FDS)
Federal Communication Commission (FCC), 7
Feed horn, 323, 324, 325
Fermat's minimum path, 229
Ferrite limiters, 396
FFT. *See* Fast Fourier transform (FFT)
FFTW library. *See* Fastest Fourier Transform in the West (FFTW) library
Field of view (FOV), 317, 334, 691
Field programmable gate arrays (FPGA), 395, 454, 474, 523
Filterbank, 537
Filtered state estimate, 719
Fine resolution. *See* High resolution
Finite impulse response (FIR) filters, 395, 460, 496, 525–526, 631
Finite pulse train, 280–283
  modulated, 281–283
FIR. *See* Finite impulse response (FIR) filters
Fire-and-forget mode, 44
Fire control radar (FCR), 37, 46–47
Firefinder, 90
First-order canceller, 631
Fixed point *vs.* floating point, 480–482
Flash point, 229
F1 layer, ionosphere, 139
F2 layer, ionosphere, 139
FLOPS, counting
  choosing efficient algorithms, 470–472
  general approach, 464–468
  mode-specific formulas, 468–470
Fluctuating targets, 247, 578–581
  Doppler spectrum of, 267–269
  vs. non-fluctuating targets, 580–581
FM. *See* Frequency modulation (FM)
FMCW. *See* Frequency modulated continuous wave (FMCW)
Focal length to diameter *(f/D)* ratio, 323–324
Fog
  attenuation and, 127–128
  types of, 127
Foliage penetration (FOPEN) SAR systems, 836
Fourier analysis, 506–522
  DFT, 514–516
  DTFT, 506–509
  fast Fourier transform, 521
  reciprocal spreading, 508
  spatial frequency, 513–514
  straddle loss, 515–516, 517–518
  summary of, 521–522
  windowing, 509–513, 516–518
Fourier spectrum, 30
Fourier transform, 187, 276–277
  matched filter and, 778–779
Fourier uncertainty principle, 784–785
Fourth-degree chi-square PDF, 254, 255, 256
FOV. *See* Field of view (FOV); Field of view (FOV)
FPGA. *See* Field programmable gate arrays (FPGA)
Fractional Doppler shift (FDS), 804
Frank codes, 824, 825–827, 829
Free-running oscillator, 352–353
Frequency
  agility, 262, 559, 580–581, 708
  dependence, of RCS, 249–250
  diversity, 581
  dividers, for exciters, 450–451
  Doppler shift as function of, 626
  downconversion
    double, 398
    and mixers, 397–399
    single, 398
  of electromagnetic waves, 6
  pulling, 379
Frequency modulated continuous wave (FMCW), 49, 361, 371, 377
Frequency modulation (FM), 371, 419
Frequency multipliers, for exciters, 450
Fresnel reflection coefficient, 148–150
Frozen precipitation, 199–200
FSS. *See* Fast switching synthesizer (FSS)
Furuno FAR2817 X-band radars, 47

G

Gain, 313. *See also* Directivity
  antenna *vs.* amplifier, 313
Gain control
  receiver dynamic range and, 408–409
Gain loss, 330, 332–333
  phase-shifter bits and, 330

Gamma density, 560, 574
Gamma PDF, 256
Gap cells, 596
  self-masking and, 604
Gases, attenuation and, 124–125
Gating, 763
  typical threshold values for, 764
Gaussian Clutter Power Spectrum
  improvement factor for, 639
Gaussian noise, 557, 775. *See also* Circular random process
Gaussian PDF, 97
Gaussian spectra, 190, 192
GCMLD. *See* Generalized censored mean level detector (GCMLD)
Gedae by Gedae, Inc., 483
Generalized censored mean level detector (GCMLD), 616
  ROC for, 619
Generalized likelihood ratio test (GLRT), 708
General-purpose processors, 474
Georgia Tech Research Institute (GTRI), 45, 177
  empirical model, 202–205
    coefficients for, 203
  sea clutter model equations, 204
Ghosts, 663
Giga, 7
Gigabit Ethernet, 477
Glint, 678, 690
Glistening surface, 152
GLRT. *See* Generalized likelihood ratio test (GLRT)
GMTI radars. *See* Ground MTI (GMTI) radars
GOCA-CFAR. *See* Greatest-of CA-CFAR (GOCA-CFAR)
GPR. *See* Ground penetration radars (GPR)
GPU. *See* Graphical processing units (GPU)
Graphical processing units (GPU), 475
Grating lobes, 330–331, 349
  element spacing and, 331, 332
Grazing angle, 144, 150, 169
  clutter surface and, 172, 173
  dependence on, land reflectivity and, 177–178
  effect of, on PDF, 193
Greatest-of CA-CFAR (GOCA-CFAR), 608–612, 618
  CFAR loss associated with, 610
  ROC associated with, 609–610, 616–617
Green's function, 217
Ground MTI (GMTI) radars, 463
Ground penetration (GPEN) SAR systems, 838
Ground penetration radars (GPR), 53
Ground plane, 143, 299, 879, 881, 887
Group delay, matched filter, 794
Gs. *See* Guard cells (Gs)
GTRI. *See* Georgia Tech Research Institute (GTRI)
Guard cells (Gs), 595, 596
  self-masking and, 604
Gunn oscillators, 360–361
Gyrotrons, 360

H

Hail, and attenuation, 128–129
Hard-tube modulator, 372
Hardware metrics, of radar signal processor, 462–464
Harry Diamond Laboratory, 177
HCE-CFAR. *See* Heterogeneous clutter estimating (HCE)-CFAR
Herley-CTI, Inc., 442, 444
Hertz, 7
Heterodyne receiver, 394–395
Heterogeneous clutter estimating (HCE)-CFAR, 618
HF. *See* High frequency (HF)
High frequency (HF), 118, 430
High-frequency optics region, 226–227
High-frequency scattering
  edge diffraction, 232–234
  end-region scattering, 231–232
  multiple-bounce scattering, 235–236
  phasor addition, 227–229
  specular scattering, 229–231
High grazing angle region, clutter surface, 174
High-order filter, 436
High Performance Computing (HPC) Challenge, 483, 484
High PRF mode, 661
High-range resolution (HRR) systems, 787
High resolution, 843
High voltages (HV), 347
Hilbert transformer, 413–414
Hittite Microwave Corp. HMC C014 mixer, 448
Homodyne receiver, 394
Horizon, defined, 134
Horizontally polarized, EM wave, 11
Hot spot, 229
HPC Challenge. *See* High Performance Computing (HPC) Challenge
HPCS Benchmarks, 484
HPEC challenge, 484
HRR systems. *See* High-range resolution (HRR) systems
Huffman coded waveform, 787
Huygen's principle, 12, 140. *See also* Diffraction
Huygen's wavelet, 217
HV. *See* High voltages (HV)

I

IBM PowerPC microprocessors, 476
IBM "Roadrunner" system, 479
IC technology. *See* Integrated circuit (IC)
Ideal (noise-free) matched filter output, 693–694
IEEE dictionary. *See* Institute of Electrical and Electronics Engineers (IEEE) dictionary
IEEE P754 32-bit format, 481
IF. *See* Intermediate frequency (IF)
IFM receivers. *See* Instantaneous frequency measurement (IFM) receivers
IID. *See* Independent and identically distributed (IID)
IIR filters, 526–527
IIR high-pass filters. *See* Infinite impulse response (IIR) high-pass filters
Imaging, 34–36
  effect of phase noise on, 428–429
Imaging, SAR, 468–469, 856–874
  ambiguity, 879–880
  CNR, 887
  coordinate systems, 856–857
  DBS, 860–870
    derivation, 860–864
    example, 866–868
    image formation, 864–865
    limitations, 868–870
  foreshortening, 880
  layover, 880
  man-made returns, 884–886
  matched filter imaging, 871–873
  NAR in, 875, 876
  PSR, 856, 865
    linear collection, 857–860
  shadowing. *See* Shadows, imaging
  SNR, 886–887
  speckle, 881–884
  survey of, 873–874
IMM estimator. *See* Interacting multiple model (IMM) estimator
Impact ionization avalanche transit time (IMPATT) diodes, 360–361
IMPATT. *See* Impact ionization avalanche transit time (IMPATT) diodes

Improvement factor, 637, 638–640
  for Gaussian Clutter Power Spectrum, 639
Incident field, 215
Independent and identically distributed (IID), 561, 591
Index of refraction, 213
InfiniBand, 477
Infinite impulse response (IIR) high-pass filters, 633
Infinite pulse train, 279–280
Initial phase, 6
Injection-lock, 353
Injection-prime, 353
In-phase (I) frequency value, 844
Instantaneous AGC, 408–409
Instantaneous frequency, *vs.* time for LFM, 790
Instantaneous frequency measurement (IFM) receivers, 392, 395
Institute of Electrical and Electronics Engineers (IEEE) dictionary, 219
  RCS definition, 219–220
Instrumentation, radar, 45–46
Integrated circuit (IC), 454
Integrated power conditioner, 369
Integrated sidelobe ratio (ISR), 794
Integration, 536–538, 548–552, 559, 570–572, 584
  binary, 550, 584
  coherent, 536–537, 549, 559
  combination of, 550–552
  gain, 536, 549
  noncoherent, 537–538, 549–550, 559, 570–572, 575, 576
    of nonfluctuating target, 570–572
Integration angle, in SAR, 845, 847–848, 849
Intel microprocessors, 478, 479
Intel's Math Kernel Library (MKL), 485
Intensity, of electromagnetic wave, 10
Interacting multiple model (IMM) estimator, 743
  implementation of
    combination of state estimates, 745–746
    mode conditioned estimates, 744
    model likelihood computations, 744–745
    mode probability update, 745
    state mixing, 744
Interference, 95
  constructive, 9
  destructive, 9
  forms of, 5
  other than noise, 105–106
  power, 598
  region, 120
Interferometric processing, 884
Interferometric SAR, 35
Interferometry, 702
Intermediate frequency (IF), 4, 360, 430, 434, 776
  oscillator, 437
  selection of, 399–400
  signal, 391
International Telecommunications Union (ITU), 7, 159
Interpulse period (IPP), 20
Intrapulse, pseudorandom biphase modulation, 437–438
Intrapulse LFM. *See* Intrapulse linear frequency modulation (LFM)
Intrapulse linear frequency modulation (LFM), 438–439
Inverse methods, image formation survey, 873
Inverse synthetic aperture radar (ISAR), 45. *See also* Synthetic aperture radar (SAR)
  image, 46
  turntable range, 46
Ionosphere, 121, 138–140
IPP. *See* Interpulse period (IPP)
ISAR. *See* Inverse synthetic aperture radar (ISAR)
Isodoppler contour, 299
Isorange contour, 299
Isotropic antenna, 310–311
  defined, 310
  power density of, 310
  radiation intensity from, 310
ISR. *See* Integrated sidelobe ratio (ISR)
ITU. *See* International Telecommunications Union (ITU)

J

Jamming signals, 5, 78
  preselection effect on rejection of, 397

K

Kalman algorithm, 728
Kalman filters, 461, 718, 719, 728–729, 737, 741
  as predictor-corrector algorithm, 729
Keller cone, 232–233, 234
Kinematic motion models, 746
  NCA motion model, 749
  NCV motion model, 747–749
  nearly constant speed motion model, 750–751
  Singer motion model, 749–750
Kinematic state estimation, 678
KLIX WSR-88D NEXRAD weather radar, 671–672
Klystrons, 348, 361–362
Knife edge effect, 141
K-point DFT, 697

L

Land clutter
  clutter frequency spectra, 190–192
  decorrelation time, 187–189
  spatial variations, 192–193
  temporal variations, 185–186
Land reflectivity
  frequency bands, 178–180
  grazing angle, 177–178
Latency, 462–463
L-band frequency, 15
Leading-edge (LE) diffraction, 233
Leading window, 596, 597
Least squares estimate (LSE), 722
LE diffraction. *See* Leading-edge (LE) diffraction
Left-handed circular (LHC), 150, 223
LEIBE MPM. *See* Microwave Propagation Model (LEIBE MPM)
LFM. *See* Linear frequency modulation (LFM)
LFM waveform. *See* Linear frequency modulated (LFM) waveform
LHC. *See* Left-handed circular (LHC)
Light, speed of, 5
Likelihood ratio (LR), 554–556
Likelihood ratio test (LRT), 554–556, 562
Linear detector, 58, 537, 563, 572–573
Linear frequency modulated (LFM) waveform, 695, 730, 774, 787–788
  ambiguity function for, 802
  vs. biphase MLS coded waveform, 823–825
  compressed response, 791–792
  match filtered response for, 792
  measurements with, 756
  NCV tracking with, 736
  peak sidelobe ratio of, 793–794, 797
  range-Doppler coupling, 803–805
    spectral interpretation of, 806–807
  Rayleigh resolution, 792–793
  sidelobe
    reduction in, 797–800
    structure of, 805
  spectrum of, 791
  summary of, 808
  Taylor weighted, straddle loss reduction associated with, 800
  time-domain description of, 789–790

time for, instantaneous frequency *versus,* 790
V-LFM, 808
Linear frequency modulation (LFM), 30, 433, 847
chirps, 627
Linearity, 523
Linearly polarized, EM wave, 11
Linear polarization, 214
scattering matrix for, 222–223
Linear shift-invariant (LSI), 523
spectral representations of, 524
Line-of-sight (LOS) region, 119–120, 558
LNA. *See* Low-noise amplifier (LNA)
LO. *See* Local oscillator (LO)
Load balancing, 482
Local oscillator (LO), 360
selection of, 399–400
signal, 23, 391
coherent system, 24
Log amplifiers, 401
Logarithmic receiver, 176
Log detector, 98
Log-likelihood ratio test, 554, 568
Log-normal distribution, 176, 256, 257
Log-video, 98
Longbow radar, 44–45
Long-pulse illumination, 222
LOS region. *See* Line-of-sight (LOS) region
Lower-power radar systems, 396
Lower sideband (LSB), 432
Low grazing angle region, clutter surface, 174
Low-noise amplifier (LNA), 339, 356, 391, 395
Low power density (LPD), 340, 341
Low PRF mode, 661
Low-voltage differential signaling (LVDS), 410
LPD. *See* Low power density (LPD)
LRT. *See* Likelihood ratio test (LRT)
LSB. *See* Lower sideband (LSB)
LSE. *See* Least squares estimate (LSE)
LSI. *See* Linear shift-invariant (LSI)
LVDS. *See* Low-voltage differential signaling (LVDS)

M

Macro component, surface boundary, 146
Magnetic field, 5, 213
Magnetic permeability, 213
Magnetron, 348, 358, 359–360
Mahalanobis distance, 763
Main beam, 13
Main lobe, 13, 782
shape of, 784
width, range resolution and, 784
Maneuver index, 731
Man-made returns, in SAR imaging, 884–886
Manual gain control, 408
Maple, 481
Marcum's Q function, 100, 564
Marcum targets, 257
Marine navigation radar, 51–52
Markovian switching coefficients, 742–743
Masking, 600, 693
mutual, 601, 604–605
self, 603–604
Massachusetts Institute of Technology Lincoln Laboratories (MIT/LL), 177
Master oscillator-power amplifier (MOPA), 353, 429
Matched filter, 530, 540–542, 567–568
for continuous-time signals, 542
defined, 774
derivation of, 779–780
form of, 776
Fourier transform and, 778–779
implementations of, 794–796
phase-coded waveform and, 812–813
point target model, 776–777
radar range equation, 775–776
relationship between radar range equation and, 780
response
for phase codes, 812–813
properties of, 782
for simple pulse, 781
response for LFM waveform, 792
SNR, radar performance and, 775
Matched filter imaging, 871–873
concept, 871
of near-range point targets, 872
MathCAD, 481
Mathematica, 481
MathWorks, Inc., the, 483
MATLAB, 101, 481, 527
Maximum length sequence (MLS), 810, 819, 822–823
vs. compressed LFM waveform, 823–824
Maximum likelihood (ML) estimators, 592, 686, 719
Maximum MSE (MMSE), 733–734
Maxwell, James Clerk, 212
Maxwell's equation, 5, 214, 216
MCAEKF. *See* Measurement covariance adaptive extended Kalman filter (MCAEKF)
MCRLB. *See* Modified Cramèr-Rao lower bound (MCRLB)
MDD. *See* Minimum detectable Doppler (MDD)
MDV. *See* Minimum detectable velocity (MDV)
Mean, 533–534
square, 533–534
Mean time between failures (MTBF), 366, 381
Measurement covariance adaptive extended Kalman filter (MCAEKF), 759
Measurement-to-track data association, 713–716, 717, 760–761
formation of new track, 762
measurement validation and gating, 762–763
nearest neighbor (NN) filter, 764
probabilistic data association filter (PDAF), 765–766
strongest neighbor (SN) filter, 764–765
Measurement validation, 763–764
Median filters, 530
Medium PRF radar, 661
in range and velocity ambiguities, 302–303
MEM. *See* Microelectro-mechanical systems (MEM)
Mercury Computer Systems, 477–478
MESFET. *See* Metal semiconductor field-effect transistor (MESFET)
Mesosphere, 120–121
Message passing interface (MPI), 477
Metal semiconductor field-effect transistor (MESFET), 364–365
Method-of-moments (MoM), 147
MFA. *See* Multiple frame assignment (MFA)
MHT. *See* Multiple hypothesis tracking (MHT)
Micro component, surface boundary, 146
Microelectro-mechanical systems (MEM), 326
Microwave filters, 435– 436
Microwave power modules (MPM), 363, 369–370
components of, 369–370
Microwave Propagation Model (LEIBE MPM), 158
Microwave tubes
crossed-field tubes, 358–359
linear beam tubes, 357–358

Millimeter wave (MMW), 347
amplitude fluctuations and, 138
clutter, 200–202
surface electrical and physical parameters at, 147
Millions of instructions per second (MIPS), 463
Minimum detectable Doppler (MDD), 658
Minimum detectable radar cross section, 72
Minimum detectable velocity (MDV), 638
Minimum mean squared error (MMSE), 534, 722, 728
Minimum peak sidelobe (MPS) codes, 774, 818–819, 820–821
lengths, and corresponding peak sidelobe levels, 819
Minimum variance (MV) estimators, 686
Minimum variance unbiased (MVU) estimator, 686
MIPS. *See* Millions of instructions per second (MIPS)
Miss, 553
Missile support radars, 46–47
MITEQAMF-4B-02000400-20-33P-LPN amplifier, 451
MITEQ Inc., 445
Miteq LNS series synthesizers, 445
MITEQ model LNS60806580 frequency synthesizer, 445, 446
phase noise of, 446
MITEQ MX2M010060 passive frequency multiplier, 450
MIT/LL. *See* Massachusetts Institute of Technology Lincoln Laboratories (MIT/LL)
Mixer output frequencies, 431–432
Mixers, 430
for exciters, 448–449
frequency downconversion and, 397–399
Mk-86 Gunfire Control system, 46
MKL. *See* Intel's Math Kernel Library (MKL)
ML. *See* Maximum likelihood (ML)
ML approach, 703, 704
ML estimators. *See* Maximum likelihood (ML) estimators
MLS. *See* Maximum length sequence (MLS)
MMIC. *See* Monolithic microwave integrated circuits (MMIC)
MMSE. *See* Minimum mean squared error (MMSE)
MNR. *See* Multiplicative noise ratio (MNR)
Modes, diffraction, 141
Modified Cramèr-Rao lower bound (MCRLB), 707, 708
Moding, 360
Modulators, 371–373
active-switch, 372–373
line-type, 371–372
M-of-n detection. *See* Binary integration
MoM. *See* Method-of-moments (MoM)
Moments, 533–534
Monolithic microwave integrated circuits (MMIC), 364
Monopulse, 27, 320–322, 409, 677, 701–703
angular position determination and, 320
AOA estimation
multiple unresolved targets, 707–708
single target, 704–707
error signal, 320–321
Monostatic antenna configurations, 18–20
Monostatic cross section, 221
Monostatic propagation, 119
Monte Carlo simulations, 730, 766–767
Moore, Gordon, 478
Moore's law, 459, 560
influence, on radar signal processor implementation, 478–480
MOPA. *See* Master oscillator-power amplifier (MOPA)
MOTR. *See* Multiple-object tracking radar (MOTR)
Moving radar, 641–642
Moving radar, range-Doppler spectrum of, 296–303
clutter
altitude line (AL), 298–299
distribution, 299–300
Doppler bandwidth, 296, 297
Doppler shift of, 296–297, 301
main lobe clutter (MLC), 298–299
range ambiguity, 300–301
sidelobe clutter (SLC), 298
specturm elements, 297–299
velocity ambiguity, 301–302
range and velocity ambiguity, 300–303
Moving target detector (MTD), 667–668
Moving target indication (MTI), 28, 283, 378, 402, 497, 629–631
blind speeds and staggered PRFs, 634–637
figures of merit, 637–640
filters. See MTI filters
from moving platform, 641–644
performance, limitations to, 640–641
processing, 424–425
pulse cancellers, 631–634
visibility factor, 640
Moving targets, pulse-Doppler detection of, 658–659
MPI. *See* Message passing interface (MPI)
MPM. *See* Microwave power modules (MPM)
MPS codes. *See* Minimum peak sidelobe (MPS) codes
MRSIM. *See* Multispectral Response Simulation (MRSIM)
MTD. *See* Moving target detector (MTD)
MTI. *See* Moving target indication (MTI)
MTI filters, 460
effect of, 630
figures, 637–640
clutter attenuation, 637, 638
improvement factor, 637, 638–640
minimum detectable velocity (MDV), 638
subclutter visibility, 637
usable Doppler space fraction (UDSF), 638
MTT. *See* Multiple target tracking (MTT)
Multicore processor, 475
Multilook processing, speckle, 882–884
Multipath
angle errors, 155
classification error, 155–156
defined, 142
divergence factor, 151–152
Fresnel reflection coefficient, 148–150
propagation paths, 143
roughness factors, 152–155
diffuse scattering, 153–154
specular scattering, 152–153
superposition, 144–145
surface boundary and, 146–148
Multiple-bounce scattering, 226, 235–236
Multiple-dwell detection principles, 108
Multiple frame assignment (MFA), 763
Multiple hypothesis tracking (MHT), 763
Multiple model filtering, 741–746
Multiple-object tracking radar (MOTR), 45
Multiple target tracking (MTT), 716
performance assessment of, 76–767
Multiplicative noise ratio (MNR), 888
Multispectral Response Simulation (MRSIM), 159
Mutual target masking, 601, 604–605
censored CFAR, 613–614
ordered statistics CFAR, 614–615
smallest-of CA-CFAR, 608, 612–613
trimmed mean-CFAR, 613–614
MV estimators. *See* Minimum variance (MV) estimators
MVU estimator. *See* Minimum variance unbiased (MVU) estimator

N

Nallatech, 474
Narrowband system, 435
NASA' Tracking and Data Relay Satellite System (TDRSS), 487
National Telecommunications and Information Administration (NTIA), 376
  spectrum efficiency (SE) and, 376
National Weather Service, 670
NATO Seasparrow AAW missiles, 43
Naval Research Laboratory (NRL), 177
Navy E2C AEW system, 46
NCA filtering. *See* Nearly constant acceleration (NCA) filtering
NCV filtering. *See* Nearly constant velocity (NCV) filtering
NCV tracking. *See* Nearly constant velocity (NCV) tracking
Nearest neighbor (NN) filter, 764
Nearly constant acceleration (NCA) filtering, 730, 739–741
  DWNA and, 749
Nearly constant speed motion model, 750–751
Nearly constant velocity (NCV) filtering
  with CWNA, 736–739
  with DWNA, 730–736
Nearly constant velocity (NCV) tracking
  with LFM waveforms, 736
  motion model, 747–749
    with CWNA, 748
    with DWNA, 748
    state vector for, 746
NEES. *See* Normalized estimation error squared (NEES)
Nested codes, 810
Neyman-Pearson criterion, 553–554, 561, 571
Neyman-Pearson (NP) detector, 591, 594–595
  threshold *vs.* CFAR threshold, 601
NLFM waveforms. *See* Nonlinear frequency modulated (NLFM) waveforms
NLSE. *See* Nonlinear least-squares estimate (NLSE)
NN filter. *See* Nearest neighbor (NN) filter
Noise, 25–26
  clutter signals *vs.,* 165–169
  defined, 64
  at Doppler shifts, 295
  effects, on precision and accuracy in radar measurements, 679–681
  figure, 65, 404–405
  power receiver, 404–406
  quantization, 505
  statistical models of, 557
  thermal receiver, 24, 64–65
  threshold detection, of signal, 26
Noise-equivalent backscatter coefficient, 887
Noncentral gamma-gamma density, 267
Noncoherent demodulation, 400–402
Noncoherent integration, 67, 537–538, 549–550, 559, 570–572, 575, 576, 684
  gain, 575, 577
Noncoherent system, 23
  vs. coherent system, 352
Noncontact radar, 49
Nonfluctuating targets, 247, 561–562, 570–572, 580, 581, 584
  defined, 561–562
  noncoherent integration of, 570–572
  vs. fluctuating targets, 580, 581
Nonlinear frequency modulated (NLFM) waveforms, 785
Nonlinear least-squares estimate (NLSE), 757–759
Nonlinear systems, shift-varying and, 530–532
Nonrecurring engineering effort (NRE), 474
No return areas (NRA), in SAR imaging, 875
Normalized estimation error squared (NEES), 725–726, 767
Normalized frequency, 507
Northrop Grumman Norden Systems, 46
Nose-view scattering center image, 239, 241
Notably biased estimate, 706
NP detector. *See* Neyman-Pearson (NP) detector
N-pulse cancellers, 634
NRE. *See* Nonrecurring engineering effort (NRE)
NRL. *See* Naval Research Laboratory (NRL)
NTIA. *See* National Telecommunications and Information Administration (NTIA)
Null hypothesis, 552
Nyquist criterion, 498
Nyquist rate, 188, 498, 501
Nyquist sampling, 24
  theorem, 496, 497–501, 795

O

Observed SNR, 685
Ogive tip diffracted field, 219
OLA method. *See* Overlap-add (OLA) method
Onboard *versus* offboard processing, 487–488
One-dimensional (1-D), CFAR window, 595
One-step smoothed estimate, 719
One-way attenuation, 122
  fog and, 127–128
  rain rate and, 125–126, 127
OpenCL, 475
Open modular architectures, of COTS processors, 476–478
Optimal detection, 552–557
Optimum detector, 564
  for coherent receiver, 563
  for nonfluctuating radar signals, 561–562
  for nonfluctuating radar signals with coherent integration, 566–570
  for nonfluctuating target with noncoherent integration, 570–572
Ordered statistics (OS)-CFAR, 614–615
  combining greatest-of with, and censored CFAR, 615–616
  ROC for, 616–617
Order statistics filter, 530
OS-CFAR. *See* Ordered statistics (OS)-CFAR
Oscillators, 359–361
  gyrotron, 360
  magnetron, 359–360
  solid-state, 360–361
OTH. *See* Over-the-horizon (OTH) radars
OTH radar. *See* Over-the-horizon (OTH) radars
Output frequencies, 435
Overlap-add (OLA) method, 471–472
Over-the-horizon (OTH) radars, 17, 121
  applications of, 40–41

P

PA. *See* Power amplifier (PA)
PAG. *See* Power-aperture-gain product (PAG)
Paired echo theory, 380–381
Parallel feed systems, 354
  corporate feeds, 354
  distributed feeds, 354
Parallel plate waveguide, 135
Parametric estimation approach, track filtering, 720–722
  constant acceleration filtering, 726–727
  constant velocity filtering, 722–726
  limiting memory in, 721
Parks-McClellan algorithm, 526, 633
Parseval's theorem, 780

Passive arrays, 353, 354
  architectures, 339, 340
    disadvantages of, 339
    vs. active array architectures, 340
Passive phased array, 339, 350
Pave Paws, 42
PCI Express, 477
PCPEM. *See* Personal Computer Parabolic Equation Model (PCPEM)
PDAF. *See* Probabilistic data association filter (PDAF)
PDF. *See* Probability density functions (PDF)
PDR. *See* Phase-derived range (PDR)
PDRO. *See* Phase-locked dielectric resonator oscillator (PDRO)
Peak detector, 400
Peak sidelobe ratio (PSR), 793
PEC. *See* Perfect electric conductor (PEC)
Pencil beam, 39
Penetration, depth of, 156–158
Perfect electric conductor (PEC), 215
"Perfect model," 721
Periodic system calibration, 682
Personal Computer Parabolic Equation Model (PCPEM), 159
PFA. *See* Polar formatting algorithm (PFA)
PFN. *See* Pulse-forming network (PFN)
Phase, of electromagnetic waves, 7–9
Phase-coded pulse, 30
Phase-coded waveforms, 788
  Barker, match filtered response for, 809
  biphase-coded waveforms, 811–812
  equivalence operations, 812
  matched filter and, 812–813
  phase codes, 810
  polyphase-coded waveforms, 811–812
  properties of, 808–810
  spectrum of, 813–816
  structure of, 808–810
Phase codes, 810
  equivalence operations, 812
  match filtered response of, 812–813
  summary of, 829–830
Phase comparison monopulse, 702–703
Phased array antenna, 14, 326–338
  advantage of, 326
  characterization of, 353–354
  issues for, 91–92
Phased array radar, 349–350
Phase-derived range (PDR), 45
Phase drift, 641
Phase errors
  on directivity pattern, 316, 317
  vs. amplitude errors, 316
Phase-locked dielectric resonator oscillator (PDRO), 442–443
Phase-locked loop (PLL) bandwidth, 443
Phase-locked oscillators (PLO), 440
  for stable local oscillator, 442–443
Phase noise, 419–421
  clutter reduction, effect on, 424–425
  imaging, effect on, 428–429
  plot, 421
  pulse-Doppler processing, effect on, 425–428
  self-coherence effect and, 422–424
  spectral characteristics, 420–421
  spectral folding effects and, 421–422
  spectrum, 440
Phase quantization error, 329, 330
  phase-shifter bits and, 330
Phase shifter, 328–330
  switched line length, 329
Phase variation, 312
  antenna size on, 312
Phasor addition, 227–229
Phasor sum, 226
PHEMT. *See* Pseudomorphic high-electron mobility transistor (PHEMT)
Pilot pulse, 796
Plane wave, 213, 311, 327
Plan position indicator (PPI) displays, 401, 838
Plateau region, surface clutter, 173, 174
PLL bandwidth. *See* Phase-locked loop (PLL) bandwidth
PLO. *See* Phase-locked oscillators (PLO)
Point scatterers, 228
Point spread response (PSR), for SAR imaging, 856, 865. *See also* Doppler beam sharpening (DBS), for SAR imaging
  linear collection, 857–860
    along-track invariance, 589
    cross-track variance, 859
    properties of, 860
Point target model, 776–777
Polar formatting algorithm (PFA), 873
Polarization, 214–215, 836
  of electromagnetic waves, 10–11
  measurement of radar, 28–29
Polarization scattering matrix (PSM), 28, 171
  for circular polarization, 223–224
  for linear polarization, 222–223
Poles, 523
Police speed measuring radars, 52–53
Polyphase-coded waveforms, 811–812
Polyphase codes, 810
  Barker, 824, 825
  Frank codes, 824, 825–827
  P1 code, 827
  P2 code, 827
  P3 code, 827–829
  P4 code, 827–829
    Doppler tolerance of, 829
Poseidon Scientific, 441, 442
Power
  PPP measurements of, 669–670
  from target, 62–64
Power amplifier (PA), 339, 347, 364
Power-aperture form, of RRE, 80
Power-aperture-gain product (PAG), 350, 368
Power-aperture product, 79, 350
Power-aperture space, 319
Power density, 10
  transmission of, 10
Power law spectra, 190, 192
Power sources, 356–370
  amplifiers, 361–364
  microwave power modules (MPM), 369–370
  oscillators, 359–361
  solid-state devices, 364–369
Power spectral density (PSD), 65, 166, 187, 427
Power spectrum, in random signal, 533–534
Power supplies, 373–375
  high voltage, 349, 373–374
  for solid-state amplifiers, 374–375
Poynting vector, 214
PPI. *See* Plan position indicator (PPI) displays
PPP. *See* Pulse pair processing (PPP)
PPS. *See* Pulses per second (PPS)
Precision and accuracy, in radar measurements, 678–681
  effects of noise on, 679–681
  error sources, 678–679, 682–683, 684
  performance considerations and, 681–683
  resolution and sampling density and, 681
  signal propagation and, 682
  in target shooting, 678
Pre-T/R tubes, 396
PRF. *See* Pulse repetition frequency (PRF)
PRI. *See* Pulse repetition interval (PRI)
Principles of Modern Radar: Basic Principles, 3
Probabilistic data association filter (PDAF), 765–766

Probability density function (PDF), 96, 166, 185, 253, 427, 532–533, 552, 583, 596, 679, 687. *See also* Swerling models
of amplitude voltage, 258
chi-square of degree 2, 254, 256
of clutter, 583
cumulative distribution function and, 186
effect of grazing angle on, 193
exponential, 253, 254, 255, 256, 257
fourth-degree chi-square, 254, 255, 256
gamma, 256
log-normal distribution, 256, 257
one-parameter, 256
Rayleigh, 253, 254, 256
two-parameter, 256
Weibull, 256, 257
Probability of detection, 96–111, 598
average, 599
Probability of false alarm, 96–111
Process control radars, 48–49
Programmable DSP module, 474
Prolate spheroid, 230
Propagation, electromagnetic waves. *See also* Multipath
atmospheric refraction. *See* Refraction
attenuation. *See* Attenuation, of electromagnetic waves
bistatic, 119
diffraction. *See* Diffraction
factor, 118–119
ionosphere, exploiting. *See* Ionosphere
monostatic, 119
paths, 119–121
regions, 119–121
turbulence. *See* Atmospheric turbulence
PSD. *See* Power spectral density (PSD)
Pseudomorphic high-electron mobility transistor (PHEMT), 365
Pseudorandom biphase modulation, intrapulse, 437–438
PSM. *See* Polarization scattering matrix (PSM)
PSR. *See* Peak sidelobe ratio (PSR); Point spread response (PSR)
Ptolemy system, 483
Pulse
biphase-coded, 30
chirp, 30
compression, 30, 74–75, 222, 371
gain, 788
waveform. *See* Pulse compression waveforms
frequency modulation of, 847
phase-coded, 30
width, 20
Pulse cancellers, MTI, 631–634
frequency response of, 632
Pulse compression waveforms
amplitude modulation, 787
defined, 773
Doppler modulation in, 807
LFM waveform, 787–788
matched filter, 774–782
phase-coded waveforms, 788
Pulse-Doppler processing, 644, 696
advantages of, 646
blind zones, 660
for detection of moving targets, 645
metrics for, 657–659
with DFT, matched filter and filterbank interpretations of, 652–653
Doppler spectrum
SNR in, 651–652
and straddle loss, sampling, 648–651
DTFT, of constant-velocity target, 646–648
dwell-to-dwell stagger, 659
effect of phase noise on, 425–428
fine doppler estimation, 653–655
modern spectral estimation in, 656–657
PRF regimes and ambiguity resolution, 660–664
transient effects, 664–665
Pulse-Doppler spectrum, 665
Pulsed radar, 348–349
Pulsed radar Doppler data acquisition
data matrix and Doppler signal model, 627–628
generic Doppler spectrum for single range bin, 628
Pulsed waveform, 20
Pulse-forming network (PFN), 371–372
Pulse-limited case, scattering coefficients, 170
Pulse pair processing (PPP), 668–673
for doppler and range rate measurements, 699
power, measurements of, 669–670
Pulse repetition frequency (PRF), 20–21, 70, 89, 190, 348, 350, 418, 468, 627
regimes, advantages and disadvantages of, 661–662
regimes and ambiguity resolution, 660–664
regimes in airborne radar, advantages and disadvantages of, 662
staggered, blind speeds and, 634–637
Pulse repetition interval (PRI), 20, 89, 486, 502, 548, 549, 627
timing, 418
Pulses per second (PPS), 21
Pulse-to-pulse decorrelation, 558, 579, 580
Pulse-to-pulse stagger, 635, 636
Pushing factors, 379
Pythagorean theorem, 857

Q

Q. *See* Quadrature (Q)
QRD. *See* Q-R decomposition (QRD)
Q-R decomposition (QRD), 485
Quadratic phase, LFM waveform and, 791
Quadratic phase error (QPE), 863
Quadrature (Q), 98
components, 420
frequency value, 844
Quadriphase codes, 810, 824
Quantization, 496, 504–506
binary representations of numeric data, 502
error, 505
noise, 505

R

RACEway (ANSI/VITA 5–1994), 477
RACEway protocols, 476
Radar(s)
air defense systems, 40
airport surveillance, 49
altimeters, 53–54
artillery locating, 48
automotive collision avoidance, 53
bands, 8
basic configurations, 18
BMD, 40–41
classification of, 376
cloud attenuation as, 16
coherent system, 23
concepts of, 4–5
datacube, 502–503
design, 557
equation, 169
exciters. *See* Exciters
functions of, 33–36
imaging of, 34–36
instrumentation of, 45–46
marine navigation, 51–52
measurements of, 27–32
military applications, 36–37
multiple functions of, 47–48
noise detection, 25–26
noncoherent system, 23
over-the-horizon search, 40–41
overview, 3–4
police speed measuring, 52–53
process control, 48–49
satellite mapping, 52
search/detect, 33

Radar(s) (*cont.*)
search mode
overview, 87–89
phased array antenna issues, 91–92
search-and-track, 93–94
search regimens, 92–93
search time, 90–91
search volume, 90
track-while-scan, 93
semiactive seekers in, 43
target identification of, 48
three-dimensional search, 39–40
track, 33–34
transmission/reception process, 5
two-dimensional search, 37–39
wake vortex detection, 51
weather, 49–51
Radar-absorbing material (RAM), 18
Radar antennas, 309–343
architectures of, 339–343
active array, 339–341
passive array, 339
subarray, 341–343
beamwidth of, 312
directivity for, 313
directivity pattern of, 312
on performance, 317–319
radiation pattern of, 311–312
angular dependence of, 311–312
sidelobes of, 313–314
Radar cross section (RCS), 10, 18, 19–20, 45, 62, 64, 94, 168, 174, 317, 425, 496, 665, 677, 775
aspect angle. *See* Aspect angle, of RCS
complex targets. *See* Radar cross section (RCS), of complex targets
customary notation, 222
decorrelation, 558–559
estimation, 699–700
IEEE RCS definition, 219–220
intuitive derivation for scattering cross section, 220–222
polarization scattering matrix (PSM)
for circular polarization, 223–224
for linear polarization, 222–223
simple target. *See* Radar cross section (RCS), of simple targets
solving for, 72
statistical models of, 558–560
extended, 560
Swerling models. *See* Swerling models
Radar cross section (RCS), of complex targets
correlation, 258, 259–263
echo amplitude, 259, 261
length, 259
decorrelation of, 259–263
frequency agility, 262
distribution, 253–258, 259
scatterers, 251–252
statistical models for, 253–263
Radar cross section (RCS), of simple targets
aspect angle of, 249–250
conducting spheres, 248
corner reflector, 248, 249
frequency dependence of, 249–250
Radar detection
algorithms, 553
scheme, 560–561
Radar measurements, 751
angle measurements
AOA estimation method. *See* AOA estimation method
monopulse technique, 701–703
sequential lobing, 700–701
coordinate systems, 709
doppler and range rate measurements, 696
DFT interpolation methods, 698–699
DFT methods, 697–698
pulse pair processing, 699
hypothesis, 552–553
with LFM waveforms, 756
overview, 677–678
parameter estimation
Cramèr-Rao lower bound (CRLB), 688–688
estimators, 685–686
Gaussian case, precision and resolution for, 688–690
signal and target variability, 690
phase measurement, 695–696
precision and accuracy, 678–681
effects of noise on, 679–681
error sources, 678–679, 682–683, 684
performance considerations and, 681–683
resolution and sampling density and, 681
signal propagation and, 682
in target shooting, 678
radar signal model, 683–685
range measurements
resolution and sampling, 690–693
split-gate and centroid, 693–695
RCS estimation, 699–700
of reduced dimension, surveillance radars with, 756–757
in sine space, 754–756
in stabilized coordinates, 752–754
Radar range equation (RRE), 26, 88, 118, 317–318
atmospheric loss, 68–69
average power, 83–84
average power form of, 73–74
clutter, 76–78
decibel form of, 72–73
as dependent variable, 72
dwell time, 83–84
energy form of, 775–776
graphical example, 75–76
matched filter and, relationship between, 780
monostatic, 119
multiple-pulse effects, 66–67
one way equation, 78–79
overview, 59–61
power-aperture form of, 80
power density at a distance R, 61–62
received power from target, 62–64
receive loss, 69
search form of, 79–80, 318–319
signal processing loss, 69–72
signal-to-noise ratio and, 66
summary of losses, 67–68
target RCS effect, 84
track form of, 80–83
for tracking, 318
transmit loss, 68
Radar receivers. *See* Receiver(s)
RADARSAT 2, resolution modes, 52
Radar signal model, 683–685
Radar signal processor
algorithm metrics, 464
counting FLOPS
choosing efficient algorithms, 470–472
general approach, 464–468
mode-specific formulas, 468–470
fixed point *versus* floating point, 480–482
hardware metrics, 462–464
implementation technology
analog-to-digital conversion, 472–473
comparison of, 474
COTS technology and modular open architectures, 476–478
Moore's law, influence of, 478–480
processor technologies, 473–475
major elements of, 459
overview, 459–460
sizing
benchmarks, 483–485
considerations in estimating timing, 482–483
data rates, 485–487

onboard *versus* offboard processing, 487–488
software tool impacts, 485
structure, 460–462
Radar signals
nonfluctuating, 561–562
detector for, 561–562, 566–570
performance for, 563–566
threshold detection of, 560–584
unknown parameters of, 561
Radar spectrum engineering criteria (RSEC), 376, 377, 378
Radar tracking algorithms
kinematic motion models, 746
NCA motion model, 749
NCV motion model, 747–749
nearly constant speed motion model, 750–751
Singer motion model, 749–750
measurements, 751
with LFM waveforms, 756
of reduced dimension, surveillance radars with, 756–757
in sine space, 754–756
in stabilized coordinates, 752–754
measurement-to-track data association, 760–761
formation of new track, 762
measurement validation and gating, 762–763
nearest neighbor (NN) filter, 764
probabilistic data association filter (PDAF), 765–766
strongest neighbor (SN) filter, 764–765
performance assessment of, 766–767
target tracking
data association, 713–716, 717
detection processing, 717
functional areas, 717–718
in modern radar systems, 716, 717
parameter estimation, 717
track filtering. *See* Track filtering
Radar video signal, 391–392
RADC. *See* Rome Air Development Center (RADC)
Radial velocity, 677
Radii of curvature, 230
Radio-acoustic sounding systems (RASS), 50
Radio frequency (RF), 4, 138, 184, 347, 349, 350, 351, 431, 434, 776
bands, 8
hardware, 682
oscillator, 347
phase shifters, 349, 350
power, 350, 351
preselection, 396–397
signal, 393
subassembly, 447
with SMT assembly technique, 447–448
Rain rate
equivalent, snow *vs.*, 129
one-way attenuation and, 125–126, 127
Rain reflectivity
atmospheric clutter and, 196–199
versus frequency band, 197
RAM. *See* Radar-absorbing material (RAM)
Ramp rate, 790
Random phase errors, 316, 317
Random processes, 532
Random signals, 532–535
autocorrelation function, 533
cross-correlation function, 533
moments and, 533–534
power spectrum, 533–534
probability density functions, 532–533
time averages, 535
white noise, 534–535
Random tracking index, 731
Random variable (RV), 679
Range, 27–28
ambiguity, 22
bins, 502
coverage, 629
frequency, 514
gates, 502, 691
measurements
resolution and sampling, 690–693
split-gate centroid method, 693–695
migration, 469, 548, 628
rate, 677
resolution. *See* Range resolution
sampling, 21–22
window, 684, 690
Range-Doppler coupling, 695, 803–805
spectral interpretation of, 806–807
Range-Doppler coupling coefficient, 736
Range-Doppler imaging, SAR as, 839–842
azimuth, 840–841
for coarse resolutions, 841
cone angle, 840
Doppler filter bandwidth, 841
Doppler filtering, 841
dwell time, 841
velocity vector, 842
Range-Doppler spectrum
for moving radar, 296–303
clutter. *See* Clutter, moving radar
range and velocity ambiguity, 300–303
for stationary radar, 295–296
Range migration algorithm (RMA), 873
Range-profiling, 439
Range resolution, 29, 690–692, 773, 782
computation of, 785
defined
by Rayleigh criterion, 783–784, 785–786
in terms of mainlobe width, 784
degraded/reduced, 784
improved/enhanced, 784
requirements of, 782–783
two point targets, 785–786
waveform bandwidth and, relationship between, 784–785
Woodward constant, 784
RASS. *See* Radio-acoustic sounding systems (RASS)
Rayleigh criterion, 783–784, 785–786
for resolution, 844
Rayleigh distributed interference, 591, 592
Rayleigh distribution, 557, 563, 569
of noise, 105
Rayleigh PDF, 98, 253, 254, 256
Rayleigh (peak-to-null) Doppler resolution, 627
Rayleigh scattering, 225
Rays
bending of, 131
path of, 131, 132
RCS. *See* Radar cross section (RCS)
Real-beam ground mapping (RBGM), 837–838
Real floating-point operations (RFLOP), 465
Received phase, 419
measurement of, 419–420
Receive loss, 69
Receiver circuits, 4
Receiver/exciter (REX), 418
Receiver operating characteristics (ROC), 96, 556, 602
for CFAR algorithms, 617–618
for generalized censored mean level detector, 619
Receiver operating curves, 102–103
Receiver(s)
analog-to-digital data conversion, 409–414
digital up/down frequency conversion, 414
direct digital coherent detection implementation, 412–414
spurious-free dynamic range, 412

Receiver(s) (*cont.*)
channelized, 395–396
crystal video, 393
demodulation
analog coherent detection
implementation and mismatch errors, 403–404
coherent, 402–403
noncoherent, 400–402
digital, 395
dynamic range, 406
coupling issues, 409
gain control, 408–409
sensitivity time control, 407–408
functions, 392
frequency downconversion and mixers, 397–399
LO and if frequencies, selection of, 399–400
receiver protection, 396
RF preselection, 396–397
homodyne, 394
instantaneous frequency measurement, 395
major elements, 392
noise power, 404–406
overview, 391–392
radar video signal, 391–392
superheterodyne, 394–395
superregenerative, 393
types, 392–396
Reciprocal spreading, 508
Reciprocity theorem, 171
Reconfigurable hardware modules, 474
Rectangular gating, 763
Rectangular window, 512
Reference oscillator, 440
Reference window, CFAR
target returns in, 603
Reference windows, 596
Reflection, of electromagnetic waves, 17–18
Reflection coefficient, 143, 144
multipath
angle errors, 155
classification error, 155–156
divergence factor, 151–152
Fresnel reflection coefficient, 148–150
roughness factors, 152–155
Reflector antennas, 322–326
parabolic, 322–326
Refraction, 118, 130
anomalous, 134–137
effective earth model and, 134
index of, 130
Snell's law and, 131, 132
standard, 130
standard atmosphere and, 131–134
Refractive index, 130
Relative phase, waves, 9
Resolution, 843–844
cross-range, 845–846
down-range resolution, 845
fine range resolutions, 846
integration angle, 845
measurement of radar, 29–32
concept of, 30
frequency of, 32
range in, 29
pulse, frequency modulation of, 847
range. *See* Range resolution
Rayleigh criterion for, 844
and sampling, 851–852
short pulses, 846
spatial, 843
stepped frequency waveform, 846–847
Resonance region, 174
Resonant region scattering, 225–226
Resource manager, 93
REX. *See* Receiver/exciter (REX)
RF. *See* Radio frequency (RF)
RFLOP. *See* Real floating-point operations (RFLOP)
RHC. *See* Right-handed circular (RHC)
Richards, M. A., 303
Rician PDF, 100, 564, 570
Right-handed circular (RHC), 150
components, 223
RMA. *See* Range migration algorithm (RMA)
Rms. *See* Root mean square (rms)
Rms bandwidth (B_{rms}), 689
RMSE. *See* Root mean squared error (RMSE)
Rms time duration, (τ_{rms}), 689
ROC. *See* Receiver operating characteristics (ROC)
Rome Air Development Center (RADC), 191
Root mean squared error (RMSE), 682, 687, 725
Root mean square (rms), 173
noise, 27
Root sum of squares (rss), 682
Roughness factors, multipath, 152–155
Rough surface theory, 173
RRE. *See* Radar range equation (RRE)
RSEC. *See* Radar spectrum engineering criteria (RSEC)
Rss. *See* Root sum of squares (rss)
RV. *See* Random variable (RV)

S

Sampled signals, vector representation of, 501
Sampling, 496
interval for, 497
nonbaseband signals, 501
Nyquist theorem, 496, 497–501
SAR. *See* Synthetic aperture radar (SAR); Specific absorption rate (SAR)
Satellite mapping radars, 52
Saturation, 505
SAW. *See* Surface acoustic wave (SAW)
S-band frequency, 15
SBC. *See* Single board computers (SBC)
SBO. *See* Shoe-box oscillator (SBO)
SBX. *See* X-Band (SBX) radar
Scalloping loss, 91, 333
Scan-to-scan decorrelation, 558, 578, 580
Scattered field, 215, 218–219
Scattered power density, 220
Scattering, 17, 224
atmospheric volumetric, 118, 121–122, 168
coefficients, 169–171
diffuse, 153–154
specular, 152–153
diffuse, 17–18
high-frequency optics region, 226–227
Rayleigh scattering, 225
resonant region scattering, 225–226
specular, 17–18
SC-cut crystal oscillator, 440
SC-cut quartz crystal oscillator. *See* Stress-compensatedcut (SC-cut) quartz crystal oscillator
Schwartz inequality, 780
SCI (IEEE 1596–1992), 477
Scintillation, 678, 690
SCR. *See* Signal-to-clutter ratio (SCR); Silicon-controlled rectifiers (SCR)
SE. *See* Spectrum efficiency (SE)
Sea clutter, 193
correlation properties, 194–195
GTRI, model equations, 204
sea spikes, 195–196
spatial variations, 196
spectral properties, 195
Search mode, radar, 33
overview, 87–89
phased array antenna issues, 91–92
search-and-track, 93–94
search regimens, 92–93
search time, 90–91
search volume, 90
track-while-scan, 93
Search radars, and low frequency, 318–319

Sea reflectivity, 180
angular and frequency dependencies, 182–183
ducting, 184–185
GTRI empirical model and, 203
physical parameters affecting, 181
Sea spikes, 195–196
Sea states, 181
physical properties for various, 147–148
Second-degree non-central chi square (NCCS2), 559
Self-coherence effect, phase noise and, 422–424
Self-masking, target, 603–604
Semiactive seekers, 43
Sensitivity time control (STC), 395
receiver dynamic range and, 407–408
Sensor measurement vector, 720
Sensor-noise only (SNO) covariance, 733
Sequential lobing, 320, 700–701
Serial RapidIO, 477
SFDR. *See* Spurious-free dynamic range (SFDR)
Shadow region, 120
Shadows, imaging, 877–879
cross-range extent of, 277
length of, 277, 878
RF energy, 275, 276
size and shape of, 878
solid objects, 276, 277
width of, 277
Sheared ridge, 804
Shift-invariant, 523
Shift register, 822
Ship-based standard missile (SM), 43
Shnidman's equation, 581–583
vs. Albersheim's equation, 582–583
Shoe-box oscillator (SBO), 442
Short-time pulse, 846
Shuttle Imaging Radar-C (SIRC), 469
Shuttle Imaging Radars (SIR), 52
Sidelobe clutter, 298
Sidelobes, 14, 226, 782, 792
performance of, LFM waveform and, 793–794
range measurement and, 693
reduction in LFM waveform, 797–800
structure of LFM waveform, 805
time, 800
Sidelooking airborne radar (SLAR), 838
Sifting property, 498
Signal-plus-noise PDF, 100–102
Signal-plus-noise-to-noise ratio, 685
Signal Processing Designer by CoWare, Inc., 483
Signal processing exercise, SAR imaging, 842–843
data acquisition, 843
image
formation, 843
synthesis, 842
Signal processing loss, 69–72
Signal processor, 4
Signal quantization, 678
Signal-to-clutter ratio (SCR), 59, 189, 422
determination of, 425–427
Signal-to-interference-plus-noise-ratio (SINR), 590
CA-CFAR performance and, 600
CFAR loss and, 601–602
Signal-to-interference ratio (SIR), 59, 189, 427, 630
Signal-to-noise ratio (SNR), 14, 25–26, 59, 189, 411, 495, 679, 717, 773
in Doppler spectrum, 651–652
loss, 566
loss associated with Taylor weighting function, 798
radar performance and, 775
radar range equation and, 66, 775–776
Signal-to-noise ratio (SNR), 318, 548, 556, 564, 886–887
Signal-to-quantization noise ratio (SQNR), 473, 505
Signal variability, 690
Signal windowing techniques, 693
Sign-magnitude, 504
encoding, 504
Silicon-controlled rectifiers (SCR), 372
Simple pulse
ambiguity surface for, 802
match filtered response for, 781
spectrum of, 785
Simulink, 483
Simultaneous lobing technique, 701
Sinc function, 278
Sinc response, 31
Sine space, 330
measurements in, 754–756
Singer motion model, 749–750
Single board computers (SBC), 462
Single canceller, 631
Single-pole single-throw (SPST) RF switch, 449
Single-pulse signal-to-noise ratio (SNR), 88
Single range bin, generic Doppler spectrum for, 628
Single-sideband frequency converter, 432–433
Single-sideband phase noise, 420
Single-transmit pulse, timing for, 453
Singular value decomposition (SVD), 485
SINR. *See* Signal-to-interference-plus-noise-ratio (SINR)
SIR. *See* Shuttle Imaging Radars (SIR); Signal-to-interference ratio (SIR)
SIRC. *See* Shuttle Imaging Radar-C (SIRC)
Sizing, signal processor
benchmarks, 483–485
considerations in estimating timing, 482–483
data rates, 485–487
onboard *versus* offboard processing, 487–488
software tool impacts, 485
Skin depth, 156–158
Skip, 17
Skirt region, 295
Skirts of filter, 435
Slant plane, 879, 881, 887
Slant range, 855
azimuth extent, 841
cross-track range and, 838
DBS image, 861, 862, 864
fine resolutions, 853
layover, 880
PSR, 857–858, 860
SNR, 886
SLAR. *See* Sidelooking airborne radar (SLAR)
Sliding window estimator, 721, 757
Slow AGC, 408
Slow time, 502
SAR data acquisition, 842
SM. *See* Ship-based standard missile (SM)
Smallest-of CA-CFAR (SOCA-CFAR), 608, 612–613, 618
CFAR loss associated with, 610, 613
ROC for, 616–617
Smoke, and attenuation, 130
SMT devices. *See* Surface mount technology (SMT) devices
Snell's law, 131, 132, 215
SN filter. *See* Strongest neighbor (SN) filter
SNO covariance. *See* Sensor-noise only (SNO) covariance
Snow, and attenuation, 128–129
versus equivalent rainfall rate, 129
types, 129
SNR. *See* Signal-to-noise ratio (SNR)
Solid-state active-aperture arrays, 368–369
high-power density, 368
low-power density, 368

Solid-state amplifiers, 364–367, 369
analog, 366
class of operation of, 366–367
Solid-state devices, 364–370
advantages of, 365–366
drawbacks of, 366
vs. vacuum tube devices, 364, 365–366
Solid-state power amplifiers (SSPA), 371
Solid-state technology, 356, 364, 367
Solid-state Transmitter/Receiver modules, 367–368
Space-based radars, 140
Space-fed feed system, 354, 355
Space-time adaptive processing (STAP), 96, 402, 460, 468, 503, 642, 717
Spatial distributions, land clutter, 192–193
Spatial Doppler signal, 627
Spatial frequency, 7, 513–514, 863–864
Spatial locality, 464
Spatial resolution, 843
SPEC. *See* Standard Performance Evaluation Corporation (SPEC) commercial benchmarks
Specific absorption rate (SAR), 384
Speckle, 881–884
multilook processing, 882–884
Speckle-reduction procedure. *See* Multilook processing, speckle
Spectral folding effects
phase noise and, 421–422
Spectral width, 50, 669, 699
Spectrum efficiency (SE), 376
Spectrum engineering, 376, 377
Spectrum management, 376
activities of, 376
Specular point, 229
Specular scattering, 17, 152–153, 226, 229–231
Specular spike, 236
Speed of light, 5
Spherical coordinates, 27, 709
Spherical waves, 213
intensity of, 10
Spikes, sea, 195–196
Split-gate centroid method, 693–695
Split-gate tracker, 693–694
Spook-9, 46
Spotlight SAR collection, 847–848
SPST RF switch. *See* Single-pole single-throw (SPST) RF switch
Spurious-free dynamic range (SFDR), 411, 412, 413
SQNR. *See* Signal-to-quantization noise ratio (SQNR)
Square law detector, 98, 400, 537, 563, 572–575, 592, 598
Neyman-Pearson, 592, 594–595
for nonfluctuating target, 573–575
SSPA. *See* Solid-state power amplifiers (SSPA)
Stabilized coordinates, measurements in, 752–754
Stable local oscillator (STALO), 441–442
phase-locked oscillators for, 442–443
phase noise, 424
Staggered PRFs, blind speeds and, 634–637
Stagger ratio, 635
STALO. *See* Stable local oscillator (STALO)
Standard missile, 43
Standard Performance Evaluation Corporation (SPEC) commercial benchmarks, 483
STAP. *See* Space-time adaptive processing (STAP)
State estimate, 719
Stationary phase point, 229
STC. *See* Sensitivity time control (STC)
Steering vector, 566
Stepped chirp, 788
Stepped frequency waveforms, 788, 846–847
Stochastic state estimation, 719, 728–730
CWNA, nearly constant velocity filtering with, 736–739
DWNA, nearly constant velocity filtering with, 730–736
LFM waveforms, nearly constant velocity tracking with, 736
multiple model filtering for highly tracking maneuvering targets, 741–746
nearly constant acceleration filtering, 739–741
Stop-and-hop model, 627, 854
“Stove pipe” aircraft, 239–243
Straddle loss, 515–516, 516–518, 684, 782, 786–787
reduction, LFM waveform and, 800
sampling Doppler spectrum and, 648–651
Stratosphere, 120
Streaming processing, 460
Stress-compensatedcut (SC-cut) quartz crystal oscillator, 433
Stretch processing, 473, 787
LFM, 847
Stripmap SAR, 14, 847
Strongest neighbor (SN) filter, 764–765
Subarray architecture, 341–343
advantage of, 341–342
analog time delay units (TDUs) in, 341–343
overlapped, 343
Subclutter visibility, 637
Subreflector, 324, 325
Subrefraction, anomalous atmosphere, 135
Sum beam, 320
Summation point, defined, 311
Superheterodyne receivers, 394–395
Superposition
of electromagnetic waves, 9–10. *See also* Interference
multipath, 144–145
Superrefraction, anomalous atmosphere, 135
Superregenerative receiver, 393
Surface
clutter. *See* Surface clutter
diffraction, 118
intervisibility, 118
multipath, 118
reflectivity, 169
wave
effects, 226
scattering, 226
Surface acoustic wave (SAW)
devices, 795
oscillators, 438, 443–444
Surface boundary, and multipath, 146–148
Surface clutter, 77, 202–205
general dependencies, 172–175
temporal and spatial variations, 175–177
value data, experimental programs
land reflectivity, 177–180
sea reflectivity, 180–185
variability of
land clutter, 185–193
sea clutter, 193–196
Surface mount technology (SMT) devices, 446–448
Survey, of image formation, 873–874
SVD. *See* Singular value decomposition (SVD)
SW0. *See* Swerling 0 (SW0)
Sweep rate, 790
Swerling models, 263–267, 559–560, 578, 579, 580
analyzing detection performance, 264
concept, 263
correlation model, 266
CPI, 264
detection performance of, 580
extension strategy, 267

PRF, 263–264
pulse-to-pulse decorrelation, 264, 266
scan-to-scan decorrelation, 264, 265, 266
Swerling 0 (SW0), 103
Swerling 2 target, 700
Swerling 1 target model, 107–108
Switches
for exciters, 449–450
Synthesizers, 444–446
Synthetic aperture radar (SAR), 28, 165, 530, 549, 596, 782, 835–837. *See also* Imaging, SAR
advantage, 836
beamwidth, 838
data collection, 852–856
geometry for, 35
as large synthetic antenna aperture, 837–839
overview, 835
processing, 402, 459
as range-Doppler imaging, 839–842
sampling requirements, 849–851
along-track sampling limit, 850
pulsed waveform, 849
SAR imaging *vs.* optical imaging, 835, 836, 837
as signal processing exercise, 842–843
Synthetic wideband waveforms, 788
System noise temperature, 405

T

Tangential signal sensitivity (TSS), 401
Target
detection, strategies for, 548–552
masking, 600
mutual, 601, 604–605
self, 603–604
motion model, 719
position, 27
RCS. *See* Radar cross section (RCS)
reflectivity. *See* Target reflectivity
tracking. *See* Target tracking
variability, 690
visibility, 640
Target motion resolution (TMR), 45
Target reflectivity
EM wave, 211
characteristics of, 212
fundamentals, 212–214
polarization, 214–215
reflection, 215–219
examples, 236–243
high-frequency scattering
edge diffraction, 232–234
end-region scattering, 231–232
multiple-bounce scattering, 235–236
phasor addition, 227–229
specular scattering, 229–231
RCS. *See* Radar cross section (RCS)
scattering regimes, 224
high-frequency optics region, 226–227
Rayleigh scattering, 225
resonant region scattering, 225–226
Target tracking
data association, 713–716, 717
detection processing, 717
functional areas, 717–718
in modern radar systems, 716, 717
parameter estimation, 717
track filtering. *See* Track filtering
Taylor aperture tapers, 316
Taylor function, 315, 321
Taylor weighting function
defined, 797
Rayleigh resolution associated with, 798
SNR loss associated with, 798
straddle loss reduction, 800
TDRSS. *See* NASA' Tracking and Data Relay Satellite System (TDRSS)
TDU. *See* Time delay units (TDU)
TE diffraction. *See* Trailing-edge (TE) diffraction
TEMPER. *See* Tropospheric Electromagnetic Parabolic Equation Routine (TEMPER)
Temporal locality, 464
Terrain Integration Rough Earth Model (TIREM), 158
Texas Instruments, Inc. (TI), 474
THAAD. *See* Theater High Altitude Air Defense (THAAD)
Theater High Altitude Air Defense (THAAD), 42, 43
Thermal noise, 25
receiver, 64–65
Thermosphere, 121
Three-dimensional search, radar, 39–40
Three-pulse canceller, 631–634
Threshold detection, 553–554, 560–584
concept of, 95–96
Thumbtack ambiguity surface, 817
Thunderstorms, 198
Time-bandwidth product, 696, 791, 793, 799–800, 804, 805, 810, 823
Time delay units (TDU), 337, 338
Time sidelobes, 800
Timing and control circuits, for exciters, 452–454
Timing jitter, 453
TIREM. *See* Terrain Integration Rough Earth Model (TIREM)
TKMOD5C, 158
TM-CFAR. *See* Trimmed mean (TM)-CFAR
TMR. *See* Target motion resolution (TMR)
Toeplitz matrices, 541
Tomographic methods, image formation survey, 873
T/R. *See* Transmit/receive (T/R) module
Track filtering, 34, 678, 713–716, 719
converted measurement filter, 760
extended Kalman filter (EKF), 759–760
nonlinear least squares estimation, 757–759
overview, 713
parametric estimation, 720–722
constant acceleration filtering, 726–727
constant velocity filtering, 722–726
limiting memory in, 721
stochastic state estimation, 728–730
CWNA, nearly constant velocity filtering with, 736–739
DWNA, nearly constant velocity filtering with, 730–736
LFM waveforms, nearly constant velocity tracking with, 736
multiple model filtering for highly tracking maneuvering targets, 741–746
nearly constant acceleration filtering, 739–741
Tracking, 678. *See also* Target tracking
radar application of, 46–47
Track partitioning, 763
Trailing-edge (TE) diffraction, 233
Transient effects, pulse-Doppler processing, 664–665
Transition region, 295
Transmission bands, 125
Transmit loss, 68
Transmit/receive (T/R) module, 350, 356
Transmit signal, 429–437
bandwidth of, 431
Transmitters, 4
configurations for, 352–356
cooling of, 382–383
design, 376–378
efficiency, 351
operation of, 381–384
safety in, 383–384
parameters of, 350–351
power source of, 356
classification of, 356
reliability of, 381–382
MTBF and, 381–382

Transmitters (*cont.*)
  on spectral purity, 378–381
    nontime-varying errors, 380–381
    time-varying errors, 379–380
Trapping, anomalous atmosphere, 135–136
Traveling waves, 226
Traveling-wave tube (TWT), 348, 363, 370, 374
  amplifiers, 433
  in microwave power modules, 370
TRF receiver. *See* Tuned radio frequency (TRF) receiver
Trihedrals, 235, 249, 885
Trimmed mean (TM)-CFAR, 613–614
T_{rms}. *See* Rms time duration, (τ_{rms})
Troposphere, 120
Tropospheric Electromagnetic Parabolic Equation Routine (TEMPER), 159
T/R tubes, 396
TSS. *See* Tangential signal sensitivity (TSS)
Tube amplifiers, 361–364
  crossed-field, 364
  linear beam, 361–364
Tuned radio frequency (TRF) receiver, 393
Turbulence, atmospheric, 118, 137–138
Two-dimensional (2-D) data set, 596
Two-dimensional (2-D) system, 37
Two-dimensional search, radar, 37–39
Two-pulse MTI canceller, 631–634
  frequency response, 636–637
Two's complement encoding, 504
TWT. *See* Traveling-wave tube (TWT)

U

UAV. *See* Unmanned aerial vehicles (UAV)
UAV radars. *See* Unmanned aerial vehicles (UAV)
UDSF. *See* Usable Doppler space fraction (UDSF)
UHF. *See* Ultra high frequency (UHF); Ultra high frequency (UHF)
Ultra high frequency (UHF), 135, 561
UMOP. *See* Unintentional modulation of pulse (UMOP)
Unambiguous Doppler coverage, 629
Unambiguous range, 23, 629
Underflow, 505
Unintentional modulation of pulse (UMOP), 378
Unmanned aerial vehicles (UAV), 370, 480
Upper sideband (USB), 432
U.S. Firefinder system, 90
U.S. Military's Common Data Link (CDL), 487
U.S. Standard Atmosphere, 131
Usable Doppler space fraction (UDSF), 638
USB. *See* Upper sideband (USB)

V

Vacuum electron device (VED), 371
Vacuum tube power booster, 370
Variance, 533–534, 685–686
Vector Signal Image Processing Library (VSIPL), 477–478, 485
VED. *See* Vacuum electron device (VED)
Velocity
  ambiguity, 301–302
  constant velocity (CV). *See* Constant velocity (CV)
  Doppler shift as function of, 626
  radial, 677
Versa module europe (VME), 476
Very high frequency (VHF), 135
VHF. *See* Very high frequency (VHF)
Video filtering, 404
Virtual sources, 140
V-LFM, 808
VME. *See* Versa module europe (VME)
VME (IEEE P1014–1987), 477
VME64 (IEEE P1014 Rev D), 477
"Voltage domain" approach, 469–470
"Voltage reflectivity," 683
Voltage standing wave ratio (VSWR), 367, 447
Volume
  clutter, 77–78
  reflectivity, 169
VSIPL. *See* Vector Signal Image Processing Library (VSIPL)
VSWR. *See* Voltage standing wave ratio (VSWR); Voltage standing wave ratio (VSWR)

W

Wake vortex detection radars, 51
Water concentration, 127
  fog *versus,* 128
Water vapor, attenuation and, 124–125
Wave
  height, 181
  impedance, 214
Waveforms
  bandwidth, range resolution and, 784–785
  biphase-coded, 788
  CW, 20
  design methods, 693
  filtered response, 782
  generation
    dwell-to-dwell frequency change, 439–440
    intrapulse, pseudorandom biphase modulation, 437–438
    intrapulse linear frequency modulation (LFM), 438–439
  Huffman coded, 787
  interpulse modulation, 787–788
  linear frequency modulated, 787–788
  modulations, 774
  phase-coded, 788
  polyphase-coded, 811–812
  pulse compression, 787–788. *See also* Pulse compression waveforms
  pulsed, 20
  stepped frequency, 788
  synthetic wideband, 788
Wavelength, of electromagnetic waves, 6
Wavelet transforms, 485
Wavenumber, 213, 513. *See also* Spatial frequency
Weather radars, 49–51
Weibull distribution, 105, 175–176, 186, 194
  PDF, 256, 257
Wenzel Associates, 440, 441
White noise, 65, 534–535, 779
  acceleration error, 730
Wideband phased array, 335–338
Wideband STAP-GMTI radar, 466–467
Windowing, 509–513
Woodward constant, range resolution, 784
WSR-88D (NEXRAD) radar, 50
WSR-88D Next-Generation Radar (NEXRAD) radar, 670

X

X-band pulsed magnetron, 359
X-band radar, 433
X-Band (SBX) radar, 42
Xilinx, 474
XPatch, 158

Z

Zeroes, 523
Zero-mean error, 687
Zero padded signal vector, 652
Zero padding, 516, 648
Zero-velocity filter, 664–667
Z transform, 522–523

Signal Analysis and Processing

Select Fourier Transforms and Properties

Continuous Time	
$x(t)$	$X(f)$
$\begin{cases} A, & -\frac{\tau}{2} \le t \le \frac{\tau}{2} \\ 0, & \text{otherwise} \end{cases}$	$A\tau \frac{\sin(\pi f \tau)}{\pi f \tau} \equiv A\tau \operatorname{sinc}(\pi f \tau)$
$\begin{cases} A\cos(2\pi f_0 t), & -\frac{\tau}{2} \le t \le \frac{\tau}{2} \\ 0, & \text{otherwise} \end{cases}$	$\frac{A\tau}{2}\operatorname{sinc}[\pi(f-f_0)\tau] + \frac{A\tau}{2}\operatorname{sinc}[\pi(f+f_0)\tau]$
$\sum_{n=-\infty}^{\infty} \delta_D(t-nT)$	$\sum_{k=-\infty}^{\infty} \delta_D(f-k\cdot PRF)$
$AB\frac{\sin(\pi Bt)}{\pi Bt} \equiv AB\operatorname{sinc}(\pi Bt)$	$\begin{cases} A, & -\frac{B}{2} \le t \le \frac{B}{2} \\ 0, & \text{otherwise} \end{cases}$
$x(t-t_0)$	$e^{-j2\pi f t_0} X(f)$
$e^{+j2\pi f_0 t} x(t)$	$X(f-f_0)$
Discrete Time	
$x[n]$	$X(\hat{f})$
$Ae^{j2\pi \hat{f}_0 n}, \quad n = 0, \ldots, N-1$	$A\frac{1-e^{-j2\pi(\hat{f}-\hat{f}_0)N}}{1-e^{-j2\pi(\hat{f}-\hat{f}_0)}} \equiv NAe^{-j\pi(\hat{f}-\hat{f}_0)(N-1)}\operatorname{asinc}(\hat{f}-\hat{f}_0, N)$
$A\delta[n-n_0]$	$e^{-j2\pi \hat{f} n_0}$
$\sum_{n=-\infty}^{\infty} \delta_D(t-nT)$	$\sum_{k=-\infty}^{\infty} \delta_D(f-k\cdot PRF)$
$A\hat{B}\frac{\sin\left[\pi \hat{B} n\right]}{\pi \hat{B} n} \equiv A\hat{B}\operatorname{sinc}\left[\pi \hat{B} n\right]$	$\begin{cases} A, & \lvert\hat{f}\rvert < \hat{B} \\ \hat{B} < \lvert\hat{f}\rvert < \pi, & \text{otherwise} \end{cases}$
$x[n-n_0]$	$e^{-j2\pi \hat{f} n_0} X\left(\hat{f}\right)$
$e^{+j2\pi \hat{f}_0 n} x[n]$	$X\left(\hat{f}-\hat{f}_0\right)$

Window Properties

Window	3 dB Mainlobe Width, relative to rectangular	Peak Sidelobe (dB)	Sidelobe Rolloff (dB per octave)	SNR Loss (dB)	Maximum Straddle Loss (dB)
Rectangular	1.0	−13.2	6	0	3.92
Hann	1.68	−31.5	18	−1.90	1.33
Hamming	1.50	−41.7	6	−1.44	1.68
Taylor, 35 dB, $\bar{n} = 5$	1.34	−35.2	0/6	−0.93	2.11
Taylor, 50 dB, $\bar{n} = 5$	1.52	−46.9	0/6	−1.49	1.64
Dolph-Chebyshev (50 dB equiripple)	1.54	−50.0	0	−1.54	1.61
Dolph-Chebyshev (70 dB equiripple)	1.78	−70.0	0	−2.21	1.19